The Genetics and Molecular Biology
of Neural Tumors

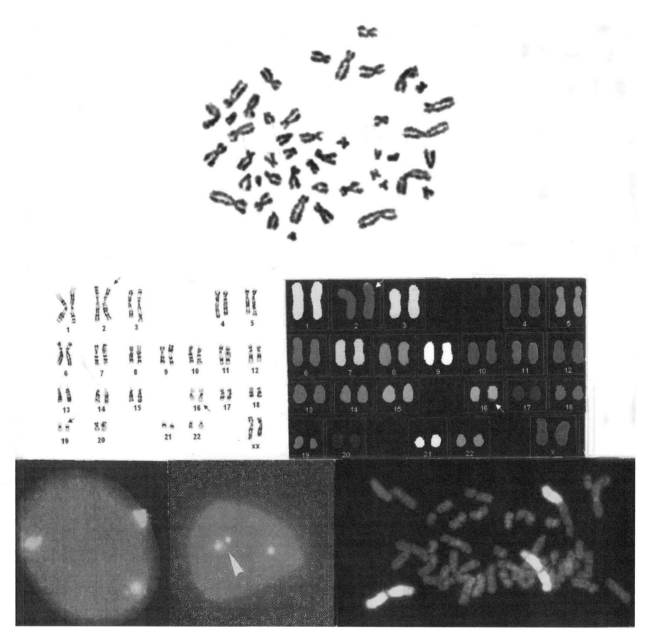

Impressions of Cytogenetic Findings. *Top*: Metaphase cell showing the normal 46 human chromosomes. This figure was part of an article that established 46 chromosomes as characteristic of human cells (Tjio, J.H. and Levan, A. [1956] The chromosome number of man. *Hereditas* **42**, 1–6). *Middle left*: G-banded abnormal karyotypes of a solitary neurofibroma containing 46 chromosomes with two unbalanced translocations, der(2)t(2;19)(p25;p13.2) and der(16)t(2;16)(p25;q24), and a del(19)(p13.2). These changes were substantiated (middle right) with SKY (arrows), a procedure in which by computer analysis the staining of each chromosome pair can be distinguished from that of all others (Nandula, S.V., Borczuk, A.G., and Murty, V.V.V.S. [2004] Unbalanced t(2;19) an t(2;16) in a neurofibroma. *Cancer Genet. Cytogenet.* **152**, 169–171). *Bottom left*: The nucleus on the left shows three fluorescence in situ hybridization (FISH) signals (instead of the normal two signals), i.e., trisomy for a centromeric probe for chromosome 8. In the nucleus on the right, the arrowhead points to the pairing of signals for a centromeric probe for chromosome 17, indicative of an iso(17q), in a medulloblastoma. A normal signal for the normal chromosome 17 is also present. *Bottom right*: FISH staining (yellow) of chromosomes 1 in a metaphase. Shown are normal chromosomes 1 and an abnormal derivative chromosome (top) containing part of a chromosome 1.

The Genetics and Molecular Biology of Neural Tumors

Avery A. Sandberg, MD, DSc
John F. Stone, PhD

Department of DNA Diagnostics,
St. Joseph's Hospital and Medical Center, Phoenix, AZ

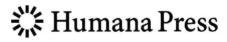 Humana Press

Avery A. Sandberg
Department of DNA Diagnostics
St. Joseph's Hospital and Medical Center
Phoenix, AZ
USA

John F. Stone
Department of DNA Diagnostics
St. Joseph's Hospital and Medical Center
Phoenix, AZ
USA

ISBN: 978-1-934115-58-9   e-ISBN: 978-1-59745-510-7

Library of Congress Control Number: 2007942034

Cover illustration: Figure 2, Chapter 8, "Medulloblastoma, Primative Neuroectodermal Tumors, and Pineal Tumors," and Bottom right image from Frontispiece.

Printed on acid-free paper

9 8 7 6 5 4 3 2 1

springer.com

We dedicate this book to our wives, Maryn Sandberg and Mary Stone

Preface

The Genetics and Molecular Biology of Neural Tumors is essentially an analectic presentation of a rapidly developing field for which information is being generated at a remarkable rate. Nevertheless, we think that this book fills a void existing currently in the tumors under consideration. In this volume, we have compiled the information available on neural tumors, presented both in text format and as tables. The latter should prove to be a ready and concentrated source of cogent information and references for the reader. By "neural tumors" we refer to those of the nervous system, exclusive of tumors solely astrocytic, glial, or neuronic (neuronal) in origin.

The information presented in the book is based on the efforts and contributions of thousands of investigators whose labors have been devoted to the cytogenetics, histopathology, embryology, molecular genetics, and molecular biology of neural tumors. The 3,500 references contained in this book are only part of the published output on the various other facets of these tumors.

The introductory segment of each chapter contains some clinical and epidemiologic information and succinct but relevant pathohistologic and immunohistochemical descriptions to facilitate the interpretation of the findings on the genetics and molecular biology of the tumors.

The fact that the genetic cause has not been rigorously established in any of the tumors presented in this book, and the emerging concept that multiple genetic factors may play an important role in the development of some tumors, have led us to a more compendial approach in presenting the available information on neural tumors, as reflected in the tables of each chapter.

An attempt has been made by us to address those topics (genetics and molecular biology) related to neural tumors of interest to clinicians (neurologists and neurosurgeons), clinical investigators, cytogeneticists, geneticists, and molecular biologists investigating neural tumors.

Despite some shortcomings in translating findings in experimental animals to human conditions and that this book is almost exclusively devoted to human neural tumors, some references from the voluminous literature on animal (primarily rodent) studies have been included.

Many of the tables in the book are based on information supplied in the publications of referenced authors, to whom we are much indebted, as we are to the authors and publishers for permission to reproduce appropriate figures.

The staff of the library at St. Joseph's Hospital, i.e., Mrs. Molly Harrington, Mrs. Jessica (Scott) Spalding, Mrs. Irma Contreras, and Mrs. Ellen Abrams, has been of incalculable help in gathering information in the literature, and Mrs. Nina E. Rosner for arranging such literature.

We owe special thanks to Ms. Jan Vaughan, who has "mothered" the volume through its various phases of birth and growth.

Avery A. Sandberg
John F. Stone
Phoenix, AZ

Contents

1

Benign Peripheral Nerve Sheath Tumors: Neurofibromas, Schwannomas, and Perineuriomas

Summary Chapter 1 deals with the genetics of molecular biology of benign peripheral nerve sheath tumors (neurofibromas, schwannomas, and perineuriomas). Besides their common association with the neurofibromatoses (NF1 and NF2) and the diversity in cellular origin of these tumors, they display diversity in their cytogenetic findings, with schwannomas (including the vestibular type) and some perineuriomas showing loss of chromosome 22 (-22) as a consistent karyotypic event. No recurrent chromosomal change has been established in neurofibromas. The relationship of gastrointestinal stromal tumors to NF1 of interest. The alterations of the genes closely associated with neurofibromas, *NF1* and its protein product neurofibromin and *NF2* and its protein product merlin, associated with schwannomas, are presented here. Also presented is the wide range of molecular and other studies on genetic and related factors possibly playing a role in the tumorigenesis and behavior of these benign nerve sheath tumors.

Keywords neurofibromas · schwannomas · perineuriomas · *NF1* (neurofibromin) · *NF2* (merlin).

Introduction

This chapter and the following two chapters deal with tumors, both benign and malignant, demonstrating a strong association with the neurofibromatoses, the *NF1* and *NF2* genes, and changes affecting chromosome 22 (1). Chapter 1 is concerned with benign tumors of neural origin, i.e., neurofibromas, schwannomas, and perineuriomas, which occur in high frequency in the neurofibromatoses (NF), although they also may be seen as sporadic tumors. Emphasis is placed on the genetic changes in these tumors, their close association with the neurofibromatosis type 1 (NF1) and NF2 (1–3), and on the genes *NF1* and *NF2* (and their protein products) and other genes in relation to their role in the tumorigenesis of neurofibromas, schwannomas, and perineuriomas.

NF1 and NF2 are disorders involving the nervous system, in which affected individuals have a high risk for developing both benign and malignant tumors (4–7, 7A). NF1 affects 1 in 3,500 individuals, and NF2 affects 1 in 30,000-42,000 individuals (8). The incidence of these syndromes is constant across all ethnic backgrounds, and there is no gender predominance.

For further clinical, pathologic, genetic, and molecular information related to the neurofibromatoses and their associated tumors, a large number of reviews and publications are available (5, 8–41).

Neurofibromatosis Type 1 (von Recklinghausen Disease)

NF1 is an autosomal dominant disorder characterized by multiple neurofibromas, malignant peripheral nerve sheath tumors (MPNST), optic nerve gliomas, astrocytomas, multiple hyperpigmented lesions (café-au-lait spots), axillary and inguinal freckling, iris hamartomas (Lisch nodules), and various osseous and vascular lesions (1). About 50% of NF1 patients have new germline mutations of the *NF1* gene. With the exception of large deletions, these spontaneous mutations occur predominantly in the paternal germline (31, 42).

In NF1, benign tumors arise in association with the peripheral nerve sheaths; the latter are composed of a mixture of schwann cells, fibroblasts, mast cells, endothelial cells, perineurial cells, and other cells (1, 43). The most common tumor associated with NF1, i.e., neurofibroma, occurs in nearly all patients with NF1. They are rarely present in childhood, but they develop during puberty and pregnancy, suggesting a hormonal influence on tumor growth. A variety of other types of tumors and leukemias may arise in NF patients (44, 45). The occurrence of NF1 features in restricted and localized areas of the body, termed segmental NF, has been shown to be due to a mutation of the *NF1* gene, and the regional distribution of manifestations in segmental NF reflects different cell clones, commensurate with the concept of somatic mosaicism (46).

Neurofibromas

Neurofibromas vary in size and number, and they are associated with small or large terminal nerves. Among the major subtypes of neurofibroma, the dermal and plexiform variants are characteristic of NF1 (34, 43). Molecular genetic investigations have indicated that NF1-associated neurofibromas are monoclonal (47). Sporadic neurofibromas are histologically identical to those occurring in NF1. The key roles played by the schwann cells in axon structure (48–50) and neoplasia of the neurofibromatoses are discussed in later sections of this chapter.

Dermal or cutaneous neurofibroma is a well-circumscribed, nonencapsulated benign tumor rich in extracellular matrix and variably composed of schwann and fibroblast-like cells, with an admixture of endothelial cells, lymphocytes, perineurial-like cells, and a large number of mast cells (34, 35, 52). These tumors usually do not occur until puberty, and they increase in frequency with age; they are present in nearly all adults with NF1. Dermal neurofibromas harbor little risk of malignant transformation, but they may carry a major cosmetic burden, particularly when numerous.

Plexiform neurofibromas (PNF), seen in about one third of NF1 cases, produce diffuse enlargement of major deep nerve trunks and their branches, and they are pathognomonic of NF1 (34, 53, 54). Histologically, PNF are composed of large hypertrophied nerves composed of spindle-shaped fibroblasts, schwann cells, and perineurial and mast cells embedded in a myxoid matrix (Figure 1.1.). PNF may occur at unusual sites,

e.g., the uterine cervix (43, 55). PNF are often elongate multinodular lesions involving either multiple trunks of a plexus or multiple fascicles of a large nerve (56, 57). In contrast to the more common dermal neurofibroma, PNF may involve multiple tissues, and they can grow to large size and cause significant morbidity by stimulating underlying bone growth or by compressing surrounding tissues (25). About 5–10% of PNF have a risk of malignant progression to MPNST, a highly aggressive and lethal tumor. It seems that MPNST arise from large PNF, and they are probably derived from Schwann cells, as reflected by many MPNST staining positively for Schwann cell markers such as S-100 (58, 59). In contrast, malignant transformation is a very rare event for other types of neurofibroma.

The difference in the mechanisms of tumorigenesis between cutaneous (dermal) neurofibromas and PNF is unclear; however, it has been suggested that the amount of *NF1* mRNA processing could play a role in determining which tumor type develops (60). In general, dermal (cutaneous) neurofibromas seem to have the lowest, PNF an intermediate, and MPNST the largest amount of mRNA editing of the *NF1* gene. These differences in mRNA processing could relate to the variability of tumor expression in NF1 patients. *NF1* promoter methylation in PNF has been reported as a possible mechanism in these tumors (61–63).

Loss of Heterozygosity (LOH) Changes in Neurofibromas

Some early studies failed to find LOH in neurofibromas (47, 64, 65), whereas LOH was found in malignant NF1-related tumors (66–68). In addition, the presence of both alleles of the *NF1* gene was described in schwann cells derived from neurofibromas (69). However, subsequent investigations reported somatic *NF1* deletions in neurofibromas (70, 71) and mutations in both copies of the *NF1* gene in a dermal neurofibroma (72). Furthermore, genetic instability of several microsatellite markers was seen in neurofibromas (73).

LOH in the *NF1* gene has been found in all types of neurofibroma (70, 71, 74–83), with both somatic and germline mutations of *NF1* expressed at the RNA level (75). The studies point to the *NF1* gene to be a tumor suppressor gene (TSG), both copies of which are altered in benign and malignant NF1-related tumors (68, 72, 75, 80, 81, 84).

It is of interest that most of the neurofibromas with LOH for the *NF1* gene (when such information was available in the publications) have been PNF (74, 78, 82), a companion finding for the higher rate and type of chromosome changes seen in PNF and their propensity for malignant change (Table 1.1).

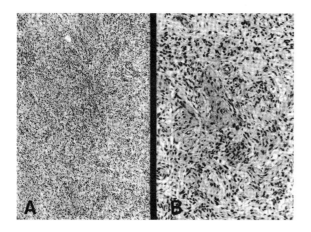

FIGURE 1.1. Low- (**A**) and high-power (**B**) histologic picture of a plexiform neurofibroma in an adult with probable NF1 (89). The tumor is composed of spindle cells arranged in thin and delicate fascicles. Cytogenetic analysis revealed the tumor to have a complex karyotype with multiple clones but each containing a t(1;22)(p23;q11) (89).

S-100 Protein

The *NF1* gene status in S-100 protein-positive and -negative cell subpopulations was evaluated (85) in archival paraffin-embedded specimens from seven PNF, two atypical PNF,

TABLE 1.1. Chromosome changes in neurofibromas.

	References
46,XX,der(2)t(2;19)(p25;p13.2),der(16)t(2;16)(p25;q24), del(19)(p13.2)	*Riccardi and Elder 1986 (94)*
47,XX,+18	*Riccardi and Elder 1986 (94)*
*46,X,t(X;2)(p22;q21),del(1)(q11),t(1;22)(p32;q11)c,der(8)t(1;8) (q11;q2?4)/46,XX,t(1;22)**c**,add(5)(q14),add(16)(p12)/ 45,XX,t(1;22)**c**,inv(8)(p23q12),der(9)t(3;9)(q12;p22), del(3)(q26 q28),der(10)t(3;10)(p11;p12),del(12)(q14)	*Rey et al 1987 (89)*
45,XX,-22	*Chadduck et al 1991-1992 (90)*
**45~46,XX,t(1;17)(q21;q21),r(3;?)(p25q29;?),add(5)(q14), der(9)t(4;?;9)(q12;?;p11),der(9)t(5;9)(q13;p13),add(15)(q22), +der(?)t(?;7)(?;q11.2),+mar	*Molenaar et al 1997 (91)*
47,XY,+der(?)t(?;12)(?;q15)	*Mertens et al 2001 (95, 96)*
*46,XY,t(1;9)(p36.3;p22)46,XX,t(2;11)(q13;q23)	*Wallace et al 2000 (92)*
*46,XX,t(2;11)(q13;q23)	*Wallace et al 2000 (92)*
*46,XX,t(1;9)(p36.1;q21.2), t(9;12)(p24;q13), t(11;17)(q?14.2;?p13)	*Wallace et al 2000 (92)*
46,XX,t(2;11)(q21;q13),t(3;19)(p12;q13.1),del(6)(q21.2), add(10)(p15),del(14)(q24),add(22)(p11.1)/ 53 58,XX,+2,add(2)(p21)x2,add(2)(q37),+6,+7,add(7)(q34), +8,add(11)(p11.2),add(11)(q23),-13,add(14)(p11.1),+16, add(16)(q12.1),+17,+19,add(19)(q13.3),+20,+21,-22	
x46,XX,der(2)t(2;19)(p25;p13.2),der(16)t(p25;q24),del(19)(p13.2)	*Nandula et al 2004 (93)*
x46,XX,t(4;9)(p31;p22)	*Sawyer et al 2005 (301)*
46,XX,der(2)t(2;17)(p23;q11-12)t(17;17)(q11-12;q25),del(17) (q11-12),der(17)t(?2;17)(p23;q11-12)	*Storlazzi et al 2005 (301A)*

*Plexiform neurofibroma.
**Diffuse neurofibroma (early malignancy?), recurrent tumors.
xSolitary neurofibroma.

one cellular/atypical PNF (all probably associated with NF1), and eight MPNST (four associated with NF1). *NF1* loss was detected in four of seven PNF and in one atypical PNF, with deletions restricted to S-100 protein-immunoreactive schwann cells. In contrast, all eight MPNST harbored *NF1* deletions, regardless of S-100 protein expression or NF1 clinical status. The results suggest that the schwann cell is the primary neoplastic component in PNF and that S-100 protein-negative cells in MPNST represent dedifferentiated schwann cells, which harbor *NF1* deletions in both NF1-associated and sporadic tumors (85).

It has been postulated that S-100–positive cells might be primarily responsible for the development of neurofibromas (86), although they do not always express myelin P0, which is a specific marker of differentiated schwann cells. S-100–positive cells from neurofibromas of NF1 patients lack *NF1* mRNA; thus, they do not express the protein neurofibromin. However, the fibroblasts from the same tumors carry at least one normal *NF1* allele; thus, they produce mRNA and neurofibromin. The lack of expression of mature schwann cell proteins (i.e., myelin P0) seen in S-100–positive cells raises the possibility that neurofibromin loss may occur in a precursor cell population or cause the differentiated schwann cell to revert to a more immature phenotype (86). It has been demonstrated that CD34+ and S-100–negative cells in neurofibromas and schwannomas are nonneoplastic supportive cells (87).

Chromosome Changes in Neurofibromas

In an early study (88), a 6.7-fold higher frequency of aneuploid mitoses was found in neurofibromas from three patients with NF1 compared with skin cultures from three healthy control-matched donors. Each of the neurofibromas contained a clone with monosomy 22 (88). Such monosomy was seen in two neurofibromas shown in Table 1.1. Cytogenetic studies of neurofibromas based on banding have been sparse (Table 1.1), with no recurrent karyotypic change being consistently present. In some neurofibromas the karyotypes may be complex (89–93), possibly indicating malignant transformation of neurofibromas to MPNST (91, 94); thus, they can be of help in arriving at a diagnosis. For example, a 62-yr-old woman presented with a solitary neurofibroma; a second recurrence showed features indicative of malignancy, but it was insufficient to establish a histologic diagnosis (91). Cytogenetic analysis revealed abnormalities compatible with MPNST, supporting the histologic suspicion of emerging malignancy.

Nine of 10 neurofibromas examined in one study (95, 96) had normal karyotypes, as did five neurofibromas (67) and seven dermal neurofibromas in other studies (92). In the latter study (92), four of six PNF tumors had karyotypic changes, although they differed from tumor to tumor, with one tumor of these having a complex karyotype, possibly reflecting malignant foci. In a study of 26 neurofibromas, 66% were found to be diploid and the remainder aneuploid (97). Sporadic neurofibromas and schwannomas seem to have cytogenetic and/or molecular genetic changes akin to the tumors associated with NF. It should be stressed that Table 1.1 contains only those neurofibromas for which complete karyotypes were available, and it does not include those tumors for which only partial karyotypes or nonclonal changes were presented.

Gain of chromosome 17 in PNF has been suggested as useful for the differential diagnosis of benign and borderline tumors (98, 99).

The *NF1* Gene

The *NF1* gene, located at 17q11.2, comprises 350-kb of genomic DNA with 60 exons and it encodes an mRNA of 11-13 kb (100–103). The localization of *NF1* at 17q11.2 was helped considerably by the finding of constitutional translocations in families with NF, i.e., t(17;22)(q11.2;q11.2) in two families (104, 105), a t(1;17)(p34.3;q11.2) (105) in one family, and an inv(17)(q11.2q25.1) (106) in another family. Appropriate studies showed involvement of the *NF1* gene in each of these translocations. The *NF1* gene product, a protein called neurofibromin, is normally expressed in neurons, oligodendrocytes, schwann cells (107), and many other cell types (92, 100, 103, 108–112). Somatic mutations in the *NF1* gene have been described in tumors unrelated to NF1 (68, 84, 113–116). An interesting observation in this regard is the transcriptional suppression of the *NF1* expression by the fusion protein resulting from t(8;21) and its fusion gene *RUNX1-MTG8* in acute myeloblastic leukemia (116A). All this evidence corroborates the hypothesis that *NF1* is a TSG (7, 74, 77, 78, 82, 85, 86, 92).

The failure to find *NF1* anomalies in some neurofibromas may be related to inactivating mutations occurring in parts of the gene not tested by the methodologies used, to the changes that are not sufficiently quantitative (85) to be detected, or to genes other than *NF1* being affected. These comments apply in particular to sporadic type of tumors. Rapid degradation of the abnormal protein and a decreased amount of mRNA transcribed from the mutant allele may contribute to failure of detection (117). The failure of mutations of *NF1* in NF1 to cluster in hot spots and their large number (>300), with only 7% observed more than once, may be another explanation.

The *NF1* gene, in part due to its large size, is associated with a spectrum of qualitative and quantitative alterations in NF1 tumors, including LOH of variable size and locations and somatic *NF1* point mutations (75, 118). NF1 is a familial tumor syndrome in which the type of germline mutation of the *NF1* gene influences the type of second-hit in the tumors (118A). Generally, the same molecular changes were seen in multiple neurofibromas (75).

The NF1 patients who develop MPNST may have any type of *NF1* germline mutation, such as a total gene deletion, a frameshift mutation, an in-frame deletion, or a missense mutation (119). No specific type of *NF1* germline mutation has been found in NF1 individuals with MPNST, but large *NF1*

gene deletions were more frequently found in this group than reported for NF1 individuals, in general.

A high rate of *NF1* mutations (76%) in 38 neurofibromas from nine NF1 patients was reported previously (119A). This change underlies reduced *NF1* expression in schwann cell cultures. Intragenic mutations were the most common, particularly frameshift mutations (41%).

Somatic mutations of the *NF1* gene have been reported in a variety of neoplasias in patients without NF1 (113).

Neurofibromin

Neurofibromin, the product of *NF1*, is present in several isoforms (112). Neurofibromin is predominantly expressed in neurons, schwann cells, oligodendrocytes, astrocytes, and leukocytes (121A). Among its functions is an association with microtubules and participation in several signaling pathways. Neurofibromin is a regulator of RAS GTPase–activating proteins, which are negative regulators of *RAS* function (120), thereby they control cell growth and differentiation (71, 92, 100, 101, 103, 108–111, 121) (Figure 1.2). The absence of neurofibromin in NF1 cells leads to increased levels of activating proteins (e.g., p21^{ras} and p13), which contribute to cellular proliferation (122–124) associated with neurofibromas, schwannomas, and MPNST (121, 125).

At the cellular level, neurofibromas have a unicellular clonal origin (47), with schwann cells being the progenitors (9, 10, 62, 126–129). Although previous studies indicated a possible crucial role of fibroblasts in the development of these benign tumors (58), mutations of *NF1* leading to neurofibroma formation were demonstrated to affect primarily the schwann cells (130), and not the fibroblasts, and a portion of the schwann cells in neurofibromas may have mutations

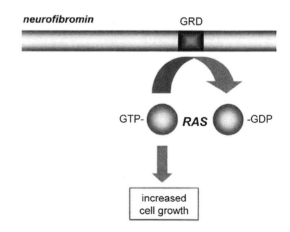

FIGURE 1.2. Schematic presentation of a possible mechanism for neurofibromin control of cell growth. Neurofibromin functions in part as a negative regulator of the *RAS* oncogene by accelerating the conversion of active guanosine triphosphate (GTP)-bound *RAS* to its inactive GDP-bound form. Active RAS, as a result of reduced neurofibromin expression, leads to increased cell growth and facilitates tumor formation (from ref. (41), with permission).

of both alleles (128). Schwann cells deficient in neurofibromin acquire tumorigenic properties, leading to neurofibroma development (62). Thus, these schwann cells acquire genetic changes, epigenetic changes, or a combination for tumor development, including angiogenic and invasive properties (125, 127).

Gastrointestinal Stromal Tumors (GIST) and NF1

GIST (Table 1.2) are mesenchymal tumors that arise throughout the gastrointestinal tract and, rarely, in the mesentery or retroperitoneum. They are characterized by strong immunohistochemical staining for CD117, and most contain activating mutations in the KIT receptor tyrosine kinase, particularly at exons 9, 11, 13, and 17. Approximately one third of tumors that are wild type at the *KIT* gene harbor mutations in the juxtamembrane and tyrosine kinase domains of a related tyrosine kinase receptor, i.e., platelet-derived growth factor receptor-α (PDGFRA). In most cases, GIST are sporadic; however, they may occur as multiple lesions in familial forms associated with *KIT* or *PDGFRA* germline mutations. In these situations, GIST arise in a background of interstitial cell of Cajal hyperplasia, and they may be associated with systemic manifestations including skin hyperpigmentation and mast cell disorders. An association between the development of multiple GIST and NF1 also has been established; however, the molecular abnormalities of GIST arising in this setting have not been elucidated, and studies assessing their molecular features are limited. One study (130A) showed that although most patients with NF1 and GIST do not have *KIT* or *PDGFRA* mutations, *KIT* germline mutations might be implicated in the pathogenesis of GIST in some patients.

The occurrence of GIST is common among NF1 patients, and study of GIST patient groups will usually reveal a substantial number with NF1 (132, 133). GIST in NF1 patients are often multicentric (134) (Table 1.2), a phenomenon rare in sporadic GIST cases, although exceptions may occur (135). Gastric GIST is common among sporadic cases, whereas the small intestine is the preferable site in NF1 patients

(136–138). Histologically, the NF1-associated GIST have a predominantly spindle-shaped cell morphology, in contrast to the epithelioid type often seen in sporadic GIST (133, 139, 140). In contrast to the non-NF1 GIST, the NF1-associated tumors lack the mutations of *KIT* (141), and this precludes their response to imatinib (142).

Mast Cells in Neurofibroma and the *KIT* Gene

The presence of mast cells in neurofibromas and schwannomas has led to investigations on the *KIT* status in these tumors (130A, 143), in view of *KIT* mutations being common in mastocytosis (144) and that normal schwann cells do not express *KIT* but schwannomas do (143, 145). Possible therapies for neurofibroma have addressed the mast cells and their *KIT* content (16, 146, 147, 162).

Most patients with NF1 have multiple GIST (134–137, 148). In a study (133) of 36 such tumor samples from nine patients with NF1, none had metastases or died of the disease, and none of the tumors showed histologic evidence of neurofibroma or schwannoma. Anomalies of *KIT* and *PDGFRA* genes were found in most of the tumors, but they differed from GIST in non-NF1 patients in their clinicopathology, phenotype, and the nature of the mutations of the *KIT* and *PDGFRA* genes. It is possible that haplodeficiency of neurofibromin promotes the growth of the interstitial cells of Cajal expressing both *KIT* and S-100 protein in GIST of the small intestine in NF1 cases. This may explain why GIST develops in NF1 patients without gain-of-function mutations of the *KIT* gene (133).

Although *KIT* germline mutations may be implicated in the pathogenesis of GIST in some NF1 patients (130A, 133), there is general agreement that the bulk of NF1-associated GIST are phenotypically and genetically distinct from sporadic GIST, indicating that different mechanisms are involved in their pathogenesis (133A–D). The response to imatinib of some patients with *KIT*- and *PDGFRA*-wild-type GIST (131) may be explained by the observations referred to above (130A).

TABLE 1.2. GIST: Features of NF1-associated and sporadic GIST.

NF1-associated GIST	Sporadic GIST
Incidence of GIST high	Relatively rare tumor
Multicentric tumors	Usually solitary tumor
Spindle-cell histology predominantly	Epithelioid histology not uncommon
Low-grade appearance and non-aggressive course	High-grade appearance and often malignant
Predominantly located in small intestine	Often located in stomach
Strong CD34 staining in >90% of cases	CD34 staining in less than 50% of tumors
KIT mutations not present	*KIT* mutations common
CD117 staining present	CD117 (50-80%) staining present
PDGFRA mutations not present	*PDGFRA* mutations present (not with *KIT* mutations)
Skenoid fibers present in almost all tumors	Only about half of tumors have skenoid fibers
S-100 expression common	S-100 expression uncommon

Interstitial neurofibromatosis is an alternate form of neurofibromatosis with neurofibromas limited to the intestine in the absence of any other features of NF1 and NF2. A study of one family (three sisters) with intestinal neurofibromatosis showed a mutation of *PDGFRA*, a condition that may be allelic to familial GIST caused by *PDGFRA* mutations (133E).

Neurofibroma and Transformation

The complexity of the genetic events leading to and associated with neurofibroma development and during transformation to MPNST was further pointed to by a study of apolipoprotein D (apoD) (149). The gene for apoD is a member of the lipocalin family of genes involved in the transport of small hydrophotic molecules (e.g., cholesterol and steroid hormones), and it is ubiquitously expressed in almost all tissues (150).

ApoD expression was studied in peripheral nerves, NF, and MPNST by quantitative reverse transcription-polymerase chain reaction (RT-PCR), in situ hybridization, and immunohistochemistry (149). Multiplex quantitative RT-PCR for messenger RNA was performed on a series of formalin-fixed and paraffin-embedded specimens that included nine MPNSTs, 12 NF (two plexiform), and four normal peripheral nerves. ApoD expression increases as NF are formed, followed by a marked decline in apoD expression as NF undergo malignant transformation into MPNST (149). These changes in apoD expression levels coincide with changes in cell cycle inhibition mediated by cyclin-dependent kinase inhibitors. The specific mechanisms underlying changes in apoD expression associated with peripheral nerve tumorigenesis and cell cycle inhibition remain to be elucidated.

Neurofibromas share at least one common feature of senescence, i.e., increased apoD expression (149). ApoD expression has been found to correlate with inhibition of the cell cycle in cellular senescence and other settings. This expression of apoD seems to parallel the pattern of CDK-1 activity in NF1-associated tumorigenesis in which CDK-1 is activated in neurofibromas (151, 152), followed by loss of cell cycle inhibitors, including p53, p16, and p27 during transformation (151–156).

Formation of neurofibromas in NF1 cases is related to constitutive activation of the *RAS* signaling pathway, resulting from inactivation of the *NF1* gene (122, 157). RAS activates the p53 and p16 tumor suppressors, thereby inducing cell cycle arrest and a phenotype indistinguishable from replicative senescence (158–160). Under other conditions, where either p16 or p53 is inactivated, *RAS* can cause transformation (158, 161).

Activated Ras guanosine triphosphate (GTP), as a result of oncogenic mutations, is common in human cancer (162). Markedly elevated levels of Ras GTP were demonstrated in NF1 neurofibromas compared with schwannoma, and they were associated with increased tumor vascularity, probably due to increased VEGF secretion (162). Normalization of the Ras GTP levels by farnesylthiosalicylic acid has been shown to be a potential approach to therapy in NF (163).

Study of gene expression in neurofibromas (nine dermal and nine plexiform) and 16 MPNST from NF1 patients revealed expression to be present only in MPNST (metalloproteinase 13) or to differ quantitatively (PDGFRA and fibronectin) (164).

Pigmented neurofibromas are a rare subtype of neurofibroma occurring in NF1 cases (165). In contrast to neurofibromas and schwannomas, pigmented neurofibromas were shown to coexpress the microphthalmia-associated transcription factor (MITF) and the *MET/HGF* oncogene (165), factors that play a role in melanocytic neoplasms.

Table 1.3 lists genetic facets, other factors, and substances and various areas that may play a role in neurofibroma genesis and biology.

Neurofibromatosis Type 2

NF2 is an autosomal dominant syndrome affecting intracranial and intraspinal nerve sheaths without peripheral manifestations, and it may be associated with other types of tumors, including brain tumors (astrocytoma, ependymoma) and meningiomas (166, 166A, 167, 168, 168A). Schwannomas are characteristic of NF2, and although sensory nerves are the preferred site of development, motor and autonomic nerves also may be affected. Visceral schwannomas are rare (34, 56). Bilateral vestibular schwannomas are characteristic of NF2.

NF2 at present has been divided into three categories. One type (Wishart) has a severe clinical presentation, with bilateral vestibular schwannomas, spinal tumors, and onset in late teens and early twenties. Another NF2 type (Gardner) has a late onset and less severe presentation, and although bilateral vestibular schwannomas are common, other tumors are not. In the third NF2 type, termed mosaic NF2, a mutation occurs during embryogenesis, rather than in the germline DNA; hence, varying portions of the patient's cells carry the mutation (168B), accounting for the variable clinical picture in these cases (unilateral vestibular schwannoma, meningioma).

Schwannomas

Schwannomas (neurilemomas, neurinomas, perineural fibroblastomas) have been subtyped on the basis of their histology, location, or both. Thus, besides conventional schwannomas, cellular, melanotic, epithelioid, and plexiform types have been described (1, 56, 132, 169–171). Because, in general, genetic alterations do not seem to vary among these tumor types, alterations are addressed to these tumors as a group.

Schwannomas are usually encapsulated benign tumors of peripheral nerve sheath composed of differentiated schwann

TABLE 1.3. NF1 and neurofibromas.

	References
Structural differences in the minimal catalytic domains of the GTPase-activating p120^{GAP} protein and neurofibromin	*Ahmadian et al 1996 (415)*
Minichromosome formation by excision of 17q in a case of NF1.	*Andersen et al 1990 (302)*
Mutations and other changes of *NF1* in tumors other than those of peripheral nerve sheaths.	*Andersen et al 1993 (303)*
	Seizinger 1993 (13)
	The et al 1993 (304)
	Koh et al 1995 (305)
	Martinsson et al 1997 (306)
	Side et al 1997 (84)
	Gran and Safali 2005 (307)
Vascularity of neurofibromas.	*Arbiser et al 1998 (308)*
Doublecortin, a microtubule-associated phosphoprotein involved in neuronal migration and differentiation, is expressed in neuro- blasts. Its expression was found in neurofibromas but not in schwannomas.	*Bernreuther et al 2006 (428)*
In NF with decreased neurofibromin protein, active RAS-GTP protein is increased.	*Bernards 1995 (120)*
	Kim et al 1995 (124)
	Bollog et al 1996 (413)
	Guha et al 1996 (121)
	Sherman et al 2000 (130)
Differential regulation of RAS-GAPase and *NF1* activity.	*Bollag and McCormick 1991 (412)*
	Bollag et al 1996 (413)
Functions of the *NF1* gene.	*Buchberg et al 1990 (157)*
	Xu et al 1990 (110,111)
Animal models for NF1.	*Cichowski et al 1999 (309)*
	Vogel et al 1999 (310)
	Weiss et al 1999 (411)
	DeClue et al 2000 (125)
Benign peripheral nerve sheath tumors are very rare in the spinal cord.	*Chi et al 2006 (430)*
Role of growth hormone receptor in neurofibroma development.	*Cunha et al 2003 (311)*
LOH at intron 38 of *NF1* in plexiform neurofibromas, but not in dermal neurofibromas.	*Däscher et al 1997 (74)*
Mammalian target of rapamycin (mTOR) signaling pathway is regulated by the neurofibromin GTPase-activating protein-related domain, and this system may represent a logical and tractable biologically based therapeutic approach for NF1 tumors (neurofibroma, astrocytoma).	*Dasgupta et al 2005 (427)*
FISH studies of 21 neurofibromas (11 plexiform, 4 cutaneous, 6 subcutaneous) failed to find somatic deletions, pointing to mutational mechanisms.	*DeLuca et al 2004 (83)*
Hedgehogs and their receptors may be involved in tumorigenesis of NF1.	*Endo et al 2003 (409)*
Pigmented (melanotic) neurofibromas must be differentiated from Bednar tumor (DFSP); the latter has more storiform growth, CD34 positivity, and S-100 protein.	*Fetsch et al 2000 (312)*
Coexpression of HGF and C-Met high in neurofibromas.	*Fukuda et al 1998 (313)*
GAP-related domain of NF1 does not play a part in tumorigenesis; it does interact with *ras* p21.	*Gomez et al 1995 (314)*
	Ballester et al 1990 (108)
	Martin et al 1990 (109)
	Xu et al 1990 (110,111)
Genetics and molecular biology of NF and surgical facets.	*Gottfried et al 2006 (408)*
NF1-associated neurofibromas have markedly elevated levels of RAS GTP and associated with increased tumor vascularity.	*Guha et al 1996 (121)*
NF1-related loci on chromosome 22, 14 and 2.	*Hulsebos et al 1996 (315)*
Growth response of neurofibromas to PDGF-BB and TGF-β1.	*Kadono et al 1994 (316)*
Interaction of PDGF and PDGF-βR and not TGF-β and CTGF may play a role in neurofibroma development.	*Kadono et al 2000 (316A)*
Aberrant splicing caused by environmental factors may reduce tumor suppressor mRNA of *NF1* and *NF2* in neurofibromas, schwannomas, and meningiomas.	*Kaufmann et al 2002 (317)*
p53 mutations rare in neurofibromas, especially in plexiform type.	*Kluwe et al 1999 (77)*
	Borman et al 1993 (318)
	Levine 1997 (319)
	Rasmussen et al 2000 (82)
LOH of *NF1* in schwann cells (*not* in fibroblasts) responsible for NF.	*Kluwe et al 1999 (78)*
Fibroblasts from cafe-au-lait lesions and from neurofibromas of NF1 patients showed increased tolerance to ionizing radiation.	*Kopelovich and Rich 1986 (320)*
Benign epithelioid peripheral nerve sheath tumors of soft tissue have to be differentiated from neurofibromas and schwannomas.	*Laskin et al 2005 (321)*
Merlin is involved in the regulation of ubiquitination and degradation of transactivation-responsive RNA-binding protein.	*Lee et al 2006 (432)*

TABLE 1.3. Continued

	References
The putative tumor suppressor GTP-binding protein that interacts with merlin in suppressing cell proliferation has been reported.	*Lee et al 2007* (429)
Enhanced levels of some growth-stimulating (midkine and stem cell factor) may contribute to diffuse tumorigenesis in NF1.	*Li et al 1992* (113)
Aberrant Notch signaling may play a role in the conversion of neurofibroma to MPNST.	*Li et al 2004* (322)
NF are not generally p53 immunoreactive as compared with MPNST in which p53 mutations may be a factor in the poor prognosis of these tumors.	*Liapis et al 1999* (410)
Microsatellite instability and promoter methylation are not major causes of *NF1* inactivation in neurofibromas, although instability in chromosome 9 has been reported.	*Luijten et al 2000* (61)
	Fargnoli et al 1997 (323)
HGF/MET role in nerve sheath tumors.	*Ma et al 2003* (324)
	Rao et al 1997 (325)
	Zhang and Vande Woude 2003 (326)
Epidermal growth factor receptor and aberrant proliferation in the CNS in children with NF1.	*Mangoura et al 2006* (407)
Estradiol binding in neurofibromas.	*Martuza et al 1981* (327)
Midkine, angiogenic factor, may play role in neurofibromas.	*Mashour et al 2001* (328)
Growth factor levels and diffuse tumorigenesis in NF1.	*Mashour et al 2004* (329)
Deletion of both copies of *NF1* in schwann cells combined with *NF1* heterozygosity in the tumor promotes NF formation in mice.	*McLaughlin and Jacks 2002* (406)
Progesterone may play a role in neurofibroma development.	*McLaughlin and Jacks 2003* (330)
Role of progesterone in neurofibroma growth.	*McLaughlin and Jacks 2003* (330)
NF1 and unexplained iron deficiency: is it related to NF growth?	*Moczygemba and Bhattacharjee* (405)
Matrix metalloproteinase expression and differences in proliferation and invasion of normal, transformed and NF1 schwann cells.	*Muir 1995* (331)
Transformation of neurofibromas in NF1 associated with *CDKN2A/p16* inactivation.	*Nielsen et al 1999* (154)
Deletion or allelic loss of varying nature of the *NF1* gene have been demonstrated in a number of different tumors other than neurofibromas (e.g. rhabdomyosarcoma, optic glioma, pilocytic astrocytoma).	*Oguzkan et al 2006* (425)
Role of p16 and p14 or p14 alone in neurofibroma.	*Petronzelli et al 2001* (332)
Possible role of 9p21 in plexiform neurofibroma	*Petty et al 1993* (333)
LARGE, located at 22q12.3-q31.1, member of glycosyltransferase family, and its possible role in neurofibroma.	*Peyard et al 1999* (334)
CD44 and schwann cells.	*Ponta et al 2003* (335)
Mutation of *HPNCC* gene may predispose to mild form of NF1.	*Raevaara et al 2004* (336)
Loss of allelic synchrony in the replication of *RB1*, *AML1* and *CMYC* genes is seen in lymphocytes of NF1 cases as compared with normals studied by FISH.	*Reish et al 2003* (337)
EGF genetic polymorphism may be associated with aggressive clinical features of NF1, but not with a malignant phenotype.	*Ribeiro et al 2007* (337A)
De novo *NF1* changes in children with *MLH1* deficiency.	*Ricciardone et al 1999* (338)
	Wang et al 1999 (339)
Cytogenetic deletion at 17q11.2 in a case of NF1 and contiguous gene syndrome.	*Riva et al 1996* (340)
Clinical and diagnostic implications of mosaicism in NF.	*Ruggieri and Husoh 2001* (28)
Merlin interacts with Ral guanine nucleotide dissociation stimulator and inhibits its activity, including its oncogenic signals.	*Ryu et al 2005* (431)
An inverse pattern of *ERBB2* expression in neurofibromin in non- neoplastic schwann cells but not in schwannomas.	*Schlegel et al 1998* (414)
Besides Ras-dependent growth inhibition, neurofibromin can exert a tumor suppressive effect via a Ras-independent pro aptotic effect.	*Shapira et al 2007* (426)
NF1 and malignancy.	*Shearer et al 1994* (44)
The plasminogen activators, urokinase (uPA) and tissue type (tPA) and their inhibitor PAI-1 were prominently expressed in the schwann cell component of neurofibromas, but not in the mast cells, macrophages, and endothelial cells of these tumors.	*Sirn et al 2006* (433)
NF1 mutations and other changes in NF1.	*Skuse et al 1989* (66)
	Ponder 1990 (341)
	Wallace et al 1990 (342)
	Hoffmeyer et al 1994 (343)
	Kehrer-Sawatzki et al 1997 (104, 344)
	Upadhyaya et al 1997 (345)
	Messiaen et al 1999, 2000 (346, 347)
	Toliat et al 2000 (348)
	Zou et al 2004 (349)
Genetics of NF1-associated peripheral nerve sheath tumors.	*Stephens 2003* (404)

TABLE 1.3. Continued

	References
Mitotic recombination resulting in a reduction to homozygosity of a germline *NF1* mutation in an NF1-associated GIST has been reported. The authors hypothesized that the LOH of *NF1* and lack of *KIT* and *PDGFRA* mutations constitute evidence of an alternative pathogenesis in NF1-associated GIST.	*Stewart et al 2007 (424)*
The interaction of merlin with cellular proteins may be a crucial factor for tumor suppression by merlin.	*Takeshima et al 1994 (168D)*
NF1 mutations in neuroblastoma.	*The et al 1993 (304)*
Neurofibromin regulation of adenyl cyclase.	*Tong et al 2002 (350)*
Clinical manifestations and management of NF1.	*Tonsgard 2006 (402)*
An interesting observation was the absence of cutaneous neurofibromas in NF1 cases with a 3-bp deletion of exon 17 of the *NF1* gene.	*Upadhyaya et al 2007 (423)*
hMLH1-deficient children and NF1 development and early onset of extracolonic cancer.	*Wang et al 1999 (339)*
NF1 is the most frequent tumor predisposition syndrome.	*Wimmer 2005 (403)*
Fornesyltransferase inhibitors block the NF1 malignant phenotype.	*Yan et al 1995 (351)*
Interaction of schwann cells with mast cells in neurofibroma formation.	*Yang et al 2003 (287)*
The genetic and molecular pathogenesis of NF1 and NF2.	*Yohay 2006 (401)*
22q12 and possible genes.	*Zucman-Rossi et al 1996 (252)*
A general overview of NF1, as related to the cellular microenvironment, including the role of mast cells and their interaction with schwann cells has been presented. This review included cogent studies in mice and a sympathetic expression and understanding of NF1 patients' vicissitudes.	*Le and Parada 2007 (455)*

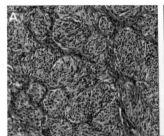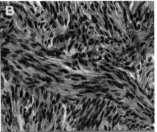

FIGURE 1.3. Histology of plexiform cellular schwannoma from the plantar subcutaneous tissue of an 8-month-old boy not affected by NF. A low-power view is shown on the left **(A)** and a high-power view is shown on the right **(B)**. The plexiform multinodular architecture of the tumor consists of fascicular proliferation of spindled cells forming interconnecting, compact nodules separated by eosinophilic extracellular material (192). The tumor demonstrated diffuse staining for S-100 and an abnormal karyotype, i.e., 47,XY,+17/48,XY,+17,+18.

cells (Figure 1.3). All ages are affected, but there is a peak incidence in the fourth to sixth decades. Generally, there is no sex predilection, the exception being a female:male ratio of 2:1 among patients with intracranial tumors. In the absence of NF2, multiple, often subcutaneous schwannomas are indicative of schwannomatosis, a rare and genetically distinct disorder (172,173).

NF2 schwannomas present at an earlier age than the sporadic type, including the diagnostic hallmark of NF2, i.e., bilateral vestibular schwannomas. Schwannomas are benign tumors, and they rarely, if ever, undergo malignant transformation (170). Unilateral vestibular schwannoma may be accompanied by other NF2-related tumors (intracranial, spinal) (170A).

In addition to schwannomas of the vestibular division of the eighth cranial nerve, other sensory nerves may be affected, including the fifth cranial nerve and spinal dorsal roots. Motor nerves such as the 12th cranial nerve also may be involved. Cutaneous schwannomas occur, and they may be plexiform in type. The incidence of sporadic forms of schwannoma is estimated as 2 in 100,000 individuals, although the incidence is being revealed to be higher partly because of the development of MRI scanning with gadolinium enhancement (174).

Of interest is the description of two cases in which a conventional schwannoma was contiguous with a deep plexiform schwannoma (174A). Cavernous malformation in a schwannoma (which may also occur in gliomas) has been described previously (174B); the possibility that loci on 7q21-q22, 7p15-p13, and 3q25-q27 may harbor cerebral cavernous malformation genes was discussed therein.

Description of clinical and pathologic aspects of schwannoma of the gastrointestinal tract has been published previously (175).

Chromosome Changes in Schwannomas

Schwannomas with karyotypic changes are shown in Table 1.4. The changes for some schwannomas, thought not to be complete in some reports, have been presented in Table 1.4 as our interpretation of the findings in the original publications (89,176–180). Except for monosomy 22 (-22), seen in various types of schwannoma, other changes have not been consistent or recurrent. Indications of loss of chromosome 22 (based on cytogenetic analysis) in schwannomas, including vestibular, were obtained in the 1970s (181, 182). Molecular studies a decade later (183, 184) established the involvement of chromosome 22 as characteristic of vestibular schwannomas.

In an early study (181), the cytogenetic findings in seven schwannomas and 1 MPNST were described; karyotypically, normal (diploid) and hypodiploid stemlines in the schwannomas were seen. Chromosomal loss was seen in the

TABLE 1.4. Chromosome changes in schwannomas.

Karyotype	References
-Y**	Rey et al 1987 (185)
-17,-22**	Rey et al 1987 (185)
-18,13q+**	Rey et al 1987 (185)
-22** (three tumors)	Rey et al 1987 (185)
44,?,t(X;1)(p11p13;q35),t(3;8)(p13;p11)	Teyssier and Ferre 1989 (177)
44,?,t(X;1)(p11p13;q35),t(3;8)(p13;p11)	Teyssier and Ferre 1989 (177)
44,X,-Y,del(3)(q12),-6*	Teyssier and Ferre 1989 (177)
45-46,?,inv(1)(p36q11)	Teyssier and Ferre 1989 (177)
46,X,-Y*	Teyssier and Ferre 1989 (177)
47,?+5*	Teyssier and Ferre 1989 (177)
44,?,-18,-22*	Rogatto and Casartelli 1989 (176)
45,?,dic(X;6),inv(9)	Rogatto and Casartelli 1989 (176)
45,?,15q+,-22*	Rogatto and Casartelli 1989 (176)
46,?,inv(9)(p13q21) *	Rogatto and Casartelli 1989 (176)
<44,X,-Y,-22*	Courturier et al 1990 (178)
<45,X,t(X;1;13),-22	Courturier et al 1990 (178)
45,XY,-22	Courturier et al 1990 (178)
45,XX,-22*	Courturier et al 1990 (178)
45,XY,-22*	Courturier et al 1990 (178)
45,XY,-22/46,XY,ipsuidic(22)	Courturier et al 1990 (178)
44-46,X,-X,del(3)(cen),-15,-22	Lodding et al 1990 (179)
¡<43-46,X,-X,-7,-15,-22	Stenman et al 1991 (180)
¡<43-46,XY,-10,-12,+20,-22	Stenman et al 1991 (180)
44-46,X,-X,-12	Stenman et al 1991 (180)
45-46,XY,del(9)(p11-p13)	Stenman et al 1991 (180)
45-46,X,-X,+5,del(6)(q21)	Stenman et al 1991 (180)
45-46,XY,-15,-21	Stenman et al 1991 (180)
45-48,XY,t(2;5)(q33-q34;q21-q22)	Stenman et al 1991 (180)
45-46,XY,-22	Stenman et al 1991 (180)
46-48,XX,+13,+20	Stenman et al 1991 (180)
46,XY/45,XY,-22/44,X,-Y,-22*	Webb and Griffin 1991 (187)
39,X,-Y,-7,r(8)(p23q24),-10,-15,-17,-19,-22/ 44,XY,-19,-22/45,XY,-22*	Bello et al 1993 (188)
43-46,XX,t(1;6)(q21;p13),t(6;13)(q25;q12),-17, del(22)(q11)	Bello et al 1993 (188)
44,XX,-17,-22*	Bello et al 1993 (188)
44,XY,-19,-22/45,XY,-22	Bello et al 1993 (188)
44,XX,-20,-22/44-45,XX,-14/46,X,-X,del(1)(q24), del(4)(q22),t(15;19)(q22;q13),-22,+2mar*	Bello et al 1993 (188)
45,X,-Y	Bello et al 1993 (188)
45,XX,ins(13;?)(p12;?),-18*	Bello et al 1993 (188)
45,XX,-14/45,XX,-22/46,XX,del(9)(q32) *	Bello et al 1993 (188)
45,XY,-16/45,XY,-17/45,XY,-22	Bello et al 1993 (188)
45,XY,-16/45,XY,-22	Bello et al 1993 (188)
45,XY,-22 (two cases)	Bello et al 1993 (188)
45,XY,-22* (three cases)	Bello et al 1993 (188)
45,XX,-22	Bello et al 1993 (188)
45,XX,-22* (five cases)	Bello et al 1993 (188)
45,XX,-22/47,XX,+7/58-106,XXXX,-10,-13,-14,-17*	Bello et al 1993 (188)
47,XX,+7/47,XX,+der(7)hsr(7)(p13)/46,XX,del(9)(q11) *	Bello et al 1993 (188)
46,Y,(X;7)(q22;q36),t(1;6)(q32;p23),t(3;16)(q11;p11),9p+	Neumann et al 1993 (353)
45,X,-Y,t(1;17)(p12;q11.2)	Rao et al 1996 (191)
45,XX,-13,-22,+mar	Rao et al 1996 (191)
46,XY,add(1)(q44),add(4)(p16),add(11)(p15),der(16) t(15;16)(q15;q24),tas(1;1)(p36.3;q44),tas(1;16)(q44;q24), tas(1;20)(q44;q13.3),tas(1;22)(q44;p13),tas(4;4)(p16;p16), tas(4;16)(p16;q24),tas(4;22)(p16;q13),tas(11;17)(p15;p13), tas(13;17)(q24;p13)*	Sawyer et al 1996 (354)
44-46,X,del(X)(q24),add(1)(q11),der(2)t(1;2)(q11;q31), der(4)t(2;4)(q11;q35),add(10)(p11)	Mertens et al 2000 (95,96)
44-46,X,-X,del(1)(p32),add(2)(p11),add(12)(p13),inc	Mertens et al 2000 (95,96)
45,X,-Y	Mertens et al 2000 (95,96)
45,XY,-19	Mertens et al 2000 (95,96)
45,XY,-22	Mertens et al 2000 (95,96)
45,XY,-22/45,X,-Y	Mertens et al 2000 (95,96)

TABLE 1.4. Continued

Karyotype	References
45,XY,-22/46,XY,del(22)(q12)	*Mertens et al 2000* (95, 96)
45,XY,-22/46,idem,+7	*Mertens et al 2000* (95, 96)
46,XX,del(15)(q22),add(22)(q12)	*Mertens et al 2000* (95, 96)
46,XX,del(22)(q11)	*Mertens et al 2000* (95, 96)
46,XY,der(12)(t(12;17)(q24;q21)/45,X,-Y	*Mertens et al 2000* (95, 96)
46,XX,der(22)t(13;22)(q14;q13)	*Mertens et al 2000* (95, 96)
46,XX,t(11;14)(q14q12)	*Mertens et al 2000* (95, 96)
46,XY,-22,+mar/47,idem,+X	*Mertens et al 2000* (95, 96)
47,XX,+r/48,idem,+r/49,idem,+2r	*Mertens et al 2000* (95, 96)
47,XY,+7	*Mertens et al 2000* (95, 96)
47,XX,+8/45,XX,add(14)(p11)	*Mertens et al 2000* (95, 96)
47,XX,+20/45,X,-X	*Mertens et al 2000* (95, 96)
47,XY,+17/48,idem,+18	*Joste et al 2004* (192)
46,XX,t(9;22)(q22;q13)	*Hameed et al 2006* (192A)

*Vestibular schwannoma.
**Study contained vestibular and spinal schwannomas, but no indication was given for as to which tumors underwent the changes.

B, C, E17-18, and G groups, and one to three marker chromosomes were present in several of these tumors. Some years later (185), three acoustic (vestibular) and four spinal schwannomas were evaluated, and karyotypically abnormal clones found in six schwannomas. Monosomy 22 was seen in four tumors. Subsequently, other investigators have described numerical losses or structural rearrangements of chromosome 22 as the predominant anomaly in schwannoma (Table 1.5).

Total loss of chromosome 22 (-22) in schwannomas has varied among studies, ranging from <20% to >50% of the cases (176, 178, 180, 185–189) (Figure 1.4). The -22 occurs in about the same frequency in the various histologic types of schwannoma, including vestibular tumors. Loss of one of the sex chromosomes, trisomy 7, or other numerical changes also may be seen in schwannoma (95, 96).

The frequency of normal karyotypes in schwannoma is difficult to evaluate accurately, because the articles generally do not report such cases. However, in studies on large numbers of schwannomas, it is not unusual for a substantial number, often >50%, to have normal karyotypes (178, 187, 188). This may be related to overgrowth of the tumor cells by normal elements or to cryptic genetic changes not detectable cytogenetically. Studies have shown that karyotypically normal cells may harbor submicroscopic deletions when examined with molecular techniques, particularly those of chromosomes 17 and 22 (178, 190).

The cytogenetic changes in *cellular* schwannoma are similar to those in other benign peripheral nerve sheath tumors (95, 96, 179, 180, 191). The cytogenetic changes of the less common *plexiform* variant of cellular schwannomas have been reported previously (192).

Cytogenetic examination of 56 schwannoma was reported in abstract form (193). Both sporadic and NF1- and NF2-associated acoustic schwannomas, spinal schwannomas, or a combination were evaluated. Monosomy of chromosome 22 (-22) was present in 21 tumors (as a sole anomaly in 15). Losses of chromosomes 14, 16, 17, 18, and Y were seen in at least one specimen each. Structural abnormalities involving chromosome 22 were identified in two samples: i(psu?)dic(22) and del(22)(q11). The authors concluded that abnormalities of chromosome 22 are characteristic of schwannoma, irrespective of location, and that chromosomal loss as opposed to structural rearrangement of 22 is the most frequent event (193).

In a subsequent study (189), cytogenetic and molecular methods were compared in 18 schwannomas for the determination of the sensitivity and consistency of each technique in detecting anomalies of chromosome 22. With the exception of a few cases, the cytogenetic and the molecular data were in agreement.

The cytogenetic findings (194) in a schwannoma and MPNST of the thigh of a 24-year-old male with NF1 showed the MPNST to have a very complex karyotype, whereas the schwannoma was diploid.

In a retroperitoneal schwannoma with many and very complex chromosome changes and intratumoral heterogeneity, 18p11 was involved in eight rearrangements with different chromosomes. Although the cytogenetic heterogeneity in this tumor included 58 chromosomal breakpoints and many numerical changes, chromosome 22 was spared relatively by these anomalies. A scheme for these unusual cytogenetic changes in this schwannoma was presented (195, 196).

Partial or complete monosomy 22 was seen by molecular studies in 22% of vestibular schwannomas and in 55% of the nonacoustic type (197). Because of the frequent loss of chromosome 22 in meningiomas, a pathogenesis similar to vestibular schwannomas has been proposed (178, 197, 198).

Seven of 17 schwannomas (various types) had abnormalities of chromosome 22 manifested as loss (-22), deletions, additions, or translocations (95, 96). Except for sex-chromosome changes, i.e., −Y (two cases), −X (two cases), +X (one case), and del(X)(q24) (one case) and +7 in two cases, the other changes were inconsistent.

TABLE 1.5. NF2 and schwannomas.

	References
Survivin, member of the inhibitor of apoptosis (IAP) gene family has multifaceted functions related to apoptosis and cell division.	*Altieri 2003* (249)
Changes of *NF2* genes in cancers not related to NF (breast, colon, brain, head, and neck).	*Arakawa et al 1994* (390)
	Ng et al 1995 (391)
	Huang et al 2002 (391A)
Genotype-phenotype correlations for NF2 tumors.	*Baser et al 2004* (296)
Deletion of 1p in 2/23 schwannomas.	*Bello et al 1995* (356)
Claudin-1 expression present in meningiomas but not in schwannomas.	*Bhattacharjee et al 2003* (357)
Mutations of the *NF2* gene are frequently found in meningiomas and occasionally in other types of tumors, e.g., mesothelioma.	*Bianchi et al 1994* (211)
	Ruttledge et al 1994 (211A)
Changes of *NF2*.	*Bijlsma et al 1992* (197)
	DeVitis et al 1996 (241)
	Hung et al 2000 (358)
Mutations of *SCH* gene in schwannomas.	*Bijlsma et al 1994* (219)
Glial growth factor may play a role in vestibular schwannoma.	*Brockes et al 1986* (359)
Heregulins/neuregulins, including glial growth factor, regulate schwann cell growth.	*Burden and Yardin 1997* (360)
	Levi et al 1995 (361)
	Kim et al 1997 (362)
	Pinkas-Kramarski et al 1997 (363, 364)
Merlin inhibits mixed-lineage kinase 3 (MLK3) regulation of β/B-Raf.	*Chadee et al 2006* (363A)
Complete inactivation of *NF2* may not be required for tumor formation, e.g., G→A transition is sufficient.	*De Klein et al 1998* (364A)
Schwannoma cells express EGFR (not present in normal schwann cells), which may play a role in tumor development.	*DeClue et al 2000* (125)
Cavernous malformation within schwannomas is rare; it has been hypothesized that it may be related to loss of chromosome 7 (7q22) affecting the *Ras* oncogene or mutation of the Ras-GAP- protein of the *NF1* gene.	*Feiz-Erfan et al 2006* (443)
Calretinin staining positive in schwannomas but not in meningiomas.	*Fine et al 2004* (365)
Several genes were overexpressed and some underexpressed in schwannomas versus normal schwann cells.	*Hanemann et al 2006* (388A)
Neuregulin expression and activation of *ERBB* receptors in vestibular schwannoma.	*Hansen and Linthicum 2004* (395)
An autocrine pathway of vestibular schwannoma growth stimula- tion involving neuroregulin and ErbB2 receptors has been described.	*Hansen et al 2006* (437)
Merlin inhibits the Rac/CDC42-dependent Scr/Thr kinase PAK1, the latter is essential for the malignant growth of *NF2*-deficient cells. PAK1-blocking agents could be potentially useful in the treatment of NF2 and related tumors.	*Hirokawa et al 2004* (440)
S-100 protein-positive granular cells containing tumors (schwannoma, MPNST) express osteopontin; the biological significance of this finding awaits elucidation.	*Hoshi et al 2005* (454)
Ploidy pattern (DNA levels) and FISH may be useful in differentiating benign from malignant schwannomas.	*Hruska et al 2004* (394)
Somatic mutations of *NF2* seen in familial and non-familial vestibular schwannomas.	*Irving et al 1994* (220)
Human schwannomas (and meningiomas) are clonal in nature and arise from a single cell.	*Jacoby et al 1990* (366)
The possibility has been raised that dendritic cell neurofibroma with pseudorosettes may be associated with *NF2* gene changes, thus representing a fourth type (in addition to neurofibroma, schwannoma and perineurioma) of peripheral nerve sheath tumor.	*Kazakov et al 2005* (434)
c-AMP-dependent protein kinase A, neuregulin/tyrosine kinase interactions.	*Kim et al 1997* (245)
Calpain-dependent proteolysis of merlin.	*Kimura et al 1998* (367)
Calpain proteolysis of merlin in schwannomas.	*Kimura et al 2000* (368)
Merlin inhibits p21-activated kinase.	*Kissil et al 2003* (369)
HGF is a mitogen for schwann cells and is present in neurofibromas.	*Krasnoselsky et al 1994* (370)
Merlin and ERM proteins interact with N-WASP and regulate its actin polymerization function. Increasing evidence has been accumulating for the role of merlin in actin cytoskeleton reorganization.	*Manchanda et al 2005* (370A)
Estradiol binding in schwannomas.	*Martuza et al 1981* (327)
Angiogenic factor, midkine, is aberrantly expressed in *NF1*-deficient schwann cells and may serve as a mitogen.	*Mashour et al 2001* (238)
Factors playing a role in apoptosis, i.e., the Fas-Fas-L system, Bcl-2 and Bax were investigated in vestibular schwannomas. None of these factors appeared to play a role in these tumors. Thus, apoptosis was not mediated via the Fas-Fas-L system, because Fas expression is deficient in vestibular schwannoms whereas Fas-L is present in the majority of these tumors.	*Mawrin et al 2002* (441)
A review of the mechanisms of merlin action and effects, including results in experimental animals, in tumor development has been published. The role possibly played by merlin in coordinating the process of growth-factor receptor signaling and cell adhesion was emphasized.	*McClatchey and Giovannini 2005* (438)
Matrix metalloproteinase expression may play a role in the proliferation and invasion of schwann cells.	*Muir et al 1995* (331)
Merlin and cell cycle control.	*Muranen et al 2005* (371)

TABLE 1.5. Continued

	References
A case with multiple cutaneous schwannomas was shown to be associated with an *NF2* gene mutation in two separate tumors, confirming the case as one of NF2 mosaicism.	*Murray et al 2006 (435)*
The significance of regulated Rac (a small GTPase of the Rho family) signaling in mediating schwann cell-exon interaction has been described. Controlling Rac activity may be a possible therapeutic approach.	*Nakai et al 2006 (439)*
Cytokines and apoptosis in schwann cells.	*Oliveira et al 2005 (392)*
Overexpression of fibroblastic growth factor 1 has a positive correlation with the incidence and growth rate of sporadic vestibular schwannomas.	*O'Reilly et al 2004 (400)*
Cell proliferation and actin organization aberrant in schwannomas.	*Pelton et al 1998 (235)*
An evaluation of the epidemiology of vestibular schwannomas indicated a possible increase in its incidence as compared to that of schwannomas at other sites.	*Propp et al 2006 (389)*
Synergistic regulation of schwann cell proliferation by heregulin and forskolin.	*Rahmatullah et al 1998 (399)*
Expression of *RAS* and nuclear oncogenes may lead to schwann cell proliferation and transformation.'	*Ridley et al 1988 (372)*
Improved culture conditions for neurofibroma-derived schwann cells make them readily available for appropriate studies.	*Rosenbaum et al 2000 (453)*
Role of *NF2* in the development of CNS tumors.	*Ruttledge and Rouleau 2005 (398)*
CGH studies showed undifferentiated GIST to be genetically different from schwannomas and leiomyomas.	*Sarlomo-Rikala et al 1998 (373)*
Merlin interacts with hepatocyte growth factor-regulated tyrosine kinase substrate.	*Scoles et al 2000 (393)*
Merlin requires HRS (tyrosine kinase substrate) interaction to be fully functional and to inhibit STAT (signal transducers and activators of transcription) activation.	*Scoles et al 2002 (375)*
Loss of gene at 22q11 in bilateral vestibular schwannoma.	*Seizinger et al 1987 (184)*
Decreased p27 and aggressive vestibular schwannoma behavior.	*Seol et al 2005 (397)*
Constitutional activation of *erbB2* may initiate immortalization and transformation of immature schwann cells.	*Sherman et al 1999 (376)*
Possible role of activated *neu/erbB2* in schwannoma cell transformation.	*Sherman et al 1999 (376)*
Structure and mutation of *NF2*.	*Shimizu et al 2002 (377)*
	Kluwe et al 2005 (378)
Schwann cells of NF proliferate in culture when exposed to a number of agents, e.g., cAMP analogs, PDGF and TGF-β.	*Sobue et al 1985 (395A)*
	Muir et al 2001 (62)
Mutations of p53, common in sarcomas and other cancers, have have been described in 4/12 malignant schwannomas.	*Stock et al 1997 (374)*
EGFR expression in vestibular schwannoma growth	*Sturgis et al 1996 (379)*
Effect of merlin (product of *NF2*) on phosphorylation of *NF2*.	*Surace et al 2004 (380)*
In addition to the actin cytoskeleton, merlin has been shown to associate with cell membrane domains and proteins of varying nature (ERM-ezrin, radixin, moesin proteins, CD44, F-actin, N-WASP, Rac1 and others).	*Takeshima et al 1994 (168D)*
	Neff et al 2006 (371A)
Pak protein in schwann cell transformation.	*Tang et al 1998 (381)*
Role of Pak protein kinases in schwann cell transformation.	*Tawk et al 2005 (382)*
Melanotic schwann expressed CD117 (*KIT*).	*Tawk et al 2005 (382)*
Altered expression of integrins and pathological ensheathment of extracellular matrix in schwannomas.	*Utemark et al 2003 (383A)*
Basal apoptosis rate of primary schwannoma cells is reduced in comparison with normal schwann cells.	*Utemark et al 2005 (383A)*
Schwannoma cells show enhanced integrin-dependent adhesion and activation of the small Rho GTPase Rac1. Also, the glial fibrillary protein is confined to the perinuclear area instead of being well spread throughout the cytoplasm. The cytoskeletal rearrangement is accompanied by changes in cell shape and increased cell motility.	*Utemark et al 2005 (383B)*
ABCA2 protein expression in vestibular schwannoma and peripheral nerve.	*Wang et al 2005 (384)*
Mutations of *NF2* have been found not only in NF2-associated tumors but also in sporadic unilateral schwannomas and cystic schwannomas.	*Welling et al 1996 (384A)*
	Welling 1998 (384B)
cDNA microarray studies of vestibular schwannomas have suggested several potentially important tumorigenic pathways associated with development of these tumors, e.g., underexpression of the apoptosis-related *LUCA-15* gene and down-regulation of ezrin, a relative of merlin.	*Welling et al 2002 (442)*
Schwannomas may form part of the tumors seen in the Carney complex.	*Wilkes et al 2005 (451)*
Histological differentiation of congenital and childhood plexiform (multinodular) cellular schwannoma from MPNST.	*Woodruff et al 2003 (385)*
A "three-hit" model appears to fit better the development of vestibular schwannoma than the "two-hit" model.	*Woods et al 2003 (452)*
p21-activated kinase (PAK-1) and merlin.	*Xiao et al 2002 (386)*
	Kissil et al 2003 (369)
Effect of merlin on actin cytoskeleton, localization of merlin with F-actin at cell membrane and along paranodal incisures in sciatic nerve.	*Xiao et al 2003 (210)*
	Gonzalez-Agosti et al 1996 (387)
	Scherer and Gutmann 1996 (388)
A study of 23 trigerminal neurinomas (schwannomas) found that the immunoglobulin-like domain 1 (LRIG-1) may	*Xiong et al 2006 (436)*

TABLE 1.5. Continued

	References
inhibit the malignant differentiation and proliferation of these tumors by a negative feedback loop of epidermal growth factor receptor (EGFR).	
Farnesyltransferase and neurofibromin interaction.	*Yan et al 1995* (351)
An array-CGH study of 21 schwannomas revealed 12 to have loss of chromosome 22.	*Mantripragada et al 2003* (370B)
Two genes, *NF2* and *Expanded*, have been shown to be part of the Hippo signaling pathway involved in growth regulation. The tumor suppressor merlin and expanded function cooperatively in cell regulation.	*Hamaratoglu et al 2006* (437A)
Chromogenic in situ hybridization demonstrated *NF2* gene deletions in 60% of meningiomas and 65% of schwannomas.	*Begnami et al 2007* (296A)
Neurofilament protein staining demonstrated intratumoral axons in many sporadic schwannomas.	*Nascimento and Fletcher 2007* (439A)

Of interest have been the results of studies (199) on 47 sporadic schwannomas, two neurofibromas (one NF1-related and one NF2-related), and three schwannomas (one NF1-related and two NF2-related). Deletions on chromosome 22 were present in 82% of the schwannomas, indicating the fundamental role of genes located on this chromosome in their development. Partial deletion of chromosome 22 was present in a substantial fraction of these tumors (27%). In addition to the *NF2* locus, the authors (199) detected multiple regions with terminal or interstitial deletions on chromosome 22, some of them probably harboring genes involved in schwannoma tumorigenesis. The findings (199) indicate a heterogeneity in the mechanisms leading to the development of schwannoma.

DNA methylation was determined in 16 tumor-related genes in schwannoma (sporadic and NF2-associated) (199A). The results indicated that DNA methylation may be involved in the process of schwannoma development and that a clearer understanding of the epigenetic mechanisms that cause transcriptional repression may contribute to the establishment of a possible relationship between the genetic changes of varying nature and the clinical and pathologic parameters in schwannoma (199B). LOH on chromosome 22 related to *NF2* was found in 23/54 (42.6%) of schwannomas (36 vestibular and

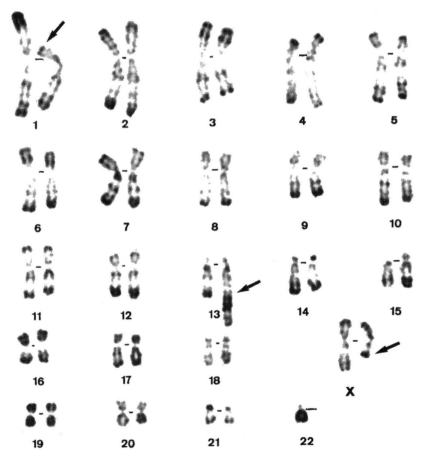

FIGURE 1.4. Karyotype of a cervical nerve schwannoma showing a t(X;1;13) (*arrows*) and -22, the latter being a common finding in schwannomas (178).

18 spinal), with LOH being significantly more frequent in vestibular than in the spinal tumors (199C). The proliferative index (Ki-67 and proliferating cell nuclear antigen) was considerably higher in the schwannomas with LOH. Using tiling path profiling, 53% of schwannomas showed heterozygous deletions of chromosome 22, predominantly as monosomy 22 (199D). The authors observed a correlation between the breakpoint position, present in tumor DNA, constitutional DNA, or both, and the location of segmental duplications. This association implicates these unstable regions in rearrangements occurring both in meiosis and mitosis (199D).

The inhibition of schwannoma cell proliferation by overexpression of the *NF2* gene was thought to operate through promotion of PDGFR degradation (200).

NF2 Gene

An elucidation of the exact function of *NF2* as a TSG and its protein product merlin has remained elusive despite the many descriptions of its interaction with and effects on an array of signaling systems and cellular pathways and other regulatory activities. The molecular and physiologic anatomy of merlin is described in an extensive literature on the subject. It is beyond the scope of this chapter to comprehensively cover the aforementioned areas; hence, descriptions are confined to facets of *NF2* and merlin that may bear more directly on tumor formation.

The *NF2* gene, located at 22q12.2, spans 110-kb (201–203) and comprises 17 exons encoding a 595-amino acid protein. A translocation, t(4;22)(q12;q12.2), in an NF2 helped to locate the *NF2* gene (204–206) at 22q12.2. The protein product of *NF2*, termed merlin (or schwannomin), has homology in the first 13 exons to the ezrin, radixin, and moesin (ERM) family of proteins, which are thought to play a role in linking cytoskeletal components to cell membrane glycoproteins, e.g., CD44 (207, 208). However, unlike other ERM proteins (209, 210) that bind strongly to the actin cytoskeleton, merlin associates weakly with actin in vitro (8).

The various characteristics and functional aspects of *NF2* and its product merlin as they are related to neural tumors (schwannomas) and to non-neural tumors (e.g., mesothelioma) (211) and other tumor types unrelated to NF2 (melanoma, breast cancer), have been reviewed previously (210). Studies have emphasized the role of merlin in cell motility and proliferation (210), highlighting the involvement of merlin in Rac signaling. Rac plays an important role in regulating actin cytoskeletal dynamics (212, 213). Merlin activity is regulated by phosphorylation in response to Rac/Pak expression. Merlin acts as a tumor suppressor, at least in part, through inhibition of Ras/Rac signaling (210) and effects on membrane ruffling (214). In addition, merlin has been shown to induce G0/G1 arrest and pRB phosphorylation (209, 215, 216). It has been demonstrated that merlin deficiency destabilizes cadherin-mediated cell–cell junctions, thus promoting tumorigenesis and metastases (217).

Mutations occur throughout the coding sequence of the *NF2* gene and at intronic sites, although they have not been described in exons 16 and 17 (190). In most cases, such mutations are accompanied by loss of the remaining wild-type allele on 22q. Still other cases demonstrate loss of 22q in the absence of detectable *NF2* gene mutations. Nonetheless, loss of merlin expression, demonstrated by Western blotting or immunohistochemistry, seems to be a universal finding in schwannomas, regardless of mutation or allelic status. This suggests that loss of merlin function is an essential step in schwannoma tumorigenesis (210).

Germline *NF2* mutations differ somewhat from somatic mutations in sporadic schwannomas and meningiomas in that they frequently are point mutations that alter splice junctions or create new stop codons, and although found in all parts of the gene, they occur preferentially in exons 1-8. A possible hot spot for these mutations seems to be position 169 in exon 2, in which a C-to-T transition at a CpG dinucleotide results in a stop at codon 57; other CpG dinucleotides in *NF2* are also commonly targets for C-to-T transitions(218).

Mutations of the *NF2* gene in schwannomas may consist of a single site mutation or larger mutations (79, 190, 219–226), mostly in the first 13 exons that lead to a truncated and nonfunctional merlin (227). Mutations that result in premature protein termination are associated with a more severe clinical course of NF2, with symptoms starting before age 20 and the development of two or more central nervous system (CNS) tumors before age 30. Single amino acid changes (missense mutations) are associated with a milder clinical disease (38, 39).

Merlin

Deficiency of merlin, secondary to changes of the *NF2* gene (228), has been reported for almost all tumors associated with NF2, and in many sporadic schwannomas and meningiomas (229–233). Regulated overexpression of merlin in schwannoma cells and viral transduction of the *NF2* gene into meningioma cells result in growth suppression in vitro and reduced tumor growth in vivo, independently of the endogenous *NF2* status (225, 234).

Loss or modifications of merlin expression in NF2-associated schwannoma cells are associated with dramatic alterations in the actin cytoskeleton (235, 236) and in cell physiology (237).

Some data suggest that merlin may integrate a number of intracellular signaling pathways (8, 238). In this regard, merlin binds to CD44, a hyaluronic acid receptor important in growth regulation, and promotes growth arrest (239). Additionally, merlin associates with the hepatocyte growth factor (HGF)-regulated tyrosine kinase substrate, HRS. HGF is one of the most powerful stimuli for schwann cell proliferation and motility, suggesting that the association of merlin with HRS and other factors might inhibit cell growth and movement (239A). Merlin binds Ral guanine nucleotide dissociation stimulator (RalGDS), a protein that acts as downstream regulator of Ras (240). This interaction could be an important pathway for regulating oncogenic transformation of cells. The findings suggest that merlin may function as a tumor suppressor by inhibiting RalGDS-mediated oncogenic signals (240B).

Merlin mediates contact inhibition of growth by suppressing recruitment of Rac to matrix adhesions (240A). Merlin suppressor signaling by the small GTPase Rac by interacting with an inhibiting the Rac target effector p21-activated kinase (PAK) (240A). Merlin may function as a tumor suppressor by inhibiting the RalGDS, a downstream molecule or Ras, oncogenic signals (240B).

A p110 subunit of the eukaryotic initiation factor 3 (elF3c) interacts with merlin and thus plays a role in cellular proliferation. High levels of elF3c were found in meningiomas with *NF2* mutations and lost merlin expression and vice versa in tumors retaining merlin. Thus, elF3c can serve as a prognostic marker in meningioma (239A).

Merlin is present in a wide range of metazoans (240C) and interferes with Ras- and Rac-dependent signal transfer; thus, it is tumor suppressive in its action (240D).

Merlin is involved in the regulation of transactivation-responsive RNA-binding protein (TRBP) level by facilitating its ubiquitination in response to such cues as cell-cell contacts (240E). Merlin affects the integrity of HEI-10, a cell cycle regulator, and thus acts as a tumor suppressor (240F).

The most consistent phenotype associated with merlin deficiency in cultured cells is loss of contact-dependent inhibition of proliferation (240G). Merlin is activated by dephosphorylations and its inhibition (by suppression of the appropriate phosphatase) causes Ras activation and cellular transformation (240H).

The high proportion of schwannomas in which inactivating mutations (nonsense, frameshift, or special donor site mutations) of *NF2* have been found indicates that the product of this gene, merlin, which is truncated and rendered nonfunctional, plays a fundamental role in controlling the growth of the schwann cells that give rise to this tumor type (190) (Figure 1.5). Complete inactivation of *NF2* occurs in >60% of schwannomas, either by mutations in both alleles or by mutation in one allele and loss of the other allele, strongly supporting the tumor suppressor model in which loss or inactivation of both copies of the *NF2* gene underlies schwannoma formation (190). This event is thought to be critical in the formation of both sporadic and inherited schwannomas (8, 95, 96, 176, 178–180, 185, 191, 192). A similar spectrum of NF2 mutations occurs in sporadic tumors and in those of patients with germline *NF2* mutations (226, 241). Merlin levels were markedly reduced in 50% of sporadic schwannomas (242).

No alterations of the *NF2* gene were seen in seven of 58 schwannomas (190). No unique or striking clinical features were evident in these seven cases. Because two of the tumors were from patients with a clinical diagnosis of NF2, it is likely

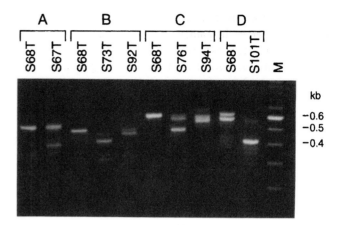

FIGURE 1.5. Molecular demonstration of mutations of the *NF2* gene in schwannomas. In each tumor, full-length cDNA was synthesized from total cytoplasmic RNA followed by nested PCR of overlapping quadrants of the cDNA, including exons 1–5 (**A**), exons 4–9 (**B**), exons 8–13 (**C**), and exons 12–17 (**D**). RNA from a spinal tumor without a mutation in an NF2 patient (S68T) served as a control in each case (190). Except for S101T, which was a schwannoma of the femoral nerve of an adult patient with NF2, all other tumors were sporadic vestibular schwannomas. The abnormalities of the *NF2* gene in each tumor are demonstrated by the different mobilities of the tumor products and/or the presence of more than one component compared with the control.

that these tumors possessed *NF2* mutations outside the exon sequences that were scanned or that they were missed by the technique used, rather than being due to mutations at an independent locus (190).

The distribution of *NF2* mutations indicates that merlin's tumor suppressor function can be eliminated by truncations beginning in any region of the protein, except possibly in the alternative COOH termini encoded by exons 16 and 17 (190). Despite the paucity of missense mutations of *NF2* in schwannomas, mutation analysis has established that inactivation of merlin is the critical event leading to the formation of both inherited and sporadic schwannomas.

Correlation between *NF2* methylation status and expression in schwannomas has been reported previously (243); the data suggested that although additional factors may be involved in the complete silencing of the *NF2* gene (by LOH at 22q, mutations, μ-calpain expression), aberrant methylation of certain CpG sites in the 5′ flanking region would significantly contribute to the reduced expression of *NF2* mRNA. Both methylation-specific and nonmethylation-specific bands in some tumors were found. *NF2* expression was not detected by Northern blot analysis, suggesting that methylation occurred in certain schwannoma tissues or that there may have been some heterogeneity in the methylation status of the CpG sites, as could be demonstrated by sequencing analysis (243).

Additional genes, in cooperation with the *NF2* gene, may be involved in the causation of schwannomas (172, 199) and point to the heterogeneity in the mechanisms leading to schwannoma development. Schwannomas (sporadic) show mutations at exons 2, 7, and 12 of *NF2*; allelic loss at 1p may contribute to the pathogenesis of a small group of these tumors; no losses at 14q were seen (244).

Stimulation of schwann cells with glial growth factor in vitro leads to increases in Ras-GTP, demonstrating a link between Ras regulation and schwann cell proliferation (238, 245). Survivin, an inhibitor of apoptosis protein (246), is expressed by benign CNS tumors (e.g., meningioma, schwannoma, pituitary adenoma) (247, 248), and in other different types of tumors (249). Survivin overexpression may be an early event in tumorigenesis, and it also could be responsible for progression of malignant and benign neoplasms, including schwannoma (250). Using a microarray approach with high-resolution and high-throughput detection of deletions in NF2 (251, 252), heterozygous deletions were found in 45% of sporadic schwannomas (251) involving a continuous 11-Mb segment of the *NF2* gene.

Various signaling systems are supported by multiple neuroregulin growth factors in the development of schwannomas (253). The ErbB receptors for the neuregulin factors are sensitive to a number of molecular inhibitors, and they deserve consideration in the treatment of schwannomas (253).

Somatostatin, a neuropeptide with a variety of effects in the central and peripheral nervous systems, has several receptors that may be expressed in tumoral tissues (254). Thus, the receptor sst_{2A} is expressed in 89% of schwannomas, in 22% of neurofibromas, and in 15% of MPNSDT. The receptor sst_4 was found in 32% of MPNST and not in the other tumors; this receptor is subject to agonist (254).

The majority of retroperitoneal schwannomas (90%) and a minority of peripheral tumors are glial fibrillary protein (GFAP) positive (255). Keratin expression parallels that of GFAP, possibly representing crossreactivity. However, keratin-positive schwannomas should not be confused with sarcomatoid carcinoma, mesotheliomas, or keratin-positive sarcomas (255).

Vestibular Schwannomas

The majority of vestibular schwannomas (>95%) are sporadic and unilateral and manifest themselves in the fourth and fifth decades of patients. When bilateral, these tumors represent a characteristic manifestation of NF2 (56, 256), a condition in which they often occur. Mutations of *NF2* and especially deletion of exon 4 are common in these tumors (257). Vestibular schwannomas (called acoustic neuromas in the past) occur at a younger age in NF2 patients than in sporadic cases (257A), they are more invasive than the sporadic tumors, and they differ in labeling index (higher in NF2-related tumors), histology (258), and morphology (259, 260). A review of the molecular biology of vestibular schwannomas presented the complex events underlying the development of these tumors (371A).

Chromosome 22 allelic loss is frequent in sporadic vestibular schwannomas, with the minimal region of LOH involving the *NF2* gene (261, 262). Some other tumor suppressor genes (*VHL, p53, NF1,* and *WT1*) do not seem to be important in the pathogenesis of these tumors.

In >50% of vestibular schwannomas, three CpG sites were methylated in a site-specific manner leading to suppressed expression of the *NF2* gene (243), and they may constitute one of the mechanisms for inactivation of this gene (263). Vestibular and peripheral schwannomas show consistent loss of merlin but not for other factors encoded by *NF2* (264). These somatic mutations, loss of merlin seen in schwannomas, or both (172, 190, 210, 221, 222, 226) contrast with the persistence of merlin in ependymomas and meningiomas (265).

Changes of *NF2* are present in bilateral vestibular schwannomas (183, 184, 220, 223, 266), although the mechanisms for *NF2* inactivation differ from those in nonfamilial tumors. Young patients with unilateral vestibular schwannoma should be suspected of having NF2; however, if neither NF2-related diagnostic clinical features are found nor a family history of NF2-related problems, it is highly unlikely that the affected patients will develop bilateral disease (267). A basis for the aforementioned statement was a study (267) in which 45 patients aged 30 years or less with unilateral vestibular schwannomas were analyzed molecularly for changes in the *NF2* gene in peripheral blood cells (constitutional changes) and in 28 vestibular schwannomas (acquired changes). No changes were identified in the blood cells of the 45 patients, whereas *NF2* point mutations and LOH were found in 21/28 tumor samples. Thus, these patients are thought unlikely to develop bilateral disease or to pass on a pathogenic *NF2* mutation to their offspring.

In a study of proliferation indices (Ki-67 and PCNA) (260, 260A), NF2-associated vestibular schwannomas were shown to have higher indices than the sporadic type. This was also true for spinal MPNST versus spinal cellular schwannomas.

The incidence of LOH and mutation rate of *NF2* in vestibular schwannomas of Chinese patients were shown to be similar to those of previous studies (268). In contrast to the germline *NF2* mutations whose nature may have an effect on the phenotype (269, 270), an examination of 91 sporadic vestibular schwannomas (262), revealed no association between the nature of the *NF2* mutation and tumor biology, pointing to other factors besides the *NF2* inactivation as determinants of tumor behavior. Similar conclusions were reached by others (251) in which array-comparative genomic hybridization (CGH) was used.

A study of families with unilateral vestibular schwannomas failed to reveal any germline mutations of the *NF2* gene, indicating that either a chance somatic mutation of *NF2* or a separate genetic locus may be responsible for familial vestibular schwannomas (270A).

Unilateral vestibular schwannoma may be associated with other NF2 tumors, and such cases should be closely followed for possible tumor development (260A).

Recurrence of vestibular schwannoma after radiosurgery may be associated with an alternative mechanism of *NF2* inactivation that may correlate with radioresistance in these tumors (270A).

Cystic vestibular schwannomas are an aggressive group of unilateral, sporadic tumors that invade the surrounding cranial nerves (270B). An age-dependent phosphorylation of p53 protein, higher in younger patients, may play a role in vestibular schwannoma development (270C). Some differences in the gene expression profiles of cystic tumors versus those of sporadic and NF2-associated vestibular schwannomas have been described (270D, 270E). Constitutive neuregulin-1 and ErbB pathways contribute to proliferation of vestibular schwannoma (271A).

Aggressive vestibular schwannomas were found to have a decreased expression of p17, a cyclin-dependent kinase inhibitor, in contrast to unremarkable alteration of other factors (p21, p53, Bcl-2, Fas-L, and caspase-3) (397). The aggressiveness of these tumors may be due to the p27 changes.

The expression of epidermal growth receptor (EGFR) and leucine-rich repeats and immunoglobulin-like domain-1 (LRIG-1) in vestibular schwannoma was investigated in relation to the biology of these tumors (270F). The high expression rate of the latter (78%) suggested that LRIG-1 may inhibit the malignant differentiation and proliferation of these tumors, possibly by a negative feedback loop of EGFR.

Recurrence of vestibular schwannomas after radiosurgery may possibly due to mechanisms of NF2 inactivation other than those described to date (270B). ABCA2, a member of the A subclass of the ATP-binding cassette (ABC) transporter superfamily, is expressed in the cytoplasm of schwann cells (384). ABCA2 was expressed in benign vestibular schwannomas, but showed phenotypic heterogeneity depending on the cellular makeup of these cells.

Genetic aspects and those of other factors and substances that may play a role in schwannoma development and biology are presented in Table 1.5.

CGH of NF Tumors

Improvements and standardization of CGH techniques (271) have led to a more reliable comparison of results between laboratories, and those on neurofibromas and schwannomas fall into this category. Note that balanced translocations, including those of chromosome 22, would not be ascertained by the CGH technique; hence, the results should be interpreted with this caveat in mind.

CGH studies have in essence shown that changes either are not present or are minimal in sporadic neurofibromas compared with NF1-associated neurofibromas and MPNST (98, 99, 272, 273) (Figure 1.6). CGH studies in a small number

of neurofibromas or schwannomas revealed either no changes or a small number of alterations (e.g., those of 3p and 4q), in contrast to the complex changes seen in MPNST (272).

CGH was used to detect changes in the relative chromosome copy number in 50 cases of peripheral nerve sheath tumors: nine MPNST, 27 neurofibromas (three plexiform), and 14 schwannomas (99). Chromosome imbalances were frequently detected in the benign and malignant tumors. In neurofibromas and schwannomas, the number of losses was higher than the number of gains, suggesting a predominant role of TSG in their tumorigenesis. Both sporadic and NF1-associated neurofibromas exhibited losses of 22q in >50% of the cases. These chromosomal regions may contain common chromosomal abnormalities characteristic of both types of neurofibromas. Gains were found in plexiform neurofibromas. The significance of the losses of chromosome 19 in these cases is not clear at present, but in NF1-associated neurofibromas, the presence of some as yet unknown TSG on chromosome 19 cannot be ruled out (98).

Figure 1.6 shows that neurofibromas associated with NF1 have losses of 17p, 19p, 17q, 19q, 22q, 4p, 16p, and chromosome 16 in the order listed, whereas significant losses are seen only at 19q and 22q of the sporadic tumors. The losses in chromosome 17 in NF1-associated neurofibromas often involved 17p11.2-p13 and 17q24-q25 (99); those in chromosome 19 involved 19p13.2-qter. Gains were seen at 4q, 5q, and 13q in the former group; no significant gains were seen in the latter group (98, 99). In another CGH study (272), gains of chromosome 7, 8q, 15q, and 17q were characteristic of MPNST, whereas neurofibromas showed no distinct pattern, although gains of 4q and 11q, not seen in MPNST, may be significant.

In a CGH study of six neurofibromas and three schwannomas, none of the former tumors showed genetic imbalances but two of the latter did (no details given in the publication) (274). In other CGH studies (273, 275), in contrast to the significant changes in MPNST, NF1-associated neurofibromas were essentially normal.

A CGH study of eight NF2-related and 13 sporadic schwannomas (276), almost all vestibular, nine (36%) showed loss of 22q harboring the *NF2* gene. In another CGH study of 76 schwannomas (66 sporadic and 10 NF2-related) (277), loss of 22q was found in 18 (24%) of the cases.

These findings are compatible with the demonstration of LOH of 22q at the *NF2* locus in other studies (183, 202, 203) and mutations of the *NF2* gene found in about half of the tumors (190, 221, 270, 278).

In a study of 17 (13 vestibular, one facial nerve and three cervical nerve) sporadic schwannomas (174) based on CGH and on fluorescence in situ hybridization (FISH) and mutation search of the *NF2* gene, the results suggested that most of these tumors (>80%) have two-hit mutations of this gene, and this seems to be the only major causative gene in the genesis of schwannomas. As part of a study of orbital benign tumor (279), two of four schwannomas were found to have losses at chromosome 16 and 22q.

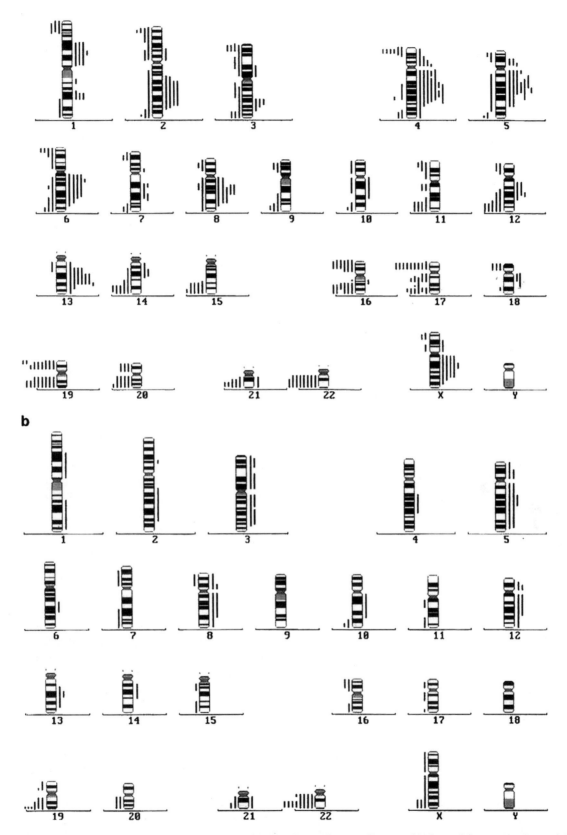

FIGURE 1.6. CGH analyses of NF1-associated neurofibromas (**a**) and of sporadic neurofibromas (**b**). Loss of chromosomal material is shown to the left of each chromosome and gains to the right. Chromosomal imbalances were more common in NF1-associated neurofibromas than in the sporadic tumors. Besides loss of 22q, losses on chromosomes 17 and 19 were frequent in the NF1-associated tumors (98).

Schwannomatosis

Schwannomatosis, either isolated or familial, without an NF2 background is a very rare and distinct condition (172, 173, 218, 260, 280, 281), and it has been proposed to be a separate form of NF, with fundamental clinical and genetic differences from NF2 (281–283). Familial schwannomatosis is thought to be a very rare occurrence (276) either as sporadic or NF2-associated. Because patients with schwannomatosis develop tumors later in life, their life expectancy is longer than that of NF2-related patients. Nor do such patients develop meningiomas, ependymomas, or astrocytomas, as observed in cases with NF2. Schwannomatosis is rarely familial, and when inherited, it commonly demonstrates incomplete penetrance and variable expressivity (283). Based on diagnostic criteria for NF2, patients with schwannomas and NF2 were readily differentiated from those with sporadic schwannomatosis or meningiomatosis, both rather rare conditions (283A). Approximately one third of patients with schwannomatosis had tumors in an anatomically limited distribution, such as a single limb, contiguous segments of the spine, or one side of the body (168C).

In schwannomatosis, the NF2 in the tumors is inactivated by truncating mutations, but in a pattern that differs from that seen in patients with NF2. Furthermore, patients with schwannomatosis do not harbor a heterozygous NF2 mutation (germline) in their normal tissues, indicating that the tumors in schwannomatosis arise as a result of acquired or somatic mutations in the cells directly involved in tumor formation (281). Examination of multiple tumors form the same patient revealed that some schwannomatosis patients were somatic mosaics for NF2 gene changes. By contrast, other individuals, particularly those with a positive family history, seemed to have an inherited predisposition to formation of tumors that carry somatic alterations of the NF2 gene (281). Patients with schwannomatosis are likely to have a variant form of NF2 and up to a 50% risk of passing on a gene predisposing to multiple schwannomas (284). Examination of samples from sporadic and familial schwannomatosis revealed genetic aberrations outside the NF2 locus on chromosome 22 (281A).

Several genetic aberrations were found in sporadic and familial schwannomatosis (281A). Rearrangements of the immunoglobulin locus were restricted to schwannomatosis/schwannoma samples. Missense mutations in the CABIN1 locus were also observed in schwannomatosis and NF2, and they make this gene a plausible candidate for contributing to the pathogenesis of these disorders (281A).

Gene Expression Studies

To gain insight into the molecular events and to determine whether peripheral nerve sheath tumors can be classified according to gene expression profiles, a study was performed on nine dermal neurofibromas, 10 PNF, 10 MPNST, and two MPNST cell lines (164, 285). All tumors (except for six sporadic MPNST) were obtained from NF1 patients. Significant differences in gene expression patterns between neurofibromas and MPNST and between dermal neurofibromas and PNF were detected. NF1-associated and sporadic MPNST could not be distinguished by their gene expression patterns. The authors (164, 285) presented a panel of discriminating genes that may assist in the subclassification of peripheral nerve sheath tumors. Studies based on cDNA array analysis and on RT-PCR (286) showed a differential expression of a number of genes in schwannomas versus control schwann cells. Thus, seven genes were overexpressed and seven genes were underexpressed in the schwannomas versus the control cells.

Of some meaning in relation to the genesis of neural tumors in NF1 is a report of GIST in NF1 patients that differed in their clinical, pathologic, phenotypic, and mutational status of KIT and PDGFRA genes from GIST in non-NF1 patients (132). Studies have demonstrated that haplodeficiency, rather than total deficiency, of neurofibromin is important for the development of neurofibromas in NF1 (287), augmenting KIT ligand signals in mast cells. It is also possible that such a state of haplodeficiency of neurofibromin promotes the growth of a specific subtype of the interstitial cells of Cajal, expressing both KIT and S-100 protein, in the small intestine. Such a hypothesis might explain why GIST develops in NF1 patients without gain-of-function mutation of the KIT gene. Further study is required to clarify the precise mechanism of KIT overexpression in NF1-associated GIST and its tumorigenesis.

Perineuriomas

Perineuriomas are exceedingly rare and benign tumors composed of neoplastic perineurial cells without a schwann cell component (288). Perineurial cells compose the connective tissue sheath surrounding each bundle of fibers in a peripheral nerve. Perineuriomas have been described according to their location as either intraneural or of soft tissue, the latter including the sclerosing type. Sclerosing perineuriomas occur as circumscribed, small (usually <2 cm), painless nodules in the fingers and hands. The tumor has a strong predilection for young adults in the third and fourth decades with a 2:1 male predominance (35). Although perineuriomas are generally benign tumors, occasional malignant examples have been described (289), but no genetic data are available on these tumors. The rare occurrence of hybrid tumors, i.e., schwannoma-perineurioma and neurofibroma-perineurioma (290), has been documented.

Immunohistochemically, the various perineuriomas show a similar picture, i.e., positivity for vimentin and epithelial antigen (EMA), laminin, and collagen IV, and negativity for S-100 protein (289), CD34, desmin, and muscle-specific actin (52). Staining for glucose transporter-1 has been reported to have diagnostic utility in sclerosing perineurioma (291). Staining for p53 may be seen in some perineuriomas (292).

Perineuriomas may arise in the intestine, principally in the colon, as polypoid intramucosal lesions with distinctive histologic features including entrapment of colonic crypts, but also more rarely as large submucosal masses. Distinguishing perineuriomas from other spindle-cell neoplasms of the gastrointestinal tract, which is important for proper clinical management, can be facilitated by immunostaining for EMA and claudin-1.

Soft tissue perineuriomas occur most commonly in the subcutis of the extremities and trunk of young to middle-aged adults, and they show a morphologic spectrum ranging from hypercellular lesions with collagenous stroma to hypocellular tumors with myxoid stroma. In addition to EMA, soft tissue perineuriomas are often positive for CD34, which may lead to difficulties in the differential diagnosis. Soft tissue perineuriomas behave in a benign fashion and rarely recur. Atypical histologic features (including scattered pleomorphic cells and infiltrative margins) seem to have no clinical significance and seem to be akin to those seen in schwannomas and atypical (bizarre) neurofibromas (293).

The histology of perineuriomas shows some distinct differences between the intraneural versus the soft tissue types, and their descriptions can be found in several sources (35, 43, 52, 56, 293, 294) (Figure 1.7).

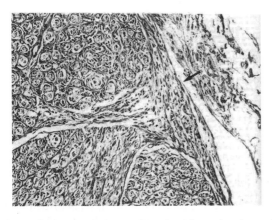

FIGURE 1.7. Histological picture of a perineurioma showing multiple fascicles to be involved by pseudo-onion bulb formation. Spindled perineurial cells form a wedge-shaped infiltrate in the extrafascicular tissue (*arrow*). This tumor was characterized by an abnormality of chromosome 22, i.e., add(22)(q11.2). (From reference 292, with permission.).

Intraneural perineuriomas typically occur in adolescence or early adulthood and show no sex predilection. Peripheral nerves of the extremities are primarily affected; cranial nerve lesions are rare. Central nervous system perineuriomas are exceedingly rare (295). Soft tissue perineuriomas occur in a variety of sites in adults of all ages, predominantly females (2:1), and they are not associated with nerves.

A case of a 30-year-old man with NF2 and bilateral vestibular schwannomas (diagnosed at age 19), parasagittal meningioma, an intraspinal ependymoma and other neoplasms, was found to have a perineurioma of the thigh (295A). The diagnosis was established by immunohistochemical and ultrastructural analysis.

Genetic Changes in Perineuriomas

The karyotypes of the six perineuriomas reported to-date are shown in Table 1.6. The neoplastic nature of perineuriomas is indicated by the clonal and some recurrent genetic changes reported in these tumors. Loss of the whole or part of chromosome 22 was established with FISH in four of five soft-tissue perineuriomas (296). Cytogenetically, three of the perineurinomas shown in Table 1.6 had loss of part or of a whole chromosome 22. These findings are in keeping with chromosome 22 loss, as seen in nerve sheath tumors and meningiomas. The possibility that the gene involved is *NF2*, located at 22q11.2, is supported by results in a cutaneous sclerosing perineurinoma of the finger in a 14-year-old girl with a karyotype: 46,XX,del(10)(q22q24),der(10),del(22)(q11-q12?)/47,idem,+der(10) (95, 96, 297). Changes of chromosome 10 seen cytogenetically have been proposed as possibly playing a role in the genesis of perineurioma in addition to that of chromosome 22 (*NF2*) (298). Using markers for the 5′ *BCR* and *NF2* loci, cryptic deletion of these loci in the abnormal metaphases was demonstrated (295). In another study, *NF2* was shown by PCR to be mutated in four of eight sclerosing and soft tissue perineuriomas (299), in three of which LOH of the *NF2* locus was also found.

Loss of chromosome 13 (-13) in a soft tissue perineurinoma in a 26-yr-old woman as the sole karyotypic changes has been reported (300). No changes in chromosome 22 could be detected by FISH.

Besides the well-defined types of tumors arising from peripheral nerves, i.e., schwannomas, neurofibroma and perineurioma, there exists a group of soft tissue tumors

TABLE 1.6. Chromosome changes in perineuriomas.

Anatomic site	Histology	Karyotype	References
Index finger	Sclerosing	46,XY,t(2;10)(p23;q24), t(der(10;10)(2p25;q24)	*Brock et al 2005* (298)
Index finger	Sclerosing	47,XX,add(3)(q23), add(6)(q21),-5,-9,-10,-22, +mar1,+mar2,+mar	*Brock et al 2005* (298)
Elbow	Intraneural	46,XX,add(2)(q11.2),add(3)(q12)	*Brock et al 2005* (298)
Intra-abdominal	Soft tissue	46,XX,t(8;9)(q13;q22)	*Brock et al 2005* (298)
Finger (left middle finger, volar)	Sclerosing	46,XX,del(10)(q22q24),der(10), del(22)(q11-12q?)/ 47,idem,+der(10)	*Mertens et al 2000* (96)
Posterior interosseous nerve	Interneural	45,XX,add(14)(p13),-22,add(22)(q11.2)	*Emory et al 1995* (292)

wherein characteristic features of perineurinoma and schwannoma or neurofibroma can occur within a single tumor (290). Such tumors may be underrecognized, probably due to the failure to recognize the perineurinomatous component. None of the patients with these tumors had NF. In one case, a point mutation of the *NF2* gene was detected (290).

Some Facets of Schwann and Mast Cells in NF Neoplasia

The key role played by schwann cells in neoplasia of peripheral nerve sheaths warrants some comments regarding the background and role of these cells. Schwann cells are the principal support cells throughout the peripheral nerve system, providing each axon with either ensheathment or a series of myelin segments. Schwann cells originate from the neural crest tissue early in embryonic development (48), migrate into the periphery along with growing axons and undergo rapid cell division (49) to accommodate an increasing number of peripheral axons (50). After their final differentiation, the schwann cells cease dividing and remain essentially quiescent for the life of the individual (49). However, the schwann cells can be reactivated to enter the cell cycle under pathological conditions, including tumor formation (51).

Schwann cell–axon interaction is characterized by processes of the schwann cells extending and migrating on developing axons before differentiation, and requiring coordinated regulation of the schwann cell cytoskeleton. This interaction is lost by the schwann cells of schwannomas due to *NF2* mutations. Because the gene is normally involved in Rac signaling, it is not surprising that schwannoma cells have elevated levels of active Rac. Thus, it seems that control of Rac signaling is essential for schwann cell–axon interaction (416).

Schwann cells, the myelin-forming cells in the peripheral nervous system, have their proliferation and differentiation controlled by intracellular signaling systems responsible for integrating stimuli from cytokines and other factors and controlling the transcriptional activities of genes regulating mitosis or differentiation. Schwann cell proliferation is under control of cyclic adenosine monophosphate signals and mitogen-activated protein kinase pathways. Differentiation of these cells is influenced by PDGF, neuregulin, and ILGF-I, who serve as ligands to receptor-type tyrosine kinase and activate common intracellular signaling cascades, mitogen-activated kinase pathways, and phosphatidylinositol-3-kinase pathways (417). The latter pathway promotes myelin formation. The balance of activities between the differentiation and proliferation pathways holds a key to schwann cell biology and abnormal growth.

Neurofibromas are benign tumors, particularly the dermal type, that have no predilection for malignant transformation. Neurofibromas are caused and their development initiated by

loss of the *NF1* gene in schwannoma cells, but their full development seems to be related to NF1-haploinsufficient cells that ultimately constitute an integral part of the make-up of these tumors. One of these cell types are the mast cells.

The literature on the cellular origin, composition, and role in neurofibroma formation has a long history that have been succinctly reviewed (418, 419). These complex tumors are composed of schwann cells, mast cells, fibroblasts, and peripheral cells embedded in collagen produced by the fibroblasts. Total loss of *NF1* in schwann cells and haploinsufficiency of *NF1* in non-neuronal cells are required for neurofibroma tumorigenesis. The latter aspect consists of a complicated series of events in which mast cells seem to play a key role (418, 420). Thus, the inciting factor for mast cell infiltration of neurofibromas is Kit ligand hypersecreted by the *NF1*-deficient schwann cells (129, 420).

The basic components of an evolving neurofibroma are *NF1*-deficient schwann cells, which act as tumorigenic instigators, and haploinsufficient mast cells acting as inducers, and haploinsufficient fibroblasts, schwann cells, perineurial cells, and endothelial cells that act as responders (62, 78, 86, 128, 419). The pathways involved in neurofibroma formation are complex (287,418,420), including the effects of the mast cells after they have migrated to the Kit ligand-secreting schwann cells. Fibroblasts and mast cells interact directly to contribute to the neurofibroma phenotype. Furthermore, the haploinsufficient mast cells secrete high concentrations of the profibrotic transforming growth factor-β (TGF-β), causing fibroblasts to proliferate and synthesize collagen in neurofibromas (420). The latter action occurs via hyperactivation of a novel Rasc-abl signaling pathway. Inhibition of c-abl (genetic or pharmacologic) reverse the fibroblasts proliferation and collagen synthesis to normal levels (420).

Although mast cells with total or partial loss of *NF1* are necessary in the pathogenesis of neurofibromas, they are not sufficient for tumor formation per se in the neural environment, pointing to the importance of the presence of other heterozygous supporting cells in the neurofibroma environment (129). This includes the haploinsufficient fibroblasts which in neurofibromas demonstrate abnormal response to cytokines, increased collagen deposition, and increased proliferation.

The complexity of the molecular events in the haploinsufficient mast cells in neurofibroma is demonstrated by the following: hyperactivation of the extracellular signal-regulated kinase (ERK) via the hematopoietic-specific Rho GTPase Rac2 directly contributes to the hyperproliferation of *NF1*-deficient mast cells in vitro and in vivo (421). Furthermore, Rac2 functions as mediator of cross-talk between phosphoinositide 3-kinase and the p21^{ras}-ERK pathway to confer distinct proliferative advantage to *NF1*-deficient mast cells (421).

Schwann cells of NF1-related neurofibromas are deficient in neurofibromin, which contains a GAP-related domain (NF1-GRD) leading to Ras activation, increased proliferation in response to certain growth stimuli, increased angiogenic potential, and altered cell morphology. Reconstitution of NF1-GRD in NF1-deficient cells led to amelioration of the changes described except for the angiogenic potential (422).

Informative reviews on the molecular, genetic, and cellular the aspects of the neurofibromatoses and pathogenesis of neurofibroma and schwannoma, are avialable (408, 444–450).

References

1. Riccardi, V.M. (2002) *Neurofibromatosis: Phenotype, Natural History and Pathogenesis*, 2nd ed. Johns Hopkins University Press, Baltimore, MD.

2. Friedman, J., Gutmann, D., MacCollin, M., and Riccardi, V. (1999) *Neurofibromatosis: Phenotype, Natural History, and Pathogenesis*. Johns Hopkins University Press, Baltimore, MD.

3. Friedman, J.M. (2002) Neurofibromatosis type 1. Clinical and diagnostic criteria. *J. Child. Neurol.* **17,** 548–554.

4. Sørensen, S.A., Mulvihill, J.J., and Nielsen, A. (1986) Long-term follow-up of von Recklinghausen neurofibromatosis. Survival and malignant neoplasms. *N. Engl. J. Med.* **314,** 1010–1015.

5. Kley, N., and Seizinger, B.R. (1995) The neurofibromatosis type 2 (NF2) tumour suppressor gene: implications beyond the hereditary tumour syndrome? *Cancer Surv.* **25,** 207–218.

6. Zöller, M.E.T., Rembeck, B., Odén, A., Samuelsson, M., and Angervall, L. (1997) Malignant and benign tumors in patients with neurofibromatosis type 1 in a defined Swedish population. *Cancer* **79,** 2125–2131.

7. Side, L.E., and Shannon, K.M. (1998) The *NF1* gene as a tumor suppressor. In *Neurofibromatosis Type 1: From Genotype to Phenotype* (Upadhyaya, M., and Cooper, D.N., eds.), BIOS Scientific Publishers Ltd., Oxford, United Kingdom, pp. 133–152.

7A. Walker, L., Thompson, D., Easton, D., Ponder, B., Ponder, M., Frayling, I., and Baralle, D. (2006) A prospective study of neurofibromatosis type 1 cancer incidence in the UK. *Br. J. Cancer* **95,** 233–238.

8. Reed, N., and Gutmann, D.H. (2001) Tumorigenesis in neurofibromatosis: new insights and potential therapies. *Trends Mol. Med.* **7,** 157–162.

9. Waggener, J.S. (1966) Ultrastructure of benign peripheral nerve sheath tumors. *Cancer* **19,** 699–709.

10. Kamata, Y. (1978) Study of the ultrastructure and acetylcholinesterase activity in von Recklinghausen's neurofibromatosis. *Acta Pathol. Jpn.* **28,** 393–410.

11. Halliday, A.L., Sobel, R.A., and Martuza, R.L. (1991) Benign spinal nerve sheath tumors: their occurrence sporadically and in neurofibromatosis types 1 and 2. *J. Neurosurg.* **74,** 248–253.

12. Hajdu, S.I., (1993) Peripheral nerve sheath tumors. Histogenesis, classification, and prognosis. *Cancer* **72,** 3549–3552.

13. Seizinger, B.R. (1993) *NF1:* a prevalent cause of tumorigenesis in human cancers? *Nat. Genet.* **3,** 97–99.

14. Huson, S.M. (1994) Neurofibromatosis 1: a clinical and genetic overview. In *The Neurofibromatoses*, 1st ed. (Huson, S.M., and Hughes, R.A.C., eds.), Chapman and Hall, London, pp. 160–203.

15. Louis, D.N., Ramesh, V., and Gusella, J.F. (1995) Neuropathology and molecular genetics of neurofibromatosis 2 and related tumors. *Brain Pathol.* **5,** 163–172.

16. Metheny, L.J., Cappione, A.J., and Skuse, G.R (1995) Genetic and epigenetic mechanisms in the pathogenesis of neurofibromatosis type 1. *J. Neuropathol. Exp. Neurol.* **54,** 753–760.

17. von Deimling, A., Krone, W., and Menon, A.G. (1995) Neurofibromatosis type 1: pathology, clinical features and molecular genetics. *Brain Pathol.* **5,** 153–162.

18. von Deimling, A., Foster, R., and Krone, W. (2000) Neurofibromatosis type 1. In *World Health Organization Classification of Tumours. Pathology and Genetics of Tumours of the Nervous System* (Kleihues, P., and Cavenee, W.K., eds.), IARCPress, Lyon, France, pp. 216–218.

19. Shen, M.H., Harper, P.S., and Upadhyaya, M. (1996) Molecular genetics of neurofibromatosis type 1 (NF1). *J. Med. Genet.* **22,** 2–17.

20. Gutmann, D.H., Aylsworth, A., Carey, J.C., Korf, B., Marks, J., Pyeritz, R.E., Rubenstein, A., and Viskochil, D. (1997) The diagnostic evaluation and multidisciplinary management of neurofibromatosis 1 and neurofibromatosis 2. *JAMA* **278,** 51–57.

21. Pollack, I.F., and Mulvihill, J.J. (1997) Neurofibromatosis 1 and 2. *Brain Pathol.* **7,** 823–836.

22. Rosenbaum, T., Petrie, K.M., and Ratner, N. (1997) Neurofibromatosis type 1: genetic and cellular mechanisms of peripheral nerve tumors formation. *Neuroscientist* **3,** 412–420.

23. Ferner, R.E. (1998) Clinical aspects of neurofibromatosis 1. In *Neurofibromatosis Type 1: From Genotype to Phenotype* (Upadhyaya, M., and Cooper, D.N., eds.), BIOS Scientific Publishers Ltd., Oxford, United Kingdom, pp. 21–38.

24. Kluwe, L., and Mautner, V.F. (1998) Mosaicism in sporadic neurofibromatosis 2 patients. *Hum. Mol. Genet.* **7,** 2051–2055.

25. Upadhyaya, M., and Cooper, D.N. (1998) *Neurofibromatosis Type 1: From Genotype to Phenotype*, 1st ed. BIOS Scientific, Oxford, United Kingdom.

26. Ruggieri, M. (1999) The different forms of neurofibromatosis. *Childs Nerv. Syst.* **15,** 295–308.

27. Angelov, L., and Guha, A. (2000) Peripheral nerve tumors. In *Neurooncology: The Essentials* (Bernstein, M., and Berger, M.S., eds.), Thieme Medical Publishers, New York, NY, pp. 434–444.

28. Ruggieri, M., and Huson, S. (2001) The clinical and diagnostic implications of mosaicism in the neurofibromatoses. *Neurology* **56,** 1433–1443.

29. Zhu, Y., and Parada, L.F. (2001) Neurofibromin, a tumor suppressor in the nervous system. *Exp. Cell Res.* **264,** 19–28.

30. Ferner, R.E., and Gutmann, D.H. (2002) International consensus statement on malignant peripheral nerve sheath tumors in neurofibromatosis 1. *Cancer Res.* **62,** 1573–1577.

31. Ferner, R.E., and O'Doherty, M.J. (2002) Neurofibroma and schwannoma. *Curr. Opin. Neurol.* **15,** 679–684.

32. Packer, R.J., Gutmann, D.H., Rubenstein, A., Viskochil, D., Zimmerman, R.A., Vezina, G., Small, J., and Korf, B. (2002) Plexiform neurofibromas in NF1. Toward biologic-based therapy. *Neurology* **58,** 1461–1470.

33. Tucker, T., and Friedman, J.M. (2002) Pathogenesis of hereditary tumors: beyond the "two-hit" hypothesis. *Clin. Genet.* **62,** 345–357.

34. Miettinen, M. (2003) Nerve sheath tumors. In *Diagnostic Soft Tissue Pathology*, Churchill Livingstone, Philadelphia, PA, pp. 343–362.

35. Miettinen, M. (2003) Perineurioma. Tumor of perineurial cells. In *Diagnostic Soft Tissue Pathology*, Churchill Livingstone, Philadelphia, PA, pp. 363–367.

36. Arun, D., and Gutmann, D.H. (2004) Recent advances in neurofibromatosis type 1. *Curr. Opin. Neurol.* **17**, 101–105.

37. Bhattacharyya, A.K., Perrin, R., and Guha, A. (2004) Peripheral nerve tumors: management strategies and molecular insights. *J. Neurooncol.* **69**, 335–349.

38. Carroll, S.L., and Stonecypher, M.S. (2004) Tumor suppressor mutations and growth factor signaling in the pathogenesis of NF–1 associated peripheral nerve sheath tumors. I. The role of tumor suppressor mutations. *J. Neuropathol. Exp. Neurol.* **63**, 1115–1123.

39. Carroll, S.L., and Stonecypher, M.S. (2005) Tumor suppressor mutations and growth factor signaling in the pathogenesis of NF-1 associated peripheral nerve sheath tumors. II. The role of dysregulated growth factor signaling. *J. Neuropathol. Exp. Neurol.* **64**, 1–9.

40. Kim, D.H., Murovic, J.A., Tiel, R.L., Moes, G., and Kline, D.G. (2005) A series of 397 peripheral neural sheath tumors: 30-year experience at Louisiana State University Health Sciences Center. *J. Neurosurg.* **102**, 246–255.

41. Ward, B.A., and Gutmann, D.H. (2005) Neurofibromatosis 1: from lab bench to clinic. *Pediatr. Neurol.* **32**, 221–228.

42. Heim, R.A., Kam-Morgan, L.N.W., Binnie, C.G., Corns, D.D., Cayouette, M.C, Farber, R.A., Ayslworth, A.S., Silverman, L.M., and Luce M.C. (1995) Distribution of 13 truncating mutations in neurofibromatosis type 1 gene. *Hum. Mol. Genet.* **4**, 975–981.

43. Weiss, S.W., and Goldblum, J.R. (2001) *Enzinger and Weiss's Soft Tissue Tumors*, 4th ed. Mosby, St. Louis, MO.

44. Shearer, P., Parham, D., Kovnar, E., Kun, L., Rao, B., Lobe, T., and Pratt, C. (1994) Neurofibromatosis type 1 and malignancy: review of 32 pediatric cases treated at a single institution. *Med. Pediatr. Oncol.* **22**, 78–83.

45. Walker, L., Thompson, D., Easton, D., Ponder, B., Ponder, M., Frayling, I., and Baralle, D. (2006) A prospective study of neurofibromatosis type 1 cancer incidence in the UK. *Br. J. Cancer* **95**, 233–238.

46. Tinschert, S., Naumann, I., Stegmann, E., Buske, A., Kaufmann, D., Thiel, G., and Jenne, D.E. (2000) Segmental neurofibromatosis is caused by somatic mutation of the neurofibromatosis type 1 (*NF1*) gene. *Eur. J. Hum. Genet.* **8**, 455–459.

47. Skuse, G.R., Kosciolek, B.A., and Rowley, P.T. (1991) The neurofibroma in von Recklinghausen neurofibromatosis has a unicellular origin. *Am. J. Hum. Genet.* **49**, 600–607.

48. Zimmer, C., and Le Dourain, N. (1993) Neural crest lineage. In *Peripheral Neuropathy*, 3rd ed. (Dyck, P.J., Thomas, P.K., Griffin, J.W., Low, P.A., and Poduslo, J.F., eds.), Saunders, Philadelphia, PA, pp. 299–316.

49. Ashbury, A.K. (1967) Schwann cell proliferation in developing mouse sciatic nerve. *J. Cel. Biol.* **34**, 735–743.

50. Lemke, G.E. (1990) Glial growth factors. *Semin. Neurosci.* **2**, 437–443.

51. De Vries, G.H. (1993) Schwann cell proliferation. In *Peripheral Neuropathy*, 3rd ed. (Dyck, P.J., Thomas, P.K., Griffin, J.W., Low, P.A., and Poduslo, J.F., eds.), Saunders, Philadelphia, pp. 290–298.

52. Scheithauer, B.W., Giannini, C., and Woodruff, J.M. (2000) Perineurioma. In *World Health Organization Classification of Tumours. Pathology and Genetics of Tumours of the Nervous System* (Kleihues, P., and Cavenee, W.K., eds.), IARC Press, Lyon, pp.169–171.

53. Korf, B.R. (1999) Plexiform neurofibromas. *Am. J. Med. Genet.* **89**, 31–37.

54. Scheithauer, B.W., Woodruff, J.M., and Erlandson, R.A., editors. (1999) *Neurofibroma Tumors of the Peripheral Nervous System. Atlas of Tumor Pathology*, third series, fascicle 24. Armed Forces Institute of Pathology, Washington, DC, pp. 177–218.

55. Wei, E.X., Albores-Saavedra, J., and Fowler, M.R., (2005) Plexiform neurofibroma of the uterine cervix. A case report and review of the literature. *Arch. Pathol. Lab. Med.* **129**, 783–786.

56. Woodruff, J.M., Kourea, H.P., Louis, D.N., and Scheithauer, B.W. (2000) Schwannoma. In *World Health Organization Classification of Tumours. Pathology and Genetics of Tumours of the Nervous System* (Kleihues, P., and Cavenee, W.K., eds.), IARC Press, Lyon, France, pp.164–166.

57. Woodruff, J.M., Kourea, H.P., Louis, D.N., and Scheithauer, B.W. (2000) Neurofibroma. In *World Health Organization Classification of Tumours. Pathology and Genetics of Tumours of the Nervous System* (Kleihues, P., and Cavenee, W.K., eds.), IARC Press, Lyon, France, pp.167–168.

58. Rosenbaum, T., Boissy, Y.L., Kombrinck, K., Brannan, C., Jenkins, N., Copeland, N.G., and Ratner, N. (1995) Neurofibromin-deficient fibroblasts fail to form perineurium in vitro. *Development* **121**, 3583–3592.

59. Morioka, N., Tsuchida, T., Etoh, T., Ishibashi, Y., and Otsuka, F. (1990) A case of neurofibrosarcoma associated with neurofibromatosis: light microscopic, ultrastructural, immunohistochemical and biochemical investigations. *J. Dermatol.* **17**, 312–316.

60. Cappione, A.J., French, B.L., and Skuse, G.R. (1997) A potential role for NF1 mRNA editing in the pathogenesis of NF1 tumors. *Am. J. Hum. Genet.* **60**, 305–312.

61. Luijten, M., Redeker, S., van Noesel, M.M., Troost, D., Westerveld, A., and Hulsebos, T.J.M. (2000) Microsatellite instability and promoter methylation as possible causes of NF1 gene inactivation in neurofibromas. *Eur. J. Hum. Genet.* **8**, 939–945.

62. Muir, D., Neubauer, D., Lim, I.T., Yachnis, A.T., and Wallace, M.R. (2001) Tumorigenic properties of neurofibromin-deficient neurofibroma Schwann cells. *Am. J. Pathol.* **158**, 501–513.

63. Fishbein, L., Eady, B., Sanek, N., Muir, D., and Wallace, M.R. (2005) Analysis of somatic *NF1* promoter methylation in plexiform neurofibromas and Schwann cells. *Cancer Genet. Cytogenet.* **157**, 181–186.

64. Menon, A.G., Anderson, K.M., Riccardi, V.M., Chung, R.Y., Whaley, J.M., Yandell, D.W., Farmer, G.E., Freiman, R.N., Lee, J.K., Li, F.P., Barker, D.F., Ledbetter, D.H., Kleicer, A., Martuza, R.L., Gusella, J.R., and Seizinger, B.R. (1990) Chromosome 17p deletions and p53 mutations associated with the formation of malignant neurofibromas in von Recklinghausen neurofibromatosis. *Proc. Natl. Acad. Sci USA* **87**, 5435–5439.

65. Lothe, R.A., Slettan, A., Saeter, G., Brøgger, A., Børresen, A.-L., and Nesland, J.M. (1995) Alterations at chromosome 17 loci in peripheral nerve sheath tumors. *J. Neuropathol. Exp. Neurol.* **54**, 65–73.

66. Skuse, G.R., Kosciolek, B.A., and Rowley, P.T. (1989) Molecular genetic analysis of tumors in von Recklinghausen neurofibromatosis: loss of heterozygosity for chromosome 17. *Genes Chromosomes Cancer* **1**, 36–41.

67. Glover, T.W., Stein, C.K., Legius, E., Andersen, L.B., Brereton, A., and Johnson, S. (1991) Molecular and cytogenetic analysis of tumors in von Recklinghausen neurofibromatosis. *Genes Chromosomes Cancer* **3**, 62–70.

68. Legius, E., Marchuk, D.A., Collins, F.S., and Glover, T.W. (1993) Somatic deletion of the neurofibromatosis type 1 gene in a neurofibrosarcoma supports a tumor suppressor gene hypothesis. *Nat. Genet.* **3**, 122–126.

69. Stark, M., Assum, G., and Krone, W. (1995) Single-cell PCR performed with neurofibroma Schwann cells reveals the presence of both alleles of the neurofibromatosis type 1 (NF1) gene. *Hum. Genet.* **96**, 619–623.

70. Colman, S.D., Williams, C.A., and Wallace, M.R. (1995) Benign neurofibromas in type 1 neurofibromatosis (NF1) show somatic deletions of the *NF1* gene. *Nat. Genet.* **11**, 90–92.

71. Serra, E., Puig, S., Otero, D., Gaona, A., Kruyer, H., Ars, E., Estivill, X., and Lázaro, C. (1997) Confirmation of a double-hit model for the *NF1* gene in benign neurofibromas. *Am. J. Hum. Genet.* **61**, 512–519.

72. Sawada, S., Florell, S., Purandare, S.M., Ota, M., Stephens, K., and Viskochil, D. (1996) Identification of *NF1* mutations in both alleles of a dermal neurofibroma. *Nat. Genet.* **14**, 110–112.

73. Ottini, L., Esposito, D.L., Richetta, A., Carlesimo, M., Palmirotta, R., Verì, M.C., Battista, P., Frati, L., Caramia, F.G., Calvieri, S., Cama, A., and Mariani-Costantini, R. (1995) Alterations of microsatellites in neurofibromas of von Recklinghauasen's disease. *Cancer Res.* **55**, 5677–5680.

74. Däschner, K., Assum, G., Eisenbarth, I., Krone, W., Hoffmeyer, S., Wortmann, S., Heymer, B., and Kehrer-Sawatzki, H. (1997) Clonal origin of tumor cells in a plexiform neurofibroma with LOH in NF1 intron 38 and in dermal neurofibromas without LOH of the NF1 gene. *Biochem. Biophys. Res. Commun.* **234**, 346–350.

75. Serra, E., Ars, E., Ravella, A., Sánchez, A., Puig, S., Rosenbaum, T., Estivill, X., and Lázaro, C. (2001) Somatic *NF1* mutational spectrum in benign neurofibromas: mRNA splice defects are common among point mutations. *Hum. Genet.* **108**, 416–429.

76. Serra, E., Rosenbaum, T., Nadal, M., Winner, U., Ars, E., Estivill, X, and Lázaro, C. (2001) Mitotic recombination effects homozygosity for *NF1* germline mutations I neurofibromas. *Nat. Genet.* **28**, 294–296.

77. Kluwe, L., Friedrich, R.E., and Mautner, V.F. (1999) Allelic loss of the *NF1* gene in NF1-associated plexiform neurofibromas. *Cancer Genet. Cytogenet.* **113**, 65–69.

78. Kluwe, L., Friedrich, R., and Mautner, V.-F. (1999) Loss of *NF1* allele in Schwann cells but not in fibroblasts derived form an NF1-associated neurofibroma. *Genes Chromosomes Cancer* **24**, 283–285.

79. Kluwe, L., Friedrich, R.E., Hagel, C., Lindenau, M., and Mautner, V.F. (2000) Mutations and allelic loss of the NF2 gene in neurofibromatosis 2-associated skin tumors. *J. Invest. Dermatol.* **114**, 1017–1021.

80. Eisenbarth, I., Beyer, K., Krone, W., and Assum, G. (2000) Toward a survey of somatic mutation of the *NF1* gene in benign neurofibromas in patients with neurofibromatosis type 1. *Am. J. Hum. Genet.* **66**, 393–401.

81. Joh, A.M., Ruggieri, M., Ferner, R., Upadhyaya, M. (2000) A search for evidence of somatic mutations in the *NF1* gene. *J. Med. Genet.* **37**, 44–49.

82. Rasmussen, S.A., Overman, J., Thomson, S.A.M., Colman, S.D., Abernathy, C.R., Trimpert, R.E., Moose, R., Virdi, G., Roux, K., Bauer, M., Rojiani, A.M., Maria, B.L., Muir, D., and Wallace, M.R. (2000) Chromosome 17 loss-of-heterozygosity studies in benign and malignant tumors of neurofibromatosis type 1. *Genes Chromosomes Cancer* **28**, 425–431.

83. De Luca, A., Bernardini, L., Ceccarini, C., Sinibaldi, L., Novelli, A., Giustini, S., Daniele, I., Calvieri, S., and Mingarelli, R. (2004) Fluorescence in situ hybridization analysis of allelic losses involving the long arm of chromosome 17 in *NF1*-aassociated neurofibromas. *Cancer Genet. Cytogenet.* **150**, 168–172.

84. Side, L., Taylor, B., Cayouette, M., Conner, E., Thompson, P., Luce, M., and Shannon, K. (1997) Homozygous inactivation of the *NF1* gene in bone marrow cells from children with neurofibromatosis type 1 and malignant myeloid disorders. *N. Engl. J. Med.* **336**, 1713–1720.

85. Perry, A., Roth, K.A., Banerjee, R., Fuller, C.E., and Gutmann, D.H. (2001) *NF1* deletions in S-100 protein-positive and negative cells of sporadic and neurofibromatosis 1 (NF1)-associated plexiform neurofibromas and malignant peripheral nerve sheath tumors. *Am. J. Pathol.* **159**, 57–61.

86. Rutkowski, J.L., Wu, K., Gutmann, D.H., Boyer, P.J., and Legius, E. (2000) Genetic and cellular defects contributing to benign tumor formation in neurofibromatosis type 1. *Hum. Mol. Genet.* **9**, 1059–1066.

87. Khalifa, M.A., Montgomery, E.A., Ismiil, N., and Azumi, N. (2000) What are the CD34+ cells in benign peripheral nerve sheath tumors? Double immunostaining study of CD34 and S-100 protein. *Am. J. Clin. Pathol.* **114**, 123–126.

88. Krone, W., and Högemann, I. (1986) Cell culture studies on neurofibromatosis (von Recklinghausen). V. Monosomy 22 and other chromosomal anomalies in cultures from peripheral neurofibromas. *Hum. Genet.* **74**, 453–455.

89. Rey, J.A., Bello, M.J., de Campos, J.M., Benítez, J., Sarasa, J.L., Boixados, J.R., and Sánchez Cascos, A. (1987) Cytogenetic clones in a recurrent neurofibroma. *Cancer Genet. Cytogenet.* **26**, 157–163.

90. Chadduck, W.M., Boop, F.A., and Sawyer, J.R. (1991–1992) Cytogenetic studies of pediatric brain and spinal cord tumors. *Pediatr. Neurosurg.* **17**, 57–65.

91. Molenaar, W.M., Dijkhuizen, T., van Echten, J., Hoekstra, H.J., and van den Berg, E. (1997) Cytogenetic support for early malignant change in a diffuse neurofibroma not associated with neurofibromatosis. *Cancer Genet. Cytogenet.* **97**, 70–72.

92. Wallace, M.R., Rasmussen, S.A., Lim, I.T., Gray, B.A., Zori, R.T., and Muir, D. (2000) Culture of cytogenetically

abnormal Schwann cells from benign and malignant NF1 tumors. *Genes Chromosomes Cancer* **27,** 117–123.

93. Nandula, S.V., Borczuk, A.C., and Murty, V.V.V.S. (2004) Unbalanced t(2;19) and t(2;16) in a neurofibroma. *Cancer Genet. Cytogenet.* **152,** 169–171.

94. Riccardi, V.M., and Elder, D.W. (1986) Multiple cytogenetic aberrations in neurofibrosarcomas complicating neurofibromatosis. *Cancer Genet. Cytogenet.* **23,** 199–209.

95. Mertens, F., Dal Cin, P., De Wever, I., Fletcher, C.D.M., Mandahl, N., Mitelman, F., Rosai, J., Rydholm, A., Sciot, R., Tallini, G., Van den Berghe, H., Vanni, R., and Willén, H. (2000) Cytogenetic characterization of peripheral nerve sheath tumors: a report of the CHAMP study group. *Cancer Genet. Cytogenet.* **123,** 86.

96. Mertens, F., Dal Cin, P., De Wever, I., Fletcher, C.D.M., Mandahl, N., Mitelman, F., Rosai, J., Rydholm, A., Sciot, R., Tallini, G., Van den Berghe, H., Vanni, R., and Willén, H. (2000) Cytogenetic characterization of peripheral nerve sheath tumours: a report of the CHAMP study group. *J. Pathol.* **190,** 31–38.

97. Scheithauer, B.W., Halling, K.C., Nascimento, A.G., Hill, E.M., Sin, F.H., and Katzmann, J.A. (1995) Neurfibroma and malignant peripheral nerve sheath tumor: a proliferation index and DNA ploidy study. *Pathol. Res. Pract.* **19,** 172–187.

98. Koga, T., Iwasaki, H., Ishiguro, M., Matsuzaki, A., and Kikuchi, M. (2002) Losses in chromosomes 17, 19 and 22q in neurofibromatosis type 1 and sporadic neurofibromas: a comparative genomic hybridization analysis. *Cancer Genet. Cytogenet.* **136,** 113–120.

99. Koga, T., Iwasaki, H., Ishiguro, M., Matsuzaki, A., and Kikuchi, M. (2002) Frequent genomic imbalances in chromosomes 17, 19, and 22q in peripheral nerve sheath tumours detected by comparative genomic hybridization analysis. *J. Pathol.* **197,** 98–107.

100. Cawthon, R.M., Weiss, R., Xu, G., Viskochil, D., Culver, M., Stevens, J., Robertson, M., Dunn, D., Gesteland, R., O'Connell, P., and White, R. (1990) A major segment of the neurofibromatosis type 1 gene. cDNA sequence, genomic structure, and point mutations. *Cell* **62,** 193–201.

101. Viskochil, D., Buchberg ,A.M., Xu, G., Cawthon, R.M., Stevens, J., Wolff, R.K., Culver, M., Carey, J.C., Copeland, N.G., Jenkins, N.A., White, R., and O'Connell, P. (1990) Deletions and a translocation interrupt a cloned gene at the neurofibromatosis type 1 locus. *Cell* **62,** 187–192.

102. Marchuk, D.A., Saulino, A.M., Tavakkol, R., Swaroop, M., Wallace, M.R., Andersen, L.B., Mitchell, A.L., Gutmann, D.H., Boguski, M., and Collins, F.S. (1991) cDNA cloning of the type 1 neurofibromatosis gene: complete sequence of the NF1 gene product. *Genomics* **11,** 931–940.

103. Li, Y., O'Connell, P., Huntsman-Breidenbach, H., Cawthon, R., Stevens, J., Xu, G., Neil, S., Robertson, M., White, R., and Viskochil, D. (1995) Genomic organization of the neurofibromatosis 1 gene. *Genomics* **25,** 9–18.

104. Kehrer-Sawatzki, H., Häussler, J., Krone, W., Bode, H., Jenne, D.E., Mehnert, K.U., Tümmers, U., and Assum, G. (1997) The second case of a t(17;22) in a family with neurofibromatosis type 1: sequence analysis of the breakpoint regions. *Hum. Genet.* **99,** 237–247.

105. Ledbetter, D.H., Rich, D.C., O'Connell, P., Leppert, M., and Carey, J.C. (1989) Precise localization of NF1 to 17q11.2 by balanced translocation. *Am. J. Hum. Genet.* **44,** 20–24.

106. Asamoah, A., North, K., Doran, S., Wagstaff, J., Ogle, R., Collins, F.S., and Korf, B.R. (1995) 17q inversion involving the neurofibromatosis type one locus in a family with neurofibromatosis type one. *Am. J. Med. Genet.* **60,** 312–316.

107. Daston, M.M., Scrable, H., Nordlund, M., Sturbaum, A.K., Nissen, L.M., and Ratner, N. (1992) The protein product of the neurofibromatosis type 1 gene is expressed at highest abundance in neurons, Schwann cells, and oligodendrocytes. *Neuron* **8,** 415–428.

108. Ballester, R., Marchuk, D., Boguski, M., Saulino, A., Letcher, R., Wigler, M., and Collins, F.S. (1990) The NF1 locus encodes a protein functionally related to mammalian GAP and yeast IRA proteins. *Cell* **63,** 851–859.

109. Martin, G.A., Viskochil, D., Bollag, G., McCabe, P.C., Crosier, W.J. Haubruck, H., Conroy, L., Clark, R., O'Connell, P., Cawthon, R.M., Innis, M.A., and McCormick, F. (1990) The GAP-related domain of the neurofibromatosis type 1 gene product interacts with *ras* p21. *Cell* **63,** 843–849.

110. Xu, G., Lin, B., Tanaka, K., Dunn., D., Wood, D., Gesteland, R., White, R., Weiss, R., and Tamanoi, F. (1990) The catalytic domain of the neurofibromatosis type 1 gene product stimulates *ras* GTPase and complements *ira* mutants of S. cerevisiae. *Cell* **63,** 835–841.

111. Xu, G., O'Connell, P., Viskochil, D., Cawthon, R., Robertson, M., Culver, M., Dunn, D., Stevens, J., Gesteland, R., White, R., and Weiss, R. (1990) The neurofibromatosis type 1 gene encodes a protein related to GAP. *Cell* **62,** 599–608.

112. Gutmann, D.H., Geist, R.T., Rose, K., and Wright, D.E. (1995) Expression of two new protein isoforms of the neurofibromatosis type 1 gene product, neurofibromin, in muscle tissues. *Dev. Dyn.* **202,** 302–311.

113. Li, Y., Bollag, G., Clark, R., Stevens, J., Conroy, L., and Fults, D., Ward, K., Friedman, E., Samowitz, W., Robertson, M., Bradley, P., McCormick, F., White, R., and Cawthon, R. (1992) Somatic mutations in the neurofibromatosis-1 gene in human tumors. *Cell* **69,** 275–281.

114. Shannon, K.M., O'Connell, P., Martin, G.A., Paderanga, D., Olson, K., Dinndorf, P., and McCormick, F. (1994) Loss of normal NF1 allele from the bone marrow of children with type 1 neurofibromatosis and malignant myeloid disorders. *N. Engl. J. Med.* **330,** 597–601.

115. Side, L.E., Emmanuel, P.D., Taylor, B., Franklin, J., Thompson, P., Castleberry, R.P., and Shannon, K.M. (1998) Mutations of the NF1 gene in children with juvenile myelomonocytic leukemia without clinical evidence of neurofibromatosis, type 1. *Blood* **92,** 267–272.

116. Öguzkan, S., Terzi, Y.K., Guler, E., Derbent, M., Isik, P., Saatci, U., and Ayter, S. (2006) NF1 cases associated with rhabdomyosarcoma of urinary bladder. A large deletion in the *NF1* gene. *Cancer Genet. Cytogenet.* **164,** 159–163.

116A. Yang, G., Khalaf, W., van de Locht, L., Jansen, J.H., Gao, M., Thompson, M.A., van der Reijden, B.A., Gutmann, D.H., Delwel, R., Clapp, D.W., and Hiebert, S.W. (2005) Transcriptional repression of the neurofibromatosis type-1 tumor suppressor by the t(8;21) fusion protein. *Mol. Cell. Biol.* **25,** 5869–5879.

117. Hoffmeyer, S., Assum, G., Griesser, J., Kaufmann, D., Nurnberg, P., and Krone, W. (1995) On unequal allelic expression

of the neurofibromin gene in neurofibromatosis type 1. *Hum. Mol. Genet.* **4**, 1267–1272.

118. Fang, L.J., Vidaud, D., Vidaud, M., and Thirion, J.P. (2001) Identification and characterization of four novel large deletions in human neurofibromatosis type 1 (NF1) gene. *Hum. Mutat.* **18**, 549–550.

118A. De Raedt, T., Maertens, O., Chmara, M., Brems, H., Heyns, I., Sciot, R., Majounie, E., Upadhyaya, M., De Schepper, S., Speleman, F., Messiaen, L., Vermeesch, J.R., and Legius, E. (2006) Somatic loss of wild type NF1 allele in neurofibromas: comparison of NF1 microdeletion and non-microdeletion patients. *Genes Chromosomes Cancer* **45**, 893–904.

119. Wu, R., López-Correa, C., Rutkowski, J.L., Baumbach, L.L., Glover, T.W., and Legius, E. (1999) Germline mutations in NF1 patients with malignancies. *Genes Chromosomes Cancer* **26**, 376–380.

119A. Maertens, O., Brems, H., Vandesompele, J., De Raedt, T., Heyns, I., Rosenbaum, T., De Schepper, S., De Paepe, A., Mortier, G., Janssens, S., Speleman, F., Legius, E., and Messiaen, L. (2006) Comprehensive NF1 screening on cultured Schwann cells from neurofibromas. *Hum. Mutat.* **27**, 1030–1040.

120. Bernards, A. (1995) Neurofibromatosis type 1 and Ras-mediated signaling: filling in the GAPs. *Biochim. Biophys. Acta* **1242**, 43–59.

121. Guha, A., Lau, N., Huvar, I., Gutmann, D., Provias, J., Pawson, T., and Boss, G. (1996) RAS-GTP levels are elevated in human NF1 peripheral nerve tumors. *Oncogene* **12**, 507–513.

121A. Trovo-Marqui A.B., and Tajara, E.H. (2006) Neurofibromin: a general outlook. *Clin. Genet.* **70**, 1–13.

122. Basu, T.N., Gutmann, D.H., Fletcher, J.A., Glover, T.W., Collins, F.S., and Downward, J. (1992) Aberrant regulation of *ras* proteins in malignant tumour cells from type 1 neurofibromatosis patients. *Nature* **356**, 713–715.

123. DeClue, J.E., Papageorge, A.G., Fletcher, J.A., Diehl, S.R., Ratner, N., Vass, W.C., and Lowy, D.R. (1992) Abnormal regulation of mammalian^{p21} ras contribuates to malignant tumor growth in von Recklinghausen (type 1) neurofibromatosis. *Cell* **69**, 265–273.

124. Kim, H.A., Rosenbaum, T., Marchionni, M.A., Ratner, N., and DeClue, J.E. (1995) Schwann cells from neurofibromin deficient mice exhibit activation of p21^{ras}, inhibition of cell proliferation and morphological changes. *Oncogene* **11**, 325–335.

125. DeClue, J.E., Heffelfinger, S., Benvenuto, G., Ling, B., Li, S., Rui, W., Vass, W.C., Viskochil, D., and Ratner, N. (2000) Epidermal growth factor receptor expression in neurofibromatosis type 1-related tumors and NF1 animal models. *J. Clin. Invest.* **105**, 1233–1241.

126. Stefansson, K., Wollmann, R., and Jerkovic, M. (1982) S-100 protein in soft tissue tumors derived form Schwann cells and melanocytes. *Am. J. Pathol.* **106**, 261–268.

127. Sheela, S., Riccardi, V.M., and Ratner, N. (1990) Angiogenic and invasive properties of neurofibroma Schwann cells. *J. Cell Biol.* **111**, 645–653.

128. Serra, E., Rosenbaum, T., Winner, U., Aledo, R., Ars, E., Estivill, X., Lenard, H.-G., and Lázaro, C. (2000) Schwann cells harbor the somatic *NF1* mutation in neurofibromas:

evidence of two different Schwann cell subpopulations. *Hum. Mol. Genet.* **9**, 3055–3064.

129. Zhu, Y., Ghosh, P., Charnay, P., Burns, D.K., and Parada, L.F. (2002) Neurofibromas in NF1: schwann cell origin and role of tumor environment. *Science* **296**, 920–922.

130. Sherman, L.S., Atit, R., Rosenbaum, T., Cox, A.D., and Ratner, N. (2000) Single cell Ras-GTP analysis reveals altered Ras activity in a subpopulation of neurofibroma Schwann cells but not fibroblasts. *J. Biol. Chem.* **275**, 30740–30745.

130A. Yantiss, R.K., Rosenberg, A.E., Sarran, L., Besmer, P., and Antonescu, C.R. (2005) Multiple gastrointestinal stromal tumors in type 1 neurofibromatosis: a pathologic and molecular study. *Mod. Pathol.* **18**, 475–484.

131. Lee, J.L., Kim, J.Y., Ryu, M.H., Kang, H.J., Chang, H.M., Kim, T.W., Lee, H., Park, J.H., Kim, H.C., Kim, J.S., and Kang, Y.K. (2006) Response to imatinib in KIT- and PDGFRA-wild type gastrointestinal stromal associated with neurofibromatosis type 1. *Dig. Dis. Sci.***51**, 1043–1046.

132. Anderson, C.E., and Salter, D.M. (2005) Schwannoma with focal smooth muscle differentiation: a potential pitfall in the interpretation of core biopsies. *Histopathology* **46**, 582–594.

133. Takazawa, Y., Sakurai, S., Sakuma, Y., Ikeda, T., Yamaguchi, J., Hashizume, Y., Yokoyama, S., Motegi, A., and Fukayama, M. (2005) Gastrointestinal stromal tumors of neurofibromatosis type 1 (von Recklinghausen's disease). *Am. J. Surg. Pathol.* **29**, 755–763.

133A. Andersson, J., Sihto, H., Meis-Kindblom, J.M., Joensuu, H., Nupponen, N., and Kindblom, L.G. (2005) NF1-associated gastrointestinal stromal tumors have unique clinical, phenotypic, and genotypic characteristics. *Am. J. Surg. Pathol.* **29**, 1170–1176.

133B. Miettinen, M., Fetsch, J.F., Sobin, L.H., and Lasota, J. (2006) Gastrointestinal stromal tumors in patients with neurofibromatosis type 1: a clinicopathologic and molecular genetic study of 45 cases. *Am. J. Surg. Pathol.* **30**, 90–96.

133C. Nemoto, H., Tate, G., Schirinzi, A., Suzuki, T., Sasaya, S., Yoshizawa, Y., Midorikawa, T., Mitsuya, T., Dallapiccola, B., and Sanada, Y. (2006) Novel *NF1* gene mutation in a Japanese patient with neurofibromatosis type 1 and a gastrointestinal stromal tumor *J. Gastroenterol.* **41**, 378–382.

133D. Stewart, D.R., Corless, C.L., Rubin, B.P., Heinrich, M.C., Messiaen, L.M., Kessler, L.J., Zhang, P.J., and Brooks, D.G. (2007) Mitotic recombination as evidence of alternative pathogenesis of gastrointestinal stromal tumours in neurofibromatosis type 1. *J. Med. Genet.* **44**, e61.

133E. De Raedt, T., Cools, J., Debiec-Rychter, M., Brems, H., Mentens, N., Sciot, R., Himpens, J., de Wever, I., Schoffski, P., Marynen, P., and Legius, E. (2006) Intestinal neurofibromatosis is a subtype of familial GIST and results from a dominant activating mutation in PDGFRA. *Gastroenterology* **131**, 1907–1912.

134. Miettinen, M., Kopczynski, J., Makhlouf, H.R., Sarlomo-Rikala, M., Gyorffy, H., Burke, A., Sobin, L.H., and Lasota, J. (2003) Gastrointestinal stromal tumors, intramural leiomyomas, and leiomyosarcomas in the duodenum. A clinicopathologic, immunohistochemical, and molecular genetic study of 167 cases. *Am. J. Surg. Pathol.* **27**, 625–641.

135. Schaldenbrand, J.D., and Appelman, H.D. (1984) Solitary solid stromal gastrointestinal tumors in von Recklinghausen's disease with minimal smooth muscle differentiation. *Hum. Pathol.* **15,** 229–232.

136. Min, K.W., and Balaton, A.J. (1993) Small intestinal stromal tumors with skeinoid fibers in neurofibromatosis: report of four cases with ultrastructural study of skeinoid fibers from paraffin blocks. *Ultrasruct. Pathol.* **17,** 307–314.

137. Ishida, T., Wada, I., Horiuchi, H., Oka, T., and Machinami, R. (1996) Multiple small intestinal stromal tumor with skeinoid fibers in association with neurofibromatosis 1 (von Recklinghausen's disease). *Pathol. Int.* **46,** 689–695.

138. Nakamura, H., Satoh, Y., Ikeda, T., Endoh, T., and Imai, K. (2000) A case of upper jejunal gastrointestinal stromal tumor (GIST) accompanied with von Recklinghausen's disease. *Nippon Shokakibyo Gakkai Zasshi.* **97,** 1385–1390.

139. Cheng, S.P., Huang, M.J., Yang, T.L., Tzen, C.Y., Liu, C.L., Liu, T.P., and Hsiao, S.C. (2004) Neurofibromatosis with gastrointestinal stromal tumors: insights into the association. *Dig. Dis. Sci.* **49,** 1165–1169.

140. Andersson, J., Sihto, H., Meis-Kindblom, J., Joensuu, H., Nupponen, N., and Kindblom, L.-G. (2005) NF1-associated gastrointestinal stromal tumors have unique clinical, phenotypic, and genotypic characteristics. *Am. J. Surg. Pathol.* **29,** 1170–1176.

141. Kinoshita, K., Hirota, S., Isozaki, K., Ohashi, A., Nishida, T., Kitamura, Y., Shinomura, Y., and Matsuzawa, Y. (2004) Absence of c-kit gene mutations in gastrointestinal stromal tumours from neurofibromatosis type 1 patients. *J. Pathol.* **202,** 80–85.

142. Sandberg, A.A., and Bridge, J.A. (2002) Updates on the cytogenetics and molecular genetics of bone and soft tissue tumors: gastrointestinal stromal tumors. *Cancer Genet. Cytogenet.* **135,** 1–22.

143. Ryan, J.J., Klein, K.A., Neuberger, T.J., Leftwich, J.A., Westin, E.H., Kauma, S., Fletcher, J.A., DeVries, G.H., and Huff, T.F. (1994) Role for the stem cell factor/KIT complex in Schwann cell neoplasia and mast cell proliferation associated with neurofibromatosis. *J. Neurosci. Res.* **37,** 415–432.

144. Yanagihori, H., Oyama, N., Nakamura, K., and Kaneko, F. (2005) c-*kit* mutations in patients with childhood-onset mastocytosis and genotype-phenotype correlation. *J. Mol. Diagn.* **7,** 252–257.

145. Badache, A., Muja, N., and De Vries, G.H. (1998) Expression of Kit in neurofibromin-deficient human Schwann cells: role in Schwann cell hyperplasia associated with type 1 neurofibromatosis. *Oncogene* **17,** 795–800.

146. Riccardi, V.M. (1987) Mast-cell stabilization to decrease neurofibroma growth. Preliminary experience with ketotifen. *Arch. Dermatol.* **123,** 1011–1016.

147. Hirota, S., Nomura, S., Asada, H., Ito, A., Morii, E., and Kitamura, Y. (1993) Possible involvement of c-kit receptor and its ligand in increase of mast cells in neurofibroma tissues. *Arch. Pathol. Lab. Med.* **117,** 996–999.

148. Bernardis, V., Sorrentino, D., Snidero, D., Avellini, C., Paduano, R., Beltrami, C.A., Digito, F., and Bartoli, E. (1999) Intestinal leiomyosarcoma and gastroparesis associated with von Recklinghausen's disease. *Digestion* **60,** 82–85.

149. Hunter, S., Weiss, S., Ou, C.-Y., Jaye, D., Young, A., Wilcox, J., Arbiser, J.L., Monson, D., Goldblum, J., Nolen, J.D., and Varma, V. (2005) Apolipoprotein D is down-regulated during malignant transformation of neurofibromas. *Hum. Pathol.* **36,** 987–993.

150. Flower, D.R. (1994) The lipocalin protein family: a role in cell regulation. *FEBS Lett.* **354,** 7–11.

151. Kourea, H.P., Orlow, I., Scheithauer, B.W., Cordon-Cardo, C., and Woodruff, J.M. (1999) Deletions of the *INK4A* gene occur in malignant peripheral nerve sheath tumors but not in neurofibromas. *Am. J. Pathol.* **155,** 1855–1860.

152. Kourea, H.P., Cordon-Cardo, C., Dudas, M., Leung, D., and Woodruff, J.M. (1999) Expression of p27(kip) and other cell cycle regulators in malignant peripheral nerve sheath tumors and neurofibromas: the emerging role of p27(kip) in malignant transformation of neurofibromas. *Am. J. Pathol.* **155,** 1885–1891.

153. Legius, E., Dierick, H., Wu, R., Hall, B.K., Marynen, P., Cassiman, J.-J., and Glover, T.W. (1994) *TP53* mutations are frequent in malignant NF1 tumors. *Genes Chromosomes Cancer* **10,** 250–255.

154. Nielsen, G.P., Stemmer-Rachamimov, A.O., Ino, Y., Møller, M.B., Rosenberg, A.E., and Louis, D.N. (1999) Malignant transformation of neufibromas in neurofibromatosis 1 is associated with *CDKN2A/p16* inactivation. *Am. J. Pathol.* **155,** 1879–1884.

155. Birindelli, S., Perrone, F., Oggionni, M., Lavarino, C., Pasini, B., Vergani, B., Ranzani, G.N., Pierotti, M.A., and Pilotti, S. (2001) Rb and TP53 pathway alterations in sporadic and NF1-related malignant peripheral nerve sheath tumors. *Lab. Invest.* **81,** 833–844.

156. Perry, A., Kunz, S.N., Fuller, C.E., Banerjee, R., Marley, E.F., Liapis, H., Watson, M.A., and Gutmann, D.H. (2002) Differential *NF1, p16,* and *EGFR* patterns by interphase cytogenetics (FISH) in malignant peripheral nerve sheath tumor (MPNST) and morphologically similar spindle cell neoplasms. *J. Neuropathol. Exp. Neruol.* **61,** 702–709.

157. Buchberg, A.M., Cleveland, L.S., Jenkins, N.A., and Copeland, N.G. (1990) Sequence homology shared by neurofibromatosis type-1 gene and *IRA-1* and *IRA-2* negative regulators of the *RAS* cyclic AMP pathway. *Nature* **347,** 291–294.

158. Serrano, M., Lin, A.W., McCurrach, M.E., Beach, D., and Lowe, S.W. (1997) Oncogenic ras provokes premature cell senescence associated with accumulation of p53 and p16INK4a. *Cell* **88,** 593–602.

159. Palmero, I., and Serrano, M. (2001) Induction of senescence by oncogenic Ras. *Methods Enzymol.* **333,** 247–256.

160. Deng, Q., Liao, R., Wu, B.L., and Sun, P. (2003) High-intensity *ras* signaling induces premature senescence by activating p38 pathway in primary human fibroblasts. *J. Biol. Chem.* **279,** 1050–1059.

161. Lin, A.W., Barradas, M., Stone, J.C., van Aelst, L., Serrano, M., and Lowe, S.W. (1998) Premature senescence involving p53 and p16 is activated in response to constitutive MEK/MAPK mitogenic signaling. *Genes Dev.* **12,** 3008–3019.

162. Guha, A. (1998) Ras activation in astrocytomas and neurofibromas. *Can. J. Neurol. Sci.* **25,** 267–281.

163. Barkan, B., Starinsky, S., Friedman, E., Stein, R., and Kloog, Y. (2006) The Ras inhibitor farnesylthiosalicylic acid as a

potential therapy for neurofibromatosis type 1. *Clin. Cancer Res.* **12,** 5533–5542.

164. Holtkamp, N., Mautner, V.F., Friedrich, R.E., Harder, A., Hartmann, C., Theallier-Janko, A., Hoffmann, K.T., and von Deimling, A. (2004) Differentially expressed genes in neurofibromatosis 1-associated neurofibromas and malignant peripheral nerve sheath tumors. *Acta Neuropathol (Berl)* **107,** 159–168.

165. Motoi, T., Ishida, T., Kawato, A., Motoi, N., and Fukayama, M. (2005) Pigmented neurofibroma: review of Japanese patients with an analysis of melanogenesis demonstrating coexpression of c-met protooncogene and microphthalmia-associated transcription factor. *Hum. Pathol.* **36,** 871–877.

166. Evans, D.G., Sainio, M., and Baser, M.F. (2000) Neurofibromatosis type 2. *J. Med. Genet.* **37,** 897–904.

166A. Bianchi, A.B., Hara, T., Ramesh, V., Gao, J., Klein-Szanto, A.J., Morin, F., Menon, A.G., Trofatter, J.A., Gusella, J.F., Seizinger, B.R., and Kley, N. (1994) Mutations in transcript isoforms of the neurofibromatosis 2 gene in multiple human tumour types. *Nat. Genet.* **6,** 185–192.

167. Louis, D.N., Stemmer-Rachamimov, and Wiestler, OD. (2000) Neurofibromatosis type 2. In *World Health Organization Classification of Tumours. Pathology and Genetics of Tumours of the Nervous System* (Kleihues, P., and Cavenee, W.K., eds.), IARC Press, Lyon, France, pp. 219–222.

168. Ruggieri, M., Iannetti, P., Polizzi, A., La Mantia, I., Spalice, A., Giliberto, O., Platania, N., Gabriele, A.L., Albanese, V., and Pavone, L. (2005) Earliest clinical manifestations and natural history of neurofibromatosis type 2 (NF2) in childhood: a study of 24 patients. *Neuropediatrics* **36,** 21–34.

168A. Kanter, W.R., Eldridge, R., Fabricant, R., Allen, J.C., and Koerber, T. (1980) Central neurofibromatosis with bilateral acoustic neuromas: genetic, clinical and biochemical distinctions from peripheral neurofibromatosis. *Neurology* **30,** 851–859.

168B. Kluwe, L., Mautner, V., and Heinrich, B. (2003) Molecular study of frequency of mosaicism in neurofibromatosis 2 patients with bilateral vestibular schwannomas. *J. Med. Genet.* **40,** 109–114.

168C. MacCollin, M., Willett, C., Heinrich, B., Jacoby, L.B., Acierno, J.S., Jr., Perry, A., and Louis, D.N. (2003) Familial schwannomatosis: exclusion of the NF2 locus as the germline event. *Neurology* **60,** 1968–1974.

168D. Takeshima, H., Izawa, I., Lee, P.S., Safdar, N., Levin, V.A., and Saya, H. (1994) Detection of cellular proteins that interact with the NF2 tumor suppressor gene product. *Oncogene* **9,** 2135–2144.

169. Ishida, T., Kuroda, M., Motoi, T., Oka, T., Imamura, T., and Machinami, R. (1998) Phenotypic diversity of neurofibromatosis 2: association with plexiform schwannoma. *Histopathology* **32,** 264–270.

170. McMenamin, M.E., and Fletcher, C.D.M. (2001) Expanding the spectrum of malignant change in schwannomas. Epithelioid malignant change, epithelioid malignant peripheral nerve sheath tumor, and epithelioid angiosarcoma: a study of 17 cases. *Am. J. Surg. Pathol.* **25,** 13–25.

170A. Aghi, M., Kluwe, L., Webster, M.T., Jacoby, L.B., Barker, F.G., II, Ojemann, R.G., Mautner, V.-F., and MacCollin, M. (2006) Unilateral vestibular schwannomas with other neurofibromatosis type 2-related tumors: clinical and molecular study of a unique phenotype. *J. Neurosurg.* **104,** 201–207.

171. Agaram, N.P., Prakash, S., and Antonescu, C.R. (2005) Deep-seated plexiform schwannoma. A pathologic study of 16 cases and comparative analysis with the superficial variety. *Am. J. Surg. Pathol.* **29,** 1042–1048.

172. Jacoby, L.B., Jones, D., Davis, K., Kronn, D., Short M.P., Gusella, J., and MacCollin, M. (1997) Molecular analysis of the NF2 tumor-suppressor gene in schwannomatosis. *Am. J. Hum. Genet.* **61,** 1293–1302.

173. Purcell, S.M., and Dixon, S.L. (1989) Schwannomatosis. An unusual variant of neurofibromatosis or a distinct clinical entity? *Arch. Dermatol.* **125,** 390–393.

174. Ikeda, T., Hashimoto, S., Fukushige, S., Ohmori, H., and Horii, A. (2005) Comparative genomic hybridization and mutation analyses of sporadic schwannomas. *J. Neurooncol.* **72,** 225–230.

174A. White, J.B., Scheithauer, B.W., Amrami, K.K., Babovic-Vuksanovic, D., and Spinner, R.J. (2006) Contiguous conventional and plexiform schwannomas. Report of two cases. *J. Neurosurg.* **104,** 319–324.

174B. Feiz-Erfan, I., Zabramski, J.M., Herrmann, L.L., and Coons, S.W. (2006) Cavernous malformation within a schwannoma: review of the literature and hypothesis of a common genetic etiology. *Acta Neurochir (Wien)* **148,** 647–652.

175. Hou, Y.Y., Tan, Y.S., Xu, J.F., Wang, X.N., Lu, S.H., Ji, Y., Wang, J., and Zhu, X.Z. (2006) Schwannomas of the gastrointestinal tract: a clinicopathological, immunohistochemical and ultrastructural study of 33 cases. *Histopathology* **48,** 536–545.

176. Rogatto, S.R., and Casartelli, C. (1989) Cytogenetic study of human neurinomas. *Cancer Genet. Cytogenet.* **41,** 278.

177. Teyssier, J.R., and Ferre, D. (1989) Frequent clonal chromosomal changes in human non-malignant tumors. *Int. J. Cancer* **44,** 828–832.

178. Couturier, J., Delattre, O., Kujas, M., Philippon, J., Peter, M., Rouleau, G., Aurias, A, and Thomas G. (1990) Assessment of chromosome 22 anomalies in neurinomas by combined karyotype and RFLP analyses. *Cancer Genet. Cytogenet.* **45,** 55–62.

179. Lodding, P., Kindblom, L.-G., Angervall, L., and Stenman, G. (1990) Cellular schwannomas. A clinicopathologic study of 29 cases. *Virchows Arch. A Pathol. Anat.* **416,** 237–248.

180. Stenman, G., Kindblom, L.-G., Johansson, M., and Angervall, L. (1991) Clonal chromosome abnormalities and in vitro growth characteristics of classical and cellular schwannomas. *Cancer Genet.C ytogenet.* **57,** 121–131.

181. Mark, J. (1972) The chromosomal findings in seven neurinomas and one neurosarcoma. *Acta Path. Microbiol. Scand. Section A.* **80,** 61–70.

182. Mark, J. (1977) Chromosomal abnormalities and their specificity in human neoplasms: an assessment of recent observations by banding techniques. *Adv. Cancer Res.* **24,** 165–222.

183. Seizinger, B.R., Martuza, R.L., and Gusella, J.F. (1986) Loss of genes on chromosome 22 in tumorigenesis of human acoustic neuroma. *Nature* **322,** 644–647.

184. Seizinger, B.R., Rouleau, G., Ozelius, L.J., Lane, A.H., St. George-Hyslop, P., >Huson, S., Gusella, J.F., and Martuza,

R.L. (1987) Common pathogenetic mechanism for three tumor types in bilateral acoustic neurofibromatosis. *Science* **236,** 317–319.

185. Rey, J.A., Bello, M.J., de Campos, J.M., Kusak, M.E., and Moreno, S. (1987) Cytogenetic analysis in human neurinomas. *Cancer Genet. Cytogenet.* **28,** 187–188.

186. Wullich, B., Kiechle-Schwarz, M., Mayfrank, L., and Schempp, W. (1989) Cytogenetic and in situ DNA-hybridization studies in intracranial tumors of a patient with central neurofibromatosis. *Hum. Genet.* **82,** 31–34.

187. Webb, H.D., and Griffin, C.A. (1991) Cytogenetic study of acoustic neuroma. *Cancer Genet. Cytogenet.* **56,** 83–84.

188. Bello, M.J., de Campos, J.M., Kusak, M.E., Vaquero, J., Sarasa, J.L., Pestaña, A., and Rey, J.A. (1993) Clonal chromosome aberrations in neurinomas. *Genes Chromosomes Cancer* **6,** 206–211.

189. Rey, J.A., Bello, M.J., de Campos, J.M., Vaquero, J., Kusak, M.E., Sarasa, J.L., and Pestaña, A. (1993) Abnormalities of chromosome 22 in human brain tumors determined by combined cytogenetic and molecular genetic approaches. *Cancer Genet. Cytogenet.* **66,** 1–10.

190. Jacoby, L.B. MacCollin, M., Barone, R., Ramesh, V., and Gusella, J.F. (1996) Frequency and distribution of *NF2* mutations in schwannomas. *Genes Chromosomes Cancer* **17,** 45–55.

191. Rao, U.N.M., Surti, U., Hoffner, L., and Yaw, K. (1996) Cytogenetic and histologic correlation of peripheral nerve sheath tumors of soft tissue. *Cancer Genet. Cytogenet.* **88,** 17–25.

192. Joste, N.E., Racz, M.I., Montgomery, K.D., Haines, S., and Pitcher, J.D. (2004) Clonal chromosome abnormalities in a plexiform cellular schwannoma. *Cancer Genet. Cytogenet.* **150,** 73–77.

192A. Hameed, M., Chen, F., Mahmet, K., Das, K., and Kim, S. (2006) Schwannoma with a reciprocal t(9;22)(q22;q13). *Cancer Genet. Cytogenet.* **164,** 92–93.

193. Rey, J.A., Couturier, J., Bello, M.J., Aurias, A., Pestaña, A. (1991) Human Gene Mapping 11. Cytogenetic analysis in a series of 56 neurinomas. *Cytogenet. Cell Genet.* **58,** 2050.

194. Örndal, C., Rydholm, A., Willén, H., Mitelman, F., and Mandahl, N. (1994) Cytogenetic intratumor heterogeneity in soft tissue tumors. *Cancer Genet. Cytogenet.* **78,** 127–137.

195. Gorunova, L., Dawiskiba, S., Andrén-Sandberg, Å., Höglund, M., and Johansson, B. (2001) Extensive cytogenetic heterogeneity in a benign retroperitoneal schwannoma. *Cancer Genet. Cytogenet.* **127,** 148–154.

196. Gorunova, L., Höglund, M., and Johansson, B. (2001) Extensive cytogenetic heterogeneity in a benign retroperitoneal schwannoma. *Cancer Genet. Cytogenet.* **129,** 72.

197. Bijlsma, E.K., Brouwer-Mladin, R., Bosch, D.A., Westerveld, A., and Hulsebos, T.J.M. (1992) Molecular characterization of chromosome 22 deletions in schwannomas. *Genes Chromosomes Cancer* **5,** 201–205.

198. Wolff, R.K., Frazer, K.A., Jackler, R.K., Lanser, M.J., Pitts, L.H., and Cox, D.R. (1992) Analysis of chromosome 22 deletions in neurofibromatosis type 2-related tumors. *Am. J. Hum. Genet.* **51,** 478–485.

199. Bruder, C.E.G., Ichimura, K., Tingby, O., Hirakawa, K., Komatsuzaki, A., Tamura, A., Yuasa, Y., Collins, V.P., and Dumanski, J.P. (1999) A group of schwannomas with interstitial deletions on 22q located outside the *NF2* locus shows no detectable mutations in the *NF2* gene. *Hum. Genet.* **104,** 418–424.

199A. Bello, M.J., Martínez-Glez, V., Franco-Hernandez, C., Pefla-Granero, C., de Campos, J.M., Isla, A., Lassaletta, L., Vaquero, J., and Rey, J.A. (2007) DNA methylation pattern in 16 tumor-related genes in schwannomas. *Cancer Genet. Cytogenet.* **172,** 84–86.

199B. Lassaletta, L., Bello, M.J., del Rio, L., Alfonso, C., Roda, J.M., Rey, J.A., and Gavilan, J. (2006) DNA methylation of multiple genes in vestibular schwannoma: relationship with clinical and radiological findings. *Otol. Neurotol.* **27,** 1180–1185.

199C. Bian, L.G., Sun, Q.F., Tirakotai, W., Zhao, W.G., Shen, J.K., Luo, Q.Z., and Bertalanffy, H. (2005) Loss of heterozygosity on chromosome 22 in sporadic schwannoma and its relation to the proliferation of tumor cells. *Chin. Med. J. (Engl.)* **118,** 1517–1524.

199D. Diaz de Stahl, T., Hansson, C.M., de Bustos, C., Mantripragada, K.K., Piotrowski, A., Benetkiewicz, M., Jarbo, C., Wiklund, L., Mathiesen, T., Nyberg, G., Collins, V.P., Evans, D.G., Ichimura, K., and Dumanski, J.P. (2005) High-resolution array-CGH profiling of germline and tumor-specific copy number alterations on chromosome 22 in patients affected with schwannomas. *Hum. Genet.* **118,** 35–44.

200. Fraenzer, J.T., Pan, H., Minimo, L., Jr., Smith, G.M., Knauer, D., and Hung, G. (2003) Overexpression of the NF2 gene inhibits schwannoma cell proliferation through promoting PDGFR degradation. *Int. J. Oncol.* **23,** 1493–1500.

201. Rouleau, G.A., Wertelecki, W., Haines, J.L., Hobbs, W.J., Trofatter, J.A., Seizinger, B.R., Martuza, R.L., Superneau, D.W., Conneally, P.M., and Gusella, J.F. (1987) Genetic linkage of bilateral acoustic neurofibromatosis to a DNA marker on chromosome 22. *Nature* **329,** 246–248.

202. Rouleau, G.A., Merel, P., Lutchman, M., Sanson, M., Zuchman, J., Marineau, C., Hoang-Xuan, K., Demczuk, S., Desmaze, C., and Plougastel, B. (1993) Alteration in a new gene encoding a putative membrane-organizing protein causes neuro-fibromatosis type 2. *Nature* **36,** 515–521.

203. Trofatter, J.A., MacCollin, M.M., Rutter, J.L., Murrell, J.R., Duyao, M.P., Parry, D.M., Eldridge, R., Kley, N., Menon, A.G., and Pulaski, K. (1993) A novel moesin-, ezrin-, radxin-like gene is a candidate for the neurofibromatosis 2 tumor suppressor. *Cell* **72,** 791–800 (Errata: 1993 *Cell* **75,** 826).

204. Arai, E., Ikeuchi, T., Karasawa, S., Tamura, A., Yamamoto, K., Kida, M., Ichimura, K., Yuasa, Y., and Tonomura, A. (1992) Constitutional translocation t(4;22)(q12;q12.2) associated with neurofibromatosis type 2. *Am. J. Med. Genet.* **44,** 163–167.

205. Arai, E., Tokino, T., Imai, T., Inazawa, J., Ikeuchi, T., Tonomura, A., and Nakamura, Y. (1993) Mapping the breakpoint of a constitutional translocation on chromosome 22 in a patient with NF2. *Genes Chromosomes Cancer* **6,** 235–238.

206. Arai, E., Ikeuchi, T., and Nakamura, Y., (1994) Characterization of the translocation breakpoint on chromosome 22q12.2 in a patient with neurofibromatosis type 2 (NF2). *Hum. Mol. Genet.* **3,** 937–939.

207. Tsukita, S., Oishi, K., Sato, N., Sagara, J., Kawai, A., and Tsukita, S. (1994) ERM family members as molecular linkers

between the cell surface glycoprotein CD44 and actin-based cytoskeletons. *J. Cell. Biol.* **126**, 391–401.

208. Sainio, M., Zhao, R., Heiska, L., Turunen, O., den Bakker, MN., Swarthoff, E., Lutchman, M., Rouleau, G.A., Jääsheläinen, J., Vaheri, A., and Carpen, O. (1997) Neurofibromatosis 2 tumor suppressor protein colocalizes ezrin and CD44 and associates with actin-containing cytoskeleton. *J. Cell Sci.* **110**, 2249–2260.

209. Sherman, L., Xu, H.-M., Geist, R.T., Saporito-Irwin, S., Howells, N., Ponta, H., Herrlich, P., and Gutmann, D.H. (1997) Interdomain binding mediates tumor growth suppression by the *NF2* gene product. *Oncogene* **15**, 2505–2509.

210. Xiao, G.-H., Chernoff, J., and Testa, J.R. (2003) *NF2*: the wizardry of merlin. *Genes Chromosomes Cancer* **38**, 389–399.

211. Bianchi, A.B., Mitsunaga, S.-I., Cheng, J.Q., Klein, W.M., Jhanwr, S.C., Seizinger, B., Kley, N., Klein-Szanto, A.J.P., and Testa, J.R. (1995) High frequency of inactivating mutations in the neurofibromatosis type 2 gene *(NF2)* in primary malignant mesotheliomas. *Proc. Natl. Acad. Sci. USA* **92**, 10854–10858.

211A. Ruttledge, M.H., Sarrazin, J., Rangaratnam, S., Phelan, C.M., Twist, E., Merel, P., Delattre, O., Thomas, G., Nordenskjold, M., Collins, V.P., Dumanski, J.P., and Rouleau, G.A. (1994) Evidence for the complete inactivation of the NF2 gene in the majority of sporadic meningiomas. *Nat. Genet.* **6**, 180–184.

212. Ridley, A.J, Paterson, H.F., Johnston, C.L., Dickmann D., and Hall, A. (1992) The small GTP-binding protein rac regulates growth factor-induced membrane ruffling. *Cell* **70**, 401–410.

213. Shaw, R.J., Pacz, J.G., Curto, M., Yaktine, A., Pruitt, W.M., Saotome, I., O'Bryan, J.P., Gupta, V., Ratner, N., Der, C.J., Jacks, T., McClatchey, A.I. (2001) The Nf2 tumor suppressor, merlin, functions in Rac-dependent signaling. *Dev. Cell* **1**, 63–72.

214. Bashour, A.M, Meng, J.J., Ip, W., MacCollin, M., and Ratner, N. (2002) The neurofibromatosis type 2 gene product, merlin, reverses the F-actin cytoskeletal defects in primary human schwannoma cells. *Mol. Cell Biol.* **22**, 1150–1157.

215. Morrison, H., Sherman, L.S., Legg, J., Banine, F., Isacke, C., Haipek, C.A., Gutmann, D.H., Ponta, H., and Herrlich, P. (2001) The NF2 tumor suppressor gene product, merlin, mediates contact inhibition of growth through interactions with CD44. *Genes Dev.* **15**, 968–980.

216. Schulze, K.M., Hanemann, C.O., Muller, H.W., and Hanenberg, H. (2002) Transduction of wild-type merlin into human schwannoma cells decreases schwannoma cell growth and induces apoptosis. *Hum. Mol. Genet.* **11**, 69–76.

217. Lallemand, D., Curto, M., Saotome, I., Giovannini, M., and McClatchey, A.I. (2003) NF2 deficiency promotes tumorigenesis and metastasis by destabilizing adherens junctions. *Genes Dev.* **17**, 1090–1100.

218. MacCollin, M., Woodfin, M., Kronn, D., and Short, M.P. (1996) Schwannomatosis: a clinical and pathologic study. *Neurology* **46**, 1072–1079.

219. Bijlsma, E.K., Merel, P., Bosch, D.A., Westerveld, A., Delattre, O., Thomas, G., and Hulsebos, T.J.M. (1994) Analysis of mutations in the *SCH* gene in schwannomas. *Genes Chromosomes Cancer* **11**, 7–14.

220. Irving, R.M., Moffat, D.A., Hardy, D.G., Barton, D.E., Xuereb, J.H., and Maher, E.R. (1994) Somatic NF2 gene mutations in familial and non-familial vestibular schwannoma. *Hum. Mol. Genet.* **2**, 347–350.

221. Jacoby, L.B., MacCollin, M., Louis, D.N., Mohney, T., Rubio, M.-P., Pulaski, K., Trofatter, J.A., Kley, N., Selzinger, B., Ramesh, V., and Gusella, J.F. (1994) Exon scanning for mutation of the *NF2* gene in schwannomas. *Hum. Mol. Genet.* **3**, 413–419.

222. Lekanne Deprez, R.H., Bianchi, A.B., Groen, N.A., Seizinger, B.R., Hagemeijer, A., van Drunen, E., Bootsma, D., Koper, J.W., Avezaat, C.J., Kley, N., and Zwarthoff, E.C. (1994) Frequent *NF2* gene transcript mutations in sporadic meningiomas and vestibular schwannomas. *Am. J. Hum. Genet.* **54**, 1022–1029.

223. Sainz, J., Baser, M.E., Ragge, N.K., Nelson, R.A., and Pulst, S.M. (1993) Loss of alleles in vestibular schwannomas. *Arch. Otolaryng. Head Neck Surg.* **119**, 1285–1288.

224. Sainz, J., Huynh, D.P., Figueroa, K., Ragge, N.K., Baser, M.E., and Pulst, S.M. (1994) Mutations of the neurofibromatosis type 2 gene and lack of the gene product in vestibular schwannomas. *Hum. Mol. Genet.* **3**, 885–891.

225. Twist, E.C., Ruttledge, M.H., Rousseau, M., Sanson, M., Papi, L., Merel, P., Delattre, O., Thomas, G., and Rouleau, G.A. (1994) The neurofibromatosis type 2 gene is inactivated in schwannomas. *Hum. Mol. Genet.* **3**, 147–151.

226. Mérel, P., Hoang-Xuan, K., Sanson, M., Moreau-Aubrey, A., Bijlsma, E.K., Lazaro, C., Moisan, J.P., Resche, F., Nishisho, I., Estivill, X., Delattre, J.Y., Poisson, M., Theillet, C., Hulsebos, T., Delattre, O., and Thomas, G. (1995) Predominant occurrence of somatic mutations of the *NF2* gene in meningiomas and schwannomas. *Genes Chromosomes Cancer* **13**, 211–216.

227. Evans, D.G.R., Trueman, L., Wallace, A., Collins, S., and Strachan, T. (1998) Genotype phenotype correlations in type 2 neurofibromatosis (NF2): evidence for more severe disease associated with truncating mutations. *J. Med. Genet.* **35**, 450–455.

228. Stemmer-Rachamimov, A.O., Ino, Y., Lim, Z.Y., Jacoby, L.B., MacCollin, M., Gusella, J.F., Ramesh, V., and Louis, D.N. (1998) Loss of the *NF2* gene and merlin occur by the tumorlet stage of schwannoma development in neurofibromatosis 2. *J. Neuropathol. Exp. Neurol.* **57**, 1164–1167.

229. Gusella, J.F., Ramesh, V., MacCollin, M., and Jacoby, L.B. (1996) Neurofibromatosis 2: loss of merlin's protective spell. *Curr. Opin. Genet. Dev.* **6**, 87–92.

230. Gutmann, D.H., Giordano, M.J., Fishback, A.S., and Guha, A. (1997) Loss of merlin expression in sporadic meningiomas, ependymomas and schwannomas. *Neurology* **49**, 267–270.

231. Gutmann, D.H., Geist, R.T., Xu, H., Kim, J.S., and Saporito-Irwin, S. (1998) Defects in neurofibromatosis 2 protein function can arise at multiple levels. *Hum. Mol. Genet.* **7**, 335–345.

232. Hitotsumatsu, T., Iwaki, T., Kitamoto, T., Mizoguchi, M., Suzuki, S.O., Hamada, Y., Fukui, M., and Tateishi, J. (1997) Expression of neurofibromatosis 2 protein in human brain tumors: an immunohistochemical study. *Acta Neuropathol.* **93**, 225–232.

233. Ueki, K., Wen-Bin, C., Narita, Y, Asai, A., and Kirino, T. (1999) Tight association of loss of merlin expression with loss of heterozygosity at chromosome 22q in sporadic meningiomas. *Cancer Res.* **59**, 5995–5998.

234. Ikeda, K., Sacki, Y., Gonzalez-Agosti, C., Ramesh, V., and Chiocca, E.A. (1999) Inhibition of NF2-negative and NF2-positive primary human meningioma cell proliferation by overexpression of merlin due to vector-mediated gene transfer. *J. Neurosrug.* **91,** 85–92.

235. Pelton, P.D., Sherman, L.S., Rizvi, T.A., Marchionni, M.A., Wood, P., Friedman, R.A., and Ratner, N. (1998) Ruffling membrane, stress fiber, cell spreading and proliferation abnormalities in human Schwannoma cells. *Oncogene* **17,** 2195–2209.

236. Gutmann, D.H., Sherman, L., Seftor, L., Haipek, C., Hoang Lu, K., and Hendrix, M. (1999) Increased expression of the NF2 tumor suppressor gene product, merlin, impairs cell motility, adhesion and spreading. *Hum. Mol. Genet.* **8,** 267–275.

237. McClatchey, A.I., Saotome, I., Mercer, K., Crowley, D., Gusella, J.F., Bronson, R.T., and Jacks, T. (1998) Mice heterozygous for a mutation at the NF2 tumor suppressor locus develop a range of highly metastatic tumors. *Genes Dev.* **12,** 1121–1133.

238. Tikoo, A., Varga, M., Ramesh, V., Gusella, J., and Maruta, H. (1994) An anti-Ras function of neurofibromatosis type 2 gene product (NF2/Merlin). *J. Biol. Chem.* **269,** 23387–22390.

239. Herrlich, P., Morrison, H., Sleeman, J., Orian-Rousseau, V., Konig, H., Weg-Remers, S., and Ponta, H. (2000) CD44 acts as a growth- and invasiveness-promoting molecule and as a tumor-suppressing cofactor. *Ann. N Y Acad. Sci.* **910,** 106–120.

239A. Scoles, D.R., Yong, W.H., Qin, Y., Wawrowsky, K., and Pulst, S.M. (2006) Schwannomin inhibits tumorigenesis through direct interaction with the eukaryotic initiation factor subunit c (eIF3c). *Hum. Mol. Genet.* **15,** 1059–1070.

240. Ryu, C.H., Kim, S.-W., Lee, K.H., Lee, J.Y., Kim, H., Lee, W.K., Choi, B.H., Lim, Y., Kim, Y.H., Lee, K.-H., Hwang, T.-K., Jun, T.-Y., and Rha, H.K. (2005) The merlin tumor suppressor interacts with Ral guanine nucleotide dissociation stimulator and inhibits its activity. *Oncogene* **24,** 5355–5364.

240A. Okada, T., Lopez-Lago, M., and Giancotti, F.G. (2005) Merlin/*NF-2* mediates contact inhibition of growth by suppressing recruitment of Rac to the plasma membrane. *J. Cell Biol.* **171,** 361–371.

240B. Ryu, C.H., Kim, S.W., Lee, K.H., Lee, J.Y., Kim, H., Lee, W.K., Choi, B.H., Lim, Y., Kim, Y.H., Lee, K.H., Hwang, T.K., Jun, T.Y., and Rha, H.K. (2005) The merlin tumor suppressor interacts with Ral guanine nucleotide dissociation stimulator and inhibits its activity. *Oncogene,* **24,** 5355–5364.

240C. Golovnina, K., Blinov, A., Akhmametyeva, E.M., Omelyanchuk, L.V., and Chang, L.-S. (2005) Evolution and origin of merlin, the product of the *Neurofibromatosis type 2* (*NF2*) tumor-suppressor gene. BMC *Evol. Biol.* **5,** 69–86.

240D. Morrison, H., Sperka, T., Manent, J., Giovannini, M., Ponta, H., and Herrlich, P. (2007) Merlin/neurofibromatosis type 2 suppresses growth by inhibiting the activation of Ras and Rac. *Cancer Res.* **67,** 520–527.

240E. Lee, J.Y., Moon, H.J., Lee, W.K., Chun, H.J., Han, C.W., Jeon, Y.W., Lim, Y., Kim, Y.H., Yao, T.P., Lee, K.H., Jun, T.Y., Rha, H.K., and Kang, J.K. (2006) Merlin facilitates ubiquitination and degradation of transactivation-responsive RNA-binding protein. *Oncogene* **25,** 1143–1152.

240F. Gronholm, M., Muranen, T., Toby, G.G., Utermark, T., Hanemann, C.O., Golemis, E.A., and Carpen, O. (2006) A functional association between merlin and HEI10, a cell cycle regulator. *Oncogene* **25,** 4389–4398.

240G. McClatchey, A.I., and Giovannini, M. (2007) Membrane organization and tumorigenesis—the NF2 tumor suppressor, Merlin. *Genes Dev.* **19,** 2265–2277.

240H. Jin, H., Sperka, T., Herrlich, P., and Morrison, H. (2006) Tumorigenic transformation by CPI-17 through inhibition of a merlin phosphatase. *Nature* **442,** 576–579.

241. De Vitis, L.R., Tedde, A., Vitelli, F., Ammannati, F., Mennonna, P., Bono, P., Grammatico, B., Grammatico, P., Radice, P., Bigozzi, U., Montali, E., and Papi, L. (1996) Analysis of the neurofibromatosis type 2 gene in different human tumors of neuroectodermal origin. *Hum. Genet.* **97,** 638–641.

242. Harwalkar, J.A., Lee, J.H., Hughes, G., Kinney, S.E., and Golubic, M. (1998) Immunoblotting analysis of schwannomin/merlin in human schwannomas. *Am. J. Otol.* **19,** 654–659.

243. Kino, T., Takeshima, H., Nakao, M., Nishi, T., Yamamoto, K., Kimura, T., Saito, Y., Kochi, M., Kuratsu, J., Saya, H., and Ushio, Y. (2001) Identification of the *cis*-acting region in the *NF2* gene promoter as a potential target for mutation and methylation-dependent silencing in schwannoma. *Genes to Cells* **6,** 441–454.

244. Leone, P.E., Bello, M.J., Mendiola, M., Kusak, M.E., de Campos, J.M., Vaquero, J., Sarasa, J.L., Pestaña, A, and Rey, J.A. (1998) Allelic status of 1p, 14q, and 22q and *NF2* gene mutations in sporadic schwannomas. *Int. Mol. Med.* **1,** 889–892.

245. Kim, H., DeClue, J.E., and Ratner, N. (1997) cAMP-dependent protein kinase A is required for Schwann cell growth: interaction between cAMP and neuregulin/tyrosine kinase pathways. *Neurosci. Res.* **49,** 236–247.

246. Ambrosini, G. Adida, C., and Altieri, D.C. (1997) A novel anti-apoptosis gene, survivin, expressed in cancer and lymphoma. *Nat. Med.* **3,** 917–921.

247. Katoh, M., Wilmotte, R., Belkouch, M.C., de Tribolet, N., Pizzolato, G., and Dietrich, P.Y. (2003) Survivin in brain tumors: an attractive target for immunotherapy. *J. Neurooncol.* **64,** 71–76.

248. Kayaselcuk, F., Zorludemir, S., Bal, N., Erdogan, B., Erdogan, S., and Erman, T. (2004) The expression survivin and Ki-67 in meningiomas: correlation with grade and clinical outcome. *J. Neurooncol.* **67,** 209–214.

249. Altieri, D.C. (2003) Survivin, versatile modulation of cell division and apoptosis in cancer. *Oncogene* **22,** 8581–8589.

250. Hassounah, M., Lach, B., Allam, A., Al-Khalaf, H., Siddiqui, Y., Pangue-Cruz, N., Al-Omeir, A., Al-Ahdal, M.N., and Aboussekhra, A. (2005) Benign tumors from the human nervous system express high levels of survivin and are resistant to spontaneous and radiation-induced apoptosis. *J. Neurooncol.* **72,** 203–208.

251. Mantripragada, K.K., Buckley, P.G., Benetkiewicz, M., De Bustos, C., Hirvela, C., Jarbo, C., Bruder, C.E., Wensman, H., Mathiesen, T., Nyberg, G., Papi, L., Collins, V.P., Ichimura, K., Evans, G., and Dumanski, J.P. (2003) High-resolution profiling of an 11 Mb segment of human chromosome 22 in sporadic schwannoma using array-CGH. *Int. J. Oncol.* **22,** 615–622.

252. Mantripragada, K.K., Tapia-Paez, I., Blennow, E., Nilsson, P., Wedell, A., and Dumanski, J.P. (2004) DNA copy-number analysis of the 22q11 deletion-syndrome region using array-CGH with genomic and PCR-based targets. *Int. J. Mol. Med.* **13,** 273–279.

253. Stonecypher, M.S., Chaudhury, A.R., Byer, S.J., and Carroll, S.L. (2006) Neuregulin growth factors and their erbB receptors form a potential signaling network for schwannoma tumorigenesis. *J. Neuropathol. Exp. Neurol.* **65,** 162–175.

254. Mawrin, C., Schulz, S., Hellwig-Patyk, A., Kirches, E., Roessner, A., Lendeckel, U., Firsching, R., Vorwerk, C.K., Keilhoff, G., Dietzmann, K., Grimm, K., Lindberg, G., Gutmann, D.H., Scheihauer, B.W., and Perry, A. (2005) Expression and function of somatostatin receptors in peripheral nerve sheath tumors. *J. Neuropathol. Exp. Neurol.* **64,** 1080–1088.

255. Fanburg–Smith, J.C., Majidi, M., Miettinen, M. (2006) Keratin expression in schwannoma; a study of 115 retroperitoneal and 22 peripheral schwannomas. *Mod. Pathol.* **19,** 115–121.

256. Yoshimoto, Y. (2005) Systematic review of the natural history of vestibular schwannoma. *J. Neurosurg.* **103,** 59–63.

257. Sainz, J., Figueroa, K., Baser, M.E., and Pulst, S.-M. (1996) Identification of three neurofibromatosis type 2 (NF2) gene mutations in vestibular schwannomas. *Hum. Genet.* **97,** 121–123.

257A. Baser, M.E., Mautner, V.-F., Parry, D.M., and Evans, D.G.R. (2005) Methodological issues in longitudinal studies: vestibular schwannoma growth rates in neurofibromatosis 2. *J. Med. Genet.* **42,** 903–906.

258. Sobel, R.A., and Wang, Y. (1993) Vestibular (acoustic) schwannomas: histologic features in neurofibromatosis 2 and in unilateral cases. *J. Neuropathol. Exp. Neurol.* **52,** 106–113.

259. Aguiar, P.H., Tatagiba, M., Samii, M., Dankoweit-Timpe, E., and Ostertag, H. (1995) The comparison between the growth fraction of bilateral vestibular schwannomas in neurofibromatosis 2 (NF2) and unilateral vestibular schwannomas using the monoclonal antibody MIB 1. *Acta Neurochir (Wien)* **134,** 40–45.

260. Antinheimo, J., Haapasalo, H., Seppälä, M., Sainio, M., Carpen, O., and Jääskeläinen, J. (1995) Proliferative potential of sporadic and neurofibromatosis 2-associated schwannomas as studied by MIB-1 (Ki-67) and PCNA labeling. *J. Neuropathol. Exp. Neurol.* **54,** 776–82.

260A. Niemczyk, K., Vaneecloo, F.M., Lecomte, M.H., Lejeune, J.P., Lemaitre, L., Skarzynski, H., Vincent, C., and Dubrulle, F. (2000) Correlation between Ki-67 index and some clinical aspects of acoustic neuromas (vestibular schwannomas). *Otolaryngol. Head Neck Surg.* **123,** 779–783.

261. Irving, R.M., Moffat, D.A., Hardy, D.G., Barton, D.E., Xuereb, J.H., and Maher, E.R. (1993) Molecular genetic analysis of the mechanism of tumorigenesis in acoustic neuroma. *Arch. Otolaryngol. Head Neck Surg.* **119,** 1222–1228.

262. Irving, R.M., Harada, T., Moffat, D.A., Hardy, D.G., Whittaker, J.L., Xuereb, J.H., and Maher, E.R. (1997) Somatic neurofibromatosis type 2 gene mutations and growth characteristics in vestibular schwannoma. *Am. J. Otolaryngol.* **18,** 754–760.

263. Gonzalez-Gomez, P., Bello, M.J., Alonson, M.E., Lomas, J., Arjona, D., de Campos, J..BM., Vaquero, J., Isla, A.,

Lassaletta, L., Gutierrez, M., Sarasa, J.L., and Rey, J.A. (2003) CpG island methylation in sporadic and neurofibromatosis type 2-associated schwannomas. *Clin. Cancer Res.* **9,** 5601–5606.

264. Stemmer-Rachamimov, A.O., Xu, L., Gonzalez-Agosti, C., Burwick, J.A., Pinney, D., Beauchamp, R., Jacoby, L.B., Gusella, J.F., Ramesh, V., and Louis, D.N. (1997) Universal absence of merlin, but not other ERM family members, in schwannomas. *Am. J. Pathol.* **151,** 1649–1654.

265. Huynh, D.P., Mautner, V., Baser, M.E., Stavrou, D., and Pulst, S.-M. (1997) Immunohistochemical detection of schwannomin and neurofibromin in vestibular schwannomas, ependymomas and meningiomas. *J. Neuropathol. Exp. Neurol.* **56,** 382–390.

266. Wu, C.L., Thakker, N., Neary, W., Black, G., Lye, R., Ramsden, R.T., Read, A.P., and Evans, D.G.R. (1998) Differential diagnosis of type 2 neurofibromatosis: molecular discrimination of NF2 and sporadic vestibular schwannomas. *J. Med. Genet.* **35,** 973–977.

267. Mohyuddin, A., Neary, W.J., Wallace, A., Wu, C.L., Purcell, S., Reid, H., Ramsden, R.T., Read, A., Black, G., and Evans, D.G.R. (2002) Molecular genetic analysis of the NF2 gene in young patients with unilateral vestibular schwannomas. *J. Med. Genet.* **39,** 315–322.

268. Bian, L.-G., Tirakotai, W., Sun, Q.-F., Zhao, W.-G., Shen, J.-K., and Luo, Q.-Z. (2005) Molecular genetics alterations and tumor behavior of sporadic vestibular schwannoma from the People's Republic of China. *J. Neurooncol.* **73,** 253–260.

269. Parry, D.M., MacCollin, M.M., Kaiser-Kupfer, M.I., Pulaski, K., Nicholson, H.S., Bolesta, M., Eldridge, R., and Gusella, J.F. (1996) Germ-line mutations in the neurofibromatosis 2 gene: correlations with disease severity and retinal abnormalities. *Am. J. Hum. Genet.* **59,** 529–539.

270. Ruttledge, M.H., Andermann, A.A., Phelan, C.M., Claudio, J.O., Han, F.Y., Chretien, N., Rangaratnam, S., MacCollin, M., Short, P., Parry, D., Michels, V., Riccardi, V.M., Weksberg, R., Kitamura, K., Bradburn, J.M., Hall, B.D., Propping, P., and Rouleau, G.A. (1996) Type of mutation in the neurofibromatosis type 2 gene (*NF2*) frequently determines severity of disease. *Am. J. Hum. Genet.* **59,** 331–342.

270A. Bikhazi, P.H., Lalwani, A.K., Kim, E.J., Bikhazi, N., Attaie, A., Slattery, W.H., Jackler, R.K., and Brackmann, D.E. (1998) Germline screening of the NF-2 gene in families with unilateral vestibular schwannoma. *Otolaryngol. Head Neck Surg.* **119,** 1–6.

270B. Lee, D.J., Maseyesva, B., Westra, W., Long, D., Niparko, J.K., and Califano, J. (2006) Microsatellite analysis of recurrent vestibular schwannoma (acoustic neuroma) following stereotactic radiosurgery. *Otol. Neurotol.* **27,** 213–219.

270C. Dayalan, A.H., Jothi, M., Keshava, R., Thomas, R., Gope, M.L., Doddaballapur, S.K., Somanna, S., Praharaj, S.S., Ashwathnarayanarao, C.B., and Gope, R. (2006) Age dependent phosphorylation and deregulation of p53 in human vestibular schwannomas. *Mol. Carcinog.* **45,** 38–46.

270D. Lasak, J.M., Welling, B., Akhmametyeva, E.M., Salloum, M., and Chang, L.-S. (2002) Retinoblastoma—cyclin-dependent kinase pathway deregulation in vestibular schwannomas. *Laryngoscope* **112,** 1555–1561.

270E. Welling, D.B., Lasak, J.M., Akhmametyeva, E.M., Ghaheri, B., Chang, L.-S. (2002) cDNA microarray analysis of vestibular schwannomas. *Otol. Neurotol.* **23,** 736–748.

270F. Xiong, Z., Cao, Y., Guo, D., Ye, F., and Lei, T. (2006) Expression of EGFR and LRIG-1 in human trigeminal neurinoma. *J. Huazhong Univ. Sci. Technolog. Med. Sci.* **26,** 86–88.

271. Kirchhoff, M., Gerdes, T., Rose, H., Maahr, J., Ottesen, A.M., and Lundsteen, C. (1998) Detection of chromosomal gains and losses in comparative genomic hybridization analysis based on standard reference intervals. *Cytometry* **31,** 163–173.

271A. Hansen, M.R., Roehm, P.C., Chatterjee, P., and Green, S.H. (2006) Constitutive neuregulin-1/ErbB signaling contributes to human vestibular schwannoma proliferation. *Glia* **53,** 593–600.

272. Schmidt, H., Taubert, H., Meye, A., Würl, P., Bache, M., Bartel, F., Holzhausen, H.-J., and Hinze, R. (2000) Gains in chromosomes 7, 8q, 15q and 17q are characteristics changes in malignant but not in benign peripheral nerve sheath tumors from patients with von Recklinghausen's disease. *Cancer Lett.* **155,** 181–190.

273. Rickert, C.H., and Paulus, W. (2004) Comparative genomic hybridization in central and peripheral nerve system tumors of childhood and adolescence. *J. Neuropathol. Exp. Neurol.* **63,** 399–417.

274. Parente, F., Grosgeroge, J., Coindre, J.-M., Terrier, P., Vilain, O., and Turc-Carel, C. (1999) Comparative genomic hybridization reveals novel chromosome deletions in 90 primary soft tissue tumors. *Cancer Genet. Cytogenet.* **115,** 89–95.

275. Mechtersheimer, G., Otaño-Joos, M., Ohl, S., Benner, A., Lehnert, T., Willeke, F., Möller, P., Otto, H.F., Lichter, P., and Joos, S. (1999) Analysis of chromosomal imbalances in sporadic and NF1-associated peripheral nerve sheath tumors by comparative genomic hybridization. *Genes Chromosomes Cancer* **25,** 362–369.

276. Antinheimo, J., Sallinen, S.L., Sallinen, P., Haapasalo, H., Helin, H., Horelli-Kuitunen, N., Wessman, M., Saino, M., Jääskeläinen, J., and Carpen, O. (2000) Genetic aberrations in sporadic and neurifibromatosis 2 (NF2)-associated schwannomas studied by comparative genomic hybridization (CGH). *Acta Neurochir. (Wien)* **142,** 163–167.

277. Warren, C., James, L.A., Ramsden, R.T., Wallace, A., Baser, M.E., Varley, J.M., and Evans, D.G. (2003) Identification of recurrent regions of chromosome loss and gain in vestibular schwannomas using comparative genomic hybridisation. *J. Med. Genet.* **40,** 802–806.

278. Zucman-Rossi, J., Legoix, P., Der Sarkissian, H., Cheret, G., Sor, F., Bernardi, A., Cazes, L., Giraud, S., Ollagnon, E., Lenoir, G., and Thomas G. (1998) NF2 gene in neurofibromatosis type 2 patients. *Hum. Mol. Genet.* **7,** 2095–2101.

279. Rodahl, E., Lybaek, H., Arnes, J., and Ness, G.O. (2005) Chromosomal imbalances in some benign orbital tumours. *Acta Ophthalmol. Scand.* **83,** 385–391.

280. Seppälä, M.T., Sainio, M.A., Johannes, M.J., Kinnunen, J.J., Setälä, K.H., and Jääskeläinen, J.E. (1998) Multiple schwannomas: schwannomatosis or neurofibromatosis type 2? *J. Neurosurg.* **89,** 36–41.

281. Wolkenstein, P., Benchikhi, H., Zeller, J., Wechsler, J., and Revuz, J. (1997) Schwannomatosis: a clinical entity distinct from neurofibromatosis type 2 *Dermatology* **195,** 228–231.

281A. Buckley, P.G., Mantripragada, K.K., Diaz de Stahl, T., Piotrowski, A., Hansson, C.M., Kiss, H., Vetrie, D., Ernberg, I.T., Nordenskjold, M., Bolund, L., Sainio, M., Rouleau, G.A., Niimura, M., Wallace, A.J., Evans, D.G., Grigelionis, G., Menzel, U., and Dumanski, J.P. (2005) Identification of genetic aberrations on chromosome 22 outside the NF2 locus in schwannomatosis and neurofibromatosis type 2. *Hum. Mutat.* **26,** 540–549.

282. Evans, D.G.R., Mason, S., Huson, S.M., Ponder, M., Harding, A.E., and Strachan, T. (1997) Spinal and cutaneous schwannomatosis is a variant form of type 2 neurofibromatosis: a clinical and molecular study. *J. Neurol. Neurosurg. Psychiatry* **62,** 361–366.

283. King, A., and Gutmann, D.H. (2000) The question of familial meningiomas and schwannomas. NF2B or not to be? *Neurology* **54,** 4–5.

283A. Antinheimo, J., Sankila, R., Carpén, O., Pukkala, E., Sainio, M., and Jääskeläinen, J. (2000) Population-based analysis of sporadic and type 2 neurofibromatotsis-associated meningiomas and schwannomas. *Neurology* **54,** 71–76.

284. Evans, D.G.R., Mason, S., Huson, S.M., Ponder, M., Harding, A.E., and Strachan, T. (1997) Spinal and cutaneous schwannomatosis is a variant form of type 2 neurofibromatosis: a clinical and molecular study. *J. Neurol. Neurosurg. Psychiatry* **62,** 361–366.

285. Holtkamp, N., Reuβ D.E., Atallah, I., Kuban, R.-J., Hartmann, C., Mautner, V.-F., Frahm, S., Friedrich, R.E., Algermissen, B., Pham, V.-A., Prietz, S., Rosenbaum, T., Estevez-Schwarz, L., and von Deimling, A. (2004) Subclassification of nerve sheath tumors by gene expression profiling. *Brain Pathol.* **14,** 258–264.

286. Hanemann, C.O., Bartelt-Kirbach, B., Diebold, R., Kämpchen, K., Langmesser, S., and Utermark, T. (2006) Differential gene expression between human schwannoma and control Schwann cells. *Neuropathol. Appl. Neurobiol.* **32,** 605–614.

287. Yang, F.-C., Ingram, D.A., Chen, S., Hingtgen, C.M., Ratner, N., Monk, K.R., Clegg, T., White, H., Mead, L., Wenning, M.J., Williams, D.A., Kapur, R., Atkinson, S.J., and Clapp, D.W. (2003) Neurofibromin-deficient Schwann cells secret a potent migratory stimulus for $Nf1^{+/-}$ mast cells. *J. Clin. Invest.* **112,** 1851–1861.

288. Tsang, W.Y.W., Chan, J.K.C., Chow, L.T.C., and Tse, C.C.H. (1992) Perineurinoma: an uncommon soft tissue neoplasm distinct from localized hypertrophic neuropathy and neurofibroma. *Am. J. Surg. Pathol.* **16,** 756–763.

289. Hirose, T., Scheithauer, B.W., and Sano, T. (1998) Perineurial malignant peripheral nerve sheath tumor (MPNST). A clinicopathologic, immunohistochemical, and ultrastructural study of seven case. *Am. J. Surg. Pathol.* **22,** 1368–1378.

290. Kazakov, D.V., Pitha, J., Sima, R., Vanecek, T., Shelekhova, K., Mukensnabl, P., and Michal, M. (2005) Hybrid peripheral nerve sheath tumors: schwannoma-perineurioma and neurofibroma-perineurioma. A report of three cases in extradigital locations. *Ann. Diagn. Pathol.* **9,** 16–23.

291. Yamaguchi, U., Hasegawa, T., Hirose, T., Fugo, K., Mitsuhashi, T., Shimizu, M., Kawai, A., Ito, Y., Chuman, H., Beppu, Y. (2003) Sclerosing perineurinoma: a clinicopathogical study of five cases and diagnostic utility of immunohistochemical staining for GLUT1. *Virchows Arch.* **443,** 159–163.

292. Emory, T.S., Scheithauer, B.W., Hirose, T., Wood, M., Onofrio, B.M., and Jenkins, R.B. (1995) Intraneural perineurioma. A clonal neoplasm associated with abnormalities of chromosome 22. *Am. J. Clin. Pathol.* **103**, 696–704.

293. Hornick, J.L., and Fletcher, C.D.M. (2005) Soft tissue perineurioma. Clinicopathologic analysis of 81 cases including those with atypical histologic features. *Am. J. Surg. Pathol.* **29**, 845–858.

294. Hornick, J.L., and Fletcher, C.D.M. (2005) Intestinal perineuriomas. Clinicopathologic definition of a new anatomic subset in a series of 10 cases. *Am. J. Surg. Pathol.* **29**, 859–865.

295. Giannini, C., Scheithauer, B.W., Steinberg, J., and Cosgrove, T.J. (1998) Intraventricular perineurioma: case report. *Neurosurgery* **43**, 1478–1482.

295A. Pitchford, C.W., Schwartz, H.S., Atkinson, J.B., and Cates, J.M.M. (2006) Soft tissue perineurioma in a patient with neurofibromatosis type 2: a tumor not previously associated with the NF2 syndrome. *Am. J. Surg. Pathol.* **30**, 1624–1629.

296. Giannini, C., Scheithauer, B.W., Jenkins, R.B., Erlandson, R.A., Perry, A., Borell, T.J., Hoda, R.S., and Woodruff, J.M. (1997) Soft-tissue perineurioma. Evidence for an abnormality of chromosome 22, criteria for diagnosis, and review of the literature. *Am. J. Surg. Pathol.* **21**, 167–173.

296A. Begnami, M.D., Palau, M., Rushing, E.J., Santi, M., and Quezado, M. (2007) Evaluation of *NF2* gene deletion in sporadic schwannomas, meningiomas, and ependymomas by chromogenic *in situ* hybridization *Hum. Pathol.* **38**, 1345–1350.

297. Sciot, R., Dal Cin, P., Hagemeijer, A., De Smet, L., Van Damme, B., and Van den Berghe, H. (1999) Cutaneous sclerosing perineurioma with cryptic *NF2* gene deletion. *Am. J. Surg. Pathol.* **23**, 849–853.

298. Brock, J.E., Perez-Atayde, A., Kozakewich, H.P.W., Richkind, K.E., Fletcher, J.A., and Vargas, S.O. (2005) Cytogenetic aberrations in perineurioma. Variation with subtype. *Am. J. Surg. Pathol.* **29**, 1164–1169.

299. Lasota, J., Fetsch, J.F., Wozniak, A., Wasag, B., Sciot, R., and Miettinen, M. (2001) The neurofibromatosis type 2 gene is mutated in perineurial cell tumors. A molecular genetic study of eight cases. *Am. J. Pathol.* **158**, 1223–1229.

300. Mott, R.T., Goodman, B.K., Burchette, J.L., and Cummings, T.J. (2005) Loss of chromosome 13 in a case of soft tissue perineurioma. *Clin. Neuropathol.* **24**, 69–76.

301. Sawyer, J.R., Parr, L.G., Gokden, N., and Nicholas, R.W. (2005) A reciprocal t(4;9)(q31;p22) in a solitary neurofibroma. *Cancer Genet. Cytogenet.* **156**, 172–174.

301A. Storlazzi, C.T., Von Steyern, F.V., Domanski, H.A., Mandahl, N., and Mertens, F. (2005) Biallelic somatic inactivation of the *NF1* gene through chromosomal translocations in a sporadic neurofibroma. *J.Cancer* **117**, 1055–1057.

302. Andersen, L.B., Tommerup, N., and Koch, J., (1990) Formation of a minichromosome by excision of the proximal region of 17q in a patient with von Recklinghausen neurofibromatosis. *Cytogenet. Cell Genet.* **53**, 206–210.

303. Andersen, L.B., Fountain, J.W., Gutmann, D.H., Tarlé, S.A., Glover, T.W., Dracopoli, N.C., Housman, D.E., and Collins, F.S. (1993) Mutations in the neurofibromatosis 1 gene in sporadic malignant melanoma cell lines. *Nat. Genet.* **3**, 118–121.

304. The, I., Murthy, A.E., Hannigan, G.E., Jacoby, L.B., Menon, A.G., Gusella, J.F., and Bernards, A. (1993) Neurofibro-

matosis type 1 gene mutations in neuroblastoma. *Nat. Genet.* **4**, 62–66.

305. Koh, T., Yokota, J., Ookawa, K., Kina, T., Koshimura, K., Miwa, S., Ariyasu, T., Yamada, H., Osaka, M., Haga, H., Hitomi, S., Sugiyama, T., and Takahashi, R. (1995) Alternative splicing of the neurofibromatosis type gene correlates with growth patterns and neuroendocrine properties of human small-cell lung-carcinoma cells. *Int. J. Cancer* **60**, 843–847.

306. Martinsson, T., Sjöberg, R.-M., Hedborg, F., and Kogner, P. (1997) Homozygous deletion of the neurofibromatosis-1 gene in the tumor of a patient with neuroblastoma. *Cancer Genet. Cytogenet.* **95**, 183–189.

307. Güran, S, and Safali, M. (2005) A case of neurofibromatosis and breast cancer: loss of heterozygosity of *NF1* in breast cancer. *Cancer Genet. Cytogenet.* **156**, 86–88.

308. Arbiser, J.L., Flynn, E., and Barnhill, R.L. (1998) Analysis of vascularity of human neurofibromas. *J. Am. Acad. Dermatol.* **39**, 950–954.

309. Cichowski, K., Shih, T.S., Schmitt, E., Santiago, S., Reilly, K., McLaughlin, M.E., Bronson, R.T., and Jacks, T. (1999) Mouse model of tumor development in neurofibromatosis type 1. *Science* **286**, 2172–2176.

310. Vogel, K.S., Klesse, L.J., Velasco-Miguel, S., Meyers, K., Rushing, E.J., and Parada, L.F. (1999) Mouse model for neurofibromatosis type 1. *Science* **286**, 2176–2179.

311. Cunha, K.S.G., Barboza, E.P., and da Fonseca, E.C. (2003) Identification of growth hormone receptor in localized neurofibromas of patients with neurofibromatosis type 1. *J. Clin. Pathol.* **56**, 758–763.

312. Fetsch, J.F., Michal, M., and Miettinen, M. (2000) Pigmented (melanotic) neurofibroma. A clinicopathologic and immunohistochemical analysis of 19 lesions from 17 patients. *Am. Surg. Pathol.* **24**, 331–343.

313. Fukuda, T., Ichimura, E., Shinozaki, T., Sano, T., Kashiwabara, K., Oyama, T., Nakajima, T., and Nakamura, T. (1998) Coexpression of HGF and c-Met/HGF receptor in human bone and soft tissue tumors. *Pathol. Int.* **48**, 757–762.

314. Gómez, L., Barrios, C., Kreicbergs, A., Zetterberg, A., Pestaña, A., and Castresana J.S. (1995) Absence of mutation at the GAP-related domain of the neurofibromatosis type 1 gene in sporadic neurofibrosarcomas and other bone and soft tissue sarcomas. *Cancer Genet. Cytogenet.* **81**, 173–174.

315. Hulsebos, T.J.M., Bijleveld, E.H., Riegman, P.H.J., Smink, L.J., and Dunham, I. (1996) Identification and characterization of *NF1*-related loci on human chromosomes 22, 14 and 2. *Hum. Genet.* **98**, 7–11.

316. Kadono, T., Soma, Y., Takehara, K., Nakagawa, H., Ishibashi, Y., and Kikuchi, K. (1994) The growth regulation of neurofibroma cells in neurofibromatosis type-1: increased responses to PDGF-BB and TGF-β1. *Biochem. Biophys. Res. Commun.* **198**, 827–834.

316A. Kadono, T., Kikuchi, K., Nakagawa, H., and Tamaki, K. (2000) Expressions of various growth factors and their receptors in tissues from neurofibroma. *Dermatology* **201**, 10–14.

317. Kaufmann, D., Leistner, W., Kruse, P., Kenner, O., Hoffmeyer, S., Hein, C., Vogel, W., Messiaen, L., and Bartelt, B. (2002) Aberrant splicing in several human tumors in the tumor suppressor genes *Neurofibromatosis Type 1*, *Neurofibromatosis Type 2* , and *Tuberous Sclerosis 2*[1]. *Cancer Res.* **62**, 1503–1509.

318. Boman, F., Peters, J., Ragge, N., and Triche, T. (1993) Infrequent mutation of the p53 gene in fibrous tumors of infancy and childhood. *Diagn. Mol. Pathol.* **2**, 14–22.

319. Levine, A.J. (1997) p53, the cellular gatekeeper for growth and division. *Cell* **88**, 323–331.

320. Kopelovich, L., and Rich, R.F. (1986) Enhanced radiotolerance to ionizing radiation is correlated with increased cancer proneness of cultured fibroblasts from precursor states in neurofibromatosis patients. *Cancer Genet. Cytogenet.* **22**, 203–210.

321. Laskin, W.B., Fetsch, J.F., Lasota, J., and Miettinen, M. (2005) Benign epithelioid peripheral nerve sheath tumors of the soft tissues. Clinicopathologic spectrum of 33 cases. *Am. J. Surg. Pathol.* **29**, 39–51.

322. Li, Y., Rao, P.K., Wen R., Song, Y., Muir, D., Wallace, P., van Horne, S.J., Tennekoon, G.I., and Kadesch, T. (2004) Notch and Schwann cell transformation. *Oncogene* **23**, 1146–1152.

323. Fargnoli, M.C., Chimenti, S., and Peris, K. (1997) Multiple microsatellite alterations on chromosome 9 in neurofibromas of NF-1 patients. *J. Invest. Dermatol.* **108**, 812–813.

324. Ma, P.C., Maulik, G., Christensen, J., and Salgia, R. (2003) c-Met: structure, functions and potential for therapeutic inhibition. *Cancer Metas. Rev.* **22**, 309–325.

325. Rao, U.N.M., Sonmez-Alpan, E., and Michalopoulos, G.K. (1997) Hepatocyte growth factor and c-MET in benign and malignant peripheral nerve sheath tumors. *Hum. Pathol.* **28**, 1066–1070.

326. Zhang, Y.-W, and Vande Woude, G.F. (2003) HGF/SF-Met signaling in the control of branching morphogenesis and invasion. *J. Cell. Biochem.* **88**, 408–417.

327. Martuza, R.L., MacLaughlin, D.T., and Ojemann, R.G. (1981) Specific estradiol binding in schwannomas, meningiomas, and neurofibromas. *Neurosurgery* **9**, 665–671.

328. Mashour, G.A., Ratner, N., Khan, G.A., Wang, H.-L., Martuza, R.L., and Kurtz, A. (2001) The angiogenic factor midkine is aberrantly expressed in *NF1*-deficient Schwann cells and is a mitogen for neurofibroma-derived cells. *Oncogene* **20**, 97–105.

329. Mashour, G.A., Hernáiz Driever, P., Hartmann, M., Drissel, S.N., Zhang, T., Scharf, B., Felderhoff-Müser, U., Sakuma, S., Friedrich, R.E., Martuza, R.L., Mautner, V.F., and Kurtz, A. (2004) Circulating growth factor levels are associated with tumorigenesis in neurofibromatosis type 1. *Clin. Cancer Res.* **10**, 5677–5683.

330. McLaughlin, M.E., and Jacks, T. (2003) Progesterone receptor expression in neurofibromas. *Cancer Res.* **63**, 752–755

331. Muir, D. (1995) Differences in proliferation and invasion by normal transformed and NF1 Schwann cell cultures are influenced by matrix metalloproteinase expression. *Clin. Exp. Metastasis* **13**, 303–314.

332. Petronzelli, F., Sollima, D., Coppola, G., Martini-Neri, M.E., Neri, G., and Genuardi, M. (2001) *CDKN2A* germline splicing mutation affecting both p16^{ink4} and p14^{arf}RNA processing in a melanoma/neurofibroma kindred. *Genes Chromosomes Cancer* **31**, 398–401.

333. Petty, E.M., Gibson, L.H., Fountain, J.W., Bolognia, J.L., Yang-Feng, T.L., Housman, D.E., and Bale, A.E. (1993) Molecular definition of a chromosome 9p21 germ-line deletion in a woman with multiple melanomas and a plexiform neurofibroma: implications for 9p tumor-suppressor gene(s). *Am. J. Hum. Genet.* **53**, 96–104.

334. Peyrard, M., Seroussi, E., Sandberg-Nordqvist, A.-C., Xie, Y.-G., Han, F.-Y., Fransson, I., Collins, J., Dunham, I., Kost-Alimova, M., Imreh, S., and Dumanski, J.P. (1999) The human *LARGE* gene from 22q12.3-q13.1 is a new, distinct member of the glycosyltransferase gene family. *Proc. Natl. Acad. Sci. USA* **96**, 598–603.

335. Ponta, H., Sherman, L., and Herrlich, P.A. (2003) CD44: from adhesion molecules to signalling regulators. *Nat. Rev. Mol. Cell Biol.* **4**, 33–45.

336. Raevaara, T.E., Gerdes, A.-M., Lönnqvist, K.E., Tybjaerg-Hansen, A., Abdel-Rahman, W.M., Kariola, R., Peltomäki, P., and Nyström-Lahti, M. (2004) HNPCC mutation *MLH1* P648S make the functional protein unstable, and homozygosity predisposes to mild neurofibromatosis type 1. *Genes Chromosomes Cancer* **40**, 261–265.

337. Reish, O., Orlovski, A., Mashevitz, M., Sher, C., Libman, V., Rosenblat, M., and Avivi, L. (2003) Modified allelic replication in lymphocytes of patients with neurofibromatosis type 1. *Cancer Genet. Cytogenet.* **143**, 133–139.

337A. Ribeiro, R., Soares, Â., Pinto, D., Catarino, R., Lopes, C., and Medeiros, R. (2007) *EGF* genetic polymorphism is associated with clinical features but not malignant phenotype in neurofibromatosis type 1 patients. *J. Neurooncol.* **81**, 225–229.

338. Ricciardone, M.D., Özçelik, T., Cevher, B., Özdag, H., Tuncer, M., Gürgey, A., Uzunalimoglu, Ö, Çetinkaya, H., Tanyeli, A., Erken, E., and Özturk, M. (1999) Human *MLH1* deficiency predisposes to hematological malignancy and neurofibromatosis type 1. *Cancer Res.* **59**, 290–293.

339. Wang, Q., Lasset, C., Desseigne, F., Frappaz, D., Bergeron, C., Navarro, C., Ruano, E., and Puisieux, A. (1999) Neurofibromatosis and early onset of cancers in h*MLH1*-deficient children. *Cancer Res.* **59**, 294–297.

340. Riva, P., Castorina, P., Manoukian, S., Dalprà, L, Doneda, L., Marini, G., den Dunnen, J., and Larizza, L. (1996) Characterization of a cytogenetic 17q11.2 deletion in an NF1 patient with a contiguous gene syndrome. *Hum. Genet.* **98**, 646–650.

341. Ponder, B. (1990) Neurofibromatosis gene cloned. *Nature* **346**, 703–704.

342. Wallace, M.R., Andersen, L.B., Fountain, J.W., Odeh, H.M., Viskochil, D., Marchuk, D.A., O'Connell, P., White, R., and Collins, F.S. (1990) A chromosome jump crosses a translocation breakpoint in the von Recklinghausen neurofibromatosis region. *Genes Chromosomes Cancer* **2**, 271–277.

343. Hoffmeyer, S., Assum, G., Kaufmann, D., and Krone, W. (1994) Unequal expression of *NF1* alleles. *Nat. Genet.* **6**, 331.

343A. Lázaro, C., Gaona, A., and Estivill, X. (1994) Two CA/GT repeat polymorphisms in intron 27 of the human neurofibromatosis type 1 (*NF1*) gene. *Hum. Genet.* **93**, 351–352.

344. Kehrer-Sawatzki, H., Schwickardt, T., Assum, G., Rocchi, M., Krone, W. (1997) A third neurofibromatosis type 1 (NF1) pseudogene at chromosome 15q11.2. *Hum. Genet.* **100**, 595–600.

345. Upadhyaya, M., Osborn, M.J., Maynard, J., Kim, M.R., Tamanoi, F., and Cooper, D.N. (1997) Mutational and functional analysis of the neurofibromatosis type 1 (*NF1*) gene. *Hum. Genet.* **99**, 88–92.

346. Messiaen, L.M., Callens, T., Roux, K.J., Mortier, G.R., De Paepe, A., Abramowicz, M., Pericak-Vance, M.A., Vance, J.M., and Wallace, M.R. (1999) Exon 10b of the *NF1* gene represents a mutational hotspot and harbors a recurrent missense mutation Y489C associated with aberrant splicing. *Genet. Med.* **1,** 248–253.

347. Messiaen, L.M. Callens, T., Mortier, G., Beysen, D., Vandenbroucke, I., Van Roy, N., Speleman, F., and De Paepe, A. (2000) Exhaustive mutation analysis of the *NF1* gene allows identification of 95% of mutations and reveals a high frequency of unusual splicing defects. Hum. *Mut.***15,** 541–555.

348. Toliat, M.R., Erdogan, F., Gewies, A., Fahsold, R., Buske, A, Tinschert, S., and Nürnberg, P. (2000) Analysis of the *NF1* gene by temperature gradient gel electrophoresis reveals a high incidence of mutations in exon 4b. *Electrophoresis* **21,** 541–544.

349. Zou, M.-X., Butcher, D.T., Sadikovic, B., Groves, T.C., Yee, S.-P., and Rodenhiser, D.I. (2004) Characterization of functional elements in the neurofibromatosis (*NF1*) proximal promoter region. *Oncogene* **23,** 330–339.

350. Tong, J., Hannan, F., Zhu, Y., Bernards, A., and Zhong Y. (2002) Neurofibromin regulates G protein-stimulated adenylyl cyclase activity. *Nat. Neurosci.* **5,** 95–96.

350A. Watanabe, T., Oda, Y., Tamiya, S., Masuda, K., and Tsuneyoshi, M. (2001) Malignant peripheral nerve sheath tumour arising within neurofibroma. An immunohistochemical analysis in the comparison between benign and malignant components. *J. Clin. Pathol.* **54,** 631–636.

351. Yan, N., Ricca, C., Fletcher, J., Glover, T., Seizinger, B.R., and Manne, V. (1995) Farnesyltransferase inhibitors block the neurofibromatosis type 1 (NF1) malignant phenotype. *Cancer Res.* **55,** 3569–3575.

352. Zucman-Rossi, J., Legoix, P., and Thomas G. (1996) Identification of new members of the Gas2 and Ras families in the 22q12 chromosome region. *Genomics* **38,** 247–254.

353. Neumann, E., Kalousek, D.K., Norman, M.G., Steinbok, P., Cochrane, D.D., and Goddard, K. (1993) Cytogenetic analysis of 109 pediatric central nervous system tumors. *Cancer Genet. Cytogenet.* **71,** 40–49.

354. Sawyer, J.R., Roloson, G.J., Bell, J.M., Thomas, J.R., Teo, C., and Chadduck, W.M. (1996) Telomeric associations in the progression of chromosome aberrations in pediatric solid tumors. *Cancer Genet. Cytogenet.* **90,** 1–13.

355. Hameed, M., Chen, F., Mahmet, K., Das, K., and Kim, S. (2006) Schwannoma with a reciprocal t(9;22)(q22;q13). *Cancer Genet. Cytogenet.* **164,** 92–93.

356. Bello, M.J., Leone, P.E., Nebreda, P., de Campos, J.M., Kusak, M.E., Vaquero, J., Sarasa, J.L., García-Miguel, P., Queizan, A., Hernández-Moneo, J.L., Pestaña, A., and Rey, J.A. (1995) Allelic status of chromosome 1 in neoplasms of the nervous system. *Cancer Genet. Cytogenet.* **83,** 160–164.

357. Bhattacharjee, M., Adesina, A.M., Goodman, C., and Powell, S. (2003) Claudin-1 expression in meningiomas and schwannomas: possible role in differential diagnosis. *J. Neuropathol. Exp. Neurol.* **62,** 581.

358. Hung, G., Faudoa, R., Baser, M.E., Xue, Z., Kluwe, L., Slattery, W., Brackman, D., and Lim, D. (2000) Neurofibromatosis type 2 phenotypes and germ-line NF2 mutations determined by an RNA mismatch method and loss of heterozygosity analysis in NF2 schwannomas. *Cancer Genet. Cytogenet.* **118,** 167–168.

359. Brockes, J.P., Breakefield, X.O., and Martuza, R.L. (1986) Glial growth factor—like activity in Schwann cell tumors. *Ann. Neurol.* **20,** 317–322.

360. Burden, S., and Yarden, Y. (1997) Neuregulins and their receptors: a versatile signaling module in organogenesis and oncogenesis. *Neuron* **18,** 847–855.

361. Levi, A.D.O., Bunge, R.P., Lofgren, J.A., Meima, L., Hefti, F., Nikolics, K., and Sliwkowski, M.X. (1995) The influence of heregulins on human schwann cell proliferation. *J. Neurosci.* **15,** 1329–1340.

362. Kim, H.A., Ling, B., and Ratner, N. (1997) *NF1*-deficient mouse Schwann cells are angiogenic and invasive and can be induced to hyperproliferate: reversion of some phenotypes by an inhibitor of farnesyl transferase. *Mod. Cell. Biol.* **177,** 862–872.

363. Pinkas-Kramarski, R., Alroy, I., and Yarden, Y. (1997) ErbB receptors and EGF-like ligands: cell lineage determination and oncogenesis through combinatorial signaling. *J. Mammary Gland Biol. Neoplasia* **2,** 97–107.

363A. Chadee, D.N., Xu, D., Hung, G., Andalibi, A., Luo, D., Gutmann, D.H., and Kyriakis, J.M. (2006) Mixed-lineage kinase 3 regulates B-Raf through maintenance of the B-Raf/Raf-1 complex and inhibition by the NF2 tumor suppressor gene. *Proc. Natl. Acad. Sci. USA* **103,** 4463–4468.

364. Pinkas-Kramarski, R., Eilam, R., Alroy, I., Levkowitz, G., Lonai, P., and Yarden, Y. (1997) Differential expression of NDF/neuregulin receptors ErbB-3 and ErbB-4 and involvement in inhibition of neuronal differentiation. *Oncogene* **15,** 2803–2815.

364A. De Klein, A., Riegman, P.H.J., Bijlsma, E.K., Heldoorn, A., Muijtjens, M., den Bakker, M.A., Avezaat, C.J.J., and Zwarthoff, E.C. (1998) A G?A transition creates a branch point sequence and activation of a cryptic exon, resulting in the hereditary disorder neurofibromatosis 2. *Hum. Mol. Genet.* **7,** 393–398.

365. Fine, S.W., McClain, S.A., and Li, M. (2004) Immunohistochemical staining for calretinin is useful for differentiating schwannomas from neurofibromas. *Am. J. Clin. Pathol.* **122,** 552–559.

366. Jacoby, L.B., Pulaski, K., Rouleau, G.A., and Martuza, R.L. (1990) Clonal analysis of human meningiomas and schwannomas. *Cancer Res.* **50,** 6783–6786.

367. Kimura, Y., Koga, H., Araki, N., Mugita, N., Fujita, N., Takeshima, H., Nishi, T., Yamashima, T., Saido, T.C., Yamasaki, T., Moritake, K., Saya, H., Nakao M. (1998) The involvement of calpain-dependent proteolysis of the tumor suppressor NF2 (merlin) in schwannomas and meningiomas. *Nat. Med.* **4,** 915–922.

368. Kimura, Y., Saya, H., and Nakao, M. (2000) Calpain-dependent proteolysis of NF2 protein: involvement in schwannomas and meningiomas. *Neuropathology* **20,** 153–160.

369. Kissil, J.L., Wilker, E.W., Johnson, K.C., Eckman, M.S., Yaffe, M.B., and Jacks, T. (2003) Merlin, the product of the Nf2 tumor suppressor gene, is an inhibitor of the p21-activated kinase, Pak1. *Mol. Cell* **12,** 841–849.

370. Krasnoselsky, A., Massay, M.J., DeFrances, M.C., Michalopoulos, G., Zarnegar, R., and Ratner, N. (1994)

Hepatocyte growth factor is a mitogen for Schwann cells and is present in neurofibromas. *J. Neurosci.* **14,** 7284–7290.

370A. Manchanda, N., Lyubimova, A., Ho, H.-Y.H., James, M.F., Gusella, J.F., Ramesh, N., Snapper, S.B., and Ramesh, V. (2005) The NF2 tumor suppressor merlin and the ERM proteins interact with N-WASP and regulate its actin polymerization function. *J. Biol. Chem.* **280,** 12517–12522.

370B. Mantripragada, K.K., Buckley, P.G., Benetkiewicz, M., De Bustos, C., Hirvelä, C., Jarbo, C., Bruder, C.E., Wensman, H., Mathiesen, T., Nyberg, G., Papi, L., Collins, V.P., Ichimura, K., Evans, G., and Dumanski, J.P. (2003) High-resolution profiling of an 11 Mb segment of human chromosome 22 in sporadic schwannoma using array-CGH. *Int. J. Oncol.* **22,** 615–622.

371. Muranen, T., Grönholm, M., Renkema, G.H., and Carpén, O. (2005) Cell cycle-dependent nucleocytoplasmic shuttling of the neurofibromatosis 2 tumour suppressor merlin. *Oncogene* **24,** 1150–1158.

371A. Neff, B.A., Welling, B., Akhmametyeva, E., and Chang, L.-S. (2006) The molecular biology of vestibular schwannomas: dissecting the pathogenic process at the molecular level. *Otol. Neurotol.* **27,** 197–208.

372. Ridley, A.J, Paterson, H.F., Noble, M., and Land, H. (1988) *ras*-mediated cell cycle arrest is altered by nuclear oncogenes to induce Schwann cell transformation. *EMBO J.* **7,** 1635–1645.

373. Sarlomo-Rikala, M., El-Rifai, W., Lahtinen, T., Andersson, L.C., Miettinen, M., and Knuutila, S. (1998) Different patterns of DNA copy number changes in gastrointestinal stromal tumors, leiomyomas, and schwannomas. *Hum. Pathol.* **29,** 476–481.

374. Schneider-Stock, R., Radig., K., Oda, Y., Mellin, W., Rys, J., Niezabitowski, A., and Roessner, A. (1997) *p53* gene mutations in soft-tissue sarcomas–correlations with p53 immunohistochemistry and DNA ploidy. *J. Cancer Res. Clin. Oncol.* **123,** 211–218.

375. Scoles, D.R., Nguyen, V.D., Qin, Y., Sun, C.-X., Morrison, H., Gutmann, D.H., and Pulst, S.-M. (2002) Neurofibromatosis 2 (NF2) tumor suppressor schwannomin and its interacting protein HRS regulate STAT signaling. *Hum. Mol. Genet.* **11,** 3179–3189.

376. Sherman, L., Sleeman, J.P., Hennigan, R.F., Herrlich, P., and Ratner, N. (1999) Overexpression of activated *neu/erb*B2 initiates immortalization and malignant transformation of immature Schwann cells *in vitro*. *Oncogene* **18,** 6692–6699.

377. Shimizu, T., Seto, A., Maita, N., Hamada, K., Tsukita, S., Tsukita, S., and Hakoshima, T. (2002) Structural basis for neurofibromatosis type 2. Crystal structure of the merlin ferm domain. *J. Biol. Chem.* **277,** 10332–10336.

378. Kluwe, L., Nygren, A.O.H. Errami, A., Heinrich, B., Matthies, C., Tatagiba, M., and Mautner, V. (2005) Screening for large mutations of the *NF2* gene. *Genes Chromosomes Cancer* **42,** 384–391.

379. Sturgis, E.M., Woll, S.S., Aydin, F., Marrogi, A.J., and Amedee, R.G. (1996) Epidermal growth factor receptor expression by acoustic neuromas. *Laryngoscope* **106,** 457–462.

380. Surace, E.I., Haipek, C.A., and Gutmann, D.H. (2004) Effect of merlin phosphorylation of neurofibromatosis type 2 (NF2) gene function. *Oncogene* **23,** 580–587.

381. Tang, Y., Marwaha, S., Rutkowski, J.L., Tennekoon, G.I., Phillips, P.C., and Field, J. (1998) A role for Pak protein kinases in Schwann cell transformation. *Proc. Natl. Acad. Sci. USA* **95,** 5139–5144.

382. Tawk, R.G., Tan, D., Mechtler, L., and Fenstermaker, R.A. (2005) Melanotic schwannoma with drop metastases to the caudal spine and high expression of CD117 (c-kit). *J. Neurooncol.* **71,** 151–156.

383. Utermark, T., Kaempchen, K., Hanemann, C.O. (2003) Pathological adhesion of primary human schwannoma cells is dependent on altered expression of integrins. *Brain Pathol.* **13,** 352–363.

383A. Utermark, T., Kaempchen, K., Antoniadis, G., Hanemann, C.O. (2005) Reduced apoptosis rates in human schwannomas. *Brain Pathol.* **15,** 17–22.

383B. Utermark, T., Schubert, S.J., and Hanemann, C. (2005) Rearrangements of the intermediate filament GFAP in primary human schwannoma cells. *Neurobiol. Dis.* **19,** 1–9.

384. Wang, Y., Yamada, K., Tanaka, Y., Ishikawa, K., and Inagaki, N. (2005) Expression of ABCA2 protein in human vestibular schwannoma and peripheral nerve. *J. Neurol. Sci.* **232,** 59–63.

384A. Welling, D.B., Guida, M., Goll, F., Pearl, D.K., Glasscock, M.E., Pappas, D.G., Linthicum, F.H., Rogers, D., and Prior, T.W. (1996) Mutational spectrum in the neurofibromatosis type 2 gene in sporadic and familial schwannomas. *Hum. Genet.* **98,** 189–193.

384B. Welling, D.B. (1998) Clinical manifestations of mutations in the neurofibromatosis type 2 gene in vestibular schwannomas (acoustic neuromas). *Laryngoscope* **108,** 178–189.

385. Woodruff, J.M., Scheithauer, B.W., Kurtkaya-Yapicier, Ö, Raffel, C., Amr, S.S., LaQuaglia, M.P., and Antonescu, C.R. (2003) Congenital and childhood plexiform (multinodular) cellular schwannoma. A troublesome mimic of malignant peripheral nerve sheath tumor. *Am. J. Surg. Pathol.* **27,** 1321–1329.

386. Xiao, G.-H., Beeser, A., Chernoff, J., and Testa, J.R. (2002) p21-activated kinase links Rac/Cdc42 signaling to merlin. *J. Biol. Chem.* **277,** 883–886.

387. Gonzalez-Agosti, C., Xu, L., Pinney, D., Beauchamp, R., Hobbs, W., Gusella, J., and Ramesh, V. (1996) The merlin tumor suppressor localizes preferentially in membrane ruffles. *Oncogene* **13,** 1239–1247.

388. Scherer, S.S., and Gutmann, D.H. (1996) Expression of the neurofibromatosis 2 tumor suppressor gene product, merlin, in Schwann cells. *J. Neurosci. Res.* **46,** 595–605.

388A. Hanemann, C.O., Bartelt-Kirbach, B., Diebold, R., Kampchen, K., Langmesser, S., and Utermark, T. (2006) Differential gene expression between human schwannoma and control Schwann cells. *Neuropathol. Appl. Neurobiol.* **32,** 605–614.

389. Propp, J.M., McCarthy, B.J., Davis, F.G., and Preston-Martin, S. (2006) Descriptive epidemiology of vestibular schwannomas. *Neurooncol.* **8,** 1–11.

390. Arakawa, H., Hayashi, N., Nagase, H., Ogawa, M., and Nakamura, Y. (1994) Alternative splicing of the NF2 gene and its mutation analysis of breast and colorectal cancers. *Hum. Mol. Genet.* **3,** 565–568.

391. Ng, H.-k, Lau, K.-m., Tse, J.Y.M., Lo, K.-w, Wong, J.H.C., Poon, W.-S., and Huang, D.P. (1995) Combined molecular

genetic studies of chromosome 22q and the neurofibromatosis type 2 gene in central nervous system tumors. *Neurosurgery* **37**, 764–773.

391A. Huang, B., Starostik, P., Kühl, J., Tonn, J.C., and Roggendorf, W. (2002) Loss of heterozygosity on chromosome 22 in human ependymomas. *Acta, Neuropathol.* **103**, 415–420.

392. Oliveira, R.B., Sampaio, E.P., Aarestrup, F., Teles, R.M.B., Silva, T.P., Oliveira, A.L., Antas, P.R.Z., and Sarno, E.N. (2005) Cytokines and *Mycobacterium leprae* induce apoptosis in human schwann cells. *J. Neuropathol. Exp. Neurol.* **64**, 882–890.

393. Scoles, D.R., Huynh, D.P., Chen, M.S., Burke, S.P., Gutmann, D.H., and Pulst, S.M. (2000) The neurofibromatosis 2 tumor suppressor protein interacts with hepatocyte growth factor-regulated tyrosine kinase substrate. *Hum. Mol. Genet.* **9**, 1567–1574.

394. Hruska, A., Bollmann, R., Kovacs, R.B., Bollmann, M., Bodi, M., and Sapi, Z. (2004) DNA ploidy and chromosome (FISH) pattern analysis of peripheral nerve sheath tumors. *Cell Oncol.* **26**, 335–345.

395. Hansen, M.R., and Linthicum, F.H., Jr. (2004) Expression of neuregulin and activation of erbB receptors in vestibular schwannomas: possible autocrine loop stimulation. *Otol. Neurotol.* **25**, 155–159

395A. Sobue, G., Sonnenfeld, K., Rubenstein, A.E., and Pleasure, D. (1985) Tissue culture studies of neurofibromatosis: effects of axolemmal fragments and cyclic adenosine 3′,5′-monophosphate analogues on proliferation of Schwann-like and fibroblast-like neurofibroma cells. *Ann. Neurol.* **18**, 68–73.

396. Baser, M.E., Kuramoto, L., Joe, H., Friedman, J.M., Wallace, A.J., Gillespie, J.E., Ramsden, R.T., and Evans, D.G. (2004) Genotype-phenotype correlations for nervous system tumors in neurofibromatosis-2: a population-based study. *Am. J. Hum. Genet.* **75**, 231–239.

397. Seol, H.J., Jung, H.-W., Park, S.-H., Hwang, S.-K., Kim, D.G., Paek, S.H., Chung, Y.-S., and Lee, C.S. (2005) Aggressive vestibular schwannomas showing postoperative rapid growth–their association with decreased p27 expression. *J. Neurooncol.* **75**, 203–207.

398. Ruttledge, M.H., and Rouleau, G.A. (2005) Role of the *neurofibromatosis Type 2* gene in the development of tumors of the nervous system. *Neurosurg. Focus* **19**, E6 1–5.

399. Rahmatullah, M., Schroering, A., Rothblum, K., Stahl, R.C., Urban, B., and Carey, D.J. (1998) Synergistic regulation of Schwann cell proliferation by heregulin and forskolin. *Mol. Cell. Biol.* **18**, 6245–6252.

400. O'Reilly, B.F., Kishore, A., Crowther, J.A., and Smith, C. (2004) Correlation of growth factor receptor expression with clinical growth in vestibular schwannomas. *Ontol. Neurotol.* **25**, 791–796.

401. Yohay, K.H., (2006) The genetic and molecular pathogenesis of NF1 and NF2. *Semin. Pediatr. Neurol.* **13**, 21–26.

402. Tonsgard, J.H. (2006) Clinical manifestations and management of neurofibromatosis type 1. *Semin. Pediatr. Neurol.* **13**, 2–7.

403. Wimmer, K. (2005) Neurofibromatosis: the most frequent hereditary tumor predisposition syndrome. *Wien Med. Wochenschr.* **155**, 273–280.

404. Stephens, K. (2003) Genetics of neurofibromatosis 1-associated peripheral nerve sheath tumors. *Cancer Invest.* **21**, 897–914.

405. Moczygemba, C., and Bhattacharjee, M. (2005) Neurofibromatosis 1 and unexplained iron deficiency in two patients: is iron depletion related to neurofibroma growth? *J. Neurooncol.* **75**, 227–228.

406. McLaughlin, M.E., and Jacks, T. (2002) Thinking beyond the tumor cell: Nf1 haploinsufficiency in the tumor environment. *Cancer Cell* **1**, 408–410.

407. Mangoura, D., Sun, Y., Li, C., Singh, D., Gutmann, D.H., Flores, A., Ahmed, A., and Vallianatos, G. (2006) Phosphorylation and neurofibromin by PKC is a possible molecular switch in EGF receptor signaling in neural cells. *Oncogene* **25**, 735–745.

408. Gottfried, O.N., Viskochil, D.H., Fults, D.W., and Couldwell, W.T. (2006) Molecular, genetic, and cellular pathogenesis of neurofibromas and surgical implications. *Neurosurgery* **58**, 1–16.

409. Endo, H., Utani, A., Matsumoto, F., Kuroki, T., Yoshimoto, S., Ichinose, M., and Shinkai, H. (2003) A possible paracrine hedgehog signaling pathway in neurofibromas from patients with neurofibromatosis type 1. *Br. J. Dermatol.* **148**, 337–341.

410. Liapis, H., Marley, E.F., Lin, Y., and Dehner, L.P. (1999) p53 and Ki-67 proliferating cell nuclear antigen in benign and malignant peripheral nerve sheath tumors in children. *Pediatr. Dev. Pathol.* **2**, 377–384.

411. Weiss, B., Bollag, G., and Shannon, K. (1999) Hyperactive Ras as a therapeutic target in neurofibromatosis type 1. *Am. J. Med. Genet.* **89**, 14–22.

412. Bollag, G., and McCormick, F. (1991) Differential regulation of rasGAP and neurofibromatosis gene product activities. *Nature* **351**, 576–579.

413. Bollag, G., Clapp D.W., Shih, S., Adler, F., Zhang, Y.Y., Thompson, P., Lange, B.J., Freedman, M.H., McCormick, F., Jacks, T., and Shannon, K. (1996) Loss of NF1 results in activation of the Ras signaling pathway and leads to aberrant growth in haematopoietic cells. *Nat. Genet.* **12**, 144–148.

414. Schlegel, J., Muenkel, K., Trenkle, T., Fauser, G., and Ruschoff, J. (1998) Expression of the ERBB2/neu and neurofibromatosis type 1 gene products in reactive and neoplastic schwann cell proliferation. *Int. J. Oncol.* **13**, 1281–1284.

415. Ahmadian, M.R., Wiesmüller, L., Lautwein, A., Bischoff, F.R., and Wittinghofer, A. (1996) Structural differences in the minimal catalytic domains of the GTPase-activating proteins p120^{GAP} and neurofibromin. *J. Biol. Chem.* **271**, 16409–16415.

416. Nakai, Y., Zheng, Y., MacCollin, M., and Ratner, N. (2006) Temporal control of Rac in Schwann cell-axon interaction is disrupted in *NF2*-mutant schwannoma cells. *J. Neurosci.* **26**, 3390–3395.

417. Ogata, T., Yamamoto, S., Nakamura, K., and Tanaka, S. (2006) Signaling axis in schwann cell proliferation and differentiation. *Mol. Neurobiol.* **33**, 51–62.

418. Viskochil, D.H. (2003) It takes two to tango: mast cell and Schwann cell interactions in neurofibromas. *J. Clin. Invest.* **112**, 1791–1793.

419. Fialkow, P.J., Sagebiel, R.W., and Gartler, S.M. (1971) Multiple cell origin of hereditary neurofibromas. *N. Engl. J. Med.* **284,** 298–300.

420. Yang, F.-C., Chen, S., Clegg, T., Li, X., Morgan, T., Estwick, S.A., Yuan, J., Khalaf, W., Burgin, S., Travers, J., Parada, L.F., Ingram, D.A., and Clapp, D.W. (2006) *Nf1+/−* mast cells induce neurofibroma like phenotypes through secreted TGF-β signaling. *Hum. Mol. Genet.* **15,** 2421–2437.

421. Ingram, D.A., Hiatt, K., King, A.J., Fisher, L., Shivakumar, R., Derstine, C., Wenning, M.J., Diaz, B., Travers, J.B., Hood, A., Marshall, M., Williams, D.A., and Clapp, D.W. (2001) Hyperactivation of p21^{ras} and the hematopoietic-specific Rho GTPase, Rac2, cooperate to alter the proliferation of neurofibromin-deficient mast cells in vivo and in vitro. *J. Exp. Med.* **194,** 57–69.

422. Thomas, S.L., Deadwyler, G.D., Tang, J., Stubbs, E.B., Jr., Muir, D., Hiatt, K.K., Clapp, D.W., and De Vries, G.H. (2006) Reconstitution of the NF1 GAP-related domain in NF1-deficient human schwann cells. *Biochem. Biophys. Res. Commun.* **348,** 971–980.

423. Upadhyaya, M., Huson, S.M., Davies, M., Thomas, N., Chuzhanova, N., Giovannini, S., Evans, D.G., Howard, E., Kerr, B., Griffith, S., Consoli, C., Side, L., Adams, D., Pierpont, M., Hachen, R., Barnicoat, A., Li., H., Wallace, P., Van Biervliet, J.P., Stevenson, D., Viskochil, D., Baralle, D., Haan, E., Riccardi, V., Turnpenny, P., Lazaro, C., and Messiaen, L. (2007) An absence of cutaneous neurofibromas associated with a 3-bp inframe deletion in exon 17 of the *NF1* gene (c.2970–2972 delAAT): evidence of a clinically significant NF1 genotype-phenotype correlation. *Am. J. Hum. Genet.* **80,** 140–151.

424. Stewart, D.R., Corless, C.L., Rubin, B.P., Heinrich, M.C., Messiaen, L.M., Kessler, L.J., Zhang, P.J., and Brooks, D.G. (2007) Mitotic recombination as evidence of alternative pathogenesis of gastrointestinal stromal tumours in neurofibromatosis type 1. *J. Med. Genet.* **44,** 61–65.

425. Oguzkan, S., Terzi Y.K., Cinbis, M., Anlar, B., Aysun, S., and Ayter, S. (2006) Molecular genetic analyses in neurofibromatosis type 1 patients with tumors. *Cancer Genet. Cytogenet.* **165,** 167–171.

426. Shapira, S., Barkan, B., Friedman, E., Kloog, Y., and Stein, R. (2007) The tumor suppressor neurofibromin confers sensitivity to apoptosis by Ras-dependent and Ras-independent pathways. *Cell Death Differ.* **14,** 895–906.

427. Dasgupta, B., Yi, Y., Chen, D.Y., Weber, J.D., and Gutmann, D.H. (2005) Proteomic analysis reveals hyperactivation of the mammalian target of rapamycin pathway in neurofibromatosis 1-associated human and mouse brain tumors. *Cancer Res.* **65,** 2755–2760.

428. Bernreuther, C., Salein, N., Matschke, J., and Hagel, C. (2006) Expression of doublecortin in tumours of the central and peripheral nervous system and in human non-neuronal tissues. *Acta Neuropathol.* **111,** 247–254.

429. Lee, H., Kim, D., Dan, H.C., Wu, E.L., Gritsko, T.M., Cao, C., Nicosia, S.V., Golemis, E.A., Liu, W., Coppola, D., Brem, S.S., Testa, J.R., and Cheng, J.Q. (2007) Identification and characterization of putative tumor suppressor NGB, a GTP-binding protein that interacts with the neurofibromatosis 2 protein. *Mol. Cell. Biol.* **27,** 2103–2119.

430. Chi, J.H., Cachola, K., and Parsa, A.T. (2006) Genetic and molecular biology of intramedullary spinal cord tumors. *Neurosurg. Clin. N. Am.* **17,** 1–5.

431. Ryu, C.H., Kim, S.-W., Lee, K.H., Lee, J.Y., Kim, H., Lee, W.K., Choi, B.H., Lim, Y., Kim, Y.H., Lee, K.-H., Hwang, T.-K., Jun, T.-Y., and Rha, H.K. (2005) The merlin tumor suppressor interacts with Ral guanine nucleotide dissociation stimulator and inhibits its activity. *Oncogene* **24,** 5355–5364.

432. Lee, J.Y., Moon, H.J., Lee, W.K., Chun, H.J., Han, C.W., Jeon, Y.-W., Lim, Y., Kim, Y.H., Yao, T.-P., Lee, K.-H., Jun, T.-Y., Rha, H.K., and Kang, J.-K. (2006) Merlin facilitates ubiquitination and degradation of transactivation-responsive RNA-binding protein. *Oncogene* **25,** 1143–1152.

433. Sirén, V., Peltonen, J., and Vaheri, A. (2006) Plasminogen activators and their inhibitor gene expression in cutaneous NF1-related neurofibromas. *Arch. Dermatol. Res.* **297,** 421–424.

434. Kazakov, D.V., Vanecek, T., Sima, R., Kutzner, H., and Michal, M. (2005) Dendritic cell neurofibroma with pseudorosettes lacks mutations in exons 1–15 of the neurofibromatosis type 2 gene. *Am. J. Dermatopathol.* **27,** 286–289.

435. Murray, A.J., Hughes, T.A.T., Neal, J.W., Howard, E., Evans, D.G.R., and Harper, P.S. (2006) A case of multiple cutaneous schwannomas; schwannomatosis or neurofibromatosis type 2? *J. Neurol. Neurosurg. Psychiatry* **77,** 269–271.

436. Xiong, Z., Cao, Y., Guo, D., Ye, F., and Lei, T. (2006) Expression of EGFR and LRIG-1 in human trigeminal neurinoma. *J. Huazhong Univ. Sci. Technolog. Med. Sci.* **26,** 86–88.

437. Hansen, M.R., Roehm., P.C., Chatterjee, P., and Green, S.H. (2006) Constitutive neuregulin-1/ErbB signaling contributes to human vestibular schwannoma proliferation. *Glia* **53,** 593–600.

437A. Hamaratoglu, F., Willecke, M., Kango-Singh, M., Nolo, R., Hyun, E., Tao, C., Jafar-Nejad, H., and Halder, G. (2006) The tumor-suppressor genes *NF2/Merlin* and *Expanded* act through Hippo signaling to regulate cell proliferation and apoptosis. *Nat. Cell Biol.* **8,** 27–36.

438. McClatchey, A.I., and Giovannini, M. (2005) Membrane organization and tumorigenesis—the NF2 tumor suppressor, merlin. *Genes Dev.* **19,** 2265–2277.

439. Nakai, Y., Zheng, Y., MacCollin, M., and Ratner, N. (2006) Temporal control of Rac in schwann cell-axon interaction is disrupted in *NF2*-mutant schwannoma cells. *J. Neurosci.* **26,** 3390–3395.

439A. Nascimento, A.F. and Fletcher, C.D.M. (2007) The controversial nosology of benign nerve sheath tumors: neurofilament protein staining demonstrates intratumoral axons in many sporadic schwannomas. *Am. J. Surg. Pathol.* **31,** 1363–1370.

440. Hirokawa, Y., Tikoo, A., Huynh, J., Utermark, T., Hanemann, C.O., Giovannini, M., Xiao, G.H., Testa, J.R., Wood, J., and Maruta, H. (2004) A clue to the therapy of neurofibromatosis type 2: NF2/merlin is a PAK1 inhibitor. *Cancer* **10,** 20–26.

441. Mawrin, C., Kirches, E., Dietzmann, K., Roessner, A., and Boltze, C. (2002) Expression pattern of apoptotic markers in vestibular schwannomas. *Pathol. Res. Pract.* **198,** 813–819.

442. Welling, D.B., Lasak, J.M., Akhmametyeva, E., Ghaheri, B., and Chang, L.-S. (2002) cDNA microarray analysis of vestibular schwannomas. *Otol. Neurotol.* **23,** 736–748.

443. Feiz-Erfan, I., Zabramski, J.M., Herrmann, L.L., and Coons, S.W. (2006) Cavernous malformation within a schwannoma: review of the literature and hypothesis of a common genetic etiology. *Acta Neurochir. (Wien)* **148,** 647–652.

444. Ferner, R.E. (2007) Neurofibromatosis 1 and neurofibromatosis 2: a twenty first century perspective. *Lancet Neurol.* **6,** 340–351.

445. Crawford, A.H., and Schorry, E.K. (2006) Neurofibromatosis update. *J. Pediatr. Orthop.* **26,** 413–423.

446. Lee, M.-J., and Stephenson, D.A. (2007) Recent developments in neurofibromatosis type 1. *Curr. Opin. Neurol.* **20,** 135–141.

447. Rubin, J.B., and Gutmann, D.H. (2005) neurofibromatosis type 1 – a model for nervous system tumour formation? *Nat. Rev. Cancer* **5,** 557–564.

448. Yohay, K. (2006) Neurofibromatosis types 1 and 2. *The Neurologist* **12,** 86–93.

449. Murovic, J.A., Kim, D.H., and Kline, D.G. (2006) Neurofibromatosis-associated nerve sheath tumors. *Neurosurg. Focus* **20,** E1.

450. Ruttledge, M.H., and Rouleau, G.A. (2005) Role of the *neurofibromatosis Type 2* gene in the development of tumors of the nervous system. *Neurosurg. Focus* **19,** E6.

451. Wilkes, D., McDermott, D.A., and Basson, C.T. (2005) Clinical phenotypes and molecular genetic mechanisms of Carney complex. *Lancet Oncol.* **6,** 501–508.

452. Woods, R., Friedman, J.M., Evans, D.G., Baser, M.E., and Joe, H. (2003) Exploring the "two-hit hypothesis" in *NF2:* tests of two-hit and three-hit models of vestibular schwannoma development. *Genet. Epidemiol.* **24,** 265–272.

453. Rosenbaum, T., Rosenbaum, C., Winner, U., Müller, H.W., Lenard, H.G., and Hanemann, C.O. (2000) Long-term culture and characterization of human neurofibroma-derived Schwann cells. *J. Neurosci. Res.* **51,** 524–532.

454. Hoshi, N., Sugino, T., and Suzuki, T. (2005) Regular expression of osteopontin in granular cell tumors: distinctive feature among Schwannian cell tumors. *Pathol. Int.* **55,** 484–490.

455. Le, L.Q., and Parada, L.F. (2007) Tumor microenvironment and neurofibromatosis type I: connecting the GAPs. *Oncogene* **26,** 4609–4616.

2

Malignant Peripheral Nerve Sheath Tumors

Summary   Chapter 2 deals with malignant peripheral nerve sheath tumors (MPNST), which are highly aggressive tumors with a poor prognosis. MPNST commonly originate in plexiform neurofibromas of patients with neurofibromatosis 1 (NF1), but they also may be sporadic. The *NF1* gene is often altered in both forms of MPNST, and it seems to be the initiating genetic event in the tumorigenesis of these tumors. The genetic molecular changes subsequent to that of the *NF1* gene have not been clearly defined. Thus, here we present studies on cell cycle genes, the *TP53* gene, the *MDM2* gene, the retinoblastoma protein pathway, and several other genes, proteins, and factors have have attempted to define the sequence of genetic alterations responsible for or affecting MPNST development and biology. Results obtained with studies on changes at 9p and 17q may reflect on important aspects of MPNST biology.

Keywords   malignant peripheral nerve sheath tumors (MPNST) · NF1 gene · plexiform neurofibroma · molecular changes

Introduction

Malignant peripheral nerve sheath tumors (MPNST), referred to also as malignant schwannomas, neurofibrosarcomas, and neurogenic sarcomas, are uncommon neoplasms, accounting for about 5% of malignant tumors of the soft tissues (1–4). Two thirds arise from neurofibromas, most often of the plexiform type (PNF) and in the setting of neurofibromatosis 1 (NF1), and seldom, if ever, from schwannomas or dermal neurofibromas (4–6). The remaining one third of MPNST arise de novo as sporadic tumors (7), thought to originate in the schwann cells of nerve sheaths. Although MPNST is very rare (1:100,000) in the general population, about 10–20% of NF1 patients develop MPNST (8–10). Very rare instances of MPNST originating in a schwannoma or ganglioneuromas have been reported previously (5, 11).

MPNST primarily occur in adults in the third to the sixth decades of life. The mean age of patients with NF1-associated MPNST is approximately a decade younger (28–36 years) than that of sporadic cases (40–44 years). Childhood and adolescent cases are uncommon, and they are rare in children under the age of 6 years (12, 13). MPNST are slightly more frequent in females. Although histologically and immunophenotypically NF1-related and sporadic MPNST are very similar, the tumors in NF1 patients tend to be multifocal (14)

and more aggressive, and they show some differences in the genetic molecular events (4).

MPNST show a wide range of biological behavior, ranging from low to high malignancy, a variety of clinical manifestations, a wide spectrum of histologic appearances (15), and differences in their etiologic background (NF1-related vs. sporadic); hence, it is not surprising to find that the genetic and molecular findings in these tumors reflect a gallimaufry of these various aspects of MPNST (16).

Large- and medium-sized nerves are more prone to involvement by MPNST than small nerves. The most common sites are the nerves of the buttock and thigh, brachial plexus and upper arm, and the paraspinal region (17, 18). The sciatic nerve is frequently affected. Cranial nerve MPNST are very uncommon, with the fifth cranial nerve being more often involved than the eighth cranial verve. With rare exceptions, MPNST of cranial nerves seem to arise de novo (3, 19). MPNST may develop at the site of prior irradiation, and radiation may accelerate the development of MPNST in patients with NF1.

There is no statistical difference in the final outcome for patients with MPNST with or without NF1 (Figure 2.1), but patients with NF1 have a high risk for developing a second MPNST (9, 20).

MPNST are highly aggressive tumors with a poor prognosis. About 60% of patients die of the disease, with an even higher mortality (80%) in individuals with paraspinal lesions

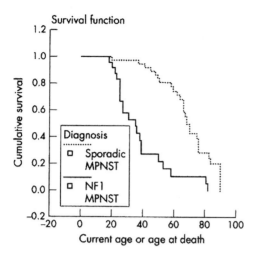

FIGURE 2.1. Kaplan–Meier analysis of survival of patients with NF1-related or sporadic MPNST ($p = 0.09$). The 5-year survival from diagnosis was 21% for NF1-related cases and 42% for sporadic cases (from ref. (9), with permission).

and those (100%) with divergent angiosarcomas (3). The local recurrence rate is high, and metastases develop in more than half the patients, most commonly to the lungs, bones, pleura, and liver; spread by the meningeal route is also possible.

Histologically, most MPNST are high-grade tumors with a high mitotic rate and necrosis. The most common histologic pattern includes a high-grade fibrosarcomatous element composed of densely packed sheets of plump but relatively uniform spindle- or oval cells (Figures 2.2 and 2.3). Many tumors show necrosis and vascular proliferation resembling that commonly seen in glioblastoma multiforme; some tumors have plump, slightly epithelioid spindle-cells. MPNST consist primarily of schwann cells, but they may include fibroblasts and mast cells (2–4).

MPNST may be difficult to differentiate with certainty from some tumors, particularly leiomyosarcoma, malignant melanoma, synovial sarcoma, and fibrosarcoma. This may account, in part, for the spurious finding of chromosome

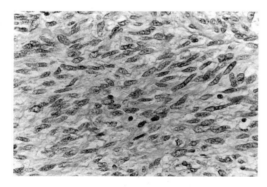

FIGURE 2.2. Histological appearance of an MPNST showing the wavy or buckle-shaped nuclei of the cells. Cytoplasmic borders are indistinct.

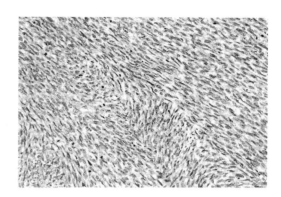

FIGURE 2.3. Histologic appearance of a malignant triton tumor (176). Shown is a spindle cell component of this tumor with a fascicular growth pattern and high cellularity with numerous mitoses.

abnormalities characteristic of some tumors, e.g., t(X;18) of synovial sarcoma in MPNST (21).

The high proliferative activity of MPNST is indicated by the Ki-67 (MIB-1) index of 5–65% versus <1% in schwannomas and neurofibromas. This index may correlate with prognosis (22). The MIB-1 index was 1 and 72% in PNF and high-grade MPNST, respectively (23).

MPNST reported to have a t(X;18), or *SYT-SSX* transcripts, changes characteristic for synovial sarcoma (24, 25), are not included in the discussion or tabulations here. This decision is supported by the failure to demonstrate the t(X;18) in the large number of MPNST (NF1-associated and sporadic) studied cytogenetically or molecularly (26,27) and by the lack of *NF1* gene changes in synovial sarcoma, a change that is seen in the preponderant number of MPNST (28). Nevertheless, exceptional and unusual cases may exist (29). Immunohistochemically, it was shown that staining for *HMGA2* can differentiate MPNST from synovial sarcomas (29A).

Sensitive and specific immunohistochemical markers are lacking for MPNST, particularly in diagnostically challenging cases, where MPNST is part of the differential diagnosis. A significant part of the findings presented here is a reflection of the continuing search for such markers.

Publications dealing with the cytogenetic and/or molecular genetics of MPNST, except for their locations and differentiating them from malignant triton tumors, have generally not indicated the subtype of the MPNST examined. Hence, this chapter is concerned with MPNST as a group of tumors without breakdown into subtypes, i.e., epithelioid (not usually associated with NF1), glandular (usually associated with NF1), plexiform, or other types (3, 4). Clinical, histologic, epidemiologic, and other aspects of MPNST may be found in previous publications (20, 30–42,42A).

We present the genetic and molecular pathways that have been shown to play a role or at least are implicated in MPNST causality and biology and, importantly, may be cogent targets for the development of successful therapy affecting MPNST. For more information on NF1, the *NF1* gene, neurofibromin, and neurofibromas, see Chapter 1.

Cytogenetics of MPNST

Complete karyotypes have been published on <100 MPNST (Table 2.1). The chromosome numbers in MPNST have ranged from hypodiploidy of <35 chromosomes to near-tetraploidy. MPNST tend to have complex karyotypes and extensive and variable cytogenetic heterogeneity (Table 2.1; Figures 2.4–2.9).

Extremely complex karyotypes in MPNST have been described previously (43–45). In >50% of MPNST transloca-tions, marker and derivative chromosomes or dicentrics were present, indicative of high frequency of structural changes associated with these tumors. In >10% of MPNST, dmin were present; ring chromosomes were rare (43, 46, 47).

Sporadic and NF1-associated MPNST have complex kary-otypic abnormalities both numerical and structural. No consis-tent karyotypic pattern characteristic of MPNST has been seen. Near-triploid or hypodiploid chromosome numbers are often present, as are chromosomal losses (46); loss of genetic material related to structural aberrations and recombinations involve almost all chromosomes.

In a study of 10 tumors (48), structural abnormalities of chromosome 17 possibly involving the *NF1* and *TP53* loci were common (49). Chromosome 22 loss also has been noted frequently (48). No specific cytogenetic differences have been observed between sporadic and NF1-associated tumors.

Frequent gains of chromosome 7, 8q, 15q, and 17q were found exclusively in MPNST and not in neurofibromas (50). Gains of 15q and 17q tend to discriminate between hered-itary and sporadic MPNST, whereas gains of 20q occur more frequently in sporadic than in NF1-associated MPNST, a finding that could be used for differentiating these tumors.

In a computer-assisted cytogenetic review of 47 MPNST and four malignant triton tumors (MTT), the most common breakpoints were 1p13, 1q21, 7p22, 9p11, 17p11, 17q11, and 22q11 (43). In another study of 20 MPNST, common break-points were 1p11, 5p15, 5q13, 6p21, 8q10, 11q13, 11q21, 14q24, 17q21, 17q25, and 20q13 (47). It would seem that 1p, 7p22, 11q13-q23, 20q13, and 22q11-q13 emerge as chro-mosomal regions with a high frequency of involvement in MPNST (45), and genes located in these regions may play a crucial role in the biology of these tumors.

In another cytogenetic study (including SKY) (45), struc-tural changes frequently seen in 19 MPNST and two MTT were 1p31-p36, 4q28-q35, 7p22, 11q22-q23, 19q13, 20q13, and 22q11-q13. Loss of chromosomal material was much more common than gain. Thus, loss involved 1p36, 3q21-qter, 9p23-pter, chromosome 10, 11q23-qter, 16/16q24, chromo-some 17, and 22/22q in about 50% of the MPNST, with gains of 7/7q and 8/8q in about 30%. These changes seemed to be of similar incidence in MPNST and MTT.

In a study of 27 MPNST, seven cases (26%) had normal karyotypes (51). Twenty cases (74%) showed clonal aberra-tions, most often with complex karyotypes. The most frequent abnormalities were losses or rearrangements of 10p, 11q, or 17q (nine cases each) and losses or rearrangements of 1p, 9p, 17p, or 22q (eight cases each). Seven cases each showed a ring chromosome, trisomy 7, or rearrangement of 11p. Only two cases showed rearrangement of 12q13-q15. Eigh-teen cases showed a variety of other less frequent aberrations affecting multiple and different chromosomes. Complex rear-rangements were more common in NF1-associated MPNST and in tumors showing heterologous differentiation. Among the cases with 17q rearrangements, three of nine had NF1. There were no evident correlations between any karyotypic abnormality and clinical parameters in these patients (51).

Although cytogenetic studies have shown complex clonal abnormalities in most MPNST (43, 47, 48), rare cases of MPNST with single abnormalities, such as +7 and +22, neither of which is diagnostically specific, have been described. Apropos the origin of MPNST from PNF is the observation (7) that schwann cells from dermal neurofibromas are diploid (cytogenetically normal), whereas most of the schwann cells originating in PNF are abnormal, some with complex karyotypes. Diploid MPNST have been reported occasionally, including one of the spinal type (17). Intratu-moral chromosomal heterogeneity in MPNST may be exten-sive (52). Triploidy or tetraploidy in MPNST, often in the presence of a high Ki-67 index is associated with a poor prog-nosis (22, 43, 46, 47, 53). The type and complexity of the chro-mosome changes seen in NF1-associated MPNST (and MTT) did not seem to differ from those in sporadic tumors.

Losses of two chromosomes may reflect involvement of two tumor suppressor genes, *NF1* located at 17q11.2 and *NF2* located at 22q12. Losses of heterozygosity (LOH) at these sites and at the loci of other tumor-associated genes seem to be involved in a complex multistep process of tumor develop-ment.

To determine meaningful cytogenetic changes observed in MPNST and to discern possible cytogenetic patterns, a computer-assisted cytogenetic analysis was performed (Table 2.2) (43). A database was constructed that permits the detection of statistically significant nonrandom chromosomal aberrations and allows direct comparison of different kary-otypes. This new approach was used to discover nonrandom changes in chromosomal material and to detect frequently involved breakpoints in MPNST. Furthermore, this approach was used to analyze cytogenetic differences between NF1-associated and sporadic MPNST.

These studies did not reveal specific structural cytoge-netic abnormalities, which could be of diagnostic importance in MPNST. However, the finding of a complex karyotype showing loss of 9p21 in combination with gain in 7q13 in a tumor histologically resembling an MPNST could be of diag-nostic importance (43). This approach indicated cytogenetic differences between NF1-associated and sporadic MPNST in chromosomal regions 1p3, 4p1, 21q−p2, and 15p1−q1. Addi-tional studies, combining different cytogenetically based tech-niques might detect differences between NF1-associated and sporadic MPNST that could be helpful in clarifying the onco-genesis of MPNST.

TABLE 2.1. Chromosome changes in MPNST.

Karyotype	References
+16	*Becher et al 1984 (171)*
−18	
del(1)(p13),del(3)(p24)	*Becher et al 1984 (171)*
Xp+,del(1)(p13),del(3)(p24),5q+	*Becher et al 1984 (171)*
4q+	
+1,+2,+7	*Riccardi and Elder 1986 (172)*
2q+,9p+	*Riccardi and Elder 1986 (172)*
?del(6)(p?)	*Riccardi and Elder 1986 (172)*
−1,−7,−11,−12,−13,−17,−19,+mar1,+mar2,+mar3,+mar4,+mar5,+mar6,+mar7,+mar8,+3mar	*Decker et al 1990 (173)*
+2,−5,+18,+18,+ x2,20,+21,−22 or +22,+mar1,+mar2,+19mar	*Glover et al 1991 (83)*
−X,+2,+5,+7,+7,−13,+mar1,+mar2,+mar3,+mar4,+mar5,+mar6,+mar7	*Fletcher et al 1991 (51)*
dic r(X:10)(Xp22.2→q26:20p13→q13),der(1)t(1:3)(p21;p24), t(2:10)(q37;q22),−3,−4,der(5) t(5:?)(q31:?),−9,der(9)t(3:9)(q21 or q13:p24 or p22),der(11)t(11:?)(q22.2:?),der(17)t(17:22:?) (q21;q13.1:?),−22,der(22)t(17:22:?)(q21:q13.1:?)	*Fletcher et al 1991 (51)*
−X,der(10)t(1;3)(p34;p21)x2,+2,der(3)t(3:?)(p21:?), der(3)t(3:?)(p23:?),der(4)t(4:?)(p15:?), del(5)(p14),+del(5)(p15),+6,del(7)(p15),inv(?7)(p11q11),+del(9) (p11)x2,+del(10)(p11)x2, −13,i(13q),−14,−16,+17,−22,+9mar	*Fletcher et al 1991 (51)*
39−45,XY,−1,−2,−3,−4,der(5)t(5:?)(p15.1:?), der(7)t(7:?)(p22:?),del(9)(p12),−11,hsr(11)(p15), −12,del(12)(p11),−16,−del(18)(p11),−16,del(18)(p11),−20,der(20)(20:?)(q13:?), −21,+der(?)t(1:?)(q24:?),+5−7mar,cx	*Fletcher et al 1991 (51)*
40−46,XX,del(7)(q22),del(22)t(22:?)(q13:?), cx	*Fletcher et al 1991 (51)*
44−89,XX,der(1)t(1:?)(q21:?),der(3)(3:?)(p21:?), der(6)t(6:?)(q13:?),der(7)t(7:?)(p21:?), der(8)t(8:?)(p12:?),der(9)(q11:p11),der(12)t(1:12)(q21:q24), der(15)t(8;15)(p11:p11), der(19)t(19:?)(q13:?),−20,del(20)(p11p12),+der(?)t(1:?)(q22:?),−der(?)t(6:?)(q21:?),cx	*Fletcher et al 1991 (51)*
46,XY,der(22)t(22:?)(q22:?)	*Fletcher et al 1991 (51)*
46,XY,psu dic(4)t(4:9)(q16:p13),cx	*Fletcher et al 1991 (51)*
54−270,X,del(X)(q26),der(1)t(1:?)(q21:?),del(1)(p11), der(3)t(3:?)(p11:?),del(3)(p13), der(4:?)(q33:?),i(5q),der(6)t(6:?)(q23:?),der(7)t(7:?)(q32:?),del(9)(p13), der(9)t(9:14) (p11:q11),der(10)t(10:?)(q26:?),del(12)(q22),der(13)t(13:?)(p11:?),del(16)(p11), del(17)(p11),cx	*Fletcher et al 1991 (51)*
58,65,X,−Y,+1,+5,+7,der(8)t(8:?)(p22:?),der(9)t(9:?)(p22:?),der(12)t(9;12)(q11;p12), der(13)t(13;17)(p11:q11.2)x2,der(15)t(15:?)(p11:?)x2,−17,−17,der(19)t(19:?)(q13.3:?), +20,+21,+22,cx	*Fletcher et al 1991 (51)*
65−79,XXYY,der(1)t(1:?)(p32:?)t(1:17)(q12:q13),del(7)(q22), der(8)t(8:?)(q24:?), psu dic(11)t(11;13)(q21:p11),der(12)t(12:?)(p13:?),cx	*Fletcher et al 1991 (51)*
44,X,−X,−7	*Valdueza et al 1991 (17)*
48,X,−Y,1p−,−3,−5,−8,−11,−14,19q,−8mar	*Valdueza et al 1991 (17)*
64−69,XY,del(1)(p21),der(6)t(6:?)(q13:?),del(8)(q22), der(9)t(9;17)(p22:q21),−17,−22	*Bello et al 1993 (17A)*
51−54,X,−X,del(1)(q23q42),del(1)(q23q42)add(1)(p36), add(9)(q34),i(12)(p10),add(12)(q24.3), add(13)(q24),t(17:22)(q11;p13),+1−3mar	*Jhanwar et al 1994 (48)*
54−56,XX,+2,−6,add(11)(p13),−13,add(14)(q32),add(15)(q22),del(17)(p11),+18, del(22)(q11.2)x2,+1−3mar,inc	*Jhanwar et al 1994 (48)*
56,XY,+5,+7,+7,+8,+10,+12,del(12)(q14),+15,del(17)(p13),+19,+20,+dmin	*Jhanwar et al 1994 (48)*
62−65<3n−>,−X,−X,−Y,add(1)(p13),+der(1)t(1:17)(p22;q11.2),−6,dup(7)(q31),der(7)t(7:7) (p22;q11.2),i(11)(q10),del(17)(p11),+1−7,mar,inc	*Jhanwar et al 1994 (48)*
64−66<3n−>,−X,−X,−Y,add(1)(p36),add(4)(q35),del(5)(q22),del(6)(q13q25), del(6)(q13),+7, del(8)(p11.2),add(8)(p11.2),−10,add(10)(p15),add(11)(q13),add(12)(q24),add(13)(p13), add(14)(q32)x2,add(14)(p13),−15,−15,−16,−17,−17,−18,−18,−19,−20,−21,−21,+22, del(22)(q11.2)x2,+1−8mar	*Jhanwar et al 1994 (48)*

TABLE 2.1. Continued

Karyotype	References
64–68<3n–>,XXY,+1,der(1)t(1;17)(p13;q11.2)x2,–2,–4,–4,+del(5)(q22),+6,+8,add(9)(p13), del(11)(q13),add(11)(p13),–12,–12,–13,add(14)(q22),–15,del(15)(q24)x2,–16,–16, add(17)(p11),–18,–18,–19,–19,–21,–21,–21,del(22)(q11.2)x2	Jhanwar et al 1994 (48)
65–74<3n–>,X,–X,–X,add(17)(q11.2),del(22)(q11.2),inc	Jhanwar et al 1994 (48)
65–71<3n–>,XX,–X,–X,+7,+12,+16,+17,del(17)(p11),–18,–19,–20,del(22)(q11.2),+1–5mar	Jhanwar et al 1994 (48)
70<3n–>,XXY,add(1)(q11),add(1)(p32),+7,add(7)(p22),t(9;10)(p11;q11), del(11)(q23), i(12)(p10),del(17)(p11),+r,+1–9mar	Jhanwar et al 1994 (48)
119–145<6n–>,XY,del(17)(p11),add(17)(q11.2),del(22)(q11.2),inc	Jhanwar et al 1994 (48)
34–36,X,–Y,del(1)(p21),del(2)(q33),–4,–5,der(7)t(4;7)(q12;p15),del(7)(q11),–8,–9,–10,–11, der(11)t(7;11)(q11;q21),der(12)t(11;12;?)(q11;p13;?),–13,–14,der(14;15) (q10;q10), –16,–17,add(19)(q13),–20,add(21)(p11q22),–22,+1–3mar/34–36,idem,add(18)(q23)	Örndal et al 1994 (52)
45,XY,+1,der(1;11)dic(1;11)(q44;q13),i(1)(q10),inv(6)(p21q12),–17	Mertens et al 1995 (175)
46,XX,t(2;7)(q32;p22)/74–79,XX,–X,–1,+2,der(3)t(1;3)(p11;q11), +4,+5,+6,der(6)t(6;14) (q11;q11)x2,+del(7)(p11p15),der(7)t(7;13)(p21;q11), dic(7;15) (p22;p13)x2, der(8)t(6;8)(p21;p11)x3,–9,der(11)t(1;11;?)(q12;p11;?)x2,–13,–16,–17,–21,–22,inc	Mertens et al 1995 (46)
58–62,X,–X,del(X)(q23),–1,–1,del(1)(q31),–3,–3,del(3)(q24),–4,t(5;7)(q35;q32),t(5;8)(q13;q23), –6,+7,add(8)(p11),add(8)(p23),+del(8)(q13),–10,del(11)(q13),–12,+13, add(14)(q24), –17,–18,add(19)(q13),add(19)(p13q13),add(20)(q12),der(20)t(10;20)(q11;q13), add(21)(q22),–22,r(?22),+1–2mar	Mertens et al 1995 (46)
61–70,X,–del(X)(p21),der(X)t(X;?;17)(q22;?;q11),–1,add(5)(p13),–6,+7,+der(7)del(7)(p11) t(6;7)(q21;q34),–8,–10,–11,–12,+der(14)t(6;14)(p11;q11),–16,–17,–18,–19,–21,–22,+2–4mar	Mertens et al 1995 (46)
61–73,X,–Y,r(X)(p22q28),der(1)t(1;8)(p22q13),+2,der(3;18)(p10;q10), add(7)(p11)x2, dic(7;?;12)(q36;?;p11),der(8)t(3;8)(q21;q24)x2,add(9)(p24), der(10)t(8;10) (q21;q26)x2, add(11)(p13),der(11;17)(p10;q10),add(12)(p13),+13, der(15;22)(q10;q10),add(15)(p11), add(19)(q13),+r,+mar,inc	Mertens et al 1995 (46)
64–70,XX,–X,+1,der(1;16)(q10;p10)add(10)(q44)x2,–2,+3,+4,+5,add(5)(p15)x2,der(6)add(6) (p23)t(2;6)(q11;q21),+7,–8,add(9)(p22)x2,–10,–10,–11,–11,der(13)t(3;?;13) (p21;?;p11)x2, –20,add(21)(p11),–22,+1–2r,+2–3mar	Mertens et al 1995 (46)
69,XX,–Y,del(1)(p21p32),t(1;3)(q44;q12)x2,+2,+6,+7,der(7)t(7;?;14) (p22;?;q11)x2,–8,–9, +10,der(12)t(12;15)(p13;q12),add(12)(q24),–13,–13,–14,der(15)t(14;?;15) (q11;?;q26), –16,+17,del(17)(q11q21)x2,–19,der(19)t(11;19)(q13;p13)x2,add(20)(q13)x2,+21, der(22)t(17;22)(q21;q13)x2,der(?;9;?)t(?;9;?)(?;q13q34;?)x2,+6mar	Mertens et al 1995 (46)
76–81,XX,–Y,add(2)(q37),del(4)(p15),add(11)(p11),del(11)(q11),del(12)(p12), add(19)(p13),inc	Mertens et al 1995 (46)
45,XX,–22	Sciot et al 1995 (176A)
47–65,XX,del(1)(p31),+del(3)(q13.3),+del(6)(q21)x2,+del(6)(q27)x2, del(7)(q21),+add(12)(q22–24),+add(16)(q24),add(17)(?q11)x2,add(20)(q11.2),–21,+mar1,+mar2, +6mar	McComb et al 1996 (176)
64,XX,del(X)(q22),+del(X)(q22),–1,–2,der(5)t(5;?9)(p13;?q13),+7,–9,+10,–11,–12,–16,–17,–18, +mar1x2	McComb et al 1996 (176)
47,XY,+7/45,X,–Y/45,X,–Y,+7,–15	Rao et al 1996 (49)
50,X,–Y,+der(1)t(1;7)(q22;q11.2),+der(1)dic(1;9)(p32;p22),+add(1)(q32), +2,+add(2)(q37), ?i(3)(q10),–5,–6,tas(7;12)(p22;p13),–8,–9,–13,add(14)(p11.2),–19,–21,+del (22)(q11.2),+1–8mar	Rao et al 1996 (49)
56,X,–X,+2,+2,+2,+5,+6,+7,+7,+15,+n17,+n19,+n20,del(22)(q?)/58–60, XX,ins(1;?)(q22;?), +2,+del(2)(q15),+5,+7,+del(7)(q22),+8,+8,+17,+19,+20,+20, del(22)(q?),+1–2mar	Rao et al 1996 (49)
45–46,XX,t(1;17)(q21;q21),der(3)t(3;?)(p25q29;?),add(5)(q14), der(9)t(4;?9)(q12;?;p11), der(9)t(5;9)(q13;p13), add(15)(q22),+der(?)t(?;7)(?;q11.2),+mar	Molenaar et al 1997 (177)
47,XY,+i(1)(q10),–18,+21/45,XY,–18	Molenaar et al 1997 (177)
59–60,XX,t(1;13)(q21;q22),add(1)(q21),der(2)t(2;17)(q36;q11)x2,+2,+6,–8,–8,–9, +der(11;13)(q10;q10),+12,+16,+add(16)(p13),–17,–18,+20,+21,+21,–22,+1–4mar	Pauwels et al 1999 (178)

TABLE 2.1. Continued

Karyotype	References
53–61,–X,–X,–Y,–1,–2,–2,–del(3)(q12),–4,–5,–5,del(5)(p1?),del(6)(q13),–7,–7,–8,–8,–9,–10,–10,–10, –11,–11,–11,–12,add(12)(p11),–13,–13,–13,–14,–14,–15,–16,–17,–17,–18,–20,–21,–21,–22,+34–40mar /75–118,idemx2 (following radiation)	*Chauveinc et al 1999 (178A)*
36–38,X,–Y,add(1)(q21),add(3)(q10),–5,–6,add(7)(q36),–8,–8,–9,der(9)add(9)(p12)t(1;9)(q23;q22), –10,add(11)(p10),add(12)(q22),–13,–13,–14,–15,–16,–16,–17,–18,dic(18;?18)(q21;?;p11), –19,–20,–21,–22,–22,+mar,+mar2,+mar3,+mar4,+mar5,+4–9mar/67–72,idemx2,–add(3)(q10), –mar2,–mar5,+6–17mar	*Plaat et al 1999 (43)*
36–43,XY,der(1)t(1:?;3)(p13;?;p24),–2,add(3)(p21),+der(4)(p11:q12),–6,–7,del(7)(q11), i(8)(q10),–9,dic(9;11)(p11),der(10)t(10:17)(q22;q11),add(12)(p11),–13,–17,–17,–18,–21,+1–5mar	*Plaat et al 1999 (43)*
42–46,XX,t(1;17)(q25;q12)	*Plaat et al 1999 (43)*
54–61,X,+X,Y,+der(1:8)(q10:q10),+der(1;14)(p13;q11),+add(2)(q11)x2, +3,+add(3)(q11),+der(3)t(3;14)(q1?2;q11),+4,+?add(4)(q11),+5,+5,+add(7) (q36)x2,+add (7)(q36)x2,+der(7)del(7)(q11p15p21)del(7)(q11.1q21)x2,+add(8)(p11), +add(8)(p2?1),der(9)t(9;?;15) (p21:?;q11),–12,–12,der(12)t(7;12)(q?21q?11)x2,–13, dup(13)(q12q34)x2,–14,–14,i(14)(q10),–15,–15,der(16)t(5:16)(q15;p13)x2,–17,–17,+psu dic(18;12)(p11.3;p11.2)x2,+r(19)(p13;q13),+2r,2–3dmin	*Plaat et al 1999 (43)*
65–70,XX,–X,add(1)(p22),add(1)(q21),–3,–4,–4,–4,add(5)(q11),+add(5)(q11),der(6)add(6)(p22) idic(6)(q14),+7,+i(7)(q10),–8,–8,–9,der(9)t(4:9)(q12:p24)x2,–10,–11,+13,–14,add(15)(q25), –17,–18,–19,add(19)(p13),–20,der(20)t(1:20)(q25:q13),–21,+der(?)t(2;?;8)(p1?;?;q13), +der(?)t(?;?;8)(?;3;13),+mar1,+mar2x5	*Plaat et al 1999 (43)*
67–73,X,–X,der(X)t(X:?;9)(q11:?;q12),der(1)t(1:15)(p13:q15), +der(1)t(1:15)(p13:q15), +2,–3,add(3)(q25),+der(4:6)(p10:p10)del(6)(p22p24)+5,–6,–6,+7,+8,+add(8)(p23), –9,inv(9)(p11q13)**c**,der(9)t(9:9)(p21:q13),–10,+10,–11,–12,der(12)t(4:12)(q11:p11), dic(12:22)(p11:q10),–13,–14,–15,–15,add(15)(p11),–16,add(16)(q11),–17,–18,+18, –19,der(19)t(17:19)(q12:q13.4),–20,–20,+21,+22,+der(?)t(?;14)(?;q11),+r1,+r2,+1–7mar/ 46,XX,inv(9)(p11q13)**c**	*Plaat et al 1999 (43)*
39,XY,add(1)(p11),–3,–4,ins(7:?)(?p11:?),–9,–9,–10,del(11)(q21q23),der(12)t(12:?;17)(p11:q11), –15,–17,–18,–19,add(19)(q11),–22,–22,+der(?)t(?;3)(?;q12), +r,+mar/<4n>,idemx2	*Mertens et al 2000 (47)*
44,XY,add(1)(p11),+5,add(13)(q14),–14,–15,del(17)(q11q21),–22	*Mertens et al 2000 (47)*
44–45,XX,del(2)(p21p23),–9,t(9:10)(q11:p11),add(11)(q21),–17,add(17)(q25),–21,+r,+2mar	*Mertens et al 2000 (47)*
45,XY,–3,–11,del(14)(q22q24),de(15)t(?14:15)(q22:p11),der(20)t(11:20)(q21:q13),+der(?)t(?:1)(?:q25)	*Mertens et al 2000 (47)*
45–48,XY,+2,t(5:14)(p15:q24),+7,+8,add(8)(q24)x2,der(9)add(9) (q22)add(9)(p13)x2, +add(9)(q11),+12,add(13)(q22),der(13)t(1:13)(q12:q34), +der(13)t(9:13) (p13:p13),–15,–21,–21,+mar	*Mertens et al 2000 (47)*
47,XX,+7	*Mertens et al 2000 (47)*
49,XY,+3,+7,+8,+i(8)(q10),del(10)(p11),–13,add(16)(q24),del(17)(q21),–22,+mar/ <4n>,XXYY,+2r,inc	*Mertens et al 2000 (47)*
49,XY,+i(8)(q10)x2,der(9:14)(q10:q10),–15,add(16)(q13),–17,–17,+r,+3mar	*Mertens et al 2000 (47)*
54,X,–Y,del(2)(p14),+4,+6,+7,+i(8)(q10),+12,+15,+19,+mar1/53–54,X,–Y,+4,+7,+7,+8, t(11:15)(q13:p11),+12,add(15)(p11),–19,+20,+21,+22,+mar1,+mar2	*Mertens et al 2000 (47)*
55,XY,+6,der(7)t(5:7)(q13:p22),+add(7)(p22),+8,add(17)(q25),+18,+18,+21,+3mar	*Mertens et al 2000 (47)*
58–83<3n>,Y,–X,add(X)(q24),+Y,–1,–4,–5,–11,–11,der(14)t(1;14)(q24;q24),der(14)t(?8;14) (q11;q32),–16,–17,der(19)t(11;19)(q13;q13),add(20)(q13),+r,+1–3mar	*Mertens et al 2000 (47)*
62–65,XX,add(X)(p22),+1,add(1)(p11)x2,+2,–4,–5,–6,–6,+8,+8,–9,add(9)(p21)x2,–10,del(10) (q22q24)x2,add(11)(q25)x2,der(11)t(11;12)(p11:q11)ins(11;?)(p11:?),–12,+13,–14,–15,–15,–16,–18,–20,–22,+mar1x2,+mar2	*Mertens et al 2000 (47)*
69–79,XXX,add(1)(q21),add(1)(q32),der(3)t(3;12)(p21:q13)x2,add(6)(q21),del(6)(q15), add(12)(q13),add(13)(p11),add(22)(q13),+der(?)t(?;5)(?;q13),inc	*Mertens et al 2000 (47)*

TABLE 2.1. Continued

Karyotype	References
69–114<4n>,XX,–X,–X,–X,add(1)(p11),–2,–6,del(6)(q21),+7,+7,+7,+7,–8,–8,–9,–9,der(9)add(9)(p11)add(9)(q34)x2,–10,–11,add(12)(q13),–15,–15,–16,–17,–17,–17,–17,–19,–21,–21,–22,+der(?)t(?;17)(?;q21)x3,+hsr(?),inc	*Mertens et al 2000 (47)*
81–100,XX,–X,–X,del(1)(p12),i(6)(p10)x2,add(10)(p13)x2,add(16)(q24)x2,+mar1x2,+mar2x2,inc	*Mertens et al 2000 (47)*
43,X,der(X)t(X:18)(p11.2;q11.2),der(1)t(X:1)(p11.2;p13),–3,add(5)(q11.2),–10,–18,der(18) t(1;18)(p32;q11.2)	*Vang et al 2000 (29)*
43,XX,–1,–1,der(3)t(3;17)(p12;p11.2)t(9;17)(p34:p13),+add(7)(q32), dup(7)(q32q36), +der(8)t(6;8)(?;p11.2)t(6;15)(?;q15),–9,der(1;14)(p10;q10),der(16)t(1;16)(p11;p11.2) del(16)(q22),der(17;21)(p10;q10),der(19)t(10;19)(q11.2p13.3)t(1;19)(q12;q13.3) t(1;17)(q44;q23),der(19)del(19)(p13.1)t(1;19)(q12;q13.3)t(1;17)(q44;q23), –21,–22,1–2dmin	*Schmidt et al 2001 (44)*
42,idem,der(10;13)(p10;q10),–13,der(14)t(3;14)(q25;q32),1–2dmin/ 43,idem,+del(3)(q11),der(10;13)(p10;q10),–13,der(14)t(3;14)(q25;q32),1–2dmin/ 61–66,XX,–X,–1,–1,+der(3)t(3;17)(p12;p11.2)t(9;17)(q34:p13)x2,del(3)(q11),+dup(7) (q32q36)x2,+der(8)t(6;8)(?;p11.2)t(6;15)(?;q15)x2,–9,–9,der(10;13)(p10;q10) x2,–13,–13, der(14)t(3;14)(q25;q32)x2,der(1;14)(p10;q10)x2,–15,–15,der(16)t(1;16)(p11;p11.2) del(16)(q22)x2,der(17;21)(p10;q10)x2,der(19)t(10;19)(q11.2;p13.3)t(1;19)(q12;q13.3)t(1;17)(q44;q23)x2,der(19)del(19)(p13.1)t(1;19)(q12;q13.3) (q11.2;p13.3) t(1;17)(q44;q23)x2,–21,–21,–22,–22.2–5dmin	*Schmidt et al 2001 (44)*
45,XX,del(12)(q11.2),+del(12)(p11.2),–14,der(19)t(8;19) (q11.2;p13.3)t(8;19)(q11.2;q13.3), der(20)t(8:20)(q11.2;q13.1),–22,1–2dmin/ 44,idem,–del(12)(q11.2),dmin/ 44,idem,–der(19)t(8;19)(q11.2;p13.3)t(8;19)(q11.2;q13.3),2dmin/ 44,idem,–21,dmin	*Schmidt et al 2001 (44)*
45,XX,del(12)(q11.2),+del(12)(p11.2),–14,der(19)t(8;19)(q11.2;p13.3), t(8;19)(q11.2;q13.3), der(20)t(8:20)(q11.2;q13.1),–22,dmin/ 44,idem,–der(19)t(8;19)(q11.2;p13.3)t(8;19)(q11.2;q13.3),1–2dmin/ 44,idem,–der(20)t(8:20)(q11.2;q13.1),dmin/ 44,idem,–9,–10/ 42,idem,–9,–10,–der(19)t(8;19)(q11.2;p13.3)t(8;19)(q11.2;q13.3),dmin/ 44,idem,–del(12)(q11.2),–10,+22	*Schmidt et al 2001 (44)*
49,X,–Y,der(5)t(5;18)(p15.3;q11.2)x2,+der(5)t(5;8)(q11.2;q22)x2, der(9)dup(9)(p13p24) t(9;15)(p24;q22)t(9;15)(q13;q26), der(10)t(10;21)(q11.2;q11.2)x2,del(10)(p11.2) del(10)(q24)x2,+12,+12,der(14;22)(q10;q10)x2,der(15)t(7;15)(q11.2;q26)x2, der(16)t(16;20)(p13.3;p13)del(16)(q22)del(20)(p11.2)x2,der(19)t(8;19)(p11.2;q13.1)x2, –20,–20,der(21)t(17;21)(q21;q11.2)x2,del(21)(q11.2)/48,idem,–der(5)t(5;8)(q11.2;q22)/ 48,idem,–12/47,idem,–der(5)t(5;8)(q11.2;q22),–12	*Schmidt et al 2001 (44)*
52,XX,+der(1;13)(q10;q10),del(3)(p21),+der(4;14)(p10;q10)x2, der(5;14)(q10;q10),+der(5)t(5;8)(q11.2;q11.2)x2,der(7)t(7;9)(q22;q22),+del(7)(p11.2)del(7)(q32), +der(8)t(8;12) (q11.2;p11.2),der(11)t(10;11)(q22;p13),+i(12)(p10),del(13)(q32),–14,–14,+del(15)(q24), +der(17;21)(p11.2;q11.2),–21,i(22)(q10),1–2dmin/ 104,idemx2/50,idem,–der(5)t(5;8)(q11.2;q11.2)x2,+der(8)t(6;8)(?;p11.2)t(6;15)(?;q15)x2,–9,–9,der(10;13) (p10;q10)x2,–13,–13,der(14)t(3;14)(q25;q32)x2,der(1;14)(p10;q10)x2,–15,–15,der(16)t(1;16) (p11;p11.2)del(16)(q22)x2, 100,idemx2,–der(5)t(5;8)(q11.2;q11.2)x2,–i(12)(p10)/ 51,idem,–der(8)t(8;12)(q11.2;p11.2)/ 50,idem,–i(12)(p10),–del(15)(q24)	*Schmidt et al 2001 (44)*
58–65,XX,–X,–1,–1,–1,del(3)(q11),der(3)t(3;17)(p12;p11.2)t(9;17) (q34:p13)x2,+der(3)t(3;17) (p12;p11.2)t(9;17)(q34:p13)x2,+dup(7) (q32q36)x2,+der(8)t(6;8) (?;p11.2)t(6;15)(?;q15)x2,–9,–9,der(10;13)(p10;q10),–13,–13, der(14)t(3;14)(q25;q32)x2,der(1;14)(p10;q10)x2,–15,–15,der(16)t(1;16)(p11;p11.2)del(16)(q22), der(17;21)(p10;q10)x2,–17,der(19)t(10;19)(q11.2;p13.3)t(1;19)(q12;q13.3)t(1;17)(q44;q23)x2,der(19)del(19)(p13.1)t(1;19)(q12;q13.3)t(1;17)(q44;q23)x2, –21,–21,–22,–22,1–4dmin/	*Schmidt et al 2001 (44)*
82–86,XX,–X,–X,–1,–1,–1,del(3)(q11)x2,der(3)t(3;17)(p12;p11.2)t(9;17)(q34:p13)x3, +dup(7)(q32q36)x3,+der(8)t(6;8)(?;p11.2)t(6;15)(?;q15)x2,–9,–9, der(10;13) (p10;q10)x2,–13,–13,der(14)t(3;14)(q25;q32)x2,der(1;14)(p10;q10)x2,–15,–15,der(16)t(1;16) (p11;p11.2)del(16)(q22)x2, der(17;21)(p10;q10)x2,–17,der(19)t(19)t(10;19)(q11.2;p13.3)t(1;19) (q12;q13.3)t(1;17)(q44;q23)x2,der(19)del(19)(p13.1)t(1;19)(q12;q13.3)t(1;17)(q44;q23)x2, –21,–21,–22,–22,1–5dmin/	*Schmidt et al 2001 (44)*

TABLE 2.1. Continued

Karyotype	References
42–44,XX,–1,–1,del(3)(q11),der(3)t(3:17)(p12:p11.2)t(9:17)(q34:p13), +der(3)t(3:17)(p12:p11.2) t(9;17)(q34:p13),+dup(7)(q32q36), +der(8)t(6:8) (?;p11.2)t(6;15)(?;q15),–9,der(10:13)(p10q10), –13,der(14)t(3:14)(q25:q32), der(1;14)(p10;q10),–15,der(16)t(1:16)(p11.2)del(16)(q22), der(17:21)(p10;q10),der(19)t(10;19)(q11.2;p13.3)t(1:19)(q12:13.3)t(1:17)(q44:q23), der(19) del(19)(p13.1)t(1;19)(q12:q13.3)t(1:17)(q44:q23),–21,–22,1–2dmin	*Frank et al 2003 (179)*
60–63<3n>,–X,add(X)(q28)x2,–1,–2,add(2)(q33~q34),–4,add(4)(p12),–5,+7,+7,–8,–9,–10,–10, +12,+14,+add(14)(p11)x2,–15,–18,–19,–22,+1–4mar/42~46,XX,del(9)(p13,del(10)(q22), add(11)(p15),add(15)(q26)	*Frank et al 2003 (179)*
61–65<3n>,XXY,+Y,add(1)(p13),–4,–6,+7,+8,–9,–10,–11,–12,–13,+14,+add(14)(q32)x2, –15,–18,–19,+20,–21,–21,+mar1,+mar1,+mar2,+mar3/62~66,idem,+4	*Frank et al 2003 (179)*
68–74<3n>,XXY,dic(1:14)(14pter→14q32:::::1p13→1qter)x2,+4,+5,+8,–9,der(10)t(10:14) (q22:q13)x2,+11,–14,–14,–14,–15,+16,+17,+19,+add(20)(p12),–22,+mar	*Frank et al 2003 (179)*
?46,Y,t(X:12)(q32:q24),t(2:4)(q35:q31)	*Gil et al 2003 (180)*
45,XY,der(1)t(1:1)(p35:q21),–18	*Van Roy et al 2003 (181)*
46,X,–X,i(1q),+2	*Van Roy et al 2003 (181)*
42,XY,–3,add(4)(q35),add(11)(p14),–12,–14,add(19)(q13.4),–21,–21,–22,–22,+mar1,+mar2, +mar3/42,XY,idem,add(1)(p36),del(7)(q31)/42,XY,del(1)(q35),del(2)(q23),add(4)(q35), +5,del(7)(q31),–9,–9,add(11)(q23),–12,–19,add(19)(q13.4),–21,–22,+mar1,+mar2	*Bridge et al 2004 (45)*
43,XY,i(1)(q10),+8,–9,–11,del(12)(p11.1),–18,–22	*Bridge et al 2004 (45)*
44,XY,del(3)(p21),del(10)(q22),–13,–16,del(20)(q13.3),add(22)(q11.2)	*Bridge et al 2004 (45)*
46,XX,add(1)(q21),–6,–10,–14,–18,+mar1,+mar2,+mar3,+mar4,inc	*Bridge et al 2004 (45)*
46,XY,t(1;3)(p36.1;q11.2),del(7)(q32),+i(8)(q10),del(9)(p13),–10,del(11)(q14),der(12)t(1:12) (q23:p12),der(14)t(10:14)(q11.2;p11.2). der(15)t(15:21)(q26:q?)?dup(21),add(20)(q13.2), –21, add(22)(q11.2),+mar1,1–3dmin	*Bridge et al 2004 (45)*
47–65,XX,del(1)(p31),+del(3)(q13.3),+del(6)(q21)x2,+del(6)(q27)x2, del(7)(q21),add(12)(q24), +add(16)(q24),add(20)(q11.2),–21,+mar1,+mar2,+6mar	*Bridge et al 2004 (45)*
58,XX,–X,–1,–3,–4,–5,–6,+7,+8,–9,–10,–11,–13,–14,+15,–16,–17,–18,–22,+mar	*Bridge et al 2004 (45)*
60–70,XY,–X,–1,add(1)(p36.3),–3,–6,–8,–9,–10,add(11)(q22),del(11)(q21),–13,–14,–15, add(15)(q24),add(16)(q21),add(16)(q24),–17,–17,–17,+add(19)(q13.2),–20,+21, add(22)(q11.2),+mar1,+mar2,+mar3,+mar4,+mar5,+mar6x2,+mar7,+mar8,+1–4mar	*Bridge et al 2004 (45)*
65,XXY,add(3)(p11.2),add(5)(q35)x2,del(7)(q11.1q34x2,add(9)(q23),–10,add(13)(p11.1), add(14)(q32)x2,–16,–19,–19,add(20)(q13.3),add(22)(q11.2)x2,1–3dmin	*Bridge et al 2004 (45)*
67,XXY,+add(1)(p35),+3,–13,add(19)(q13),–20,–20,–21	*Bridge et al 2004 (45)*
71–75,XY,–X, or –Y,–3,dic(3;9)(q11,p11)x2,–4,del(4)(q28q35),der(7)del(7)(q22)t(1:17:7) (?;?:p22)x2,der(7)t(7:7)(p22:?),+der(7)t(7:7)(p22:?),–8,add(8)(q24.3)x2,–9,–9,+11, del(11)(q23)x2,del(12)(q24),+der(14)t(7:14)(?;p11),–15,–16,–16,–17,der(17) del(17)(p13)t(17:17)(q25:?)x2,+18,der(19)t(13:19:21)(q?:?p11q11:q?)x2,der(20)t(12:20) (?q13:p13)x2,der(21)t(17:21)(?:q22)x2,der(22)t(8:17:22)(?;?:p11), +mar1,+mar2,+mar3x2,+5–9mar	*Bridge et al 2004 (45)*
71–82,XXX,del(1)(q21)x2,add(2)(q?),del(4)(p13),–5,–5,+del(7)(p11),–8,–8,–9,–9,–10,–10, del(12)(p11),+14,add(16)(q24),add(20)(q11),+mar1,+mar2,inc	*Bridge et al 2004 (45)*
72,XXX,+1,add(1)(p31)x2,del(2)(p13p23),del(4)(p14p16)x2, del(6)(q13q26)x2,+7, add(7)(q36)x2,+8,–9,–10,del(11)(q23),+12,–14,–15,–16,–17,–18,+add(19)(q13.3), +20,+21,+mar1,+mar2,+mar3,1–5dmin	*Bridge et al 2004 (45)*

TABLE 2.1. Continued

Karyotype	References
58–59,XX,der(1;2)(1qter?1p36::12q11?12qter),+dup(2)(q21q31), +3,+4,+i(5)(q10),+6,+i(7)(q10),der(8)t(8;8)(q24:q24), +12,−13,−13,der(15)t(15;17)(p11:q11),+16,−17,−17,+18,+20,+21,+22,+3~8mar	*Aoki et al 2006 (203)*
73–82,XXY,−Y,der(1)t(1;5)(p32;q13),add(2)(q35),add(2)(q21), add(3)(q21),add(4)(q31), add(5)(p15.3),del(6)(q21),+del(6)(q21),+8,+9,+10,add(11)(q21),−12,−14,add(16)(p13). +add(16)(q13),+add(17)(q23),+18,der(19)t(2;19)(q12;p13),der(19)t(2;19)(q12;p13),+20,+21,−22,+der(?)t(5:?)(q13:?),der(?)t(18:?)(q11:?)	*Aoki et al 2006 (203)*
72–76,XXX,+X,+2.add(3)(q29),+5,−6,+7.i(8)(q10),add(9)(p22),+add(9)(p22),−10, der(10;15)t(q10;q10),−11,+add(12)(p13),−13,−14,−15,−16,−16,+17,−18,dup(18)(q23q11). ins(19;14)(p13;q3?q11),+ins(19;14)(p13;q3?q11),+20,+20,der(22)t(9;22)(q10;p12),+der(22)t(9:22)(q10;p12),+der(?)t(21:?)(q11:?)x2,+4~6mar	*Aoki et al 2006 (203)*
40−XY,−3,−4,−5,add(6)(q11.2),+i(8)(q10),−1,add(11)(p15),−14,−17,−17,−18,add(19)(p13),−20,−21,−22,+4mar/79,idemx2,−15	*Kobayashi et al 2006 (53)*
44,−X,add(5)(q22),add(8)(p23),add(9)(p11),add(10)(p11),−13,−16,−21,−22,+3mar	*Kobayashi et al 2006 (53)*
46,X,add(X)(p11.2),t(5;10)(q13;q24)/46,X,add(X)(p11.2),add(4)(q21),−6,add(8)(q11.2),add(10)(p11.2),add(12)(q24.1),add(12)(q13),−13,add(18)(p11.2),+2mar	*Kobayashi et al 2006 (53)*
47,XX,add(8)(q22),del(13)(q32)add(14)(q32),+mar/47,XX,del(1)(p34.1), add(2)(q11.2), der(3)ins(3:?)(p21:?)add(3)(q12),add(12)(q24.1),−13,ins(14;?)(q22:?),add(17)(p11.2), −18,−22,+4mar	*Kobayashi et al 2006 (53)*
47,XX,+7	*Kobayashi et al 2006 (53)*
58,add(X)(q24),−Y,+der(1)add(1)(p11.2)add(1)(q32),+2,+del(3)(p11.2), add(3)(q21),+add(5)(q11.2),+?der(7)add(7) (p22)add(7)(q11.2)x2,+add(8)(p11.2),−9,der(9) add(9)(p24)add(9)(q11),der(11)ins(11:?)(q23:?)t(7;11)(q11.2:q23)x2,add(13)(q21),+14,+14, add(15)(p11.2),add(15)(p11.2),del(16)(p11.2)x2,−17,add(17)(q25),−18,+add(19)(q13.3)x2, add(19)(q13.1),add(21)(p11.2)x2,del(22)(q13),+3mar	*Kobayashi et al 2006 (53)*
76,−Y,add(X)(p11.2),+add(X)(p22.3),+del(1)(q21),+add(1)(p13), +add(1)(p13),+2,+2, +add(3)(q21)x2,+add(4)(p16)x2,+dup(5)(q31q35)x2,+7,+7,+8,+8, der(9)t(1:9)(p13:p11.2) x2, +add(9)(p11.2)x2,+10,+10,+11,add(12)(p11.2),+add(12)(p11.2),+13,+add(13)(q32),+14, +16,+16,−17,−17,+18,+18,add(19)(p13.1)x2−20,−20,+6mar	*Kobayashi et al 2006 (53)*
78,X,del(X)(q22),+1,+2,+2,+2,+3,+4,+5,+5,+6,+6,+add(7)(p22)x4, +9,+9,+add(9)(p22), +del(9)(p22)x2,+10,+11,+add(11)(p15),+12,+12,+12,+13,+13,+14,add(14)(q32)x2, −16,+17,+18,+19,add(20)(q13.1)x2,add(21)(p11.2)x2,−22,del(22)(q13)	*Kobayashi et al 2006 (53)*
33−35,XY,−1,der(2)(4?→4?::6?→6?::1?→1?::8?→8?::20?::13?→13?::2p14→ 2q37::7?→7?::1q22::1qter),−3,der(4)(4pter→4q31::3q12→3qter),del(4)(pter→q11:), del(5)(pter→q11:),−6,der(7)(7pter→q11::12q11→12q15::15q11→15qter),der(8) (8pter→8q24.3::17?→17?::12?→12?::17?→17?::8→8?::21?→21?::20?→ 20?::5?→5?),der(8)(p21→q10::4q11→4qter),−9,−10,der(10)(17?→17?::10p15→10qter), der(11)(11pter→11q22::13?→13?),−12,−13,−14,−15,der(15)(15pter→15q24::14q11→14qter), −16,der(17)(17pter→17q11.1::9q13→9q21),−19,−20,der(21)(5?→5?::21p11.2→21qter),+mar	*Ishiguro et al 2006 (215)*
61~75,XX,−X,der(1)add(1)(p13)add(1)(q43), +der(1)add(1)(p13)add(1)(q43),+2,add(2)(q31), +3,+4,+5,−5,i(5)(q10),+6,del(6)(q21),+del(6)(q21),+7,+8,−9,−10,−11,add(11)(q23),+12,−13, +14,+15,−16,−17,−18,add(18)(p11.2)x2,+19,−20,+21,+22,+mar1,+mar2,+mar3,+mar4,+mar4	*Kleinschmidt-DeMasters 2007 (216)*

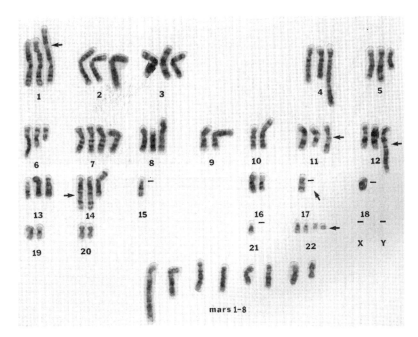

FIGURE 2.4. Complex karyotype of an MPNST containing 63 chromosomes, including 8 unidentified markers (mars 1-8), as well as a number of abnormal chromosomes (some pointed to by horizontal arrows) and a missing chromosome 17 (upward pointing arrow) (48).

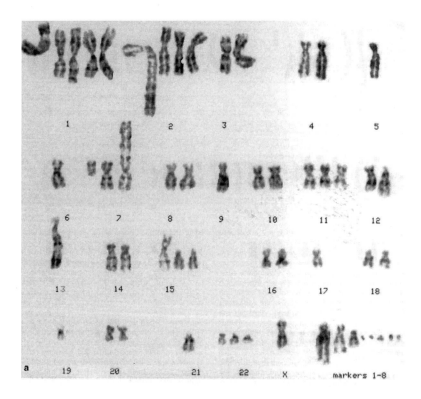

FIGURE 2.5. Complex karyotype of an MPNST of the thigh containing a −17, a change possibly affecting the *NF1*, and/or *TP53* genes (49).

Comparative Genomic Hybridization (CGH) Studies in MPNST

Similar to chromosome banding analysis, CGH has revealed a broad spectrum of chromosomal imbalances in MPNST (54,54A). Despite this complexity, several nonrandom chromosomal changes, predominantly representing DNA gains, were identified. It should be pointed out that the CGH tech-nique is probably less sensitive to losses, especially in poly-ploid tumors, which MPNST often are; a high proportion of these are intermingled with diploid stromal cells, which among other factors may also play a role (44). It should be reiterated that CGH offers a global view of the en masse chromosomal changes in all the cells of the tumor, and the method is not capable of revealing balanced changes, such as translo-cations. An early CGH study of two MPNST showed gain of 1q21−q22 in one tumor and gain of 12q14 in the other tumor

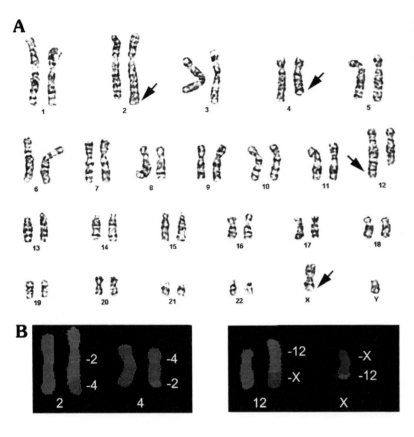

FIGURE 2.6. Karyotype of an MPNST (A) of the skull base showing two translocations as the only anomalies (180). The translocations (arrows) are (2;4)(q35;q31) and (X;12)(q22;q24). Neither translocation is specific for or common in MPNST, as is true of other changes in this type of tumor. Fish staining of the two translocations is shown in B.

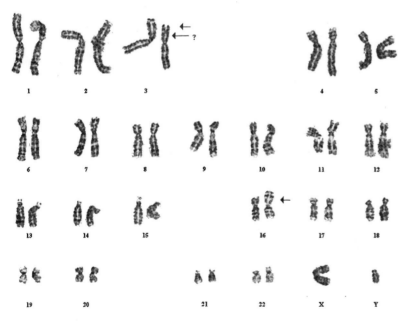

FIGURE 2.7. G-banded karyotype of an MPNST with a t(3;16)(p2?;p13.3) (arrows) as the only cytogenetic anomaly (courtesy or Dr. Kathleen E. Richkind).

(55). Except for high-level amplifications that were confined to occasional tumors, no differences between sporadic and NF1-associated MPNST were found (54). In both, the number of gains was significantly higher than the number of losses, suggesting a predominant role of proto-oncogene activation during MPNST progression. Candidate regions with potentially relevant proto-oncogenes included chromosomal bands or regions: 17q24−q25, 7p11−p13, 5p15, 8q22−q24, and 12q21−q24; those with putative tumor suppressor genes (TSG) were 9p21−p24, 13q14−q22, and 1p. High-level amplifications were restricted to sporadic tumors and affected eight different chromosomal subregions. In three of these MPNST, identical subregions on chromosomal arms 5p and 12q were coamplified.

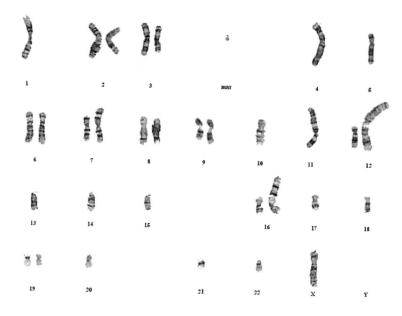

FIGURE 2.8. G-banded karyotype of an MPNST from a 22-year-old male with a hypodiploid number of 33 chromosomes. A composite karyotype consisted of 33,X,−Y,add(2)(q31),add(3)(q21),−4,add(4)(p16), dic(5;12)(p13;p13),add(7)(p22),−10,−11,add(11) (p15),−13,−14,−15,add(16)(p13.3),−17,−18,−20, −21,−22 (courtesy or Dr. Kathleen E. Richkind). Comparison of the karyotypes in Figures 2.4–2.9 shows the range of chromosomal changes that may be seen in MPNST, i.e., from pseudodiploidy (see Figure 2.7) with only one cytogenetic event to the complex and many changes seen in the severe hypodiploid karyotype (this figure).

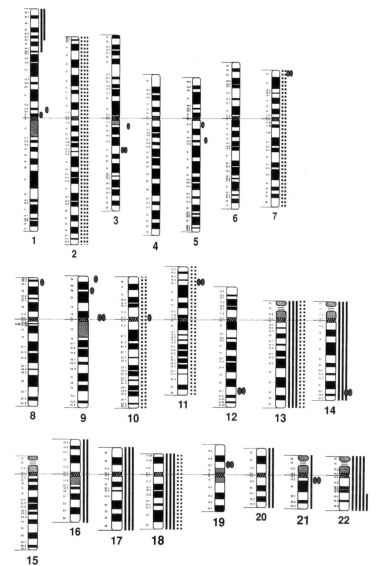

FIGURE 2.9. Karyotypic alterations in MPNST. Solid lines indicate loss, broken lines indicate gain, and black oval symbols indicate breaks (53). Alhough the findings were variable, loss of chromosomes 22, 14, and 17 are to be noted.

TABLE 2.2. Computer-assisted cytogenetic analysis of MPNST[a].

Significant losses:	9p2, 11p1, 11q2, 18p1
Other losses:	1p3, 9p1, 11q1, 12q2, 17p1, 18q1−q2, 19p1, 22q1, X,22q1, X, Y
	Loss of 17q1 on which the *NF1* gene is located (17q11.2) is not a common cytogenetic finding in NF1-related MPNST.
Sporadic MPNST vs. NF1-related MPNST:	Relative loss in NF1-related MPNST: 1p3, 4p1, 21p1−q2 and relative gain in 15p1−q1
	The differences between NF1-related and sporadic MPNST might reflect different oncogenetic pathways.
	9p2 losses and 7q1 gains could be of oncogenetic importance in MPNST
Gains:	chromosome 7 (especially 7q1)
Chromosomal breakpoints:	1p21−p22 (28% of NF1 vs. 0% of sporadic)
	1p32−p34 (17% of NF1 vs. 0% of sporadic)
	8p11−p12 (7% of NF1 vs. 27% of sporadic)
	17q10−q12 (24% of NF1 vs. 7% of sporadic)
Most involved breakpoints:	1p13, 1q21, 7p22, 9p11, 17p11, 17q11, 22q11

[a]Based on data of Plaat et al. (43).

One study (56) investigated 31 MPNST from 23 patients, including nine with NF1, by means of CGH (Figure 2.10). The MPNST in 21 of the 23 patients revealed changes, with a mean value of 11 aberrations per sample (range 2–29). The minimal common regions of the most frequent gains were 8q23−q24.1 (12 cases), 5p14 (11 cases), and 6p22-pter, 7p15-p21, 7q32-q35, 8q21.1-q22, 8q24.2-qter, and 17q22-qter (10 cases each). Seventeen high-level amplifications were detected in 8 of the 21 samples. In three cases, the high-level amplifica-tions involved 8q24.1-qter, and in 2 cases each regions 5p14, 7p14-pter, 8q21.1-q23, and 13q32-q33. The minimal common region of frequent losses was 14q24.3-qter (five cases). The gain of 8q as a single common anomaly in a primary tumor, its recurrence, and metastasis suggests that this aberration is an early change in the tumorigenesis of MPNST. Compa-rable aberrations were observed in separate tumors of the same patients affected by NF1, indicating a limited number of random secondary changes. In sporadic MPNST, the most frequent gains were narrowed down predominantly to 5p, chromosome 6, 8q, and 20q, whereas in MPNST from patients with NF1, gains in 7q, 8q, 15q, and 17q were frequent. The occurrence of gain of both 7p15-p21 and 17q22-qter was asso-ciated with a statistically significant poor survival rate.

Metaphase-CGH and microarray-CGH was used to deline-ate the changes in eight MPNST and eight schwannomas by the former and in five MPNST in the latter (56A). Gains were found at 3q13-qq26, 5p13-p14, and 12q11-q23, and losses were found at 1p31, 10p, 11q24-qter, 16, and 17. Microarray-CGH revealed gains of the *EGFR*, *DAB2*, *MSH2*, *KCNK12*, *DDX15*, *CDK6*, and *LAMA3* genes and losses of *CDH1*, *GLTSCR2*, *EGR1*, *CTSB*, *GATA3*, and *SULT2A1* genes, pointing to the complexity of the genetic changes in MPNST.

In a CGH study of MPNST (57), gains of chromo-somal material exceeded losses, suggesting oncogene activa-tion during tumor progression. Thus, gains of 17q and the X-chromosome were found in two of four of NF1-associated MPNST and of 4q in three of five sporadic tumors. This contrasts with a higher rate of losses in neurofibromas and schwannomas indicative of TSG playing a key role in these tumors. Losses of 17p were rare in these benign tumors as based on CGH (54, 56), only 3/52 showing this change. It seems that genomic alterations at 8q, primarily gains, in NF1-related MPNST is a recurrent finding (47, 50, 54, 58).

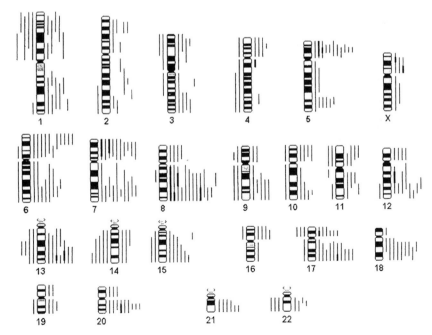

FIGURE 2.10. Gains (shown by lines to the right of the chromosomes) and losses (to the left) of chromosomal material in MPNST determined by CGH. The findings of primarily gains reflect the hyperdiploidy often seen in MPNST. High-level amplifications are shown by thick lines. Gains at 7p and 17q were stressed by the authors (from ref. (56), with permission).

Molecular Genetics of MPNST

The complexity of the genetic changes and their mechanisms of action in MPNST are recurring themes in this chapter, and they are partially reflected in the range of studies presented. The information constitutes at least some of the pieces in the puzzle of the stepwise process in MPNST development and progression. However, it is still unknown where these pieces fit into the puzzle. Undoubtedly, some of the pieces are still missing; nevertheless, an attempt has been made here to present as many of the puzzle pieces as are available in the literature.

In keeping with the diversity of the cytogenetic changes, it is not surprising that the results of molecular studies in MPNST have shown similar variability. Also, this variability reflects the range of the histology and biology of these tumors, the different methodologies and approaches used, and the state of the tumors examined (e.g., archival vs. nonpreserved tissue). Other factors, e.g., intratumoral heterogeneity, anatomic site of the tumor, and previous therapy, may play a role in the nature and diversity of the findings obtained. Nevertheless, a consensus on the molecular findings in MPNST may help in the understanding of the biology of these tumors.

MPNST, like most tumors, require an orchestrated and successive chain of genetic events for full development and progression. These genetic pathways may not be similar among MPNST, and they may account for the biologic and clinical variability among these tumors. Thus, the molecular events responsible for and associated with MPNST pathogenesis are presented for NF1-related MPNST and the sporadic types of these tumors, whenever such data are available.

Based on FISH studies, *NF1* deletions and losses, often due to -17, are common in MPNST whether of NF1 origin or sporadic; these changes are less common in PNF, and they are not seen in cellular schwannomas (28). Changes occurring at 17p are next in significance in that they involve the *TP53* gene located on that chromosome arm. However, stress must also be put on changes at 9p21, a region that harbors a number of genes (*INK4B*, *ARF*, *INK4A*, and *CDKN2A*) controlling the cell cycle and often affected in MPNST. It is not surprising that some of the molecular changes seen in NF1-related MPNST, e.g., those of *NF1*, p53, p16, and others, are seen in some neurofibromas, especially of the plexiform type, because they may serve as precursors of MPNST (28, 58–61).

The NF1 Gene in MPNST

The key gene in MPNST tumorigenesis is the *NF1* gene located at 17q11.2. MPNST originating in NF1 patients all have alterations of the *NF1* gene (62–66); most sporadic MPNST probably start out with *NF1* changes as well (58). For example, no detectable levels of *NF1* transcripts were found in 26 NF1-related and 18 sporadic MPNST (77). These *NF1*

alterations per se seem not to be sufficient for the progression and full development of MPNST, which require an array of additional genetic changes, many of which have been examined (see below); however, these have not revealed a consistency within either of the aforementioned categories of MPNST. A search for possible genes involved (other than *NF1*) in the development of MPNST (sporadic and NF1-related) has shown some genes to be involved often and others to play a minor (if any) role, but none has been demonstrated to be consistently involved (67).

The *NF1* gene has been discussed in the previous chapter in relation to neurofibroma pathogenesis. *NF1* is a TSG negatively regulating $p21^{ras}$ and *RAS*, which probably play a key growth role in MPNST development (68). The *NF1* product, neurofibromin, has a GTPase-like enzymatic activity whose inactivation leads to constitutive activation of the *RAS* oncogene and its effects in the pathway in the development of MPNST (68–71), especially of NF1-related tumors (72). The increased levels of active Ras proteins lead to the activation of numerous Ras-dependent signal transduction pathways including the microtubule-associated protein kinase (MAPK) pathway. An additional role in MPNST development is the inactivation of the *CDKN2A/p16* pathway (73, 74) located at 9p21. Reflecting this aspect of NF1 was the finding of highly elevated Ras~GTP levels in NF1-related MPNST and to a lesser extent in NF1-neurofibromas (70), but not in schwannomas. Neurofibromin was absent in the MPNST. There is evidence that malignant progression in MPNST is related to alterations of genes controlling cell cycle regulation (58–61). A clearly implicated gene is *TP53*. Both *TP53* gene mutations and alterations of protein expression have been found in MPNST (75, 76).

In NF1-related MPNST, inactivation of both *NF1* alleles has been demonstrated, implicating changes of this gene as an initial event in MPNST tumorigenesis (64); microsatellite results are consistent with the occurrence of somatic inactivation by LOH of the second *NF1* allele (71). Furthermore, sporadic MPNST also show alterations at the *NF1* locus (58, 77). *NF1* gene inactivations are involved in the early stages of nerve sheath tumorigenesis, i.e., in the formation of neurofibromas, which are precursors of MPNST.

A possible explanation for the lack of LOH or mutations of *NF1* in MPNST and the heterogeneity of molecular results could be the presence of normal cells, clonal heterogeneity of the tumors per se, or both. For example, these factors may account for the results in a study of MPNST of NF1 cases (78), in which LOH for *NF1* was found in only two of seven tumors (no hypermethylation present in any of these tumors), microsatellite instability was observed in five of 11 MPNST (not present in all 70 neurofibromas), and no subtle mutations of *TP53* and *CDKN2A* in seven MPNST, although LOH of these genes was shown in four of 11 tumors. The results led the authors to state that NF1 tumorigenesis is a complex multistep process involving a variety of different types of genetic defects at multiple loci (78).

The demonstration of *NF1* inactivation or alteration in neurofibromas (58–61, 79–82), especially of the plexiform type from which MPNST originate, made the search for similar changes in MPNST logical. In an early study (65), LOH at the locus of *NF1* was demonstrated in three of six MPNST, findings which were confirmed by other investigators (63, 76, 83). The results of these studies did not necessarily reflect an inactivation of the normal *NF1* allele (in NF1 patients), because LOH is often confined to 17p only (76, 83). An indication of inactivation or alteration of the *NF1* gene in MPNST was obtained with somatic cell hybrids in an established cell line containing two cytogenetically identical der(13)t(13;17)(p11.2;q11.2) (84). A specific deletion of the *NF1* was demonstrated subsequently in a MPNST from an NF1 case (75). Another study (63) also pointed to inactivation of *NF1* from a patient from NF1. Mutations of the *NF1* gene were found in 75% of NF1-associated MPNST (75). These authors (75) stated that inactivation of the *NF1* gene initiates MPNST tumorigenesis but that tumor progression occurs by an accumulation of additional genetic abnormalities.

The prevalence of *NF1* deletions in MPNST (NF1-related and sporadic), regardless of S-100 protein expression or NF1 clinical status, has been noted (58). The interpretation is that S-100 protein-negative tumor cells within MPNST (sporadic and NF1-related) represent dedifferentiated schwann cells that still harbor *NF1* deletions, i.e., the loss of *NF1* represents an early tumorigenic event that is still detectable in high-grade neoplastic clones no longer manifesting immunohistochemical evidence of schwann cell differentiation (23). The finding of divergent epithelial differentiation, mesenchymal differentiation, or both in some MPNST (e.g., MTT) and complete lack of S-100 protein expression in others would further support this dedifferentiation hypothesis (58). The decreased CD34 reactivity found in these cells parallels that of S-100, and it is related to loss of fibroblast-like cells in high-grade MPNST (23, 85). LOH studies of MPNST have been largely limited to examples from NF1 patients (59, 60, 64–66, 79, 81, 82), where *NF1* loss has been common. Monosomy 22 was identified in one sporadic MPNST (49), suggesting that *NF1* may be implicated in some of these cases as well. Another study found no mutations in nine sporadic MPNST within the GAP-related domain by a polymerase chain reaction (PCR)/single-strand conformational polymorphism (86). It also was shown (64) that results obtained by molecular DNA analysis and interphase cytogenetic supplement each other when interpreting genetic events during tumor development. The variability in *NF1* involvement in sporadic MPNST points to a possible heterogeneity in the genetic pathways leading to the development of some of these tumors (62, 86), although in general *NF1* changes are present in most MPNST of sporadic origin.

The mechanisms of the effects, interactions, and components of the pathways of p53, retinoblastoma protein (pRb), and other cell cycle controllers are discussed in Chapter 9.

Here, we present the changes of some members of these pathways in MPNST.

Although NF1-related and sporadic MPNST showed equivalent involvement of the *NF1* gene (58, 63, 77), some studies (86) pointed to the possibility of different genetic pathways leading to MPNST. In NF1-associated MPNST, microsatellite results are consistent with somatic inactivation by LOH of the second allele of *NF1* in MPNST (71). The disagreement regarding lack of LOH of *NF1* in neurofibromas found by some workers (71, 76, 83), but found by others (79, 80, 82, 87), may be due to the type of neurofibromas examined (plexiform vs. others), because LOH of p53 may be an early event in some of these tumors destined to become MPNST. LOH of 17q (where *NF1* is located) is consistent with a somatic inactivation of the second NF1 allele in MPNST (65, 71, 72). Cytogenetic and LOH studies have shown loss of 17p in MPNST (46, 64, 75, 76). In a CGH study, gain of material at 17q24-qter was found (87A). Only a subpopulation of schwann cells in neurofibromas, most often of the plexiform type, exhibited LOH at the *NF1* gene (59), supporting the hypothesis that schwann cells are the progenitor cells of these neurofibromas.

Cell Cycle Genes and Their Proteins in MPNST

The genes and their protein products affecting the cell cycle at various stages and studied in MPNST are involved in hierarchical pathways consisting of an array of components. The specific functions (at least of some) of these components have not been entirely deciphered and their causal or contributory roles in MPNST are still in the process of being clarified. Presented here are those genes and their proteins that have been studied in MPNST. These findings underline the importance of disturbed cell cycle control as a critical step in MPNST tumorigenesis (88).

To a large extent, molecular studies in MPNST have reflected and expanded those obtained by cytogenetics, FISH, and CGH. This expansion has been primarily due to the higher sensitivity of the molecular techniques used in the studies. The bulk of these investigations have addressed the changes in NF1-associated MPNST versus those obtained in sporadic MPNST. In the former tumors, the *NF1* gene is almost invariably involved by mutations; thus, it is an initial causal event in MPNST tumorigenesis. In sporadic MPNST, changes affecting the *NF1* gene are common (approx 50%), but other genetic alterations seem to play a key role in initiating tumorigenesis. Thus, the exact genes involved remain unknown. The initial genetic event in MPNST, i.e., *NF1* inactivation, is insufficient for the full development of MPNST, which then requires a series of additional changes often (but not always) involving genes and their products operative in the control of cell cycle events. Together, the findings on cell cycle controllers indicate their frequent, but not absolute, involvement in MPNST biology and point to other possible pathways in the causation, progression, or both of MPNST tumorigenesis (23, 89–91). Examples of the results of some of these studies are presented.

TP53 Gene and Its Protein p53 in MPNST

Under normal conditions, the low detectable levels of p53 in cellular nuclei are due to its short half-life. These levels rise during cellular DNA damage and in tumors, including MPNST, and they may be due to gene mutations leading to stabilization and inactivation of p53 (92). The many functions of *TP53* and its protein product p53 are listed in Chapter 9. A global map of p53 transcription binding sites in the human genome has been published previously (92A).

Results related to *TP53* mutations, p53 expression, or both in MPNST are subject to methodological vicissitudes, aside from the variability of the histology, biology, and the state of the tumors examined. This is particularly true of the immunohistochemical evaluation of p53 status with antibodies that lack specificity and may account for some of the divergent results obtained (16, 93–95).

Thus, the p53 changes in MPNST have ranged from lack of p53 mutations to LOH in 50–75% of these tumors, both sporadic and NF1-associated (62, 63, 66, 71, 88, 96–102).

The involvement of p53 and its gene *TP53* is the most common change, next to that of *NF1*, described in MPNST (93, 100). A frequent manifestation is overexpression or accumulation of p53 in the nuclei of these tumors. In general, overexpression of p53 in MPNST, usually determined immunohistochemically and confirmed by molecular means (67), was found to be much more frequent than mutations of *TP53*, indicating other causes for the accumulation of nuclear p53 than mutations. This overexpression has been associated in a small percentage of MPNST with mutations of *TP53*, leading to the production of an abnormal p53, which apparently is not metabolized as readily as normal p53. Other possible causes for p53 accumulation may be a faulty system for its metabolic breakdown, possible excessive or abnormal mRNA produced by *TP53* resulting in accumulation of p53, or a block in the use of p53 by the cell. Other mechanisms may be responsible for stabilization and accumulation of p53 in MPNST, e.g., binding to viral or cellular oncoproteins (103).

In early studies on p53 and *TP53* in MPNST, mutations or LOH were found in the few tumors (NF1-associated) examined (75, 76, 96, 104, 105). These changes were not seen in the neurofibromas of the affected patients. Subsequent investigations on larger series of MPNST, both NF1-associated and sporadic, indicated that mutations in these tumors are relatively rare. In fact, biallelic inactivation of the *TP53* rarely contributes to MPNST development (98).

Expression of p53 was shown in 29–100% of MPNST, whereas p53 levels were low or absent in neurofibromas (71, 90, 91, 95). Using a more sensitive method than immunochemistry, i.e., Western blot, 14 of 15 MPNST were shown to have a moderate expression of p53 (67). As mentioned, a few mutations of p53 have been reported previously (71, 75, 76, 96, 106), although biallelic inactivation of p53 is rare (98). The presence of p53 overexpression or mutation lends support to the notion that p53 alterations play a role in the development of MPNST from NF1-associated neurofibromas, but their late appearance precludes their use as a predictive marker of malignant transformation (98A).

Two MTT (one sporadic and one NF1-related) showed strong nuclear reactivity for p53 and high Ki-67 indices. LOH for the p53 locus was shown in the sporadic MTT. The sporadic tumor did not contain any changes of the *NF1* gene, whereas it was lost in the NF1-related MPNST (102).

The number of firmly established *TP53* mutations in MPNST is very small, with several different sequence changes having been reported. p53 mutations and 17p deletions were found in MPNST with *NF1* mutations (76, 96).

After the involvement of *NF1* and its protein product neurofibromin, in MPNST, that of *TP53* (located at 17p13) and its protein product p53 seem to be next in frequency. In general, p53 changes in MPNST have ranged from lack of p53 mutations to LOH in 50–75% of these tumors, both sporadic and NF1-associated (62, 63, 66, 71, 73, 88, 96–102).

An example of studies showing a high incidence of *TP53* changes in MPNST is a study (71) in which *TP53* mutations coupled with LOH at the 17p13 locus were found in 43 versus 9% in sporadic versus NF1-associated MPNST; p53 overexpressions not associated with *TP53* mutations or mdm2 overexpression were present in 71 versus 25% in sporadic versus NF1-associated MPNST (71). No LOH of *TP53* was found in the neurofibromas of the MPNST of NF1 patients or in the one neurofibroma of a patient with sporadic MPNST. Reflections on the variability of *TP53* and p53 results in MPNST, for example, is the lack of mutations in one study of NF1-associated and sporadic MPNST (98) and results obtained by others (71).

The p53 reaction was found to be more common in NF1-associated MPNST versus sporadic tumors, i.e., 88 versus 43% (23). This contrasts with the findings of others whose results were opposite of the ones just cited (6, 71, 95).

In contrast to MPNST, neurofibromas, including PNF, show no accumulation of p53 (Figure 2.11) (6, 23).

In 15 neurofibromas and seven MPNST from nine individuals (seven with NF1, six of whom developed MPNST) (64), genetic alterations at nine polymorphic loci on chromosome 17 were examined. Allelic imbalance was detected only in the malignant tumors from NF1 patients (four of six MPNST). Complete LOH of 17q loci was found in three of these tumors, including loci within the *NF1* gene. Two of the malignant tumors also showed deletions of 17p. No mutations were detected within exon 5–8 of the *TP53* gene in any MPNST; none of the tumors was p53-positive. The number of chromosomes 17 present in each tumor was evaluated by FISH on interphase nuclei with a centromere-specific probe. A deviation from the disomic status of chromosome 17 was

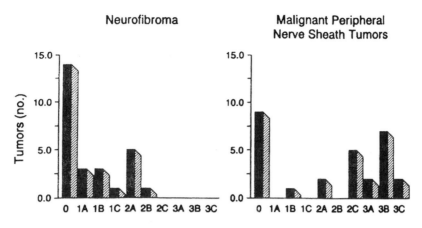

FIGURE 2.11. Results of a study on p53 in neurofibromas and MPNST (95). Intensity and extent of p53 protein immunoreactivity was graded as absent (0), weak (1), moderate (2), or strong (3). Percentage of nuclei stained was graded as A (<20%), B (20–60%), or C (>60%). The results show much higher intensity of p53 staining in MPNST vs. neurofibromas. It is not surprising that some of the molecular changes seen in NF1-related MPNST, e.g., those of *NF1*, p53, p16, and others, are seen in some neurofibromas, especially of the plexiform type, because they may serve as precursors of MPNST (28, 58, 59, 61).

observed in two of the MPNST from NF1 patients. These results support the hypothesis of inactivation of both *NF1* gene alleles during development of MPNST in patients with NF1. No evidence for a homozygous mutated condition of these tumors was found (64).

To assess the presence of p53 accumulation in MPNST and its correlation with clinical and pathologic features, 12 neurofibromas and 10 MPNST of pediatric patients were studied (6). Six MPNST were associated with NF1, all of which developed within PNF. Nuclear p53 staining detected by immunohistochemistry was present in 60% of MPNST, regardless of their origin (6). In contrast, the neurofibromas were p53 immunonegative. Rare p53-positive nuclei were detected in the transitional zone in two of six MPNST arising in PNF. Half of the NF1 patients with p53-positive MPNST developed recurrence or metastases or a second malignancy within 2 years of diagnosis, whereas patients with p53-positive sporadic MPNST were free of disease 1 to 7 years later. Mutations and accumulation of p53 were more frequent in NF1-associated MPNST (6) and may be factors accounting for the poor prognosis in these cases. In contrast to neurofibromas, there was a correlation between p53 immunoreactivity and Ki-67 expression in high-grade MPNST.

A high Ki-67 or MIB-1 proliferation index, present in almost all MPNST, p53 positivity in >60% of these tumors and changes in other genes, tended to distinguish MPNST of even low-grade from neurofibromas (6). The p53 expression in high-grade MPNST is associated with concomitant expression of the proliferation marker Mib-1, and it seems to be more frequent in NF1-related MPNST than in the sporadic type. However, the lack of significant alterations of the cell cycle regulators in PNF adjacent to high-grade MPNST, or in the majority of low-grade MPNST, indicates that these alterations are unlikely to be initial events in the malignant process.

The occurrence of *TP53* and *p16*^{INK4A} gene deregulation and the presence of microsatellite alterations at markers located at 17p, 17q, 9p21, 22q, 11q, 1p, or 2q loci in MPNST either related (14 cases) or unrelated (14 cases) to NF1 was investigated (71). The results indicate that, in MPNST, *p16*^{INK4A} inactivation almost equally affects both groups. However, *TP53* mutations and LOH involving the TP53 locus (43 vs. 9%), and p53 wild-type overexpression, related or not to mdm2 overexpression (71 vs. 25%), seem to mainly be restricted to sporadic MPNST.

LOH for all chromosome 17 polymorphisms tested was found in a MPNST (62). On the remaining chromosome 17 homolog, a 200-kb tumor-specific deletion of *NF1* was demonstrated.

Mutations of the *CDKN2A* and *TP53* genes were not seen in a study of 7 MPNST(78). Four of these tumors had LOH of either *CDKN2A* or *TP53*.

MDM2 Pathway in MPNST

MDM2 transcription is activated by functional *TP53*. *MDM2* is one of the central components of the negative regulation of p53 in cells, and it interacts with and prevents p53 from further stimulating transcription of downstream genes and targets p53 for degradation (107). MDM2 activity is inhibited by p14^{ARF}, one of the two transcripts of the *CDKN2A* locus (108–110).

Changes in *CDKN2A*, as seen in most MPNST (67), probably alter the expression of p14 and the absence of p14 would lead to high levels of MDM2. However, MDM2 is only expressed in a minority of MPNST (67, 71, 111), indicating that p14 expression is not absent in these tumors.

Immunoreactivity for MDM2 was found in 13/49 MPNST, especially in tumors of the extremities. No correlation with prognosis or nuclear p53 was found. Coexpression of p53 and

MDM2 was seen in 11 of 49 of the tumors, with no correlation to survival (16). No *MDM2* amplification was found in two MPNST studied (106). In contrast to other sarcomas (111A, 112) coexpression of p53 and MDM2 did not have a prognostic significance in MPNST (16).

Although no amplification of *MDM2* was encountered in one study (106), 33 of 49 MPNST showed expression of mdm2 by immunoreactivity (22), but they carried no prognostic value. The overexpression of MDM2/p53 was thought to carry a poor prognosis (111A,112), but others disagree (22).

The pRB Pathway in MPNST

The *p53* and *RB* tumor suppressor genes play an important role in the development of the malignant phenotype in a significant subset of both sporadic and NF1-associated MPNST, whereas alterations in the *PTEN* tumor suppressor gene seem not to play a significant role in these tumors. The disturbances in the p53 and pRb pathways resulting in the inhibition of apoptosis in MPNST should be taken into account in the development and planning of therapy for these tumors (113). These authors (113) stated that current therapy of MPNST by DNA-damaging drug-based schemes, which disregard the alterations of the genetic profiles of MPNST, could make these drugs ineffective, because they depend on apoptosis-mediated cell killing mechanisms.

There seems to be an inverse relationship between pRb and p16^{INK4A}, i.e., cells (presumably including those of MPNST) showing loss of p16^{INK4A} expression tend to retain pRb (wild-type) expression, whereas cells without pRb expression generally express wild-type p16 (114,114A). Both by immunohistochemistry (71, 90, 91, 111, 115) and molecular means (67), MPNST have been shown to express pRb, even though some of the MPNST examined showed 13q-, a chromosomal region where the *RB1* gene is located.

In a study of 12 MPNST, it was found that disruption of the *RB1* pathway was common in MPNST and that alteration of *CDKN2A* was particularly frequent (89). *CDKN2A* encodes p16^{INK4A} and p14^{ARF}, which have an impact on the *RB1* and *TP53* pathways, respectively. Homozygous deletions of exon 2 of *CDKN2A*, thus affecting p16^{INK4A} and p14^{ARF}, have been identified in MPNST (74, 90, 91). These results are in line with the finding that none of six soft tissue sarcomas with *TP53* and *RB1* alterations had mutations in *CDKN2A* (116), findings also reflected in some neuroblastomas (117, 118).

Although only a few recurrent cytogenetic changes of chromosome 9 in MPNST have been reported, particularly rearrangements of 9p, FISH studies with 9p21-p23 markers in nine familial and three sporadic MPNST (89) showed interstitial deletions that supported *CDKN2A* as a possible target gene (74, 89, 119). Nine MPNST showed aberrations of *CDKN2A* by Southern blot analyses as well, and in four of these MPST, expression of *CDKN2A* could not be detected by Northern blot analysis. Neither mutations of *CDKN2A*

were identified by sequencing of the coding region nor gene inactivation by promoter methylation. Analyses of additional markers excluded mismatch repair deficiency as an important mechanism in the genesis of these tumors. Almost all of the MPNST showed DNA changes in one or more of the cell cycle-associated genes, i.e., *CDKN2A*, *CDKN2B*, *RB1*, *CDK4*, *MDM2*, and *CCND2*. Hence, disruption of the pRB pathway is common in MPNST, and the reduction of *CDKN2A* expression is particularly frequent and probably contributes to MPNST development (89).

LOH for the *RB* gene was found (88) in a substantial proportion (44%) of MPNST, including two cases with NF1. This alteration was associated with loss of expression of the RB protein in one case; in another case, loss of pRb occurred without LOH. These findings point to the important role of *RB* gene products in MPNST tumorigenesis. In another study, loss of immunohistochemical expression of pRb in 11% of MPNST was described (90, 91); this loss of pRb was found to be significantly associated with tumor recurrence. It was demonstrated that frequent occurrence of homozygous deletions of the *CDKN2A* gene in these tumors may precede the transformation from benign neurofibroma to MPNST in NF1 (74). Because the *CDKN2A* gene and its gene product p16 are intimately linked to the *RB* pathway by preventing pRb protein phosphorylation (and subsequent inactivation allowing cell division) via inactivating cyclin-dependent kinases 4 and 6, a loss of pRb function either due to allelic or protein loss or both may further distort cell cycle deregulation resulting in highly malignant phenotype of MPNST. Thus, frequent aberrations in the *CDKN2A* and *RB* genes occur in MPNST, and they suggest that disruption of the RB pathway may be common in MPNST (89).

The phosphorylation status of the pRb protein has been shown to be controlled in part by the *PTEN* tumor suppressor gene; the *PTEN* gene is mutated in various human cancers, but no mutations have been found within the complete coding region of this gene in MPNST (88).

Even if all MPNST share a common initiating genomic event, it is likely that individual tumors may progress via alterations along several different molecular pathways. This is reflected in the heterogeneity of the cytogenetic and molecular changes among MPNST, for example, in the EGFR positivity in only 35% of tumors (both NF1-related and sporadic) (28, 77). The molecular diversity within MPNST seems to characterize several subgroups (77). Thus, a small subset of MPNST of nonaggressive nature expresses markers of neurological and schwann cell differentiation, e.g., *S-100* and *L1CAM*, while lacking expression of *Ki67*, *CCNB1*, and *CCNB2* (77).

Inactivation of p16^{INK4A} is sufficient for abnormal stimulation of the cell cycle and MPNST tumorigenesis, and altered expression of other central components of the cell cycle is usually not necessary for the development of MPNST (67). Thus, only 2/15 MPNST examined (67) had p16 expression, as shown by Western blotting, with NF1-related and sporadic tumors showing equally the lack of p16 expression. This is in agreement with previous studies based on other methodologies (23,71,74). Apparently, methylation of p16 does not seem to be a mechanism for its inactivation in MPNST (67, 89). Neither has methylation of the *CDKN2A* locus been observed in the 12 MPNST studied (89). Although methylation of the genes of p14 and p16 was observed in a few MPNST (113), it would seem that such methylation is not a common mechanism for p14 and p16 inactivation. The inactivation of p16 is associated with gross gene alterations of the *CDKN2A* locus (67), including homozygous deletions in 15% of MPNST (74, 89–91). These findings were confirmed by I-FISH and PCR (28, 113). Because neurofibromas, including the plexiform type, express p16, the alterations of p16 seen in MPNST must play a role in malignant transformation.

Inactivation of p16^{INK4A} was found equally in NF1-associated and sporadic MPNST (71). Hemizygous or homozygous p16 deletion was present in 75% of MPNST, but it was not observed in neurofibromas (28).

Studies on MPNST of p27, p53, pRb, CD1, p21, cyclin D, p15, p16, p14, *CDKN2A*, *INK4A*, and *INK4B* have been presented previously (23,67,68,71,88–91,114,119,120).

Studies on genes or their products involved in cell cycle control (p53, mdm2, pRb, p16, cyclin D, and cdk4) in series of soft tissue sarcomas (STS), including MPNST, showed significant expression of cdk4 (93%), none for p16 and significant levels for the others (approx 30%) (111). In a previous study, all six MPNST examined showed p53 nuclear accumulation (93).

In another study (113), p14, p15, and p16 (all located at 9p21), and all cell cycle regulators were frequently inactivated in MPNST (NF1-assoiated and sporadic) and led to impaired (through LOH or promoter methylation) apoptosis. The close relation of p14^{ARF} and *TP53*, and of p15^{INK4B} and p16^{INK4A} and *Rb*, point to the key role of the inactivation of the p53 and pRb pathways in the pathogenesis of MPNST (113). Similar observations and conclusions have been made in other studies of a similar nature (67,71,88).

Frequent alterations of cell cycle regulators p53, p16, and p27 are present in high-grade MPNST (23). These alterations plus decreased S-100 protein, Leu 7, and CD34 immunoreactivity patterns, and increased proliferation markers Mib-1 and TopoIIα in high-grade MPNST, support the concept that these altered cell cycle regulators influence tumor progression in the PNF to MPNST sequence. However, the lack of significant alterations of these cell cycle regulators in PNF adjacent to high-grade MPNST, or in the majority of low-grade MPNST, indicates that these alterations are unlikely to be early events in malignant transformation. Epithelial growth factor receptor (EGFR), which regulates the cell cycle through both pRB-dependent and pRB-independent pathways, also might play an important role in NF1 tumorigenesis and schwann cell transformation (121). The question remains as to whether there are as yet unidentified genetic events that initiate progression from PNF to MPNST. A study provided genetic evidence that the *NF1* haploinsufficient state of the somatic tissues surrounding peripheral nerve heath tumors, including fibroblasts and mast cells, provides a functional contribution to tumor formation, either through initiation or progression of tumorigenesis (23).

INK4A, p16, and p19 in MPNST

The *INK4A* gene, a candidate tumor suppressor gene located on 9p21, encodes two protein products, p16^{INK4A} and p19^{ARF}; p16 is a negative cell cycle regulator capable of arresting cells in the G1 phase by inhibiting cyclin-dependent kinases CDK 4 and CDK6, thereby preventing pRb phosphorylation, and p19^{ARF} prevents mdm2-mediated neutralization of p53. Loss of *INK4A* is a frequent molecular alteration involved in the genesis of a number of neoplasms, including tumors of neuroectodermal origin, and it has been observed in 60–75% of MPNST (both NF1-related and sporadic) (28, 71, 89–91, 116, 119). The absence of cyclin-dependent kinase inhibitor of p16^{INK4A} and alterations in the *CDKN2A* gene have been described in NF1-related MPNST (67). The absence of p16 activates the cell cycle in most MPNST (67). Normal expression of *RB1*, *TP53*, cyclin D3, *CDK2*, *CDK4*, p21^{CIP1}, and p27^{KIP1} were found in these tumors. Thus, *INK4A* deletions are frequent events in MPNST, and they may participate in tumor progression. Silencing of p16 by methylation, which occurs often in several tumor types, was uncommon in MPNST (90,91).

Deletions of the gene *INK4A*, whose protein products are p16^{INK4A} and p19^{ARF}, located at 9p21, of a homozygous nature have been reported in MPNST (91, 116). These deletions involved exon 2 of the gene, thus affecting both p16 and p19, and they were present in most of the tumor samples, both NF1-related and sporadic. The elimination of p16 and p19 activities leads to a curtailing of both pRb and p53 pathways and a role in MPNST tumorigenesis (91). These changes, in concert with those of p27 underexpression and cyclin-E overexpression, facilitate G1- and S-phase progression in the cell cycle and participate in the progression from neurofibroma to MPNST (91).

Homozygous deletions of the *CDKN2A* gene, which encodes the p16 cell cycle inhibitory molecules, occur in the progression of neurofibromas to MPNST, being found in 50% of MPNST but not in neurofibromas (74). Mutations or methylation of this gene were not observed in three studies (74, 90, 91). It seems likely that additional oncogenes and tumor suppressor genes probably play a role in malignant transformation of neurofibromas. The karyotypic complexity seen in MPNST lends support to this notion (64). Other studies have shown that alterations of the *CDKN2A* gene do not contribute to the oncogenesis of the majority of soft tissue tumors (STS) (122).

EGFR in MPNST

The epidermal growth factor receptor (EGFR), which regulates the cell cycle through both pRb-dependent and pRb-independent pathways, may play an important role in NF1 tumorigenesis and schwann cell transformation (121,122A,BC). The question remains as to whether there are yet unidentified genetic events that initiate progression from PNF to MPNST. A study provided genetic evidence that the *NF1* haplo-insufficient state of the somatic tissue surrounding peripheral nerve sheath tumors, including fibroblasts and mast cells, provides a functional contribution to tumor formation, either through initiation or progression of tumorigenesis (123). Studies are needed on genetic factors that, if detected, could provide insight into critical events that initiate tumor progression and malignant transformation. In the meantime, histologic criteria, although imperfect, remain the basis for clinically distinguishing PNF from MPNST, and determination of p53 reactivity may help to discriminate low-grade and high-grade MPNST in a small biopsy (23).

EGFR expression is associated with the development of the schwann cell-derived tumors characteristic of NF1 and in animal models of this disease (121). This is somewhat of a paradox, since schwann cells normally lack EGFR and respond to ligands other than epidermal growth factor (EGF). Nevertheless, immunoblotting, Northern analysis, and immunohistochemistry revealed that each of three MPNST cell lines from NF1 patients expressed EGFR, as did seven of seven sporadic MPNST, a non-NF1 MPNST cell line, and the S-100 cells from each of nine benign neurofibromas. All of the cells or cell lines expressing EGFR responded to EGF by activation of downstream signaling pathways. Thus, EGFR expression may play an important role in NF1-related MPNST.

Cyclins-D in MPNST

Cyclin-D1 (a labile factor) was weakly expressed in 11/15 MPNST by Western blot (67). Conversely, cyclin-D3 was highly expressed in all 15 MPNST examined. Findings with immunochemistry with antibodies to cyclin-D1 were similar to those of the molecular studies mentioned above (90, 91, 111). Cyclin-D2 was found to be amplified in 1/15 MPNST (89). No amplification of cyclins D1 and D3 was found. This may possibly be related to lack of specificity and sensitivity of the antibodies used (67).

D3 is the most widely expressed cyclin. It may have roles other than cell cycle regulation, e.g., differentiation (124).

CDK4 and CDK2 Proteins in MPNST

Using Western blot, CDK4 and CDK2 proteins were expressed in 15 of 15 MPNST (67). Similar findings had been reported by others (111). In contrast, another study detected CDK4 expression in only one of 26 MPNST (71). An overexpression of CDK2 or CDK4 could be an indicator of an accelerated cell cycle. In addition to their association with corresponding cyclins, these kinases need the proper phosphorylations and dephosphorylations at conserved threonine residues to accelerate the cell cycle (125). Compared with CDK2, amplification of CDK4 is rare in MPNST (89). Overexpression or amplification of CDK2 and CDK4 may not be necessary for tumorigenesis if p16 is nonfunctional (67).

Deletions of the *CDKN2A* gene (74, 90, 91) together with alterations of the expression of p27 (KIP1) (90) are additional evidence of an altered cell cycle control of MPNST (88). Furthermore, the close relationship in the activities between p53 and p14^{ARF} and between p15^{INK4B}/p16^{INK4A} and pRb is a further indication of the interactions of the various pathways in cell regulation.

CDK Inhibitors p21^{CIP1} and p27^{KIP1} in MPNST

Both p21 and p27 were found to be highly expressed in all 15 MPNST examined by Western blot (67). It has been suggested that p27 is involved in malignant transformation of neurofibromas, based on the observation of decreased immunostaining of nuclear p27 in high-grade MPNST versus neurofibromas, and a strong cytoplasmic p27 staining in the MPNST (23, 90, 91). Increased expression of p27 may protect cells from untoward effects of high levels of cyclins E and D1 (126); high levels have been found for p27 and cyclin E in MPNST (90, 91). These events are a reflection of the inhibition of cyclins D, E, and A exhibited by p21^{CIP1} and p27^{KIP1} (62, 65, 79).

Because p27 is sequestered in the cytoplasm, it cannot inhibit cyclin E/CDK2 located in the nucleus of the cancer cells (126). Neurofibromas express nuclear p27 and lack cyclin E; MPNST have no nuclear p27, acquire cytoplasmic p27 expression and high levels of cyclin E. Thus, dysregulation of the G1 transition occurs and contributes to the neurofibroma to MPNST progression. The presence of cytoplasmic p27 is associated with a poor prognosis (90, 91).

Transcription of p21 is activated by p53 in response to cellular DNA damage or stress (114A). Immunohistochemically, 16/35 MPNST were found to express p21 (90, 91). Using Western blot, 15 of 15 MPNST expressed p21, eight of these tumors showing a moderate to high expression (67). Some p21 is necessary for the association of CDK4 and cyclin D and target the complex to the nucleus; high p21 activity inhibits the CDK4/cyclin D activity and arrests the cell cycle (127). Levels of expression of p53, p16, and p27 are not reliable indices of MPNST progression (23), nor is the expression of p21^{WAF1}/cip1 (22,68). However, p27 expression (cytoplasmic) was found to be associated with a poor prognosis of MPNST (91).

TABLE 2.3. Immunohistochemical expression of cell cycle regulators and Ki-67 indices in MPNST (35 tumors) and neurofibromas (16 tumors). Data shown as percentages of tumors involved.[a]

Parameter examined	p53	pRb	p21	p27N	p27C	Cyclin D	Cyclin E	Ki-67
MPNST	29	89	46	9	54	29	42	59
NF	0	94	69	94	75	31	31	0

N, nuclear expression; C, cytoplasmic expression.
[a]Based on Kourea et al. (91).

Increased p53 expression and decreased expression of p16 and p27 may be involved in progression of PNF to MPNST (23). No differences in the expression of p16 (located at 9p21) and p27 (located at 12p13) were seen in NF1-related versus sporadic MPNST (23, 71). Findings related to some of the pathways discussed above are shown in Table 2.3 for one study in which neurofibromas and MPNST were analyzed (91). Table 2.4 lists genetic parameters related to MPNST that either have not been explored or whose role in MPNST biology is still uncertain.

Gene Expression in MPNST

Gene expression studies based on cDNA arrays in MPNST showed these tumors to differ significantly from neurofibromas; of the latter, PNF, the precursor for MPNST, had a pattern that was distinctly different from that of dermal neurofibromas (128). The gene expression profiles of NF1-related and sporadic MPNST were similar. In another expression profiling study of MPNST (77), by using several previously reported statistical approaches, the authors were unable to identify a molecular signature for distinguishing NF1-asasociated from sporadic MPNST. However, based on a somewhat different approach, nine of 42 tumors showed a relative overexpression of transcripts associated with differentiation and down-regulation of proliferation and growth factor-associated transcripts. All of these nine tumors lacked expression of EGFR, and they were mostly from NF1 patients. No correlation was found with a number of histologic and clinical parameters (77).

It was shown (77A) that although MPNST cell lines are heterogeneous in cellular growth, a common gene expression profile in MPNST cell lines and primary MPNST distinguishes them from normal schwann cells. The MPNST cell lines thus provide a valuable resource for generating in vitro evidence to support preclinical and clinical trials. Overexpression of *TWIST1* is constant across all tested MPNST cell lines and primary tumors, and it is necessary for cell migration. Further investigation into the functional roles of *TWIST1* and its transcriptional targets may help to uncover novel biomarkers or drug targets for improved diagnosis and treatment of MPNST.

Changes at 9p21 Locus in MPNST

A significant percentage (40%) of MPNST has been found to show changes at 9p cytogenetically (53); chromosomal instability played a more important role than microsatellite instability in MPNST. The inactivation of the 9p21 chromosomal locus is a frequent and unique hallmark of the genetic profile of MPNST, both NF1-related and sporadic (46, 89, 113, 129). Such inactivation usually affects the p14^{ARF} and p15 ^{NK4B}/p16^{INK4A} pathways, with the former having a close relationship with p53 and the latter with pRb (113). Inactivation of the 9p21-related pathways occurs in 75% of MPNST, equally distributed among NF1-related and sporadic tumors. This inactivation of 9p21 is achieved through homozygous deletions (46% of tumors) and promoter methylation (18% of tumors) (113).

PCR analysis (113) indicated that the inactivation of 9p21 gene cluster in 77% of MPNST represents the most frequent hallmark of NF1-related and sporadic MPNST gene profile. This inactivation is mainly achieved through homozygous deletion (46%) and to a lesser extent through promoter methylation (18%), and it is additionally supported by loss of expression (80%) of one or more of the three tumor suppressor genes *p15^{INK4b}*, *p14^{ARF}*, and *p16^{INK4a}* located at 9p21 (113). Because of the close relationship between *p14^{ARF}* and *TP53*, and between *p15 ^{INK4b}*, *p16^{INK4a}*, and *pRb*, the DNA analysis results and the mRNA expression data strongly support a coinactivation of the *TP53* and *Rb* pathways in these tumors (75%). A failure of a balanced and coordinated function of p53 and pRb in cell growth control and apoptosis is thus expected in most MPNST and, in particular, in *TP53* mutation carrying sporadic MPNST (71).

Using real-time quantitative RT-PCR, 14 plexiform NF and nine MPNST were examined for expression of 489 genes (113A). The expression of 28 (5.7%) of the genes was significantly different between MPNST and plexiform neurofibromas; 16 genes were up-regulated and 12 were down-regulated in MPNST. The altered genes were mainly involved in cell proliferation, senescence, apoptosis, and extracellular matrix remodeling. Other genes were involved in the Ras signaling pathway and the Hedgehog-Gli signaling pathway. Several of the down-regulated genes were schwann cell- or mast cell-specific, pointing to a depletion, dedifferentiation, or both of schwann and mast cells during malignant transformation of plexiform neurofibromas (113A).

It was shown that although MPNST cell lines are heterogeneous in cellular growth, a common gene expression profile in these cell lines and primary MPNST distinguishes them from normal schwann cells (113B). Overexpression of *TWIST1* was constant for all tested cell lines and primary MPNST and seems to be necessary for cell migration (113B).

TABLE 2.4. Continued

	References
NF1 patients with subcutaneous neurofibromas are three times more likely to have internal plexiform neurofibromas or MPNST than patients without subcutaneous tumors. Also, individuals with internal plexiform subcutaneous neurofibromas were 20 times more likely to have MPNST than those without such neurofibromas.	*Tucker et al 2005* (223)
Examination of the lymphocyte DNA of patients with MPNST and NF1 revealed an array of *NF1* gene mutations, which did not bear a relationship between the mutation type (nonsense, missense, frameshift, splice site mutation and multi-exonic deletions), gene location of the mutations and the clinical manifestation of MPNST.	*Upadhyaya et al 2006* (213)
Genes located at distal 1p may play a role in MPNST.	*Van Roy et al 2003* (181)
Ki-67 proliferation index is a reliable marker for progression in MPNST.	*Watanabe et al 2001***A** (22)
Survival after diagnosis of MPNST is less than 5 years.	*Ducatman et al 1986* (166)
	White 1971 (166A)
	Ghosh et al 1973 (166B))
	Wong et al 1998 (166C)
No specific germline mutations of *NF1* were found in patients with MPNST, though large deletions were more frequently found in this group than in others.	*Wu et al 1999* (224)
Tumor cell line from a spontaneous MPNST in rats.	*Yamate et al 2003* (190)

Growth Factors and Their Receptors in MPNST

Hepatocyte Growth Factor Activator (HGFA), c-Met, and CD44

It seems from the findings presented above that TSG mutations and alterations of oncogenes play an important role in MPNST pathogenesis; nevertheless, it is likely that dysregulated signaling by as yet unidentified growth factors and their receptors also contribute to the formation of these tumors. Some of these changes will be presented in the following paragraphs.

Although numerous genes have been implicated in aberrant cell growth and metastasis in MPNST (6, 36, 74, 76, 90, 91, 95, 121, 129, 130), their specific contribution to MPNST progression is not clear. The combined expression of two such genes, the receptor tyrosine kinase c-Met and its ligand, HGF, also called scatter factor, has been implicated in the growth and metastasis of many cancers (131). Normal rodent schwann cells express c-Met but not HGF; also, HGF promotes schwann cell mitogenesis (131). However, HGF and c-Met are concomitantly expressed in MPNST (49), in neurofibromas (16, 49, 133), and in schwannomas (134). These findings suggest that HGF and c-Met together form an autocrine signaling loop that promotes the phenotypes of MPNST and other peripheral nerve sheath tumors. This autocrine loop does not contribute to cell proliferation, but it does contribute to cell invasion (metastases) in MPNST (135).

HGF is a heparin-binding protein that is normally produced by a variety of mesenchymal cells, and it acts as a paracrine mitogen for epithelial cells that express c-Met (135A). Cells secrete HGF as an inactive zymogen that is activated by proteolytic cleavage at residues Arg494-Val495 (136, 136A, 137). A specific serine protease (urokinase) that induces HGF activation, called HGFA, was purified from serum (137–139).

Thus, tumor cells with HGF/c-Met autocrine activity would need either to express HGFA or to be in the vicinity of cells that express HGFA or a related protease. In this regard, the activation of c-Met in colorectal carcinomas seems to depend on the aberrant expression of HGFA by the carcinoma cells themselves (140). Eight neurofibromas were negative for HGF. Reactivity for c-Met was higher in MPNST than in neurofibromas, and it may help to distinguish low-grade MPNST from atypical neurofibromas (49). Expression of CD44 is also tied to MPNST cell invasion (141). Because CD44 (a transmembrane glycoprotein) does not influence c-Met phosphorylation (135), other mechanisms for the CD44 effects probably exist. In contrast to CD44, an HGF-neutralizing antibody that blocked c-Met phosphorylation almost completely blocked cell invasion in MPNST (135). These results differ from those of previous studies (142, 143). At present, the influence on and inactivation of CD44 with c-Met in MPNST cells are not clear.

HGF, c-Met, and CD44 are coexpressed in MPNST, but their localization did not correlate with increased cell proliferation (135). A MPNST cell line that expressed all of these proteins, converted pro-HGF to active HGF, and it exhibited constitutive c-Met phosphorylation. Blocking c-Met activity or expression inhibits the invasive behavior of these cells but not their proliferation. Although a CD44 splice variant contributes to MPNST cell invasion and interacts with c-Met and HGF, it is not required for c-Met activation. These data indicate that the HGF/c-Met autocrine loop can promote MPNST invasion through a CD44-independent mechanism and suggest that c-Met, HGFA, and HGF are potential molecular targets for agents to inhibit MPNST metastasis (135).

The cellular effects of HGF are dependent on its binding to the transmembrane tyrosine kinase receptor encoded by the proto-oncogene *C-MET*, which is expressed predominantly in epithelial cells. HGF is secreted by mesenchymal cells, and it has pleiotropic biological activities on several cell types.

HGF has been shown to be present in schwann cell tumors (132); eight of 14 MPNST were positive for both HGF and c-Met, possibly indicative of an autocrine-mediated signal transduction, and it may be involved in tumor progression. The HGF/c-Met autocrine loop could serve as a target for therapies aimed at preventing metastatic spread of MPNST (135). For example, a combination of monoclonal antibodies against HGF inhibited the growth of tumor lines in mice (144), and other approaches inactivating HGF and c-Met showed tumor inhibition (145).

Another protein that has been implicated in promoting c-Met activation by HGF is the CD44 transmembrane glycoprotein. CD44 plays roles in cell–cell and cell–matrix adhesion, activation of high-affinity growth factor receptors, and tumor growth and metastasis (146). Several signaling molecules, such as c-Src (147, 148) and *ras* (149) have been shown to be part of the HGF/c-Met signaling pathways. Src kinase has no influence on MPNST cell proliferation. However, in MPNST cell lines elevated CD44 expression and cell invasion were dependent on Src kinase activity, but independent of MAPK and MEK (141). Thus, Src and CD44 could be putative targets for therapy against MPNST invasion or metastases. Src kinase inhibitors may block MPNST invasion through several pathways that include lowering CD44 expression, preventing HGF synthesis and blocking downstream signaling by c-Met (135). Tumor cells, including MPNST (130), express a large number of distinct CD44 proteins as a result of alternative RNA splicing. It is possible that HGHF/c-Met autocrine signaling in tumors also may require the presence of particular CD44 splice variant-encoded proteins.

Platelet-Derived Growth Factor (PDGF)

PDGF BB, and to a lesser extent fibroblast growth factor 2 (FGF2), were mitogenic for two MPNST-derived schwann cell lines, but not for a schwann cell line derived from a schwannoma (from a non-NF1 patient) or for transformed rat schwann cells (150). Levels of expression of PDGF receptors α and β, to which PDGF BB binds, were significantly increased in the two MPNST-derived cell lines compared with the non-NF1 schwann cell lines. The level of tyrosyl-phosphorylated PDGF receptor β is strongly increased upon stimulation by PDGF BB. In comparison, only modest levels of tyrosyl-phosphorylated PDGF receptor α were observed, upon stimulation by PDGF AA or PDGF BB. Hence, it would seem that PDGF AA is a weak mitogen for the MPNST-derived cells by comparison with PDGF BB. These results indicate that the mitogenic effect of PDGF BB for the MPNST-derived schwann cell lines is primarily transduced by a PDGF receptor β. New differentiation factor (NDF) β, a potent mitogen for normal schwann cells, was unable to stimulated proliferation of the transformed schwann cell lines, due to a dramatic downregulation of the erbB3 receptor. Therefore, aberrant expression of growth factor receptors by schwann cells, such as the PDGF receptors, could represent

an important step in the process leading to schwann cell hyperplasia. The activation of signal transduction pathways by PDGF-α and -β receptors has been reviewed previously (150A).

A study (151) characterized PDGF-β receptor expression levels and signal transduction pathways in NF1-related MPNST cell lines and compared these levels with the expression of PDGF-β receptors in normal human schwann cells. Based on Western blotting, PDGF-β receptor expression levels were similar in schwann cells and NF1-related MPNST cell lines. MAPK and Akt also were phosphorylated in both cell types to a similar degree in response to PDGF-β chains (PDGF BB). However, increased intracellular calcium (Ca^{2+}) levels in response to PDGF BB were observed only in the NF1-related MPNST cell lines; schwann cells did not show any increase in intracellular calcium when stimulated with PDGF BB. Calmodulin kinase I, II (CAMKII) is phosphorylated in response to PDGF BB in the NF1-related MPNST cell lines, whereas no phosphorylation of CAMKII was observed in schwann cells. The decreased growth of NF1-related MPNST cell lines after treatment with CAMKII inhibitor is consistent with the view that aberrant activation of the calcium-signaling pathway by PDGF BB contributes to the formation of MPNST in NF1 patients.

Overexpression of growth factors, their receptors, or both is thought to play an important role in cellular transformation. Neurofibromas, which fail to grow when implanted under the skin of nude mice, do grow when implanted into peripheral nerves, implicating nerve growth factors in their growth (151A). However, the growth factors and their receptors promoting neurofibroma growth, MPNST growth, or both remain unknown. It has been shown that, in contrast to normal schwann cells or non-NF1–transformed schwann cells, MPNST-derived schwann cell lines expressed high levels of c-Kit, the tyrosine kinase receptor for stem cell factor (SCF) (152,152A). Moreover, SCF was able to stimulate the proliferation of MPNST-derived cell lines. Because the effect of SCF on MPNST-derived schwann cells was only moderate, the possibility was investigated (150) that in a manner similar to what has been described in a hematopoietic system (153), SCF has a modest mitogenic effect on its own, but it may act synergistically with other factors to induce cell proliferation. Several potential co-factors for SCF were tested, including NDFβ, a member of a family of growth factors called neuregulins, which are potent mitogens for both human and rat schwann cells (154, 155,155A); PDGF-BB, and FGF2, and forskolin, an agent that increases the levels of intracellular cAMP, known to potentiate the mitogenicity of FGF2, PDGF BB, and NDFβ, by inducing increased expression of their receptors (156).

SCF had no synergistic effect on cell proliferation when combined with either NDFβ, PDGF BB, FGF2, or forskolin (150). However, PDGF BB and to a lesser extent FGF2 seemed to be strongly mitogenic on their own. NDFβ, the most potent mitogen for normal schwann cells, had no effect on MPNST-derived schwann cell proliferation. Further studies (150) revealed drastic changes in the pattern of growth factor receptors expressed by the MPNST-derived schwann cells. Results (150) revealed that growth factors, PDGF BB in particular, may play a significant role in the aberrant growth of schwann cells that leads to tumor formation in NF1.

Schwann cell lines derived from NF1-related MPNST have numerous growth-factor receptors that are aberrantly regulated. It was reported that MPNST cell lines derived from NF1 patients overexpressed c-Kit, the tyrosine kinase receptor from stem cell factor, and PDGF-β receptors (150, 152). Schwann cells derived from MPNST overexpress PDGF receptors, and they can be induced to proliferate by PDGF BB (150, 151), but not by PDGF AA (157). Schwann cells from NF1-related MPNST, and thus deficient in neurofibromin, can be induced to proliferate by the c-Kit ligand (152). In addition, EGFR are highly expressed in NF1-related MPNST cell lines (28, 121). Furthermore, it has been reported that transforming growth factor (TGF)-β receptors and HGFA levels were elevated in cells derived from NF1-related MPNST (16).

In addition to aberrant growth factor receptor expression, the gain-of-function of several signal transduction pathways has been linked with the conversion of benign neurofibromas to MPNST. In some cases, these activated pathways are Ras-dependent. For example, one study (141) suggested that hyperactivated Ras increases CD44 expression, which contributes to the invasive behavior of MPNST. Other pathways involved in the development of MPNST are Ras-independent. A study demonstrated that cells derived form NF1-related MPNST signal through the Notch pathway, contributing to the lethal phenotype of the tumors (158, 159). Other studies indicated that increased activities of the G protein-stimulated adenylyl cyclase and cAMP pathway might also contribute to the progression of the disease (160–162). Clearly, several pathways are abnormally activated and contribute to the formation of NF1-related MPNST.

Topoisomerase-II (TopoII)

TopoIIα is a nuclear enzyme whose activity has been linked with cellular dedifferentiation and a potentially aggressive tumor phenotype in epithelial neoplasms (23). TopoIIα has been shown to be present in 16 of 27 MPNST, with all 16 positive tumors being high grade; although negative results may be seen in an occasional high-grade MPNST, such negativity is usually associated with low-grade tumors. The MIB-1 results paralleled those of TopoIIα.

Apolipoprotein D (ApoD)

ApoD, a member of the lipocalin superfamily of protein expression, known to be associated with cell cycle inhibi-

tion, increases as neurofibromas are formed, followed by a marked decline in neurofibromas undergoing malignant transformation into MPNST. These changes coincide with those in cell cycle inhibition mediated by cyclin-dependent kinase inhibitors (163). The specific mechanisms remain to be elucidated.

Protein Gene Product 9.5 (PGP9.5)

PGP9.5 is a broad neural marker expressed in nerve fibers and neurons of both peripheral nerves and central nervous system. It is a more sensitive marker than S-100 for MPNST (94 vs. 38%), including epithelioid and conventional MPNST ("spindle-cell sarcoma") (164).

Checkpoint with Forkhead-Associated Domain and Ring Finger (CHFR)

CHFR has been identified as defining an early mitotic checkpoint (164A). Abnormalities of CHFR and related substances may result in mitotic abnormalities leading to chromosomal instability and aneuploidy. An investigation of CHFR levels in 96 MPNST showed 66% to have low levels, which correlated with a high mitotic count; a high Ki-67 labeling index; and a poor prognosis (164A). MPNST with a normal karyotype showed strong CHFR activity.

Neuregulin-1 (NRG-1)

It is hypothesized that NRG-1 family of growth and differentiation factors contribute to the pathogenesis of peripheral nerve sheath tumors. NRG-1 proteins stimulate the proliferation, survival, and migration of schwann cells, and activating mutations of the NRG-1 receptor subunit ErbB2 promote MPNST formation (164B).

Malignant Triton Tumors

MTT is the designation for MPNST with rhabdomyosarcomatous differentiation (3, 4, 165, 166). The name is derived from triton salamander, in which development of both myoid and neural elements from transplanted sciatic nerve were experimentally demonstrated.

Triton tumors are clinically similar to MPNST, except that they are more often associated with NF1 (60–70%). MTT also occur at a younger age (mean, 33 years) when associated with NF1 compared with patients without NF1 (mean, 42 years). MTT present a variety of locations, with the head and neck and trunk being among the most common. Clinicopathologic series show a highly malignant behavior, with the 5-year survival being 10–29%.

Characteristic of MTT tumor is focal rhabdomyosarcomatous differentiation occurring in the background of a high-grade spindle-cell or pleomorphic MPNST. Well-differentiated rhabdomyoblastic components are scattered between the spindle-cells or seen in clusters. MTT are positive for desmin, myoglobin, and MyoD1. S-100 protein positivity is variable, and remnants of neurofibroma can often be demonstrated. Electron microscopy is also useful in verifying the skeletal muscle differentiation.

In MTT arising outside of the clinical setting of NF1, other genetic mechanisms may be operative. A molecular analysis of a non-NF1 triton tumor revealed retention of *NF1* gene expression, suggesting that loss of *NF1* expression is not required for the development of this tumor (102). Unfortunately, LOH analysis was uninformative; therefore, allelic loss of *NF1* expression could not be shown (102). NF1 protein was identified in protein extracts form this tumor (102), complementing the *NF1* mRNA findings. As with other MPNST, mutant *p53* accumulation likely heralds or is associated with aggressive MTT.

Nuclear accumulation of p53 protein and Ki-67 immunoreactivity in two MTT (one not associated with NF1) was noted (102). This abnormal expression of p53 protein, confirmed by p53 LOH in a non-NF1 tumor, is in agreement with findings in other histologic types of MPNST. Ki-67 immunoreactivity in both MTT in parallel with *p53* nuclear staining further supports the proposed association of *p53* mutations with increased cellular proliferation in vivo. As with other MPNST, mutant *p53* accumulation likely heralds or is associated with aggressive tumors.

Cytogenetic studies of MTT (Table 2.5; Figures 2.12–2.14) have revealed complex karyotypes in most cases, akin to those already presented for MPNST, varying from case to case without recurrent changes. However, it has been suggested (167) that genes located at 7p22 and 11p15 and perhaps those at 7q36, 12p13, 13p11.2, 17q11.2, and 19p13.1 may play a role in a stepwise process of involvement of a succession of these genes in the genesis of MTT. MTT and p53-positive MPNST in children were high-grade tumors (6).

TABLE 2.5. Chromosome changes in MTT.

Karyotype	References
(1;2)(p13;p25),t(3;14)(p12;q31),ins(4;?5)(q28;?q22q32),-5,add(9)(p24),add(15)(p26),add(22)(q13),+mar1,+mar2/47,XX,add(6)(p25),t(7;8)(p11;p11),t(12;21)(q21;q21),+22/47,XX,t(2;10)(q31;p15),t[(3,(der7)][p13;q36]],der(7)t(7;?7)(p22;?q?)+mar	*Travis et al 1994* (191)
63,XX,-X,+dic(1;22)(p36;p10),+del(2)(q33),t(2;?12)(p25;?),-3,add(4)(p34),+5,dic(7;?11)(p22;?),-10,-11,der(110t(5;11)(q11;q23),-12,-12,-14,add(14)(p11),-15,-16,add(16)(p13),-17,-18,-19,-20,del(20)(q13),?add(21)(q22),+mar1,+mar2,+mar3,+mar4, +2-3dmin	*Travis et al 1994* (191)
*64,XX,del(X)(q22),+del(X)(q22),-1,-2,der(5)t(5;?9)(p13;?q13),+7,-9,+10,-11,-12,-16,-17, -18,+mar1x2	*McComb et al 1996* (176)
*69,XX,-X,del(1)(p21),-3,der(5)t(5;?9)(p13;q13),+der(5)t(5;?9)(p13;?q13),+der(5)t(5;?9)(p13;?q13),+7,+8,+8,-9,-9,-9,-10,-10,dup(11)(p11.2p15),del(12)(q15),del(12)(q15), -13,-14,-16,-17,-18,+21,+22,+mar1x2,+mar2,+mar3,+3mar	*McComb et al 1996* (176)
59-60,XX,t(1;13)(q21;q22),add(1)(q21),der(2)t(2;17)(q36;q11)x2,+2,+6,-8,-8,-9,+der(11;13)(q10;q10),+12,+16,+add(16)(p13),-17,-18,+20,+20,+21,+21,-22,+1-4mar	*Pauwels et al 1999* (178)
*59-72,X,-X,add(X)(q23),-1,add(1)(p11),add(1)(p23),del(1)(q21),der(1)t(1;22)(p13;q11),add(2)(q21),add(2)(q31),+del(2)(q14-q21),+3,-4,-4,-4,+5,der(6)t(6;17)(p11;q21),+der(6)t(6;17)(p11;q21),+add(7)(p22)x2,+add(8)(p11),+i(8)(q10),+der(9)t(1;9)(q12;p13),-10,add(10)(p12),+add(10)(p12),-11,der(12)t(4;12)(q13;p12),+der(12)t(4;12)(q13;p12),+13,-14,-14,+15,+15,+add(16)(q23),-17,-17,+18,-19,-20,add(20)(p12),-21,-21,-22,-22,-22,+der(?)t(?;1)(?;p22),+1-5mar	*Plaat et al 1999* (43)
46,XX,i(8)(q10),der(13)t(1;13)(q10;q10)	*Hennig et al 2000* (192)
48-55,XY,der(7)add,7(p?)dup(7),der(7)t(7;20)(p22;?)ins(20;19),der(7)ins(8;7)(?;p22q36)t(3;8)t(8;20),-8,-8,r(8)dup(8),+der(8)r(8;22),-9, der(11)t(11;20)(p15;?)ins(20;19)der(12)t(8;12)(q21;p13),der(13)t(13;13)(q25;p11),-17,-19,der(19)t(17;19)(q11.2;q13.1) -20,-22,+4-7r	*Haddadin et al 2003* (167)
44,XX,-2,-3,del(3)(p13),-4,add(6)(q27),der(7)t(7;?7)(p22;?q?),-10,der(12)t(12;21)(q24;q21),add(13)(q34),add(15)(q26),-16,-17,-21,+mar1,+mar2,+mar3,+mar4,+mar5/45,XX,-2,-3,del(3)(p13),ins(4;?5)(q28;?q22q32),add(6)(q27),der(7)t(7;?7)(p22;?q?),add(13)(q34),add(15)(q26),-16,-17,-21,+mar1,+mar2,+mar3,+mar4	*Bridge et al 2004* (45)
64,XX,-X,+1,der(1)t(1;15)(p11;q11.2)x2,+2,-3,-4,+5,-6,+7,+8,-9,-10,-11,-12,-13,-14,-15,-15,-15,-16,+17,+18,+20,+21	*Bridge et al 2004* (45)
*49,XY,der(14;15)(q10q10),+i(8q)x4	*Magrini et al 2004* (193)
46,XY,t(7;9)(q11.2),der(16)t(1;16)(q23;q13)	*Velagaleti et al 2004* (194)

*NF1-associated..

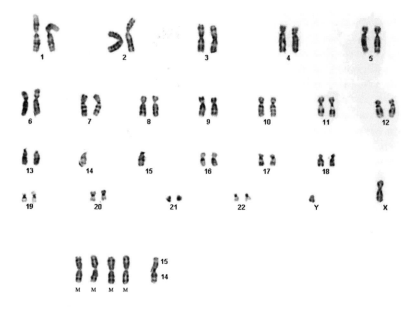

FIGURE 2.12. Karyotype of a malignant triton tumor (MTT) with four i(8q) chromosomes as an outstanding anomaly (193). A t(14;15) was also present. However, no consistent or characteristic cytogenetic anomaly has been found to characterize MTT.

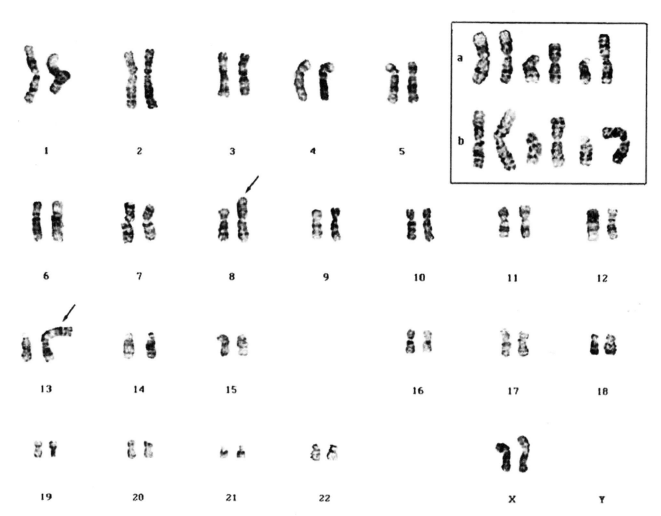

FIGURE 2.13. Pseudodiploid karyotype with 46 chromosomes in a malignant triton tumor (MTT) with an i(8q)(q10) and der(13)t(1;13)(q10;q10). Inset shows enlarged normal and derivative chromosomes 1, 8 and 13 in this MTT (192).

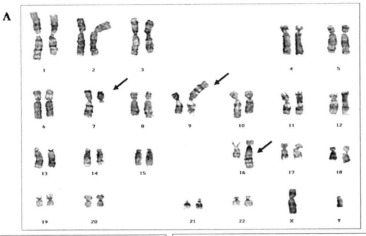

FIGURE 2.14. Karyotype (A) of a malignant triton tumor (MTT) (194). Shown is at t(7;9)(q11.2;q24) and an unbalanced der(16)t(1;16)(q23;q13). In (B) is shown a partial karyotype of the derivative chromosome 16 with breakpoints and in (C) a partial karyotype of the balanced t(7;9) with the breakpoints.

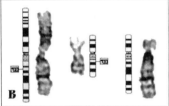

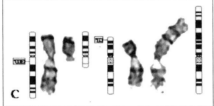

Summarizing Comments: Neurofibromas, *NF1* Gene, Schwann Cells, and MPNST/MTT

Neurofibromin is a Ras-GTPase-activating protein, i.e., absence of neurofibromin leads to increased levels of Ras-GTPase, these being high in MPNST, moderate in NF1-related neurofibromas, and absent in non-NF1 schwannomas. Thus, *RAS* plays an important negative regulatory role in the pathogenesis of peripheral nerve tumors (70).

Inactivation of tumor-related genes is an essential mechanism of tumorigenesis and most likely includes the malignant transformation of neurofibromas to MPNST. Such inactivation, in combination with aberrant growth factor recep-

TABLE 2.6. Some features of and studies on schwann cells.

	References
Factors influencing schwann cell development and apoptosis include NRG, TGF-β, FGF, PDGF, and insulin-like growth factor-I, MAPK, and a kinase of MAPK (MEK). Schwann cells express receptors for these trophic factors.	*Badache et al 1998* (152)
PDGF and FGF are mitogenic in rat schwann cells.	*Davis and Stroobant 1990* (195)
Insulin-like growth factor-I (IGF-I), rescues schwann cells from apoptosis via PI3 kinase.	*Delaney et al 1999* (196)
Glial growth factor (GGF), one of the products of the neuregulin gene, is mitogenic for schwann cells. Neuregulin and its receptors regulate schwann cell apoptosis.	*Grinspan et al 1996* (197)
Based on studies in neurofibromin-deficient mice, though Ras-GTP is regulated by neurofibromin, mutations of *NF1* by itself is unlikely to explain the hyperplasia observed in schwann cell tumors in NF1.	*Kim et al 1995* (170)
Cyclin-D1 mediates response of schwann cells to the proliferative effects of cAMP and *NF1*.	*K im et al 2001* (161)
HGF is mitogenic for schwann cells and is present in neurofibromas.	*Krasnoselsky et al 1994* (132)
Forskolin and heregulin/glial growth factor (GGF) are stimulators of schwann cell growth.	*Levi et al 1995* (155)
Heregulins may affect schwann cell proliferation through p185erbB2, an EGF-like receptor tyrosine kinase that is encoded by the *erbB2* proto-oncogene.	*Levi et al 1995* (155)
	Morrissey et al 1995 (198)
Gas6, a ligand for protein tyrosine kinase receptors of the Axl/Rse family, supports schwann cell proliferation.	*Li et al 1996* (199)
Notch intracellular domain signaling may contribute to the trans- formation of neurofibromas to MPNST.	*Li et al 2004* (159)
An inhibitor of MAPK was shown to be effective against MPNST, but not against normal schwann cell proliferation.	*Mattingly et al 2006* (208)
The role and mechanisms of signaling involving erbB receptors and EGF-like ligands has been discussed.	*Pinkas-Kramarski et al 1997* (200)
Endogenous neuregulins may function in an autocrine/paracrine loop or in a juxtacrine (cell-to-cell)-mediated signal.	*Raabe et al 1996* (201)
CD44, a transmembrane glycoprotein, enhanced neuregulin Signaling in schwann cells.	*Sherman et al 2000* (202)
Activation of the neuregulin-1/ErbB signaling pathway promotes the proliferation of schwann cells in MPNST.	*Stonecypher et al 2005* (164B)

tors expression levels and increased signal transduction activities, plays critical roles in the tumor progression from benign neurofibromas to MPNST in NF1 patients. Hypermethylation, which generally occurs at CpG islands located at the promoter/exon 1 regions of tumor-related genes, results in their silencing (168), and it also may play a role in progression of neurofibroma to MPNST (169).

Nevertheless, the actual mechanisms leading to the formation of neurofibromas and MPNST remain unclear. Both epigenetic events and multiple genetic changes have been proposed to contribute to the formation of the peripheral nerve sheath tumors in NF1 (149A). Mechanisms by which benign neurofibromas transform into MPNST involve several genetic alterations. The process requires the inactivation of the *NF1* gene, but additional genetic mutational inactivations, such as of *TP53* and *INK4A* genes and other cell cycle controllers, also may be involved (71, 76, 170). However, animal models with the loss of normal *Nf1* and *p53* alleles developed soft tissue sarcomas but did not develop MPNST (117, 118, 123). These studies suggest that a combination of genetic mutations in concert with other variable abnormalities leads to the development of different types of NF1 malignancies.

Although MPNST may contain fibroblasts and mast cell (38), the tumors consist primarily of cells derived from schwann cells (31,149A). Schwann cells derived from NF1-related MPNST have numerous growth-factor receptors that are aberrantly regulated. Thus, these cells may overexpress c-Kit, the tyrosine kinase receptor for a stem-cell factor, and PDGF-β receptors (150, 152), and EGFR (28, 121). Also, the transforming growth factor-β receptors and HGF-α receptors are elevated in NF1-related schwann cells derived from MPNST (16). Schwann cells have to be reactivated from their dormant state to reenter the cell cycle on their route to MPNST development. Some of the factors and substances listed in Table 2.6 may be involved in the transformation of neurofibromas to MPNST. Table 2.6 also presents some aspects of schwann cells of relevance not only to MPNST but also to neurofibromas and meningiomas.

References

1. Lewis, J.J., and Brennan, M.F. (1996) Soft tissue sarcomas. *Curr. Prob. Surg.* **33**, 817–872.
2. Weiss, S.W., and Goldblum, J.R. (2001) *Enzinger and Weiss's Soft Tissue Tumors*, 4th ed. Mosby, St. Louis, MO.
3. Scheithauer, B.W., Woodruff, J.M., and Erlandson, R.A. (1999) *Tumors of the Peripheral Nervous System.* Armed Forces Institute of Pathology, Washington, DC.
4. Miettinen, M. (2003) MPNST. In *Pathology and Genetics of Tumours of the Nervous System* (Kleihues, P., and Cavenee, W.K., eds.), IARC Press, Lyon, France, pp. 367–370.
5. Woodruff, J.M., Selig, A.M., Crowley, K., and Allen, P.W. (1994) Schwannoma (neurilemoma) with malignant transformation. A rare, distinctive peripheral nerve tumor. *Am. J. Surg. Pathol.* **18**, 882–895.

6. Liapis, H., Marley, E.F., Lin, Y., and Dehner, L.P. (1999) p53 and Ki-67 proliferating cell nuclear antigen in benign and malignant peripheral nerve sheath tumors in children. *Pediatr. Dev. Pathol.* **2**, 377–384.
7. Wallace, M.R., Rasmussen, S.A., Lim, I.T., Gray, B.A., Zori, R.T., and Muir, D. (2000) Culture of cytogenetically abnormal Schwann cells from benign and malignant NF1 tumors. *Genes Chromosomes Cancer* **27**, 117–123.
8. Poyhonen, M., Niemela, S., and Herva, R. (1997) Risk of malignancy and death in neurofibromatosis. *Arch. Pathol. Lab. Med.* **121**, 139–143.
9. Evans, D.G.R., Baser, M.E., McGaughran, J., Sharif, S., Howard, E., and Moran, A. (2002) Malignant peripheral nerve sheath tumours in neurofibromatosis 1. *J. Med. Genet.* **39**, 311–314.
10. Woodruff, J.M. (1996) Pathology of major peripheral nerve sheath tumors. In *Soft Tissue Tumors*, International Academy of Pathology Monograph (Weiss, S.W., and Brooks, J.S.J., eds.), Williams and Wilkins, Baltimore, MD, pp. 129–161.
11. Ricci, A., Jr., Parham, D.M., Woodruff, J.M., Callihan, T., Green, A., and Erlandson, R.A. (1984) Malignant peripheral nerve sheath tumors arising from ganglioneuromas. *Am. J. Surg. Pathol.* **8**, 19–29.
12. Ducatman, B.S., Scheithauer, B.W., Piepgras, D.G., and Reiman, H.M. (1984) Malignant peripheral nerve sheath tumors in childhood. *J. Neurooncol.* **2**, 241–248.
13. Meis, J.M., Enzinger, F.M., Martz, K.L., and Neal, J.A. (1992) Malignant peripheral nerve sheath tumors (malignant schwannomas) in children. *Am. J. Surg. Pathol.* **16**, 694–707.
14. Molenaar, W.M., Ladde, B.E., Schraffordt Koops, H., and Dam-Meiring, A. (1989) Two epithelioid malignant schwannomas in a patient with neurofibromatosis. Cytology, histology and DNA-flow-cytometry. *Pathol. Res. Pract.* **184**, 529–534.
15. Sangüeza, O.P., and Requena, L. (1998) Neoplasms with neural differentiation: a review. Part II: malignant neoplasms. *Am. J. Dermatopathol.* **20**, 89–102.
16. Watanabe, T., Oda, Y., Tamiya, S., Masuda, K., and Tsuneyoshi, M. (2001) Malignant peripheral nerve sheath tumour arising within neurofibroma. An immunohistochemical analysis in the comparison between benign and malignant components. *J. Clin. Pathol.* **54**, 631–636.
17. Valdeuza, J.M., Hagel, C., Westphal, M., Hänsel, M., and Herrmann, H.-D. (1991) Primary spinal malignant schwannoma: clinical, histological and cytogenetic findings. *Neurosurg. Rev.* **14**, 283–291.
17A. Bello, M.J., De Campos, J.M., Kusak, M.E., Vaquero, J., Sarasa, J.L., Pestana, A., and Rey, J.A. (1993) Clonal chromosome aberrations in neurinomas. *Genes Chromosomes Cancer* **6**, 206–211.
18. Dorsi, M.J., and Belzberg, A.J. (2004) Paraspinal nerve sheath tumors. *Neurosurg. Clin. N. Am.* **15**, 217–222.
19. Bhattacharyya, A.K., Perrin, R., and Guha. A. (2004) Peripheral nerve tumors: management strategies and molecular insights. *J. Neurooncol.* **69**, 335–349.
20. Doorn, P.F., Molenaar, W.M., Buter, J., and Hoekstra, H.J. (1995) Malignant peripheral nerve sheath tumors in patients with and without neurofibromatosis. *Eur. J. Surg. Oncol.* **21**, 78–82.

21. O'Sullivan, M.J., Kyriakos, M., Zhu, X., Wick, M.R., Swanson, P.E., Dehner, L.P., Humphrey, P.A., and Pfeifer, J.D. (2000) Malignant peripheral nerve sheath tumors with t(X;18). A pathologic and molecular genetic study. *Mod. Pathol.* **13**, 1253–1263.

22. Watanabe, T., Oda, Y., Tamiya, S., Kinukawa, N., Masuda, K., and Tsuneyoshi, M. (2001**A**) Malignant peripheral nerve sheath tumours: high Ki67 labelling index is the significant prognostic indicator. *Histopathology* **39**, 187–197.

23. Zhou, H., Coffin, C.M., Perkins, S.L., Tripp, S.R., Liew, M., and Viskochil, D.H. (2003) Malignant peripheral nerve sheath tumor. A comparison of grade, immunophenotype, and cell cycle/growth activation marker expression in sporadic and neurofibromatosis 1-related lesions. *Am. J. Surg. Pathol.* **27**, 1337–1345.

24. Sandberg, A.A., and Bridge, J.A. (1996) *The Cytogenetics of Bone and Soft Tissue Tumors.* R.G. Landes Company, Austin/Georgetown.

25. Sandberg, A.A., and Bridge, J.A. (2002) Updates on the cytogenetics and molecular genetics of bone and soft tissue tumors. Synovial sarcoma. *Cancer Genet. Cytogenet.* **133**, 1–23.

26. Coindre, J.-M., Hostein, I., Benhattar, J., Lussan, C., Rivel, J., and Guillou, L. (2002) Malignant peripheral nerve sheath tumors are t(X;18)-negative sarcomas. Molecular analysis of 25 cases occurring in neurofibromatosis type 1 patients, using two different RT-PCR-based methods of detection. *Mod. Pathol.* **15**, 589–592.

27. Tamborini, E., Agus, V., Perrone, F., Papini, D., Romanò, R., Pasini, B., Gronchi, A., Colecchia, M., Rosai, J., Pierotti, M.A., and Pilotto, S. (2002) Lack of SYT-SSX fusion transcripts in malignant peripheral nerve sheath tumors on RT-PCR analysis of 34 archival cases. *Lab. Invest.* **82**, 609–618.

28. Perry, A., Kunz, S.N., Fuller, C.E., Banerjee, R., Marley, E.F., Liapis, H., Watson, M.A., and Gutmann, D.H. (2002) Differential *NF1*, *p16*, and *EGFR* patterns by interphase cytogenetics (FISH) in malignant peripheral nerve sheath tumor (MPNST) and morphologically similar spindle cell neoplasms. *J. Neuropathol. Exp. Neurol.* **61**, 702–709.

29. Vang, R., Biddle, D.A., Harrison, W.R., Heck, K., and Cooley, L.D. (2000) Malignant peripheral nerve sheath tumor with a t(X;18). A synovial sarcoma variant? *Arch. Pathol. Lab. Med.* **124**, 864–867.

29A. Hui, P., Li, N., Johnson, C., De Wever, I., Sciot, R., Manfioletti, G., and Tallini, G. (2005) HMGA proteins in malignant peripheral nerve sheath tumor and synovial sarcoma: preferential expression of *HMGA2* in malignant peripheral nerve sheath tumor. *Mod. Pathol.* **18**, 1519–1526.

30. Sordillo, P.P., Helson, L., Hajdu, S.I., Magill, G.B., Kosloff, C., Golbey, R.B., and Beattie, E.J. (1981) Malignant schwannoma—clinical characteristics, survival, and response to therapy. *Cancer* **47**, 2503–2509.

31. Morioka, N., Tsuchida, T., Etoh, T., Ishibashi, Y., and Otsuka, F. (1990) A case of neurofibrosarcoma associated with neurofibromatosis: light microscopic, ultrastructural, immunohistochemical and biochemical investigations. *J. Dermatol.* **17**, 312–316.

32. Hajdu, S.I. (1993) Peripheral nerve sheath tumors. Histogenesis, classification, and prognosis. *Cancer* **72**, 3549–3552.

33. Woodruff, J.M., and Christensen, W.N. (1993) Glandular peripheral nerve sheath tumors. *Cancer* **72**, 3618–3628.

34. Swanson, P.E., Scheithauer, B.W., and Wick, M.R. (1995) Peripheral nerve sheath neoplasms. Clinicopathologic and immunochemical observations. In *Pathology Annual* (Rosen, P.P., and Fechner, R.E., eds.), Appleton & Lange, Stamford, CT, pp. 1–82.

35. Ordóñez, N.G., and Tornos, C. (1997) Malignant peripheral nerve sheath tumor of the pleura with epithelial and rhabdomyoblastic differentiation. Report of a case clinically simulating mesothelioma. *Am. J. Surg. Pathol.* **21**, 1515–1521.

36. McCarron, K.F., and Goldblum, J.R. (1998) Plexiform neurofibroma with and without associated malignant peripheral nerve sheath tumor: a clinicopathologic and immunohistochemical analysis of 54 cases. *Mod. Pathol.* **11**, 612–617.

37. McMenamin, M.E., and Fletcher, C.D.M. (2001) Expanding the spectrum of malignant change in schwannomas. Epithelioid malignant change, epithelioid malignant peripheral nerve sheath tumor, and epithelioid angiosarcoma: a study of 17 cases. *Am. J. Surg. Pathol.* **25**, 13–25.

38. Takeuchi, A., and Ushigome, S. (2001) Diverse differentiation in malignant peripheral nerve sheath tumours associated with neurofibromatosis-1: an immunohistochemical and ultrastructural study. *Histopathology* **39**, 298–309.

39. Ferner, R.E., and Gutmann, D.H. (2002) International Consensus Statement on malignant peripheral nerve sheath tumors in neurofibromatosis 1. *Cancer Res.* **62**, 1573–1577.

40. Klijanienko, J., Caillaud, J.-M., Lagacé, R., and Vielh, P. (2002) Cytohistologic correlations of 24 malignant peripheral nerve sheath tumor (MPNST) in 17 patients: the Institut Curie experience. *Diagn. Cytopathol.* **27**, 103–108.

41. Perrin, R.G., and Guha, A. (2004) Malignant peripheral nerve sheath tumors. *Neurosurg. Clin. N. Am.* **15**, 203–216.

42. Kim, D.H., Murovic, J.A., Tiel, R.L., Moes, G., and Kline, D.G. (2005) A series of 397 peripheral neural sheath tumors: 30-year experience at Louisiana State University Health Sciences Center. *J. Neurosurg.* **102**, 246–255.

42A. Friedrich, R.E., Kluwe, L., Funsterer, C., and Mautner, V.F. (2005) Malignant peripheral nerve sheath tumors (MPNST) in neurofibromatosis type 1 (NF1): diagnostic findings on magnetic resonance images and mutation analysis of the *NF1* gene. *Anticancer Res.* **25**, 1699–1702.

43. Plaat, B.E.C., Molenaar, W.M., Mastik, M.F., Hoekstra, H.J., te Meerman, G.J., and van den Berg, E. (1999) Computer-assisted cytogenetic analysis of 51 malignant peripheral-nerve-sheath tumors: sporadic *vs.* neurofibromatosis-type-1-aassociated malignant schwannomas. *Int. J. Cancer* **83**, 171–178.

44. Schmidt, H., Taubert, H., Würl, P., Bache, M., Bartel, F., Holzhausen, H.-J., and Hinze, R. (2001) Cytogenetic characterization of six malignant peripheral nerve sheath tumors: comparison of karyotyping and comparative genomic hybridization. *Cancer Genet. Cytogenet.* **128**, 14–23.

45. Bridge, R.S., Jr., Bridge, J.A., Neff, J.R., Naumann, S., Althof, P., and Bruch, L.A. (2004) Recurrent chromosomal imbalances and structurally abnormal breakpoints within complex karyotypes of malignant peripheral nerve sheath tumour and malignant triton tumour: a cytogenetic and molecular cytogenetic study. *J. Clin. Pathol.* **57**, 1172–1178.

46. Mertens, F., Rydholm, A., Bauer, H.F.C., Limon, J., Nedoszytko, B., Szadowska, A., Willén, H., Heim, S., Mitelman, F., and Mandahl, N. (1995) Cytogenetic findings in malignant peripheral nerve sheath tumors. *Int. J. Cancer* **61**, 793–798.

47. Mertens, F., Dal Cin, P., De Wever, I., Fletcher, C.D.M., Mandahl, N., Mitelman, F., Rosai, J., Rydholm, A., Sciot, R., Tallini, G., van den Berghe, H., Vanni, R., and Willén, H. (2000) Cytogenetic characterization of peripheral nerve sheath tumours: a report of the CHAMP study group. *J. Pathol.* **190**, 31–38.

48. Jhanwar, S.C., Chen, Q., Li, F.P., Brennant, M.F., and Woodruff, J.M. (1994) Cytogenetic analysis of soft tissue sarcomas. Recurrent chromosome abnormalities in malignant peripheral nerve sheath tumors (MPNST). *Cancer Genet. Cytogenet.* **78**, 138–144.

49. Rao, U.N.M., Surti, U., Hoffner, L., and Yaw, K. (1996) Cytogenetic and histologic correlation of peripheral nerve sheath tumors of soft tissue. *Cancer Genet. Cytogenet.* **88**, 17–25.

50. Schmidt, H., Taubert, H., Meye, A., Würl, P., Bache, M., Bartel, F., Holzhausen, H.-J., and Hinze, R. (2000) Gains in chromosomes 7, 8q, 15q and 17q are characteristic changes in malignant but not in benign peripheral nerve sheath tumors from patients with Recklinghausen's disease. *Cancer Lett.* **155**, 181–190.

51. Fletcher, J.A., Kozakewich, H.P., Hoffer, F.A., Lage, J.M., Weidner, N., Tepper, R., Pinkus, G.S., Morton, C.C., and Corson, J.M. (1991) Diagnostic relevance of clonal cytogenetic aberrations in malignant soft-tissue tumors. *N. Engl. J. Med.* **324**, 436–443.

52. Örndal, C., Rydholm, A., Willén, H., Mitelman, F., and Mandahl, N. (1994) Cytogenetic intratumor heterogeneity in soft tissue tumors. *Cancer Genet. Cytogenet.* **78**, 127–137.

53. Kobayashi, C., Oda, Y., Takahira, T., Izumi, T., Kawaguchi, K., Yamamoto, H., Tamiya, S., Yamada, T., Oda, S., Tanaka, K., Matsuda, S., Iwamoto, Y., and Tsuneyoshi, M. (2006) Chromosomal aberrations and microsatellite instability of malignant peripheral nerve sheath tumors: a study of 10 tumors from nine patients. *Cancer Genet. Cytogenet.* **165**, 98–105.

54. Mechtersheimer, G., Otaño-Joos, M., Ohl, S., Benner, A., Lehnert, T., Willeke, F., Möller, P., Otto, H.F., Lichter, P., and Joos, S. (1999) Analysis of chromosomal imbalances in sporadic and NF1-associated peripheral nerve sheath tumors by comparative genomic hybridization. *Genes Chromosomes Cancer* **25**, 362–369.

54A. Rickert, C.H., and Paulus, W. (2004) Comparative genomic hybridization in central and peripheral nerve system tumors of childhood and adolescence. *J. Neuropathol. Exp. Neurol.* **63**, 399–417.

55. Forus, A., Olde Weghuis, D., Smeets, D., Fodstad Ø., Myklebost, O., and Geurts van Kessel, A. (1995) Comparative genomic hybridization analysis of human sarcomas: I. Occurrence of genomic imbalances and identification of a novel major amplicon at 1q21-q22 in soft tissue sarcomas. *Genes Chromosomes Cancer* **14**, 8–14.

56. Schmidt, H., Würl, P., Taubert, H., Meye, A., Bache, M., Holzhausen, H.-J., and Hinze, R. (1999) Genomic imbalances of 7p and 17q in malignant peripheral nerve sheath tumors are clinically relevant. *Genes Chromosomes Cancer* **25**, 205–211.

56A. Nakagawa, Y., Yoshida, A., Numoto, K., Kunisada, T., Wai, D., Ohata, N., Takeda, K., Kawai, A., and Ozaki, T. (2006) Chromosomal imbalances in malignant peripheral nerve sheath tumor detected by metaphase and microarray comparative genomic hybridization. *Oncol. Rep.* **15**, 297–303.

57. Koga, T., Iwasaki, H., Ishiguro, M., Matsuzaki, A., and Kikuchi, M. (2002) Frequent genomic imbalances in chromosomes 17, 19, and 22q in peripheral nerve sheath tumours detected by comparative genomic hybridization analysis. *J. Pathol.* **197**, 98–107.

58. Perry, A., Roth, K.A., Banerjee, R., Fuller, C.E., and Gutmann, D.H. (2001) *NF1* deletions in S-100 protein-positive and negative cells of sporadic and neurofibromatosis 1 (NF1)-associated plexiform neurofibromas and malignant peripheral nerve sheath tumors. *Am. J. Pathol.* **159**, 57–61.

59. Kluwe, L., Friedrich, R.E., and Mautner, V.F. (1999) Allelic loss of the *NF1* gene in NF1-associated plexiform neurofibromas. *Cancer Genet. Cytogenet.* **113**, 65–69.

60. Kluwe, L., Friedrich, R., and Mautner, V.-F. (1999) Loss of *NF1* allele in Schwann cells but not in fibroblasts derived form an NF1-associated neurofibroma. *Genes Chromosomes Cancer* **24**, 283–285.

61. Serra, E., Rosenbaum, T., Winner, U., Aledo, R., Ars, E., Estivill, X., Lenard, H.-G., and Lázaro, C. (2000) Schwann cells harbor the somatic *NF1* mutation in neurofibromas: evidence of two different Schwann cell subpopulations. *Hum. Mol. Genet.* **9**, 3055–3064.

62. Legius, E., Marchuk, D.A., Collins, F.S., and Glover, T.W. (1993) Somatic deletion of the neurofibromatosis type 1 gene in a neurofibrosarcoma supports a tumor suppressor gene hypothesis. *Nature Genet.* **3**, 122–126.

63. Lothe, R.A., Saeter, G., Danielsen, H.E., Stenwig, A.E., Høyheim, B., O'Connell, P., and Børresen, A.-L. (1993) Genetic alterations in a malignant schwannoma from a patient with neurofibromatosis (NF1). *Pathol. Res. Pract.* **189**, 465–471.

64. Lothe, R.A., Slettan, A., Saeter, G., Brøgger, A., Børresen, A.-L., and Nesland, J.M. (1995) Alterations at chromosome 17 loci in peripheral nerve sheath tumors. *J. Neuropathol. Exp. Neurol.* **54**, 65–73.

65. Skuse, G.R., Kosciolek, B.A., and Rowley, P.T. (1989) Molecular genetic analysis of tumors in von Recklinghausen neurofibromatosis: loss of heterozygosity for chromosome 17. *Genes Chromosomes Cancer* **1**, 36–41.

66. Rasmussen, S.A., Overman, J., Thomson, S.A., Colman, S.D., Abernathy, C.R., Trimpert, R.E., Moose, R., Virdi, G., Roux, K., Bauer, M., Rojiani, A.M., Maria, B.L., Muir, D., and Wallace, M.R. (2000) Chromosome 17 loss-of-heterozygosity studies in benign and malignant tumors in neurofibromatosis type 1. *Genes Chromosomes Cancer* **28**, 425–431.

67. Ågesen, T.H., Flørenes, V.A., Molenaar, W.M., Lind, G.E., Berner, J.-M., Plaat, B.E.C., Komdeur, R., Myklebost, O., van den Berg, E., and Lothe, R.A. (2005) Expression patterns of cell cycle components in sporadic and neurofibromatosis type 1-related malignant peripheral nerve sheath tumors. *J. Neuropathol. Exp. Neurol.* **64**, 74–81.

68. DeClue, J.E., Papageorge, A.G., Fletcher, J.A., Diehl, S.R., Ratner, N., Vass, W.C., and Lowry, D.R. (1992) Abnormal

regulation of mammalian p21^{ras} contributes to malignant tumor growth in von Recklinghausen (type 1) neurofibromatosis. *Cell* **69**, 265–273.

69. Basu, T.N., Gutmann, D.H., Fletcher, J.A., Glover, T.W., Collins, F.S., and Downward, J. (1992) Aberrant regulation of *ras* proteins in malignant tumour cells from type 1 neurofibromatosis patients. *Nature* **356**, 713–715.

70. Guha, A., Lau, N., Huvar, I., Gutmann, D., Provias, J., Pawson, T., and Boss, G. (1996) Ras-GTP levels are elevated in human NF1 peripheral nerve tumors. *Oncogene* **12**, 507–513.

71. Birindelli, S., Perrone, F., Oggionni, M., Lavarino, C., Pasini, B., Vergani, B., Ranzani, G.N., Pierotti, M.A., and Pilotti, S. (2001) Rb and TP53 pathway alterations in sporadic and NF1-related malignant peripheral nerve sheath tumors. *Lab. Invest.* **81**, 833–844.

72. Shen, M.H., Harper, P.S., and Upadhyaya, M. (1996) Molecular genetics of neurofibromatosis type 1 (*NF1*). *J. Med. Genet.* **33**, 2–17.

73. Schneider-Stock, R., Radig, K., and Roessner, A. (1997) Loss of heterozygosity on chromosome 9q21 (p16 gene) uncommon in soft-tissue sarcomas. *Mol. Carcinog.* **18**, 63–65.

74. Nielsen, G.P., Stemmer-Rachamimov, A.O., Ino, Y., Møller, M.B., Rosenberg, A.E., and Louis, D.N. (1999) Malignant transformation of neurofibromas in neurofibromatosis type 1 is associated with *CDKN2A/p16* inactivation. *Am. J. Pathol.* **155**, 1879–1884.

75. Legius, E., Dierick, H., Wu, R., Hall, B.K., Marynen, P., Cassiman, J.-J., and Glover, T.W. (1994) *TP53* mutations are frequent in malignant NF1 tumors. *Genes Chromosomes Cancer* **10**, 250–255.

76. Menon, A.G., Anderson, K.M., Riccardi, V.M., Chung, R.Y., Whaley, J.M., Yandell, D.W., Farmer, G.E., Freiman, R.N., Lee, J.K., Li, F.P., Barker, D.F., Ledbetter, D.H., Kleider, A., Martuza, R.L., Gusella, J.F., and Seizinger, B.R. (1990) Chromosome 17p deletions and p53 gene mutations associated with the formation of malignant neurofibrosarcomas in von Recklinghausen neurofibromatosis. *Proc. Natl. Acad. Sci. USA* **87**, 5435–5439.

77. Watson, M.A., Perry, A., Tihan, T., Prayson, R.A., Guha, A., Bridge, J., Ferner, R., and Gutmann, D.H. (2004) Gene expression profiling reveals unique molecular subtypes of neurofibromatosis type 1-associated and sporadic malignant peripheral nerve sheath tumors. *Brain Pathol.* **14**, 297–303.

77A. Miller, S.J., Rangwala, F., Williams, J., Ackerman, P., Kong, S., Jegga, A.G., Kaiser, S., Aronow, B.J., Frahm, S., Kluwe, L., Mautner, V., Upadhyaya, M., Muir, D., Wallace, M., Hagen, J., Quelle, D.E., Watson, M.A., Perry, A., Gutmann, D.H., and Ratner, N. (2006) Large-scale molecular comparison of human schwann cells to malignant peripheral nerve sheath tumor cell lines and tissues. *Cancer Res.* **66**, 2584–2591.

78. Upadhyaya, M., Han, S., Consoli, C., Majounie, E., Horan, M., Thomas, N.S., Potts, C., Griffiths,S., Ruggieri, M., von Deimling, A., and Cooper, D.N. (2004) Characterization of the somatic mutational spectrum of the neurofibromatosis type 1 (NF1) gene in neurofibromatosis patients with benign and malignant tumors. *Hum. Mutat.* **23**, 134–146.

79. Colman, S.D., Williams, C.A., and Wallace, M.R. (1995) Benign neurofibromas in type 1 neurofibromatosis (NF1) show somatic deletions of the *NF1* gene. *Nat. Genet.* **11**, 90–92.

80. Sawada, S., Florell, S., Purandare, S.M., Ota, M., Stephens, K., and Viskochil, D. (1996) Identification of *NF1* mutations in both alleles of a dermal neurofibroma. *Nat. Genet.* **14**, 110–112.

81. Däschner, K., Assum, G., Eisenbarth, I., Krone, W., Hoffmeyer, S., Wortmann, S., Heymer, B., and Kehrer-Sawatzki, H. (1997) Clonal origin of tumor cells in a plexiform neurofibroma with LOH in NF1 intron 38 and in dermal neurofibromas without LOH of the NF1 gene. *Biochem. Biophys. Res. Commun.* **234**, 346–350.

82. Serra, E., Puig, S., Otero, D., Gaona, A., Kruyer, H., Ars, E., Estivill, X., and Lázaro, C. (1997) Confirmation of a double-hit model for the *NF1* gene in benign neurofibromas. *Am. J. Hum. Genet.* **61**, 512–519.

83. Glover, T.W., Stein, C.K., Legius, E., Andersen, L.B., Brereton, A., and Johnson, S. (1991) Molecular and cytogenetic analysis of tumors in von Recklinghausen neurofibromatosis. *Genes Chromosomes Cancer* **3**, 62–70.

84. Reynolds, J.E., Fletcher, J.A., Lytle, C.H., Nie, L., Morton, C.C., and Diehl, S.R. (1992) Molecular characterization of a 17q11.2 translocation in a malignant schwannoma cell line. *Hum. Genet.* **90**, 450–460.

85. Weiss, S.W., and Nickoloff, B.J. (1993) CD-34 is expressed by a distinctive cell population in peripheral nerve, nerve sheath tumors, and related lesions. *Am. J. Surg. Pathol.* **17**, 1039–1045.

86. Gómez, L., Barrios, C., Kreicbergs, A., Zetterberg, A., Pestaña, A., and Castresana, J.S. (1995) Absence of mutation at the GAP-related domain of the neurofibromatosis type 1 gene in sporadic neurofibrosarcomas and other bone and soft tissue tumors. *Cancer Genet. Cytogenet.* **81**, 173–174.

87. Ottini, L., Esposito, D.L., Richetta, A., Carlesimo, M., Palmirotta, R., Verì, M.C., Battista, P., Frati, L., Caramia, F.G., Calvieri, S., Cama, A., and Mariani-Costantini, R. (1995) Alterations of microsatellites in neurofibromas of von Recklinghauasen's disease. *Cancer Res.* **55**, 5677–5680.

87A. Lothe, R.A., Karhu, R., Mandahl, N., Mertens, F., Saeter, G., Heim, S., Børresen-Dale, A.-L., and Kallioniemi, O.-P. (1996) Gain of 17q24-qter detected by comparative genomic hybridization in malignant tumors from patients with von Recklinghausen's neurofibromatosis. *Cancer Res.* **56**, 4778–4781.

88. Mawrin, C., Kirches, E., Boltze, C., Dietzmann, K., Roessner, A., and Schneider-Stock, R. (2002) Immunohistochemical and molecular analysis of p53, RB, and PTEN in malignant peripheral nerve sheath tumors. *Virchows Arch.* **440**, 610–615.

89. Berner, J.-M., Sørlie, T., Mertens, F., Henriksen, J., Saeter, G., Mandahl, N., Brøgger, A., Myklebost, O., and Lothe, R.A. (1999) Chromosome band 9p21 is frequently altered in malignant peripheral nerve sheath tumors: studies of *CDKN2A* and other genes of the pRB pathway. *Genes Chromosomes Cancer* **26**, 151–160.

90. Kourea, H.P., Orlow, I., Scheithauer, B.W., Cordon-Cardo, C., and Woodruff, J.M. (1999A) Deletions of the *INK4A* gene

occur in malignant peripheral nerve sheath tumors but not in neurofibromas. *Am. J. Pathol.* **155,** 1855–1860.

91. Kourea, H.P., Cordon-Cardo, C., Dudas, M., Leung, D., and Woodruff, J.M. (1999) Expression of p27^{kip} and other cell cycle regulators in malignant peripheral nerve sheath tumors and neurofibromas. The emerging role of p27^{kip} in malignant transformation of neurofibromas. *Am. J. Pathol.* **155,** 1885–1891.

92. Levine, A.J. (1997) p53, the cellular gatekeeper for growth and division. *Cell* **88,** 232–311.

92A. Wei, C.-L., Wu, Q., Vega, V.G., Chiu, K.P., Ng, P., Zhang, T., Shahab, A., Yong, H.C., Fu, Y.T., Weng, Z., Liu, J.J., Zhao, X.D., Chew, J.-L., Lee, Y.L., Kuznetsov, V.A., Sung, W.-K., Miller, L.D., Lim, B., Liu, E.T., Yu, Q., Ng, H.-H., and Ruan, Y. (2006) A global map of p53 transcription-factor binding sites in the human genome. *Cell* **124,** 207–219.

93. Kawai, A., Noguchi, M., Beppu, Y., Yokoyama, R., Mukai, K., Hirohashi, S., Inoue, H., and Fukuma, H. (1994) Nuclear immunoreaction of p53 protein in soft tissue sarcomas. *Cancer* **73,** 2499–2505.

94. Baas, I.O., van den Berg, F.M., Mulder, J.W., Clement, M.J., Slebos, R.J., Hamilton, S.R, and Offerhaus, G.J. (1996) Potential false-positive results with antigen enhancement for immunohistochemistry of the p53 gene product in colorectal neoplasms. *J. Pathol.* **178,** 264–267.

95. Halling, K.C., Scheithauer, B.W., Halling, A.C., Nascimento, A.G., Zeismer, S.C., Roche, P.C., and Wollan, P.C. (1996) p53 expression in neurofibroma and malignant peripheral nerve sheath tumor. An immunohistochemical study of sporadic and NF1-associated tumors. *Am. J. Clin. Pathol.* **106,** 282–288.

96. Nigro, J.M., Baker, S.J., Preisinger, A.C., Jessup, J.M., Hostetter, R., Cleary, K., Bigner, S.H., Davidson, N., Baylin, S., Devilee, P., Glover, T., Collins, F.S., Weston, A., Modali, R., Harris, C.C., and Vogelstein, B. (1989) Mutations in the *p53* gene occur in diverse human tumour types. *Nature* **342,** 705–708.

97. Toguchida, J., Yamaguchi, T., Ritchie, B., Beauchamp, R.L., Dayton, S.H., Herrera, G.E., Yamamuro, T., Kotoura, Y., Sasaki, M.S., and Little, J.B. (1992) Mutation spectrum of the p53 gene in bone and soft tissue sarcomas. *Cancer Res.* **52,** 6194–6199.

98. Lothe, R.A., Smith-Sørensen, B., Hektoen, M., Stenwig, A.E., Mandahl, N., Saeter, G., and Mertens, F. (2001) Biallelic inactivation of *TP53* rarely contributes to the development of malignant peripheral nerve sheath tumors. *Genes Chromosomes Cancer* **30,** 202–206.

98A. Leroy, K., Dumas, V., Martin-Garcia, N., Falzone, M.-C., Voisin, M.-C., Wechsler, J., Revuz, J., Créange, A., Levy, E., Lantieri, L., Zeller, J., and Wolkenstein, P. (2001) Malignant peripheral nerve sheath tumors associated with neurofibromatosis type 1. A clinicopathologic and molecular study of 17 patients. *Arch. Dermatol.* **137,** 908–913.

99. Wadayama, B., Toguchida, J., Yamaguchi, T., Sasaki, M.S., and Yamamuro, T. (1993) p53 expression and its relationship to DNA alterations in bone and soft tissue sarcomas. *Br. J. Cancer* **68,** 1134–1139.

100. Weiss, S.W. (1996) p53 gene alterations in benign and malignant nerve sheath tumors. *Am. J. Clin. Pathol.* **106,** 271–272.

101. Schneider-Stock, R., Radig, K., Oda, Y., Mellin, W., Rys, J., Niezabitowski, A., and Roessner, A (1997) p53 gene mutations in soft-tissue sarcomas—correlations with p53 immunohistochemistry and DNA ploidy. *J. Cancer Res. Clin. Oncol.* **123,** 211–218.

102. Strauss, B.L., Gutmann, D.H., Dehner, L.P., Hirbe, A., Zhu, X., Marley, E.F., and Liapis, H. (1999) Molecular analysis of malignant triton tumors. *Hum. Pathol.* **30,** 984–988.

103. Blagosklonny, M.V. (1997) Loss of function and p53 protein stabilization. *Oncogene* **15,** 1889–1893.

104. Andreassen Å, Øyjord, T., Hovig, E., Holm, R., Flórenes, V.A., Nesland, J.M., Myklebost, O., Høie, J., Bruland, Ø.S., Børresen, A.-L., and Fodstad, Ø. (1993) *p53* abnormalities in different subtypes of human sarcomas. *Cancer Res.* **53,** 468–471.

105. Castresana, J.S., Rubio, M.P., Gomez, L., Kreicbergs, A., Zetterberg, A., and Barrios, C. (1995) Detection of TP53 gene mutations in human sarcomas. *Eur. J. Cancer* **31A,** 735–358.

106. Kindblom, L.G., Ahlden, M., Meis-Kindblom, J.M., and Stenman, G. (1995) Immunohistochemical and molecular analysis of p53, MDM2, proliferating cell nuclear antigen and Ki67 in benign and malignant peripheral nerve sheath tumours. *Virchows Arch.* **427,** 19–26.

107. Woods, D.B., and Vousden, K.H. (2001) Regulation of p53 function. *Exp. Cell Res.* **264,** 56–66.

108. Pomerantz, J., Schreiber-Agus, N., Liégeois, N.J., Silverman, A., Alland, L., Chin, L., Potes, J., Chen, K., Orlow, I., Lee, H.-W., Cordon-Cardo, C., and DePinho, R.A. (1998) The *Ink4a* tumor suppressor gene product, p19^{Arf}, interacts with MDM2 and neutralizes MDM2's inhibition of p53. *Cell* **92,** 713–723.

109. Zhang, Y., Xiong, Y., and Yarbrough, W.G. (1998) ARF promotes MDM2 degradation and stabilizes p53: ARF-INK4a locus deletion impairs both the Rb and p53 tumor suppression pathways. *Cell* **92,** 725–734.

110. Funk, J.O. (1999) Cancer cell cycle control. *Anticancer Res.* **19,** 4772–4780.

111A. Cordon-Cardo, C., Latres, E., Drobnjak, M., Oliva, M.R., Pollack, D., Woodruff, J.M., Marechal, V., Chen, J., Brennan, M.F., and Levine, A.J. (1994) Molecular abnormalities of *mdm2* and *p53* genes in adult soft tissue sarcomas. *Cancer Res.* **54,** 794–799.

111. Yoo, J., Park, S.Y., Kang, S.J., Shim, S.I., and Kim, B.K. (2002) Altered expression of G1 regulatory proteins in human soft tissue sarcomas. *Arch. Pathol. Lab. Med.* **126,** 567–573.

112. Würl, P., Meye, A., Schmidt, H., Lautenschläger, C., Kalthoff, H., Rath, F.-W., and Taubert, H. (1998) High prognostic significance of Mdm2/p53 co-overexpression in soft tissue sarcomas of the extremities. *Oncogene* **16,** 1183 1185.

113. Perrone, F., Tabano, S., Colombo, F., Degrada, G., Birindelli, S., Gronchi, A., Colecchia, M., Pierotti, M.A., and Pilotti, S. (2003) *p15*^{INK4b}, *p14*^{ARF}, and *p16*^{INK4a} inactivation in sporadic and neurofibromatosis type 1-related malignant peripheral nerve sheath tumors. *Clin. Cancer Res.* **9,** 4132–4138.

113A. Lévy, P., Vidaud, D., Leroy, K., Laurendeau, I., Wechsler, J., Bolasco, G., Parfait, B., Wolkenstein, P., Vidaud, M., and Bièche, I. (2004) Molecular profiling of malignant peripheral nerve sheath tumors associated with neurofibromatosis type 1, based in large-scale real-time RT-PCR. *Mol. Cancer* **3,** 20–32.

113B. Miller, S.J., Rangwala, F., Williams, J., Ackerman, P., Kong, S., Jegga, A.G., Kaiser, S., Aronow, B.J., Frahm, S., Kluwe, L., Mautner, V., Upadhyaya, M., Muir, D., Wallace, M., Hagen, J., Quelle, D.E., Watson, M.A., Perry, A., Gutmann, D.H., and Ratner, N. (2006) Large-scale molecular comparison of human schwann cells to malignant peripheral nerve sheath tumor cell lines and tissues. *Cancer Res.* **66**, 2584–2591.

114. Sherr, C.J., and Roberts, J.M. (1995) Inhibitors of mammalian G1 cyclin-dependent kinases. *Genes Dev.* **9**, 1149–1163.

114A. Sherr, C.J. (1996) Cancer cell cycles. *Science* **274**, 1672–1677.

115. Karpeh, M.S., Brennan, M.F., Cance, W.G. Woodruff, J.M., Pollack, D., Casper, E.S., Dudas, M.E., Latres, E., Drobnjak, M., and Cordon-Cardo, C. (1995) Altered patterns of retinoblastoma gene product expression in adult soft-tissue sarcomas. *Br. J. Cancer* **72**, 986–991.

116. Orlow, I., Drobnjak, M., Zhang, Z.-F., Lewis, J., Woodruff, J.M., Brennan, M.F., and Cordon-Cardo, C. (1999) Alterations of the INK4A and INK4B genes in adult soft tissue sarcomas: effect on survival. *J. Natl. Cancer Inst.* **91**, 73–79.

117. Beltinger, C.P., White, P.W., Sulman, E.P., Maris, J.M., and Brodeur, G.M. (1995) No CDKN2 mutations in neuroblastomas. *Cancer Res.* **55**, 2053–2055.

118. Kawamata, N., Seriu, T., Koeffler, H.P., and Bartram, C.R. (1996) Molecular analysis of the cyclin-dependent kinase inhibitor family (CDKN2/MTS1/INK4A), p18 (INK4C) and p27 (Kip1) genes in neuroblastomas. *Cancer* **77**, 570–575.

119. Serrano, M., Hannon, G.J., and Beach, D. (1993) A new regulatory motif in cell-cycle control causing specific inhibition of cyclin D/CDK4. *Nature* **366**, 704–707.

120. Pindzola, J.A., Palazzo, J.P., Kovatich, A.J., Tuma, B., and Michael, N. (1998) Expression of p21$^{WAF1/CIP1}$ in soft tissue sarcomas: a comparative immunohistochemical study with p53 and Ki-67. *Pathol. Res. Pract.* **194**, 685–691.

121. DeClue, J.E., Heffelfinger, S., Benvenuto, G., Ling, B., Li, S., Rui, W., Vass, W.C., Viskochil, D., and Ratner, N. (2000) Epidermal growth factor receptor expression in neurofibromatosis type 1-related tumors and NF1 animal models. *J. Clin. Invest.* **105**, 1233–1241.

122. Schneider-Stock, R., Walter, H., Haeckel, C., Radig, K., Rys, J., and Roessner, A. (1998) Gene alterations at the CDKN2A (p16/MTS1) locus in soft tissue tumors. *Int. J. Oncol.* **13**, 325–329.

122A. Merlino, G.T., Xu, Y.H., Ishii, S., Clark, A.J., Semba, K., Toyoshima, K., Yamamoto, T., and Pastan, I. (1984) Amplification and enhanced expression of the epidermal growth factor receptor gene in A431 human carcinoma cells. *Science* **224**, 417–419.

122B. Velu, T.J., Beguinot, L., Vass, W.C., Zhang, K., Pastan, I., and Lowy, D.R. (1989) Retroviruses expressing different levels of the normal epidermal growth factor receptor: biological properties and new bioassay. *J. Cell Biochem.* **39**, 153–166.

122C. Peng, D., Fan, Z., Lu, Y., DeBlasio, T., Scher, H., and Mendelsohn, J. (1996) Anti-epidermal growth factor receptor monoclonal antibody 225 up-regulates p27KIP1 and induces G1 arrest in prostatic cancer cell line DU145. *Cancer Res.* **56**, 3666–3669.

123. Zhu, Y., Ghosh, P., Charnay, P., Burns, D.K., and Parada, L.F. (2002) Neurofibromas in NF1: schwann cell origin and role of tumor environment. *Science* **296**, 920–922.

124. Bartkova, J., Lukas, J., Strauss, M., and Bartek, J. (1998) Cyclin D3: requirement for G1/S transition and high abundance in quiescent tissues suggest a dual role in proliferation and differentiation. *Oncogene* **17**, 1027–1037.

125. Morgan, D.O. (1995) Principles of CDK regulation. *Nature* **374**, 131–134.

126. Weinstein, I.B. (2000) Disorders of cell circuitry during multistage carcinogenesis: the role of homeostasis. *Carcinogenesis* **21**, 857–864.

127. LaBaer, J., Garrett, M.D., Stevenson, L.F., Slingerland, J.M., Sandhu, C., Chou, H.S., Fattaey, A., and Harlow, E. (1997) New functional activities for the p21 family of CDK inhibitors. *Genes Dev.* **11**, 847–862.

128. Holtkamp, N., Reuß D.E., Atallah, I., Kuban, R.-J., Hartmann, C., Mautner, V.-F., Frahm, S., Friedrich, R.E., Algermissen, B., Pham, V.-A., Prietz, S., Rosenbaum, T., Estevez-Schwarz, L., and von Deimling, A. (2004) Subclassification of nerve sheath tumors by gene expression profiling. *Brain Pathol.* **14**, 258–264.

129. Lévy, P., Ripoche, H., Laurendeau, I., Lazar, V., Ortonne, N., Parfait, B., Leroy, K., Wechsler, J., Salmon, I., Wolkenstein, P., Dessen, P., Vidaud, M., Vidaud, D., and Bièche, I. (2007) Microarray-based identification of *Tenascin C* and *Tenascin XB*, genes possibly involved in tumorigenesis associated with neurofibromatosis type 1. *Clin. Cancer Res.* **13**, 398–407.

130. Sherman, L., Jacoby, L.B., Lampe, J., Pelton, P., Aguzzi, A., Herrlich, P., and Ponta, H. (1997) CD44 expression is aberrant in benign Schwann cell tumors possessing mutations in the neurofibromatosis type 2, but not type 1, gene. *Cancer Res.* **57**, 4889–4897.

131. Jiang, W., Hiscox, S., Matsumoto, K., and Nakamura, T. (1999) Hepatocyte growth factor/scatter factor, its molecular, cellular and clinical implications in cancer. *Crit. Rev. Oncol. Hematol.* **29**, 209–248.

132. Krasnoselsky, A., Massay, M.J., DeFrances, M.C., Michalopoulos, G., Zarnegar, R., and Ratner, N. (1994) Hepatocyte growth factor is a mitogen for Schwann cells and is present in neurofibromas. *J. Neurosci.* **14**, 7284–7290.

133. Fukuda, T., Ichimura, E., Shinozaki, T., Sano, T., Kashiwabara, K., Oyama, T., Nakajima, T., and Nakamura, T. (1996) Coexpression of HGF and c-Met/HGF receptor in human bone and soft tissue tumors. *Pathol. Intl.* **48**, 759–762.

134. Moriyama, T., Kataoka, H., Kawano, H., Yokogami, K., Nakano, S., Goya, T., Uchino, H., Koono, M., and Wakisaka, S. (1998) Comparative analysis of expression of hepatocyte growth factor and its receptor, c-Met, in gliomas, meningiomas and schwannomas in humans. *Cancer Lett.* **124**, 149–155.

135. Su, W., Gutmann, D.H., Perry, A., Abounader, R., Laterra, J., and Sherman, L.S. (2004) CD44-independent hepatocyte growth factor/c-Met autocrine loop promotes malignant peripheral nerve sheath tumor cell invasion in vitro. *GLIA* **45**, 297–306.

135A. Zhang, Y.W., and Vande Woude, G.F. (2003) HGF/SF-met signaling in the control of branching morphogenesis and invasion. *J. Cell Biochem.* **88**, 408–417.

136. Lokker, N.A., Mark, M.R., Luis, E.A., Bennett, G.L., Robbins, K.A., Baker, J.B., and Godowski, P.J. (1992) Structure-function analysis of hepatocyte growth factor: identification of variants that lack mitogenic activity yet retain high affinity receptor binding. *EMBO J.* **11**, 2503–2510.

136A. Naka, D., Ishii, T., Yoshiyama, Y., Miyazawa, K., Hara, H., Hishida, T., and Kidamura, N. (1992) Activation of hepatocyte growth factor by proteolytic conversion of a single chain form to a heterodimer. *J. Biol. Chem.* **267**, 20114–20119.

137. Naldini, L., Tamagnone, L., Vigna, E., Sachs, M., Hartmann, G., Birchmeier, W., Daikuhara, Y., Tsubouchi, H., Blasi, F., and Comoglio, P.M. (1992) Extracellular proteolytic cleavage by urokinase is required for activation of hepatocyte growth factor/scatter factor. *EMBO J.* **11**, 4825–4833.

138. Miyazawa, K., Shimomura, T., Kitamura, A., Kondo, J., Morimoto, Y., and Kitamura, N. (1993) Molecular cloning and sequence analysis of the cDNA for a human serine protease responsible for activation of hepatocyte growth factor. *J. Biol. Chem.* **268**, 10024–10028.

139. Miyazawa, K., Shimomura, T., and Kitamura, N. (1996) Activation of hepatocyte growth factor in the injured tissues is mediated by hepatocyte growth factor activator. *J. Biol. Chem.* **271**, 3615–3618.

140. Kataoka, H., Hamasuna, R., Itoh, H., Kitamura, N., and Koono, M. (2000) Activation of hepatocyte growth factor/scatter factor in colorectal carcinoma. *Cancer Res.* **60**, 6148–6159.

141. Su, W., Sin, M., Darrow, A., and Sherman, L.S. (2003) Malignant peripheral nerve sheath tumor cell invasion is facilitated by Src and aberrant CD44 expression. *Glia* **42**, 350–358.

142. van der Voort, R., Taher, T.E.I., Wielenga, V.J.M., Spaargaren, M., Prevo, R., Smit, L., David, G., Hartmann, G., Gherardi, E., and Pals, S.T. (1999) Heparan sulfate-modified CD44 promotes hepatocyte growth factor/scatter factor-induced signal transduction through the receptor tyrosine kinase c-Met. *J. Biol. Chem.* **274**, 6499–6506.

143. Orian-Rousseau, V., Chen, L., Sleeman, J.P., Herrlich, P., and Ponta, H. (2002) CD44 is required for two consecutive steps in HGF/c-Met signaling. *Genes Dev.* **16**, 3074–3086.

144. Cao, B., Su, Y., Oskarsson, M., Zhao, P., Kort, E.J., Fisher, R.J., Wang, L.-M., and Vande Woude, G.F. (2001) Neutralizing monoclonal antibodies to hepatocyte growth factor/scatter factor (HGF/SF) display antitumor activity in animal models. *Proc. Natl. Acad. Sci. USA* **98**, 7443–7448.

145. Abounader, R., Lal, B., Luddy, C., Koe, G., Davidson, B., Rosen, E.M., and Laterra, J. (2002) In vivo targeting of SF/HGF and c-met expression via U1snRNA ribozymes inhibits glioma growth and angiogenesis and promotes apoptosis. *FASEB J.* **16**, 108–110.

146. Ponta, H., Sherman, L., and Herrlich, P.A. (2003) CD44: from adhesion molecules to signaling regulators. *Nat. Rev. Mol. Cell Biol.* **4**, 33–45.

147. Rahimi, N., Hung, W., Tremblay, E., Saulnier, R., and Elliott, B. (1998) c-Src kinase activity is required for hepatocyte growth factor-induced motility and anchorage-independent growth of mammary carcinoma cells. *J. Biol. Chem.* **273**, 33714–33721.

148. Hung, W., and Elliott, B. (2001) Co-operative effect of c-Src tyrosine kinase and Stat3 in activation of hepatocyte growth

149. Irby, R.B., and Yeatman, T.J. (2000) Role of Src expression and activation in human cancer. *Oncogene* **19**, 5636–5642.

149A. Rosenbaum, T., Patrie, K.M., and Ratner, N. (1997) Neurofibromatosis type 1: Genetic and cellular mechanisms of peripheral nerve tumor formation. *Neuroscientist* **3**, 412–420.

150. Badache, A., and De Vries, G.H. (1998) Neurofibrosarcoma-derived schwann cells overexpress platelet-derived growth factor (PDGF) receptors and are induced to proliferate by PDGF BB. *J. Cell. Physiol.* **177**, 334–342.

150A. Eriksson, A., Siegbahn, A., Westermark, B., Heldin, C.-H., and Claesson-Welsh, L. (1992) PDGF α- and β-receptors activate unique and common signal transduction pathways. *EMBO J.* **11**, 543–550.

150B. Allison, K.H., Patel, R.M., Goldblum, J.R., and Rubin, B.P. (2006) Superficial malignant peripheral nerve sheath tumor. A rare and challenging diagnosis. *Am. J. Clin. Pathol.* **124**, 685–692.

151. Dang, I., and DeVries, G.H. (2005) Schwann cell lines derived from malignant peripheral nerve sheath tumors respond abnormally to platelet-derived growth factor-BB. *J. Neurosci Res.* **79**, 318–328.

151A. Lee, J.K., Sobel, R.A., Ciocca, E.A., Kim, T.S., and Martuza, R.L. (1992) Growth of human acoustic neuromas, neurofibromas and schwannomas in the subrenal capsule and sciatic nerve of the nude mouse. *J. Neurooncol.* **14**, 101–112.

152. Badache, A., Muja, N., and De Vries, G.H. (1998) Expression of Kit in neurofibromin-deficient Schwann cells: role in Schwann cell hyperplasia associated with type 1 neurofibromatosis. *Oncogene* **17**, 795–800.

152A. Ryan, J.J., Klein, K.A., Neuberger, T.J., Leftwich, J.A., Westin, E.H., Kauma, S., Fletcher, J.A., DeVries, G.H., and Huff, T.F. (1994) Role for the stem cell factor/KIT complex in Schwann cell neoplasia associated with neurofibromatosis. *J. Neurosci.* **37**, 415–432.

153. McNiece, I.K., Langley, K.E., and Zsebo, K.M. (1991) Recombinant human stem cell factor synergises with GM-CSF, G-CSF, IL-3 and epo to stimulate human progenitor cells of the myeloid and erythroid lineages. *Exp. Hematol.* **19**, 226–231.

154. Marchionni, M.A., Goodearl, A.D.J., Chen, M.S., Bermingham-McDonogh, O., Kirk, C., Hendricks, M., Danehy, F., Misumi, D., Sudhalter, J., Kobayashi, K., Wroblewski, D., Lynch, C., Baldassare, M., Hiles, I., Davis, J.B., Hsuan, J.J., Totty, N.F., Otsu, M., McBurney, R.N., Waterfield, M.D., Stroobant, P., and Gwynne. D. (1993) Glial growth factors are alternatively spliced erbB2 ligands expressed in the nervous system. *Nature* **362**, 312–318.

155. Levi, A.D.O., Bunge, R.P., Lofgren, J.A., Meima, L., Hefti, F., Nikolics, K., and Sliwkowski, M.X. (1995) The influence of heregulins on human Schwann cell proliferation. *J. Neurosci.* **15**, 1329–1340.

155A. Rutkowski, J.L., Kirk, C.J., Lerner, M.A., and Tennekoon, G.I. (1995) Purification and expansion of human schwann cells *in vitro*. *Nat. Med.* **1**, 80–83.

156. Weinmaster, G., and Lemke, G. (1990) Cell-specific cyclic AMP-mediated induction of the PDGF factor. *EMBO J.* **9**, 915–920.

factor expression in mammary carcinoma cells. *J. Biol. Chem.* **276**, 12395–12403.

157. Hardy, M., Reddy, U.R., and Pleasure, D. (1992) Platelet-derived growth factor and regulation of Schwann cell proliferation in vivo. *J. Neurosci. Res.* **31**, 254–262.

158. Li, H., Velasco-Miguel, S., Vass, W.C., Parada, L.F., and DeClue, J.E. (2002) Epidermal growth factor receptor signaling pathways are associated with tumorigenesis in the *Nf1:p53* mouse tumor model. *Cancer Res.* **62**, 4507–4513.

159. Li, Y., Rao, P.K., Wen, R., Song, Y., Muir, D., Wallace, P., van Horne, S.J., Tennekoon, G.I., and Kadesch, T. (2004) Notch and Schwann cell transformation. *Oncogene* **23**, 1146–1152.

160. Guo, H.-F., Tong, J., Hannan, F., Luo, L., and Zhong, Y. (2000) A neurofibromatosis-1-regulated pathway is required for learning in *Drosophila*. *Nature* **403**, 895–898.

161. Kim, H.A., Ratner, N., Roberts, T.M., and Stiles, C.D. (2001) Schwann cell proliferative responses to cAMP and *Nf1* are mediated by cyclin D1. *J. Neurosci.* **21**, 1110–1116.

162. Tong, J., Hannan, F., Zhu, Y., Bernards, A., and Zhong, Y. (2002) Neurofibromin regulates G protein-stimulated adenylyl cyclase activity. *Nature Neurosci.* **5**, 95–96.

163. Hunter, S., Weiss, S., Ou, C.-Y., Jaye, D., Young, A., Wilcox, J., Arbiser, J.L., Monson, D., Goldblum, J., Nolen, J.D., and Varma, V. (2005) Apolipoprotein D is downregulated during malignant transformation of neurofibromas. *Hum. Pathol* **36**, 987–993.

164. Hoang, M.P., Sinkre, P., and Albores-Saavedra, J. (2001) Expression of protein gene product 9.5 in epithelioid and conventional malignant peripheral nerve sheath tumors. *Arch. Pathol. Lab. Med.* **125**, 1321–1325.

164A. Kobayashi, C., Oda, Y., Takahira, T., Izumi, T., Kawaguchi, K., Yamamoto, H., Tamiya, S., Yamada, T., Iwamoto, Y., and Tsuneyoshi, M. (2006) Aberrant expression of CHFR in malignant peripheral nerve sheath tumors. *Mod. Pathol.* **19**, 524–532.

164B. Stonecypher, M.S., Byer, S.J., Grizzle, W.E., and Carroll, S.L. (2005) Activation of the neuregulin-1/ErbB signaling pathway promotes the proliferation of neoplastic Schwann cells in human malignant peripheral nerve sheath tumors. *Oncogene* **24**, 5589–5605.

165. Daimaru, Y., Hashimoto, H., and Enjoji, M. (1984) Malignant 'triton' tumors: a clinicopathologic and immunohistochemical study of nine cases. *Hum. Pathol.* **15**, 768–778.

166. Ducatman, B.S., Scheithauer, B.W., Piepgras, D.G., Reiman, H.M., and Ilstrup, D.M. (1986) Malignant peripheral nerve sheath tumors: a clinicopathologic study of 120 cases. *Cancer* **57**, 2006–2021.

166A. White, H.R. (1971) Survival in malignant schwannoma: an 18-year study. *Cancer* **27**, 720–729.

166B. Ghosh, B.C., Ghosh, L., Huvos, A.G., and Fortner, J.G. (1973) Malignant schwannoma: a clinicopathologic study. *Cancer* **31**, 184–190.

166C. Wong, W.W., Hirose, T., Scheithauer, B.W., Schild, S.E., and Gunderson, L.L. (1998) Malignant peripheral nerve sheath tumor: analysis of treatment outcome. *Int. J. Radiat. Oncol.* **42**, 351–360.

167. Haddadin, M.H., Hawkins, A.L., Long, P., Morsberger, L.A., Depew, D., Epstein, J.I., and Griffin, C.A. (2003) Cytogenetic study of malignant triton tumor: a case report. *Cancer Genet. Cytogenet.* **144**, 100–105.

168. Esteller, M., and Herman, J.G. (2002) Cancer as an epigenetic disease: DNA methylation and chromatin alterations in human tumours. *J. Pathol.* **196**, 1–7.

169. Gonzalez-Gomez, P., Bello, M.J., Arjona, D., Alonso, M.E., Lomas, J., de Campos, J.M., Kusak, M.E., Gutierrez, M., Sarasa, J.L., and Rey, J.A. (2003) Aberrant CpG island methylation in neurofibromas and neurofibrosarcomas. *Oncol. Rep.* **10**, 1519–1523.

170. Kim, H.A., Rosenbaum, T., Marchionni, M.A., Ratner, N., and DeClue, J.E. (1995) Schwann cells from neurofibromin deficient mice exhibit activation of p21^{ras}, inhibition of cell proliferation and morphological changes. *Oncogene* **11**, 325–335.

171. Becher, R., Wake, N., Gibas, Z., Ochi, H., and Sandberg, A.A. (1984) Chromosome changes in soft tissue sarcomas. *J. Natl. Cancer Inst.* **72**, 823–831.

172. Riccardi, V.M., and Elder, D.W. (1986) Multiple cytogenetic aberrations in neurofibrosarcoma complicating neurofibromatosis. *Cancer Genet. Cytogenet.* **23**, 199–209.

173. Decker, H.J.H., Cannizzaro, L.A., Mendez, M.J., Leong, S.P.L., Bixenman, H., Berger, C., and Sandberg, A.A. (1990) Chromosomes 17 and 22 involved in marker formation in neurofibrosarcoma in von Recklinghausen disease. *Hum. Genet.* **85**, 337–342.

174. Rey, J.A., Bello, M.J., Kusak, M.E., de Campos, J.M., and Pestaña, A. (1993) Involvement of 22q12 in a neurofibrosarcoma in neurofibromatosis type 1. *Cancer Genet. Cytogenet.* **66**, 28–32.

175. Mertens, F., Heim, S., Kullendorff, C.-M., Donnér, M., Hägerstrand, I., Mitelman, F., and Mandahl F. (1995) Clonal karyotypic evolution in a pediatric neurofibrosarcoma. *Cancer Genet. Cytogenet.* **81**, 135–138.

176. McComb, E.N., McComb, R.D., DeBoer, J.M., Neff, J.R., and Bridge, J.A. (1996) Cytogenetic analysis of a malignant triton tumor and a malignant peripheral nerve sheath tumor and a review of the literature. *Cancer Genet. Cytogenet.* **91**, 8–12.

176A. Sciot, R., Dal Cin, P., Fletcher, C.D.M., De Wever, I., De Vos, R., Van Damme, B., and Van den Berghe, H. (1995) Monosomy 22 in a malignant peripheral nerve sheath tumour of the kidney in childhood: a genetic link with other malignant paediatric renal neoplasms. *Histopathology* **27**, 373–376.

177. Molenaar, W.M., Dijkhuizen, T., van Echten, J., Hoekstra, H.J., and van den Berg, E. (1997) Cytogenetic support for early malignant change in a diffuse neurofibroma not associated with neurofibromatosis. *Cancer Genet. Cytogenet.* **97**, 70–72.

178. Pauwels, P., Dal Cin, P., Sciot, R., Lammens, M., Penn, O., van Nes, E., Kwee, W.S., and van den Berghe, H. (1999) Primary malignant peripheral nerve sheath tumour of the heart. *Histopathology* **34**, 56–59.

178A. Chauveinc, L., Dutrillaux, A.M., Validire, P., Padoy, E., Sabatier, L., Couturier, J., and Dutrillaux, B. (1999) Cytogenetic study of eight new cases of radiation-induced solid tumors. *Cancer Genet. Cytogenet.* **114**, 1–8.

179. Frank, D., Gunawan, B., Holtrup, M., and Füzesi, L. (2003) Cytogenetic characterization of three malignant peripheral nerve sheath tumors. *Cancer Genet. Cytogenet.* **144**, 18–22.

180. Gil, Z., Fliss, D.M., Voskoboimik, N., Trejo-Leider, L., Khafif, A., Yaron, Y., and Orr-Urtreger, A. (2003) Two novel translocations, t(2;4)(q35;q31) and t(X;12)(q22;q24), as the

only karyotypic abnormalities in a malignant peripheral nerve sheath tumor of the skull base. *Cancer Genet. Cytogenet.* **145,** 139–143.

181. Van Roy, N., Van Gele, M., Vandesompele, J., Messiaen, L., Van Belle, S., Sciot, R., Mortéle, K., Gyselinck, J., Michiels, E., Forsyth, R., Van Marck, E., De Paepe, A., and Speleman, F. (2003) Evidence for involvement of a tumor suppressor gene on 1p in malignant peripheral nerve sheath tumors. *Cancer Genet. Cytogenet.* **143,** 120–124.

182. Frahm, S., Mautner, V.F., Brems, H., Legius, E., Debiec-Rychter, M., Friedrich, R.E., Knofel, W.T., Peiper, M., and Kluwe, L. (2004) Genetic and phenotypic characterization of tumor cells derived from malignant peripheral nerve sheath tumors of neurofibromatosis type 1 patients. *Neurobiol. Dis.* **16,** 85–91.

183. Hirose, T., Tani, T., Shimada, T., Ishizawa, K., Shimada, S., and Sano, T. (2003) Immunohistochemical demonstration of EMA/Glut1-positive perineurial cells and CD34-positive fibroblastic cells in peripheral nerve sheath tumors. *Mod. Pathol.* **16,** 293–298.

184. Kawaguchi, K., Oda, Y., Saito, T., Takahira, T., Yamamoto, H., Tamiya, S., Iwamoto, Y., and Tsuneyoshi, M. (2005) Genetic and epigenetic alterations of the *PTEN* gene in soft tissue sarcomas. *Hum. Pathol.* **36,** 357–363.

185. Leroy, X., Aubert, S., Leteurtre, E., and Gosselin, B. (2003) Expression of CD117 in a malignant peripheral nerve sheath tumor arising in a patient with type 1 neurofibromatosis. *Histopathology* **42,** 511–513.

185A. Meye, A., Würl, P., Hinze, R., Berger, D., Bache, M., Schmidt, H., Rath, F.-W., and Taubert, H. (1998) No p16^{INK4A}/CDKN2/MTS1 mutations independent of p53 status in soft tissue sarcomas. *J. Pathol.* **184,** 14–17.

186. Miettinen, M., and Cupo, W. (1993) Neural cell adhesion molecule distribution in soft tissue tumors. *Hum. Pathol.* **24,** 62–66.

187. Oda, Y., Saito, T., Tateishi, N., Ohishi, Y., Tamiya, S., Yamamoto, H., Yokoyama, R., Uchiumi, T., Iwamoto, Y., Kuwano, M., and Tsuneyoshi, M. (2005) ATP-binding cassette superfamily transporter gene expression in human soft tissue sarcomas. *Int. J. Cancer* **114,** 854–862.

188. Saito, T., Oda, Y., Kawaguchi, K., Takahira, T., Yamamoto, H., Tamiya, S., Tanaka, K., Matsuda, S., Sakamoto, A., Iwamoto, Y., and Tsuneyoshi, M. (2003) PTEN/MMAC1 gene mutation is a rare event in soft tissue sarcomas without specific balanced translocations. *Int. J. Cancer* **104,** 175–178.

189. Salomon, D.S., Brandt, R., Ciardiello, F., and Normanno, N. (1995) Epidermal growth factor-related peptides and their receptors in human malignancies. *Crit. Rev. Oncol. Hematol.* **19,** 183–232.

190. Yamate, J., Yasui, H., Benn, S.J., Tsukamoto, Y., Kuwamura, M., Kumagai, D., Sakuma, S., and LaMarre, J. (2003) Characterization of newly established tumor lines from a spontaneous malignant schwannoma in F344 rats: nerve growth factor production, growth inhibition by transforming growth factor-β1, and macrophage-like phenotype expression. *Acta Neuropathol.* **106,** 221–233.

190A. Nagashima, Y., Ohaki, Y., Tanaka, Y., Sumino, K., Funabiki, T., Okuyama, T., Watanabe, S., Umeda, M., anf Misugi, K. (1990) Establishment of an epithelioid malignant

schwannoma cell line (YST-1). *Virchows Arch. B Cell Pathol. Incl. Mol. Pathol.* **59,** 321–327.

191. Travis, J.A., Sandberg, A.A., Neff, J.R., and Bridge, J.A. (1994) Cytogenetic findings in malignant triton tumor. *Genes Chromosomes Cancer* **9,** 1–7.

192. Hennig, Y., Löschke, S., Katenkamp, D., Bartnitzke, S., and Bullerdiek, J. (2000) A malignant triton tumor with an unbalanced translocation (1;13)(q10;q10) and an isochromosome (8)(q10) as the sole karyotypic abnormalities. *Cancer Genet. Cytogenet.* **118,** 80–82.

193. Magrini, E., Pragliola, A., Fantasia, D., Calabrese, G., Gaiba, A., Farnedi, A., Collina, G., and Pession, A. (2004) Acquisition of i(8q) as an early event in malignant triton tumors. *Cancer Genet. Cytogenet.* **154,** 150–155.

194. Velagaleti, G.V.N., Miettinen, M., and Gatalica, Z. (2004) Malignant peripheral nerve sheath tumor with rhabdomyoblastic differentiation (malignant triton tumor) with balanced t(7;9)(q11.2;p24) and unbalanced translocation der(16)t(1;16)(q23;q13). *Cancer Genet. Cytogenet.* **149,** 23–27.

195. Davis, J.B., and Stroobant, P. (1990) Platelet-derived growth factors and fibroblast growth factors are mitogens for rat Schwann cells. *J. Cell Biol.* **110,** 1353–1360.

196. Delaney, C.L., Cheng, H.-L., and Feldman, E.L. (1999) Insulin-like growth factor-1 prevents caspase-mediated apoptosis in Schwann cells. *J Neurobiol.* **41,** 540–548.

197. Grinspan, J.B., Marchionni, M.A., Reeves, M., Coulaloglou, M., and Scherer, S.S. (1996) Axonal interactions regulate Schwann cell apoptosis in developing peripheral nerve: neuregulin receptors and the role of neuregulins. *J. Neurosci.* **16,** 6107–6118.

198. Morrissey, T.K., Levi, A.D.O., Nuijens, A., Sliwkowski, M.X., and Bunge, R.P. (1995) Axon-induced mitogenesis of human Schwann cell involves heregulin and p185^{erbB2}. *Proc. Natl. Acad. Sci. USA* **92,** 1431–1435.

199. Li, R., Chen, J., Hammonds, G., Phillips, H., Armanini, M., Wood, P., Bunge, R., Godowski, P.J., Sliwkowski, M.X., and Mather, J.P. (1996) Identification of Gas6 as a growth factor for human Schwann cells. *J. Neurosci.* **16,** 2012–2019.

200. Pinkas-Kramarski, R., Alroy, I., and Yarden, Y. (1997) ErbB receptors and EGF-like ligands: cell lineage determination and oncogenesis through combinatorial signaling. *J. Mammary Gland Biol. Neoplasia* **2,** 97–107.

201. Raabe, T.D., Clive, D.R., Neuberger, T.J., Wen, D., and DeVries, G.H. (1996) Cultured neonatal Schwann cells contain and secrete neuregulins. *J. Neurosci. Res.* **46,** 263–270.

202. Sherman, L.S., Rizvi, T.A., Karyala, S., and Ratner, N. (2000) CD44 enhances neuregulin signaling by Schwann cells. *J. Cell Biol.* **150,** 1071–1083.

203. Aoki, M., Nabeshima, K., Nishio, J., Ishiguro, M., Fujita, C., Koga, K., Hamasaki, M., Kaneko, Y., and Iwasaki, H. (2006) Establishment of three malignant peripheral nerve sheath tumor cell lines, FU-SFT8611, 8710 and 9817: Conventional and molecular cytogenetic characterization. *Int. J. Oncol.* **29,** 1421–1428.

204. Dang, I., Nelson, J.K., and DeVries, G.H. (2005) c-Kit receptor expression in normal human Schwann cells and Schwann cell lines derived from neurofibromatosis type 1 tumors. *J. Neurosci. Res.* **82,** 465–471.

205. Friedrich, C., Holtkamp, N., Cinatl, J., Jr., Sakuma, S., Mautner, V.-F., Wellman, S., Michaelis, M., Henze, G., Kurtz, A., and Driever, P.H. (2005) Overexpression of Midkine in malignant peripheral nerve sheath tumor cells inhibits apoptosis and increases angiogenic potency. *Int. J. Oncol.* **27**, 1433–1440.

206. Holtkamp, N., Okuducu, A.F., Mucha, J., Afanasieva, A., Hartmann, C., Atallah, I., Estevez-Schwarz, L., Mawrin, C., Friedrich, R.E., Mautner, V.-F., and von Deimling, A. (2006) Mutation and expression of *PDGFRA* and *KIT* in malignant peripheral nerve sheath tumors, and its implications for imatinib sensitivity. *Carcinogenesis* **27**, 664–671.

207. Karube, K., Nabeshima, K., Ishiguro, M., Harada, M., and Iwasaki, H. (2006) cDNA microarray analysis of cancer associated gene expression profiles in malignant peripheral nerve sheath tumours. *J. Clin. Pathol.* **59**, 160–165.

208. Mattingly, R.R., Kraniak, J.M., Dilworth, J.T., Mathieu, P., Bealmear, B., Nowak, J.E., Benjamins, J.A., Tainsky, M.A., and Reiners, J.J., Jr. (2006) The mitogen-activated protein kinase/extracellular signal-regulated kinase kinase inhibitor PD184352 (Cl-1040) selectively induces apoptosis in malignant schwannoma cell lines. *J. Pharmacol. Exp. Ther.* **316**, 456–465.

209. Mawrin, C., Schulz, S., Hellwig-Patyk, A., Kirches, E., Roessner, A., Lendeckel, U., Firsching, R., Vorwerk, C.K., Keilhoff, G., Dietzmann, K., Grimm, K., Lindberg, G., Gutmann, D.H., Scheithauer, B.W., and Perry, A. (2005) Expression and function of somatostatin receptors in malignant peripheral nerve sheath tumors. *J. Neuropathol. Exp. Neurol.* **64**, 1080–1088.

210. Sabah, M., Cummins, R., Leader, M., and Kay, E. (2006) Loss of p16 (INK4A) expression is associated with allelic imbalance/loss of heterozygosity of chromosome 9p21 in microdissected malignant peripheral nerve sheath tumors. *Appl. Immunohistochem. Mol. Morphol.* **14**, 97–102.

211. Hagel, C., Zils, U., Peiper, M., Kluwe, L., Gotthard, S., Friedrich, R.E., Zurakowski, D., von Deimling, A., and Mautner, V.F. (2007) Histopathology and clinical outcome of NF1-associated vs. sporadic malignant peripheral nerve sheath tumors. *J. Neurooncol.* **82**, 187–192.

212. Storlazzi, C.T., Brekke, H.R., Mandahl, N., Brosjö, O., Smeland, S., Lothe, R.A., and Mertens, F. (2006) Identification of a novel amplicon at distal 17q containing the *BIRC5/SURVIVIN* gene in malignant peripheral nerve sheath tumours. *J. Pathol.* **209**, 492–500.

213. Upadhyaya, M., Spurlock, G., Majounie, E., Griffiths, S., Forrester, N., Baser, M., Huson, S.M., Gareth Evans, D., and Ferner, R. (2006) The heterogeneous nature of germline mutations in NF1 patients with malignant peripheral nerve sheath tumours (MPNSTs). *Hum. Mutat.* **27**, 716.

214. Lee, P.R., Cohen, J.E., and Fields, R.D. (2006) Immune system evasion by peripheral nerve sheath tumor. *Neurosci. Lett.* **297**, 126–129.

215. Ishiguro, M., Iwasaki, H., Takeshita, M., Hirose, Y., and Kaneko, Y. (2006) A cytogenetic analysis in two cases of malignant peripheral nerve sheath tumor showing hypodiploid karyotype. *Oncol. Rpts.* **16**, 225–232.

216. Kleinschmidt-DeMasters, B.K., Lovell, M.A., Donson, A.M., Wilkinson, C.C., Madden, J.R., Addo-Yobo, S.O., Lillehei, K.O., and Foreman, N.K. (2007) Molecular array analyses of 51 pediatric tumors shows overlap between malignant intracranial ectomesenchymoma and MPNST but not medulloblastoma or atypical teratoid rhabdoid tumor. *Acta Neuropathol.* **113**, 695–703.

217. Rodriguez, F.J., Scheithauer, B.W., Abell-Aleff, P.C., Elamin, E., and Erlandson, R.A. (2007) Low grade malignant peripheral nerve sheath tumor with smooth muscle differentiation. *Acta Neuropathol.* **113**, 705–709.

218. Miller, S.J., Rangwala, F., Williams, J., Ackerman, P., Kong, S., Jegga, A.G., Kaiser, S., Aronow, B.J., Frahm, S., Kluwe, L., Mautner, V., Upadhyaya, M., Muir, D.,. Wallace, M., Hagen, J., Quelle, D.E., Watson, M.A., Perry, A., Gutmann, D.H., and Ratner, N. (2006) Large-scale molecular comparison of human Schwann cells to malignant peripheral nerve sheath tumor cell lines and tissues. *Cancer Res.* **66**, 2584–2591.

219. Mattingly, R.R., Kraniak, J.M., Dilworth, J.T., Mathieu, P., Bealmear, B., Nowak, J.E., Benjamins, J.A., Tainsky, M.A., and Reiners, J.J. (2006) The mitogen-activated protein kinase/extracellular signal-regulated kinase kinase inhibitor PD184352 (CI-1040) selectively induces apoptosis in malignant schwannoma cell lines. *J. Pharmacol. Exp. Ther.* **316**, 456–465.

220. Robanus-Maandag, E., Giovannini, M., van der Valk, M., Niwa-Kawakita, M., Abramowski, V., Antonescu, C., Thomas, G., and Berns, A. (2004) Synergy of *Nf2* and *p53* mutations in development of malignant tumours of neural crest origin. *Oncogene* **23**, 6541–6547.

221. Kluwe, L., Friedrich, R.E., Peiper, M., Friedman, J., and Mautner, V.F. (2003) Constitutional NF1 mutations in neurofibromatosis 1 patients with malignant peripheral nerve sheath tumors. *Hum. Mutat.* **22**, 420.

222. Su, W., Sin, M., Darrow, A., and Sherman, L.S. (2003) Malignant peripheral nerve sheath tumor cell invasion is facilitated by Src and aberrant CD44 expression. *Glia.* **42**, 350–358.

223. Tucker, T., Wolkenstein, P., Revuz, J., Zeller, J., and Friedman, J.M. (2005) Association between benign and malignant peripheral nerve sheath tumors in NF1. *Neurology* **65**, 205–211.

224. Wu, R., López-Correa, C., Rutkowski, J.L., Baumbach, L.L., Glover, T.W., and Legius, E. (1999) Germline mutations in NF1 patients with malignancies. *Genes Chromosomes Cancer* **26**, 376–380.

3
Meningioma

Summary   Meningioma is the subject matter of Chapter 3. Cytogenetically, meningioma is characterized by loss of chromosome 22 (-22) or less often by changes of the long arm of this chromosome (22q). The changes of chromosome 22 affect the gene *NF2*, which probably constitute the original tumorigenic events leading to meningioma development. In addition to changes of chromosome 22, those of chromosomes 1, 14, 10, 9, 17, and 18 (shown in the order of frequency) may play a role in full tumor genesis and biology. Comparative genomic hybridization and loss of heterozygosity studies have generally confirmed those of cytogenetics. The alterations of the *NF2* gene, already mentioned as a key molecular genetic event in meningiomas, and its protein product merlin (schwannomin) are presented in Chapter 3. Other areas presented are those of pediatric, radiation-induced, multiple and familial meningiomas. Molecular facets that have been investigated in meningiomas are included in Chapter 3.

Keywords   meningiomas · loss of chromosome 22 (-22) · *NF2* gene · radiation-induced meningiomas.

Introduction

Meningiomas are neoplasms that arise from the leptomeningeal covering of the brain and spinal cord, and they account for 15–20% of all central nervous system (CNS) tumors (1). Meningiomas constitute 18–25% of all primary intracranial tumors and 25% of intraspinal tumors; >80% of meningiomas are benign in nature. About 12% of all meningiomas recur within 5 years and 19% within 20 years of surgical treatment (2, 3). The current World Health Organization (WHO) grading system comprises three grades. The most common meningiomas (80–90%) are slow-growing and considered benign tumors, and they correspond to grade I (typical meningiomas). About 5–15% are classified as grade II (atypical) and 1–3% as grade III (anaplastic/malignant) tumors. Grades II and II tumors exhibit more aggressive clinical behavior, with a higher risk of recurrence compared with typical grade I meningiomas (Table 3.1) (4, 5). Brain invasion may constitute one of the criteria for the diagnosis of atypical meningioma (MII) (6). Meningiomas are usually sporadic, but a few families have been described with multiple tumors inherited in an autosomal dominant manner (8–10). Meningiomas occur in as many as half of the patients with the dominantly inherited familial neurofibromatosis-2 (NF2) syndrome (11). Most NF2 patients who develop meningiomas do so earlier in life than the patients with sporadic tumors, and the meningiomas are multiple in 1–8% of the cases (9,10) and mostly fibroblastic in histology (12). The *NF2* gene, responsible for NF2, is located at 22q12 (13, 14). Germline mutations of this gene are present in almost all meningiomas in NF2 patients and somatic mutations of *NF2* in about 30% of sporadic meningiomas (15–24). These findings indicate that inactivation of the *NF2* gene is important in the development of a significant number of sporadic meningiomas, which also display loss of heterozygosity (LOH) for markers located on chromosome 22 (25). Thus, *NF2* acts as a tumor suppressor gene (TSG) in at least a significant subgroup of sporadic meningiomas. A classification of meningiomas according to histologic type and grade is shown in Table 3.2 and Figure 3.1. The frequency of various grades of meningioma and their recurrence rates are shown in Table 3.3.

Rhabdoid meningiomas are rare (Figure 3.1b) but aggressive in nature (26, 27). They are usually grade III tumors and may have to be differentiated from malignant rhabdoid tumors and atypical teratoid-rhabdoid tumors (see Chapter 9). The latter tumors show consistent loss of INI1 nuclear expression immunohistochemically, whereas composite rhabdoid meningioma retain such expression (28). The authors (28) concluded that determining the INI1 status in meningiomas by immunohistochemistry is a relatively simple, sensitive, and specific technique for distinguishing rhabdoid meningioma from other

TABLE 3.4. Karyotypes of meningioma with structural chromosome changes.

Karyotype	References
43,XX,−1,−6,−7,−11,−22,+2mar/42,idem,−9	*Mark et al 1972 (58)*
44,XY,−1,−17,−22,+mar/43,idem,−8,−19,+mar	*Mark et al 1972 (58)*
45,XX,−22/44,idem,−8/44,XX,−8,der(12)t(12;21)(p?;q?),−21	*Mark et al 1972 (58)*
47,XX,+12,del(22)(q11)/46,idem,−8	*Mark et al 1972 (58)*
41−42,XY,−1,−5,−8,add(9)(q?),−14,−18,−22,+r	*Mark 1971, 1973 (60, 66)*
45,XY,−22/45,idem,del(12)(q?)	*Mark 1973 (59, 396)*
45,XX,−22/45,XX,−8,del(22)(q?)	*Mark 1973 (59, 396)*
45,XX,−22/48,XX,+6,+14,del(22)(q?)	*Mark 1973 (59, 396)*
38,X,−Y,del(1)(p?),−4,−9,−10,−13,−14,−17,−21,−22,+mar	*Zankl et al 1975 (397)*
41,X,−X,del(1)(p?),−10,−15,−18,−22	*Zankl et al 1975 (397)*
44,Y,−X,−14,−22,+mar/43,idem,−10,−11,−15,+2mar	*Zankl et al 1975 (397)*
45,XX,del(20)(q11),−22/44,XX,−20,−22	*Zankl et al 1975 (397)*
45,XX,del(1)(p?),del(2)(p?),del(4)(p?),−15,−22,+mar	*Zankl et al 1979 (398)*
45,XX,1p−,2p−,4p−,−15,−22,+mar	*Zankl et al 1982 (115A)*
41−43,XX,−1,2−,3−,4−,6,−8,−9,−15,−19,−22,+5−7mar	*Battersby et al 1986 (8)*
40,XY,−1,−6,−12,der(14;15)(q10;q10),−16,−22	*Katsuyama et al 1986 (398A)*
41,XX,−1,−10,der(11)t(1;11)(p11;p15),−14,−18,−22	*Katsuyama et al 1986 (398A)*
42,XY,−1,−6,−8,−14,del(22)(q?)	*Katsuyama et al 1986 (398A)*
38−39,X,−X,−3,−13,−14,−18,−21,−22,+3−4mar	*Al Saadi et al 1987 (71)*
41,X,−X,−1,−22,inc/46,XX,−1,−22,+r,+mar,dmin	*Al Saadi et al 1987 (71)*
43,XX,−14,−19,−22/44,idem,+r	*Al Saadi et al 1987 (71)*
43−45,XX,t(9;11),add(20)(p?),−22,+2−5mar	*Al Saadi et al 1987 (71)*
44,XX,del(1)(p?),−7,−18, −20,t(22;22)	*Al Saadi et al 1987 (71)*
44,XX,t(1;22),t(15;17)/43,idem,t(14;15)/42,idem,t(6;8), t(6;15;17), t(14;15)	*Al Saadi et al 1987 (71)*
44,X,−X,−6,−7,−11,−12,−12,−13,−13, −17,−18,+19,+20,+22,+5mar	*Al Saadi et al 1987 (71)*
44,XX,−7,−22,dmin	*Al Saadi et al 1987 (71)*
45,XY,t(4;14),−22	*Al Saadi et al 1987 (71)*
46,XY,add(7)(q?)	*Al Saadi et al 1987 (71)*
46,XY,del(2)(q24)	*Al Saadi et al 1987 (71)*
46,XX,t(1;3)(q?;q?)	*Al Saadi et al 1987 (71)*
46,XX,t(3;7)(p13;q36)	*Al Saadi et al 1987 (71)*
48,XX,+17,+mar	*Al Saadi et al 1987 (71)*
41,XX,−1,−2,−4,−9,−15,−19,−22,+der(1)t(?1;15)(?::1q42→p11::15q11→qter),del(3)(p13),+der(9)t(9;?)(p11;?)	*Maltby et al 1998 (72)*
41−43,XX,der(1)t(1;15)(p11;q11)add(1)(q42),add(2)(p23),del(3)(p13),der(4)t(4;8)(p11;q11),−6,−8,add(9)(p11), −15,dic(19;?),−22	*Maltby et al 1998 (72)*
42,XY,del(1)(p34),del(3)(p11), add(6)(q21), dic(14;?),−17,?der(19)t(17;19)(q21;q12),−20,−21,−22,?add(22)(q11)	*Maltby et al 1998 (72)*
44,XY,der(1)t(1;?;19)(p22;?q12), add(7)(p11),−19,−22	*Maltby et al 1988 (72)*
44,XX,dic(1;9)(p?;q?),−22	*Maltby et al 1988 (72)*
45,XX,add(1)(p11),−22/44,idem,dic(9;14)(q34;p11)	*Maltby et al 1988 (72)*
45,XY,der(1)t(1;7)(q32;p13),del(6)(q21),der(15)t(15;17)(q22;q21),−17,der(19)t(15;19)(q22;q13)	*Maltby et al 1988 (72)*
45,XX,−22/45,idem,add(12)(q11)/46,XX,t(17;22)(q23;q12)	*Maltby et al 1988 (72)*
40,XX,der(1;11)(q10;q10),−8,−13,−15,−18,−22	*Rey et al 1988 (73)*
42,XY,del(1)(p22),add(3)(q27),del(11)(p13),−13,−14,−18,−21,−22,+mar	*Rey et al 1988 (73)*
43,X,−Y,der(1)t(1;1)(p34;q21),−14,−22	*Rey et al 1988 (73)*
44,XX,del(1)(p32),−4,−5,−14,−22,+2mar	*Rey et al 1988 (73)*
45,XY,add(1),−22	*Rey et al 1988 (73)*
45,XY,add(8)(p12),−22	*Rey et al 1988 (73)*
45,XX,dic(19;22)(p13;q11)/45,XX,der(19)dic(19;22)del(19)(p13),−22/45,XX,del(1)(p34),der(3)t(1;3)(q12;q29), −19,−22,+mar/45,XX,r(1)(p34q43),−19,−22,+mar/44,XX,r(1),−11,−15,−19,−22,+2mar	*Rey et al 1988 (399)*
45,XY,+1,der(1;6)(q10;p10)×2,t(3;10)(q21;p13),+6,−22	*Rey et al 1988 (399)*
45,XX,−1,+r(1)(p34q43)M7,−19,−22,+M1 or M2/44,XX,−1,+r(1)(dupM7)M8,−11,−15,−19,−22, +M1 or M2,+markers involving 11p(Marker9−Marker10)	*Rey et al 1988 (399)*
45,XX,r(19),−22	*Rey et al 1988 (399)*
45−46,Y,dic(X;1)(q28;p11),+1,del(1)(p11)×−2,−4,der(6;11)(p10;q10),−14,−22,+r,+mar	*Rey et al 1988 (399)*
49,XX,inv(1)(p12−13q24)1,del(2)(q33)2,del(6)(q21)3,+7,−8,+11,+12,del(13)(q21)4,−16,−17,+19,+20, −22,+2mar	*Rey et al 1988 (399)*
44,X,−Y,del(4)(q31q35),del(5)(q?),−22	*Rogatto and Casartelli 1988 (400)*
45,XX,del(12)(p12p13),del(19)(p13),−22	*Rogatto and Casartelli 1988 (400)*
78−82,XX?,del(22)(q?),inc	*Rogatto and Casartelli 1988 (400)*
32,X,+del(3)(p13),+5,+8,+9,+del(13)(q21),+15,+19,+22, +mar	*Rönne et al 1988 (401)*
41−45,X,−Y,+del(1)(p32),del(2)(q31), dic(6;22)(p12;q13), inv(6)(p21q12),−13,−14,−18,−20,−21,−22	*Casartelli et al 1989 (75)*
40,X,−Y,−10,−12,−13,−14,−22/39,idem,der(17)t(17;22)(p13;q11), −22	*López−Ginés et al 1989 (402)*
45,XX,−22/44,idem,del(X)(q26),+4,−9,−14	*Poulsgård et al 1989 (403)*
42,XY,der(1)t(1;?)(p12;?),-6,der(12;15)(q10;q10),-18,-22	*Woo et al 2008 (620)*

TABLE 3.4. Continued

Karyotype	References
45,XX,−22/44,idem,der(1)del(1)(p33)del(1)(q43),+der(1)t(1;1)(p33;q44)dup(1)(q24q31),−4,−6,−8,+19	*Poulsgård et al 1989* (403)
44,X,+der(3)(p?),−13,+t(13;15),−15,−22	*Strachan et al 1989* (131)
44,XX,−6,+add(7),−16,−22	*Strachen et al 1989* (131)
45,XY,−22/46,XY,del(22)(q?)	*Strachan et al 1989* (131)
46,X,?der(4)t(4;15)(q28;q11),t(4;14)(q22;q32)	*Teyssier and Ferre 1989* (404)
44−45,XX,−1,del(7)(q11),−22,−22	*Westphal et al 1989* (132)
45,XX,−22/45,idem,t(21;22)	*Westphal et al 1989* (132)
46,XY,del(4)(q24),t(4;7)(q24;q36)	*Westphal et al 1989* (132)
46,XX,+del(14)(q22q32),−22	*Wullich et al 1989* (405)
52,XX,+8,+9,+12,+15,+16,+20,+21,−22	*Wullich et al 1989* (405)
44,XX,del(1)(p21),−14,−19,−22,+2mar	*Casalone et al 1990* (74)
44,XY,−14,−22/44,idem,del(1)(p2?2)	*Casalone et al 1990* (74)
44,X,−Y,−22/44,idem,add(9)(q34)	*Casalone et al 1990* (74)
46,XY,add(1)(p?)	*Casalone et al 1990* (74)
46,XX,del(1)(p2?1)	*Casalone et al 1990* (74)
46,XX,del(7)(q22q32)	*Casalone et al 1990* (74)
46,XY,t(1;7;14)(q25;q32;q22),t(18;22)(q12;q11)	*Casalone et al 1990* (74)
46,XY,t(2;7;14)(q23;q36;q22)	*Casalone et al 1990* (74)
??,X?,add(7)(q31),add(9)(p?),add(22)(q11)	*Casalone et al 1990* (74)
??,X?,der(X)(p22),+2mar	*Casalone et al 1990* (74)
41,X,−Y,del(1)(p?),der(6),−14,−15,−22	*Logan et al 1990* (406)
43,XY,−10,−14,−22/43,idem,del(1)(p?)	*Logan et al 1990* (406)
44,XY,−6,t(6;19),−19,−22	*Logan et al 1990* (406)
45,XY,−22/45,idem,add(1)(q?)/44,idem, add(1),−22	*Logan et al 1990* (406)
45,X,−Y/45,XX,inv(X)(p11q12)	*Berra et al 1991* (133)
45,XX,−22/45,idem,der(22)t(1;22)(q12;p12)	*Berra et al 1991* (133)
45,XY,−22/46,idem,+7/45,idem,add(11)(p15)	*Berra et al 1991* (133)
46,XX,+i(4)(p10) or i(5)(p10),−22/45,idem,−X	*Berra et al 1991* (133)
46,XY,t(Y;1)(q12;q31)/46,X,der(Y)t(Y;1)	*Berra et al 1991* (133)
47−50,XY,+Y,+5,+8,+12,+13,der(17),−18,+20,−22	*Chadduck et al 1991* (407)
43,XX,−9,−22/41,idem,r(?1),−3,−11,+mar	*Chio et al 1991* (408)
45,XX,−22/46,idem,+mar	*Chio et al 1991* (408)
45,XY,−1,t(1;22)(p11;q11),t(4;22)(p16;q11)/44, idem,add(13)(p?), −14/44,idem,−10,+12,add(13)(p?),−14/45, idem,−10,+12,add(13)(p?),−14,+16	*Lekanne Deprez et al 1991* (237)
46,XX,add(1)(p36),add(2)(q31),inv(2)(q22q33)	*Meloni et al 1991* (409)
46,X,−Y,der(1),+3,+7,+8, −9,−17,+19,−21,−22,+mar	*Vagner−Capodano et al 1991* (410)
45,XY,t(15;22)(p?;q?)	*Cho et al 1992* (411)
44,XY,add(1)(p13)x2,add(2)(p11), del(4)(q31), del(6)(q21),−14,−16,del(17)(q21),−22,+mar	*Karnes et al 1992* (412)
46,XX,del(22)(q11q13)/45,XX,−22	*Tonk et al 1992* (413)
42−44,XY,del(1)(p13p22),tas(4;21)(p16;p13), tas(8;22)(p23;p13), inc	*Vagner−Capodano et al 1992* (414)
45,XX,tas(5;12)(p15;p13),tas(12;13)(p13;p13),−22	*Vagner−Capodano et al 1992* (414)
46,XY,t(1;3)(q42;q21),t(2;16)(p23;p13),inv(9)(p13q34)	*Bello et al 1993* (415)
46,XX,t(2;3)(q35;q11)	*Bello et al 1993* (415)
48,X,−Y,t(1;4)(q21;q15),+7,+20,+21	*Bello et al 1993* (415)
50−52,X,−X,+5,+6,+12,−14,−16,+18, +19,+20,+21,+2mar	*Bello et al 1993* (415)
36,X,−Y,+2,del(2)(p14)x2,−5,−6,−7,add(8)(q?),+add(9)(p?), −10,−10,−11,+14,+14,−16,add(17)(q?), −20,−20,−21,−22,−22/41,XY,del(2),dic(3;?)(q26;?),−8,add(10)(p?),−13,−17,add(17),+18,−19,−21, −22/43,XY,del(2),add(8)(p?),add(10),i(12)(q10),add(14)(q?),−17,add(17),−19,−20,−22,+mar/ 45,XY,−1,+3,−7,−8,add(8)(q?),−17,add(17),−19,+21,+2mar/71,XY,−X,+Y,del(2),dic(2;?)(q37;?),+4,+7, −8,add(8)(p?),+9,add(10)(p?),+11,+12,add(14),−15,−16,add(17),−18,−18,+21,−22,+2mar	*Doco−Fenzy et al 1993* (106)
39,X,−Y,−2,−10,add(11)(p?),+add(15)(q?),−19,−20,−21,−22/42,XY,−1,der(3)t(1;3)(q11;q19), add(11),−18,−19,−22/43,XY,der(1)t(1;11)(q11;p11),−10,add(11),−19,−22//43,XY,−1,add(11), −19,−22/44,XY,−1,−11,add(11)	*Doco−Fenzy et al 1993* (106)
43,X,−X,−13,−22/43,X,−X,t(3;7),del(4)(q35),add(5)(q?),t(8;11)(p21;p12),der(9)(q31),−11,−22/ 43,X,−X,der(1)(p?),+del(4)(q2?),dic(7;15)(p11;p11),−15,−21,−22	*Doco−Fenzy et al 1993* (106)
44,XX,t(2;9)(p12;q?),−12,−19/45,XX,−13/45,XX,−14/92,XXXX,add(15)(p?)	*Doco−Fenzy et al 1993* (106)
44,XY,−1,t(1;5;6)(p12;q23;q25),−14,−22,+r/43,idem,−18/40,idem,+2,−3,−4,−18,−18,−r/82,idemx2,−5, −9,−15,−17,−18	*Doco−Fenzy et al 1993* (106)
44,X,−X,−22/43,idem,−15/43,idem,−10/44,idem,t(4;10)(p13;q26)/44,idem,t(4;14)(q13;q31)/42,idem,−9, −20/42,idem,−13,−16/43,idem,−13/43,idem,−4/43,idem,−12	*Doco−Fenzy et al 1993* (106)
45,XX,der(1)t(1;2)(q43;q12),−22/79,idem,+1,+3,+4,+5,+11,+12, +14,+16,−17,+18,+19,+20,+21	*Doco−Fenzy et al 1993* (106)
45,XY,−11/45,XY,t(6;8)(q?;p?),−8	*Doco−Fenzy et al 1993* (106)
45,XX,−19/45,XX,−22/45,XX,del(11)(p14),−17	*Doco−Fenzy et al 1993* (106)

TABLE 3.4. Continued

Karyotype	References
45,XX,−22/43,idem,t(8;19)(p23;q13),−19,t(19;20)(q13;p13),−20,add(22)(q?)/43,idem,t(4;19)(q35;q13), −8,t(14;21),add(22)/44,idem,t(4;17)(q25;q25),−17,add(22)/44,idem,+t(2;19)(p25;q13)t(14;21)(q?;q?),−19, −21,add(22),45,idem,add(22)	*Doco−Fenzy et al 1993* (106)
45,XX,−22/45,idem,r(19)/45,idem,+8,−10,r(19)/45,idem,add(6)(q?),+8,−10	*Doco-Fenzy et al 1993* (106)
46,XX,add(5)(q?)/45,XX,−9/45,XX,−21	*Doco-Fenzy et al 1993* (106)
46,XY,del(2)(q22)	*Doco-Fenzy et al 1993* (106)
46,XX,del(4)(q?),t(4;15)(q28;q11),der(14)t(4;14)(q22;q32)	*Doco-Fenzy et al 1993* (106)
46,XY,del(22)(q12)/44,idem,−8,−13	*Doco-Fenzy et al 1993* (106)
46,XY,t(1;2)(q42;q13)/46, XY,t(1;13)(q22;q22)/46,XY,add(3)(q?)	*Doco-Fenzy et al 1993* (106)
57,XX,+3,+4,+8,+9,+12,+14,+15,+15,+17,+19,+20/58, idem,+3,+10,+14,−19, +21/58,idem,+3, +8,−19,−21, +mar/59,idem, −X,+3,+5,+7, +10,−12,−19,+21	*Doco-Fenzy et al 1993* (106)
47,XY,del(11)(q?13q?21),+r	*Dressler et al 1993* (416)
46,XX,del(1)(p32)	*López-Ginés et al 1993* (417)
45,XX,t(1;15),t(7;11),−22	*Pagni et al 1993* (256)
46,XX,t(1;19)(q21;p13) (2 cases)	*Prempree et al 1993* (418)
39,XX,−1,−4,−6,+der(7),−8,−10,−12,+der(14),−17,−18,−22	*Vagner-Capodano et al 1993* (77)
40,X,−Y,der(1),+3,−5,−8,−9,−12,−14,−18,+19,der(21),−22	*Vagner-Capodano et al 1993* (77)
42−44,XY,del(1)(p13),−22,−22	*Vagner-Capodano et al 1993* (77)
44,X,−X,−1,−22,+mar	*Vagner-Capodano et al 1993* (77)
46,X,der(X),t(3;7)(q11;q11)	*Vagner-Capodano et al 1993* (77)
46,XX,inv(2)	*Vagner-Capodano et al 1993* (77)
46,XX,t(1;5)(q22;q35)	*Vagner-Capodano et al 1993* (77)
47,XX,+mar	*Vagner-Capodano et al 1993* (77)
47,XY,−4,−4,+5,+12,−16,+der(19),+mar	*Vagner-Capodano et al 1993* (77)
50,XX,del(1)(q25),+3,+17,+2mar	*Vagner-Capodano et al 1993* (77)
50,XX,+3,+5,der(9),+12,+20	*Vagner-Capodano et al 1993* (77)
44−46,XX,t(1;5)(p13;q?),add(2)(q31),der(2)t(2;14)(p24;q24),t(5;9)(q11;q32),del(6)(q13q26), t(7;8)(p21;q22),−14,−15,add(20)(q13),add(22)(q13), del(22)(q13),+2mar	*Albrecht et al 1994* (419)
42,X,−Y,−1,−6,−8,−18,+20,−22,+r	*Biegel et al 1994* (420)
45,XY,der(2)t(2;3)(q37;p23),dic(3;22)(p14−p21;q13),del(22)(q?)	*Biegel et al 1994* (420)
45,XX,dic(1;6)(p31;q14−q16),der(19)add(19)(p13)add(19)(q13),add(22)(q13)	*Biegel et al 1994* (420)
45,XY,r(6),add(22)(q11),add(22)(q13)	*Biegel et al 1994* (420)
42−44,XY,del(1)(p13p32),−22	*Figarella-Branger et al 1994* (421)
47,XY,+mar	*Figarella-Branger et al 1994* (421)
44,XX,add(3)(q26),der(17)t(17;21)(p11;q11),add(18)(p11), −19,−21,−22,+mar	*Gollin and Janecka 1994* (422)
45,XX,−22/46,idem,+mar	*Gollin und Janecka 1994* (422)
39,X,−Y,add(1)(p32),−10,−14,−16,−18	*Griffin et al 1994* (134)
39−43,XX,del(1)(p31),+del(1)(q22),−6,der(11)t(6;11)(q11;p14),−18,add(19)(q13.4),−21,−22	*Griffin et al 1994* (134)
40,X,−X,der(1)t(1;2)(p11;p11),−2,der(6)t(1;6)(p11;p11),der(10)t(2;10)(q21;q21),+der(17) t(17;19)(p11;p11),−19,−22,inc	*Griffin et al 1994* (134)
41−42,XX,del(11)(q13q21),−22,+mar,inc	*Griffin et al 1994* (134)
43−44,X,−Y,del(1)(p12),der(7)t(7;12)(q11;p13), inv(22)(q11q13), inc	*Griffin et al 1994* (134)
45,XX,del(2)(p16),add(11)(q24),−22	*Griffin et al 1994* (134)
45,XX,−1,−6,−7,der(10)t(1;10)(q25;q26),der(11)t(11;22)(p15;p11),t(15;**22**)(q10;q10)	*Griffin et al 1994* (134)
45,XX,−22,46,XX,del(22)(q12)	*Griffin et al 1994* (134)
46,XY,add(11)(p15)	*Griffin et al 1994* (134)
46,XX,ins(5;?)(?q11;?),r(21)	*Griffin et al 1994* (134)
60−63,XXX,−1,−2,+5,−6,−7,inv(9),−11,−15,+21,−22,inc	*Griffin et al 1994* (134)
41,X,-X,del(1)(p22),-4,6q+,8q+,10q+,11p+, der(12)t(1;12)(q11;q24)×2,-14,-15,-22, del(22)(q11)/44,XX,-14,-22	*Yamada et al 1994* (135)
45,XX,-22/idem,r(11)(p15q25)	*Yamada et al 1994* (135)
50,XY,del(1)(p32),+3,+13,+13,+15,-22,+mar	*Yamada et al 1994* (135)
56,XY,+3,+4,+5,+8,+9,+14,+15,+17,+18,+20/58, XY,+X,+3,+5,+7,+8,+10,+14,+14,+15, +17,+19/59,XY,+X,+3,+4,+5,+8,+9,+13, +14,+15,+17, +18,+19,+20	*Yamada et al 1994* (135)
70,XY,-X,-4,-7,+9,+15,+17,+18,+19,+20,+21,-22/70, idem, add(19)(p13)	*Agamanolis and Malone 1995* (423)
42,XY,del(1)(p12), add(6)(q12),-14,-16, add(16)(q24),-19,-22	*Henn et al 1995* (424)
45,X,-Y/46,XY,add(13)(q34)	*Henn et al 1995* (424)
45,XY,del(1)(p12),t(14;19)(q22;q13),-22/44,XY,del(1)(p12),-14,-22/44,del(1)(p?),-14,-22	*Henn et al 1995* (424)
45,XY,del(1)(p13),-22	*Henn et al 1995* (424)
45,XX,-1,+t(1;4)(p11;q11),+t(3;7)(p12;q21),-4,-6,-7,+mar	*Henn et al 1995* (424)
45,XX,-22/47,XX,+der(X)	*Henn et al 1995* (424)
60-77,XXX,del(1)(p12),2-3dmin,inc	*Henn et al 1995* (424)
39,X,-X,del(1)(p?),-6,-7,del(8)(q?),add(9)(p?),add(11)(p?)×2,-13,-14,add(20)(p?),-21,-22	*Lekanne Deprez et al 1995* (78)

TABLE 3.4. Continued

Karyotype	References
41,XX,add(1)(p12),der(1;2)(p10;q10),-3,-4,der(6)t(1;6)(q12;q14),der(8)t(8;?13)(q24;?q14),del(11)(q11), -13,der(16)t(4;16)(q21;p13),der(18)t(11;18)(q12;q12),der(19)t(3;19)(q11;q12),-22/42,idem,+r/84,idem×2,+2r	*Lekanne Deprez et al 1995 (78)*
42,XY,i(1)(q10),-4,-14,-18,-22	*Lekanne Deprez et al 1995 (78)*
42,X,-X,-1,-12,-14,-17,-18,-19,-20,-22,+r	*Lekanne Deprez et al 1995 (78)*
42-43,XY,dic(1;19)(p12;p13),del(11)(p12), dic(11;21)(p11;p12),tas(11;21)(p15;p13),-14, tas(19;21)(q13;p13),-21,-22	*Lekanne Deprez et al 1995 (78)*
43,XX,der(1)t(1;16)(p11;q11),dic(6;16)(p12;q11) or dic(6;22)(p12;q11),-14,-16,-22	*Lekanne Deprez et al 1995 (78)*
43,XX,-14,-19,-22/42,idem,tas(7;19)(p22;q13)	*Lekanne Deprez et al 1995 (78)*
44,X,del(X)(q13),trc(1;22;19)(p12;q11;q10),inv(7)(p14-p15q34-q35),der(15),der(16)t(16;22) (q24;q?),i(18)(p10),inc	*Lekanne Deprez et al 1995 (78)*
44,Y,der(X),del(1)(p?),t(1;6)(p32;p21),del(2)(q?),-4,del(5)(q?),add(6)(p?),+del(6)(q?), der(7),add(10)(q?),del(12)(q?),der(12;17)(q10;q10),-15,-19,del(22)(q?),der(22)t(2;22)(q11;q11)./ 44,dem,add(9)(q?) or -9,add(11)(p?) or -11,-21	*Lekanne Deprez et al 1995 (78)*
44,XX,dic(1;19)(q10;p13),add(2)(p21), der(2)t(2;8)(q1?4;q21), add(4)(q21),der(8)t(2;8)(p21;q21), del(12)(p12),t(14;16)(q11;q11-12),der(18)t(2;18)(q14;p11),-22	*Lekanne Deprez et al 1995 (78)*
45,XX,del(1)(q?),add(3)(q?),-10/44,XX,del(2)(q?),-6,-7,add(10)(p?),add(11)(p?),add(13)(q?),add(19)(q?), -22,add(22)(q?)/45,XX,del(2),-6,-7,add(10),add(11),-13,add(19),-22,add(22),+r	*Lekanne Deprez et al 1995 (78)*
45,XX,dic(1;19)(p13;q11),del(13)(q13q21),+20,-22/46,idem,+20./45,idem,+20,-21	*Lekanne Deprez et al 1995 (78)*
45,XX,dic(19;22)(p1?2;q1?1)/42-44,idem,del(1)(p?),del(6)(q?),add(11)(p?),-21,-22	*Lekanne Deprez et al 1995 (78)*
45,XX,-22/44,idem,dic(13;22)(p11;p11)	*Lekanne Deprez et al 1995 (78)*
45,XX,-22/45,idem,r(17)	*Lekanne Deprez et al 1995 (78)*
46,XX,add(2)(q22)	*Lekanne Deprez et al 1995 (78)*
46,XX,t(3;7)(1?2;q35)	*Lekanne Deprez et al 1995 (78)*
46,XX,t(7;14)(q25;q11-q12)	*Lekanne Deprez et al 1995 (78)*
47,XX,+?del(13)(q14q22)	*Lekanne Deprez et al 1995 (78)*
47,XX,+7,+21,-22/47,idem,add(9)(q?)	*Lekanne Deprez et al 1995 (78)*
54-59,X,-X,+1,del(1)(p?)×2,+3,+5, add(6)(q?), +13,+17,+18,+29,+20,+22,+markers	*Lekanne Deprez et al 1995 (78)*
??,XY,t(1;2)(q4?1;q2?2),-19,add(19)(q?),add(22)(q?),+hsr(?),inc	*Lekanne Deprez et al 1995 (78)*
46,XX,-15,+mar	*López-Ginés et al 1995 (67)*
37-41,X,add(Y)(p11),add(1)(q23),-6,-8,add(11)(p13),-14,-22,+mar/45,X,-Y	*Perry et al 1996 (80)*
40-44,X,add(X)(p?21),der(1)add(1)(p13)dup(1)(q23q42),-2,-3-4,-6,add(9)(q14),-10,dic(11;?)(p11;?), -15,-17,add(17)(q2?5),-18,add(19)(p13),+r,+5-6mar	*Perry et al 1996 (80)*
41,X,-X,der(1)t(1;12)(p?13;q1?1),dic(3;7)(p12;p11),dic(6;22)(q12;q11),-18/40,idem,-13,-21,+mar/41, idem,-4,add(19)(p13),add(20)(q13),+mar	*Perry et al 1996 (80)*
42,XY,del(1)(p13),t(1;3)(q44;q12),-3,-4,add(6)(q13),add(11)(p15),-14,add(16)(p13),-18,-22,+r	*Perry et al 1996 (80)*
44,XX,der(1)t(1;10)(p22;q11),dic(3;10)(p12;q11),dic(6;13)(q13;p11),del(7)(p13p15)	*Perry et al 1996 (80)*
42,XY,add(1)(p13),-6,-8,-14,-22/42, idem,add(17)(q23),add(18)(q12)	*Bhattacharjee et al 1997 (425)*
46,X,-X,r(1),t(3;10)(p14;q22),der(16)add(16)(p13),add(16)(q23),add(17)(q23),add(20)(q13)add(22)(q11), add(22)(q12),+r/46,idem,-r(1),+der(1)(add(1)(p31)add(1)(q41)	*Bhattacharjee et al 1997 (425)*
46,X,inv(X)(p22q26),t(3;5;17) (p22q13; q11;q22)/46,XX,t(7;11)(p15;q14,add(9)(q31)/46,XX,del(6)(q22), add(8)(q22),t(12;17)(q14;q23)/46,XX,t(12;13;13)(q24;p1?;q33;q14)	*Chauveinc et al 1997 (108)*
44,XY,t(3;9)(q13;q22)**c**,-14,-22/50,idem,+3,+5, +der(9), t(3;9),+19,+20,+mar/66,XY,-X,+1,+2,+der(3)t(3;9), t(3;9)**c**,-4,+7,-8,-13,-14,-14,i(15)(q10),+17,-19,-20,-22,-22,+2mar/85,XXYY,der(3),t(3;9)**c**,-5,-10,-14,-14, -15,i(15)(q10),-16,-18,-18,-20,-22,+3mar	*Niazi et al 1998 (426)*
34-59,XY,del(1)(p22p33),t(10;11)(p15;p15), del(11)(p13), -19,-22	*Zattara-Cannoni et al 1998 (427)*
35-44,XY,-1,add(9)(p24),i(11)(q10),add(15)(q26),-22	*Zattara-Cannoni et al 1998 (427)*
39,X,-Y,del(1)(p13),-4,-6,-13,-14,+17,-22,+3mar,dmin	*Zattara-Cannoni et al 1998 (427)*
39-43,XX,-14,-18,-22,+mar	*Zattara-Cannoni et al 1998 (427)*
39-44,X,-Y,t(1;7)(p11;q11),-22	*Zattara-Cannoni et al 1998 (427)*
40,X,-Y,-1,-4,-10,-14,-22,+2mar	*Zattara-Cannoni et al 1998 (427)*
40-45,XX,t(1;5)(p12;p15)	*Zattara-Cannoni et al 1998 (427)*
41,X,-Y,del(1)(p13),-4,-14,-16,+17,-19,-20,add(21)(p13)	*Zattara-Cannoni et al 1998 (427)*
42,X,-Y,del(2)(p12),+5,del(11)(q14),10-,15,-22	*Zattara-Cannoni et al 1998 (427)*
43,XX,-4,-5,-12,-17,-22,+2mar	*Zattara-Cannoni et al 1998 (427)*
44,XY,del(1)(p13),-5	*Zattara-Cannoni et al 1998 (427)*
44,XY,-7,-10,-18,-22,+2mar/45,XY,del(1)(p12),del(7)(p11),t(10;10)(p15;p15)	*Zattara-Cannoni et al 1998 (427)*
44-45,XX,+mar	*Zattara-Cannoni et al 1998 (427)*
45,X,-Y/41,X,-Y,del(1)(p13),-4,-14,-16,+17,-19,-20,add(21)(p13),-22/41,X,-Y,del(1),-4,-14, -16,+17,-19,-20,add(21),-22,dmin,inc	*Zattara-Cannoni et al 1998 (427)*
45,XX,del(1)(q32),+7,-10,-11,t(22;22),+mar	*Zattara-Cannoni et al 1998 (427)*
45,XX,del(5)(q23q34),-22/52-90,XX,inc	*Zattara-Cannoni et al 1998 (427)*
45,XX,-1,del(7)(p12)/45,idem,dmin	*Zattara-Cannoni et al 1998 (427)*
45,XY,-1,t(2;19)(p11;p13)	*Zattara-Cannoni et al 1998 (427)*
45,XX,+7,-10,tas(14;22)/44,XX,-1,-2,-3,-14,-15,-21,+mar	*Zattara-Cannoni et al 1998 (427)*
45,XX,+7,-11,der(11)t(11;11)(p11;q12)	*Zattara-Cannoni et al 1998 (427)*

Tᴀʙʟᴇ 3.4. Continued

Karyotype	References
45,XX,-22/40-42,idem,-1,-2,add(11)(p15)	*Zattara-Cannoni et al 1998* (427)
45,XX,-22/46,idem,+mar	*Zattara-Cannoni et al 1998* (427)
46,XX,add(11)(p15)	*Zattara-Cannoni et al 1998* (427)
46,XX,del(1)(q32),der(11)	*Zattara-Cannoni et al 1998* (427)
46,XY,del(11)(q14)	*Zattara-Cannoni et al 1998* (427)
46,XY,+der(1),-22	*Zattara-Cannoni et al 1998* (427)
46,XX,dmin	*Zattara-Cannoni et al 1998* (427)
46,XY,t(3;7)(p11;q36)/47-94,XY,inc	*Zattara-Cannoni et al 1998* (427)
47,XY,+mar	*Zattara-Cannoni et al 1998* (427)
48,XX,+2mar	*Zattara-Cannoni et al 1998* (427)
39-44,XY,add(1)(p34),der(1;2)(p10;q10),-5,add(6)(q16),inv(7)(p12q21),-8,-13,del(19)(q13), -22,r(22),+2mar	*Chauveinc et al 1999* (257)
40-45,XX,t(3;4)(q21;q25),del(5)(p11),-14,add(22)(q11)×2	*Debiec-Rychter et al 1999* (428)
41-43,XY,-1,-10,-14,-18,der(20)t(1;20)(q21;q13),-22	*Debiec-Rychter et al 1999* (428)
42-44,X,-Y,-6,der(15;22)(q10;q10)	*Debiec-Rychter et al 1999* (428)
44-45,XY,-11,add(11)(p13),-21,t(21;21)(p11;p11),-22	*Debiec-Rychter et al 1999* (428)
45,XY,add(1)(p32),-10,-18,+mar,inc	*Debiec-Rychter et al 1999* (428)
45,XX,-21,-22,+mar	*Debiec-Rychter et al 1999* (428)
45,XX,-22/46,idem,r(12)(p13q24)	*Debiec-Rychter et al 1999* (428)
45-46,XX,del(2)(p11),der(11)t(6;11)(q11;p14),-21	*Debiec-Rychter et al 1999* (428)
45-46,XX,del(10)(q22),+del(11)(p11),dic(13;22)(p11;p11),-22	*Debiec-Rychter et al 1999* (428)
45-46,X,-Y,+8,del(10)(q23),-14,+2mar	*Debiec-Rychter et al 1999* (428)
48-57,XY,+3,+5,+7,+der(7)t(1;7)(p13;q12),del(9)(p13),add(17)(p11), add(19)(q13),+20,+20	*Debiec-Rychter et al 1999* (428)
55-118,XXX,der(1)t(1;11)(p11;q11),+12,+17,+18	*Debiec-Rychter et al 1999* (428)
78-96,XXYY,del(1)(q21),-2,-2,-6,-8,add(8)(p21)×2,-10,add(11)(p11)×2,del(12)(p11)×2,-14, -15,del(17)(q25)×2,del(18)(p11)×2,der(18)t(12;18)(p11;p11)×2,add(19)(q13)×2, del(22)(q11),+4mar	*Debiec-Rychter et al 1999* (428)
81-90,XY,-X,-Y,-1,der(1)t(1;1)(p13;q21),-2,add(2)(p16),-8,-10,-14,-18,-19,add(19)(q13),dic(21;22)(q11;q11)×2	*Debiec-Rychter et al 1999* (428)
82-94,XXYY,+3,+5,+7,-8,+9,+13,-14,-15,-17,-18, -19,+20,-22,del(22)(q11)×2,+2mar	*Debiec-Rychter et al 1999* (428)
40,XX,der(1)t(1;3)(p12-13;q11),-3,-6,-10,-11,-18	*Steilen-Gimbel et al 1999* (429)
45,XY,der(1)t(1;3)(p12-13;q11),-3	*Steilen-Gimbel et al 1999* (429)
45,XY,der(1)t(1;3)(p12-13;q11),-3,-22	*Steilen-Gimbel et al 1999* (429)
40,X,-Y,-10,-12,-13,-14,-22/39,idem,dic(17;22)(q25;p11)	*Cerdá-Nicolás et al 2000* (104)
41,XY,t(1;14)(p11;p13),-5,+dic(6;?)(q22;?),del(7)(p11),-9,-10,-12,tas(14;22)(p13;p13),-16,-18, +r,+mar	*Cerdá-Nicolás et al 2000* (104)
43,XX,der(1)t(1;14)(p11;q11),-6,add(7)(p11),-14,-22	*Cerdá-Nicolás et al 2000* (104)
44,X,-Y,der(2)del(2)(p24)del(2)(q33),-22	*Cerdá-Nicolás et al 2000* (104)
44,XX,-7,der(14)t(7;14)(p11;p11),-17,der(18)t(7;18)(q11;q23)	*Cerdá-Nicolás et al 2000* (104)
45,XX,del(1)(p13),-22	*Cerdá-Nicolás et al 2000* (104)
45,XX, der(1)t(1;4)(p22;q12), der(4;5)(q10;q10), der(4)t(4;5)(q31;q11),-17,+18	*Cerdá-Nicolás et al 2000* (104)
45,XX,-14,+mar,inc	*Cerdá-Nicolás et al 2000* (104)
46,XX,del(14)(q22)	*Cerdá-Nicolás et al 2000* (104)
46,XY,der(4)t(4;6)(p16;p21),der(17)t(17;17)(p11;q11q22),del(22)(q13), +mar	*Cerdá-Nicolás et al 2000* (104)
46,XX,-22,+mar	*Cerdá-Nicolás et al 2000* (104)
47,XY,del(1)(q11),-7,+8,+9,-10,+13,-15,-19,-22,+2mar	*Cerdá-Nicolás et al 2000* (104)
46,XX,t(1;4)(q44;q21)	*Go et al 2000* (430)
38-42,X,-X,der(1;8)(q10;q10),der(1)add(1)(q32)del(1)(p13p21),der(1)t(1;14)(p36;q11),-2,-3,-4,-5, der(6)t(6;7)(q11;q11),+?inv(6)(p25q13),der(7)t(2;7)(q?11;q11),+add(9)(q11),der(10)t(X;10)(q11;q26), -11,-11,-12,?del(13)(q13q31),-14,add(16)(q22),-17,der(17)t(12;17)(q24;p11),-18,add(19)(q13), der(19)t(1;19)(q12;q13),add(22)(q12),+r,+2mar,dmin/47,XX,del(11)(q12),t(3;15)(p11;p11),add(19)(q13.4), inv(20)(p13q11.2),+r/46-47,X,-X,add(3)(q27),add(7)(q22),del(11)(q22q23),del(13)(q14q31)/46,XX, der(3)t(3;?14)(p21;?q24),add(5)(q35)	*Sawyer et al 2000* (107)
39,X,-Y,der(1)t(1;1)(p22;q21),-8,t(9;22)(q12;q12),-10,add(11)(p11),-11,-16,-18,-22	*Sawyer et al 2000* (107)
39-40,X,-X,der(1;2)(q10;q10),-2,-3,der(7)t(3;7)(q?12;p14),-22/46,XX,t(9;12)(p21;q23)	*Sawyer et al 2000* (107)
40,X,der(Y)t(Y;1)(q21;q10),-1,-10,-14,-15,-18,-22/40,idem,tas(7;20)(p22;q13)	*Sawyer et al 2000* (107)
40,XY,der(1;2)(q10;q10),-4,-8,-13,-14,-22/80,idem×2	*Sawyer et al 2000* (107)
40,XY,-1,der(1)t(1;13)(q44;q14),der(1;16)(q10;q10),-6,-8,r(8),der(9)t(9;12)(p11;q11),-14, -16,-18,-22	*Sawyer et al 2000* (107)
40,X,-X,-1,-6,-14,-18,der(19)t(1;19)(q10;p13.3),-22/39-40,XX,der(19)t(1;19)(q10;p13.3),add(19)(q13.4)/39-40, idem,der(19)(q10;p13.3),dic(19;20)(p12;q13)	*Sawyer et al 2000* (107)
40,X,-Y,-1,-10,add(11)(p15),-14,-18,-22,der(22)t(1;22)(q10;p11)	*Sawyer et al 2000* (107)
40-42,XY,der(1) t(1;14)(p32;q11) del(14)(q13q22), t(1;13)(p32;p12),-4 , der(6) t(4;6)(q12;q13), -10,-11, tas(11;20)(p15;q13.3),der(12)t(1;12)(p32;p13),del(14)(q13q22),der(14)t(11;14) (q13;q24),dic(17;22)(p11;q13),del(22)(q12)/46,XY,t(1;7)(q25;q32),t(1;13)(p32;p12), add(5)(q35),del(5)(q15q21),del(13)(q12q22),del(17)(q12q21)	*Sawyer et al 2000* (107)

Table 3.4. Continued

Karyotype	References
41-46,XY,del(1)(p32p34),t(1;15)(q10;q25),?t(2;15)(p25;q25),del(3)(p23),del(5)(q22q31),der(10)t(10;11)(q22;q13),-10,del911)(q21q23),+15,-16,?del(16)(q22),-19,del(20)(p11),-22,add(22)(q13),+r	*Sawyer et al 2000* (107)
41-47,XY,der(1)t(1;11)(q10;p15),del(6)(q21),del(10)(q25),-14,add(19)(p13.3),+20,add(20)(q13),-22	*Sawyer et al 2000* (107)
43-44,XY,add(1)(p11),-4,-6,-17,der(22)t(1;22)(q10;q12)/43-44,idem,der(11)t(1;11)(q10;q25)	*Sawyer et al 2000* (107)
44,XX,der(1)t(1;18)(p13;?;q11),del(6)(q13),-18,-22/45,X,-X	*Sawyer et al 2000* (107)
44,Y,t(X;12)(p10;p10),der(1;17)t(1;17)(p32;p13)del(1)(p13p21),add(2)(q24),?del(3)(q?27),del(4)(q32),t(5;6;7)(p15;p16;q11),ins(8;2)(q11;q13q37),add(15)(q25),del(16)(q22),add(18)(q23),dic(19;22)(p11;q11),-20,add(20)(q11)	*Sawyer et al 2000* (107)
44,XX,-1,inv(3)(q13q23),t(4;7)(q31;q31),t(8;22)(q22;q11),tas(11;13)(p15;p11),-19,der(19)t(1;19)(q10;p13)/41-43,idem,der(4)t(4;7)(q31;q31),-7,-18	*Sawyer et al 2000* (107)
44-45,XY,der(1)add(1)(p22)del(1)(q42),add(2)(q33),der(6)add(6)(p23)t(6;16)(q13;q11.2),-8,add(9)(q22),-12,-14,-16,tas(17;22)(p13;p13),add(19)(p13.3),del(22)(q11),der(22)t(14;22)(q13;q13)ins(22;?)(q13;?)add(14)(q32),+mar1,+mar2	*Sawyer et al 2000* (107)
44-45,XX,der(1)t(1;3)(p11;p21),del(3)(p21),add(4)(p11),der(10)t(4;10)(q11;q11.2),-19,add(21)(q22)	*Sawyer et al 2000* (107)
44-45,XX,der(1)t(1;4)(p13;p12),-4,der(8)t(8;15)(q24;q22),-9,tas(14;22)(p13;p13),-15,der(22)t(3;22)(q11;q13),+r	*Sawyer et al 2000* (107)
45,X,dic(X;22)(q11;q12)/45,idem,del(6)(q15q23)	*Sawyer et al 2000* (107)
45,XY,dic(19;22)(p13;q11)/45,XY,dic(19;22)(p11;q11)	*Sawyer et al 2000* (107)
45,XX,tas(1;11)(p36;p15),tas(11;16)(p15;q24),-22/45,XX,der(8)t(8;11)(p23;q12),tas(8;16)(p23;p13),-22/45,XX,tas(8;16)(p23;q24),tas(11;19)(p15;p13)/45,XX,tas(8;12)(p23;p13),r(16),-22	*Sawyer et al 2000* (107)
45,XX,tas(2;20)(q37;q13),tas(5;20)(q35;q13),-22	*Sawyer et al 2000* (107)
46,XY,der(1)inv(1)(p21p?36)inv(1)(p?36p31),t(6;7)(p?23;q?23),del(7)(p12p12)/46,Y,t(X;17)(q13;q21),t(2;6)(p21;q13),del(7)(q34),inv(7)(p22;q11.2),t(15;17)(q10;q10),t(18;19)(q11.2;q13.3)	*Sawyer et al 2000* (107)
46,XX,der(2)t(2;12)(q33;q13)	*Sawyer et al 2000* (107)
46,XX,inv(4)(p15q12),t(4;11)(q21;p15),del(5)(q13q31),der(11)t(11;14)(p15;q11),add(14)(q13)×2,t(16;20)(p11;q13),add(18)(?p11)/46,XX,t(14;20)(q24;q13.3)/46,X,?del(X)(p11p11),t(7;13)(q32;q34)	*Sawyer et al 2000* (107)
46,XX,tas(19;22)(q13;q13)/46,XX,del(22)(q12)/45,XX,-22	*Sawyer et al 2000* (107)
46,XX,t(2;7)(q24;p14)/46,idem,add(20)(q13)	*Sawyer et al 2000* (107)
46,XX,t(2;14)(q33;q22)	*Sawyer et al 2000* (107)
46,XX,-1,+5,der(10)t(1;10)(q10;q26),-14,+20/46,idem,der(21)t(1;21)(q10;p11)	*Sawyer et al 2000* (107)
52-55,XX,der(1;3)(q10;p10),+5,+5,tas(9;11)(q34;p15),+11,+12,+19,+20,+21,+21,-22	*Sawyer et al 2000* (107)
46,XX,del(1)(p32)	*López-Ginés et al 2001* (68)
38-75,X,-Y,der(1)t(1;22)(p11;q12),add(13)(q34),-22	*Zattara-Cannoni et al 2001* (258)
43-47,XY,der(1)t(1;22)(p11;q12),-10,-18,-22	*Zattara-Cannoni et al 2001* (258)
44-45,XY,der(1)t(1;22)(p11;q12),del(7)(p12),-22	*Zattara-Cannoni et al 2001* (258)
44-45,XY,der(1)t(1;22)(p11;q12),-22,+mar	*Zattara-Cannoni et al 2001* (258)
45-66,XX,der(1)t(1;22)(p11;q12),add(6)(p25),-22	*Zattara-Cannoni et al 2001* (258)
45-84,XX,der(1)t(1;22)(p11;q12),-7,-11,-22,+2mar	*Zattara-Cannoni et al 2001* (258)
44-45,XX,t(1;19)(q21;q13),tas(3;9)(p26;p24),dic(18;22)(p11;p11),tas(3;9)(p26;p24)	*Sawyer et al 2002* (431)
38,X,-Y,del(1)(p12),der(4)t(4;7)(p11;q11),-6,-7,-9,-10,der(11)t(6;11)(p11;p11),-13,-14,-17,-18,-22,+3mar	*Henn et al 2003* (102)
40-42,XX,del(1)(p34),+dic(1;9)(p12;q34),-5,i(6)(p10),del(7)(q32),-9,-10,der(12),-14,-16,-22	*Henn et al 2003* (102)
41,X,-Y, der(1)t(1;7)(p32;p12),der(5)(p14),-7,-14,-18,-22	*Henn et al 2003* (102)
41,XX,-1, der(7) t(1;7)(q11;p11),-10,der(11),-14,-19,-22	*Henn et al 2003* (102)
43,XX,-1,der(7)t(1;7)(q11;p11),der(4),der(6),-13,-14,-15,der(18),+20,+20,-21,-22	*Henn et al 2003* (102)
43,XY,-1,der(7)t(1;7)(q11;p11),der(11)t(11;21)(p11;q11),del(18)(q21),-22	*Henn et al 2003* (102)
44,XX,der(1)t(1;3)(p32;q11),-3,der(7;11)(q10;q10)	*Henn et al 2003* (102)
45,XX,del(7)(p12),-14,-22,+mar/45,idem, del(1)(p32)	*Henn et al 2003* (102)
45,XX,dic(7;22)(p11;q11)/46,idem,+mar	*Henn et al 2003* (102)
45,XX,-1,del(7)(q11),+der(7)t(1;7)(q11;p11),-22	*Henn et al 2003* (102)
45,XX,-1,der(7)t(1;7)(q11;p11),-2,+mar	*Henn et al 2003* (102)
46,XY,dup(7)(q11q36)	*Henn et al 2003* (102)
46,XY,i(7)(q10)	*Henn et al 2003* (102)
46,XY,inv(7)(p22q21)	*Henn et al 2003* (102)
43,XX,der(1)t(1;11)(p11;q11), del(2)(p12), del(6)(q28), dic(7;22)(p12;q11), -11,-13,-22	*Ketter et al 2003* (432)
38-41,XX,der(1;7)(q10;q10),-4,del(6)(q15q23),t(6;11)(p?23;p11),inv(9)(p11q13),-10,-14,der(19)t(11;19)(q11;p13.3),-22	*Sawyer et al 2003* (99)
39-40,X,-X,dic(1;12)(p11;?p12),-10,inv(13)(q22q26),-14,-18,-22	*Sawyer et al 2003* (99)
39-42,X,-X,dic(1;12)(p13;p11),der(2)add(2)(q11),add(2)(q?31),?t(3;5)(q?25;q35),t(4;15)(q?23;q?26),?del(6)(q23q23),add(8)(q?13),t(8;10)(q24;q?23),t(9;19)(p?13;q?13),t(10;16)(q11;p13),del(13)(14q22),-22	*Sawyer et al 2003* (99)
40-46,X,t(X;1)(q22;q21),add(1)(p?22),der(1)t(1;6)(p?22;p12),add(2)(p11),t(2;22)(p11.2;q11.2),-6,dic(7;22)(p11;p11.2),del(11)(q14),der(12)t(6;12)(p11.1;p13),-16,-19,add(20)(p13)	*Sawyer et al 2003* (99)
41-43,X,-Y,dic(1;4)(p13;p11),-13,-14,-22	*Sawyer et al 2003* (99)
42,X,-Y,del(1)(p13),der(2)t(2;4)(p11;?),-4,t(6;9)(q15;?),-18,-22	*Sawyer et al 2003* (99)
42,X,-X,dic(1;7)(p13;p11),der(2;4)(q10;q10),-6,+9,der(9;15)(q10;q10),dic(22;22)(q11;q11)	*Sawyer et al 2003* (99)

Table 3.4. Continued

Karyotype	References
42-44,XX,-8,-22/44,idem,r(11)(p15q25)/41-43,idem,dic(1;11)(p32;p13),-11/41-43, idem,dic(1;11)(p32;p13),tas(9;16)(p24;p13.3)/42,idem,dic(1;11)(p13;p13), dic(9;16)(p24;q11.2)/46,XX,tas(9;16)(p24;p13)	*Sawyer et al 2003* (99)
43,X,-Y,der(3)t(3;5)(q29;p13),dup(5)(p12p13),der(12)t(5;12)(p13;p13),-14,-22	*Sawyer et al 2003* (99)
43-45,XY,der(1)t(1;16)(p13;q13),-10,?add(14)(q?12),-16,add(16)(q12.1),del(22)(q13), der(22)t(16;22)(q11;q13)	*Sawyer et al 2003* (99)
43-45,XY,dic(1;22)(p11,q12-q13),-18,add(22)(q11.2),+mar/43-45,idem,tas(2;12)(p11;p11),43-45,idem,del(2)(p11)	*Sawyer et al 2003* (99)
43-45,XY,dic(19;22)(p13;q11)/43-45,idem,tas(1;19)(p36;q13/43-44,idem,dic(1;19)(p22-p32;q13.3)/43-44, idem,dic(1;19)(p11;q13.1)/43-44,idem,dic(1;19)(p11;q11)/43-44,idem,dic(1;4)(p11;q21),der(19)t(4;19)(q12;q13.4)	*Sawyer et al 2003* (99)
44,XY,der(1;9)(q10;q10),dic(4;9)(p?13;q11),-22/44,XY,der(1;4)(q10;q10),-22/44,XY,dic(4;18)(p?13;q11.1),-22	*Sawyer et al 2003* (99)
44-45,Y,?X;10)(p22.1;q25),-1,?t(5;13)(q15;q14),-6,der(7)t(7;7)(p12;q32),der(7)del(7) (p11.2p15)del(7)(q11.2q22),del(8)(q13),?t(18;22)(q11.2;q13),-22,+mar1,+mar2	*Sawyer et al 2003* (99)
45,X,der(X)t(X;22)(q21;q12),add(5)(q?34),del(8)(p11.1p12),add(12)(p11.2),-22,add(22)(q11.2)	*Sawyer et al 2003* (99)
45,XY,add(1)(p36.3),inv(1)(p22q25),der(5)t(5;?12)(p14;p13)der(5)(q35),add(6)(p22),?add(8)(p21), der(12)t(5;12)(p14;p13),del(17)(p12),?dic(19;22)(q13.1;q11.2)/44,idem,tas(19;inv1)(q13;p22)	*Sawyer et al 2003* (99)
45,XY,der(3;7)(q10;q10)	*Sawyer et al 2003* (99)
45,XX,der(4)t(4;22)(q28;q11)ins(4;?)(?q28),der(7)t(7;22)(p12;p11),der(12)t(7;12)(p13;p12.2)del(7)(p21), der(22)t(7;22)(p21;q13.2),-22/45, idem,-der(7), +i(7)(q10)	*Sawyer et al 2003* (99)
45,XX,dic(1;22)(p34-35;q13)/45,XX,dic(1;22)(p31-32;q12-13)/45,XX,(p11;q12-13)	*Sawyer et al 2003* (99)
45,XY,?dic(3;12)(q12;q13),der(4)t(4;12)(p16;q13),add(7)(q36),t(10;11)(q24;p15)	*Sawyer et al 2003* (99)
45,XX,i(1)(q10),der(3)(q29;p11),?inv(2)(p15q37),der(3)t(3;6)(q29;p11),-4,add(6)(q11), add(11)(q23),-22,del(22)(q12),+mar	*Sawyer et al 2003* (99)
45,X,tas(Y;22)(q12;p11),dic(1;11)(p13;p13),+5,-7,+15,+20,-22	*Sawyer et al 2003* (99)
45,t(X;12)(q?13;q?13),-Y,t(1;1)(p?13;q?13),t(3;22)(p26;q?11),t(10;11)(p10;p10)	*Sawyer et al 2003* (99)
45,XY,t(7;10)(p10;p10),t(12;19)(p13;q13),dic(19;19)(p13.3;q13.3),add(12)(p13)/43, idem,der(1)t(1;2)(p11;p13),-2,-dic(19;19),+trc(19;19;22)(p13.3;q13p13;q12.2)	*Sawyer et al 2003* (99)
45,XX,-22/45,idem,r(X)/44,idem,-X/45,idem,tas(19;22)(q13;p11)	*Sawyer et al 2003* (99)
45,XX,-22/45,idem,tas(1;11)(p36.3;p15.5)/45,idem,dic(1;11)(p32;p15.5)/45,idem,del(1)(p22)/45,idem, del(1)(p13)/45,idem,tas(1;22)(p36.3;q13)/45,idem,tas(1;14)(p36.3.q13)/45,idem,tas(1;13)(p36.3;q34)/45, idem,tas(1;19)(p36.3;q13.4)	*Sawyer et al 2003* (99)
45,XX,-22/45,idem,tas(1;19)(p36.3;p24),tas(10;19)(p15;q13.4),tas(15;19)(p11.2;q13.4),tas(17;18)(p13;q23), tas(17;19)((p13;p13.3)	*Sawyer et al 2003* (99)
45,XX,-22/45,idem,tas(8;22)(?;p13)	*Sawyer et al 2003* (99)
45-46,XX,der(1)t(1;9)(q44;q11)/46,XX,dic(1;12)(q44;p11)	*Sawyer et al 2003* (99)
46,XY,der(2)t(2;9)((p21;p22),del(14)(q22),dic(17;22)(p11.1;q11.2),inv(19)(p13.3q13.4),der(22)t(2;22)(p23;q11)/46, idem,dic(17;22)×2	*Sawyer et al 2003* (99)
46,XX,del(17)(q11.1)/46,XX, dup(17)(q23-q24)/46,XX,der(11)t(11;17)(p15/q11.1)/46,XX, tas(11;19)(p15;q13.4)/46,XX,dic(11;19)(p11;q11)	*Sawyer et al 2003* (99)
46,XX,der(1)t(1;12)(p36;q12),t(2;4)(q35;q21)	*Sawyer et al 2003* (99)
46,XX,der(7;16),t(7;11)(q11;q11),del(11)(q11),t(15;22),t(16;22)	*Sawyer et al 2003* (99)
46,XX,der(8)t(6;8)(?;p23),der(12)t(11;12)(q13;p13)	*Sawyer et al 2003* (99)
46,XX,der(22)t(7;22)(q11;q11)	*Sawyer et al 2003* (99)
46,XY,dic(3;7)(p11;p?13),+r(7),der(14)t(7;14)(q11;q32)	*Sawyer et al 2003* (99)
46,XY,tas(1;8)(q44;p23)/46,XY,add(1)(q44)/46,XY,del(1)(q11)	*Sawyer et al 2003* (99)
46,XX,t(2;12)(p25),del(4)(p16),?add(6)(q27),ider(22)(q10)del(22(q11)/46,idem,t(7;12)(q10;q10),-12, dic(19;19)(p13;q11),add(22)(q13)	Sawyer et al 2003 (99)
47,XY,+12,-14,+15,+20,-22/46,idem,dic(7;19)(p11;p11)	*Sawyer et al 2003* (99)
41,XY,der(1)t(1;14)(p11;q11),-10,-14,-18,-22	*López-Ginés et al 2004* (96)
43,X?,del(1)(p13),del(10)(q21),-13,-14,+21,-22	*López-Ginés et al 2004* (96)
43,XY,del(1)(p21),-14,-18,-22	*López-Ginés et al 2004* (96)
43,XY,del(1)(p32),i(3)(q10),-15, 18,-20, del(22)(q12-q13)	*López-Ginés et al 2004* (96)
43,XY,der(1) t(1;11)(p22;q24),der(6)t(6;9)(q15;q22),der(9)t(9;13)(q13;q12),?del(11)(p11),del(11)(q14),-13,-14	*López-Ginés et al 2004* (96)
45,XX,add(1)(q21),der(4)t(1;4)(?;q21),der(6)t(1;6)(?;q23),-14,+15,-22	*López-Ginés et al 2004* (96)
45,XY,del(1)(p13),-22	*López-Ginés et al 2004* (96)
45,XX,der(1)t(1;22)(p32;?),del(4)(q25),t(8;11)(q24;q21),der(14)t(14;22)(q32;q11)-22,del(22)(q11)	*López-Ginés et al 2004* (96)
46,XX,der(1)t(1;3)(p32;p21),der(3)t(3;7)(p13;q11),add(7)(q11)	*López-Ginés et al 2004* (96)
42,XY,del(1)(p13),-4,t(4;7)(q12;q11),-6,t(6;7)(p12;p12),del(7),-14,der(17),-22	*van Tilborg et al 2005* (101)
42,X,-X,dic(1;22)(p11;p11),der(6)t(6;13)(q13;q13),-13,-14,der(16)t(13;16)(?;p13),-22	*van Tilborg et al 2005* (101)
42,X,-X,-14,-19,-22,many tas	*van Tilborg et al 2005* (101)
42-43,XY,dic(1;19)(p21;p13),del(11)(p12),tas(11;21)(pter;pter),-14,-21,-22	*van Tilborg et al 2005* (101)
42,der(X),Y,1p-,t(1;6)(p32;p21),2q-,-4,-5,6q+,der(7),10q+,12q-,der(12)t(12;17)(q10;q10),-15,-19,22q-	*van Tilborg et al 2005* (101)
44,X,del(X)(p13),t(1;22;19?)(q534→p12::cen(22)→q11::cen(19)→qter)	*van Tilborg et al 2005* (101)
44,XX,der(1)t(1;6)t(2;3)(p24;q32),del(6)(q16),add(7)(q36)add(8)(p12),del(11)(q13),-11,del(17)(q21q24),-18,-20,-22	*van Tilborg et al 2005* (101)

TABLE 3.4. Continued

Karyotype	References
44,XX,i(1)(q10),der(6)t(6;9)(q21;q13),der(15)t(6;15)(p11;p12),-18,-22	*van Tilborg et al 2005* (101)
+44,XY,-1,del(7)(p14p21),add(22)(q12)	*van Tilborg et al 2005* (101)
+44,XX,-22,markers	*van Tilborg et al 2005* (101)
45,XX,-1,add(6)(q10),-16,der(22),t(1;22),+mar	*van Tilborg et al 2005* (101)
45,XY,del(2)(p23),-4,add(5)(q34),-14,add(19)(p12),add(22)(q11),+mar	*van Tilborg et al 2005* (101)
46,XX,t(2;7)(q32;q35)	*van Tilborg et al 2005* (101)
46,XX,t(7;14)(q35;q12)	*van Tilborg et al 2005* (101)
92< n4>,XXYY,-1,-1,-1,add(1)(q)×3,-2,-2,add(2)(p),del(5)(p13)×2,-9,-9,-9,add(9)(p13)×2, -14,-14,del(14)(q2)×2,+4mar/62-69<3>,XYY,-1,-2,-2,dic(2;?)(q2?1;?),-3,-3,+del(3)(p)×2,-4,add(4)(q)×2,-6,-6,-6,+7,+add(8)(q),-9,-9,i(9)(q10),+11,+12,-13,-14,-15,-16,-16,-17,i(17)(q10)×2,-18,-19,-22,-22,del(22)(q11),+10-19mar	*Pelz et al 2007* (103A)
50,XX,+3,+5,+12,+17,+20,-22	*Ketter et al 1987* (587)
50,XX,+4,+12,+17,+20	Ketter et al 1987 (587)
51,XX,+3,+5,+12,+13,+mar	Ketter et al 1987 (587)
51,XX,+12,+13,+16,+17,+20	Ketter et al 1987 (587)
52,XX,+3,+5,+9,+10,+12,+17,-22	Ketter et al 1987 (587)
52,XX,+5,+10,+12,+15,+17,+20	Ketter et al 1987 (587)
53,XX,+3,+5,+8,+9,+12,+13,+20	Ketter et al 1987 (587)
53,XX,+5,+7,+10,+10,+12,+16,+17	Ketter et al 1987 (587)
54,X,-Y,t(4;5)(q35;q12),+5,+7,+9,+11,+12,+13,+17,+18,+20	Ketter et al 1987 (587)
54,XX,+3,+4,+5,+9,+12,+13,+15,+20	Ketter et al 1987 (587)
54,XX,+3,+8,+9,+12,+13,+15,+20,+20	Ketter et al 1987 (587)
56,XY,+3,+4,-7,+10,+12,-13,+16,+19,+20,+20,+4mar	Ketter et al 1987 (587)
56,X,-X,+4,+5,+8,+12,+14,+15,+17,+18,+19,+20	Ketter et al 1987 (587)
56,XXX,+3,+4,+5,+8,+9,+12,-13,-16,+dic(13;16)(p13;p13),+17,+20,+20	Ketter et al 1987 (587)
59,XX,+4,+5,+5,+6,+7,+11,+12,+13,+16,+17,+18,+19,+20	Ketter et al 1987 (587)
62,XX,+5,+5,+6,+7,+8,+9,+11,+13,+13,+16,+17,+18,+18,+19,+20,+22	*Ketter et al 1987* (587)

Not included in this table are meningiomas with only numerical changes (including -22) or those tumors with i(22) or del(22q)(414A, 414B)*.
+Corrected karyotypes.

structural changes (deletions, additions, and translocations) involving chromosome 22. The number given may be higher than that given, based on the possibility that marker chromosomes in some of the karyotypes may have harbored abnormal chromosomes 22.

Of interest is loss of sex chromosomes observed in meningioma. Nearly 15% of meningiomas in male patients showed loss of the Y (-Y), with 5% of the tumors showing -Y as the sole anomaly. In one series of 70 meningiomas, FISH studies revealed loss of the Y in >45% of the male cases

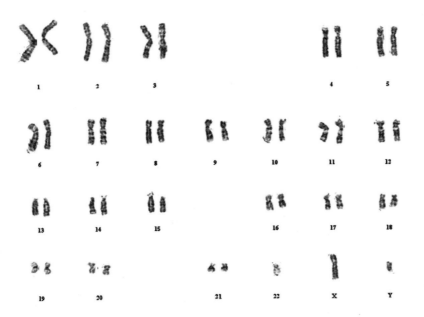

FIGURE 3.2. The karyotype is that of a meningioma from a 63-year-old male containing as the sole anomaly the most common change in meningioma, i.e., a lost chromosome 22 (-22) (karyotype supplied by Dr. Kathleen E. Richkind).

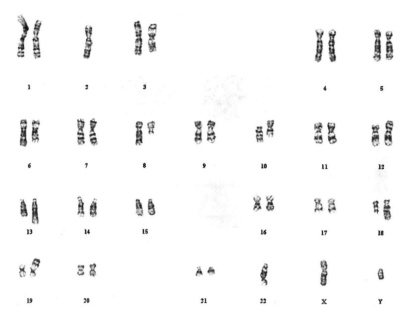

FIGURE 3.3. A complex karyotype in a meningioma of a 30-year-old male showing: 44,XY,t(1;13) (p32;q32),-2,?del(2),del(6)(q21),t(8;18) (q13;q23),del(10)(p11.2) or add(10)(p11.2), del(10)(q24),add(12)(p13),add(19)(p13.3),-22,add(22)(p11.2). The karyotype in this tumor had a chromosome number ranging from 43 to 46, with cells showing variable losses: -1, -7, -18, -19, and two to three markers (karyotype courtesy of Dr. Kathleen E. Richkind).

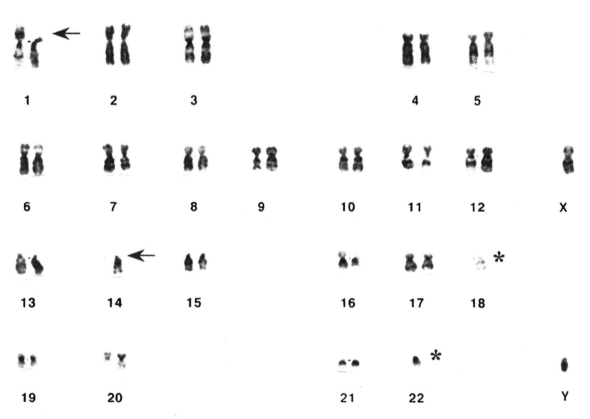

FIGURE 3.4. Karyotype of a meningioma from a 54-year-old male showing three of the most common cytogenetic changes in addition to -22, i.e., 1p-, -14, and -18 (96). Other changes involve chromosomes 4, 8, and 11. The meningioma was grade II and the chromosome changes reflect the atypical nature of the tumor.

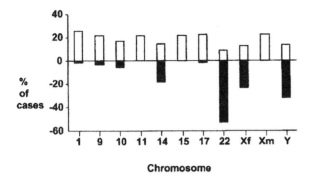

FIGURE 3.5. Numerical chromosome changes in 70 meningiomas. Loss is shown by a black bar and gains by an white bar. Xf, X-chromosome in females; Xm, X-chromosome in males. Of note are losses of chromosomes 22, 14, and sex chromosomes, frequent cytogenetic anomalies in meningioma (from ref. 88, with permission).

(Figure 3.5) (88). Loss of the Y chromosome in meningiomas did not seem to be age related (89). About 5% of meningiomas in female patients showed loss of an X-chromosome (-X), which was usually associated with other karyotypic anomalies. In about 18% of meningiomas, changes other than those mentioned above were encountered, but none was recurrent or specific.

The loss or involvement of chromosome 22 thus seems to be a key event associated with a preponderant percentage of meningiomas. It is possible that tumors without apparent involvement of chromosome 22 could contain changes in this chromosome not discernible cytogenetically and require molecular approaches. Even the latter may be in some cases insufficient for the establishment of subtle changes in chromosome 22.

Some meningiomas are associated with complex karyotypic changes, usually in addition to -22 or structural changes of that chromosome (Figure 3.6). These cases may represent atypical or aggressive meningiomas (grades II and III). The presence of a complex karyotype in benign meningiomas was thought by the authors (90) to be indicative of an intrinsically malignant potential.

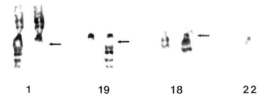

1	19	18	22

FIGURE 3.6. Partial karyotype of chromosome changes in a recurrent transitional meningioma in a 61-year-old female (431). The complete karyotype of this case is shown in Table 3.4. The two changes shown in this figure are t(1;19)(q21;q13.3) and dic(18;22)(p11;p11). Although this meningioma did not show loss of chromosome 22 or its long arm, changes seen in most meningiomas, the short arm of chromosome 1 (1p) was involved in a translocation with chromosome 19 and the short arm of chromosome 22 was translocated to the short arm of chromosome 18.

In the progression from typical to atypical and then malignant meningiomas, the karyotypic changes may be witness to, if not the cause of, these progressions (91). Thus, it is not surprising to find that the most complex karyotypes were often associated with atypical or malignant meningiomas (80, 90, 92).

As mentioned previously, in addition to the crucial role played by chromosome 22 changes in meningioma, allelic loss of material at 1p, 14q, and 10q has been established with CGH analyses as recurrent events in the genetic pathway of meningioma development and progression (68, 84, 93–97). However, some meningiomas may show these changes without an apparent involvement of chromosome 22 (68, 98). A role for telomeric fusion in the formation of deletions, dicentrics and unbalanced translocations of 1p in meningiomas has been advocated previously (99).

Even though it is often difficult to ascertain whether normal (diploid) karyotypes in meningioma cultures are the result of overgrowth by normal cells or are truly representative of the tumors, the latter is a possibility as reflected by the rather frequent finding of such cytogenetically normal cells in primary meningiomas (59, 63, 71, 100). In fact, the percentage of such cells has ranged from 20 to 65% among the studies cited. It is possible that molecular genetic changes, particularly of the NF2 gene, may be present in some of these cells, although studies to that effect have not been published to date.

A study apropos diploid meningiomas is one in which 55 meningiomas of varying anatomic sites were successfully karyotyped (101). The meningiomas were divided into two groups: those with -22, LOH, or both for the NF2 gene and others without either anomaly. Of the latter group, 23 of 27 tumors had normal karyotypes; two had -Y; and two had translocations, one with a t(2;7)(q32;q35) and the other with a t(7;14)(q35;q12), as the only changes. In the former group of 28 meningiomas with either -22, LOH, or both of NF2, eight tumors had -22 as the sole change, one tumor had changes in addition to -22, and one tumor with LOH of NF2 was not associated with changes of chromosome 22 (46,X,-X,+7). Translocations and other structural changes affecting chromosome 7 in meningioma are shown in Figure 3.7 (102).

Although no family histories were presented, including that for the nine vestibular meningiomas, the findings in the aforementioned study (101) are of interest because half of the meningiomas had no changes of chromosome 22 or to NF2, indicating that other mechanisms were responsible for their genesis. Even though the findings in this study do not exclude the possibility of cryptic changes of chromosome 22 and NF2 being responsible for the development of meningiomas, they strongly point, as have others, to a variety of genetic mechanisms underlying meningioma development.

Trisomy or tetrasomy of chromosome 22 is relatively rare in meningioma (103), and it almost always is associated with hyperdiploidy and complex cytogenetic changes (Figure 3.8). A malignant meningioma with pseudotetraploid and pseudotriploid clones with very complex karyotypes has

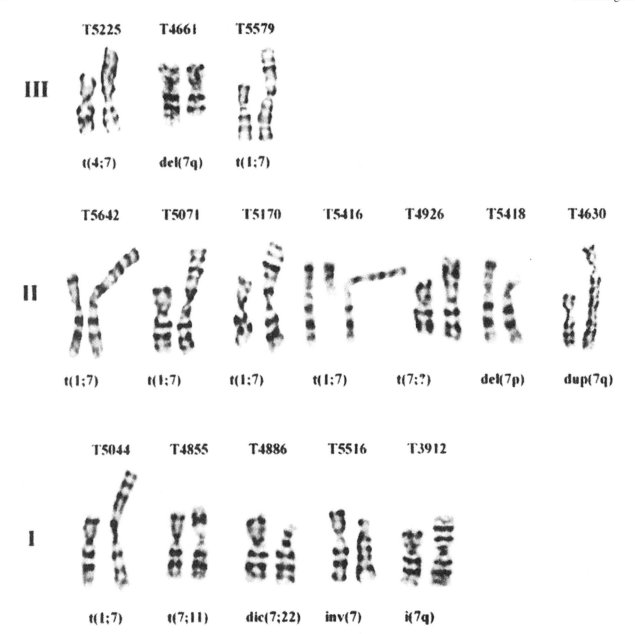

FIGURE 3.7. Changes of chromosome 7 associated with meningioma progression (102). Each partial karyotype (with the normal chromosome shown on the right) represents a tumor with its grade indicated by I–III. Although changes of chromosome 7 shown in this Figure are not as frequent as other cytogenetic changes in meningioma, they seem to play a role in the progression of some meningiomas. Note that the changes are represented by translocations, deletions, and other structural changes of chromosome 7.

been described previously (103A). FISH and CGH studies confirmed the complexity of the changes.

Chromosome changes in addition to -22 are usually associated with atypical features and increased aggressiveness of meningiomas (104). Complex karyotypes were observed in 34% of grade I, 45% of grade II, and in 70% of grade III meningiomas, associated with losses of chromosomes 10, 14, 18, and 22 and structural changes affecting chromosomes 1, 4, 7, 14, and 22 (104). In rare tumors telomere association and dmin were seen.

The participation of more than two chromosomes in translocations in meningiomas has been described in six cases, i.e., t(1;7;14)(q25;q32;q22), t(2;7;14)(q23;q36;q22) (74, 105), t(1;5;6)(p12;q23;q25) (106), t(5;6;7)(p15;p16;q11) and der(1)t(1;6;7;22) (107), and in one radiation-induced meningioma with t(3;5;17)(q22q13;q11;q22) (108).

The presence of marker chromosomes, whose origin could not be determined with conventional cytogenetic techniques, has been described in at least 65 meningiomas. Such markers may occur singly or as many as 15 in rare tumors. The origin of the markers should be resolved by newer approaches,

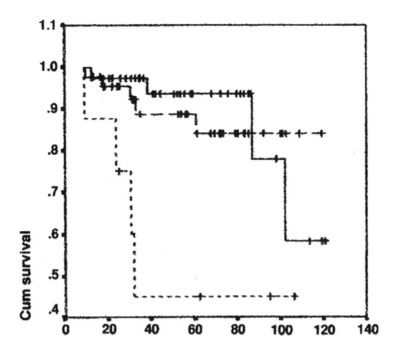

Disease free survival: time in months from surgery to relapse

FIGURE 3.8. Cumulative survival of 88 patients with meningioma after surgery as related to chromosome 22 status in the tumors. Solid line, no loss of 22 (43% of patients); broken line, loss of 22 (49% of patients), and dotted line, polysomy of 22 (8% of patients). The poor survival in patients with trisomy or tetrasomy of chromosome 22 was related by the authors to the hyperdiploid karyotypes present in these cases (from ref. 103, with permission).

e.g., SKY, but no such studies have been published to date. Ring chromosomes are relatively rare in meningiomas; only a dozen tumors containing rings have been described. Although they have been described in only a dozen meningiomas, dmin or hsr are usually indicative of gene amplification. A small number of meningiomas contained telomere associations (78, 99, 109).

About one fourth of meningiomas contained either balanced or unbalanced translocations, but none was sufficiently recurrent or characteristic; some meningiomas contained more than one translocation.

Nearly all chromosomes previously thought to be lost by cytogenetic or CGH studies of grade III meningiomas were shown by SKY to be part of marker chromosomes, thus presenting a more balanced karyotype than previously thought (110). The majority of meningiomas of all grades showed loss of one chromosome 22 as determined by SKY as well as involvement of chromosomes 8 in a number of unbalanced translocations with different partners (110).

Meningiomas are genetically heterogeneous and display different patterns of chromosomal changes (determined by multiplex-FISH), with the presence of more than one tumor cell clone detected in about half of cases, including atypical/anaplastic cases (111). The pathways of intratumoral clonal evolution observed in the benign tumors were different from those observed in atypical/anaplastic meningiomas, suggesting that the latter tumors might not always represent a more advanced stage of histologically benign meningiomas.

A correlation between the chromosome changes in meningioma and recurrence has been reported (112), including the observation that such recurrence is particularly high for those tumors with deletion of 1p.

Nucleolar organizer regions of chromosomes have received some attention in the past (115A, 113–115) as prognostic indicators, but no recent studies with newer techniques have been published. The cytogenetic classification and findings in meningioma can serve as criteria for clinical management (116).

Regional genetic heterogeneity within meningiomas, as is true of other tumors, was studied for chromosome 1, 14, and 22 in 77 meningioma samples from 72 patients (117). Table 3.5 and Figure 3.7 show the percentages of single and combined chromosomal aberrations as they relate to the grade of meningioma. Combined changes were more frequent in grades II and II meningiomas, with grade III being devoid of diploid cells. In addition, these authors (117) demonstrated that molecular genetics may be an important adjunct to standard pathology. The appearance of multiple chromosomal aberrations in different regions of meningiomas showed the importance of regional heterogeneity in the biological behavior of meningiomas. When detecting chromosomal aberrations in subtotally resected meningiomas, it is reasonable to assume that residual portions likely possess

TABLE 3.5. Percentages of single and combined chromosomal aberrations determined by interphase-FISH in 77 meningiomas.

Chromosomal aberrations (homogeneous or heterogeneous)[a]	Grade 1 (n = 59)[b]	Grade II (n = 13)[b]	Grade III (n = 5)[b]
1p deletion alone	8.5	0	0
14q deletion alone	6.8	0	0
22q deletion alone	8.5	0	0
1p deletion + 14q deletion	6.8	23.1	0
1p deletion + 22q deletion	5.1	0	0
1p deletion + 14q deletion + 22q deletion	13.6	46.2	20
1p deletion + 14q deletion + 22q trisomy	1.7	23.1	80

[a]Deletion of 22q alone, deletion of 1p + trisomy of 22q, deletion of 14q + deletion of 22q, and deletion of 14q + trisomy of 22q were not found in any cases.
[b]Chromosomal aberrations were found in 51% of grade I, 92.4% of grade II, and 100% of grade III meningiomas.
From Pfisterer et al 2004 (117).

TABLE 3.6. Chromosome and other changes (%) in meningioma.

NF2 mutations	Grade I	Grade II	Grade III	Total
	24	35	44	27
Loss of 22	48	64	58	52
Loss of 1p	14.7	62	77	24
Loss of 14q	19.7	50	57	38
Loss of 10q	5	29	49	22
Loss of 9p	11	37	48	31
Del p16	13	12	33	16
Amplification of 17q	8	20	51	31
Telomerase changes	8	68	90	28

Based on Dezamis and Sanson 2003 (241).

TABLE 3.7. Changes (%) affecting 22q and NF2 in meningiomas according to grade and histology.

Histology	LOH at 22q	No. of cases with NF2 mutations
WHO grade I tumors	24/61 (39)	11/61 (18)
WHO grade II tumors	16/24 (67)	8/24 (33)
WHO grade III tumors	3/3 (100)	2/3 (67)
Transitional tumors	23/57 (40)	10/57 (18)
Meningotheliomatous tumors	15/22 (68)	9/22 (41)
Other histological subtypes	5/9 (56)	2/9 (22)

Based on Lomas et al. 2005 (165).

chromosomal aberrations. Thus, it would seem helpful for genetic studies of a subtotally resected meningioma to be done using tissue obtained as close as possible to the margin, remaining tumor mass, or areas of suspected tumor. Because of the high incidence of regional heterogeneity in the distribution of genetic aberrations in low-grade meningiomas, well-designed FISH studies need to incorporate studies of regional heterogeneity.

If, as indicated in some publications (76, 118–122), with rare exceptions (123, 124) meningiomas are of monoclonal origin (whether single, multiple, recurrent, or as a result of irradiation), the karyotypic heterogeneity pointing to an epigenetic effect as a possibly responsible cause of this heterogeneity. Some features related to the most common chromosomal changes in meningioma (Tables 3.5 and 3.6) and some of the consequences of these changes are presented in a number of the following sections.

Chromosome 22 Changes in Meningioma

As mentioned, chromosome 22 plays a crucial role in meningioma tumorigenesis, and its anomalies hold the key to deciphering the genesis and biology of this tumor. Loss of chromosome 22 (Table 3.7), including the NF2 gene at 22q12, in approximately 50% of sporadic meningiomas has been reported previously (15, 16, 125–127). LOH studies in meningiomas were in agreement regarding chromosome 22 changes

(128), although discrepancies were observed in several cases. Isochromosomes 22 may be present in meningiomas and although they may be involved in tumor progression, they do not seem to play a part in the genesis of these tumors (129). Indications that the gene for meningiomas is on chromosome 22 were obtained in a patient with multiple tumors (130).

Monosomy 22 (-22) may be accompanied in some meningiomas by changes affecting the other chromosome 22, i.e., deletions, translocations, and telomeric associations (60, 72, 99, 131–135). These changes may be indicative of the vulnerability of chromosome 22 to alterations, a pathway to its loss from the meningioma genome, or both (136, 137). The ultimate result is loss of merlin expression in meningiomas.

Merlin (schwannomin) the protein product of NF2, is a member of the 4.1 structural protein family (including moesin, radixin, and ezrin) that links cytoplasmic membrane proteins to the cytoskeleton (138–140). Merlin is encoded by 17 exons of the NF2 gene with alternate splicing of the 16th exon resulting in two highly similar isoforms with comparable functions. No inactivating mutations have been observed at codon 16 (138). The mechanisms by which merlin exerts a tumor-suppressive activity and inhibits meningeal cell proliferation are not completely established (141), but they may include protein interactions that effect Ras signaling, phosphatitidylinositol 3-kinase activation and cyclin D1 expression (138). The disruption of the signaling cascade that leads to cytoskeletal reorganization is considered to be critical to tumor formation. Overexpression of merlin in both

NF2-negative and *NF2*-positive human meningiomas significantly inhibited their proliferation in vitro, which provides further evidence for a role of merlin as a negative regulator of tumor growth (142). Merlin functions as a tumor and metastasis suppressor by controlling cadherin-mediated cell–cell contact; merlin deficiency promotes tumorigenesis and metastasis by destabilizing adherens junctions (143). More details regarding merlin may be found in the section on schwannomas in Chapter 1. Mutations in *NF2* and loss of merlin may or may not be associated with increased mitotic activity and aggressive tumor growth (144–146). Mutations of *NF2* play an important role in the development of sporadic meningiomas as well as indicating a different tumorigenesis of meningioma variants (147). It was hoped that discovery of the *NF2* gene might provide major clues about the pathogenesis of sporadic meningiomas as well. However, it is now recognized that deletions of the putative tumor suppressor *NF2* gene at 22q12.2 with loss of merlin expression occur primarily in benign, fibrous, or transitional subtypes of meningioma (148, 152). In contrast, in sporadic meningiomas, which are more commonly meningothelial, inactivating mutations of the *NF2* gene occur in only 27% of the tumors (148, 152), and complete loss of merlin immunoreactivity occurs in only 60% of cases (137). Thus, in sporadic meningiomas, the importance of *NF2* mutations is less certain and does not seem to be essential for development of some tumor variants.

The changes affecting chromosome 22 in meningiomas are closely tied to those of the *NF2* gene (22). Thus, truncating mutations of the *NF2*, represent the most frequent gene alteration in meningioma. LOH at 22q12.2 has been described in 40–70% of all NF2-associated and sporadic meningiomas (1, 144, 149–151), and in 60% of sporadic tumors (15, 25, 152, 153). The protein product of *NF2* is merlin, a negative regulator of tumor growth (138, 142), and its inactivation by mutations of *NF2* seems to be important for meningioma development. Thus, the majority of meningiomas show absent or reduced immunoreactivity to merlin (154–156), which is strongly associated with LOH of 22q (25). In patients with NF2, almost all of the meningiomas have deletions of 22q (146, 157).

Molecular genetic and protein studies showed that the frequency of *NF2*/merlin alterations differs among the three most common benign meningioma variants. Whereas fibroblastic and transitional meningiomas harbor *NF2* mutations in approximately 70–80% of cases, meningothelial meningiomas carry mutations in only 25% of cases (16, 148). A close correlation between the fibroblastic variant and LOH at 22q also was found (15). As a consequence, reduced merlin expression was observed in the majority of fibroblastic and transitional meningiomas but rarely in meningothelial tumors (154, 156), suggesting that the genetic origin of these latter tumors is largely independent of *NF2* gene alterations. Moreover, the similar frequency of *NF2* gene mutations in atypical and anaplastic, and in benign fibroblastic and transitional meningiomas, suggests that *NF2* mutations are an early alter-

ation involved in the formation of most benign meningiomas, but other genetic events, reflected in additional chromosome changes, are involved in the tumorigenesis of or progression to higher grade meningiomas (33, 83). Quantitative analysis of *NF2* mutation transcripts in meningiomas led the authors (152) to conclude that molecular differences existed among the histologic subtypes of these tumors and that there is likely an NF2-independent pathogenesis for meningiomas of meningothelial histology. Thus, loci other than that of *NF2* on chromosome 22 may play a role in meningioma tumorigenesis, as has been amply demonstrated previously (1, 158–162).

NF2 mutations or LOH often associated with loss of merlin seem to represent an early alteration in most meningiomas (21, 23, 136, 137, 163–165), with NF2-associated tumors developing along the same genetic pathway as sporadic meningiomas (146), which includes inactivation of the *NF2* gene followed by LOH of 1p, 10q, 6q, and 14q. LOH of 19q and 17p is very rare in either type of meningioma. The *NF2* changes in familial meningioma resemble those seen in sporadic cases (166). Proliferative activity is similar in both types of meningioma.

In some meningiomas, no *NF2* mutation or LOH of 22q could be detected despite a lack of merlin. An alternative mechanism that involves merlin degradation by the protease μ-calpain, a calcium-dependent neutral cysteine protease, was proposed to account for this phenomenon (163, 314). Studies have demonstrated activation of μ-calpain in >50% of meningiomas (314), although no association between activation of the μ-calpain and merlin status could be established (25, 156). Instead, concordance of LOH of 22q and merlin loss suggested that other mechanisms, such as homozygous deletions or methylation as well as undetected *NF2* mutations, may account for the loss of merlin (25).

Because the frequency of LOH on chromosome 22 exceeds that of *NF2* mutations in meningiomas, and because interstitial deletions of 22q did not include the *NF2* locus in some tumors (15), the possible role of other genes in meningioma genesis has been addressed (33). A large number of candidate genes located on 22q has been investigated in regard to their role in meningioma pathogenesis. These genes include *ADTB1*, *RPP22*, and *GA22* (located at 22q12.2) (167), *MN1* (located outside of 22q12.1) (78, 159), *hSNF5/INI1* (located at 22q11.23) (168), *CLTCL1/CLH-22* (located at 22q11.21) (160), *BAM22* (located at 22q12) (169), and *LARGE* (located at 22q12.3) (162). Although some of the candidate genes may be involved in meningioma genesis (33), the findings to date do not offer an unequivocal role for any of these genes.

Among sporadic tumors, 11 of 14 schwannomas, three of eight ependymomas, and 16 of 19 meningiomas were shown to have significantly reduced or absent merlin expression, suggesting that *NF2* may be involved in the pathogenesis of these sporadic tumors (136).

In a study of sporadic meningiomas, the *NF2* status and that of 22q were evaluated in relation to the tumor site (170). In keeping with previous results, meningothelial meningiomas

TABLE 3.8. Salient genetic and molecular changes in meningioma tumorigenesis.

Benign meningioma (grade I)
Inactivation or loss of *NF2* (merlin), -22 and/or 22q-
   Loss of proteins 4.1B (DAL-1) and 41.R
   Gain of PR

Recurrent meningioma (grade I)
-14 or 14q-, loss of *NDRG2*

Atypical meningioma (grade II)
1p-, 10q-, -14 or 14q-, loss of *NDRG2*, 9p- (loss of p16),
   loss of PR, loss of TSLC-1

Anaplastic (malignant) meningioma (grade III)
6q-, mutations of *TP53* (rare), mutations of *PTEN* (rare),
   17q+, 9p- (loss of p16), loss of p14^{ARF}, -14 or 14q-,
   loss of *NDRG2*, loss of PR

The changes shown in the table are cumulative. Some of the changes, e.g., -14 or 14q-, are shown for more than one grade of meningioma, because they may appear either at an early or late stage of the tumors. For details, see text.

showed almost no *NF2* mutation and intact 22q in contrast to a high frequency of changes in transitional and fibrous tumors. Because meningothelial meningiomas are often located in the anterior skull base, this localization may be related to the histogenetic and genetic findings in these tumors. A high frequency (74%) of large alterations (deletions and duplications) in the *NF2* gene was found in sporadic and familial meningioma using multiplex ligation-dependent probe amplification (MLPA), a technique that has proven to be accurate and simple for detecting alterations that escape other screening methods. In the specific case of tumor samples, in which the cytogenetic and molecular alterations are usually abundant, it was confirmed that the MLPA technique can discriminate between various possible losses present simultaneously in both alleles of a gene (171).

Next to those of chromosomes 22, the changes of chromosomes 1, 14, 9, 10, 17, and 18 (shown in order of their frequency) and parameters associated with these changes (Tables 3.5 and 3.6) may play a key role in tumor progression (31, 82, 87, 94, 95, 172–174). Some of these changes are presented here. An example of the experience of one group of investigators with chromosome 22 and the *NF2* gene in meningioma is shown in Table 3.5.

NF2-associated meningiomas are relatively rare; the majority of these tumors occur as isolated and sporadic events. Nevertheless, deletions of chromosome 22 are found in all NF2-associated meningiomas and in 54–78% of sporadic tumors (19–21, 32, 56, 62, 78, 129, 148, 150, 151, 175–177). Table 3.8 shows the frequent cytogenetic changes and the associated molecular alterations in various grades of meningioma.

Chromosome 1 Changes in Meningioma

Next to those of chromosome 22, the changes affecting chromosome 1, especially 1p, are the most common in meningioma (Figures 3.9 and 3.10) (178). Deletions in 23 of 50 of

1p36 and in 33 of 50 of 22qter were observed in meningiomas by using FISH (179). Deletions or alterations of 1p possibly affecting genes at 1p22 and 1p21.1-p13 are events associated with meningioma progression (31, 68, 82, 84, 87, 94, 95, 97, 172–174, 180, 181), with the deletions spanning 1p34-pter (172). Telomeric fusion may be another mechanism for loss of 1p in meningiomas (99).

Deletions of 1p and 3p determined by CGH may contribute to meningioma tumorigenesis, and they have been suggested as an alternative to loss of chromosome 22 (98). The possibility of a gene on 3p, e.g., *RASSF1A*, located at 3p21.3, playing a role in meningioma development in some cases has been indicated in studies (98, 182).

Deletion of 1p32 is not associated with involvement of *RAD54* (98, 182), p73 (185, 186), and p18 (125, 188) and other genes (189, 190). In contrast, loss of 1p seen in 30% of sporadic meningiomas was associated with inactivation of *NF2* in most of these cases (87, 126, 164, 187, 191–194).

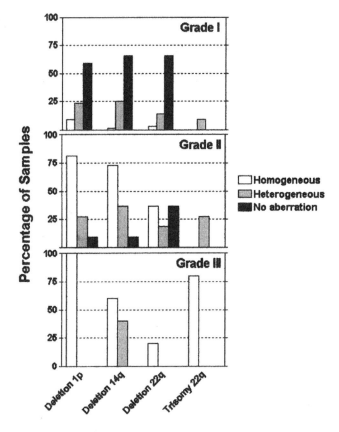

FIGURE 3.9. The results reveal several karyotypic features in meningioma, i.e., loss of 1p, although present in a small percentage of grade I tumors, increases with progression so that 100% of the grade III tumors contain this change. In parallel with those of 1p, are losses of 14q and chromosome 22. The presence of trisomy 22q in >75% of grade III meningiomas is probably related to the hyperploidy commonly encountered in these tumors (from ref. 117, with permission).

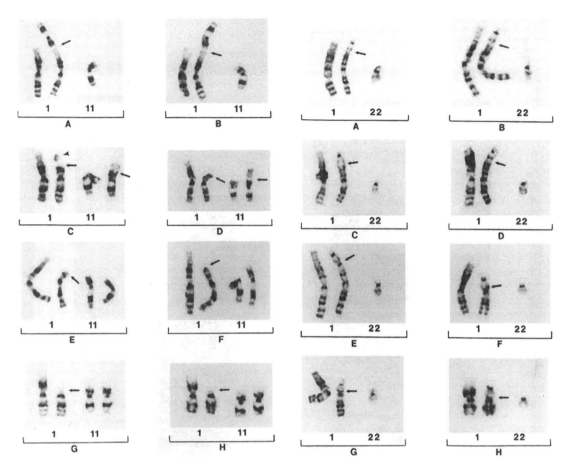

FIGURE 3.10. Telomeric fusions involving chromosome 1 (1p) in meningioma (99). These fusions may play a role in the formation of clonal deletions, dicentrics and unbalanced translocations involving 1p in meningiomas.

Together, the analyses of individual genes on 1p do not support a significant meningioma suppressor function of any of the genes investigated (33). A report (195) provided comprehensive ordering and identification of transcripts in the smallest region of overlapping deletion at 1p34 in meningioma and presented 17 likely candidate genes. This may serve as an important step toward the eventual elucidation of the 1p TSG.

Deletion at 1p36.1-p34, a locus for the alkaline phosphatase (*ALPL*) gene, leading to loss of activity of this enzyme, may play a role in meningioma biology (196, 491), especially in recurrence (197).

LOH of 1p, observed in >25% of meningiomas (82, 84, 189), was associated with deletions of chromosome 22 and with changes in *NF2* in meningiomas (189). LOH of 1p may serve as an important indicator based on its ability to predict a greater likelihood of recurrence than other indicators (189). Schemes for the clonal evolution of meningiomas based on chromosome changes, i.e., the presence or absence of 1p-, have been presented (32) along with mechanisms involving chromosome 22.

Allelic loss at 1p and 22q or at 1p alone may lead to DNA methylation of multiple promoter-associated CpG islands in meningioma corresponding to genes *THBS1*, *TIMP3*, $p16^{INK4A}$, $p14^{ARF}$, and *p73* (198). This methylation status may be related to the dichotomous *TP53* expression in meningiomas, i.e., suppressed or reduced expression as a result of CpG methylation in grades I and II tumors and enhanced expression in some grade III meningiomas (199).

Using tiling path microarray, the authors (200) found an association between the presence of segmental duplications and deletion breakpoints in 1p in meningiomas, suggesting their role in the generation of tumor-specific aberrations.

Monosomy 7p in meningiomas, seen in tumor progression, is associated with loss of 1p. Hence, it represents a cofactor rather than an independent feature of meningioma progression (102).

The search for a TSG at 22q12 in addition to the *NF2* gene led to the study of two genes, *GAR22* and *RRP22*, but neither was found to be mutated in 12 meningiomas. Half of the meningiomas had LOH at 22q12-q22, with none showing *NF2* mutations (167).

Chromosome 14 Changes in Meningioma

The third most common cytogenetic change affecting meningiomas consists of loss, alterations (32), or deletions of chromosome 14 (31, 32, 79, 84, 85, 87, 94, 95, 173, 201, 202). Based on the collective findings in meningioma (Table 3.4), >100 of these tumors had monosomy 14 (-14), with a much lesser number of meningiomas containing structural changes (deletions or translocations) affecting this chromosome. In 20 meningiomas, -14 was the only other change in tumors with -22 and in some with -22 and -Y/-X. Only one meningioma had monosomy 14 (-14) as the sole cytogenetic anomaly and another a del(14)(q22). About six meningiomas with +14 or add(14) have been reported previously (Table 3.4). No particular TSG or consistent and commonly deleted region in chromosome 14 has been established, although they seem to be related to meningioma progression and prognosis (94, 95, 203).

Based on cytogenetic findings (96), it seemed that the presence of deletions of 1p and alterations of chromosome 14 were associated with aggressive meningiomas. Of interest was the presence of -22 (or other abnormalities of 22) in these cases, i.e., only one of six meningiomas had -22 and 1p changes, whereas all eight tumors with chromosomes 1 and 14 changes also had -22 or other changes of chromosome 22.

In a study based on the numerical changes of eight chromosomes in meningioma, it seemed that those of chromosome 14 (especially gains) were an adverse prognostic factor (203).

Based on interphase FISH and CGH studies (204), 40.3 and 14.5% of 124 meningiomas showed loss or gain of 14q32, respectively. As in other studies, monosomy 14 was associated with an adverse prognosis. Other clinical correlations between chromosome 14 changes were presented in the study (204). The question raised some years ago (131) as to whether the results of chromosome analysis of meningiomas would predict recurrence has been partially answered by -14 and 14q- being such changes.

The possible role of the gene N-myc downstream-regulated gene 2 (NDRG2) located at 14q11.2 in meningioma progression has received attention (205). Inactivation of this gene may result from loss of 14q, a cytogenetic event not uncommon in the transformation of benign meningiomas to the more aggressive grades II and III. In aggressive meningiomas without such chromosomal loss, hypermethylation of the CpG island within the NDRG2 promoter region may be a likely mechanism of inactivation in the majority of meningiomas with loss of NDRG2 expression (205). Using high-density oligonucleotide microarrays, NDRG2 lacked detectable mRNA expression in all five anaplastic meningiomas examined (205). NDRG2 is normally expressed in brain, heart, and muscle cells, and it is one of four members of the NRDRG family (206, 207). The precise functions of this protein family are unknown. NDRG2 has been implicated in cell growth (208), differentiation (209), apoptosis (210), and response to mini-alocorticoid stimuli in the kidney (211). NDRG2 is up-regulated in the

brain in Alzheimer disease (212), suggesting a role in both cell growth and neurodegeneration (47). Further studies on this gene should clarify its role in meningioma biology.

The findings regarding chromosome 14 changes in meningioma are particularly apropos, indicating that a tumor suppressor gene, NDRG2, located at 14q11.2 is frequently inactivated as determined by gene expression microarray and immunohistochemical assays in aggressive meningiomas (primarily grade III) (205). This gene may play a key role in meningioma progression and serve as a useful and functionally relevant biomarker in predicting the behavior of meningiomas (205). Strong promoter hypermethylation may be one mechanism of transcriptional downregulation of NDRG2 in aggressive meningiomas (205).

Chromosome 10 Changes in Meningioma

Deletions of chromosome 10 (10q-) are not infrequent in meningioma (Table 3.4), and they are seen in more than half of grade III tumors and may not be necessarily related to tumor progression (31, 81, 84, 213–216). The PTEN and DMBT1 genes, located on 10q, are not affected by these deletions (213, 217, 218). The allelic losses of chromosome 10 thus seem to be complex, and LOH of this chromosome has an unfavorable clinical aspect (215, 216). No mutations of the DMBT-1 gene, located at 10q26.1-q26.2, in atypical or malignant meningiomas were encountered (214).

Chromosome 9 Changes in Meningioma and Their Consequences

Changes of chromosome 9 occur primarily in grade III meningiomas (Table 3.4) (31, 174). These changes, e.g., -9 or del(9p21), are more likely to be revealed by FISH (219) than by other methodologies. Genes located on 9p (CDKN2A, p14^{ARF} and CDKN2B) often show aberrations in grade III meningiomas, but they are infrequent in grade I meningiomas (173, 220). The same applies to the pRB and p53 pathways. Alterations of CDKN2A are associated with a relatively short survival (219).

Chromosome 17 Changes in Meningioma

The TP53 gene located on 17p (17p13.1) is rarely involved in meningioma (30, 173, 174, 218, 221–225), even in malignant tumors (224), although some studies indicated a role of p53 mutations in meningioma aggressiveness (226). Methylation may play a part in the TP53 status in meningioma (199). CGH studies have shown amplification at 17q (q22-q23) in 42% of anaplastic meningiomas, but no such amplification was seen in grade I meningiomas (31). A study using 17q microsatellite markers identified copy number increases as follows: 14% in grade I, 21% in grade II, and 61% in grade (31, 227). The *PS6K* gene and possibly others located at the site of 17q amplification may play a role in tumor progression (33, 227, 228). Allelic imbalance of 17q is a frequent event in anaplastic meningioma (228). The role played by the *erbB-2* gene, located at 17q12-q21.32 (229), in meningioma has not been elucidated.

Chromosome 18 Changes in Meningioma

The differentially expressed in adenocarcinomas of the lung-1 (DAL-1) protein, whose gene is located at 18p11.3, shares homology with merlin as part of the 4.1 family of membrane-associated proteins (230). Expression of DAL-1 is lost in about three fourths of meningiomas, and it seems to be an early event (231, 232). LOH at 18p11.3 was seen in 71% of meningiomas (233), with combined loss of merlin and DAL-1 in 50%. Loss of 18q determined by CGH was encountered in the more advanced tumors (31, 233). A study (126) based on CGH of sporadic meningiomas showed that LOH of 18p was not as common in these tumors as reported previously (231), acting more as a progression event than an early event in meningioma formation. An analysis of TSG genes located on the long arm of chromosome 18, i.e., *MADH2*, *MADH4*, *APM-1*, and *DCC*, showed that they are not inactivated in meningiomas (233).

CGH Studies in Meningioma

Several CGH analyses of meningiomas have been performed with the results not too divergent among these studies (Figures 3.11 and 3.12) (31, 32, 98, 164, 181, 192–194, 204, 234, 235). One of the first CGH studies limited itself to changes in chromosomes 1 and 3 (98), which was performed on meningiomas without chromosome 22 deletions, based on LOH

analysis (98). In two of 25 tumors, deletions of 1p and 3p were demonstrated. The authors (98) concluded that these findings may represent changes that are an alternate to those of chromosome 22 in the genesis of some meningiomas. The next CGH analysis addressed the full chromosome complement of meningiomas from 62 patients (31). An overview of the results is shown in Table 3.9. LOH studies of 17 loci confirmed the CGH results in 95% of the cases (31), including all changes at 22q. Not shown in Table 3.9 are the changes affecting a larger number of chromosome arms in grades II and III meningiomas, i.e., losses in <25% and gains in <40% of the tumors. Some of these unlisted changes may, in fact, play an important role in meningioma biology.

The results of other CGH studies are essentially variations on the themes presented in Table 3.8, i.e., loss of chromosome 22 or 22q is seen in about 50% of meningiomas; loss of 1p increases with tumor grade, as does loss of 14q and 18q; and in grade III loss of chromosome 10 becomes significant. Gains of chromosomes and their amplification probably reflect the hyperploidy of grades II and III meningiomas. CGH studies in radiation-induced meningiomas (RIM) resembled those in other types of meningiomas (234, 235).

CGH was applied (164) to 20 meningiomas (13 typical, four atypical, and three anaplastic) to investigate the genetic pathways underlying their development. Typical meningiomas displayed only a few genetic changes, i.e., monosomy 22. Anaplastic meningiomas manifested more aberrations than typical meningiomas, frequently exhibiting losses of 1p, 2p, 6q, chromosome 10, and 14q, and gain of 20q, in addition to monosomy 22. The average number of alterations increased significantly with grade progression. These CGH findings suggest that losses of 1p, 2p, 6q, chromosome 10, and 14q and gain of 20q are genetic changes implicated in the malignant progression of meningioma.

A CGH study (32) of 35 meningiomas showed frequent loss at 1p, 6p, and chromosomes 14, 18, and 22. Loss of the Y and to a lesser extent an X also was noticed. The results reflect those of others.

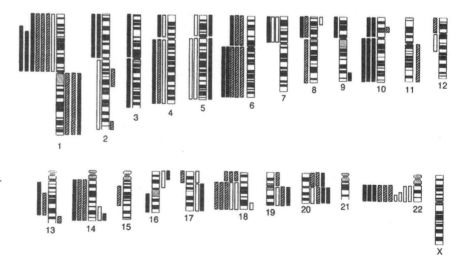

FIGURE 3.11. Results of CGH analysis of chromosome losses and gains in meningioma. Losses are shown to the left and gains to the right of each chromosome. Shading of the bars indicates grade of the tumors: white, grade I; hatched, grade II; and black, grade III. Losses of 22q, 1p, chromosome 6 and 18 are the most common (from ref. 164, with permission).

FIGURE 3.12. Possibility of recurrence based on the alkaline phosphatase (Pal) reaction in meningiomas. In the normal group Pal expression (as determined histochemically) was uniform in all tumor cells, whereas in the abnormal group such expression was heterogenous or completely lacking (from ref. 197, with permission).

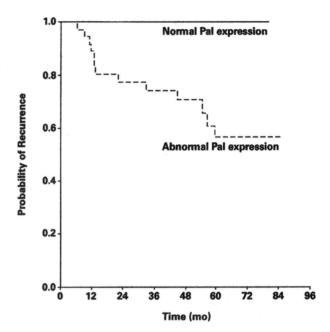

In a study of meningioma in Turkish patients (193, 194), 15 grade I, seven grade II, and three grade III tumors were examined with CGH. The most frequent abnormality in grade I tumors was loss of 22q (47%) followed by loss of 1p (33%). In grade III tumors, loss of 1p was present in three of three tumors, with losses of 1p, 10q, 14q, 15q, 18q, and 22q and gains of 12q, 15q, and 18p also present. Combined loss of 1p and 14q was seen in two of 15 grade I, three of seven grade II, and two of three grade III tumors. Amplification of 17q was seen in grades II and III meningiomas. A CGH study of changes of chromosome 14 in two meningiomas confirmed changes observed by FISH (204).

In a CGH analysis (31) of sporadic meningiomas, the number of genomic aberrations was 2.9 ± 0.7 for grade I (19 tumors), 9.2 ± 1.2 for grade II (21 tumors), and 13.3 ± 1.9 for grade III (19 tumors) meningiomas, indicating that tumor progression is associated with an increase in the genomic changes. The most frequent alteration in grade I meningiomas was loss of 22q (58% of tumors), 71% in grade II, and 42% in grade III tumors.

Based on array-CGH results, the number of chromosome aberrations per tumor was higher in invasive meningiomas than in noninvasive tumors, i.e., 67.4 versus 40.5 (395A). Furthermore, loss of 1p, 6q, and 14q and gain of 15q and chromosome 20 were frequent in invasive tumors, findings previously described in similar studies.

Array-CGH of the original tumor showed loss of chromosomes 4 and 17, 9p24-p21, 11q23-qter, and 13q12-q21. One of the cell lines had similar changes, indicating origin from at least part of the original tumor.

LOH Studies in Meningioma

LOH methodologies, being more sensitive than CGH, have revealed a somewhat higher percentage of allelic losses in meningiomas, but qualitatively the changes with both techniques have been similar (Tables 3.10–3.13). Thus, the commonest LOH was encountered for chromosome arms 22q,

TABLE 3.9. CGH results (%) in meningioma.

Grade I (19 tumors)	Loss	Gain
(68% with abnormalities)	22q (53)	
(2.9 ± 0.7)*	1p (26)	
(Ki-67 index 1.3 ± 0.4)	14q (21)	
Grade II (21 tumors)		
(94% with abnormalities)	1p (76)	20q (48)
(9.2 ± 1.2)*	22q (57)	12q (43)
(Ki-67 index 5.9 ± 1.0)	14q (43)	15q (43)
	18q (43)	
	10q (29)	
	18p (29)	
Grade III (19 tumors)		
(100% with abnormalities)		
(13.3 ± 1.9)*	1p (79)	17q (63)
(Ki-67 index 9.5 ± 1.6)	14q (63)	20q (58)
	10q (58)	17p (47)
	6q (53)	1q (42)
	10p (47)	12q (42)
	18q (47)	15q (42)
	22q (42)	
	18p (37)	
	9p (32)	
	6p (26)	
	X (26)	
	4p (21)	
	11p (21)	

This table contains only chromosome arm losses of > 0% of the tumors and gains of >40% of the tumors.
*Mean number of genomic aberrations per tumor.
Based essentially on findings of Weber et al. (31).

TABLE 3.10. Examples of results (%) on allelic changes in meningioma.

	Grade I	Grade II	Grade III
LOH			
1p-	11	40	70
10q-	12	27	40
14q-	0	47	55
Simon et al 1995 (84)			
LOH			
1p- (29/51)	14	83.3	100%
9p- (10/51)	7	29	33
10q- (14/51)	9.5	24	100
14q- (15/51)	29	25	100
Leuraud et al 2004 (187)			
CGH			
1p-	26	76	79
9p-	5	10	32
10q-	5	38	58
14q- (15/51)	21	43	63
Weber et al 1997 (31)			

1p, 10q, 14q, and 9p in the order shown, with the percentage of tumors involved by changes of 1p, 10q, 14q, and 9p being progressively higher with increasing grades of meningiomas or recurrence. Generally, LOH or other changes of 22q served as background for the appearance of anomalies for the aforementioned chromosomes. LOH of chromosome 2 is common in benign meningiomas of the elderly (236).

The discrepancy between the higher incidence of chromosome 22 LOH and the lower frequency of *NF2* gene mutations has led to the search for other TSG on chromosome 22 in proximity to but distinct from the *NF2* gene (15, 16, 130, 144, 158, 176, 237–239). Other possible TSG candidates have been proposed, including *BAM22*, *LARGE*, *MN1*, and *INI1* genes (159, 162, 168, 169).

In a comprehensive allelotype and genetic analysis (214), using the determination of LOH with a large number of microsatellite markers, of the 86 meningiomas examined 58% had mutations of *NF2*. Common regions of LOH were seen at 1p, 10q, 14q, and 22q. Those of 10q and 14q were seen more frequently in grades II–III meningiomas and LOH of 6q, 13q, and 18q were common in grade III tumors.

Studies on LOH of various chromosomal alleles in meningioma have revealed some variability, possibly related to differences in the methodologies used, the nature of the tumors examined, and the purposes of the studies. However, in general the findings on LOH have been in agreement with, and often expanded, the results obtained by CGH, cytogenetics, and FISH. Tables 3.10–3.13 contain a representative listing of the LOH findings and other changes in meningioma.

Observations on LOH and those with CGH have demonstrated an array of changes in meningioma of all three grades (126, 192, 194). In these studies, losses of chromosomal material outnumbered gains.

LOH at 22q12.2 was found in 40–70% of meningiomas (150, 151). Truncating mutations of *NF2* and associated LOH

at 22q have been encountered in up to 60% of sporadic meningiomas (15, 16, 25, 148, 176), findings indicative that inactivation of merlin, the protein product of *NF2*, plays an important role in meningioma development. This is reflected in the absent or reduced immunoreactivity to merlin seen in the majority of meningiomas (138, 142, 154–156).

LOH of chromosome 22 exceeds the number of *NF2* mutations in meningioma, possibly indicating that LOH may affect loci that do not include that of *NF2* (15, 16, 153). Several other genes with locations on 22q12.2 have been investigated as to their possible role in meningioma development, with the results being either negative or equivocal (78, 159, 160, 162, 167–169). Relatively early LOH studies (151, 175, 238) did not show the nonrandom changes seen in subsequent analyses (31, 82, 84, 97, 146, 174, 187).

Table 3.8 shows some of the salient events that occur after *NF2* inactivation and probably responsible in some measure for the biologic and clinical progression of meningiomas. Included in these events are loss of chromosomal material at 1p, 14, 10q, and 9p, reactivation of telomerase, inactivation of the p16/*CDKN2A* gene, changes particularly prominent in grades II and III meningiomas (33, 187, 240, 241). LOH study of meningiomas was found to be a useful method of predicting atypia and anaplasia in these tumors (242). Besides confirming previously described LOH on 1p, 9p, 10q, and 14q, these authors (242) also found LOH on 11q, possibly involving the multiple endocrine neoplasia type 1 gene. A study has shown that meningiomas are not an infrequent occurrence in MEN1, and loss of function of the *MEN1* gene may play a role in meningioma pathogenesis (242A).

Pediatric Meningiomas

Pediatric NF2-associated meningiomas are uncommon and poorly characterized in comparison with sporadic meningiomas in adult cases. Rhabdoid meningioma, a subtype of malignant meningioma, is apt to occur in children (27). To elucidate their molecular features, MIB-1, progesterone receptor (PR), *NF2*, merlin, *DAL-1*, DAL-1 protein, and chromosomal arms 1p and 14q in 53 meningiomas from 40 pediatric patients were analyzed (157) by using immunohistochemistry and dual-color FISH. Fourteen pediatric (42%) patients and the seven adult cases had NF2. Meningioma grading revealed 21 benign (40%), 26 atypical (49%), and six anaplastic (11%) examples. Other aggressive findings included a high mitotic index (32%), a high MIB-1 LI (37%), aggressive variant histology (e.g., papillary, clear cell) (25%), brain invasion (17%), recurrence (39%), and patient death (17%). FISH analysis demonstrated deletions of *NF2* in 82%, *DAL-1* in 82%, 1p in 60%, and 14q in 66%. NF2-associated meningiomas did not differ from sporadic pediatric tumors except for a higher frequency of merlin loss in the former and a higher frequency of brain invasion in the latter. Thus, although pediatric NF2-associated meningiomas share the

TABLE 3.11. LOH and other studies in meningiomas.

	References
LOH of 1p and 22q in 1/33 and 9/33 grade I meningiomas; 10/15 and 12/15 grade II meningiomas; 2/2 and 2/2 grade III meningiomas, 1q not involved	*Bello et al 1994 (82)*
Specific chromosomal regions or bands involved in meningioma, shown by LOH, CGH, or other methodologies have shown variability and these are listed for some of the chromosomes:	
1p: 1p36, 1p32-p35	*Bello et al 1995 (191)*
1p34-pter	*Boström et al 1998 (217)*
1p32	*Sulman et al 1998 (189)*
1p22-p13	*Leone et al 1999 (87)*
1p36.21-p23	*Murakami et al 2003 (97)*
14q: 14q24.3-q31, 14q32.1-q32.2	*Simon et al 1995 (84)*
	Menon et al 1997 (85)
	Tse et al 1997 (202)
14q21	*Weber et al 1997 (31)*
14q22-q24, 14q32	*Leone et al 1999 (87)*
Allelic gain and amplification of 17q22-q23; loss of 18q21	*Büschges et al 2002 (227)*
11/36 meningiomas with changes in chromosomes: 1, 2, 3, 6, 7, 10, 11, 14, 17, 18, X	*Dumanski et al 1987 (150)*
LOH at 17p and 22q correlates with aggressive and malignant behavior of meningiomas	*Kim et al 1998 (434)*
LOH of five aggressive, recurrent meningiomas revealed loss of 22q and ultimately 1p- in all five tumors; loss of 14q was not observed, but 10q- was seen in two.	*Lamszus et al 1999 (174)*
LOH study of 22 NF2-associated meningiomas showed 30/30 to have anomalies of 1p, 10q, 6q, 14q, chromosome 18, and 9p in the frequency shown. These changes are similar to those in sporadic tumors.	*Lamszus et al 2000 (146)*
LOH of 10q or 10q- shown in meningioma,	*Rempel et al 1993 (81))*
	Mihaila et al 2003 (215)
	Leuraud et al 2004 (187)
although the *PTEN* located on 10q is not involved in meningioma	*Boström et al 1998 (217)*
10q- may be associated with a poor prognosis	*Mihaila et al 2003 (215)*
Allelic loss (not statistically significant) at 1p, 3p, 5p, 5q, 11p, 13p, 17p	*Schneider et al 1992 (238)*
8/40 meningiomas had LOH at chromosomes 4, 10, 11, 13, 14, 19: No changes at chromosomes 1, 12, 17, 18, 21	*Seizinger et al 1987 (151)*
High rates (40–60% in grades II and III meningiomas) of LOH at 14q32.3?	*Simon et al 1995 (84)*
	Menon et al 1997 (85)
	Tse et al 1997 (202)
	Weber et al 1997 (31)
	Leone et al 1999 (87)
LOH changes at 1p, 10q, and 14q in grade I meningiomas could be indicative of recurrence or progression	*Simon et al 1995 (84)*
LOH of 1p was seen in 50% of sporadic meningiomas with chromosome 22 changes, whereas only 10% had 1p changes in the absence of chromosome alterations	*Sulman et al 1998 (189)*
LOH of 9p (also shown by CGH), containing a number of genes involved in cell cycle regulation, may be necessary for meningioma progression from grade II to III and is found in 32% of sporadic grade III tumors and only rarely in grades I and II.	*Weber et al 1997 (31)*
In sporadic tumors, LOH or loss of 22q has been shown to be present in at least 65% of meningiomas,	*Zang and Singer 1967 (56)*
	Westphal et al 1989 (132)
	Ruttledge et al 1994 (15, 16)
with 1p- loss being next in frequency.	*Jiménez-Lara et al 1992 (178)*
	Bello et al 1995 (191)
	Simon et al 1995 (84)
	Weber et al 1997 (31)
	Leone et al 1999 (87)
These changes are followed by those of 10q- and 14q-	*Rempel et al 1993 (81)*
	Simon et al 1995 (84)
	Weber et al 1997 (31)
	Boström et al 1998 (217)
	Peters et al 1998 (213)
	Leone et al 1999 (87)
and then 6q- and 18q-.	*Weber et al 1997 (31)*

TABLE 3.12. Some facets of LOH changes in meningioma.

		References
1.	LOH changes and MIB-index are similar in NF2-related and sporadic meningiomas.	*Lamszus et al 2000* (146)
2.	Order in which chromosome arms are associated with LOH in NF2-related and sporadic meningiomas is as follows: 22q > 1p > 10q > 14q > 6q > 18q > 9p.	*Lamszus et al 2000* (146)
3.	Other chromosome arms may be involved by LOH, but at a lower frequency than that of the chromosome arms shown above.	*Lamszus 2004* (33)
4.	Progression of meningiomas to higher grades may be specifically associated with LOH of 9p, amplification or allelic gain of 17q22-q23 present in anaplastic meningioma, LOH involving both 22q and 17p correlates with aggressive and malignant behavior.	*Weber et al 1997* (31) *Kim et al 1998* (434) *Büschges et al 2002* (227)
5.	The specific allelic losses in meningioma appear to be cell lineage-related rather than linked to the loss of the NF2 gene function.	*Weber et al 1997* (31)
6.	About 50% of sporadic meningiomas with loss chromosome 22 have a deletion of 1p, in contrast to only 10% in tumors with normal chromosomes 22.	*Sulman et al 1998* (189)
7.	Changes in addition to those of NF2 may be necessary for progression from grade I to grade II meningiomas, the so-called stepwise process of tumor development.	*Weber et al 1997* (31)
8.	LOH shown at 18q21 but gene involved at that locus remains unknown.	*Büschges et al 2001* (233)
9.	Meningiomas with LOH for 1p had a MIB-1 index of 8.6 ± 8.0, whereas those without such a change had an index of 2.0 ± 2.1.	*Lamszus 2004* (33)
10.	Generally, LOH of 1p is seen in approx 5–10% of grade I meningiomas, 40–60% of grade II and 80–100% of grade III.	*Lamszus 2004* (33)
11.	Allelic loss of 22q in sporadic meningiomas has ranged from 30% to >80% with a mean figure of 50%. Meningiomas associated with NF2 had allelic loss in 100% of the tumors.	*Lamszus 2004* (33)
12.	Telomerase activity correlated significantly with allelic loss of 1p and 10q, but not with that of 22q.	*Leuraud et al 2004* (187)
13.	Progression for survival correlated with allelic loss of 1p, 9p, 10q, and telomerase activity.	*Leuraud et al 2004* (187)
14.	Allelic loss or mutations of NF2 are much more common in fibrous, transitional, atypical and anaplastic sporadic meningiomas than in meningothelial tumors. NF2 change necessary for progression of transitional tumors to grades II and II.	*Lamszus et al 1999* (174)
15.	Allelic loss of 14q has been stressed as an important event in meningioma (present in 40–60%) of grades II and II tumors; others have not found loss of 14q to be frequent. The same applies to allelic loss of 10q.	

TABLE 3.13. Some LOH and other molecular studies of loss of NF2 and/or chromosome 22 or 22q.

	References
LOH of 22q: 9/33 in grade I, 12/15 in grade II, and 2/2 in grade III meningiomas	*Bello et al 1994* (82)
LOH confirmed -22 or 22q- shown by cytogenetics; no LOH of chromosome 22 seen in tumors with a normal karyotype	*Dumanski et al 1987* (150) *Meese et al 1987* (175)
LOH of NF2 in 2/7 multiple meningiomas; truncating mutations of NF2 in 3/7	*Heinrich et al 2003* (326)
LOH in 5/5 recurrent meningiomas	*Lamszus et al 1999* (174)
LOH of 22q in 30/30 NF2-related meningiomas	*Lamszus et al 2000* (146)
In sporadic meningiomas 2/3 of somatic mutations occur in exons 1–7 of NF2, 1/3 in exons 8–13, and none in exons 14 and 15	*Lekanne Deprez et al 1994* (17) *Ruttledge et al 1994* (15) *Joseph et al 1995* (433) *Wellenreuther et al 1995* (148) *De Vitis et al 1996* (22)
Sporadic meningiomas with LOH of chromosome 22, often had loss of 1p and 14q; this was especially true of grades II and III tumors	*Leone et al 1999* (87)
LOH of 22q in 37/51 meningiomas: 48% in grade I, 87.5% in grade II, and 100% in grade III	*Leuraud et al 2004* (187)
In germline mutations of the NF2 gene in patients with NF2- related meningiomas, the mutations are spread over the entire gene	*Merel et al 1995* (19, 20) *Perry et al 1996* (80)
LOH at 22q is sole reason for loss of merlin in meningiomas	*Ueki et al 1999* (25)
LOH of 22q in 11/17 sporadic meningiomas	*Weber et al 1997* (31)
LOH or mutations of NF2 in 70–83% of grade I and 60–100% in grade III meningiomas	*Wellenreuther et al 1995* (148)
Mutation rates of NF2 similar in all histologic types of meningiomas, although a changed NF2 is necessary for progression of transitional tumors to grades II and III meningiomas	*Wellenreuther et al 1995* (148)
LOH of 22q in at least 65% of sporadic meningiomas	*Zang and Singer 1967* (56) *Westphal et al 1989* (132) *Ruttledge et al 1994* (15, 16)
Extracranial metastatic meningiomas were shown to have LOH at 1p, 9p, 10q, 14q and 22q.	*Gladin et al 2007* (189A)

TABLE 3.14. Some characteristics of pediatric meningiomas.

Uncommon
Occur mostly in males
Associated with NF2
Aggressive, brain invasion
High mitotic and MIB-1 indexes
Papillary or clear-cell histology
Deletions of *NF2* in 82% of meningiomas
Loss of *DAL-1* in 82% of meningiomas
Loss of 1p in 60% of meningiomas
Loss of 14q in 66% of meningiomas
High frequency of merlin loss

Based on Perry et al. (157).

common molecular alterations of their adult, sporadic counterparts, a higher fraction is genotypically and phenotypically aggressive. Given the high frequency of undiagnosed NF2 in the pediatric cases, a careful search for other features of this disease is warranted in any child presenting with a meningioma.

By the end of follow-up (157), 42% of pediatric patients qualified for the diagnosis of NF2 and in five of these 14 patients (36%), meningioma was the presenting manifestation. Another study (243) similarly found that three of 22 children presenting with meningioma later developed typical features of NF2.

An overview of some clinical and genetic findings in pediatric meningiomas is presented in Table 3.14. It has been stressed (157,243,244) that any child presenting with a meningioma should be carefully evaluated for other features of NF2 because of the important relationship between pediatric meningiomas and NF2 (245). In young children with meningioma, the diagnosis of NF2 may be established only years later (246). Emphasis has been put on the aggressive nature of these tumors (157), as reflected in some of the findings in Table 3.14. Somatic mutations of *NF2* have been found in sporadic pediatric meningiomas, akin to findings in adult sporadic tumors. Two meningioma from a child showed monosomy 22 (-22), even though there was an excess of other chromosomes (247).

Radiation-Induced Meningiomas

RIM are encountered after radiation for such conditions as tinea capitis, pituitary tumors, and other primary brain tumors or after dental X-rays (248–255). Loss of chromosome 22 has been described in RIM in several studies (108,256,257). CGH studies of RIM, although showing some variability of findings between laboratories, have generally revealed changes similar to those of sporadic meningiomas (234,235,258).

Most of the knowledge about the association of ionizing radiation and meningioma was obtained from immigrants to Israel who had been treated with low-dose cranial irradiation for tinea capitis between 1948 and 1960 (259,260). Thus,

patients who were treated during childhood developed meningiomas after a latency period of 20–40 years (248,261,262). RIM are often clinically and histologically more aggressive (corresponding to grade II or grade III), are often multiple, and they have a higher proliferative activity than their sporadic counterparts (1). Examination of 16 RIM revealed them to be aggressive tumors with high recurrence rate, to be of high grade, and almost always associated with complex cytogenetic changes particularly involving 1p, 6q, and 22 (92).

A few studies addressed the genetic mechanisms involved in RIM development (256). Mutation analyses of RIM showed that *NF2* mutations are relatively rare and occur in less than 25% of these tumors, compared with >50% in sporadic meningiomas (218,255,263). Likewise, allelic losses on 22q were only detected in 2/7 RIM, and a CGH study found monosomy for chromosome 22 in only one of five cases (235). Thus, *NF2* alterations seem to play a less important role in RIM than in sporadic meningiomas.

Several other genes, including some that are prone to developing radiation-associated mutations, also were analyzed in RIM. Mutations in the *TP53* and *PTEN* genes were confined to single cases, and no mutations were observed in the *HRAS*, *KRAS*, and *NRAS* genes (218).

In a CGH study (234) of 16 radiation-related and 17 nonradiation-related meningiomas, losses of chromosomes 22 (in approx 50%) and chromosome 1 (35–37%) were the most frequent, and they showed no significant differences as to radiation background. Although these data would seem to favor the view that the tumorigenic pathways involved in meningioma genesis as detected at the level of chromosome aberrations obtained with CGH are similar, regardless of the nature of the previous radiation, in another study (263) of seven radiation-related and eight nonradiation-related meningiomas, the *NF2* gene was not mutated or underexpressed in any of the former tumors versus 50% of the latter tumors. Allelic losses at chromosome 22 were detected only in two of seven radiation-related tumors (263). In a subsequent study (235), on six RIM and one sporadic meningioma, CGH analysis revealed loss of 1p (in five of seven) and 7p (in four of seven) of the tumors, with 6q loss in three of five. Only one of five RIM had loss of chromosome 22.

Ironically, radiation represents one of the accepted treatments for meningiomas that are recurrent, clinically aggressive, or that have failed surgical treatment. It has been estimated that the relative risk for the development of a subsequent meningioma in children receiving low-dose cranial irradiation is nearly 10-fold over those without such exposure (250,260), suggesting that there may be a critical window of susceptibility during childhood for neoplastic transformation of meningothelial cells by radiation. Support for this hypothesis comes from the previously mentioned reports of a significant increase in the incidence of meningiomas in Israel after the widespread use of low-dose scalp irradiation to treat children with tinea capitis in the 1950s (250,260).

There is some debate regarding whether RIM are more prone to be malignant. RIM present at an earlier age arise within the prior irradiation field by definition, and they are more likely to be multifocal. Histologic findings in RIM include high cellularity, marked pleomorphism/atypia with numerous giant cells, vacuolated nuclei, vascular hyalinization, and increased mitotic activity (249). However, none of these features are specific and any or all may be encountered in meningiomas unassociated with prior irradiation.

Genetic studies have shown that the *NF2* gene is less often implicated in radiation-induced than in sporadic meningiomas (218, 263). Instead, there are often complex structural and numerical chromosomal abnormalities (258). A specific genetic signature in RIM has not been identified. However, in one study (258) a characteristic derivative chromosome 1 in six RIM was observed, suggesting that a region on 1p13 may be critical to the development of these meningiomas.

In a study of seven RIM and eight nonradiation-related tumors using several techniques (single strand conformational polymorphism, immunoblotting, and LOH based on microsatellite analysis), the *NF2* gene was not mutated or underexpressed in any of the RIM versus 50% in the latter tumors (263). Allelic losses at chromosome 22 were detected in two of seven RIM.

Telomerase

Telomeres consist of stretches of repetitive DNA sequences that maintain chromosomal stability. Telomeric DNA is progressively shortened with each mitosis. When telomeres reach a critical length, the cells ultimately become senescent. Moreover, shortened telomeres can initiate aberrant fusion or recombination of chromosome ends; thus, they may contribute to cancer development. Telomerase is a reverse transcriptase that stabilizes telomere length, and it is necessary for unlimited cell proliferation. In contrast to germline cells and most embryonic cells, telomerase is not active in most normal adult tissues. However, it is frequently reactivated in cancer (264). Telomerase activity may help to identify benign meningiomas that are likely to grow or recur (265). Telomerase activation in meningiomas may be a critical step in the pathogenesis of malignant or atypical meningiomas (266). Elongation of the telomeres is associated with malignant potential.

Telomerase activity has been detected in 3–21% of grade I, 58–92% of grade II, and 100% of grade III meningiomas (265, 266, 266A–C). Such activity correlated significantly with the presence of 1p- and 10q- in meningiomas, but not with 22q- (187).

The expression of one essential component of telomerase, a reverse transcriptase subunit (hTERT), may be a more sensitive index than total telomerase, i.e., all telomerase-positive meningiomas also expressed hTERT, but not vice versa (266); recurrent meningioma may express hTERT but not telomerase. These findings suggest that hTERT may serve as a more sensitive index for an aggressive course of meningiomas than telomerase (33). Incidentally, telomeric association may be seen in meningiomas (107, 109), and it accounts for some of the cytogenetic changes in these tumors. In a study of the gene expression of the two core components of telomerase, hTERT and the RNA subunit (hTR), such expression was related to the proliferation index (Ki-67) (267). Telomerase activity was found in two of 20 meningiomas, both atypical, with the remaining 18 tumors being benign. hTERT mRNA transcripts were consistently low or absent in meningiomas. The absence of a positive correlation between telomerase activity and hTERT mRNA could not be attributed to the presence of hTERT spliced variants. Telomerase activity scores correlated with the Ki-67 index. A positive association was also seen between the Ki-67 index and the degree of hTERT mRNA expression.

Human telomerase reverse transcriptase expression is an early event in meningioma tumorigenesis in contrast to telomerase activity of tumors (268).

The mean proliferating cell nuclear antigen (PCNA), Ki-67, and telomerase reverse transcriptase indices were higher in relapsing residual meningiomas than in a stable group (269).

Hormones and Their Receptors in Meningioma

The expression of several hormone receptors, e.g., those for androgens (AR), estrogens (ER), and PR, has been investigated in meningiomas. References to these studies may be found in previous publications (34, 270–282).

The higher incidence of meningiomas in females (1, 282A), the aggravation of symptoms by high progesterone levels (e.g., pregnancy and menstruation (283, 284)), and nonrandom coincidence with breast cancer (285–288), lent cogency to the study of certain hormones and their receptors in meningioma and the genesis of a sizable literature on this area (280). However, a study has indicated that endogenous hormones (pregnancy and indirectly smoking) may have protective effects for meningioma development in premenopausal women (288A). Estrogen receptor levels are low in meningioma and occur in about 10% of the tumors (43, 274, 284, 289). PR and AR are present in about two thirds of meningiomas, and the latter are more frequently expressed in the tumors of women than men (272, 273, 290–293). An inverse relationship between PR and bcl-2 related to prognosis in meningioma has been described previously (294). Meningiomas that are progesterone receptor-positive are less likely to recur (295). Progesterone, locally synthesized, exerts its actions through PR, and it may be involved in the regulation of cell growth and development in neurogenic tumors, including meningiomas (296). The expression of PR alone is a favorable prognostic sign in meningiomas. Absence of both PR and ER or the presence of only ER [separated meningiomas into different types according to their aggressiveness and abnormal karyotypes, especially involvement of chromosomes 14 and 22] (297). Higher frequency and levels of PR have been found in meningiomas of the optic nerve sheath than in other benign meningiomas (298). Immunohistochemical localization of PR and heat shock protein 27 and absence of ER and hormone-induced protein PS2 have been reported in meningiomas (299). An inverse relation has been found between the cell proliferation index (Ki-67) and expression of PR (300, 301). Autonomous expression of PR in meningioma may be due to ER mutants with loss of exon 4 are rendered unable to bind heat shock protein 90 (303). Although some hope was held for the therapeutic effectiveness of agents affecting the hormonal parameters of meningiomas, to date that has not been the case. Meaningful conclusions from the clinical application of hormone antagonists in the treatment of meningioma await appropriate trials and evaluation (34, 292, 305).

A relationship between exogenous hormone exposure and meningioma formation is not strong (306). Immunohistochemical detection of PR and correlation with Ki-67 labeling indices in paraffin-embedded sections of meningioma has been reported previously (307, 308).

PR possibly leads to peritumoral edema of intracranial meningiomas by secretion of some substances (e.g., prostaglandins and biogenic amines) that result in vagogenic edema (309). Growth hormone receptor expression and function in meningiomas have been discussed, and a possible adjuvant to therapy with receptor-specific antagonists raised (310). Nevertheless, it seems that the many studies on hormones and their receptors in meningioma in the 1980s and 1990s, although yielding information of some interest, with some exceptions (34), have not found realistic applicability therapeutically.

DAL-1 and -4.1 Protein in Meningioma

DAL-1/4.1B is a member of the protein 4.1 superfamily, which includes the product of the *NF2* gene (merlin) and encompasses structural proteins that play an important role in membrane processes via interactions with actin, spectrin, and the cytoplasmic domains of integral membrane proteins. DAL-1/4.1B localizes within chromosomal region 18p11.3, which is affected by LOH in various tumors.

The protein 4.1B has been shown to be a tumor suppressor involved in the molecular pathogenesis of sporadic meningiomas (157, 231, 232). Deletions and loss of protein expression of the 4.1B gene are common in meningioma, regardless of histologic grade, suggesting that the 4.1 protein loss, like *NF2* inactivation, is an early genetic event in meningioma tumorigenesis (315). Similar to merlin, the reexpression of the 4.1B protein in deficient meningioma cell lines results in reduced cell proliferation (316). The 4.1B region required for meningioma cell growth suppression is contained within a 503-amino acid fragment, termed DAL-1 (315, 317). The

low mutational frequency in meningiomas (318) discounts sequence variations in DAL-1/4.1B as the main mechanism underlying participation of this gene in the neoplastic transformation of meningiomas, and it suggests that other inactivating mechanisms, such as epigenetic changes, may participate in DAL-1/4.1B silencing.

Protein 4.1B interacts with tumor suppressor in lung cancer-1 (TSLC-1), suggesting a common signaling pathway (315). However, TSLC-1 loss was more frequent in high-grade meningiomas, and there was no correlation between TSLC-1 loss and 4.1B protein expression (315).

It would seem that merlin and the 4.1B protein may be functionally distinct proteins with different mechanism of action (316). It is of interest that early molecular studies (144) pointed to a locus different from that of *NF2* in the genesis of meningioma. Another protein 4.1R, may play a role as a tumor suppressor in meningioma pathogenesis (317). Loss of this protein was observed in six of 15 meningiomas and two cell lines.

Loss of expression of protein 4.1B, to which merlin is structurally related, is an early event in the tumorigenesis of sporadic meningiomas (315). Because protein 4.1B interacts with the TSLC-1 protein expressed in leptomeninges, whose gene is located at 11q23, the status of the latter protein was investigated in meningiomas (315). In a series of 123 meningiomas, TSLC-1 expression was absent in 48% of grade I, 69% of grade II, and 85% of grade III meningiomas. Furthermore, TSLC-1 loss was associated with high proliferative indices and decreased survival. The authors (315) concluded that TSLC-1 plays an important role in meningioma pathogenesis.

Multiple and Familial Meningiomas

Multiple meningiomas (meningiomatosis) occur in 4–10% of patients with meningioma (10, 37, 318–321, 321A), usually through subarachnoid spread via the CSF (119–122). This mode of spread is supported by the identical *NF2* gene mutations in multiple meningiomas (322) and pattern of X-chromosome inactivation. Familial meningiomas are very rare (323); they occur in NF2 subjects (321), although rare non-NF2 familial cases have been reported (239). In both types of families, the meningiomas are usually multiple.

Alternatively, *NF2* mosaicism could underlie some cases. *NF2* mutations and clonal origin were more frequently found in patients with a larger number of tumors than in patients with only two tumors, which is compatible with either mosaicism or with an increased ability of liquorigenic seeding caused by a mutant *NF2* gene.

The introduction of new techniques for visualization of tumors, such as computed tomography, has afforded a reliable assessment of the presence of multiple meningiomas (37, 324). In fact, the question has been raised whether solitary meningiomas exist, in view of the finding of additional such tumors in cases with seemingly solitary tumors (324).

Any child presenting with a meningioma should be carefully examined for other features of NF2. The child also should be closely followed-up for NF2 if no evidence is found at the time of meningioma diagnosis (157).

A family with predisposition to meningioma development was found to have a deletion of an Alu sequence in a c-*sis* allele (325). This would create a constitutive expression of platelet-derived growth factor (PDGF)-β. The authors (325) hypothesized that the change in c-*sis* could be pathogenetically important in the development of meningiomas. An identical deletion was found in one of three of sporadic meningiomas. Changes of c-*sis* in another family with meningiomas and a constitutional t(14;22) have been described previously (325A). In a family with inherited multiple meningiomas (8), cytogenetic analysis (Table 3.4) revealed a hypodiploid karyotype with -22 and several markers.

A study focused specifically on differences between multiple meningiomas in sporadic versus familial non-NF2 cases (326). All *NF2* mutations that were detected occurred in tumors from patients with no affected relatives, suggesting that the *NF2* inactivation is frequently involved in multiple sporadic meningiomas but that it is rare in multiple meningioma kindreds. In agreement with this observation, linkage analysis of a family with multiple meningiomas showed no segregation with the *NF2* locus (239). In another multiple meningioma kindred, immunoreactivity for merlin was detected, implying that the *NF2* gene was not inactivated (166). Interestingly, most non-NF2 meningioma family members develop meningothelial meningiomas, which is in line with the observation that this variant seems to arise independently of *NF2* inactivation.

Examination of multiple and recurrent meningiomas in an adult patient showed -22 to be an early event followed by mutations of *NF2* and subsequently loss of one X-chromosome and other changes (322). In a publication listing risk factors for meningioma recurrence cytogenetic patterns are mentioned, but no discussion was presented (327).

Meningioma and Meningioangiomatosis

Meningiomatosis are rare seizure-associated, and enigmatic cortical and leptomeningeal lesions, encountered either sporadically (328, 329) or in the setting of NF2 (330, 331). It is thought to be hamartomatous or reactive in nature and characterized by a perivascular spindle-cell proliferation of presumed meningothelial origin, based on the presence of psammoma bodies, occasional epithelial membrane antigen (EMA) immunoreactivity, and the coexistence of an adjacent meningioma in some cases (332–335). It has been speculated that such meningiomas arise as a result of neoplastic transformation in a perivascular meningothelial cell within the meningioangiomatosis. However, a case was encountered

(336) with identical genetic alterations in both the meningioangiomatosis and meningioma components, suggesting the alternate possibility that meningiomas may occasionally spread extensively along perivascular spaces, thus mimicking the architectural pattern of meningioangiomatosis. In fact, the resemblance of meningioangiomatosis to brain invasion by meningioma has been emphasized (337). Based on a study of merlin levels, 4.1B protein, PR, MIB-1, and *NF2* in meningioangiomatosis with and without meningioma (330), the authors suggested that the meningioangiomatic part is neoplastic, representing a perivascular spread form a meningioma (leptomeningeal or intracerebral), rather than an underlying hamartoma. The cells of "pure" meningioangiomatosis (i.e., perivascular spindle-cells) are genetically and immunohistochemically similar to non-neoplastic meningothelial cells (330).

The rare condition known as meningothelial hyperplasia may have overlapping features with meningioma, which presents difficulties in the differential diagnosis (330A). A study showed that the cells in meningothelial hyperplasia retain both merlin and protein 4.1B expression, and FISH showed no deletions of *NF2* or 4.1B, with some cases having polysomy of 22q and chromosome 18 and possible polyploidy, changes very different from those seen in meningioma (330A).

Proliferation Index in Meningioma

The possible value of proliferation indices (e.g., Ki-67, MIB-1, and PCNA) (338) in the prognosis of meningiomas has received much attention (30, 301, 302, 339–346), although others indicate that these indices may be of minor value (337).

The possible recurrence of intracranial meningioma was evaluated on the basis of their MIB-1 index and p53 expression (304). There was a statistically significant correlation between the MIB-1 index and tumor grade. The majority (76%) of meningiomas with a low index did not recur. Positive p53 expression was present in 26 of 45 meningiomas, and although such expression was higher in atypical and anaplastic meningiomas, no significant correlation was found between the p53 expression and recurrence. Thus, quantitative MIB-1 labeling is a useful technique in the diagnostic assessment of meningiomas (304).

In the experience of some authors, an increased MIB-1 labeling index may be indicative of recurrence of meningiomas (341, 348, 349). This index is also higher in atypical and anaplastic meningiomas a compared to simple tumors (350).

The MIB-1 and Ki-67 indices show a significant increase from benign (mean 3.8%), to atypical (mean 7.2%) to anaplastic (14.7%) meningioma (351). MIB-1 indices may vary considerably among anaplastic meningiomas (1.3–24.2%; mean 11.7%) (352). One study found that the Ki-67 index was a crucial criterion of differentiating anaplastic (mean 11%) from atypical (mean 2.1%) and simple (0.7%) meningioma (352). In Tables 3.15 and 3.16 are shown the results of studies on the Ki-67 (MIB-1) index and PCNA index in meningioma.

MIB-1 labeling index, expression of bcl-2 and mitosis count did not correlate with recurrence of meningioma. Apoptosis count was significantly higher in the initial resection specimens than in recurrent tumors vs. nonrecurrent tumors

TABLE 3.15. Proliferation indices (%) in recurrent meningiomas.

	Nonrecurrent meningioma	Recurrent meningioma	Reference
Ki-67[a]	1.0	7.9	*Madsen and Schrøder 1997* (342)
Ki-67	0.1–4.56	0.6–6.26	*Takeuchi et al 1997* (344)
MIB-1[b]	1.9	10.9	*Ohkoudo et al 1998* (438A)
PCNA[c]	<2.0–4.0	>7.0–38	*Cobb et al 1996* (439)
PCNA	0.1–5.44	0.77–7.60	*Takeuchi et al 1997* (344)

[a] Ki-67 is a monoclonal antibody recognizes a cell cycle stage-related protein selectively expressed in cells that have entered the cell cycle other than the G0 phase (338). The monoclonal antibody recognizes both Ki-67 and PCNA.
[b] MIB-1 is a monoclonal antibody that recognizes the nonhistone nuclear Ki-67. Ki-67 is absent during G0 and early G1 phases of the cell cycle; reaches a maximum during the G2 and M phases.
[c] PCNA, a cell cycle-regulated protein is a cofactor of DNA polymerase-δ, an enzyme required for DNA replication; two- to three-fold increases of PCNA are present between early G1 and early S phases and a plateau through the G2 phase.

TABLE 3.16. Proliferation indices in meningioma.

Grade I	Grade II	Grade III	References
Ki-67 index			
3.4	6.6	11.8	*Langford et al 1996* (340)
0.99	8.5		*Striepecke et al 1996* (343)
1.0	6.9	12.7	*Madsen and Schrøder 1997* (342)
0.6	1.6	3.1	*Möller and Braendstrup 1997* (347)
0.75	3.2	6.04	*Hsu et al 1998* (345)
1.6	7.4	14.7	*Debiec-Rychter et al 1999* (428)
<2	12.00	11.75	*Rushing et al 1999* (435)
2.0	8.0	15.4	*Murakami et al 2003* (97)
MIB-1 index			
2.47			*Karamitopoulou et al 1994* (225)
0.73	2.08	10.98	*Kolles et al 1995* (352)
1.06	2.75	10.90	*Nakasu et al 1995* (436)
2.69	10.58	19.1	*Ohkoudo et al 1996* (348)
3.80	7.17	14.71	*Maier et al 1997* (437)
2.26	4.49		*Antinheimo et al 1997* (145)
1.74	1.89		*Antinheimo et al 1997* (145)
1.30	9.30	15.00	*Karamitopoulou et al 1998* (438)
1.41	2.13		*Lamszus et al 2000* (146)
PCNA index			
1.36%	2.52		*Striepecke et al 1996* (343)
2.6	9.7	19.9	*Møller and Braendstrup 1997* (347)
1.23[b]	13.38	16.6	*Hsu et al 1998* (345)

The numbers represent the percentages of Ki-67-, MIB-1-, or PCNA positive tumors.
Based on Lamszus et al (146).

TABLE 3.17. Histologic and molecular findings in meningioma.

	No. of cases (%)
Total number of tumors	88 (100)
WHO grade I tumors	61 (70)
WHO grade II tumors	24 (27)
WHO grade III tumors	3 (3)
Transitional tumors	57 (65)
Meningotheliomatous tumors	22 (25)
Other subtypes	9 (10)
LOH 22q	43 (49)
NF2 mutation	21 (24)
NF2 methylation	23 (26)
LOH 22q and *NF2* mutation	16 (18)
LOH 22q and *NF2* methylation	5 (6)
NF2 mutation and *NF2* methylation	1 (1)
LOH 22q and *NF2* mutation and *NF2* methylation	2 (2)
LOH 22q (sole alteration)	20 (23)
NF2 mutation (sole alteration)	2 (2)
NF2 methylation (sole alteration)	15 (17)
No alterations	27 (31)

Based on Rey et al 1993 (128), Leone et al 1999 (87), and Lomas et al 2005 (165).

(353). Mutations of bcl-2 correlated with higher grade of meningiomas.

Evaluation of cell proliferation index Ki-67, DNA ploidy, and AgNOR staining, integrated with standard histopathology can provide better information for grading meningiomas (354). The correlation between increasing grades of meningioma, survivin, and Ki-67 index has been described (355). This also applies to survival. An improved method for Ki-67 determination in meningiomas was presented (356). Ki-67 index has no advantage over counting mitoses in meningioma as far as prognosis is concerned. Thus, mitotic activity justifies its role in meningioma grading (357).

It seems that the proliferative indices, e.g., MIB-1, do not show significant differences in NF2-related versus sporadic meningiomas (146). An analysis of Tables 3.15 and 3.16 supports this conclusion. A summary of some of the clinical and laboratory findings in meningioma is shown in Table 3.17.

Miscellaneous Aspects of Meningioma

A symposium on meningioma in 1995, followed by publication in 1996 of presentations at the symposium, dealt primarily with the clinical aspects of meningioma and served as an excellent summary of the status of this tumor at that time (358). The pathology, genetics, and biology of meningioma were reviewed (33) 9 years after the aforementioned symposium, containing discussions and references to the very large body of the molecular and genetic events associated with meningioma.

Hemangiopericytomatous meningiomas or hemangiopericytomas (HPC) of the meninges have presented a diagnostic problem to pathologists and clinicians. For example, the six

TABLE 3.18. Genetic, molecular, metabolic, and miscellaneous aspects of meningiomas.

	References
Use of dermosomal proteins in the differential diagnosis of meningioma versus other brain tumors.	*Akat et al 2003 (440)*
The tumor suppressors p53/p21 signaling pathway and *PTEN* play important roles in the development of benign meningiomas.	*Al-Khalaf et al 2007 (591)*
The p73 protein, which has a high sequence homology with *p53* and similar structure, is not involved in meningioma tumorigenesis.	*Alonso et al 2001 (441)* *Lomas et al 2001 (185)*
Pediatric (non-NF2) meningioma.	*Amirjamshidi et al 2000 (442)*
Aberrant methylation of the $p14^{ARF}$ gene in grades II and III meningiomas may be a mechanism for the accumulation of wild-type p53 protein in these tumors and this deregulation of the p14-MDM2-p53 pathway may contribute to the progression of meningiomas.	*Amatya et al 2004 (443)*
High expression of EGFR and ERB2-4, epidermal growth factor receptors, in meningiomas.	*Andersson et al 2004 (444)* *Carroll et al 1997 (365)* *Johnson et al 1994 (445)*
NF2-related meningiomas showed more mitotic figures and nuclear pleomorphism than sporadic meningiomas. The proliferation potential was much higher in the former than the latter and may be related to earlier onset, multiplicity and aggressiveness of NF2-related meningiomas than sporadic ones.	*Antinheimo et al 1997 (145)*
Somatostatin receptor mRNA in meningiomas.	*Arena et al 2004 (445A)*
Connexin proteins (26 and 43) can be useful in differentiating various types of meningiomas (positive) from hemangiopericytoma (negative).	*Arishima et al 2002 (446)*
Interphase cytogenetic studies in meningioma.	*Arnoldus et al 1991, 1992 (447,448)*
Loss of chromosome 22 in 50% or more in meningioma by interphase cytogenetics; loss of Y or X rare.	*Arnoldus et al 1991, 1992 (447,448)*
The establishment of 3 immortalized meningioma cell lines preserving their meningothelial features and overcoming their senescence has been described.	*Baia et al 2006 (395B)*
No mutations of p16 in meningiomas.	*Barker et al 1997 (449)* *Simon et al 1997 (450)*
Age-associated increase in the prevalence of 22q LOH in subsets of benign meningiomas has been described.	*Baser and Poussaint 2006 (584)*
Nearly 3/4 of pediatric meningiomas had absent or minimal merlin expression based on in situ hybridization studies.	*Begnami et al 2007 (589)*
No loss of chromosome 22; other mechanism? only 6 cases (one with -Y, one with -X).	*Bello et al 1993 (415)*
Hypermethylation of the DNA repair gene *MGMT* was found in 14/73 meningiomas without significant *TP53* mutations. The role of gene hypermethylation in cancer has been reviewed. Aberrant CpG island hypermethylation is associated with atypical and anaplastic meningioma.	*Bello et al 2004 (198,311)* *Esteller et al 2001 (312)* *Liu et al 2005 (313)*
MGMT silencing through promoter hypermethylation may lead to *TP53* mutations in some meningiomas	*Bello et al 2004 (311)*
Claudin-1 and EMA expression can help to differentiate meningiomas from schwannomas and other tumors.	*Bhattacharjee et al 2003 (451)* *Hahn et al 2006 (452)*
Role of pial blood supply in peritumoral edema in meningiomas.	*Bitzer et al 1997 (361)*
A possible role for PDGF in meningioma growth has been proposed.	*Black et al 1994 (366)* *Johnson et al 2002 (368)*
Coexpression of PDGF and its receptor in meningiomas indicates an autocrine or paracrine stimulation of meningioma growth. Following receptor binding, the signals are transduced intracellularly via phosphorylation of members of the mitogen-activated protein kinase (MAPK) cascade, resulting in enhanced meningioma cell proliferation. Studies have shown that beside the MAPK signaling pathway, another route involving the phosphinositol-3 kinase (P13K)/Akt pathway seems involved in the control of meningioma cell proliferation in response to transforming growth factor-β.	*Black et al 1994 (366)* *Johnson et al 2001 (367)* *Johnson et al 2002 (368)*
VEGF-B, placenta growth factor, scatter factor/hepatocyte growth factor, fibroblast growth factor-2, EGF, PDGR, and ILGF have been investigated as to their role in angiogenesis and malignancy in meningiomas with equivocal results.	*Black et al 1994, 1996 (274,366)* *Sanson and Cornu 2000 (289)*
Multiplicity of meningioma development.	*Borovich et al 1988 (324)*
Optic nerve meningiomas have a strong association with NF2, paralleling the occurrence of optic nerve gliomas in NF1.	*Bosch et al 2006 (565)*
PTEN (*MMAC1*) mutations not present in meningiomas with monosomy 10 (-10).	*Boström et al 1998 (217)*
Loss of 6_1-S phase regulators (CDKN2A and CKN2B and p14^{ARF}) leads to aggressive meningiomas.	*Boström et al 2001 (173)*
PTEN gene mutations involved in low-grade meningiomas.	*Peters et al 1998 (213)*
The lack of membranous β-catenin and/or E-cadherin in meningiomas may indicate an altered interaction between meningioma cells independent of loss of NF2, -22 and tumor grade	*Brunner et al 2006 (375)*
The lack of membranous β-catenin and/or membranous E-cadherin in meningiomas may indicate an altered interaction between meningioma cells independent of loss of *NF2* and independent of tumor grade and chromosome changes, including -22.	*Brunner et al 2006 (375)*
The lack of membranous beta-catenin and/or membranous E-cadherin in meningiomas may indicate an altered interaction between meningioma cells independent of loss of *NF2* and of tumor grade.	*Brunner et al 2006 (592)*
NF2 gene expression was studied in different histotypic meningiomas. Such expression was higher in meningothelial meningiomas than in other types but was not grade-related.	*Buccoliero et al 2007 (590)*
PS6K (a putative oncogene at 17q23) amplification seen in a small subset of aggressive meningiomas.	*Büschges et al 2002 (227)* *Cai et al 2001 (94)*

TABLE 3.18. Continued

	References
Multiple meningiomas in the same patient may differ histologically and cytogenetically.	*Butti et al 1989* (9)
	Rönne and Poulsgård 1990 (123)
Response of tumor cells to growth factors and other mitogens is mediated by specific receptors, including protein tyrosine kinase and G protein-coupled receptors. In response to stimulation, these receptors are activated and initiate intracellular signaling events. Growth factor receptors, such as epidermal growth factor receptor (EGFR), are known to be overexpressed in human meningiomas.	*Carroll et al 1997* (365)
Estrogen receptors in meningioma.	*Carroll et al 1999* (453)
Dicentric chromosomes were relatively common in meningiomas, as was telomere association. Telomerase activity was absent in these tumors.	*Carroll et al 1999* (593)
KRAS expression is elevated in meningiomas with a normal karyotype or only –22 (376). No amplification or major rearrangements of *KRAS* were found.	*Carstens et al 1988* (376)
Higher p53 expression in atypical vs. typical meningiomas; inverse relationship in *erb*B-2 expression.	*Chozick et al 1996* (454)
Indices of gliomas and meningiomas in Denmark 1943–1997; in Rochester, Minnesota.	*Christiensen et al 2003* (455)
	Radhakrishnan et al 1995 (456)
Secretory meningiomas.	*Çolakŏglu et al 2003* (457)
Five novel immunogenic antigens and their genes were identified in meningiomas. Three of the genes reside on chromosome 6 at q23 and one each on chromosomes 3 and 17, respectively. These antigens may be useful for diagnosis and possibly therapy of meningiomas.	*Comtesse et al 2001* (566)
MGEA6, overexpressed potential gene in meningioma.	*Comtesse et al 2001* (458)
Notch signaling pathway may be critical in meningioma progression.	*Cuevas et al 2005* (459)
Expression of extracellular matrix markers in meningioma.	*Das et al 2003* (335A)
Truncated NF2 proteins are unstable and undergo accelerated degradation in meningiomas (and schwannomas).	*den Bakker et al 2001* (460)
The aggressive behavior of a pediatric meningosarcoma has been attributed to its genetic instability, deletion of 17 p13 possibly involving p53 and the complexity of the karyotype.	*de Mesa et al 2005* (567)
Platelet-activating factor is present in meningiomas, as it is in other tumors, and might act as a tumor-altering factor by affecting angiogenic and/or cytokine networks.	*Denizot et al 2006* (377)
The expression of a number of proto-oncogenes was studied in meningioma and the results indicate a possible role for *MYC* and *FOS* in these tumors. The authors proposed that the nuclear transcription-regulating genes *MYC* and *FOS* are usually under the control of TSG, which are lost in meningiomas. This is supported by the more than 70% occurrence of proto-oncogene messenger RNA expression for *MYC* and *FOS* in meningiomas.	*Detta et al 1993* (378)
High mutation rates (>50%) of *NF2* gene in sporadic meningioma have been reported.	*Dumanski et al 1990* (144)
	Trofatter et al 1993 (14)
	Merel et al 1995 (19,20)
	Ruttledge et al 1994 (16)
	Wellenreuther et al 1995 (148)
	Leone et al 1999 (87)
In some meningiomas the NF2 and meningioma locus differ and in anaplastic cases may be outside chromosome 22.	*Dumanski et al 1990* (144)
The development of recurrence of meningiomas after resection is usually due to regrowth of the primary tumor and rarely to the emergence of an unrelated meningioma. Such recurrences usually acquire cytogenetic changes in addition to those seen in the primary tumors.	*Lindblom et al 1994* (83)
	Espinosa et al 2006 (568)
Multiple meningiomas in adults may have other causes than *NF2* mutations; the latter are a common cause in children with familial meningiomas.	*Evans et al 2005* (581)
Disturbance of balanced expression of molecules that promote opposing functions important in the development of the meningioma phenotype.	*Fathallah-Shaykh et al 2003* (461)
Familial meningiomas (non-NF2).	*Ferrante et al 1987* (462)
	Heinrich et al 2003 (326)
	Maxwell et al 1998 (166)
	McDowell 1990 (463)
	Pulst et al 1993 (239)
Neural cell-adhesion molecule isoforms and epithelial cadherin adhesion molecules play a role in the morphogenesis and histogenesis of meningiomas.	*Figarella-Branger et al 1994A* (379)
Loss of chromosome 22 in sporadic meningiomas shows no evidence of genomic imprinting; equal parental origin of -22.	*Fontaine et al 1990* (464)
Review of the molecular diagnostics of CNS tumors, including meningioma.	*Fuller and Perry 2005* (380)
Neither *BCL2*/bcl2 nor *ROS1* expression could be correlated with nor are they concurrent in any of the histologic types of meningioma	*Girish et al 2005* (381)
Tissue plasminogen activator expression not abnormal in meningioma, in contrast to glioblastoma.	*Goh et al 2005* (382)
The vascular endothelial growth factor-A (VEGF-A) has been found to be expressed in meningioma and related to the extent of peritumoral edema.	*Goldman et al 1997* (359)
	Provias et al 1997 (360)
	Bitzer et al 1998 (361)
	Yoshioka et al 1999 (**362**)

TABLE 3.18. Continued

	References
Expression of VEGF-A may be related to recurrence of benign meningiomas and their vascularity, though this was not confirmed in another study.	*Yamasaki et al 2000* (363) *Provias et al 1997* (359) *Samoto et al 1995* (364) *Lamszus et al 2000* (364A)
CpG island methylation and mutation of *RB1* gene in meningioma.	*Gonzalelz-Gomez et al 2003* (465)
Genes investigated individually in meningiomas: *DAL-1/4.1B, p18, TP53, PTEN, KRAS, NRAS, HRAS, CDKN2A (p16), CDKN2B (p15), p14^ARF CDKN2C.*	*Gutmann et al 2000* (231) *Boström et al 2001* (173) *Joachim et al 2001* (218) *Nunes et al 2005* (126)
S100A5: a marker of recurrence in grade I meningiomas.	*Hancq et al 2004* (466)
Based on the rate of *NF2* mutations in meningiomas, it is possible to divide these tumors into subgroups which overlap with histological variants.	*Hartmann et al 2006* (569)
Survivin was present in meningiomas, as it is in high levels of malignant tumors, reducing the tumors' capacity to undergo spontaneous and radiation-induced apoptosis.	*Hassounah et al 2005* (383)
The protein p62 binds to ubiquitin and thus may affect a number of cellular events. By using p62 labeling, it was shown that signs of a functioning proteosomal system exist in nonrecurrent meningiomas.	*Karja et al 2006* (385)
The involvement of a novel hyaluronidase gene in meningioma development is proposed.	*Heckel et al 1998* (384)
1p loss and enhanced glucose metabolism in meningiomas.	*Henn et al 1995* (424)
Mitochondrial and cytoplasmic metabolic energy pathways in meningioma.	*Herting et al 2003* (467)
Phosphofructokinase and lactate dehydrogenase were significantly increased in anaplastic vs. benign meningiomas.	*Herting et al 2003* (467)
Family history of cancer and meningioma risk.	*Hill et al 2004* (468)
Platelet-activating factor, which may arise from infiltrating leukocytes, is important in the development of peritumoral edema in meningioma.	*Hirashima et al 1998* (570)
Osteopontin in meningiomas.	*Hirota et al 1995* (469)
Progesterone and estrogen receptors in meningiomas; but not in high-grade tumors.	*Hsu et al 1997* (282) *Verhage et al 1995* (470) *Jacobs et al 1999* (471)
Rac2 gene, though underexpressed in some meningiomas, does not seem to play a role in tumorigenesis.	*Hwang et al 2005 (472,472A)*
Transforming growth factor-β in meningioma development.	*Johnson et al 2004* (473)
NF2 changes distinguish hemangiopericytomas from meningiomas.	*Joseph et al 1995* (433)
p53 and p21^WAF1/CIP1 in meningioma progression and recurrence.	*Kamei et al 2000* (474)
Meningioma in VHL patients and *VHL* gene changes.	*Kanno et al 2003* (475)
Most immunohistochemical markers have shown not to have diagnostic or prognostic relevance in meningioma, though prostaglandin D synthesase holds promise.	*Kawashima et al 2001* (476)
High expression of *sis* and *c-myc* in 6/15 and 12/15 sporadic meningiomas and mutations in *sis*. No other information on area since these early publications.	*Kazumoto et al 1990* (477) *Schmidt et al 1990* (325)
ECM (extracellular matrix)-associated proteins (SPARC, tenascin, stromelysin-3) and invasiveness of meningiomas. Sclerosing meningioma.	*Kilic et al 2002* (478A) *Rempel et al 1999* (370) *Kim et al 2004* (478)
Calpain proteolysis of NF2 in meningioma.	*Kimura et al 1998* (163)
Epigenetic (e.g., methylation) silencing of merlin and 4.1R	*Kino et al 2001* (479)
Shifting trend in female:male ratio in meningioma development in patients in the 35–59 year group in Scandinavian countries may be due to increasing hormone use.	*Klaeboe et al 2005* (480)
ODC mRNA as a predictor of meningioma recurrence.	*Klekner et al 2001* (481)
Apoptosis and meningioma recurrence.	*Konstantinidou et al 2001* (482)
Proliferation rate of meningioma as reflected in MIB-1, DNA topoisomerase II or cyclin A immunostaining or other indices predicts prognosis.	*Korshunov et al 2002* (483) *Cobb et al 1996* (439) *Hsu et al 1998* (345) *Shibuya et al 1992* (484)
Risk factors and frequency of meningioma in elderly.	*Krampla et al 2004* (485)
Clonal spread in multiple or recurrent meningiomas.	*Larson et al 1995* (119) *Stangl et al 1997* (122) *von Deimling et al 1999* (121)
An established cell line from a malignant meningioma had 45–65 chromosomes without -22 or loss of a sex chromosome.	*Lee 1990* (486)
Possible involvement of the gene *MN1* in a meningioma with a t(4;22)(p16;q11).	*Lekanne Deprez et al 1995* (159)
NF2 mutations observed in 32% of sporadic meningiomas, 5% of sporadic vestibular schwannomas and in 100% of meningiomas in NF patients.	*Lekanne-Lopez et al 1994* (17)
RAD54L polymorphism as risk factor in meningioma.	*Leone et al 2003* (184)
COX-2 and tumor grade in meningioma.	*Lin et al 2003* (487)
Aberrant CpG island hypermethylation in atypical and anaplastic meningioma.	*Liu et al 2005* (313)

TABLE 3.18. Continued

	References
Study of *TP53* in meningiomas showed reduced expression (as a result of CpG methylation) in grades I and II and increased expression in grade III tumors.	*Lomas et al 2004* (199)
HER2 protein overexpressed in 30% of meningiomas with epithelial differentiation.	*Loussouarn et al 2006* (485A)
Large alterations of the *NF2* gene were found in sporadic and familial meningiomas using multiplex ligation-dependent probe amplification (MLPA), a method thought to be accurate and simple for detecting alterations that escape other methods.	*Martínez-Glez et al 2007* (171)
p53 gene mutations are rare, but do occur, in meningiomas.	*Mashiyama et al 1991* (488)
It was shown that both MAPK and P13K/Akt are activated at different levels in benign and malignant meningiomas. Such activation contributes to the aggressive behavior of malignant meningiomas, where MAPK activation is involved in both proliferation and apoptosis.	*Mawrin et al 2005* (369)
Nuclear DNA content and recurrence in meningioma.	*Meixensberger et al 1996* (489)
	Kudoh et al 1995 (339)
Role of Ki-67 and PCNA in predicting recurrence of meningioma still uncertain.	*Möller and Braendstrup 1997* (347)
Reduced tumor alkaline phosphatase as indicator of meningioma recurrence (**Fig. 3.14**).	*Murakami et al 1993* (490)
	Niedermayer et al 1997 (196)
	Müller et al 1999 (491)
	Bouvier et al 2005 (197)
Meningioma cell lines.	*Murphy et al 1991* (492)
	Lee et al 1990 (242)
	Tanaka et al 1989 (493)
Cyclin A expression in meningiomas provides significant clinical information, especially as an independent prognostic factor.	*Nakabayashi et al 2003* (387)
Incidental meningiomas, including at autopsies.	*Nakasu et al 1987* (320)
	Wood et al 1957 (495)
MIB-1 and Ki-67 indices.	*Nakasu et al 2001* (?)
	Sandberg et al 2001 (494A)
	Perry et al 1998 (30)
Eicosanoids are lipidmediators produced by cyclooxygenases (COX-1 and COX-2) and lipoxygenases (LO) and found to be overexpressed in gliomas and meningiomas. Inhibition of eicosanoids cascade may be a useful approach to the therapy of these tumors.	*Nathoo et al 2004* (471A)
Decrease in somastatin receptor as an indicator of response to gamma knife surgery of meningiomas.	*Nicolato et al 2005* (496)
p73 expression and tumor grade in meningiomas.	*Nozaki et al 2001* (186)
Rhabdoid meningioma in a 12-yr-old female with Turner syndrome expressing desmin. *INI* gene was not changed, whereas *NF2* had exon 9 deleted.	*Nozza et al 2005* (497)
Meningiomas possess cholecystokinin and its receptors, the activation of which may increase cell growth via an autocrine/paracrine mechanism.	*Oikonomou et al 2005* (498)
Meningioma recurrence and metalloproteinases.	*Okada et al 2004* (499)
Proteomic analysis of meningiomas revealed protein expression patterns unique to the various grades in these tumors.	*Okamoto et al 2006* (572)
A link between *NAT2* (N-acetyltransferase) polymorphism and meningioma development has been postulated.	*Olivera et al 2006* (386)
Peritumoral edema and VEGF and its receptors in meningioma.	*Otsuka et al 2004* (36)
	Paek et al 2002 (500)
Cell line behavior (phenotypic change).	*Pistolesi et al 2002* (35)
Invasive nature of meningiomas tied to overexpression of matrix metalloproteins (MMP9 and MMP2).	*Pallini et al 2000* (501)
	Perrett et al 2002 (502)
	Nordqvist et al 2001 (503)
	Siddique et al 2003 (504)
Growth factors and receptors (PDGF-BB and PDGFR-β; EGF and EGFR; TGF-α; IGF-II, VEGF) and their role in meningioma.	*Perry et al 2004* (48)
	Johnson et al 2001 (367)
	Yang et al 2000 (505)
	Torp et al 1992 (506)
	Carroll et al 1997 (365)
	Halper et al 1999 (507)
	Nordqvist et al 1997 (508)
	Yamasaki et al 2000 (363)
	Harland et al 1998 (509)
	Pagotto et al 1995 (510)
Loss of merlin and tumors in mice; merlin activity.	*Perry et al 2004* (48)
PTEN gene in meningiomas.	*Peters et al 1998* (213)
BAM22, a gene of the β-adaptin family (located at 22q12), was inactivated in sporadic meningiomas.	*Peyrard et al 1994* (511)
Other studies pointed in that direction.	*Dumanski et al 1990* (144)
	Lekanne Deprez et al 1991 (237)

TABLE 3.18. Continued

	References
	Sanson et al 1993 (512)
	Ng et al 1995 (177)
LARGE gene at 22q12.3-q13.1: possible role in meningioma.	*Peyrard et al 1999* (162)
The *GADD45A* gene does not play a role in meningioma tumorigenesis; *EPB41* (4.1R protein) does.	*Piaskowski et al 2005* (513)
Peritumoral edema in meningioma is related to VEGF and	*Pistolesi et al 2002* (35)
VEGF-R expression in these tumors	*Otsuka et al 2004* (36)
IL-4 in meningioma.	*Puri et al 2005* (514)
Meningiomas express IL-4R and its receptors; possible targets	*Puri et al 2005* (514)
for cytotoxic therapy.	
Ets-1 expression may be involved in meningioma recurrence by upregulating matrix metalloproteinases and	*Okuducu et al 2006* (388)
may serve as an indicator of high risk of recurrence. Ets-1 expression in typical and atypical meningiomas	*Okuducu et al 2006* (389)
has been reported.	
Genetic instability in meningioma.	*Pykett et al 1994* (69)
No elevated expression of *CDK4* and *MDM2* in meningiomas.	*Pykett et al 1997* (514A)
	Weber et al 1997 (31)
No elevated *p53* expression in meningiomas.	*Pykett et al 1997* (514A)
Predictors of meningioma recurrence: tumor location, atypical or malignant histology and high Ki-67 or MIB-1	*Ragel and Jensen 2003* (34)
index.	
Potential systemic therapies for meningioma.	*Ragel and Jensen 2003* (34)
An occupational risk for meningioma (and acoustic neuromas) among a number of occupations has been	*Rajaraman et al 2005* (515)
reported. Further, polymorphism of the gene for the enzyme δ-aminolevulinic. acid dehydratase (*ALAD*),	
which is inhibited by several chemicals, including lead, was postulated to possibly be the background to	
meningioma development.	
Occupations associated with an increased risk of meningioma (or acoustic schwannoma) have been pointed out,	*Rajaraman et al 2005* (574)
although studies on larger numbers of cases and controls are necessary.	
Merlin interacts with a number of important proteins:	
sodium-hydrogen exchange regulator factor (NHE-RF)	*Reczek et al 1997* (516)
BII-spectrin (fodrin)	*Murthy et al 1998* (517)
hepatocyte growth factor-regulated tyrosine	*Scoles et al 1998* (518)
kinase substrate (HRS)	*Sun et al 2002* (519)
schwannomin interacting protein-1 (SCHIP-1)	*Scoles et al 2000* (520)
paxillin	*Goutebroze et al 2000* (521)
other ERM proteins	*Fernandez-Valle et al 2002* (522)
β1-integrin	*Obremski et al 1998* (523)
CD44	*Gronholm et al 1999* (524)
	Morrison et al 2001 (525)
	Tsukita et al 1994 (526)
The gene *SPARC* (also known as BM-40 and osteonectin/orteonectin) is a developmentally regulated gene	*Rempel et al 1999* (370)
whose protein product, SPARC, is a secreted, extracellular matrix-associated protein implicated in the	
modulation of cell adhesion and migration. SPARC was not expressed in grade I meningiomas, but was	
highly expressed in invasive tumors, regardless of the grade. However, there has been no follow-up on this	
marker in subsequent years.	
Phospholipid modifications in the functional properties of meningioma have been described.	*Riboni et al 1984* (390)
INI-1 gene (integrase interactor-1) does not play an important role in meningioma, though it may not be so	*Rieske et al 2003* (527)
	Schmitz et al 2001 (168)
A possible Li-Fraumeni syndrome (MFH, glioblastoma and an atypical meningioma) in a 40-yr-old male. The	*Rieske et al 2005* (528)
meningioma showed *TP53* mutations and LOH at 22q.	
Prognostic value of proliferation index (MIB-1) and telomerase RNA localization in papillary meningioma	*Rushing et al 1999* (435)
Single nucleotide polymorphisms within the *KRAS* and *ERCC2* genes increase the risk for meningioma,	*Sadetzki et al 2005* (391)
whereas those of *cyclin D1* and *p16* modify the risk to develop meningiomas when comparing irradiated and	
non-irradiated populations.	
IGFBP-6 important in brain edema and invasion by meningiomas	*Sandberg Nordqvist and Mathiesen 2002* (529)
Meningioma diversity and brain invasion may be related to expression of IGFBP-6, though IGF-II and IGFB-2	*Sandberg Nordqvist and Mathiesen 2002* (529)
and -5 may play a role	
Allelic loss of chromosome 22 and histopathologic predictors of meningiomas recurrence	*Sanson et al 1992* (530)
Hormone receptors in meningioma (androgen, somatostatin, growth hormone, prolactin)	*Sanson and Cornu 2000* (289)
	Dutour et al 1998 (531)
	Konstantinidou et al 2003 (293)
	Carroll et al 1996 (532)
	Muccioli et al 1997 (533)
p18^{INK4c} in meningiomas	*Santarius et al 2000* (188)
	Simon et al 2001 (220)

TABLE 3.18. Continued

	References
Caution has been advised in comparing gene expression profile results performed on tumor tissue vs. those based on primary cultures of meningioma. Differences in the number and nature of up- and down-regulated genes notable between the two sources of cells.	*Sasaki et al 2003* (583)
p15, p16, CDK4 and cyclin D1 changes infrequent in meningioma	*Sato et al 1996* (534)
Expression of merlin in tissues.	*Scherer and Gutmann 1996* (535)
	Stemmer-Rachamimov et al 1997 (536)
	den Bakker et al 1999 (537)
E-cadherin loss, as determined by immunoreactivity, was observed in malignant meningiomas. E-Cadherin is closely related to the differentiation and organogenesis of meningioma cells.	*Schwechheimer et al 1998* (392)
Merlin directly interacts with and inhibits cellular proliferation mediated by the expression of both PRBP and eIFc3 subunit p110. The latter appears to be involved in NF2 pathogenesis and NF2-related tumors.	*Tohma et al 1991* (393)
	Scoles et al 2006 (575)
Cranial and spinal meningiomas have been described in identical twin boys.	*Sedzimir et al 1973* (564)
Loss of merlin associated with defect of cell growth and motility.	*Shaw et al 2001* (537A)
	Lallemand et al 2003 (143)
Re-expression of wild-type merlin in meningioma cell lines reduces growth in vitro and in vivo and cell mobility.	*Sherman et al 1997* (?)
	Gutmann et al 1999 (539)
	Morrison et al 2001 (525)
	Ikeda et al 1999 (142)
Similar to adult sporadic meningioma, some meningiomas may results from mutations of the *NF2* gene. *NF2* did not seem to be involved in pediatric ependymomas or rhabdoid tumors.	*Slavc et al 1995* (582)
Deletion of Alu sequences in c-*sis* were found in individuals with predisposition to meningioma development.	*Smidt et al 1990* (576)
Ornithine decarboxylase as diagnostic tool for meningioma.	*Stenzel et al 2004* (540)
S6K overexpression and meningioma progression.	*Surace et al 2004* (315)
Loss of tumor suppressor in lung cancer-1 (*TSLC1*) expression in meningiomas was associated with a poor survival, particularly of atypical tumors.	*Surace et al 2004* (573A)
Secretory meningiomas are characterized by unique epithelial differentiation with glandular lumina containing secretory globules (hyaline bodies). These tumors are relatively rare. The presence of an increased number of mast cells has been described in secretory meningioma.	*Tirakotai et al 2006* (577)
Inositol phosphate turnover and response to mitogenic stimulation. by EGF in meningiomas.	*Todo and Fahlbusch 1994* (541)
Abnormalities of pRb and CDK4 rare in meningiomas.	*Tse et al 1998* (542)
A cytogenetic approach to chromosome changes in meningiomas indicated that loss of chromosome 22 (-22) was present as a sole change in some of these tumors, whereas others had additional abnormalities.	*Vagner-Capodano et al 1992* (410)
Meningiomas without *NF2* gene involvement have a normal karyotype and no obvious genetic or epigenetic aberrations, suggesting that genes involved in the pathogenesis of these tumors are altered by events beyond the detection of the methods used.	*van Tilborg et al 2006* (578)
Changes of *TP53* are not likely to play a role in meningioma pathogenesis.	*Verheijen et al 2002* (588)
Subarachnoid spread of meningioma.	*von Deimling et al 1995* (120)
Expression of metalloproteinases and cathepsin D were not useful diagnostically or prognostically.	*von Randow et al 2006* (579)
DNA microarray studies in meningiomas revealed not only the loss of chromosome 22, but also other frequent changes, i.e. amplification of MSH2, deletions of GSCL and HIRA in a significant number of these tumors.	*Wada et al 2005* (580)
TP53 mutations in meningioma.	*Wang et al 1995* (222)
Homozygous deletions of *CDKN2A* in 4/16 grade III meningiomas, amplification of *CDK4* and *MDM2*, 17q23 amplified in 8/6 grade III tumors.	*Weber et al 1997* (31)
Model for genomic alterations in benign, atypical and anaplastic meningiomas.	*Weber et al 1997* (31)
Fas-APO1 (CD95) upregulated in grade II meningiomas leading to increased apoptotic indices.	*Weisberg et al 2001* (543)
EGF binding in meningiomas.	*Weisman et al 1987* (544)
p53 deletions may be involved in the malignant progression of meningiomas.	*Yakut et al 2002* (545)
14-3-3 proteins and their possible role in meningioma.	*Yu et al 2002* (546)
	Perry et al 2004 (48)
Hyperdiploidy in meningiomas, especially those tumors with 50 or more chromosomes, is quite rare (Table 3.4). In a study in which 16 hyperdiploid meningiomas were selected from the collection of the authors and constituting 2.4% of the tumors, it was found that hyperdiploid meningiomas showed an aggressive biology when compared to that of the majority of common-type meningiomas. Incidentally, only 2/16 hyperdiploid meningiomas showed a -22.	*Zankl et al 1971* (585)
	Scholz et al 1996 (586)
	Ketter et al 2007 (587)
Cytogenetic findings in meningeal hemangiopericytoma differentiate it from meningioma (no loss o chromosome 22).	*Zattara-Cannoni et al 1996* (371)
The oncogene *ROS1* for tyrosine receptor kinase is commonly expressed in adult meningiomas and may play a role in the etiology of these tumors.	*Zhou and Sharma 1995* (394)
Elevated MI and Ki-67 LI and ER negativity are *predictive* factors of recurrence of benign and completely resected meningiomas, especially those associatd with Bcl-2 positivity.	*Maiuri et al 2007* (594)
An association of risk and protective haplotypes of ATM (ataxia telangiectasia mutated) gene in meningioma has been described.	*Malmer et al 2007* (595)

TABLE 3.18. Continued

	References
A study of angiogenesis in meningiomas revealed its dependence on the types of VEGF isoforms involved, i.e., grades II and III meningiomas (presenting numerous microvessels) expressed the 121 and 125 isoforms, whereas grade I tumors (with few microvessels) expressed the 189 isoform.	*Pistolesi et al 2004* (596)
This is keeping with the observation that atypical meningiomas had the highest number of blood vessels, when compared to benign tumors General VEGF expression levels were not different among these tumors. However, overexpression of VEGF and increased angiogenic potential have been reported in meningiomas with malignant transformation	*Lewy-Trenda et al 2003* (597)
	Shono et al 2000 (598)
In meningiomas the density of vessels simultaneously expressing VPF/VEGF and its receptor flt-1 correlates strongly with the amount of microcystic pattern. Thus, it is likely that vessel permeability and the angiogenic effects of VPF/VEGF are responsible for the histologic appearance of microcystic meningiomas.	*Christov et al 1999* (599)
Based on studies with the specific marker for neo-angiogenesis, endoglin (CD105), it was shown that it may serve as a target for therapeutic approaches blocking blood supply to meningiomas.	*Barresi et al 2007* (600)
A strong link between VEGF mRNA expression and peritumoral edema was found.	*Kalkanis et al 1996* (601)
The expression of several genes linked to cell cycle regulation and cellular proliferation is upregulated in atypical and anaplastic meningiomas, when compared with benign tumors, based on micro- array gene expression profiles.	*Wrobel et al 2005* (602)
The caveolin-1 protein shows an increased expression in aggressive meningiomas, as reflected by its association with atypical subtypes, high histologic grade and growth fraction.	*Barresi et al 2006* (603)
Immunostaining for the mitosis-specific antibody anti-phospho- histone H3 (PHH3 mitotic index, MI) was a more reliable method as a predictor of meningioma recurrence or fatal outcome than evaluation of mitotic index or Ki-67 labeling index.	*Kim et al 2007* (604)
Hypoxia, as reflected in the expression of carbonic anhydrase-9 in the tumor microenvironment, is a component of grade 3 meningiomas and may be an indicator of a poor prognosis.	*Yoo et al 2007* (605)
Functional differences in various types of meningeal cells may be due to differences in prostaglandin D synthase (β-Trace) expression. These differences may involve CSF absorption, tumorigenesis and other processes.	*Yamashima et al 1997* (606)
In vitro effects of steroids (progesterone, antiprogestogens and 17β-estradiol) on growth of meningiomas in cell culture were examined and found not to alter such growth. In contrast, androgen antagonists inhibited the growth.	*Adams et al 1990* (607)
Expression of the levels of mini-chromosome maintenance-2 protein (Mcm2) in meningiomas may facilitate the identification of patients with a high risk of recurrence.	*Hunt et al 2002* (608)
Low or absent expression of progesterone receptor expression was found in meningiomas with a high risk of recurrence.	*Strik et al 2002* (619)
Mitosin, a cellular proliferation-associated molecule, may be used as a recurrence marker for meningioma.	*Konstantinidou et al 2003* (610)
Attention in this and other chapters has been given to proliferation markers (Ki-67, MIB, PCNA, mitotic index), not because they reflect a specific molecular or genetic event, but because of the correlation made in such studies of these markers with diagnostic and prognostic aspects of meningiomas, as well as with molecular and genetic changes in these tumors.	*Ohta et al 1994* (611) *Abramovich and Prayson 1999* (612) *Nakaguchi et al 1999* (613)
The prognostic aspects, particularly that of recurrence, of meningiomas have received much attention, especially the histology of these tumors. Because of the many clinical, topographical, radiological and surgical factors, histology is not solely decisive for the prognosis. Mitotic activity, cellular pleomorphism with prominent nucleoli and micronecrosis and focally-raised cell density have been discussed as indicators of an unfavorable program.	
A study addressed the possibility, already applied to other tumors, of the identification of genetic prognostic markers that may reliably reflect tumor biology.	*Ketter et al 2007* (614) *Niedermayer et al 1996* (615) *Ildan et al 1999* (616) *Mantle et al 1999* (617) *Rahnenführer et al 2005 et al 2005* (618)
An oncogenetic tree model that estimates the most likely cyto-genetic pathways in more than 650 meningiomas was calculated and a genetic progression score (GPS) developed. High GPS values correlated highly with early recurrence, regardless of the WHO grade. As a quantitative measure, the GPS allows a more precise assessment of the prognosis of meningiomas than prof-fered cytogenetic markers based on single chromosome aberrations.	*Ketter et al 2007* (614)
The role of tumor angiogenesis, i.e., is it a cause or consequence of cancer has not been clearly defined. Nevertheless, findings on angiogenesis in meningiomas reveal cogent information regarding these tumors.	*Shchors and Evan 2007* (619)
In a study comparing the clinical course, cytogenetics and cytokinetics, the results led the authors to conclude that *de novo* malignant meningiomas and meningiomas with malignant transformation may represent subgroups of atypical and anaplastic meningiomas. Of the 28 meningiomas examined, half had -22, 11 had -14, 11 had chromosome 1 changes and 5 loss of chromosome 18. Ten of these tumors had -14 and -22.	*Krayenbühl et al 2007* (619A)
Using arrayCGH, a study demonstrated a higher frequency of biallelic *NF2* inactivation in fibroblastic (52%) vs. meningothelial (18%) meningiomas, with a similar picture reflected in macro-mutations of *NF2*. Promoter methylation of *NF2* did not appear to play a part in meningioma development.	*Hnnsson et al 2007* (621)
Detection of the allelic status of lp by FISH may be of help in predicting the prognosis in meningiomas, including the malignant progression of these tumors.	*Ishino et al 1998* (622)

cases reported in one study (371) have not been included in Table 3.4, particularly because the cytogenetic changes, all showing preponderantly gain of chromosomes and of a similar nature in all six of the tumors, do not reflect those of meningioma, e.g., no tumor had a -22. Nor are the changes compatible with those seen in HPC (372).

High-throughput microarray immunochemistry may be an accurate and efficient way of distinguishing hemangiopericytomas from meningiomas and simple from aggressive meningiomas, by using EMA, E-cadherin, and PDGFR-β as markers (373).

Even within the benign meningiomas, there is a wide heterogeneity in the outcomes, which cannot be accounted for by clinical or pathologic variables. To overcome this shortcoming, studies have used gene expression profiling to iden-

tify genes that are differentially expressed between benign and malignant meningiomas (374).

A review of new approaches for the treatment of refractory meningiomas (34) indicated that hydroxyurea has been the best-studied agent, but that complete tumor regression is not likely to be achieved with this chemotherapeutic drug. The authors (34) cover a range of factors and drugs as approaches to nonsurgical meningioma therapy (e.g., angiogenesis inhibition, somatostatin agonist, growth hormone inhibitors, growth factor receptor inhibitors, signal transduction pathway inhibition, mitogen-associated protein kinase inhibition, estrogen and progesterone receptor blocking effects, and gene therapy), and although some aspects of these approaches may find utility in the therapy for refractory meningiomas (34), it is difficult to deduct an optimistic view for the development of

TABLE 3.19. Signal transduction and other pathways in meningioma.

	References
PDGF pathways	*Maxwell et al 1990 (549)*
	Wang et al 1990 (550)
	Adams et al 1991 (547)
	Nister et al 1994 (550A)
	Todo et al 1996 (548)
	Shamah et al 1997 (538A)
RAS-RAF1-MEK1-MAPK/ERK pathway	
Lovastatin is a potent inhibitor of MEK1-MAPK/ERK pathway	*Johnson et al 2002 (571)*
	Pronk and Bos 1994 (395C)
	Lewis et al 1998 (551)
	Johnson et al 2004 (473)
	Johnson and Toms 2005 (395)
	Mawrin et al 2005 (369)
PI3K and MAPK kinases pathway	*Yart et al 2001 (552)*
	Conway et al 1999 (553)
	Johnson and Toms 2005 (395)
PKB/Akt and p70^{SGK}	*Walker et al 1998 (369A)*
	Johnson et al 2002 (368)
	Mawrin et al 2005 (369)
	Song et al 2005 (554)
EGFR pathway (EGF and TGF-α)	*Weisman et al 1987 (544)*
	Jones et al 1990 (555)
	Johnson et al 1994 (445)
	Linggood et al 1995 (556)
	Carroll et al 1997 (365)
	Hsu et al 1998 (345)
	Van Stetten et al 1999 (556A)
IGF pathway	*Nordqvist et al 1997 (508)*
	Johnson and Toms 2005 (395)
	Watson et al 2002 (374)
H-*RAS* mutations and activation	*Shu et al 1999 (557)*
	Gire et al 2000 (557A)
COX-2 and prostaglandins	*Castelli et al 1989 (558)*
	Gaetani et al 1991 (559)
	Ragel et al 2005 (559A)
PLC-γ	*Wahl et al 1989 (559B)*
	Johnson et al 1994, 2004 (445, 473)
	Wang et al 1998 (560A)
	Buckley et al 2004 (560)
TGF-β	*Johnson et al 1992 (561, 562)*
MGMT (DNA repair enzyme)	*Yin et al 2003 (563)*

agents capable of inducing complete tumor regression, at least in the near future.

Expression by meningiomas of type II interleukin receptors may serve as a target for cytotoxin/immunotoxin therapy in patients with meningioma who are not amenable to surgery or for recurrent tumors (394A).

Signal transduction and other pathways and their interactions in meningioma have been comprehensively reviewed and appropriate references may be found in that review (395). Although these pathways probably do not play a key role in the genesis of meningioma, the changes in these pathways may be a reflection of the genetic events associated with full development and progression of these tumors. These pathways involve genetic elements and their products that may bear upon the biology of meningiomas susceptible to therapeutic approaches to these tumors.

In a review of mitogenic signal transduction pathways operative in meningioma (395), the authors stated that the identification and mapping of multiple pathways affecting meningioma growth and apoptosis offer new opportunities for treatment approaches (and possibly prevention) of these tumors. Expression or activation of elements in these pathways in meningioma may be associated with invasiveness and malignant progression of these tumors, because some of these pathways are intimately associated with the control of cell proliferation and apoptosis. The interactions among these pathways are complex and their effects in meningioma have not been fully elucidated. Table 3.19 lists signal transduction pathways that have been studied in meningioma.

Tables 3.18 and 3.19 list genetic and hereditary aspects, factors, and substances, and other cogent areas bearing on meningioma development, progression, and biology, not discussed or expanded upon in the text.

References

1. Louis, D.N., Scheithauer, B.W., Budka, H., von Deimling, A., and Kepes, J.J. (2000) Meningiomas. In *World Health Organization Classification of Tumours. Pathology and Genetics of Tumours of the Nervous System* (Kleihues, P., and Cavenee, W.K., eds.), IARC Press, Lyon, France, pp. 176–184.
2. Jääskeläinen, J. (1986) Seemingly complete removal of histologically benign intracranial meningioma: late recurrence rate and factors predicting recurrence in 657 patients. A multivariate analysis. *Surg. Neurol.* **26,** 461–469.
3. Stafford, S.L., Perry, A., Suman, V.J., Meyer, F.B., Scheithauer, B.W., Lohse, C.M., and Shaw, E.G. (1998) Primarily resected meningiomas: outcome and prognostic factors in 581 Mayo Clinic patients, 1978 through 1988. *Mayo Clin. Proc.* **73,** 936–942.
4. Kleihues, P., Burger, P.C., and Scheithauer, B.W. (1993) The new WHO classification of brain tumors. *Brain Pathol.* **3,** 255–268.
5. Kleihues, P., Louis, D.N., Scheithauer, B.W., Rorke, L.B., Reifenberger, G., Burger, P.C., and Cavenee, W.K. (2002) The WHO classification of tumors of the nervous system. *J. Neuropathol. Exp. Neurol.* **61,** 215–225.
6. Perry, A., Scheithauer, B.W., Stafford, S.L, Lohse, C.M., and Wollan, P.C. (1999) "Malignancy" in meningiomas. A clinicopathologic study of 116 patients, with grading implications. *Cancer* **85,** 2046–2056.
7. Memon, M.Y. (1980) Multiple and familial meningiomas without evidence of neurofibromatosis. *Neurosurgery* **7,** 262–264.
8. Battersby, R.D.E., Ironside, J.W., and Maltby, E.L. (1986) Inherited multiple meningiomas: a clinical, pathological and cytogenetic study of an affected family. *J. Neurol Neurosurg. Psychiatry* **49,** 362–368.
9. Butti, G., Assietti, R., Casalone, R., and Paoletti, P. (1989) Multiple meningiomas: a clinical, surgical, and cytogenetic analysis. *Surg. Neurol.* **31,** 255–260.
10. Domenicucci, M., Santoro, A., D'Osvaldo, D.H., Delfini, R., Cantore, G.P., and Guidetti, B. (1989) Multiple intracranial meningiomas. *J. Neurosurg.* **70,** 41–44.
11. Martuza, R.L., and Eldridge, R. (1988) Neurofibromatosis 2. (Bilateral acoustic neurofibromatosis.) *N. Engl. J. Med.* **318,** 684–688.
12. McLendon, R.E., and Tien, R.D. (1998) Genetic syndromes associated with tumors and/or hamartomas. In *Russel and Rubinstein's Pathology of Tumors of the Nervous System* (Bigner, D.D., McLendon, R.E., and Bruner, J.M., eds.), Arnold Publishers, London, UK, pp. 371–418.
13. Rouleau, G.A., Mérel, P., Lutchman, M., Sanson, M., Zucman, J., Marineau C., Hoang-Zuang, K., Demczuk, S., Desmaze, C., Plougastel, B., Pulst, S.M., Lenoir, G., Bijlsma, E., Fashold, R., Dumanski, J., de Jong, P., Parry, D., Eldridge, R., Auraias, A., Delattre, O., and Thomas, G. (1993) Alteration of a new gene encoding a putative membrane-organizing protein causes neuro-fibromatosis type 2. *Nature* **363,** 515–521.
14. Trofatter, J.A., MacCollin, M.M., Rutter, J.L., Murell, J.R., Duyao, M.P., Parry, D.M., Eldridge, R., Kley, N., Menon, A., Pulski, K., Haase, V.H., Ambrose, C.M., Nunroe, D., Bove, C., Haines, J., Martua, R.L., MacDonald, M., Seizinger, B.R., Short, M.P., Bucklert, A.J., and Gusella, J.F. (1993) A novel moesin-, ezrin, radixin-like gene is a candidate for the neurofibromatosis 2 tumor suppressor. *Cell* **72,** 791–800.
15. Ruttledge, M.H., Xie, Y.-G., Han, F.-Y., Peyrard, M., Collins, V.P., Nordenskjöld, M., and Dumanski, J.P. (1994) Deletions on chromosome 22 in sporadic meningioma. *Genes Chromosomes Cancer* **10,** 122–130.
16. Ruttledge, M.H., Sarrazin, J., Rangaratnam, S., Phelan, C.M., Twist, E., Merel, P., Delattre, O., Thomas, G., Nordenskjöld, M., Collins, V.P., Dumanski, J.P., and Rouleau, G.A. (1994B) Evidence for complete inactivation of the NF2 gene in the majority of sporadic meningiomas. *Nat. Genet.* **6,** 180–184.
17. Lekanne Deprez, R.H., Bianchi, A.B., Groen, N.A., Seizinger, B.R., Hagemeijer, A., van Drunen, E., Bootsma, D., Koper, J.W., Avezaat, C.J., Kley, N., and Zwarthoff, E.C. (1994A) Frequent NF2 gene transcript mutations in meningiomas and vestibular schwannomas. *Am. J. Hum. Genet.* **54,** 1022–1029.
18. Louis, D.N., Ramesh, V., and Gusella, J.F. (1995) Neuropathology and molecular genetics of neurofibromatosis 2 and related tumors. *Brain Pathol.* **5,** 163–172.

19. Mérel, P., Hoang-Xuan, K., Sanson, M., Bijlsma, E., Rouleau, G., Laurent-Puig, P., Pulst, S., Baser, M., Lenoir, G., Sterkers, J.M., Philippon, J., Resche, F., Mautner, V.F., Fischer, G., Hulsebos, T., Aurias, A., Delattre, O., and Thomas, G. (1995) Screening for germ-line mutations in the NF2 gene. *Genes Chromosomes Cancer* **12,** 117–127.

20. Mérel, P., Hoang-Xuan, K., Sanson, M., Moreau-Aubrey, A., Bijlsma, E.K., Lazaro, C., Moisan, J.P., Resche, F., Nishisho, I., Estivill, X., Delattre, J.Y., Poisson, M., Theillet, C., Hulsebos, T., Delattre, O., and Thomas G. (1995) Predominant occurrence of somatic mutations of the NF2 gene in meningiomas and schwannomas. *Genes Chromosomes Cancer* **13,** 211–216.

21. Papi, L., De Vitis, L.R., Vitelli, F., Ammannati, F., Mennonna, P., Montali, E., and Bigozzi, U. (1995) Somatic mutations in the neurofibromatosis type 2 gene in sporadic meningiomas. *Hum. Genet.* **95,** 347–351.

22. De Vitis, L.R., Vitelli, A.T.F., Mennonna, F.A.P., Montali, U.B.E., and Papi, L. (1996) Screening for mutations in the neurofibromatosis type 2 (NF2) gene in sporadic meningiomas. *Hum.Genet.* **97,** 632–637.

23. Harada, T., Irving, R.M., Xuereb, J.H., Barton, D.E., Hardy, D.G., Moffat, D.A., and Maher, E.R. (1996) Molecular genetic investigation of the neurofibromatosis type 2 tumor suppressor gene in sporadic meningioma. *J. Neurosurg.* **84,** 847–851.

24. Sanson, M. (1996) A new tumor suppressor gene responsible for type 2 neurofibromatosis is inactivated in neurinoma and meningioma. *Rev. Neurol.* (Paris) **152,** 1–10.

25. Ueki, K., Wen-Bin, C., Narita, Y., Asai, A., and Kirino, T. (1999) Tight association of loss of merlin expression with loss of heterozygosity at chromosome 22q in sporadic meningiomas. *Cancer Res.* **59,** 5995–5998.

26. Kepes, J.J., Moral, L.A., Wilkinson, S.B., Abdullah, A., and Llena, J.F. (1998) Rhabdoid transformaiton of tumor cells in meningiomas: a histologic indication of increased proliferative activity: report of four cases. *Am. J. Surg. Pathol.* **22,** 231–238.

27. Martínez-Lage, J.F., Ñiguez, B.F., Sola, J., Pérez-Espejo, M.A., Ros de San Pedro, J., and Fernandez-Cornejo, V. (2006) Rhabdoid meningioma: a new subtype of malignant meningioma also apt to occur in children. *Childs Nerv. Syst.* **22,** 325–329.

28. Perry, A., Fuller, C.E., Judkins, A.R., Dehner, L.P., and Biegel, J.A. (2005) INI1 expression is retained in composite rhabdoid tumors, including rhabdoid meningiomas. *Mod. Pathol.* **18,** 951–958.

29. Bannykh, S.I., Perry, A., Powell, H.C., Hill, A., and Hansen, L.A. (2002) Malignant rhabdoid meningioma arising in the setting of preexisting ganglioglioma: a diagnosis supported by fluorescence in situ hybridization. *J. Neurosurg.* **97,** 1450–1455.

30. Perry, A., Stafford, S.L., Scheithauer, B.W., Suman, V.J., and Lohse, C.M. (1998) The prognostic significance of MIB-1, p53, and DNA flow cytometry in completely resected primary meningiomas. *Cancer* **82,** 2262–2269.

31. Weber, R.G., Boström, J., Wolter, M., Baudis, M., Collins, V.P., Reifenberger, G., and Lichter, P. (1997) Analysis of genomic alterations in benign, atypical, and anaplastic meningiomas: toward a genetic model of meningioma progression. *Proc. Natl. Acad. Sci. USA* **94,** 14719–14724.

32. Zang, K.D. (2001) Meningioma: a cytogenetic model of a complex benign human tumor, including data on 394 karyotyped cases. *Cytogenet. Cell Genet.* **93,** 207–220.

33. Lamszus, K. (2004) Meningioma pathology, genetics, and biology. *J. Neuropathol. Exp. Neurol.* **63,** 275–286.

34. Ragel, B., and Jensen, R.L. (2003) New approaches for the treatment of refractory meningiomas. *Cancer Control* **10,** 148–158.

35. Pistolesi, S., Fontanini, G., Camacci, T., De Ieso, K., Boldrini, L., Lupi, G., Padolecchia, R., Pingitore, R., and Parenti, G. (2002) Meningioma-associated brain oedema: the role of angiogenic factors and pial blood supply. *J. Neurooncol.* **60,** 159–164.

36. Otsuka, S., Tamiya, T., Ono, Y., Michiue, H., Kurozumi, K., Daido, S., Kambara, H., Date, I., and Ohmoto, T. (2004) The relationship between peritumoral brain edema and the expression of vascular endothelial growth factor and its receptors in intracranial meningiomas. *J. Neurooncol.* **70,** 349–357.

37. Borovich, B., and Doron, Y. (1986) Recurrence of intracranial meningiomas: the role played by regional multicentricity. *J. Neurosurg.* **64,** 58–63.

38. Collins, V.P., Nordenskjöld, M., and Dumanski, J.P. (1990) The molecular genetics of meningiomas. *Brain Pathol.* **1,** 19–24.

39. Baumgartner, J.E., and Sorenson, J.M. (1996) Meningioma in the pediatric population. *J. Neurooncol.* **29,** 223–228.

40. Bondy, M., and Ligon, B.L. (1996) Epidemiology and etiology of intracranial meningiomas: a review. *J. Neurooncol.* **29,** 197–205.

41. Kyritsis, A.P. (1996) Chemotherapy for meningiomas. *J. Neurooncol.* **29,** 269–272.

42. Langford, L.A. (1996) Pathology of meningiomas. *J. Neurooncol.* **29,** 217–221.

43. McCutcheon, I.E. (1996) The biology of meningiomas. *J. Neurooncol.* **29,** 207–216.

44. Sandberg, A.A. (1990) *The Chromosomes in Human Cancer and Leukemia.* 2nd ed. Elsevier, New York.

45. Radner, H., Katenkamp, D., Reifenberger, G., Deckert, M., Pietsch, T., and Wiestler, O.D. (2001) New developments in the pathology of skull base tumors. *Virchows Arch.* **438,** 321–335.

46. Weiss, S.W., and Goldblum, J.R. (2001) *Enzinger and Weiss's Soft Tissue Tumors,* 4th ed. Mosby, St. Louis, MO.

47. Lusis, E., and Gutmann, D.H. (2004) Meningioma: an update. *Curr. Opin. Neurol.* **17,** 687–692.

48. Perry, A., Gutmann, D.H., and Reifenberger, G. (2004) Molecular pathogenesis of meningiomas. *J. Neurooncol.* **70,** 183–202.

49. Claus, E.B., Bondy, M.L., Schildkraut, J.M., Wiemels, J.L., Wrensch, M., and Black, P.M. (2005) Epidemiology of intracranial meningioma. *Neurosurgery* **57,** 1088–1095.

50. Ng, H.K. (2005) Molecular genetics of meningiomas. *Asian J. Surg.* **28,** 11–12.

51. Ragel, B.T., and Jensen, R.L. (2005) Molecular genetics of meningiomas. *Neurosurg. Focus* **19** (E9), 1–8.

52. Willis, J., Smith, C., Ironside, J.W., Erridge, S., Whittle, I.R., and Everington, D. (2005) The accuracy of meningioma grading: a 10-year retrospective audit. *Neuropathol. Appl. Neurobiol.* **31,** 141–149.

53. Hartmann, C., Boström, J., and Simon, M. (2006) Diagnostic and molecular pathology of meningiomas. *Expert Rev. Neurother.* **6**, 1671–1683.

54. Riemenschneider, M.J., Perry, A., and Reifenberger, G. (2006) Histological classification and molecular genetics of meningiomas. *Lancet Neurol.* **5**, 1045–1054.

55. Daumas-Duport, C., (2002) Commentary on the WHO classification of tumors of the nervous system. *J. Neuropathol. Exp. Neurol.* **61**, 226–229.

56. Zang, K.D., and Singer, H. (1967) Chromosomal constitution of meningiomas. *Nature* **216**, 84–85.

57. Singer, H., and Zang, D.K. (1970) Cytologische und cytogenetische Untersuchungen an Hirntumoren. I. Die Chromosomenpathologie des menschlichen Meningeoms. *Hum. Genet.* **9**, 172–184.

58. Mark, J., Levan, G., and Mitelman, F. (1972) Identification by fluorescence of the G chromosome lost in human meningiomas. *Hereditas* **71**, 163–168.

59. Mark, J. (1973) Karyotype patterns in human meningiomas. A comparison between studies with G- and Q-banding techniques. *Hereditas* **75**, 213–220.

60. Mark, J. (1973) Origin of the ring chromosome in a human recurrent meningioma studied with G-band technique. *Acta Pathol. Microbiol. Scand. Sect.* **81**, 588–590.

61. Zankl, H., and Zang, K.D. (1972) Cytological and cytogenetical studies on brain tumors. IV. Identification of the missing G chromosome in human meningiomas as no. 22 by fluorescence technique. *Humangenetik* **14**, 167–169.

62. Zankl, H., and Zang, K.D. (1980) Correlation between clinical and cytogenetical data in 180 human meningiomas. *Cancer Genet. Cytogenet.* **1**, 351–356.

63. Zang, K.D. (1982) Cytological and cytogenetical studies on human meningioma. *Cancer Genet. Cytogenet.* **6**, 249–274.

64. Rey, J.A., Pestaña, A., and Bello, M.J. (1992) Cytogenetic and molecular genetics of nervous system tumors. *Oncol. Res.* **4**, 321–331.

65. Mark, J. (1971) Chromosomal aberrations and their relation to malignancy in meningiomas: a meningioma with ring chromosomes. *Acta Pathol. Microbiol. Scand. A* **79**, 193–200.

66. Mark, J. (1977) Chromosomal abnormalities and their specificity in human neoplasms: an assessment of recent observations by banding techniques. *Adv. Cancer Res.* **24**, 165–222.

67. López-Ginés, C., Cerdá-Nicolás, M., Barcia-Salorio, J.L., and Llombart-Bosch, A. (1995) Cytogenetical findings in recurrent meningiomas. A study of 10 tumors. *Cancer Genet. Cytogenet.* **85**, 113–117.

68. López-Ginés, C., Cerdá-Nicolás, M., Gil-Benso, R., Barcia-Salorio, J.L., and Llombart-Bosch, A. (2001) Loss of 1p in recurrent meningiomas: a comparative study in successive recurrences by cytogenetics and fluorescence in situ hybridization. *Cancer Genet. Cytogenet.* **125**, 119–124.

69. Pykett, M.J., Murphy, M., Harnish, P.R., and George, D.L. (1994) Identification of a microsatellite instability phenotype in meningiomas. *Cancer Res.* **54**, 6340–6343.

70. Prowald, A., Wemmert, S., Biehl, C., Storck, S., Martin, T., Henn, W., Ketter, R., Meese, E., Zang, K.D., Steudel, W.I., and Urbschat, S. (2005) Interstitial loss and gain of sequences on chromosome 22 in meningiomas with normal karyotype. *Int. J. Oncol.* **26**, 385–393.

71. Al Saadi, A., Latimer, F., Madercic, M., and Robbin, T. (1987) Cytogenetic studies of human brain tumors and their clinical significance. II. Meningioma. *Cancer Genet. Cytogenet.* **26**, 127–141.

72. Maltby, E.L., Ironside, J.W., and Battersby, R.D.E. (1988) Cytogenetic studies in 50 meningiomas. *Cancer Genet. Cytogenet.* **31**, 199–210.

73. Rey, J.A., Bello, M.J., de Campos, J.M., Kusak, E., and Moreno, S. (1988) Chromosomal involvement secondary to -22 in human meningiomas. *Cancer Genet. Cytogenet.* **33**, 275–290.

74. Casalone, R., Simi, P., Granata, P., Minelli, E., Giudici, A., Butti, G., and Solero, C.L. (1990) Correlation between cytogenetic and histopathological findings in 65 human meningiomas. *Cancer Genet. Cytogenet.* **45**, 237–243.

75. Casartelli, C., Rogatto, S.R., and Neto, J.B. (1989) Karyotypic evolution of human meningioma. Progression through malignancy. *Cancer Genet. Cytogenet.* **40**, 33–45.

76. Poulsgård, L., Rönne, M., and Schmidek, H.H. (1993) The cytogenetic and molecular genetic analysis of meningiomas. In *Molecular Genetics of Nervous System Tumors* (Levine, A.J., and Schmidek, H.H., eds.), Wiley-Liss, New York, pp. 249–254.

77. Vagner-Capodano, A.M., Grisoli, F., Gambarelli, D., Sedan, R., Pellet, W., and De Victor, B. (1993) Correlation between cytogenetic and histopathological findings in 75 human meningiomas. *Neurosurgery* **32**, 892–900.

78. Lekanne Deprez, R.H., Riegman, P.H., van Drunen, E., Warringa, U.L., Groen, N.A., Stefanko, S.Z., Koper, J.W., Avezaat, C.J.J., Mulder, P.G.H., Zwarthoff, E.C., and Hagemeijer, A. (1995) Cytogenetic, molecular genetic and pathological analyses in 126 meningiomas. *J. Neuropathol. Exp. Neurol.* **54**, 224–235.

79. Schneider, B.F., Shashi, V., von Kap-herr, C., and Golden, W.L. (1995) Loss of chromosomes 22 and 14 in the malignant progression of meningiomas. A comparative study of fluorescence in situ hybridization (FISH) and standard cytogenetic analysis. *Cancer Genet. Cytogenet.* **85**, 101–104.

80. Perry, A., Jenkins, R.B., Dahl, R.J., Moertel, C.A., and Scheithauer, B.W. (1996) Cytogenetic analysis of aggressive meningiomas. Possible diagnostic and prognostic implications. *Cancer* **77**, 2567–2573.

81. Rempel, S.A., Schwechheimer, K., Davis, R.L., Cavenee, W.K., and Rosenblum, M.L. (1993) Loss of heterozygosity for loci on chromosome 10 is associated with morphologically malignant meningioma progression. *Cancer Res.* **53**, 2386–2392.

82. Bello, M.J., de Campos, J.M., Kusak, M.E., Vaquero, J., Sarasa, J.L., Pestaña, A., and Rey, J.A. (1994) Allelic loss at 1p is associated with tumor progression in meningiomas. *Genes Chromosomes Cancer* **9**, 296–298.

83. Lindblom, A. Ruttledge, M., Collins, V.P., Nordenskjöld, M., and Dumanski, J.P. (1994) Chromosomal deletions in anaplastic meningiomas suggest multiple regions outside chromosome 22 as important in tumor progression. *Int. J. Cancer* **56**, 354–357.

84. Simon, M., von Deimling, A., Larson, J.J., Wellenreuther, R., Kaskel, P., Waha, A., Warnick, R.E., Tew, J.M., Jr., and Menon, A.G. (1995) Allelic losses on chromosomes 14, 10, and 1 in atypical and malignant meningiomas: a

genetic model of meningioma progression. *Cancer Res.* **55**, 4696–4701.

85. Menon, A.G., Rutter, J.L., von Sattel, J.P. , Snyder, H., Murdoch, C., Blumenfeld, A., Martuza, R.L., von Deimling, A., Gusella, J.F., and Houseal, T.W. (1997) Frequent loss of chromosome 14 in atypical and malignant meningioma: identification of a putative "tumor progression" locus. *Oncogene* **14**, 611–616.

86. Vogelstein, B., and Kinzler, K.W. (1993) The multistep nature of cancer. *Trends Genet.* **9**, 138–141.

87. Leone, P.E., Bello, M.J., de Campos, J.M., Vaquero, J., Sarasa, J.L., Pestaña, A., and Rey, J.A. (1999) *NF2* gene mutations and allelic status of 1p, 14q an 22q in sporadic meningiomas. *Oncogene* **18**, 2231–2239.

88. Sayagués, J.M., Tabernero, M.D., Maíllo, A., Díaz, P., Basillo, A., Bortoluci, A., Gomez-Moreta, J., Santos-Briz, A., Morales, F., Orfao, A. (2002) Incidence of numerical chromosome aberrations in meningioma tumors as revealed by fluorescence in situ hybridization using 10 chromosome-specific probes. *Cytometry (Clin. Cytometry)* **50**, 153–159.

89. Zankl, H., Weiss, A.F., and Zang, K.D. (1975). Cytological and cytogenetical studies on brain tumors. VI. No evidence for a translocation in 22-monosomic meningiomas. *Humangenetik* **30**, 343–348.

90. Al-Mefty, O., Kadri, P.A.S., Pravdenkova, S., Sawyer, J.R., Stangeby, C., and Husain, M. (2004) Malignant progression in meningioma: documentation of a series and analysis of cytogenetic findings. *J. Neurosurg.* **101**, 210–218.

91. Cerdá-Nicolás, M., López-Ginés, C., Barcia-Salorio, J.L., and Llombart-Bosch, A. (1998) Evolution to malignant in a recurrent meningioma: morphological and cytogenetic findings. *Clin. Neuropathol.* **17**, 210–215.

92. Al-Mefty, O., Topsakal, C., Pravdenkova, S., Sawyer, J.R., and Harrison, M.J. (2004) Radiation-induced meningiomas: clinical, pathological, cytokinetic and cytogenetic characteristics. *J. Neurosurg.* **100**, 1002–1013.

93. Simon, M., Kokkino, A.J., Warnick, R.E., Tew, J.M., Jr., von Deimling, A., and Menon, A.G. (1996) Role of genomic instability in meningioma progression. *Genes Chromosomes Cancer* **16**, 265–269.

94. Cai, D.X., James, C.D., Scheithauer, B.W., Couch, F.J., and Perry, A. (2001) *PS6K* amplification characterizes a small subset of anaplastic meningiomas. *Am. J. Clin. Pathol.* **115**, 213–218.

95. Cai, D.X., Banerjee, R., Scheithauer, B.W., Lohse, C.M., Kleinschmidt-DeMasters, B.K., and Perry, A. (2001) Chromosome 1p and 14q FISH analysis in clinicopathologic subsets of meningioma: diagnostic and prognostic implications. *J. Neuropathol. Exp. Neurol.* **60**, 628–636.

96. López-Ginés, C., Cerdá-Nicolás, M., Gil-Benso, R., Callaghan, R., Collado, M., Roldan, P., and Llombart-Bosch, A. (2004) Association of loss of 1p and alterations of chromosome 14 in meningioma progression. *Cancer Genet. Cytogenet.* **148**, 123–128.

97. Murakami, M., Hashimoto, N., Takahashi, Y., Hosokawa, Y., Inazawa, J., Mineura, K. (2003) A consistent region of deletion on 1p36 in meningiomas: identification and relation to malignant progression. *Cancer Genet. Cytogenet.* **140**, 99–106.

98. Carlson, K.M., Bruder C., Nordenskjöld, M., Dumanski, J.P. (1997) 1p and 3p deletions in meningiomas without detectable aberrations of chromosome 22 identified by comparative genomic hybridization. *Genes Chromosomes Cancer* **20**, 419–424.

99. Sawyer, J.R., Husain, M., Lukacs, J.L., Stangeby, C., Lichti Binz, R., and Al-Mefty, O. (2003) Telomeric fusion as a mechanism for the loss of 1p in meningioma. *Cancer Genet. Cytogenet.* **145**, 38–48.

100. Yamada, K., Kondo, T., Yoshioka, M., and Oami, H. (1980) Cytogenetic studies in twenty human brain tumors: association of No. 22 chromosome abnormalities with tumors of the brain. *Cancer Genet. Cytogenet.* **2**, 293–307.

101. van Tilborg, A.A.G., Al Allak, B., Velthuizen, S.C.J.M., de Vries, A., Kros, J.M., Avezaat, C.J.J., de Klein, A., Beverloo, H.B., and Zwarthoff, E.C. (2005) Chromosomal instability in meningiomas. *J. Neuropathol. Exp. Neurol.* **64**, 312–322.

102. Henn, W., Niedermayer, I., Ketter, R., Reichardt, S., Freiler, A., and Zang, K.D. (2003) Monosomy 7p in meningiomas: a rare constituent of tumor progression. *Cancer Genet. Cytogenet.* **144**, 65–68.

103. Maíllo, A., Díaz, P., Sayagués, J.M., Blanco, A., Tabernero, M.D., Ciudad, J., López, A., Gonçalves, J.M., and Orfao, A. (2001) Gains of chromosome 22 by fluorescence in situ hybridization in the context of an hyperdiploid karyotype are associated with aggressive clinical features in meningioma patients. *Cancer* **92**, 377–385.

103A. Pelz, A.-F., Klawunde, P., Skalej, M., Wieacker, P., Kirches, E., Schneider, T., and Mawrin, C. (2007) Novel chromosomal aberrations in a recurrent malignant meningioma. *Cancer Genet. Cytogenet.* **174**, 48–53.

104. Cerdá-Nicolás, M., López-Ginés, C., Perez-Bacete, M., Barcia-Salorio, J.L., and Llombart-Bosch, A. (2000) Histopathological and cytogenetic findings in benign, atypical and anaplastic human meningiomas: a study of 60 tumors. *Clin. Neuropathol.* **19**, 259–267.

105. Casalone, R., Granata, P., Simi, P., Tarantino, E., Butti, G., Buonaguidi, R., Faggionato, F., Knerich, R., and and Solero, C.L. (1987) Recessive cancer genes in meningiomas? An analysis of 31 cases. *Cancer Genet. Cytogenet.* **27**, 145–159.

106. Doco-Fenzy, M., Cornillet, P., Scherpereel B., Depernet, B., Bisiau-Leconte, S., Ferre, D., Pluot, M., Graftiaux, J.P., and Teyssier, J.R. (1993) Cytogenetic changes in 67 cranial and spinal meningiomas: relation to histopathological and clinical pattern. *Anticancer Res.* **13**, 845–850.

107. Sawyer, J.R., Husain, M., Pravdenkova, S., Krisht, A., and Al-Mefty, O. (2000) A role for telomeric and centromeric instability in the progression of chromosome aberrations in meningioma patients. *Cancer* **88**, 440–453.

108. Chauveinc, L., Ricoul, M., Sabatier, L., Gaboriaud, G., Srour, A., Bertagna, X., and Dutrillaux, B. (1997) Dosimetric and cytogenetic studies of multiple radiation-induced meningiomas for a single patient. *Radiother. Oncol.* **43**, 285–288.

109. Vagner-Capodano, A.M., Grisoli, F., Gambarelli, D., Figarella, D., and Pellissier, J.F. (1992) Telomeric association of chromosomes in human meningiomas. *Ann. Genet.* **35**, 69–74.

110. Storck, S., Henn, W., Freiler, A., Lindemann, U., Reichardt, S., Zang, K.D., and Martin, T. (2001) Analysis of chromosomal aberrations in meningiomas by spectral karyotyping. *Cancer Genet. Cytogenet.* **128**, 70.

111. Sayagués, J.M., Tabernero, M.D., Maíllo, A., Espinosa, A., Rasillo, A., Díaz, P., Ciudad, J., López, A., Merino, M., Gonçalves, J.M., Santos-Briz, A., Morales, F., Orfao, A. (2004) Intratumoral patterns of clonal evolution in meningiomas as defined by multicolor interphase fluorescence *in situ* hybridization (FISH). Is there a relationship between histopathologically benign and atypical/anaplastic lesions? *J. Mol. Diagn.* **6**, 316–325.

112. Steudel, W.I., Feld, R., Henn, W., and Zang, K.D. (1996) Correlation between cytogenetic and clinical findings in 215 human meningiomas. *Acta. Neurochir.* [Suppl.] (Wien) **65**, 73–76.

113. Plate, K.H., Ruschoff, J., and Mennel, H.D. (1990) Nucleolar organizer regions in meningiomas. Correlation with histopathologic malignancy grading, DNA cytometry and clinical outcome. *Anal. Quant. Cytol. Histol.* **12**, 429–438.

114. Kunishio, K., Ohmoto, T., Matsuhisa, T., Maeshiro, T., Furuta, T., and Matsumoto, K. (1994) The significance of nucleolar organizer region (AgNOR) score in predicting meningioma recurrence. *Cancer* **73**, 2200–2205.

115. De Stefano, V., Salvatore, G., Monticelli, A., Riccio, P., Cappabianca, P., and Bucciero, A. (1996) Prognostic significance of nucleolar organizer regions in meningiomas. *J. Neurosurg. Sci.* **40**, 89–92.

115A. Zankl., H., Huwer, H., and Zang, K.D. (1982) Cytogenetic studies on the nucleolar organizer region (NOR) activity in meningioma cells with normal and hypodiploid karyotypes. *Cancer Genet. Cytogenet.* **6**, 47–53.

116. Ketter, R., Henn, W., Niedermayer, I., Steilen-Gimbel, H., König, J., Zang, K.D., and Steudel, W.-I. (2001) Predictive value of progression-associated chromosomal aberrations for the prognosis of meningiomas: a retrospective study of 198 cases. *J. Neurosurg.* **95**, 601–607.

117. Pfisterer, W.K., Hank, N.C., Preul, M.C., Hendricks, W.P., Pueschel, J., Coons, S.W., and Scheck, A.C. (2004) Diagnostic and prognostic significance of genetic regional heterogeneity in meningiomas. *J. Neurooncol.* **6**, 290–299.

118. Jacoby, L.B., Pulaski, K., Rouleau, G.A., and Martuza, R.L. (1990) Clonal analysis of human meningiomas and schwannomas. *Cancer Res.* **50**, 6783–6786.

119. Larson, J.J., Tew, J.M., Jr., Simon, M., and Menon, A.G. (1995) Evidence for clonal spread in the development of multiple meningiomas. *J. Neurosurg.* **83**, 705–709.

120. von Deimling, A., Kraus, J.A., Stangl, A.P., Wellenreuther, R., Lenartz, D., Schramm, J., Louis, D.N., Ramesh, V., Gusella, J.F., and Wiestler, O.D. (1995) Evidence for subarachnoid spread in the development of multiple meningiomas. *Brain Pathol.* **5**, 11–14.

121. von Deimling, A., Larson, J., Wellenreuther, R., Stangl, A.P., van Velthoven, V., Warnick, R., Tew, J., Jr., Balko, G., and Menon, A.G. (1999) Clonal origin of recurrent meningiomas. *Brain Pathol.* **9**, 645–650.

122. Stangl, A.P., Wellenreuther, R., Lenartz, D., Kraus, J.A., Menon, A.G., Schramm, J., Wiestler, O.D., and von Deimling, A. (1997) Clonality of multiple meningiomas. *J. Neurosurg.* **86**, 853–858.

123. Rönne, M., and Poulsgård, L. (1990) A case of meningioma with multiclonal origin. *Anticancer Res.* **10**, 539–542.

124. Zhu, J., Frosch, M.P., Busque, L., Beggs, A.H., Dashner, K., Gilliland, D.G., and Black, P.M. (1995) Analysis of meningiomas by methylation- and transcription-based clonality assays. *Cancer Res.* **55**, 3865–3872.

125. Leuraud, P., Marie, Y., Robin, E., Huguet, S., He, J., Mokhtari, K., Cornu, P., Hoang-Xuan, K., and Sanson, M. (2000) Frequent loss of 1p32 region but no mutation of the *p18* tumor suppressor gene in meningiomas. *J. Neurooncol.* **50**, 207–213.

126. Nunes, F., Shen, Y., Niida, Y., Beauchamp, R., Stemmer-Rachamimov, A.O., Ramesh, V., Gusella, J., and MacCollin, M. (2005) Inactivation patterns of *NF2* and *DAL-1/4.1B* (*EPB41L3*) in sporadic meningioma. *Cancer Genet. Cytogenet.* **162**, 135–139.

127. Ruttledge, M.H., and Rouleau, G.A. (2005) Role of the neurofibromatosis type 2 gene in the development of tumors of the nervous system. *Neurosurg. Focus* **19**, E6.

128. Rey, J.A., Bello, M.J., de Campos, J.M., Vaquero, J., Kusak, M.E., Sarasa, J.L., and Pestaña, A. (1993) Abnormalities of chromosome 22 in human brain tumors determined by combined cytogenetic and molecular genetic approaches. *Cancer Genet. Cytogenet* **66**, 1–10.

129. Casalone, R., Minelli, E., Granata, P., and Giudici, A. (1990) Pseudodicentric isochromosome(22) in meningiomas. *Cancer Genet. Cytogenet.* **45**, 273–275.

130. Lekanne Deprez, R.H., Groen, N.A., Louz, D., Hagemeijer, A., van Drunen, E., Koper, J.W., Avezaat, C.J.J., Bootsma, D., van der Kwast, T.H., and Zwarthoff, E.C. (1994) Constitutional DNA-level aberrations in chromosome 22 in a patient with multiple meningiomas. *Genes Chromosomes Cancer* **9**, 124–128.

131. Strachan, R., Clarke, C., Nurbhai, M., and Marks, S. (1989) Will chromosome karyotyping of meningiomas aid prediction of tumour recurrence? *Br. J. Neurosurg.* **3**, 583–590.

132. Westphal, M., Hånsel, M., Kunzman, R., Hölzel, F., and Hermann, H.-D. (1989) Spectrum of karyotypic aberrations in cultured human meningiomas. *Cytogenet.Cell Genet.* **52**, 45–49.

133. Berra, B., Papi, L., Bigozzi, U., Serino, D., Morichi, R., Mennonna, P., Rapelli, S., Cogliati, T., and Montali. E. (1991) Correlation between cytogenetic data and ganglioside pattern in human meningiomas. *Int. J. Cancer* **47**, 329–333.

134. Griffin, C.A., Hruban, R.H., Long, P.P., Miller, N., Volz, P., Carson, B., and Brem, H. (1994) Chromosome abnormalities in meningeal neoplasms: do they correlate with histology? *Cancer Genet. Cytogenet.* **78**, 46–52.

135. Yamada, K., Kasama, M., Kondo, T., Shinoura, N., and Yoshioka, M. (1994) Chromosome studies in 70 brain tumors with special attention to sex chromosome loss and single autosomal trisomy. *Cancer Genet. Cytogenet.* **73**, 46–52.

136. Gutmann, D.H., Giordano, M.J., Fishback, A.S., and Guha, A. (1997) Loss of merlin expression in sporadic meningiomas, ependymomas and schwannomas. *Neurology* **49**, 267–270.

137. Lee, J.H., Sundaram, V., Stein, D.J., Kinney, S.E., Stacey, D.W., and Golubic, M. (1997) Reduced expression of schwannomin/merlin in human sporadic meningiomas. *Neurosurgery* **40**, 578–587.

138. Gusella, J.F., Ramesh, V., MacCollin, M., and Jacoby, L.B. (1999) Merlin: the neurofibromatosis 2 tumor suppressor. *Biochim. Biophys. Acta* **1423**, M29–M36.

139. Hovens, C.M., and Kaye, A.H. (2001) The tumour suppressor protein NF2/merlin: the puzzle continues. *J. Clin. Neurosi.* **8**, 4–7.

140. Xiao, G.H., Chernoff, J., and Testa, J.R. (2003) NF2: the wizardry of merlin. *Genes Chromosomes Cancer* **38**, 389–399.

141. Zwarthoff, E.D. (1996) Neurofibromatosis and associated tumor suppressor genes. *Pathol. Res. Pract.* **192**, 647–657.

142. Ikeda, K., Saeki, Y., Gonzalez-Agosti, C., Ramesh, V., and Chiocca, E.A. (1999) Inhibition of *NF2*-negative and *NF2*-positive primary human meningioma cell proliferation by overexpression of merlin due to vector-mediated gene transfer. *J. Neurosurg.* **91**, 85–92.

143. Lallemand, D., Curto, M., Saotome, I., Giovannini, M., and McClatchey, A.I. (2003) *NF2* deficiency promotes tumorigenesis and metastasis by destabilizing adherens junctions. *Genes Dev.* **17**, 1090–1100.

144. Dumanski, J.P., Rouleau, G.A., Nordenskjöld, M., and Collins, V.P. (1990) Molecular genetic analysis of chromosome 22 in 81 cases of meningioma. *Cancer Res.* **50**, 5863–5867.

145. Antinheimo, J., Haapasalo, H., Haltia, M., Tatagiba, M., Thomas, S., Brandis, A., Sainio, M., Carpen, O., Samii, M., and Jääskeläinen, J. (1997) Proliferation potential and histological features in neurofibromatosis 2-associated and sporadic meningiomas. *J. Neurol.* **87**, 610–614.

146. Lamszus, K., Vahldiek, F., Mautner, V.-F., Schichor, C., Tonn, J., Stavrou, D., Fillbrandt, R., Westphal, M., and Kluwe, L. (2000) Allelic losses in neurofibromatosis 2-associated meningiomas. *J. Neuropathol. Exp. Neurol.* **59**, 504–512.

147. Kim, J.H., Kim, I.S., Kwon, S.Y., Jang, B.C., Suh, S.I., Shin, D.H., Jeon, C.H., Son, E.I., and Kim, S.P. (2006) Mutational analysis of the *NF2* gene in sporadic meningiomas by denaturing high-performance liquid chromatography. *Int. J. Mol. Med.* **18**, 27–32.

148. Wellenreuther, R., Kraus, J.A., Lenartz, D., Menon, A.G., Schramm, J., Louis, D.N., Ramesh, V., Gusella, J.F., Wiestler, O.D., and von Deimling, A. (1995) Analysis of the neurofibromatosis 2 gene reveals molecular variants of meningioma. *Am. J. Pathol.* **146**, 827–832.

149. Mark, J. (1970) Chromosomal patterns in human meningiomas. *Eur. J. Cancer* **6**, 489–498.

150. Dumanski, J.P., Carlbom, E., Collins, V.P., and Nordenskjöld, M. (1987) Deletion mapping of a locus on human chromosome 22 involved in the oncogenesis of meningioma. *Proc. Natl. Acad. Sci.USA* **84**, 9275–9279.

151. Seizinger, B.R., de la Monte, S., Atkins, L., Gusella, J.F., and Martuza, R.L. (1987) Molecular genetic approach to human meningioma: loss of genes on chromosome 22. *Proc. Natl. Acad. Sci. USA* **84**, 5419–5423.

152. Wellenreuther, R., Waha, A., Vogel, Y., Lenartz, D., Schramm, J., Wiestler, O.D., and von Deimling, A. (1997) Quantitative analysis of neurofibromatosis type 2 gene transcripts in meningiomas supports the concept of distinct molecular variants. *Lab. Invest.* **77**, 601–606.

153. Lomas, J., Bello, M.J., Alonso, M.E., Gonzalez-Gomez, P., Arjona, D., Kuask, M.E., de Campos, J.M., Sarasa, J.L., and Rey, J.A. (2002) Loss of chromosome 22 and absence of NF2 gene mutation in a case of multiple meningiomas. *Hum. Pathol.* **33**, 375–378.

154. Hitotsumatsu, T., Iwaki, T., Kitamoto, T., Mizoguchi, M., Suzuki, S.O., Hamada, Y., Fukui, M., and Tateishi, J. (1997) Expression of neurofibromatosis 2 protein in human brain tumors: an immunohistochemical study. *Acta. Neuropathol. (Berl.)* **93**, 225–232.

155. Huynh, D.P., Mautner, V., Baser, M.E., Stavrou, D., and Pulst, S.-M. (1997) Immunohistochemical detection of schwannomin and neurofibromin in vestibular schwannomas, ependymomas and meningiomas. *J. Neuropathol. Exp. Neurol.* **56**, 382–390.

156. Evans, J.J., Jeun, S.-S., Lee, J.H., Harwalkar, J.A., Shoshan, Y., Cowell, J.K., and Golubic, M. (2001) Molecular alterations in the *Neurofibromatosis Type 2* gene and its protein rarely occurring in meningothelial meningiomas. *J. Neurosurg.* **94**, 111–117.

157. Perry, A., Giannini, C., Raghavan, R., Scheithauer, B.W., Banerjee, R., Margraf, L., Bowers, D.C., Lytle, R.A., Newsham, I.F., and Gutmann, D.H. (2001) Aggressive phenotypic and genotypic features in pediatric and NF2-associated meningiomas: a clinicopathologic study of 53 cases. *J. Neuropathol. Exp. Neurol.* **60**, 994–1003.

158. Akagi, K., Kurahashi, H., Arita, N., Hayakawa, T., Monden, M., Mori, T., Takai, S., and Nishisho, I. (1995) Deletion mapping of the long arm of chromosome 22 in human meningiomas. *Int. J. Cancer* **60**, 178–182.

159. Lekanne Deprez, R.H., Riegman, P.H.J., Groen, N.A., Warringa, U.L., van Biezen, N.A., Molijn, A.C., Bootsma, D., de Jong, P.J., Menon, A.G., Kley, N.A., Seizinger, B.R., and Zwarthoff, E.C. (1995) Cloning and characterization of *MN1*, a gene from chromosome 22q11, which is disrupted by a balanced translocation in meningioma. *Oncogene* **10**, 1521–1528.

160. Kedra, D., Peyrard, M., Fransson, I., Collins, J.E., Dunham, I., Roe, B.A., and Dumanski, J.P. (1996) Characterization of a second human clathrin heavy chain polypeptide gene (*CLH-22*) from chromosome 22q11. *Hum. Mol. Genet.* **5**, 625–631.

161. Durmaz, R., Arslantas, A., Artan, S., Ozon, Y.H., Isiksoy, S., Basaran, N., and Tel, E. (1998) The deletion of 22q13 region in both intracranial and spinal meningiomas in a patient (case report). *Clin. Neurol. Neurosurg.* **100**, 219–223.

162. Peyrard, M., Seroussi, E., Sandberg-Nordqvist, A.C., Xie, Y.G., Han, F.Y., Fransson, I., Collins, J., Dunham, I., Kost-Alimova, M., Imreh, S., Dumanski, J.P. (1999) The human LARGE gene from 22q12.3-q13.1 is a new, distinct member of the glycosyltransferase gene family. *Proc. Natl. Acad. Sci. USA* **96**, 598–603.

163. Kimura, Y., Koga, H., Araki, N., Mugita, N., Fujita, N., Takeshima, H., Nishi, T., Yamashima, T., Saido, T.C., Yamasaki, T., Moritake, K., Saya, H., and Nakao, M. (1998) The involvement of calpain-dependent proteolysis of the tumor suppressor NF2 (merlin) in schwannomas and meningiomas. *Nat. Med.* **4**, 915–922.

164. Ozaki, S., Nishizaki, T., Ito, H., and Sasaki, K. (1999) Comparative genomic hybridization analysis of genetic alterations associated with malignant progression of meningioma. *J. Neurooncol.* **41**, 167–174.

165. Lomas, J., Bello, M.J., Arjona, D., Alonso, M.E., Martinez-Glez, V., Lopez-Marin, I., Amiñoso, C., de Campos, J.M., Isla, A., Vaquero, J., and Rey, J.A. (2005) Genetic and epigenetic alteration of the *NF2* gene in sporadic meningiomas. *Genes Chromosomes Cancer* **42**, 314–319.

166. Maxwell, M., Shih, S.D., Galanopoulos, T., Hedley-Whyte, T., and Cosgrove, G.R. (1998) Familial meningioma: analysis of expression of neurofibromatosis 2 protein Merlin. *J. Neurosurg.* **88**, 562–569.

167. Zucman-Rossi, J., Legoix, P., and Thomas, G. (1996) Identification of new members of the Gas2 and Ras families in the 22q12 chromosome region. *Genomics* **38**, 247–254.

168. Schmitz, U., Mueller, W., Weber, M., Sévenet, N., Delattre, O., and von Deimling, A. (2001) INI1 mutations in meningiomas at a potential hotspot in exon 9. *Br. J. Cancer* **84**, 199–201.

169. Peyrard, M., Pan, H.-Q., Kedra, D., Fransson, I., Swahn, S., Hartman, K., Clifton, S.W., Roe, B.A., and Dumanski, J.P. (1996) Structure of the promoter and genomic organization of the human β-adaptin gene (BAM22) from chromosome 22q12. *Genomics* **36**, 112–117.

170. Kros, J., de Greve, K., van Tilborg, A., Hop, W., Pieterman, H., Avezaat, C., Lekanne Deprez, R., and Swarthoff, E. (2001) NF2 status of meningiomas is associated with tumor localization and histology. *J. Pathol.* **194**, 367–372.

171. Martínez-Glez, V., Franco-Hernández, C., Lomas, J., Peña-Granero, C., de Campos, J.M., Isla, A., and Rey, J.A. (2007) Multiplex ligation-dependent probe amplification (MLPA) screening in meningioma. *Cancer Genet. Cytogenet.* **173**, 170–172.

172. Boström, J., Mühlbauer, A., and Reifenberger, G. (1997) Deletion mapping of the short arm of chromosome 1 identifies a common region of deletion distal to *D1S496* in human meningiomas. *Acta. Neuropathol.* **94**, 479–485.

173. Boström, J., Meyer-Puttlitz, B., Wolter, M., Blaschke, B., Weber, R.G., Lichter, P., Ichimura, K., Collins, V.P., and Reifenberger, G. (2001) Alterations of the tumor suppressor genes *CDKN2A (p16^{INK4a})*, *p14^{ARF}*, *CDKN2B (p15^{INK4b})* in atypical and anaplastic meningiomas. *Am. J. Pathol.* **159**, 661–669.

174. Lamszus, K., Kluwe, L., Matschke, J., Meissner, H., Laas, R., and Westphal, M. (1999) Allelic losses at 1p, 9q, 10q, 14q, and 22q in the progression of aggressive meningiomas and undifferentiated meningeal sarcomas. *Cancer Genet. Cytogenet.* **110**, 103–110.

175. Meese, E., Blin, N., and Zang, K.D. (1987) Loss of heterozygosity and the origin of meningiomas. *Hum. Genet.* **77**, 349–351.

176. Cogen, P.H., Daneshvar, L., Bowcock, A.M., Metzger, A.K., and Cavalli-Sforza, L.L. (1991) Loss of heterozygosity for chromosome 22 DNA sequences in human meningioma. *Cancer Genet. Cytogenet.* **53**, 273–277.

177. Ng, H.K., Lau, K.M., Tse, J.Y., Lo, K.W., Wong, J.H., Poon, W.S., and Huang, D.P. (1995) Combined molecular genetic studies of chromosome 22q and the neurofibromatosis type 2 gene in central nervous system tumors. *Neurosurgery* **37**, 764–773.

178. Jiménez-Lara, A.M., Rey, J.A., Bello, M.J., de Campos, J.M., Vaquero, J., Kusak, M.E., Pestaña, A. (1992) Cytogenetics of human meningiomas: an analysis of 125 cases. *Cancer Genet. Cytogenet.* **63**, 174.

179. Yilmaz, Z., Sahin, F.I., Atalay, B., Özen, Ö., Caner, H., Bavbek, M., Demirhan, B., Altinörs, N. (2005) Chromosome 1p36 and 22qter deletions in paraffin block sections of intracranial meningiomas. *Pathol. Oncol. Res.* **11**, 224–228.

180. Ishino, S., Hashimoto, N., Fushiki, S., Date, K., Mori, T., Fujimoto, M., Nakagawa, Y., Ueda, S., Abe, T., and Inazawa, J. (1998) Loss of material from chromosome arm 1p during malignant progression of meningioma revealed by fluorescent in situ hybridization. *Cancer* **83**, 360–366.

181. Maruno, M., Yoshimine, T., Muhammad, A.K., Ninomiya, H., and Hayakawa, T. (1998) Chromosomal losses and gains in meningiomas: comparative genomic hybridization (CGH) study in the whole genome. *Neurol. Res.* **20**, 612–616.

182. Horiguchi, K., Tomizawa, Y., Tosaka, M., Ishiuchi, S., Kurihara, H., Mori, M., and Saito, N. (2003) Epigenetic inactivation of *RASSF1A* candidate tumor suppressor gene at 3p21.3 in brain tumors. *Oncogene* **22**, 7862–7865.

183. Mendiola, M., Bello, M.J., Alonso, J., Leone, P.E., Vaquero, J., Sarasa, J.L., Kusak, M.E., de Campos, J.M., Pestaña, A., and Rey, J.A. (1999) Search for mutations of the *hRAD54* gene in sporadic meningiomas with deletion at 1p32. *Mol. Carcinog.* **24**, 300–304.

184. Leone, P.E., Mendiola, M., Alonso, J., Paz-y-Miño, C., and Pestaña, A. (2003) Implications of a RAD54L polymorphism (2290C/T) in human meningiomas as a risk factor and/or a genetic marker. *BMC Cancer* **3**, 6–14.

185. Lomas, J., Bello, M.J., Arjona, D., Gonzalez-Gomez, P., Alonso, M.E., de Campos, J.M., Vaquero, J., Ruiz-Barnes, P., Sarasa, J.L., Casartelli, C., and Rey, J.A. (2001) Analysis of p73 gene in meningioma with deletion at 1p. *Cancer Genet. Cytogenet.* **129**, 88–91.

186. Nozaki, M., Tada, M., Kashiwazaki, H., Hamou, M.-F., Diserens, A.-C., Shinohe, Y., Sawamura, Y., Iwasaki, Y., de Tribolet, N., and Hegi, M.E. (2001) p73 is not mutated in meningiomas as determined with a functional yeast assay but p73 expression increases with tumor grade. *Brain Pathol.* **11**, 296–305.

187. Leuraud, P., Dezamis, E., Aguirre-Cruz, L., Taillibert, S., Lejeune, J., Robin, E., Mokhtari, K., Boch, A.-L., Cornu, P., Delatrre, J.-Y., and Sanson, M. (2004) Prognostic value of allelic losses and telomerase activity in meningiomas. *J. Neurosurg.* **100**, 303–309.

188. Santarius, T., Kirsch, M., Nikas, D.C., Imitola, J., and Black, P.M. (2000) Molecular analysis of alterations of the p18^{INK4c} gene in human meningiomas. *Neuropathol. Appl. Neurobiol.* **26**, 67–75.

189. Sulman, E.P., Dumanski, J.P., White, P.S., Zhao, H., Maris, J.M., Mathiesen, T., Bruder, C., Cnaan, A., and Brodeur, G.M. (1998) Identification of a consistent region of allelic loss on 1p32 in meningiomas: correlation with increased morbidity? *Cancer Res.* **58**, 3226–3230.

189A. Gladin, C.R., Salsano, E., Menghi, F., Grisoli, M., Ghielmetti, F., Milanesi, I., Pollo, B., Brock, S., Cusin, A., Minati, L., Finocchiaro, G., and Bruzzone, M.G. (2007) Loss of heterozygosity studies in extracranial metastatic meningiomas. *J. Neurooncol.* **85**, 81–85.

190. Bello, M.J., de Campos, J.M., Vaquero, J., Kusak, M.E., Sarasa, J.L., and Rey, J.A. (2000) High-resolution analysis

of chromosome arm 1p alterations in meningioma. *Cancer Genet. Cytogenet.* **120**, 30–36.

191. Bello, M.J., Leone, P.E., Nebreda, P., de Campos, J.M., Kusak, M.E., Vaquero, J., Sarasa, J.L., García-Miguel, P., Queizan, A., Hernández-Moneo, J.L., Pestaña, A., and Rey, J.A. (1995) Allelic status of chromosome 1 in neoplasms of the nervous system. *Cancer Genet. Cytogenet.* **83**, 160–164.

192. Khan, J., Parsa, N.Z., Harada, T., Meltzer, P.S., and Carter, N.P. (1998) Detection of gains and losses in 18 meningioma by comparative genomic hybridization. *Cancer Genet. Cytogenet.* **103**, 95–100.

193. Arslantas, A., Artan, S., Öner, Ü., Durmaz, R., Müslümanoglu, H., Atasoy, M.A., Basaran, N., and Tel, E. (2002) Comparative genomic hybridization analysis of genomic alterations in benign, atypical and anaplastic meningiomas by. *Acta Neurol. Belg.* **102**, 53–62.

194. Arslantas, A., Artan, S., Öner, Ü., Durmaz, R., Müslümanoglu, H., Atasoy, M.A., Basaran, N., and Tel, E. (2003) Detection of chromosomal imbalances in spinal meningiomas by comparative genomic hybridization. *Neurol. Med. Chir. (Tokyo)* **43**, 12–19.

195. Sulman, E.P., White, P.S., and Brodeur, G.M. (2004) Genomic annotation of the meningioma tumor suppressor locus on chromosome 1p34. *Oncogene* **23**, 1014–1020.

196. Niedermayer, I., Feiden, W., Henn, W., Steilen-Gimbel, H., Steudel, W.-I., and Zang, K.D. (1997) Loss of alkaline phosphatase activity in meningiomas: a rapid histochemical technique indicating progression-associated deletion of a putative tumor suppressor gene on the distal part of the short arm of chromosome 1. *J. Neuropathol. Exp. Neurol.* **56**, 879–886.

197. Bouvier, C., Liprandi, A., Colin, C., Giorgi, R., Quilichini, B., Metellus, P., and Figarella-Branger, D. (2005) Lack of alkaline phosphatase activity predicts meningioma recurrence. *Am. J. Clin. Pathol.* **124**, 252–258.

198. Bello, M.J., Amiñoso, C., Lopez-Marin, I., Arjona, D., Gonzalez-Gomez, P., Alonso, M.E., Lomas, J., de Campos, J.M., Kusak, M.E., Vaquero, J., Isla, A., Gutierrez, M., Sarasa, J.L., and Rey, J.A. (2004) DNA methylation of multiple promoter-associated CpG islands in meningiomas: relationship with the allelic status at 1p and 22q. *Acta Neuropathol.* **108**, 413–421.

199. Lomas, J., Amiñoso, C., Gonzalez-Gomez, P., Alonso, M.E., Arjona, D., Lopez-Marin, I., de Campos, J.M., Isla, A. Vaquero, J., Gutierrez, M., Sarasa, J.L., Bello, M.J., and Rey, J.A. (2004) Methylation status of *TP73* in meningiomas. *Cancer Genet. Cytogenet.* **148**, 148–151.

200. Buckley, P.G., Jarbo, C., Menzel, U., Mathiesen, T., Scott, C., Gregory, S.G., Langford, C.F., and Dumanski, J.P. (2005) Comprehensive DNA copy number profiling of meningioma using a chromosome 1 tiling path microarray identifies novel candidate tumor suppressor loci. *Cancer Res.* **65**, 2653–2661.

201. Zankl, H., and Huwer, H. (1978) Are NOR's easily translocated to deleted chromosomes? *Hum. Genet.* **42**, 137–142.

202. Tse, J.Y.M., Ng, H.-K., Lau, K.-M., Lo, K.-W., Poon, W.-S., and Huang, D.P. (1997) Loss of heterozygosity of chromosome 14q in low- and high-grade meningiomas. *Hum. Pathol.* **28**, 779–785.

203. Maíllo, A., Orfao, A., Sayagués, J.M., Díaz, P., Gómez-Moreta, J.A., Caballero, M., Santamarta, D., Santos-Briz, A.,

Morales, F., and Tabernero, M.D. (2003) New classification scheme for the prognostic stratification of meningioma on the basis of chromosome 14 abnormalities, patient age, and tumor histopathology. *J. Clin. Oncol.* **21**, 3285–3295.

204. Tabernero, M.D., Espinosa, A.B., Maíllo, A., Sayagués, J.M., del Carmen Alguero, M., Lumbreras, E., Díaz, P., Gonçalves, J.M., Onzain, I., Merino, M., Morales, F., and Orfao, A. (2005) Characterization of chromosome 14 abnormalities by interphase in situ hybridization and comparative genomic hybridization in 124 meningiomas. *Am. J. Clin. Pathol.* **123**, 744–751.

205. Lusis, E.A., Watson, M.A., Chicoine, M.R., Lyman, M., Roerig, P., Reifenberger, G., Gutmann, D.H., and Perry, A. (2005) Integrative genomic analysis identifies *NDRG2* as a candidate tumor suppressor gene frequently inactivated in clinically aggressive meningioma. *Cancer Res.* **65**, 7121–7126.

206. Shimono, A., Okuda, T., and Kondoh, H. (1999) N-myc-dependent repression of ndr1, a gene identified by direct subtraction of whole mouse embryo cDNAs between wile type and N-myc mutant. *Mech. Dev.* **83**, 39–52.

207. Qu, X., Zhai, Y., Wei, H., Zhang, C., Xing, G., Yu, Y., and He, F. (2002) Characterization and expression of three novel differentiation-related genes belong to the human NDRG gene family. *Mol. Cell Biochem.* **229**, 35–44.

208. Deng, Y., Yao, L., Chau, L., Ng, S.S., Peng, Y., Liu, X., Au, W.S., Wang, J., Li, F., Ji, S., Han, H., Nie, X., Li, Q., Kung, H.F., Leung, S.Y., and Lin, M.C. (2003) N-Myc downstream-regulated gene 2 (NDRG2) inhibits glioblastoma cell proliferation. *Int. J. Cancer* **106**, 984 (erratum).

209. Choi, S.C., Kim, K.D., Kim, J.T., Kim, J.W., Yoon, D.Y., Choe, Y.K., Chang, Y.S., Paik, S.G., and Lim, J.S. (2003) Expression and regulation of NDRG2 (N-myc downstream regulated gene 2) during the differentiation of dendritic cells. *FEBS Lett.* **553**, 413–418.

210. Wu, G.Q., Liu, X.P., Wang, L.F., Zhang, W.H., Zhang, J., Li, K.Z., Dou, K.F., Zhang, X.F., and Yao, L.B. (2003) Induction of apoptosis of HepG2 cells by NDRG2. *Xi Bao Yu Fen Zi Mian Yi Xue Za Zhi* **19**, 357–360.

211. Boulkroun, S., Fay, M., Zennaro, M.-C., Escoubet, B., Jaisser, F., Blot-Chabaud, M., Farman, N., and Courtois-Coutry, N. (2002) Characterization of Rat *NDRG2* (N-Myc downstream regulated gene 2), a novel early mineralocorticoid-specific induced gene. *J. Biol. Chem.* **277**, 31506–31515.

212. Mitchelmore, C., Buchmann-Moller, S., Rask, L., West, M.J., Troncoso, J.C., and Jensen, N.A. (2004) NDRG2: a novel Alzheimer's disease associated protein. *Neurobiol. Dis.* **16**, 48–58.

213. Peters, N., Wellenreuther, R., Rollbrocker, B., Hayashi, Y., Meyer-Puttlitz, B., Duerr, E.-M., Lenartz, D., Marsh, D.J., Schramm, J., Wiestler, O.D., Parsons, R., Eng, C., and von Deimling, A. (1998) Analysis of the PTEN gene in human meningiomas. *Neuropathol. Appl. Neurobiol.* **24**, 3–8.

214. von Deimling, A., Fimmers, R., Schmidt, M.C., Bender, B., Fassbender, F., Nagel, J., Jahnke, R., Kaskel, P., Duerr, E.-M., Koopmann, J., Maintz, D., Steinbeck, S., Wick, W., Platten, M., Müller, D.J., Przkora, R., Waha, A., Blümcke, B., Wellenreuther, R., Meyer-Puttlitz, B., Schmidt, O., Mollenhauer, J., Poustka, A., Stangl, A.P., Lenartz, D., von Ammon, K., Henson, J.W., Schramm, J., Louis, D.N., and Wiestler, O.D.

(2000) Comprehensive allelotype and genetic analysis of 466 human nervous system tumors. *J. Neuropathol. Exp. Neurol.* **59**, 544–558.

215. Mihaila, D., Gutiérrez, J.A., Rosenblum, M.L., Newsham, I.F., Bögler, O., Rempel, S.A. on behalf of NABTT CNS Consortium. (2003) Meningiomas: analysis of loss of heterozygosity on chromosome 10 in tumor progression and the delineation of four regions of chromosomal deletion in common with other cancers. *Clin. Cancer Res.* **9**, 4435–4442.

216. Mihaila, D., Jankowski, M., Gutiérrez, J.A., Rosenblum, M.L., Newsham, I.F., Bögler, O., Rempel, S.A. on behalf of NABTT CNS Consortium. (2003) Meningiomas: loss of heterozygosity on chromosome 10 and marker-specific correlations with grade, recurrence, and survival. *Clin. Cancer Res.* **9**, 4443–4451.

217. Boström, J., Cobbers, J.M.J.L., Wolter, M., Tabatabai, G., Weber, R.G., Lichter, P., Collins, V.P., and Reifenberger, G. (1998) Mutation of the *PTEN* (*MMAC1*) tumor suppressor gene in a subset of glioblastomas but not in meningiomas with loss of chromosome arm 10q. *Cancer Res.* **58**, 29–33.

218. Joachim, T., Ram, Z., Rappaport, Z.H., Simon, M., Schramm, J., Wiestler, O.D., and von Deimling, A. (2001) Comparative analysis of the *NF2, TP53, PTEN, KRAS, NRAS* and *HRAS* genes in sporadic and radiation-induced human meningiomas. *Int. J. Cancer* **94**, 218–221.

219. Perry, A., Banerjee, R., Lohse, C.M., Kleinschmidt-DeMasters, B.K., and Scheithauer, B.W. (2002) A role for chromosome 9p21 deletions in the malignant progression of meningiomas and the prognosis of anaplastic meningiomas. *Brain Pathol.* **12**, 183–190.

220. Simon, M., Park, T.-W., Köster, G., Mahlberg, R., Hackenbroch, M., Boström, J., Löning, T., and Schramm, J. (2001) Alterations of the $INK4a^{p16-p14ARF}/INK4b^{p15}$ expression and telomerase activation in meningioma progression. *J. Neurooncol.* **55**, 149–158.

221. Ohgaki, H., Eibl, R.H., Schwab, M., Reichel, M.B., Mariani, L., Gehring, M., Petersen, I., Höll, T., Wiestler, O.D., and Kleihues, P. (1993) Mutations of the *p53* tumor suppressor gene in neoplasms of the human nervous system. *Mol. Carcinog.* **8**, 74–80.

222. Wang, J.L., Zhang, Z.J., Hartman, M., Smits, A., Westermark, B., Muhr, C., and Nister, M. (1995) Detection of TP53 gene mutation in human meningiomas: a study using immunohistochemistry, polymerase chain reaction/single-strand conformation polymorphism and DNA sequencing techniques on paraffin-embedded samples. *Int. J. Cancer* **64**, 223–228.

223. Matsuno, A., Nagashima, T., Matsuura, R., Tanaka, H., Hirakawa, M., Murakami, M., Tamura, A., and Kirino, T. (1996) Correlation between MIB-1 staining index and the immunoreactivity of p53 protein in recurrent and non-recurrent meningiomas. *Am. J. Clin. Pathol.* **106**, 776–781.

224. Prayson, R.A. (1996) Malignant meningioma: a clinicopathologic study of 23 patients including MIB1 and p53 immunohistochemistry. *Am. J. Clin. Pathol.* **105**, 719–726.

225. Karamitopoulou, E., Perentes, E., Diamantis, P., and Maraziotis, T. (1994) Ki-67 immunoreactivity in human central nervous system tumors: a study with MIB 1 monoclonal antibody on archival material. *Acta Neuropathol.* **87**, 47–54.

226. Cho, H., Ha, S.Y., Park, S.H., Park, K., and Chae, Y.S. (1999) Role of p53 gene mutation in tumor aggressiveness of intracranial meningiomas. *J. Korean Med. Sci.* **14**, 199–205.

227. Büschges, R., Ichimura, K., Weber, R.G., Reifenberger, G., and Collins, V.P. (2002) Allelic gain and amplification on the long arm of chromosome 17 in anaplastic meningiomas. *Brain Pathol.* **12**, 145–153.

228. Bueschges, R., Ichimura, K., Reifenberger, G., and Collins, V.P. (2001) Anaplastic meningiomas exhibit high frequencies of allelic imbalance of the long arm of chromosome 17. *Cancer Genet. Cytogenet.* **128**, 69.

229. Popescu, N.C., King, C.R., and Kraus, M.H. (1989) Localization of the human erbB-2 gene on normal and rearranged chromosomes 17 to bands q12–21.32. *Genomics* **4**, 362–366.

230. Kittiniyom, K., Mastronardi, M., Roemer, M., Wells, W.A., Greenberg, E.R., Titus-Ernstoff, L., and Newsham, I.F. (2004) Allele-specific loss of heterozygosity at the DAL-1/4.1B (EPB41L3) tumor-suppressor gene locus in the absence of mutation. *Genes Chromosomes Cancer* **40**, 190–203.

231. Gutmann, D.H., Donahoe, J., Perry, A., Lemke, N., Gorse, K., Kittiniyom, K., Rempel, S.A., Gutierrez, J.A., and Newsham, I.F. (2000) Loss of DAL-1, a protein 4.1-related tumor suppressor, is an important early event in the pathogenesis of meningiomas. *Hum. Mol. Genet.* **9**, 1495–1500.

232. Perry, A., Cai, D.X., Scheithauer, B.W., Swanson, P.E., Lohse, C.M., Newsham, I.F., Weaver, A., and Gutmann, D.H. (2000) Merlin, DAL-1, and progesterone receptor expression in clinicopathologic subsets of meningioma: a correlative immunohistochemical study of 175 cases. *J. Neuropathol. Exp. Neurol.* **59**, 872–879.

233. Büschges, R., Boström, J., Wolter, M., Blaschke, B., Weber, R.G., Lichter, P., Collins, V.P., and Reifenberger, G. (2001) Analysis of human meningiomas for aberrations of the *MADH2, MADH4, APM-1* and *DCC* tumor suppressor genes on the long arm of chromosome 18. *Int. J. Cancer* **92**, 551–554.

234. Rienstein, S., Loven, D., Israeli, O., Ram, Z., Rappaport, Z.H., Barkai, G., Goldman, B., Aviram-Goldring, A., and Friedman, E. (2001) Comparative genomic hybridization analysis of radiation-associated and sporadic meningiomas. *Cancer Genet. Cytogenet.* **135**, 135–140.

235. Rajcan-Separovic, E., Maguire, J., Loukianova, T., Nisha, M., Kalousek, D. (2003) Loss of 1p and 7p in radiation-induced meningiomas identified by comparative genomic hybridization. *Cancer Genet. Cytogenet.* **144**, 6–11.

236. Baser, M.E., and Poussaint, T.Y. (2006) Age associated increase in the prevalence of chromosome 22q loss of heterozygosity in histological subsets of benign meningioma. *J. Med. Genet.* **43**, 285–287.

237. Lekanne Deprez, R.H., Groen, N.A., van Biezen, N.A., Hagemeijer, A., van Drunen, E., Koper, J.W., Avezaat, C.J.J., Bootsma, D., and Zwarthoff, E.C. (1991) A t(4;22) in a meningioma points to the localization of a putative tumor-suppressor gene. *Am. J. Hum. Genet.* **48**, 783–790.

238. Schneider, G., Lutz, S., Henn, W., Zang, K.D., and Blin, N. (1992) Search for putative suppressor genes in meningioma: significance of chromosome 22. *Hum. Genet.* **88**, 579–582.

239. Pulst, S.-M., Rouleau, G.A., Marineau, C., Fain, P., and Sieb, J.P. (1993) Familial meningioma is not allelic to neurofibromatosis 2. *Neurology* **43**, 2096–2098.

240. Figarella-Branger, D., Bouvier-Labit, C., Liprandi, A., and Pellissier, J.-F. (2000) Prognostic factors in meningiomas. *Ann. Pathol.* **20**, 438–447.

241. Dezamis, E., and Sanson, M. (2003) Génétique moléculaire des meningiomas et correlations genotype/phenotype. *Rev. Neurol.* (Paris) **159**, 727–738.

242. Lee, J.Y.K., Finkelstein, S., Hamilton, R.L., Rekha, R., King, J.T., Jr., and Omalu, B. (2004) Loss of heterozygosity analysis of benign, atypical, and anaplastic meningiomas. *Neurosurgery* **55**, 1163–1173.

242A. Asgharian, B., Chen, Y.-J., Patronas, N.J., Peghini, P.L., Reynolds, J.C., Vortmeyer, A., Zhuang, Z., Venzon, D.J., Gibril, F., and Jensen, R.T. (2004) Meningiomas may be a component tumor of multiple endocrine neoplasia type 1. *Clin. Cancer Res.* **10**, 869–880.

243. Evans, D.G.R., Birch, J.M., and Ramsden, R.T. (1999) Pediatric presentation of type 2 neurofibromatosis. *Arch. Dis. Child.* **81**, 496–499.

244. Evans, D.G.R., Watson, C., King, A., Wallace, A.J., and Baser, M.E. (2005) Multiple meningiomas: differential involvement of the *NF2* gene in children and adults. *J. Med. Genet.* **42**, 45–48.

245. Erdincler, P., Lena, G., Sarioglu, A.C., Kuday, C., and Choux, M. (1998) Intracranial meningiomas in children: review of 29 cases. *Surg. Neurol.* **49**, 136–141.

246. Deen, H.G., Scheithauer, B.W., and Ebersold, M.J. (1982) Clinical and pathological study of meningiomas of the first two decades of life. *J. Neurosurg.* **56**, 317–322.

247. Wullich, B. Mayfrank, L., Schwechheimer, K., Finke, J., and Schempp, W. (1990) Chromosome abnormalities in multiple meningiomas of a child. *Genes Chromosomes Cancer* **2**, 166–168.

248. Modan, B., Baidatz, D., Mart, H., Steinitz, R., and Levin, S.G. (1974) Radiation-induced head and neck tumours. *Lancet* **1**, 277–279.

249. Rubinstein, A.B., Shalit, M.N., Cohen, M.L., Zandbank, U., and Reichenthal, E. (1984) Radiation-induced cerebral meningioma: a recognizable entity. *J. Neurosurg.* **61**, 996–971.

250. Ron, E., Modan, B., Boice, J.D., Alfandary, E., Stovall, M., Chetrit, A., and Katz, L. (1988) Tumors of the brain and nervous system after radiotherapy in childhood. *N. Engl. J. Med.* **319**, 1033–1039.

251. Harrison, M.J., Wolfe, D.E., Lau, T.-S., Mitnick, R.J., and Sachdev, V.P. (1991) Radiation-induced human meningiomas: experience at the Mount Sinai Hospital and review of the literature. *J. Neurosurg.* **75**, 564–574.

252. Salvati, M., Cervoni, L., Puzzilli, F., Bristot, R., Delfini, R., and Gagliardi, F.M. (1997) High-dose radiation-induced meningiomas. *Surg. Neurol.* **47**, 435–442.

253. Gosztonyi, G., Slowik, F., and Pásztor, E. (2004) Intracranial meningiomas developing at long intervals following low-dose X-ray irradiation of the head. *J. Neurooncol.* **70**, 59–65.

254. Longstreth, W.T., Jr., Phillips, L.E., Drangsholt, M., Koepsell, T.D., Custer, B.S., Gehrels, J.-A., and van Belle, G. (2004) Dental X-rays and the risk of intracranial meningioma. A population-based case-control study. *Cancer* **100**, 1026–1034.

255. De Tommasi, A., Occhiogrosso, M., De Tommasi, C., Cimmino, A., Sanguedolce, F., and Vailati, G. (2005)

Radiation-induced intracranial meningiomas: review of six operated cases. *Neurosurg. Rev.* **28**, 104–114.

256. Pagni, C.A., Canavero, S., Fiocchi, F., and Ponzio, G. (1993) Chromosome 22 monosomy in a radiation-induced meningioma. *Ital. J. Neurol. Sci.* **14**, 377–379.

257. Chauveinc, L., Dutrillaux, A.M., Validire, P., Padoy, E., Sabatier, L., Couturier, J., and Dutrillaux, B. (1999) Cytogenetic study of eight new cases of radiation-induced solid tumors. *Cancer Genet. Cytogenet.* **114**, 1–8.

258. Zattara-Cannoni, H., Roll, P., Figarella-Branger, D., Lena, G., Dufour, H., Grisoli, F., and Vagner-Capodano, A.-M. (2001) Cytogenetic study of six cases of radiation-induced meningiomas. *Cancer Genet. Cytogenet.* **126**, 81–84.

259. Strojan, P., Popovic, M., and Jereb, B. (2000) Secondary intracranial meningiomas after high-dose cranial irradiation: report of five cases and review of the literature. *Int. J. Radiat. Oncol. Biol. Phys.* **48**, 65–73.

260. Sadetzki, S., Flint-Richter, P., Ben-Tal, T., and Nass, D. (2002) Radiation-induced meningioma: a descriptive study of 253 cases. *J. Neurosurg.* **97**, 1078–1082.

261. Iacono, R.P., Apuzzo, M.L.J., Davis, R.L., and Tsai, F.Y. (1981) Multiple meningiomas following radiation therapy for medulloblastoma. *J. Neurosurg.* **55**, 282–286.

262. Soffer, D., Pittaluga, S., Feiner, M., and Beller, A.J. (1983) Intracranial meningiomas following low-dose irradiation to the head. *J. Neurosurg.* **59**, 1048–1053.

263. Shoshan, Y., Chernova, O., Jeun, S.-S., Somerville, R.P., Israel, Z., Barnett, G.H., and Cowell, J.K. (2000) Radiation-induced meningioma: a distinct molecular genetic pattern? *J. Neuropathol. Exp. Neurol.* **59**, 614–620.

264. Falchetti, M.L., Pallini, R., Larocca, L.M., Verna, R., and D'Ambrosio, E. (1999) Telomerase in intracranial tumours: prognostic potential for malignant gliomas and meningiomas. *J. Clin. Pathol.* **52**, 234–236.

265. Langford, L.A., Piatyszek, M.A., Xu, R., Schold, S.C., Wright, W.E., and Shay, J.W. (1997) Telomerase activity in ordinary meningiomas predicts poor outcome. *Hum. Pathol.* **28**, 416–420.

266. Chen, H.J., Liang, C.L., Lu, K., Lin, J.W., and Cho, C.L. (2000) Implication of telomerase activity and alterations of telomere length in the histologic characteristics of intracranial meningiomas. *Cancer* **89**, 2092–2098.

266A. Falchetti, M.L., Larocca, L.M., and Pallini, R. (2002) Telomerase in brain tumors. *Child's Nerv. Syst.* **18**, 112–117.

266B. Boldrini, L., Pistolesi, S., Gisfredi, S., Ursino, S., Lupi, G., Caniglia, M., Pingitore, R., Basolo, F., Parenti, G., and Fontanini, G. (2003) Telomerase in intracranial meningiomas *Int. J. Mol. Med.* **12**, 943–947.

266C. Simon, M., Park, T.-W., Leuenroth, S., Hans, V.H.J., Löning, T., and Schramm, J. (2000) Telomerase activity and expression of the telomerase catalytic subunit, hTERT, in meningioma progression. *J. Neurosurg.* **92**, 832–840.

267. Cabuy, E., and de Ridder, L. (2001) Telomerase activity and expression of telomerase reverse transcriptase correlated with cell proliferation in meningiomas and malignant brain tumors in vivo. *Virchows Arch.* **439**, 176–184.

268. Maes, L., Kalala, J.P., Cornelissen, R., and de Ridder, L. (2006) Telomerase activity and hTERT protein expression in meningiomas: an analysis in vivo versus in vitro. *Anticancer Res.* **26**, 2295–2300.

269. Maes, L., Kalala, J.P., Cornelissen, R., and de Ridder, L. (2006) PCNA, Ki-67 and hTERT in residual benign meningiomas. *In Vivo* **20**, 271–275.

270. Piquer, J., Cerda, M., Lluch, A., Barcia Salorio, J.L., and Garcia-Conde, J. (1991) Correlations of female steroid hormone receptors with histologic features in meningiomas. *Acta Neurochir* (Wien) **110**, 38–43.

271. Brandis, A., Mirzai, S., Tatagiba, M., Walter, G.F., Samii, M., and Ostertag, H. (1993) Immunohistochemical detection of female sex hormone receptors in meningiomas: correlation with clinical and histological features. *Neurosurgery* **33**, 212–218.

272. Carroll, R.S., Glowacka, D., Dashner, K., and Black, P.M. (1993) Progesterone receptor expression in meningiomas. *Cancer Res.* **53**, 1312–1316.

273. Carroll, R.S., Zhang, J., Dashner, K., Sar M., Wilson, E.M., and Black, P.M. (1995) Androgen receptor expression in meningiomas. *J. Neurosurg.* **82**, 453–460.

274. Black, P., Carroll, R., and Zhang, J. (1996) The molecular biology of hormone and growth factor receptors in meningiomas. *Acta Neurochir. Suppl.* (Wien) **65**, 50–53.

275. Bouillot, P., Pellissier, J.-F., Devictor, B., Graziani, N., Bianco, N., Grisoli, F., and Figarella-Branger, D. (1994) Quantitative imaging of estrogen and progesterone receptors, estrogen-regulated protein, and growth fraction: immunocytochemical assays in 52 meningiomas. *J. Neurosurg.* **81**, 765–773.

276. Rubinstein, A.B., Loven, D., Geier, A., Reichenthal, E., and Gadoth, N. (1994) Hormone receptors in initially excised versus recurrent intracranial meningiomas. *J. Neurosurg.* **81**, 184–187.

277. Speirs, V., Boyle-Walsh, E., and Fraser, W.D. (1997) Constitutive co-expression of estrogen and progesterone receptor mRNA in human meningiomas by RT-PCR and response of in vitro cell cultures to steroid hormones. *Int. J. Cancer* **72**, 714–719.

278. Blankenstein, M.A., Verheijen, F.M., Jacobs, J.M., Donker, T.H., van Duijnhoven, M.W., and Thijssen, J.H. (2000) Occurrence, regulation, and significance of progesterone receptors in human meningioma. *Steroids* **65**, 795–800.

279. Taddei, G.L., Caldarella, A., Raspollini, M.R., Taddei, A., and Buccoliero, A.M. (2002) Estrogen and progesterone receptors in meningiomas: immunohistochemical (Mib-1, p53) and clinico-morphological correlations. *Pathologica* **94**, 10–15.

280. Wolfsberger, S., Doostkam, S., Boecher-Schwarz, H.-G., Roessler, K., van Trotsenburg, M., Hainfellner, J.A., and Knosp., E. (2004) Progesterone-receptor index in meningiomas: correlation with clinico-pathological parameters and review of the literature. *Neurosurg. Rev.* **27**, 238–245.

281. Roser, F., Nakamura, M., Bellinzona, M., Rosahl, S.K., Ostertag, H., and Samii,, M. (2004) The prognostic value of progesterone receptor status in meningiomas. *J. Clin. Pathol.* **57**, 1033–1037.

282. Bozzetti, C., Camisa, R., Nizzoli, R., Manotti, L., Guazzi, A., Naldi, N., Mazza, S., Nizzoli, V., and Cocconi, G. (1995) Estrogen and progesterone receptors in human meningiomas: biochemical and immunocytochemical evaluation. *Surg. Neurol.* **43**, 230–233.

282A. Korhonen, K., Salminen, T., Raitanen, J., Auvinen, A., Isola, J., and Haapasalo, H. (2006) Female predominance in meningiomas can not be explained by differences in progesterone, estrogen, or androgen receptor expression. *J. Neurooncol.* **80**, 1–7.

283. Bickerstaff, E.R., Small, J.M., and Guest, I.A. (1958) The relapsing course of certain meningiomas in relation to pregnancy and menstruation. *J. Neurol. Neurosurg. Psychiatry* **21**, 89–91.

284. Black, P.M. (1993) Meningiomas. *Neurosurgery* **32**, 643–657.

285. Bonito, D., Giarelli, L., Falconieri, G., Bonifacio-Gori, D., Tomasic, G., and Vielh, P. (1993) Association of breast cancer and meningioma. Report of 12 new cases and review of the literature. *Pathol. Res. Pract.* **189**, 399–404.

286. Miller, R.E. (1986) Breast cancer and meningioma. *J. Surg. Oncol.* **31**, 182–183.

287. Lieu, A.S., Hwang, S.L., and Howng, S.L. (2003) Intracranial meningioma and breast cancer. *J. Clin. Neurosci.* **10**, 553–556.

288. Kubo, M., Fukutomi, T., Akashi-Tanaka, S., and Hasegawa, T. (2001) Association of breast cancer with meningioma: report of a case and review of the literature. *Jpn. J. Clin. Oncol.* **31**, 510–513.

288A. Lee, E., Grutsch, J., Persky, V., Glick, R., Mendes, J., and Davis, F. (2006) Association of meningioma with reproductive factors. *Int. J. Cancer* **119**, 1152–1157.

289. Sanson, M., and Cornu, P. (2000) Biology of meningiomas. *Acta Neurochir. (Wien)* **142**, 493–505.

290. Martuza, R.L., MacLaughlin, D.T., and Ojemann, R.G. (1981) Specific estradiol binding in schwannomas, meningiomas, and neurofibromas. *Neurosurgery* **9**, 665–671.

291. Tonn, J.C., Ott, M.M., Paulus, W., Meixensberger, J., and Roosen, K. (1996) Progesterone receptors in tumor fragment spheroids of human meningiomas. *Acta Neurochir.* **65**, 105–107.

292. Hsu, D.W., Efird, J.T., and Hedley-Whyte, E.T. (1997) Progesterone and estrogen receptors in meningiomas: prognostic considerations. *J. Neurosurg.* **86**, 113–120.

293. Konstantinidou, A.E., Korkolopoulou, P., Mahera, H., Kotsiakis, X., Hranioti, S., Eftychladis, C., and Patsouris, E. (2003) Hormone receptors in non-malignant meningiomas correlate with apoptosis, cell proliferation and recurrence-free survival. *Histopathology* **43**, 280–290.

294. Verheijen, F.M., Donker, G.H., Viera, C.S., Sprong, M., Jacobs, H.M., Blaauw, G., Thijssen, J.H.H., and Blankenstein, M.A. (2002) Progesterone receptor, bcl-2 and Bax expression in meningiomas. *J. Neurooncol.* **56**, 35–41.

295. Fewings, P.E., Battersby, R.D.E., and Timperley, W.R. (2000) Long-term follow up of progesterone receptor status in benign meningioma: a prognostic indicator of recurrence? *J. Neurosurg.* **92**, 401–405.

296. Inoue, T., Akahira, J.-I., Suzuki, T., Darnel, A.D., Kaneko, C., Takahashi, K., Hatori, M., Shirane, R., Kumabe, T., Kurokawa, Y., Satomi, S., and Sasano, H. (2002) Progesterone production and actions in the human central nervous system and neurogenic tumors. *J. Clin. Endocrinol. Metab.* **87**, 5325–5331.

297. Pravdenkova, S., Al-Mefty, O., Sawyer, J., and Husain, M. (2006) Progesterone and estrogen receptors: opposing prognostic indicators in meningiomas. *J. Neurosurg.* **105**, 163–173.

298. Thom, M., and Martinian, L. (2002) Progesterone receptors are expressed with higher frequency by optic nerve sheath meningiomas. *Clin. Neuropathol.* **21,** 5–8.

299. Assimakopoulou, M. (2000) Immunohistochemical localization of progesterone receptor and heat shock protein 27 and absence of estrogen receptor and PS2. *Cancer Detect. Prev.* **24,** 163–168.

300. Tonn, J.-C., Ott, M.M., Bouterfa, H., Kerkau, S., Kapp, M., Müller-Hermelink, H.K., and Roosen, K. (1997) Inverse correlation of cell proliferation and expression of progesterone receptors in tumor spheroids and monolayer cultures of human meningiomas. *Neurosurgery* **41,** 1152–1159.

301. Ho, D.M.-T., Hsu, C.-Y., Ting, L.-T., and Chiang, H. (2002) Histopathology and MIB-1 labeling index predicted recurrence of meningiomas. A proposal of diagnostic criteria for patients with atypical meningioma. *Cancer* **94,** 1538–1547.

302. Hsu, D.W., Pardo, F.S., Efird, J.T., Linggood, R.M., and Hedley-Whyte, E.T. (1994) Prognostic significance of proliferative indices in meningioma. *J. Neuropathol. Exp. Neurol.* **53,** 247–255.

303. Koehorst, S.G., Jacobs, H.M., Thijssen, J.H., and Blankenstein, M.A. (1993) Detection of an oestrogen receptor-like protein in human meningiomas by band shift assay using a synthetic oestrogen responsive element (ERE). *Br. J. Cancer* **68,** 290–294.

304. Lanzafame, S., Torrisi, A., Barbagallo, G., Emmanuele, C., Alberio, N., and Albanese, V. (2000) Correlation between histological grade, MIB-1, p53, and recurrence in 69 completely resected primary intracranial meningiomas with a 6 year mean follow-up. *Pathol. Res. Pract.* **196,** 483–488.

305. Schrell, U.M.H., Nomikos, P., Schrauzer, Th., Anders, M., Marschalek, R., Adams, E.F., and Fahlbusch, R. (1996) Hormonal dependency of cerebral meningiomas. *Acta Neurochir.* **65,** 54–57.

306. Custer, B., Longstreth, W.T., Jr., Phillips, L.E., Koepsell, T.D., and Van Belle, G. (2006) Hormonal exposures and the risk of intracranial meningioma in women: a population-based case-control study. *BMC Cancer* **6,** 152–160.{\}

307. Nagashima, G., Aoyagi, M., Wakimoto, H., Tamaki, M., Ohno, K., and Hirakawa, K. (1995) Immunohistochemical detection of progesterone receptors and the correlation with Ki-67 labeling indices in paraffin-embedded sections of meningiomas. *Neurosurgery* **37,** 478–483.

308. Gursan, N., Gundogdu, C., Albayrak, A., and Kabalar, M.E. (2002) Immunohistochemical detection of progesterone receptors and the correlation with Ki-67 labeling indices in paraffin-embedded sections of meningiomas. *Int. J. Neurosci.* **112,** 463–470.

309. Maiuri, F., Montagnani, S., Iaconetta, G., Gallicchio, B., Bernardo, A., and Signorelli, F. (1994) Correlation between sex hormone receptors and peritumoral edema in intracranial meningiomas. *J. Neurosurg. Sci.* **38,** 29–33.

310. Friend, K.E., Radinsky, R., and McCutcheon, I.E. (1999) Growth hormone receptor expression and function in meningiomas: effect of a specific receptor antagonist. *J. Neurosurg.* **91,** 93–99.

311. Bello, M.J., Alonso, M.E., Amiñoso, C., Anselmo, N.P., Arjona, D., Gonzalez-Gomez, P., Lopez-Marin, I., de Campos, J.M., Gutierrez, M., Isla, A., Kusak, M.E., Lassaletta, L., Sarasa, J.L., Vaquero, J., Casartelli, C., and Rey, J.A. (2004) Hypermethylation of the DNA repair gene *MGMT*: association with *TP53* G:C to A:T transitions in a aseries of 469 nervous system tumors. *Mutat. Res.* **554,** 23–32.

312. Esteller, M., Corn, P.G., Baylin, S.B., and Herman, J.G. (2001) A gene hypermethylation profile of human cancer. *Cancer Res.* **61,** 3225–3229.

313. Liu, Y., Pang, J.C., Dong, S., Mao, B., Poon, W.S., and Ng, H. (2005) Aberrant CpG island hypermethylation profile is associated with atypical and anaplastic meningiomas. *Hum. Pathol.* **36,** 416–425.

314. Kimura, Y., Saya, H., and Nakao, M. (2000) Calpain-dependent proteolysis of NF2 protein: involvement in schwannomas and meningiomas. *Neuropathology* **20,** 153–160.

315. Surace, E.I., Lusis, E., Haipek, C.A., and Gutmann, D.H. (2004) Functional significance of S6K overexpression in meningioma progression. *Ann. Neurol.* **56,** 295–298.

316. Gutmann, D.H., Hirbe, A.C., Huang, Z.Y., and Haipek, C.A. (2001) The Protein 4.1 tumor suppressor, DAL-1, impairs cell motility, but regulates proliferation in a cell type-specific fashion. *Neurobiol. Dis.* **8,** 266–278.

317. Robb, V.A., Li., W., Gascard, P., Perry, A., Mohandas, N., and Gutmann, D.H. (2003) Identification of a third Protein 4.1 tumor suppressor, Protein 4.1R, in meningioma pathogenesis. *Neurobiol. Dis.* **13,** 191–202.

318. Martínez-Glez, V., Bello, M.J., Franco-Hernández, C., de Campos, J.M., Isla, A., Vaquero, J., and Rey, J.A. (2005) Mutational analysis of the Dal-1/4.1B tumour-suppressor gene locus in meningiomas. *Int. J. Mol. Med.* **16,** 771–774.

319. Nasher, H.C., Grote, W., Lohr, E., and Gerhard, L. (1981) Multiple meningiomas. Clinical and computer tomographic observations. *Neuroradiology* **21,** 259–263.

320. Nakasu, S., Hirano, A., Shimura, T., and Llena, J.F. (1987) Incidental meningiomas in autopsy study. *Surg. Neurol.* **27,** 319–32.

321. Antinheimo, J., Sankila, R., Carpen, O., Pukkala, E., Sainio, M., Jääskeläinen, J. (2000) Population-based analysis of sporadic and type 2 neurofibromatosis-associated meningiomas and schwannomas. *Neurology* **54,** 71–76.

321A. Lusins, J.O., and Nakagawa, H. (1981) Multiple meningiomas evaluated by computed tomography. *Neurosurgery* **9,** 137–141.

322. Zhu, J.J., Maruyama, T., Jacoby, L.B., Hermann, J.G., Gusella, J.F., Black, P.M., and Wu, J.K. (1999) Clonal analysis of a case of multiple meningiomas using multiple molecular genetic approaches: pathology case report. *Neurosurgery* **45,** 409–416.

323. King, A., and Gutmann, D.H. (2000) The question of familial meningiomas and schwannomas: NF2B or not to be? *Neurology* **54,** 4–5.

324. Borovich, B., Doron, Y., Braun, J., Feinsod, M., Goldsher, D., Gruszkiewicz, J., Guilburd, J.N., Zaaroor, M., Levi, L., and Soustiel, J.F. Lemberger, A. (1988) The incidence of multiple meningiomas—do solitary meningiomas exist? *Acta. Neurochir.(Wien)* **90,** 15–22.

325. Schmidt, M., Kirsch, I., and Ratner, L. (1990) Deletion of Alu sequences in the fifth c-*sis* intron in individuals with meningiomas *J. Clin. Invest.* **86,** 1151–1157.

325A. Bolger, G.B., Stamberg, J., Kirsch, I.R., Hollis, G.F., Schwarz, D.F., and Thomas, G.H. (1985) Chromosome translocation t(14;22) and oncogene (c-sis) variant in a pedigree with familial meningioma. *N. Engl. J. Med.* **312**, 564–567.

326. Heinrich, B., Hartmann, C., Stemmer-Rachamimov, A.O., Louis, D.N., and MacCollin, M. (2003) Multiple meningiomas: investigating the molecular basis of sporadic and familial forms. *Int. J. Cancer* **103**, 483–488.

327. Ayerbe, J., Lobato, R.D., de la Cruz, J., Alday, R., Rivas, J.J., Gómez, P.A., and Cabrera, A. (1999) Risk factors predicting recurrence in patients operated on for intracranial meningioma. A multivariate analysis. *Acta. Neurochir. (Wien)* **141**, 921–932.

328. Jallo, G.I., Silvera, V.M., and Abbott, I.R. (2000) Meningioangiomatosis. *Pediatr. Neurosurg.* **32**, 220–221.

329. Jallo, G.I., Kothbauer, K., Mehta, V., Abbott, R., and Epstein, F. (2005) Meningioangiomatosis without neurofibromatosis: a clinical analysis. *J. Neurosurg. (Pediatrics 4)* **103**, 319–324.

330. Perry, A., Kurtkaya-Yapicier, Ö., Scheithauer, B.W., Robinson, S., Prayson, R.A., Kleinschmidt-DeMasters, B.K., Stemmer-Rachamimov, A.O., and Gutmann, D.H. (2005) Insights into meningioangiomatosis with and without meningioma: a clinicopathologic and genetic series of 24 cases with review of the literature. *Brain Pathol.* **5**, 55–65.

330A. Perry, A., Lusis, E.A., and Gutmann, D.H. (2005) Meningothelial hyperplasia: a detailed clinicopathologic, immunohistochemical and genetic study of 11 cases. *Brain Pathol.* **15**, 109–115.

331. Omeis, I., Hillard, V.H., Braun, A., Benzil, D.L., Murali, R., and Harter, D.H. (2006) Meningioangiomatosis associated with neurofibromatosis: report of 2 cases in a single family and review of the literature. *Surg. Neurol.* **65**, 595–603.

332. Wiebe, S., Muñoz, D.G., Smith, S., and Lee, D.H. (1999) Meningioangiomatosis. A comprehensive analysis of clinical and laboratory features. *Brain* **122**, 709–726.

333. Kirn, N.R., Choe, G., Shin, S.-H., Wang, K.-C., Cho, B.K., Choi, K.S., and Chi, J.G. (2002) Childhood meningiomas associated with meningioangiomatosis: report of five cases and literature review. *Neuropathol. Appl. Neurobiol.* **28**, 48–56.

334. Perry, A., and Dehner, L.P. (2003) Meningeal tumors of childhood and infancy. An update and literature review. *Brian Pathol.* **13**, 386–408.

335. Deb, P., Gupta, A., Sharma, M.C., Gaikwad, S., Singh, V.P., and Sarkar, C. (2006) Meningioangiomatosis with meningioma: an uncommon association of a rare entity—report of a case and review of the literature. *Childs Nerv. Syst.* **22**, 78–83.

335A. Das, A., Tan, W.-L., and Smith, D.R. (2003) Expression of extracellular matrix markers in benign meningiomas. *Neuropathology* **23**, 275–281.

336. Sinkre, P., Perry, A., Cai, D., Raghavan, R., Watson, M., Wilson, K., and Barton Rogers, B. (2001) Deletion of the NF2 region in both meningioma and juxtaposed meningioangiomatosis: case report supporting a neoplastic relationship. *Ped. Develop. Pathol.* **4**, 568–572.

337. Giangaspero, F., Guiducci, A., Lenz, F.A., Mastronardi, L., and Burger, P.C. (1999) Meningioma with meningioangiomatosis: a condition mimicking invasive meningiomas in children and young adults. Report of two cases and review of the literature. *Am. J. Surg. Pathol.* **23**, 872–875.

338. Gerdes, J., Li, L., Schlueter, C., Duchrow, M., Wohlenberg, C., Gerlach, C., Stahmer, I., Kloth, S., Brandt, E., and Flad, H.-D. (1991) Immunobiochemical and molecular biologic characterization of the cell proliferation-associated nuclear antigen that is defined by monoclonal antibody Ki-67. *Am. J. Pathol.* **138**, 867–873.

339. Kudoh, C., Sugiura, K., Yoshimizu, N., and Detta, A. (1995) Rapidly growing histologically benign meningiomas: cell kinetic and deoxyribonucleic acid ploidy features: report of three cases. *Neurosurgery* **37**, 998–1001.

340. Langford, L.A., Cooksley, C.S., and DeMonte, F. (1996) Comparison of MIB-1 (Ki-67) antigen and bromodeoxyuridine proliferation indices in meningiomas. *Hum. Pathol.* **27**, 350–354.

341. Matsuno, A., Fujimaki, T., Sasaki, T., Nagashima, T., Ide, T., Asai, A., Matsuura, R., Utsunomiya, H., and Kirino, T. (1996) Clinical and histopathologic analysis of proliferative potentials of recurrent an non-recurrent meningiomas. *Acta Neuropathol.* **91**, 504–510.

342. Madsen, C., and Schröder, H.D. (1997) Ki-67 immunoreactivity in meningiomas—determination of the proliferative potential of meningiomas using monoclonal antibody Ki-67. *Clin. Neuropathol.* **16**, 137–142.

343. Striepecke, E., Handt, S., Weis, J., Koch, A., Cremerius, U., Reineke, T., Bull, U., Schröder, J.M., Zang, K.D., and Bocking, A. (1996) Correlation of histology, cytogenetics and proliferation fraction (Ki-67 and PCNA) quantitated by image analysis in meningiomas. *Pathol. Res. Pract.* **192**, 816–824.

344. Takeuchi, H., Kubota, T., Kabuto, M., Kitai, R., Nozaki, J., and Yamashita, J. (1997) Prediction of recurrence in histologically benign meningiomas: proliferating clel nuclear antigen and Ki-67 immunohistochemical study. *Surg. Neurol.* **48**, 501–506.

345. Hsu, D.W., Efird, J.T., and Hedley-Whyte, E.T. (1998) MIB-1 (Ki-67) index and transforming growth factor-alpha (TGFα) immunoreactivity are significant prognostic predictors for meningiomas. *Neuropathol. Appl. Neurobiol.* **24**, 441–452.

346. Ahmed, R., Soomro, I.N., Aziz, S.A., and Hasan, S.H. (1999) p53 and PCNA expression in benign, atypical and malignant meningiomas. *J. Pak. Med. Assoc.* **49**, 241–243.

347. Mǿller, M.-L., and Braendstrup, O. (1997) No prediction of recurrence of meningiomas by PCNA and Ki-67 immunohistochemistry. *J. Neurooncol.* **34**, 241–246.

348. Ohkoudo, M., Sawa, H., Shiina, Y., Sato, H., Kamata, K., Iijima, J., Yamamoto, H., Fujii, M., and Saito, I. (1996) Morphometrical analysis of nucleolin immunohistochemistry in meningiomas. *Acta Neuropathol.* **92**, 1–7.

349. Schiffer, D., Ghimenti, C., and Fiano, V. (2005) Absence of histological signs of tumor progression in recurrences of completely resected meningiomas. *J. Neurooncol.* **73**, 125–130.

350. Amatya, V.J., Takeshima, Y., Sugiyama, K., Kurisu, K., Nishisaka, T., Fukuhara, T., and Inai, K. (2001) Immunohistochemical study of Ki-67 (MIB-1), p53 protein, p21WAF1, and p27KIP1 expression in benign, atypical, and anaplastic meningiomas. *Hum. Pathol.* **32**, 970–975.

351. Maier, H., Öfner, D., Hittmair, A., Kitz, K.,, and Budka, H. (1992) Classic, atypical, and anaplastic meningioma: three histopathological subtypes of clinical relevance. *J. Neurosurg.* **77**, 616–623.

352. Kolles, H., Niedermayer, I., Schmitt, C., Henn, W., Feld, R., Steudel, W.I., Zang, K.D., and Feiden, W. (1995) Triple approach for diagnosis and grading of human meningiomas: histology, morphometry of Ki-67/Feulgen stainings, and cytogenetics. *Acta Neurochir. (Wien)* **137**, 174–181.

353. Abramovich, C.M., and Prayson, R.A. (2000) Apoptotic activity and bcl-2 immunoreactivity in meningiomas. Association with grade and outcome. *Am. J. Clin. Pathol.* **114**, 84–92.

354. Ferraraccio, F., Accardo, M., Giangaspero, F., and Cuccurullo, L. (2003) Recurrent and atypical meningiomas—a multiparametric study using Ki67 labelling index, AgNOR and DNA Feulgen staining. *Clin. Neuropathol.* **22**, 187–192.

355. Kayaselcuk, F., Zorlundemir, S., Bal, N., Erdogan, B., Erdogan, S., and Erman, T. (2004) The expression of survivin and Ki-67 in meningiomas: correlation with grade and clinical outcome. *J. Neurooncol.* **67**, 209–214.

356. Kim, Y.J., Romeike, B.F., Uszkoreit, J., and Feiden, W. (2006) Automated nuclear segmentation in the determination of the Ki-67 labeling index in meningiomas. *Clin. Neuropathol.* **25**, 67–73.

357. Torp, S.H., Lindboe, C.F., Granli, U.S., Moen, T.M., and Nordtomme, T. (2001) Comparative investigation of proliferation markers and their prognostic relevance in human meningiomas. *Clin. Neuropathol.* **20**, 190–195.

358. De Monte, F., Ligon, L., and Sawaya, R. (editors) (1996) Meningiomas Symposiuim, University of Texas M.D. Anderson Cancer Center, February 25, 1995. *J. Neurooncol.* **29**, 195–272.

359. Goldman, C.K., Bharara S, Palmer, C.A., Vitek, J., Tsai, J.-C., Weiss, H.L., and Gillespie, G.Y. (1997) Brain edema in meningiomas is associated with increased vascular endothelial growth factor expression. *Neurosrugery* **40**, 1269–1277.

360. Provias, J., Claffey, K., delAguila, L., Lau, N., Feldkamp, M., and Guha, A. (1997) Meningiomas: role of vascular endothelial growth factor/vascular permeability factor in angiogenesis and peritumoral edema. *Neurosurgery* **40**, 1016–1026.

361. Bitzer, M., Opitz, H., Popp, J., Morgalla, M., Gruber, A., Heiss, E., and Voigt, K. (1998) Angiogenesis and brain edema in intracranial meningiomas: influence of vascular endothelial growth factor. *Acta Neurochir. (Wien)* **140**, 333–340.

362. Yoshioka, H., Hama, S., Taniguchi, E., Sugiyama, K., Arita, K., and Kurisu, K. (1999) Peritumoral brain edema associate with meningiomas: influence of vascular endothelial growth factor expression and vascular blood supply. *Cancer* **85**, 936–944.

363. Yamasaki, F.., Yoshioka, H., Hama, S., Sugiyama, K., Arita, K., and Kurisu, K. (2000) Recurrence of meningiomas. Influence of vascular endothelial growth factor expression. *Cancer* **89**, 1102–1110.

364. Samoto, K., Ikezaki, K., Ono, M., Shono, T., Kohno, K., Kuwano, M., and Fukui, M. (1995) Expression of vascular endothelial growth factor and its possible relation with neovascularization in human brain tumors. *Cancer Res.* **55**, 1189–1193.

364A. Lamszus, K., Lengler, U., Schmidt, N.O., Stavrou, D., Ergun, S., and Westphal, M. (2000) Vascular endothelial growth factor, hepatocyte growth factor/scatter factor, basic fibroblast growth factor, and placenta growth factor in human meningiomas and their relation to angiogenesis and malignancy. *Neurosurgery* **46**, 938–947.

365. Carroll, R.S., Black, P.M., Zhang, J., Kirsch, M., Percec, I., Lau, N., and Guha A. (1997) Expression and activation of epidermal growth factor receptors in meningiomas. *J. Neurosurg.* **87**, 315–323.

366. Black, P.M., Carroll, R., Glowacka, D., Riley, K., and Dashner, K. (1994) Platelet-derived growth factor expression and stimulation in human meningiomas. *J. Neurosurg.* **81**, 388–393.

367. Johnson, M.D., Woodard, A., Kim, P., and Frexes-Steed, M. (2001) Evidence for a mitogen-associated protein kinase activation and transduction of mitogenic signals by platelet-derived growth factor in human meningioma cells. *J. Neurosurg.* **94**, 293–300.

368. Johnson, M.D., Okediji, E.J., Woodard, A., Toms, S.A., and Allen, G.S. (2002) Evidence for phosphatidylinositol 3-kinase-Akt-p70^{S6K} pathway activation and transduction of mitogenic signals by platelet-derived growth factor in meningioma cells. *J. Neurosurg.* **97**, 668–675.

369. Mawrin, C., Sasse, T., Kirches, E., Kropf, S., Schneider, T., Grimm, C., Pambor, C., Vorwerk, C.K., Firsching, R., Lendeckel, U., and Dietzmann, K. (2005) Different activation of mitogen-activated protein kinase and Akt signaling is associated with aggressive phenotype of human meningioma. *Clin. Cancer Res.* **11**, 4074–4082.

369A. Walker, T.R., Moore, S.M., Lawson, M.F., Panettieri, R.A., Jr., and Chilvers, E.R. (1998) Platelet-derived growth factor-BB and thrombin activate phosphoinositide 3-kinase and protein kinase B: role in mediating airway smooth muscle proliferation. *Mol. Pharmacol.* **54**, 1007–1015.

370. Rempel, S.A., Ge, S., and Gutierrez, J.A. (1999) SPARC: a potential diagnostic marker of invasive meningiomas. *Clin. Cancer Res.* **5**, 237–241.

371. Zattara-Cannoni, H., North, M.O., Gambarelli, D., Figarella-Branger, D., Graziani, N., Grisoli, F., and Vagner-Capodano, A. M. (1996) The contribution of cytogenetics to the histogenesis of meningeal hemangiopericytoma. *J. Neurooncol.* **29**, 137–142.

372. Bello, M.J., Rey, J.A., Pestaña, A., de Campos, J.M., Sarasa, J.L., Kusak, M.E., Vaquero, J. (1992) Meningeal hermangiopericytoma or hemangiopericytic meningioma? A cytogenetic and molecular analysis. *Cancer Genet. Cytogenet.* **63**, 78–80.

373. Lusis, E.A., Chicoine, M.R., and Perry, A. (2005) High throughput screening of meningioma biomarkers using a tissue microarray. *J. Neurooncol.* **73**, 219–223.

374. Watson. M.A., Gutmann, D.H., Peterson, K., Chicoine, M.R., Kleinschmidt-DeMasters, B.K., Brown, H.G., and Perry, A. (2002) Molecular characterization of human meningiomas by gene expression profiling using high-density oligonucleotide microarrays. *Am. J. Pathol.* **161**, 665–672.

375. Brunner, E.C., Romeike, B.F.M., Jun;g, M., Comtesse, N., and Meese, E. (2006) Altered expression of β-catenin/E-cadherin in meningiomas. *Histopathology* **49**, 178–187.

376. Carstens, C., Meese, E., Zang, K.D., and Blin, N. (1988) Human KRAS oncogene expression in meningioma. *Cancer Lett.* **43**, 37–41.

377. Denizot, Y., De Armas, R., Caire, F., Pommepuy, I., Truffinet, V., and Labrousse, F. (2006) Platelet-activating factor and human meningiomas. *Neuropathol. Appl. Neurobiol.* **32**, 674–678.

378. Detta, A., Kenny, B.G., Smith, C., Logan, A., and Hitchcock, E. (1993) Correlation of proto-oncogene expression and proliferation in meningiomas. *Neurosurgery* **33**, 1065–1074

379. Figarella-Branger, D., Pellissier, J.F., Bouillot, P., Bianco, N., Mayan, M., Grisoli, F., and Rougon, G. (1994) Expression of neural cell-adhesion molecule isoforms and epithelial cadherin adhesion molecules in 47 human meningiomas: correlation with clinical and morphological data. *Mod. Pathol.* **7**, 752–761.

380. Fuller, C.E., and Perry, A. (2005) Molecular diagnostics in central nervous system tumors. *Adv. Anat. Pathol.* **12**, 180–194.

381. Girish, V., Sachdeva, N., Minz, R.W., Radotra, B., Mathuria, S.N., and Arora, S.K. (2005) Bcl2 and ROS1 expression in human meningiomas: an analysis with respect to histological subtype. *Indian J. Pathol. Microbiol.* **48**, 325–330.

382. Goh, K.Y.C., Poon, W.S., Chan, D.T.M., and Ip, C.P. (2005) Tissue plasminogen activator expression in meningiomas and glioblastomas. *Clin. Neurol. Neurosurg.* **107**, 296–300.

383. Hassounah, M., Lach, B., Allam, A., Al-Khalaf, H., Siddiqui, Y., Pangue-Cruz, N., Al-Omeir, A., Al-Ahdal, M.N., and Aboussekhra, A. (2005) Benign tumors from the human nervous system express high levels of survivin and are resistant to spontaneous and radiation-induced apoptosis. *J. Neurooncol.* **72**, 203–208.

384. Heckel, D., Comtesse, N., Brass, N., Blin, N., Zang, K.D., and Meese, E. (1998) Novel immunogenic antigen homologous to hyaluronidase in meningioma. *Hum. Mol. Genet.* **7**, 1859–1872.

385. Karja, V., and Alafuzoff, I. (2006) Protein p62 common in investigations in benign meningiomas—a possible predictor of malignancy. *Clin. Neuropathol.* **25**, 37–43.

386. Olivera, M., Martínez, C., Molina, J.A., Alonso-Navarro, H., Jiménez-Jiménez, F.J., García-Martín, E., Benítez, J., and Agúndez, J.A.G. (2006) Increased frequency of rapid acetylator genotypes in patients with brain astrocytoma and meningioma. *Acta Neurol. Scand.* **113**, 322–326.

387. Nakabayashi, H., Shimizu, K., and Hara, M. (2003) Prognostic significance of cyclin A expression in meningiomas. *Appl. Immunohistochem. Mol. Morphol.* **11**, 9–14.

388. Okuducu, A.F., Zils, U., Michaelis, S.A., Mawrin, C., and von Deimling, A. (2006) Increased expression of avian erythroblastosis virus E26 oncogene homolog 1 in World Health Organization grade 1 meningiomas is associated with an elevated risk of recurrence and is correlated with the expression of its target genes matrix metalloproteinase-2 and MMP-9. *Cancer* **107**, 1365–1372.

389. Okuducu, A.F., Zils, U., Michaelis, S.A., Michaelides, S., and von Deimling, A. (2006) Ets-1 is up-regulated together with its target gene products matrix metalloproteinase-2 and matrix metalloproteinase-9 in atypical and anaplastic meningioma. *Histopathology* **48**, 836–845.

390. Riboni, L., Ghidoni, R., Sonnino, S., Omodeo-Sale, F., Gaini, S.M., and Berra, B. (1984) Phospholipid content and composition of human meningiomas. *Neurochem. Pathol.* **2**, 171–188.

391. Sadetzki, S., Flint-Richter, P., Starinsky, S., Novikov, I., Lerman, Y., Goldman, B., and Friedman, E. (2005) Genotyping of patients with sporadic and radiation-associated meningiomas. *Cancer Epidemiol. Biomarkers Prev.* **14**, 969–976.

392. Schwechheimer, K., Zhou, L., and Birchmeier, W. (1998) E-cadherin in human brain tumours: loss of immunoreactivity in malignant meningiomas. *Virchows Arch.* **432**, 163–167.

393. Tohma, Y., Yamashima, T., and Yamashita, J. (1992) Immunohistochemical localization of cell adhesion molecule epithelial cadherin in human arachnoid villi and meningiomas. *Cancer Res.* **52**, 1981–1987.

394. Zhou, J.F., and Sharma, S. (1995) Expression of the ROS1 oncogene for tyrosine receptor kinase in adult human meningiomas. *Cancer Genet. Cytogenet.* **83**, 148–154.

394A. Puri, S., Joshi, B.H., Sarkar, C., Mahapatra, A.K., Hussain, E., and Sinha, S. (2005) Expression and structure of interleukin 4 receptors in primary meningeal tumors. *Cancer* **103**, 2132–2142.

395. Johnson, M., and Toms, S. (2005) Mitogenic signal transduction pathways in meningiomas: novel targets for meningioma chemotherapy? *J. Neuropathol. Exp. Neurol.* **64**, 1029–1036.

395A. Korshunov, A., Cherekaev, V., Bekyashev, A., and Sycheva, R. (2007) Recurrent cytogenetic aberrations in histologically benign, invasive meningiomas of the sphenoid region. *J. Neurooncol.* **81**, 131–137.

395B. Baia, G.S., Slocum, A.L., Hyer, J.D., Misra, A., Sehati, N., VandenBerg, S.R, Feuerstein, B.G., Deen, D.F., McDermott, M.W., and Lal, A. (2006) A genetic strategy to overcome the senescence of primary meningioma cell cultures. *J. Neurooncol.* **78**, 113–121.

395C. Pronk, G.J., and Bos, J.L. (1994) The role of p21ras in receptor tyrosine kinase signaling. *Biochim. Biophys. Acta* **1198**, 131–147.

396. Mark, J. (1973) The fluorescence karyotypes of three human meningiomas with hyperdiploid-hypotriploid stemlines. *Acta Neuropathol.* **25**, 46–53.

397. Zankl, H., Seidel, H., and Zang, K.D. (1975). Cytological and cytogenetical studies on brain tumors. V. Preferential loss of sex chromosome in human meningiomas. *Humangenetik* **27**, 119–128.

398. Zankl., H., Ludwig., B., May, G., and Zang, K.D. (1979) Karyotypic variations in human meningioma cell cultures under different in vitro conditions. *J. Cancer Res. Clin. Oncol.* **93**, 165–172.

398A. Katsuyama, J., Papenhausen, P.R., Herz, F., Gazivoda, P., Hirano, A., and Koss, L.G. (1986) Chromosome abnormalities in meningiomas. *Cancer Genet. Cytogenet.* **22**, 63–68.

399. Rey, J.A., Bello, M.J., de Campos, J.M., and Kusak, M.E. (1988) Incidence and origin of dicentric chromosomes in cultured meningiomas. *Cancer Genet. Cytogenet* **35**, 55–60.

400. Rogatto, S.R., and Casartelli, C. (1988) Cytogenetic analysis of human meningiomas. *Brazil. J. Genet.* **11**, 729–744.

401. Rönne, M., Poulsgård, L., and Elberg, J.J. (1988) A case of meningioma with frequent relapses and a hyperhaploid stemline. *Anticancer Res.* **8**, 545–548.

402. López-Ginés, C., Piquer, J., Cerdá-Nicolás, M., Barcia-Salorio, J.L., and Barcia-Marino, C. (1989) Meningiomas: karyotypes and histological patterns. *Clin. Neuropathol.* **3**, 130–133.

403. Poulsgård, L., Rönne, M., and Schröder, H.D. (1989) Cytogenetic studies of 19 meningiomas and their clinical significance. I. *Anticancer Res.* **9**, 109–112.

404. Teyssier, J.R., and Ferre, D. (1989) Frequent clonal chromosomal changes in human non-malignant tumors. *Int. J. Cancer* **44**, 828–832.

405. Wullich, B., Kiechle-Schwarz, M., Mayfrank, L., and Schempp, W. (1989) Cytogenetic and in situ DNA-hybridization studies in intracranial tumors of a patient with central neurofibromatosis. *Hum. Genet.* **82**, 31–34.

406. Logan, J.A., Seizinger, B.R., Atkins, L., and Martuza, R.L. (1990) Loss of the Y chromosome in meningiomas. a molecular genetic approach. *Cancer Genet. Cytogenet.* **45**, 41–47.

407. Chadduck, W.M., Boop, F.A., and Sawyer, J.R. (1991) Cytogenetic studies of pediatric brain and spinal cord tumors. *Pediatr. Neurosurg.* **92**, 57–65.

408. Chio, C.-C., Lin, S.-J., Yang, S.-H., and Cheng, Y.-C. (1991) Cytogenetic study of twenty-two intracranial tumors. *J. Formosan Med. Assoc.* **90**, 504–508.

409. Meloni, A., Morgan, R., Bridge, J., Erling, M.A., Lewin, R.J., and Sandberg, A.A. (1991) Cytogenetic findings in typical and atypical meningiomas. *Cancer Genet. Cytogenet.* **51**, 35–39.

410. Vagner-Capodano, A.M., Hairion, D., Gambarelli, D., Perez-Castillo, A.M., and Grisoli, F. (1991) A new approach of brain tumors: the cytogenetic study. *J. Neuroradiol.* **18**, 107–121.

411. Cho, J.H., Gong, G.Y., Yu, E.S., Whang, C.J., Jee, K.J., and Lee, I. (1992) Cytogenetic analysis of meningiomas. *J. Korean Med. Sci.* **7**, 162–166.

412. Karnes, P.S., Tran, T.N., Cui, M.Y., Raffel, C., Gilles, F.H., Barranger, J.A., and Ying, K.I. (1992) Cytogenetic analysis of 39 pediatric central nervous system tumors. *Cancer Genet. Cytogenet.* **59**, 12–19.

413. Tonk, V., Osella, P., Delasmorenas, A., Wyandt, H.E., and Milunsky, A. (1992) Abnormalities of chromosome 22 in meningiomas and confirmation of the origin of a dicentric 22 by in situ hybridization. *Cancer Genet. Cytogenet.* **64**, 65–68.

414. Vagner-Capodano, A.J., Gentet, J.C., Gambarelli, D., Pellissier, J.F., Gouzien, M., Lena, G., Genitori, L., Choux, M., and Raybaud, C. (1992) Cytogenetic studies in 45 pediatric brain tumors. *Pediatr. Hematol. Oncol.* **9**, 223–235.

414A. Weiss, A.F., Portmann, R., Fischer, H., Simon, J., and Zang, K.D. (1975) Simian virus 40-related antigens in three human meningiomas with defined chromosome loss. *Proc. Natl. Acad. Sci. USA* **72**, 609–613.

414B. Thangavelu, M., Turina, J., Tomita, T., and Chou, P.M., (1998) Clonal cytogenetic abnormalities in pediatric brain tumors: cytogenetic analysis and clinical correlation. *Pediatr. Neurosurg.* **28**, 15–20.

415. Bello, M.J., de Campos, J.M., Vaquero, J., Kusak, M.E., Sarasa, J.L., Rey, J.A., and Pestaña, A. (1993) Chromosome 22 heterozygosity is retained in most hyperdiploid and pseudodiploid meningiomas. *Cancer Genet. Cytogenet.* **66**, 117–119.

416. Dressler, L.G., Duncan, M.H., Varsa, E.E., and McConnell, T.S. (1993) DNA content measurement can be obtained using archival material for DNA flow cytometry. A comparison with cytogenetic analysis in 56 pediatric solid tumors. *Cancer* **72**, 2033–2041.

417. López-Ginés, C., Cerdá-Nicolás, M., Peydro-Olaya, A., and Llombart-Bosch, A. (1993) Case of meningioma with del(1)(p32) as sole anomaly. *Cancer Genet. Cytogenet.* **70**, 74–76.

418. Prempree, T., Amornmarn, R., Faillace, W.J., Arce, C.A., and Nguyen, T.Q. (1993) 1;19 translocation in human meningioma. *Cancer* **71**, 2306–2311.

419. Albrecht, S., Goodman, J.C., Rajagopolan, S., Levy, M., Cech, D.A., Cooley, L.D. (1994) Malignant meningioma in Gorlin's syndrome: cytogenetic and p53 gene analysis. *J. Neurosurg.* **81**, 466–471.

420. Biegel, J.A., Parmiter, A.H., Sutton, L.N., Rorke, L.B., and Emanuel, B.S. (1994) Abnormalities of chromosome 22 in pediatric meningiomas. *Genes Chromosomes Cancer* **9**, 81–87.

421. Figarella-Branger, D., Vagner-Capodano, A.M., Bouillot, P., Graziani, N., Gambarelli, D., Devictor, B., Zattara-Cannonni, H., Bianco, N., Grisoli, F., and Pellissier, J.F. (1994) Platelet-derived growth factor (PDGF) and receptor (PDGFR) expression in human meningiomas: correlations with clinicopathological features and cytogenetic analysis. *Neuropathol. Appl. Neurobiol.* **20**, 439–447.

422. Gollin, S.M., and Janecka, I.P. (1994) Cytogenetics of cranial base tumors. *J. Neurooncol.* **20**, 241–254.

423. Agamanolis, D.P., and Malone, J.M. (1995) Chromosomal abnormalities in 47 pediatric brain tumors. *Cancer Genet. Cytogenet.* **81**, 125–134.

424. Henn, W., Cremerius, U., Heide, G., Lippitz, B., Schröder, J.M., Gilsbach, J.M., Büll U., and Zang, K.D. (1995) Monosomy 1p is correlated with enhanced in vivo glucose metabolism in meningiomas. *Cancer Genet. Cytogenet.* **79**, 144–148.

425. Bhattacharjee, M.B., Armstrong, D.D., Vogel, II., and Cooley, L.D. (1997) Cytogenetic analysis of 120 primary pediatric brain tumors and literature review. *Cancer Genet. Cytogenet.* **97**, 39–53.

426. Nazi, M., van Dijken, P.J., and Al Moutaery, K. (1998) A patient with meningioma showing multiple cytogenetic abnormalities and a constitutional translocation (3;9)(q13.3;q22). *Cancer Genet. Cytogenet.* **105**, 11–13.

427. Zattara-Cannoni, H., Gambarelli, D., Dufour, H., Figarella, D., Vollot, F., Grisoli, F., and Vagner-Capodano, A. M. (1998) Contribution of cytogenetics and FISH in the diagnosis of meningiomas: a study of 189 tumors. *Ann. Genet.* **41**, 164–175.

428. Debiec-Rychter, M., Biernat, W., Limon, J., Kordek, R., Izycka, E., Borowska-Lehman, J., Imielinski, B., and Liberski, P.P. (1999) Cytogenetic and proliferative potentials in meningiomas. *Pol. J. Pathol.* **50**, 243–248.

429. Steilen-Gimbel, H., Niedermayer, I., Feiden, W., Freiler, A., Steudel, W.I., Zang, K.D., and Henn, W. (1999) Unbalanced translocation t(1;3)(p12–13;q11) in meningiomas as the unique feature of chordoid differentiation. *Genes Chromosomes Cancer* **26**, 270–272.

430. Go, Y., Ohjimi, Y., Iwasaki, H., Oka, K., Ishiguro, M., Kaneko, Y., Tsuchimochi, H., Tomonaga, M., and

Kikuchi, M. (2000) A case of papillary meningioma with a t(1;4)(q44;q21). *Cancer Genet. Cytogenet.* **119,** 37–41.

431. Sawyer, J.R., Thomas, E.L., and Al-Mefty, O. (2002) Translocation (1;19)(q21;q13.3) is a recurrent reciprocal translocation in meningioma. *Cancer Genet. Cytogenet.* **134,** 88–90.

432. Ketter, R., Henn, W., Feiden, W., Prowald, A., Sittel, C., Steudel, W.-I., and Strowitzki, M. (2003) Nasoethmoidal meningioma with cytogenetic features of tumor aggressiveness in a 16-year-old child. *Pediatr. Neurosurg.* **39,** 190–194.

433. Joseph, J.T., Lisle, D.K., Jacoby, L.B., Paulus, W., Barone, R., Cohen, M.L., Roggendorf, W.H., Bruner, J.M., Gusella, J.F., and Louis, D.L. (1995) *NF2* gene analysis distinguishes hemangiopericytoma from meningioma. *Am. J. Pathol.* **147,** 1450–1455.

434. Kim, J.H., Lee, S.H., Rhee, C.H., Park, S.Y., and Lee, J.H. (1998) Loss of heterozygosity on chromosome 22q and 17p correlates with aggressiveness of meningiomas. *J. Neurooncol.* **40,** 101–106.

435. Rushing, E.J., Colvin, S.M., Gazdar, A., Miura, N., White, C.L., III, Coimbra, C., and Burns, D.K. (1999) Prognostic value of proliferation index and expression of the RNA component of human telomerase (hTR) in papillary meningiomas. *J. Neurooncol.* **45,** 199–207.

436. Nakasu, S., Nakajima, M., Matsumura, K., Nakasu, Y., and Handa, J. (1995) Meningioma: proliferating potential and clinicoradiological features. *Neurosurgery* **37,** 1049–1055.

437. Maier, H., Wanschitz, J., Sedivy, R., Rössler, K., Öfner, D., and Budka, H. (1997) Proliferation and DNA fragmentation in meningioma subtypes. *Neuropathol. Appl. Neurobiol.* **23,** 496–506.

438. Karamitopoulou, E., Perentes, E., Tolnay, M., and Probst, A. (1998) Prognostic significance of MIB-1, p53, and bcl-2 immunoreactivity in meningiomas. *Hum. Pathol.* **29,** 140–145.

438A. Ohkoudo, M., Sawa, H., Hara, M., Saruta, K. Aiso, T., Ohki, R., Yamamoto, H., Maemura, E., Shiina, Y., Fujii, M., and Saito, I. (1998) Expression of p53, MDM2 protein and Ki-67 antigen in recurrent meningiomas. *J. Neurooncol.* **38,** 41–49.

439. Cobb, M.A., Husain, M., Andersen, B.J., and Al-Mefty, O. (1996) Significance of proliferating cell nuclear antigen in predicting recurrence of intracranial meningioma. *J. Neurosurg.* **84,** 85–90.

440. Akat, K., Mennel, H.-D., Kremer, P., Gassler, N., Bleck, C.K.E., and Kartenbeck, J. (2003) Moleuclar characterization of desmosomes in meningiomas and arachnoidal tissue. *Acta Neuropathol.* **106,** 337–347.

441. Alonso, M.E., Bello, M.J., Gonzalez-Gomez, P., Lomas, J., Arjona, D., de Campos, J.M., Kusak, M.E., Sarasa, J.L., Isla, A., and Rey, J.A. (2001) Mutation analysis of the p73 gene in nonastrocytic brain tumours. *Br. J. Cancer* **85,** 204–208.

442. Amirjamshidi, A., Mehrazin, M., and Abbaddioun, K. (2000) Meningiomas of the central nervous system occurring below the age of 17: report of 24 cases not associated with neurofibromatosis and review of the literature. *Childs Nerv. Syst.* **16,** 406–416.

443. Amatya, V.J., Takeshima, Y., Inai, K. (2004) Methylation of p14(ARF) gene in meningiomas and its correlation to the p53 expression and mutation. *Mod. Pathol.* **17,** 705–710.

444. Andersson, U., Guo, D., Malmer, B., Bergenheim, A.T., Brännström, T., Hedman, H., and Henriksson, R. (2004) Epidermal growth factor receptor family (EGFR, ErbB2–4) in gliomas and meningiomas. *Acta. Neuropathol.* **108,** 135–142.

445. Johnson, M.D., Horiba, M., Winnier, A.R., and Arteaga, C.L. (1994) The epidermal growth factor receptor is associated with phospholipase C-γ1 in meningiomas. *Hum. Pathol.* **25,** 146–153.

445A. Arena, S., Barbieri, F., Thellung, S., Pirani, P., Corsaro, A., Villa, V., Dadati, P., Dorcaratto, A., Lapertosa, G., Ravetti, J.-L., Spaziante, R., Schettini, G., and Florio, T. (2004) Expression of somatostatin receptor mRNA in human meningiomas and their implication in *in vitro* antiproliferative activity. *J. Neurooncol.* **66,** 155–166.

446. Arishima, H., Sato, K., and Kubota, T. (2002) Immunohistochemical and ultrastructural study of Gap junction proteins connexin26 and 43 in human arachnoid villi and meningeal tumors. *J. Neuropathol. Exp. Neurol.* **61,** 1048–1055.

447. Arnoldus, E.P.J., Noordermeer I.A., Boudewijn Peters, A.C., Voormolen, J.H.C., Bots, G.T.A.M., Raap, A.K., and van der Ploeg, M. (1991) Interphase cytogenetics of brain tumors. *Genes Chromosome Cancer* **3,** 101–107.

448. Arnoldus, E.P., Wolters, L.B., Voormolen, J.H., van Duinen, S.G., Raap, A.K., van der Ploeg, M., and Peters, A.C. (1992) Interphase cytogenetics: a new tool for the study of genetic changes in brain tumors. *J. Neurosurg.* **76,** 997–1003.

449. Barker, F.G., Chen, P., Furman, F., Aldape, K.D., Edwards, M..S., and Israel, M.A. (1997) P16 deletion and mutation analysis in human brain tumors. *J. Neurooncol.* **31,** 17–23.

450. Simon, M., von Deimling, A., and Menon, A.G. (1997) Absence of p16 mutations in malignant meningioma progression. *Zentralbl. Neurochir. Suppl.* **6(A).**

451. Bhattacharjee, M., Adesina, A.M., Goodman, C., and Powell, S. (2003) Claudin-1 expression in meningiomas and schwannomas: possible role in differential diagnosis. *J. Neuropathol. Exp. Neurol.* **62,** 581.

452. Hahn, H.P., Bundock, E.A., and Hornick, J.L. (2006) Immunohistochemical staining for claudin-1 can help distinguish meningiomas from histologic mimics. *Am. J. Clin. Pathol.* **125,** 203–208.

453. Carroll, R.S., Zhang, J., and Black, P.M. (1999) Expression of estrogen receptors alpha and beta in human meningiomas *J. Neurooncol.* **42,** 109–116.

454. Chozick, B.S., Benzil, D.L., Stopa, E.G., Pezzullo, J.C., Knuckey, N.W., Epstein, M.H., Finkelstein, S.D., and Finch, P.W. (1996) Immunohistochemical evaluation of erbB-2 and p53 protein expression in benign and atypical human meningiomas. *J. Neurooncol.* **27,** 117–126.

455. Christensen, H.C., Kosteljanetz, M., and Johansen, C. (2003) Indices of gliomas and meningiomas in Denmark, 1943 to 1997. *Neurosurgery* **52,** 1327–1334.

456. Radhakrishnan, K., Mokri, B., Parisi, J.E., O'Fallon, W.M., Sunku, J., and Kurland, L.T. (1995) The trends in incidence of primary brain tumors in the population of Rochester, Minnesota. *Ann. Neurol.* **37,** 67–73.

457. Çolakoğlu, N., Demirtas, E., Oktar, N., Yüntem, N., Islekel, S., and Özdamar, N. (2003) Secretory meningiomas. Report of clinical, immunohistochemical findings in 12 cases and review of the literature. *J. Neurooncol.* **62,** 233–241.

458. Comtesse, N., Niedermayer, I., Glass, B., Maldener, E., Nastainczyk, W., Feiden, W., and Meese, E. (2001) MGEA6, first overexpressed and immunogenic antigen found in human primary intracranial tumors. *Cancer Genet. Cytogenet.* **128,** 69.

459. Cuevas, I.C. Slocum, A.L., Jun, P., Costello, J.F., Bollen, A.W., Riggins, G.J., McDermott, M.W., and Lal, A. (2005) Meningioma transcript profiles reveal deregulated notch signaling pathway. *Cancer Res.* **65,** 5070–5075.

460. den Bakker, M.A., van Tilborg, A.A., Kros, J.M., and Zwarthoff, E.C. (2001) Truncated NF2 proteins are not detected in meningiomas and scahwannomas. *Neuropathology* **21,** 168–173.

461. Fathallah-Shaykh, H.M., He, B., Zhao, L.-J., Engelhard, H.H., Cerullo, L., Lichtor, T., Byrne, R., Munoz, L., Von Roenn, K., Rosseau, G.L., Glick, R., Sherman, C., and Farooq, H. (2003) Genomic expression discovery predicts pathways and opposing functions behind phenotypes. *J. Biol. Chem.* **278,** 23830–23833.

462. Ferrante, L., Acqui, M., Artico, M., Mastronardi, L., and Nucci, F. (1987) Familial meningiomas: report of two cases. *J. Neurosurg.* **31,** 145–151.

463. McDowell, J.R. (1990) Familial meningioma. *Neurology* **40,** 312–314.

464. Fontaine, B., Rouleau, G.A., Seizinger, B., Jewell, A.F., Hanson, M.P., Martuza, R.I., and Gusella, J.F. (1990) Equal parental origin of chromosome 22 losses in human sporadic meningioma: no evidence for genomic imprinting. *Am. J. Hum. Genet.* **47,** 823–827.

465. Gonzalez-Gomez, P., Bello, M.J., Alonso, M.E., Arjona, D., Lomas, J., de Campos, J.M., Vaquero, J., Isla, A., and Rey, J.A. (2003) CpG island methylation status and mutation analysis of the RB1 gene essential promoter region and protein-binding pocket domain in nervous system tumors. *Br. J. Cancer* **88,** 109–114.

466. Hancq, S., Salmon, I., Brotchi, J., De Witte, O., Gabius, H.-J., Heizmann, C.W., Kiss, R., and Decaestecker, C. (2004) S100A5: a marker of recurrence in WHO grade I meningiomas. *Neuropathol. Appl. Neurobiiol.* **30,** 178–187.

467. Herting, B., Meixensberger, J., Roggendorf, W., and Reichmann, H. (2003) Metabolic patterns in meningiomas. *J. Neurooncol.* **65,** 119–123.

468. Hill, D.A., Linet, M.S., Black, P.M., Fine, H.A., Selker, R.G., Shapiro, W.R., and Inskip, P.D. (2004) Meningioma and schwannoma risk in adults in relation to family history of cancer. *Neurooncol.* **6,** 274–280.

469. Hirota, S., Nakajima, Y., Yoshimine, T., Kohri, K., Nomura, S., Taneda, M., Hayakawa, T., and Kitamura, Y. (1995) Expression of bone-related protein messenger RNA in human meningiomas: possible involvement of osteopontin and development of psammoma bodies. *J. Neuropathol. Exp. Neurol.* **54,** 698–703.

470. Verhage, A., Go, K.G., Visser, G.M., Blankenstein, M.A., and Vaalburg. W. (1995) The presence of progesterone receptors I arachnoid granulations and in the lining of arachnoid cysts: its relevance to expression of progesterone receptors in meningiomas. *Br. J. Neurosurg.* **9,** 47–50.

471. Jacobs, H.M., van Spriel, A.B., and Koehorst, S.G.A. (1999) The truncated estrogen receptor alpha variant lacking exon 5 is not involved in progesterone receptor expression in meningiomas. *J. Steroid Biochem. Mol. Biol.* **71,** 167–172.

471A. Nathoo, N., Barnett, G.H., and Golubic, M. (2004) The eicosanoids cascade: possible role in gliomas and meningiomas. *J. Clin. Pathol.* **57,** 6–13.

472. Hwang, S.-L., Lieu, A.-S., Chang, J.-H., Cheng, T.-S., Cheng, C.-Y., Lee, K.-S., Lin, C.-L., Howng, S.-L., and Hong, Y.-R. (2005) Rac2 expression and mutation in human brain tumors. *Acta Neurochir. (Wien)* **147,** 551–554.

472A. Hwang, S.L., Chang, J.H., Cheng, T.S., Sy, W.D., Lieu, A.S., Lin, C.L., Lee, K.S., Howng, S.L., and Hong, Y.R. (2005) Expression of Rac3 in human brain tumors. *J. Clin. Neurosci.* **12,** 571–574.

473. Johnson, M.D., Okediji, E., and Woodard, A. (2004) Transforming growth factor-β effects on meningioma cell proliferation and signal transduction pathways. *J. Neurooncol.* **66,** 9–16.

474. Kamei, Y., Watanabe, M., Nakayama, T., Kanamaru, K., Waga, S., and Shiraishi, T. (2000) Prognostic significance of p53 and p21$^{WAF1/CIP1}$immunoreactivity and tumor micronecrosis for recurrence of meningiomas. *J. Neurooncol.* **46,** 205–213.

475. Kanno, H., Yamamoto, I., Yoshida, M., and Kitamura, H. (2003) Meningioma showing *VHL* gene inactivation in a patient with von Hippel-Lindau disease. *Neurology* **60,** 1197–1199.

476. Kawashima, M., Suzuki, S.O., Yamashima, T., fukui, M., and Iwaki, T. (2001) Prostaglandin D synthase (β-trace) in meningeal hemangiopericytoma. *Mod. Pathol.* **14,** 197–201.

477. Kazumoto, K., Tamura, M., Hoshino, H., and Yuasa, Y. (1990) Enhanced expression of the *sis* and c-*myc* oncogenes in human meningiomas. *J. Neurosurg.* **72,** 786–791.

478. Kim, N.R., Im, S.-H., Chung, C.K., Suh, Y.-L., Choe, G., and Chi, J.G. (2004) Sclerosing meningioma: immunohistochemical analysis of five cases. *Neuropathol. Appl. Neurobiol.* **30,** 126–135.

478A. Kilic, T., Bayri, Y., Ozduman, K., Acar, M., Diren, S., Kurtkaya, O., Ekinci, G., Bugra, K., Sav, A., Ozek, M.M., and Pamir, M.N. (2002) Tenascin in meningioma: expression is correlated with anaplasia, vascular endothelial growth factor expression and peritumoral edema but not with tumor border shape. *Neurosurgery* **51,** 183–193.

479. Kino, T., Takeshima, H., Nakao, M., Nishi, T., Yamamoto, K., Kimura, T., Saito, Y., Kochi, M., Kuratsu, J., Saya, H., and Ushio, Y. (2001) Identification of the cis-acting region in the NF2 gene promoter as a potential target for mutation and methylation-dependent silencing in schwannomas. *Genes Cell* **6,** 441–454.

480. Klaeboe, L., Lonn, S., Scheie, D., Auvinen, A., Christensen, H.C., Feychting, M., Johansen, C., Salminen, T., and Tynes, T. (2005) Incidence of intracranial meningiomas in Denmark, Finland, Norway and Sweden, 1968–1997. *Int. J. Cancer* **117,** 996–1001.

481. Klekner, A., Röhn, G., Schillinger, G., Schröder, R., and Klug, N. (2001) ODC mRNA as a prognostic factor for predicting recurrence in meningiomas. *J. Neurooncol.* **53,** 67–75.

482. Konstantinidou, A., Korkolopoulou, P., Patsouris, E., Mahera, H., Hranioti, S., Kotsiakis, X., and Davaris, P. (2001) Apoptosis detected with monoclonal antibody to single-

stranded DNA is a predictor of recurrence in meningiomas. *J. Neurooncol.* **55,** 1–9.

483. Korshunov, A., Shishkina, L., and Golanov, A. (2002) DNA topoisomerase II-α and cyclin A immunoexpression in meningiomas and its significance. *Arch. Pathol. Lab. Med.* **126,** 1079–1086.

484. Shibuya, M., Hoshino, T., Ito, S., Wacker, M.R., Prados, M.D., Davis, R.L., and Wilson, C.B. (1992) Meningiomas: clinical implications of a high proliferative potential determined by bromodeoxyuridine labeling. *Neurosurgery* **30,** 494–498.

485. Krampla, W., Newrkla, S., Pfisterer, W., Jungwirth, S., Fischer, P., Leitha, T., Hruby, W., and Tragl, K.H. (2004) Frequency and risk factors for meningioma in clinically healthy 75-year-old patients. Results of the Transdanube Ageing Study (VITA). *Cancer* **100,** 1208–1212.

485A. Loussouarn, D., Brunon, J., Avet-Loiseau, H., Campone, M., and Mosnier, J.-F. (2006) Prognostic value of HER2 expression in meningiomas: an immunohistochemical and fluorescence in situ hybridization study. *Hum. Pathol.* **37,** 415–421.

486. Lee, W.H. (1990) Characterization of a newly established malignant meningioma cell line of the human brain: IOMM-Lee. *Neurosurgery* **27,** 389–395.

487. Lin, C.-C.D., Kenyon, L., Hyslop, T., Hammond, E., Andrews, D.W., Curran, W.J., Jr., and Dicker, A.P. (2003) Cyclooxygenase-2 (COX-2) expression in human meningioma as a function of tumor grade. *Am. J. Clin. Oncol.* **26,** S98–S102.

488. Mashiyama, S., Murakami, Y., Yoshimoto, T., Sekiya, T., and Hayashi, K. (1991) Detection of p53 gene mutations in human brain tumors by single-strand conformation polymorphism analysis of polymerase chain reaction products. *Oncogene* **6,** 1313–1318.

489. Meixensberger, J., Janka, M., Zellner, A., Roggendorf, W., and Roosen, K. (1996) Prognostic significance of nuclear DNA content in human meningiomas: a prospective study. *Acta Neurochir.* **65,** 70–72.

490. Murakami, M., Kuratsu, J.-I., Mihara, Y., Matsuno, K., and Ushio, Y. (1993) Histochemical study of alkaline phosphatase in primary human brain tumors: diagnostic implications for meningiomas and neurinomas. *Neurosurgery* **32,** 180–184.

491. Müller, P., Henn, W., Niedermayer, I., Ketter, R., Feiden, W., Steudel, W.-I., Zang, K.D., and Steilen-Gimbel, H. (1999) Deletion of chromosome 1p and loss of expression of alkaline phosphatase indicate progression of meningiomas. *Clin. Cancer Res.* **5,** 3569–3577.

492. Murphy, M., Chen, J.N., and George, D.L. (1991) Establishment and characterization of a human leptomeningeal cell line. *J. Neurosci. Res.* **30,** 475–483.

493. Tanaka, K., Sato, C., Maeda, Y., Koike, M., Matsutani, M., Yamada, K., and Miyaki, M. (1989) Establishment of a human malignant meningioma cell line with amplified c-myc oncogene. *Cancer* **64,** 2243–2249.

494. Nakasu, S., Li, D.H., Okabe, H., Nakajima, M., and Matsuda, M. (2001) Significance of MIB-1 staining indices in meningiomas. *Am. J. Surg. Pathol.* **25,** 472–478.

494A. Sandberg, D.I., Edgar, M.A., Resch, L., Rutka, J.T., Becker, L.E., and Souweidane, M.M. (2001) MIB-1 staining index of pediatric meningiomas. *Neurosurgery* **48,** 590–597.

495. Wood, M.W., White, R.J., and Kernohan, J.W. (1957) One hundred intracranial meningiomas found incidentally at necropsy. *J. Neuropathol. Exp. Neurol.* **16,** 337–340.

496. Nicolato, A., Giorgetti, P., Foroni, R., Grigolato, D., Pasquin, I.P., Zuffante, M., Soda, C., Tommasini, A., and Gerosa, M. (2005) Gamma knife radiosurgery in skull base meningiomas: a possible relationship between somatostatin receptor decrease and early neurological improvement without tumor shrinkage at short term imaging follow-up. *Acta. Neurochir. (Wien)* **147,** 367–375.

497. Nozza, P., Raso, A., Rossi, A., Milanaccio, C., Pezzolo, A., Capra, V., Gambini, C., and Pietsch, T. (2005) Rhabdoid meningioma of the tentorium with expression of desmin in a 12-year-old Turner syndrome patient. *Acta Neuropathol.* **110,** 205–206.

498. Oikonomou, E., Machado, A.L., Buchfelder, M., and Adams, E.F. (2005) Meningiomas expressing and responding to cholecystokinin (CCK). *J. Neurooncol.* **73,** 199–204.

499. Okada, M., Miyake, K., Matsumoto, Y., Kawai, N., Kunishio, K., and Nagao, S. (2004) Matrix metalloproteinase-2 and matrix metalloproteinase-9 expressions correlate with the recurrence of intracranial meningiomas. *J. Neurooncol.* **66,** 29–37.

500. Paek, S.H., Kim, C.-Y., Kim, Y.Y., Park, I.A., Kim, M.S., Kim, D.G., and Jung, H.-W. (2002) Correlation of clinical and biological parameters with peritumoral edema in meningioma. *J. Neurooncol.* **60,** 235–245.

501. Pallini, R., Casalbore, P., Mercanti, D., Maggiano, N., and Larocca, L.M. (2000) Phenotypic change of human cultured meningioma cells. *J. Neurooncol.* **49,** 9–17.

502. Perret, A.G., Duthel, R., Fotso, M.J., Brunon, J., and Mosnier, J.F. (2002) Stromelysin-3 is expressed by aggressive meningiomas. *Cancer* **94,** 765–772.

503. Nordqvist, A.C.S., Smurawa, H., and Mathiesen, T. (2001) Expression of matrix metalloproteinases 2 and 9 in meningiomas associated with different degrees of brain invasiveness and edema. *J. Neurosurg.* **95,** 839–844.

504. Siddique, K., Yanamandra, N., Gujrati, M., Dinh, D., Rao, J.S., and Olivero, W. (2003) Expression of matrix metalloproteinases, their inhibitors, and urokinase plasminogen activator in human meningiomas. *Int. J. Oncol.* **22,** 289–294.

505. Yang, S.-Y., and Xu, G.-M. (2000) Expression of PDGF and its receptor as well as their relationship to proliferating activity and apoptosis of meningiomas in human meningiomas. *J. Clin. Neurosci (Suppl. 1)* **8,** 49–53.

506. Torp, S.H., Helseth, E., Dalen, A., and Unsgaarad, G. (1992) Expression of epidermal growth factor receptor in human meningiomas and meningeal tissue. *APMIS* **100,** 797–802.

507. Halper, J., Jung, C., Perry, A., Suliman, H., Hill, M.P., and Scheithauer, B. (1999) Expression of TGF-α In meningiomas. *J. Neurooncol.* **45,** 127–134.

508. Nordqvist, A.C., Peyrard, M., Pettersson, H., Mathiesen, T., Collins, V.P., Dumanski, J.P., and Schalling, M. (1997) A high ratio of insulin-like growth factor II insulin-like growth factor binding protein 2 messenger RNA as a marker for anaplasia in meningiomas. *Cancer Res.* **57,** 2611–2614.

509. Harland, S.P., Kuc, R.E., Pickard, J.D., and Davenport, A.P. (1998) Expression of endothelin (A) receptors in human

gliomas and meningiomas, with high affinity for the selective antagonist PD156707. *Neurosurgery* **43**, 890–898.

510. Pagotto, U., Arzberger, T., Hopfner, U., Sauer, J., Renner, U., Newton, C.J., Lange, M., Uhl, E., Weindl, A., and Stalla, G.K. (1995) Expression and localization of endothelin-1 and endothelin receptors in human meningiomas: evidence for a role in tumoral growth. *J. Clin. Invest.* **96**, 2017–2025.

511. Peyrard, M., Fransson, I., Xie, Y.-G., Han, F.-Y., Ruttledge, M.H., Swahn, S., Collins, J.E., Dunham, I., Collins, V.P., and Dumanski, J.P. (1994) Characterization of a new member of the human β-adaptin gene family from chromosome 22q12, a candidate meningioma gene. *Hum. Mol. Genet.* **3**, 1393–1399.

512. Sanson, M., Marineau, C., Desmaze, C., Lutchman, M., Ruttledge, M., Baron, C., Narod, S., Delattre, O., Lenoir, G., Thomas, G., Aurias, A., and Rouleau, G.A. (1993) Germline deletion in neurofibromatosis type 2 kindred inactivates the NF2 gene and a candidate meningioma locus. *Hum. Mol. Genet.* **2**, 1215–1220.

513. Piaskowski, S., Rieske, P., Szybka, M., Woniak, K., Bednarek, A., Pluciennik, E., Jaskolski, D., Sikorska, B., and Liberski, P.P. (2005) *GADD45A* and *EPB41* as tumor suppressor genes in meningioma pathogenesis. *Cancer Genet. Cytogenet.* **162**, 63–67.

514. Puri, S., Joshi, B.H., Sarkar, C., Mahapatra, A.K., Hussain, E., and Sinha, S. (2005) Expression and structure of interleukin 4 receptors in primary meningeal tumors. *Cancer* **103**, 2132–2142.

514A. Pykett, M.J., Landers, J., and George, D.L. (1997) Expression patterns of the p53 tumor suppressor gene and mdm2 proto-oncogene in human meningiomas. *J. Neurooncol.* **32**, 39–44.

515. Rajaraman, P., Schwartz, B.S., Rothman, N., Yeager, M., Fine, H.A., Shapiro, W.R., Selker, R.G., Black, P.M., and Inskip, P.D. (2005) δ-aminolevulinic acid dehydratase polymorphism and risk of brain tumors in adults. *Environ. Health Perspect.* **113**, 1209–1211.

516. Reczek, D., Berryman, M., and Bretscher, A. (1997) Identification of EBP50: a PDZ-containing phosphorprotein that associates with members of the ezrin-radixin-moesin family. *J. Cell Biol.* **139**, 169–179.

517. Murthy, A., Gonzalez-Agosti, C., Cordero, E., Pinney, D., Candia, C., Solomon, F., Gusella, J., and Ramesh, V. (1998) NHE-RF, a regulatory cofactor for Na$^+$H$^+$ exchange, is a common interactor for merlin and ERM (MERM) proteins. *J. Biol. Chem.* **273**, 1273–1276.

518. Scoles, D.R., Huynh, D.P., Morcos, P.A., Coulsell,, E.R., Robison, N.G.G., Tamanoi, F., and Pulst, S.M. (1998) Neurofibromatosis 2 tumour suppressor schwannomin interacts with beta II-spectrin. *Nat. Genet.* **18**, 354–359.

519. Sun, C.X., Haipek, C., Scoles, D.R., Pulst, S.M., Giovannini, M., Komada, M., and Gutmann, D.H. (2002) Functional analysis of the relationship between the neurofibromatosis 2 (NF2) tumor suppressor and its binding partner, hepatocyte growth factor-regulated tyrosine kinase substrate (HRS HGS). *Hum. Mol. Genet.* **11**, 3167–3178.

520. Scoles, D.R., Huynh, D.P., Chen, M.S., Burke, S.P., Gutmann, D.H., and Pulst, S.M. (2000) The neurofibromatosis 2 (NF2) tumor suppressor protein interacts with hepatocyte growth factor-regulated tyrosine kinase substrate, HRS. *Hum. Mol. Genet.* **9**, 1567–1574.

521. Goutebroze, L., Brault, E., Muchardt, C., Camonis, J., and Thomas, G. (2000) Cloning and characterization of SCHIP-1, a novel protein interacting specifically with spliced isoforms and naturally occurring mutant NF2 proteins. *Mol. Cell Biol.* **20**, 1699–1712.

522. Fernandez-Valle, C., Tang, Y., Ricard, J., Rodenas-Ruano, A., Taylor, A., Hackler, E., Biggerstaff, J., and Iacovelli, J. (2002) Paxillin binds schwannomin and regulates its density-dependent localization and effect on cell morphology. *Nat. Genet.* **31**, 354–362.

523. Obremski, V.J., Hall, A.M., and Fernandez-Valle, C. (1998) Merlin, the neurofibromatosis 2 tumor suppressor gene product, and betal integrin associate in isolated and differentiating Schwann cells. *J. Neurobiol.* **37**, 487–501.

524. Gronholm, M., Sainio, M., Zhao, F., Heiska, L., Vaheri, A., and Carpen, O. (1999) Homotypic and heterotypic interaction of the neurofibromatosis 2 tumor suppressor protein merlin and the ERM protein ezrin. *J. Cell Sci.* **112**, 895–904.

525. Morrison, H., Sherman, L.S., Legg, J., Banine, F., Isacke, C., Haipek, C.A., Gutmann, D.H., Ponta, H., and Herrlich, P. (2001) The NF2 tumor suppressor gene product, merlin, mediates contact inhibition of growth through interactions with CD44. *Genes Dev.* **15**, 968–980.

526. Tsukita, S., Oishi, K., Sato, N., Sagara, J., Kawai, A., and Tsukita, S. (1994) ERM family members as molecular linkers between the cell surface glycoprotein CD44 and actin-based cytoskeletons. *J. Cell Biol.* **126**, 391–401.

527. Rieske, P., Zakrzewska, M., Piaskowski, S., Jaskólski, D., Sikorska, B., Papierz, W., Zakrzewski, K., and Liberski, P.P. (2003) Molecular heterogeneity of meningioma with INI1 mutation. *J. Clin. Pathol. Mol. Pathol.* **56**, 299–301.

528. Rieske, P., Zakrzewska, M., Biernat, W., Bartkowiak, J., Zimmermann, A., and Liberski, P.P. (2005) Atypical molecular background of glioblastoma and meningioma developed in a patient with Li-Fraumeni syndrome. *J. Neurooncol.* **71**, 27–30.

529. Sandberg Nordqvist, A.-C., and Mathiesen, T. (2002) Expression of IGF-II, IGFBP-2, -5, and -6 in meningiomas with different brain invasiveness. *J. Neurooncol.* **57**, 19–26.

530. Sanson, M., Richard, S., Delattre, O., Poliwka, M., Mikol, J., Philippon, J., and Thomas, G. (1992) Allelic loss on chromosome 22 correlates with histopathological predictors of recurrence of meningiomas. *Int. J. Cancer* **50**, 391–394.

531. Dutour, A., Kumar, U., Panetta, R., Ouafik, L., Fina, F., Sasi, R., and Patel, Y.C. (1998) Expression of somatostatin receptor subtypes in human brain tumors. *Int. J. Cancaer* **76**, 620–628.

532. Carroll, R.S., Schrell, U.M., Zhang, J., Dashner, K., Nomikos, P., Fahlbusch, R., and Black, P.M. (1996) Dopamine D1, dopamine D2, and prolactin receptor messenger ribonucleic acid expression by the polymerase chain reaction in human meningiomas. *Neurosurgery* **38**, 367–375.

533. Muccioli, G., Ghe, C., Faccani, G., Lanotte, M., Forni, M., and Ciccarelli, E. (1997) Prolactin receptors in human meningiomas: characterization and biological role. *J. Endocrinol.* **153**, 365–371.

534. Sato, K., Schauble, B., Kleihues, P., and Ohgaki, H. (1996) Infrequent alterations of the p15, p16, CDK4 and cyclin D1 genes in non-astrocytic human brain tumors. *Int. J. Cancer* **66**, 305–308.

535. Scherer, S.S., and Gutmann, D.H. (1996) Expression of the neurofibromatosis 2 tumor suppressor gene product, merlin, in Schwann cells. *J. Neurosci. Res.* **46**, 595–605.

536. Stemmer-Rachamimov, A.O., Gonzalez-Agosti, C., Xu, L. Burwick, J.A., Beauchamp, R., Pinney, D., Louis, D.N., and Ramesh, V. (1997) Expression of NF2-encoded merlin and related ERM family proteins in the human central nervous system. *J. Neuropathol. Exp. Neurol.* **46**, 735–742.

537. den Bakker, M.A., Vissers, K.J., Molijn, A.C., Kros, J.M., Zwarthoff, E.C., and van der Kwast, T.H. (1999) Expression of the neurofibromatosis type 2 gene in human tissues. *J. Histochem. Cytochem.* **47**, 1471–1480.

537A. Shaw, R.J., Paez, J.G., Curto, M., Yaktine, A., Pruitt, W.M., Saotome, I., O'Bryan, J.P., Gupta, V., Ratner, N., Der, C.J., Jacks, T., and McClatchey, A.I. (2001) The NF2 tumor suppressor, merlin, functions in Rac-dependent signaling. *Dev. Cell* **1**, 63–72.

538. Sherman, L., Xu, H.M., Geist, R.T., Saporito-Irwin, S., Howells, N., Ponta, H., Herrlich, P., and Gutmann, D.H. (1997) Interdomain binding mediates tumor growth suppression by the NF2 gene product. *Oncogene* **15**, 2505–2509.

538A. Shamah, S.M., Alberta, J.A., Giannobile, W.V., Guha, A., Kwon, Y.K., Carroll, R.S., Black, P.M., and Stiles, C.D. (1997) Detection of activated platelet-derived growth factor receptors in human meningioma. *Cancer Res.* **57**, 4141–4147.

539. Gutmann, D.H., Sherman, L., Seftor, L., Haipek, C., Lu, K.-H., and Hendrix. M. (1999) Increased expression of the Nf2 suppressor gene product, merlin, impairs cell motility, adhesion and spreading. *Hum. Mol. Genet.* **8**, 267–276.

540. Stenzel, W., Röhn, G., Miletic, H., Radner, H., Deckert, M., and Ernestus, R.-I. (2004) Diagnostic impact of ornithine decarboxylase in meningiomas. *J. Neurooncol.* **66**, 59–64.

541. Todo, T., and Fahlbusch, R. (1994) Accumulation of inositol phosphates in low-passage human meningioma cells following treatment with epidermal growth factor. *J. Neurosurg.* **80**, 890–896.

542. Tse, J.Y., Ng, H.K., Lo, K.W., Chong, E.Y., Lam, P.Y., Ng, E.K., Poon, W.S., and Huang, D.P. (1998) Analysis of cell cycle regulators: p16INK4A, pRb, and CDK4 in low- and high-grade meningiomas. *Hum. Pathol.* **29**, 1200–1207.

543. Weisberg, S., Ashkenazi, E., Israel, Z., Attia, M., Shoshan, Y., Umansky, F., and Brodie, C. (2001) Anaplastic and atypical meningiomas express high levels of Fas and undergo apoptosis in response to Fas ligation. *Am. J. Pathol.* **159**, 1193–1197.

544. Weisman, A.S., Raguet, S.S., and Kelly, P.A. (1987) Characterization of the epidermal growth factor receptor in human meningioma. *Cancer Res.* **47**, 2172–2176.

545. Yakut, T., Bekar, A., Doygun, M., Acar, H., Egeli, U., and Ogul, E. (2002) Evaluation of relationship between chromosome 22 and p53 gene alterations and the subtype of meningiomas by the interphase-FISH technique. *Teratog. Carcinog. Mutagen.* **22**, 217–225.

546. Yu, T., Robb, V.A., Singh, V., Gutmann, D.H., and Newsham, I.F. (2002) The 4.1/ezrin/radixin/moesin domain of the DAL-1/Protein 4.1B tumour suppressor interacts with 14-3-3-proteins. *Biochem. J.* **365**, 783–789.

547. Adams, E.F., Todo, T., Schrell, U.M., Thierauf, P., White, M.C., and Fahlbusch, R. (1991) Autocrine control of human meningioma proliferation: secretion of platelet-derived growth-factor-like molecules. *Int. J. Cancer* **49**, 398–402.

548. Todo, T., Adams, E.F., Fahlbusch, R., Dingermann, T., and Wener, H. (1996) Autocrine growth stimulation of human meningioma cells by platelet-derived growth factor. *J. Neurosurg.* **84**, 852–859.

549. Maxwell, M., Galanopoulos, T., Hedley-Whyte, E.T., Black, P.M., and Antoniades, H.N. (1990) Human meningiomas co-express platelet-derived growth factor (PDGF) and PDGF-receptor genes and their protein products. *Int. J. Cancer* **46**, 16–21.

550. Wang, J.L., Nister, M., Hermansson, M., Westermark, B., and Ponten, J. (1990) Expression of PDGF beta-receptors in human meningioma cells. *Int. J. Cancer* **46**, 772–778.

550A. Nister, M., Enblad, P., Backstrom, G., Soderman, T., Persson, L., Heidin, C.H., and Westermark, B. (1994) Platelet-derived growth factor (PDGF) in neoplastic and non-neoplastic cystic lesions of the central nervous system and in the cerebrospinal fluid. *Br. J. Cancer* **69**, 952–956.

551. Lewis, T.S., Shapiro, P.S., and Ahn, N.G. (1998) Signal transduction through MAP kinase cascades. *Adv. Cancer Res.* **74**, 49–139.

552. Yart, A., Laffargue, M., Mayeux, P., Chretien, S., Peres, C., Tonks, N., Roche, S., Payrastre, B., Chap, H., and Raynal, P. (2001) A critical role for phosphoinositide 3-kinase upstream of Gab1 and SHP2 in the activation of Ras and mitogen-activated protein kinases by epidermal growth factor. *J. Biol. Chem.* **276**, 8856–8864.

553. Conway, A.-M., Rakhit, S., Pyne, S., and Pyne, N.J. (1999) Platelet-derived-growth-factor stimulation of the p42/p44 mitogen-activated protein kinase pathway in airway smooth muscle: role of pertussis-toxin-sensitive G-proteins, c-Src tyrosine kinases and phosphoinositide 3-kinase. *Biochem. J.* **337**, 171–177.

554. Song, G., Ouyang, G., and Bao, S. (2005) The activation of Akt/PKB signaling pathway and cell survival. *J. Cell. Mol. Med.* **9**, 59–71.

555. Jones, N.R., Rossi, M.L., Gregoriou, M., and Hughes, J.T. (1990) Epidermal growth factor receptor expression in 72 meningiomas. *Cancer* **66**, 152–155.

556. Linggood, R.M., Hsu, D.W., Efird, J.T., and Pardo, F.S. (1995) TGF α expression in meningioma–tumor progression and therapeutic response. *J. Neurooncol.* **26**, 45–51.

556A. Van Setten, G.B., Edstrom, L., Stibler, H., Rasmussen, S., Schultz, G. (1999) Levels of transforming growth factor alpha (TGF-alpha) in human cerebrospinal fluid. *Int. J. Dev. Neurosci.* **17**, 131–134.

557. Shu, J., Lee, J.H., Harwalkar, J.A., Oh-Siskovic, S., Stacey, D.W., and Golubic, M. (1999) Adenovirus-mediated gene transfer of dominant negative Ha-Ras inhibits proliferation of primary meningioma cells. *Neurosurgery* **44**, 579–588.

557A. Gire, V., Marshall, C., and Wynford-Thomas, D. (2000) PI-3-kinase is an essential anti-apoptotic effector in the proliferative response of primary human epithelial cells to mutant RAS. *Oncogene* **19**, 2269–2276.

558. Castelli, M.G., Chiabrando, C., Fanelli, R., Martelli, L., Butti, G., Gaetani, P., and Paoletti, P. (1989) Prostaglandin and thromboxane synthesis by human intracranial tumors. *Cancer Res.* **49**, 1505–1508.

559. Gaetani, P., Butti, G., Chiabrando, C., Danova, M., Castelli, M.G., Riccardi, A., Assietti, R., and Paoletti, P. (1991) A study on the biological behavior of human brain tumors. *J. Neurooncol.* **10**, 233–240.

559A. Ragel, B.T., Jensen, R.L., Gillespie, D.L., Prescott, S.M., and Couldwell, W.T. (2005) Ubiquitous expression of cyclooxygenase-2 in meningiomas and decrease in cell growth following in vitro treatment with the inhibitor celecoxib: potential therapeutic application. *J. Neurosurg.* **103**, 508–517.

559B. Wahl, M.I., Nishibe, S., Suh, P.G., Rhee, S.G., and Carpenter, G. (1989) Epidermal growth factor stimulates phosphorylation of phospholipase C-II independently of receptor internalization and exatracellular calcium. *Proc. Natl. Acad. Sci. USA* **86**, 1568–1572.

560. Buckley, C.T., Sekiya, F., Kim, Y.J., Rhee, S.G., and Caldwell, K.K. (2004) Identification of phospholipase C-γ1 as a mitogen-activated protein kinase substrate. *J. Biol. Chem.* **279**, 41807–41814.

560A. Wang, Z., Gluck, S., Zhang, L., and Moran, M.F. (1998) Requirement for phospholipase C-gamma1 enzymatic activity in growth factor-induced mitogenesis. *Mol. Cell. Biol.* **18**, 590–597.

561. Johnson, M.D., Federspiel, C.F., Gold, L.I., and Moses, H.L. (1992) Transforming growth factor-β and transforming growth factor β-receptor expression in human meningioma cells. *Am. J. Pathol.* **141**, 633–642.

562. Johnson, M.D., Gold, L.I., and Moses, H.L. (1992) Evidence for transforming growth factor-beta expression in human leptomeningeal cells and transforming growth factor-beta-like activity in human cerebrospinal fluid. *Lab. Invest.* **67**, 360–368.

563. Yin, D., Xie, D., Hofmann, W.-K., Zhang, W., Asotra, K., Wong, R., Black, K.L., and Koeffler, H.P. (2003) DNA repair gene O^6-methylguanine-DNA methyltransferase: promoter hypermethylation associated with decreased expression and G:C to A:T mutations of *p53* in brain tumors. *Mol. Carcinog.* **36**, 23–31.

564. Sedzimir, C.B., Frazer, A.K., and Roberts, J.R. (1973) Cranial and spinal meningiomas in a pair of identical twin boys. *J. Neurol. Neurosurg.* **36**, 368–376.

565. Bosch, M.M., Wichmann, W.W., Boltshauser, E., and Landau, K. (2006) Optic nerve sheath meningiomas in patients with neurofibromatosis type 2. *Arch Ophthalmol.* **124**, 379–385.

566. Comtesse, N., Heckel, D., Rácz, A., Brass, N., Glass, B., and Meese, E. (1999) Five novel immunogenic antigens in meningioma: cloning, expression analysis, and chromosomal mapping. *Clin. Cancer Res.* **5**, 3560–3568.

567. de Mesa, R.L., Sierrasesumaga, L., de Cerain, A.L., Calasanz, M.J., and Patino-Garcia, A. (2005) Pediatric meningiosarcoma: clinical evolution and genetic instability. *Pediatr. Neurol.* **32**, 352–354.

568. Espinosa, A.B., Tabernero, M.D., Maíllo, A., Sayagués, J.M., Ciudad, J., Merino, M., Alguero, M.C., Lubombo, A.M., Sousa, P., Santos-Briz, A., and Orfao, A. (2006) The cytogenetic relationship between primary and recurrent meningiomas points to the need for new treatment strategies in cases at high risk of relapse. *Clin. Cancer Res.* **12**, 772–780.

569. Hartmann, C., Sieberns, J., Gehlhaar, C., Simon, M., Paulus, W., and von Deimling, A. (2006A) *NF2* mutations in secretory and other rare variants of meningiomas. *Brain Pathol.* **16**, 15–19.

570. Hirashima, Y., Hayashi, N., Fukuda, O., Ito, H., Endo, S., and Takaku, A. (1998) Platelet-activating factor and edema surrounding meningiomas. *J. Neurosurg.* **88**, 304–307.

571. Johnson, M.D., Woodard, A., Okediji, E.J., Toms, S.A., and Allen, G.S. (2002) Lovastatin is a potent inhibitor of meningioma cell proliferation: evidence for inhibition of a mitogen associated protein kinase. *J. Neurooncol.* **56**, 133–142.

572. Okamoto, H., Li, J., Vortmeyer, A.O., Jaffe, H., Lee, Y.S., Glasker, S., Sohn, T.S., Zeng, W., Ikejiri, B., Proescholdt, M.A., Mayer, C., Weil, R.J., Oldfield, E.H., and Zhuang, Z. (2006) Comparative proteomic profiles of meningioma subtypes. *Cancer Res.* **66**, 10199–10204.

573. Olivera, M., Martínez, C., Molina, J.A., Alonso-Navarro, H., Jiménez-Jiménez, F.J., García-Martín, E., Benítez, J., and Agúndez, J.A.G. (2006) Increased frequency of rapid acetylator genotypes in patients with brain astrocytoma and meningioma. *Acta Neurol. Scand.* **113**, 322–326.

573A. Surace, E.I., Lusis, E., Murakami, Y., Scheithauer, B.W., Perry, A., and Gutmann, D.H. (2004) Loss of tumor suppressor in lung cancer-1 (TSLC1) expression in meningioma correlates with increased malignancy grade and reduced patient survival. *J. Neuropathol. Exp. Neurol.* **63**, 1015–1027.

574. Rajaraman, P., De Roos, A.J., Stewart, P.A., Linet, M.S., Fine, H.A., Shapiro, W.R., Selker, R.G., Black, P.M., and Inskip, P.D. (2004) Occupation and risk of meningioma and acoustic neuromas in the United States. *Am. J. Ind. Med.* **45**, 395–407.

575. Scoles, D.R., Yong, W.H., Qin, Y., Wawrowsky, K., and Pulst, S.M. (2006) Schwannomin inhibits tumorigenesis through direct interaction with the eukaryotic initiation factor subunit c (eIF3c). *Hum. Mol. Genet.* **15**, 1059–1070.

576. Smidt, M., Kirsch, I., and Ratner, L. (1990) Deletion of Alu sequences in the fifth c-*sis* intron in individuals with meningiomas. *J. Clin. Invest.* **86**, 1151–1157.

577. Tirakotai, W., Mennel, H.-D., Celik, I., Hellwig, D., Bertalanffy, H., and Riegel, T. (2006) Secretory meningioma: immunohistochemical findings and evaluation of mast cell infiltration. *Neurosurg. Rev.* **29**, 41–48.

578. van Tilborg, A.A., Morolli, B., Giphart-Gassler, M., de Vries, A., van Geenen, D.A., Lurkin, I., Kros, J.M., and Zwarthoff, E.C. (2006) Lack of genetic and epigenertic changes in meningiomas without NF2 loss. *J. Pathol.* **208**, 564–573.

579. von Randow, A.J., Schindler, S., and Tews, D.S. (2006) Expression of extracellular matrix-degrading proteins in classic, atypical, and anaplastic meningiomas. *Pathol. Res. Pract.* **202**, 365–372.

580. Wada, K., Maruno, M., Suzuki, T., Kagawa, N., Hashiba, T., Fujimoto, Y., Hashimoto, N., Izumoto, S., and Yoshimine, T. (2005) Chromosomal and genetic abnormalities in benign and malignant meningiomas using DNA microarray. *Neurol. Res.* **27**, 747–754.

581. Evans, D.G.R., Watson, C., King., A., Wallace, A.J., and Baser, M.E. (2005) Multiple meningiomas: differential involvement of the *NF2* gene and its protein rarely occurring in meningothelial meningiomas. *J. Neurosurg.* **94**, 111–117.

582. Slavc, I., MacCollin, M.M., Dunn, M., Jones, S., Sutton, L., Gusella, J.F., and Biegel, J.A. (1995) Exon scanning for mutations of the NF2 gene in pediatric ependymomas, rhabdoid tumors and meningiomas. *Int. J. Cancer* **64**, 243–247.

583. Sasaki, T., Hankins, G.R., and Helm, G.A. (2003) Comparison of gene expression profiles between frozen original meningiomas and primary cultures of the meningiomas by GeneChip. *Neurosurgery* **52**, 892–899.

584. Baser, M.E., and Poussaint, T.Y. (2006) Age associated increase in the prevalence of chromosome 22q loss of heterozygosity in histological subsets of benign meningioma. *J. Med. Genet.* **43**, 285–287.

585. Zankl, H., Singer, H., and Zang, K.D. (1971) Cytological and cytogenetical studies on brain tumors. II. Hyperdiploidy, a rare event in human primary meningiomas. *Humangenetik.* **11**, 253–257.

586. Scholz, M., Gottschalk, J., Striepecke, E., Firsching, R., Harders, A., and Füzesi, L. (1996) Intratumorous heterogeneity of chromosomes 10 and 17 in meningiomas using non-radioactive in situ hybridization. *J. Neurosurg.* **40**, 17–23.

587. Ketter, R., Kim, Y.-J., Storck, S., Rahnenführer, J., Romeike, B.F.M., Steudel, W.-I., Zang, K.D., and Henn, W. (2007) Hyperdiplity defines a distinct cytogenetic entity of meningiomas. *J. Neurooncol.* **83**, 213–221.

588. Verheijen, F.M., Sprong, M., Kloosterman, J.M., Blaauw, G., Thijssen, J.H., and Blankenstein, M.A. (2002) TP53 mutations in human meningiomas. *Int. J. Biol. Markers* **17**, 42–48.

589. Begnami, M.D., Rushing, E.J., Santi, M., and Quezado, M. (2007) Evaluation of *NF2* gene deletion in pediatric meningiomas using chromogenic in situ hybridization. *Int. J. Surg. Pathol.* **15**, 110–115.

590. Buccoliero, A.M., Castiglione, F., Degl'Innocenti, D.R., Gheri, C.F., Garbini, F., Taddei, A., Ammannati, F., Mennonna, P., and Taddei, G.L. (2007) *NF2* gene expression in sporadic meningiomas: relation to grades or histotypes real-time-pCR study. *Neuropathology* **27**, 36–42.

591. Al-Khalaf, H.H., Lach, B., Allam, A., AlKhani, A., Alrokayan, S.A., and Aboussekhra, A. (2007) The p53/p21 DNA damage-signaling pathway is defective in most meningioma cells. *J. Neurooncol.* **83**, 9–15.

592. Brunner, E.C., Romeike, B.F., Jung, M., Comtesse, N., and Meese, E. (2006) Altered expression of beta-catenin/E-cadherin in meningiomas. *Histopathology* **49**, 178–187.

593. Carroll, T., Maltby, E., Brock, I., Royds, J., Timperley, W., and Jellinek, D. (1999) Meningiomas, dicentric chromosomes, gliomas, and telomerase activity. *J. Pathol.* **188**, 395–399.

594. Maiuri, F., Del Basso De Caro, M., Esposito, F., Cappabianca, P., Strazzullo, V., Pettinato, G., and de Divitiis, E. (2007) Recurrences of meningiomas: predictive value of pathological features and hormonal and growth factors. *J. Neurooncol.* **82**, 63–68.

595. Malmer, B.S., Feychting, M., Lönn, S., Lindström, S., Grönberg, H., Ahlbom, A., Schwartzbaum, J., Auvinen, A., Collatz-Christensen, H., Johansen, C., Kiuru, A., Mudie, N., Salminen, T., Schoemaker, M.J., Swerdlow, A.J., and Henriksson, R. (2007) Genetic variation in p53 and ATM haplotypes and risk of glioma and meningioma. *J. Neurooncol.* **82**, 229–237.

596. Pistolesi, S., Boldrini, L., Gisfredi, S., De Ieso, K., Camacci, T., Caniglia, M., Lupi, G., Leocata, P., Basolo, F., Pingitore, R., Parenti, G., and Fontanini, G. (2004) Angiogenesis in intracranial meningiomas: immunohistochemical and molecular study. *Neuropathol. Appl. Neurobiol.* **30**, 118–125.

597. Lewy-Trenda, I., Omulecka, A., Janczukowicz, J., and Papierz, W. (2003) The morphological analysis of vasculature and angiogenic potential in meningiomas: immunoexpression of CD31 and VEGF antibodies. *Folia Neuropathol.* **41**, 149–153.

598. Shono, T., Inamura, T., Torisu, M., Suzuki, S.O., and Fukui, M. (2000) Vascular endothelial growth factor and malignant transformation of a meningioma: case report. *Neurol. Res.* **22**, 189–193.

599. Christov, C., Lechapt-Zalcman, E., Adle-Biassette, H., Nachev, S., and Gherardi, R.K. (1999) Vascular permeability factor/vascular endothelial growth factor (VPF/VEGF) and its receptor flt-1 in microcystic meningiomas. *Acta Neuropathol.* **98**, 414–420.

600. Barresi, V., Cerasoli, S., Vitarelli, E., and Tuccari, G. (2007) Density of microvessels positive for CD105 (endoglin) is related to prognosis in meningiomas. *Acta Neuropathol.* **114**, 147–156.

601. Kalkanis, S.N., Carroll, R.S., Zhang, J., Zamani, A.A., and Black, P.McL. (1996) Correlation of vascular endothelial growth factor messenger RNA expression with peritumoral vasogenic cerebral edema in meningiomas. *J. Neurosurg.* **85**, 1095–1101.

602. Wrobel, G., Roerig, P., Kokocinski, F., Neben, K., Hahn, M., Reifenberger, G., and Lichter, P. (2005) Microarray-based gene expression profiling of benign, atypical and anaplastic meningiomas identifies novel genes associated with meningioma progression. *Int. J. Cancer* **114**, 249–256.

603. Barresi, V., Cerasoli, S., Paioli, G., Vitarelli, E., Giuffrè, G., Guiducci, G., Tuccari, G., and Barresi, G. (2006) Caveolin-1 in meningiomas: expression and clinico-pathological correlations. *Acta Neuropathol.* **112**, 617–626.

604. Kim, Y.-J., Ketter, R., Steudel, W.-I., and Feiden, W. (2007) Prognostic significance of the mitotic index using the mitosis marker anti-phosphohistone H3 in meningioma. *Am. J. Clin. Pathol.* **128**, 118–125.

605. Yoo, H., Baia, G.S., Smith, J.S., McDermott, M.W., Bollen, A.W., VandenBerg, S.R., Lamborn, K.R., and Lai, A. (2007) Expression of the hypoxia marker carbonic anhydrase 9 is associated with anaplastic phenotypes in meningiomas. *Clin. Cancer Res.* **13**, 68–75.

606. Yamashima, T., Sakuda, K., Tohma, Y., Yamashita, J., Oda, H., Irikura, D., Eguchi, N., Beuckmann, C.T., Kanaoka, Y., Urade, Y., and Hayaishi, O. (1997) Prostaglandin D synthase (b-trace) in human arachnoid and meningioma cells: roles as a cell marker or in cerebrospinal fluid absorption, tumorigenesis, and calcification process. *J. Neurosci.* **17**, 2376–2382.

607. Adams, E.F., Schrell, U.M.H., Fahlbusch, R., and Thierauf, P. (1990) Hormonal dependency of cerebral meningiomas. Part 2: *In vitro* effects of steroids, bromocriptine, and epidermal growth factor on growth in meningiomas. *J. Neurosurg.* **73**, 750–755.

608. Hunt, D.P., Freeman, A., Morris, L.S., Burnet, N.G., Bird, K., Davies, T.W., Laskey, R.A., and Coleman, N. (2002) Early recurrence of benign meningioma correlates with expression of mini-chromosome maintenance-2 protein. *Br. J. Neurosurg.* **16**, 10–15.

609. Strik, H.M., Strobelt, I., Pietsch-Breitfeld, B., Iglesias-Rozas, J.R., Will, B., Meyermann, R. (2002) The impact of proges-

terone receptor expression on relapse in the long-term clinical course of 93 benign meningiomas. *In Vivo* **16,** 265–270.

610. Konstantinidou, A.E., Korkolopoulou, P., Kavatzas, N., Mahera, H., Thymara, I., Kotsiakis, X., Perdiki, M., Patsouris, E., and Davaris, P. (2003) Mitosin, a novel marker of cell proliferation and early recurrence in intracranial meningiomas. *Histol. Histopathol.* **18,** 67–74.

611. Ohta, M., Iwaki, T., Kitamoto, T., Takeshita, I., Tateishi, J., and Fukui, M. (1994) MIB1 staining index and scoring of histologic features in meningioma. *Cancer* **74,** 3176–3189.

612. Abramovich, C.M., and Prayson, R.A. (1999) Histopathologic features and MIB-1 labeling indices in recurrent and nonrecurrent meningiomas. *Arch. Pathol. Lab. Med.* **123,** 793–800.

613. Nakaguchi, H., Fujimaki, T., Matsuno, A., Matsuura, R., Asai, A., Suzuki, I., Sasaki, T., and Kirino, T. (1999) Postoperative residual tumor growth of meningioma can be predicted by MIB-1 immunohistochemistry. *Cancer* **85,** 2249–2254.

614. Ketter, R., Urbschat, S., Henn, W., Feiden, W., Beerenwinkel, N., Lengauer, T., Steudel, W.-I., Zang, K.D., and Rahnenführer, J. (2007) Application of oncogenetic trees mixtures as a biostatistical model of the lconal cytogenetic evolution of meningiomas. *Int. J. Cancer* **121,** 1473–1480.

615. Niedermayer, I., Kolles, H., Zang, K.D., and Feiden, W. (1996) Characterization of intermediate type (WHO "atypical") meningiomas. *Clin. Neuropathol.* **15,** 330–336.

616. Ildan, F., Tuna, M., Göçer, A.I, Boyar, B., Bagdatoglu, H., Sen, O., Haciyakupoglu, S., and Burgut, H.R. (1999) Correlation of the relationships of brain-tumor interfaces, magnetic resonance imaging, and angiographic findings to predict cleavage of meningiomas. *J. Neurosurg.* **91,** 384–390.

617. Mantle, R.E., Lach, B., Delgado, M.R., Baeesa, S., and Bélanger, G. (1999) Predicting the probability of meningioma recurrence based on the quantity of peritumoral brain edema on computerized tomography scanning. *J. Neurosurg.* **91,** 375–383.

618. Rahnenführer, J., Beerenwinkel, N., Schulz, W.A., Hartmann, C., von Deimling, A., Wullich, B., and Lengauer, T. (2005) Estimating cancer survival and clinical outcome based on genetic tumor progression scores. *Bioinformatics* **21,** 2438–2446.

619. Shchors, K., and Evan, G. (2007) Tumor angiogenesis: cause or consequence of cancer? *Cancer Res.* **67,** 7059–7061.

619A. Krayenbühl, N., Pravdenkova, S., and Al-Mefty, O. (2007) De novo versus transformed atypical and anaplastic meningiomas: comparisons of clinical course, cytogenetics, cytokinetics, and outcome. *Neurosurgery* **61,** 495–504.

620. Woo, K.-S., Sung, K.-S., Kim, K.-U., Shaffer, L.G., and Han, J.-Y. (2008) Characterization of complex chromosome aberrations in a recurrent meningioma combining standard cytogenetic and array comparative genomic hybridization techniques. *Cancer Genet. Cytogenet.* **180,** 56–59.

621. Hansson, C.M., Buckley, P.G., Grigelioniene, G., Piotrowski, A., Hellström, A.R., Mantripragada, K., Jarbo, C., Mathiesen, T., and Dumanski, J.P. (2007) Comprehensive genetic and epigenetic analysis of sporadic meningioma for macromutations on 22q and micromutations within the *NF2* locus. *BMC Genomics* **8,** 16–31.

622. Ishino, S., Hashimoto, N., Fushiki, S., Date, K., Mori, T., Fujimoto, M., Nakagawa, Y., Ueda, S., Abe, T., and Inazawa, J. (1998) Loss of material from chromosome arm 1 p during malignant progression of meningioma revealed by fluorescent in situ hybridization. *Cancer* **83,** 360–366.

4

Hemangioblastoma of the Central Nervous System

Summary Chapter 4 deals with hemangioblastoma of the central nervous system. Although these tumors are benign, they may be associated with serious complications due to tumor edema and recurrence. Hemangioblastomas are the most common tumors in von Hippel-Lindau (VHL) disease and hence carry the changes in the *VHL* gene seen in this disease. A high proportion of sporadic hemangioblastomas show alterations of the *VHL* gene. Thus, the *VHL* gene and its protein product pVHL play a crucial role in hemangioblastoma tumorigenesis. Cytogenetic changes involving 3p and 6q have been seen in hemangioblastoma, but these changes are not specific for these tumors. Similar results have been obtained with comparative genomic hybridization and loss of heterozygosity studies. Molecular studies of areas other than the VHL pathway on hemangioblastoma, although relatively few, are also presented.

Keywords hemangioblastoma · *VHL* gene · pVHL · 3p and 6q alterations.

Introduction

Although capillary hemangioblastomas may occur at anatomic sites other than the central nervous system (CNS), in this chapter the term "hemangioblastoma" refers only to those of the CNS. Hemangioblastoma are benign, highly vascularized neoplasms which occur as sporadic tumors (75%) or as a manifestation of the von Hippel-Lindau (VHL) disease (25%) (1,2) and they account for 2% of all intracranial tumors, for 7–10% of primary posterior fossa tumors (3–6), and for 2% of primary spinal cord tumors (7–11). About 50% of hemangioblastomas occur in the spinal cord, 37% in the cerebellum and 10% in the brainstem (12–15). The mean age of onset of hemangioblastoma and range of its frequency at various anatomic sites are shown in Table 4.1.

Sporadic hemangioblastomas occur predominantly in the cerebellum, whereas VHL-associated hemangioblastomas are localized not only in the cerebellum but also in the brain stem and spinal cord; supratentorial lesions are rare. Multiple tumors at various sites are often and almost exclusively found in VHL patients (Figure 4.1) (16). Hemangioblastomas may be cystic (65%) or present as solid (35%) tumors.

Hemangioblastomas are slow-growing, benign tumors frequently associated with cysts in the cerebellum, brain stem, or spinal cord. Symptoms generally arise from impaired cerebral spinal fluid flow, due to a cyst or solid tumor mass, resulting in an increase of intracranial pressure. Hemangioblastomas may produce erythropoietin (17), and thus cause secondary polycythemia.

Hemangioblastomas are characterized histologically by two main components: large vacuolated stromal cells and an abundant capillary network (Figure 4.2). Cellular (high recurrence type) and reticular variants (recurrence-free) are distinguished on the basis of the abundance of the stromal cell component (18). Histologically, hemangioblastomas consist of a rich vascular plexus that is surrounded by polygonal stromal cells (Figure 4.3). The stromal cells constitute the neoplastic cells of hemangioblastoma (19, 21, 21A), and their nuclei may vary in size, with occasional atypical and hyperchromatic forms. However, their most striking morphological feature consists of numerous lipid-containing vacuoles, resulting in the typical "clear cell" morphology of hemangioblastomas. This feature can sometimes lead to difficult diagnostic problems, i.e., differentiating between capillary hemangioblastomas and metastatic renal cell carcinoma (RCC) (22). The stromal and capillary endothelial cells differ significantly in their antigen expression patterns. Stromal cells lack endothelial cell markers, such as von Willebrand factor and CD34, and they do not express endothelium-associated adhesion molecules, such as CD31 (platelet/endothelial cell adhesion molecule 1) (23). Unlike endothelial cells, stromal cells express neuron-specific enolase, neural cell adhesion molecule, and ezrin (23, 24).

TABLE 4.1. Frequency and age at onset of tumors in VHL disease.

	Mean (range) age of onset (years)	Frequency in patients (%)
CNS		
Craniospinal hemangioblastomas		
Cerebellum	33 (9–78)	44–72
Brain stem	32 (12–46)	10–25
Spinal cord	33 (12–66)	13–50
Lumbosacral nerve roots	Unknown	<1
Supratentorial	Unknown	<1
Retinal hemangioblastomas	25 (1–67)	25–60
Endolymphatic sac tumors	22 (12–50)	10
Visceral		
Renal cell carcinoma or cysts	39 (16–67)	25–60
Pheochromocytomas	30 (5–58)	10–20
Pancreatic tumor or cyst	36 (5–70)	35–70
Epididymal cystadenoma	Unknown	25–60
Broad ligament cystadenoma	Unknown (16–46)	Unknown

Based on Lonser et al. 2003 (14).

Vimentin is the major intermediate filament expressed by stromal cells, which usually do not express glial fibrillary acidic protein (24). Some characteristics of the stromal cells and endothelial cells of hemangioblastoma are shown in Table 4.2.

The stromal cells express high levels of mRNA and protein for epidermal growth factor receptor (EGFR), but the *EGFR* gene is not amplified in hemangioblastomas (26, 27). A subpopulation of the stromal cells also expresses transforming growth factor (TGF)-α, an EGFR ligand, which suggests an autocrine or paracrine TGF-α loop (26).

Vascular endothelial growth factor (VEGF), a prime regulator of physiological and pathological angiogenesis (28, 29), is highly expressed in the stromal cells, with corresponding endothelial expression of its receptors, VEGFR-1 and -2 (30), and the endothelial cell receptor Tie-1 (31). A subpopulation of stromal cells also may express high levels of the endothe-

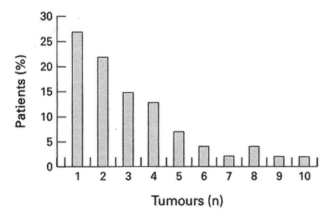

FIGURE 4.1. Incidence of multiple hemangioblastomas in familial (VHL) settings. Even though most of the patients had multiple hemangioblastomas, 27% had single tumors (from ref. 62, with permission).

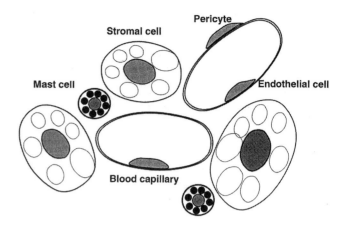

FIGURE 4.2. Schematic representation of hemangioblastomas: the stromal cells are thought to constitute the major neoplastic component with the high vascularity of these tumors being due to proliferation of the blood capillaries stimulated by VEG and VEGR (from ref. 41, with permission).

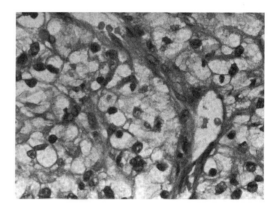

FIGURE 4.3. Histological view of hemangioblastoma showing the vascular framework and plump, clear stromal cells (from ref. 142, with permission).

lial receptors, suggesting paracrine and autocrine growth factor-receptor loops (31). Most hemangioblastomas contain numerous mast cells; these are of interest because VEGF binds to heparin. The endothelial cells of hemangioblastomas

TABLE 4.2. Some findings in stromal cells (SC) and endothelial cells (EC) of hemangioblastoma.

	References
Transthyretin and transferrin in SC	*Bleistein et al 2002* (148)
Ezrin expression in SC	*Böhling et al 1996* (25)
Expression of adhesion molecules in SC vs. EC	*Böhling et al 1996* (23)
Localization of γ-enolase in SC	*Feldenzer et al 1987* (149)
Angiogenic histogenesis in SC	*Lach et al 1999* (150)
Immunohistochemical and ultrastructural studies of SC	*Omulecka et al 1995* (151)
Immunoperoxidase staining in SC	*Tanimura et al 1984* (152)

also express receptors for other angiogenic growth factors, including platelet-derived growth factors (27). In addition to VEGF, erythropoietin and the transcription factor hypoxia-inducible factor (HIF)-2α are up-regulated in the stromal cells (32, 33). Aquaporins, factors involved in the passage of water across cell membranes, were shown to be overexpressed in the stromal, but not the endothelial cells of hemangioblastomas, particularly the cystic group of tumors (34).

Although hemangioblastomas are infrequent tumors, they are an important clinical entity because the morbidity and mortality associated with them can be reduced significantly, if these lesions are appropriately diagnosed and treated (6, 7, 12); surgical outcomes are generally favorable, although relapse or multifocal occurrence may still produce serious complications. Molecular genetic analysis is a safe and specific approach in confirming or excluding VHL disease in patients with hemangioblastoma (35). Early identification of *VHL* mutation gene carriers is important in reducing disease morbidity and mortality.

Hemangioblastomas can be life-threatening due to the associated edema of these tumors, their large cystic components when located in the posterior fossa or paraplegia if localized in the spinal canal (15). An early diagnosis, however, is the key for adequate management of such patients and their relatives. Since the *VHL* tumor suppressor gene was identified (36), molecular genetic testing for disease-predisposing mutations is possible. Because there are many therapeutic options for the broad range of lesions associated with VHL disease, molecular genetic testing should be offered to patients at risk. The slow growth of hemangioblastoma is reflected in the low mitotic rate and the low number of tumor cells entering the cell cycle, with a Ki-67 index of <1%.

VHL Disease

The attention and space given in this chapter to VHL disease, and particularly to changes in its responsible gene, *VHL*, are a reflection of the germline mutations of this gene causing not only the VHL disease and its tumors, including hemangioblastomas and RCC, but also to loss of heterozygosity (LOH) of *VHL* in the tumors per se being found to be responsible for the development of a significant percentage of sporadic hemangioblastomas (37–39).

VHL disease is inherited through an autosomal dominant trait and characterized by the development of capillary hemangioblastomas of the CNS and retina, clear cell renal carcinoma (RCC), pheochromocytoma (PCC), pancreatic islet and inner ear tumors, and cysts and cystadenomas in the kidney, pancreas, and epididymis (16,40–42). The syndrome is caused by germline mutations of the *VHL* tumor suppressor gene, located on 3p25 (36,43), with 90% penetrance by 65 years of age (39A). VHL is estimated to occur at a rate of 1:36,000 individuals (1,44,45). The clinical picture varies among VHL patients and several subtypes of VHL have been formulated (Table 4.3) (14,16).

Hemangioblastomas are the most common tumors in VHL, affecting at least 60–80% of all patients (15, 46–48). The average age of presentation for hemangioblastoma in VHL disease is 33 years (Figure 4.4) (15). The tumors arise anywhere along the craniospinal axis and are often associated with edema or cysts (cysts occur in 30–80% of hemangioblastomas), or both. A small percentage of hemangioblastomas may show aggressive behavior, even malignant facets, and some of the characteristics of these tumors are listed in Table 4.4.

The clinical diagnosis of VHL disease is based on the presence of capillary hemangioblastomas in the CNS or retina, the presence of one of the typical VHL-associated tumors, or a previous family history (Table 4.1). Retinal hemangioblastomas are typically the first manifestations of VHL disease (49–53).

In a retrospective series of 40 patients with hemangioblastomas (48), various clinical aspects were compared in patients with sporadic hemangioblastomas versus those with VHL disease (39%) (Table 4.5). Eight patients had multiple lesions, and five of these and five others had spinal cord hemangioblastomas. This percentage of VHL patients (38%) is higher than reported by others (54–57), which may be due to referral bias in the authors' institution. The diagnosis of VHL was clinically based. The authors (48) concluded that although surgical outcomes for hemangioblastomas are generally favorable, VHL patients nevertheless require lifelong surveillance to detect these lesions before they become symptomatic, in agreement with findings in the literature (9, 56, 58), reemphasizing the necessity of long-term follow-up in patients diagnosed with VHL disease. In this series of cases (48), 67% of patients with VHL disease

TABLE 4.3. Association of VHL disease types with risk of VHL-associated tumors and with types of germline mutations in the *VHL* gene.

VHL disease type	Risk of tumor types observed in VHL families			Germline VHL mutation types
	HB	RCC	PCC	
1	High	High	Low	Full gene deletions, partial gene deletions, nonsense mutations, and splice acceptor mutations
2A	High	Low	High	Missense mutations
2B	High	High	High	Partial gene deletions, nonsense mutations, and missense mutations
2C	No	No	High	Missense mutations

HB, hemangioblastoma (from Bryant et al 2003 (153)).

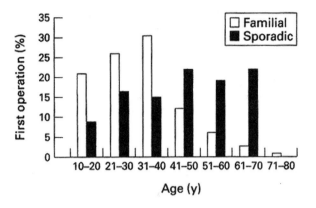

FIGURE 4.4. Age at surgical presentation of patients with familial (VHL) or sporadic hemangioblastoma. The familial cases develop and require surgery for hemangioblastoma at a much earlier age than patients with sporadic tumors (from ref. 62, with permission).

TABLE 4.4. Hemangioblastomas with aggressive behavior.

	Reference
Recurrent with glial differentiation	*Adams and Hilton 2002* (154)
Cerebellar with extensive dissemination	*Hande and Nagpal 1996* (155)
Cerebellar with dissemination in the cranium, brain stem and spinal cord	*Kato et al 2005* (156)
Malignant spread	*Mohan et al 1976* (157)
Natural history of cerebellar hemangioblastoma	*Slater et al 2003* (158)
Disseminated hemangioblastoma of CNS without VHL	*Weil et al 2002* (159)

developed new CNS lesions, with a calculated rate of one new lesion developing every 2.1 years. Similar to previously published series, patients with VHL disease presented with hemangioblastomas at a younger age than those with sporadic disease (38 vs. 45 years) (Figure 4.5), often with multiple lesions (53%), or spinal hemangioblastomas (47 vs. 12%), or with new lesions throughout the follow-up period. Screening with magnetic resonance imaging identified >75% of presymptomatic new lesions, leading to appropriate treatment. The authors (48) found no difference in surgical outcome between patients with sporadic disease compared with those with VHL disease, and additional surgical operations required in VHL patients were not associated with an increased risk of serious complications. These findings may serve as important guidelines for neurosurgeons treating the neoplastic aspects of patients with VHL disease (59).

Cryptic germline mutations of the *VHL* gene may be present in rare patients with hemangioblastomas as the only manifestation, as shown in a study in which three of 69 patients had such a germline mutation (60). Germline mutations were found in two of four cases with multiple hemangioblastomas (60). These results reiterate the value of genetic testing in patients with hemangioblastoma, even in the absence of VHL disease (61).

TABLE 4.5. Some characteristics of patients with sporadic hemangioblastoma versus VHL disease.

	Sporadic disease	VHL syndrome
No. of patients (% of total)	25 (62)	15 (38)
No. of males	14 (56)	7 (47)
No. of female	11 (44)	8 (53)
Mean age at presentation of neurological symptoms	45 years	38 years
No. of patients with each lesion		
Cerebellar hemangioblastoma	17 (68)	8 (53)
Medullary hemangioblastoma	4 (16)	1 (7)
Spinal hemangioblastoma	3 (12)	7 (47)
Supratentorial hemangioblastoma	1 (4)	2 (13)

Based on Conway et al. 2001 (48).

Molecular genetic analysis for *VHL* gene mutations has been reported to be superior to clinical information in the diagnosis of VHL disease (48, 62). This analysis provides an essential and reliable diagnosis and enables screening for other manifestations of VHL disease, such as RCC; in addition, the analysis provides prognostic information. VHL families may be characterized by the presence or absence of PCC (Table 4.3). In contrast to type 2 families, who have been

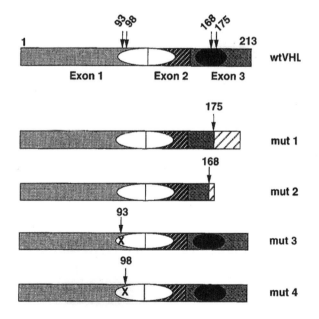

FIGURE 4.5. Schematic representation of the domain structure wild-type pVHL (wtVHL) and some of its mutated forms (mut 1–4). The three exons of *VHL* are shown by different shadowing, with the ovals signifying binding domains (*black*, elongin). The numbers above give the amino acid residues. The *arrows* indicate residues involved in mutated pVHL: mut 1 is a 1-bp deletion creating a frameshift after residue 175; mut 2 is a 4-bp insertion creating a frameshift after residue 168; mut 3 and mut 4 are missense mutations, Gly-93-Asp and Tyr-98-His, respectively. The mut 1 and mut 2 are devoid of elongin binding domain; mut 3 and mut 4 have intact functional elongin binding domains (from *ref.* 96, with permission).

found to have missense mutations, most type 1 families are affected by deletions or premature terminal mutations (42). It is possible that differences in the nature of the mutations in VHL disease account for differences in tumor phenotype.

Few characteristics of patients harboring hemangioblastomas were predictive of the VHL disease. Studies mentioned previously (48, 56) have demonstrated that patients with VHL disease tend to present with hemangioblastomas at a younger age than do patients with sporadic disease. A high degree of suspicion for VHL disease should be raised in young patients with hemangioblastomas, especially in those with spinal hemangioblastomas. In addition, multiple lesions are essentially diagnostic of VHL.

Although the percentage of VHL disease-associated hemangioblastomas decreases after the fourth decade of life and is infrequent in patients without other symptomatic lesions and a negative family history, it is recommended that every patient with hemangioblastoma be screened for *VHL* germline mutations. This family history provides key information and enables screening for extraneurological tumors in these patients and investigation of a patient's family to ameliorate and manage the VHL disease.

VHL Gene and Its Protein pVHL

As with most genetic and metabolic pathways affecting various cellular events and functions, those related to the *VHL* gene, its protein product pVHL and associated factors, are no exception in their complexity. In agreement with the saying that "a picture tells more than a thousand words," the *VHL* pathway and others involved in hemangioblastoma and RCC development have been presented graphically in several figures in this chapter, by using information published by authors intimately involved with various aspects of the field. The topics covered are listed in Tables 4.6 and 4.7.

Because almost all patients with VHL disease develop hemangioblastomas and a high proportion are affected with RCC, and the frequent involvement of the *VHL* gene in sporadic hemangioblastomas, the *VHL* pathway leading to the development of these tumors is given much attention here.

The *VHL* gene is expressed in a variety of human tissues, e.g., the epithelial cells of the skin; the gastrointestinal,

TABLE 4.6. Some pathways for hemangioblastoma development.

1. Germline mutations of the *VHL* gene cause VHL disease, predisposing the individuals to the development of multiple tumors, most frequently hemangioblastoma.
2. Alterations involving the *VHL* gene are frequently detected by mutation, SCCP or LOH analysis in 10–50% of sporadic hemangioblastomas.
3. Hypermethylation of a normally unmethylated CpG island of the *VHL* gene may inactivate the gene, without *VHL* mutations and lead to hemangioblastoma formation.
4. Possible involvement of tumor suppressor genes in the genesis of sporadic hemangioblastoma.

TABLE 4.7. Some functions of the *VHL* gene and pVHL.

1. Down-regulation of hypoxia-inducible mRNAs	*Maxwell et al 1999* (91)
2. Proper assembly of the extracellular fibronectin matrix	*Ohh et al 1998* (98)
3. Regulation of exit from the cell cycle	*Pause et al 1998* (93)
4. Regulation of expression of carbonic anhydrases 9 and 12	*Ivanov et al 1998* (96)
5. Transcription elongation	*Duan et al 1995* (99)
6. Regulation of VEGF and angiogenesis	*Pause et al 1999* (160)
	Vortmeyer et al 2003 (161)

respiratory, and urogenital tracts and endocrine and exocrine organs (63, 64). In the CNS, immunoreactivity for pVHL is prominent in neurons, including the Purkinje cells of the cerebellum (65, 66).

Mutational inactivation of the *VHL* gene (Figures 4.5 and 4.6) in affected family members is responsible for their genetic susceptibility to RCC and hemangioblastomas, but the mechanisms by which pVHL causes neoplastic transformation, has not been totally elucidated. Several signaling pathways seem to be involved (67), one of which points to a role of pVHL in protein degradation. The domain of the *VHL* gene involved in the binding to elongins is frequently mutated in VHL-associated neoplasms (68) (see below).

A large spectrum of germline mutations spread over all three exons of the *VHL* tumor suppressor gene has been detected in VHL patients (69–72). Missense mutations are most common, but nonsense mutations, microdeletions/insertions, splice site mutations, and large deletions also occur (72–74). The spectrum of clinical manifestations of the VHL disease (Table 4.3) probably reflects the type of germline mutation. Phenotypes are based on the absence (type 1) or presence (type 2) of PCC. VHL type 2 is usually associated with missense mutations and subdivided on the presence (type 2A) or absence (type 2B) of RCC (1, 75–77).

According to its function as a tumor suppressor gene, *VHL* gene mutations are also common in sporadic hemangioblastomas and RCC (69, 70).

Germline mutations within the *VHL* gene have been identified in up to 100% of affected families with VHL disease. The initiation of tumor formation in VHL disease is generally thought to require the biallelic inactivation of the *VHL* gene by two genetic alterations to lose its tumor suppressor activity, the so-called two-hit events in tumor development (Figure 4.7). In VHL disease, the first hit is an inherited germline mutation in the *VHL* gene, which is present in one allele in each cell of the body. The second hit is a somatic DNA alteration (e.g., deletion or mutation of the remaining wild-type allele), which is acquired during the patient's lifetime and is present only in the tumor cells. Known somatic inactivating mechanisms include recombination events resulting in a loss of heterozygosity (LOH of the *VHL* gene), and to a lesser extent intragenic point mutations.

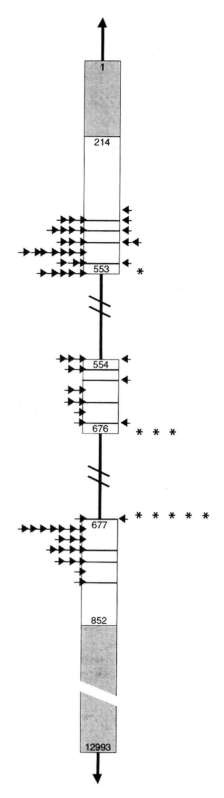

Hypermethylation of normally unmethylated sites of the promoter that are rich in 5'-CG-3' dinucleotides, so called CpG islands, has been suggested as an epigenetic mechanism of tumor suppressor gene inactivation (78, 79). Genome-wide changes in methylation pattern are known to occur in all forms of neoplasia. Although nonisland CpGs become hypomethylated, certain CpG islands become densely hypermethylated. Furthermore, in normal tissues, extensive methylation of promoter region CpG islands is associated with transcriptional silencing. This is well known for imprinted alleles and genes on the inactive X chromosome (80). The role of tumor suppressor gene hypermethylation, however, is still unclear in many tumors, including hemangioblastoma.

Investigations on hypermethylation of the *VHL* gene in hemangioblastomas have yielded diverse results, and they were usually based on small series of cases. In one study (81), hypermethylation in four of eight investigated VHL-related hemangioblastomas was present. By contrast, another study (82) investigated eight (three VHL-associated and five sporadic) hemangioblastomas, and it did not detect hypermethylation of the *VHL* gene in any of the tumors. In agreement with the second study, another study (83) could not find any hypermethylation of the *VHL* gene promoter region. All groups (81–83) used methods that were similar. However, the particular enzymes used were different (SmaI and Not1 I vs. EheI) (83). The different results could be explained by different levels of methodological sensitivity. Quantitative analyses of methylation might help to clarify this controversial issue.

The β-domain of pVHL interacts with the α subunits of HIF-1, which mediates cellular responses to hypoxia (84–89). Under normal conditions and in the presence of functional pVHL, the α subunits are rapidly degraded. Under hypoxic conditions and in pVHL-deficient cells, the subunits are stabilized, with a concomitant induction of hypoxia-regulated genes, including that of VEGF (90, 91). Constitutive overexpression of VEGF through this signaling pathway (92) could explain the extraordinary and rich capillary component of VHL-associated neoplasms (91). The capillaries of hemangioblastomas are recruited through overexpression of VEGF and related angiogenic factors secreted by stromal cells, which are considered nonneoplastic.

Additional functions of pVHL protein may contribute to malignant transformation and the evolution of the phenotype of VHL-associated lesions. Studies in RCC cell lines suggested that pVHL is involved in the control of cell cycle exit, i.e., the transition from the G2 into the quiescent G0 phase, possibly by preventing accumulation of the cyclin-dependent kinase inhibitor, p27 (93). Another study showed that only wild-type but not tumor-derived pVHL binds to fibronectin. As a consequence, VHL-associated RCC cells showed a defective assembly of an extracellular fibronectin matrix (94). Through a down-regulation of the response of cells to hepatocyte growth factor/scatter factor and reduced levels of tissue inhibitor of metalloproteinase 2,

FIGURE 4.6. Another schematic representation of mutations of the *VHL* gene in familial (VHL) and sporadic hemangioblastomas. *Mottled areas* designate untranslated regions of the gene. *Arrows* pointing to the right indicate substitutions, and *arrows* pointing to the left indicate insertions. Deletions are shown by horizontal lines and splice mutations, by asterisks (from *ref.* 16, with permission).

FIGURE 4.7. Schematic representation of the genetic events leading to LOH in VHL-related and some sporadic hemangioblastomas. Germline mutations result in a faulty allele in all body cells, only requiring LOH of the other allele for tumor development. In contrast, sporadic hemangioblastoma develop through somatic mutations of the appropriate cells followed by LOH of the second allele. However, there may be pathways for hemangioblastoma development that do not involve the *VHL* gene (see text) (from ref. 83, with permission).

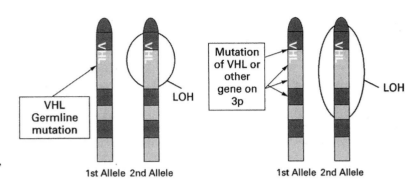

pVHL-deficient tumor cells exhibit a high capacity for invasion (95). Also, inactivated pVHL causes an overexpression of transmembrane carbonic anhydrases that are involved in extracellular pH regulation (96), but the biological significance of this dysregulation remains to be determined.

The molecular aspects and mechanisms of action of the *VHL* gene were included in timely and informative reviews (14,97). The following paragraphs contain some of the salient points of these reviews, and they incorporate some of the information presented here about the functions of the *VHL* gene (Tables 4.8 and 4.9 and Figure 4.8).

The *VHL* gene has three exons that encode the protein product pVHL (36), which is a tumor suppressor protein

and component of a ubiquitin ligase, localized in the nucleus or cytoplasm, the extent to which being dependent on cell density (67, 94, 98). The pVHL forms a complex with other proteins, including elongin B, elongin C, and Cullin 2 (CUL2) to form the VCB–CUL2 complex (99–104), and negatively regulating hypoxia-inducible messenger ribonucleic acid (67). This protein complex controls and determines ubiquitin-dependent proteolysis of large cellular proteins. Under normal oxygen levels, the VCB–CUL2 complex binds to and targets the α subunits of HIF-1 and -2 for ubiquitin-mediated degradation (14, 104–107).

Abnormal or absent pVHL function may disrupt tumor suppression indirectly through HIF-mediated effects or

TABLE 4.8. Functions and loss of functions of *VHL*.

Functions of *VHL* Gene under Normal Cellular Conditions
VHL gene
↓
pVHL
↓
VCB–CUL2 complex: elongin C, elongin B, CUL2, Rbx1
(102, 162, 163)
↓
Functions as a ubiquitin ligase 3
(111, 160, 163)
↓
Ubiquinates HIFα (part of HIF transcription factor)
in the presence of oxygen
(91, 164, 165)
↓
Interaction of proline-dependent hydroxylation of a conserved proline residue in HIFα
(114, 115)
Loss of Functions of *VHL*
Loss of *VHL* gene
↓
Loss of pVHL functions
↓
Accumulation of HIFα irrespective of oxygen concentration
(100, 164–168)
↓
Deregulated activation of HIF target genes: VEGF, TGFα, erythropoietin,
PDGFβ, glucosin transfer-1, and other hypoxia-inducible genes
↓
VEGF important target of HIF in hemangioblastoma because of its major role
in tumor angiogenesis
(169)

TABLE 4.9. Description of some vascular growth factors and receptors involved in hemangioblastoma development.

1. Tie is an endothelial cell-specific receptor and is a tyrosine kinase with immunoglobulin and epidermal factor homology domains. Tie is upregulated during normal development, wound healing and ovulation, as well as in the vasculature of some tumors including hemangioblastoma.
2. FLT-4 (*fms*-like tyrosine kinase) belongs to the endothelial cell receptor subfamily, and is characterized by seven immunoglobulin-like loops. FLT-4, like other endothelial cell receptors, regulates angiogenesis in tumors.
3. The endothelial cell-specific angiogenic VEGF or vascular permeability factor (VPF) and PLGF are secreted glycoproteins.
4. VEGF is up-regulated in many tumors.
5. PLGF is up-regulated in a number of tumors including meningiomas, but down-regulated in thyroid tumors.
6. VEGF has two endothelial cell-specific tyrosine kinase receptors, VEGFR-1 (FLT-1) and VEGFR-2 (KDR). PLGF potentiates the mitogenic activity of low concentrations of VEGF; it binds to VEGFR-1 but not to VEGFR-2.
7. Hemangioblastomas have a subpopulation of granular mast cells that show immunoreactivity for erythropoietin and secrete heparin. VEGF is a heparin binding growth factor requiring cell surface-associated heparin-like molecules for interaction with its receptor. Certain forms of PLGF bind heparin.
8. The release of heparin by mast cell deregulation may contribute to the abundant vascular proliferation in hemangioblastoma, possibly through the interaction with VEGF and PLGF.

Based on Hatva et al. 1996 (31).

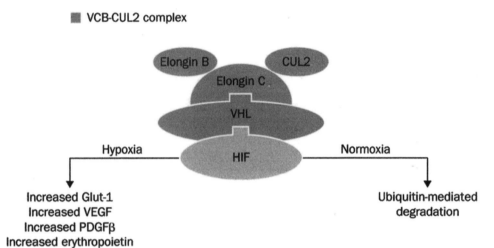

FIGURE 4.8. Interaction of pVHL with other proteins including elongin B, elongin C, and CUL2 to form the VCB–CUL2 complex. See text for details involved in this VHL pathway (from ref. 14, with permission).

directly through pVHL-mediated effects, or both. Many of the tumor suppressive effects of pVHL could result from the degradation of HIF (109–111). Under normal circumstances, HIF can coordinate the cell's response to hypoxia. Through transcriptional regulation, HIF enhances glucose uptake and increases expression of angiogenic, growth, and mitogenic factors, including VEGF, platelet-derived growth factor β polypeptide (PDGFβ), erythropoietin (which, as mentioned, can cause polycythemia that occasionally arises in VHL disease), and transforming growth factor α (14, 88, 104).

Subsequently, disruption of pVHL-mediated degradation of HIF could contribute to tumor formation through multiple mechanisms. If pVHL function were absent or abnormal, HIF could stimulate angiogenesis, which is critical for persistence of tumors associated with VHL disease. HIF-mediated angiogenesis could result from increased levels of VEGF or PDGFβ, or both, which are known to be important for proliferation of endothelial cells and pericytes. This might explain the highly vascular nature of tumors associated with VHL disease, especially hemangioblastoma and RCC. Moreover, high vascular permeability of the tumor vessels, resulting from increased VEGF levels, might underlie the peritumoral

edema and cysts generally present in this disorder. Another potential mechanism of HIF-mediated carcinogenesis is over-production of TGFα. Besides being a potent mitogenic factor, raised levels of TGFα can stimulate cellular overexpression of epidermal growth factor receptors (receptors for TGFα), creating an autocrine loop (41, 88, 104, 107, 112–115).

Cells defective for pVHL cannot degrade HIF α-subunits, perhaps explaining the overproduction of HIF target genes such as *VEGF* in these defective cells (91). Hydroxylation of a conserved proline residue at the core of the HIF1-α subunit is required for pVHL binding (91, 114). This hydroxylation has been found to be regulated by an enzyme requiring molecular oxygen and iron, suggesting that the enzyme may function as a cellular oxygen sensor (114, 115). Mutations associated with type 2C (pheochromocytoma, PCC only) phenotype VHL disease show normal HIF regulation in vitro, suggesting that pVHL-mediated effects other than HIF dysregulation may be necessary for the development of VHL-associated PCC (112). Specifically, the pVHL mutants with an increased risk of PCC-only phenotype can down-regulate HIF but cannot promote fibronectin matrix assembly (113). This suggests that after pVHL inactivation, abnormal fibronectin assembly may contribute to PCC pathogenesis in VHL disease (113).

This study (113) of *VHL* mutations commonly found in PCC-only families suggested that after *VHL* inactivation, abnormal fibronectin assembly may play a role in PCC pathogenesis (116). This also indicated that pVHL functions may be tissue-specific.

In a study of RCC and the *VHL* gene, *VHL* substrate recognition was shown to be essential to the tumor suppressor function of the gene (117). Furthermore, normoxic stabilization of HIF-1α alone, although capable of mimicking some aspects of *VHL* gene loss, was not sufficient to reproduce tumorigenesis, indicating that HIF-1α is not the critical oncogenic substrate of the *VHL* gene. How these findings relate to hemangioblastoma tumorigenesis has yet to be clarified.

pVHL, through its oxygen-dependent polyubiquitylation of HIF, plays a central role in the oxygen-sensing pathway. The interaction between pVHL and HIF is governed by posttranslational prolyl hydroxylation of HIF by a conserved family of Egl-nine (EGLN) enzymes (97,117A). In the absence of pVHL, HIF becomes stabilized and induces the expression of its target genes, many of which regulate angiogenesis, cell growth, and survival. Furthermore, HIF may play a role in tumor formation when pVHL is defective. This may apply to the tumorigenesis of sporadic hemangioblastomas, which frequently lack pVHL functions, and clear cell renal carcinoma (97).

Other possible mechanisms of tumorigenesis caused by absent or abnormal pVHL, independently of HIF, include disruption of the normal cell cycle, increased angiogenesis, and abnormalities in the extracellular matrix (14, 21). The inability to leave the cell cycle (i.e., to enter G0) is seen in cells lacking pVHL (115). This event might take place early in tumorigenesis. Furthermore, mutations of pVHL itself could increase VEGF translational regulation (118, 119). These mutations might augment the angiogenic effects mediated by HIF and further increase tumor vessel permeability.

Finally, although cells without pVHL can secrete fibronectin, they cannot properly assemble a fibronectin extracellular matrix, which could contribute to carcinogenesis (114). Overall, HIF-mediated, direct pVHL-mediated, and unknown effects of abnormal or absent pVHL probably interact to induce formation of the various tumors in VHL disease (14).

Sporadic Hemangioblastomas

The genetic pathways for VHL-associated and those for sporadic hemangioblastoma have much in common. However, a significant portion of the sporadic tumors follow a pathway (or pathways) hitherto not completely established, as illustrated in the following paragraphs.

Somatic inactivation of the *VHL* gene has been found in varying percentages of sporadic hemangioblastomas (69–71,

74, 82) and sporadic RCC (38, 120–122), and it has been reported in up to 50% of hemangioblastomas (9, 58, 62, 67); LOH including the *VHL* region has reported in the stromal cell component of 50% of hemangioblastomas (71).

In the past, studies on mutation analysis of the *VHL* gene in VHL-related and sporadic hemangioblastomas have yielded a broad range of results (21,37,69–71,81,82). This phenomenon might be the consequence of the small number of cases in some series and the incomplete analysis of the *VHL* gene inactivating mechanisms.

To obtain an informative picture of *VHL* gene function in hemangioblastoma pathogenesis, a series of 29 VHL disease-associated and 13 sporadic hemangioblastomas were investigated for all suggested mechanisms of *VHL* inactivation (83). Using a method shown to have a sensitivity of 86% in detecting *VHL* germline mutations, only three somatic *VHL* gene mutations were found (83), all of which occurred in sporadic hemangioblastomas. Only one of these tumors showed biallelic *VHL* gene inactivation due to LOH of the *VHL* gene region at 3p. The relatively low frequencies of somatic *VHL* gene mutations in sporadic hemangioblastomas and biallelic *VHL* gene inactivation agree with those of some of the previous studies, in which somatic mutations had been reported with an average of 18% (seven of 38 cases) (69–71, 82). Thus, biallelic *VHL* gene inactivation is not a common mechanism in the tumorigenesis of sporadic hemangioblastoma (83). According to the authors (83), this is very interesting, as in the past, the two-hit mechanism was viewed as dogma for the *VHL* gene in both VHL disease-associated and sporadic hemangioblastoma (69–71). Furthermore, that five of 10 sporadic tumors showed 3p LOH, but four of them lacked structural *VHL* gene alterations, implies that the *VHL* gene plays a minor role in these sporadic tumors and that there are other genes on 3p involved in hemangioblastoma tumorigenesis (83). This had already been suggested for RCC, where mutations in tumor suppressor genes at 3p14-p21 seem to have a primary role in tumorigenesis in tumors with 3p LOH, but without *VHL* gene inactivation (123).

In summary, data (83) suggest that the genetic pathways involved in pathogenesis in VHL-associated hemangioblastoma versus sporadic hemangioblastoma may be distinct. Further studies on molecular events at 3p in hemangioblastomas might clarify this hypothesis.

Cytogenetic and Comparative Genomic Hybridization (CGH) Studies in Hemangioblastoma

There has been a paucity of cytogenetic studies in hemangioblastoma. In one study (124), three hemangioblastomas in patients with VHL disease had normal karyotypes. In another study (124A) of three similar cases the results were equivocal, as were those in four hemangioblastomas in patients without VHL disease. The indications of possible involvement of tumor suppressor genes (other than *VHL*) in the genesis of sporadic hemangioblastomas may be found in the results of CGH and LOH studies of these tumors (Figures 4.9–4.11).

The most frequent chromosomal anomaly in hemangioblastoma (sporadic and VHL-related) established by CGH is either loss of the whole chromosome 3 (-3) or of 3p (125–128). These findings are not uncommon in human tumors, in general (129,129A).

Four regions in 3p have been reported to harbor a (putative) tumor suppressor gene, i.e., 3p12, 3p14.2, 3p21.3, and 3p25. The *VHL* gene has been identified at 3p25 (36), and it is involved in hemangioblastoma oncogenesis. CGH did not detect a loss involving chromosome 3 in one hemangioblastoma (125); however, small losses undetectable by CGH might be present or other mechanisms such as hypermethylation might affect genes on chromosome 3, thus initiating hemangioblastoma oncogenesis. Interestingly, this tumor was the only one containing a loss of chromosome 8 on which the

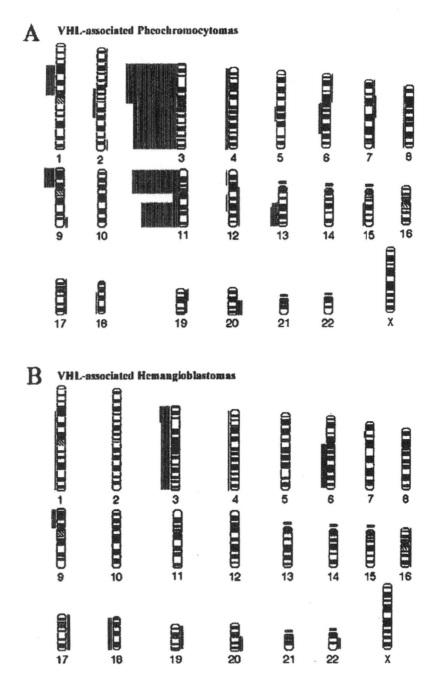

FIGURE 4.9. CGH studies of VHL-associated PCC and hemangioblastomas. Losses of chromosomes 3 and 11 are frequent in PCC compared with hemangioblastomas. Loss of chromosome 11 was not seen in hemangioblastomas (from ref. 131, with permission).

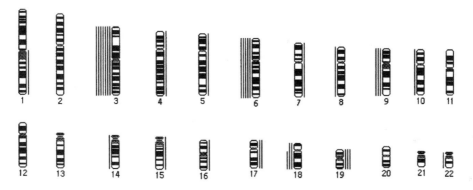

FIGURE 4.10. CGH analysis of sporadic hemangioblastoma showing loss of chromosomes 3 and 6 and gain of chromosome 19 (from ref. 125, with permission).

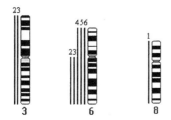

FIGURE 4.11. CGH study of hemangioblastoma showing loss of chromosome 6 (127) seen in five of 22 tumors. Such loss may involve both VHL-related and sporadic hemangioblastoma.

gene for elongin C is located. As mentioned already, pVHL binds to elongin C, which additionally binds to elongin B and CUL2, forming the VCB–complex. This complex induces degradation of their substrates, and it is involved in transcription elongation (130). Aberrations of either the *VHL* gene or elongin C might result in destabilization of the VCB–complex. In a study comparing chromosome loss in VHL-related PCC and hemangioblastomas by CGH (131), loss of chromosome 3 was the commonest finding in both types of tumors (94 and 70%, respectively), whereas loss of chromosome 11 found in 86% of the PCC was not seen in the hemangioblastomas. Because mutations of the *VHL* gene were present in all of these patients (131), divergent genetic pathways may be operative in these tumors (125, 126).

Next to chromosome 3 loss, sporadic hemangioblastomas lose the whole chromosome 6. Concomitant loss of 3p and 6q is also not uncommon in hemangioblastomas (27, 128). Loss of 6q has been frequently reported in other neoplasms (129). *TSG* genes have been suggested to be located at 6p21.2, 6q23-q25, and 6q24-q35 (129). Besides *TSG* genes, others such as *VEGF* (located at 6p12) and *EZRIN* (located at 6q25-q26) might be important in hemangioblastoma pathogenesis. Up-regulation of VEGF (transcript and protein) has been reported in the stromal cells of hemangioblastomas and of the corresponding receptors in the tumor endothelium, suggesting a paracrine mechanism responsible for the unusual vascularization in these tumors (30, 132). Normally, the VCB–complex induces degradation of a transcription factor for HIF, and

destabilization of this complex results in the up-regulation of *VEGF*. Frequent loss of chromosome 6 in sporadic hemangioblastomas might suggest that the remaining *VEGF* allele is sufficient for up-regulation of *VEGF* (125). The other gene, *EZRIN*, may be important in hemangioblastomas, because the ezrin protein has a diffuse cytoplasmic expression in the stromal cells of hemangioblastomas, in contrast to its normal localization close to the plasma membrane (25). This could be caused by overexpression of the protein or by expression of an aberrant ezrin protein with deficient binding capability to the membrane or cytoskeleton (25).

Losses of 18q and chromosome 9, detected in 30% of the sporadic hemangioblastomas by CGH, have been reported in ≥40% of human neoplasms and several known or putative TSG have been located on these chromosomes (129). A gain of chromosome 19, detected by CGH in 30% of the sporadic hemangioblastomas (125, 126), is seen in 2–36% of a variety of tumors (129A).

Two hemangioblastomas did not contain aberrations detectable by CGH, possibly indicating that no aberrations larger than 2 Mb were present in these samples (125). However, because it was reported that hemangioblastomas consist of both neoplastic stromal cells and nonneoplastic cells, including vessels, macrophages, and reactive astrocytes (21, 71), it is also possible that aberrations were not detected due to the high percentage of normal DNA.

A CGH analysis of cellular and reticular variants of hemangioblastoma showed that loss of chromosome 6 is limited to and significantly associated with the cellular variant (133). In contrast, loss of chromosome 19 characterized the reticular variant. The cellular variant also showed loss of 22q and chromosome 19 and gain of chromosome 4. The reticular variant showed loss of 22q and chromosome 3. The data may point toward different genetic pathways in the pathogenesis of the two histologic subtypes of capillary hemangioblastoma (133).

Because patients with germline mutations in the *VHL* gene are predisposed to multiple tumors, including hemangioblastomas, RCC and PCC, a close correlation between the oncogenesis of these tumors is to be expected. The overview of the frequency of aberrations detected by CGH in these tumors

(125) showed that although hemangioblastomas contain a complete loss of chromosome 3, RCC most frequently lost only 3p, whereas PCC lost only 3q. Losses of chromosomes 6 and 9, frequently occur in RCC but not in PCC. Conversely, a gain of chromosome 19 is frequently reported for PCC, but less frequently for RCC. Loss of chromosome 18 is rare in both PCC and RCC. The results (125) indicate that RCC and PCC demonstrate different frequencies of chromosomal imbalances implying different oncogenic pathways. In hemangioblastomas chromosomal imbalances characteristic for either RCC or PCC were detected. Because hemangioblastomas do not harbor either loss of 3p (seen in RCC) or loss of 3q (seen in PCC) but a loss of the complete chromosome 3, there seems to be one pathway for the oncogenesis of hemangioblastomas that showed similarities to both the pathways for RCC and PCC (125).

In summary, losses of chromosome 3 and 6 were most frequently detected imbalances in sporadic hemangioblastomas (125), and the results suggest that there is a characteristic pathway of chromosomal imbalances in which loss of chromosome 3 (harboring the *VHL* gene) and loss of chromosome 6 occur as sequential events, followed by a loss of chromosome 9 and/or a loss of 18q and/or a gain of chromosome 19. This pathway shows some similarities to those of RCC and PCC.

LOH in Hemangioblastoma

Reported frequencies of LOH on 3p in VHL-related hemangioblastomas have ranged from 14 to 100% (21, 37, 81, 82). One study showed 3p LOH in 13 of 21 informative VHL-related hemangioblastomas (62%) investigated (83), indicating that a two-hit inactivation of the *VHL* gene is a common mechanism in VHL disease-associated hemangioblastomas. In sporadic tumors, LOH of the telomeric 3p region was detected in 50% of cases (83). These results are in good agreement with two previous studies, which found allelic loss on 3p in one of two (82) and 10 of 19 (71) sporadic hemangioblastomas.

Mutations leading to LOH of the tumor suppressor gene *VHL* in sporadic hemangioblastomas have been described previously. In one study of 13 sporadic hemangioblastomas, three were shown to have mutations of the *VHL* gene. In four other tumors, the authors were unable to do sequencing due to the poor quality of the DNA, although these hemangioblastomas had abnormal single strand conformational polymorphism patterns (69). In a subsequent study (70), of 20 hemangioblastomas, 10 were found to have mutations of the *VHL* gene. In two patients germline mutations were present, the mutations being identical in leukocyte and tumor DNA. From these studies, it can be inferred that various types of *VHL* gene mutations may be responsible for sporadic hemangioblastomas in a significant percentage of these cases.

LOH at 22q13 has been demonstrated in sporadic hemangioblastomas (134). Although a number of tumor suppressor genes are located in the affected region, including *NF2*, the authors (134) discuss the unlikelihood of *NF2* gene involvement. In three of eight informative hemangioblastomas, LOH near the *VHL* gene was detected. The possibility exists that an as yet unidentified gene at 22q13.2, in addition to *VHL*, may play a role in the pathogenesis of hemangioblastoma (134).

Molecular Studies in Hemangioblastoma

A possible role for cyclin D1 in hemangioblastoma has been advocated. pVHL regulates the expression of cyclin D, which plays a key role in cell cycle regulation and carcinogenesis (135–137). Results indicate that the *VHL* gene is required for the down-regulation of cyclin D1 through HIF at high cell density (Figure 4.12). Suppression of cyclin D1 expression at high cell density is impaired by the loss of functional pVHL and the deregulated cyclin D1 expression causes the high-level phosphorylation of pRB even at high cell density (137).

Although cyclins D may play a role in various neoplasias (138, 139), in a study of 17 sporadic hemangioblastomas the results related to cyclin D1 were too diverse to be statistically significant (139A) of an unequivocal correlation between

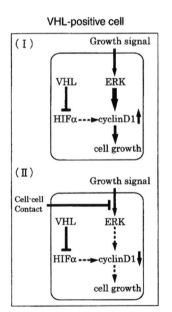
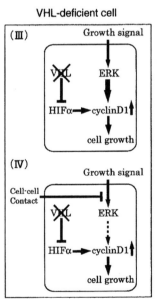

FIGURE 4.12. Schematic representation of cyclin D1 regulation by pVHL at different cell densities. At low cell density (**I**), all growth remains at a high level through an extracellular growth signal inducing cyclin D1 expression via the activation of extracellular signal-regulated kinase. HIF((or HIF-1() is degraded by pVHL. In dense cell populations (**II**), the growth signal is inhibited (including ERF) and cyclin D1 levels are reduced. Cells with mutated VHL show high levels (**III** and **IV**) of cyclin D1, regardless of cell density (from ref. 137, with permission).

cyclin D1 genotype and the occurrence of sporadic hemangioblastomas. The authors indicated that despite the high expression of cyclin D1 in some hemangioblastomas, the *CCND1* genotype is unlikely to be an important genetic modifier in the oncogenesis of sporadic hemangioblastomas and that other factors may be responsible for the induction of these tumors. A possible mechanism of tumorigenesis caused by absent or abnormal VHL protein independent of HIF is the up-regulation of cyclin D1 alone (135–137).

Hemangioblastoma being essentially a vascular tumor is regulated and contributed to by a variety of growth factor-receptor interactions (Figure 4.13 and Table 4.9) (31). VEGF mRNA was highly expressed in all seven hemangioblastomas (two from VHL cases) (31). Some hemangioblastoma tumor cells expressed high levels of placental growth factor (PLGF). In addition, significantly elevated levels of Tie mRNA, Tie protein, *VEGFR1*, and *VEGFR2* but not *FLT4* mRNA were observed in the endothelia of hemangioblastomas. A subpopulation of stromal cells highly expressed the receptors. Upregulation of the endothelial growth factors and receptors may result in autocrine or paracrine stimulation of endothelial cells and their precursors involved in the genesis of hemangioblastoma.

The strong expression of the aforementioned factors in hemangioblastoma stromal ells and in endothelial cells may indicate that the tumor cells retain the potential to differentiate into endothelial cells forming the typical vessels

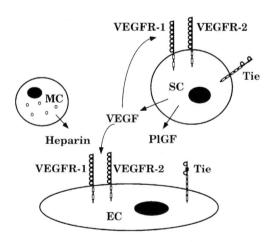

FIGURE 4.13. Relevant features of vascular endothelial growth factor receptor expression by hemangioblastomas. The endothelial cell-specific receptors Tie, VEGFR-1 and VEGFR-2 are known to be early markers for endothelial cell progenitors. Their strong expression in hemangioblastoma stromal cells and in endothelial cells may indicate that the tumor cells, like angioblasts in early embryos, retain the potential to differentiate into endothelial cells forming the typical vessels of hemangioblastomas. In addition, the same cells express genes encoding growth factors binding to the endothelial receptor tyrosine kinases, thus possibly establishing autocrine and paracrine signal transduction for the stimulation of growth of these tumors (31). SC, stromal cell; EC, endothelial cell; MC, mast cell; VEGFR, VEGF receptor; see Table 4.9 (from ref. 31, with permission).

of hemangioblastoma (31). In addition, these cells express genes encoding growth factors binding to the endothelial receptor tyrosine kinase, thus possibly establishing autocrine and paracrine signal transduction for the stimulation of growth of hemangioblastoma (31). A tumor suppressor gene (TSG), *ZAC1*, has been shown to be inactivated in hemangioblastoma through loss of the expressed (paternal) allele, pointing to its possible role in non-VHL cases with hemangioblastoma (137A). The gene *ZAC1* (*LOT1* or *PLAG1*), located at 6q24-q25, encodes a finger protein akin to p53, with tumor suppressor activity through induction of apoptosis and cell cycle arrest at G1.

Hemangioblastoma versus Metastatic RCC

The problem of differentiating hemangioblastoma from metastatic RCC to the brain has received much attention. The histopathological differential diagnosis of hemangioblastoma from metastatic RCC is made difficult by the facts that 1) there is considerable morphologic overlap between the two lesions; 2) both tumors are common in VHL patients; and 3) RCC frequently metastasizes to the CNS, both as sporadic tumors and in the context of VHL disease. Most of the approaches to this problem have used immunohistochemical techniques (140). Thus immunohistochemically, the use of cytokeratin, epithelial membrane antigen (e.g., immunoperoxidase) (141), pan-epithelial antigen, and a panel of antisera (142) as markers for RCC has been advocated, because hemangioblastomas react negatively. The possible use of steroid receptors, especially for progesterone, to differentiate hemangioblastoma from metastatic RCC, was thwarted by the presence of such receptors in both types of tumors (143). However, CNS positivity was seen only in metastatic RCC and not in CNS hemangioblastoma (144), possibly serving as a means of differentiating these lesions. More recently, D2-40, a monoclonal antibody that recognizes an oncofetal antigen (M2A), was shown to react positively only with hemangioblastoma (both sporadic and VHL-associated) and not with metastatic RCC (145). Staining for ezrin may be helpful in the differential diagnosis of capillary hemangioblastoma from other brain tumors (25). Because CD10 is expressed in clear cell RCC, whereas inhibin α is expressed in hemangioblastoma, CD10 and inhibin α can be useful in the distinction of these two entities (146). In total, 22 cases of cerebellar hemangioblastoma, five cases of metastatic clear cell RCC to the CNS, and 16 primary cases of clear cell RCC were studied with immunohistochemical staining of both CD10 and inhibin α. All 22 cases of hemangioblastoma were immunonegative for CD10 in the stromal cells. In contrast, all five cases of metastatic clear cell RCC and 16 cases of primary clear cell RCC showed positive CD10 membranous staining. In all, 20 cases of hemangioblastoma (20 of 22; 91%) expressed inhibin α in the stromal cells. Two cases of primary clear cell RCC (two of 16; 13%) and three cases of metastatic clear cell RCC (three of five; 60%) showed immunopositivity for inhibin α. In conclusion, in addition to the immunostaining of inhibin α, CD10 is a useful marker for distinguishing between hemangioblastoma and metastatic clear cell RCC. Similar results with inhibin α were obtained in other studies (147).

Metastasis of RCC to cerebellar hemangioblastoma in a patient with VHL disease has been reported previously (147A). Their case was the third to be reported.

References

1. Neumann, H.P., and Wiestler, O.D. (1991) Clustering of features of von Hippel-Lindau syndrome: evidence for a complex genetic locus. *Lancet* **337**, 1052–1054.
2. Neumann, H.P., and Wiestler, O.D. (1994) von Hippel-Lindau disease: a syndrome providing insights into growth control and tumorigenesis. *Nephrol. Dial. Transplant* **9**, 1832–1833.
3. Cushing, H., and Bailey, B. (1928) *Tumors Arising from the Blood-Vessels of the Brain: Angiomatous Malformation and Hemangioblastomas.* Charles C. Thomas, Springfield, IL.
4. Olivecrona, H. (1952) The cerebellar angioreticulomas. *J. Neurosurg.* **9**, 317–330.
5. Obrador, S., and Martin-Rodriguez, J.G. (1977) Biological factors involved in the clinical features and surgical management of cerebellar hemangioblastomas. *Surg. Neurol.* **7**, 79–85.
6. Constans, J.P., Meder, F., Maiuri, F., Donzelli, R., Spaziante, R., and de Divitiis, E. (1986) Posterior fossa hemangioblastomas. *Surg. Neurol.* **25**, 269–275.
7. Hurth, M., Andre, J.M., Djindjian, R., Escourolle, R., Houdart, R., Poirer, J., and Rey, A. (1975) Intraspinal hemangioblastomas. *Neurochirurgie* **21 (Suppl 1)**, 1–136.
8. Murota, T., and Symon, L. (1989) Surgical management of hemangioblastomas of the spinal cord: a report of 18 cases. *Neurosurgery* **25**, 699–708.
9. Symon, L., Murota, T., Pell, M., and Bordi, L. (1993) Surgical management of haemangioblastoma of the posterior foss. *Acta Neurochir. (Wien)* **120**, 103–110.
10. Wolf, A. (1941) Tumors of the spinal cord, nerve roots, and membranes in surgical diseases. In *Surgical Diseases of the Spinal Cord, Membrane, and Nerve Roots: Symptoms, Diagnosis and Treatment* (Elsberg, C.A., ed.), P.B. Hoeger, Inc., New York, pp. 231–364.
11. Yasargil, M.G., Antic, J., Laciga, R., de Preux, J., Fideler, R.W., and Boone, S.C. (1976) The microsurgical removal of intermedullary spinal hemangioblastomas: report of twelve cases and a review of the literature. *Surg. Neurol.* **6**, 141–148.
12. Silver, M.L., and Hennigar, G. (1952) Cerebellar hemangiomas (hemangioblastoma): a clinicopathological review of 40 cases. *J. Neurosurg.* **9**, 484–494.
13. Lonser, R.R., Weil, R.J., Wanebo, J.E., De*Vroom*, H.L., and Oldfield, D.H. (2003) Surgical management of spinal cord hemangioblastomas in patients with von Hippel-Lindau disease. *J. Neurosurg.* **98**, 106–116.
14. Lonser, R.R., Glenn, G.M., Walther, M., Chew, E.Y., Libutti, S.K., Linehan, W.M., and Oldfield, E.H. (2003) von Hippel-Lindau disease. *Lancet* **361**, 2059–2067.
15. Wanebo, J.E., Lonser, R.R., Glenn, G.M., and Oldfield, E.H. (2003) The natural history of hemangioblastomas of the central nervous system in patients with von Hippel-Lindau disease. *J. Neurosurg.* **98**, 82–94.
16. Böhling, T., Plate, K.H., Haltia, M.J., Alitalo, K., and Heumann, H.P.H. . (2000) von Hippel-Lindau disease and capillary hemangioblastoma. In *World Health Organization Classification of Tumours. Pathology and Genetics of Tumours of the Nervous System* (Kleihues, P., and Cavenee, W.K., eds.), IARC Press, Lyon, France, pp. 223–226.
17. Tachibana, O., Yamashima, T., Yamashita, J., and Takinami, K. (1994) Immunohistochemical study of basic fibroblast growth factor and erythropoietin in cerebellar hemangioblastomas. *Noshuyo Byori* **11**, 169–172.
18. Hasselblatt, M., Jeibmann, A., Gerβ J., Behrens, C., Rama, B., Wassmann, H., and Paulus, W. (2005) Cellular and reticular variants of haemangioblastoma revisited: a clinicopathologic study of 88 cases. *Neuropathol. Appl. Neurobiol.* **31**, 618–622.
19. Becker, I., Paulus, W., and Roggendorf, W. (1989) Histogenesis of stromal cells in cerebellar hemangioblastomas. An immunohistochemical study. *Am. J. Pathol.* **134**, 271–275.
20. Berkman, R.A., Merrill, M.J., Reinhold, W.C., Monacci, W.T., Saxena, A., Clark, W.C., Robertson, J.T., Ali, I.U., and Oldfield, E.H. (1993) Expression of the vascular permeability factor/vascular endothelial growth factor gene in central nervous system neoplasms. *J. Clin. Invest.* **91**, 153–159.
21. Vortmeyer, A.O., Gnarra, J.R., Emmert-Buck, M.R., Katz, D., Linehan, W.M., Oldfield, E.H., and Zhuang, Z. (1997) von Hippel-Lindau gene deletion detected in the stromal cell component of a cerebellar hemangioblastoma associated with von Hippel-Lindau disease. *Hum. Pathol.* **28**, 540–543.
21A. Ding, X.-H., Zhou, L.-F., Tan, Y.-Z., Zhao, Y., and Zhu, J.-J. (2007) Histologic and histogenetic investigations of intracranial hemangioblastomas. *Surg. Neurol.* **67**, 239–245.
22. Mills, S.E., Ross, G.W., Perentes, E., Nakagawa, Y., and Scheithauer, B.W. (1990) Cerebellar hemangioblastoma: immunohistochemical distinction from metastatic renal cell carcinoma. *Surg. Pathol.* **3**, 121–132.
23. Wizigmann-Voos, S., and Plate, K.H. (1996) Pathology, genetics and cell biology of hemangioblastomas. *Histol. Histopathol.* **11**, 1049–1061.
24. Böhling, T., Mäenpää, A., Timonen, T., Vantunen, L., Paetau, A., and Haltia, M. (1996) Different expression of adhesion molecules on stromal cells and endothelial cells of capillary hemangioblastoma. *Acta Neuropathol.* **92**, 461–466.
25. Böhling, T., Turunen, O., Jääskeläinen, J., Carpen, O., Saino, M., Wahlström, T., Vaheri, A., Haltia, M. (1996) Ezrin expression in stromal cells of capillary hemangioblastoma. *Am. J. Pathol.* **148**, 367–373.
26. Reifenberger, G., Reifenberger, J., Bilzer, T., Wechsler, W., and Collins, V.P. (1995) Coexpression of transforming growth factor-α and epidermal growth factor receptor in capillary hemangioblastomas of the central nervous system. *Am. J. Pathol.* **147**, 245–250.
27. Böhling, T., Hatva, E., Kujala, M., Claesson-Welsh, L., Alitalo, K, and Haltia, M (1996) Expression and growth

factors and growth factor receptors in capillary hemangioblastoma. *J. Neuropathol. Exp. Neurol.* **55**, 522–527.

28. Beckner, M.E. (1999) Factors promoting tumor angiogenesis. *Cancer Invest.* **17**, 594–623.

29. Machein, M.R., and Plate, K.H. (2000) VEGF in brain tumors. *J. Neurooncol.* **50**, 109–120.

30. Wizigmann-Voos, S., Breier, G., Risau, W., and Plate, K.H. (1995) Up-regulation of vascular endothelial growth factor and its receptors in von Hippel-Lindau disease-associated and sporadic hemangioblastomas. *Cancer Res.* **55**, 1358–1364.

31. Hatva, E., Böhling, T., Jääskeläinen, J., Persico, M.G., Haltia, M., and Alitalo, K. (1996) Vascular growth factors and receptors in capillary hemangioblastomas and hemangiopericytomas. *Am. J. Pathol.* **148**, 763–775.

32. Flamme, I., Krieg, M., and Plate, K.H. (1998) Up-regulation of vascular endothelial growth factor in stromal cells of hemangioblastomas is correlated with up-regulation of the transcription factor HRF/HIF-2?. *Am. J. Pathol.* **153**, 25–29.

33. Krieg, M., Marti, H.H., and Plate, K.H. (1998) Coexpression of erythropoietin and vascular endothelial growth factor in nervous system tumors associated with von Hippel-Lindau tumor suppressor gene loss of function. *Blood* **92**, 3388–3393.

34. Chen, Y., Tachibana, O., Oda, M., Xu, R., Hamada, J., Yamashita, J., Hashimoto, N., and Takahashi, J.A. (2006) Increased expression of aquaporin 1 in human hemangioblastomas and its correlation with cyst formation. *J. Neurooncol.* **80**, 219–225.

35. Catapano, D., Muscarella, L.A., Gaurnieri, V., Zelante, L., D'Angelo, V.A., and D'Agruma, L. (2005) Hemangioblastomas of central nervous system: molecular genetic analysis and clinical management. *Neurosurgery* **56**, 1215–1221.

36. Latif, F., Tory, K., Gnarra, J.R., Yoa, M., Duh, F.M., Orcutt, M.L., Stackhouse, T., Kuzmin, I., Modi, W., Geil., L., Schmidt, L., Zhou, F., Li, H., Wei, M.H., Chen, F., Glenn, G., Choyke, P., Walther, M.M., Weng, Y., Duan, D.R., Dean, M., Glavac, D., Richards, F.M., Crossey, P.A., Ferguson-Schmidt, M.A., Le Paslier, D., Chumakov, I., Cohen, D., Chinault, A.C., Maher, E.R., Linehan, W.M., Zbar, B., and Lerman, M.I. (1993) Identification of the von Hippel-Lindau disease tumor suppressor gene. *Science* **260**, 1317–1320.

37. Crossey, P.A., Foster, K., Richards, F.M., Phipps, M.E., Latif, F., Tory, K., Jones, M.H., Bentley, E., Kumar, R., Lerman, M.I., Zbar, B., Affara, N.A., Ferguson-Smith, M.A., and Maher, E.R.. (1994) Molecular genetic investigations of the mechanism of tumourigenesis in von Hippel-Lindau disease: analysis of allele loss in VHL tumours. *Hum. Genet.* **93**, 53–58.

38. Gnarra, J.R., Tory, K., Weng, Y., Schmidt, L., Wei, M.H., Li, H., Latif, F., Liu, S., Chen, F., Duh, F.M., Lubensky, I., Duan, D.R., Florence, C., Pozatti, R., Walther, M.M., Bander, N.H., Grossman, H.B., Brauch, H., Pomer, S., Brooks, J.D., Isaacs, W.B., Lerman, M.I., Zbar, B., and Linehan, W.M. (1994) Mutations of the VHL tumor suppressor gene in renal carcinoma. *Nat. Genet.* **7**, 85–90.

39. Maher, E.R., Yates, J.R.W., and Ferguson-Smith, M.A. (1990) Statistical analysis of the two stage mutation model in von Hippel-Lindau disease, and in sporadic cerebellar haemangioblastoma and renal cell carcinoma. *J. Med. Genet.* **27**, 311–314.

39A. Maher, E.R., Yates, J.R., Harries, R., Benjamin, C., Harris, R., Moore, A.T., and Ferguson-Smith, M.A. (1990) Clinical features and natural history of von Hippel-Lindau disease. *Q. J. Med.* **77**, 1151–1163.

40. Maher, E.R., and Kaelin, W.G., Jr. (1997) von Hippel-Lindau disease. *Medicine (Baltimore)* **76**, 381–391.

41. Richard, S., Campello, C., Taillandier, L., Parker, F., and Resche, F. (1998) Haemangioblastoma of the central nervous system in von Hippel-Lindau disease. French VHL Study Group. *Intern. Med.* **243**, 547–553.

42. Friedrich, C.A. (1999) von Hippel-Lindau syndrome: a pleomorphic condition. *Cancer* **86**, 2478–2482.

43. Stolle, C., Glenn, G., Zbar, B., Humphrey, J.S., Choyke, P., Walther, M., Pack, S., Hurley, K., Andrey, C., Klausner, R., and Linehan, W.M. (1998) Improved detection of germline mutations in the von Hippel-Lindau disease tumor suppressor gene. *Hum. Mutat.* **12**, 417–423.

44. Maher, E.R., Iselius, L., Yates, J.R., Littler, M., Benjamin, C., Harris, R., Sampson, J., Williams, A., Ferguson-Smith, M.A., and Morton, N. (1991) Von Hippel-Lindau disease: a genetic study. *J Med. Genet.* **28**, 443–447.

45. Maddock, I.R., Moran, A., Maher, E.R., Teare, M.D., Norman, A., Payne, S.J., Whitehouse, R., Dodd, C., Lavin, M., Hartley, N., Super, M., and Evans, D.G.R. (1996) A genetic register for von Hippel-Lindau disease. *J. Med. Genet.* **33**, 120–127.

46. Filling-Katz, M.R., Choyke, P.L., Oldfield, E., Charnas, L., Patronas, N.J., Glenn, G.M., Gorin, M.B., Morgan, J.K., Linehan, W.M., Seizinger, B.R., and Zbar. B (1991) Central nervous system involvement in von Hippel-Lindau disease. *Neurology* **41**, 41–46.

47. Neumann, H.P.H., Eggert, H.R., Scheremet, R., Schumacher, M., Mohadjer, M., Wakhloo, A.K., Volk, B., Hettmannsperger, U., Riegler, P., Schollmeyer, P. , and Wiestler, O. (1992) Central nervous system lesions in von Hippel-Lindau syndrome. *J. Neurol. Neurosurg. Psychiatry* **55**, 898–901.

48. Conway, J.E., Chou, D., Clatterbuck, R.E., Brem, H., Long, D.M., and Rigamonti, D. (2001) Hemangioblastomas of the central nervous system in von Hippel-Lindau syndrome and sporadic disease. *Neurosurgery* **48**, 55–63.

49. Choyke, P.L., Glenn, G.M., Walther, M.M., Patronas, N.J., Linehan, W.M., and Zbar, B. (1995) von Hippel-Lindau disease: genetic, clinical and imaging features. *Radiology* **194**, 629–642.

50. Horton, W.A., Wong, V., and Eldridge, R. (1976) Von Hippel-Lindau disease: clinical and pathological manifestations in nine families with 50 affected members. *Arch. Intern. Med.* **136**, 769–777.

51. Manski, T.J., Heffner, D.K., Glenn, G.M., Patronas, N.J., Pikus, A.T., Katz, D., Lebovics, R., Sledjeski, K., Choyke, P.L., Zbar, B., Linehan, W.M., and Oldfield, E.H. (1997) Endolymphatic sac tumors: a source of morbid hearing loss in von Hippel-Lindau disease. *JAMA* **277**, 1461–1466.

52. Melmon, K.L., and Rosen, S.W. (1964) Lindau's disease: review of the literature and study of a large kindred. *Am. J. Med.* **36**, 595–617.

53. Neumann, H.P. (1987) Basic criteria for clinical diagnosis and genetic counseling in von Hippel-Lindau syndrome. *VASA* **16**, 220–226.

54. Cramer, F., and Kimsey, M.W. (1952) The cerebellar hemangioblastomas: review of fifty-three cases, with special reference to cerebellar cysts and the association with polycythemia. *Arch. Neurol. Psychiatry* **67**, 237–252.

55. Palmer, J.J. (1972) Hemangioblastomas: r review of 81 cases. *Acta Neurochir. (Wien)* **27**, 125–148.

56. Neumann, H.P.H., Eggert, H.R., Weigel, K., Friedburg, H., Wiestler, O.D., and Schollmeyer, P. (1989) Hemangioblastomas of the central nervous system. *J. Neurosurg.* **70**, 24–30.

57. Seizinger, B.R., Rouleau, G.A., Ozelius, L.J., Lane, A.H., Farmer, G.E., Lamiell, J.M., Haines, J., Yuen, J.W., Collins, D., Majoor-Krakauer, D., Bonner, T., Mathew, C., Rubenstein, A., Halperin, J., McConkie-Rosell, A., Green, J.S., Trofatter, J.A., Ponder, B.A., Eierman, L., Bowmer, M.I., Schimke, R., Oostra, B., Aronin, N., Smith, D.I., Drabkin, H., Waziri, M.H., Hobbs, W.J., Martuza, R.L., Conneally, P.M., Hsia, Y.E., and Gusella, J.F. (1988) Von Hippel-Lindau disease maps to the region of chromosome 3 associated with renal cell carcinoma. *Nature* **332**, 268–269.

58. Niemelä, M., Lemeta, S., Summanen, P., Böhling, T., Sainio, M., Kere, J., Poussa, K., Sankila, R., Haapasalo, H., Kääriäinen, H., Pukkala, E., and, Jääskeläinen, J. (1999) Long-term prognosis of haemangioblastoma of the CNS: impact of von Hippel-Lindau disease. *Acta Neurochir (Wien)* **141**, 1147–1156.

59. Gunel, M., and Awad, I.A. (2001) Comments concerning to von Hippel-Lindau (VHL) disease is caused by loss-of-function of the *VHL* gene located on chromosome 3p25. *Neurosurgery* **48**, 62.

60. Hes, F.J., McKee, S., Taphoorn, M.J.B., Rehal, P., van der Luijt, R.B., McMahon, R., van der Smagt, J.J., Dow, D., Zewald, R.A., Whittaker, J., Lips, C.J.M., MacDonald, F., Pearson, P.L., and Maher, E.R. (2000) Cryptic von Hippel-Lindau disease: germline mutations with haemangioblastoma only. *J. Med. Genet.* **37**, 939–943.

61. Richards, F.M., Webster, A.R., McMahon, R., Woodward, E.R., Rose, S., and Maher, E.R. (1998) Molecular genetic analysis of Hippel-Lindau disease *J. Intern. Med.* **243**, 527–533.

62. Gläskar, S., Bender, B.U., Apel, T.W., Natt, E., van Velthoven, V., Scheremet, R., Zentner, J., and Neumann, H.P.H. (1999) The impact of molecular genetic analysis of the VHL gene in patients with haemangioblastomas of the central nervous system. *J. Neurol. Neurosurg. Psychiatry* **67**, 758–762.

63. Corless, C.L., Kibel, A.S., Iliopoulos, O., and Kaelin, W.G., Jr. (1997) Immunostaining of the von Hippel-Lindau gene product in normal and neoplastic human tissues. *Hum. Pathol.* **28**, 459–464.

64. Sakashita, N., Takeya, M., Kishida, T., Stackhouse, T.M., Zbar, B., and Takahashi, K. (1999) Expression of von Hippel-Lindau protein in normal and pathological human tissues. *Histochem. J.* **31**, 133–144.

65. Los, M., Jansen, G.H., Kaelin, W.G., Lips, C.J., Blijham, G.H., and Voest, E.E. (1996) Expression pattern of the von Hippel-Lindau protein in human tissues. *Lab. Invest.* **75**, 231–238.

66. Nagashima, Y., Miyagi, Y., Udagawa, K., Taki, A., Misugi, K., Sakai, N., Kondo, K., Kaneko, S., Yao, M., and Shuin, T. (1996) von Hippel-Lindau tumour suppressor gene. Localization of expression by in situ hybridization. *J. Pathol.* **180**, 271–274.

67. Ohh, M., and Kaelin, W.G., Jr. (1999) The von Hippel-Lindau tumour suppressor protein: new perspectives. *Mol. Med. Today* **5**, 257–263.

68. Stebbins, C.E., Kaelin, W.G., Jr., and Pavletich, N.P. (1999) Structure of the VHL-ElonginC-ElonginB complex: implications for VHL tumor suppressor function. *Science* **284**, 455–461.

69. Kanno, H., Kondo, K., Ito, S., Yamamoto, I., Fujii, S., Torigoe, S., Sakai, N., Hosaka, M., and Yao, M. (1994) Somatic mutations of the von Hippel-Lindau tumor suppressor gene in sporadic central nervous system hemangioblastomas. *Cancer Res.* **54**, 4845–4847.

70. Oberstrass, J., Reifenberger, G., Reifenberger, J., Wechsler, W., and Collins, V.P. (1996) Mutation of the von Hippel-Lindau tumor suppressor gene in capillary haemangioblastomas of the central nervous system. *J. Pathol.* **179**, 151–156.

71. Lee, J., Dong, S., Park, W., Yoo, N., Kim, C., Jang, J., Chi, J., Zbar, B.L.I.A., Linehan., W.M., Vortmeyer, A.O., and Zhuang, Z. (1998) Loss of heterozygosity and somatic mutations of the VHL tumor suppressor gene in sporadic cerebellar hemangioblastomas. *Cancer Res.* **58**, 504–508.

72. Zbar, B., Kishida, T., Chen, F., Schmidt, L, Maher, E.R., Richards, F.M., Crossey, P.A., Webster, A.R., Affara, N.A., Ferguson-Smith, M.A., Brauch, H., Glavac, D., Neumann, H.P., Tisherman, S., Mulvihill, J.J., Gross, D.J., Shuin, T., Whaley, J., Seizinger, B., Kley, N., Olschwang, S., Boisson, C., Richard, S., Lips, C.H., Linehan, W.M., and Lerman, M. (1996) Germline mutation in the von Hippel-Lindau disease (VHL) gene in families from North America, Europe, and Japan. *Hum. Mutat.* **8**, 348–357.

73. Neumann, H.P.H., and Bender, B.U. (1998) Genotype-phenotype correlations in von Hippel-Lindau disease. *J. Intern. Med.* **243**, 541–545.

74. Olschwang, S., Richard, S., Boisson, C., Giraud, S., Laurent-Puig, P., Resche, F., and Thomas, G. (1998) Germline mutation profile of the VHL gene in von Hippel-Lindau disease and in sporadic haemangioblastoma. *Hum. Mutat.* **12**, 424–430.

75. Chen, F., Slife, L., Kishida, T., Mulvihill, J., Tisherman, S.E., and Zbar, B. (1996) Genotype-phenotype correlation in von Hippel-Lindau disease: identification of a mutation associated with VHL type 2A. *J. Med. Genet.* **33**, 716–717.

76. Glavac, D., Neumann, H.P., Wittke, C., Jaenig, H., Masek, O., Streicher, T., Pausch, F., Engelhardt, D., Plate, K.H., Hofler, H., Chen, F., Zbar, B., and Brauch, H. (1996) Mutations in the VHL tumor suppressor gene and associated lesions in families with von Hippel-Lindau disease from central Europe. *Hum. Genet.* **98**, 271–280.

77. Atuk, N.O., Stolle, C., Owen, J.A., Jr., Carpenter, J.T., and Vance, M.L. (1998) Pheochromocytoma in von Hippel-Lindau disease: clinical presentation and mutation analysis in a large, multigenerational kindred. *J. Clin. Endocrinol. Metab.* **83**, 117–120.

78. Herman, J.G., Latif, F., Weng, Y., Lerman, M.I., Zbar, B., Liu, S., Samid, D., Duan, D.R., Gnarra, J.R., Linehan, W.M., and Baylin, S.B. (1995) Silencing of the VHL tumor suppressor gene by DNA methylation in renal cell carcinoma. *Proc. Natl. Acad. Sci. USA* **91**, 9700–9704.

79. Jones, P.A., and Laird, P.W. (1999) Cancer epigenetics comes of age. *Nat. Genet.* **21**, 163–167.

80. Baylin, S.B., Herman, J.G., Graff, J.R., Vertino, P.M., and Issa, J.P. (1998) Alterations in DNA methylation: a fundamental aspect of neoplasia. *Adv. Cancer Res.* **72**, 141–196.

81. Prowse, A.H., Webster, A.R., Richards, F.M., Richard, S., Olschwang, S., Resche, F., Affara, N.A., and Maher, E.R. (1997) Somatic inactivation of the VHL gene in von Hippel-Lindau disease tumors. *Am. J. Hum. Genet.* **60**, 765–771.

82. Tse, J.Y.M., Wong, J.H.C., Lo, K., Poon, W., Huang, D., and Ng, H. (1997) Molecular genetic analysis of the von Hippel-Lindau disease tumor suppressor gene in familial and sporadic cerebellar haemangioblastomas. *Am. J. Clin. Pathol.* **107**, 459–466.

83. Gläskar, S., Bender, B.U., Apel, T.W., van Velthoven, V., Mulligan, L.M., Zentner, J., and Neumann, H.P.H. (2001) Reconsideration of biallelic inactivation of the *VHL* tumour suppressor gene in hemangioblastomas of the central nervous system. *J. Neurol. Neurosurg. Psychiatry* **70**, 644–648.

84. Zagzag, D., Zhong, H., Scalzitti, J.M., Laughner, E., Simons, J.W., and Semenza, G.L. (2000) Expression of hypoxia-inducible factor 1α in brain tumors. Association with angiogenesis, invasion, and progression. *Cancer* **88**, 2606–2618.

85. Zagzag, D., Krishnamachary, B., Yee, H., Okuyama, H., Chiriboga, L., Aktar Ali, M., Melamed, J., and Semenza, G.L. (2005) Stromal cell-derived factor-1α and CXCR4 expression in hemangioblastoma and clear cell-renal cell carcinoma: von Hippel-Lindau loss-of-function induces expression of a ligand and its receptor. *Cancer Res.* **65**, 6178–6188.

86. Kaelin, W.G., Iliopoulos, O., Lonergan, K.M., and Ohh, M. (1998) Functions of the von Hippel-Lindau tumour suppressor protein. *J. Intern. Med.* **243**, 535–539.

87. Kondo, K., and Kaelin, W.G., Jr. (2001) The von Hippel-Lindau tumor suppressor gene. *Exp. Cell Res.* **264**, 117–125.

88. Kaelin, W.G., Jr. (2002) Molecular basis of the VHL hereditary cancer syndrome. *Nat. Rev. Cancer* **2**, 673–682.

89. Kwon, S.J., Song, J.J., and Lee, Y.J. (2005) Signal pathway of hypoxia-inducible factor-1α phosphorylation and its interaction with von Hippel-Lindau tumor suppressor protein during ischemia in MiaPaCa-2 pancreatic cancer cells. *Clin. Cancer Res.* **11**, 7607–7613.

90. Chan, C.-C., Vortmeyer, A.O., Chew, E.Y., Green, W.R., Matteson, D.M., Shen, De F., Linehan, W.M., Lubensky, I.A., and Zhuang, Z. (1999) *VHL* gene deletion and enhanced *VEGF* gene expression detected in the stromal cells of retinal angioma. *Arch. Ophthalmol.* **117**, 625–630.

91. Maxwell, P.H., Wiesener, M.S., Chang, G.-W., Clifford, S.C., Vaux, E.C., Cockman, M.E., Wykoff, C.C., Pugh, C.W., Maher, E.R., and Ratcliffe, P.J. (1999) The tumour suppressor protein VHL targets hypoxia-inducible factors for oxygen-dependent proteolysis. *Nature* **399**, 271–275.

92. Fischer, I., Gagner, J.P., Law, M., Newcomb, E.W., and Zagzag, D. (2005) Angiogenesis in gliomas: biology and molecular pathophysiology. *Brain Pathol.* **15**, 297–310.

93. Pause, A., Lee, S., Lonergan, K.M., and Klausner, R.D. (1998) The von Hippel-Lindau tumor suppressor gene is required for cell cycle exit upon serum withdrawal. *Proc. Natl. Acad. Sci. USA* **95**, 993–998.

94. Ohh, M., Yauch, R.L., Lonergan, K.M., Whaley, J.M., Stemmer-Rachamimov, A.O., Louis, D.N., Gavin, B.J., Kley, N., Kaelin, W.G., Jr., and Iliopoulos, O. (1998) The von Hippel-Lindau tumor suppressor protein is required for proper assembly of an extracellular fibronectin matrix. *Mol. Cell* **1**, 959–968.

95. Koochekpour, S., Jeffers, M., Wang, P.H., Gong, C., Taylor, G.A., Roessler, L.M., Stearman, R., Vasselli, J.R., Stetler-Stevenson, W.G., Kaelin, W.G., Jr., Linehan, W.M., Klausner, R.D., Gnarra, J.R., and Vande Woude, G.F. (1999) The von Hippel-Lindau tumor suppressor gene inhibits hepatocyte growth factor/scatter factor-induced invasion and branching morphogenesis in renal carcinoma cells. *Mol. Cell. Biol.* **19**, 5902–5912.

96. Ivanov, S.V., Kuzman, I., Wei, M.-H., Pack, S., Geil, L., Johnson, B.E., Stanbridge, E.J., and Lerman, M.I. (1998) Down-regulation of transmembrane carbonic anhydrases in renal cell carcinoma by wild-type von Hippel-Lindau transgenes. *Proc. Natl. Acad. Sci. USA* **95**, 12596–12601.

97. Kim, W.Y., and Kaelin, W.G. (2004) Role of *VHL* gene mutation in human cancer. *J. Clin. Oncol.* **22**, 4991–5004.

98. Lee, S., Chen, D.Y.T., Humphrey, J.S., Gnarra, J.R., Linehan, W.M., and Klausner, R.D. (1996) Nuclear/cytoplasmic localization of the von Hippel-Lindau tumor suppressor gene product is determined by cell diversity. *Proc. Natl. Acad. Sci. USA* **93**, 1770–1775.

99. Duan, D.R., Pause, A., Burgess, W.H., Aso, T., Chen, D.Y., Garrett K.P., Conaway, R.C., Conaway, J.W., Linehan, W.M., and Klausner, R.D. (1995) Inhibition of transcription elongation by the VHL tumor suppressor protein. *Science* **269**, 1402–1406.

100. Iliopoulos, O., Levy, A.P., Jiang, C., Kaelin, W.G., Jr., and Goldberg, M.A. (1996) Negative regulation of hypoxia-inducible genes by the von Hippel-Lindau protein. *Proc. Natl. Acad. Sci. USA* **93**, 10595–10599.

101. Iwai, K., Yamanaka, K., Kamura, T., Minato, N., Conaway, R.C., Conaway, J.W., Klausner, R.D., and Pause, A. (1999) Identification of the von Hippel-Lindau tumor-suppressor protein as part of an active E3 ubiquitin ligase complex. *Proc. Natl. Acad. Sci. USA* **96**, 12436–12441.

102. Pause, A., Lee, S., Worrell, R.A., Chen, D.Y.T., Burgess, W.H., Linehan, W.M., and Klausner, R.D. (1997) The von Hippel-Lindau tumor-suppressor gene product forms a stable complex with human CUL-2, a member of the Cdc53 family of proteins. *Proc. Natl. Acad. Sci. USA* **94**, 2156–2161.

103. Feldman, D.E., Thulasiraman, V., Ferreyra, R.G., Frydmann, J. (1999) Formation of the VHL-elongin BC tumor suppressor complex is mediated by the chaperonin TriC. *Mol. Cell.* **4**, 1051–1061.

104. Iliopoulos, O. (2001) von Hippel-Lindau disease: genetic and clinical observations. *Front. Horm. Res.* **28**, 131–166.

105. Carmeliet, P., Dor, Y., Herbert, J.-M., Fukumura, D., Brusselmans, K., Dewerchin, M., Neeman, M., Bono, F.,

Abramovitch, R., Maxwell, P., Koch, C.J., Ratcliffe, P., Moons, L., Jain, R.K., Collen, D., and Keshet, E. (1998) Role of HIF-1α in hypoxia-mediated apoptosis, cell proliferation and tumour angiogenesis. *Nature* **394**, 485–490.

106. Cockman, M.E., Masson, N., Mole, D.R., Jaakkola, P., Chang, G.-W., Clifford, S.C., Maher, E.R., Pugh, C.W., Ratcliffe, P.J., and Maxwell, P.H. (2000) Hypoxia inducible factor-α binding and ubiquitylation by the von Hippel-Lindau tumor suppressor protein. *J. Biol. Chem.* **275**, 25733–25741.

107. Seagroves, T., and Johnson, R.S. (2002) HIF-2α overexpression directly contributes to renal clear cell tumorigenesis: evidence for HIF as a tumor promoter. *Cell* **1**, 403.

108. Lonergan, K.M., Iliopoulos, O., Ohh, M., Kamura, T., Conaway, R.C., Conaway, J.W., and Kaelin, W.G., Jr. (1998) Regulation of hypoxia-inducible mRNAs by the von Hippel-Lindau tumor suppressor protein requires binding to complexes containing elongins B/C and Cul2. *Mol. Cell Biol.* **18**, 732–741.

109. Ohh, M., Park, C.W., Ivan, M., Hoffman, M.A., Kim, T.Y., Huang, L.E., Pavletich, N., Chau, V., and Kaelin, W.G. (2000) Ubiquitination of hypoxia-inducible factor requires direct binding to the beta-domain of the von Hippel-Lindau protein. *Nat. Cell Biol.* **2**, 423–427.

110. Wykoff, C.C., Pugh, C.W., Maxwell, P.H., Harris, A.L., and Ratcliffe, P.J. (2000) Identification of novel hypoxia dependent and independent target genes of the von Hippel Lindau (VHL) tumour suppressor by mRNA differential expression profiling. *Oncogene* **19**, 6297–6305.

111. Tanimoto, K., Makino, Y., Pereira, T., and Poellinger, L. (2000) Mechanism of regulation of the hypoxia-inducible factor-1 alpha by the von Hippel-Lindau tumor suppressor protein. *EMBO J.* **19**, 4298–4309.

112. Clifford, S.C., Cockman, M.E., Smallwood, A.C., Mole, D.R., Woodward, E.R., Maxwell, P.H., Ratcliffe, P.J., and Maher, E.R. (2001) Contrasting effects on HIF-1α regulation by disease-causing pVHL mutations correlate with patterns of tumourigenesis in von Hippel-Lindau disease. *Hum. Mol. Genet.* **10**, 1029–1038.

113. Hoffman, M.A., Ohh, M., Yang, H., Klco, J.M., Ivan, M., and Kaelin, W.G., Jr. (2001) von Hippel-Lindau protein mutants linked to type 2C VHL disease preserve the ability to downregulated HIF. *Hum. Mol. Genet.* **10**, 1019–1027.

114. Ivan, M., Kondo, K., Yang, H., Kim, W., Valiando, J., Ohh, M., Salic, A., Asara, J.M., Lane, W.S., and Kaelin, W.G., Jr. (2001) HIFα targeted for VHL-mediated destruction by proline hydroxylation: implications for O2 sensing. *Science* **292**, 464–468.

115. Jääkkola, P., Mole, D.R., Tian, Y.M., Wilson, M.I., Gielbert, J., Gaskell, S.J., Kriegsheim, Av., Hebestreit, H.F., Mukherji, M., Schofield, C.J., Maxwell, P.H., Pugh, C.W., and Ratcliffe, P.J. (2001) Targeting of HIF-α to the von Hippel-Lindau ubiquitylation complex by O2-regulated prolyl hydroxylation. *Science* **292**, 468–472.

116. Benn, D.E., Marsh, D.J., and Robinson, B.G. (2002) Genetics of pheochromocytoma and paraganglioma. *Curr. Opin. Endocrinol. Diabetes* **9**, 79–86.

117. Maranchie, J.K., Vasselli, J.R., Riss, J., Bonifacino, J.S., Linehan, W.M., and Klausner, R.D. (2002) The contribution of VHL substrate binding and HIF1-α to the phenotype of VHL loss in renal cell carcinoma. *Cancer Cell* **1**, 247–255.

117A. Kim, W.Y., and Kaelin, W.G., Jr. (2003) The von Hippel-Lindau tumor suppressor protein: new insights into oxygen sensing and cancer. *Curr. Opin. Genet. Dev.* **13**, 55–60.

118. Gnarra, J.R., Zhou, S., Merrill, M.J., Wagner, J.R., Krumm, A., Papavassiliou, E., Oldfield, E.H., Klausner, R.D., and Linehan, W.M. (1996) Post-transcriptional regulation of vascular endothelial growth factor mRNA by the product of the *VHL* tumor suppressor gene. *Proc. Natl. Acad. Sci. USA* **93**, 10589–10594.

119. Mukhopadhyay, D., Knebelmann, B., Cohen, H.T., Ananth, S., and Sukhatme, V.P. (1997) The von Hippel-Lindau tumor suppressor gene product interacts with Sp1 to repress vascular endothelial growth factor promoter activity. *Mol. Cell. Biol.* **17**, 5629–5639.

120. Brieger, J., Weidt, E.J., Schirmacher, P., Störkel, S., Huber, C., and Decker, H.J. (1999) Inverse regulation of vascular endothelial growth factor and VHL tumour suppressor gene in sporadic renal cell carcinomas is correlated with vascular growth: an in vivo study on 29 tumours. *J. Mol. Med.* **77**, 505–510.

121. Gallou, C., Joly, D., Mejean, A., Staroz, F., Martin, N., Tarlet, G., Orfanelli, M.T., Bouvier, R., Droz, D., Chretien, Y., Marechal, J.M., Richard, S., Junien, C., and Beroud, C. (1999) Mutations of the VHL gene in sporadic renal cell carcinoma: definition of a risk factor for VHL patients to develop an RCC. *Hum. Mutat.* **13**, 464–475.

122. Kondo, K., Klco, J., Nakamura, E., Lechpammer, M., and Kaelin, W.G., Jr. (2002) Inhibition of HIF is necessary for tumor suppression by the von Hippel-Lindau tumor suppressor protein. *Cancer Cell* **1**, 237–246.

123. Clifford, S.C., Prowse, A.H., Affara, N.A., Buys, C.H.C.M., and Maher, E.R. (1998) Inactivation of the von Hippel-Lindau (*VHL*) tumour suppressor gene and allelic losses at chromosome arm 3p in primary renal cell carcinoma: evidence for a *VHL*-independent pathway in clear cell renal tumorigenesis. *Genes Chromosomes Cancer* **22**, 200–209.

124. Jordan, D.K., Patil, S.R., Divelbiss, J.E., Vemuganti, S., Headley, C., Waziri, M.H., and Gurll, N.J. (1989) Cytogenetic abnormalities in tumors of patients with von Hippel-Lindau disease. *Cancer Genet. Cytogenet.* **42**, 227–241.

124A. Decker, H.-J., Klauck, S.M., Lawrence, J.B., McNeil, J., Smith, D., Gemmill, R.M., Sandberg, A.A., Neumann, H.H.P., Simon, B., Green, J., and Seizinger, B.R. (1994) Cytogenetic and fluorescence in situ hybridization studies on sporadic and hereditary tumors associated with von Hippel-Lindau syndrome (VHL). *Cancer Genet. Cytogenet.* **77**, 1–13.

125. Sprenger, S.H.E., Gijtenbeek, J.M.M., Wesseling, P., Sciot, R., van Calenbergh, F., Lammens, M., and Jeuken, J.W.M. (2001) Characteristic chromosomal aberrations in sporadic cerebellar hemangioblastomas revealed by comparative genomic hybridization. *J.Neurooncol.* **52**, 241–247.

126. Gijtenbeek, J.M.M., Jacobs, B., Sprenger, S.H.E., Eleveld, M.J., Geurts van Kessel, A., Kros, J.M., Sciot, R., van Callenbergh, F., Wesseling, P., and Jeuken, J.W.M. (2002) Analysis of von Hippel-Lindau mutations with comparative genomic hybridization in sporadic and hereditary hemangioblastomas: possible genetic heterogeneity. *J. Neurosurg.* **97**, 977–982.

127. Lemeta, S., Aalto, Y., Niemelä, M., Jääskeläinen, J., Sainio, M., Kere, J., Knuutila, S., and Böhling, T. (2002) Recurrent

DNA sequence copy losses on chromosomal arm 6q in capillary hemangioblastoma. *Cancer Genet. Cytogenet.* **133,** 174–178.

128. Lemeta, S., Pylkkänen, L., Saino, M., Niemelä, M., Saarikoski, S., Husgafvel-Pursiainen, K., and Böhling, T. (2004) Loss of heterozygosity at 6q is frequent and concurrent with 3p loss in sporadic and familial capillary hemangioblastomas. *J. Neuropathol. Exp. Neurol.* **63,** 1072–1079.

129. Knuutila, S., Aalto, Y., Autio, K., Bjorkqvist, A.M., El-Rafai, W., Hemmer, S., Huhta, T., Kettunen, E., Kiuru-Kuhlefelt, S., Larramendy, M.L., Lushnikova, T., Monni, O., Pere, H., Tapper, J., Tarkkanen, M., Varis, A., Wasenius, V.M., Wolf, M., and Zhu, Y. (1999) DNA copy number losses in human neoplasms. *Am. J. Pathol.* **155,** 683–694.

129A. Knuutila, S., Bjorkqvist, A.M., Autio, K., Tarkkanen, M., Wolf, M., Monni, O., Szymanska, J., Larramendy, M.L., Tapper, J., Pere, H., El-Rafai, W., Hemmer, S., Wasenius, V.M., Vidgrenn, V., and Zhu, Y. (1998) DNA copy number amplifications in human neoplasms. *Am. J. Pathol.* **152,** 1107–1123.

130. Richard, S., David, P., Marsot-Dupuch, K., Giraud, S., Beroud, C., and Resche, F. (2000) Central nervous system haemangioblastomas, endolymphatic sac tumors, and von Hippel-Lindau disease. *Neurosurg. Rev.* **23,** 1–22.

131. Lui, W.O., Chen, J., Gläsker, S., Bender, B.U., Madura, C., Khoo, S.K., Kort, E, Larsson, C., Neumann, H.P.H., and Teh, B.T. (2002) Selective loss of chromosome 11 in pheochromocytomas associated with the VHL syndrome. *Oncogene* **21,** 1117–1122.

132. Stratmann, R., Krieg, M., Haas, R., and Plate, K.H. (2000) Putative control of angiogenesis in hemangioblastomas by the von Hippel-Lindau tumor suppressor gene. *J. Neuropathol. Exp. Neurol.* **56,** 1242–1252.

133. Rickert, C.H., Hasselblatt, M., Jeibmann, A., and Paulus, W. (2006) Cellular and reticular variants of hemangioblastoma differ in their cytogenetic profiles. *Hum. Pathol.* **37,** 1452–1457.

134. Beckner, M.E., Sasatomi, E., Swalsky, P.A., Hamilton, R.L., Pollack, I.F., and Finkelstein, S.D. (2004) Loss of heterozygosity reveals non-*VHL* hemangioblastomas at 22q13. *Hum. Pathol.* **35,** 1105–1111.

135. Bindra, R.S., Vasselli, J.R., Stearman, R., Linehan, W.M., and Kalusner, R.D. (2002) VHL-mediated hyoxia regulation of cyclin D1 in renal carcinoma cells. *Cancer Res.* **62,** 3014–3019.

136. Zatyka, M., Fernandes da Silva, N., Clifford, S.C., Morris, M.R., Wiesener, M.S., Eckardt, K.-U., Houlston, R.S., Richards, F.M., Latif, F., and Maher, E.R. (2002) Identification of cyclin D1 and other novel targets for the von Hippel-Lindau tumor suppressor gene by expression array analysis and investigation of cyclin D1 genotype as a modifier in von Hippel-Lindau disease. *Cancer Res.* **62,** 3803–3811.

137. Baba, M., Hirai, S., Yamada-Okabe, H., Hamada, K., Tabuchi, H., Kobayashi, K., Kondo, K., Yoshida, M., Yamashita, A., Kishida, T., Nakaigawa, N., Nagashima, Y., Kubota, Y., Yao, M., and Ohno. S. (2003) Loss of von Hippel-Lindau protein causes cell density dependent deregulation

of cyclinD1 expression through hypoxia-inducible factor. *Oncogene* **22,** 2728–2738.

137A. Lemeta, S., Jarmalaite, S., Pylkkänen, L., Böhling, T., Husgafvel-Pursiainen, K. (2007) Preferential loss of the nonimprinted allele for the *ZACl* tumor suppressor gene in human capillary hemangioblastoma. *J. Neuropathol. Exp. Neurol.* **66,** 860–867.

138. Sherr, C.J. (1995) D-type cyclins. *Trends Biochem. Sci.* **20,** 187–190.

139. Donnellan, R., and Chetty, R. (1998) Cyclin D1 and human neoplasia. *J. Clin. Pathol:Mol. Pathol.* **51,** 1–7.

139A. Gijtenbeek, J.M.M., Boots-Sprenger, S.H.E., Franke, B., Wesseling, P., and Jeuken, J.W.M. (2005) Cyclin D1 genotype and expression in sporadic hemangioblastomas. *J. Neurooncol.* **74,** 261–266.

140. Miller, C.R., and Perry, A. (2004) Immunohistochemical differentiation of hemangioblastoma from metastatic clear cell renal carcinoma: an update. *Adv. Anat. Pathol.* **11,** 325–326.

141. Hufnagel, T.J., Kim, J.H., True, L.D., and Manuelidis, E.E. (1989) Immunohistochemistry of capillary hemangioblastoma. Immunoperoxidase-labeled antibody staining resolves the differential diagnosis with metastatic renal cell carcinoma, but does not explain the histogenesis of the capillary hemangioblastoma. *Am. J. Surg. Pathol.* **13,** 207–216.

142. Gouldesbrough, D.R., Bell, J.E., and Gordon, A. (1988) Use of immunohistochemical methods in the differential diagnosis between primary cerebellar haemangioblastoma and metastatic renal carcinoma. *J. Clin. Pathol.* **41,** 861–865.

143. Brown, D.F., Dababo, M.A., Hladik, C.L., Eagan, K.P., White, C.L., 3rd., and Rushing, E.J. (1998) Hormone receptor immunoreactivity in hemangioblastomas and clear cell renal cell carcinomas. *Mod. Pathol.* **11,** 55–59.

144. Andrew, S.M., and Gradwell, E. (1986) Immunoperoxidase labelled antibody staining in differential diagnosis of central nervous system haemangioblastomas and central nervous system metastases of renal carcinomas. *J. Clin. Pathol.* **39,** 917–919.

145. Roy, S., Chu, A., Trojanowski, J.Q., and Zhang, P.J. (2005) D2-40, a novel monoclonal antibody against the M2A antigen as a marker to distinguish hcmangioblastomas from renal call carcinomas. *Acta Neuropathol.* **109,** 497–502.

146. Jung, S.-M., and Kuo, T.T. (2005) Immunoreactivity of CD10 and inhibin alpha in differentiating hemangioblastoma of central nervous system from metastatic clear cell renal cell carcinoma. *Mod. Pathol.* **18,** 788–794.

147. Hoang, M.P., and Amirkhan, R.H. (2003) Inhibin alpha distinguishes hemangioblastoma from clear cell renal cell carcinoma. *Am. J. Surg. Pathol.* **27,** 1152–1156.

147A. Mottolese, C., Stan, H., Giordano, F., Frappaz, D., Alexei, D., and Streichenberger, N. (2001) Metastasis of clear-cell renal carcinoma to cerebellar hemangioblastoma in von Hippel Lindau disease: rare or not investigated? *Acta Neurochir. (Wien)* **143,** 1059–1063.

148. Bleistein, M., Geiger, K., Franz, K., Stoldt, P., and Schlote, W. (2000) Transthyretin and transferrin in hemangioblastoma stromal cells. *Pathol. Res. Pract.* **196,** 675–681.

149. Feldenzer, J.A., and McKeever, P.E. (1987) Selective localization of γ-enolase in stromal cells of cerebellar hemangioblastomas. *Acta Neuropathol.* **72,** 281–285.

150. Lach, B., Gregor, A., Rippstein, P., and Omulecka, A. (1999) Angiogenic histogenesis of stromal cells in hemangioblastoma: ultrastructural and immunohistochemical study. *Ultrastruct. Pathol.* **23,** 299–310.

151. Omulecka, A., Lach, B., Alwasiak, J., and Gregor, A. (1995) Immunohistochemical and ultrastructural studies of stromal cells in hemangioblastoma. *Folia Neuropathol.* **33,** 41–50.

152. Tanimura, A., Nakamura, Y., Hachisuka, H., Tanimura, Y., and Fukumura, A. (1984) Hemangioblastoma of the central nervous system: nature of the stromal cells as studied by the immunoperoxidase technique. *Hum. Pathol.* **15,** 866–869.

153. Bryant, J., Farmer, J., Kessler, L.J., Townsend, R.R., and Nathanson, K.L. (2003) Pheochromocytoma: the expanding genetic differential diagnosis. *J. Natl. Cancer Inst.* **95,** 1196–1204.

154. Adams, S.A., and Hilton, D.A. (2002) Recurrent haemangioblastoma with glial differentiation. *Neuropathol. Appl. Neurobiol.* **28,** 142–146.

155. Hande, A.M., and Nagpal, R.D. (1996) Cerebellar haemangioblastoma with extensive dissemination. *Br. J. Neurosurg.* **10,** 507–511.

156. Kato, M., Ohe, N., Okumura, A., Shinoda, J., Nomura, A., Shuin, T., and Sakai, N. (2005) Hemangioblastomatosis of the central nervous system without Hippel-Lindau disease: a case report. *J. Neurooncol.* **72,** 267–270.

157. Mohan, J., Brownell, B., and Oppenheimer, D.R. (1976) Malignant spread of haemangioblastoma: report on two cases. *J. Neurol. Neurosurg. Psychiatry* **39,** 515–525.

158. Slater, A., Moore, N.R., and Huson, S.M. (2003) The natural history of cerebellar hemangioblastomas in von Hippel-Lindau disease. *Am. J. Neuroradiol.* **24,** 1570–1574.

159. Weil, R.J., Vortmeyer, A.O., Zhuang, Z., Pack, S.D., Theodore, N., Erickson, R.K., and Oldfield, E.H. (2002) Clinical and molecular analysis of disseminated hemangioblastomatosis of the central nervous system in patients without von Hippel-Lindau disease. *J. Neurosurg.* **96,** 775–787.

160. Pause, A., Peterson, B., Schaffar, G., Stearman, R., and Klausner, R.D. (1999) Studying interactions of four proteins in the yeast two-hybrid system: structural resemblance of the pVHL/elongin BC/hCUL-2 complex with the ubiquitin ligase complex SKP1/cullin/F-box protein. *Proc. Natl. Acad. Sci. USA* **96,** 9533–9538.

161. Vortmeyer, A.O., Frank, S., Jeong, S.-Y., Yuan, K., Ikejiri, B., Lee, Y.-S., Bhowmick, D., Lonser, R.R., Smith, R., Rodgers, G., Oldfield, E.H., and Zhuang, Z. (2003) Developmental arrest of angioblastic lineage initiates tumorigenesis in von Hippel-Lindau disease. *Cancer Res.* **63,** 7051–7055.

162. Kishida, T., Stackhouse, T.M., Chen, F., Lerman, M.I., and Zbar, B. (1995) Cellular proteins that bind the von Hippel-Lindau disease gene product: mapping of binding domains and the effect of missense mutations. *Cancer Res.* **55,** 4544–4548.

163. Kamura, T., Koepp, D.M., Conrad, M.N., Skowyra, D., Moreland, R.J., Iliopoulos, O., Lane, W.S., Kaelin, W.G., Jr., Elledge, S.J., Conaway, R.C., Harper, J.W., and Conaway, J.W. (1999) Rbx1, a component of the VHL tumor suppressor complex and SCF ubiquitin ligase. *Science* **284,** 657–661.

164. Semenza, G.L. (2000) HIF-1 and human disease: one highly involved factor. *Genes Dev.* **14,** 1983–1991.

165. Semenza, G.L. (2000) Hypoxia, clonal selection, and the role of HIF-1 in tumor progression. *Crit. Rev. Biochem. Mol. Biol.* **35,** 71–103.

166. Knebelmann, B., Ananth, S., Cohen, H.T., and Sukhatme, V.P. (1998) Transformation growth factor α is a target for the von Hippel-Lindau tumor suppressor. *Cancer Res.* **58,** 226–231.

167. Zhong, H., De Marzo, A.M., Laughner, E., Lim, M., Hilton, D.A., Zagzag, D., Buechler, P., Isaacs, W.B., Semenza, G.L., and Simons, J.W. (1999) Overexpression of hypoxia-inducible factor 1α in common human cancers and their metastases. *Cancer Res.* **59,** 5830–5835.

168. Zhu, H., and Bunn, H.F. (2001) Signal transduction. How do cells sense oxygen? *Science* **292,** 449–451.

169. Ferrara, N. (2002) VEGF and the quest for tumour angiogenesis factors. *Nat. Rev. Cancer* **2,** 795–803.

5
Paraganglioma and Pheochromocytoma

Summary   Paragangliomas (PGL) are rare tumors, with those of the upper body (head and neck, and especially of the carotid body) being of parasympathetic origin, whereas abdominal PGL originate in the sympathoadrenal neuroendocrine system. The latter is the origin of pheochromocytomas (PCC). Chapter 5 deals with the changes affecting the mitochondrial complex II genes controlling the succinate dehydrogenase enzyme system (*SDHB*, *SDHC*, and *SDHD*) found in familial PGL. However, the preponderant number of PGL is of sporadic origin, and their genetic aspects have not been clearly elucidated. The relation to and association of PCC with some conditions and the role played by their gene such as von Hippel-Lindau (VHL) disease, multiple endocrine neoplasia 2, and neurofibromatosis 1 are presented in Chapter 5. The somewhat limited cytogenetic findings in PGL and PCC are discussed. The results of loss of heterozygosity and comparative genomic hybridization studies for these tumors are presented, and of the molecular aspects of *RET*, and other genes. Hypotheses on the role of the oxygen status in the development of PGL of the carotid body are included.

Keywords   paraganglioma · pheochromocytoma · SDH enzyme system · VHL · MEN-2 genes.

Introduction

Paragangliomas (PGL) and pheochromocytomas (PCC) are embryologically related tumors, sharing a neural-crest origin, several clinical features, and an overlapping genetic profile, although they do show variability as to site, histology, and biology. The occurrence of these tumors in familial settings, their association with hereditary syndromes, and their genetic alterations have been the subject of a sizeable literature (1–15, 15A, 15B).

The exact incidence of PGL or PCC is difficult to establish because of the rarity of these tumors and the fact that they may be either hereditary or sporadic. From the literature, the range seems to be <1 case per 100,000 to <1 case per 2,0000,000, this range probably depending on the geographic location of the patient population, the presence of a familial element and the bases for diagnosis.

Paraganglioma

Because the term "glomus" has been applied to some of the aggregations of paraganglial cells, e.g., glomus tympanicum and glomus jugulare, the term "glomus tumor" rather than "paraganglioma" has persisted in some publications (e.g., 16–19). True glomus tumors (glomangiomas) are small, benign mesenchymal neoplasms, the majority of which arise in the dermis or subcutis of the extremities, where PGL almost never occur. The most common site for glomangiomas is the subungual region of the fingers, followed by the palm, wrist, forearm, and foot, none of them a site of PGL (20). Glomangiomas arise from neuromyoarterial cells surrounding cutaneous arteriovenous anastomoses that serve as temperature regulators. These tumors are primarily sporadic in nature and familial cases are very rare. In one such family, no linkage to 11q23 was found (in contrast to PGL) (20). PGL are derived from paraganglia, a diffuse neuroendocrine system dispersed from the skull base to the pelvic floor.

Several patients with multiple PGL have been reported previously (18), with most of them being bilateral carotid body (a small oxygen-sensing organ located at the bifurcation of the carotid artery in the head and neck) PGL and sporadic in nature. A significant number of these patients had three or more PGL, involving in addition to the carotid body the paraganglia tympanicum, vagale, and jugulare (18). Malignant PGL of the carotid body can be catecholamine-secreting (20A) and metastasize (21, 22). Diagnostic and surgical aspects of treating multiple or malignant PGL have been discussed previously (23, 24).

Although PGL are very rare tumors, their hereditary nature in a significant percentage of cases, up to 30% (25,26) and the deciphering of the specific genes associated with these tumors has been a fascinating and unique development.

PGL are highly vascularized, slow-growing and generally benign tumors arising mainly in the neural-crest cells (paraganglionic chemoreceptors) of the parasympathetic system of the head and neck (80% of cases) or the intra-abdominal paraganglia (17%) of the sympathoadrenal neuroendocrine system (27). PGL of the central nervous system (CNS) are very rare and occur almost exclusively in the cauda equina (28). PGL located in the pituitary fossa has been described (29). PGL also may occur at unusual locations, e.g., gallbladder (30), spine (31), and in the abdomen associated with cutaneous angiolipoma (32). Most of PGL occur sporadically, and an uncertain percentage of PGL are associated with hereditary conditions, e.g., von Hippel-Lindau (VHL) syndrome, multiple endocrine neoplasia-2 (MEN-2), and familial PGL and gastrointestinal stromal tumors (GIST).

The carotid body is the most common tumor site for PGL (33). Although the carotid bodies are very small (<4 mm in diameter and combined weight <15 mg) (27), even when not malignant, PGL of these bodies may grow to be of sufficient size to encroach upon cranial nerves or spread by embolization causing, for example, facial palsy (34, 34A, 34B) or metastasize when malignant (21).

Catecholamine-secreting PGL account for approximately 20% of all catecholamine-secreting tumors (35), and although at one time it was thought that such PGL were associated with a higher rate of malignancy than PCC, other studies have described an equivalent rate (36, 37). Malignant changes (38) are more common in sporadic PGL (12%) than in the familial type (2.5%).

The presence of both PGL and PCC in some patients has been reported, e.g., approximately 5% of Dutch patients with head and neck PGL also had PCC (39–43).

It is rare for head and neck PGL of parasympathetic origin to be secreting tumors (44), whereas PCC and abdominal PGL both of sympathoprogenitor origin are frequently functional (syndromic). Moreover, most patients with head and neck PGL develop a slow-growing asymptomatic vascular mass, whereas patients with PCC suffer from hypertensive crises and paroxysmal symptoms related to the secretion of catecholamines such as norepinephrine and epinephrine (45).

PGL also can occur in the parasympathetic paraganglia adjacent to the aortic arch, neck, and skull base as local "nonfunctioning" masses, commonly referred to as head and neck PGL. Less common are jugulotympanic and vagal PGL (46).

A concise and informative review of head and neck PGL has been published (47), including references to the early literature. In fact, a whole issue of a journal was devoted to PGL of the head and neck in which a wide range of areas related to these tumors was presented (47). A short but informative

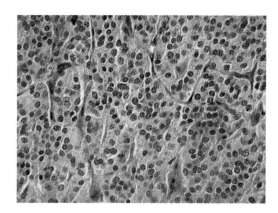

FIGURE 5.1. Histologic presentation of a PGL. The tumor is well differentiated and resembles a normal paraganglion, composed of chief cells arranged in nests surrounded by sparse and inconspicuous single layer of sustentacular cells (from ref. (293), with permission).

history of PGL, the nomenclature related to it, and incidence and malignant transformation were presented (47).

PGL of the head and neck consist of large, oval-shaped cells or chief cells of neuroectodermal origin that are usually arranged in clusters surrounded by flat, elongated satellite or sustentacular cells (27) (Figure 5.1). The chief cells are considered responsible for chemoreceptor activity. In approximately 1% of PGL, the chief cells contain neurosecretory granules that can release catecholamines akin to those in PCC. The sustentacular cells lack these granules and neither secrete nor contain catecholamines (48), and they are considered supportive in nature. It is difficult to predict PGL tumor behavior histologically, but vascular invasion and tumor necrosis have been suggested as features correlating with malignant behavior (46). The extreme vascularity of PGL may be related to the expression of angiogenic growth factors, i.e., vascular endothelial growth factor (VEGF) and platelet-derived-VEGF (48A). Experiences of the clinical, genetic counseling, diagnostic, and other aspects (49) of PGL have been described previously (50–57).

Pheochromocytoma

PCC is preponderantly a tumor of the adrenal medulla; in rare cases, PCC may originate in abdominal paraganglia, especially in the organ of Zuckerkandl (58) or its remnants, the urinary bladder, and the thorax (27, 44, 58A, 59, 59A). About 10% of PCC may be associated with VHL disease, the MEN-2 syndrome, neurofibromatosis-1 (NF1), and rarely with other conditions (e.g., Carney triad, Sturge–Weber syndrome) (Tables 5.1 and 5.2).

PCC, considered by some as a PGL of the adrenal medulla, is a rare neuroendocrine chromaffin-staining tumor of neural-crest origin that usually causes hypertension and clinical symptoms by oversecretion of catecholamines (27, 60–62, 83A). Very rarely, PCC also may arise at extra-adrenal

TABLE 5.1. Hereditary syndromes associated with pheochromocytoma.[a]

Syndrome gene	Gene locus	Clinical phenotype	Risk of pheochromocytoma (%)	Mutated germline
MEN-2A	10q11.2	Medullary carcinoma of the thyroid, hyperparathyroidism	50–70	*RET*
MEN-2B	10q11.2	Medullary carcinoma of the thyroid, multiple mucosal neuromas, marfanoid habitus, hyperparathyroidism, skeletal abnormalities	50–70	*RET*
Neurofibromatosis type 1	17q11.2	Neurofibromas and schwannomas, MPNST of peripheral nerves, café au lait spots	1–5	*NF1*
von Hippel-Lindau disease	3p25	Retinal angiomas, CNS hemangioblastoma, renal cell carcinoma, pancreatic and renal cysts	10–20	*VHL*

[a] Based on Dluhy 2002 (66) and Thompson 2005 (296).

TABLE 5.2. Features of PCC in MEN-2A and sporadic PCC.

MEN-2A	Sporadic
Age 38 ± 11 yrs	Age 47 ± 16 years
52% asymptomatic	Nearly 100% asymptomatic (hypertension)
Usually bilateral	Unilateral

Based on Pomares et al. 1998 (307).

sites, e.g., the meninges (63) or the cerebellopontine angle (64).

The adrenal medulla and by extension PCC is keyed to respond to stress (e.g., cold, hypoglycemia, hemorrhage, immobilization, and hypoxia) by secreting catecholamines (e.g., adrenalin, noradrenalin, and dopamine).

Extra-adrenal paraganglial cell clusters and organs of sympathetic nervous system origin include that of Zuckerkandl, and prevertebral and paravertebral thoraco-abdominal and pelvic paraganglia. Other sites include the ovary, testis, vagina, urethra, prostate, bladder, and liver (27, 65). These sites seem to respond to stress in a manner similar to that of the adrenal medulla.

The endocrine (chromaffin) cells of the sympathioadrenal neuroendocrine system, from which PCC originate, synthesize and secrete catecholamines, and they exhibit a characteristic histochemical (chromaffin) reaction when treated with oxidizing agents (27, 44). When stored catecholamines are released by a PCC, patients experience a paroxysm of episodic headache, sweating and palpitations, besides the sustained hypertension found in these patients. The unrecognized pres-

ence of this tumor may result in a lethal hypertensine crisis, cardiac arrhythmia, or myocardial infarction (66).

Histologically, PCC typically have an "organoid" structure composed of cells arranged in compartments separated by thin vascular septa (Figure 5.2). The cell size varies from small to large, and shapes vary from round to epithelioid, and they are occasionally spindled. The cytoplasm varies from eosinophilic to basophilic; the latter variants often have a granular cytoplasm. Nuclear pleomorphism is not uncommon, and some tumors have nuclear pseudoinclusions (cytoplasmic invaginations). Eosinophilic cytoplasmic globules occur in some cases, and a minority have lipofuscin (neuromelanin) pigmentation. Extensive hemorrhage or stromal fibrosis may be present (62). PCC cells may be chromaffin-positive (27, 44). Distinct differences in the histopathologic phenotypes of PCC associated with VHL disease versus PCC associated with MEN-2 have been described previously (67).

Past concepts regarding PCC as 10% tumors, i.e., that 10% of PCC are malignant, 10% are bilateral, 10% are extra-adrenal (of which 10% are extraabdominal), 10% not associated with hypertension, and 10% as hereditary tumors, have been challenged, if not abrogated, by the results of the studies presented in this chapter, especially those obtained after 2000 (26, 68).

A presentation dealing with the genetics and diagnostic aspects of PCC, especially the changes and levels of biochemical substances (e.g., catecholamine, metanephrine, and vanillymandelic acid), and the therapeutic approaches for the treatment of PCC has been published (69). The frequency

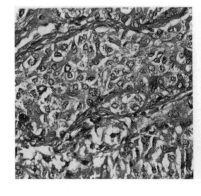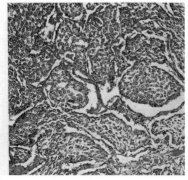

FIGURE 5.2. Pheochromocytoma of the adrenal gland showing (right) typical "Zellballen" configuration and (left) higher power showing the abundant granular cytoplasm of the tumor cells (from ref. (493), with permission).

TABLE 5.3. Baseline characteristics of 314 patients with PCC.

Characteristic	All patients (no.)	FS patients		SP patients	
		No.	%	No.	%
No.	314	56	17.8	258	82.2
Sex					
Male	140	28		112	
Female	174	28		146	
Age at diagnosis, years					
Mean	41.3		30.2		43.6
Range	7–80		10–69		7–80
Adrenal tumors	264	49	87.5	215	83.3
Bilateral tumors	41	31	55.4	10	3.9
Extra-adrenal tumors	58	10	17.9	48	18.6
Malignant tumors	52	5	8.9	47	18.2
Tumor diameter, mm					
Mean	59.2		52.2		60.3
SD	34.7		28.5		35.5

FS, patients with a family history and/or a syndromic presentation; SP, patients with an apparently sporadic presentation; SD, standard abbreviation.
Based on Amar et al. 2005 (73).

TABLE 5.4. Summary of LOH results (%) according to diagnosis and chromosomal region.

Diagnosis	LOH in region		
	1p36.2~pter	1p32	1cen~p13
Pheochromocytomas (PCC)	22/42 (52)	28/42 (67)	28/42 (67)
Benign sporadic	15/28	19/27	20/27
Malignant	3/5	4/6	4/6
Benign hereditary	4/9	5/9	4/9
MEN-2A	3/5	3/5	3/5
VHL	0/3	1/3	0/3
NF1	1/1	1/1	1/1
Paragangliomas (PGL)	5/5 (100)	5/7 (71)	5/7 (71)
Benign	2/2	1/3	2/3
Malignant	3/3	4/4	3/4
All	27/47 (57)	33/49 (67)	33/49 (67)

Based on Edström Elder et al. 2002 (72).

of malignancy in all PCC has been reported to range from 13 to 26% (70). Other facets of malignant PGL and PCC have been described previosly (35, 71). In Tables 5.3 and 5.4 are shown clinical and genetic aspects of PGL and PCC as observed in two different clinics (72, 73).

Clinical screening and diagnostic aspects of PCC have been discussed in a number of publications (74–83, 83A).

Cytogenetic and Comparative Genomic Hybridization (CGH) Studies in PGL and PCC

No cytogenetic studies have been reported in PGL, and the information related to chromosomal changes in these tumors has been primarily supplied by CGH studies.

Although the number of PCC karyotypically characterized is relatively small, several aspects are worthy of note (Figure 5.3 and Table 5.5). The finding of trisomy 7 (+7)

in most PCC analyzed cytogenetically, including its presence with other cytogenetic abnormalities, may be significant and constitute an important change in the formation, progression, or both of PCC (84). Of interest is a PCC with the karyotype 46,XX,der(16)t(1;16)(q31;p13), and other changes (Table 5.5) in a 47-year-old woman, whose tumor was diagnosed after the finding of a PCC and other stigmata of VHL in her 20-year-old son. The tumor in the latter patient did not have chromosome changes. Both tumors were thought not be malignant (85). The other two complex karyotypes in Table 5.5 (86) were established in two malignant PCC. It would seem that PCC may be associated with rather simple karyotypic changes, although with progression complex karyotypes, which may play a role in the biology of these tumors, may develop.

It should be pointed out that Table 5.5 contains only those PCC in which cytogenetic abnormalities were found; normal karyotypes, not infrequently seen in PCC, including a case with VHL, have not been included (85–87).

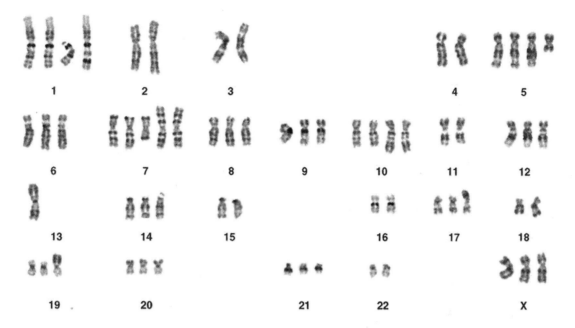

FIGURE 5.3. Banded complex karyotype of an advanced sporadic pheochromocytoma containing 64 chromosomes (86). Note the five chromosomes 7 (three structurally abnormal) and besides the numerical changes, structural ones affecting chromosomes 5, 10, 13, 17, 19, and 20.

TABLE 5.5. Chromosome changes in pheochromocytoma.

Karyotype	References
47,XX,+7*	*Jordan et al 1989* (87)
47,XX,+7*	*Kiechle-Schwarz et al 1989* (85)
46,XX,der(16)t(1;16)(q31;p13)/46,XX,der(8)t(3;8)(p21;p23)/ 46,XX,dic(3;4)(p16;p26)del(3)(q13)/46,XX,der(4)t(3;4)(p21;p16)*	*Kiechle-Schwarz et al 1989* (85)
47,XX,+7* (two cases)	*Decker et al 1994* (308)
46,XX,del(1)(p31-p33),del(4)(q31),del(6)(q26?),del(8q)(q21.2 or q21.3), del(13)(q32)**	*Pfragner et al 1998* (89)
47,X,der(X),add(3)(q26 or q27),+7**	*Pfragner et al 1998* (89)
59-64,XY,der(X)t(X;22)(p11;p11),add(3)(p12),-4,+7,add(7)(q21)x2, -8,add(8)(q24),-9,add(9)(p23),add(11)(p15),-12,add(12)(q24),-13, add(14)(p11)x2,-15,-16,-17,-18,+20,-21,-22,add(22)(p11),+mar	*Gunawan et al 2004* (86)
60-68,XX,-Y,+X,del(1)(p13),+dic(1;16)(q22;p13),-2,-4,+i(5)(p10), +7,+7,i(7)(p10),i(7)(q10)x2,+10,der(10)t(3;10)(q25;q26)x2,-11, -13,-13,i(13)(q10),-15,-16,add(17)(p12),-18,add(19)(p13),-22	*Gunawan et al 2004* (86)

*Associated with VHL.
**Cell lines.

Fluorescence in situ hybridization (FISH) studies with alphoid probes in archival tissues of 23 PCC (18 primary and five metastatic) revealed overrepresentation of chromosomes 1 and 7 and loss of chromosome 15 (88).

A continuous cell line of a human sporadic PCC was shown to contain a hypodiploid clone with -19, -17, -21, or -22. Deletions of 1p and 4q were more frequent than mutations of the *RET* gene in this cell line (89).

In CGH analyses of 29 PCC (four with family history) and 24 PGL (five with family history) (90), the PCC had a homogeneous CGH profile, each containing more than four genomic imbalances, whereas the CGH profile of the PGL revealed fewer alterations (Figure 5.4). The most frequent rearrangement found in the PGL was total or partial loss of chromosome 1 (67%), mainly involving 1p21-p32, 1q32, and

1q42-qter. Loss of 1q was found only in PGL, whereas in PCC 1p was frequently lost (69%), constituting the most common change in sporadic PCC (91, 92). Other studies (90) found that 72% of the PCC, versus only 37.5% of the PGL, showed partial or total loss of chromosome 3. Loss of chromosomal arm 3q has been reported as frequent in PCC (Figure 5.5) (91) and in abdominal PGL (Figure 5.6) (92). The changes of chromosomes 1 and 3 probably play a key role in the pathogenesis of both chromaffin and nonchromaffin tumors (Figures 5.4–5.6).

Loss involving 8p (62% of cases), gains of 17q and 12q (48 and 45% of cases, respectively), and losses of 18p and 9p (34 and 31% of cases, respectively) were found by CGH in PCC (90). Gain of the short arm of chromosome 19 was

A

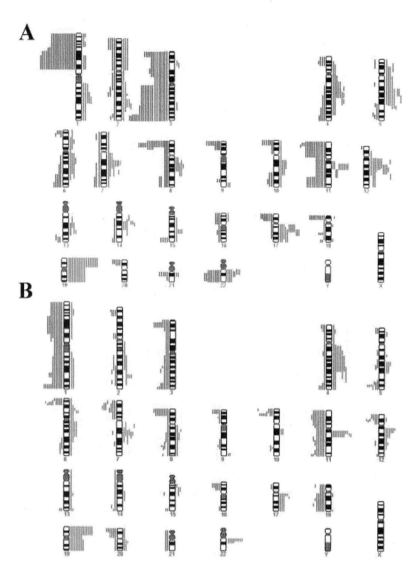

B

FIGURE 5.4. DNA copy number changes identified by CGH in 29 PCC and 24 PGL (20 of head and neck) (90). Note prominent losses of 1p, chromosome 3, especially 3q, and chromosome 11 and gain of chromosome 19 in the PCC. Similar but not as frequent changes were seen in PGL.

frequent in PGL (54%) and in PCC (59%) (90). Partial loss of 6p, 7p, 8p, and 10p and or chromosome 18 was found along with a gain of 12q. At least 50% of parasympathetic PGL were found to have loss of all or part of chromosome 11 (Figure 5.7) (90, 91). The differences of the CGH profiles of PCC versus PGL, e.g., loss of 3q in PCC, loss of 1q only in PGL, and loss of 8p22-p23 in PCC and PGL, suggest that genes related to specific tumor development may be located in distinct chromosomal regions (90). Also, gain of 11q13 may be significantly associated with malignant PCC and PGL.

In general, the frequency of chromosomal anomalies in the CGH study just discussed (90) was greater than in previous studies, possibly related to technical advances in CGH software that have enhanced the capability of detecting losses (93). Because the PGL studied (90) were mainly located in the head and neck region, it is not surprising that the CGH alterations of these parasympathetic PGL do not correspond to those found in PCC. Furthermore, it has been reported that parasympathetic PGL are structurally and func-

tionally different from those belonging to the sympathoadrenal neuroendocrine system (27).

PCC is rare in children, and it is associated with a poor prognosis (94). A CGH study of 14 such tumors showed combined loss of 3p and 11p *or* whole chromosomes in 10 of 14 PCC. All 10 cases with 3p and 11p loss carried *VHL* mutations (94), nine constitutive and one somatic restricted to the tumor. Of the four remaining cases without *VHL* involvement, two had other familial syndromes (one of *NF1* and of *SDHD*) and two were of unknown etiology. Thus, true sporadic PCC is rare in children, and such cases should be checked for a related predisposing gene (94, 95).

A sporadic retroperitoneal PGL in a 16-year-old girl with a metastasis to the first lumbar vertebra was examined by CGH revealing loss of 1p and chromosome 3 and gains of 1q, chromosomes 4 and 5, 11q, and 13q (95A). No telomerase activity was found in the tumor.

Loss of 8p22-p23 in the five PCC with signs of malignancy, along with gain of 11cen-q13, is of interest in that these

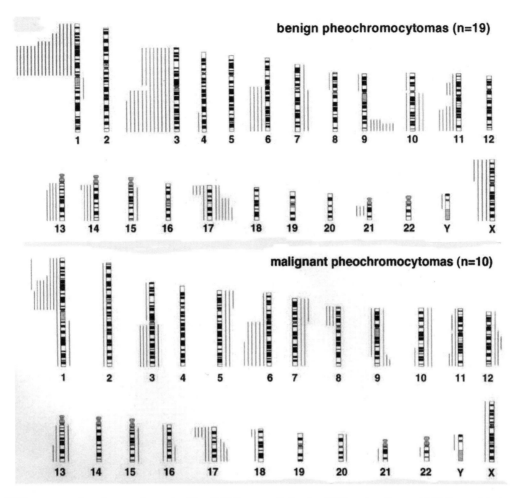

FIGURE 5.5. DNA copy number changes determined by CGH in sporadic benign PCC and malignant PCC. Loss of 1p is prominent in both groups of PCC, whereas loss of chromosome 3 or 3q is more evident in benign tumors than in the malignant ones. Note 6q loss in both groups (from ref. 91, with permission).

tumors showed more accumulated alterations than usually seen in benign tumors (90). Half of the malignant PGL (three of six) showed both loss of 8p and gain of 11q13, suggesting these alterations could be markers of malignancy. However, both alterations were seen together in tumors without signs of malignancy, although in a much smaller proportion of tumor types (three of 23 PCC and four of 19 PGL). A CGH analysis (96) characterized the genetic profiles of 36 VHL-related PCC, the results of which were compared with those of sporadic and five MEN-2–related PCC. In the 36 VHL-related tumors, loss of chromosomes 3 and 11 was found in 34 tumors (94%) and 31 tumors (86%), respectively. There was significant concordance of deletions in chromosomes 3 and 11, suggesting that they are involved in two different but necessary and complementary genetic pathways. The loss of chromosome 11 seemed to be specific for VHL-related PCC, because it was not present in any of the 10 VHL-related CNS hemangioblastomas studied, and it was significantly less common in sporadic and MEN-2–related PCC. The authors (96) concluded that sporadic and MEN-2–related PCC have a distinct genetic pathway versus that of VHL-related cases. Loss of 1p was very common in the former two types of PCC but not in VHL-related cases. Similar observations had been made (103) in sporadic PCC in which 61% had loss of 1p and in 80% of MEN-2–related PCC. No 1p loss was found in the 2 VHL-related PCC. Based on array-CGH studies, sporadic PCC were found to have as the most common change loss of chromosome 22 (44% of 66 tumors), either as monosomy or terminal deletion of 22q (97). Another common change was loss of 1p.

Microarray-based CGH of mouse PCC cell lines showed changes similar to those of human PCC (98).

Hybridization performed by FISH, with a probe containing the *FBXO25* gene (90) confirmed that, with one exception, almost all the evaluable PCC with 8p22-p23 deletions by CGH showed genomic deletion involving this gene. However, analysis of the *FBXO25* gene by sequencing its eight exons did not find any alteration in the 15 tumors with loss of 8p22-p23 and thus indicates that this gene probably does not play an important role in PCC or PGL development (90).

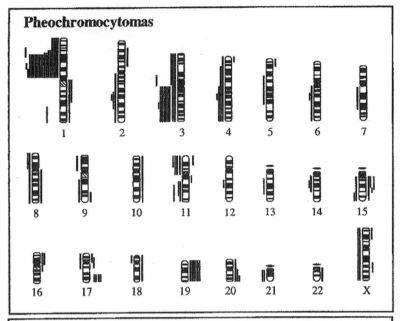

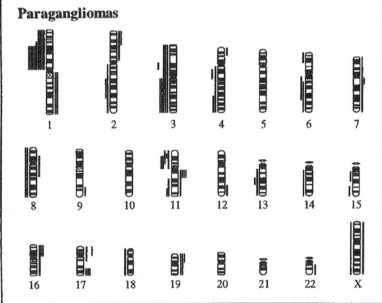

FIGURE 5.6. DNA copy number alterations detected by CGH in 23 PCC (seven sporadic, five in MEN-2, four malignant, six unknown, and one in NF) and 11 abdominal PGL (three benign and eight malignant). Each line represents one change detected in a single tumor with losses shown on the left and gain on the right of each chromosome. Prominent are the losses of 1p and 3q in PCC and to lesser extent in PGL. Loss of 11q is to be noted (from ref. (92), with permission).

Loss of Heterozygosity (LOH) in PGL and PCC

Loss of heterozygosity or LOH on specific chromosomes or chromosomal regions have been extensively studied in hereditary and sporadic PGL and PCC (Table 5.6).

Early studies (99) indicated LOH in PCC (sporadic and familial) at 1p, 22q, 17p, and 3p, in the order of their frequency. LOH at 3p25.5 (locus of *VHL* gene) and 3p21 was reported in PCC associated with VHL, and some sporadic PCC (100, 101). In one study, LOH at 1p34-p36 was found in 5/11 (45%) of PCC, but not in PGL (102); LOH at 3p25 was present in five or nine (45%) of sporadic PCC and in none of

the PGL, although LOH at 3p21 was seen in two of four PGL. Subsequent LOH studies of PCC and PGL confirmed and extended, e.g., LOH of chromosome 11, the involvement of the chromosomes listed above in PCC and PGL, although the percentage of cases affected by these changes varied among studies (92, 96, 103–105). These findings are indicative that multiple genes are involved in the formation, progression, or both of PCC and PGL. Especially frequent are LOH of 1p, primarily in sporadic PCC and MEN-2–associated tumors, a chromosomal area thought to harbor key genes in PCC and PGL tumorigenesis.

Loss of 3p has been reported in both familial and sporadic PCC (103). In one study, 91% of PCC associated with VHL had LOH at 3p and 21 and 15% of 22q and 1p (106), whereas

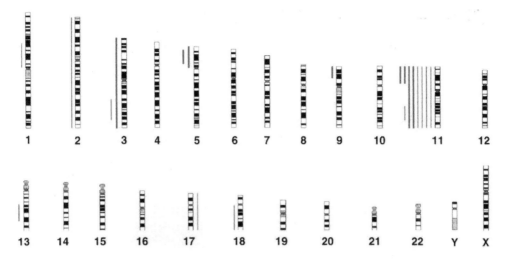

FIGURE 5.7. DNA copy number alterations detected by CGH in 16 PGL (nine sporadic and seven familial), primarily of head and neck. Note loss of chromosome 11 or 11p in eight cases, the most prominent finding in this study. The difference in the findings of this study and those of Figure 5.6 may be related to the fact that in the latter study only abdominal PGL were investigated (from ref. 105, with permission).

TABLE 5.6. Published Studies on Loss of Chromosomal Material in PCC and PGL.

Chromosome arm	PCC or PGL	Sporadic or Familial		References
1p	PCC	+	+	*Benn et al 2000 (103)*
11q	PGL	+		*Bikhazi et al 2000 (104)*
1p and 3q	PCC	+		*Dannenberg et al 2000 (91)*
11q, 11p, 5p	PGL	+	+	*Dannenberg et al 2001 (105)*
1p	PGL and PCC	+	+	*Edström Elder et al 2002 (72)*
1p and 11q	PCC and PGL	+	+	*Edström et al 2000 (92)*
1p, 3p, 10q, 17q, 22q	PCC	+	+	*Khosla et al 1991 (99)*
3p and 11p	PCC	+		*Lui et al 2002 (96)*
1p	PCC	+	+	*Moley et al 1992 (107)*
1p	PCC	+	+	*Opocher et al 2003 (114)*
1p and 22q	PCC	+	+	*Shin et al 1993 (120)*
22q	PCC	+	+	*Takai et al 1987 (117)*
				Tanaka et al 1992 (119)
3p	PCC	+	+	*Tory et al 1989 (100)*
				Zeiger et al 1995 (101)
1p and 11p	PCC	+		*Tsutsumi et al 1989 (309)*
				Yokogoshi et al 1990 (310)
1p and 3p	PCC and PGL	+	+	*Vargas et al 1997 (102)*
1p, 3p, 17p, 22q	PCC	+	+	*Pfranger et al 1998 (89)*
1p, 3p, 22q	PCC	+	+	*Bender et al 2000 (106)*

sporadic PCC had LOH of 3p in 71% of the tumors, 53% at 22q and 24% at 1p. In both types of PCC, no intragenic mutations of *VHL* or *RET* were found (106). Genes other than *VHL*, especially on 1p, are significant for sporadic PCC development, and they suggest that the genetic pathways involved in sporadic versus VHL PCC genesis are distinct (106).

Loss of 1p is present in 80% of MEN-2–associated and sporadic PCC and only in approx15% of VHL-associated PCC, indicative of at least two different genetic pathways for PCC (96, 106). Three commonly involved regions on 1p have been identified: 1p36 harboring genes for PCC (72, 102, 103, 107), medullary thyroid carcinoma 99, 108, 108A), parathyroid adenoma (109, 109A), and neuroblastoma (110–112). At

least three tumor suppressor genes are present within 1pter-1p34 (103), with *SDHB* located at 1p36.13 (113). The second and third target regions on 1p, shown primarily by CGH, were 1p22-p32 and 1cen-1p31 (72, 91, 92, 114). Consensus regions of deletion were defined in sporadic and MEN-2 PCC in a study using microarray-based CGH with BAC spanning 1p (115), i.e., 1cen-1p21.1, 1p21.3-1p31.3, and 1p34.3-1p36.33.

A range of LOH of 3p has been found in sporadic PCC, being as high as 75% (101) and as low as 15% (99), although the authors (102) of another study of 3p LOH in PCC (55% of the cases) attributed the low number to methodological aspects. LOH of 1p (1p34-p36) and 3p (3p21 and 3p25) in 12 sporadic PCC and five extra-adrenal PGL without hereditary history was investigated (102). LOH of the 1p locus was found in five of 11 PCC (45%) and in none of PGL. LOH at 3p25 (locus of the *VHL* gene) was found in five of nine (55%) PCC and none in the three PGL examined. LOH at 3p21 was found in two or four of the PGL. The allelic deletions at 1p and 3p seemed to be separate events. The (102) results suggest that extra-adrenal PGL may have a different molecular mechanism of tumorigenesis than PCC. LOH of 1p was seen in >40% of the sporadic PCC in the studies just mentioned. LOH studies in sporadic, MEN-2-, NF1-, and VHL-associated PCC have frequently shown loss of 1p, 3p, 3q, 17p, 17q, and 22q (99, 101, 107, 108, 116–121). This suggests that these areas (especially 1p and 3p) may contain tumor suppressor genes important for the development or progression of these tumors. Frequent LOH of 6p in PCC has been described (17 sporadic- and one VHL-associated (122)).

LOH studies of PGL have not been as numerous as those of PCC (Table 5.6). A study of 5 sporadic extra adrenal PGL showed no LOH of 1p or of 3p25 in three of these tumors. However, LOH at 3p21 was found in 2/4 PGL examined. The results of PCC and PGL suggest that extra-adrenal PGL may have a different molecular mechanism of tumorigenesis than PCC (102).

LOH at 11q13 and 11q22-q23, loci for PGL2 and PGL1 respectively, was seen in three of eight sporadic PGL of the head and neck, one of the former locus and two of the latter (104). The finding of complete LOH at 11q22-q23 in the aneuploid fraction of sporadic and hereditary head and neck PGL (123) indicates that sporadic and familial PGL share a similar molecular pathogenesis. LOH for 1p in PGL was common in malignant tumors (four of four) and in two of three in other sporadic PGL (72). Genetic testing of members of families with PGL has been advocated, using mapping of the loci on 11q13 and 11q23 (124, 125).

Gains of 11q13, as part of gain of region 11cen-q13, seem to characterize abdominal PGL showing malignant features or are overtly malignant, and malignant PCC (90, 92, 126).

Molecular Genetics of PGL and PCC

Both PGL and PCC are associated with complex and heterogeneous genetics, related not only to the occurrence of hereditary forms and sporadic types of these tumors but also to the complexity of the genetic pathways and as yet undeciphered and unknown genes of causal relationship to these tumors (127). It has already been indicated that the incidence of hereditary PGL and PCC is underestimated (25), possibly due to a number of factors.

Why *RET*, *VHL*, *SDHB*, and *SDHD* mutations should be a frequent cause of hereditary PGL and PCC, whereas somatic mutations are rare in sporadic tumors (128–133), is unclear. Possibly, mutations in these genes only promote tumorigenesis if they are present at a specific stage of organ development (90). It is also likely that other genes may be responsible for these tumors (134).

For example, the amplification of the normal allele in heterozygous patients may not allow the detection of rare mutations such as large deletions and chromosomal rearrangements (135, 136). Thus, the number of patients with germline mutations may be slightly higher (137). About 10% of sporadic PCC are hereditary. Young age at diagnosis and extra-adrenal or bilateral PCC are predictive of a germline mutation (73, 137, 138).

Genetic testing is proposed for all cases with PCC or functional PGL as depicted in an algorithm suggested by the authors (Figure 5.8) (137) and as reflected in some of the pedigrees in the literature (Figure 5.9).

In a recent study using all coding exons for *SDHB*, *SDHC*, *SDHD*, and *VHL*, and six of *RET* exons (137), only one causative mutation per patient (PGL or PCC) was found. *SDHB* mutations were found in 15 of 18 patients with malignant tumors (137), a higher number than reported by other (138A). The former authors (73) discuss possible factors responsible for the difference (e.g., clinical evaluation, follow-up, and source of recruitment). Disease associated with *SDHB* mutations showed more aggressiveness and a poor prognosis than those with other changes (73). Further fundamental and therapeutic research is needed to discover new therapeutic

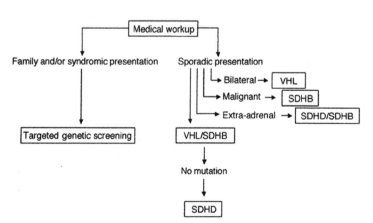

FIGURE 5.8. Suggested protocol for genetic testing in PCC and functional PGL after a thorough evaluation of affected patients (from ref. 137, with permission).

FIGURE 5.9. Examples of familial PCC in three families. The ages of the patients at PCC diagnosis are shown underneath the affected individuals. In none of the families was MEN-2–associated *RET* mutated; missense *VHL* mutations were identified in two of these kindreds (family A and family 386), with no clinical evidence of VHL disease (from ref. 3, with permission).

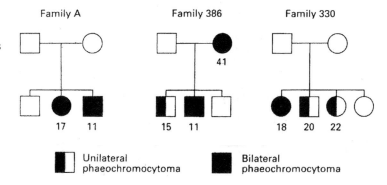

targets and to determine curative and preventive protocols for malignant PCC (71, 73).

Type 2C VHL mutants maintain normal regulation of hypoxia-inducible factor (HIF)α, suggesting that distinct VHL functions might be affected in 2C tumors (139, 140).

SDHB mutations seem to be associated with a higher risk of extra-adrenal PCC, whereas those of *SDHD* often have head and neck PGL (12, 138A); the genotype-phenotype of *SDH* mutations seem to be broader than initially described, e.g., the occurrence of renal cell carcinoma (RCC) and medullary thyroid carcinoma (138A, 141).

In addition to the hypoxic response, an oxydoreductase imbalance involving both PCC with dysfunction of SDH and certain tumors with *VHL* mutations has been revealed by expression profiling (142).

Using an integrated genomics approach involving lineage, expression profiling, and genome-wide deletion mapping, two novel loci that concurrently increase PCC predisposition were identified (142).These familial PCC loci were found at 2q (acting as a tumor suppressor gene [TSG]) and another on 16p, coinciding with a locus for familial neuroblastoma.

Molecular Genetics of PGL

Mutations of the enzyme complex II (SDH) and LOH on chromosome 11 result in a small fraction of sporadic and in almost all familial forms of PGL (41–43, 143).

A search for imprinting effects in hereditary PGL at 11q23 has been reported (144).

Mapping of the hereditary paraganglioma (PGL) region at 11q23 has been partially accomplished (41–43, 143).

The development of PGL in diverse anatomical locations in subjects with *SDHB*, *SDHC*, and *SDHD* germline mutations indicates that the paraganglionic system throughout the body is a target for PGL (45, 150). Thus, the possibility of mitochondrial complex II germline mutations should be raised in the differential diagnosis of all PGL, including PCC. Whether certain subunit mutations are more strongly associated with a given anatomical location, hormonal activity, malignancy, age at onset, tumor multiplicity, and size remains to be established. Other genetic loci as indicated by the LOH and CGC

studies, and environmental factors also may affect phenotypic expression of PGL.

Hereditary PGL is closely tied in with germline mutations affecting genes encoding for the succinate dehydrogenase (SDH) enzyme system (Tables 5.7–5.9), which is a heterotetrameric complex with functions in the Krebs cycle related to the oxidation of succinate to fumarate leading to ATP production (151, 152). SDH constitutes a moiety of complex II of mitochondria (150, 153) (Tables 5.7 and 5.8). The catalytic subunits of SDH are encoded by the genes *SDHD*, *SDHB*, *SDHC*, and *SDHA* (Table 5.7). *SDHD* encodes the smallest subunit of SDH. *SDHA* and *SDHB* are anchored to the mitochondrial inner membrane through membrane-spanning subunits encoded by *SDHC* and *SDHD* (154, 155). A large number of mutations (approx 30) in hereditary PGL have been described for *SDHD* and *SDHB* each (45, 156).

Germline mutations in complex II genes are associated with the development of PGL in diverse anatomical locations, including PCC, a finding that has important implications for the clinical management of patients and genetic counseling of families. Consequently, patients with PGL, including PCC, and a complex II germline mutation should be diagnosed as potentially hereditary, regardless of family history, anatomical location, or multiplicity of tumors.

SDHB in PGL

A large number of different germline mutations have been described in the *SDHB* gene (located at 11q13), which are associated with PGL (and PCC) (Figure 5.10) (145, 157–159) (Table 5.10). *SDHB*-associated tumors (approx. 10% of familial PGL) (35, 35A) are often syndromic (secrete catecholamines) of which approximately 10% are malignant (160) and approximately 10% are familial. The malignancy rate in PGL is similar to that seen in PCC (36, 37).

The preponderant number of PGL (approx 90%) are of sporadic nature, and their genetic bases have not been established rigorously; the remaining cases (approx 10%) are hereditary (familial) in which mutations and alterations of genes encoding for the SDH moiety of the mitochondrial complex II play a causal role (153). These genes, i.e., *SDHD* in 50%, *SDHB* in 20%, and *SDHC* in 10% of the familial

TABLE 5.7. Paraganglioma loci and their genes.

Region	References
PGL1 locus for *SDHD* at 11q23 for familial cases of carotid PGL	*Baysal et al 1997,2000* (41,145,146)
	Hirawake et al 1997 (162)
PGL1 locus associated with germline mutations of the *SDHD*	*Heutink et al 1992, 1994* (2,181)
Gene and a founder effect and maternal imprinting	*Baysal et al 1997, 1999* (145,146,311)
	Milunsky et al et al 1997, 2001 (190,312)
	Täschner et al 2001 (43)
	Lee et al 2003 (313)
	Riemann et al 2004 (314)
LOH at *PGL1* locus in the aneuploid fraction of PGL (hereditary and sporadic)	*van Schothorst et al 1996, 1998* (39,40,123)
	Bikhazi et al 2000 (104)
SDHD has 4 exons and a protein product of 159 amino acids	*Baysal 2003* (153)
PGL2 locus at 11q13 and mutation of unknown gene linked to PGL in a family	*Mariman et al 1993, 1995* (124,315)
LOH of this locus has been seen in carotid body PGL	*Bikhazi et al 2000* (104)
SDHB has 8 exons and a protein product of 280 amino acids	*Baysal 2003* (153)
PGL3 locus at 1q23.3 associated with germline mutations of the *SDHC* gene in a family with PGL	*Niemann and Muller 2000* (163)
	Niemann et al 1999, 2001 (161, 161A)
SDHC has 6 exons and a protein product of 169 amino acids	*Baysal 2003* (153)
PGL4 locus at 1p36 associated with germline mutations of the	*Astuti et al* 2002 (113)
SDHB gene in two families with PCC and PGL; also seen in 1/24 sporadic PCC	

Mutations of the gene *SDHA*, located at 5p15, do not lead to PGL; mutations of *SDHA* lead to optic atrophy, ataxia, myopathy, neuropathy, encephalopathy, and Leigh syndrome based on Bourgeon et al. 1995 (316), Birch-Machin et al. 2000 (317), Parfait et al. 2000 (318), and Baysal et al. 2001 (319).

TABLE 5.8. Mitochondrial enzyme system affected by mutations in PGL and PCC.

The enzyme system succinate dehydrogenase (succinate-ubiquinone oxydoreductase) or SDH is part of the mitochondrial complex II involved in the generation of ATP (153, 320, 321, 324) through an aerobic electron transfer chain and the Krebs cycle (159).
The gene for SDH consists of four parts: *SDHA, SDHB, SDHC*, and *SDHD* (153).
The complex II activity is selectively and complete lost in PGL with *SDHD, SDHB* and *SDHC* mutations (170, 202). A database for these mutations has been established. The subsequent overexpression of several hypoxia-inducible genes in PGL (48A) is in harmony with the hypothesis that loss of complex II mimics hypoxic stimulation, which leads to adaptive proliferation of the paraganglial chief cells resulting in PGL (150).
An overview of known *SDH* mutations and the associated tumor types can be found at the SDH Mutation Database at the Leiden Open Variation Database at http://chromim.liacs.nl/lovd_sdh/.

TABLE 5.9. Germline mutations in PGL.

Familial	Sporadic
SDHD mutations 50%	5%
SDHB mutations 20%	3%
SDHC mutations 10%	3%

Based on Baysal et al. 2002, 2004 (135, 167).

cases, each has unique features and a catena of molecular events in their causation of PGL.

An interesting case is that of a 32-year-old man who was found to have three PGL in the abdomen and who 7 years later had developed lytic lesions of the left femur (35). The tumors secreted catecholamines. This patient's 27-yr-old son was found to have locally metastatic abdominal catecholamine-secreting PGL (35). In both patients, the PGL had *SDHB* mutations.

The more aggressive and other features of *SDHB*-related PGL of sympathetic origin in the abdomen or thoracic cavity have received some special attention (137A). The clinical presentation, biochemical phenotypes, and genotype–

phenotype correlations in patients with *SDHB*-associated PCC and PGL have been recently addressed (137A).

The typical PGL associated with *SDHB* mutations originates from the extra-adrenal abdominal or thoracic paraganglia. In contrast to *SDHC* and *SDHD*-associated PGL, up to 70% of *SDHB*-related abdominal and thoracic PGL develop into metastatic disease (137).

Recognizing *SDHB*-related disease in patients may be delayed by incomplete and age-dependent penetrance, despite an autosomal dominant pattern of inheritance (137A). Many patients with *SDHB*-related tumors have no family history of PGL. About 4–7% of patients with apparently sporadic abdominal or thoracic PGL or PCC have been shown to carry *SDHB* mutations (12, 137). In addition, the diagnosis may be delayed due to lack of symptoms related to catecholamine excess (137B–D).

An evaluation of 29 patients (16 males, 13 females) with *SDHB*-associated abdominal or thoracic PGL, led the authors to the following conclusions (137A). *SDHB*-associated PCC and PGL is characterized by a high malignant potency, warranting aggressive therapy, strict follow-up, and family screening. The diagnosis may be delayed by a negative family history and an atypical clinical presentation with signs and

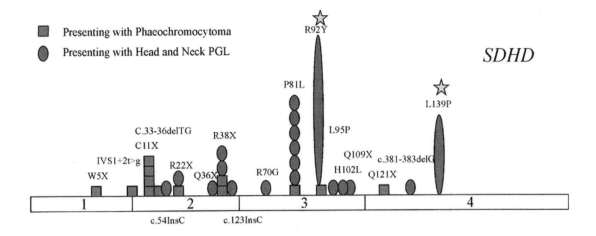

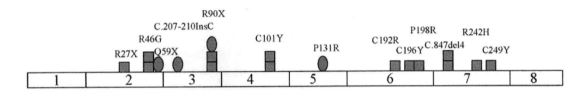

FIGURE 5.10. Schematic presentation of germline mutations of *SDHD* and *SDHB* genes in a Dutch patient population associated with PGL and/or PCC development. The stars have been placed over the 2 documented cases with founder mutations. Of note is the occurrence of PGL and PCC in these cases, with the number of cases of PGL outnumbering those of PCC (from ref. 11, with permission).

TABLE 5.10. *SDHB* germline mutations (14 loci) have been seen in the following tumors.

	References
Familial catecholamine secreting PGL	*Astuti et al 2001* (157)
Familial catecholamine-secreting PGL, PGL of head and neck, and familial PCC	*Astuti et al 2001* (157)
Familial head and neck PGL	*Baysal et al 2002* (167)
Nonfamilial head and neck PGL	*Baysal et al 2002* (167)
Nonfamilial catecholamine-secreting tumors	*Neumann et al 2002* (12)
Nonfamilial catecholamine-secreting tumor and familial catecholamine-secreting PGL	*Young et al 2002* (35)

Based on Young et al. 2002 (35).

symptoms that are predominantly related to tumor growth rather than to catecholamine excess. The biochemical phenotype usually consists of hypersecretion of norepinephrine, dopamine, or both, but 10% of tumors are biochemically silent. The clinical expression of these tumors in individual patients cannot be predicted by the type and location of the *SDHB* gene mutation.

SDHC in PGL

The locus for the *SDHC* gene has been shown to be at 1q23.3 (161), rather than the previously reported 1q21 (162).

Mutations of the *SDHC* gene in one family with syndromic PGL (163) and in one isolated case of malignant PGL (164–166) have been reported. In both instances, transmission was through the mother. However, no mutations of *SDHC* were found in four other series of patients with familial PGL (or PCC) (151, 167–169).

In a study (170) using a number of approaches (45, 171), a family with head and neck PGL was analyzed and discovered to have a 8.37-kb *SDHC* deletion, which spanned two AluY elements and removed exon 6. The deletion caused PGL3 following both maternal and paternal transmissions in the pedigree and also was detected in an unrelated sporadic case that showed allele sharing with the familial cases at seven polymorphic markers near *SDHC*, suggesting a common ancestral origin.

The presence (170) of a large deletion in a complex II gene confirmed the role of *SDHC* in familial and sporadic PGL. The observation of both paternal and maternal disease transmissions in PGL3, together with earlier findings, suggests that imprinted transmission in hereditary PGL is restricted to *SDHD* among complex II genes.

On the basis of *SDHC*'s high Alu content and the demonstration by special techniques (170) of an Alu-mediated deletion in *SDHC* and the failure to find conventional mutations in this gene in several PGL series mentioned above, genomic deletions in *SDHC* should be looked for in appropriate familial and sporadic PGL in which *SDHD* and *SDHB* mutations are absent. Because the *SDHC* deletion caused tumors through both paternal and maternal transmissions, it is unlikely that an absolute parent of origin effect operates at the *SDHC* locus. This observation, together with the earlier results on the transmission of *SDHB* mutations, strongly suggests that the parent of origin effect on the *SDHD* gene is not a functional consequence of complex II mutations but a locus-specific epigenetic phenomenon operating exclusively on the *SDHD* gene at chromosomal region 11q23. A mouse cell line with a mutated *SDHC* provided evidence that oxygen production from mitochondria results in oxidative stress which may affect apoptosis and tumorigenesis (172).

SDHD in PGL

Much of the genetics of hereditary head and neck PGL (and of PCC) has been elucidated by the demonstration of a group of genes encoding for the complex II enzyme system of mitochondria. The first to be characterized was the *SDHD* gene (located at 11q23) associated with the PGL-1 locus (173).

In a study of four kindreds with familial PGL (174), the usefulness of *SDHD* mutation analysis was confirmed in families with more than one member exhibiting sympathetic or parasympathetic PGL and indicated that genetic testing of *SDHD* and other SDH genes may be an important tool for the identification of the syndrome in cases with multiple or bilateral PGL. This study (174) also provided information on the geographic variability of the mutations.

In a study of 57 unselected patients with parasympathetic PGL (19 cases with family history), germline mutations of the *SDHD* gene were shown to be present in all patients with familial PGL and in 13 of 38 patients with apparently sporadic PGL (13). All were missense mutations in highly conserved regions of the *SDHD* gene and were not seen in 200 controls. Somatic mutations of *SDHD* do not play a significant role in sporadic PGL.

Most *SDHD* mutations in PCC patients are nonsense or splicing mutations (12, 143). Missense mutations are more likely to cause PGL (Figure 5.10). The *SDHD* gene is subject to polymorphisms (175), some of which may be difficult to differentiate from mutations with low penetrance (176, 177).

The PGL1 (SDHD) phenotype is transmitted through fathers and not by mothers (45). The presence of *SDHD* founder mutations clearly indicated that this sex-specific transmission effect operates over multiple generations and therefore is reversible in gametes; *SDHD* is subject to genomic imprinting. PGL1 associated with *SDHD* mutations are inherited almost exclusively via the paternal line, a finding inconsistent with autosomal dominant transmission (17). These findings can be explained in terms of the genomic imprinting hypothesis (178, 179), i.e., the maternally derived gene is inactivated during oncogenesis and can be reactivated only during spermatogenesis.

The genomic imprinting effect of *SDHD* locus may have helped the spread of the founder mutations (91, 156, 180) by effectively halving the overall penetrance of the mutant alleles (applies to The Netherlands cases) (43).

SDHD mutations are transmitted from mothers to their progeny unaltered (41, 42), as would be expected from an autosomal dominant inheritance. Thus, the stable transmission of *SDHD* mutations from both sexes suggests that the parent-of-origin effects at the PGL1 locus may be caused by differential epigenetic alterations around the *SDHD* gene during sex-specific gametogenesis, consistent with genomic imprinting (170). In contrast mutations of *SDHB* and *SDHC* cause PGL after maternal (163, 167) transmission.

Parent-of-origin effect has been demonstrated only for the PGL1 locus, suggesting the presence of an epigenetic mechanism at 11q23 that is important for the expression of *SDHD*, perhaps in a tissue-specific manner (170). Approximately 36% of nonfamilial cases in The Netherlands carry the two founder *SDHD* mutations (43). The transmission pattern of *SDHD* mutations shows strict parent-of-origin effects (17, 145, 181).

Because the genetic component of hereditary PGL can often be obscured by imprinting and low-gene penetrance, all at-risk relatives of an affected individual, regardless of the presence of a positive family history, should be tested for *SDHD* mutations (45, 143). Loss of other alleles of *SDHD* or others in tumors indicates their function as tumor suppressor genes (42, 159, 163, 183).

In summary, germline mutations in *SDHB*, *SDHC*, and *SDHD* account for the majority of PGL with a positive familial history. *SDHD* seems to be the most commonly mutated gene. Approximately 8% of all nonfamilial PGL also harbor occult germline mutations in the complex II genes. Multiple PGL in nonfamilial cases, regardless of their anatomical distribution, and, because of the rarity of PGL, the presence of two or more first- or second-degree affected relatives should herald the possibility of complex II mutations. However, more studies are required to determine whether subjects with solitary PGL and no family history are likely to have complex II germline mutations. In a study, 45 of 66 (68%) nonfamilial PCC patients who had germline mutations in *VHL*, *RET*, *SDHD*, or *SDHB* presented with solitary tumors (12). The high combined rate of germline mutations in these genes in nonfamilial PCC may therefore justify gene testing in subjects presenting with a single PCC. However, complex II gene mutations were discovered only in one of 32 (~3%) subjects with nonfamilial head and neck PGL presenting with solitary tumors (167). Thus, a recommendation for gene testing seems to be currently premature for sporadic patients presenting with a single head and neck PGL.

PGL and Atmospheric Oxygen

An interesting and somewhat provocative interpretation and extrapolation of the higher incidence of carotid body PGL at high altitudes in some clinical conditions have been advanced (170). Stressed by the author (170) is the role of chronic hypoxia as a risk factor for carotid body PGL in subjects (human and animal) at high altitudes (33, 183–186) and the high incidence of carotid body PGL in patients with chronic cyanotic disease, cirrhosis of the liver, and lung disorders (186–188). The incidence of PGL of the carotid body is 10-fold higher at high altitudes than at sea level (185, 189). PGL at high altitudes are only 1% familial; hence, they are associated with a low rate of *SDHD* mutations (170).

Because of the phenotypic similarity of carotid body PGL due to chronic hypoxia to that of PGL1, it has been hypothesized by analogy that *SDHD* mutations may impair oxygen sensing in the carotid body (41, 42). Furthermore, environmental oxygen levels that are primarily determined by atmospheric pressure (elevation) have a possible effect on penetrance, expression, and prevalence of *SDHD* mutations (156).

Needing a cogent explanation is the presence of multiple founder mutations in PGL in The Netherlands versus the USA (41, 42, 167, 190), Australia (191), England (192), Germany, Poland (12), and Spain (180), countries without multiple founder mutations, and the 3.4- to 8.5-fold frequency of *SDHD* mutations in cases of PGL in The Netherlands versus that in the United States.

Also needing an explanation is tumor multiplicity of PGL of 201 versus 453 in The Netherlands versus the United States, respectively, and much higher prevalence of *SDHD* mutations in The Netherlands versus the United States (156).

Several modifying effects of altitude on the phenotypic severity of *SDHD* mutations and on some facets of PGL bear directly on the areas just mentioned:

1. Subjects with multiple PGL (or with PCC) are more likely to reside in higher altitudes than those with single tumors or without PCC.
2. The higher rate of germline mutation carriers among sporadic PGL cases in The Netherlands than in the United States seem to be due to reduced gene penetrance in The Netherlands possibly related to the unusually low altitude (see below).
3. Clustering of three of four *SDHD* founder mutations in The Netherlands (13, 43) may be partially explained by reduced penetrance of the *SDHD* mutation leading to relaxation of natural selection.
4. Genomic imprinting and environment may act jointly to reduce overall penetrance of *SDHD* mutations and allow the founder mutations to increase in frequency over generations through genetic drift (170).

The role of the HIFs in the sensing of oxygen levels and the control effected by prolyl and asparaginyl hydroxylations have been reviewed previously (170A). An explanation for the possible causes underlying the aforemtnioned observations has been advanced (167, 170), in which the difference of approximately 3% in atmospheric pressure between The Netherlands and the United States, although seemingly minor, suggests that the *SDHD* gene product may interact with molecular oxygen and that defective oxygen sensing may be the ultimate cause of PGL. However, the exact molecular mechanisms remain unknown (170).

The reports dealing with PGL and high altitudes have generally ignored the high hemoglobin levels of residents living at such altitudes. Although it is true that the partial oxygen pressure decreases with increased geographic elevation, the amount of oxygen delivered to tissues by the elevated hemoglobin levels at high altitudes compensates for the decreased partial pressure of oxygen. This also would apply to paraganglional tissues. Thus, rather than implicate hypoxia, which residents at high altitude do not suffer from, in PGL development, it may be more accurate to state that organs such as the carotid body are sensitive to the changes in partial pressure of oxygen as reflected in the reduced hemoglobin saturation with oxygen.

Increased erythropoietin production in chronic mountain sickness of Andean natives is associated with polycythemia and is related to ventilating inefficiency, rather than to altered sensitivity to hypoxia, cobalt levels, or sleep abnormalities (170B).

Inheritance of PGL

Haplotype analysis suggested that most Dutch families with PGL type 1 descended from a single individual who was presumably the first ancestral carrier of the gene mutation (i.e., founder effect) (39). Although PGL of the head and neck were prominent in the Dutch patients, an increased frequency of PCC also was noted (40).

Early studies based on linkage mapping of the *PGL1* locus indicated that the prevalence of PGL was very high in the Netherlands. Subsequently, after the establishment of the association of the *SDHD* gene with *PGL1*, three Dutch founder mutations of this gene were established (13, 43) in 82 unrelated subjects and families. A founder *SDHD* mutation was detected in several U.S. families (150, 167, 170).

In The Netherlands, 33 of 93 sporadic PGL cases carried *SDHD* germline mutations; only two of 37 in the United States. Genomic imprinting alone is unlikely to explain this difference, because genomic imprinting operates in all *SDHD*-linked pedigrees thus far reported (167, 170).

Some subjects develop PGL tumors as early as age eight, whereas others develop them at very advanced ages (145, 146). Some cases present with multiple tumors at early ages, whereas others develop only a single tumor in their lifetime. Some subjects develop head and neck PGL (167), others develop PCC (12), and others develop both types of tumors during their lifetime (192, 193). Because SDHD mutations were first detected in families with head and neck PGL (41, 42), and *SDHB* mutations in families with PCC (192, 193), there are probably anatomical preferences of distinct subunits. In fact, mutation screening of unselected patients suggests that *SDHD* mutations are more likely to be found in subjects with head and neck PGL (167), whereas *SDHB* mutations are more likely to be found in those with PCC (158).

Mitochondria and PGL

Mitochondria play a primary causal role in the development of certain tumors (148, 194), but in none as clearly as in hereditary PGL and less commonly in PCC. Mitochondrial complex II (succinate dehydrogenase; succinate:ubiquinone oxydoreductase), composed of SDHD, SDHA, SDHB, and SDHC, all participate in aerobic electron transport and the Krebs tricarboxylic acid cycle. Dysregulation of the hypoxia-responsive genes (SDH group) leads to impairment of some mitochondrial functions. Because the carotid body is the most common site of head and neck PGL and it is linked to oxygen sensing, the genetic changes affecting complex II (normally involved in hypoxia and oxidation-reduction) may play a role in the predisposition to these tumors (45). These considerations may account for the frequency of these tumors in residents at high altitudes (167).

Apoptosis in PGL

Evasion of apoptosis is one of the acquired hallmarks of most cancers (195–198). Because mitochondria play an essential role in the apoptotic process (199, 200), the effects of the various gene mutations on the SDH system in mitochondria remain uncertain (201), as does the contribution of apoptosis to the generation of PGL and PCC (202).

Carotid Body and PGL

The carotid body is the most common site for PGL, followed by the jugular bulb, ganglia in the neck, middle ear, vagal system, tympanic tissues, and aortic arch. At least 20 distinct sites contain paraganglia.

The carotid body is a small organ (<15 mg) (27) located at the medial aspect of the bifurcation of the carotid artery within its adventitia. It plays a major role in the acute adaptation to hypoxia (203) by stimulating the cardiopulmonary system, resulting in increased (ventilation) respiratory and heart rates by stimulating the central respiratory centers via efferent glossopharyngeal nerves (204, 205).

It has been proposed that the ultimate cause of PGL is a defect in sensing environmental oxygen levels (42, 153) by the carotid body caused by *SDHD* and *SDHB* subunit mutations (159, 202). SDH seems to be crucial for the regulation of cellular reactive oxygen species (206). The complete loss of *SDHB*, *SDHC*, or *SDHD* would lead to a constitutive activation of the hypoxic stimulation pathways and PGL development, mimicking the high frequency of sporadic occurrence

of PGL in individuals who live at high altitude (33, 42, 167, 191, 207). However, how defective mitochondrial complex II translates into a mitogenic signal remains unknown (8). A possible role of reactive oxygen species levels or the cytoplasmic redox ratio (208) associated with a cascade of cellular events triggered by these changes has been mentioned (8). It is possible that the mutations of the SDH genes resulting in defective mitochondrial complex II functions may be responsible for the hypertrophy, then hyperplasia, and ultimately the development of PGL (8). The role of HIF-1 (209), important in chronic hypoxia as a factor involved in the carotid body response to acute hypoxia, remains unknown (8, 209).

The cellular response to hypoxia involves the inhibition of HIF-1α hydroxylases activity, which then present the VHL-depending targeting the HIF-1α to the ubiquitin-proteasome pathway (153, 210).

Molecular Genetics of PCC

PCC in VHL, MEN-2, and NF1

Many of the genes overexpressed in VHL-associated PCC were linked to hypoxia-driven angiogenic pathways activated in VHL tumorigenesis and included besides *VEGF*, *VEGFR*, and a large number of other genes. Overexpression of the hypoxia-inducible transcription factor *HIF-2α*, in norepinephrine-predominant VHL tumors, indicates that the overexpression of this gene depends on the noradrenergic biochemical phenotype (211, 211A).

Pheochromocytomas are highly vascular tumors that arise from mutations in a diverse and apparently unrelated group of TSG and oncogenes. The authors (212) show that three of the genes that cause hereditary PCC have a common function. Specifically, these genes, *VHL*, *SDHB*, and *SDHD*, encode proteins that regulate a transcription factor known as HIFα, which helps cells adapt to hypoxia (low oxygen levels) (213). *VHL* is named after its role in VHL disease, an inherited disorder that predisposes individuals to PCC and other tumors. Previous studies showed that when cells lack *VHL*, HIFα is not degraded, resulting in a signal that resembles hypoxia. The authors (212) found that loss of two genes that cause two distinct PCC syndromes (the genes *SDHB* and *SDHD*, which encode the subunits B and D of the succinate dehydrogenase, a component enzyme of the energy and respiratory system in mitochondria) also triggers a HIFα response. The researchers further discovered that high HIFα levels can suppress *SDHB*. This suggests a regulatory loop that further enhances the "hypoxia" profile of tumors. This finding provides a rational explanation for the shared features of these distinct syndromes and may be relevant for other cancers with a prominent hypoxic pattern. The link between hypoxia signals (via *VHL*) and mitochondrial signals (via *SDH*) in PCC is mediated by HIF. These features explain the shared mutations and suggest an additional mechanism for increased HIFα activity in PCC (212).

The interrelationships between the HIFα regulatory loop linking with hypoxia and mitochondrial signals and the changes in the *SDH* and other genes in PCC and PGL have been presented and reviewed previously (212–218).

Although the preponderant number (90%) of PCC (15, 17, 219) are sporadic in nature, the familial form of PCC and those associated with specific syndromes (e.g., VHL, NF1, and MEN-2) have generated a more substantial literature than the sporadic cases of PCC (Table 5.11). Based on the genetic basis of PCC, the choice and method of screening patients with these tumors have been discussed previoulsy (220).

Of 271 unrelated and nonsyndromic PCC without a family history, 66 (24%) were found (12) to have germline mutations in the *VHL* (207), *RET* (221), *SDHB* (222), or *SDHD* (163) genes. Although young age of onset was associated with a germline mutation, 35% of those found to have germline mutations presented after the age of 30 years and 8% after the age of 40 years. Extra-adrenal tumors were found to be associated with heredity, and, in particular, with germline mutations in the *SDHD* gene. The relatively high frequency of apparently sporadic disease in *SDHD* mutation carriers might be attributed to maternal imprinting. No evidence of genomic imprinting was obtained for the *SDHB* gene (12, 167, 192, 193). These data suggest that all presentations of PCC and PGL, irrespective of family history, syndromic manifestations, or age at diagnosis, should be subjected to clinical genetic testing for some of these four genes (*RET*, *VHL*, *SDHB*, and *SDHD*).

Atlhough only a small subset of sporadic PCC has mutations of *RET* or *VHL* (130, 223–225), dominantly acting activating mutations in the *RET* oncogene and inactivating mutations of *VHL*, *NF1*, and other genes, along with concomitant LOH of the second allele and other chromosomal abnormalities might play an important role in PCC tumorigenesis (89). However, most sporadic PCC are probably initiated by other mechanisms, i.e., cumulative and possibly hierarchical genetic changes (89). In previous studies (226) *RET* was expressed in PCC.

The incidence of PCC is sufficiently high in VHL disease and MEN-2 to warrant screening of patients with these syndromes for PCC, even if they are asymptomatic (227).

Patients with PCC associated with MEN-2 were more symptomatic and had a higher incidence of hypertension and higher concentrations of metanephrines (epinephrine metabolite) but lower total plasma concentrations of catecholamines than patients with PCC associated with VHL disease (228). The latter patients had elevated plasma levels of norepinephrine metabolite, normethanephrine.

PCC, VHL Disease, and *VHL* Gene

Some features of VHL-associated PCC and PGL are shown in Tables 5.12 and 5.13. PCC and hemangioblastoma constitute an integral part of VHL disease (229) and for further information, including molecular genetic mechanisms related to the *VHL* gene and its protein, see Chapter 4 on hemangioblastoma.

That PCC constitutes an integral part of VHL disease has been well established (11, 160, 230–236). PCC may be the original and sole manifestation of VHL disease (237, 238); hence, mutational analysis of the VHL gene, a tumor suppressor gene located at 3p25, in these patients may lead to the establishment of the diagnosis of VHL disease and identification of asymptomatic individuals in the family at-risk for the disease (239). VHL disease, an autosomal dominant syndrome, has many manifestations in which

TABLE 5.11. Germline mutations in PCC.

Gene	All patients (n = 314)		FS patients (n = 56)		SP patients (n =258)	
	No.	%	No.	%	No.	%
VHL	25	8.0	16	28.6	9	3.5
SDHB	21	6.7	3	5.4	18	7
RET	16	5.1	15	26.8	1	0.4
SDHD	11	3.5	9	16.1	2	0.8
SDHC	0	0	0	0	0	0
NF1	13	4.1	13	23.2	0	0
Total	86	27.4	56	100	30	11.6

FS, patients with a family history and/or a syndromic presentation; SP, patients with an apparently sporadic presentation.
Based on Amar et al. 2005 (137).

TABLE 5.12. VHL Disease, *VHL* Gene, PCC and PGL.

	References
VHL gene mutation (germline)	*Abbott et al 2006 (335)*
V84L manifests as bilateral PCC Population studies	*Brauch et al 1995 (322)*
VHL gene as a tumor suppressor gene	*Chen et al 1995 (323)*
	Kaelin and Maher 1998 (222)
	Latif et al 1993 (325)
	Walther et al 1999 (233,276)
VHL gene mutations	*Crossey et al 1994 (327)*
VHL gene and its protein	*Friedrich 1999 (328)*
VHL gene located at 3p25	*Hosoe et al 1990 (329)*
10–15% of VHL patients develop PCC	*Manger and Gifford 1996 (160)*
	Mantero et al 2000 (330)
VHL gene and hypoxia	*Maxwell et al 1999 (211A)*
Errors in mitosis may predispose to tumor formation	*Neumann et al 1988 (332)*
VHL gene, HIF-1 and cancer	*Pugh and Ratcliff 2003 (210)*
PCC may be the first manifestation of VHL	*Richard et al 1994 (237)*
VHL and PGL	*Schimke et al 1998 (333)*
	Zanelli and van der Walt 1996 (334)
VHL gene changes in sporadic PCC	*Walther et al 1999 (233)*

For more information on the *VHL* gene and its protein product pVHL, see Chapter 4 on hemangioblastoma.

TABLE 5.13. Comparison of Sporadic Pheochromocytoma and Pheochromocytoma Associated with von Hippel-Lindau disease.

Patient characteristic	von Hippel-Lindau pheochromocytoma	Sporadic pheochromocytoma	P value
Age (years)	29.9	39.7	0.0034
Symptoms	6/37	23/26	<0.001
Hypertension	3/37	24/26	<0.001
Diagnostic studies	23/37	26/26	<0.001
Volume (cm^3)	4.2	35.4	<0.001
Urine epinephrine level, g/24 h	6.5	14.1	<0.001
Urine metanephrine level, μmol/day	6.59	26.36	<0.001
Vanillylmandelic acid level, μmol/day	38	97	<0.001

Based partially on material in (233).

hemangioblastoma (see Chapter 4), renal cell carcinoma of the kidney and PCC are the most frequent tumors (234, 240). VHL consists of several clinical types that vary in their phenotypic manifestations of the *VHL* gene, i.e., type 1 is not associated with PCC, and type 2A is associated with PCC and RCC and type 2B is associated with PCC but not with RCC. The known association of PCC development with two hereditary syndromes, i.e., VHL and MEN-2 (whose responsible proto-oncogene *RET* is located at 10q11.2), has led to an examination of changes of these genes in sporadic PCC. Not surprisingly, the results have not been consistent, in part due to failure to rule out or establish possible germline mutations in the affected individuals to be certain of the sporadic nature of the PCC. The germline mutations in VHL that predispose families to PCC in particular, do not impair HIF-1α ubiquitylation (140, 221).

Alterations of the *VHL* gene is sporadic PCC (48 benign, 24 malignant) were investigated with several techniques (221A). The germline mutations of *VHL* were identified in one patient with a benign tumor and in one with a malignant tumor. Tumor-specific *VHL* mutations and accompanying LOH were found in two of 47 benign and in four of 23 malignant PCC. Expression of the VHL protein was present in all tumors.

The frequent and significant association of PCC with VHL disease (10–15% of patients) and commonly in MEN-2 (50–70% of cases) and much more rarely in NF1 (approx 3% of cases), Sturge–Weber syndrome, and tuberous sclerosis, has generated investigations addressing a number of genetic parameters.

The functions and their mechanisms, changes, and effects of the *VHL* tumor suppressor gene and its protein product in familial and nonfamilial settings have received much attention and have been reviewed previously (11). The variability of the phenotypes of tumors associated with VHL may be related to the contrasting effects on HIF-1α regulation by different mutations affecting pVHL (221). Reviews on various aspects of MEN-2 and the *RET* oncogene have been published previously (9, 241–245).

PCC and MEN-2 Syndromes

Some differences in MEN-2–associated PCC versus sporadic PCC are shown in Table 5.2, along with features of MEN-2 PCC (Table 5.14). PCC develop in 50% of patients with MEN-2 (9); hence, some discussion of the *RET* proto-oncogene (whose mutations are the cause of MEN-2) and its effects is presented here.

Activating germline mutations of the *RET* proto-oncogene cause multiple endocrine neoplasia type 2 (MEN-2) (246–249) (Figure 5.11). Specific mutations of the *RET* gene are related to the disease phenotype in MEN-2 (250). MEN-2A is characterized by medullary thyroid carcinoma (MTC), which generally appears first (251), followed by PCC in 50% of cases, in addition to parathyroid hyperplasia or adenoma in 15–30% of cases. In families with MEN-2A and FMTC, 96% have a *RET* mutation in any one of five cysteine residues (243, 247, 252–254), and a strong association of a codon 634 mutations with the presence of PCC. Similarly, MTC and frequently bilateral PCC are features of MEN-2B, but hyperparathyroidism is rare. Other features of MEN-2B include mucosal neuromas and a Marfanoid habitus. In MEN-2B families, the disease-causing factor is a single germline *RET* mutation in 95% of cases (243, 249, 252–254, 254A).

RET encodes for a receptor tyrosine kinase (expressed in tissues and tumor derived form the neural crest) with four identified ligands: glial cell-line-derived neurotrophic factor (GDNF) (243A), persephin, neurturin, and artemin; and four associated coreceptors: GDNF receptor α-1-4 (GFRα-1-4) (243, 255). Each GDNF family member has a high affinity for and interacts with a specific member of the GFRα family (243A, 256, 256A). Activation of RET takes place either through normal ligand coreceptor binding or by MEN-2–related mutations leading to ligand-independent dimerization (257) or altered substrate specificity (258). The observation is that in MEN-2A PCC, either the mutant *RET* allele is duplicated or the wild-type allele is lost (259). However, imbalances of chromosome 10 (where *RET* is located) at 10q11.2 copy number imbalances as determined by CGH are rare events in PCC (91).

TABLE 5.14. MEN-2, *MEN-2* gene, and PCC.

	References
Allelic loss of 1p (1p34-p36) in 29-45% of sporadic and in 100% of PCC in MEN-2	*Baysal 2002 (45)*
Gonosomal aneuploidy in lymphocytes of MEN-2 patients	*Behmel et al 1997 (336)*
RET gene and its mutations in MEN-2A	*Eng et al 1994, 1996 (251,252)*
	Qiao et al 2001 (337)
	Williamson et al 1997 (109)
	Zedenius et al 1994 (270)
A case of MEN-2A with a malignant PCC and ganglioneuroma.	*Gullu et al 2005 (344)*
MEN-2, PCC and hyperparathyroidism	*Howe et al 1993 (338)*
Cases of medullary thyroid carcinoma with germline mutation of codon 892 of the *RET* gene and MEN-2 should be screened for PCC	*Jimenez et al 2004 (343)*
Deletions of chromosome 1 in MEN-2	*Mathew et al 1987 (116)*
Incidence of PCC in MEN-2 in Europe	*Modigliani et al 1995 (339)*
PCC seen in both MEN-2A and MEN-2B	*Nguyen et al 2001 (340)*
	Ponder 2001 (242)
	Solcia et al 2000 (44)
Variability of phenotypes with *RET* mutations in *MEN-2*	*Ponder 1999 (241)*
MEN-2 gene located at 10q11	*Ponder 2001 (242)*
RET codon 634 mutations in MEN-2 may have a direct impact on tumor aggressiveness.	*Puñales et al 2003 (341)*
Unique *RET* mutation (C609S) in MEN-2 associated with PCC and MTC.	*Kinlaw et al 2005 (342)*
Loss of mutation of a gene on chromosome 22 in 2/8 PCC in MEN-2A	*Takai et al 1987 (117)*

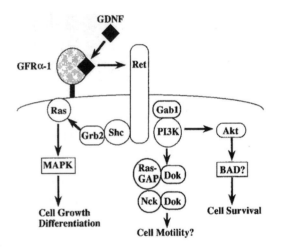

FIGURE 5.11. Schematic illustration of the intracellular pathway through activated *RET* as seen in MEN-2. For factors related to PCC pathogenesis in MEN-2, see text (from ref. 264, with permission).

The function of RET is mediated by multiple interactions in downstream signaling pathways (260). The mitogen-activated protein kinase (MAPK) pathway is stimulated in a RAS-mediated activation by RET in PCC cells (261). Proliferation induced by MEN-2A–mutated RET in vitro is enhanced by signal transducer and activator of transcription 3, a downstream target and transcription factor involved in G1 to S phase cell cycle transition (262). Also, the involvement of phosphatidylinositol 3-kinase (PI3K) downstream effectors in MEN-2A (263) and MEN-2B (264) further elucidates the mechanisms by which the mutant RET exerts its oncogenic ability. More specifically, a novel mechanism for cyclin adenosine monophosphate activation of PI3K in chromaffin cells has been proposed as playing a role in the development and progression of PCC (265).

An example is a study in which mutations of both the MEN-2A and -2B types were reported in occasional and supposedly sporadic PCC, although germline mutations were not excluded in the cases, except in one case where a new MEN-2B–like mutation was established (251).

In a study of three families with PCC without additional features (or history) (3), no *RET* mutations were encountered, but in two families missense *VHL* mutations were identified. The authors recommended, as others have (227,238,266), that patients with familial, multiple or early onset PCC be investigated for germline *VHL*, *MEN-2*, and *RET* mutations.

Six PCC from MEN-2A patients all contained mutations within exons 10 and 11 of *RET*, whereas none of seven sporadic PCC showed that change (128). The homodimers of the RET protein were detected in PCC from MEN-2A patients but not in a sporadic PCC (267).

Studies on *RET* changes in sporadic PCC have shown considerable variability, i.e., no changes in some studies (see above), an occasional case in others (226, 268, 416A), and as high as 10–15% of sporadic PCC (253).

In a study (269) of sporadic PCC, *RET* mutations were seen in six of 28 tumors (three each from MEN-2A and -2B). Five of the six PCC had missense mutations. None of the patients were shown to have germline (i.e., constitutional) *RET* mutations. In two separate studies (270, 271) on a total of 39 patients with sporadic PCC (three with *RET* mutations), none was found to have an unsuspected germline *RET* mutation. The recommendation (269) that all patients with PCC be tested for germline *RET* mutations has been criticized (272) in view that zero of 67 cases cited above showed positive results. This was supported by results of subsequent studies on the *RET* and *VHL* genes in patients with sporadic PCC (130, 223). In sporadic PCC, none showed mutations for *MEN-2* or *RET* and only one had a *VHL* mutation (not present in constitutional DNA) (134). The authors suggested that other genes must be responsible for the development of sporadic PCC. Mutations of the *VHL* gene were found in two of 17 sporadic PCC (273). These two mutations were of the same nature and occurred in a CpG island, which is a mutational hotspot in type 2A VHL families (273). No mention was made by the authors whether the two cases were tested for germline status.

In summary, sporadic PCC may be associated with germline mutations in the *VHL* (8.5% of cases) or *RET* (3.6% of cases) genes without a family history of VHL, MEN-2, or NF1 and approximately 3.5% of nonfamilial PCC have *SDHD* mutations. Mutations of *SDHB* in 4.5–8% of nonfamilial PCC may harbor occult germline mutations. The occurrence of head and neck PGL in these cases or family members indicates mutation in the mitochondrial complex II.

PCC and Mitochondrial Complex II Genes

A study of 20 PCC (19 sporadic with four malignant, one familial) and of 10 familial PCC failed to reveal any mutations in the *SDHD* gene (176, 274), although 15% of the samples had an intronic single nucleotide polymorphism, the significance of which is uncertain.

Four kindreds with familial PCC were investigated for germline *SDHD* mutations and one found in a two-generation family consisting of four children with PCC (192). None of the members of these kindreds had germline mutations of *VHL* or *RET*. No such mutations were seen in 24 sporadic PCC.

Mutations of the *SDHB* gene were investigated in families with PCC (some with PGL) (157). Four germline mutations in eight families were found, i.e., in two of five kindreds with familial PCC, in two of three kindreds with PCC and PGL, and one of 24 cases with sporadic PCC. No *SDHC* mutations were seen.

In another evaluation of the role of the *SDHB* gene in familial PCC (158), six of seven probands (four with PCC and PGL, one with PCC only, two with extra-adrenal PCC) with a family history of PCC were found to have germline mutations of *SDHB*. Such mutations have been reported to occur in sporadic PCC that were extra-adrenal in their location or malignant in character (159).

It has been suggested that germline mutations of *SDHB* occur more frequently in familial PCC than do those of *SDHD* (158), with or without associated PGL. This study had a number of interesting aspects. Three probands with *SDHB* mutations had parents (one mother and two fathers) with such mutations but without any disease. In one case, the proband's father had two siblings with PCC. Paternal inheritance of the *SDHB* mutation was found in three families and maternal in one (158). In none of 14 sporadic PCC were *SDHB* mutations found in the retained allele (158). The authors suggested that the *SDHB* gene may be variably penetrant with its expression, possibly being modified by other genetic or epigenetic factors.

A sporadic malignant PCC in a 55-year-old female without a family history was shown to be induced by a germline mutation of the *SDHB* gene (275). Within the tumor, LOH at 1p led to a null *SDHB* allele and complete loss of complex II activity. High expression of hypoxic-angiogenic responsive genes, i.e., *EPAS1* (endothelial PAS domain protein 1) and VEGF in the tumor cells and of VEGF receptors and EPAS1 in the endothelial cells were present. The authors (275) attributed the "spectacular tumor vascularization and tumor growth" to the genetic changes.

The aforementioned studies and others indicate that familial PCC may be caused by mutations of the *SDHD* and *SDHB* genes or of the *VHL* gene (12, 167). These changes occur in 8% of sporadic PCC without a family history or syndromic manifestations.

In some of the *SDHD* and *SDHB* families with PCC, head and neck PGL were present, with identical mutations inducing both tumor types. Only three PCC were seen in 63 Dutch familial head and neck PGL cases, suggesting that *SDHD* mutations (founder, Dutch) are infrequently associated with abdominal paraganglial involvement.

The broad genetic heterogeneity of PCC is underscored by the demonstration of two susceptibility loci on 2q and 16p, consistent with a recessive form of this tumor (142). This digenic trait was found in a large kindred of Brazilian-Portuguese origin. PCC (including two bilateral tumors) were detected in six siblings.

Levels of HIF-1α, HIF-2α/EPAS1, VEGF, and its receptor VEGF-R1 were shown to be increased in the *SDHD*-related PCC.

PCC and NF1

Neurofibromatosis type 1 is an autosomal dominant disorder caused by acquired mutations in the *NF1* gene (located at 17q11.2) (see Chapter 1). PCC occur in approximately 5% of patients with NF1 with mean patient age of 42 years at presentation with PCC (276).

In a study of 33 PCC (five from NF1 cases, three familial, and one with MEN-2B), 10 thought to be malignant, the MIB-labeling index was higher in the malignant tumors, and these patients tended to be young males (277). None of the malignant PCC were from NF1 patients. LOH of *NF1* was found in all seven PCC from patients with NF1, as well as, loss of neurofibromin (121). In seven of 20 PCC from non-NF1 patients reduced or absent *NF1* expression was found (278).

A 20-year-old woman with NF1 who developed a PCC, pulmonary PGL and a jugular glomus tumor may be indicative of a possible link between NF1 and neural crest tumors (279). A number of aspects of NF1 and PCC, including molecular and clinical descriptions, have been reported (280–287).

Carney Triad

Carney triad is defined by pulmonary chondromas, GIST, and extra-adrenal PGL (10,288). Less than 25% of patients exhibit all three tumors, and additional components have been recognized including adrenal cortical adenomas, duodenal stromal tumors, and possibly esophageal leiomyomas (288,289). The possibility that the triad might be familial has been raised, because there are now three reports of Carney triad patients with siblings having bilateral PGL (288, 290). Some PGL-GIST diads may be familial and inherited in an apparent autosomal dominant manner, but differ from the Carney triad by the absence of female predilection and predominance of PGL. A report described PGL in one member of monozygotic twins and the other with gastric GIST (290A).

Tables 5.15–5.19 list factors and areas related to PGL and PCC either not covered in the text or for easy search and identification by the reader. Descriptions of the pathology and clinical aspects of PGL and PCC can be found in a number of sources (25, 46, 60, 62, 125, 291–295).

For concise yet informative descriptions of the pathologic, clinical, and genetic aspects of benign and malignant PCC, composite PCC and PGL, and parasympathetic and sympathetic PGL, and information on the inherited disorders associated with these tumors (MEN-2, VHL disease, familial PGL and RCC, and MF1), chapters in the WHO Classification of Tumours of Endocrine Organs may be consulted (296–306).

The Proceedings of the First International Symposium on Pheochromocytoma, which was held in October 2005, was published in book form in 2006 (306A) and contains a wide range of information dealing with almost every facet of pheochromocytoma and some of PGL. A second symposium is scheduled for 2008.

TABLE 5.15. Some features of PGL.

	References
Clinical aspects of PGL	*Amar et al 2005 (73)*
SDHD mutations impair oxygen-sensing; high altitude associated with phenotypic "severity" in PGL1	*Aström et al 2003 (156)*
SDHD (located at 11q23) mutations in familial (hereditary) PGL (mitochondrial complex II gene)	*Badenhop et al 2001 (191)*
Multicentric PGL may occur in approx 10% of cases of PGL	*Balatsouras et al 1992 (18)*
The germline mutation rate in patients with PGL is at least 20% in Western Europe and, hence, systematic testing for germline mutations of the complex II genes (*SHDH*, *SDHB*, and *SDHC*) has been recommended	*Bauters et al 2003 (169)*
Maternal imprinting is present in *SDHD* mutation-positive familial head and neck PGL	*Baysal et al 2000 (41,42)*
Bilateral head and neck PGL may occur in 10% of cases without a family history	*Baysal et al 2002 (167)*
Epidemiology, incidence, and clinical facets of PGL	*Baysal et al 2002 (167)*
Familial PGL show mutations of *SDHD* and *SDHB* in 70% of cases; such mutations were seen in only 8% of sporadic cases	*Baysal et al 2002 (167)*
Positive family history in 10-50% of head and neck PGL	*Baysal et al 2002 (167)*
PGL of head and neck usually do not secrete catecholamines (are nonsyndromic) and clinically recognized by regional pressure and/or embolic effects	*Baysal 2003 (153)*
Variation exists in the prevalence, penetrance and expressivity of SDH subunit mutations in PGL	*Baysal 2003 (153)*
SDHB and *SDHD* mutations are not fully penetrant and thus some at-risk carriers may remain clinically unaffected	*Baysal 2003 (153)*
No genomic imprinting at the PGL1 (*SDHD*) locus; hence, parent-of-origin effects in PGL1 may be caused by other mechanisms	*Baysal 2004 (170)*
Possible epigenetic mechanism at 11q23 may be important for the expression of *SDHD*, perhaps in a tissue-specific manner	*Baysal 2004 (170)*
Familial PGL: multiple, bilateral, early age; autosomal dominant with genomic imprinting of the paternal allele; non-affected members of family should be examined for defect	*Bikhazi et al 1999 (125)*
Sporadic head and neck PGL: LOH at 11q23 and 11q22-q23	*Bikhazi et al 2000 (104)*
PGL of the organ of Zuckerkandl and GIST (gastric) in mono-zygotic twins, respectively. Carney syndrome?	*Boccon-Gibod et al 2004 (290A)*
In the Carney triad and familial-related syndrome (PGL and GIST), the PGL are usually multicentric and involve a range of ganglia, both sympathetic and parasympathetic, including the organ of Zuckerkandl. PCC may occur in a rare case.	*Carney 1999 (288)* *Carney and Stratakis 2002 (10)*
Familial and bilateral PGL of carotid body.	*Chase 1933 (364)*
Multiple PGL in NF	*DeAngelis et al 1987 (279)*
The *RET* oncogene is expressed but not mutated in sporadic PGL (52 PGL in 44 patients).	*de Krijger et al 2000 (131)*
Allelic imbalance is confined to 11q in PGL at the site of the PGL loci	*Devilee et al 1994 (345)* *Koreth et al 1999 (346)*
Head and neck PGL: 26 tumors from 22 patients (18 with family history), two also had PCC: allelic imbalance at 11q	*Devilee et al 1994 (345)*
Of 36 sympathoadrenal PGL, 61% showed nondiploid patterns (including tetraploid and peritetraploid cases) without necessarily indicating malignancy.	*Garcia-Escudero et al 2001 (452)*
Prognosis of PGL is not readily reflected in DNA changes, although they correlate with tumor location which is per se an indicator of prognosis.	*Gonzalez-Campora et al 1993 (453)*

TABLE 5.15. Continued

	References
PGL infiltrate and destroy adjacent tissue and may be malignant, most frequently in carotid body PGL	*Gospeth et al 1998* (347)
	Maier et al 1999 (348)
Positive family history in approximately 10% of PGL in the USA vs. 50% in The Netherlands; more recently, a higher incidence (approximately 25%) reported in the United States 2001	*Grufferman et al 1980* (219)
	van der Mey et al 1989 (17)
	Drovdlic et al (49)
90% of PGL are sporadic in nature;	*Grufferman et al 1980* (219)
<5% of PGL are malignant	*Baysal et al 2002* (167)
PGL may release catecholamines or other amines	*Hirano et al 1998* (349)
Characterization of the *SDHD* gene subunits. Other aspects of the succinate dehydrogenase system have been studied.	*Hirawake et al 1999* (474)
	Leckschat et al 1993 (448)
Metastatic carotid body PGL in a VHL patient	*Hull et al 1982* (350)
PGL in MEN-2	*Kennedy and Nager1986* (16)
Formation of carotid body tumors in patients with mutations (germline) in SDH genes (mitochondrial complex II)	*Koch et al 2002* (67)
Subsets of PGL are familial and can produce catecholamines in patients with germline mutations in genes associated with the mitochondrial complex II	*Koch et al 2002* (67)
Hereditary deficiencies of clotting factors in patients with carotid body PGL	*Kroll et al 1964* (351)
80% of PGL arise in the head and neck;	*Lack 1997* (27)
17% are intraabdominal and originate in the sympathoadrenal neuroendocrine system	
IGF-II (insulin growth factor) expressed in 9/9 PGL	*Li et al 1998* (352)
Insulin-like growth factor (IGF-II) is expressed in cancers of the prostate, breast, bladder, and PGL	Li et al 1998 (352)
Management of PGL described; early removal is advocated to prevent metastases and local invasion	*Patetsios et al 2002* (451)
RET or *VHL* mutations not associated with sporadic PGL of head and neck, though very rare cases of PGL may be seen in VHL, MEN-2 and NF1. No pure PGL family has been described with a mutation of *VHL*, *RET* or *NF1*	*Ponder 2001* (242)
	Baysal 2002 (45)
Inherited *SDHD* mutations may cause rare PGL of spinal cord	*Masuoka et al 2001* (353)
Counseling based on predictive DNA diagnosis in hereditary PGL	*Oosterwijk et al 1996* (19)
Incidence of PGL of carotid body is 10-fold in Andean mountain residents than in those at sea level	*Pacheco-Ojeda et al 1988* (189)
Reactive oxygen species (ROS) is generated by mitochondrial complex II response to hypoxia	*Paddenberg et al 2003* (354)
Novel *SDHD* mutation causing skipping of exon 3 in familial PGL	*Renard et al 2003* (356)
Family with PGL but without *RET* or *VHL* changes	*Sköldberg et al 1998* (357)
DNA flow cytometry: 37% of PGL had abnormalities (familial and sporadic)	*van der Mey et al 1991* (358)
Symptoms of PGL: bilateral and alveolar hypoventilation	*Roncoroni et al 1993* (359)
SDH and disease	*Rustin et al 2002* (360)
Examples of PGL at locations other than head and neck: cauda equina, cerebellopontine angle, pituitary fossa, sella, and meninges	*Salame et al 2001* (29)
	Deb et al 2005 (64)
	Mercuri et al 2005 (63)
	Yang et al 2005 (28)
	Zorlu et al 2005 (361)
Scintigraphy with octreotide, a somatostatin analog with high affinity for type 2 somatostatin receptors, is a reliable means of detecting PGL	*Telischi et al 2000* (362)
BCL2 expression may play a role in carotid body tumorigenesis	*Wang et al 1997* (363)
Somatostatin receptors in PGL	*Zak and Lawson 1982* (65)
	Reubi et al 1992 (118)
An investigation of patients with Carney syndrome and their tumors, (including PGL) revealed no association with SDH- inactivation, including subunit A. Loss of 1p was seen in two of five PGL examined by CGH.	*Baysal et al 2004* (499)
	Matyakhina et al 2007 (500)

TABLE 5.16. Some features of PCC.

	References
Higher catecholamines in young patients with PCC in a VHL family with a new mutation of *VHL* gene.	*Atuk et al 1998* (365)
Although not as frequent as in neuroblastoma, CpG island methylation in the promoter region of the gene *RASSF1A* was found in 5/23 sporadic PCC. This also contrasts with allelic loss at 3p in 46% of PCC (38.5% at 3p21.3).	*Astuti et al 2001* (193)
Regardless of tumor weight, PCC with a "malignant histology" are highly prone to metastasize when more than 5% of the nuclei are MIB-1-positive or CD44 immunostaining is negative or both. PCC with more than 10 chromosome copy number changes determined by CGH should be considered malignant.	*August et al 2004* (491)
Expression of telomeric profile discriminates between benign and malignant PCC.	*Boltze et al 2003* (366)
The MIB-1 nuclear proliferation index was elevated only in malignant PCC; a diploid DNA pattern did not necessarily predict benign behavior of PCC.	*Brown et al 1999* (460)
	van der Harst 2000 (460A)
Frequent screening of patients with medullary thyroid carcinoma cases associated with MEN-2A for the possible presence of PCC has been advocated.	*Casanova et al 1993* (458)

TABLE 5.16. Continued

	References
In 15 cases of PCC (only 1 familial and 5 bilateral) three germline mutations were demonstrated (1 bilateral and 2 with unilateral).	*Cascón et al 2002 (168)*
In a study of 33 PCC (19 men and 14 women; with a mean age of 45), which included 10 malignant tumors, five with NF1, 3 familial and one with MEN-2, indices of malignancy were addressed. High MIB-1 index, extra-adrenal location, tumor weight, and young age were indices of malignancy.	*Clarke et al 1998 (277)*
Hypoxia and increase of transcription and stability of tyrosine hydroxylases mRNA in PCC cells.	*Czyzyk-Krezeska et al 1994 (367)*
TP53, p16 and *VHL*-associated *CUL2* gene do not play a role in pathogenesis of PCC	*Dahia et al 1995 (367A)*
	Herfarth et al 1997 (368)
	Aguiar et al 1996 (369)
	Duerr et al 1999 (370)
Glial cell line-derived neurotrophic factor (GDNF) does not play a role in the genesis of sporadic PCC (18 benign, 4 malignant).	*Dahia et al 1997 (371)*
In a study of 48 benign and 24 malignant PCC, *VHL* mutations and LOH were found in two of the benign and in four of the malignant tumors.	*Dannenberg et al 2003 (461)*
Malignant PCC showed a higher frequency of p53 and bcl-2 expression as compared to benign tumors. Overexpression of c-erbB-2 was associated with familial PCC.	*de Krijger et al 1999 (462)*
Plasma normethanephrine and metanephrine levels are a sensitive way of diagnosing PCC (familial).	*Eisenhofer et al 1999 (372)*
The usefulness of determining the levels of various catecholamine metabolites in the diagnosis and follo;w-up of PCC, and their metabolism by these tumors, can be found in a number of reports and studies.	*Eisenhofer 2001 (485)*
	Eisenhofer et al 1998 (486)
	Eisenhofer et al 2003 (487)
	Isobe et al 2000 (488)
	Isobe et al 1998 (489)
AKT activated in PCC (sporadic and MEN 2-related), but not in benign adrenocortical adenomas.	*Fassnacht et al 2006 (373)*
Angiogenesis in PCC.	*Favier et al 2002 (374)*
Family with germline mutation of *SDHD* showed loss of remaining allele leading to abrogation of mitochondrial complex II transfer activity in PCC. Sporadic PCC did not show any loss of activity or mutations of *SDHD* and *SDHB*.	*Gimenez-Roqueplo et al 2003 (159)*
Observations on *SDHD* and *SDHB* in sporadic and familial PCC.	*Gimm et al 2000 (143)*
	Agiuar et al 2001 (274)
	Astuti et al 2001 (192)
	Kytolä et al 2002 (176)
Multiple mucosal neuromas, medullary carcinoma of the thyroid and PCC; a syndrome or variation of MEN-2?	*Gorlin et al 1968 (375)*
Familial PCC with novel mutation of *VHL*	*Gross et al 1996 (376)*
Multiple PGL occur in only 6% of PGL cases. In sporadic PGL, <5% have bilateral tumors in the head and neck, whereas 32% of familial PGL show bilateral tumors.	*Grufferman et al 1980 (219)*
Management of "incidentaloma" of the adrenal gland.	*Grumbach et al 2003 (377)*
Pituitary adenylate cyclase-activating polypeptide (PACAP) exerts trophic effects on the differentiation, proliferation and survival of neuronal cells and could serve as a target for therapeutic effects in PCC.	*Grumolato et al 2003 (483)*
Overexpression of topoisomerase II-α and MIB-1 and loss of RB protein were common in malignant PCC, whereas p53, E-cadherin and HER-2/neu did not have a diagnostic utility in tumor behavior.	*Gupta et al 2000 (463)*
Loss of *NF1* expression in PCC not in NF1 patients.	*Gutmann et al 1995 (278)*
Neuropeptide Y expression in benign and malignant PCC.	*Helman et al 1989 (378)*
VHL and PCC	*Hes et al 2003 (236)*
	Lonser et al 2003 (234)
	Gijtenbeek et al 2005 (240)
Algorithms for screening and localization of PCC.	*Ilias and Pacak 2004 (78)*
	Sawka et al 2004 (80)
Imaging of benign and malignant PCC.	*Ilias et al 2003 (75)*
	Kann et al 2004 (78)
RET germline mutations and genotype/phenotype correlation in PCC and Medullary thyroid carcinoma.	*Jimenez et al 2004 (343)*
RAS and nitric oxide in PCC cells.	*Jeong et al 2002 (380)*
High urinary dopamine levels, extra-adrenal location, high tumor weight, elevated tumor dopamine concentration, and postoperative hypertension are all indices of malignancy in PCC.	*John et al 2000 (464)*
Neurofibromin insufficiency may induce schwann cell proliferation in composite PCC.	*Kimura et al 2002 (465)*
Therapy for malignant PCC.	*Lam et al 2005 (81)*
RET is a target for RNA splicing deregulation at its 5'-end in familial and sporadic PCC.	*Le Hir et al 2002 (450)*
Best tests of biochemical diagnosis of PCC.	*Lenders et al 2002 (58A)*
	Eisenhofer 2003 (381)
	Eisenhofer et al 2003 (74)
	Sawka et al 2003 (382)
Review of PCC and its various features.	*Lenders et al 2005 (389)*
Mutations of p53 may play a role in the tumorigenesis of benign and functional PCC.	*Lin et al 1994 (457)*
Codon-specific development of PCC in MEN-2. <10% of PCC patients have a positive family history	*Machens et al 2006 (383)*
	Maher and Eng 2002 (11)

Table 5.16. Continued

	References
Clinical and diagnostic aspects of PCC.	*Manger 2003 (76)*
	Amar et al 2005 (73)
Diagnosis and management of PCC.	*Manger and Eisenhofer 2004 (479)*
Although RET may play a limited role in normal adult tissues, high levels of expression may have an association with PCC development.	*Matias-Guiu et al 1995 (449)*
Intronic single nucleotide polymorphisms in *RET* and sporadic PCC subset.	*McWhinney et al 2003 (384)*
PCC occur in both MEN-2A (associated with *RET* gene located at 10q11.2 defects in the cysteine-rich domain encoded by exons 10 and 11) and MEN-2B (associated with *RET* defects in the tyrosine kinase domain encoded by exon 16), the former associated with medullary thyroid carcinoma and parathyroid abnormalities and the latter with changes of enteric ganglia. Malignancy seems to develop in these PCC as they get larger.	*Modigliani et al 1995 (339)* *Ponder 1999, 2001 (241, 242)* *Puñales et al 2003 (341)*
ACTH synthesis was more likely to occur in familial PCC than in sporadic tumors, whereas the reverse was true for vasoactive intestinal peptide. The expression of other neuropeptides was less informative, though ACTH was more likely to be expressed in malignant PCC and neuron-specific enolase in benign forms.	*Moreno et al 1999 (459)*
Familial PCC and cerebellar hemangioblastoma in a case and review of literature.	*Mulholland et al 1969 (385)*
Incidence of PCC approximately 1 in 300,000 to 1,000,000 90% of PCC are sporadic 10% of PCC are extra-adrenal <10% of PCC are malignant, though higher rates (13–26%) have been reported.	*Mundschenk and Lehnert 1998 (70)* *Baysal et al 2002 (45)* *Mundschenk and Lehnert 1998 (70)* *Baysal et al 2002 (45)*
Somatostatin receptor subtypes in PCC and octreotide scintigraphy	*Mundschenk et al 2003 (77)*
Malignancy in PCC is more common in nondiploid cases than in diploid cases. No biochemical marker was a reliable index of malignancy.	*Nativ et al 1992 (456)* *Tormey et al 2000 (455)*
Apparently 3–8% of sporadic PCC may be the sole manifestation of hereditary disease and associated with germline *VHL* mutations in patients.	*Neumann et al 1993 (227)* *Bar et al 1997 (130)* *Brauch et al 1997 (223)* *van der Harst 1998 (232)*
A high MIB-1 (Ki-67) score in PCC suggests a malignant facet.	*Ohji et al 2001 (276A)*
Genetics of PCC in VHL and NF1.	*Opocher et al 2005 (478)*
Biochemical diagnosis, localization and management of PCC: sporadic and MEN-2-related.	*Pacak et al 2005 (477)*
Effects of NGF (nerve growth factor) on catecholamines in 2 PCC cultures.	*Pfragner et al 1984 (386)*
A continuous human cell line from a benign sporadic PCC has been established and characterized. Earlier attempts to cultivate human PCC resulted in cell lines with limited life spans.	*Pfragner et al 1984 (89)* *Pfragner and Walser 1980 (445)* *Pfragner et al 1984 (386)* *Evers et al 1992 (446)*
In 2/17 sporadic PCC the same mutation of the *VHL* was present	*Iyengar et al 1995 (273)*
Somatostatin receptors in PCC.	*Reubi et al 1992 (118)* *Mundschenk et al 2003 (77)*
Stress, release of adrenal catecholamines and the catecholamine-ergic systems	*Sabban and Kvetnanský et al 2001 (387)*
Enhanced expression of tenascin, an extracellular matrix glyco-protein, is present in PCC, especially in malignant tumors.	*Salmenkivi et al 2001 (454)*
Loss of inhibin/activin-betaB-subunit expression may be an indicator of malignant potential in PCC.	*Salmenkivi et al 2001 (454B)*
Composite PCC in a patient with NF1.	*Satake et al 2001 (466)*
Neutrophilia in PCC. (Letter and Reply)	*Sevastos et al 2005 (388)* *Sawka et al 2005 (389)*
Estimates of literature indicate that ~8.5% of sporadic PCC have *VHL* mutations and ~3.6% *RET* mutations.	*Takaya et al 1996 (390)* *Baysal 2002 (45)*
Clinicopathologic and immunophenotypic review of PCC of the adrenal gland.	*Thompson 2002 (59)*
Erythropoietin and its receptor expressed in VHL-associated PCC; only the receptor was expressed in MEN-2 PCC.	*Vogel et al 2005 (391)*
NF1 and PCC.	*Walther et al 1999 (326)*
GDNF is a natural ligand for *RET*.	*Woodward et al 1997 (133)*
No mutations of GDNF in familial PCC. Seen in one sporadic PCC.	
Nerve growth factor (NGF) activities mediated by the TrkA/p75 neurotrophin receptor (NTR) complex tyrosine kinase receptor. Proapoptotic effects of 75 NTR on PC12 (PCC) cells related to up-regulation of cholesterol biosynthetic enzymes and consequent cholesterol biosynthesis.	*Yan et al 2005 (393)*
Peptide EM66 a marker for benign vs. malignant PCC.	*Yon et al 2003 (392)*
A relatively high incidence of p53 gene mutations or intronic sequence alterations in multiple and malignant PCC was described. Such changes were not seen in benign tumors.	*Yoshimoto et al 1998 (467)*
Studies showes that NF1-associated PCC patients have germline *NF1* mutations that favor the cysteine-serine rich domain over the RAS GTPase activating protein domain. These genetic data might help direct functional evaluation of neurofibromin in its role in both heritable and sporadic PCC and PGL tumors.	*Bausch et al 2007 (501)*
A plan for preoperative management of PCC patients has been proposed.	*Pacak 2007 (501 A)*

TABLE 5.17. Genetic Changes in Paraganglioma (PGL).

	References
PCC have increased levels of IGF-II which are not reflected in peripheral levels. Receptors for IGF-I and IGF-II were found in these tumors.	*Gelato and Vassalotti 1990* (484)
SDHA is not associated with tumor development.	*Bourgeron et al 1995* (316)
PGL (carotid body) in VHL: a rare association.	*Zanelli and van der Walt 1996* (334)
Functioning carotid PGL in VHL disease.	*Schimke et al 1998* (333)
All four peptides are encoded in the nucleus, then imported into the mitochondria where they are modified, folded and assembled.	*Ackrell 2002* (417)
The occurrence of malignancy in PGL of *SDHC* mutation carriers has been pointed out.	*Niemann et al 2003* (166) *Niemann 2006* (166)
The association of PCC with *SDHB* mutations in childhood may require more detailed investigation.	*Beck et al 2004* (418)
Study of 7 German cases with carotid body PGL revealed one novel mutation of *SDHD*, no Dutch founder mutation and no evidence of G12S polymorphism.	*Leube et al 2004* (480)
Screening of *SDHB*, *SDHC* and *SDHD* for mutations was performed in familial and sporadic head and neck PGL. The 14 sporadic cases showed no changes, in contrast to the familial cases (2/3).	*Mhatre et al 2004* (419)
Germline mutations of *SDHB* in heritable PGL may be associated with early-onset RCC.	*Vanharanta et al 2004* (141)
Germline missense, nonsense and frameshift mutations across 4 exons of *SDHD* in PGL has been demonstrated, as well as intronic splice site mutations and 96-kilobase deletion spanning *SDHD* in a family with head and neck PGL.	*Bertherat et al 2005* (420) *McWhinney et al 2004* (136)
Study of 17 sporadic PGL and the patients revealed only one *SDHB* and one *SDHD* mutation, indicating that inactivation of these genes is not a major factor in sporadic PGL. Active succinate dehydrogenase systems in most of the tumors is also an indication of the conclusion. LOH and FISH studies showed frequent loss of regions on chromosome 11.	*Braun et al 2005* (421)
Multifocal PGL tumors are common in *SDHD*-associated disease.	*Dannenberg et al 2005* (422)
Mutations of *SDHD* described in a Greek family with carotid body PGL.	*Liapis et al 2005* (423)
Four kindreds with PGL (familial) of head and neck with *SDHD* mutations; tumors may be bilateral or multiple.	*Velasco et al 2005* (174)
A study of 44 consecutive patients with malignant PGL revealed *SDHB* mutations in 13. Half of the tumors were extra-adrenal and catecholamine-producing.	*Brouwers et al 2006* (424)
Various types of mutations (missense, nonsense, frameshift, splice site) of *SDHB* are associated with about half of all malignant PGL.	*Brouwers et al 2006* (424)
A novel mutation of *SDHB* was described in a case of familial PCC associated with PGL of the para-aortic area and carotid body.	*Cascón et al 2006* (416)
Gross *SDHB* deletions found in 3/24 patients with sporadic PGL who were negative for point mutations. The authors discuss the fact that this is the first description of gross deletions in patients with sporadic PGL. The extra-adrenal location of the tumors seems to constitute a determining factor for whether to include these cases in genetic testing for gross deletions of the *SDHB* gene.	*Cascón et al 2006* (416)
A case of a 16-year-old girl with idiopathic hypotension, polycythemia and multiple PGL of the renal hilum has been described.	*Dionne et al 2006* (425)
Sporadic abdominal PGL in a 68-year-old male with a germline mutation of *SDHB*.	*Elston et al 2006* (426)
A case of carotid PGL (bilateral) had increased urinary dopamine excretion. Other catecholamines were excreted within the normal range. Familial cases were more likely to have bilateral PGL than non-familial cases.	*Jeffery et al 2006* (427)
A PGL and a myxopapillary ependymoma in the cauda equina in a 38-year-old woman has been described. Another such case was also reported.	*Keith et al 2006* (428) *Yang et al 2005* (28)
Resection of a PGL of the organ of Zuckerkandl in a patient with a carotid body PGL has been described.	*Mithani et al 2006* (429)
A novel alteration in the *SDHD* gene in a pedigree with familial PGL has been described.	*Ogawa et al 2006* (430)
Missense dysfunction of SHDH may be important in familial *SDHD* –associated PGL, as well as expressional silencing of the wildtype allele.	*Ogawa et al 2006* (430)
Functioning PGL and GIST of the jejunum in three women (unrelated). No changes in *SDH* genes or *KIT* were present. Is it a new syndrome?	*Perry et al 2006* (431)
Sclerosing PGL in 19 cases described. May be mistaken for an aggressive malignant neoplasm. They were located in the carotid body, parapharyngeal region and mediastinum.	*Plaza et al 2006* (432)
Loss of chromosome 11 regions, including *PGL1* and *PGL2* loci, may result in a complicated phenotype, e.g., multiple tumors; one such patient had 6 primary PGL.	*Riemann et al 2004* (314)
No PCC associated with *SDHC* has been reported, though a PGL (malignant) of the carotid bifuncation producing catecholamines has been reported.	*Schiavi et al 2005* (433) *Muller et al 2005* (434)
Patients with head and neck PGL may harbor *SDHC* mutations, in addition to those of *SDHB* and *SDHD*.	*Schiavi et al 2005* (433)
PGL associated with *SDHC* mutations are benign and seldom multifocal. It was recommended that an analysis of germline mutations for *SDHC* be performed in apparently sporadic cases to identify risk of inheritance.	*Schiavi et al 2005* (433)
In a study of mutations of *SDH* genes in 34 patients with PGL, 79% were found to have mutations of *SDHD* and the remainder mutations of *SDHB*. No mutations of *SDHC* were encountered.	*Badenhop et al 2004* (152)
The enhanced expression of *FGFR1* and concomitant expression of the ligand hFGF in PGL may result in a mitogenic signal and possible anti-apoptotic effects that could contribute to the development of these tumors of high vascularity.	*Douwes Dekker et al 2007* (435)

TABLE 5.18. Genetic changes in PCC.

	References
Basic fibroblast growth factor (bFGF) was shown to be present in PCC and to possibly exert an autocrine and paracrine function in the growth and neovascularization of PCC.	*Statuto et al 1993* (405)
The product of the *VHL* gene, pVHL, is a component of a ubiquitin ligase, which targets the transcription factor HIF subunits HIF-1α and HIF-2α for ubiquitination and proteolytic degradation in the presence of oxygen, by forming ubiquitin-ligase complex with elongin B and C, and with Cullin-2 and Rbx1.	*Iliopoulos et al 1996* (406) *Kamura et al 1999* (408)
An isolated familial PCC as a variant of VHL disease has been described.	*Ritter et al 1996* (409)
Alterations of the *RET* oncogene in familial and sporadic PCC.	*Rodien et al 1997* (416A)
Contrary to medullary thyroid carcinoma, *RET* mutations are not associated with aggressive behavior of sporadic PCC.	*van der Harst et al 1998* (475)
Chronic stimulation of the *RET* signaling pathway may be involved in the pathogenesis of some sporadic PCC.	*Le Hir et al 2000* (481)
Contrasting effects on HIF1α by mutations of *VHL* relate to patterns of tumorigenesis in VHL disease, i.e., hemangioblastomas and RCC occur in type 1 VHL but PCC do not.	*Clifford et al 2001* (139)
PCC in VHL disease and MEN2 display distinct biochemical and clinical phenotypes, e.g., the tendency for PCC in the former to be associated with norepinephrine secretion vs. epinephrine in the latter.	*Eisenhofer et al 2001* (228)
Carriers of *VHL* germline mutations can present with a form fruste of the VHL disease presenting initially with unilateral PCC and hence mutation analysis should be performed.	*Frenzel et al 2001* (410)
HIF deregulation may play a role in PCC pathogenesis in the setting of VHL disease.	*Hoffman et al 2001* (140) *Semenza 2002* (140A)
Patients with VHL type 2 (with PCC) harbor missense mutations in contrast to type 1 cases (without PCC) who more frequently exhibit deletions or premature termination mutations.	*Iliopoulos et al 2001* (411)
SDHB gene mutation and its consequences in a sporadic PCC.	*Gimenez-Roqueplo et al 2002* (275)
Somatic mutations of *VHL*, *RET*, *SDHB* and *SDHD* are rare in sporadic PCC.	*Benn et al 2003* (158) *Cascón et al 2004* (412)
The common expression of hTERT and telomerase may identify PCC of more aggressive nature.	*Boltze et al 2003* (366)
Intronic single nucleotide polymorphisms in the *RET* gene are associated with a subset of apparently sporadic PCC and may modulate age of onset.	*McWhinney et al 2003* (384)
The higher concentrations of EM66, a novel secretogranin II- derived peptide present in chromaffin cells of the adrenal gland, in benign PCC vs. low levels in malignant tumors, can be used diagnostically.	*Yon et al 2003* (392)
Familial related PCC with a *VHL* mutation (new).	*Sanso et al 2004* (476)
A *RET* mutation with decreased penetrance in the family of a patient with sporadic (?) PCC has been described.	*Arum et al 2005* (395)
Association of non-functioning PCC, ganglioneuroma, adrenal cortical adenoma and vertebral hemangioma in a case with a variant of the *VHL* gene has been described.	*Bernini et al 2005* (400)
Catecholamines play a role in internal remodeling of the carotid body.	*Bernini et al 2006* (400A)
Low molecular weight proteomic results distinguish metastatic from benign PCC.	*Brouwers et al 2005* (396)
Hypermethylation of CpG island promoters is associated with transcriptional inactivation of TSG in neoplasia, e.g., inactivation of *p16* and *PTEN* has been related to the development of PCC. In a study of MEN-2 and sporadic PCC, hypermethylation was detected in 48% of PCC for *RASSF1A*, 24% for *p16*, 36% for *MSH2*, 16% for *CDH1* and 8% for *PTEN*. No methylation of *VHL*, *MLH1* or *TIMP3* was observed. In the MEN-2-related tumors, *p16* inactivation was seen in 42 vs. 8% in sporadic PCC and *RASSF1* in 58 vs. 38%, respectively. Combined methylation of *p16* and *RASSF1* was seen only in *MEN2*- associated tumors.	*Dammann et al 2005* (490)
The AKT/protein kinase-β pathway is a major one involved in the regulation of apoptosis and cell survival. Increased activation of AKT in PCC has been described, both sporadic and MEN2 related. This activation is not associated with LOH of the *PTEN* gene and, therefore, most likely is caused by signaling dysregulation upstream of PI3K.	*Fassnacht et al 2005* (373)
SDHB and *SDHD* hasve been implicated in the pathogenesis of PCC, as well as the *RET* and *VHL* genes.	*Gimm 2005* (482)
Different expression of catecholamine transporters in VHL-PCC vs. MEN-2-PcC.	*Huynh et al 2005* (399)
Novel *VHL* mutations in a Japanese family with PGL and hepatic hemangioma.	*Takahashi et al 2005* (401A)
Synchronous mediastinal ganglioneuroma and retroperitoneal PCC.	*Takeda et al 2005* (413)
Catecholamine excess (37.5%) and PCC (20%) in a well-defined Dutch population with *SDHD*-linked head and neck PGL.	*Van Houtum et al 2005* (414)
MEN-2-associated PCC may be derived from a different cell than in VHL disease, based on the differential expression of erythropoietin.	*Vogel et al 2005* (391)
Patients with PCC and *SDHB* mutations have a later age of onset, extra-adrenal (abdominal or thoracic) tumors and a higher rate of malignancy than those with tumors without this mutation. On the other hand, patients with PCC with *SDHD* mutations had a propensity to develop PGL and multiple tumors with infrequent malignant changes and extra-adrenal PCC.	*Benn et al 2006* (415)
LOH/single nucleotide polymorphism involved *TP53* in 30%, *RB1* in 21%, *WT1* in 27% and *NF1* in 40% of PCC (95 sporadic, 48 MEN-2–associated). Changes in malignant PCC described.	*Blanes et al 2006* (394)
Hypermethylation of CpG islands is not part of sporadic PCC and other genes must be involved in the genetics of these tumors.	*Cascón et al 2006* (416)
Insulin-like growth factor receptor is overexpressed in PCC.	*Fottner et al 2006* (397)
Circulating EM66 is a highly sensitive marker for the diagnosis and follow-up of PCC.	*Guillemot et al 2006* (399)
Cases of MEN2 should be screened for *RET* mutations of codons 918, 634, and 630 from age 10 and for the remainder codons at age 20 for possible PCC risk.	*Machens et al 2006* (383)
Composite PCC and adrenal neuroblastoma in an infant.	*Tatekawa et al 2006* (402)
Occurrence of *SDHB* gene mutations in PCC, tumors which are often extra-adrenal and malignant.	*van Nederveen et al 2006* (326)
PTEN mutations or inactivation play a minor role in malignant PCC.	*van Nederveen et al 2006* (403)
Malignant PCC and Bcl-2 expression.	*Wakasugi et al 2006* (404)

TABLE 5.19. Genetic changes in PGL and PCC.

	References
Occasional members of families with PGL may also have PCC	*Pritchett 1982* (355)
S-100 protein staining in cells morphologically similar to the sustacular cells of normal paraganglia and adrenal medulla were found in all PGL and in benign and aggressive PCC.	*Schroder and Johannsen 1986* (471)
A 21-year-old male with familial PGL of the carotid body and multiple extra-adrenal PCC.	*Jensen et al 1991* (379)
Insulin-like growth factor (IGF)-II immunoreactivity was found in normal and neoplastic paraganglionic tissues (PCC and PGL), although its biological significance has not been determined.	*Suzuki et al 1992* (492)
Non-diploid PCC and PGL are more likely to show aggressive behavior than diploid tumors and hence the former should be carefully monitored.	*Pang and Tsao 1993* (468)
DNA levels were measured in PGL and PCC. Of the 56 PGL, 59% were diploid and 41% aneuploid. Of carotid body PGL 13/14 were diploid. Of PGL in other sites 7/12 were aneuploid. DNA levels can be used as predictive indices of PGL and PCC behavior.	*Held et al 1997* (472)
Overexpression of p53 (thought not frequently) was more common in bilateral PCC and PGL than in single tumors; loss of Rb expression was often seen in PCC, whereas mdm2 overexpression was common in PGL.	Lam et al 2001 (470)
The differential diagnosis of PCC, PGL and composite tumors is discussed.	*McNichol 2002* (469)
In seven PGL cases (three with PCC) and two with a family history, novel *SDHD* mutations were shown in two cases (one with PCC).	*Cascón et al 2002* (180)
A review of some of the conditions and genes associated with PCC (*VHL*, *NF1*, *MEM2*, *SDHD* , *p53*, and *RET*).	*Koch et al 2002* (494)
Germline mutations of *SDHD* in familial head and neck PGL involve the 3′ part of the gene (5′ in familial PCC). Mutations involve exons 2 and 3 and exons 2–7 of *SDHB* in PGL.	*Maher and Eng 2002* (11)
Germline mutations of *SDHD* in familial PCC involve the 5′ part of the gene (3′ in familial PGL). Mutations of exons 2–7 of *SDHB* in PGL.	*Maher and Eng 2002* (11)
A 9-year-old boy with several extra-adrenal PCC-secreting catecholamines and -producing symptoms, developed a PGL 7 months following removal of the PCC. A novel germline mutation of the *VHL* gene was found in the boy but not in the mother.	*Reichardt et al 2002* (473)
Review of the various HIF factors, especially *HIF1* in cancer, and possible relevance to PGL and PCC.	*Semenza 2002* (140A)
SDHB mutation-positive PCC in PGL families do not display genomic imprinting	*Benn et al 2003* (158)
The combined use of Ki-67/MIB-1 and hTERT expression, in addition to histopathology, provides a highly specific means of identifying benign PCC and abdominal PGL that are not at risk of developing recurrent or metastatic disease.	*Edström-Elder et al 2003* (447)
Complex II may be considered as one of the primary factors in the regulation of mitochondrial respiratory function by responding to the cellular iron level, thereby influencing cellular growth.	*Yoon et al 2003* (437)
A study demonstrated the complete loss of the maternal chromosome 11 in *SDHD*-linked PGL and PCC, suggesting that the combined loss of the wildtype *SDHD* allele and maternal 11p region is essential for tumorigenesis. Thus, the exclusive paternal transmission of the disease can be explained by a somatic genetic mechanism targeting both the *SDHD* gene on 11q23 and a paternally imprinted gene at 11p15.5, rather than imprinting of *SDHD*.	*Hensen et al 2004* (438)
Review of the clinical presentations of sporadic and familial PCC and PGL.	*Kaltsas et al 2004* (14)
Novel mutation (K40E) of *SDHB* in an Australian family with with PCC and PGL.	*McDonnell et al 2004* (439)
SDH mutations database for PGL and PCC.	*Bayley et al 2005* (436)
Another layer of control between VHL and SDH pathways was suggested by the finding that high HIFα levels seem to repress *SDHB*.	*Dahia et al 2005* (212)
Update on mitochondrial tumor suppressors (*SDHB*, *SDHD*, and fumarate hydratase).	*Gottlieb and Tomlinson 2005* (440)
Histological grading of PCC and extra-adrenal sympathetic PGL.	*Kimura et al 2005* (15A)
A study has shown that mutations of genes (*VHL*, *RET*, *NF1*, *SDH*) giving rise to the development of PCC and PGL may do so by decreasing the activity of 2-oxoglutarate-dependent oxygenase (EgIN3 prolyl hydroxylase) through a number of different mechanisms, leading to reduced apoptosis of neural crest cells during development.	*Lee et al 2005* (214) *Maxwell 2005* (331)
Novel germline mutations of *SDHB* and *SDHD* in sporadic head and neck PGL (minor role) and familial PGL and/or PCC (major role) were described	*Bayley et al 2006* (441)
GIST in *SDHB* mutation-associated family with PCC/PGL has been reported.	*Bolland et al 2006* (442)
Psammomatous melanotic schwannomas are part of the Carney complex and may be its first manifestation.	*Carrasco et al 2006* (497) *Mosunjac et al 2007* (498)
Review of evolving concepts on PGL and PCC tumorigenesis.	*Dahia 2006* (218)
Although inactivation of *VHL* and *SDHB* and *SDHD* may disrupt similar HIF-dependent and HIF-independent signaling pathways, their effects on target gene expression were not identical and may explain the observed clinical differences in PCC and PGL seen with germline *VHL* vs. *SDHB* and *SDHD* mutations.	*Pollard et al 2006* (443)
A single founder *SDHD* mutation was present in six families in an area of central Italy and was associated with widely variable interfamilial and intrafamilial expressivity.	*Simi et al 2006* (444)
Frequent losses of 1p and 3q occur in benign PCC of Japanese as determined by CGH.	*Kino et al 2007* (496)
Superiority of fluorodeoxyglucose positron emission tomography to other functional imaging techniques in the evaluation of meta-static *SDHB*-associated PCC and PGL had been reported.	*Timmers et al 2007* (495)

References

1. Parry, D.M., Li, F.P., Strong, L.C., Carney, J.A., Schottenfeld, D., Reimer, R.R., and Grufferman, S. (1982) Carotid body tumors in humans: genetics and epidemiology. *J. Natl. Cancer Inst.* **68**, 573–578.

2. Heutink, P., van Schothorst, E.M., van der Mey, A.G.L., Bardoel, A., Breedveld, G., Pertijs, J., Sandkuijl, L.A., van Ommen, G.-J.B., Cornelisse, C.J., Oostra, B.A., and Devilee, P. (1994) Further localization of the gene for hereditary paragangliomas and evidence for linkage in unrelated families. *Eur. J. Hum. Genet.* **2**, 148–158.

3. Crossey, P.A., Eng, C., Ginalska-Malinoswska, M., Lennard, T.W.J., Wheeler, D., Ponder, B.A.J., and Maher, E.R. (1995) Molecular genetic diagnosis of von-Hippel-Lindau disease in familial phaeochromocytoma. *J. Med. Genet.* **32**, 885–886.

4. Nilsson, O., Tissell, L.E., Jansson, S., Ahlman, H., Gimm, O., and Eng, C. (1999) Adrenal and extra adrenal pheochromocytomas in a family with germline *RET* V804L mutation. *JAMA* **281**, 1587–1588.

5. van Baars, F.M., Cremers, C.W., van den Broek, P., Geerts, S., and Veldman, J.F. (1982) Genetic aspects of nonchromaffin paraganglioma. *Hum. Genet.* **60**, 305–309.

6. Hes, F., Zewald, R., Peeters, T., Sijmons, R., Links, T., Verheij, J., Matthijs, G., Legius, E., Mortier, G., van der Torren, K., Rosman, M., Lips, C., Pearson, P., and van der Luijt, R. (2000) Genotype-phenotype correlations in families with deletions in the von Hippel-Lindau (VHL) gene. *Hum. Genet.* **106**, 425–431.

7. Petropoulos, A.E., Luetje, C.M., Camarata, P.J., Whittaker, K., Lee, G., and Baysal, B.E. (2000) Genetic analysis in the diagnosis of familial paragangliomas. *Laryngoscope* **110**, 1225–1229.

8. Baysal, B.E., and Myers, E.N. (2002) Etiopathogenesis and clinical presentation of carotid body tumors. *Microsc. Res. Tech.* **59**, 256–261.

9. Benn, D.E., Marsh, D.J., and Robinson, B.G. (2002) Genetics of pheochromocytoma and paraganglioma. *Curr. Opin. Endocrinol. Diabetes* **9**, 79–86.

10. Carney, J.A., and Stratakis, C.A. (2002) Familial paraganglioma and gastric stromal sarcoma: a new syndrome distinct form the Carney triad. *Am. J. Med. Genet.* **108**, 132–139.

11. Maher, E.R., and Eng, C. (2002) The pressure rises: update on the genetics of phaeochromocytoma. *Hum. Mol. Genet.* **11**, 2347–2354.

12. 12. Neumann, H.P.H., Bausch, B., McWhinney, S.R., Bender, B.U., Gimm, O., Franke, G., Schipper, J., Klisch, J., Altehöher, C., Zerres, K., Januszewicz, A., and Eng, C. (2002) Germ-line mutations in nonsyndromic pheochromocytoma. *N. Engl. J. Med.* **346**, 1459–1466.

13. Dannenberg, H., Dinjens, W.N.M., Abbou, M., van Urk, H., Pauw, B.K.H., Mouwen, D., Mooi, W.J., and de Krijger, R.R. (2002) Frequent germ-line succinate dehydrogenase subunit D gene mutations in patients with apparently sporadic parasympathetic paraganglioma. *Clin. Cancer Res.* **8**, 2061–2066.

14. Kaltsas, G.A., Papadogias, D., and Grossman, A.B. (2004) The clinical presentation (symptoms and signs) of sporadic and familial chromaffin cell tumours (phaeochromocytomas and paragangliomas). *Front. Horm. Res.* **31**, 61–75.

15. Gimm, O., Koch, C.A., Januszewicz, A., Opocher, G., and Neumann, H.P. (2004) The genetic basis of pheochromocytoma. *Front. Horm. Res.* **31**, 45–60.

15A. Kimura, N., Watanabe, T., Noshiro, T., Shizawa, S., and Miura, Y. (2005) Histological grading of adrenal and extra-adrenal pheochromocytomas and relationship to prognosis: a clinicopathological analysis of 116 adrenal pheochromocytomas and 30 extra-adrenal sympathetic paragangliomas including 38 malignant tumors. *Endocr. Pathol.* **16**, 23–32.

15B. Suarez, C., Rodrigo, J.P., Ferlito, A., Cabanillas, R., Shaha, A.R., and Rinaldo, A. (2006) Tumours of familial origin in the head and neck. *Oral Oncol.* **42**, 965–978.

16. Kennedy, D.W., and Nager, G.T. (1986) Glomus tumor and multiple endocrine neoplasia. *Otolaryngol. Head Neck Surg.* **94**, 644–648.

17. van der Mey, A.G., Maaswinkel-Mooy, P.D., Cornelisse, C.J., Schmidt, P.H., and Van de Kamp, J.J. (1989) Genomic imprinting in hereditary glomus tumours: evidence for new genetic theory. *Lancet* **2**, 1291–1294.

18. Balatsouras, D.G., Eliopoulos, P.N., and Economou, C.N. (1992) Multiple glomus tumours. *J. Laryngol. Otol.* **106**, 538–543.

19. Oosterwijk, J.C., Jansen, J.C., Van Schothorst, E.M., Oosterhof, A.W., Devilee, P., Bakker, E., Zoeteweij, M.W., and Van der Mey, A.G. (1996) First experiences with genetic counselling based on predictive DNA diagnosis in hereditary glomus tumours (paragangliomas). *J. Med. Genet.* **33**, 379–383.

20. Blume-Peytavi, U., Adler, Y.D., Geilen, C.C., Ahmad, W., Christiano, A., Goerdt, S., and Orfanos, C.E. (2000) Multiple familial cutaneous glomangioma: a pedigree of 4 generations and critical analysis of histologic and genetic differences of glomus tumors. *Am. Acad. Dermatol.* **42**, 633–639.

20A. Strauss, M., Nicholas, G.G., Abt, A.B., Harrison, T.S., and Seaton, J.F. (1983) Malignant catecholamine-secreting carotid body paraganglioma. *Otolaryngol. Head Neck Surg.* **91**, 315–321.

21. Zbaren, P., and Lehmann, W. (1985) Carotid body paraganglioma with metastases. *Laryngoscope* **95**, 450–454.

22. Walsh, R.M., Leen, E.J., Gleeson, M.J., and Shaheen, O.H. (1997) Malignant vagal paraganglioma. *J. Laryngol. Otol.* **111**, 83–88.

23. Maher, E.R., Webster, A.R., Richards, F.M., Green, J.S., Crossey, P.A., Payne, S.J., and Moore, A.T. (1996) Phenotypic expression in von Hippel-Lindau disease: correlations with germline VHL gene mutations. *J. Med. Genet.* **33**, 328–332.

24. Dalainas, I., Nano, G., Casana, R., Bianchi, P., Stegher, S., Malacrida, G., and Tealdi, D.G. (2006) Carotid body tumours. A 20-year single-institution experience. *Chir. Ital.* **58**, 631–635.

25. Hegarty, J.L., and Lalwani, A.K. (2000) Paragangliomas of the head and neck: implications of molecular genetics in clinical medicine. *Curr. Opin. Otolaryngol. Head Neck Surg.* **8**, 384–390.

26. Benn, D.E., and Robinson, B.G. (2006) Genetic basis of phaeochromocytoma and paraganglioma. *Best Pract. Res. Clin. Endocrinol. Metab.* **20**, 435–450.

27. Lack, E.E. (1997) *Atlas of tumor pathology. Tumors of The Adrenal Gland and Adrenal Paraganglia*, 3rd ed. Armed Forces Institute of Pathology, Washington, DC, p. 405.

28. Yang, S.-Y., Jin, Y.J., Park, S.H., Jahng, T.A., Kim, H.J., and Chung, C.K. (2005) Paragangliomas in the cauda equina region. clinicopathoradiologic findings in four cases. *J. Neurooncol.* **72,** 49–55.

29. Salame, K., Ouaknine, G.E.R., Yossipov, J., and Rochkind, S. (2001) Paraganglioma of the pituitary fossa: diagnosis and management. *J. Neurooncol.* **54,** 49–52.

30. Mehra, S., and Chung-Park, M. (2005) Gallbladder paraganglioma. A case report with review of the literature. *Arch. Pathol. Lab. Med.* **129,** 523–526.

31. Moran, C.A., Rush, W., and Mena, H. (1997) Primary spinal paragangliomas: a clinicopathological and immunohistochemical study of 30 cases. *Histopathology* **31,** 167–173.

32. Lee, S.P., Nicholson, G.I., and Hitchcock, G. (1977) Familial abdominal chemodectomas with associated cutaneous angiolipomas. *Pathology* **9,** 173–177.

33. Arias-Stella. J., and Bustos, F. (1976) Chronic hypoxia and chemodectomas in bovines at high altitudes. *Arch. Pathol. Lab. Med.* **100,** 636–639.

34. Valavanis, A. (1986) Preoperative embolization of the head and neck: Indications, patient selection, goals, and precautions. *Am. J. Neuroradiol.* **7,** 943–952.

34A. Herdman, R.C.D., Gillespie, J.E., and Ramsden, R.T. (1993) Radiology in focus. Facial palsy after glomus tumour embolization. *J. Laryngol. Otol.* **107,** 963–966.

34B. Marangos, N., and Schumacher, M. (1999) Radiology in focus. Facial palsy after glomus jugulare tumour embolization. *J. Laryngol. Otol.* **113,** 268–270.

35. Young, A.L., Baysal, B.E., Deb, A., and Young, W.F., Jr. (2002) Clinical Case Seminar. Familial malignant catecholamine-secreting paraganglioma with prolonged survival associated with mutation of the succinate dehydrogenase B gene. *J. Clin. Endocrinol. Metab.* **87,** 4101–4105.

35A. Maier-Woelfle, M., Brändle, M., Komminoth, P., Saremaslani, P., Schmid, S., Locher, T., Heitz, P.U., Krull, I., Galeazzi, R.L., Schmid, C., and Perren, A. (2004) A novel succinate dehydrogenase subunit B gene mutation, H132P, causes familial malignant sympathetic extraadrenal paragangliomas. *J. Clin. Endocrinol. Metab.* **89,** 362–367.

36. Pommier, R.F., Vetto, J.T., Billingsly, K., Woltering, E.A., and Brennan, M.F. (1993) Comparison of adrenal and extraadrenal pheochromocytomas. *Surgery* **114,** 1160–1166.

37. Goldstein, R.E., O'Neill, J.A., Jr., Holcomb., G.W., 3rd, Morgan, W.M., 3rd, Neblett, W.W., 3rd, Oates, J.A., Brown, N., Nadeau, J., Smith, B., Page, D.L., Abumrad, N.N., Scott, H.W., Jr. (1999) Clinical experience over 48 years with pheochromocytoma. *Ann. Surg.* **229,** 755–764.

38. Linnoila, R.I., Keiser, H.R., Steinberg, S.M., and Lack, E.E. (1990) Histopathology of benign versus malignant sympathoadrenal paragangliomas: Clinicopathologic study of 120 cases including unusual histologic features. *Hum. Pathol.* **21,** 1168–1180.

39. van Schothorst, E.M., Jansen, J.C., Grooters, E., Prins, D.E.,Wiersinga, J.J., van der Mey, A.G., van Ommen, G.J., Devilee, P., and Cornelisse, C.J. (1998) Founder effect at PGL1 in hereditary head and neck paraganglioma families from the Netherlands. *Am. J. Hum. Genet.* **63,** 468–473.

40. van Schothorst, E.M., Beekman, M., Torremans, P., Kuipers-Dijkshoorn, N.J., Wessels, H.W., Bardoel, A.F.J, van der Mey, A.G.L., van der Vijver, M.J., van Ommen, G.J.B., Devilee, P., and Cornelisse, C.J. (1998) Paragangliomas of the head and neck region show complete loss of heterozygosity at 11q22-q23 in chief cells and the flow-sorted DNA aneuploid fraction. *Hum. Pathol.* **29,** 1045–1049.

41. Baysal, B.E., Willett-Brozick, J.E., Lawrence, E.C., Drovdlic, C.M., Myssiorek, D., Ferrell, R.E., Myers, E.N., and Rubinstein, W.S. (2000) Genetic heterogeneity in hereditary paraganglioma (PGL): SDHD is the primary locus in imprinted PGL pedigrees. *Am. J. Hum. Genet.* **67(Suppl. 2),** A83.

42. Baysal, B.E., Ferrell, R.E., Willett-Brozick, J.E., Lawrence, E.C., Myssiorek, D., Bosch, A., van der Mey, A., Täschner, P.E.M., Rubinstein, W.S., Myers, E.N., Richard, C.W., III, Cornelisse, C.J., Devilee, P., and Devlin, B. (2000) Mutations in *SDHD*, a mitochondrial complex II gene, in hereditary paraganglioma. *Science* **287,** 848–851.

43. Täschner, P.E.M., Jansen, J.C., Baysal, B.E., Bosch, A., Rosenberg, E.H., Bröcker-Vriends, A.H.J.T., van der Mey, A.G.L., van Ommen, G.-J.B., Cornelisse, C.J., and Devilee, P. (2001) Nearly all hereditary paragangliomas in The Netherlands are caused by two founder mutations in the *SDHD* gene. *Genes Chromosomes Cancer* **31,** 274–281.

44. Solcia, E., Klöppel, G., and Sobin, L.H. (2000) *Histological Typing of Endocrine Tumours*, 2nd. ed. Springer, New York, p. 160.

45. Baysal, B.E. (2002) Hereditary paraganglioma targets diverse paraganglia. *J. Med. Genet.* **39,** 617–622.

46. Miettinen, M. (2003) Paraganglioma. In *Diagnostic Soft Tissue Pathology*, Churchill Livingstone, Philadelphia, PA, pp. 388–389.

47. Myssiorek, D. (2001) Head and neck paragangliomas. An overview. In *The Otolaryngologic Clinics of North America*, vol. 34 (Myssiorek, D. Guest ed.), W.B. Saunders Company, Philadelphia, PA, pp. 829–836.

48. Genner, G.G., and Grimley, P.M. (1974) Tumors of the extra-adrenal paraganglion system (including chemoreceptors). In *Atlas of Tumor Pathology*, Series 2, Fascicle 9, Armed Forces Institute of Pathology, Washington, DC.

48A. Jyung, R.W., LeClair, E.E., Bernat, R.A., Kang, T.S., Ung, F., McKenna, M.J., and Tuan, R.S. (2000) Expression of angiogenic growth factors in paragangliomas. *Laryngoscope* **110,** 161–167.

49. Drovdlic, C.M., Myers, E.N., Peters, J.A., Baysal, B.E., Brackmann, D.E., Slattery, W.H., III, and Rubinstein, W.S. (2001) Proportion of heritable PGL cases and associated clinical characteristics. *Laryngoscope* **111,** 1822–1827.

50. Pellitteri, P.K., Rinaldo, A, Myssiorek, D., Gary Jackson, C., Bradley, P.J., Devaney, K.O., Shaha, A.R., Netterville, J.L., Manni, J.J., and Ferlito, A. (2004) Paragangliomas of the head and neck. *Oral Oncol.***40,** 563–575.

51. Dundee, P., Clancy, B., Wagstaff, S., and Briggs, R. (2005) Paraganglioma: the role of genetic counseling and radiological screening. *J. Clin. Neurosci.* **12,** 464–466.

52. Gujrathi, C.S., and Donald, P.J. (2005) Current trends in the diagnosis and management of head and neck paragangliomas. *Curr. Opin. Otolaryngol. Head Neck Surg.* **13,** 339–342.

53. Sevastos, N., Theodossiades, G., Malaktari, S., and Archimandritis, A.J. (2005) Persistent neutrophilia as a preceding symptom of pheochromocytoma. *J. Clin. Endocrinol. Metab.* **90,** 2472–2473.

54. Antonitsis, P., Saratzis, N., Velissaris, I., Lazaridis, I., Melas, N., Ginis, G., Giavroglou, C., and Kiskinis, D. (2006) Management of cervical paragangliomas: review of a 15-year experience. *Langenbecks Arch. Surg.* **391**, 396–402.

55. Cunningham, S.C., Suh, H.S., Winter, J.M., Montgomery, E., Schulick, R.D., Cameron, J.L., and Yeo, C.J. (2006) Retroperitoneal paraganglioma: single-institution experience and review of the literature. *J. Gastrointest. Surg.* **10**, 1156–1163.

56. Killert, M., Minovi, A., Mangold, R., Hendus, J., Draf, W., and Bockmuhl, U. (2006) Paraganglioma of the head and neck—tumor control, functional results and quality of life. *Laryngorhinootologie* **85**, 649–656.

57. Young, W.F., Jr., and Abboud, A.L. (2006) Editorial: Paraganglioma—All in the family. *J. Clin. Endocrinol. Metab.* **91**, 790–792.

58. Subramanian, A., and Maker, V.K. (2006) Organs of Zuckerkandl: their surgical significance and a review of a century of literature. *Am. J. Surg.* **192**, 224–234.

58A. Lenders, J.W.M., Pacak, K., Walther, M.M., Linehan, W.M., Mannelli, M., Friberg, P., Keiser, H.R., Goldstein, D.S., and Eisenhofer, G. (2002) Biochemical diagnosis of pheochromocytoma. Which test is best? *JAMA* **287**, 1427–1434.

59. Thompson, L.D. (2002) Pheochromocytoma of the Adrenal gland Scaled Score (PASS) to separate benign from malignant neoplasms: a clinicopathologic and immunophenotypic study of 100 cases. *Am. J. Surg. Pathol.* **26**, 551–566.

59A. van der Harst, E., de Herder, W.W., de Krijger, R.R., Bruining, H.A., Bonjer, H.J., Lamberts, S.W., van den Meiracker, A.H., Stijnen, T.H., and Boomsma, F. (2002) The value of plasma markers for the clinical behaviour of phaeochromocytomas. *Eur. J. Endocrinol.* **147**, 85–94.

60. Koch, C.A., Vortmeyer, A.O., Huang, S.C., Alesci, S., Zhuang, Z., and Pacak, K. (2001) Genetic aspects of phcochromocytoma. *Endocr. Regul.* **35**, 43–52.

61. Maher, E.R., and Eng, C. (2000) Genetics of phaeochromocytoma. In *Genetics of Endocrine and Metabolic Disorders* (Thakker, R., ed.), Chapman & Hall, New York.

62. Miettinen, M. (2003) Pheochromocytoma (paraganglioma) of the adrenals. In *Diagnostic Soft Tissue Pathology,*. Churchill Livingstone, Philadelphia, PA, 389–395.

63. Mercuri, S., Gazzeri, R., Galarza, M., Esposito, S., and Giordano, M. (2005) Primary meningeal pheochromocytoma: case report. *J. Neurooncol.* **73**, 169–172.

64. Deb, P., Sharma, M.C., Gaikwad, S., Gupta, A., Mehta, V.S., and Sarkar, C. (2005) Cerebellopontine angle paraganglioma–report of a case and review of the literature. *J. Neurooncol.* **74**, 65–69.

65. Zak, F.G., and Lawson, W. (1982) Anatomy and topography. In *The Paraganglionic Chemoreceptor System*, 1st ed. Springer-Verlag, New York, pp. 15–49.

66. Dluhy, R.G. (2002) Pheochromocytoma?death of an axiom. *N. Engl. J. Med.* **346**, 1486–1488.

67. Koch, C.A., Mauro, D., Walther, M.M., Linehan, W.M., Vortmeyer, A.O., Jaffe, R., Pacak, K., Chrousos, G.P., Zhuang, Z., and Lubensky, I.A. (2002) Pheochromocytoma in von Hippel-Lindau disease: distinct histopathologic phenotype compared to pheochromocytoma in multiple endocrine neoplasia type 2. *Endocr. Pathol.* **13**, 17–27.

68. Edström Elder, E., Elder, G., and Larsson, C. (2005) Pheochromocytoma and functional paraganglioma syndrome: no longer the 10% tumor. *J. Surg. Oncol.* **89**, 193–201.

69. Pacak, K., Lindhan, W.M., Eisenhofer, G., Walther, M.M., and Goldstein, D.S. (2001) Recent advances in genetics, diagnosis, localization, and treatment of pheochromocytoma. *Ann. Intern. Med.* **134**, 315–329.

70. Mundschenk, J., and Lehnert, H. (1998) Malignant pheochromocytoma. *Exp. Clin. Endocrinol. Diabetes* **106**, 373–376.

71. Eisenhofer, G., Bornstein, S.R., Brouwers, F.M., Cheung, N.-K.V., Dahia, P.L., de Krijger, R.R., Giordano, T.J., Greene, L.A., Goldstein, D.S., Lehnert, H., Manger, W.M., Maris, J.M., Neumann, H.P.H., Pacak, K., Shulkin, B.L., Smith, D.I., Tischler, A.S., and Young, W.F., Jr. (2004) Malignant pheochromocytoma: current status and initiatives for future progress. *Endocr. Relat. Cancer* **11**, 423–436.

72. Edström Elder, E., Nord, B., Carling, T., Juhlin, C., Bäckdahl, M., Höög, A., and Larsson, C. (2002) Loss of heterozygosity on the short arm of chromosome 1 in pheochromocytoma and abdominal paraganglioma. *World J. Surg.* **26**, 965–971.

73. Amar, L., Servais, A., Gimenez-Roqueplo, A.-P., Zinzindohoue, F., Chatellier, G., and Plouin, P.-F. (2005) Year of diagnosis, features at presentation and risk of recurrence in patients with pheochromocytoma or secreting paraganglioma. *J. Clin. Endocrinol. Metab.* **90**, 2110–2116.

74. Eisenhofer, G., Goldstein, D.S., Walther, M.M., Friberg, P., Lenders, J.W.M., Keiser, H.R., and Pacak, K. (2003) Biochemical diagnosis of pheochromocytoma: How to distinguish true- from false-positive test results. *J. Clin. Endocrinol. Metab.* **88**, 2656–2666.

75. Ilias, I., Yu, J., Carrasquillo, J.A., Chen, C.C., Eisenhofer, G., Whatley, M., McElroy, B., and Pacak, K. (2003) Superiority of 6-[^{18}F]-fluorodopamine positron emission tomography *versus* [^{131}I]-metaiodobenzylguanidine scintigraphy in the localization of pheochromocytoma. *J. Clin. Endocrinol. Metab.* **88**, 4083–4087.

76. Manger, W.M. (2003) Editorial: in search of pheochromocytomas. *J. Clin. Endocrinol. Metab.* **88**, 4080–4082.

77. Mundschenk, J., Unger, N., Schulz, S., Höllt, V., Schulz, S., Steinke, R., and Lehnert, H. (2003) Somatostain receptor subtypes in human pheochromocytoma: Subcellular expression pattern and functional relevance for octreotide scintigraphy. *J. Clin. Endocrinol. Metab.* **88**, 5150–5157.

78. Ilias, I., and Pacak, K. (2004) Current approaches and recommended algorithm for the diagnostic localization of pheochromocytoma. *J. Clin. Endocrinol. Metab.* **89**, 479–491.

79. Kann, P.H., Wirkus, B., Behr, T., Klose, K.-J., and Meyer, S. (2004) Endosonographic imaging of benign and malignant of pheochromocytoma. *J. Clin. Endocrinol. Metab.* **89**, 1694–1697.

80. Sawka, A.M., Gafni, A., Thabane, L., and Young, W.F., Jr. (2004) The economic implications of three biochemical screening algorithms for pheochromocytoma. *J. Clin. Endocrinol. Metab.* **89**, 2859–2866.

81. Lam, M.G.E.H., Lips, C.J.M., Jager, P.L., Dullaart, R.P.F., Lentjes, E.G.W.M., van Rijk, P.P., and de Klerk, J.M.H. (2005) Repeated [^{131}I]metaiodobenzylguanidine therapy in two patients with malignant pheochromocytoma. *J. Clin. Endocrinol. Metab.* **90**, 5888–5895.

82. Brain, K.L., Kay, J., and Shine, B. (2006) Measurement of urinary metanephrines to screen for pheochromocytoma in an unselected hospital referral population. *Clin. Chem.* **52,** 2060–2064.

83. Lumachi, F., Polistina, F., Favia, G., and D'Amico, D.F. (1998) Extraadrenal and multiple pheochromocytomas. Are there really any differences in pathophysiology and outcome? *J. Exp. Clin. Cancer Res.* **17,** 303–305.

83A. Bryant, J., Farmer, J., Kessler, L.J., Townsend, R.R., and Nathanson, K.L. (2003) Pheochromocytoma: The expanding genetic differential analysis. *J. Natl. Cancer Inst.* **95,** 1196–1204.

84. Decker, H.J.-H., Walter, T.A., Neumann, H.P.H., and Sandberg, A.A. (1988) Cytogenetics of familial pheochromocytoma: Importance of trisomy 7 in tumor progression? *Blut* **57,** 270.

85. Kiechle-Schwarz, M., Neumann, H.P., Decker, H.J., Dietrich, C., Wullich, B., and Schempp, W. (1989) Cytogenetic studies on three pheochromocytomas derived from patients with von Hippel-Lindau syndrome. *Hum. Genet.* **82,** 127–130.

86. Gunawan, B., Schlomm, T., Schulten, H.-J., Seseke, F., Ringert, R.-H., and Füzesi, L. (2004) Cytogenetic characterization of 5 pheochromocytomas. *Cancer Genet. Cytogenet.* **154,** 163–166.

87. Jordan, D.K., Patil, S.R., Divelbiss, J.E., Vemuganti, S., Headley, C., Waziri, M.H., and Gurll, N.J. (1989) Cytogenetic abnormalities in tumors of patients with von Hippel-Lindau disease. *Cancer Genet. Cytogenet.* **42,** 227–241.

88. Van Dekken, H., Bosman, F.T., Teijgeman, R., Vissers, C.J., Tersteeg, T.A., Kerstens, H.M., Vooijs, G.P., and Verhofstad, A.A. (1993) Identification of numerical chromosome aberrations in archival tumours by in situ hybridization in routine paraffin sections: evaluations of 23 phaeochromocytomas. *J. Pathol.* **171,** 161–171.

89. Pfragner, R., Behmel, A., Smith, D.P., Ponder, B.A., Wirnsberger, G., Rinner, I., Porta, S., Henn, T., and Niederle, B. (1998) First continuous pheochromocytoma cell line: KNA. Biological, cytogenetic and molecular characterization of KNA cells. *J. Neurocytol.* **27,** 175–186.

90. Cascón, A., Ruiz-Llorente, S., Rodríguez-Perales, S., Honrado, E., Martínez-Ramírez, Á., Letón, R., Montero-Conde, C., Benítez, J., Dopazo, J., Cigudosa, J.C., and Robledo, M. (2005) A novel candidate region linked to development of both pheochromocytomas and head/neck paraganglioma. *Genes Chromosomes Cancer* **42,** 260–268.

91. Dannenberg, H., Speel, E.J.M., Zhao, J., Saremaslani, P., van der Harst, E., Roth, J., Heitz, P.U., Bonjer, H.J., Dinjens, W.N.M., Mooi, W.J., Komminoth, P., and de Krijger, R.R. (2000) Losses of chromosomes 1p and 3q are early genetic events in the development of sporadic pheochromocytomas. *Am. J. Pathol.* **157,** 353–359.

92. Edström, E., Mahlamäki, E., Nord, B., Kjellman, M., Karhu, R., Höög, A., Goncharov, N., Teh, B.T., Bäckdahl, M., and Larsson, C. (2000) Comparative genomic hybridization reveals frequent losses of chromosomes 1p and 3q in pheochromocytomas and abdominal paragangliomas, suggesting a common genetic etiology. *Am. J. Pathol.* **156,** 651–659.

93. Kirchhoff, M., Gerdes, T., Rose, H., Maahr, J., Ottesen, A.M., and Lundsteen, C. (1998) Detection of chromosomal gains and losses in comparative genomic hybridization analysis based on standard reference intervals. *Cytometry* **31,** 163–173.

94. Hering, A., Guratowska, M., Bucsky, P., Claussen, U., Decker, J., Ernst, G., Hoeppner, W., Michel, S., Neumann, H., Parlowsky, T., and Loncarevic, I. (2006) Characteristic genomic imbalances in pediatric pheochromocytoma. *Genes Chromosomes Cancer* **45,** 603–607.

95. Pozo, J., Munoz, M.T., Martos, G., and Argente, J. (2005) Sporadic phaeochromocytoma in childhood: Clinical and molecular variability. *J. Pediatr. Endocrinol. Metab.* **18,** 527–532.

95A. Blasius, S., Brinkschmidt, C., Poremba, C., Terpe, H.J., Halm, H., Schleef, J., Ritter, J., Wortler, K., Bocker, W., and Dockhorn-Dworniczak, B. (1998) Metastatic retroperitoneal paraganglioma in a 16-year-old girl. Case report, molecular pathological and cytogenetic findings. *Pathol. Res. Pract.* **194,** 439–444.

96. Lui, W.O., Chen, J., Gläsker, S., Bender, B.U., Madura, C., Khoo, S.K., Kort, E., Larsson, C., Neumann, H.P.H., and The, B.T. (2002) Selective loss of chromosome 11 in pheochromocytomas associated with the VHL syndrome. *Oncogene* **21,** 1117–1122.

97. Jarbo, C., Buckley, P.G., Piotrowski, A., Mantripragada, K.K., Benetkiewicz, M., Diaz de Ståhl, T., Langford, C.F., Gregory, S.G., Dralle, H., Gimm, O., Bäckdahl, M., Geli, J., Larsson, C., Westin, G., Åkerström, G., and Dumanski, P. (2006) Detailed assessment of chromosome 22 aberrations in sporadic pheochromocytoma using array-CGH. *Int. J. Cancer* **118,** 1159–1164.

98. Powers, J.F., Tischler, A.S., Mohammed, M., and Naeem, R. (2005) Microarray-based comparative genomic hybridization of pheochromocytoma cell lines from neurofibromatosis knockout mice reveals genetic alterations similar to those in human pheochromocytomas. *Cancer Genet. Cytogenet.* **159,** 27–31.

99. Khosla, S., Patel, V.M., Hay, I.D., Schaid, D.J., Grant, C.S., van Heerden, J.A., and Thibodeau, S.N. (1991) Loss of heterozygosity suggests multiple genetic alterations in pheochromocytomas and medullary thyroid carcinomas. *J. Clin. Invest.* **87,** 1691–1699.

100. Tory, K., Brauch, H., Linehan, M., Barba, D., Oldfield, E., Filling-Katz, M., Seizinger, B., Nakamura, Y., White, R., Marshall, F.F., Lerman, M.I., and Zbar, B. (1989) Specific genetic change in tumor associated with von Hippel-Lindau disease. *J. Natl. Cancer Inst.* **81,** 1097–1101.

101. Zeiger, M.A., Zbar, B., Keiser, H., Linehan, W.M., and Gnarra, J.R. (1995) Loss of heterozygosity on the short arm of chromosome 3 in sporadic, von Hippel-Lindau disease-associated, and familial pheochromocytoma. *Genes Chromosomes Cancer* **13,** 151–156.

102. Vargas, M.P., Zhuang, Z., Wang, C., Vortmeyer, A., Linehan, W.M., and Merino, M.J. (1997) Loss of heterozygosity on the short arm of chromosomes 1 and 3 in sporadic pheochromocytoma and extra-adrenal paraganglioma. *Hum. Pathol.* **28,** 411–415.

103. Benn, D.E., Dwight, T., Richardson, A.L., Delbridge, L., Bambach, C.P., Stowasser, M., Gordon, R.D., Marsh, D.J., and Robinson, B.G. (2000) Sporadic and familial pheochromocytomas are associated with loss of at least two discrete intervals on chromosome 1p. *Cancer Res.* **60,** 7048–7051.

104. Bikhazi, P.H., Messina, L., Mhatre, A.N., Goldstein, J.A., and Lalwani, A.K. (2000) Molecular pathogenesis in sporadic head and heck paraganglioma. *Laryngoscope* **110,** 1346–1348.

105. Dannenberg, H., de Krijger, R.R., Zhao, J., Speel, E.J.M., Saremaslani, P., Dinjens, W.N.M., Mooi, W.J., Roth, J., Heitz, P.U., and Komminoth, P. (2001) Differential loss of chromosome 11q in familial and sporadic parasympathetic paragangliomas detected by comparative genomic hybridization. *Am. J. Pathol.* **158,** 1937–1942.

106. Bender, B.U., Gutsche, M., Glasker, S., Muller, B., Kirste, G., Eng, C., and Neumann, H.P. (2000) Differential genetic alterations in von Hippel-Lindau syndrome-associated an sporadic pheochromocytomas. *J. Clin. Endocrinol. Metab.* **85,** 4568–4574.

107. Moley, J.F., Brother, M.B., Fong, C.T., White, P.S., Baylin, S.B., Nelkin, B., Wells, S.A., and Brodeur, G.M. (1992) Consistent association of 1p loss of heterozygosity with pheochromocytomas from patients with multiple endocrine neoplasia type 2 syndromes. *Cancer Res.* **52,** 770–774.

108. Mulligan, L.M., Gardner, E., Smith, B.A., Mathew, C.G.P., and Ponder, B.A.J. (1993) Genetic events in tumour initiation and progression in multiple endocrine neoplasia type 2. *Genes Chromosomes Cancer* **6,** 166–177.

108A. Marsh, D.J., Theodosopoulos, G., Martin-Schulte, K., Richardson, A.L., Philips, J., Roher, H.D., Delbridge, L., and Robinson, B.G. (2003) Genome-wide copy number imbalances identified in familial and sporadic medullary thyroid carcinoma. *J. Clin. Endocrinol. Metab.* **88,** 1866–1872.

109. Williamson, C., Pannett, A.A.J., Pang, J.T., Wooding, C., McCarthy, M., Sheppard, M.N., Monson, J., Clayton, R.N., and Thakker, R.V. (1997) Localisation of a gene causing endocrine neoplasia to a 4 cM region on chromosome 1p35-p36. *J. Med. Genet.* **34,** 617–619.

109A. Valimaki, S., Forsberg, L., Farnebo, L.O., and Larsson, C. (2002) Distinct target regions for chromosome 1p deletions in parathyroid adenomas and carcinomas. *Int. J. Oncol.* **21,** 727–735.

110. Bauer, A., Savelyeva, L., Claas, A., Praml, C., Berthold, F., and Schwab, M. (2001) Smallest region of overlapping deletion in 1p36 in human neuroblastoma: A 1 Mbp cosmid and PAC contig. *Genes Chromosomes* Cancer **31,** 228–239.

111. Ejeskär, K. Abel, F., Sjöberg, R., Backstrom, J., Kogner, P., and Martinsson, T. (2000) Fine mapping of the human preprocortistatin gene (CORT) to neuroblastoma consensus deletion region 1p36.3→p36.2, but absence of mutations in primary tumors. *Cytogenet. Cell Genet.* **89,** 62–66.

112. White, P.S., Maris, J.M., Beltinger, C., Sulman, E., Marshall, H.N., Fujimori, M., Kaufman, B.A., Biegel, J.A., Allen, C., Hilliard, C., Valentine, M.B., Look, A.T., Enomoto, H., Sakiyama, S., and Brodeur, G.M. (1995) A region of consistent deletion in neuroblastoma maps within human chromosome 1p36.2–36.3. *Proc. Natl. Acad. Sci USA* **92,** 5520–5524.

113. Astuti, D., Latif, F., Dallol, A., Dahia, P.L.M., Douglas, F., George, E., Sköldberg, F., Husebye, E.S., Eng, C., and Maher, E.R. (2002) Gene mutations in the succinate dehydrogenase subunit SDHB cause susceptibility to familial pheochromocytoma and to familial paraganglioma. Erratum to Astuti et al. (2001) *Am. J. Hum. Genet.* **69,** 49–54. *Am. J. Hum. Genet.* **70,** 565.

114. Opocher, G., Schiavi, F., Vettori, A., Pampinella, F., Vitiello, L., Calderan, A., Vianello, B., Murgia, A., Martella, M., Taccaliti, A., Mantero, F., and Mostacciuolo, M.L. (2003) Fine analysis of the short arm of chromosome 1 in sporadic and familial pheochromocytoma. *Clin. Endocrinol (Oxford)* **59,** 707–715.

115. Aarts, M., Dannenberg, H., deLeeuw, R.J., van Nederveen, F.H., Verhofstad, A.A., Lenders, J.W., Dinjens, W.N.M., Speel, E.J.M., Lam, W.L., and de Krijger, R.R. (2006) Microarray-based CGH of sporadic and syndrome-related pheochromocytomas using a 0.1–0.2 Mb bacterial artificial chromosome array spanning arm 1p. *Genes Chromosomes Cancer* **45,** 83–93.

116. Mathew, C.G.P., Smith, B.A., Thorpe, K., Wong, Z., Royle, N.J., Jeffreys, A.J., and Ponder, B.A.J. (1987) Deletion of genes on chromosome 1 in endocrine neoplasia. *Nature* **328,** 524–526.

117. Takai, S., Tateishi, H., Nishisho, I., Miki, T., Motomura, K., Miyauchi, A., Kato, M., Ikeuchi, T., Yamamoto, K., Okazaki, M., Yamamoto, M., Honjo, T., Kumahara, Y., and Mori, T. (1987) Loss of genes on chromosome 22 in medullary thyroid carcinoma and pheochromocytoma. *Jpn. J. Cancer Res.* **78,** 894–898.

118. Reubi, J.C., Waser, B., Khosla, S., Kvols, L., Goellner, J.R., Krenning, E., and Lamberts, S. (1992) In vitro and in vivo detection of somatostatin receptors in phaechromocytomas and paragangliomas. *J. Clin. Endocrinol. Metab.* **74,** 1082–1089.

119. Tanaka, N., Nishisho, I., Yamamoto, M., Miya, A., Shin, E., Karakawa, K., Fujita, S., Kobayashi, T., Rouleau, G.A., Mori, T., and Takai, S. (1992) Loss of heterozygosity on the long arm of chromosome 22 in pheochromocytoma. *Genes Chromosomes Cancer* **5,** 399–403.

120. Shin, E., Fujita, S., Takami, K., Kurahashi, H., Kurita, Y., Kobayashi, T., Mori, T., Nishisho, I., and Takai, S. (1993) Deletion mapping of chromosome 1p and 22q in pheochromocytoma. *Jpn J. Cancer Res.* **84,** 402–408.

121. Gutmann, D.H., Cole, J.L., Stone, W.J., Ponder, B.A.J., and Collins, F.S. (1994) Loss of neurofibromin in adrenal gland tumors from patients with neurofibromatosis type I. *Genes Chromosomes Cancer* **10,** 55–58.

122. Lementa, S., Salmenkivi, K., Pylkkänen, L., Sainio, M., Saarikoski, S.T., Arola, J., Heikkilä, P., Haglund, C., Husgafvel-Pursiainen, K., and Böhling, T. (2006) Frequent loss of heterozygosity at 6q in pheochromocytoma. *Hum. Pathol.* **37,** 749–754.

123. van Schothorst, E.M., Jansen, J.C., Bardoel, A.F.J., van der Mey, A.G.L., James, M.J., Sobol, H., Weissenbach, J., van Ommen, G.-J.B., Cornelisse, C.J., and Devilee, P. (1996) Confinement of PGL, an imprinted gene causing hereditary paragangliomas, to a 2-cM interval on 11q22-q23 and exclusion of DRD2 and NCAM as candidate genes. *Eur. J. Hum. Genet.* **4,** 267–273.

124. Mariman, E.C., van Beersum, S.E., Cremers, C.W., Struycken, P.M., and Ropers, H.H. (1995) Fine mapping of a putatively imprinted gene for familial non-chromaffin paragangliomas to chromosome 11q13.1: evidence for genetic heterogeneity. *Hum. Genet.* **95,** 56–62.

125. Bikhazi, P.H., Roeder, E., Attaie, A., and Lalwani, A.K. (1999) Familial paragangliomas: the emerging impact of

molecular genetics on evaluation and management. *Am. J. Oto.* **20**, 639–643.

126. Bockmuhl, U., Schluns, K., Schmidt, S., Matthias, S., and Petersen, I. (2002) Chromosomal alterations during metastasis formation of head and neck squamous cell carcinoma. *Genes Chromosomes Cancer* **33**, 29–35.

127. Dannenberg, H., Komminoth, P., Dinjens, W.N., Speel, E.J., and de Krijger, R.R. (2003) Molecular genetic alterations in adrenal and extra-adrenal pheochromocytomas and paragangliomas. *Endocr. Pathol.* **14**, 329–350.

128. Komminoth, P., Kunz, E., Hiort, O., Schröder, S., Matias-Guiu, X., Christiansen, G., Roth, J., and Heitz, P.U. (1994) Detection of *RET* proto-oncogene point mutations in paraffin-embedded pheochromocytoma specimens by nonradioactive single-strand conformation polymorphism analysis and direct sequencing. *Am. J. Pathol.* **145**, 922–929.

129. Eng, C., Crossey, P.A., Mulligan, L.M., Healey, C.S., Houghton, C., Prowse, A., Chew, S.L., Dahia, P.L., O'Riordan, J.L.H, Toledo, S.P.A., Smith, D.P., Maher, E.R., and Ponder, B.A.J. (1995) Mutations of the *RET* proto-oncogene and the von Hippel-Lindau disease tumour suppressor gene in sporadic and syndromic phaeochromocytoma. *J. Med. Genet.* **32**, 934–937.

130. Bar, M., Friedman, E., Jakobovitz, O., Leibowitz, G., Lerer, I., Abeliovich, D., and Gross, D.J. (1997) Sporadic phaeochromocytomas are rarely associated with germline mutations in the von Hippel-Lindau and *RET* genes. *Clin. Endocrinol.* **47**, 707–712.

131. de Krijger, R.R., van der Harst, E., Muletta-Feurer, S., Bruining, H.A., Lamberts, S.W.J., Dinjens, W.N.M., Roth, J., Heitz, P.U., and Komminoth, P. (2000) *RET* is expressed but not mutated in extra-adrenal paragangliomas. *J. Pathol.* **191**, 264–268.

132. Pawlu, C., Bausch, B., Reisch, N., and Neumann, H.P.H. (2005) Genetic testing for pheochromocytoma-associated syndromes. *Ann. Endocrinol. (Paris)* **66**, 178–185.

133. Woodward, E.R., Eng, C., McMahon, R.,,. Affara, N.A., Ponder, B.A.J., and Maher, E.R. (1997) Genetic predisposition to phaeochromocytoma: analysis of candidate genes *GDNF*, *RET* and *VHL*. *Hum. Mol. Genet.* **6**, 1051–1056.

134. Hofstra, R.M.W., Stelwagen, T., Stulp, R.P., de Jong, D., Hulsbeek, M., Kamsteeg, E.J., van den Berg, A., Landsvater, R.M., Vermey, A., Molenaara, W.M., Lips, C.J.M., and Buys, C.H.C.M. (1996) Extensive mutation scanning of *RET* in sporadic medullary thyroid carcinoma and of *RET* and *VHL* in sporadic pheochromocytoma reveals involvement of these genes in only a minority of cases. *J. Clin. Endocrinol. Metab.* **81**, 2881–2884.

135. 135. Baysal, B.E., Willett-Brozick, J.E., Filho, P.A.A., Lawrence, E.C., Myers, E.N., and Ferrell, R.E. (2004) An Alu-mediated partial *SDHC* deletion causes familial and sporadic paraganglioma. *J. Med. Genet.* **41**, 703–709.

136. McWhinney, S.R., Pilarski, R.T., Forrester, S.R., Schneider, M.C., Sarquis, M.M., Dias, E.P., and Eng, C. (2004) Large germline deletions of mitochondrial complex II subunits *SDHB* and *SDHB* in hereditary paraganglioma. *J. Clin. Endocrinol, Metab.* **89**, 5694–5699.

137. Amar, L., Bertherat, J., Baudin, E., Ajzenberg, C., Bressac-de Paillerets, B., Chabre, O., Chamontin, B., Delemer, B., Giraud, S., Murat, A., Nicoli-Sire, P., Richard, S., Rohmer, V., Sadoul, J.-L., Strompf, L., Schlumberger, M., Bertagna, X., Plouin, P.-F., Jeunemaitre, X., and Gimenez-Roqueplo, A.-P. (2005) Genetic testing in pheochromocytoma or functional paraganglioma. *J. Clin. Oncol.* **23**, 8812–8818.

137A. Timmers, H.J.L.M., Kozupa, A., Eisenhofer, G., Raygada, M., Adams, K.T., Solis, D., Lenders, J.W.M., and Pacak, K. (2007) Clinical presentations, biochemical phenotypes, and genotype-phenotype correlations in patients with *succinate dehydrogenase subunit B*-associated pheochromocytomas and paragangliomas. *J. Clin. Endocrinol. Metab.* **92**, 779–786.

137B. Eisenhofer, G., Goldstein, D.S., Sullivan, P., Csako, G., Brouwers, F.M., Lai, E.W., Adams, K.T., and Pacak, K. (2005) Biochemical and clinical manifestations of dopamine-producing paragangliomas: utility of plasma methoxytyramine. *J. Clin. Endocrinool. Metab.* **90**, 2068–2075.

137C. Bonnet, S., Durand, X., Baton, O., Gimenez-Roqueplo, A.P., Baudin, E., Visset, J., Algayres, J.P., and Baranger, B. (2006) Malignant hereditary paraganglioma: problems raised by nonfunctional forms management. *Ann. Chir.* **131**, 626–630.

137D. Proye, C., Fossati, P., Fontaine, P., Lefebvre, J., Decoulx, M., Wemeau, J.L., Dewailly, D., Rwamasirabo, E., and Cecat, P. (1986) Dopamine-secreting pheochromocytoma: an unrecognized entity? Classification of pheochromocytomas according to their type of secretion. *Surgery* **100**, 1154–1162.

138. Bravo, E.L., and Tagle, R. (2003) Pheochromocytoma: state-of-the-art and future prospects. *Endocrine Rev.* **24**, 539–553.

138A. Neumann, H.P.H., Pawlu, C., Peçzkowska, M., Bausch, B., McWhinney, S.R., Muresan, M., Buchta, M., Franke, G., Klisch, J., Bley, T.A., Hoegerle, S., Boedeker, C.C., Opocher, G., Schipper, J., Januszewicz, A., and Eng, C. for the European-American Paraganglioma Study Group. (2004) Distinct clinical features of paraganglioma syndromes associated with *SDHB* and *SDHD* gene mutations. *JAMA* **292**, 943–951.

139. Clifford, S.C., Cockman, M.E., Smallwood, A.C., Mole, D.R., Woodward, E.R., Maxwell, P.H., Ratcliffe, P.J., and Maher, E.R. (2001) Contrasting effects on HIF-1α regulation by disease-causing pVHL mutations correlate with patterns of tumourigenesis in von Hippel-Lindau disease. *Hum. Mol. Genet.* **10**, 1029–1038.

140. Hoffman, M.A., Ohh, M., Yang, H., Kleo, J.M., Ivan, M., and Kaelin, W.G. (2001) von Hippel-Lindau protein mutants linked to type 2C VHL disease preserve the ability to down-regulate HIF. *Hum. Mol. Genet.* **10**, 1019–1027.

140A. Semenza, G.L. (2002) Involvement of hypoxia-inducible factor 1 in human cancer. *Intern. Med.* **41**, 79–83.

141. Vanharanta, S., Buchta, M., McWhinney, S.R., Virta, S.K., Peçzkowska, M., Morrison, C.D., Lehtonen, R., Januszewicz, A., Järvinen, H., Juhola, M., Mecklin, J.-P., Pukkala, E., Herva, R., Kiuru, M., Nupponen, N.N., Aaltonen, L.A., Neumann, H.P.H., and Eng, C. (2004) Early-onset renal cell carcinoma as a novel extraparaganglial component of *SDHB*-associated heritable paraganglioma. *Am. J. Hum. Genet.* **74**, 153–159.

142. Dahlia, P.L.M., Hao, K., Rogus, J., Colin, C., Pujana, M.A.G., Ross, K., Magoffin, D., Aronin, N., Casón, A., Hayashida, C.Y., Li, C., Toledo, S.P.A., and Stiles, C.D. for the Familial Pheochromocytoma Consortium. (2005) Novel pheochromocytoma susceptibility loci identified by integrative genomics. *Cancer Res.* **65**, 9651–9658.

195. Green, D.R., and Reed, J.C. (1998) Mitochondria and apoptosis. *Science* **281**, 1309–1312.

196. Hanahan, D., and Weinberg, R.A. (2000) The hallmarks of cancer. *Cell* **100**,57–70.

197. Mandavilli, B.S., Santos, J.H., and Van Houten, B. (2005) Mitochondrial DNA repair and aging. *Mutat. Res.* **509**, 127–151.

198. Kroemer, G. (2003) Mitochondrial control of apoptosis: an introduction. *Biochem. Biophys. Res. Commun.* **304**, 433–435.

199. Ravagnan, L., Roumier, T., and Kroemer, G. (2002) Mitochondria, the killer organelles and their weapons. *J. Cell Physiol.* **192**, 131–137.

200. Kim, J.-S., He, L., and Lemasters, J.J. (2003) Mitochondrial permeability transition: a common pathway to necrosis and apoptosis. *Biochem. Biophys. Res. Commun.* **304**, 463–470.

201. Cavalli, L.R., and Liang, B.C. (1998) Mutagenesis, tumorigenicity, and apoptosis: are the mitochondria involved? *Mutat. Res.* **398**, 19–26.

202. Gimenez-Roqueplo, A.-P., Favier, J., Rustin, P., Mourad, J.-J., Plouin, P.-F., Corvol, P., Rötig, A., and Jeunemaitre, X. (2001) The R22X mutation of the *SDHD* gene in hereditary paraganglioma abolishes the enzymatic activity of complex II in the mitochondrial respiratory chain and activates the hypoxia pathway. *Am. J. Hum. Genet.* **69**, 1186–1197.

203. Nurse, C.A., and Vollmer, C. (1997) Role of basic FGF and oxygen in control of proliferation, survival, and neuronal differentiation in carotid body chromaffin cells. *Dev. Biol.* **184**, 197–206.

204. Heath, D. (1991) The human carotid body in health and disease. *J. Pathol.* **164**, 1–8.

205. Gonzalez, C., Almaraz, L., Obeso, A., and Rigual, R. (1994) Carotid body chemoreceptors: from natural stimuli to sensory discharges. *Physiol. Rev.* **74**, 829–898.

206. Ishii, N., Fujii, M., Hartman, P.S., Tsuda, M., Yasuda, K., Senoo-Matsuda, N., Yanase, S., Ayusawa, D., and Suzuki, K. (1998) A mutation in succinate dehydrogenase cytochrome*b* causes oxidative stress and ageing in nematodes. *Nature* **394**, 694–697.

207. Maxwell, P.H., and Ratcliffe, P.J. (2002) Oxygen sensors and angiogenesis. *Semin. Dev. Biol.* **3**, 29–37.

208. Lahiri, S., Rozanov, C., Roy, A., Storey, B., and Buerk, D.G. (2001) Regulation of oxygen sensing in peripheral arterial chemoreceptors. *Int. J. Biochem. Cell Biol.* **33**, 755–774.

209. Semenza, G.L. (2000) HIF-1 and human disease: one highly involved factor. *Genes Dev.* **14**, 1983–1991.

210. Pugh, C.W., and Ratcliff, P.J. (2003) The von Hippel-Lindau tumor suppressor, hypoxia-inducible factor-1 (HIF-1) degradation, and cancer pathogenesis. *Semin. Cancer Biol.* **13**, 83–89.

211. Eisenhofer, G., Huynh, T.-T., Pacak, K., Brouwers, F.M., Walther, M.M., Linehan, W.M., Munson, P.J., Manelli, M., Goldstein, D.S., and Elkahloun, A.G. (2004) Distinct gene expression profiles in norepinephrine- and epinephrine-producing hereditary and sporadic pheochromocytomas: activation of hypoxia-driven angiogenic pathways in von Hippel-Lindau syndrome. *Endoc. Relat. Cancer* **11**, 897–911.

211A. Maxwell, P.H., Wiesener, M.S., Chang, G.-W., Clifford, S.C., Vaux, E.C., Cockman, M.E., Wykoff, C.C., Pugh, C.W., Maher, E.R., and Ratcliffe, P.J. (1999) The tumour suppressor protein VHL targets hypoxia-inducible factors for oxygen-dependent proteolysis. *Nature* **399**, 271–274.

212. Dahlia, P.L.M., Ross, K.N., Wright, M.E., Hayashida, C.Y., Santagata, S., Barontini, M., Kung, A.L., Sanso, G., Powers, J.F., Tischler, A.S., Hodin, R.., Heitritter, S., Moore, F., Jr., Dluhy, R., Sosa, J.A., Ocal, I.T., Benn, D.E., Marsh, D.J., Robinson, B.G., Schneider, K., Garber, J., Arum, S.M., Korbonits, M., Grossman, A., Pigny, P., Toledo, S.P.A., Nosé, V., Li, C., and Stiles, C.D. (2005) HIF-1α regulatory loop links hypoxia and mitochondrial signals in pheochromocytomas. *PloS Genet.* **1**, 72–80.

213. Brière, J.-J., Favier, J., Bénit, P., El Ghouzzi, V., Lorenzato, A., Rabier, D., Di Renzo, M.F., Gimenez-Roqueplo, A.-P., and Rustin, P. (2005) Mitochondrial succinate is instrumental for HIF1α nuclear translocation in SDHA-mutant fibroblasts under normoxic conditions. *Hum. Mol. Genet.* **14**, 3263–3269.

213A. Acker, T., Diez-Juan, A., Aragones, J., Tjwa, M., Brusselmans, K., Moons, L., Fukumura, D., Moreno-Murciano, M.P., Herbert, J.-M., Burger, A., Riedel, J., Elvert, G., Flamme, I., Maxwell, P.H., Collen, D., Dewerchin, M., Jain, R.K., Plate, K.H., and Carmelie, P. (2005) Genetic evidence for a tumor suppressor role of HIF-2α. *Cancer Cell* **8**, 131–141.

214. Lee, S., Nakamura, E., Yang, H., Wei, W., Linggi, M.S., Sajan, M.P., Farese, R.V., Freeman, R.S., Carter, B.D., Kaelin, W.G., Jr., and Schlisio, S. (2005) Neuronal apoptosis linked to EgIN3 prolyl hydroxylase and familial pheochromocytoma genes: Developmental culling and cancer. *Cancer Cell* **8**, 155–167.

215. Pollard, P.J., Brière, J.J., Alam, N.A., Barwell, J., Barclay, E., Wortham, N.C., Hunt, T., Mitchell, M., Olpin, S., Moat, S.J., Hargreaves, I.P., Heales, S.J., Chung, Y.L., Griffiths, J.R., Dalgleish, A., McGrath, J.A., Gleeson, M.J., Hodgson, S.V., Poulsom, R., Rustin, P., and Tomlinson, I.P.MN. (2005) Accumulation of Krebs cycle intermediates and over-expression of HIF1α in tumours which result from germline *FH* and *SDH* mutations. *Hum. Mol. Genet.* **14**, 2231–2239.

216. Selak, M.A., Armour, S.M., MacKenzie, E.D., Boulahbel, H., Watson, D.G., Mansfield, K.D., Pan, Y., Simon, M.C., Thompson, C.B., and Gottlieb, E. (2005) Succinate links TCA cycle dysfunction to oncogenesis by inhibiting HIF-α prolyl hydroxylase. *Cancer Cell* **7**, 77–85.

217. Temes, E., Martín-Puig, S., Acosta-Iborra, B., Castellanos, M.C., Feijoo-Cuaresma, M., Olmos, G., Aragonés, J., and Landazuri, M.O. (2005) Activation of HIF-prolyl hydroxylases by R59949, an inhibitor of the diacylglycerol kinase. *J. Biol. Chem.* **280**, 24238–24244.

218. Dahia, P.L.M. (2006) Evolving concepts in pheochromocytoma and paraganglioma. *Curr. Opin. Oncol.* **18**, 1–8.

219. Grufferman, S., Gillman, M.W., Pasternak, L.R., Peterson, C.L., and Young, W.G., Jr. (1980) Familial carotid body tumors: case report and epidemiologic review. *Cancer* **46**, 2116–2122.

220. Plouin, P.F., and Gimenez-Roqueplo, A.P. (2006) The genetic basis of pheochromocytoma: who to screen and how? *Nat. Clin. Pract. Endocrinol. Metab.* **2**, 60–61.

221. Clifford, S.C., and Maher, E.R. (2001) von Hippel-Lindau disease: clinical and molecular perspectives. *Adv. Cancer Res.* **82**, 85–105.

221A. Dannenberg, H., De Krijger, R.R., van der Harst, E., Abbou, M., IJzendoorn, Y., Komminoth, P., and Dinjens, W.N. (2003) Von Hippel-Lindau gene alterations in sporadic benign and malignant pheochromocytomas. *Int. J. Cancer* **105**, 190–195.

222. Kaelin, W.G., and Maher, E.R. (1998) The *VHL* tumour suppressor gene paradigm. *Trends Genet.* **14**, 423–426.

223. Brauch, H., Hoeppner, W., Jahnig, H., Wohl, T., Engelhardt, D., Spelsberg, F., and Ritter, M.M. (1997) Sporadic pheochromocytomas are rarely associated with germline mutations in the vhl tumor suppressor gene or the ret protooncogene. *J. Clin. Endocrinol. Metab.* **82**, 4101–4104.

224. Le Hir, H., Colucci-D'Amato, L.G., Charlet-Berguerand, N., Plouin, P.-F., Bertagna, X., de Franciscis, V., and Thermes, C. (2000) High levels of tyrosine phosphorylated proto-Ret in sporadic pheochromocytomas. *Cancer Res.* **60**, 1365–1370.

225. Le Hir, H., Charlet-Berguerand, N., Gimenez-Roqueplo, A., Mannelli, M., Plouin, P., de Franciscis, V., and Thermes, C. (2000) Relative expression of the RET9 and RET51 isoforms in human pheochromocytomas. *Oncology* **58**, 311–318.

226. Santoro, M., Rosati, R., Grieco, M., Berlingieri, M.T., D'Amato, G.L., de Franciscis, V., and Fusco, A. (1990) The ret proto-oncogene is consistently expressed in human pheochromocytomas and thyroid medullary carcinomas. *Oncogene* **5**, 1595–1598.

227. Neumann, H.P.H., Berger, D.P., Sigmund, G., Blum, U., Schmidt, D., Parmer, R.J., Volk, B., and Kirste, G. (1993) Pheochromocytomas, multiple endocrine neoplasia type 2, and Von Hippel-Lindau disease. *N. Engl. J. Med.* **329**, 1531–1538.

228. Eisenhofer, G., Walther, M.M., Huynh, T.-T., Li, S.-T., Bornstein, S.R., Vortmeyer, A., Mannelli, M., Goldstein, D.S., Linehan, W.M., Lenders, J.W.M., and Pacak, K. (2001) Pheochromocytomas in von Hippel-Lindau syndrome and multiple endocrine neoplasia type 2 display distinct biochemical and clinical phenotypes. *J. Clin. Endocrinol. Metab.* **86**, 1999–2008.

229. Böhling, T., Plate, K.H., Haltia, M.J., Alitalo, K., and Neumann, H.P.H. (2000) Von Hippel-Lindau disease and capillary haemangioblastoma. In *World Health Organization Classification of Tumours. Pathology and Genetics of Tumours of the Nervous* System (Kleihues, P., and Cavenee, W.K., eds.), IARC Press, Lyon, France, pp. 223–226.

230. Carman, C.T., and Brashear, R.E. (1960) Pheochromocytoma as an inherited abnormality: report of the tenth affected kindred and review of the literature. *N. Engl. J. Med.* **263**, 419–423.

231. Tisherman, S.E., Tisherman, B.E., Tisherman, S.A., Dunmire, S., Levey, G.S., and Mulvihill, J.J. (1993) Three-decade investigation of familial pheochromocytoma. An allele of von Hippel-Lindau disease? *Arch. Intern. Med.* **153**, 2550–2556.

232. van der Harst, H.E., de Krijger, R.R., Dinjens, W.N., Weeks, L.E., Bonjer, H.J., Bruining, H.A., Lamberts, S.W., and Koper, J.W. (1998) Germline mutations in the vhl gene in patients presenting with phaeochromocytomas. *Int. J. Cancer* **77**, 337–340.

233. Walther, M.M., Reiter, R., Keiser, H.R., Choyke, P.L., Venzon, D., Hurley, K., Gnarra, J.R., Reynolds, J.C., Glenn, G.M., Zbar, B., and Linehan, W.M. (1999) Clinical and genetic characterization of pheochromocytoma in von Hippel-Lindau families: comparison with sporadic pheochromocytoma gives insight into natural history of pheochromocytoma. *J. Urol.* **162**, 659–664.

234. Lonser, R.R., Glenn, G.M., Walther, M., Chew, E.Y., Libutti, S.K., Linehan,W.M., and Oldfield, E.H. (2003) von Hippel-Lindau disease. *Lancet* **361**, 2059–2067.

235. Kang, H.C., Kim, I.J., Park, J.H., Shin, Y., Jang, S.G., Ahn, S.A., Park, H.W., Lim, S.K., Oh, S.K., Kim, D.J., Lee, K.W., Choi, Y.S., Park, Y.J., Lee, M.R., Kim, D.W., and Park, J.G. (2005) Three novel *VHL* germline mutations in Korean patients with von Hippel-Lindau disease and pheochromocytomas. *Oncol. Rep.* **14**, 879–883.

236. Hes, F. J., Höppener, J.W.M., and Lips, C.J.M. (2003) Pheochromocytoma in von Hippel-Lindau disease. *J. Clin. Endocrinol. Metab.* **88**, 969–974.

237. Richard, S., Beigelman, C., Duclos, J.-M., Fendler, J.-P., Plauchu, H., Plouin, P.-F., Resche, F., Schlumberger, M., Vermesse, B., and Proye C. (1994) Pheochromocytoma as the first manifestation of von Hippel-Lindau disease. *Surgery* **116**, 1076–1081.

238. Walther, M.M., and Linehan, W.M. (1996) von Hippel-Lindau disease and pheochromocytoma. *JAMA* **275**, 839–840.

239. Garcia, A., Matias-Guiu, X., Cabezas, R., Chico, A., Prat, J., Baiget, M., and De Leiva, A. (1997) Molecular diagnosis of von Hippel-Lindau disease in a kindred with a predominance of familial phaeochromocytoma. *Clin. Endocrinol.* **46**, 359–363.

240. Gijtenbeek, J., Jacobs, B., Boots-Sprenger, S., Bonne, A., Lenders, J., Küsters, B., Wesseling, P., and Jeuken, J. (2005) Molecular analysis as a tool in the differential diagnosis of VHL disease-related tumors. *Diagn. Mol. Pathol.* **14**, 115–120.

241. Ponder, B.A.J. (1999) The phenotypes associated with *ret* mutations in the multiple endocrine neoplasia type 2 syndrome. *Cancer Res.* **59**, 1736s–1742s.

242. Ponder, B.A.J. (2001) Multiple endocrine neoplasia type 2. In *The Metabolic & Molecular Bases of Inherited Disease*, 8th edition, vol. 1 (Scriver, C.R., Beaudet, A.L., Valle, D., and Sly, W.S., eds.), McGraw-Hill, New York, pp. 931–942.

243. Hansford, J.R., and Mulligan, L.M. (2000) Multiple endocrine neoplasia type 2 and *RET*: from neoplasia to neurogenesis. *J. Med. Genet.* **37**, 817–827.

243A. Treanor, J.J.S., Goodman, L., de Sauvage, F., Stone, D.M., Poulsen, K.T., Beck, C.D., Gray, C., Armanini, M.P., Pollock, R.A., Hefti, F., Phillips, H.S., Goddard, A., Moore, M.W., Buj-Bello, A., Davies, A.M., Asai, N., Takahashi, M., Vandlen, R., Henderson, C.E., and Rosenthal, A. (1996) Characterization of a multicomponent receptor for GDNF. *Nature* **382**, 80–83.

244. Wautot, V., Vercherat, C., Lespinasse, J., Chambe, B., Lenoir, G.M., Zhang, C.X., Porchet, N., Cordier, M., Beroud, C., and Calender, A. (2002) Germline mutations of MEN1 in multiple endocrine neoplasia type 1: search for correlation between phenotype and the functional domains of the MEN1 protein. *Hum. Mutat.* **20**, 35–47.

245. Weber, F., and Eng, C. (2005) Editoria;: Germline variants within RET: Clinical utility or scientific playtoy? *J. Cliln. Endocrinol. Metab.* **90**, 6334–6336.

246. Santoro, M., Melillo, R.M., Carlomagno, F., Visconti, R., De Vita, G., Salvatore, G., Lupoli, G., Fusco, A., and Vecchio, G. (1998) Moleuclar biology of the *MEN2* gene. *J. Intern. Med.* **243**, 505–508.

247. Takahashi, M., Asai, N., Iwashita, T., Murakami, H., and Ito, S. (1998) Molecular mechanisms of development of multiple endocrine neoplasia 2 by *RET* mutations. *J. Intern. Med.* **243**, 509–513.

248. Mulligan, L.M., Kwok, J.B.J., Healey, C.S., Elsdon, M.J., Eng, C., Gardner, E., Love, D.R., Mole, S.E., Moore, J.K., Papi, L., Ponder, M.A., Telenius, H., Tunnacliffe, A., and Ponder, B.A.J.. (1993) Germline mutations of the *RET* proto-oncogene in multiple endocrine neoplasia type 2A. *Nature* **363,** 458–460.

249. Mulligan, L.M., Marsh, D.J., Robinson, B.G., Schuffenecker, I., Zedenius, J., Lips, C.J.M., Gagel, R.F., Takai, S.-I., Noll, W.W., Fink, M., Raue, F., Lacroix, A., Thibodeau, S.N., Frilling, A., Ponder, B.A.J., and Eng, C., for the International *RET* Mutation Consortium. (1995) Genotype-phenotype correlation in multiple endocrine neoplasia type 2: report of the International *RET* Mutation Consortium. *J. Intern. Med.* **238,** 343–346.

250. Mulligan, L.M., Eng, C., Healey, C.S., Clayton, D., Kwok, J.B.J., Gardner, E., Ponder, M.A., Frilling, A., Jackson, C.E., Lehnert, H., Neumann, H.P.H., Thibodeau, S.N., and Ponder, B.A.J. (1994) Specific mutations of the *RET* proto-oncogene are related to disease phenotype in MEN 2A and FMTC. *Nat. Genet.* **6,** 70–74.

251. Eng, C., Smith, D.P., Mulligan, L.M., Nagai, M.A., Healey, C.S., Ponder, M.A., Gardner, E., Scheumann, G.F.W., Jackson, C.E., Tunnacliffe, A., and Ponder, B.A.J. (1994) Point mutation within the tyrosine kinase domain of the *RET* proto-oncogene in multiple endocrine neoplasia type 2B and related sporadic tumours. *Hum. Mol. Genet.* **3,** 237–241.

252. Eng, C., Clayton, D., Schuffenecker, I., Lenoir, G., Cote, G., Gagel, R.F., Ploos van Amstel, H.K., Lips, C.J.M., Nishisho, I., Takai, S.-I., Marsh, D.J., Robinson, B.G., Frank-Raue, K., Raue, F., Xue, F., Noll, W.W., Romei, C., Pacini, F., Fink, M., Niederle, B., Zadenius, J., Nordenskjold, M., Komminoth, P., Handy, G.H., Gharib, H., Thibodeau, S.N., Lacroix, A., Frilling, A., Ponder, B.A.J., and Mulligan, L.M. (1996) The relationship between specific *RET* proto-oncogene mutations and disease phenotype in multiple endocrine neoplasia type 2: International *RET* Mutation Consortium analysis. *JAMA* **276,** 1575–1579.

253. Eng, C., and Mulligan, L.M. (1997) Mutations of the RET proto-oncogene in the multiple endocrine neoplasia type 2 syndromes, related sporadic tumours, and Hirschsprung disease. *Hum. Mutat.* **9,** 97–109.

254. Eng, C. (1999) *RET* proto-oncogene in the development of human cancer. *J. Clin. Oncol.* **17,** 380–393.

254A. Gimm, O., Marsh, D.J., Andrew, S.D., Frilling, A., Dahia, P.L., Mulligan, L.M., Zajac, J.D., Robinson, B.G., and Eng, C. (1997) Germline dinucleotide mutation in codon 883 of the RET proto-oncogene in multiple endocrine neoplasia type 2B without codon 918 mutation. *J. Clin. Endocrinol. Metab.* **82,** 3902–3904.

255. Mulligan, L.M. (2001) Multiple endocrine neoplasia type 2: Molecular aspects. In *Genetic Disorders of Endocrine Neoplasia: Frontiers in Hormone Research* (Dahia, P.L.M., and Eng, C., eds.), Karger, Basel, Switzerland, pp. 81–102.

256. Jing, S., Wen, D., Yu, Y., Holst, P.L., Luo, Y., Fang, M., Tamir, R., Antonio, L., Hu, Z., Cupples, R., Louis, J.-C., Hu, S., Altrock, B.W., and Fox, G.M. (1996) GDNF-induced activation of the Ret protein tyrosine kinase is mediated by GDNFR-α, a novel receptor for GDNF. *Cell* **85,** 1113–1124.

256A. Sanicola, M., Hession, C., Worley, D., Carmillo, P., Ehrenfels, C., Walus, L., Robinson, S., Jaworski, G., Wei, H., Tizard,

R., Whitty, A., Pepinsky, R.B., and Cate, R.L. (1997) Glial cell line-derived neurotrophic factor-dependent RET activation can be mediated by two different cell-surface accessory proteins. *Proc. Natl. Acad. Sci. USA* **94,** 6238–6243.

257. Santoro, M., Carlomagno, F., Romano, A., Bottaro, D.P., Dathan, N.A., Grieco, M., Fusco, A., Vecchio, G., Matškova, B., Kraus, M.H., and Di Fiore, P.P. (1995) Activation of *RET* as a dominant transforming gene by germline mutations of MEN 2A and MEN2B. *Science* **267,** 381–383.

258. Songyang, Z., Carraway, K.L., III, Eck, M.J., Harrison, S.C., Feldman, R.A., Mohammadi, M., Schlessinger, J., Hubbard, S.R., Smith, D.P., Eng, C., Lorenzo, M.J., Ponder, B.A.J., Mayer, B.J., and Cantley, L.C. (1995) Catalytic specificity of protein-tyrosine kinases is critical for selective signaling. *Nature* **373,** 536–539.

259. Huang, S.C., Koch, C.A., Vortmeyer, A.O., Pack, S.D., Lichtenauer, U.D., Mannan, P., Lubensky, I.A., Chrousos, G.P., Gagel, R.F., Pacak, K., and Zhuang, Z. (2000) Duplication of the mutant *RET* allele in trisomy 10 or loss of the wild-type allele in multiple endocrine neoplasia type 2-associated pheochromocytomas. *Cancer Res.* **60,** 6223–6226.

260. Donis-Keller, H. (1995) The RET proto-oncogene and cancer. *J. Int. Med.* **238,** 319–325.

261. Califano, D., Rizzo, C., D'Alessio, A., Colucci-D'Amato, G.L., Calì, G., Cannada Bartoli, P., Santelli, G., Vecchio, G., and de Franciscis, V. (2000) Signaling through Ras is essential for *ret* oncogene-induced cell differentiation in PC12 cells. *J. Biol. Chem.* **275,** 19297–19305.

262. Schuringa, J.J., Wojtachnio, K., Hagens, W., Vellenga, E., Buys, C.H.C.M., Hofstra, R., and Kruijer, W. (2001) MEN2-A-RET-induced cellular transformation by activation of STAT3. *Oncogene* **20,** 5350–5358.

263. Segouffin-Cariou, C., and Billaud, M. (2000) Transforming ability of MEN2A-RET requires activation of the phosphatidylinositol 3-kinase/AKT signaling pathway. *J. Biol. Chem.* **275,** 3568–3576.

264. Murakami, H., Iwashita, T., Asai, N., Shimono, Y., Iwata, Y., Kawai, K., and Takahashi, M. (1999) Enhanced phosphatidylinositol 3-kinase activity and high phosphorylation state of its downstream signaling molecules mediated by Ret with the MEN 2B mutation. *Biochem. Biophys. Res. Commun.* **262,** 68–75.

265. Powers, J.F., Misra, S., Schelling, K., Varticovski, L., and Tischler, A.S. (2001) Mitogenic signaling by cyclic adenosine monophosphate in chromaffin cells involves phosphatidylinositol 3-kinase activation. *J. Cell Biochem. Suppl.* **36,** 89–98.

266. Neumann, H.P.H., Eng, C., and Mulligan, L.M. (1996) Reply to: von Hippel-Lindau disease and pheochromocytoma. *J. Natl. Cancer Inst.* **275,** 840.

267. Wada, M., Asai, N., Tsuzuki, T., Maruyama, S., Ohiwa, M., Imai, T., Funahashi, H., Takagi, H., and Takahashi, M. (1996) Detection of Ret homodimers in MEN 2A-associated pheochromocytomas. *Biochem. Biophys. Res. Commun.* **218,** 606–609.

268. Takaya, K., Yoshimasa, T., Arai, H., Tamura, N., Miyamoto, Y., Itoh, H., and Nakao, K. (1996) Expression of the RET proto-oncogene in normal human tissues, pheochromocytomas, and other tumors of neural crest origin. *J. Mol. Mech.* **74,** 617–621.

269. Beldjord, C., Desclaux-Arramond, F., Raffin-Sanson, M., Corvol, J.-C., de Keyzer, Y., Luton, J.-P., Plouin, P.-F., and Bertagna, X. (1995) The *RET* protooncogene in sporadic pheochromocytomas: frequent MEN 2-like mutations and new molecular defects. *J. Clin. Endocrinol. Metab.* **80**, 2063–2068.

270. Zedenius, J., Wallin, G., Hamberger, B., Nordenskjöld, M., Weber, G., and Larsson, C. (1994) Somatic and MEN 2A de novo mutations identified in the RET proto-oncogene by screening of sporadic MTC:s. *Hum. Mol. Genet.* **3**, 1259–1262.

271. Lindor, N.M., Honchel, R., Khosla, S., and Thibodeau, S.N. (1995) Mutations in the RET protooncogene in sporadic pheochromocytomas. *J. Clin. Endocrinol. Metab.* **80**, 627–629.

272. Arnold, A. (1996) *RET* mutation screening in sporadic pheochromocytoma. *J. Clin. Endocrinol. Metab.* **81**, 430.

273. Iyengar, S., Tallini, G., Sirugo, G., Bale, A.E., and Kidd, K.K. (1995) Mutation analysis of the *VHL* gene in individuals with sporadic and familial pheochromocytoma. *Am. J. Hum. Genet.* **57 (Suppl.) A67**, 357.

274. Aguiar, R.C.T., Cox, G., Pomeroy, S.L., and Dahia, P.L.M. (2001) Analysis of the *SDHD* gene, the susceptibility gene for familial paraganglioma syndrome (PGL1), in pheochromocytoma. *J. Clin. Endocrinol. Metab.* **86**, 2890–2894.

275. Gimenez-Roqueplo, A.-P., Favier, J., Rustin, P., Rieubland, C., Kerlan, V., Plouin, P.-F., Rötig, A., and Jeunemaitre, X. (2002) Functional consequences of a *SDHB* gene mutation I an apparently sporadic pheochromocytoma. *J. Clin. Endocrinol. Metab.* **87**, 4771–4774.

276. Walther, M.M., Herring, J., Enquist, E., Keiser, H.R., and Linehan, W.M. (1999) von Recklinghausen's disease and pheochromocytomas. *J. Urol.* **162**, 1582–1586.

276A. Ohji, H., Sasagawa, I., Iciyanagi, O., Suzuki, Y., and Nakada, T. (2001) Tumour angiogenesis and Ki-67 expression in pheochromocytoma. *BJU Int.* **87**, 381–385.

277. Clarke, M.R., Weyant, R.J., Watson, C.G., and Cary, S.E. (1998) Prognostic markers in pheochromocytoma. *Hum. Pathol.* **29**, 522–526

278. Gutmann, D.H., Geist, R.T., Rose, K., Wallin, G., and Moley, J.F. (1995) Loss of neurofibromatosis type I (*NF1*) gene expression in pheochromocytomas from patients without NF1. *Genes Chromosomes Cancer* **13**, 104–109.

279. DeAngelis, L.M., Kelleher, M.B., Post, K.D., and Fetell, M.R. (1987) Multiple paragangliomas in neurofibromatosis: A new neuroendocrine neoplasia. *Neurology* **37**, 129–133.

280. Huson, S.M., Harper, P.S., and Compston, D.A. (1988) Von Recklinghausen neurofibromatosis. A clinical and population study in south-east Wales. *Brain* **111 (Pt 6)**, 1355–1381.

281. Bausch, B., Koschker, A.-C., Fassnacht, M., Stoevesandt, J., Hoffmann, M.M., Eng, C., Allolio, B., and Neumann, H.P.H. (2006) Comprehensive mutation scanning of *NF1* in apparently sporadic cases of pheochromocytoma. *J. Clin. Endocrinol. Metab.* **91**, 3478–3481.

282. Bausch, B., Borozdin, W., and Neumann, H.P.H. (2006) Clinical and genetic characteristics of patients with neurofibromatosis type 1 and pheochromocytoma. *N. Engl. J. Med.* **354**, 2729–2731.

283. Caiazzo, R., Mariette, C., Piessen, G., Jany, I., Carnaille, B., and Triboulet, J.P. (2006) Type 1 neurofibromatosis, pheochromocytoma and somatostatinoma of the ampulla. Literature review. *Ann. Chir.* **131**, 393–397.

284. Gursoy, A., and Erdogan, M.F. (2006) Severe pulmonary involvement and pheochromocytoma in atypical patient with neurofibromatosis type 1. *Endocr. Pract.* **12**, 469–471.

285. Katechis, D., Makaryus, A.N., Spatz, A., Freeman, J., and Diamond, J.A. (2005) Acute myocardial infarction in a patient with pheochromocytoma and neurofibromatosis. *J. Invasive Cardiol.* **17**, 331–333.

286. Lew, J.I., Jacome, F.J., and Solorzano, C.C. (2006) Neurofibromatosis-associated pheochromocytoma. *J. Am. Coll. Surg.* **202**, 550–551.

287. Xu, W., Mulligan, L.M., Ponder, M.A., Liu, L., Smith, B.A., Mathew, C.G., and Ponder, B.A. (1992) Loss of NF1 alleles in phaeochromocytomas from patients with type I neurofibromatosis. *Genes Chromosomes Cancer* **4**, 337–342.

288. Carney, J.A. (1999) Gastric stromal sarcoma, pulmonary chondroma, and extra-adrenal paraganglioma (Carney Triad): Natural history, adrenocortical component, and possible familial occurrence. *Mayo Clin. Proc.* **74**, 543–552.

289. Appelman, H.D. (1999) The Carney triad: A lesson in observation, creativity, and perseverance. *Mayo Clin. Proc.* **74**, 638–640.

290. Colwell, A.S. D'Cunha, J., and Maddaus, M.A. (2001) Carney's triad paragangliomas. *J. Thorac Cardiovasc. Surg.* **121**, 1011–1012.

290A. Boccon-Gibod, L., Boman, F., Boudjemaa, S., Fabre, M., Leverger, G., and Carney, J.A. (2004) Separate occurrence of extra-adrenal paraganglioma and gastrointestinal stromal tumor in monozygotic twins: probable familial Carney syndrome. *Pediatr. Dev. Pathol.* **7**, 380–384.

291. Brodeur, G.M., and Shimada, H. (1998) Pheochromocytomas and paragangliomas, vol. 2. In *Russell and Rubinstein's Pathology of Tumors of the Nervous System:* (Bigner, D.D., McLendon, R.E., and Bruner, J.M., eds.)., Arnold, London, pp. 535–560.

292. Erickson, D., Kudva, Y.C., Ebersold, M.J., Thompson, G.B., Grant, C.S., van Heerden, J.A., and Young, W.F. (2001) Benign paragangliomas: clinical presentation and treatment outcomes in 236 patients. *J. Clin. Endocrinol. Metab.* **86**, 5210–5216.

293. Soffer, D., and Scheithauer, B.W. (2000) Paraganglioma. In *Pathology and Genetics of Tumours of the Nervous* System (Kleihues, P., and Cavenee, W.K., eds.), IARC Press, France, Lyon, pp. 112–114.

294. Weiss, S.W., and Goldblum, J.R. (2001) *Enzinger and Weiss's Soft Tissue Tumors*, 4th ed. Mosby, St. Louis, MO.

295. Edström Elder, E., Hjelm Skog, A.L., Hoog, A., and Hamberger, B. (2003) The management of benign and malignant pheochromocytoma and abdominal paraganglioma. *Eur. J. Surg. Oncol.* **29**, 278–283.

296. Thompson, L.D.R., Young, W.F., Jr., Kawashima, A., Komminoth, P., and Tischler, A.S. (2004) Malignant adrenal phaeochromocytoma. In *Pathology and Genetics of Tumours of Endocrine Organs: World Health Organization Classification of Tumors* (DeLellis, R.A., Lloyd, R.V., Heitz, P.U., and Eng, C., eds), Kliehues, P., Sobin, L.H., Series Eds., IARC Press, Lyon, France, pp. 147–150.

297. McNicol, A.M., Young, W.F., Jr., Kawashima, A., Komminoth, P., and Tischler, A.S. (2004) Benign

phaeochromocytoma. In *Pathology and Genetics of Tumours of Endocrine Organs: World Health Organization Classification of Tumors* (DeLellis, R.A., Lloyd, R.V., Heitz, P.U., and Eng, C., eds), Kliehues, P., Sobin, L.H., Series Eds., IARC Press, Lyon, France, pp. 151–155.

298. Tischler, A.S., Kimura, N., Lloyd, R.V., and Komminoth, P. (2004) Composite phaeochromocytoma or paraganglioma. In *Pathology and Genetics of Tumours of Endocrine Organs: World Health Organization Classification of Tumors* (DeLellis, R.A., Lloyd, R.V., Heitz, P.U., and Eng, C., eds.), Kliehues, P., Sobin, L.H., Series Eds., IARC Press, Lyon, France, pp. 156–158.

299. Kimura, N., Chetty, R., Capella, C., Young, W.F., Jr., Koch, C.A., Lam, K.Y., DeLellis, R.A., Kawashima, A., Komminoth, P., and Tischler, A.S. (2004) Extra-adrenal paraganglioma: Carotid body, jugulotympanic, vagal, laryngeal, aortico-pulmonary. In *Pathology and Genetics of Tumours of Endocrine Organs: World Health Organization Classification of Tumors* (DeLellis, R.A., Lloyd, R.V., Heitz, P.U., and Eng, C., eds.), Kliehues, P., Sobin, L.H., Series Eds., IARC Press, Lyon, France, pp. 159–161.

300. Tischler, A.S., Komminoth, P., Kimura, N., Young, W.F., Jr., Chetty, R., Albores-Saavedra, J., and Kleihues, P. (2004) Extra-adrenal paraganglioma: Gangliocytic, cauda equina, orbital, nasopharyngeal. In *Pathology and Genetics of Tumours of Endocrine Organs: World Health Organization Classification of Tumors* (DeLellis, R.A., Lloyd, R.V., Heitz, P.U., and Eng, C., eds.), Kliehues, P., Sobin, L.H., Series Eds., IARC Press, Lyon, France, pp. 162–163.

301. Kimura, N., Capella, C., De Krijger, R.R., Thompson, L.D.R., Lam, K.Y., Komminoth, P., Tischler, A.S., and Young, W.F., Jr. (2004) Extra-adrenal sympathetic paraganglioma: Superior and inferior paraaortic. In *Pathology and Genetics of Tumours of Endocrine Organs: World Health Organization Classification of Tumors* (DeLellis, R.A., Lloyd, R.V., Heitz, P.U., and Eng, C., eds.), Kliehues, P., Sobin, L.H., Series Eds., IARC Press, Lyon, France, pp. 164–165.

302. Tischler, A.S., and Komminoth, P. (2004) Extra-adrenal sympathetic paraganglioma: Cervical, intrathoracic and urinary bladder. In *Pathology and Genetics of Tumours of Endocrine Organs: World Health Organization Classification of Tumors* (DeLellis, R.A., Lloyd, R.V., Heitz, P.U., and Eng, C., eds.), Kliehues, P., Sobin, L.H., Series Eds., IARC Press, Lyon, France, pp. 165–166.

303. Gimm, O., Morrison, C.D., Suster, S., Komminoth, P., Mulligan, L., and Sweet, K.M. (2004) Multiple endocrine neoplasia type 2. In *Pathology and Genetics of Tumours of Endocrine Organs: World Health Organization Classification of Tumors* (DeLellis, R.A., Lloyd, R.V., Heitz, P.U., and Eng, C., eds.), Kliehues, P., Sobin, L.H., Series Eds., IARC Press, Lyon, France, pp. 211–217.

304. Maher, E.R., Nathanson, K., Komminoth, P., Neumann, H.P.H., Plate, K.H., Bohling, T., and Schneider, K. (2004) Von Hippel-Lindau syndrome (VHL). In *Pathology and Genetics of Tumours of Endocrine Organs: World Health Organization Classification of Tumors* (DeLellis, R.A., Lloyd, R.V., Heitz, P.U., and Eng, C., eds.), Kliehues, P., Sobin, L.H., Series Eds., IARC Press, Lyon, France, pp. 230–237.

305. Nathanson, K., Baysal, B.E., Drovdlic, C., Komminoth, P., and Neumann, H.P. (2004) Familial paraganglioma-phaeochromocytoma syndromes caused by *SDHB*, *SDHC* and *SDHD* mutations. In *Pathology and Genetics of Tumours of Endocrine Organs: World Health Organization Classification of Tumors* (DeLellis, R.A., Lloyd, R.V., Heitz, P.U., and Eng, C., eds.), Kliehues, P., Sobin, L.H., Series Eds., IARC Press, Lyon, France, pp. 238–242.

306. Evans, D.G.R., Komminoth, P., Scheithauer, B.W., and Peltonen, J. (2004) Neurofibromatosis type 1. In *Pathology and Genetics of Tumours of Endocrine Organs: World Health Organization Classification of Tumors* (DeLellis, R.A., Lloyd, R.V., Heitz, P.U., and Eng, C., eds), Kliehues, P., Sobin, L.H., Series Eds., IARC Press, Lyon, France, pp. 243–248.

306A. Pacak, K., and Eisenhofer, G., eds. (2006) Pheochromocytoma. First International Symposium. In *Annals of the New York Academy of Sciences*, Vol. 1073, Blackwell Publishing, Boston, MA.

307. Pomares, F.J., Canas, R., Rodriguez, J.M., Hernandez, A.M., Parrilla, P., and Tebar, F.J. (1998) Differences between sporadic and multiple endocrine neoplasia type 2A phaeochromocytoma. *Clin. Endocrinol.* **48,** 195–200.

308. Decker, H.-J., Klauck, S.M., Lawrence, J.B., McNeil, J., Smith, D., Gemmill, R.M., Sandberg, A.A., Neumann, H.H.P., Simon, B., Green, J., and Seizinger, B.R. (1994) Cytogenetic and fluorescence in situ hybridization studies on sporadic and hereditary tumors associated with von Hippel-Lindau syndrome (VHL). *Cancer Genet. Cytogenet.* **77,** 1–13.

309. Tsutsumi, M., Yokota, J., Kakizoe, T., Koiso, K., Sugimura, T., and Terada, M. (1989) Loss of heterozygosity on chromosomes 1p and 11p in sporadic pheochromocytoma. *J. Natl. Cancer Inst.* **81,** 367–370.

310. Yokogoshi, Y., Yoshimoto, K., Saito, S. (1990) Loss of heterozygosity on chromosomes 1 and 11 in sporadic pheochromocytomas. *Jpn. J. Cancer Res.* **81,** 632–638.

311. Baysal, B.E., van Schothorst, E.M., Farr, J.E., Grashof, P., Myssiorek, D., Rubinstein, W.S., Täschner, P., Cornelisse, C.J., Devlin, B., Devilee, P., and Richard, C.W., 3rd. (1999) Repositioning the hereditary paraganglioma critical region on chromosome band 11q23. *Hum. Genet.* **104,** 219–225.

312. Milunsky, J., DeStefano, A.L., Huang, X.-L., Baldwin, C.T., Michels, V.V., Jako, G., and Milunsky, A. (1997) Familial paragangliomas: linkage to chromosome 11q23 and clinical implications. *Am. J. Med. Genet.* **72,** 66–70.

313. Lee, S.-C., Chionh, S.-B., Chong, S.-M., and Täschner, P.E.M. (2003) Hereditary paraganglioma due to the SDHD M1I mutation in a second Chinese family: A founder effect? *Laryngoscope* **113,** 1055–1058.

314. Riemann, K., Sotlar, K., Kupka, S., Braun, S., Zenner, H.-P., Preyer, S., Pfister, M., Pusch, C.M., Blin, N. (2004) Chromosome 11 monosomy in conjunction with a mutated *SDHD* initiation codon in nonfamilial paraganglioma cases. *Cancer Genet. Cytogenet.* **150,** 128–135.

315. Mariman, E.C.M., van Beersum, S.E.C., Cremers, C.W.R.J., van Baars, F.M., and Ropers, H.H. (1993) Analysis of a second family with hereditary non-chromaffin paragangliomas locates the underlying gene at the proximal region of chromosome 11q. *Hum. Genet.* **91,** 357–361.

316. Bourgeron, T., Rustin, P., Chretien, D., Birch-Machin, M., Bourgeois, M., Viegas-Pequignot, E., Munnich, A., and Rotig. A. (1995) Mutation of a nuclear succinate dehydrogenase

gene results in mitochondrial respiratory chain deficiency. *Nat. Genet.* **11,** 144–149.

317. Birch-Machin, M.A., Taylor, R.W., Cochran, B., Ackrell, B.A.C., and Turnbull, D.M. (2000) Late-onset optic atrophy, ataxia, and myopathy associated with a mutation of a complex II gene. *Ann. Neurol.* **48,** 330–335.

318. Parfait, B., Chretien, D., Rotig, A., Marsac, C., Munnich, A., and Rustin, P. (2000) Compound heterozygous mutations in the flavoprotein gene of the respiratory chain complex II in a patient with Leigh syndrome. *Hum. Genet.* **106,** 236–243.

319. Baysal, B.E., Rubinstein, W.S., and Täschner, P.E. (2001) Phenotypic dichotomy in mitochondrial complex II genetic disorders. *J. Mol. Med.* **79,** 495–503.

320. Scheffler, I.E. (1998) Molecular genetics of succinate:quinone oxidoreductase in eukaryotes. *Prog. Nucleic Acids Res. Mol. Biol.* **60,** 267–315.

321. Hederstedt, L. (2003) Complex II is complex too. *Science* **299,** 671–672.

322. Brauch, H., Kishida, T., Glavac, D., Chen, F., Pausch, F., Hofler, H., Latif, F., Lerman, M.I., Zbar, B., and Neumann, H.P.H. (1995) von Hippel-Lindau (VHL) diseases with pheochromocytoma in the Black Forest region of Germany: evidence for a founder effect. *Hum. Genet.* **95,** 551–556.

323. Chen, F., Kishida, T., Yao, M., Hustad, T., Glavac, D., Dean, M., Gnarra, J.R., Orcutt, M.L., Duh., F.M., Glenn, G., Green, J., Hsia, E., Lamiell, J., Li, H., Wei, M.H., Schmidt, L., Tory, K., Kuzmin, I., Stackhouse, T., Latif, F., Linehan, W.M., Lerman, M., and Zbar, B. (1995) Germline mutations in the von Hippel-Lindau disease tumor suppressor gene: Correlations with phenotype. *Hum. Mutat.* **5,** 66–75.

324. Favier, J., Briere, J.J., Strompf, L., Amar, L., Filali, M., Jeunemaitre, X., Rustin, P., Gimenez-Roqueplo, A.P.; PGL.NET Network. (2005) Hereditary paraganglioma/ pheochromocytoma and inherited succinate dehydrogenase deficiency. *Horm. Res.* **63,** 171–179.

325. Latif, F., Tory, K., Gnarra, J., Yao, M., Duh., F.-M., Orcutt, M.L., Stackhouse, T., Kuzmin, I., Modi, W., Geil, L., Schmidt, L., Zhou, F., Li, H., Wei, M.H., Chen, F., Glenn, F., Choyke, P., Walther, M.M., Weng, Y., Duan, D.-S.,R., Dean, M., Glavac, D., Richards, F.M., Crossey, P.A., Ferguson-Smith, M.A., Le Paslier, D., Chumakov, I., Cohen, D., Chinault, A.C., Maher, E.R., Linehan, W.M., Zbar, B., and Lerman, M.I. (1993) Identification of the von Hippel-Lindau disease tumor suppressor gene. *Science* **260,** 1317–1320.

326. van Nederveen, F.H., Dinjens, W.N.M., Korpershoek, E., and de Krijger, R.R. (2006) The occurrence of SDHB gene mutations in pheochromocytoma. *Ann. N Y Acad. Sci.* **1073,** 177–182.

327. Crossey, P.A., Richards, F.M., Foster, K., Green, J.S., Prowse, A., Latif, F., Lerman, M.I., Zbar, B., Affara, N.A., Ferguson-Smith, M.A., and Maher, E.R. (1994) Identification of intragenic mutations in the von Hippel-Lindau disease tumour suppressor gene and correlation with disease phenotype. *Hum. Mol. Genet.* **3,** 1303–1308.

328. Friedrich, C.A. (1999) Von Hippel-Lindau syndrome. A pleomorphic condition. *Cancer* **86,** 2478–2482.

329. Hosoe, S., Brauch, H., Latif, F., Glenn, G., Daniel, G., Bale, S., Choyke, P., Gorin, M., Oldfield, E., Berman, A., Goodman, J., Orcutt, M.L., Hampsch, K., Dllisio, J., Modi, W., McBride, W., Anglard, P., Weiss, G., Walther, M.M., Linehan, W.M.,

Lerman, M.I., and Zbar, B. (1990) Localization of the von Hippel-Lindau disease gene to a small region of chromosome 3. *Genomics* **8,** 634–640.

330. Mantero, F., Terzolo, M., Arnaldi, G., Osella, G., Masini, A.M., Alì, A., Giovagnetti, M., Opocher, G., and Angeli, A., On Behalf of the Study Group on Adrenal Tumors of the Italian Society of Endocrinology. (2000) A survey on adrenal incidentaloma in Italy. *J. Clin. Endocrinol. Metab.* **85,** 637–644.

331. Maxwell, P.H. (2005) A common pathway for genetic events leading to pheochromocytoma. *Cancer Cell* **8,** 91–93.

332. Neumann, H.P.H., Schempp, W., and Wienker, T.F. (1988) High-resolution chromosome banding and fragile sites in von Hippel-Lindau syndrome. *Cancer Genet. Cytogenet.* **31,** 41–46.

333. Schimke, R.N., Collins, D.L., and Rothberg, P.G. (1998) Functioning carotid paraganglioma in the von Hippel-Lindau syndrome. *Am. J. Med. Genet.* **80,** 533–534.

334. Zanelli, M., and van der Walt, J.D. (1996) Carotid body paraganglioma in von Hippel-Lindau disease: a rare association. *Histopathology* **29,** 178–181.

335. Abbott, M.-A., Nathanson, K.L., Nightingale, S., Maher, E.R., and Greenstein, R.M. (2006) The von Hippel-Lindau (*VHL*) germline mutation V84L manifests as early-onset bilateral pheochromocytoma. *Am. J. Med. Genet.* **140A,** 685–690.

336. Behmel, A., Reichl, E., Pfragner, R., and Niederle, B. (1997) High incidence of gonosomal aneuploidy in lymphocytes of patients with sporadic, MEN1 and MEN2 associated neuroendocrine tumors: age-related X chromosome instability versus genuine mosaicism. *Medizinische Genetik.* **44,** 1–3.

337. Qiao, S., Iwashita, T., Furukawa, T., Yamamoto, M., Sobue, G., and Takahashi, M. (2001) Differential effects of leukocyte common antigen-related protein on biochemical and biological activities of RET-MEN2A and RET-MEN2B mutant proteins. *J. Biol. Chem.* **276,** 9460–9467.

338. Howe, J.R., Norton, J.A., and Wells, S.A., Jr. (1993) Prevalence of pheochromocytoma and hyperparathyroidism in multiple endocrine neoplasia type 2A. results of long-term follow-up. *Surgery* **114,** 1070–1077.

339. Modigliani, E., Vasen, H.M., Raue, K., Dralle, H., Frilling, A., Gheri, R.G., Brandi, M.L., Limbert, E., Niederle, B., Forgas, L., Rosenberg-Bourgin, M., Calmettes, C., and the Euromen Study Group. (1995) Pheochromocytoma in multiple endocrine neoplasia type 2: European study. *J. Intern. Med.* **238,** 363–367.

340. Nguyen, L., Niccoli-Sire, P., Caron, P., Bastie, D., Maes, B., Chabrier, G., Chabre, O., Rohmer, V., Lecomte, P., Henry, J.F., Conte-Devolx, B and the French Calcitonin Tumors Study Group (2001) Pheochromocytoma in multiple endocrine neoplasia type 2: a prospective study. *Eur. J. Endocrinol.* **144,** 37–44.

341. Puñales, M.K., Graf, H., Gross, J.L., and Maia, A.L. (2003) *RET* codon 634 mutations in multiple endocrine neoplasia type 2: variable clinical features and clinical outcome. *J. Clin. Endocrinol. Metab.* **86,** 2644–2649.

342. Kinlaw, W.B., Scott, S.M., Maue, R.A., Memoli, V.A., Harris, R.D., Daniels, G.H., Porter, D.M., Belloni, D.R., Spooner, E.T., Ernesti, M.M., and Noll, W.W. (2005) Multiple endocrine neoplasia 2A due to a unique C609S *RET* mutation presents with pheochromocytoma and reduced penetrance of

medullary thyroid carcinoma. *Clin. Endocrinol. (Oxford)* **63**, 676–682.

343. Jimenez, C., Habra, M.A., Huang, S.-C.E., El-Naggar, A., Shapiro, S.E., Evans, D.B., Cote, G., and Gagel, R.F. (2004) Pheochromocytoma and medullary thyroid carcinoma: a new genotype-phenotype correlation of the *RET* protooncogene 891 germline mutation. *J. Clin. Endocrinol. Metab.* **89**, 4142–4145.

344. Gullu, S., Gursoy, A., Erdogan, M.F., Dizbaysak, S., Erdogan, G., and Kamel, N. (2005) Multiple endocrine neoplasia type 2A/localized cutaneous lichen amyloidosis associated with malignant pheochromocytoma and ganglioneuroma. *J. Endocrinol. Invest.* **28**,734–737.

345. Devilee, P., van Schothorst, E.M., Bardoel,, A.F.J, Bonsing, B., Kuipers-Dijkshoorn, N.J., James, M.R., Fleuren, G., van der Mey, A.G.L, and Cornelisse, C.J. (1994) Allelotype of head and neck paragangliomas: allelic imbalance is confined to the long arm of chromosome 11, the site of the predisposing locus PGL. *Genes Chromosomes Cancer* **11**, 71–78.

346. Koreth, J., Bakkenist, C.J., and McGee, J.O. (1999) Chromosomes, 11Q and cancer: a review. *J. Pathol.* **187**, 28–38.

347. Gospeth, J., Welkoborsky, H.J., and Mann, W. (1998) Biology and growth velocity of tumors of the globus jugulotympanicum and glomus caroticum. *Laryngorhinootologie***77**, 429–433.

348. Maier, W., Marangos, N., and Laszig, R. (1999) Paraganglioma as a systemic syndrome: pitfalls and strategies. *J. Laryngol. Otol.* **113**, 978–982.

349. Hirano, S., Shoji, K., Kojima, H., and Omori, K. (1998) Dopamine-secreting carotid body tumor. *Am. J. Otolaryngol.* **19**, 412–416.

350. Hull, M.T., Roth, L.M., Glover, J.L., and Walker, P.D. (1982) Metastatic carotid body paraganglioma in von Hippel-Lindau disease. An electron microscopic study *Arch. Pathol. Lab. Med.* **106**, 235–239.

351. Kroll, A.J., Alexander, B., Cochios, F., and Pechet, L. (1964) Hereditary deficiencies of clotting factors VII and X associated with carotid-body tumors. *N. Engl. J. Med.* **270**, 6–13.

352. Li, S.L., Goko, H., Xu, Z.D., Kimura, G., Sun, Y., Kawachi, M.H., Wilson, T.G., Wilczynski, S., and Fujita-Yamaguchi, Y. (1998) Expression of insulin-like growth factor (IGF)-II in human prostate, breast, bladder, and paraganglioma tumors. *Cell Tissue Res.* **291**, 469–479.

353. Masuoka, J., Brandner, S., Paulus, W., Soffer, D., Vital, A., Chimelli, L., Jouvet, A., Yonekawa, Y., Kleihues, P., and Ohgaki, H. (2001) Germline SDHD mutation in paraganglioma of the spinal cord. *Oncogene* **20**, 5084–5086.

354. Paddenberg, R., Ishaq, B., Goldenberg, A., Faulhammer, P., Rose, F., Weissmann, N., Braun-Dullaeus, R.C., and Kummer, W. (2003) Essential role of complex II of the respiratory chain in hypoxia-induced ROS generation in the pulmonary vasculature. *Am. J. Physiol.* **284**, L710–L719.

355. Pritchett, J.W. (1982) Familial occurrence of carotid body tumor and pheochromocytoma. *Cancer* **49**, 2578–2579.

356. Renard, L., Godfraind, C., Boon, L.M., and Vikkula, M. (2003) A novel mutation in the *SDHD* gene in a family with inherited paragangliomas—implications of genetic diagnosis for follow up and treatment. *Head Neck* **25**, 146–151.

357. Sköldberg, F., Grimelius, L., Woodward, E.R., Rorsman, F., Van Schothorst, E.W., Winqvist, O., Karlsson, F.A.,

Åkerström, G., Kämpe, O., and Husebye, E.S. (1998) A family with hereditary extra-adrenal paragangliomas without evidence for mutations in the von Hippel-Lindau disease or *ret* genes. *Clin. Endocrinol* **48**, 11–16.

358. van der Mey, A.G., Cornelisse, C.J., Hermans, J., Terpstra, J.L., Schmidt, P.H., and Fleuren, G.J. (1991) DNA flow cytometry of hereditary and sporadic paragangliomas (glomus tumours). *Br. J. Cancer* **63**, 298–302.

359. Roncoroni, A.J., Montiel, G.C., and Semeniuk, G.B. (1993) Bilateral carotid body paraganglioma and central alveolar hypoventilation. *Respiration* **60**, 243–246.

360. Rustin, P., Munnich, A., and Rotig, A. (2002) Succinate dehydrogenase and human diseases: new insights into a well-known enzyme. *Eur. J. Hum. Genet.* **10**, 289–291.

361. Zorlu, F., Selek, U., Ulger, S., Donmez, T., and Erden, E. (2005) Paraganglioma in sella. *J. Neurooncol.* **73**, 265–267.

362. Telischi, F.F., Bustillo, A., Whiteman, M.L.H., Serafini, A.N., Reisberg, M.J., Gomez-Marin, O., Civantos, F.J., and Balkany, T.J. (2000) Octreotide scintigraphy for the detection of paragangliomas. *Head Neck Surg.* **122**, 358–362.

363. Wang, D.G., Johnston, C.F., Barros D'Sa, A.A., and Buchanan, K.D. (1997) Expression of apoptosis-suppressing gene bcl-2 in human carotid body tumours. *J. Pathol.* **183**, 218–221.

364. Chase, W.H. (1933) Familial and bilateral tumors of the carotid body. *J. Pathol. Bacteriol.* **36**, 1–12.

365. Atuk, N.O., Stolle, C., Owen, J.A., Jr., Carpenter, J.T., and Vance, M.L. (1998) Pheochromocytoma in von Hippel-Lindau disease: clinical presentation and mutation analysis in a large, multigenerational kindred.. *J. Clin. Endocrinol. Metab.* **83**, 117–120.

366. Boltze, C., Mundschenk, J., Unger, N., Schneider-Stock, R., Peters, B., Mawrin, C., Hoang-Vu., C., Roessner, A., and Lehnert, H. (2003) Expression profile of the telomeric complex discriminates between benign and malignant pheochromocytoma. *J. Clin. Endocrinol. Metab.* **88**, 4280–4286.

367. Czyzyk-Krezeska, M.F., Furnari, B.A., Lawson, E.E., and Millhorn, D.E. (1994) Hypoxia increases rate of transcription and stability of tyrosine hydroxylases mRNA in pheochromocytoma (PC12) cells. *J. Biol. Chem.* **269**, 760–764.

367A. Dahia, P.L., Aguiar, R.C., Tsanaclis, A.M., Bendit, I., Bydlowski, S.P., Abelin, N.M., and Toledo, S.P. (1995) Molecular and immunohistochemical analysis of P53 in phaeochromocytoma. *Br. J. Cancer* **72**, 1211–1213.

368. Herfarth, K.K., Wick, M.R., Marshall, H.N., Gartner, E., Lum, S., and Moley, J.F. (1997) Absence of TP53 alterations in pheochromocytomas and medullary thyroid carcinomas. *Genes Chromosomes Cancer* **20**, 24–29.

369. Aguiar, R.C., Dahia, P.L, Sill, H., Toledo, S.P., Goldman, J.M., and Cross, N.C. (1996) Deletion analysis of the p16 tumour suppressor gene in phaeochromocytomas. *Clin. Endocrinol.* **45**, 93–96.

370. Duerr, E.-M., Gimm, O., Neuberg, D.S., Kum, J.B., Clifford, S.C., Toledo, S.P.A., Maher, E.R., Dahia, P.L.M., and Eng, C. (1999) Differences in allelic distribution of two polymorphisms in the VHL-associated gene *CUL2* in pheochromocytoma patients without somatic *CUL2* mutations. *J. Clin. Endocrinol. Metab.* **84**, 3207–3211.

371. Dahia, P.L., Aguiar, R.C., Tsanaclis, A.M., Bendit, I., Bydlowski, S.P., Abelin, N.M., and Toledo, S.P. (1995) Molecular and immunohistochemical analysis of P53 in phaeochromocytoma. *Br. J. Cancer* **72**, 1211–1213.

372. Eisenhofer, G., Lenders, J.W.M., Linehan, W.M., Walther, M.M., Goldstein, D.S., and Keiser, H.R. (1999) Plasma normethanephrine and metanephrine for detecting pheochromocytoma in Hippel-Lindau disease and multiple endocrine neoplasia type 2. *N. Engl. J. Med.* **340**, 1872–1879.

373. Fassnacht, M., Weismann, D., Ebert, S., Adam, P., Zink, M., Beuschlein, F., Hahner, S., and Allolio, B. (2006) AKT is highly phosphorylated in pheochromocytomas but not in benign adrenocortical tumors. *J. Clin. Endocrinol. Metab.* **90**, 4366–4370.

374. Favier, J., Plouin, P.-F., Corvol, P., and Gasc, J.-M. (2002) Angiogenesis and vascular architecture in pheochromocytomas. Distinctive traits in malignant tumors. *Am. J. Pathol.* **161**, 1235–1246.

375. Gorlin, R.J., Sedano, H.O., Vickers, R.A., and Cervenka, J. (1968) Multiple mucosal neurinomas, pheochromocytoma and medullary carcinoma of the thyroid: A syndrome. *Cancer* **22**, 293–299.

376. Gross, D.J., Avishai, N., Meiner, V., Filon, D., Zbar, B., and Abeliovich, D. (1996) Familial pheochromocytoma associated with a novel mutation in the von Hippel-Lindau gene. *J. Clin. Endocrinol. Metab.* **81**, 147–149.

377. Grumbach, M.M., Biller, B.M.K., Braunstein, G.D., Campbell, K.K., Carney, J.A., Godley, P.A., Harris, E.L., Lee, J.K.T., Oertel, Y.C., Posner, M.C., Schlechte, J.A.., and Wieand, H.S. (2003) Management of the clinically inapparent adrenal mass ("incidentaloma"). *Ann. Intern. Med.* **138**, 424–429.

378. Helman, L.J., Cohen, P.S., Averbuch, S.D., Cooper, M.J., Keiser, H.R., and Israel, M.A. (1989) Neuropeptide Y expression distinguishes malignant from benign pheochromocytoma. *J. Clin. Oncol.* **7**, 1720–1725.

379. Jensen, J.C., Choyke, P.L., Rosenfeld, M., Pass, H.I., Keiser, H., White, B., Travis, W., and Linehan, W.M. (1991) A report of familial carotid body and multiple extra-adrenal pheochromocytomas. *J. Urol.* **145**, 1040–1042.

380. Jeong, H.S., Kim, S.W., Baek, K.J., Lee, H.S., Kwon, N.S., Kim, Y.-M., and Yun, H.-Y. (2002) Involvement of Ras in survival responsiveness to nitric oxide toxicity in pheochromocytoma cells. *J. Neurooncol.* **60**, 97–107.

381. Eisenhofer, G. (2003) Editorial: Biochemical diagnosis of pheochromocytoma—Is it time to switch to plasma-free metanephrines? *J. Clin. Endocrinol. Metab.* **88**, 550–552.

382. Sawka, A.M., Jaeschke, R., Singh, R.J., and Young, W.F., Jr. (2003) A comparison of biochemical tests for pheochromocytoma: Measurement of fractionated plasma metanephrines compared with the combination of 24-hour urinary metanephrines and catecholamines. *J. Clin. Endocrinol. Metab.* **88**, 553–558.

383. Machens, A., Brauchkhoff, M., Holzhausen, H.-J., Thanh, P.N., Lehnert, H., and Dralle, H. (2006) Codon-specific development of pheochromocytoma in multiple endocrine neoplasia type 2. *J. Clin. Endocrinol. Metab.* **90**, 3999–4003.

384. McWhinney, S.R., Boru, G., Binkley, P.K., Pęczkowska, M., Januszewicz, A.A., Neumann, H.P.H., and Eng, C. (2003) Intronic single nucleotide polymorphisms in the *RET* protooncogene are associated with a subset of apparently sporadic pheochromocytoma and may modulate age of onset. *J. Clin. Endocrinol. Metab.* **88**, 4911–4916.

385. lland, S.G., Atuk, N.O., and Walzak, M.P. (1969) Familial pheochromocytoma associated with cerebellar hemangioblastoma. A case history and review of the literature. *JAMA* **207**, 1709–1711.

386. Pfragner, R., Sadjak, A., and Walser, V. (1984) The effect of nerve growth factor (NGF) on the catecholamine contents of two human pheochromocytomas in tissue culture. *Exp. Pathol.* **26**, 21–31.

387. Sabban, E.L., and Kvetnanský, R. (2001) Stress-triggered activation of gene expression in catecholaminergic systems: dynamics of transcriptional events. *TRENDS Neurosci.* **24**, 91–98.

388. Sevastos, N., Theodossiades, G., Malaktari, S., and Archimandritis, A.J. (2005) Persistent neutrophilia as a preceding symptom of pheochromocytoma. *J. Clin. Endocrinol. Metab.* **90**, 2472–2473.

389. Lenders, J.W., Eisenhofer, G., Mannelli, M., and Pacak, K. (2005) Phaeochromocytoma. *Lancet* **366**, 665–675.

390. Takaya, K., Yoshimasa, T., Arai, H., Tamura, N., Miyamoto, Y., Itoh, H., and Nakao, K. (1996) The RET protooncogene in sporadic pheochromocytomas. *Intern. Med.* **35**, 449–452.

391. Vogel, T.W.A., Brouwers, F.M., Lubensky, I.A., Vortmeyer, A.O., Weil, R.J., Walther, M.M., Oldfield, E.H., Linehan, W.M., Pacak, K., and Zhuang, Z. (2005) Differential expression of erythropoietin and its receptor in von Hippel-Lindau-associated and multiple endocrine neoplasia type 2-associated pheochromocytomas. *J. Clin. Endocrinol. Metab.* **90**, 3747–3751.

392. Yon, L., Guillemot, J., Montero-Hadjadje, M., Grumolato, L., Leprince, J., Lefebvre, H., Contessse, V., Plouin, P.-F., Vaudry, H., and Anouar, Y. (2003) Identification of the secretogranin II-derived peptide EM66 in pheochromocytomas as a potential marker for discriminating benign *versus* malignant tumors. *J. Clin. Endocrinol. Metab.* **88**, 2579–2585.

393. Yan, C., Korade Mirnics, Z., Portugal, C.F., Liang, Y., Nylander, K.D., Rudzinski, M., Zaccaro, C., Saragovi, H.U., and Schor, N.F. (2005) Cholesterol biosynthesis and the pro-apoptotic effects of the p75 nerve growth factor receptor in PC12 pheochromocytoma cells. *Mol. Brain Res.* **139**, 225–234.

394. Blanes, A., Sanchez-Carrillo, J.J., and Diaz-Cano, S.J. (2006) Topographic molecular profile of pheochromocytomas: Role of somatic down-regulation of mismatch repair. *J. Clin. Endocrinol. Metab.* **91**, 1150–1158.

395. Arum, S.M., Dahia, P.L., Schneider, K., and Braverman, L.E. (2005) A *RET* mutation with decreased penetrance in the family of a patient with a "sporadic" pheochromocytoma. *Endocrine* **28**, 193–198.

396. Brouwers, F.M., Petricoin, E.F., 3rd, Ksinantova, L., Breza, J., Rajapakse, V., Ross, S., Johann, D., Mannelli, M., Shulkin, B.L., Kvetnansky, R., Eisenhofer, G., Walther, M.M., Hitt, B.A., Conrads, T.P., Veenstra, T.D., Mannion, D.P., Wall, M.R., Wolfe, G.M., Fusaro, V.A., Liotta, L.A., and Pacak, K. (2005) Low molecular weight proteomic information distinguishes metastatic from benign pheochromocytoma. *Endocr. Relat. Cancer* **12**, 263–272.

397. Fottner, C., Minnemann, T., Kalmbach, S., and Weber, M.M. (2006) Overexpression of the insulin-like growth factor 1 receptor in human pheochromocytomas. *J. Mol. Endocrinol.* **36,** 279–287.

398. Guillemot, J., Anouar, Y., Montero-Hadjadje, M., Grouzmann, E., Grumolato, L., Roshmaninho-Salgado, J., Turquier, V., Duparc, C., Defebvre, H., H., Plouin, P.-F., Klein, M., Muresan, M., Chow, B.K.C., Vaudry, H., and Yon, L. (2006) Circulating EM66 is a highly sensitive marker for the diagnosis and follow-up of pheochromocytoma. *Int. J. Cancer* **118,** 2003–2012.

399. Huynh, T.-T., Pacak, K., Brouwers, F.M., Abu-Asab, M.S., Worrell, R.A., Walther, M.M., Elkahloun, A.G., Goldstein, D.S., Cleary, S., and Eisenhofer, G. (2005) Different expression of catecholamine transporters in phaeochromocytomas from patients with von Hippel-Lindau syndrome and multiple endocrine neoplasia type 2. *Eur. J. Endocrinol.* **153,** 551–563.

400. Bernini, G.P., Moretti, A., Mannelli, M., Ercolino, T., Bardini, M., Caramella, D., Taurino, C., and Salvetti, A. (2005) Unique association of non-functioning pheochromocytoma, ganglioneuroma, adrenal cortical adenoma, hepatic and vertebral hemangiomas in a patient with a new intronic variant in the *VHL* gene. *J. Endocrinol Invest.* **28,** 1032–1037.

400A. Bernini, G., Franzoni, F., Galetta, F., Moretti, A., Taurino, C., Bardini, M., Santoro, G., Ghiadoni, L., Bernini, M., and Salvetti, A. (2006) Carotid vascular remodeling in patients with pheochromocytoma. *J. Clin. Endocrinol. Metab.* **91,** 1754–1760.

401. Takahashi, M., Yang, X.J., McWhinney, S., Sano, N., Eng, C., Kagawa, S., Teh, B.T., and Kanayama, H.-O. (2005) cDNA microarray analysis assists in diagnosis of malignant intrarenal pheochromocytoma originally masquerading as a renal cell carcinoma. *Am. J. Med. Genet.* **42,** e48.

401A. Takahashi, K., Iida, K., Okimura, Y., Takahashi, Y., Naito, J., Nishikawa, S., Kadowaki, S., Iguchi, G., Kaji, H., and Chihara, K. (2005) A novel mutation in the von Hippel-Lindau tumor suppressor gene identified in a Japanese family with pheochromocytoma and hepatic hemangioma. *Intern. Med.* **45,** 265–269.

402. Tatekawa, Y., Muraji, T., Nishijima, E., Yoshida, M., and Tsugawa, C. (2006) Composite pheochromocytoma associated with adrenal neuroblastoma in an infant: a case report. *J. Pediatr. Surg.* **41,** 443–445.

403. van Nederveen, F.H., Perren, A., Dannenberg, H., Petri, B.J., Dinjens, W.N., Komminoth, P., and de Krijger, R.R. (2006) *PTEN* gene loss, but not mutation, in benign and malignant phaeochromocytomas. *J. Pathol.* **209,** 274–280.

404. Wakasugi, S., Kinouchi, T., Taniguchi, H., Yokoyama, K., Fukuchi, K., Noguchi, A., Takeshita, M., and Hashizume, T. (2006) A case of malignant pheochromocytoma with early intense uptake and immediate rapid washout of 99mTc-tetrofosmin characterizing the overexpression of anti-apoptotic Bcl-2. *Ann. Nucl. Med.* **20,** 325–328.

405. Statuto, M., Ennas, M.G., Zamboni, G., Bonetti, F., Pea, M., Bernardello, F., Pozzi, A., Rusnati, M., Gualandris, A., and Presta, M. (1993) Basic fibroblast growth factor in human pheochromocytoma: a biochemical and immunohistochemical study. *Int. J. Cancer* **53,** 5–10.

406. Illiopoulos, O., Levy, A.P., Jiang, C., Kaelin, W.G., Jr., and Goldberg, M.A. (1996) Negative regulation of hypoxia-inducible genes by the von Hippel-Lindau protein. *Proc. Natl. Acad. Sci. USA* **93,** 10595–10599.

407. Iwai, K., Yamanaka, K., Kamura, T., Minato, N., Conaway, R.C., Conaway, J.W., Klausner, R.D., and Pause, A. (1999) Identification of the von Hippel-Lindau tumor-suppressor protein as part of an active E3 ubiquitin ligase complex. *Proc. Natl. Acad. Sci. USA* **96,** 12436–12441.

408. Kamura, T., Koepp, D.M., Conrad, M.N., Skowyra, D., Moreland, R.J., Iliopoulos, O., Lane, W.S., Kaelin, W.G., Jr., Elledge, S.J., Conaway, R.C., Harper, J.W., and Conaway, J.W. (1999) Rbx1, a component of the VHL tumor suppressor complex and SCF ubiquitin ligase. *Science* **284,** 657–661.

409. Ritter, M.M., Frilling, A., Crossey, P.A., Hoppner, W., Maher, E.R., Mulligan, L., Ponder, B.A., and Engelhardt, D. (1996) Isolated familial phaechromocytoma as a variant of von Hippel-Lindau syndrome. *J. Clin. Endocrinol. Metab.* **81,** 1035–1037.

410. Frenzel, S., Apel, T.W., Heidemann, P.H., Zerres, K., Neumann, H.P., and Dorr, H.G. (2001) Phaeochromocytoma associated with a de novo VHL mutation as form fruste of von Hippel-Lindau disease. *Eur. J. Pediatr.* **160,** 421–424.

411. Illiopoulos, O. (2001) von Hippel-Lindau disease: genetic and clinical observations. In *Genetic Disorders of Endocrine Neoplasia. Frontiers of Hormone Research Series* (Dahia, P.L.M., and Eng, C., eds.) Karger, Basel (Switzerland), pp. 131–166.

412. Cascón, A., Ruiz-Llorente, S., Fraga, M.F., Letón, R., Tellería, D., Sastre, J., Diez, J.J., Martínez-Diaz-Guerra, G., Diaz Perez, J.A., Benítez, J., Esteller, M., and Robledo, M. (2004) Genetic and epigenetic profile of sporadic pheochromocytomas. *J. Med. Genet.* **41,** e30.

413. Takeda, S., Minami, M., Inoue, Y., and Matsuda, H. (2005) Synchronous mediastinal ganglioneuroma and retroperitoneal pheochromocytoma. *Ann. Thorac. Surg.* **80,** 1525–1527.

414. van Houtum, W.H., Corssmit, E.P., Douwes Dekker, P.B., Jansen, J.C., van der Mey, A.G., Brocker-Vriends, A.H., Taschner, P.E., Losekoot, M., Frolich, M., Stokkel, M.P., Cornelisse, C.J., and Romijn, J.A. (2005) Increased prevalence of catecholamine excess and phaeochormocytomas in a well-defined Dutch population with *SDHD*-linked head and neck paragangliomas. *Eur. J. Endocrinol.* **152,** 87–94.

415. Benn, D.E., Gimenez-Roqueplo, A.-P., Reilly, J.R., Bertherat, J., Burgess, J., Byth, K., Croxson, M., Dahia, P.L.M., Elston, M., Gimm, O., Henley, D., Herman, P., Murday, V., Niccoili-Sire, P., Pasieka, J.L., Rohmer, V., Tucker, K., Jeunemaitre, X., Marsh, D.J., Plouin, P.-F., and Robinson, B.G. for the International SDH Consortium. (2006) Clinical presentation and penetrance of pheochromocytoma/paraganglioma syndromes. *J. Clin. Endocrinol. Metab.* **91,** 827–836.

416. Cascón, A., Montero-Conde, C., Ruiz-Llorente, S., Mercadillo, F., Letón, R., Rodríguez-Antona, C., Martínez-Delgado, B., Delgado, M., Díez, A., Rovira, A., Díaz, J.Á., and Robledo, M. (2006) Gross *SDHB* deletions in patients with paraganglioma detected by multiplex PCR: A possible hot spot? *Genes Chromosomes Cancer* **45,** 213–219.

416A. Rodien, P., Jeunemaitre, X., Dumont, C., Beldjord, C., and Plouin, P.F. (1997) Genetic alterations in the RET proto-oncogene in familial and sporadic pheochromocytomas. *Horm. Res* **47,** 263–268.

417. Ackrell, B.A. (2002) Cytopathies involving mitochondrial complex II. *Mol. Aspects Med.* **23**, 369–384.

418. Beck, O., Fassbender, W.J., Beyer, P., Kriener, S., Neumann, H.P., Klingebiel, T., and Lehrnbecher, T. (2004) Pheochromocytoma in childhood: implication for further diagnostic procedures. *Acta Paediatr.* **93**, 1630–1634.

419. Mhatre, A.N., Li, Y., Feng, L., Gasperin, A., and Lalwani, A.K. (2004) *SDHB, SDHC,* and *SDHD* mutation screen in sporadic and familial head and neck paraganglioma. *Clin. Genet.* **66**, 461–466.

420. Bertherat, J., and Gimenez-Roqueplo, A.-P. (2005) New insights in the genetics of adrenocortical tumors, pheochromocytomas and paragangliomas. *Horm. Metab. Res.* **37**, 384–390.

421. Braun, S., Riemann, K., Kupka, S., Leistenschneider, P., Sotlar, K., Schmid, H., and Blin, N. (2005) Active succinate dehydrogenase (SDH) and lack of *SDHD* mutations in sporadic paragangliomas. *AnticancerRes.* **25**, 2809–2814.

422. Dannenberg, H., van Nederveen, F.H., Abbou, M., Verhofstad, A.A., Komminoth, P., de Krijger, R.R., and Dinjens, W.N.M. (2005) Clinical characteristics of pheochromocytoma patients with germline mutations in *SDHD. J. Clin. Oncol.* **23**, 1894–1901.

423. Liapis, C.D., Bellos, J.K., Halapas, A., Lembessis, P., Koutsilieris, M., and Kostakis, A. (2005) Carotid body paraganglioma and *SDHD* mutation in a Greek family. *Anticancer Res.* **25**, 2249–2452.

424. Brouwers, F.M., Eisenhofer, G., Tao, J.J., Kant, J.A., Adams, K.T., Linehan, W.M., and Pacak, K. (2006) High frequency of *SDHB* germline mutations in patients with malignant catecholamine-producing paragangliomas: Implications for genetic testing. *J. Clin. Endocrinol. Metab.* **91**, 4505–4509.

425. Dionne, J.M., Wu, J.K., Heran, M., Murphy, J.J., Jevon, G., and White, C.T. (2006) Malignant hypertension, polycythemia, and paragangliomas. *J. Pediatr.* **148**, 540–545.

426. Elston, M.S., Benn, D., Robinson, B.G., and Conaglen, J.V. (2006) An apparently sporadic paraganglioma with an *SDHB* gene germline mutation presenting at age 68 years. *Int. Med. J.* **36**, 129–131.

427. Jeffery, J., Devendra, D., Farrugia, J., Gardner, D., Murphy, M.J., Williams, R., Ayling, R.M., and Wilkin, T.J. (2006) Increased urinary dopamine excretion in association with bilateral carotid body tumours—clinical, biochemical and genetic findings. *Ann. Clin. Biochem.* **43**, 156–160.

428. Keith, J., Lownie, S., and Ang, L.-C. (2006) Co-existence of paraganglioma and myxopapillary ependymoma of the cauda equina. *Acta. Neuropathol.* **111**, 617–618.

429. Mithani, S.K., Marohn, M.R., Freischlag, J.A., Dackiw, A.P., and Zeiger, M.A. (2006) Laparoscopic resection of a paraganglioma of the organ of Zuckerkandl in a patient with a carotid body tumor. *Am. Surg.* **72**, 55–59.

430. Ogawa, K., Shiga, K., Saijo, S., Ogawa, T., Kimura, N., and Horii, A. (2006) A novel G106D alteration of the *SDHD* gene in a pedigree with familial paraganglioma. *Am. J. Med. Genet.* **140A**, 2441–2446.

431. Perry, C.G., Young, W.F., Jr., McWhinney, S.R., Bei, T., Stergiopoulos, S., Knudson, R.A., Ketterling, R.P., Eng, C., Stratakis, C.A., and Carney, J.A. (2006) Functioning paraganglioma and gastrointestinal stromal tumor of the jejunum in three women. Syndrome or coincidence? *Am. J. Surg. Pathol.* **30**, 42–49.

432. Plaza, J.A., Wakely, P.E., Jr., Moran, C., Fletcher, C.D.M., and Suster, S. (2006) Sclerosing paraganglioma. Report of 19 cases of an unusual variant of neuroendocrine tumor that may be mistaken for an aggressive malignant neoplasm. *Am. J. Surg. Pathol.* **30**, 7–12.

433. Schiavi, F., Boedeker, C., Bausch, B., Peçzkowska, M., Fuentes Gomez, C., Strassburg, T., Pawlu, C., Buchta, M., Salzmann, M., Hoffmann, M.M., Berlis, A., Brink, I., Cybulla, M., Muresan, M., Walter, M.A., Forrer, F., Välimäki., M., Kawecki, A., Szutkowski, Z., Schipper, J., Walz, M.K., Pigny, P., Bauters, C., Willet-Brozick, J.E., Baysal, B.E., Januszewicz, A., Eng, C., Opocher, G., and Neumann, H.P.H. for the European-American Paraganglioma Study Group. (2005) Predictors of prevalence of paraganglioma syndrome associated with mutations of the *SDHC* gene. *JAMA* **294**, 2057–2063.

434. Muller, U., Troidl, C., and Niemann, S. (2005) *SDHC* mutations in hereditary paraganglioma/pheochromocytoma. *Fam. Cancer* **4**, 9–12.

435. Douwes Dekker, P.B., Kuipers-Dijkshoorn, N.J., Baelde, H.J., van der Mey, A.G.L., Hogendoorn, P.C.W., and Cornelisse, C.J. (2007) Basic fibroblast growth factor and fibroblastic growth factor receptor-1 may contribute to head and neck paraganglioma development by an autocrine or paracrine mechanism. *Hum. Pathol.* **38**, 79–85.

436. Bayley, J.-P., Devilee, P., and Taschner, P.E.M. (2005) The SDH mutation database: an online resource for succinate dehydrogenase sequence variants involved in pheochromocytoma, paraganglioma and mitochondrial complex II deficiency. *BMC Med. Genet.* **6**, 39–44.

437. Yoon, Y.-S., Byun, H.-O., Cho, H., Kim, B.-K., and Yoon, G. (2003) Complex II defect via down-regulation of iron-sulfur subunit induces mitochondrial dysfunction and cell cycle delay in iron chelation-induced senescence-associated growth arrest. *J. Biol. Chem.* **278**, 51577–51586.

438. Hensen, E.F., Jordanova, E.S., van Minderhout, I.J., Hogendoorn, P.C.W., Taschner, P.E.M., van der Mey, A.G.L., Devilee, P., and Cornelisse, C.J. (2004) Somatic loss of maternal chromosome 11 causes parent-of-origin-dependent inheritance in *SDHD*-linked paraganglioma and phaeochromocytoma families. *Oncogene* **23**, 4076–4083.

439. McDonnell, C.M., Benn, D.E., Marsh, D.J., Robinson, B.G., and Zacharin, M.R. (2004) K40E: a novel succinate dehydrogenase (SDH)B mutation causing familial phaeochromocytoma and paraganglioma. *Clin. Endocrinol. (Oxford)* **61**, 510–514.

440. Gottlieb, E., and Tomlinson, I.P. (2005) Mitochondrial tumour suppressors: a genetic and biochemical update. *Nat. Rev. Cancer* **5**, 857–866.

441. Bayley, J.-P., van Minderhout, I., Weiss, M.M., Jansen, J.C., Oomen, P.H.N. Menko, F.H., Pasini, B., Ferrando, B., Wong, N., Alpert, L.C., Williams, R., Blair, E., Devilee, P., and Taschner, P.E.M. (2006) Mutation analysis of SDHB and SDHC: novel germline mutations in sporadic head and neck paraganglioma and familial paraganglioma and/or pheochromocytoma. *BMC Med. Genet.* **7**, 1–10.

442. Bolland, M., Benn, D., Croxson, M., McCall, J., Shaw, J.F., Baillie, T., and Robinson, B. (2006) Gastrointestinal stromal

tumour in succinate dehydrogenase subunit B mutation-associated familial phaeochromocytoma/paraganglioma. *ANZ J. Surg.* **76,** 763–764.

443. Pollard, P.J., El-Bahrawy, M., Poulsom, R., Elia, G., Killick, P., Kelly, G., Hunt, T., Jeffery, R., Seedhar, P., Barwell, J., Latif, F., Gleeson, M.J., Hodgson, S.V., Stamp, G.W., Tomlinson, I.P.M., and Maher, E.R. (2006) Expression of HIF-1α, HIF-2α (EPAS1), and their target genes in paraganglioma and phaeochromocytoma with *VHL* and *SDH* mutations. *J. Clin. Endocrinol. Metab.* **91,** 4593–4598.

444. Simi, L., Sestini, R., Ferruzzi, P., Gaglianò, M.S., Gensini, F., Mascalchi, M., Guerrini, L., Pratesi, C., Pinzani, P., Nesi, G., Ercolino, T., Genuardi, M., and Mannelli, M. (2006) Phenotype variability of neural crest derived tumours in six Italian families segregating the same founder *SDHD* mutation Q109X. *Am. J. Med. Genet..* **42,** e52.

445. Pfragner, R., and Walser, V. (1980) Long-term tissue culture of pheochromocytomas. *Exp. Pathol.(Jena)* **18,** 423–429.

446. Evers, B.M., Rady, P.L., Stephen, K., Tyring, K., Sanchez, R.I., Rajamaran, S., Townsend, M., Jr., and Thompson, J.C. (1992) Amplification of the HER-2/neu proto-oncogene in human endocrine tumors. *Surgery* **112,** 211–218.

447. Edström Elder, E., Xu, D., Hoog, A., Enberg, U., Hou, M., Pisa, P., Gruber, A., Larsson, C., and Backdahl, M. (2003) Ki-67 and hTERT expression can aid in the distinction between malignant and benign pheochromocytoma and paraganglioma. *Mod. Pathol.* **16,** 246–255.

448. Leckschat, S., Ream-Robinson, D., and Scheffler, I.E. (1993) The gene for the iron sulfur protein of succinate dehydrogenase (SDH-IP) maps to human chromosome 1p35–36.1. *Somat. Cell Mol. Genet.* **19,** 505–511.

449. Matias-Guiu, X., Colomer, A., Mato, E., Cuatrecasas, M., Komminoth, P., Prat, J., and Wolfe, H. (1995) Expression of the ret proto-oncogene in phaeochromocytoma. An in situ hybridization and Northern blot study. *J. Pathol.* **176,** 63–68.

450. Le Hir, H., Charlet-Berguerand, N., de Franciscis, V., and Thermes, C. (2002) 5'-end RET splicing: absence of variants in normal tissues and intron retention in pheochromocytomas. *Oncology* **63,** 84–91.

451. Patetsios, P., Gable, D.R., Garrett, W.V., Lamont, J.P., Kuhn, J.A., Shutze, W.P., Kourlis, H., Grimsley, B., Pearl, G.J., Smith, B.L., Talkington, C.M., and Thompson, J.E. (2002) Management of carotid body paragangliomas and review of a 30-year experience. *Ann. Vasc. Surg.* **16,** 331–338.

452. Garcia-Escudero, A., de Miguel-Rodriguez, M., Moreno-Fernandez, A., Navarro-Bustos, G., Galera-Ruiz, H., and Galera-Davidson, H. (2001) Prognostic value of DNA flow cytometry in sympathoadrenal paragangliomas. *Anal. Quant. Cytol.. Histol.* **23,** 238–244.

453. Gonzalez-Campora, R., Diaz Cano, S., Lerma-Puertas, E., Rios Martin, J.J., Salguero Villadiego, M., Villar Rodriguez, J.L., Bibbo, M., and Davidson, H.G. (1993) Paragangliomas. Static cytometric studies of nuclear DNA patterns. *Cancer* **71,** 820–824.

454. Salmenkivi, K., Haglund, C., Arola, J., and Heikkilä, P. (2001) Increased expression of tenascin in pheochromocytomas correlates with malignancy. *Am. J. Surg. Pathol.* **25,** 1419–1423.

454B. Salmenkivi, K., Arola, J., Voutilainen, R., Ilvesmaki, V., Haglund, C., Kahri, A.I., Heikkilä, P., and Liu, J. (2001) Inhibin/activin betaB-subunit expression in pheochromocytomas favors benign diagnosis. *J. Clin. Endocrinol. Metab.* **86,** 2231–2235.

455. Tormey, W.P., Fitzgerald, R.J., Thomas, G., Kay, E.W., and Leader, M.B. (2000) Catecholamine secretion and ploidy in phaeochromocytoma. *Int. J. Clin. Pract.* **54,** 520–523.

456. Nativ, O., Grant, C.S., Sheps, S.G., O'Fallon, J.R., Farrow, G.M., van Heerden, J.A., and Lieber, M.M. (1992) The clinical significance of nuclear DNA ploidy pattern in 184 patients with pheochromocytoma. *Cancer* **69,** 2683–2687.

457. Lin, S.R., Lee, Y.J., and Tsai, J.H. (1994) Mutations of the p53 gene in human functional adrenal neoplasms. *J. Clin. Endocrinol. Metab.* **78,** 483–491.

458. Casanova, S., Rosenberg-Bourgin, M., Farkas, D., Calmettes, C., Feingold, N., Heshmati, H.M., Cohen, R., Conte-Devoix, B., Guillausseau, P.J., Houdent, C., et al. (1993) Phaeochromocytoma in multiple endocrine neoplasia type 2 A: survey of 100 cases. *Clin. Endocrinol. (Oxford)* **38,** 531–537.

459. Moreno, A.M., Castilla-Guerra, L., Martinez-Torres, M.C., Torres-Olivera, F., Fernandez, E., and Galera-Davidson, H. (1999) Expression of neuropeptides and other neuroendocrine markers in human phaeochromocytoma. *Neuropeptides* **33,** 159–163.

460. Brown, H.M., Komorowski, R.A., Wilson, S.D., Demeure, M.J., and Zhu, Y.r. (1999) Predicting metastasis of pheochromocytomas using DNA flow cytometry and immunohistochemical markers of cell proliferation: a positive correlation between MIB-1 staining and malignant tumor behavior. *Cancer* **86,** 1583–1589.

460A. van der Harst, E., Bruining, H.A., Jaap Bonjer, H., van der Ham, F., Dinjens, W.N., Lamberts, S.W., de Herder, W.W., Koper, J.W., Stijnen, T., Proye, C., Lecomte-Houcke, M., Bosman, F.T., and de Krijger, R.R. (2000) Proliferative index in phaeochromocytomas: does it predict the occurrence of metastases? *J. Pathol.* **191,** 175–180.

461. Dannenberg, H., de Krijger, R.R., van der Harst, E., Abbou, M., IJzendoorn, Y., Komminoth, P., and Dinjens, W.N. (2003) Von Hippel-Lindau gene alterations in sporadic benign and malignant pheochromocytomas. *Int. J. Cancer* **105,** 190–195.

462. de Krijger, R.R., van der Harst, E., van der Ham, F., Stijnen, T., Dinjens, W.N. Koper, J.W., Bruining, H.A., Lamberts, S.W., and Bosman, F.T. (1999) Prognostic value of p53, bcl-2, and c-erbB-2 protein expression in phaeochromocytomas. *J. Pathol.* **188,** 51–55.

463. Gupta, D., Shidham, V., Holden, J., and Layfield, L. (2000) Prognostic value of immunohistochemical expression of topoisomerase alpha II, MIB-2, p53, E-cadherin, retinoblastoma gene protein product, and HER-2/neu in adrenal and extra-adrenal pheochromocytomas. *Appl. Immunohistochem. Mol. Morphol.* **8,** 267–274.

464. John, H., Ziegler, W.H., Hauri, D., and Jaeger, P. (1999) Pheochromocytomas: can malignant potential be predicted? *Urology* **53,** 679–683.

465. Kimura, N., Watanabe, T., Fukase, M., Wakita, A., Noshiro, T., and Kimura, I. (2002) Neurofibromin and *NF1* gene analysis in composite pheochromocytoma and tumors associated with von Recklinghausen's disease. *Mod. Pathol.* **15,** 183–188.

466. Satake, H., Inoue, K., Kamada, M., Watanabe, H., Furihata, M., and Shuin, T. (2001) Malignant composite pheochromocytoma of the adrenal gland in a patient with von Recklinghausen's disease. *J. Urol.* **165,** 1199–1200.

467. Yoshimoto, T., Naruse, M., Zeng, Z., Nishikawa, T., Kasajima, T., Toma, H., Yamamori, S., Matsumoto, H., Tanabe, A., Naruse, K., and Demura, H. (1998) The relatively high frequency of p53 gene mutations in multiple and malignant phaeochromocytomas. *J. Endocrinol.* **159,** 247–255.

468. Pang, L.C., and Tsao, K.C. (1993) Flow cytometric DNA analysis for the determination of malignant potential in adrenal and extra-adrenal pheochromocytomas or paragangliomas. *Arch. Pathol. Lab. Med.* **117,** 1142–1147.

469. McNichol, A.M. (2001) Differential diagnosis of pheochromocytomas and paragangliomas. *Endocr. Pathol.* **12,** 407–415.

470. Lam, K.Y., Lo, C.Y., Wat, N.M., Luk, J.M., and Lam, K.S. (2001) The clinicopathological features and importance of p53, Rb, and mdm2 expression in phaeochromocytomas and paragangliomas. *J. Clin. Pathol.* **54,** 443–448.

471. Schroder, H.D., and Johannsen, L. (1986) Demonstration of S-100 protein in sustentacular cells of phaeochromocytomas and paragangliomas. *Histopathology* **10,** 1023–1033.

472. Held, E.L., Gal, A.A., DeRose, P.B., and Cohen, C. (1997) Image cytometric nuclear DNA quantitation of paragangliomas in tissue seactions. Prognostic significance. *Anal. Quant. Cytol. Histol.* **19,** 501–506.

473. Reichardt, P., Apel, T.W., Domula, M., Trös, R.-B., Krause, I., Bierbach, U., Neumann, H.P.H., and Kiess, W. (2002) Recurrent polytopic chromaffin paragangliomas in a 9-year-old boy resulting from a novel germline mjtation in the von Hippel-Lindau gene. *J. Pediatr. Hematol. Oncol.* **24,** 145–148.

474. Hirawake, H., Taniwaki, M., Tamura, A., Amino, H., Tomitsuka, E., and Kita, K. (1999) Characterization of the human SDHD gene encoding the small subunit of cytochrome b (cybS) in mitochondrial succinate-ubiquinone oxydoreductase. *Biochim. Biophys. Acta.* **1412,** 295–300.

475. van der Harst, E., de Krijger, R.R., Bruining, H.A., Lamberts, S.W., Jaap Bonjer, H., Dinjens, W.N., Proye, C., Koper, J.W., Bosman, F.T., Roth, J., Heitz, P.U., and Komminoth, P. (1998) Prognostic value of RET proto-oncogene point mutations in malignant and benign, sporadic phaeochromocytomas. *Int. J. Cancer* **79,** 537–540.

476. Sanso, G., Rudaz, M.C., Levin, G., and Barotini, M. (2004) Familial isolated pheochromocytoma presenting a new mutation in the von Hippel-Lindau gene. *Am. J. Hypertens.* **17,** 1107–1011.

477. Pacak, K., Ilias, I., Adams, K.T., and Eisenhofer, G. (2005) Biochemical diagnosis, localization and management of pheochromocytoma: focus on multiple endocrine neoplasia type 2 in relation to other hereditary syndromes and sporadic forms of the tumour. *J. Intern. Med.* **257,** 60–68.

478. Opocher, G., Conton, P., Schiavi, F., Macino, B., and Mantero, F. (2005) Pheochromocytoma in von Hippel-Lindau disease and neurofibromatosis type 1.*Fam. Cancer* **4,** 13–16.

479. Manger, W.M., and Eisenhofer, G. (2004) Pheochromocytoma: diagnosis and management update. *Curr. Hypertens. Rep.* **6,** 477–484.

480. Leube, B., Huber, R., Goecke, T.O., Sandman, W., and Royer-Pokora, B. (2004) *SDHD* mutation analysis in seven German patients with sporadic carotid body paraganglioma: one novel mutation, no Dutch founder mutation and further evidence that G12S is a polymorphism. *Clin. Genet.* **65,** 61–63.

481. Le Hir, H., Colucci-D'Amato, L.G., Charlet-Berguerand, N., Plouin, P.F., Bertagna, X., de Franciscis, V., and Thermes, C. (2000) High levels of tyrosine phosphorylated proto-ret in sporadic pheochromocytoma. *Cancer Res.* **60,** 1365–1370.

482. Gimm, O. (2005) Pheochromocytoma-associated syndromes: genes, proteins and functions of RET, VHL and SDHx. *Fam. Cancer* **4,** 17–23.

483. Grumolato, L., Elkahloun, A.G., Ghzili, H., Alexandre, D., Coulouarn, C., Yon, L., Salier, J.P., Eiden, L.E., Fournier, A., Vaudry, H., and Anouar, Y. (2003) Microarray and suppression subtractive hybridization analyses of gene expression in pheochromocytoma cells reveal pleiotropoic effects of pituitary adenylate cyclase-activating polypeptide on cell proliferation, survival, and adhesion. *Endocrinology* **144,** 2368–2379.

484. Gelato, M.C., and Vassalotti, J. (1990) Insulin-like growth factor-II: possible local growth factor in pheochromocytoma. *J. Clin. Endocrinol. Metab.***71,** 1168–1174.

485. Eisenhofer, G. (2001) Free or total metanephrines for diagnosis of pheochromocytoma: what is the difference? *Clin. Chem.* **47,** 988–989.

486. Eisenhofer, G., Keiser, H., Friberg, P., Mezey, E., Huynh, T.T., Hiremagalur, B., Ellingson, T., Duddempudi, S., Eijsbouts, A., and Lenders, J.W. (1998) Plasma metanephrines are markers of pheochromocytoma produced by catechol-O-methyltransferase within tumors. *J. Clin. Endocrinol. Metab.* **83,** 2175–2185.

487. Eisenhoer, G., Goldstein, D.S., Kopin, I.J., and Crout, J.R. (2003) Pheochromocytoma: rediscovery of catecholamine-metabolizing tumor. *Endocr. Pathol.* **14,** 193–212.

488. Isobe, K., Nakai, T., Yashiro, T., Nanmoku, T., Yukimasa, N., Ikezawa, T., Suzuki, E., Takekoshi, K., and Nomura, F. (2000) Enhanced expression of mRNA coding for the adrenaline-synthesizing enzyme phenylethanolamine-N-methyl transferase in adrenaline-secreting pheochromocytomas. *J. Urol.* **163,** 357–362.

489. Isobe, K., Nakai, T., Yukimasa, N., Nanmoku, T., Takekoshi, K., and Nomura, F. (1998) Expression of mRNA coding for four catecholamine-synthesizing enzymes in human adrenal pheochromocytomas. *Eur. J. Endocrinol.* **138,** 383–387.

490. Dammann, R., Schagdarsurengin, U., Seidel, C., Trumpler, C., Hoang-Vu, C., Gimm, O., Dralle, H., Pfeifer, G.P., and Brauckhoff, M. (2005) Frequent promoter methylation of tumor-related genes in sporadic and men2-associated pheochromocytomas. *Exp. Clin. Endocrinol. Diabetes* **113,** 1–7.

491. August, C., August, K., Schroeder, S., Bahn, H., Hinze, R., Baba, H.A., Kersting, C., and Buerger, H. (2004) CGH and CD44/MIB-1 immunohistochemistry are helpful to distinguish metastasized form nonmetastasized sporadic pheochromocytomas. *Mod. Pathol.* **17,** 1119–1128.

492. Suzuki, T., Watanabe, K., Sugino, T., Tanigawa, T., and Satoh, S. (1992) Immunocytochemical demonstration of IGF-II immunoreactivity in human phaeochromocytoma and extra-adrenal abdominal paraganglioma. *J. Pathol.* **167,** 199–203.

493. Rosai, J. (2004) *Rosai and Ackerman's Surgical Pathology, Ninth ed., Vol. 1.* Mosby, Edinburgh, London, New York, Oxford, Philadelphia, St. Louis, Sydney, Toronto, p. 1137.

494. Koch, C.A., Pacak, K., and Chrousos, G.P. (2002) Genetics of endocrine disease. The molecular pathogenesis of hereditary

and sporadic adrenocortical and adrenomedullary tumors. **87,** 5367–5384.

495. Timmers, H.J.L.M., Kozupa, A., Chen, C.C., Carrasquillo, J.A., Ling, A., Eisenhofer, G., Adams, K.T., Solis, D., Lenders, J.W.M., and Pacak, K. (2007) Superiority of fluorodeoxyglucose positron emission tomography to other functional imaging techniques in the evaluation of metastatic *SDHB*-associated pheochromocytoma and paraganglioma. *J. Clin. Oncol.* **25,** 2262–2269.

496. Kino, M., Suzuki, H., Naya, Y., Komiya, A., Imamoto, T., Ichikawa, T., Tatsuno, I., Ishida, H., Shido, T., and Seki, N. (2007) Comparative genomic hybridization reveals frequent losses of 1p and 3q in benign pheochromocytomas of Japanese patients. *Cancer Genet. Cytogenet.* **28,** 169–172.

497. Carrasco, C.A., Rojas-Salazar, D., Chiorino, R., Venega, J.C., and Wohlik, N. (2006) Melanotic nonpsammomatous trigeminal schwannoma as the first manifestation of Carney complex. *Neurosurgery* **59,** E1334–E1339.

498. Mosunjac, M.B., Johnston, E.I., and Mosunjac, M.I. (2007) Fine-needle aspiration cytologic diagnosis of metastatic melanotic schwannoma: familial case of a mother and daughter with Carney's complex and literature review. *Diagn. Cytopathol.* **35,** 130–134.

499. Baysal, B.E., Lawrence, E.C., Willett-Brozick, J.E., and Ferrell, R.E. (2004) Sequence analysis of succinate dehydrogenase subunit A gene (*SDHA*) for paraganglioma tumor susceptibility. *Am. J. Hum. Genet.* **75 (Suppl.) 91,** 384.

500. Matyakhina, L., Bei, T.A., McWhinney, S.R., Pasini, B., Cameron, S., Gunawan, B., Stergiopoulos, S.G., Boikos, S., Muchow, M., Dutra, A., Pak, E., Campo, E., Cid, M.C., Gomez, F., Gaillard, R.C., Assie, G., Füzesi, L., Baysal, B.E., Eng, C., Carney, J.A., and Stratakis, C.A. (2007) Genetics of Carney triad: recurrent losses at chromosome 1 but lack of germline mutations in genes associated with paragangliomas and gastrointestinal stromal tumors. *J. Clin. Endocrinol. Metab.* **92,** 2938–2943.

501. Bausch, B., Borozdin, W., Mautner, V.F., Hoffmann, M.M., Boehm, D., Robledo, M., Cascon, A., Harenberg, T., Schiavi, F., Pawlu, C., Peczkowska, M., Letizia, C., Calvieri, S., Arnaldi, G., Klingenberg-Noftz, R.D., Reisch, N., Fassina, A., Brunaud, L., Walter, M.A., Mannelli, M., MacGregor, G., Fausto Palazzo, F., Barontini, M., Walz, M.K., Kremens, B., Brabant, G., Pfaffle, R., Koschker, A.-C., Lohoefner, F., Mohaupt, M., Gimm, O., Jarzab, B., McWhinney, S.R., Opocher, G., Januszewicz, A., Kohlhase, J., Eng, C., and Neumann, H.P.H., for the European-American Phaeochromocytoma Registry and Study Group (2007) Germline *NF1* mutational spectra and loss-of-heterozygosity analyses in patients with pheochromocytoma and neurofibromatosis type 1. *J. Clin. Endocrinol. Metab.* **92,** 2784–2792.

501A. Pacak, K. (2007) Approach to the patient. Preoperative management of the pheochromocytoma patient. *J. Clin. Endocrinol. Metab.* **92,** 4069–4079.

6
Atypical Teratoid/Rhabdoid Tumors of the Central Nervous System

Summary   Chapter 6 deals with atypical teratoid/rhabdoid tumors (AT/RT) of the central nervous system. Almost all of these tumors have monosomy 22 (-22) or a deletion of this chromosome. The gene *hSNF5/INI1* involved in AT/RT maps to chromosome band 22q11.2, and it is consistently mutated in these tumors. The relationship of the *INI1* gene, as a functional center of the ATP-dependent chromatin-remodeling complex SWI/SNF, to other molecular events probably playing a role in AT/RT development and biology is presented. These events may include cyclin D1, a factor critical in cell cycle control, and p16^{INK4A} and retinoblastoma. The nature and genetics of rhabdoid tumors are discussed, including those of composite tumors. Chapter 6 also covers familial AT/RT, rhabdoid tumor predisposition syndrome, and tumors (other than AT/RT) with inactivation of the *INI1* gene.

Keywords   AT/RT · *INI1* gene · monosomy 22 (-22) · nature of rhabdoid tumors.

Introduction

Atypical teratoid/rhabdoid tumors (AT/RT) of the central nervous system (CNS) were described as an entity about two decades ago (1), and the actual term for and characteristics of these tumors were subsequently introduced and defined (2–4). The term AT/RT theoretically unified the concepts of CNS malignant rhabdoid tumors (MRT) and of atypical teratoid tumors of infancy. Histologically, AT/RT included a unique combination of peripheral neuroectodermal tumor (PNET)-like epithelial and mesenchymal features in addition to rhabdoid cells. The AT/RT concept quickly gained popularity and was included in the World Health Organization (WHO) classification of CNS tumors in 2002 (4A) as a malignant embryonal CNS tumor of children that was composed of rhabdoid cells, with or without cells resembling a typical PNET, primitive neuroepithelial cells, epithelial tissue, and neoplastic mesenchyme (Figure 6.1) (4A). Reports on the histologic background, clinical aspects, immunochemistry, and differential diagnosis of AT/RT have been published previously (2–11, 12, 12A). Further details and information on AT/RT may be found in Tables 6.1 and 6.2.

AT/RT occur predominantly in infants and children, some of whom also may have a primary renal MRT (13–15), but the majority of CNS AT/RT are not associated with a renal neoplasm. More than 90% of AT/RT occur in children under 5 years of age (3), with the majority manifesting in infants less than 2 years of age (3). There is a slight male preponderance. About 50% of AT/RT arise in the posterior fossa, with a predilection for the cerebello-pontine angle. Approximately one third of AT/RT have cerebrospinal fluid spread at presentation. The prognosis is grim: the majority of the patients die in the first year after diagnosis, in contrast to children with medulloblastoma who currently have a 5-year survival rate of 60–80% (Figure 6.2). AT/RT are quite rare in adults, and some changes in cases published to-date are shown in Table 6.3.

Some AT/RT are composed entirely of rhabdoid cells, whereas others show a combination of rhabdoid cells and areas resembling PNET/medulloblastoma (9). Approximately one third of tumors contain malignant mesenchymal elements, most often spindle cells, and one quarter exhibit a mature epithelial population that may be glandular or squamous (4,7). Any or all cell types or patterns may be seen in a given tumor in combination with the rhabdoid component. Necrosis and abundant mitotic activity are common. Immunohistochemical features are complex and vary depending upon the cellular composition of the tumor. In general, rhabdoid cells express vimentin and EMA, and less often smooth muscle actin (SMA). A number show neurofilament, glial fibrillary acid protein (GFAP), and keratin positivity, but not germ cell markers or desmin (4, 7, 10). The primitive neuroectodermal cells may express neurofilament protein, GFAP, or desmin, but some are immunohistochemically undifferentiated. The

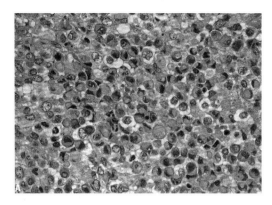

FIGURE 6.1. Histological view of a typical AT/RT consisting almost entirely of rhabdoid cells with eccentric nuclei and prominent nucleoli.

epithelial component expresses keratin but sometimes EMA and vimentin as well. Mesenchymal cells express vimentin, sometimes SMA, but rarely desmin. These reactions have been advocated as a means of differentiating AT/RT from other confusing tumors (16–19). However, these reactions are not unique, and they are certainly not specific for AT/RT and other tumors (e.g., choroid plexus tumors) may show reactivity not dissimilar from that of AT/RT for the substances mentioned (20). AT/RT are rapidly growing tumors that have MIB-1 labeling indices of 50–80%.

Childhood AT/RT are distributed anatomically as follows: posterior fossa, 50–60% (cerebellum, cerebello-pontine angle, and/or brainstem); supratentorial, 30–40% (cerebral or suprasellar); pineal, 5%; multifocal, 2% and spinal, 2%. Tumors in adults are all located in the cerebrum. The tendency to arise in the cerebello-pontine angle with invasion of surrounding structures is a distinct feature of this tumor (4).

TABLE 6.1. References to pathology, clinical aspects, and immunohistochemistry of AT/RT.

	References
Clinicopathology	Weinblatt and Kochen 1992 (113)
	Muller et al 1994 (114)
	Wick et al 1995 (115)
	Bergmann et al 1997 (116)
	Bhattacharjee et al 1997 (18,19)
	Korones et al 1999 (117)
	Oka and Scheithauer 1999 (118)
	Oulab et al 1999 (119)
	Bambakidis et al 2002 (120)
Therapy	Cossu et al 1993 (16)
	Hilden et al 1998, 2004 (8,121)
	Weiss et al 1998 (122)
	Hirth et al 2003 (123)
	Howes et al 2005 (124)
	Zimmerman et al 2005 (125)
Immunohistochemistry	Behring et al 1996 (6)
	Judkins et al 2004, 2005 (127,128)
	Bufford et al 2004 (129)

The typical rhabdoid cell is medium-sized, round to oval, with an eccentric nucleus that commonly has a prominent nucleolus. Cytoplasm has a fine granular homogeneous character or may contain a poorly defined denser pink body resembling an inclusion. Cell borders are typically distinct. Mitotic figures are usually abundant. Small rhabdoid cells may have a tapering cytoplasmic tail, whereas others may be huge, bizarre forms with more than one nucleus (4).

Cytogenetic and Fluorescence in Situ Hybridization (FISH) Changes in AT/RT

A unique genetic alteration has been documented in 90% of AT/RT. Those arising in the CNS (and in the kidney) demonstrate monosomy or a deletion of chromosome 22 by cytogenetics, FISH, or loss of heterozygosity (LOH). The gene involved, hSNF5/INI1, maps to chromosome band 22q11.2, and it is consistently mutated in AT/RT (21–23). Somatic INI1 mutations have only rarely been reported in other CNS tumors, making mutation analysis a valuable diagnostic adjunct in difficult cases (22) (Figure 6.3).

The karyotypic changes in about 60 AT/RT have been reported (Table 6.4). Of these, approximately 25% had normal karyotypes (probably due to overgrowth by normal cells), and about 50% had anomalies of chromosome 22, i.e., -22, del(22), or r(22) as the sole change; the remaining cases had more complicated karyotypes with half of them showing involvement of chromosome 22 (deletions, translocations). Altogether, 65% of the karyotypes with abnormalities had involvement of chromosome 22. No cases with dmin were observed, in contrast to their presence in medulloblastoma and PNET. However, in a tumor cell line (24), a hsr was noted. Translocations or derivative chromosomes were present in more than one half of the karyotypes published (Table 6.4). Translocations involving 22q22 with a variety of other chromosomes have been described in CNS, renal, and other AT/RT tumors, e.g., chromosomes 1, 8, 9, 11, and 18 (25).

Except for one AT/RT with a wide range of chromosome numbers, i.e., 32–78, and two in the near-triploid range, most of the AT/RT had near-diploid karyotypes, i.e., 45–47 chromosomes.

Using probes for 22q11.2 for FISH analyses on archival paraffin-embedded tumor tissues, 22q deletions were identified in six of eight AT/RT, but in none of 12 medulloblastomas (26). Of four tumors that were difficult to classify on histologic grounds, three had deletions of 22q and review of their morphology and immunophenotype resulted in their being classified as AT/RT. The remaining tumor was negative for 22q deletion and was thought to be a large-cell medulloblastoma. This study demonstrates the utility of interphase FISH in the differential diagnosis of confusing tumors (27).

TABLE 6.2. Miscellaneous aspects of AT/RT.

	References
AT/RT in a 4-year-old girl with constitutional r(22).	*Rubio 1997 (130)*
Extrarenal rhabdoid tumors tend to carry a poor prognosis.	*Fanburg-Smith et al 1998 (131)*
	Schofield 2002 (132)
Familial cases of AT/RT.	*Biegel et al 1999 (22)*
	Sévenet et al 1999 (90)
	Taylor et al 2000 (107)
	Janson et al 2006 (111)
AT/RT in two siblings.	*Proust et al 1999 (133)*
Incidence of AT/RT possibly 2.1% of CNS tumors.	*Rorke and Biegel 2000 (4)*
Lung metastases in AT/RT (6-year-old).	*Güler et al 2001 (134)*
Monozygotic twins (2.5–3.0 months old), one with a posterior fossa tumor (malignant rhabdoid) and the other with a tumor mimicking a medulloblastoma have been described. Both tumors were shown to have inactivation of *INI1* through deletion of exon 7 in one allele and a nonsense mutation in the other. Constitutional DNA showed a nonsense mutation of one allele.	*Fernandez et al 2002 (135)*
Despite an extensive literature on the characteristics and clinical course of AT/RT, the exact incidence remains difficult to determine. One report indicated that in children with malignant brain tumors and less than 5 years of age, 10–15% had AT/RT.	*Packer et al 2002 (10)*
80% of children with AT/RT die within 1 year.	*Packer et al 2002 (10)*
AT/RT in fourth ventricle with metastases in pineal area, hypothalamic septum and thalamic area.	*Fenton and Foreman 2003 (136)*
Gefitinib antitumor activity in MRT demonstrated in vitro and in vivo.	*Kuwahara et al 2004 (137)*
AT/RT must be differentiated from medulloblastoma, PNET, choroid plexus carcinoma and germline tumors, with which they are often confused.	*Biegel et al 1990 (138)*
	Rorke et al 1996 (3)
Therapy for AT/RT (surgery, radiotherapy and chemotherapy) has serious limitations because of sequelae and is rarely curative, with <20% of the children surviving 2 years.	*Rorke et al 1996 (3)*
	Burger et al 1998 (7)
AT/RT and rhabdoid tumors in general occur in very young children <2 years old.	*Burger et al 1998 (7)*
	Ho et al 2000 (139)
A cell line originating from a 3-year-old girl with AT/RT has been described and immunophenotype and neoplastic properties delineated. This established cell line had significant aneuploidy (chromosome number of 65–120) with many structural anomalies., including hsr on 5p. No -22 was noted.	*Yachnis et al 1998 (24)*
Rab-family gene *(Rab36)* located at 22q11.2 and deleted in MRT, but is not a TSG.	*Mori et al 1999 (140)*
In two AT/RT examined by immunohistochemistry, the following gave positive results: insulin-like growth factor 2 (IGF-2), IGFR-1, cathepsin D, and Ki-67 (1.4 and 10%, respectively).	*Ogino et al 1999 (141)*
Inactivating mutations of *BRG1* have been identified in human tumors and cell lines and the heterozygous inactivation of *BRG1* is associated with a tumor-prone phenotype in mice.	*Bultman et al 2000 (142)*
	Wong et al 2000 (143)
	Reisman et al 2003 (144)
Haploinsufficiency of SNF5 (integrase interactor 1) predisposes mice to rhabdoid tumors.	*Roberts et al 2000 (38)*
High expression of abnormal p16 and pRB	*Berrak et al 2002 (145)*
Low expression of abnormal p53 and PTEN	*Berrak et al 2002 (145)*
One AT/RT studied immunohistochemically showed diffuse cytoplasmic positivity for IGF-II (insulin-like growth factor II). All medulloblastomas and cerebral neuroblastomas were negative.	*Ogino et al 2001 (146)*
Alpha-internexin, an intermediate protein expressed in normal neurons, was found to be expressed in AT/RT and medulloblastoma, indicating that these primitive tumors exhibit neuronal differentiation.	*Kaya et al 2003 (147)*
Osteopontin expression was found to be up-regulated in AT/RT, with levels also elevated in plasma and CNS. Osteopontin might be a potential marker for recurrence of AT/RT.	*Kao et al 2005 (148, 149)*
No amplification of overexpression of EGFR in AT/RT.	*Jeibmann et al 2006 (150)*
Homozygous deletions were most often seen in AT/RT and extrarenal rhabdoid tumors, whereas truncating mutations were detected in a high percentage of AT/RT and kidney tumors.	*Biegel et al 2002 (32)*
An AT/RT in a 5-month-old girl and a PNET in the spinal canal of the father have been described. Although changes in *TP53* were shown to be present in the PNET, neither tumor showed changes of *INI1* or LOH abnormalities.	*Izycka-Swierzewska et al 2003 (53)*
Overexpression of the osteopontin gene may serve as a diagnostic marker in AT/RT.	*Kao et al 2005 (150A)*

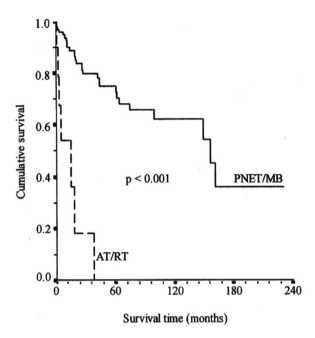

FIGURE 6.2. Kaplan–Meir survival curve for patients with AT/RT and those with PNET/medulloblastoma. The poor survival of the AT/RT patients is related to the extreme biologic aggressiveness of these tumors, especially the occurrence of early metastases (from ref. 139, with permission).

The *INI1* Gene and SWI/SNF Complex in AT/RT

The *INI1* (*SMARCB1, hSNF5*) gene contains nine exons, and it has a coding sequence of approximately 1.2 kb (28). This gene is ubiquitously expressed. A cryptic splice donor site in exon 2 results in an alternate transcript expressed in all tissues.

TABLE 6.3. Adult patients with AT/RT.

Age (years)	Gender	Alterations of 22q	References
18	Male	Not reported	*Cossu et al 1993* (16)
18	Male	Monosomy 22	*Wyatt-Ashmead et al 2001* (102)
20	Male	Not reported	*Arrazola et al 2000* (151)
20	Female (pregnant)	No abnormalities by FISH for *INI1*	*Erickson et al 2005* (152)
21	Female	22q deletion	*Bruch et al 2001* (26) *Fuller et al 2001* (45)
21	Male	Not reported	*Horn et al 1993* (153)
27	Male	Not reported	*Sugita et al 1999* (154)
	Female (pregnant)	Not reported	*Lutterbach et al 2001* (155)
31	Female	Not reported	*Pimentel et al 2003* (156)
32	Male	Not reported	*Fisher et al 1996* (157)
34	Female	*BCR* deletion	*Bruch et al 2001* (26) *Fuller et al 2001* (45)
34	Male	Not reported	*Ashraf et al 1997* (158)
35	Male	Not reported	*Byram et al 1999* (159)

Based on Erickson et al. 2005 (152).

The INI1 protein is a component of the mammalian SW1/SNF complex, which functions in an ATP-dependent manner to alter chromatin structure (29).

Somatic mutations or intragenic deletions of *INI1* have been documented in a high proportion of AT/RT, most of which create a novel stop codon (30). Although mutations have been observed throughout the coding sequence, there seem to be two potential hot spots, i.e., deletions/mutations in exons 5 and 9 (Figure 6.4). Germline *INI1* mutations have been detected in five patients with AT/RT and renal rhabdoid tumors (Table 6.5) (22). The nature of the changes affecting *INI1* in AT/RT are shown in Table 6.6.

AT/RT demonstrate monosomy 22 (-22) at a higher frequency than renal or extrarenal rhabdoid tumors, whereas extrarenal rhabdoid tumors have a very high rate of homozygous deletions of the entire *INI1* gene (31, 32). The loss of one allele, with concomitant mutation in the remaining copy of *INI1*, was observed in 55% of AT/RT. Compound heterozygous mutations or apparent splicing mutations were detected in a few. The majority of AT/RT had single base-pair nonsense mutations or frameshift mutations that are predicted to result in the coding of a novel stop codon, and premature truncation of the protein (31). Approximately 10% of AT/RT cases do not have coding sequence mutations in any of the nine exons of *INI1*, and yet they have decreased expression levels of *INI1* by reverse transcription-polymerase chain reaction (PCR) analysis or undetectable levels of the protein by Western blot analysis. Approximately 15% of AT/RT show no alterations of the *INI1* gene at the DNA, RNA, or protein levels. One mechanism proposed for the loss of expression of *INI1* in tumors without coding sequence mutations is hypermethylation of the *INI1* promoter (31). However, bisulfite modification and sequence analysis of DNA from primary rhabdoid tumor did not demonstrate hypermethylation of the *INI1* without coding sequence mutations.

Given that mutations cannot be identified in 100% of cases with AT/RT with typical morphology and immunohistochemical patterns, if the histologic features and immunostaining are consistent with AT/RT, the absence of an *INI1* mutation would not be sufficient to change the diagnosis (31). Alternatively, the presence of an *INI1* mutation, in a tumor with features suggestive of medulloblastoma or PNET, should be sufficient to change the diagnosis to AT/RT and treat the patient accordingly.

Mechanisms of *INI1*-mediated tumor suppression include induction of G1 arrest and control of the mitotic spindle checkpoint (33–37).

It has been shown that *INI1* participates in forming the functional center of the ATP-dependent chromatin-remodeling complex SWI/SNF (Figures 6.5 and 6.6 and Tables 6.7 and 6.8). This complex affects transcription by reorganizing nucleosomes associated with specific eukaryotic cellular genes. *INI1* is thought to interact with *MYC*, which recognizes the E-box motif of the promoter region. The *MYC* family acts as sequence-specific transactivators, repressing

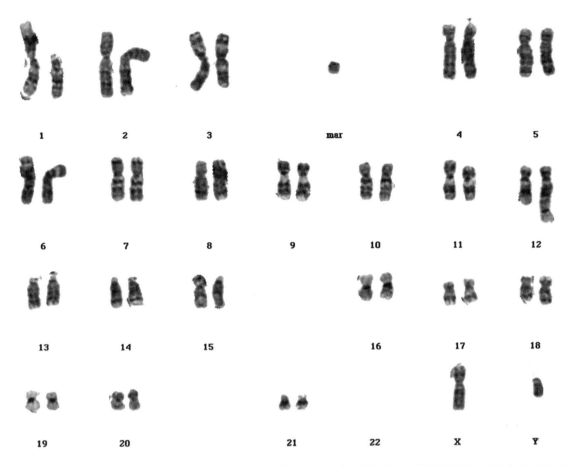

FIGURE 6.3. Karyotype of an AT/RT in a boy nearly 4-years-old showing the following: 45,XY,del(1) (p13),dic(12;22)(q24.2;p13), add(18)(p11.2),-22,+mar. The loss of a chromosome 22 is a common cytogenetic finding in AT/RT.

transcription by recruiting negative regulators and possibly activating transcription by recruiting SWI/SNF. In addition, *INI1* interacts with other diverse proteins affecting apoptosis, cellular differentiation, replication, transcription, and tumor suppression (31, 31A). For example, the SWI/SNF complex has been shown to interact with C/EBP-β, HRX, steroid hormone receptors, and erythroid Kruppel-like factor. Animal studies (38–40) have demonstrated that the *INI1* component of SWI/SNF plays a critical role in the development of various tissue types by regulating cell growth and proliferation (41, 42). *INI1* encodes a subunit of SWI/SNF chromatin remodeling complex, which is one of several evolutionarily conserved multiprotein complexes that disrupt nucleosomes in vitro in an ATP-dependent manner (43) and to transcriptional repression through the actions of histone deacetylases. It has been demonstrated that (35) wild-type ectopic *INI1* expression inhibits cell entry in to the S phase and causes cell cycle arrest by down-regulating E2F targets such as *cyclin A*, *E2F1*, and *CDC6*. Similarly, (44) adenovirus transduction of MRT cell lines with different isoforms of *INI1* suppressed cell growth and up-regulated senescence proteins.

In addition to the characterization of *INI1* as a tumor suppressor gene (TSG), some of the specific mutations and deletions of *INI1* have been clarified. One study (31) used

interphase FISH, PCR-based microsatellite heteroduplex, and sequence analysis to analyze the *INI1* locus of 100 primary malignant rhabdoid tumors from different sites. It was found (31) that a variety of abnormalities, including germline mutations, deletions, or insertions leading to novel stop codons, single base-pair nonsense mutations, and intragenic deletions may involve *INI1*. Interestingly, 25% of the histologically confirmed rhabdoid tumors did not show *INI1* deletions or mutations (32). This finding is consistent with other studies (31,45,46), which reported detectable chromosome 22 defects in only 75–85% of cases. Although *INI1* plays an important role in the preponderant number of rhabdoid tumors, it is evident that not all tumors with a rhabdoid phenotype have monosomy 22 (-22) or *INI1* gene mutation. Likewise, not all tumors with this genetic loss display histologic rhabdoid transformation. Some authors have proposed that *INI1* defects uniquely characterize true malignant rhabdoid tumors such as AT/RT; however, the presence of a genetic abnormality is not presently required for diagnosis.

An immunohistochemical study of 12 AT/RT revealed an absence of *INI1* expression in all of the tumors (47). The results were compared with those of other brain tumors; unfortunately, no medulloblastomas or PNET, tumors with which

TABLE 6.4. Chromosome abnormalities in AT/RT of CNS.

Karyotype	References
CNS	
Normal karyotypes (13 cases)	*Griffin et al 1988* (160), *Kaiserling et al 1996* (161), *Rorke et al 1996* (3), *Bhattacharjee et al 1997 (17)*, *Biegel et al 1999* (22), *Proust et al 1999 (133)*, *Lee et al 2002* (9)
Karyotypes with only monosomy 22, or partial, del or r(22) (28 cases)	*Biegel et al 1989, 1990, 1999* (22, 138, 162); *Douglass et al 1990* (163); *Fort et al 1994* (164); *Hasserjian et al 1995* (165); *Muller et al 1994, 1995*(114, 166) *Rorke et al 1996* (3); *Rubio 1997* (130); *Burger et al 1998* (7); *Savla et al 2000* (101); *Lee et al 2002* (9)
Abnormal karyotypes	
32-78<4n>,XXYY,add(1)(q32)x2,-2,-3,add(3)(p22),-5, -6,add(6)(q23)x2,+7,+8,-9,-9,-10,-11,-12 or add(12)(p11.1),-13,-14,-14,t(14;14)(q10;q10),-17,-17, -18,-21,-22,+mar	*Bhaattacharjee et al 1997* (17, 18)
45,XY,der(13;14)(q10;q10)	*Biegel et al 2000* (49)
45,XY,-22,der(22)t(22;22)dic(22;22)(p12;p12)add(22) (q13)(qter~p12::p12~q13::?)	*Bhaattacharjee et al 1997* (18)
46,XY,add(22)(p11),del(22)(q11)	*Hasserjian et al 1995* (165)
46,XY,der(2),t(2;?)(p11;?),del(16)(q21.1), ins(17) (pter→cen→q21::?::q21→qter),der(19), t(1;19)(q25;q13.4),del(20)(q13.1)	*Karnes et al 1992* (167)
46,XX,der(6),t(6;11)(p22;p15),der(11), r(6;11)(p25p22;p15q5)	*López-Ginés et al 2000* (52)
46,XX,ins(4;1)(p16.3;p34.3p13.3),del(22)(q13.1)	*Agamanolis et al 1995* (168)
46,XY,t(11;22)(p15.5;q11.23)	*Ota et al 1993* (67)
46,XY,t(12;22)(q24;q11.2)	*Huisman et al 2000* (169)
46,XY,t(12;22)(q24.3;q11.2)	*Sawyer et al 1998* (170)
46,XX,-8,+der(8)t(8;22)(p11;q?12)x2,-22, del(22)(q12q?13)/46,idem,-del(22)(q12q?13),+r(22)	*Klopfenstein et al 1997* (171)
46,XX,-9,-22,+i(1q),+der(22),t(9;22)(p13;q11)/ 45,XX,-9,-10,-22,+i(1q),+der(22),t(9;22)(p13;q11), -6,-13,del(17p), +2mar	*Biegel et al 1992* (172)
	Rorke et al 1996 (3)
46,XX,+?20,-22,del(22)(q11)	*Douglass et al 1990* (163)
47,XY,+der(1),t(1;1)(p12;q22)	*Hasserjian et al 1995* (165)
61,XX,+1,+1,+7,+7,+11,+11,+14,+14,+add(16q?),+19, +add(19)(q13),+20,+20,+21,+2mar	*Biegel et al 1999* (22)
65-120,XX,with hsr	*Yachnis et al 1998 (24)*
46,XX,t(15;17)(q23;q24),-20,del(22(q11q13),+mar	*Thangavelu et al 1998* (173)
46,XY,-8,t(18;22)(q21;p11.2),+mar	*Handgretinger et al* (174)

AT/RT is often confused, were included in the study. However, the results are in agreement with molecular and other studies indicating the very high incidence of *INI1* mutations in AT/RT. Fresh and archival AT/RT tissues have been examined by a number of FISH techniques, which revealed alterations of 22q indicative of changes affecting the *INI1* genes, thus expanding and supporting the cytogenetic findings in these tumors (7, 23, 31, 32, 48–53).

Cyclin D and *INI1* in AT/RT

Epigenetic inactivating mechanisms as such as hypermethylation have not been reported in AT/RT (54). Although the molecular mechanism downstream of *INI1* inactivation is not fully understood, several lines of evidence have accumulated.

Cyclin D1, a critical protein controlling the G1 stage of the cell cycle, is a potential direct target of the *INI1* gene in cultured cells. Transfection of the *INI1* gene induced G1 arrest and directly suppressed *cyclin D1* transcription, but the arrest was reverted by the coexpression of *cyclin D1* (34, 35, 54). This pathway can be a potential molecular target for AT/RT treatment. However, the behavior of *cyclin D1* in primary tumors remains unclear. It has been shown (55) that *cyclin D1* overexpression is associated with *INI1* inactivation in AT/RT.

Cyclin D1 is a critical downstream effector of INI1; INI1 directly represses *cyclin D1* transcription. *Cyclin D1* expression from a heterologous promoter is sufficient to overcome INI1-mediated cell cycle arrest (56). *Cyclin D1* is depressed in human and mouse genetically based MRT (56,57). In addition, genetic ablation of *cyclin D1* is sufficient to abrogate the genesis of MRT in vivo mouse models (57). These studies

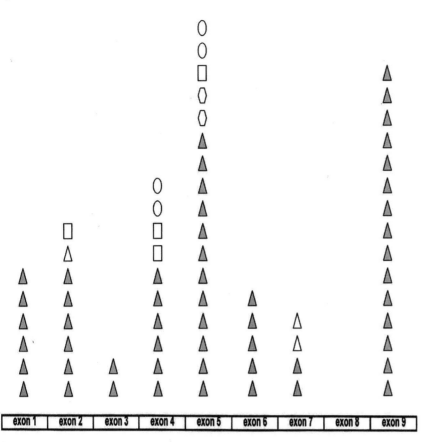

INI-1 Gene

▲ **Somatic Mutation**   ☐ **Medulloblastoma**   ◊ **Choroid Plexus**
   AT/RT

△ **Germline Mutation**   ⬡ **OPNET**
   AT/RT

FIGURE 6.4. Distribution of mutations in the 9 exons of the *INI1* gene in AT/RT and other tumors. Exons 5 and 9 show the highest frequency. Homozygous deletions of exons 1–9 were seen in 9 AT/RT. Some of the non-AT/RT tumors were diagnosed as AT/RT subsequent to the study of the *INI1* gene.

suggest that MRT are critically dependent on the presence of *cyclin D1* for genesis, survival, or both.

Cyclin D1 has multiple roles during the cell cycle, cell propagation, and tumorigenesis (58). It associates with cyclin-dependent kinases cdk4 and cdk6, leading to the phosphorylation and inactivation of the retinoblastoma (Rb) protein. It sequesters p27, releasing cyclin E to further inactivate Rb, resulting in G1-to-S transition. Cyclin D1 modulates the activity of several transcription factors in a cdk-indpendent manner, which may be important for its function in tumor cells. There are two isoforms of cyclin D1, D1a and D1b, which confer differential transformation ability to mammalian cells.

Based on the dependence of AT/RT on cyclin D1, it was hypothesized that inhibition of cyclin D1, its activity, or its pathway may be an appropriate therapeutic strategy for treat-

ment of children with AT/RT (59), an area that has not been fully explored.

Further information on the interaction of *INI1* with cyclin D1, p16[INK4A], and Rb and its role in the cell cycle can be found in appropriate publications (34,55) and on the interplay of SWI/SNF with *BRG1* (brahma-related gene 1), histone deacetylase (HDAC), and Rb in the control of the cell cycle (56,60).

AT/RT can be misdiagnosed since a variety of brain tumors, including ependymoma, potentially harboring LOH at 22q may be confused in the differential diagnosis. However, *hSNF5/INI1* inactivation can serve to distinguish AT/RT from other brain tumors. Because this inactivation is associated with cyclin D1 overexpression in AT/RT, cyclin D1 could serve as a target for *hSNF5/INI1* in primary tumors (59A).

TABLE 6.5. Germline mutations in patients with kidney and brain AT/RT.

Case	Age/sex	Germline mutations	Somatic change in MRT of kidney	Somatic change in AT/RT	References
1.	5 months male	Nonsense mutation at codon 201 (de novo)	Deletion of 22q11.2 region	AT/RT: deletion of 22q11.2 region	*Sévenet et al 1999* (84)
2.	4 months male	7-bp deletion at codons 297–299	Deletion of 22q11.2 region	AT/RT: loss of entire chromosome 22	*Biegel et al 2000* (49)
3.	6 months male	1-bp deletion at codon 91	Nonsense mutation at codon 260	CNS: loss of entire chromosome 22	*Savla et al 2000* (101)
4.	8 months male	Nonsense mutation at codon 198	Deletion of 22q11.2 region	CNS: loss of entire chromosome 22	*Savla et al 2000* (101)
5.	Unknown	Nonsense mutation at codon 257	Deletion of 22q11 region	AT/RT: ND	*Biegel et al 2002* (32)
6.	Unknown	Nonsense mutation at codon 198	ND	AT/RT: ND	*Biegel et al 2002* (32)
7.	Unknown	Nonsense mutation at codon 66	Deletion of 22q11 region	AT/RT: ND	*Biegel et al 2002* (32)
8.	Newborn female	Deletion of entire *hSNF5/INI1* gene (de novo)	Deletion of 22q11 region	AT/RT: deletion of 22q11.2 region	*Kusafuka et al 2004* (175)

ND, not done; except for cases 3 and 4 (reported as "CNS tumors"); all others had AT/RT.
Reports of the occurrence of kidney and CNS rhabdoid tumors but without supplying cytogenetic or molecular information have appeared (13, 14, 72).

TABLE 6.6. Changes affecting the *INI1* gene leading to its homozygous inactivation in AT/RT.

	References
Germline mutations, then loss of wild-type allele, e.g., mutation of other homolog	*Rousseau-Merck et al 1999, 2005 (23, 50)*
deletions	*Biegel et al 2002 (31, 32)*
insertions leading to novel stop codons	
single base-pair nonsense mutations	
intragenic deletions	
compound homozygous deletions or mutations	
mitotic recombinations	

Some Genetic Aspects of Rhabdoid Tumors

An issue to be resolved in the characterization of rhabdoid malignancy is whether a rhabdoid tumor represents a unique entity or a nonspecific phenotypic change. It remains unclear whether adult and childhood rhabdoid tumors share a common cell of origin, pathogenesis, and set of genetic alterations, or conversely, are the common result of advanced dedifferentiation in tumor progression. A study (45) attempted to clarify the nature of various tumors with rhabdoid features by using FISH analysis to test for deletions of the 22q11.2 locus in AT/RT, extrarenal malignant rhabdoid tumors, and composite extrarenal rhabdoid tumors, including sarcomas, carcinomas, or melanomas with rhabdoid features. It was found that 77% of AT/RT and 75% of MRT demonstrated *INI1* loss, whereas only 13% of composite tumors had this finding. These results show that tumors other than MRT are capable of developing a rhabdoid-like phenotype without acquiring the *INI1* gene mutation (45). In a comparative genomic hybridization (CGH) study on three AT/RT cases (61), it was found that one case showed a loss of 8p instead of the expected *INI1* mutations, suggesting that other genetic abnormalities may be involved in rhabdoid tumor development.

Rhabdoid cells characteristically have an eccentric nucleus displaced by a cytoplasmic inclusion body consisting of whorled bundles of 10-nm intermediate filaments. These inclusions possibly represent failure of normal cytoskeleton formation with resultant disorganization of vimentin and cytokeratin. A study (62, 63) analyzed rhabdoid tumors and epithelioid sarcomas using immunohistochemistry. It was demonstrated that although epithelioid sarcomas are diffusely positive for many cytokeratin subunits, rhabdoid cell inclusion bodies were immunopositive for only vimentin, CK8, and CK18. Further evaluation by three-dimensional confocal imaging showed that vimentin is diffusely distributed, but CK8 and CK18 focally form conglomerates consistent with the bundles of whorled intermediate filaments seen by electron microscopy (64). Direct sequencing of the CK8 gene (*KRT8*) in seven malignant rhabdoid tumors showed missense mutations in all of them. The genetic defects involved regions known to participate in protofilament-to-protofilament linkage and a phosphorylation site known to affect filament organization (64).

MRT are rare but highly aggressive neoplasms and constitute a phenotypically heterogenous group of tumors of extrarenal and extra-CNS locations (65, 66). MRT and their cell lines are often characterized by the same cytogenetic and molecular changes seen in renal rhabdoid tumors and AT/RT, i.e., inactivation of *INI1* and loss of chromosome 22 or its long-arm (22q-) (21, 48, 51, 67–71).

The association of MRT of the kidney with brain tumors has been known for some years (72); the brain tumors included medulloblastomas, a pineoblastoma, a giant-cell astrocytoma, and a PNET. There is a likelihood that some of these tumors were, in fact, AT/RT, considering that problems in the differential diagnosis of these tumors are not rare; no genetic data in this study were provided (72).

The association of germline mutations of *INI1* with the development of MRT of the kidney *and* brain tumors (most of them AT/RT) is shown for the eight published cases in

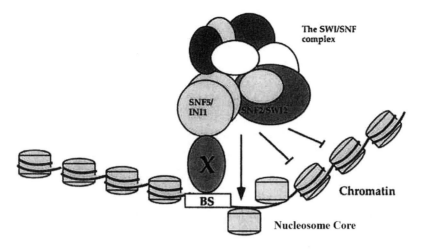

FIGURE 6.5. Schematic presentation of the recruitment of the SWI/SNF complex to specific target sites by the inactivation of *INI1* with sequence-specific regulatory proteins. BS is a binding site with a specific sequence and X is any of sequence-specific regulatory proteins that can bind to it. The repeating units represent nucleosomes in the chromatin which are formed by ~200 bp of DNA double helix wrapped around a core particle composed of histones. Nucleosomes are connected by linker DNA and associated with their neighbors to form a helical fiber of ~30 nm in diameter (31).

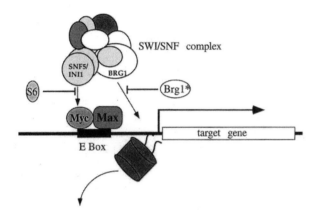

FIGURE 6.6. Schematic presentation of another effect of *INI1*, i.e., activation of transcription of MYC-MAX heterodimers mediated by the recruitment of SWI/SNF complex via the direct interaction of *INI1* with *MYC* (from ref. 31, with permission).

Table 6.5. It should be stressed that germline mutations of the *INI1* gene have been described in cases of AT/RT without kidney tumors and vice versa (22).

Composite Rhabdoid Tumors

MRT exist as a specific entity, e.g., malignant rhabdoid tumor of the kidney and AT/RT (Table 6.9). However, there exists a group of tumors with a secondary rhabdoid morphologic phenotype encountered within a wide array of tumor types, including some of the brain displaying cytologic anaplasia and aggressive behavior (73), and grouped under the label of "composite rhabdoid tumors." These tumors can be differentiated from AT/RT by the demonstration of 22q deletions

by FISH on fresh or archival tissues (3, 26, 45). Table 6.9 describes some features of these two types of tumors.

The retention of *INI1* expression in "composite" rhabdoid tumors indicates that these tumors are genetically distinct from AT/RT and other tumors with 22q deletions or loss of chromosome 22, e.g., meningiomas, medulloblastomas (73,74). These authors suggested that the immunohistochemical demonstration of *INI1* expression is a relatively simple, sensitive, and specific technique for distinguishing AT/RT from composite rhabdoid tumors of the brain, such as meningioma (75–78).

Tumor necrosis factor-related apoptosis-inducing ligand (TRAIL/Apo2L) or Apo2 ligand is a potent inducer of apoptosis in cancer cells but not in normal cells (79, 80). TRAIL/Apo2L is a potent inducer of apoptosis in MRT cells, an effect that can be enhanced by doxorubicin (81). Furthermore, the inhibition of nuclear factor-β and of phosphatidylinositol 3-kinase/Akt signaling also enhances the sensitivity of MRT cells to TRAIL, providing a basis for therapeutic application to AT/RT (81).

Tumors, Other Than AT/RT, with Inactivation of *INI1*

The demonstration of *INI1* inactivation of brain tumors other than AT/RT, e.g., medulloblastoma and those of the choroid plexus, raised questions regarding the meaning and implications of these changes. In the case of medulloblastomas (49), subsequent evaluation indicated that these *INI1*-positive tumors were most likely misdiagnosed and represented AT/RT, with prominent primitive neuroectodermal-like components (82). In fact, no mutations of *INI1* were encountered in a large series of medulloblastomas (83).

TABLE 6.7. *INI1* gene in AT/RT.

	References
The *INI1* (Integrase interactor 1) gene has been isolated.	*Kalpana et al 1994 (28)*
Inactivating biallelic alterations (in the majority of AT/RT) of the *INI1* gene. *INI1* is a TSG and a component of chromatin remodeling SWI/SNF complex	*Versteege et al 1998 (21)*
Loss of *INI1* gene occurs in virtually 100% of AT/RT and germ- line mutations predispose individuals to these tumors.	*Versteege et al 1998 (21)* *Biegel et al 1999, 2000, 2002 (22, 31, 32, 49)* *Taylor et al 2000 (107)*
INI1 protein interacts directly with *MYC* and is necessary for *MYC*-mediated transactivation through the SWI/SNF complex.	*Change et al 1999 (108)*
INI1 is a TSG involved in key biochemical pathways implicated in cell growth and development.	*Cheng et al 1999 (108)*
A number of cancers, including brain tumors, such as astrocytomas, ependymomas, and neurinomas, with LOH of 22q have been tested for mutations or deletions of *INI1*. Negative results were obtained.	*Sévenet et al 1999 (84)* *Savla et al 2000 (101)*
75–85% of AT/RT have *INI1* changes; Not all tumors with rhabdoid phenotype have -22; Tumors with changes may not show rhabdoid phenotype.	*Biegel et al 2000, 2002 (31, 49)* *Fuller et al 2001 (45)* *Uno et al 2002 (46)* *Wharton et al 2003 (61)*
Evidence for *INI1* as a TSG has been established in mouse models.	*Guidi et al 2001 (40)* *Roberts et al 2000, 2002 (38, 39)* *Klochendler-Yeivin et al 2000 (176)*
Generation of several murine *INI1* knockout models in which heterozygous mice developed MRT of soft tissue, has been described. *INI1*-null mice are embryonically lethal.	*Klochendler-Yeivin et al 2000 (176)* *Roberts et al 2000 (39)* *Guidi et al 2001 (40)*
Family with germline mutations of *INI1* with various brain tumors (AT/RT and choroid plexus).	*Taylor et al 2000 (107)*
Cell cycle arrest shown by *INI1* due to recruitment of HDAC (histone deacetylase) activity to the cyclin D1 promoter, thereby causing its repression and G0-G1 arrest. Repression of *cyclin D1* gene expression may serve as a useful goal of therapy for AT/RT.	*H.S. Zhang et al 2000 (36)*
Gene at 22q11 involved in ependymal tumors but different from *INI1*.	*Kraus et al 2001 (83)*
INI1 may be involved in choroid plexus carcinoma but not in other brain tumors.	*Weber et al 2001 (86)*
Actions of *INI1* include induction of G1 arrest and control of mitotic spindle checkpoint.	*Ae et al 2002 (33)* *Betz et al 2002 (34)* *Versteege et al 2002 (35)* *Zhang et al 2002 (36)*
Re-expression of *INI1* in human rhabdoid tumor cell lines causes a G1 cell cycle arrest, flattened cell morphology, activation of senescence-associated proteins, down-regulation of a subset of E2F target genes, activation of RB and up-regulation of $p16^{INK4A}$	*Betz et al 2002 (34)* *Versteege et al 2002 (35)* *Reincke et al 2003 (44)* *Vries et al 2005 (37)*
INI1 is a potent regulator of the entry of a cell into the S phase, an effect that may account for its TSG.	*Versteege et al 2002 (35)*
In rhabdoid tumors without genomic changes of 22q or *INI1*, abnormal methylation of promoter regions could not be demonstrated.	*F. Zhang et al 2002 (54)*
Novel germline mutation of *INI1* in a child with RPS and absence of *MYC* amplification in pineal AT/RT and renal MRT. Mutation present in both tumors and normal renal tissue. LOH of 22q.	*Fujisawa et al 2003, 2005 (55, 109)*
INI1 modulates cell growth and actin cytoskeleton organization through regulation of pRB-E2F and Rho pathways.	*Medjkane et al 2004 (43)*
Cyclin D1 is overexpressed in AT/RT with inactivated *INI1*.	*Fujisawa et al 2005 (55)*
INI1 expression retained by immunohistochemistry in composite rhabdoid tumors, including meningiomas with 22q deletions.	*Perry et al 2005 (73)*
Inactivation of *INI1* in rhabdoid tumors of the human vs. the mouse differs in its mechanisms, i.e., mitotic recombination acting in mouse rhabdoid tumors regardless of anatomic location, whereas in the human rhabdoid tumorigenesis is related to specific chromosome 22 rearrangements leading to preferential anatomic tumor location.	*Rousseau-Merck et al 2005 (50)*
An embryonal tumor of the cerebellum with rhabdoid cells was distinguished from AT/RT by histology and absence of *INI1* changes and EMA staining.	*Sasaki et al 2005 (177)*
AT/RT arising in a ganglioneuromas had LOH at 22q11.2 and mutation of *INI1*.	*Allen et al 2006 (178)*
Loss of *INI2* may influence the regulation of multiple CDK inhibitors involved in cell line replicative senescence.	*Chai et al 2005 (179)*
Rhabdoid cells undergo replicative senescence after *INI1* re- expression, including G1 cell cycle arrest, a flattened morphology and $p16^{INK4A}$ and $p21^{CIP1/WAF1}$ inactivation.	*Chai et al 2005 (179)*
A family with predisposition to rhabdoid tumors (including AT/RT) showed no linkage in *INI1* pointing to involvement of other loci.	*Frühwald et al 2006 (110)*
Loss of INI1 protein in all 18 AT/RT examined by immuno-histochemistry.	*Haberler et al 2006 (112)*

TABLE 6.8. SWI/SNF in AT/RT.

	References
Many accessory proteins are part of SWI/SNI2 multiprotein complexes that regulate transcription both positively and negatively by modifying chromatin at regulatory sites.	*Kingston et al 1996* (29) *Morozov et al 1998* (180) *Cheng et al 1999* (108)
The positive regulators include the SWI/SNF complex of which *INI1* is a part. The ATPase activity of the SWI/SNI2 homolog is essential for chromatin remodeling activity of all SWI/SNF complexes.	*Khavari et al 1993* (60) *Laurent et al 1993* (181) *Muchardt and Yaniv1999, 2001* (41, 42) *Phelan et al 1999* (182)
Components of the SWI/SNF are variable except for *INI1*.	*Peterson 1998* (183) *Schnitzler et al 1998* (185, 186) *Muchardt and Yaniv1999* (41)
SWI/SNF converts nucleosomes to a stable state. *MYC* interacts with *INI1* and requires SWI/SNF complex for transactivation function. Interaction of single components of SWI/SNF with pRB and *BRCA1* has been described.	*Schnitzler et al 1998* (185, 186) *Cheng et al 1999* (108) *Dunaief et al 1999* (186) *Bochar et al 2000* (187) *Klochendlerl-Yeivin et al 2002* (188)
SWI/SNF activity is required for *RB*-mediated transcriptional repression and subsequent inhibition of proliferation.	*Strobeck et al 2000* (189) *Wong et al 2000* (143) *Zhang et al 2000* (54)
SWI/SNF may play a role in oncogenesis. BRG1/*BRG1* and HBRM interact and collaborate with pRB and HDAC to repress E2F1/*E2F1* activity.	*Zhang et al 2000* (36)

The presence of changes of the *INI1* gene in choroid plexus tumors has been shown to affect a significant percentage of these tumors (carcinomas and papillomas) (84–88). The meaning of these findings and the enigma regarding the pathogenesis and cellular origin of these *INI1*-positive tumors have been discussed previously (82).

In view of the experience with medulloblastomas and the wide spectrum of histological presentations of AT/RT, some CNS tumors with inactivated *INI1* may be in fact, AT/RT. As experience to date has shown, the genotype of a tumor takes precedence over the phenotype in favoring (if not establishing) a precise diagnosis. Further molecular and genetic studies should clarify the patho- and histogenesis of these confusing entities. At the molecular level, there are already some indications that choroid plexus tumors may show changes that are not usually seen in AT/RT, e.g., mutations of p53, methylation of *RASSF1A* and *TRAIL*-related genes (88,89).

Constitutional (germline) mutations of the *INI1* gene in families with a variety of cancers and brain tumors have been documented (84, 90). The occurrence of choroid plexus carcinoma and AT/RT in two families is of direct interest to this section of the chapter. The findings point to a role for the *INI1* gene in the development of several tumors (MRT of both renal and soft tissue, and choroid plexus tumors) that may be related to their tumorigenetic process.

No reports of AT/RT in Li-Fraumeni syndrome families with p53 germline mutations have been documented, whereas choroid plexus tumors have been described in such families (91).

An examination of the cytogenetic changes reported in choroid plexus tumors revealed them to be pseudodiploid to hyperdiploid (92–97) without characteristic chromosome changes and only a few containing anomalies of chromosome 22 (97, 98). Thus, papillomas show hyperdiploidy with gain of chromosomes 7, 9, 12, 15, 17, and 18, whereas carcinomas show hyperdiploidy with rearrangements of 7p11-p12, 9q11-q12, 15q22, and 19q13.4 (99). Gain of 9p has been described in hyperplasia of the choroid plexus, a finding not observed in AT/RT (100).

Mutations or homozygous deletions have been detected in atypical papillary tumors or carcinomas of the choroid plexus, but not in papillomas (84,86,101). Phenotypic overlap between AT/RT and these tumors (102) is indicative of a

TABLE 6.9. Features of MRT or AT/RT versus composite rhabdoid tumor.

Parameter	MRT or AT/RT	Composite rhabdoid tumor
Patient age	Child (<3 years)	Adult
Location of tumor	Kidney, brain, soft tissue	Extrarenal visceral organs, dura, skin
Histology	Primitive neuroectodermal, Mesenchymal, and epithelial cells	Recognizable parent neoplasm
Immunohistochemistry	Polyphenotypic profile, loss of *INI1* expression	Single lineage profile, retain *INI1*
Ultrastructure	Lack of differentiation	Differentiation consistent with parent neoplasm
Genetics	22q deletions common as are *INI1* mutations and homozygous deletions	22q deletions uncommon; genetic features represent those of the parent neoplasm

common if not identical pathogenesis. Three choroid plexus papillomas had normal karyotypes, whereas of four atypical papillomas, three were abnormal (hyperdiploid) with only numerical chromosome changes. Two of these atypical papillomas had +22 and one of these atypical papillomas had dmin (17). A choroid plexus carcinoma had a hypodiploid karyotype with -22.

A malignant choroid plexus papilloma with a hypodiploid karyotype of 33–35 chromosomes retained two copies of chromosome 22 (103), whereas in another hypodiploid choroid plexus carcinoma with 33 chromosomes, monosomy 22 (-22), and several structural anomalies were found (104).

The occurrence of severe hypodiploidy in a significant portion of choroid plexus tumors examined cytogenetically is a finding not seen in AT/RT. Choroid plexus tumors with rhabdoid phenotype displayed -22 in three and del(22)(q12) in one phenotypic and genotypic overlap in AT/RT (102). A CGH study of 34 choroid plexus papillomas and 15 carcinomas (105) found chromosomal imbalances in all tumors except for two papillomas. A high incidence ($\geq$50% of the tumors) of gains was found for 7q, 5q, 7p, 5p, and 9p in the papillomas and for 12p, 12q, 20p, chromosome 1, 4q, and 20q in the carcinomas. Losses affected 10q in papillomas and 22q in the carcinomas. Loss of 22q in papillomas (47%) and carcinomas (73%) is not reflected in the cytogenetic findings of these tumors, and the disparity may be due to the differences in the types of data obtained cytogenetically versus those demonstrated by CGH.

Inactivation of *INI1* has been demonstrated in epithelioid sarcoma, a tumor affecting the distant extremities of young adults (106). Some of these tumors have involvement of 22q, including translocations. In a study of 11 such tumors (six proximal and five conventional types) by using spectral karyotyping, FISH, and array CGH, along with immunohistochemistry, Western blotting and reverse transcription-PCR, six of 11 (five were proximal) of the tumors showed inactivation of *INI1*, pointing to possible involvement of this gene in the genesis, progression, or both of these tumors (106). The authors (106) presented several arguments against epithelioid sarcoma being part of the MRT spectrum, including differences in histological aspects, restriction of *INI1* inactivation to deletions in epithelioid sarcomas versus the high frequency of point mutations in MRT and the heterogeneous distribution of these deletions, which in some cases may represent relevant secondary changes and not the primary genetic event.

Familial AT/RT and Rhabdoid Predisposition Syndrome (RPS)

RPS (107) was described in a family in which the proband had an AT/RT, her uncle a choroid plexus carcinoma, whereas the child's mother was asymptomatic, although she carried a germline *INI1* mutation. The pattern of inheritance was autosomal dominant with incomplete penetrance (107). The

RPS may account for cases of Li-Fraumeni syndrome lacking the p53 germline mutations (84). The mutation of *INI1* gene induced a splicing error, resulting in the truncation of the protein but preserved reactivity for *MYC* (107, 108). A RPS case has been described (109) who had a germline mutation of *INI1* but with absence of *MYC* amplification. RPS is highly penetrant (21, 84, 90).

A study of a family in which one child had a rhabdoid tumor of the kidney and another AT/RT showed no evidence of involvement of *INI1* gene in these tumors (110). These two cases were investigated by CGH and array CGH, FISH, gene dosage analysis by denaturing high-performance liquid chromatography, and DNA sequencing. Immunohistochemistry showed normal expression for *INI1* within the nuclei of the tumors. The affected children inherited different paternal and maternal *INI1* alleles determined by haplotype analysis.

In a study of a three-generation family in which two half-brothers were diagnosed with AT/RT at 2 and 17 months, respectively, a germline insertion mutation in exon 4 of the *INI1* gene inherited from their healthy mother was demonstrated (111). A maternal uncle died in childhood from a brain tumor and a MRT of the kidney and probably carried the same germline mutation. Because the mother and uncle had different fathers, the grandmother is also an obligate carrier of the mutation. The identification of two unaffected carriers in a family segregating a germline mutation and rhabdoid tumor supports the hypothesis that there may be variable risks of development of rhabdoid tumors in the context of a germline mutation (111). There may be a developmental window in which most rhabdoid tumors occur. The findings in this family highlight the importance of mutation analysis in all patients with a suspected rhabdoid tumor (111).

Immunohistochemical expression of INI1 protein was not found in tumors displaying characteristic morphologic features of AT/RT and in fraction of PNET (112). Furthermore, a certain number of tumors without rhabdoid phenotype but lacking the INI1 protein were detected by immunohistochemical analysis. These cases represent most likely AT/RT lacking rhabdoid features due to a sampling problem. Follow-up data suggested an aggressive biologic behavior of tumor with lack of INI1 protein, and poor outcome if treated by conventional therapy regimens. Therefore, immunohistochemical INI1 protein analysis should be routinely performed in all embryonal pediatric CNS tumors to detect cases with lack of INI1 protein, because patients with such tumors are likely to benefit from intensified treatment (112).

References

1. Lefkowitz, I.B., Rorke, L.B., and Packer, R.J. (1987) Atypical teratoid tumor of infancy: definition of an entity. *Ann. Neurol.* **22**, 448–449.
2. Rorke, L.B., Packer, R., and Biegel, J. (1995) Central nervous system atypical teratoid/rhabdoid tumors of infancy and childhood. *J. Neurooncol..* **24**, 21–28.

3. Rorke, L.B., Packer, R.J., and Biegel, J.A. (1996) Central nervous system atypical teratoid/rhabdoid tumors of infancy and childhood: definition of an entity. *J. Neurosurg.* **85,** 56–65.

4. Rorke, L.B., and Biegel, J.A. (2000) Atypical teratoid/rhabdoid tumour. In *World Health Organization Classification of Tumours. Pathology and Genetics of Tumours of the Nervous System* (Kleihues, P., and Cavenee, W.K., eds.), IARC Press, Lyon, France, pp. 145–148.

4A. Kleihues, P., Louis, D.N., Scheithauer, B.W., Rorke, L.B., Reifenberger, G., Burger, P.C., and Cavenee, W.K. (2002) The WHO classification of tumors of the nervous system. *J. Neuropathol. Exp. Neurol.* **61,** 215–225.

5. Rorke, L.B. (1997) Atypical teratoid/rhabdoid tumours. In *Pathology and Genetics–Tumours of the Nervous System* (Kleihues, P., and Cavenee, W.K., eds.), IARC Press, Lyon, France, pp. 110–111.

6. Behring, B., Brück, W., Goebel, H.H., Behnke, J., Pekrun, A., Christen, H.-J., and Kretzschmar, H.A. (1996) Immunohistochemistry of primary central nervous system malignant rhabdoid tumors: report of five cases and review of the literature. *Acta Neuropathol.* **91,** 578–586.

7. Burger, P.C., Yu, I.-T., Tihan, T., Friedman, H.S., Strother, D.R., Kepner, J.L., Duffner, P.K., Kun, L.E., and Perlman, E.J. (1998) Atypical teratoid/rhabdoid tumor of the central nervous system: a highly malignant tumor of infancy and childhood frequently mistaken for medulloblastoma. A Pediatric Oncology Group Study. *Am. J. Surg. Pathol.* **22,** 1083–1092.

8. Hilden, J.M., Watterson, J., Longee, D.C., Moertel, C.L., Dunn, M.E., Kurtzberg, J., and Scheithauer, B.W. (1998) Central nervous system atypical teratoid/rhabdoid tumor: response to intensive therapy and review of the literature. *J. Neurooncol.* **49,** 265–275.

9. Lee, M.-C., Park, S.-K., Lim, J.-S., Jung, S., Kim, J.-H., Woo, Y.-J., Lee, J.-S., Kim, H.-I., Jeong, M.-J., and Cho, H.-Y. (2002) Atypical teratoid/rhabdoid tumor of the central nervous system: clinico-pathological study. *Neuropathology* **22,** 252–260.

10. Packer, R.J., Biegel, J.A., Blaney, S., Finlay, J., Geyer, J.R., Heideman, R., Hilden, J., Janss, A.J., Kun, L., Vezina, G., Rorke, L.B., and Smith, M. (2002) Atypical teratoid/rhabdoid tumor of the central nervous system: report on workshop. *J. Pediatr. Hematol./Oncol.* **24,** 337–342.

11. Miettinen, M. (2003) Extra renal rhabdoid tumors. In *World Health Organization Classification of Tumours. Pathology and Genetics of Tumours of the Nervous System* (Kleihues, P., and Cavenee, W.K., eds.), IARC Press, Lyon, France, pp. 449–455.

12. Tekkök, I.H., and Sav, A. (2005) Primary malignant rhabdoid tumor of the central nervous system–a comprehensive review. *J. Neurooncol.* **73,** 241–252.

12A. Strother, D. (2005) Atypical teratoid rhabdoid tumors of childhood: diagnosis, treatment and challenges. *Expert Rev. Anticancer Ther.* **5,** 907–915.

13. Cohn, R.D., Frank, Y., Stanek, A.E., and Kalina, P. (1995) Malignant rhabdoid tumor of the brain and kidney in a child: clinical and pathologic features. *Pediatr. Neurol.* **13,** 65–68.

14. Parellada, J.A., Gupta, P., and Rose, W. (1998) Rhabdoid tumor of the kidney with primitive neuroectodermal tumor of the central nervous system. *Eur. J. Radiol.* **28,** 219–221.

15. Perilongo, G., Sutton, L., Czaykowski, D., Gusnard, D., and Biegel, J. (1991) Rhabdoid tumor of the central nervous system. *Med. Pediatr. Oncol.* **19,** 310–317.

16. Cossu, A., Massarelli, G., Manetto, V., Viale, G., Tanda, F., Bosincu, L., Iuzzolino, P., Cossu, S., Padovani, R., and Eusebi, V. (1993) Rhabdoid tumours of the central nervous system. *Virchows Archiv. A. Pathol. Anat.* **422,** 81–85.

17. Bhattacharjee, M.B., Armstrong, D.D., Vogel, H., and Cooley, L.D. (1997) Cytogenetic analysis of 120 primary pediatric brain tumors and literature review. *Cancer Genet. Cytogenet.* **97,** 39–53.

18. Bhattacharjee, M., Hicks, J., Dauser, R., Strother, D., Chintagumpala, M., Horowitz, M., Cooley, L., and Vogel, H. (1997) Primary malignant rhabdoid tumor of the central nervous system. *Ultrastruct. Pathol.* **21,** 361–368.

19. Bhattacharjee, M., Hicks, J., Langford, L., Dauser, R., Strother, D., Chintagumpala, M., Horowitz, M., Cooley, L., and Vogel, H. (1997) Central nervous system atypical teratoid/rhabdoid tumors of infancy and childhood. *Ultrastruct. Pathol.* **21,** 369–3378.

20. Aguzzi, A., Brandner, S., and Paulus, W. (2000) Choroid plexus tumours. In *World Health Organization Classification of Tumours. Pathology and Genetics of Tumours of the Nervous System* (Kleihues, P., and Cavenee, W.K., eds.), IARC Press, Lyon, France, pp. 84–86.

21. Versteege, I., Sévenet, N., Lange, J., Rousseau-Merck, M.-F., Ambros, P., Handgretinger, R., Aurias, A., and Delattre, O. (1998) Truncating mutations of hSNF5/INI1 in aggressive paediatric cancer. *Nature* **394,** 203–206.

22. Biegel, J.A., Zhou, J.-Y., Rorke, L.B., Stenstrom, C., Wainwright, L.M., and Fogelgren, B. (1999) Germ-line and acquired mutations of *INI1* in atypical teratoid and rhabdoid tumors. *Cancer Res.* **59,** 74–79.

23. Rousseau-Merck, M.-F., Versteege, I., Legrand, I., Couturier, J., Mairal, A., Delattre, O., and Aurias, A. (1999) *hSNF5/INI1* inactivation is mainly associated with homozygous deletions and mitotic recombinations in rhabdoid tumors. *Cancer Res.* **59,** 3152–3156.

24. Yachnis, A.T., Neubauer, D., and Muir, D. (1998) Characterization of a primary central nervous system atypical teratoid/rhabdoid tumor and derivative cell line: immunophenotype and neoplastic properties. *J. Neuropathol. Exp. Neurol.* **57,** 961–971.

25. Rosty, C., Peter, M., Zucman, J., Validire, P., Delattre, O., and Aurias, A. (1998) Cytogenetic and molecular analysis of a t(1;22)(p36;q11.2) in a rhabdoid tumor with a putative homozygous deletion of chromosome 22. *Genes Chromosomes Cancer* **21,** 82–89.

26. Bruch, L.A., Hill, A., Cai, D.X., Levy, B.K., Dehner, L.P., and Perry, A. (2001) A role for fluorescence in situ hybridization detection of chromosome 22q dosage in distinguishing atypical teratoid/rhabdoid tumors from medulloblastoma/central primitive neuroectodermal tumors. *Hum. Pathol.* **32,** 156–162.

27. Raisanen, J., Biegel, J.A., Hatanpaa, K.J., Judkins, A., White, C.L., III, and Perry, A. (2005) Chromosome 22q deletions in atypical teratoid/rhabdoid tumors in adults. *Brain Pathol.* **15,** 23–28.

28. Kalpana, G.V., Marmon, S., Wang, W., Crabtree, G.R., and Goff, S.P. (1994) Binding and stimulation of HIV-1

integrase by a human homolog of yeast transcription factor SNF5. *Science* **266,** 2002–2006.

29. Kingston, R.E., Bunker, C.A., and Imbalzano, A.N. (1996) Repression and activation by multiprotein complexes that alter chromatin structure. *Genes Dev.* **10,** 905–920.

30. Biegel, J.A. (2006) Molecular genetics of atypical teratoid/rhabdoid tumor. *Neurosurg. Focus* **20,** E11.

31. Biegel, J.A., Kalpana, G., Knudsen, E.S., Packer, R.J., Roberts, C.W.M., Thiele, C.J., Weissman, B., and Smith, M. (2002) The role of *INI1* and the SWI/SNF complex in the development of rhabdoid tumors: meeting summary from the Workshop on Childhood Atypical Teratoid/Rhabdoid Tumors. *Cancer Res.* **62,** 323–328.

31A. Biegel, J.A. (2006) Molecular genetics of atypical teratoid/rhabdoid tumor. *Neurosurg. Focus* **20,** E11.

32. Biegel, J.A., Tan, L., Zhang, F., Wainwright, L, Russo, P., and Rorke, L.B. (2002) Alterations of the *hSNF5/INI1* gene in central nervous system atypical teratoid/rhabdoid tumors and renal and extrarenal rhabdoid tumors. *Clin. Cancer Res.* **8,** 3461–3467.

33. Ae, K., Kobayashi, N., Sakuma, R., Ogata, T., Kuroda, H., Kawaguchi, N., Shinomiya, K., and Kitamura, Y. (2002) Chromatin remodeling factor encoded by ini1 induces G1 arrest and apoptosis in ini1-deficient cells. *Oncogene* **21,** 3112–3120.

34. Betz, B.L., Strobeck, M.W., Reisman, D.N., Knudsen, E.S., and Weissman, B.E. (2002) Re-expression of hSNF5/INI1/BAF47 in pediatric tumor cells leads to G$_1$ arrest associated with induction of p16ink4a and activation of RB. *Oncogene* **21,** 5193–5203.

35. Versteege, I., Medjkane, S., Rouillard, D., and Delattre, O. (2002) A key role of the hSNF5/INI1 tumour suppressor in the control of the G1-S transition of the cell cycle. *Oncogene* **21,** 6403–6412.

36. Zhang, H.S., Gavin, M., Dahiya, A.., Postigo, A.A., Ma., D., Luo, R.X., Harbour, J.W., and Dean, D.C. (2000) Exit from G$_1$ and S phase of the cell cycle is regulated by repressor complexes containing HDAC-Rb-hSWI/SNF and Rb-hSWI/SNF. *Cell* **101,** 79–89.

37. Vries, R.G.J., Bezrookove, V., Zuijderduijn, L.M.P., Kheradmand Kia, S., Houweling, A., Oruetxebarria, I., Raap, A.K., and Verrijzer, C.P. (2005) Cancer-associated mutations in chromatin remodeler hSNF5 promote chromosomal instability by compromising the mitotic checkpoint. *Genes Dev.* **19,** 665–670.

38. Roberts, C.W.M., Galusha, S.A., McMenamin, M.E., Fletcher, C.D.M., and Orkin, S.H. (2000) Haploinsufficiency of Snf5 (integrase interactor 1) predisposes to malignant rhabdoid tumors in mice. *Proc. Natl. Acad. Sci. USA* **97,** 13796–13800.

39. Roberts, C.W., Leroux, M.M., Fleming, M.D., AND Orkin, S.H. (2002) Highly penetrant, rapid tumorigenesis through conditional inversion of the tumor suppressor gene Snf5. *Cancer Cell* **2,** 415–425.

40. Guidi, C.J., Sands, A.T., Zambrowicz, B.P., Turner, T.K., Demers, D.A., Webster, W., Smith, T.W., Imbalzano, A.N., and Jones, S.N., (2001) Disruption of Ini1 leads to peri-implantation lethality and tumorigenesis in mice. *Mol. Cell. Biol.* **21,** 3598–3603.

41. Muchardt, C., and Yaniv, M. (1999) The mammalian SWI/SNF complex and the control of cell growth. *Semin. Cell Dev. Biol.* **10,** 165–167.

42. Muchardt, C., and Yaniv, M. (2001) When the SWI/SNF complex remodels the cell cycle. *Oncogene* **20,** 3067–3075.

43. Medjkane, S., Novikov, E., Versteege, I., and Delattre, O. (2004) The tumor suppressor hSNF5/INI1 modulates cell growth and actin cytoskeleton organization. *Cancer Res.* **64,** 3406–3413.

44. Reincke, B.S., Rosson, G.B., Oswald, B.W., and Wright, C.F. (2003) INI1 expression induces cell cycle arrest and markers of senescence in malignant rhabdoid tumor cells. *J. Cell. Physiol.* **194,** 303–313.

45. Fuller, C.E., Pfeifer, J., Humphrey, P., Bruch, L.A., Dehner, L.P., and Perry, A. (2001) Chromosome 22q dosage in composite extrarenal rhabdoid tumors: clonal evolution or a phenotype mimic? *Hum. Pathol.* **32,** 1102–1108.

46. Uno, K., Takita, J., Yokomori, K., Tanaka, Y., Ohta, S., Shimada, H., Gilles, F.H., Sugita, K., Abe, S., Sako, M., Hashizume, K., and Hayashi, Y. (2002) Aberrations of the *hSNF5/INI1* gene are restricted to malignant rhabdoid tumors or atypical teratoid/rhabdoid tumors in pediatric solid tumors. *Genes Chromosomes Cancer* **34,** 33–41.

47. Sigauke, E., Rakheja, D., Maddox, D.L., Hladik, C.L., White, C.L, III, Timmons, C.F., and Raisanen, J. (2006) Absence of expression of *SMARCB1/INI1* in malignant rhabdoid tumors of the central nervous system, kidneys and soft tissue: an immunohistochemical study with implications for diagnosis. *Mod. Pathol.* **19,** 717–725.

48. Biegel, J.A., Allen, C.S., Kawasaki, K., Shimizu, N., Budarf, M.L., and Bell, C.J. (1996) Narrowing the critical region for a rhabdoid tumor locus in 22q11. *Genes Chromosomes Cancer* **16,** 94–105.

49. Biegel, J.A., Fogelgren, B., Wainwright, L.M., Zhou, J.-Y., Bevan, H., and Rorke, L.B. (2000) Germline *INI1* mutation in a patient with a central nervous system atypical teratoid tumor and renal rhabdoid tumor. *Genes Chromosomes Cancer* **28,** 31–37.

50. Rousseau-Merck, M.-F., Fiette, L., Klochendler-Yeivin, A., Delattre, O., and Aurias, A. (2005) Chromosome mechanisms and *INI1* inactivation in human and mouse rhabdoid tumors. *Cancer Genet. Cytogenet.* **157,** 127–133.

51. Simons, J., Teshima, I., Zielenska, M., Edwards, V., Taylor, G., Squire, J., and Thorner, P. (1999) Analysis of chromosome 22q as an aid to the diagnosis of rhabdoid tumor. A case report. *Am. J. Surg. Pathol.* **23,** 982–988.

52. Lòpez-Ginés, C., Cerda-Nicolas, M., Kepes, J., Donat, J., Gil-Benso, R., and Llombart-Bosch, A. (2000) Complex rearrangement of chromosomes 6 and 11 as the sole anomaly in atypical teratoid/rhabdoid tumors of the central nervous system. *Cancer Genet. Cytogenet.* **122,** 149–152.

53. Izycka-Swieszewska, E., Debiec-Rychter, M., Wasag, B., Wozniak, A., Gasecki, D., Plata-Nazar, K., Bartkowiak, J., Lasota, J., and Limon, J. (2003) A unique occurrence of a cerebral atypical teratoid/rhabdoid tumor in an infant and a spinal canal primitive neuroectodermal tumor in her father. *J. Neurooncol.* **61,** 219–225.

54. Zhang, F., Tan, L., Wainwright, L.M., Bartolomei, M.S., and Biegel, J.A. (2002) No evidence for hypermethylation of the

hSNF5/INI1 promoter in pediatric rhabdoid tumors. *Genes Chromosomes Cancer* **34**, 398–405.

55. Fujiisawa, H., Misaki, K., Takabatake, Y., Hasegawa, M., and Yamashita, J. (2005) Cyclin D1 is overexpressed in atypical teratoid/rhabdoid tumor with *hSNF5/INI1* gene inactivation. *J. Neurooncol.* **73**, 117–124.

56. Zhang, Z.-K., Davies, K.P., Allen, J., Zhu, L., Pestell, R.G., Zagzag, D., and Kalpana, G.V. (2002) Cell cycle arrest and repression of cyclin D1 transcription by INI1/hSNF5. *Mol. Cell. Biol.* **22**, 5975–5988.

57. Tsikitis, M., Zhang, Z., Edelman, W., Zagzag, D., and Kalpana, G.V. (2005) Genetic ablation of *cyclin D1* abrogates genesis of rhabdoid tumors resulting from *Ini1* loss. *Proc. Natl. Acad. Sci. USA* **102**, 12129–12134.

58. Fu, M., Wang, C., Li, Z., Sakamaki, T., and Pestell, R.G. (2004) Minireview: cyclin D1: normal and abnormal functions. *Endocrinology* **145**, 5439–5447.

59. Alarcon-Vargas, D., Zhang, Z., Agarwal, B., Challagulla, K., Mani, S., and Kalpana, G.V. (2006) Targeting cyclin D1, a downstream effector of INI1/hSNF5, in rhabdoid tumors. *Oncogene* **25**, 722–734.

59A. Fujisawa, H., Misaki, K., Takabatake, Y., Hasegawa, M., and Yamashita, J. (2005) Cyclin D1 is overexpressed in atypical teratoid/rhabdoid tumor with *hSNF5/INI1* gene inactivation. *J. Neurooncol.* **73**, 117–124.

60. Khavari, P.A., Peterson, C.L., Tamkun, J.W., Mendel, D.B., and Crabtree, G.R. (1993) BRG1 contains a conserved domain of the SWI2/SNF2 family necessary for normal mitotic growth and transcription. *Nature* **366**, 170–174.

61. Wharton, S.B., Wardle, C., Ironside, J.W., Wallace, W.H., Royds, J.A., and Hammond, D.W. (2003) Comparative genomic hybridization and pathological findings in atypical teratoid/rhabdoid tumour of the central nervous system. *Neuropathol. Appl. Neurobiol.* **29**, 254–261.

62. Shiratsuchi, H., Oshiro, Y., Saito, T., Itakura, E., Kinoshita, Y., Tamiya, S., Oda, Y., Komiyama, S., and Suneyhosi, M. (2001) Cytokeratin subunits of inclusion bodies in rhabdoid cells: immunohistochemical and clinicopathological study of malignant rhabdoid tumor and epithelioid sarcoma. *Int. J. Surg. Pathol.* **9**, 37–48.

63. Shiratsuchi, H., Saito, T., Sakamoto, A., Itakura, E., Tamiya, S., Oshiro, Y., Oda, Y., Toh, S., Komiyama, S., and Tsuneyoshi, M. (2002) Mutation analysis of human cytokeratin 8 gene in malignant rhabdoid tumor: a possible association with intracytoplasmic inclusion body formation. *Mod. Pathol.* **15**, 146–153.

64. Itakura, E., Tamiya, S., Morita, K., Shiratsuchi, H., Kinoshita, Y., Oshiro, Y., Oda, Y., Ohta, S., Furue, M., and Tsuneyoshi, M. (2001) Subcellular distribution of cytokeratin and vimentin in malignant rhabdoid tumor: three-dimensional imaging with confocal laser scanning microscopy and double immunofluorescence. *Mod. Pathol.* **14**, 854–861.

65. Horie, H., Etoh, T., and Maie, M. (1992) Cytogenetic characteristics of a malignant rhabdoid tumor arising from the paravertebral region. A case report. *Acta Pathol. Jpn.* **42**, 460–465.

66. Ogino, S., Ro, J.Y., and Redline, R.W. (2000) Malignant rhabdoid tumor: a phenotype? An entity?—a controversy revisited. *Adv. Anat. Pathol.* **7**, 181–190.

67. Ota, S., Crabbe, D.C.G., Tran, T.N., Triche, T.J., and Shimada, H. (1993) Malignant rhabdoid tumor. A study with two established cell lines. *Cancer* **71**, 2862–2872.

68. Suzuki, A, Ohta, S., and Shimada, M. (1997) Gene expression of malignant rhabdoid tumor cell lines by reverse transcriptase-polymerase chain reaction. *Diagn. Mol. Pathol.* **6**, 326–332.

69. Rosson, G.B., Hazen-Martin, D.J., Biegel, J.A., Willingham, M.C., Garvin, A.J., Oswald, B.W., Wainwright, L., Brownlee, N.A., and Wright, C.F. (1998) Establishment and molecular characterization of five cell lines derived form renal and extrarenal malignant rhabdoid tumors. *Mod. Pathol.* **11**, 1228–1237.

70. Sugimoto, T., Hosoi, H., Horii, Y., Ishida, H., Mine, H., Takahashi, K., Abe, T., Ohta, S, and Sawada, T. (1999) Malignant rhabdoid-tumor cell line showing neural and smooth-muscle-cell phenotypes. *Int. J. Cancer* **82**, 678–686.

71. White, F.V., Dehner, L.P., Belchis, D.A., Conard, K., Davis, M.M., Stocker, J.T., Zuppan, C.W., Biegel, J.A., and Perlman, E.J. (1999) Congenital disseminated malignant rhabdoid tumor. a distinct clinicopathologic entity demonstrating abnormalities of chromosome 22q11. *Am. J. Surg. Pathol.* **23**, 249–256.

72. Bonnin, J.M., Rubinstein, L.J., Palmer, N.F., and Beckwith, J.B. (1984) The association of embryonal tumors originating in the kidney and in the brain. A report of seven cases. *Cancer* **54**, 2137–2146.

73. Perry, A., Fuller, C.E., Judkins, A.R., Dehner, L.P., and Biegel, J.A. (2005) INI1 expression is retained in composite rhabdoid tumors, including rhabdoid meningiomas. *Mod. Pathol.* **18**, 951–958.

74. Bannykh, S.I., Perry, A., Powell, H.C., Hill, A., and Hansen, L.A. (2002) Malignant rhabdoid meningioma arising in the setting of preexisting ganglioglioma: a diagnosis supported by fluorescence in situ hybridization. *J. Neurosurg.* **97**, 1450–1455.

75. Hauser, P., Slowik, F., Bognar, L., Babosa, M., Fazekas, I., and Schuler, D. (2001) Atypical teratoid/rhabdoid tumor or medulloblastoma? *Med. Pediatr. Oncol.* **36**, 644–648.

76. Hojo, H., and Abe, M. (2001) Rhabdoid papillary meningioma. *Am. J. Surg. Pathol.* **25**, 964–969.

77. Rittierodt, M., Tschernig, T., Samii, M., Walter, G.F., and Stan, A.C. (2001) Evidence of recurrent atypical meningioma with rhabdoid transformation and expression of pyrogenic cytokines in a child presenting with a marked acute-phase response: case report and review of the literature. *J. Neuroimmunol.* **120**, 129–137.

78. Wyatt-Ashmead, J., Kleinschmidt-DeMasters, B.K., Hill, D.A., Mierau, G.W., McGavran, L., Thompson, S.J., and Foreman, N.K. (2001) Rhabdoid glioblastoma. *C.lin. Neuropathol.* **20**, 248–255.

79. Ashkenazai, A., and Dixit, V.M. (1999) Apoptosis control by death and decoy receptors. *Curr. Opin. Cell Biol.* **11**, 225–260.

80. Walczak, H., Miller, R.E., Ariail, K., Gliniak, B., Griffith, T.S., Kubin, M., Chin, W., Jones, J., Woodward, A., Le, T., Smith, C., Smolak, P., Goodwin, R.G., Rauch, C.T., Schuh, J.C., and Lynch, D.H. (1999) Tumiricidal activity of tumor necrosis factor-related apoptosis-inducing ligand *in vivo*. *Nat. Med.* **5**, 157–163.

81. Yoshida, S., Narita, T., Koshida, S., Ohta, S., and Takeuchi, Y. (2003) TRAIL/Apo2L ligands induce apoptosis in malignant rhabdoid tumor cell lines. *Pediatr. Res.* **54,** 709–717.

82. Gessi, M., Giangaspero, F., and Pietsch, T. (2003) Atypical teratoid/rhabdoid tumors and choroids plexus tumors: when genetics "surprise" pathology. *Brain Pathol.* **13,** 409–414.

83. Kraus, J.A., de Millas, W., Sörensen, N., Herbold, C., Schichor, C., Tonn, J.C., Wiestler, O.D., von Deimling, A., and Pietsch, T. (2001) Indications for a tumor suppressor gene at 22q11 involved in the pathogenesis of ependymal tumors and distinct from *hSNF5/INI1*. *Acta Neuropathol.* **102,** 69–74.

84. Sévenet, N., Lellouch-Tubiana, A., Schofield, D., Hoang-Xuan, K., Gessler, M., Birnbaum, D., Jeanpierre, C., Jouvet, A., and Delattre, O. (1999) Spectrum of *hSNF5/INI1* somatic mutations in human cancer and genotype-phenotype correlations. *Hum. Mol. Genet.* **8,** 2359–2368.

85. Chen, T., Maitra, A., Salva, J., Wyatt-Ashmead, J., and Tomlinson, G. (2001) Mutational analysis of hSNF5/INI1 in choroid plexus carcinoma and rhabdoid tumors. *Proc. Am. Assoc. Cancer Res.* **42,** 860.

86. Weber, M., Stockhammer, F., Schmitz, U., and von Deimling, A. (2001) Mutational analysis of *INI1* in sporadic human brain tumors. *Acta Neuropathol.* **101,** 479–482.

87. Gessi, M., Koch, A., Milde, U., Wiestler, O.D., and Pietsch, T. (2002) Analysis of hSNF5/INI-1 and TP53 in chodoir plexus tumor. *Proc. Am. Assoc. Cancer Res.* **43,** 1138.

88. Zakrzewska, M., Wojcik, I., Zakrzewski, K., Polis, L., Grajkowska, W., Roszkowski, M., Augelli, B.J., Liberski, P.P., and Rieske, P. (2005) Mutational analysis of *hSNF5/INI1* and *TP53* gene in choroid plexus carcinomas. *Cancer Genet. Cytogenet.* **156,** 179–182.

89. Michalowski, M.B., de Fraipont, F., Michelland, S., Entz-Werle, N., Grill, J., Pasquier, B., Favrot, M.-C., and Plantaz, D. (2006) Methylation of *RASSF1A* and TRAIL pathway-related genes is frequent in childhood intracranial ependymomas and benign choroid plexus papilloma. *Cancer Genet. Cytogenet.* **166,** 74–81.

90. Sévenet, N., Sheridan, E., Amram, D., Schneider, P., Handgretinger, R., and Delattre, O. (1999) Constitutional mutations of the *hSNF5/INI1* gene predisposes to a variety of cancers. *Am. J. Hum. Genet.* **65,** 1342–1348.

91. Vital, A., Bringuier, P.-P., Huang, H., San Galli, F., Rivel, J., Ansoborlo, S., Cazauran, J.-M., Taillandier, L., Kleihues, P., and Ohgaki, H. (1998) Astrocytomas and choroid plexus tumors in two families with identical *p53* germline mutations. *J. Neuropathol. Exp. Neurol.* **57,** 1061–1069.

92. Chadduck, W.M., Boop, F.A., and Sawyer, J.R. (1991–92) Cytogenetic studies of pediatric brain and spinal cord tumors. *Pediatr. Neurosurg.* **17,** 57–65.

93. Donovan, M.J., Yunis, E.J., DeGirolami, U., Fletcher, J.A., and Schofield, D.E. (1994) Chromosome aberrations in choroid plexus carcinoma. *Genes Chromosomes Cancer* **11,** 267–270.

94. Punnett, H.H., Tomczak, E.Z., de Chadarevian, J.-P., and Kanev, P.M. (1994) Cytogenetic analysis of a choroid plexus papilloma. *Genes Chromosomes Cancer* **10,** 282–285.

95. Mertens, F.., Heim, S., Mandahl, N., Mitelman, F., Brun, A., Strömblad, L.-G., Kullendorff, C.-M., and Donner, M. (1995) Recurrent chromosomal imbalances in choroid plexus tumors. *Cancer Genet. Cytogenet.* **80,** 83–84.

96. Roland, B., and Pinto, A. (1996) Hyperdiploid karyotype in a choroid plexus papilloma. *Cancer Genet. Cytogenet.* **90,** 130–131.

97. Dobin, S.M., and Donner, L.R. (1997) Pigmented choroid plexus carcinoma: a cytogenetic and ultrastructural study. *Cancer Genet. Cytogenet.* **96,** 37–41.

98. Li, Y.S., Fan, Y.-S., and Armstrong, R.F. (1996) Endoreduplication and telomeric association in a choroid plexus carcinoma. *Cancer Genet. Cytogenet.* **87,** 7–10.

99. Rickert, C.H., and Paulus, W. (2001) Tumor of the choroid plexus. *Microsc. Res. Tech.* **52,** 104–111.

100. Norman, M.G., Harrison, K.J., Poskitt, K.J., and Kalousek, D.K. (1995) Duplication of 9p and hyperplasia of the choroid plexus: a pathologic, radiologic and molecular cytogenetics study. *Pediatr. Pathol. Lab. Med.* **15,** 109–120.

101. Savla, J., Chen, T.T.-Y., Schneider, N.R., Timmons, C.F., Delattre, O., and Tomlinson, G.E. (2000) Mutations of the hSNF5/INI1 gene in renal rhabdoid tumors with second primary brain tumors. *J. Natl. Cancer Inst.* **92,** 648–650.

102. Wyatt-Ashmead, J., Kleinschmidt-DeMasters, B., Mierau, G.W., Malkin, D., Orsini, E., McGavran, L., and Foreman, N.K. (2001) Choroid plexus carcinomas and rhabdoid tumors: phenotypic and genotypic overlap. *Pediatr. Dev. Pathol.* **4,** 545–549.

103. Petersen, S.E., Frederiksen, P., and Friedrich, U. (1981) Cytogenetic analysis and flow cytometric DNA measurement of a human tumor with pronounced hypodiploidy. *Cancer Genet.Cytogenet.* **4,** 1–9.

104. Neumann, E., Kalousek, D.K., Norman, M.G., Steinbok, P., Cochrane, D.D., and Goddard, K. (1993) Cytogenetic analysis of 109 pediatric central nervous system tumors. *Cancer Genet. Cytogenet.* **71,** 40–49.

105. Rickert, C.H., Wiestler, O.D., and Paulus, W. (2002) Chromosomal imbalances in choroid plexus tumors. *Am. J. Pathol.* **160,** 1105–1113.

106. Modena, P., Lualdi, E., Facchinetti, F., Galli, L., Teixeira, M.R., Pilotti, S., and Sozzi, G. (2005) *SMARCB1/*INI1 tumor suppressor gene is frequently inactivated in epithelioid sarcomas. *Cancer Res.* **65,** 4012–4019.

107. Taylor, M.D., Gokgoz, N., Andrulis, I.L., Mainprize, T.G., Drake, J.M., and Rutka, J.T. (2000) Familial posterior fossa brain tumors of infancy secondary to germline mutation of the *hSNF5* gene. *Am. J. Hum. Genet.* **66,** 1403–1406.

108. Cheng, S.-W.G., Davies, K.P., Yung, E., Beltran, R.J., Yu, J., and Kalpana, G.V. (1999) c-MYC interacts with INI1/hSNF5 and requires the SWI/SNF complex for transactivation function. *Nat. Genet.* **22,** 102–105.

109. Fujisawa, H., Takabatake, Y., Fukusato, T., Tachibana, O., Tsuchiya, Y., and Yamashita, J. (2003) Molecular analysis of the rhabdoid predisposition syndrome in a child: a novel germline *hSNF5/INI1* mutation and absence of *c-myc* amplification. *J. Neurooncol.* **63,** 257–262.

110. Frühwald, M.C., Hasselblatt, M., Wirth, S., Kohler, G., Schneppenheim, R., Subero, J.I., Siebert, R., and Kordes, U., Jurgens, H., and Vormoor, J. (2006) Non-linkage of familial rhabdoid tumors to *SMARCB1* implies a second locus for the rhabdoid tumor predisposition syndrome. *Pediatr. Blood Cancer* **47,** 273–278.

111. Janson, K., Nedzi, L.A., David, O., Schorin, M., Walsh, J.W., Bhattacharjee, M., Pridjian, G., Tan, L., Judkins, A.R., and

Biegel, J.A. (2006) Predisposition to atypical teratoid/rhabdoid tumor due to an inherited *INI1* mutation. *Pediatr. Blood Cancer* **47**, 279–284.

112. Haberler, C., Laggner, U., Slavc, I., Czech, T., Ambros, I.M., Ambros, P.F., Budka, H., and Hainfellner, J.A. (2006) Immuno-histochemical analysis of INI1 protein in malignant pediatric CNS tumors: lack of INI1 in atypical teratoid/rhabdoid tumors and in a fraction of primitive neuroectodermal tumors without rhabdoid phenotype. *Am. J. Surg. Pathol.* **30**, 1462–1468.

113. Weinblatt, M., and Kochen, J. (1992) Rhabdoid tumor of the central nervous system. *Med. Pediatr. Oncol.* **20**, 258.

114. Muller, M., Hubbard, S.L., Provias, J., Greenberger, M., Becker, L.E., and Rutka, J.T. (1994) Malignant rhabdoid tumour of the pineal region. *Can. J. Neurol. Sci.* **21**, 273–277.

115. Wick, M.R., Ritter, J.H., and Dehner, L.P. (1995) Malignant rhabdoid tumors: a clinicopathologic review and conceptual discussion. *Semin. Diagn. Pathol.* **12**, 233–248.

116. Bergmann, M., Spaar, H.J., Ebhard, G., Masini, T., Edel, G., Gullotta, F., and Meyer, H. (1997) Primary malignant rhab-doid tumours of the central nervous system: an immunohisto-chemical and ultrastructural study. *Acta Neurochir. (Wien)* **139**, 961–969.

117. Korones, D.N., Meyers, S.P., Rubio, A., Torres, C., and Cons-tine, L.S. (1999) A 4-year-old girl with a ventriculoperitoneal shunt metastasis of a central nervous system atypical tera-toid/rhabdoid tumor. *Med. Peidatr. Oncol.* **32**, 389–391.

118. Oka, H., and Scheithauer, B.W. (1999) Clinicopathological characteristics of atypical teratoid/rhabdoid tumor. *Neurol. Med. Chir. (Tokyo)* **39**, 510–517.

119. Oulab Ben Taib, N., Nagy, N., Salmon, I., David, P., Brotchi, J., and De Witte, O. (1999) Malignant rhabdoid tumors of the central nervous system. Report of a case. *Neuro-Chirurgie* **45**, 237–242.

120. Bambakidis, N.C., Robinson, S., Cohen, M., and Cohen, A.R. (2002) Atypical teratoid/rhabdoid tumors of the central nervous system: clinical, radiographic and pathologic features. *Pediatr. Neurosurg.* **37**, 64–70.

121. Hilden, J.M., Meerbaum, S., Burger, P., Finlay, J., Janss, A., Scheithauer, B.W., Walter, A.W., Rorke, L.B., and Biegel, J.A. (2004) Central nervous system atypical teratoid/rhabdoid tumor: results of therapy in children enrolled in a registry. *J. Clin. Oncol.* **22**, 2877–2884.

122. Weiss, E., Behring, B., Behnke, J., Christen, H.J., Pekrun, A., and Hess, C.F. (1998) Treatment of primary malignant rhab-doid tumor the brain: report of three cases and review of the literature. *Int. J. Radiat. Oncol. Biol. Phys.* **41**, 1013–1019.

123. Hirth, A., Pedersen, P.H., Wester, K., Mork, S., and Hilgestad, J. (2003) Cerebral atypical teratoid/rhabdoid tumor of infancy: long-term survival after multimodal treatment, also including triple intrathecal chemotherapy and gamma knife radiosurgery—case report. *Pediatr. Hematol. Oncol.* **20**, 327–332.

124. Howes, T.L., Buatti, J.M., O'Dorisio, M.S., Kirby, P.A., and Ryken, T.C. (2005) Atypical teratoid/rhabdoid tumor case report: treatment with surgical excision, radiation therapy, and alternative medicine. *J. Neurooncol.* **72**, 85–88.

125. Zimmerman, M.A., Goumnerova, L.C., Proctor, M., Scott, R.M., Marcus, K., Pomeroy, S.L., Turner, C.D., Chi, S.N., Chordas, C., and Kieran, M.W. (2005) Continuous remission of newly diagnosed and relapsed central nervous

system atypical teratoid/rhabdoid tumor. *J. Neurooncol.* **72**, 77–84.

126. Reddy, A.T. (2005) Atypical teratoid/rhabdoid tumors of the central nervous system. *J. Neurooncol.* **75**, 309–313.

127. Judkins, A.R., Mauger, J., Rorke, L.B., and Biegel, J.A. (2004) Immunohistochemical analysis of hSNF5/INI1 in pediatric CNS neoplasms. *Am. J. Surg. Pathol.* **28**, 644–650.

128. Judkins, A.R., Burger, P.C., Hamilton, R.L., Kleinschmidt-DeMasters, B., Perry, A., Pomeroy, S.L., Rosenblum, M.K., Yachnis, A.T., Zhou, H., Rorke, L.B., and Biegel, J.A. (2005) INI1 protein expression distinguishes atypical tera-toid/rhabdoid tumor from choroid plexus carcinoma. *J. Neuropathol. Exp. Neurol.* **64**, 391–397.

129. Bouffard, J.-P., Sandberg, G.D., Golden, J.A., and Rorke, L.B. (2004) Double immunolabeling of central nervous system atypical teratoid/rhabdoid tumors. *Mod. Pathol.* **17**, 679–683.

130. Rubio, A. (1997) Four year old girl with ring chromosome 22 and brain tumor. *Brain Pathol.* **7**, 1027–1028.

131. Fanburg-Smith, J.C., Hengge, M., Hengge, U.R., Smith, J.S., Jr., and Miettinen, M. (1998) Extrarenal rhabdoid tumors of soft tissue: a clinicopathologic and immuno-histochemical study of 18 cases. *Ann. Diagn. Pathol.* **2**, 351–362.

132. Schofield, D. (2002) Extrarenal rhabdoid tumour. In *World Health Organization Classification of Tumours. Pathology and Genetics of Tumours of Soft Tissue and Bone* (Fletcher, C.D.M., Unni, K.K., and Mertens, F., eds.), IARC Press, Lyon, France, pp. 219–220.

133. Proust, F., Laquerriere, A., Costantin, B., Ruchoux, M.M., Vannier, J.P., and Fréger, P. (1999) Simultaneous presentation of atypical teratoid/rhabdoid tumor in siblings. *J. Neurooncol.* **43**, 63–70.

134. Güler, E., Varan, A., Söylemezoglu, F., Çaglar, K., Basaran Demirkazik, F., and Büyükpamukçu, M. (2001) Extraneural metastasis in a child with atypical teratoid rhabdoid tumor of the central nervous system. *J. Neurooncol.* **54**, 53–56.

135. Fernandez, C., Bouvier, C., Sévenet, N., Liprandi, A., Coze, C., Lena, G., and Figarella-Branger, D. (2002) Congen-ital disseminated malignant rhabdoid tumor and cerebellar tumor mimicking medulloblastoma in monozygotic twins. Pathologic and molecular diagnosis. *Am. J. Surg. Pathol.* **26**, 266–270.

136. Fenton, L.Z., and Foreman, N.K. (2003) Atypical tera-toid/rhabdoid tumor of the central nervous system in children: an atypical series and review. *Pediatr. Radiol.* **33**, 554–558.

137. Kuwahara, Y., Hosoi, H., Osone, S., Kita, M., Iehara, T., Kuroda, H., and Sugimoto, T. (2004) Antitumor activity of Gefitinib in malignant rhabdoid tumor cells *in vitro* and *in vivo*. *Clin. Cancer Res.* **10**, 5940–5948.

138. Biegel, J.A., Rorke, L.B., Packer, R.J., and Emanuel B.S. (1990) Monosomy 22 in rhabdoid or atypical tumors of the brain. *J. Neurosurg.* **73**, 710–714.

139. Ho, D.M.-T., Hsu, C.-Y., Wong, T.-T., Ting, L.-T., and Chiang, H. (2000) Atypical teratoid/rhabdoid tumor of the central nervous system: a comparative study with primitive neuroectodermal tumor/medulloblastoma. *Acta Neuropathol.* **99**, 482–488.

140. Mori, T., Fukuda, Y., Kuroda, H., Matsumura, T., Ota, S., Sugi-moto, T., Nakamura, Y., and Inazawa, J. (1999) Cloning and characterization of a novel *Rab*-family gene, *Rab36*, within

the region at 22q11.2 that is homozygously deleted in malignant rhabdoid tumors. *Biochem. Biophys. Res. Commun.* **254,** 594–600.

141. Ogino, S., Cohen, M.L., and Abdul-Karim, F.W. (1999) Atypical teratoid/rhabdoid tumor of the CNS: cytopathology and immunohistochemistry of insulin-like growth factor-II, insulin-like growth factor receptor type 1, cathepsin D, and Ki-67. *Mod. Pathol.* **12,** 379–385.

142. Bultman, S., Gebuhr, T., Yee, D., La Mantia, C., Nicholson, J., Gilliam, A., Randazzo, F., Metzger, D., Chambon, P., Crabtree, G., and Magnuson, T. (2000) A Brg1 null mutation in the mouse reveals functional differences among mammalian SWI/SNF complex. *Mol. Cell* **6,** 1287–1295.

143. Wong, A.K., Shanahan, F., Chen, Y., Lian, L., Ha, P., Hendricks, K., Ghaffari, S., Iliev, D., Penn, B., Woodland, A.M., Smith, R., Salada, G., Carillo, A., Laity, K., Gupte, J., Swedlund, B., Tavtigian, S.V., Teng, D.H., and Lees, E. (2000) BRG1, a component of the SWI-SNF complex is mutated in multiple human tumor cell lines. *Cancer Res.* **60,** 6171–6177.

144. Reisman, D.N., Sciarrotta, J., Wang, W., Funkhouser, W.K., and Weissman, B.E. (2003) Loss of BRG1/BRM in human lung cancer cell lines and primary lung cancers: correlation with poor prognosis. *Cancer Res.* **63,** 560–566.

145. Berrak, S.G., Özek, M.M, Canpolat, C., Dagçinar, A., Sav, A., El-Naggar, A., and Langford, L.A. (2002) Association between DNA content and tumor suppressor gene expression and aggressiveness of atypical teratoid/rhabdoid tumors. *Childs. Nerv. Syst.* **18,** 485–491.

146. Ogino, S., Kubo, S., Abdul-Karim, F.W., and Cohen, M.L. (2001) Comparative immunohistochemical study of insulin-like growth factor II and insulin-like growth factor receptor type 1 in pediatric brain tumors. *Pediatr. Dev. Pathol.* **4,** 23–31.

147. Kaya, B., Mena, H., Miettinen, M., and Rushing, E.J. (2003) Alpha-internexin in medulloblastoma and atypical teratoid-rhabdoid tumors. *Clin. Neuropathol.* **22,** 215–221.

148. Kao, C.-L., Chiou, S.-H., Chen, Y.-J., Singh, S., Lin, H.-T., Liu, R.-S., Lo, C.-W., Yang, C.-C., Chi, C.-W., Lee, C.-h., Wong, T.-T. (2005) Increased expression of osteopontin gene in atypical teratoid/rhabdoid tumor of the central nervous system. *Mod. Pathol.* **18,** 769–778.

149. Kao, C.-L., Chiou, S.-H., Ho, D.M.-T., Chen, Y.-J., Liu, R.-S., Lo, C.-W., Tsai, F.-T., Lin, C.-H., Ku, H.-H., Yu, S.-M., Wong, T.-T. (2005) Elevation of plasma and cerebrospinal fluid osteopontin levels in patients with atypical teratoid/rhabdoid tumor. *Am. J. Clin. Pathol.* **123,** 297–304.

150. Jeibmann, A., Buerger, H., Frühwald, M., and Hasselblatt, M. (2006) No evidence for epidermal growth factor receptor amplification and overexpression in atypical teratoid-rhabdoid tumors. *Acta Neuropathol.* **112,** 513–514.

150A. Kao, C.-L., Chiou, S.-H., Chen, Y.-J., Singh, S., Lin, H.-T., Liu, R.-S., Lo, C.-W., Yang, C.-C., Chi, C.-W., Lee, C.-H., and Wong, T.-T. (2005) Increased expression of osteopontin gene in atypical teratoid/rhabdoid tumor of the central nervous system. *Mod. Pathol.* **18,** 769–778.

151. Arrazola, J., Pedrosa, I., Méndez, R., Saldaña, C., Scheithauer, B.W., and Martínez, A. (2000) Primary malignant rhabdoid tumour of the brain in an adult. *Neuroradiology* **42,** 363–367.

152. Erickson, M.L., Johnson, R., Bannykh, S.I., de Lotbiniere, A., and Kim, J.H. (2005) Malignant rhabdoid tumor in a pregnant adult female: literature review of central nervous system rhabdoid tumors. *J. Neurooncol.* **74,** 311–319.

153. Horn, M., Schlote, W., Lerch, K.D., Steudel, W.I., Harms, D., and Thomas, E. (1993) Malignant rhabdoid tumor: primary intracranial manifestation in an adult. *Acta Neuropathol.* **83,** 445–448.

154. Sugita, Y., Takahashi, Y., Hayashi, I., Morimatsu, M., Okamoto, K., and Shigemori, M. (1999) Pineal malignant rhabdoid tumor with chondroid formation in an adult. *Pathol. Int.* **49,** 1114–1118.

155. Lutterbach, J., Liegibel, J., Koch, D., Madlinger, A., Frommhold, H., and Pagenstecher, A. (2001) A typical teratoid/rhabdoid tumors in adult patients: case report and review of the literature. *J. Neurooncol.* **52,** 49–56.

156. Pimentel, J., Silva, R., and Pimental T. (2003) Primary malignant rhabdoid tumors of the central nervous system: considerations about two cases of adult presentation. *J. Neurooncol.* **61,** 121–126.

157. Fisher, B.J., Siddiqui, J., Macdonald, D., Cairney, A.E., Ramsey, D., Munoz, D., and Del Maestro, R. (1996) Malignant rhabdoid tumor of brain: an aggressive clinical entity. *Can. J. Neurol. Sci.* **23,** 257–263.

158. Ashraf, R., Bentley, R.C., Awan, A.N., McLendon, R.E., and Ragozzino, M.W. (1997) Implantation metastasis of primary malignant rhabdoid tumor of the brain in an adult (one case report). *Med. Pediatr. Oncol.* **28,** 223–227.

159. Byram, D. (1999) Comment, Int. J. Radiol. Oncol. Biol. Phys. **45,** 247 on Weiss, E., Behring, B., Behnke, J., Christen, H.J., Pekrun, A., and Hess, C.F. (1998) Treatment of primary malignant rhabdoid tumor of the brain: report of three cases and review of the literature. *Int. J. Radiol. Oncol. Biol. Phys.* **41,** 1013–1019.

160. Griffin, C.A., Hawkins, A.L., Packer, R.J., Rorke, L.B., and Emanuel, B.S. (1988) Chromosome abnormalities in pediatric brain tumors. *Cancer Res.* **48,** 175–180.

161. Kaiserling, E., Ruck, P., Handgretinger, R., Leipoldt, M., and Hipfel, R. (1996) Immunohistochemical and cytogenetic findings in malignant rhabdoid tumor. *Gen. Diagn. Pathol.* **141,** 327–337.

162. Biegel, J.A., Rorke, L.B., and Emanuel B.S. (1989) Monosomy 22 in rhabdoid or atypical teratoid tumors of the brain. *N. Engl. J. Med.* **321,** 906.

163. Douglass, E.C., Valentine, M., Rowe, S.T., Parham, D.M., Williams, J.A., Sanders, J.M., and Houghton, P.J. (1990) Malignant rhabdoid tumor: a highly malignant childhood tumor with minimal karyotypic changes. *Genes Chromosomes Cancer* **2,** 210–216.

164. Fort, D.W., Tonk, V.S., Tomlinson, G.E., Timmons, C.F., and Schneider, N.R. (1994) Rhabdoid tumor of the kidney with primitive neuroectodermal tumor of the central nervous system: associated tumors with different histologic, cytogenetic, and molecular findings. *Genes Chromosomes Cancer* **11,** 146–152.

165. Hasserjian, R.P., Folkerth, R.D., Scott, R.M., and Schofield, D.E. (1995) Clinicopathologic and cytogenetic analysis of malignant rhabdoid tumor of the central nervous system *J. Neurooncol.* **25,** 193–203.

166. Muller, M., Hubbard, S.L., Fukuyama, K., Dirks, P., Matsuzawa, K., and Rutka, J.T. (1995) Characterization of a pineal region malignant rhabdoid tumor. *Pediatr. Neurosurg.* **22**, 204–209.

167. Karnes, P.S., Tran, T.N., Cui, M.Y., Raffel, D., Gilles, F.H., Barranger, J.A., and Ying, K.L. (1992) Cytogenetic analysis of 39 pediatric central nervous system tumors. *Cancer Genet. Cytogenet.* **59**, 12–19.

168. Agamanolis, D.P., and Malone, J.M. (1995) Chromosomal abnormalities in 47 pediatric brain tumors. *Cancer Genet. Cytogenet.* **81**, 125–134.

169. Huisman, T.A.G.M., Brandner, S., Niggli, F., Betts, D.R., Boltshauser, E., and Martin, E. (2000) Malignant rhabdoid tumor of the brain: quantitative ^{1}H MR-spectroscopy and cytogenetics. *Neuropediatrics* **31**, 159–161.

170. Sawyer, J.R., Goosen, L.S., Swanson, C.M., Tomita, T., and de León, G.A. (1998) A new reciprocal translocation (12;22)(q24.3;q11.2–12) in a malignant rhabdoid tumor of the brain. *Cancer Genet. Cytogenet.* **101**, 62–67.

171. Klopfenstein, K., Soukup, S., Blough, R., Mazewski, C., Ballard, E., Gotwals, B., and Lampkin, B. (1997) Chromosome analyses in a rhabdoid tumor of the brain. *Cancer Genet. Cytogenet.* **93**, 152–156.

172. Biegel, J.A., Burk, C.D., Parmiter, A.H., and Emanuel, B.S. (1992) Molecular analysis of a partial deletion of 22q in a central nervous system rhabdoid tumor. *Genes Chromosomes Cancer* **5**, 104–108.

173. Thangavelu, M., Turina, J., Tomita, T., and Chou, P.M. (1998) Clonal cytogenetic abnormalities in pediatric brain tumors: cytogenetic analysis and clinical correlation. *Pediatr. Neurosurg.* **28**, 15–20.

174. Handgretinger, R., Kimmig, A., Koscielnak, E., Schmidt, D., Rudolph, G., Wolburg, H., Paulus, W., Schilbach-Stueckle, K., Ottenlinger, C., Menrad, A., Sproll, M., Bruchelt, G., Dopfer, R., Treuner, J., and Niethammer, D. (1990) Establishment and characterization of a cell line (Wa-2) derived from an extrarenal rhabdoid tumor. *Cancer Res.* **50**, 2177–2182.

175. Kusufuka, T., Miao, J., Yoneda, A., Kuroda, S., and Fukuzawa, M. (2004) Novel germ-line deletion of *SNF5/INI1/SMARCB1* gene in neonate presenting with congenital malignant rhabdoid tumor of kidney and brain primitive neuroectodermal tumor. *Genes Chromosomes Cancer* **40**, 133–139.

176. Klochendler-Yeivin, A., Fiette, L., Barra, J., Muchardt, C., Babinet, C., and Yaniv, M. (2000) The murine SNF5/INI1 chromatin remodeling factor is essential for embryonic development and tumor suppression. *EMBO Rep.* **1**, 500–506.

177. Sasaki, A., Kurihara, H., Ishiuchi, S., Hirato, J., Saito, N., and Nakazato, Y. (2005) Pediatric embryonal tumor of the cerebellum with rhabdoid cells and novel intracytoplasmic inclusions: distinction from atypical teratoid/rhabdoid tumors. *Acta Neuropathol.* **110**, 69–76.

178. Allen, J.C., Judkins, A.R., Rosenblum, M.K., and Biegel, J.A. (2006) Atypical teratoid/rhabdoid tumor evolving from an optic pathway ganglioglioma: case study. *Neurooncol.* **8**, 79–82.

179. 179. Chai, J., Charboneau, A.L., Betz, B.L., and Weissman, B.E. (2005) Loss of the *hSNF5* gene concomitantly inactivates p21$^{CIP/WAF1}$ and p16^{INK4a} activity associated with replicative senescence in A204 rhabdoid tumor cells. *Cancer Res.* **65**, 10192–10198.

180. Morozov, A., Yung, E., and Kalpana, G.V. (1998) Structure-function analysis of integrase interactor 1/hSNF5L1 reveals differential properties of two repeat motifs present in the highly conserved region. *Proc. Natl. Acad. Sci. USA* **95**, 1120–1125.

181. Laurent, B.C., Treich, I., and Carlson, M. (1993) The yeast SNF2/SWI2 protein has DNA-stimulated ATPase activity required for transcriptional activation. *Genes Dev.* **7**, 583–591.

182. Phelan, M.L., Sif, S., Narlikar, G.J., and Kingston, R.E. (1999) Reconstitution of a core chromatin remodeling complex from SWI/SNF subunits. *Mol. Cell* **3**, 247–253.

183. Peterson, C.L. (1998) SWI/SNF complex: dissection of a chromatin remodeling cycle. *Cold Spring Harb. Symp. Quant. Biol.* **63**, 545–552.

184. Schnitzler, G., Sif, S., and Kingston, R.E. (1998) Human SWI/SNF interconverts a nucleosome between its base state and a stable remodeled state. *Cell* **94**, 17–27.

185. Schnitzler, G., Sif, S., and Kingston, R.E. (1998) A model for chromatin remodeling by the SWI/SNF family. *Cold Spring Harb. Symp. Quant. Biol.* **63**, 535–543.

186. Dunaief, J.L., Strober, B.E., Guha, S., Khavari, P.A., Alin, K., Luban, J., Begemann, M., Crabtree, G.R., and Goff, S.P. (1994) The retinoblastoma protein and BRG1 form a complex and cooperate to induce cell cycle arrest. *Cell* **79**, 119–130.

187. Bochar, D.A., Wang, L., Beniya, H., Kinev, A., Xue, Y., Lane, W.S., Wang, W., Kashanchi, F., and Shiekhattar, R. (2000) BRCA1 is associated with a human SWI/SNF-related complex: linking chromatin remodeling to breast cancer. *Cell* **102**, 257–265.

188. Klochendler-Yeivin, A., Muchardt, C., and Yaniv, M. (2002) SWI/SNF chromatin remodeling and cancer. *Curr. Opin. Genet. Dev.* **12**, 73–79.

189. Strobeck, M.W., Knudsen, K.E., Fribourg, A.F., DeCristofaro, M.F., Weissman, B.E., Imbalzano, A.N., and Knudsen, E.S. (2000) BRG-1 is required for RB-mediated cell cycle arrest. *Procl Natl. Acad. Sci. USA* **97**, 7748–7753.

7
Neuroblastoma and Related Tumors

Summary   Chapter 7 deals with neuroblastoma (NB) and related tumors. The specific genetic events responsible for NB development have not been established, even though several changes occur with frequency. Among these changes the amplification of the oncogene *MYCN* is present in about one third of NB cases and carries with it a poor clinical outcome. This amplification is associated often with the presence of double minutes, homogeneously staining regions, or both, in the karyotypes of NB cells. Cytogenetically, some changes in the karyotypes of NB seem to occur in a significant number of these tumors and carry with them unfavorable prognostic aspects. Thus, deletions of 1p and 11q and gain of 17q have been observed in a significant percentage of NB; alterations of a number of other chromosomes also have been described. The cytogenetic changes in NB are reflected in comparative genomic hybridization studies and in loss of heterozygosity findings. The role of apoptosis and telomerase activity in NB biology are described. The epigenetic changes involving caspases and tyrosine kinase (Trk) receptors are discussed. The findings in related tumors, i.e., olfactory NB, glioneuroblastoma, ganglioneuroma, and ganglioglioma, are presented, and those for confusing tumors such as central neurocytoma and cerebellar liponeurocytoma. Genetic and hereditary susceptibility to NB and the intriguing spontaneous cures or regressions of NB are presented. Factors, such as angiogenesis, p16 pathway, p53 pathway, H-*ras* gene, CD44, and interference RNA also are included in Chapter 7.

Keywords   neuroblastoma and related tumors · *MYCN* amplification · deletions of 1p and 11q and gain of 17q.

General Information on Neuroblastoma (NB)

Many reviews and comprehensive articles on NB, including a book devoted to the multidisciplinary aspects of the disease (1,1A), have been published and these resources may be consulted for further information on NB (2–12).

NB and related tumors, ganglioneuroblastoma (GNB) and ganglioneuromas, are derived from primordial neural crest cells (13) that migrate from the mantle layer of the developing spinal cord and populate the primordia of the sympathetic ganglia and adrenal medulla (14). These tumors can be conceptualized as three different maturational manifestations of a common neoplasm: NB, the least differentiated, resembles the fetal adrenal medulla, and it is made up of primitive neuroblasts. Ganglioneuroblastoma (differentiating NB) possesses primitive neuroblasts along with maturing ganglion cells; the number and arrangement of the cells vary so that the tumor may assume a wide range of appearances. Ganglioneuroma, a fully differentiated tumor, is characterized by a mixture of mature schwann cells and ganglion cells (15). NB and ganglioneuroblastoma are considered malignant, and in most clinical studies patients with these two tumors are treated similarly, on the presumption that the amount of ganglionic differentiation plays a less significant role than other factors, such as age of the patient and stage of the disease. In contrast, ganglioneuromas are benign tumors requiring only conservative therapy.

NB is the third most common childhood malignancy (16); it occurs at a rate of about 1 in 10,000 live births (8). NB is the most common tumor in children under 1 year of age and accounts for about 7–10% of childhood cancers, with an incidence in the United States of 1/100,000 children under the age of 15 years, and it is the second most common neural tumor after central nervous system (CNS) tumors (17). About one fourth of NB are congenital. Half are diagnosed by the age of 2 years, 90% are diagnosed by the age of 5 years, and only sporadic cases occur during adolescence or adult life. The peak age at the time of presentation is about 18 months. In most large series, there is a slight male predominance. The distribution of NB and ganglioneuroblastomas generally follows the distribution of the sympathetic ganglia; hence, they are found in a paramidline position at any point between the base of the skull and pelvis, in addition to the adrenal medulla and organ of Zuckerkandl (8). About 65% of NB occur within the retroperitoneum, 15% in the mediastinum, 2% in the cervical region, 2% in the sacral region, and the remainder in several different areas. About half of all retroperitoneal tumors arise in the adrenal gland, the most common site of NB origin, and they typically present as an abdominal mass.

The majority of patients with NB over 1 year of age develop locally aggressive disease, metastatic disease, or both, which is usually fatal. Patient age and stage of disease, and other clinical aspects, are important factors in determining outcome in NB (18–20). Infants often experience complete regression of their disease, whereas the majority of children over 1.5 years of age have metastatic spread at diagnosis and a poor prognosis despite intensive therapy. Identification of recurrent genomic abnormalities has allowed the recognition of distinct NB subtypes that are predictive of clinical behavior. For example, tumors that have a high propensity for spontaneous regression usually exhibit only whole chromosome gains and losses, whereas metastatic NB is characterized by numerous recurrent structural chromosome abnormalities (21).

NB is characterized by heterogeneity at several levels, i.e., clinical, biologic, genetic, and molecular (22–24). For example, clinically NB is a remarkably heterogeneous disease, ranging from the rare occurrence of spontaneous cures and regressions and differentiation into benign ganglioneuromas to tumors with a very aggressive behavior in the majority of the cases (3, 10) and reflected in the characterization of NB subgroups by recurrent genetic or molecular findings, although the identification of the responsible genes has remained elusive.

NB is associated with the production of catecholamines, their precursors and by-products (10, 25), products of a metabolic pathway leading to the conversion of phenylalanine to norepinephrine and epinephrine. Catecholamine secretion is a key diagnostic feature of NB, and the concentrations of the catecholites nomovanillic acid (HVA) and vanillylmandelic acid (VMA) in the urine before and after excision of these tumors can serve as reliable indices of complete excision or recurrence of the NB. Furthermore, the ratio of HVA to VMA correlates with degree of tumor differentiation and survival.

The microscopic pattern of NB varies depending on the degree of differentiation (8). The most primitive tumors resemble the anlage of the developing sympathetic nervous system and adrenal medulla and are composed of sheets of small rounded cells, which are divided into small lobules by delicate fibrovascular stroma (Figure 7.1). These cells, which are almost devoid of cytoplasm, have round-to-polygonal deeply staining nuclei. The diagnosis of poorly differentiated NB is sometimes suggested by the presence of calcification or morula-like clusters of cells that represent the earliest form of rosette formation. Differentiated tumors have ganglion cells. Ganglionic differentiation is heralded by enlargement of the

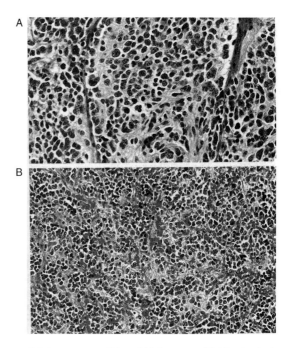

FIGURE 7.1. Low-power (**A**) and high-power (**B**) histologic features of NB of poorly-differentiated nature composed of a sheet-like proliferation of monotonous small round cells with scant cytoplasm (from ref. 8, with permission).

cells with acquisition of a discernible rim of eosinophilic cytoplasm (8).

One characteristic feature of NB is the tendency of neuroblasts to undergo varying degrees of cell maturation, characterized by the acquisition of an increased amount of cytoplasm with Nissl substance and the vacuolation of the nuclei with increased prominence of nucleoli (10). Tumors with this feature are known as "differentiating neuroblastomas." This process continues to the point of creation of fully mature ganglion cells, indistinguishable from nonneoplastic neurons, at which point the combination of neuroblasts and mature ganglion cells is termed a "ganglioneuroblastoma."

A grading scheme for NB is known as the Shimada classification. This classification is independently predictive of clinical behavior, and it is now used as an internationally accepted standard (15, 26–28) (Table 7.1). Evaluations and modifications of the Shimada classification have been published (10, 29–32, 32A). Nevertheless, the Shimada classification remains an essentially reproducible and applicable system useful clinically and prognostically accurate (10, 44). Table 7.2 presents features used in the staging of NB, noticeably including genetic and molecular facets of these tumors (33, 34).

Whereas patients in the low-risk subgroup have an overall survival rate of >95%, patients in the high-risk subgroup have a rate of long-term survival of <40% despite dose-intensive, multimodal therapy (35–38). These differences reflect the heterogeneity of NB. For example, many high-risk tumors have *MYCN* amplification, but >60% do not (39), suggesting

TABLE 7.1. International neuroblastoma staging system.

Stage	Definition
1	Localized tumor confined to the area of origin; complete gross resection, with or without microscopic residual disease; representative ipsilateral lymph nodes negative for tumor.
2A	Localized tumor with incomplete gross excision; representative ipsilateral nonadherent lymph nodes negative for tumor.
2B	Localized tumor with or without complete gross excision, with ipsilateral nonadherent lymph nodes positive for tumor.
3	Unresectable unilateral tumor infiltrating across the midline, with or without regional lymph node involvement; *or* localized unilateral tumor with contralateral regional lymph node involvement; *or* midline tumor with bilateral extension by infiltration or by lymph node involvement.
4	Any primary tumor with dissemination to distant lymph nodes, bone, bone marrow, liver, skin, and/or organs except as defined for stage 4S.
4S	Localized primary tumor (as defined for stages 1, 2A or 2B), with dissemination limited to skin, liver, and/or bone marrow.

Based on Brodeur et al. 1993 (26).

TABLE 7.2. Some features of various stages of NB.

Stage 1
   Patients <1 year of age
   Triploid DNA tumor content
   Single copy *MYCN*
   Few genomic changes
   Tumor regresses spontaneously, although slowly
Stage 4S
   Initially grows and may metastasize and then regresses
   High TrkA expression
   Triploid DNA tumor content
   No *MYCN* amplification
Advanced stages
   Patients >1 year of age
   Metastases to bone and lymph nodes
   TrkA expression downregulated
   Diploid DNA and near-diploid or tetraploid karyotype, often with dmin or hsr
   MYCN amplified, loss of 1p, 17q+, 11q-
   Marked genetic heterogeneity

that there are other genetic pathways in the development of NB (40).

Not infrequently, poorly differentiated NB may be confused with related tumors such as peripheral neuroepithelioma or other primitive neuroectodermal tumors, other round-cell sarcomas, or lymphomas (8). This problem can be readily solved now through cytogenetic analysis that usually reveals a t(11;22) in the former tumors and is not present in NB (41,42). The area of genomic medicine and its role in NB have been addressed previously (43).

NB is a cancer of early childhood in which genomic changes in the tumor correlate with its behavior (3,44, 44A). A particularly aggressive subtype of NB is genetically char-

acterized by amplification of the *MYCN* oncogene and of 1p deletions (45). Gain of parts of the long arm of 17q (17q+) has also been shown to be a characteristic feature of the aggressive subtype (46–49). The frequent 1p-deletion in clinically advanced tumors implies that the 1p region contains at least one gene (tumor suppressor gene, TSG) that is prone to inactivation in NB tumors. Amplification of *MYCN*, which plays a critical part in neurodevelopment, is amplified in about 20% of cases of NB, and it was one of the first tumor-derived genetic markers shown to be of clinical value; it continues to provide important prognostic information (50).

Genetic Aspects of NB

Cytogenetics

The published cases of NB with abnormal and complete karyotypes or with cogent cytogenetic information are shown in Table 7.3. In most cytogenetic studies, >60% of the NB tumors examined had karyotypic changes. The question is whether the tumors with normal karyotypes may represent normal cells that were in division at the expense of the NB cells, or tumor cells with subtle chromosomal abnormalities (as revealed by comparative genomic hybridization [CGH] and fluorescence in situ hybridization [FISH] in some cases).

Table 7.3 reveals that approximately 60% of the abnormal karyotypes of NB contain chromosome numbers in the diploid range, with 20% showing near-triploid and another 20% near-tetraploid numbers. Translocations were present in 17%, del(1p) in 20%, double minutes (dmin) in 27%, and changes affecting chromosome 1 (e.g., deletions, additions, duplications, isochromosome, and involvement in translocations) in 60% of the karyotypes.

A cytogenetic feature that is related to clinical outcome of NB is aneuploidy, based on cytogenetic analysis of metaphase chromosomes or cellular DNA content (51–56). Hyperdiploidy is usually indicative of poor outcome in adult NB cases, but in infants with NB it is a favorable finding. In contrast, NB with near-diploidy or high ploidy (e.g., tetraploidy) are associated with a poor prognosis. Hyperdiploidy with nonamplified *MYCN* has been reported to have a favorable prognosis in children with nondisseminated NB (57). Children with NB and tumor hyperdiploidy but with the absence of an abnormal chromosome 1 have a more favorable outcome than those cases that have the chromosome 1 anomaly (58).

As mentioned, chromosome 1, particularly its short arm (1p), was involved by deletions (Figure 7.2) translocations and additions or duplications in 60% of the karyotypes (Table 7.3). The translocations with 1p most often involved chromosome 17 (>50% of the translocations), followed by chromosomes 11 (13%), 7 (10%), and 3 (<5%).

Even though del(1p) is a very common karyotypic change in NB, it does occur in several other cancers and tumor

TABLE 7.3. Chromosome changes in neuroblastoma.

Karyotype	References
46,XX,del(1)(p31)	*Brodeur et al 1977 (163)*
46,XX,del(1)(p31),add(22)(p?)	*Brodeur et al 1977 (163)*
46,XX,del(22)(q13)	*Douglass et al 1980 (718)*
45,XX,del(1)(p32),dup(1)(p13p31),del(6)(q23),del(11)(q23),+14,+15,-18	*Brodeur et al 1981 (85)*
45,XY,t(1;14)(p32;q31),add(8)(q24),dup(14)(q11q24),-21	*Brodeur et al 1981 (85)*
46,XY,del(6)(q25),dmin	*Brodeur et al 1981 (85)*
53,XX,+dup(1)(p13p32),t(3;15)(q13:q26),+del(4)(q31),t(7;11)(q36:q13),+15,+20,+3mar	*Brodeur et al 1981 (85)*
82,XXY,-Y,del(1)(p22)x2,-3,-7,del(10)(q22)x2,-11,-12,-15,del(17)(p11)x2,-19,-19,-20,-21,-21,-22,+2mar	*Brodeur et al 1981 (85)*
47,XY,1p-,+der(5),-9,+der(9),+17,-22,dmin	*Haag et al 1981 (719)*
47,XY,1p-,+1p-,-9,der(9),+17,-22,dmin	*Haag et al 1981 (719)*
46,XX,add(1)(p32),add(11)(p?),add(17)(q?),dmin	*Gilbert et al 1982 (164)*
46,XY,add(1)(p?),add(19)(q?),add(20)(q?),-22,inc	*Gilbert et al 1982 (164)*
46,XX,add(1)(p32),add(6)(q?),add(13)(q?),add(16)(q?),dmin	*Gilbert et al 1982 (164)*
46,XX,hsr(13)(p?)	*Gilbert et al 1982 (164)*
46,XY,t(1;11)(p?:p?),dmin	*Gilbert et al 1982 (164)*
47,XY,?hsr(13),inc	*Gilbert et al 1982 (164)*
46,XY,+der(1)(p?),-17,dmin	*Gilbert et al 1984 (219)*
48,XY,+der(1)t(1;11)(p13;q11),+5,add(8)(p33),add(14)(q32)	*Gilbert et al 1984 (219)*
49,XX,+del(1)(p31)x3,t(9;22)(q34:q11)	*Gilbert et al 1984 (219)*
85,XXY,-Y,del(1)(p22)x2,+del(1),-3,-6,+7,-8,-8,-10,-12,-13,+14,+17,-21,-21,-22	*Gilbert et al 1984 (219)*
92,XXYY,add(1)(p22)x2,del(3)(q12)x2,dmin	*Gilbert et al 1984 (219)*
83,XXY,-Y,t(1;7)(p?:q?)x2,-2,der(2),der(3),-6,-8,-9,-10,-11,t(13;15)(q34:q23)cx2,-16,-16, +mar	*Sanger et al 1984 (220)*
46,XX,t(1;11)(p32:p15),del(2)(p21),add(7)(p?),dmin	*Franke et al 1985,1986 (222)*
45,XY,del(1)(p31),-8,+9,?add(13)(q?),-17,dmin	*Hafez et al 1985 (223)*
48,XY,del(1)(p31),del(13)(q?),+8,+mar	*Hafez et al 1985 (223)*
49,X,-X,del(1)(p31),+6,+9,+10,+18	*Hafez et al 1985 (223)*
45,X,-X,del(1)t(1;13),del(3)(p21),del(15)(q22),add(19)(q13)x2	*Kaneko et al 1985 (224)*
45,XX,add(1)(p22),add(2)(p23),t(4:6)(q31:q25),hsr(15)(p?),hsr(19)(q?),dmin,inc	*Franke et al 1986 (222)*
46,XY,add(1)(p36),add(1)(p32),add(1)(p22),add(3)(q?),add(9)(p?),add(11)(q?),add(12)(p?), add(18)(q?)	*Franke et al 1986 (222)*
46,XX,del(1)(p3?),add(11)(p?),dmin	*Franke et al 1986 (222)*
46,XX,t(1;1)(p32:q21),inv(2)(p13p23),add(8)(p?),dmin	*Franke et al 1986 (222)*
46,XX,t(1;2)(p21:q36),add(2)(p23),hsr(4)(p16),add(6)(q15),add(7)(p?),add(10)(q24),hsr(11) (q13),?i(17)(q10),add(20)(p?)	*Franke et al 1986 (222)*
46,XY,t(7;21)(p22:p13),add(19)(p13)	*Franke et al 1986 (222)*
57,XY,del(1)(p32),add(13)(q34),inc	*Franke et al 1986 (222)*
61,XY,t(X;6)(q13:q21),idic(1)(p13),hsr(6)(q21),hsr(11)(q25),?i(17)(q10),dmin,inc	*Franke et al 1986 (222)*
63-84,XY,add(1)(p32),t(5;11)(q13:q23),del(5)(q11q13),del(10)(p11), del(12)(q13), hsr(20)(q?),inc	*Franke et al 1986 (222)*
70,XY?,add(19)(p13),inc	*Franke et al 1986 (222)*
84,XY?,t(1;13:?)(p32:q34:?),der(5)inv(5)(p15q11)del(5)(q13),dmin,inc	*Franke et al 1986 (222)*
46,XY,dmin (2 cases)	*O'Malley et al 1986 (720)*
45,XY,add(1)(p22),add(6)(q27),add(8)(p23),del(9)(p22),-10,dmin	*Kaneko et al 1987 (54)*
45,XY,del(1)(p22),add(9)(q32),+12,-13,der(17)t(13;17)(q11:p13),-22	*Kaneko et al 1987 (54)*
46,X,der(X:1)(q28:q12),add(1)(p22),-2,add(5)(q33),add(13)(q22),add(16)(q22),add(19)(q12), add(21)(q22),-22,+2mar,dmin	*Kaneko et al 1987 (54)*
46,XX,add(1)(p22)x2,+del(1)(p31),add(3)(p25),add(4)(q35),-5,hsr(6)(p21),-8,add(11)(q21), -16,-18,+3mar,dmin	*Kaneko et al 1987 (54)*
46,XY,add(1)(p22),del(3)(p23),add(19)(p13),-21,der(21)t(21;21)(q22:q21),-22,+2mar,dmin	*Kaneko et al 1987 (54)*
46,XX,add(1)(p32),del(3)(p23),del(11)(p13)	*Kaneko et al 1987 (54)*
46,XY,add(1)(p32),-10,der(10)t(10;10)(q26:q21),+mar,dmin	*Kaneko et al 1987 (54)*
46,X,-Y,add(1)(p11),+add(1)(q12),add(12)(p13),add(17)(p13),dmin	*Kaneko et al 1987 (54)*

TABLE 7.3. Continued

Karyotype	References
46,X,-X,add(1)(p34),+8,del(10)(q24),dmin	Kaneko et al 1987 (54)
46,XX,add(2)(q37),del(9)(p13),hsr(21)(p11)	Kaneko et al 1987 (54)
46,XY,-4,del(6)(p21),t(9;?)(q34;?),-17,add(17)(q25),+2mar	Kaneko et al 1987 (54)
47,XX,add(5)(p15),-7,-11,+3mar	Kaneko et al 1987 (54)
48,XY,+add(1)(p21)x2,add(11)(q22)	Kaneko et al 1987 (54)
51,XY,t(1;11)(p34;q25),+del(6)(q21),+7,+8,+9,-10,-11,-12,+18,+21,del(22)(q13),+2mar	Kaneko et al 1987 (54)
61,XXX,-3,-4,-6,-6,-11,-15,-16,-21	Kaneko et al 1987 (54)
61,XY,-X,-4,-5,-6,-10,-12,-16,+17,-19,-19,-20,-22,-22,+4mar	Kaneko et al 1987 (54)
66,XY,-X,+3,-4,-7,-9,-11,-15,-20,+21,+2mar	Kaneko et al 1987 (54)
68,XX,-X,-3,-4,-8,-9,+12,+13,+13,+15,+16,+18,-19,+22	Kaneko et al 1987 (54)
70,XXY,+Y,-4,-6,+12,+12,+13,-15,-16,+18,+18,-22	Kaneko et al 1987 (54)
74,XX,-X,add(1)(p22),-4,-4,-5,-7,+8,+9,+10,+11,+11,-14,-15,-17,+22,+22,+6mar	Kaneko et al 1987 (54)
74,XX,-X,-4,+8,+8,+8,+12,+14,+17,-19,+?22	Kaneko et al 1987 (54)
81,XX,-Y,-Y,add(1)(p13),add(3)(q29),del(3)(p12),add(6)(q23),add(7)(q36),-8,-8,-9,add(9)(q34),
-10,-11,-12,-13,-14,-15,-16,-21,-22,der(22)t(22;22)(p11;q11),+3mar	Kaneko et al 1987 (54)
82,XX,-Y,del(1)(q42),del(3)(p21),inc	Kaneko et al 1987 (54)
84,XX,-Y,del(1)(q32),+2,-4,-5,-6,add(6)(q23),add(7)(p11),+8,-9,-10,-11,-13,-14,-15,
add(16)(q24)x2,+add(16),-17,add(17)(q25),hsr(17)(q25),-19,-19,-20,-20,add(21)(q22),-22,add(22)(q13)x2,+6mar	Kaneko et al 1987 (54)
85,XX,-X,-X,del(1)(p13)x2,+3,del(3)(p21)x3,-4,-8,-9,-11,-13,-14,-15,del(15)(q22),+16,+17,-19,add(19)(q13)
add(19;?)(q13;?)x3,+21,-22	Kaneko et al 1987 (54)
88,XXYY,del(1)(p31p34)x2,-10,-10,-16,-17,dmin	Kaneko et al 1987 (54)
100,XXX,-X,+1,+2,+3,-4,+6,+del(6)(q23),+7,-9,-11,+12,-13,+16,+17,+18,+18,+18,-19,+21	Kaneko et al 1987 (54)
47,XX,der(1)(p?),hsr(9)(q22),t(12;17)(q14;q21),add(16)(p13),inc	Christiansen and Lampert 1988 ,1989 (60, 721)
45,XY,del(1)(p32),t(1;11)(q43;q14),-4,del(7)(q31),-10,+mar,dmin	Donti et al 1988 (232)
50,XY,+5,+8,+9,+16	Hayashi et al 1988 (53)
60,XY,-X,-4,-7,-11,-13,-14,-15,-16,+17,-21,-22	Hayashi et al 1988 (53)
61,XY,-X,-1,-3,-4,-6,-11,-14,-15,+16,+17,-21,-22	Hayashi et al 1988 (53)
61,XY,-X,+Y,-2,-3,-4,-5,-8,+11,+11,-14,-17,-19,-20,-20,-21,+mar	Hayashi et al 1988 (53)
64,XX,-X,-1,-3,-4,-7,-10,-14,-15,+17,+20,+22	Hayashi et al 1988 (53)
67,XY,-X,-1,-3,+4,+5,-6,+9,-11,+13,-17,-21,+22	Hayashi et al 1988 (53)
67,XY,-X,-2,-3,-4,-5,-6,+9,+10,+11,+12,-15,-16,-17,-18,+19,+20,+21,+22	Hayashi et al 1988 (53)
68,XX,-X,+2,-3,-4,-6,-8,+9,+11,-13,+14,-15,+16,+17,+18,-19,-20,+22	Hayashi et al 1988 (53)
70,XXY,+1,-2,+5,-6,-8,+9,-10,+11,+13,-16,+17,+18,-22	Hayashi et al 1988 (53)
71,XX,-Y,-2,-4,+5,+9,+10,+14,+16,+17,+18	Hayashi et al 1988 (53)
72,XY,-X,+Y,+3,-4,-5,-6,+7,+9,+10,+12,-15,+18,+22	Hayashi et al 1988 (53)
74,XY,-X,+1,-2,+3,-4,+8,+9,+11,+14,+16,+17,-18,+19,-20,+21	Hayashi et al 1988 (53)
75,XY,-X,+Y,+9,+10,+14,+18,+20,+21	Hayashi et al 1988 (53)
76,XY,-X,+Y,+3,-4,+8,+9,+10,+11,+12,+19,+21	Hayashi et al 1988 (53)
77,XXY,-2,+4,+6,+7,+8,+9,+inv(9)x2,+10,+11,+12,-13,+15,-16,+17,+18,-19,+20,+21,-22	Hayashi et al 1988 (53)
46,XX,add(1)(p22),del(4)(q21)	Pearson et al 1988 (722)
46,XY,del(1)(p22)	Pearson et al 1988 (722)
46,XX,dmin,inc	Pearson et al 1988 (722)
47-48,XY,+11,t(14;14)(q32;q12),+add(19)(q12),+mar/90-110,XY,t(14;14),add(17)(p13), add(19),+6-8mar	Pearson et al 1988 (722)
49,XY,del(1)(q21),dup(1)(q21q32),+10,add(11)(p13),+12,add(14)(q32),-20,-22,+4mar	Pearson et al 1988 (722)
66-68,XXX,-1,del(1)(p31),-2,-4,-6,-7,-10,-12,-13,-15,-16,-17,-19,-20,-22,+5-8mar	Pearson et al 1988 (722)
76,XXY,+X,+1,del(1)(p22)x2,-3,-3,-4,-7,-14,-17,-18,-21,-21,+9mar	Pearson et al 1988 (722)
81,XY,+del(3)(p12),del(3)(p14),+4,+4,-5,+7,-11,-13,-16,-18,-19,-20,add(21)(q22)x2,-22, del(22)(q13),+5mar	Pearson et al 1988 (722)

TABLE 7.3. Continued

Karyotype	References
45,XY,del(1)(p32),-2,der(2)t(2;2)(p24:q14),der(15)t(2;15)(p11;p11),dmin	*Vernole and Melino 1988 (723)*
42,X?,del(1)(p32),add(3)(p25),add(11)(q25),+hsr(?),inc	*Hayashi et al 1989 (724)*
45,X?,add(1)(p34),inv(16)(p13q22)	*Hayashi et al 1989 (724)*
45,X,-X,del(1)(p13) (2 cases)	*Hayashi et al 1989 (724)*
46,XX,add(1)(p22),inc	*Hayashi et al 1989 (724)*
46,XX,add(1)(p32),dmin	*Hayashi et al 1989 (724)*
46,XX,add(1)(p34)	*Hayashi et al 1989 (724)*
46,X?,add(1)(p12),del(11)(p13),dmin	*Hayashi et al 1989 (724)*
46,X,-X,add(1)(p32),add(20)(q13),+mar,dmin	*Hayashi et al 1989 (724)*
46,X,-X,del(1)(p13),t(1;5)(q21;q33),add(2)(q37),add(11)(q25),inc	*Hayashi et al 1989 (724)*
47,XY,add(1)(p12),add(5)(q34),add(12)(q24)	*Hayashi et al 1989 (724)*
54,XX,+1,+3,+12,+15,+17,+17,+21,+22	*Hayashi et al 1989 (724)*
74,XXY,+3,-4,+5,+6,-8,+9,+10,-11,+13,+17,+18,-19,+20	*Hayashi et al 1989 (724)*
80,XX?,del(1)(p32),dmin,inc	*Hayashi et al 1989 (724)*
82,XXYY,del(1)(p32),del(20)(q21),inc	*Hayashi et al 1989 (724)*
84,XXXX,add(1)(p34),inc	*Hayashi et al 1989 (724)*
48,X,-Y,del(1)(p22-p31),+t(1;18)(p22-p31;q11-q12),+7,+14	*Petkovic et al 1989 (341)*
46,XY,t(1;10)(p32:q24)/46,idem,dmin	*Skinnider et al 1990 (725)*
46,X,-Y,+X,add(7)(q?),add(14)(q?)	*Carlsen et al 1991 (726)*
46,XX,del(1)(p22),dup(1)(p22q11),dmin	*Carlsen et al 1991 (726)*
46,X,-X,hsr(2)(q?),add(4)(q?),add(5) (q?),add(6)(q?),t(6;11)(q?;q?),+7,hsr(8)(q?),hsr(?9)(p?)	*Carlsen et al 1991 (726)*
42,X,-Y,del(1)(p32p36),-9,der(12)t(8;12)(q11:p13),der(14)t(14;18)(p11:q11),-18,-18, add(19)(q13),-21,dmin	*Fletcher et al 1991 (727)*
44,X,-Y,del(1)(p32),der(6)t(3;6)(q13:q12-q15),-11,dmin	*Fletcher et al 1991 (727)*
46,X,-1,i(1)(q10),+der(?12)t(1;12)(p11-p13:q11-q13),dmin	*Fletcher et al 1991 (727)*
46,XX,del(1)(p31),der(1)del(1)(p31)dup(1)(q21q32),der(4)x2,hsr(6)(q22),hsr(8)(q24), del(14)(q24),dmin	*Petkovicc and Cepulic 1991 (342)*
46,XY,+idic(1)(p13),+11,t(11;17)(q13;q11),-19,+f,+r,dmin	*McRobert et al 1992 (233)*
46,X,t(X;11)(q22:p15),?der(8)(?)46,XX,t(3;5)(p13;q22)	*Dressler et al 1993 (728)*
49,XY,add(1)(p34),add(12)(p13),dmin	*Sainati et al 1992 (729)*
49,XX,+1,add(1)(p34)x2,+i(8)(q10),-17,+20,+mar,dmin	*Sainati et al 1992 (729)*
46,XY,t(2;7)(p16:p22)	*Dressler et al 1993 (728)*
51,XY,+dup(1)(q21q32),der(2)t(?1;2)(?q25:?q37),+4,+4,+5,-8,der(9)t(?6;9)(?p11:?p11),-10, +11,-14,-16,+17,+17,+17,-20,+mar	*Dressler et al 1993 (728)*
92,XXXX,der(1)x2,del(3)(q?)x2,der(4)x2,der(11)x2	*Neumann et al 1993 (730)*
59,XY,-X,i(1)(q10),-3,-4,-6,-8,-9,-10,-11,del(11)(p12),+12,-13,add(14)(q?),-15,+17,+17, -19,-20,-22	*Petkovic et al 1993 (343)*
88,XX,-X,-X,del(1)(p22)x2,+del(1)(q21),t(1;5)(p22:q13)x2,-3,add(7)(q?)x2,+add(7)x2, -9,-10,+11,+12,der(14;14)(q10;q10),-15,-16,-16,+19,+19,-22	*Petkovic et al 1993 (343)*
46,XY,der(1)t(1;11)(p22;q12),inv(2)(p13p23),der(22)t(7;22)(p13q13),dmin	*Srivatsan et al 1993 (262)*
46,XX,add(1)(p36),del(6)(q12)	*Bown et al 1994 (731)*
46,XX,del(1)(p31-p32),r(2),hsr(13)(q?),-17,+mar,dmin	*Bown et al 1994 (731)*
46,XX,del(3)(p?),del(4)(p1?),+2mar	*Bown et al 1994 (731)*
46,XY,der(1:21)(q10:q10)	*Bown et al 1994 (731)*
46,XY,t(1:5)(p32:q31)	*Bown et al 1994 (731)*
??X?,der(1)t(1;17)(p34-p36;q12-q21),inc (5 cases)	*Caron et al 1994 (225)*
30-83,XY,der(1)(p?),inc	*Z. Chen et al 1994 (732)*
31-88,XX,del(6)(q23),inc	*Z. Chen et al 1994 (732)*
34-88,XX,der(1)(p?),+hsr(?),inc	*Z. Chen et al 1994 (732)*
46,XX,del(3)(p21),del(11)(p11),inc	*Z. Chen et al 1994 (732)*
46,XX,del(11)(q13q21)/64,XX?,del(11),i(12)(q10),add(13)(p11),inc	*Z. Chen et al 1994 (732)*
46-64,XY,add(1)(p36),dup(12)(q13q14),inc	*Z. Chen et al 1994 (732)*

TABLE 7.3. Continued

Karyotype	References
60,XY?,der(1)(q?),der(5)(p?),add(6)(q25),inc	Z. Chen et al 1994 (732)
73,XY?,add(1)(p22),dmin,inc	Z. Chen et al 1994 (732)
80-108,XY?,der(1)(p?),inc	Z. Chen et al 1994 (732)
43-44,XY,+2,-6,der(10)t(10:11)(q26:q21),del(11)(q21),-12,-13,+mar/84-90,XXYY,der(10)x2, del(11)x2, +2mar	Fujii et al 1994 (733)
52,XX,+3,+4,+5,+7,+12,+13,-17,+21,17-72dmin/52,idem,del(11)(q23)	Fujii et al 1994 (733)
45,X,-Y,add(1)(p34),der(15)t(Y:15)(q11:p13),dmin/	Lo et al 1994 (734)
46,X,-Y,add(1)(p34)x2,hsr(8)(q22),der(15)t(Y:15)(p11:p13)/46,X-Y,add(p34)x2,der(15)t(Y:15), +hsr(19)(p12)/46,X-Y,add(1)(p34)x2,+hsr(3)(p23),der(15)t(Y:15)	
48-54,XX,+add(1)(q41),+2,+7,+7,inv(9),+17,+mar	Lo et al 1994 (734)
49,XY,+1,der(1)t(1:17)(p36;q11-q12)x2,+2,+7,-15,+20,dmin	Van Roy et al 1994 (220)
82,XY?,der(1)t(1:17)(p35-p36;q11-q12),dmin,inc	Van Roy et al 1994 (220)
hypotetraploid(~80 chromosomes),del(1p),+markers	Z. Chen et al 1994 (732)
82,XX,+X,+Y,+del(1p)(p32p36)x2,+2,+3,+4,+4,+6,+7,+7,+8,+der(9q),+10,+10,+11,+12,+13,+13,+14,+15,+der(16p),+17,+18,+18,+19,+20, +21,+der(22)x2,+2mar,dmin	Z. Chen et al 1994 (732)
44-47,X?,dmin,inc	Avet-Loiseau et al 1995 (234)
46,XY,del(6)(q23)	Avet-Loiseau et al 1995 (234)
46,XX,der(1)t(1:7)(p32:q11),der(5)t(5:17)(q35:q12-q21),7-129dmin	Avet-Loiseau et al 1995 (234)
47,XY,add(1)(p35),add(4)(p16),del(11)(q1?3),+add(17)(q12-q21)	Avet-Loiseau et al 1995 (234)
47-50,X?,del(1)(q11),t(4:?17)(p15:q11),add(5)(q35),-11,add(11)(p13),add(15)(q26),+mar	Avet-Loiseau et al 1995 (234)
53,XX,?del(1)(p36),+3,+7,+8,+13,+20	Avet-Loiseau et al 1995 (234)
68-76,X?,?del(1)(p35),inc	Avet-Loiseau et al 1995 (234)
75-92,XX,del(X)(p21)x2,i(1)(q10),-3,-5,-10,-11,-14,-15,-15,+?16,i(17)(q10)x2,der(19)x2,-21/75-92,idem,hsr(5)(q31),der(12)t(11:12)(q14:q24)	Avet-Loiseau et al 1995 (234)
77-82,XXYY,del(1)(p33)x2,del(4)(p13p15)x2,-6,-9,-10,-11,-13,-14,-15,der(16)t(?11:16) (q13:q24),-17,del(17)(p12)x2,add(19)(p13),-20,-21,-21	Avet-Loiseau et al 1995 (234)
83-89,XXYY,add(1)(p31)x3,del(2)(p24)x2,t(3:8)(q11:p1?)x2,-4,-5,+7,der(11)t(11:?17) (q11-q12)x2,+18,+add(19)(p13)	Avet-Loiseau et al 1995 (234)
84-90,X?,del(1)(p35),inc	Avet-Loiseau et al 1995 (234)
87-93,XXYY,add(1)(p11),+14,add(19)(p13)	Avet-Loiseau et al 1995 (234)
92,XX,-X,der(X)t(X:?5)(q11:q23),del(17)(p22)x2,t(3:8)(p21:p22),del(5)(q15-q21)x2, dup(7)(q31q36)x2,del(15)(q14)x2,dmin	Avet-Loiseau et al 1995 (234)
61,XX,-X,-3,-4,+7,-10,-11,-14,-16,-19,-20	Hattinger et al 1996 (735)
65-68,XXXX,del(1)(p32),t(1:4)(q41:q22),add(2)(q32-q36)x2,del(6)(q22),add(9)(p21), del(9)(p21),-10,-17,-19,+20,-22,+mar,dmin	Hattinger et al 1996 (735)
46,XY,der(1)t(1:17)(p34:q21),dmin	Mitchell et al 1996 (736)
41-43,XX,der(5)t(2:5)(p11:p11),der(6)t(6:8)(p23:q11),der(9:?:10) tas(9:?:10)(p11:?:q26)hsr(?),add(14)(q31),add(16)(p12),-17,-18,der(18)t(9:18)(?:p11),der(21)t(17:21)(q11:p11),der(22) t(18:22)(q11:p12)	Sawyer et al 1996 (737)
40-43,XY,der(3)t(3:17)(p26:q21),-7,der(11)t(11:12)(p15:q13),der(15)t(9:15) (q13:q26),-17, der(17)t(7:17)(p11:p13)dup(7)(p11p22)add(7)(p22),dic(18:18) (q11:q11),der(19)t(7:19) (q11:q13),add(20)(q13),dic(20:21)(p13:?),der(21)t(7:21)(p11:q22),-22,+r	Cowan et al 1997 (738)
45,XX,der(1)t(1:11)(p32:q13),dup(2)(p1?p2?),der(9)t(9:17)(p1?3:q21),-13	Lastowska et al 1997 (228)
45,XY,der(1)t(1:17)(p36:q11-q12),add(3)(p25),add(7)(q11),-10,del(11) (q23),add(12)(p13), -13,-14,add(16)(p11-p12),del(19)(p13),+2mar	Lastowska et al 1997 (228)
46,XY,add(1)(p3?6),del(3)(p14),add(5)(p15),der(11)t(11:17)(p15:q21),+der(11)t(11:17) del(11)(q24)	Lastowska et al 1997 (228)
46,XX,del(1)(p22-p32),der(2)t(2:2)(p23:q1?)hsr(2)(q1?),der(14)t(14:17)(q32:q21)	Lastowska et al 1997 (228)
46,XX,der(1)t(1:17)(p22:q21),dmin	Lastowska et al 1997 (228)
46,XY,der(1)t(1:17)(p32:q11),add(2)(p23-p25),dmin	Lastowska et al 1997 (228)
46,XY,der(1)t(1:17)(p35-p36:q11-q12)	Lastowska et al 1997 (228)
47,XY,add(1)(p22),del(10)(q22),der(14:17)(q10:q10),+17,+18,dmin	Lastowska et al 1997 (228)
47,XX,add(1)(p36),+3,add(6)(q21),der(10)t(10:17)(q22:q21)	Lastowska et al 1997 (228)

TABLE 7.3. Continued

Karyotype	References
47,XX,+3,trp(17)(q21q25),add(19)(q13)	*Lastowska et al 1997 (228)*
49-50,X,-Y,der(2)t(1;2)(q11;p13),-10,add(11)(q22),+12,der(16)t(16;17)(q?11-q12;q21), +add(17)(q23),+18,add(21)(q22),-22,+3mar	*Lastowska et al 1997 (228)*
53,XX,+der(1)t(1;17)(p1?;q11-q12),+4,+6,+7,+7,+12,der(13;14)(q10;q10),+14,+17	*Lastowska et al 1997 (228)*
62,XXY,+1,add(1)(p13)x2,del(3)(q12),-4,-6,-9,-13,-14,-14,-15,hsr(17)(q11q12),-19,-20,+21	*Gotoh et al 1998 (245)*
44,X,-Y,der(1)t(1;7)(p32;q11),der(17)t(1;17)(q21;p11),add(19)(p13),-20/89,idemx2,+20	*Kaneko and Knudson 2000 (739)*
67-74,XX,-X,+del(1)(p13)x2,+3,+4,+7,add(7)(q36)x2,-8,-10,+11,+13,-14,+15,+18,+18,-190,+21,+22,add(22)(q13)x2,dmin	*Kaneko and Knudson 2000 (739)*
69,XX,-X,-1,-3,+6,+7,+9,+12,-13,-15,+16,der(16)t(1;16)(q21;q22)x2,+17,+18,+20,-21,-22,-22	*Kaneko and Knudson 2000 (739)*
73-78,XXXX,der(1)inv(1)(p11q24)del(1)(p34)x2,+hsr(1)(q42),+2,-3,+5,del(7)(p11),-9,-10, add(10)(p11),+13,+14,add(14)(q24)x2,+16,+18,-19,+20,+del(22)(q13),+4mar	*Kaneko and Knudson 2000 (739)*
77-78,XXXX,+1,add(1)(p13)x2,del(4)(p14),der(4)add(4)(p16)add(4) (q31)x2,+5,+6,+7,+8,add(8)(q2)x2,+add(11)(q25),+12,+13,-15,+18	*Kaneko and Knudson 2000 (739)*
100,XXXY,der(1)t(1;17)(p36;q21)x2,+5,+9,+12,+13,+16,+18,+18,dmin	*Kaneko and Knudson 2000 (739)*
83-87,XY,-X,-Y,add(1)(p32),-2,-11,-22	*Kaneko and Knudson 2000 (739)*
88-89,X?,add(1)(p13)x2,inc	*Kaneko and Knudson 2000 (739)*
88,XYY,-X,?der(1)t(1;17)(p32;q12)x2,+add(5)(p14),-8,-8,-14,-16,-22,+mar,dmin	*Zhang et al 2000 (143)*
46,XY,del(1)(p32p36),dmin/90-93,idemx2,+12	*Zhang et al 2000 (143)*
46,XX,dmin	*Zhang et al 2000 (143)*
46,XX,dmin,inc	*Cohen et al 2001 (740)*
47,XX,der(1)t(1;12)(p13;q11-q12)x2,+der(1)t(1;6)(q4?;?),der(16)t(16;17) (q22;q11),der(19)t(8;19)(?;q13)/47,idem,der(10)t(4;10)(q22;q26)	*Udayakumar et al 2001 (741)*
46,XY,del(1)(p13),t(9;12)(q34;q22),-19,+mar	*Udayakumar et al 2001 (741)*
48,XX,dup(1)(q31q44),del(2)(q31),del(3)(p21),+2mar	*Udayakumar et al 2001 (741)*
48-50,XXY,del(1)(p22),+6,+8,+2mar	*Udayakumar et al 2001 (741)*
65,XY,del(1)(q21),-5,+6,+13,+15,+16,+16,-21,+2mar	*Udayakumar et al 2001 (741)*
70-80,XY,del(1)(p22)x3,i(1)(q10)x2,+6,+8mar,5-6dmin	*Udayakumar et al 2001 (741)*
72-75,XY,+Y,+add(1)(p32)x3,+5-7mar	*Udayakumar et al 2001 (741)*
73,XY,-X,i(1)(q10),+4,+7,+7,+11,+11,+12,+13,+15,+19,+20,+20,+3mar	*Udayakumar et al 2001 (741)*
87-89,XYY,-X,add(1)(p31)x2,add(1)(q42),-2,del(3)(q11),-4,add(4)(q35),-5,-7,-8,-9,-11,-12,-13,+add(16)(q12),-17,+18,-20,+6mar,dmin	*Udayakumar et al 2001 (741)*
45,X,-Y,del(3)(p21),der(3)t(3;17)(p22;q23)/44,X,-Y,der(3)t(3;17)(p1?4;q22),der(4) t(3;4)(p22;p13),+7,-9,der(11)t(11;17),-17 (p15;q24),der(17)t(8;17)(p12;q23)/44,X,-Y,der(3)t(3;17) (q22;q11)	*Stark et al 2002 (67)*
45,X,-Y,der(5)t(5;6)t(5;12),del(6)(q12),del(9)(q22), der(14)t(1;14)(q31;q23),der(19)t(9;19)(q22;q11)	*Stark et al 2002 (67)*
46,XY,del(1)(p21),der(5)t(5;17)(p13;q22),dmin/92,idemx2	*Stark et al 2002 (67)*
46,XX,der(1)t(1;17)(p3?5;q21),del(2)(p21),dmin	*Stark et al 2002 (67)*
46,X,-X,der(2)(2;15)(p23;q22),der(3)t(3;6)(p13;p22),-6,der(15) t(6;15)(p11;q13),+der(19)t(1;19)(q21;q13),-21,dmin	*Stark et al 2002 (67)*
67,X,del(X)(p11),-X,del(1)(p36),dup(1)(q25),+7,-8,der(11)t(3;11)(?;p14),-11,der(12)t(X;12) (p?;q22),der(12)t(12;22)(q?;q22),+17,der(18)t(1;18) (q?;q22),der(20)t(X;20)(p?;q13),-21	*Stark et al 2002 (67)*
86,XXYY,-Y,der(2)t(2;3)(q37;q22)x2,der(5)t(5;7)(p15;q?)x2,-8,-9,-10,del(11)(q?),der(11) del(11)(p?)t(11;17)(q21;q23),+der(11)t(11;17)x2,dic(12;22)(p12;q13)x2,-15,-16,+18,+18,-21,-21	*Stark et al 2002 (67)*
87,XXX,-X,-3,-4,+7,+8,-10,-11,+13,-14,-16,+17,-17,-19,der(20)t(5;20)(?;q13),-21,-21,-22	*Stark et al 2002 (67)*
90,XX,-X,-X,-2,der(3)t(3;7)(p?;q?)x2,der(4)t(4;13)(p11;q13)x2,-5,dup(7)(q?)x2,+der(11) t(11;17)(q14;q21)x2,der(21)t(7;21)(?;q11)x2,der(22)t(17;22)(p12;q22)x2/	*Stark et al 2002 (67)*

TABLE 7.3. Continued

Karyotype	References
82,XX,-X,-X,-2,der(3)t(3;9)(p26;q32)x2,del(3)(p13),der(3)t(3;7)(p13;q?),der(4)t(4;13) (p11;q13)x2,-5,-9,der(9)t(2;9)(p?;q?)x2,-11,der(11)t(11;17),der(13)t(7;13)(?;p11), -15,-15,+17,+17,-18,der(18)t(18;22)(p11;q11),der(21)t(7;21)x2,der(21)t(18;21)(q11;p11),-21,-22	Stark et al 2002 (67)
94,XXYY,+11,+11,der(11)t(11;17)(q12;q12)x4	Stark et al 2002 (67)
95,XXYY,dup(1)(q?),der(3)t(3;17)(p22;q21)x2,+der(3)t(3;18)(p25;q12),+4,der(9)t(9;18) (p2?3;q21)x2,der(11)t(1;11)(q?;q14),der(11)t(7;11)(q?;q14),+17,dmin	Leon et al 2004 (742)
41-43,X-Y,-1,dup(1)(q?),+2,add(4)(q?),add(5)(q?),+7,add(11)(p?),+15,+20,inc	Betts et al 2005 (743)
34-39,X,-X,der(1)t(1;17)(p32;q21),del(2)(p21),-3,-8,-9,add(10)(p15),-11,der(16)t(16;17) (p13;q21),-17,der(19)t(17;19)(q21;p13),-21,-22/34-39,idem,-6,-12,-13,-14,-18,+1-2mar	Betts et al 2005 (743)
44,XX,der(1)t(1;17)(p32;q21),-8,-17,100dmin/45,XX,+1,der(1)t(1;17)x2,-8,-17,dmin	Betts et al 2005 (743)
44-45,XY,add(1)(p22),add(5)(q31),add(9)(p12),-10,der(16)t(12;16)(q11;p11),dmin	Betts et al 2005 (743)
45,XY,der(1)t(1;11)(p34;q13),add(2)(q35),add(8)(q24),del(15)(q22),del(16)(q22),add(17) (p1?),+1-3mar/82-87,X?,der(1)t(1;11)(p34;q13)x2,inc	Betts et al 2005 (743)
45,X,-X,der(1)t(1;17)(p13;q21),add(9)(p12),add(12)(p11),dmin/	Betts et al 2005 (743)
44-45,X,-X,der(1)t(1;17),add(9),add(12),-17,add(22)(p11),dmin/84,idemx2	Betts et al 2005 (743)
46,-X,der(X)t(X;7)(q22;q11),der(1)t(1;2)(p34;?),der(4)t(4;17)(p12;q21),der(16)t(X;16) (q26;q22),+17,der(21)t(19;21)(q13;p11)/46,idem,del(11)(q23)/47,idem,+18	Betts et al 2005 (743)
46,XY,der(1)dup(1)(p13p32)del(1)(p32p36)t(1;8)(p36;q11)del(1)(q21q25)add(1)(q32), der(8)t(1;8)(p36;q11),dmin	Betts et al 2005 (743)
46,XY,der(1)t(1;7)(p36;?),der(6)t(6;14)(q21;q11),der(7)t(6;7)(?p21;p22),der(16)t(16;17;9) (p13;q21q25;q21)t(9;17)(q21;q25),der(17)t(4;17)(q26-q28;p11),20-100dmin/92,idemx2	Betts et al 2005 (743)
46,XX,der(1)t(1;17)(p32;q21),der(1)t(1;17)(q33;q21),100dmin	Betts et al 2005 (743)
46,XY,der(1)t(1;17)(p36;q21),der(22)t(17;22)(q21;p12),50-100dmin/46,idem,del(11)(q23q23)	Betts et al 2005 (743)
46,XY,der(1)t(1;?2)(p21;?),der(19)t(17;19)(q21;p13),100-200dmin/46,idem,add(4)(p16)	Betts et al 2005 (743)
46,XX,der(3)t(3;17)(p13;q23),10-100dmin	Betts et al 2005 (743)
46,XX,dup(2)(p16) or inv(2)(p16p?)/46,idem,del(11)(q23-q24)	Betts et al 2005 (743)
47,XY,der(1)t(1;17)(p34;q21),+18,100-200dmin/47,idem,+mar/82-87,XXYY,der(1)t(1;17)x2,inc	Betts et al 2005 (743)
47,XX,der(3)t(3;17)(p21;q21),del(4)(p12),-6,+7,+8,inv(9)(p11q11)c,add(10)(q11),add(11)(q13)/ 47,idem,+1,dic(1;5)(p11;p11),-13,der(15)t(15;17)(q24;q21),+mar	Betts et al 2005 (743)
47,XY,t(4;22)(p16;q11),+7,dic(7;7)(q24;q10),del(11)(q22q25),add(17)(p12),+der(17)t(X;17) (q13;p12)/94,idemx2	Betts et al 2005 (743)
47,XX,+17/47,XX,+2,+17,-18	Betts et al 2005 (743)
47-48,XX,del(1)(q41),add(2)(p21),dup(3)(?q23q25),add(7)(p21),+14,+1-2mar/46-49,idem,+del(1),-11,del(15)(q13),del(17)(q23),+18,add(21)(p13)	Betts et al 2005 (743)
47-48,X,-Y,del(3)(p21),add(4)(p16),del(10)(q22),del(11)(13q22),add(13)(q34),add(14)(p11), add(15)(p11),add(17)(p11),add(18)(p11),+der(?)t(?;17)(?;q21),+2-4mar/47-48,idem, +del(2)(q14),-del(10),+add(13)(p11),-add(18),add(21)(p11),add(22)(q13)	Betts et al 2005 (743)
48,XY,+6,+17	Betts et al 2005 (743)
48-49,XY,+add(1)(q21),+add(3)(q27-q29),add(4)(p16),-10,+del(11)(q22q25),del(14)(q24), der(15)t(1;15)(q12;p11),add(19)(q13),der(19)(q13;q13),+2-3mar	Betts et al 2005 (743)

The presence of hsr and dmin, particularly the former, has been stated to be high (approx 50%) in established NB cell cultures (75, 76). Table 7.4, containing the published karyotypes of NB cell cultures, reveals that 25% of the cell lines had hsr and 12.5% had dmin, in contrast to the relative scarcity of hsr in NB tumors specimens, and the presence of dmin in 20% of these specimens (Table 7.3).

The dmin are often seen as a separate phenomenon from the chromosomes in direct metaphase preparations from primary NB tumors, whereas hsr are seldom seen in such preparations, but they are evident in approximately 50% of NB cell lines (76). The hsr are integrated as extra segments into abnormal chromosomes in which they can reach up to 80 Mbp (77, 78), and their integration sites vary in different NB cell lines (75, 76, 79, 80).

A study of the outcome in children with NB with hsr showed no difference from those with dmin (81); both are associated with a poor prognosis. The higher incidence of hsr in NB cell lines compared with primary tumors may be a compensatory mechanism to ensure cell survival in vitro and does not per se contribute to increased tumorigenicity (81). Dmin harboring amplified *MYCN* can be spontaneously eliminated as micronuclei in cell lines and tumors (82, 83). This propensity for dmin elimination may be the reason for the switch from dmin to hsr in cell lines. Other explanations for this switch are possible (81). Elimination in vivo of dmin containing amplified *MYCN* through the formation of micronuclei of NB cells has been described previously (83).

Either dmin or hsr (or both in small subsets of cells) occur in about 90% of NB cell lines (41). Evidence suggests that there is selection in vitro for cells derived from tumors that have preexisting dmin or hsr, and no evidence exists that these abnormalities develop de novo with time in culture of NB cells. Identification of cells containing dmin or hsr in NB can be established by interphase FISH (84).

As mentioned, the dmin are short, bipartite chromosome fragments that seem separate from the major chromosomes in metaphase spreads, and they are usually multiple. Hsr are uniformly staining elongations of the genome that are inserted into chromosomes in a spanner manner. Both of these phenomena were noted on cytogenetic studies of NB explants that were immortalized into cell lines (85). It became apparent that those tumors that could easily be transferred to an in vitro existence were associated with poor clinical outcome in vivo, and cytogenetic studies of these cells were noteworthy for the presence of both dmin and hsr. Cloning experiments of dmin and hsr revealed that they were composed of amplified segments of *MYCN*, a gene located on 2p24 and having homology to the *MYCN* proto-oncogene (45). Thus, the immortalization of cells in vitro and the aggressive clinical behavior in vivo result from multiplied segments of a tumor-related gene, with high levels of production of the resultant protein. Amplification of *MYCN* typically ranges from 50 to 400 copies of this usually single segment (10).

The large body of knowledge about amplified sequences in tumor cells contrasts with a nearly complete lack of knowledge about the nuclear morphology and topology of the aberrant chromatin structures (dmin and hsr) that harbor these sequences (86). Such knowledge is likely to be essential for a better understanding of the function of dmin and hsr in the context of nuclear organization. A study (86) on the question whether hsr and dmin occupy a distinct nuclear compartment clearly separated from surrounding chromosome territories or whether chromatin loops from hsr and dmin and chromosome territories intermingle with each other, led the authors (86) to hypothesize that dmin are located within the interchromosomal domain space and that stretches of hsr-chromatin align along this space. Such a topology could facilitate access of amplified genes to transcription and splicing complexes that are assumed to localize in this space.

Each interphase chromosome occupies a distinct territory in the cell nucleus (87–90), and the organization of chromosome territories is considered an important factor affecting nuclear functions (88, 91, 92). The compartmentalized structure of the nucleus ensures that molecules necessary for certain biological processes, such as transcription and splicing, are present in the right place and at the right time (91, 93, 94, 94A). Although the topology of chromatin in tumor and normal cell nuclei may differ profoundly, a detailed analysis of these differences is still lacking (86).

CGH and Array-CGH Studies in NB

Many CGH studies in NB tumors and cell lines have benn published (46, 95–103). These studies confirmed recurrent genomic changes established previously by other techniques (e.g., cytogenetics, FISH, and LOH), i.e., loss of 1p and 11q, gain of 17q, and amplification of 2p23 (*MYCN*) and less frequently loss of 3p, 4p, 9p, and 14q, although each study contained some findings additional to or different from others (Figure 7.4). For example, one study (104) described a subset of NB of stages 3 and 4 without *MYCN* amplification or 1p- and in another study (105) gain of 1q21-q25 was found in eight of 16 NB with stage 4 disease.

Essentially, most of the findings in NB obtained with cytogenetics (except for balanced changes such as translocations and inversions not detectable by CGH) and FISH, especially 17q gain, were corroborated and extended by CGH studies. The essential use of combined approaches (e.g., cytogenetics, FISH, SKY, or multiplex [M]-FISH and CGH) in deciphering and establishing the genomic changes in NB cell lines (and applicable also to primary tumors) was stressed previously (66). One CGH study (66) of seven cell lines with previously well-characterized genetic aberrations not only demonstrated 1p loss but also showed one or more extra copies of 17q due to unbalanced translocations with chromosome 1 or other partner chromosomes.

A CGH study of NB in which changes of 1p were not impressive, whereas gains of 2p, whole chromosomes 7 and

TABLE 7.4. Some features of chromosome 1 changes in NB.

	References
Analysis of 1p region of patient with constitutional t(1;17) (p36;q12-q21), which disrupts genes affecting nuclear and tRNA	*Laureys et al 1990 (174)*
1p deletions and mapping of 1p36.2-3.	*van der Drift et al 1995 (205)* *Van Roy et al 1995 (240)* *Shapira et al 1997 (744)* *Ejeskär et al 2001 (173)* *Slavotinek et al 1999 (745)*
Cases of NB with constitutional deletions within 1p36.2-p36.3 have been described	*Biegel et al 1993 (175)* *White et al 2001 (212)*
NB with *MYCN* amplification had deletions proximal to 1p36; NB with single copy *MYCN* often had small terminal deletions of 1p36 only. *MYCN* amplified tumors showed no parental preference of the deleted 1p, whereas single-copy *MYCN* NB preferentially deleted the maternal-derived allele. The authors hypothesized that at least two tumor suppressor genes are located within 1p35-p36: an imprinted gene distal to 1p36 defined by the smallest region of overlap common to all distal 1p-deleted NB and a nonimprinted 1p35-p36.1 locus that is deleted in *MYCN*- amplified aggressive tumors.	*Caron et al 1993 (177)*
The occurrence of 1p alterations in early and advanced stages of NB detracts somewhat from its prognostic value.	*Avigad et al 1994 (746)*
DAN gene at 1p36.1-p36.13.	*Enomoto et al 1994 (747)*
PITSLRE (protein kinase gene complex), located at 1p36, alterations in NB	*Lahti et al 1994 (748)*
Interstitial deletions at 1p32 in three NB point to possible additional TSG in proximal 1p	*Schleiermacher et al 1994 (182)*
NB with large terminal deletions of 1p had amplified *MYCN* and poor outcome, whereas NB with interstitial deletions of 1p had single copy *MYCN* and a favorable outcome.	*Takeda et al 1994 (200)*
Studies of possible genes involved in the t(1;15)(p36.2;q24) in the NGP cell line.	*Amler et al 1995, 2000 (198,749)*
Candidate genes at 1p35-p36 in cell lines of NB, one of which may be associated with *MYCN* amplification.	*Cheng et al 1995 (750)*
Some methodologies for determining 1p status (some for *MYCN* amplification also) include: microsatellite instability determined with short-tandem repeat polymorphisms, cDNA m-array CGH and FISH.	*Martinsson et al 1995 (190)* *Scaruffi et al 2004 (751)* *Strehl and Ambros 1993 (752)* *Combaret et al 1995 (753)*
RT-PCR, in archival samples.	*Peter et al 1992 (754)*
Constitutional t(1;10)(p22;q21) in a patient with NB.	*Anderson et al 2006 (755)*
1p36, its subbands and TSG.	*Leong et al 1993 (756)*
LOH at 1p in familial NB.	*Mead and Cowell 1995 (757)*
DR3 at 1p36; deleted in NB.	*Roberts et al 1998 (758)*
Mass screening revealed loss of 1p to be more common in young children with NB than in infants with this disease.	*Versteeg et al 1995 (759)* *Tonini et al 1997 (605)* *Grenet et al 1998 (760)*
Deletion of 1p36 present in the tumors of two siblings.	*Kaneko et al 1999 (761)*
Genomic imprinting, if it plays a role in some of the anomalies of NB (*MYCN* amplification, 17q gain, loss of 4p or involvement of 4p) might affect distal 1p but not the others.	*Lo Cunsolo et al 1999 (575)* *Caron and Hogarty 2000 (665)*
Changes of *CORT* gene in NB and 1p36.3→p36.2 deletion; no mutations of *CORT* found.	*Ejeskär et al 2000 (1203)*
Patched2 (at 1p32-p34) not mutated in NB.	*Jogi et al 2000 (762)*
LOH of 1p22 and 11q23 may be a more useful prognostic marker than *MYCN* amplification, elevated serum LDH or age in stage 4 NB.	*Mora et al 2000 (763)*

TABLE 7.4. Continued

	References
LOH of 1p is seen in all categories of NB but not in ganglion-neuromas. LOH in two separate regions of 1p: 1p36 and 1p22 Stage 4 tumors were diploid with LOH of these 1p regions. Local tumors were hyperdiploid with 1p36 deletion. Stage 4A NB that regressed spontaneously were hyperdiploid with 1p22 and 1p36 deletions	Mora et al 2000 (131)
Homozygously deleted region of 1p36.2-1p36.3 was identified in two cell lines.	Nakagawara et al 2000 (764) Ohira et al 2000 (206)
Hybrid map of 1p36.1 and three rearrangements in this region in NB.	Spieker et al 2000, 2001 (195,196)
Possible TSG at 1p and 14q	White and Versteeg 2000 (211)
Intratumoral heterogeneity of 1p deletions and MYCN amplification in NB.	Ambros et al 2001 (765)
Homozygous deletions in NB cell line at 1p36 defined by high-density STS map spanning	Y.Z. Chen et al 2001 (766)
Possible gene at 1p36.2-p36.3 for NB.	De Toledo et al 2001 (767)
Biological characteristics of NB with partial deletion of 1p.	Hiyama et al 2001 (768)
No evidence for an imprinted TSG at 1p36 in NB.	Hogarty et al 2001, 2002 (179,769)
EPB4.1 gene at 1p36 in NB.	Huang et al 2001 (1127)
The *KIF1B* gene at 1p36.2 is homozygously deleted in NB, but not considered a TSG.	Yang et al 2001 (770)
Two classes of alterations detected by FISH at 1p36.	Spitz et al 2002 (186)
KIF1Bα gene at 1p36.2 not involved in NB.	Y.Y. Chen et al 2003 (771)
LOH at 1p common in NB, including one ganglioneuroma.	Gonzalez-Gomez et al 2003 (772)
FISH analysis of high-risk NB revealed alterations of 1p36 and 11q23 (29%), 3p36 (19%) and *MYCN* amplification (19%): these cases define different groups of NB with an increased risk or progression.	Spitz et al 2003 (271)
Gene expression profiling of genes at 1p35-p36 in NB.	Janoueix-Lerosey et al 2004 (210)
Since no evidence for inactivation of a major TSG has been found in NB, 1p deletion may result in haploinsufficiency of several genes critical to initiation and/or progression of NB.	Janoueix-Lerosey et al 2004 (210)
APITD1, a possible TSG for NB at 1p36.2, with a putative binding domain for p53; shows low expression in NB.	Krona et al 2004 (773)
EXTL1 at 1p36.1 not involved in NB.	Mathysen et al 2004 (774)
LOH of 1p36 was present in 209/898 (23%) of NB cases and often associated with high-risk tumors and *MYCN* amplification.	Attiyeh et al 2005 (50)
A number of genes at 1p36.22 were down-regulated in NB with a poor prognosis, but not due to methylation of these genes.	Carén et al 2005 (775)
ENO1 (α-enolase) gene at 1p36.2; reduces cell death in NB.	Ejeskär et al 2005 (776)
Reduced expression of *CAMTA1* (at 1p36.3) correlates with adverse outcome in NB.	Henrich et al 2006 (777A)
In a study utilizing an oligonucleotide-based microarray, LOH of 1p and 11q of NB was associated with decreased expression of 61% an 27% of the genes mapping to 1p35-p36 and 11q, respectively. The data suggested that multiple genes may be targeted for LOH in NB.	Wang et al 2006 (777A)
Based on mRNA expression of a number of genes mapped to 1p36.2, decreases and much less commonly increases, in the expression of some genes differentiated NB with favorable biology from stage 4 NB. Tumors with deleted 1p showed significant differences in expression vs. NB with intact 1p. A complete loss of expression was not found for any of the NB.	Fransson et al 2007 (777B)
It was shown that miRNA-34a (located at 1p 36.23) is generally expressed at lower levels in unfavorable NB and cell lines relative to normal adrenal tissue.	Welch et al 2007 (777C)

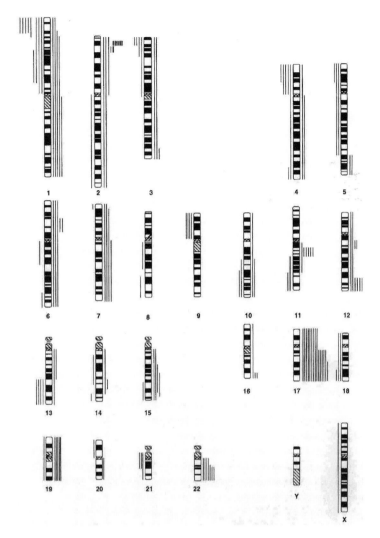

FIGURE 7.4. CGH results on 20 NB showing loss of chromosomal material (*left lines*) or gain of such material (*right lines*). Amplification of 2p (associated with *MYCN* amplification) is shown by *thick lines*. The most common changes were loss of 1p and gain of 17q. Gains of whole chromosomes 1, 17, and 22 also were present (from ref. 98, with permission).

17 and 17q, and loss of 11q were prominent has been reported previously (106).

A CGH study of 11q changes (107) showed loss of 11q without *MYCN* amplification in 59% of the 50 NB with stage 4 disease and in only 15% of those with amplification. Loss of 11q correlated with loss of 3p. Loss of 11q did correlate with prognosis, and it was not significantly associated with amplification of *MYCN* or loss of 1p.

Differences in NB genomic changes in infantile versus older patients were demonstrated by CGH (108) in that the incidence of 17q+ were lower in the former, as were losses of 11q and 14q.

Examination of NB cell lines with CGH showed a prevalence of nonreciprocal translocations (109), the most common being der(1)t(1;17). The authors stated that NB cell lines are characterized by changes predominantly involving chromosome fragments replicating early in the S phase.

A CGH study showed that long-term survivors with NB had primarily numerical chromosomal changes in their tumors (gains of 1, 7, 17, and 19), whereas those with aggressive disease were associated with structural genomic alterations (loss of 1p, 9p, 11q, and 18q and gain of 12q). Unbalanced gain of 17q was present in at least 50% of these NB (103). Similar results were obtained in other studies (102,111). Loss of 1p was found in some cases only with FISH.

cDNA microarray CGH is based on hybridization of tumor DNA to genomic bacterial artificial chromosomes, which provides a way to analyze genomic changes in tumor tissue with high sensitivity (112), detecting single-copy number aberrations in tissues in the presence of ≤60% contamination by normal cells. Furthermore, array CGH is an approach that can be applied to paraffin-embedded tissues (113,114).

Studies using cDNA microarray CGH have shown an association between gain of 1q21-q25 with prognosis (115), that 15 genes were up-regulated in stage 4 NB, including genes for cell adhesion molecules and cytoskeleton proteins (116), that the pattern of changes in murine NB resemble those of the human tumors (117), and that the 4S grade NB can be divided into three groups, i.e., one group with no *MYCN* amplification, oen group with a small *MYCN* amplicon as the only detectable imbalance, and one group with a large *MYCN* amplicon associated with coamplification of other genes (118).

A report of unique amplification of 1p34.2 and a more complex one at p36.3 in NB based on CGH and confirmed with cDNA microarray CGH has appeared (119).

In another cDNA microarray CGH study (120, 121), no genomic alterations common to all stages of NB were found. However, gain of 7q32, 17q21 and 17q23-q24 and loss of 3p21 were common to stages 1 and 4. The most common specific changes in stage 4+ tumors were loss of 1p36 and gain of 2p24-p25 and such tumors had fewer genomic alterations than stages 1 or 4-, indicating that this subgroup of poor risk NB requires a lesser number of genomic changes for malignant development.

In another array-CGH study, loss of whole chromosomes 3, 4, 14, and X and gain of 7 and 17 was found in low-grade NB (122), whereas deletions of 1p, 3p, 11q, and gain of 17q and *MYCN* amplification were seen in advanced grade NB with a poor outcome. Amplifications of 2p24 (*MYCN*) and other segments of chromosome 2, and of 6q, 12q, and 17q were described previously.

MYCN amplification, loss of 3p and 11q, and gain of 17q, findings common in NB obtained by other methodologies, were confirmed by cDNA microarray CGH (123). A review of genome-wide measurement of DNA copy number changes in NB (124), stressed the analytical power of array CGH as a rapid method for detecting alterations, i.e., amplifications, deletions and gains, and, in conjunction with cytogenetic data, allowing high-resolution mapping of unbalanced translocations.

LOH in NB

Inactivation of tumor suppressor genes (TSG) and activation of proto-oncogenes occur during multistep carcinogenesis of NB. Proto-oncogenes are activated by promoter CpG demethylation, point mutations, chromosomal translocations, and gene amplifications. Examples of the latter are the *MYCN* and *NAG* genes at 2p24, which are coamplified in NB (125–127) and an array of genes on 17q23.2, which are also coamplified in NB (128).

TSG are inactivated by promoter CpG hypermethylation, point mutations, chromosomal translocations, and deletions. TSG of NB may be located at 1p36, 4p16, 11q23, and 14q32, based on deletion analyses (129–132).

LOH of several specific genomic regions is frequently observed in NB tumors and cell lines, but biallelic inactivation of TSG, crucial for the genesis and evolution of neoplastic cells, through homozygous deletions (HD) (133, 134) are rare, and no NB TSG has yet been identified. A systematic search

for HD on 1p36 in a panel of 46 NB cell lines did not reveal such HD (135). HD was detected only of the *CDKN2A* ($p16^{INK4a}/p14^{ARF}$) gene at 9p21 and was observed in four of 46 cell lines (135). HD was observed in corresponding primary tumor samples for two of the cell lines, but it was not present in constitutional samples. These results suggest that for NB, large HD do not occur within 1p36, that most known TSG are not homozygously deleted, and that biallelic inactivation of *CDKN2A* may contribute to tumorigenicity in a subset of cases.

Deletions of 3p occur coincidentally with 11q deletions and define a subset of NB with aggressive disease without the oncogenic influence of deregulated *MYCN* (102, 136). HD of other chromosomes of possible relevance to NB include 4p, 9p, 14q, and 19q (137–141), although in general HD are rare in NB (141). This complicates the use of available approaches for identifying the TSG that could be targeted by the HD of several chromosomes.

Telomerase in NB

The telomerase complex includes telomerase RNA (hTR), human telomerase reverse transcriptase (hTERT), and several associated proteins. hTR acts as a template for the telomeric repeat synthesis, and hTERT (mapped to 5p15.33) has reverse transcriptase activity (142). hTR is sometimes expressed in cells without telomerase activity; in contrast, the degree of telomerase activity correlates with the expression of hTERT. Most normal somatic cells lack telomerase activity, because they do not express the *hTERT* gene, whereas most cancer cells, including NB, express *hTERT* (143, 144). Therefore, detection of *hTERT* mRNA expression by reverse transcription-polymerase chain reaction (RT-PCR) analysis may be useful for NB diagnosis.

There are some practical aspects related to measuring telomerase status in NB tumor tissue. The determination of telomerase activity requires fresh tissue, which may not be always available. In contrast, assay of the mRNA of the telomerase catalytic subunit hTERT can be performed on archival material by RT-PCR. The clinical utility of telomerase activity or hTERT levels as a prognostic indicator in NB has been stressed by a number of authors (145–150, 150A).

Telomerase is an enzyme that synthesizes the TTAGGG telomeric repeats for maintaining the integrity of chromosomal ends. This enzyme is expressed in germline but not in normal somatic cells. Increased telomerase activity is detectable in most cancer cells and seems to be a prerequisite for malignant transformation (151). Telomerase activity has been detected in most NB, high levels being present in advanced-stages and associated with *MYCN* amplification or 1p deletion with poor clinical outcome (97, 145, 152–158). The level of expression of the RNA template component of hTR is lower in localized and stage 4S NB than in advanced stage tumors (154). Low-level hTR expression and telomerase activity may limit the growth potential of some NB, whereas high expression of hTR and telomerase activity may contribute to aggressive tumor behavior by allowing unlimited cell proliferation (44, 159). Therefore, although elevated telomerase expression may simply be a marker of escape from cellular senescence, markedly increased levels may be associated with genomic instability and an increased likelihood of additional mutational events.

Telomerase activity was not detectable in ganglioneuromas or normal adrenal tissue (145).

Patients whose tumors were negative for telomerase activity before and after chemotherapy uniformly did well, even though all 40 cases studied were unfavorable by the Shimada classification (149). In two of seven cases where the tumors were positive before and negative or weakly telomerase-positive after treatment, the patients did well clinically. In contrast, in cases that were telomerase-positive after chemotherapy, all had a poor outcome. Thus, telomerase activity (measured in these studies by PCR-enzyme-linked immunosorbent assay) seemed to be a prognostic factor for NB (149). The correlation between telomerase activity and NB tumor aggressiveness is a mechanism that seems to be separate from cellular differentiation (160). Telomere length and telomere repeat binding factors as prognostic markers in solid tumors, including NB, have been discussed previously (161). Telomerase activity and TrkA expression can serve as significant prognostic indices in NB (156).

Changes of Chromosome 1 (Especially of 1p) in NB

Cytogeneticists first identified alterations of the distal short arm of chromosome 1 (1p) in primary NB and tumor-derived cell lines >20 years ago (162–164). Molecular analyses have subsequently confirmed and extended these findings (165–167). Deletions within 1p occur in ~35% of primary NB, and these deletions tend to be large and terminal, often extending from ~1p32 to the telomere (168–170). Interstitial deletions are infrequent (171) , and a smallest region of deletion (SRD) has been identified consistently within 1p36.3 (130, 172, 173). Constitutional genomic alterations within this region have also been observed in patients with NB, changes that may predispose them to the genesis of this tumor (174, 175). Transfer of 1p material into a NB-derived cell line has induced phenotypic changes consistent with loss of malignant potential; furthermore, reversion to tumorigenicity correlated with loss of the transferred 1p homologue (176). Together, these data strongly support the presence of one or more NB suppressor genes within chromosome subband 1p36.3. However, tumor-specific biallelic inactivation has not been demonstrated for any candidate 1p36 suppressor gene in NB, despite intensive investigations. Thus, alternate models of tumor suppressor inactivation, such as genomic imprinting, somatic epigenetic silencing, haploinsufficiency, and dominant inhibitor effect may be operative.

Reports documenting preferential deletion of the maternally inherited 1p in NB without *MYCN* amplification led to the hypothesis that an imprinted suppressor gene is targeted for disruption in this subset of NB (177, 178), supported by the observation that 1p deletions in NB without *MYCN* amplification tend to be smaller and more telomeric than those in NB with *MYCN* amplification (178). This hypothesis was tested (179) in a large series of NB, both with and without *MYCN* amplification, and showed that no parent-of-origin effect for 1p36 deletion was apparent to support the presence of an imprinted NB suppressor gene in this region.

LOH of 1p is strongly associated with high-risk features of NB, i.e., age greater than one year at diagnosis (Figure 7.5), the presence of metastatic disease and *MYCN* amplification (138, 180–182), although there is no general agreement concerning the independent prognostic significance of 1p LOH (45, 183). LOH of 1p36 may predict a decreased event-free survival, but not overall survival (184) and an increased risk of disease relapse in patients with localized tumors (138, 180, 185, 186). Whether there is more than one TSG distal to the 1p36 and whether the retained allele is inactivated by genetic or epigenetic events are unknown (187).

Lack of *MYCN* amplification in tumors with 11q23 LOH contrasts with the findings regarding 1p36 LOH. Although 1p36 LOH was highly associated with a decreased probability of survival, there was a statistically significant overlap between tumors with 1p36 LOH and those with *MYCN* amplification. The association between 1p36 LOH and *MYCN* amplification may partially explain why 1p36 LOH was not independently associated with survival after the adjustment for *MYCN* amplification in multivariate analyses (50). After the multivariate model was restricted to the low-risk and intermediate-risk groups, which are made up almost entirely of patients with tumors that do not have *MYCN* amplification, 1p36 LOH was independently associated with progression-free survival, confirming a previous report (188).

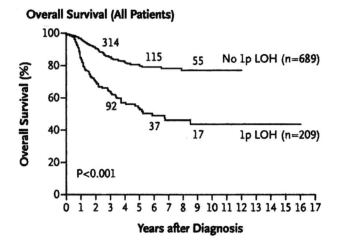

FIGURE 7.5. Overall survival of NB patients with and without LOH of 1p, showing a statistically significant difference between the two groups (50).

It was shown (50) that 1p36 LOH and unbalanced 11q LOH are strongly associated with outcome in patients with NB. The addition of these indices to the currently used prognostic variables may allow for more precise treatment recommendations.

A large number of studies have been performed in an attempt to define the region on 1p likely to contain a gene or genes responsible for NB development and/or affecting the biology of these tumors (130,170,173,174,178,181,189–197). In early studies, several loci on 1p possibly containing putative NB suppressor genes were identified by FISH and LOH, i.e., 1p36.1-pter (170, 190, 191, 194, 198), 1p36.1-p35 (199), and 1p34-p32 (182, 200). From these studies, a region of (193), a finding subsequently confirmed by others (130,131,171,191,196).

Comprehensive and large-scale molecular genetic approaches were used to further characterize the initially defined SRD at 1p36.2-1p36.3 (130, 170, 171, 190, 191, 196, 201), with the objective of further narrowing this region and number of candidate genes (193). Allelic loss studies of 737 primary NB and genotype analysis of 46 NB cell lines were performed (193). Together, the results defined a single region within 1p36.3 that was consistently deleted in 25% of tumors and 87% of cell lines. Two NB patients had constitutional deletions of distal 1p36 that overlapped with the tumor-defined region. The tumor- and constitutionally derived deletions together defined an SRD. The 1p36.3 SRD was deleted in all but one of the 184 tumors with 1p loss. Physical mapping and DNA sequencing determined that the SRD minimally spans an estimated 729-kb region. Genomic content and sequence analysis of the SRD identified 15 characterized, nine uncharacterized, and six predicted genes in the region. The physical mapping, sequencing, sequence analysis, and transcriptional profiles of genes within the defined SRD also were described (193).

However, progress in further narrowing the region and defining candidate TSG within 1p36 has been slow, due in part to most 1p deletions being large. In addition, whereas several NB cell lines and constitutional chromosomal rearrangements involving 1p36 have been identified, the affected chromosomal regions do not tightly cluster (196, 202–208). Moreover, linkage analysis of familial cases has excluded 1p36 as containing a predisposition locus (209). Thus, the gene or genes responsible for NB development have remained elusive (44, 45).

Interestingly, several genes differentially expressed between 1p-normal and 1pLOH samples encode proteins possibly involved in neural differentiation, cell cycle regulation, and apoptosis (210). It has been established that 1p deletion is frequently associated with 17q gain, as a result of unbalanced translocations. Therefore, in addition to the gene dosage effect for 1p genes, gene dosage imbalances secondary to segmental trisomy of 17q also might disrupt critical stoichiometric relationships in cell growth and physiology, which may directly contribute to tumor development or progression (210).

Cytogenetic, molecular genetic, and functional analysis of primary NB tumors and cell lines have identified several genomic regions frequently exhibiting hemizygous deletions (211). Deletion of 1p36 correlates strongly with advanced disease, and it is the most well-characterized region of deletion in NB (181,188,212). The existence of one or more additional TSG located elsewhere within 1p36 (174, 189, 194) is also a possibility. Despite the characterization of numerous candidates, no 1p36 TSG has yet been identified, largely because of the paucity of tumors with localized 1p36 rearrangements (211).

Furthermore, HD of known tumor suppressor loci has been reported only rarely in NB (191, 213, 214). The lack of evidence for genetic alterations in the genes encoding the DNA damage sensor *p53* and the cyclin-dependent kinase inhibitory protein *p16^{INK4a}* is notable, as these two genes are commonly disrupted in most malignancies (215, 216).

The most common structural chromosome abnormalities in NB tumors and cell lines are associated with aggressive clinical behavior and include loss of the distal part of 1p (1p-) and 11q (11q-), *MYCN* amplification and unbalanced gain of the long of chromosome 17 (17q+) (3, 46, 102, 104, 107, 111, 189, 217). Unbalanced gain of 17q material is the most common abnormality detected in advanced stage NB, occurring in 80–90% of cases irrespective of *MYCN* amplification or 11q loss status (46, 96, 99). Table 7.4 presents information on chromosome 1 changes in NB, most of which are not included in the text. The table also shows some of the genes localized to 1p or thought to be involved in NB.

Changes of 17q in NB

The recurrent nature of 17q changes in NB were first pointed to >20 years ago (218, 219), and it has been recognized that gain of that arm (17q+) is perhaps the most common genetic abnormality in NB, including the frequent occurrence of unbalanced translocations (213, 220–223). Chromosome 17 gain may be due to the formation of isochromosomes of 17q (98). Cytogenetic, CGH, and other methodologies have shown that gain at 17q21-qter occurs in 50–75% of primary NB (47,48, 95–99, 111, 224), an incidence much higher than that for 1p- and *MYCN* amplification. Unbalanced 17q gain is associated with an unfavorable clinical outcome (Figure 7.6) (46, 48, 49, 189, 225, 226). Opinions to the contrary have been expressed (227), in which loss of 1p and *MYCN* amplification were thought to be more reliable indices of outcome.

In NB, frequent genetic alterations consist of unbalanced translocations between chromosome 17 and various partners (228, 229), most frequently 1p, 3p, and 11q, leading to deletion of part of the recipient chromosome, and to gain of chromosome arm 17q (3, 222, 230, 231). At the cytogenetic level, 17q11 to 17q21 were the most common sites of the translocation breakpoints, although 17q23 may be involved in translocations with 11q (Tables 7.5 and 7.6). Several gains of 17q and loss of 11q in NB occur as a consequence of unbalanced

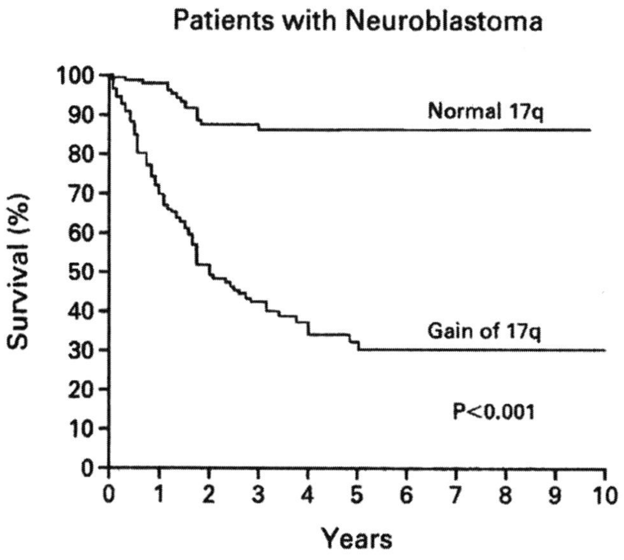

Patients with Neuroblastoma

FIGURE 7.6. Survival of patients with NB with normal 17q and those with gain of 17q. The unfavorable outcome for the latter group of patients is significantly worse than for the former group. In this study, the presence of single-copy *MYCN* or 1p deletions did not significantly affect survival of NB cases with either a normal 17q or gain of 17q. This contrasts with the poor survival in cases with *MYCN* amplification *and* 1p deletions (from ref. 48, with permission).

or more unidentified genes located in this region provides a selective advantage for NB growth (236). Overexpressed *BIRC5* (survivin), *NME1* (NM23), and *PPM1D* may be involved in this gain of genomic material (128,237,238). It is likely that multiple 17q genes are differentially up-regulated to provide an oncogenic effect in high-risk NB.

The complexity that may exist in the der(1)t(1;17) translocations in NB is illustrated by a study of a NB cell line with this translocation (239). The 17q breakpoint was mapped by FISH. Subsequently, a rearranged fragment was identified by Southern analysis, cloned in lambda vector, and sequenced. The chromosome rearrangement was more complex than expected due to the presence of an interstitial 4p telomeric sequence between 1p and 17q. Three different genes, which may play a role in NB development, were disrupted by the translocation. The 3′ untranslated region of the *PIP5K2B* gene on 17q was directly fused to the (TTAGGG)n repeat of the 4p telomere, and the (1;4) fusion disrupted the microtubule-actin crosslinking factor 1 (*MACF1*) and *POLN* genes. Interestingly, the (1;4) fusion was present at diagnosis and at relapse, whereas the (4;17) fusion was detected at relapse only, leading to a secondary 17q gain confirmed by array CGH and indicating that 17q gain may not be a primary event in NB. The results indicate that the deletion of chromosome 1 may precede the gain of chromosome 17q, which may, therefore, be a late rather than an initiating event in NB development. Screening of a panel of NB cell lines identified interstitial telomeric sequences in three, suggesting that this may be a recurrent mechanism leading to unbalanced translocations in NB (239).

The breakpoints on 17q occur between 17q21 to the telomere, but they consistently include 17q23→qter (203,229, 240, 241). Gains with breakpoints proximal to 17q21 correlated with a better survival than gains with more distal breakpoints (229). Gain of 17q is associated with advanced stage

translocations between the long arms of these chromosomes (68, 98, 220, 228, 229, 232–234). The extent of the variation of the breakpoints involved in the t(11;17), however, is unknown. It is possible that this variation is a consequence of the many different chromosomal mechanisms leading to these genomic imbalances and that the breakpoints in t(11;17) probably occur over a limited region (235).

The 17q breakpoints vary, but preferential gain of a region distal to 17q22-17q21 suggests that a dosage effect of one

TABLE 7.5. Some aspects of 17q involvement in NB.

	References
Recurrent t(1;17) show nonhomologous mitotic recombination during the S/G2 phase as a mechanism for LOH in NB.	*Caron et al 1994* (225)
Molecular cytogenetic analysis of t(1;17) in NB.	*Van Roy et al 1995* (240)
Poor prognosis is associated with 17q gain, but does not involve the *SSTR2* gene in NB.	*Abel et al 1999* (47)
Survivin (at 17q25) is associated with a poor prognosis and promoter cell survival in NB.	*Islam et al 2000* (237,238)
Molecular analysis of chromosome 17 gain in NB.	*Janoueix-Lerosey et al 2000* (253)
	Lastowska et al 2001, 2002 (256,257)
Physical closeness of amplified *MYCN* and 17q in NB.	*O'Neill et al 2001* (778)
nm23 gene on 17q and *MYCN* and *MYC* downstream pathways.	*Godfried et al 2002* (779)
I-FISH studies of 17q gain in NB.	*Trakhtenbrot et al 2002* (243)
Further molecular studies of constitutional t(1;17) in a patient with NB.	*Van Roy et al 2002* (241)
Detection of single-copy 17q gain with RT-PCR.	*Morowitz et al 2003* (780)
PPM1D as target in 17q gain in NB.	*Saito-Ohara et al 2003* (128)
Gain of 17q may be seen in NB with a favorable prognosis.	*Tomioka et al 2003* (705)
Regions syntenic with human 17q are gained in mouse and rat NB.	*Lastowska et al 2004* (258)
Nature, variety and complexity of chromosome 17 translocations in NB.	*Schleiermacher et al 2004, 2005* (236,239)
Topo-IIα gene in t(1;17) in NB.	*Yoon et al 2006* (781)
The gain on the distal segment of 17q, not uncommon in NB, was shown to contain critical regions that may reflect on the course and therapy of the disease.	*Vandesompele et al 2008* (781A)

TABLE 7.6. 17q rearrangements in neuroblastoma tumors and cell lines.

17q Changes	Cell Line	References
Primary tumors		
der(1)t(1;17)(p12:q11)x3		Betts et al 2005 (743)
der(1)t(1:17)(p13:q11)		McRobert et al 1992 (233)
der(1)t(1:17)(p13:q21)		Betts et al 2005 (743)
der(1)t(1:17)(p1?:q11-q21)		Lastkowska et al 1997 (228)
der(1)t(1:17)(p22:q21)		Lastkowska et al 1997 (228)
der(1)t(1:17)(p32:q11.2)		Lastkowska et al 1997 (228)
der(1)t(1:17)(p32:q12)		Kaneko and Knudson 2000 (739)
der(1)t(1:17)(p32:q21) (3 cases)		Betts et al 2005 (743)
der(1)t(1:17)(p32:q21),der(1)t(1:17)(q33:q21), der(16)t(16:17)(p13:q21),der(19)t(17:19)(q21:p13)		Betts et al 2005 (743)
der(1)t(1:17)(p3?:q21)		Stark et al 2002 (67)
der(1)t(1:?17)(p3?:q21)		Betts et al 2005 (743)
der(1)t(1:17)(p34:q21)		Mitchell et al 1996 (736)
der(1)t(1:17)(p34:q21)x2,der(13)t(13:17)(q24:q21)		Betts et al 2005 (743)
der(1)t(1:17)(p34-p36:q12-q21)		Caron et al 1994 (225)
der(1)t(1:17)(p35-p36:q11-q12)		Van Roy et al 1994 (220)
der(1)t(1:17)(p36:q11-q12)		Van Roy et al 1994 (220)
der(1)t(1:17)(p36:q11-q12)		Lastkowska et al 1997 (228)
der(1)t(1:17)(p36:q21)x2		Kaneko and Knudson 2000 (739)
der(1)t(1:17)(p36:q21),der(22)t(17:22)(q21:p12)		Betts et al 2005 (743)
der(1)t(1:17)(p36:q21),der(22)t(17:22)(q11:q13)x2		Betts et al 2005 (743)
der(1)t(1:17)(p36:?),der(16)t(16:17:9)(p13:q21q25:q21), der(17)t(4:17)(q26-q28:p11)		Betts et al 2005 (743)
der(1)t(1:17)(p3?6:q11-q12)		Lastkowska et al 1997 (228)
der(3)t(3:17)(p13:q23)		Betts et al 2005 (743)
der(3)t(3:17)(p22:q21)x2		Stark et al 2002 (67)
der(3)t(3:17)(p22:q22),der(11)t(11:17)(q14:q21), der(17)t(8:17)(p12:q23)		Stark et al 2002 (67)
der(3)t(3:17)(p26:q21),der(17)t(17:17)(p11:q13)		Sawyer et al 1996 (737)
der(3)t(3:17)(q21:q21),der(15)t(15:17)(q24:q21)		Betts et al 2005 (743)
der(4)t(4:17)(p12:q21)		Betts et al 2005 (743)
der(5)t(5:17)(p13:q22)		Stark et al 2002 (67)
der(5)t(5:17)(q35:q12-q21)		Avet-Loiseau et al 1995 (234)
der(8)t(8:17)(p21:q21)		Betts et al 2005 (743)
der(9)t(9:17)(p1?3:q21)		Lastkowska et al 1997 (228)
der(10)t(10:17)(q22:q21)		Lastkowska et al 1997 (228)
der(11)t(11:17)(p15:q21)		Lastkowska et al 1997 (228)
der(11)t(11:17?)(q11-q13:q11:q11-q12)		Avet-Loiseau et al 1995 (234)
der(11)t(11:17)(q11-q23:q11) (7 cases)		Lawce et al 2000 (226)
der(11)t(11:17)(q12:q12)x4		Stark et al 2003 (67)
der(11)t(11:17)(q13:q11)		McRobert et al 1992 (233)
der(11)t(11:17)(q21:q23)x2,der(22)t(17:22)(p12:q22)x2		Stark et al 2002 (67)
der(12)t(12:17)(q14:q21)		Christiansen and Lampert et al 1988 (60)
der(14)t(14:17)(q32.3:q21)		Lastkowska et al 1997 (228)
der(14:17)(q10:q10)		Lastkowska et al 1997 (228)
der(16)t(16:17)(q?11-?q12:q21)		Lastkowska et al 1997 (228)
der(16)t(16:17)(q22:q11)		Cohen et al 2001 (740)
der(17)t(X:17)(q13):p12)		Betts et al 2005 (743)

TABLE 7.6. Continued

17q Changes	Cell Line	References
der(19)t(17;19)(q12;p13)		Betts et al 2005 (743)
der(21)t(11;17)(q12;q12)x4		Stark et al 2002 (67)
der(21)t(17;21)(q11;p11)		Sawyer et al 1996 (737)
der(22)t(17;22)(p12;q22)x2		Betts et al 2005 (743)
der(?)t(?;17)(?;q12)		Betts et al 2005 (743)
der(?)t(?;17)(?;q21)		Betts et al 2005 (743)
i(17)(q10) (2 cases)		Franke et al 1986 (222B)
i(17)(q10)		Christiansen and Lampert et al 1988 (60)
i(17)(q10)x2		Avet-Loiseau et al 1995 (234)
t(4;?17)(p15;q11)		Avet-Loiseau et al 1995 (234)
t(11;17)(q11-q23;q11)x7		Lawce et al 2000 (226)
trp(17)q21q25		Lastkowska et al 1997 (228)
der(1)t(1;17)(p22;q21)	HD-N-16	Amler and Schwab et al 1989 (77)
der(1)t(1;17)(p32;q12)	IMR-32	Tumilowicz et al 1970 (632)
		Kim et al 2001 (783)
der(1)t(1;17)(p32;q21)	NLF	Brodeur et al 1977 (163)
der(1)t(1;17)(p33-p34;q11)	SJNB-12	Van Roy et al 1995 (240)
der(1)t(1;17)(p33-p34;q12)	SJNB-8	Van Roy et al 1995 (240)
der(1)t(1;17)(p34.1;q21),	NB-12	Shapira et al 1997 (744)
der(5)t(5;17)(q13;q21),der(17)t(6;17)(q21;q21)		
der(1)t(1;17)(p34.3;q21)	NMB	Brodeur et al 1977 (163)
der(1)t(1;17)(p34.3;q21)	MHH-HB-11	Pietsch et al 1988 (650)
der(1)t(1;17)(p35;q21)	IGR-N-337	Valent et al 1999 (784)
der(1)t(1;17)(p36;q11)	TR-14	Cowell and Rupniak 1983 (643)
der(1)t(1;17)(p36;q11.2), der(11)t(11;17)(q13;q11.2)	GI-ME-N	Donti et al 1988 (232)
		Van Roy et al 2001 (66)
der(1)t(1;17)(p36.1-2;q21)	LA-N-5	Seeger et at 1977 (647)
der(2)t(2;10;17)(p14;q11;q22), der(9)t(9;17)(q34;q21),der(16)t(16;17)(q24;q22)	IGR-N-91	Valent et al 1999 (784)
der(2)t(2;17)(q13;q23), der(6)t(6;17)(q25;q11.2),der(8)t(8;17)(q21;q12)	Vi-8-56	Savelyeva et al 1994 (203)
der(3)t(3;17)(p13;q21), der(11)t(11;17)(q13;q11)	SK-N-AC	Van Roy et al 1994 (220)
		Kim et al 2001 (783)
der(5)t(5;17)(q35;q21)	IGR-N-331	Valent et al 1999 (784)
der(6)t(6;17)(q27;q21)	NGP	Brodeur et al 1977 (163)
der(11)t(11;17)(p15;q12), t(11;17)(q22;q12)	GICIN-1	Panarello et al 2000 (785)
der(11)t(11;17)(q13;q11-q12) (2 cases)	N206	Van Roy et al 1994 (220)
der(11)t(11;17)(q23;q11-q12)	GI-ME-N	Donti et al 1988 (232)
der(11)t(11;17)(q13;q21)	BCH-N-JW	McConville et al 2001 (786)
der(19)t(17;19)(q11-q12;q13)	SKNF1	Van Roy et al 1994 (220)
der(20)t(17;20)(q21;q13)	SMS-KCNR	Reynolds et al 1986 (648)
		Kim et al 2001 (783)
der(22)t(17;22)(q21;q13)	SH-EP	Ross et al 1983 (787)
der(22)t(17;22)(q21;q13)	SKNSH	Biedler and Spengler 1976 (73)
der(22)t(17;22)(q21;q13)	SH-SY5Y	Kim et al 2001 (783)

disease, 1p loss, *MYCN* amplification, age >1 year, and di-tetraploid chromosome number. Gain of 17q may serve as an independent prognostic marker (48).

The mechanisms leading to the translocations with 17q are not well understood. They are thought to arise in the S/G2 phase, and a particular occurrence within early replicating chromosome regions has been observed (109, 225). However, no study has investigated the molecular structure of the breakpoints. Such an analysis may provide insight into the mechanisms leading to these translocations, and it may possibly highlight an underlying defect of a chromosome maintenance pathway (242).

Chromosome 17 alteration occurring at the time of relapse has been described in a cytogenetic analysis (60). Furthermore, clonal variations of 17q copy number have been detected in FISH studies, suggesting that these alterations may be secondary events in NB (243). In one study (239), the der(1)t(1;4;17) may either have existed at the time of diagnosis as a rare subclone or may have occurred later, possibly induced by antineoplastic treatment.

The full recognition of chromosome 17 involvement in NB as being the most frequent genetic change, followed by those of 1p loss, the presence of dmin and/or hsr and amplification of *MYCN*, came with the introduction of FISH techniques that revealed gain of 17q, often through unbalanced translocations, to be a key and frequent event in NB tumors and cell lines (189, 203, 220, 224, 225, 228, 240). The gain of 17q in these unbalanced translocations was manifested by its presence in addition to two normal copies of chromosomes 17. The translocation partners most often were 1p and 11q, the former being more frequently involved than the latter. Although less frequent, a number of other chromosomes have been shown to be involved in 17q translocations. In addition, 17q sequences may have an affinity for amplified *MYCN* (230, 244, 245). Most of the data on chromosome 17 changes generated on NB by cytogenetics (73, 219, 246, 247) were clarified and subsequently enlarged by FISH analyses (203, 220, 240), and in particular gain of 17q, was further corroborated by CGH studies (46, 95–99, 104, 229, 248, 249), which showed frequent gain of chromosome 17 in >70% of stages 1 and 2 tumors and unbalanced 17q gain in >70% of stage 4 tumors, usually associated with amplified *MYCN* and loss of 1p or in NB with loss of 11q but without *MYCN* amplification. In one study (107), the genetic profile of stage 4 NB was explored and showed that >50% of NB without *MYCN* amplification had loss of 11q and 86% had 17q gain.

Studies on large series of NB have shown the following: most NB had 17q gain, 1p loss and *MYCN* amplification simultaneously; the next largest group of NB had 17q gain as the sole feature; *MYCN* amplification rarely occurs in the absence of 17q gain or 1p loss or both, pointing to *MYCN* amplification as occurring subsequent to the other changes (189, 224, 230). A relatively large though variable segment may be involved in the 17q gain. These findings support the conclusion that a gene dosage effect, as a consequence of the unbalanced translocations, rather than the involvement of a

specific gene on 17q, probably plays a role in NB biology (230). Gain of whole chromosome 17 (or no gain) is associated with NB with a favorable outcome.

Studies of a patient with a constitutional t(1;17)(p36.1;q12-q21) with NB (174) were incomplete, because the cytogenetic status of the tumor cells could not be determined. Cloning of the breakpoint sequences on 1p36 has been unsuccessful (205, 221, 250–252). Study of the breakpoint on 17q has been narrowed down within a 25-kb segment located between the *ACCN1* and *TLK2* genes and near the distal breakpoints of two microdeletions.

In NB, the positions of the breakpoints along chromosome 17 are highly heterogeneous (253–256), and several distinct 17q translocations may occur within the same tumor (109, 224, 257, 258). This strongly suggests that the supposed oncogenic role of these translocations may rely on an increased dosage of genes located in the distal part of 17q, together with the haploinsufficiency of genes encoded by the deleted arm of the recipient chromosome.

The relationship of 17q gain to survival in NB has been investigated in several studies with some variability in conclusions, although the general consensus was that 17q gain identifies a larger proportion of high-risk NB and with greater predictive power than any other clinical or tumor genetic factor (46, 48, 49, 189, 229, 259, 259A).

In a study (48), a strong association was found between *MYCN* amplification, 1p deletion and gain of 17q. The discrepancy between these data (48) and those of others (110) may be due to the limited number of samples in the latter study. Nevertheless, the combined data support a role of 17q in the development and progression of NB and clearly define a subset of longer-term survival patients whose tumors do not show gross 17q abnormalities.

Table 7.5 presents features related to 17q+ in NB, most of which have not been discussed in the text, and contains information cogent to the possible role of 17q+ in NB. Table 7.6 shows rearrangements of chromosome 17, primarily unbalanced translocations involving 17q, in NB cell lines and primary tumors.

Changes of Chromosome 11 in NB

Karyotype analysis of NB tumors revealed that 11q loss was present in 15% of cases (260); CGH-based studies showed such loss in 10–41% of NB tumors (95–99, 101, 104), with the distal half of 11q being the most frequently affected by the loss and a high frequency of 11q deletions in stage 4 NB lacking *MYCN* amplification (99, 261). Microarray analysis of gene expression in NB confirmed such tumors to be associated with 11q loss (21).

LOH studies have shown 11q loss in up to 1/3 of NB tumors (168, 262, 263) Introduction of a normal chromosome 11 into a NB cell line with loss of 11q resulted in morphological changes similar to those observed with chromosome 1 in cell lines with 1p loss (176), thus supporting the possible presence of a TSG on 11q.

Allelic loss of 11q is present in 35–45% of NB (99, 129, 136). Although not seen in NB with *MYCN* amplification or 1p loss, loss of 11q is observed in advanced stage NB, in older patients and with unfavorable pathology (44). Thus, 11q loss appears to be an independent predictor of poor outcome in NB, especially of disease relapse (136).

A small number of 11q constitutional rearrangements in patients with NB have been described, including deletion of 11q23-qter (2, 264, 265), translocations involving 11q21 and 11q22 (265A), and an inversion of 11q21q23 (266). No tumors were analyzed in these cases; hence, the possible role played by the changes in 11q in NB development remains unknown.

Aggressive disease ultimately develops in a subgroup of NB patients with loss of 11q within the low-risk and intermediate-risk groups despite the lack of *MYCN* amplification (267). Because 11q23 LOH occurs almost exclusively in tumors without *MYCN* amplification, it was postulated (50) that it may be a useful markers for tumors that are aggressive but lack *MYCN* amplification (Figure 7.7).

There are some indications that 11p may harbor TSG at 11p13-11p15 and 11p15.3 possibly related to NB (268). Evidence for this includes occasional deletions of 11p (Table 7.1) (260) and reported cases of NB with LOH restricted to 11p (262). A small number of individuals with Beckwith–Wiedemann syndrome (BWS) have been reported to have NB (211). This is of interest because a gene at 11p15 is implicated in BWS, although the imprinting changes seen in these cases could not be found in 15 NB analyzed (269).

LOH at chromosome arms 1p and 11q occurs frequently in NB (129, 136, 188, 193, 270, 271). Previous studies have suggested that there is an association between LOH at 1p36 or 11q23 and features of high-risk NB (138, 272). Whereas 1p36 LOH was found to be associated with *MYCN* amplification, 11q23 LOH was rarely observed in tumors with this ampli-

fication (129, 272). Given that 11q23 LOH occurs primarily in tumors without *MYCN* amplification, it was hypothesized (50) that 11q23 LOH could be a useful prognostic marker, especially in cases associated with low or intermediate risk. Therefore, the allelic status at 1p and 11q in a large series of accrued NB cases was determined (50). It was shown (50) that 1p36 LOH is present in about 23% of primary NB, is highly associated with a poor outcome, and is independently predictive of an unfavorable progression-free survival in low-risk and intermediate-risk patients. It also was shown that unbalanced LOH of 11q (with retention of 11p) is present in about 17% of primary NB, predominantly in those tumors without *MYCN* amplification, and that this LOH is an independently significant marker of decreased event-free and progression-free survival. The clinical usefulness of the identification of 1p36 LOH and unbalanced 11q LOH is applicable to patients with localized disease whose tumors do now show *MYCN* amplification (50).

NB exhibiting deletion of 11q23 represents a genetic subtype distinct from those exhibiting *MYCN* amplification or 1p36 deletion (129, 273, 274). *MFRP* and *RNF26* genes located within the commonly deleted region of NB at 11q23.2 were cloned and characterized (275). Table 7.7 presents information on 11q changes in NB, most of which have not been discussed in the text.

Changes of MYCN in NB

The *MYCN* gene (located at 2p24) was cloned in 1983 as an amplified DNA sequence in NB cell lines with dmin and/or hsr (75, 276). Amplified material at 2p24 is transformed to dmin during the process of amplification (276), and the dmin may then be linearly integrated into random chromosome regions as hsr (45, 77, 78), a phenomenon much more common in cell lines of NB (approx 50%) than in primary tumors. The number of dmin may vary from a few to hundreds, with *MYCN* having usually 50–400 copies per cell, with correspondingly high levels of MycN protein expression (277).

MYCN amplification (>10 copies per haploid karyotype) occurs in a significant subset (approx 30%) of NB and correlates highly with advanced stage of disease and is usually present at diagnosis (96, 152, 217, 278, 279), and it is associated with rapid progression with a poor prognosis independent of tumor stage or patient age or biologic parameters (Figures 7.8–7.10) (45, 280–282). The prognostic impact of *MYCN* amplification is especially evident in infants with stage 4 disease (283–285). *MYCN* amplification in NB without metastases ultimately is associated with tumor progression, most clearly demonstrated in cases with nodular ganglioneuroblastoma (286–288).

The prognostic correlation of *MYCN* status in NB has been shown to be related to overexpression of this gene regardless of *MYCN* amplification status (289, 290). *MYCN* amplification was not related to clinical outcome in NB in the absence of its overexpression. Others have found contrary results (291, 292).

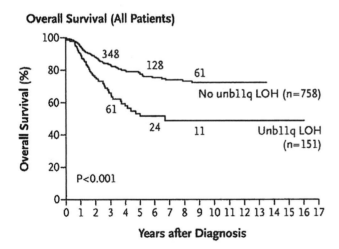

FIGURE 7.7. Overall survival of NB patients with and without LOH of 11q, showing a statistically significant difference between the two groups (from ref. 50, with permission).

TABLE 7.7. Some features of deletions and other changes of 11q in NB.

	References
Loss of 11q and gain 17q often due to an unbalanced t(11;17); breakpoints were not clearly established.	*Donti et al 1988* (232)
	McRobert et al 1992 (233)
	Van Roy et al 1994 (220)
	Avet-Loiseau et al 1995 (234)
	Lastowska et al 1997, 1998 (228, 229)
	Stark et al 2003 (68)
About 1/3 of NB have loss of 11q molecularly.	*Srivatsan et al 1993* (262)
	Takita et al 1995 (263)
	Brinkschmidt et al 1997 (96)
	Lastowska et al 1997 (98)
	Plantaz et al 2001 (107)
The allelic status 1p and 11p in NB and ganglioneuromas has been reported.	*Kurahashi et al 1995* (527)
Loss of 11q in approximately 30% of NB of advanced stages with poor outcome	*Mertens et al 1997* (260)
	Luttikhuis et al 2001 (270)
	Spitz et al 2003,, 2006 (227, 271, 788)
	Attiyeh et al 2005 (50)
defines a distinct subtype almost always with unbalanced 17q+ and other chromosomal changes (e.g., 3p-).	*Stallings et al 2004* (235)
In t(11;17) the breakpoints on both chromosomes vary.	*Vandesompele et al 1998* (104)
	Breen et al 2000 (102)
	Spitz et al 2003 (227, 271)
Deletion of 3p is nonrandomly associated with deletion of 11q; no statistically significant association between 3p and 11q loss and more aggressive NB.	*Breen et al 2000* (102)
Allelic deletions of 11q seen in 44% of NB.	*Guo et al 1999, 2000* (129, 273)
LOH mapped to 11q23 indicated the possible location of a TSG In contrast to 1p LOH, there was an inverse relation between 11q23 LOH and *MYCN* amplification.	*Maris et al 2001* (272)
	Maris and Matthay 1999 (45)
Allelic deletion at 11q23 is common in *MYCN* single-copy NB, but with an adverse prognosis.	*Guo et al 2000* (273)
	Luttikhuis et al 2001 (270)
LOH of 11q in 129/295 NB (44%) at 11q14-q23 (possible TSG in that area). No correlation with adverse prognostic variables, but an inverse relationship of 11q LOH with *MYCN* amplification; 11q inactivated during progression.	*Maris et al 2001* (272)
Unbalanced gain of 11p occurs at a relatively high frequency in NB with 11q-.	*Stallings et al 2003* (274)
NB with loss of 11q have a very high frequency (approx 90%) of unbalanced t(11;17).	*Stallings et al 2004* (235)
A balanced t(11;17) was seen in one NB with *MYCN* amplification, a different genetic subtype from the 11q- tumors.	*Stallings et al 2004* (235)
RT-PCR for 11q- in stratifying cases.	*Boensch et al 2005* (789)
NB differentiation gene on chromosome 11.	*De Preter et al 2005* (268)
High-density single nucleotide polymorphism arrays showed LOH of 11q in 15/22 NB which was highly associated with LOH of 3p, both present in 9/22 samples. LOH of 1p was seen in 1/3 of the cases. These events were usually associated by copy number loss, indicating hemizygous deletions within these regions. Gain of copy number most frequently involved 17q (21/22 cases).	*George et al 2007* (789A)

For example, high levels of *MYCN* amplification result in a benign phenotype by inducing apoptosis and enhancing favorable gene expression (293). Whether relative overexpression of *MYCN* RNA, MYCN protein, or both, in the absence of genomic amplification, is biologically or clinically significant continues to be controversial (294), as demonstrated by contradictory publications on this subject (295–298).

MYCN is normally expressed in the developing nervous system. The gene product is a DNA-binding protein. It contains a helix-loop-helix/leucine zipper motif that mediates its interactions with DNA and related proteins Max and Mad (299). Amplification of *MYCN* causes an imbalance between its product and the proteins and DNA sequences with which it interacts, resulting in unrestrained cell proliferation. Forced overexpression of *MYCN* in transgenic mice causes neurob-

lastic tumors, further indicating the relationship of this gene and tumorigenesis of NB (45, 300).

Amplification of *MYCN* is used as an index for treatment stratification and for clinical correlations in NB, reflecting the utility of *MYCN* amplification as a tumor-specific genetic alteration (45). Testing generally requires fresh tissue, hence portions of tumors should be frozen in all suspected NB, before their submersion in fixatives. Initially, Southwestern blots were used to detect this phenomenon, but interphase FISH is generally used today (301). By FISH analysis, *MYCN* amplification is apparent in cells with dmin as multiple, randomly distributed sites of fluorescence, whereas hsr are indicated by the presence of long, contiguous strands of fluorescence.

A large number of genes and genetic pathways are affected by amplification of *MYCN* (302–304). Serial analysis of

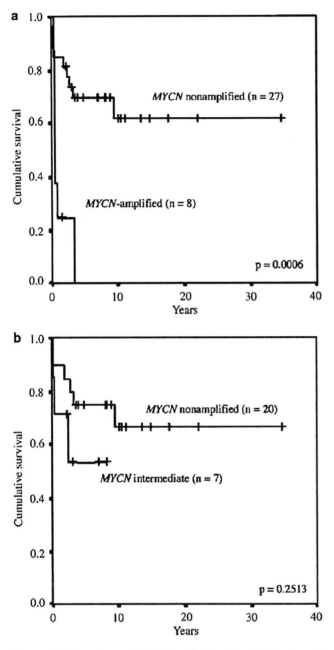

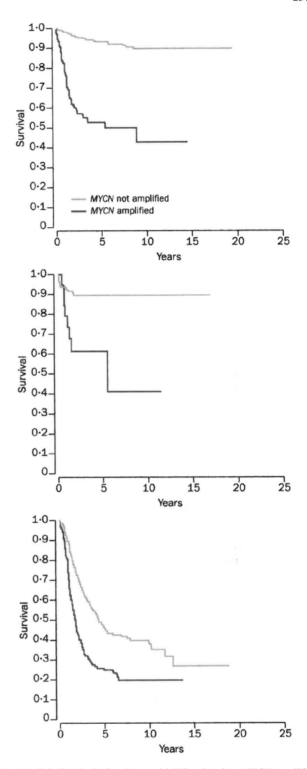

FIGURE 7.8. Overall survival of NB patients with and without amplified *MYCN*, showing the poor outcome of the latter group of patients (**a**); for a comparison of survival of those NB patients with nonamplified *MYCN* tumors versus those with intermediate or low amplification, see (**b**) (from ref. 858, with permission).

FIGURE 7.9. Survival of patients with NB related to *MYCN* amplification and stage of disease. (*Top*) Cases of stages 1–3. (*Middle*) Stage 4S. (*Bottom*) Stage 4. The statistically significant differences indicate the poor outcome inherent in *MYCN* amplification in all stages of NB (from ref. 7, with permission).

gene expression (SAGE) of *MYCN* amplification in NB has revealed that *MYCN* enhances the expression of a number of genes involved with protein synthesis (303); other SAGE studies have revealed that the *MEIS1* (located at 2p14) oncogene is overexpressed in 25% of NB (305). Gene expression showed many genes to be coexpressed and coamplified with *MYCN* overexpression in NB, suggesting pathways that contribute to the malignant behavior of these tumors (304).

The *ID2* (target of Rb protein gene) is critical for cellular proliferation, and it is the oncogene effector of *MYCN* (306).

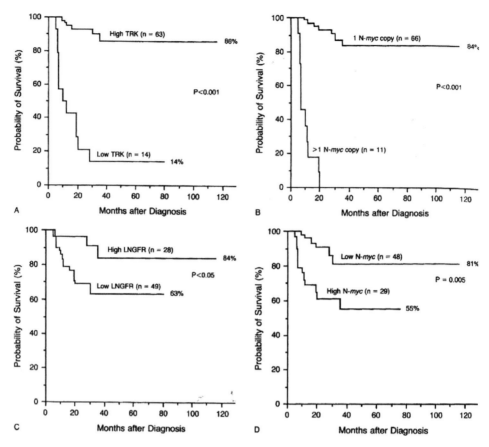

FIGURE 7.10. Probability of survival as related to the level of expression of Trk (*top left*), LNGFR (low-affinity nerve growth factor receptor, p75^{LNGFR}) (*bottom left*), and *MYCN* (*bottom right*). The relation of *MYCN* amplification to survival (*top right*) (from ref. 431, with permission).

ID2 expression is independent of *MYCN* expression and lacks prognostic value (290, 307, 308), although *ID2* expression as a predictor of poor outcome in NB has been suggested (306).

High levels of *MYCN* expression were not prognostic of adverse outcome in advanced-stage NB with nonamplified *MYCN*. A trend for improved outcome was observed with high *MYCN* expression (292). *MYCN* amplification remains diagnostically unique among other indices (309).

Although nearly all NB express *MYCN*, the highest levels of expression were seen in tumors (or NB cell lines) with *MYCN* amplification (310). Without *MYCN* amplification, expression of *MYCN* did not correlate with disease progression. Not all NB that failed to respond to therapy had *MYCN* amplification (310). The DNA index in NB did not correlate with *MYCN* expression or amplification.

Five-year survival of patients with NB is related to *MYCN* expression when considered in conjunction with TrkA expression, i.e., the best survival (>88%) was seen in cases with increased *MYCN* expression (but no amplification) and TrkA expression, in approximately 64% of cases when *MYCN* and TrkA expressions were low and *MYCN* was not amplified and in only approximately 9% when *MYCN* expression was associated with *MYCN* amplification and low expression of TrkA (293). Table 7.8 shows a representative study of *MYCN*

changes in NB (311). A correlation of these changes with some clinical parameters is presented in Table 7.8.

Table 7.9 presents some features of the *MYCN* changes in NB, most of which have not been discussed in the text. The array of these changes is reflective of the role played by *MYCN* in NB biology.

Changes of Chromosomes (Other Than 1, 11, and 17) in NB

The following sections present changes (e.g., allelic deletions and amplifications) affecting various chromosomal regions in NB (211). Such changes are not unique to NB, and they may be seen in many other cancers. Here, however, we emphasize changes and their effects in NB.

Chromosome 2

A gain at 2p24 was found by CGH in five NB without *MYCN* gene amplification. The *DDX1* gene, located at 2p24, although far from the *MYCN* gene (400 kbp), has been found to be coamplified in NB cell lines and tumors (312,313). Thus, gain at 2p24 region could be associated with the presence of extra copies and amplified DNA sequences of some genes (314).

TABLE 7.8. Example of findings seen in a study of NB.

MYCN amplification or gain present in approximately 25% of cases with NB and associated with advanced-stage disease, older patient age, loss of 1p and additional chromosome aberrations.

MYCN amplification predicts a poor outcome of NB of all stages and age.

MYCN gain tumors correlated with 11q alterations and showed no increased *MYCN* mRNA.

MYCN status discriminated between a favorable or poor prognosis in NB, independent of stage, age, and the degree of *MYCN* amplification.

NB with amplified *MYCN* showed a more favorable outcome when combined with nonstage 4 disease and patient age less than 1 year.

MYCN gain in NB cases had a poor event-free survival in stages 1–3 and 4S, probably related to genetic alterations other than *MYCN*.

MYCN gains seen in advanced-stage NB and older patients, but is not associated with higher *MYCN* expression and prediction of outcome.

Spitz et al. 2004 (311).

TABLE 7.9. Some features of *MYCN* in NB.

	References
MYCN down-regulates MHC class 1 antigen expression in NB.	*Bernards et al 1986* (790)
Characterization of *MYCN* protein.	*Slamon et al 1986* (791)
FISH analysis of *MYCN* status in NB is now considered the optimal method,	*Cohen et al 1988* (792)
	Shapiro et al 1993 (301)
	Misra et al 1995 (793)
including status in archival tissues.	*Hachitanda et al 1997* (794)
Other methodologies have been described.	*Noguchi et al 1988* (795)
MYCN expression in NB, in contrast to amplification, did not reflect prognosis.	*Nisen et al 1988* (796)
MYCN copy numbers were the same in metastases as in the primary NB.	*Tsuda et al 1988* (797)
MDR1 gene expression in NB correlates with previous therapy, regardless of *MYCN* status.	*Bourhis et al 1989* (798)
Suppression of MYC expression by *MYCN* in NB.	*Breit and Schwab 1989* (799)
MYCN amplification frequently associated with loss of 1p, low expression of TrkA and undifferentiated and stroma-poor histology in NB.	*Fong et al 1989* (168)
	Nakagawara et al 1993 (431)
	Suzuki et al 1993 (458)
	Shimada et al 1995 (287)
Prolonged half-life of *MYCN* in a NB cell line without *MYCN* amplification had been described.	*Cohn et al 1990* (800)
	Wada et al 1988 (801)
Inverse relation between *MYCN* amplification or MDR expression and catecholamine excretion in NB.	*Nakagawara et al 1990* (802, 803)
MYCN amplification and deletion of a possible TSG in NB.	*Schwab 1990* (1220)
NB express both *MYCN* and *MYC*; but, the level of *MYCN* expression is highest in NB with *MYCN* amplification. DNA index did not correlate with *MYCN* expression.	*Slavc et al 1990* (310)
DNA ploidy is a good prognostic index in NB in addition to *MYCN* amplification.	*Bourhis et al 1991* (804)
High levels of p19/nm23 protein associated with advanced stage disease and *MYCN* amplification.	*Hailat et al 1991* (1205)
MYCN antisense DNA results in differentiation and growth arrest in NB cell lines.	*Negroni et al 1991* (805)
	Schmidt et al 1994 (806)
Reverse relation between *MYCN* and helix-loop-helix gene in NB.	*Ellmeier et al 1992* (807)
NB with *MYCN* amplification had deletions proximal to 1p36;NB with single copy *MYCN* more often had small terminal deletions of 1p36 only. *MYCN* amplified tumors showed no parental preference of the deleted 1p, whereas single-copy *MYCN* NB preferentially deleted the maternal-derived allele. The authors hypothesized that at least two tumor suppressor genes are located within 1p35-p36: an imprinted gene distal to 1p36 defined by the smallest region of overlap common to all distal 1p-deleted NB and a nonimprinted 1p35-p36.1 locus that is deleted in *MYCN*-amplified aggressive tumors.	*Caron et al 1993* (177)
MYCN amplification determined by differential polymerase chain reaction.	*Gilbert et al 1993* (1204)
MRP gene expression correlates with *MYCN* amplification and overexpression in NB.	*Bordow et al 1994* (1194)
MYCN down-regulated during NGF-induced differentiation in NB.	*J. Chen et al 1994* (808)
CD44 plays a role in *MYCN* amplification and cell differentiation in NB.	*Gross et al 1994, 1997* (809, 810)
	Christiansen et al 1995 (811)
Duplication of *MYCN* at 2p24 may be an alternative mechanism to amplification in NB.	*Corvi et al 1995* (812)
The *MYCN* gene may be retained in single copy during amplification in NB.	*Corvi et al 1995* (813)
	Schwab et al 1995 (814)
Nonsyntenic amplification of MDM2 gene and *MYCN* in NB has been described.	*Corvi et al 1995* (1200)
DDX1 amplification may occur concomitant with that of *MYCN*, but may not be directly correlated.	*Squire et al 1995* (1222)

TABLE 7.9. Continued

	References
MYCN increases expression of α-prothymosis and ornithine decarboxylase and accelerates progression into the S-phase of NB cells.	*Noguchi et al 1996 (1215)*
	Lutz et al 1996 (627)
Mapping of a *MYCN* amplicon in NB.	*Reiter and Brodeur 1996, 1998 (1217, 1218)*
MYCN amplification is associated with unbalanced gain of 17q in NB.	*Brinkschmidt et al 1997 (96)*
	Lastowska et al 1997 (46)
	Plantaz et al 1997, 2001 (99, 107)
A physical association of *MYCN* with 17q has been reported, i.e., hsr on 4q and chromosome 16 were flanked by 17q material.	*O'Neill et al 2001 (778)*
This was further illustrated by the presence of hsr at 17q in primary tumors, and hsr in 1p flanked at each end by chromosome 17 material in a NB cell line.	*Kaneko et al 1987 (54)*
	Rudduck et al 1992 (244)
	Van Roy et al 1994 (220)
	Savelyeva et al 1994 (203)
	Van Roy et al 1994 (220)
MYCN plays a causative role in the invasive phenotype of NB, including metastatic spread of these tumors.	*Goodman et al 1997 (1238).*
Integrin expression is down-regulated by *MYCN*. Activin A (but not other members of the transforming growth factor-β superfamily) is also down-regulated by *MYCN*, thus depriving NB of a growth inhibiting signal.	*Judware and Culp 1997 (815)*
Expression of *MYCN* may be independent of copy number in NB.	*Lutz and Schwab 1997 (816)*
Some cases of NB without *MYCN* amplification are associated with loss of 1p, 3p, 14q, or 11q. In particular, there is an inverse relation between loss of 11q and *MYCN* amplification.	*Plantaz et al 1997 (99)*
	Guo et al 1999 (129)
	Spitz et al 2003 (271)
Prognosis in localized NB is favorable in the bulk of cases, except those with *MYCN* amplification. Prognostic factors in NB include patient age, tumor stage (poor for stages II and III) and *MYCN* amplification (poor outcome).	*Rubie et al 1997 (183, 817)*
	Matthay et al 1998 (818)
Cellular effects and mechanism of action of *MYCN* in NB.	*Spengler et al 1997 (819)*
	Fulda et al 1999 (820)
	Freeman-Edward et al 2000 (821)
	Schmidt et al 2000 (284)
	Ribatti et al 2002 (822)
	Rössler et al 2001 (823)
	Wartiovaara et al 2002 (824)
	Edsjö et al 2004 (825)
	Kim and Carroll 2004 (826)
	Ren et al 2004 (827)
	Sugihara et al 2004 (828)
	Giannini et al 2005 (829)
	Jung et al 2005 (821A)
NMI protein interacts with regions that differ in *MYCN* and *MYC* and is localized to the cytoplasm of NB, in contrast to the nuclear localization of *MYCN*.	*Bannasch et al 1999 (1190)*
Role of *MYCN* in the differentiation and apoptosis in NB.	*Galderisi et al 1999 (830)*
MYCN amplification as an independent adverse prognostic factor in NB.	*Gallego et al 1999 (831)*
	George et al 2001 (832)
Some forms of chemotherapy may be effective in NB with *MYCN* amplification and the latter can serve as a drug development target for NB.	*Kawa et al 1999 (833)*
	Lu et al 2003 (834)
Lack of inactivated p73 in NB with single-copy *MYCN*.	*Kong et al 1999 (1209)*
HuD, a neuronal-specific RNA-binding protein, regulates *MYCN* stability in NB.	*Lazarova et al 1999 (1211)*
Evaluations of the role of *MYCN* in neurogenesis and other neuronal effects and in human neoplastic disease have been presented.	*Nesbit et al 1999 (1236)*
	Knoepfler et al 2002 (1237)
Expression of *MYCN* and *MYC* is found in 90–100% and 80–85% of NB, respectively.	*Zhe et al 1999 (835)*
MYCN down-regulates angiogenesis inhibtor activin A'.	*Breit et al 2000 (1195, 1196)*
Structure of *MYCN* amplicon.	*George and Squire 2000 (836)*
MYCN induction stimulated by ILGF-1.	*Misawa et al 2000 (1213)*
LOH of 1p seen in all categories of NB but not in ganglion- neuromas. LOH in two separate regions of 1p: 1p36 and 1p22. Stage 4 tumors were diploid with LOH of these 1p regions. Local tumors were hyperdiploid with 1p36 deletion.	*Mora et al 2000 (131)*
Stage 4A NB that regressed spontaneously were hyperdiploid with 1p22 and 1p36 deletions	*Mora et al 2000 (131)*
Rapid FISH analysis for *MYCN* amplification and 1p- in NB.	*Taylor et al 2000 (837)*

TABLE 7.9. Continued

	References
NCOG is coamplified with *MYCN* in aggressive childhood NB.	*Anglesio et al 2001* (1189)
Expression of p27 is prognostic and independent of *MYCN* amplification in NB.	*Bergmann et al 2001* (1192)
A rearranged *MYCN* in a case of NB did not contribute to the aggressive behavior of the tumor, as much as *MYCN* overexpression of the germline gene.	*B. Chen et al 2001* (838)
MYCN status of NB is a prognostic indicator regardless of patient age.	*Goto et al 2001* (286)
CpG islands frequently coamplified with *MYCN* in NB.	*Wimmer et al 2001* (1230)
FYN pathway (differentiation and growth) target of *MYCN*.	*Berwanger et al 2002* (383)
Circulating *MYCN* DNA has been proposed as a marker for NB.	*Combaret et al 2002, 2005* (839–841)
MYCN overexpression down-regulates IL-6: the latter inhibits angiogenesis and suppresses NB growth.	*Hatzi et al 2002* (1206)
MYCN and Bcl-2 co-overexpressions induce MMP-2 secretion and activation in NB.	*Noujaim et al 2002* (1216)
Microchrome maintenance protein (MCM7) expression and *MYCN* amplification interrelated in NB.	*Shohet et al 2002* (1221)
	Tsai et al 2004 (1225)
Several NB cases with *MYCN* gene gain (3-4 copies of the gene in diploid cells) without amplification and subsequent overexpression (3- to 7-fold) of the gene have been described.	*Valent et al 2002* (842)
The most frequent mechanism of *MYCN* gene gain in these NB was an unbalanced translocation of 2p (with the *MYCN* gene)	*Van Roy et al 2000* (843)
with various partner chromosomes. Subsequent studies on NB with simultaneous amplification and gain of *MYCN* led to proposed pathways by which *MYCN* gene copy-number can be increased.	*Valent et al 2004* (844)
MYCN enhances MDR1 gene expression in metastatic NB model.	*Blanc et al 2003* (1193)
The relationship of TrkB expression to *MYCN* expression and retinoiod acid therapy has been presented.	*Edsjö et al 2003* (1203)
BIN1 induces apoptosis in NB cell lines with *MYCN* amplification.	*Hogarty et al 2003* (1207)
MYCN gain and amplification in a 4S NB with progression, in contrast to the usual favorable prognosis in these tumors with nonamplified *MYCN*.	*Noguera et al 2003* (845)
	Tonini et al 1997 (846)
	Schmidt et al 2005 (285)
FISH analysis of high-risk NB revealed alterations of 1p36 and 11q23 (29%), 3p36 (19%) and *MYCN* amplification (19%): these cases define different groups of NB with an increased risk or progression.	*Spitz et al 2003* (271)
Reduced NIN1 expression (through *MYCN* action) seen in unfavorable NB.	*Tajiri et al 1999, 2003* (1223, 1224)
Apoptosis and *MYCN* in NB.	*van Golen et al 2003* (847)
	van Noesel et al 2003 (848)
	Cui et al 2005 (849)
c-KIT is preferentially expressed in *MYCN* amplified NB.	*Vitali et al 2003* (1227)
MYCN important in NB development in mice.	*Hansford et al 2004* (850)
MYCN regulates MRP1 promoters in NB.	*Manohar et al 2004* (1212)
MYCN amplification is an independent prognostic indicator in NB in contrast to *MYCN* expression.	*Tanaka et al 2004* (851)
Htwist and *MYCN* cooperate in overriding failsafe programs in cancer cells.	*Valsesia-Wittmann et al 2004* (1226)
Amplification of DDX1 may reflect an improved survival in *MYCN* amplified NB.	*Weber et al 2004* (1229)
NB with hyperdiploidy or triploidy and nonamplified *MYCN* have a favorable outcome.	*Bagatell et al 2005* (852)
	George et al 2006 (853)
	Spitz et al 2006 (854)
MYCN DNA in the circulation predicts amplification of this gene in NB.	*Combaret et al 2005* (841)
	Gotoh et al 2005 (855)
No correlation between DDX1 amplification and that of *MYCN* in NB.	*De Preter et al 2005* (1201)
HMGA1: target of *MYCN* in NB.	*Giannini et al 2005* (829, 856)
Pro-apoptotic activity of *MYCN* in activation-induced cell death in microglia.	*Jung et al 2005* (821A)
Enlarged nucleoli may be indicative of *MYCN* amplification in NB.	*Kobayashi et al 2005* (857)
Chromogenic in situ hybridization detects *MYCN* hotspot amplification associated with Ki-67 expression and inversely related to nestin expression.	*Korja et al 2005* (858)
The cause, effects and significance of *MYCN* amplification in NB have been the subjects of numerous laboratory and clinical studies, as shown in this table.	*Maris 2005* (859)
MYCN directly modulates MDM2 levels leading to inhibition of apoptosis in NB.	*Slack et al 2005* (860A)
Role of p27, Rb, and *MYCN* in the induction of cell arrest has discussed.	*Wallick et al 2005* (1228)
Disruption of cooperation between Ras and *MYCN* promotes growth arrest in NB.	*Yaari et al 2005* (1231)
Retinoid acid-binding and protein II are a target of *MYCN* in NB.	*Gupta et al 2006* (861)
Inhibition of c-MYC in NB by small interfering RNA.	*Kabilova et al 2006* (862)
Patterns of regulation of nuclear *MYCN* and *MYB* in retinoic acid treated NB cells have been reported.	*Thiele 1991* (862A)
MYCN amplification is preferentially initiated at some consensus sites flanking the gene and followed by rearrangements with a random loss of domain sequences resulting in the formation of amplicons.	*Akiyama et al l994* (862B)
At least one gene is co-amplified with *MYCN* in NB, possibly *DDX1*.	*George et al l997* (862C)
Senescence was induced by hydroxyurea in *MYCN* amplified NB cell lines.	*Narath et al 2007* (862D)

Chromosome 2 harbors the locus (2p34) for the oncogene *MYCN*, which is amplified in about one third of cases of NB. This amplification, as it is related to 2p, has been investigated regarding its composition and origin (75,315).

Low level gain of distal 2p in NB as a risk factor for high-level *MYCN* amplification has been reported (316). Partial trisomy 2p in NB may be related to this amplification (317).

Changes in 2p resulting in four fusion transcripts have been described in the NB cell line IMR (318). Amplification, high expression, or both of *MEIS1* gene (at 2p24) has been described in NB and NB cell lines (319,320), although much remains to be ascertained about the role of *MEIS1* (a homeobox gene) in NB (321,322).

Allelic loss of 2q, among losses in other chromosomes, was reported to be present in NB (263). Subsequent examination of 82 NB revealed allelic loss of 2q in 32% of the tumors (323), especially of 2q33. This chromosomal band includes the genes for caspase-8 and caspase-10, which were found not to be expressed in 70% and 7% of 27 NB cell lines, respectively (323), possibly through CpG island methylation.

Chromosome 3

LOH at 3p as a possible site for a gene related to NB development had been suggested (324,325). This was based on a study of 59 NB in which nine (15.3%) showed LOH of 3p (325). Of interest is that the cases with total loss of chromosome 3 were long-term survivors, whereas those cases with LOH of 3p had a poor outcome (325). The possible relation of loss of 3p to other parameters in NB was pointed to in studies in which 3p deletions coincided with 11q deletions in a subset of patients with aggressive disease without the oncogenic influence of deregulated *MYCN* (102,136).

The region 3p21.3 harbors several genes, including the *BLU* TSG, which has been found to be hypermethylated in 86% of NB cell lines (326). No correlation was found between hypermethylation of *RASSF1A* (frequent in NB) and *BLU* promoter region CpG islands in NB, indicating that they are unrelated events. The *SEMA3B* gene, located on 3p, may be methylated in NB and it may play a role in the pathogenesis of this tumor (327). *VHL* gene (located at 3p25) expression has prognostic value in NB, i.e., low levels are associated with poor survival (328).

Critical regions of involvement on 3p in NB (primary tumors and cell lines) have been delineated, i.e., 3p21.31-3p21.2 as a minimal region of loss and the shortest regions of overlap as 3p22.3-3p22.2 and 3p25.3-3p25.1) (329).

Chromosome 4

LOH of distal 4p in NB in approx 25% of cases was established by several groups of investigators (132,137). Similar findings were obtained with CGH in sporadic NB (104,249), and in familial NB (95,132).

Chromosome 5

Examination by RT-PCR of 24 NB showed nine (37.5%) to have allelic imbalance on 5q (330). The 5q change was found only in tumors containing a single copy of *MYCN* and exhibiting hyperdiploidy and to be associated with a favorable outcome (long-term survival).

Chromosome 7

Gain of an entire chromosome 7 or its long arm (7q) is a common abnormality in NB. Thus, in one study (274), 17 of 43 NB (40%) had gain of an entire chromosome 7 and five of 43 (12%) gain of 7q. These gains were present in tumors of all stages. An unbalanced t(3;7)(p21;q11.2) led to 7q gain and loss of 11q. These results, in general, are similar to those of others (99,107). It has been hypothesized (274) that the gain of chromosome 7 may be an early event in NB, because it was the sole change seen in two NB by CGH, and gain of 7q may play a role in tumor progression, because it was seen in tumors of higher stages. Amplification of 7q21 was reported to be associated with resistance to doxorubicin therapy (274A).

Chromosome 9

Losses of 9p were detected in 50% of advanced tumors and in 14% of localized NB, which confirm previous observations on the defect of this chromosome in advanced neuroblastoma (263,331,332). However, a higher frequency of 9p deletions in advanced tumors suggests that the *p16^INK4A* gene, which usually shows allelic loss in 9-deleted NB, is not essential to tumor progression but may participate in the deregulation of the cell cycle.

LOH was detected in 16.4% of 177 NB examined for LOH at 9p21 (140). This genomic change was associated with a favorable outcome in stage 4 patients. Hemizygous deletion of the region harboring the *CDKN2A* and *CDKN2B* loci was demonstrated in two tumors by FISH; *MTAP* expression was present by immunostaining in all but one tumor (140). The authors suggested a role for genes located telomeric of 9p21 in good-risk NB. In a previous study (333), it was shown that *CDKN2*, which encodes p16, a cell cycle control protein commonly inactivated in cancer, was not mutated in NB. In a previous study (334), regions of deletion distal to 9p21 were described.

A correlation of 9p loss with 1p loss has been noticed (335). Two distinct regions of 9p21 were observed that might be loci for genes important in NB development, apart from those for *p16* and *p15*. Initial studies for mutations of the p16 gene *(CDKN2* or *MTS1)* located at 9p21, in NB cell lines, were negative (134). In contrast, others (331,336,337)\ besides showing deletions of 9p21 (LOH in 36% of NB), also found that *p16* was not expressed in 72% (13 of 19) of NB and in five of 13 cell lines due to methylation of CpG island surrounding the first exon of the *p16* gene. The authors (331,347) suggested that *p16* may be a candidate TSG for NB and that its inactivation may contribute to the progression of these tumors.

Chromosome 10

LOH of 10q is not uncommon in cancers other than NB; in the latter, deletions of 10q are rare, though loss or gain of the whole chromosome 10 may be seen not infrequently in NB with complex karyotypes. Translocations involving 10q are more common than deletions (Table 7.3). Of genes located on 10q, *PTEN* (at 10q23.3) and *DMBT1* (at 10q23.3-q26.1) may contribute to NB development, the latter gene through methylation of a positive promoter (338, 339).

A study of 26 NB and 12 NB cell lines, LOH at 10q was detected in 18% of tumors and microsatellite instability in 14% (124). Promoter hypermethylation of the *MGMT* gene was present in 8% of the tumors and in 25% of the cell lines. These data suggest that 10q alterations may be implicated in the development of a small number of NB (124).

11p

Gain of 11p (11p11.2→p14) occurs in a varying percentage (possibly as high as 20%) of NB (107, 274) but only in those tumors that have loss of 11q, and it is probably related to the mechanism leading to 11q loss (274). 11p loss is associated with a poor prognosis.

Chromosome 12

In a gene expression study of 95 NB, a subset of tumors overexpressed several contiguous genes located at 12q13-q15 (340). Amplification of genes at 12q13-q15 was demonstrated in some NB.

Chromosome 14

Cytogenetically, loss of 14q in NB is only rarely seen (85, 341–343). This may be due to the deletions being too small for microscopic visualization or that the mechanism of LOH involves homologous mitotic recombination (45). A common region of deletion has been shown to be 14q23→qter (141, 344–346). LOH for 14q was first identified in six of 12 NB studied (347) by using a single restriction fragment length polymorphism (RFLP) marker within 14q32. Subsequent studies have shown collectively that 49 of 180 informative tumors (27%) have allelic loss at 14q32 (85, 262, 263, 348–350). A study (141) of 372 primary NB with markers evenly spaced along 14q showed LOH in 22%, with a common region of deletion within 14q23-qter. LOH for 14q was highly correlated with 11q LOH, inversely related to *MYCN* amplification and present in all clinical risk groups, indicating that this abnormality may occur early in tumor development. No correlation existed with clinicopathological features, patient age, tumor stage, *MYCN* status and ploidy.

In a study of 54 NB, 31% (17/54) showed LOH at 14q32 (345, 346). There was no statistical correlation between LOH and such parameters as age, stage of tumor, *MYCN* amplification and ploidy. The authors (345, 346) proposed the existence of a putative tumor-related gene at 14q32 important in the tumorigenesis of NB.

RFLP and microsatellite analysis showed 19/108 NB with allelic loss (LOH) of 14q, primarily at 14q32, but also at 14q23 and 14q12. These changes did not seem to carry a prognostic impact (344). The possible presence of at least two TSG has been suggested to exist on 14q (344). LOH of 1p and 14q occur independently in NB, the former usually being associated with *MYCN* amplification, whereas the latter is not (348).

Chromosome 18

Allelic loss of 18q that includes the *DCC* locus has been described in about 20% of NB, with 18q21.1 being the most commonly involved region (110, 263, 351); the *deleted in colorectal carcinoma* gene (*DCC*) has TSG functions and loss of *DCC* expression in NB may contribute to the dissemination of NB cells, perhaps through alterations in growth and differentiation pathways distinct from those regulated by *MYCN*, as suggested by the authors of one study (352). This was based on a demonstration that 38% of primary 62 NB and 50% of 12 cell lines showed undetectable *DCC* expression (352).

In another study of the *DCC* gene in NB, reduced or undetectable expression in 12 of 25 (48%) of primary tumors and in seen of 16 (44%) cell lines was found (353). However, the *DCC* and *DPC4* genes, located at 18q, a region frequently deleted in NB, were shown not to be involved by inactivating mutations in NB (352, 353).

Chromosome 19

The heterozygous deletion of 19q13 has been described in NB (139), and it is characteristic of aggressive NB, especially in local-regional recurrent NB. In a search for a candidate gene related to NB tumorigenesis, the expression profile of 89 NB was compared with that of 10 benign ganglioneuromas (354). A candidate TSG was thought to be the epithelial membrane protein 3 (*EMP3*) gene located at 19q13.3 and epigenetically silenced by hypermethylation. This gene also may play a crucial role in gliomas (354, 355). The *PTPRD* gene, located on 19q, has been suggested as a TSG in NB (356).

Gain of an entire chromosome 19 in NB has been reported, but its role in tumor progression remains unclear (110), because no structural anomalies of this chromosome were encountered. Hypermethylation-silencing of the myelin-related gene *EMP3* has been shown to play a role in the development of NB (and glioma) (356A). *EMP3* exhibits features of a TSG and is located at 19q13 exhibiting LOH in NB. The presence of *EMP3* hypermethylation is a poor prognostic factor in NB (*356B*).

Molecular Biology of NB

Gene Expression Studies and Profiles in NB

Gene expression studies have shown distinct differences in the genes expressed in spontaneously regressing NB versus progressing or metastatic tumors (357, 358). Similar approaches showed that NB are more frequently associated with indices of aggressiveness (e.g., loss of 1p and *MYCN* amplification) than GNB (359). Furthermore, the schwann cells had a less aggressive phenotype than the neuroblastic cells; thus, they may play a role in tumor maturation.

Gene expression and various array studies of NB and NB cell lines have included clinical correlations, methodological considerations, mathematical interpretation, prediction of outcome based on discriminating genes, and mass screening of NB (80, 360–372), and such studies are likely to yield information applicable to the diagnosis, biology, and the development of appropriate therapies of the various forms of NB (373) and to supply accurate identification of NB subgroups with applicability to forms of therapy (374).

The first microarray-based gene expression profiling study of NB (375) demonstrated that different small round blue-cell tumors, including NB, rhabdomyosarcoma, non-Hodgkin lymphoma, and Ewing tumors, could be distinguished on the basis of their patterns of gene expression. Subsequent microarray studies have facilitated the separation of differentiating NB from poorly differentiated tumors (116) and high-risk from low-risk tumors (376–378). Expression profiles for a number of genes mapped to 1p36.2 were examined in 55 NB. Five genes were found to have a decreased expression and one with an increased expression (379), when comparing tumors with favorable biology to stage 4 NB. When NB with loss of 1p were compared with those with an intact 1p, significant differences were noted in a number of genes. Complete loss of expression was not seen for any gene. Although some of the genes with diminished expression in unfavorable and 1p- deleted NB could contribute to tumor development, a combination of lowly expressed genes at 1p could possibly cause the unfavorable outcome associated with 1p- deletion in NB (379).

One study (374) demonstrated that gene expression profiles of clinically indistinguishable, high-risk metastatic NB that lack *MYCN* gene amplification provide new definitions of high and low risks of disease progression. Molecular risk classification at diagnosis may ultimately improve risk stratification for patients who are older than 12 months at diagnosis, and it is especially important for patients who are older than 18 months at diagnosis because their outcomes currently cannot be estimated by any other means. Current therapeutic strategies were successful for patients (374) who were classified clinically as having high-risk disease, but whose tumors had gene expression profiles associated with progression-free survival (molecular low-risk tumors). However, therapeutic modifications aimed at reducing toxicities should be consid-ered for this group. Gene expression profiling studies showed the involvement of cell cycle genes in aggressive NB (380), patterns for high-risk NB (381) and outcome prediction in NB (381A).

Studies of NB cell lines based on single-nucleotide polymorphism arrays (SNP) (382) confirmed previously described allelic imbalances in these lines, i.e., loss of 1p, gain of 17q, *MYCN* amplification and 11q and 14q deletions. In addition, SNP revealed changes not detected by CGH or karyotyping, e.g., gain of 8q21.1~q24.3 and chromosome 12 and loss of 16p12.3~q13.2 and LOH at 11q14.3~q23.3 in one cell line, and loss at 8p21.2~p23.3 and 9p21.3~p22.1 with corresponding LOH in another. Thus, SNP seems to be a powerful tool in deciphering allelic imbalances in NB (382).

Differentiating stages IVS and IV of NB by gene expression were not informative (372, 383). However, when gene expression patterns of stage IV and IVS tumors were investigated using SAGE and quantitative RT-PCR to uncover transcripts related to spontaneous regression and progression of disseminated NB (358), the results showed a statistically significant divergence of genes involved in these stages.

Apoptosis

Apoptosis (physiologic or programmed cell death) is a very complex mechanism involving many molecular pathways and genes (384). During normal fetal and postnatal development of the nervous system, a large number of the neurons die via apoptosis, a phenomenon akin to the spontaneous and quantitative regression of favorable NB (385). Thus, NB cells present a paradox in that they may be susceptible to apoptotic signals or that the apoptotic mechanism may be defective and thus contribute to tumor formation.

The subject of apoptosis in NB was addressed at length and its significance in the biology of NB, as related to complete or partial regression, was discussed previously (385). Some facets of apoptosis and NB are shown in Table 7.10, and a few key areas discussed in the following paragraphs.

Morphologically, apoptosis is characterized by condensation and fragmentation of the cell, leading to the formation of apoptotic bodies that are ultimately phagocytosed and digested by nearby cells. Terminal deoxynucleotidyl transferase dUTP nick-end labeling (TUNEL) stains the nuclei with DNA fragmentation in primary tissue samples and TUNEL-positive cells aggregate and locate around calcifications in some tumors. This suggests that massive apoptosis of NB cells occurs and may be related to regression of NB (386, 387). Immature tumor cells adjacent to thin fibrovascular stroma proliferate and often express Bcl-2 that inhibits the process of apoptosis. At an increasing distance from the stroma, tumor cells differentiate and cease proliferation, whereas Bcl-2 expression is decreased.

The Bcl-2 family of intracellular proteins is responsible for relaying of the apoptotic signal. *BCL2* has been found to be highly expressed in NB cell lines (388–390) and in

TABLE 7.10. Some features of apoptosis in NB.

	References
Gene expression including Bcl-2 and and apoptosis in NB.	*Hoehner et al 1995, 1997 (387, 392)*
Therapy-related drug resistance occurs via an increase in Bcl-2 expression and alterations in apoptosis.	*Lasorella et al 1995 (402)*
The process of apoptosis and the factors involved in it have been reviewed.	*Weinreb et al 1995 (863)*
	Hengartner 2000 (864)
	Hempstead 2002 (865)
DNA fragmentation, Bcl-2 and apoptosis.	*Que et al 1996 (396)*
Apoptosis at onset and relapse of NB	*Tonini et al 1997 (867)*
Drug-induced apoptosis in NB.	*Fulda et al 1998 (868)*
Mitochondria and apoptosis in NB.	*Fulda et al 1998 (869)*
Histopathological evaluation of apoptosis.	*Soini et al 1998 (870)*
MYCN sensitization of NB to drug-induced apoptosis.	*Fulda et al 1999 (820)*
MYCN product and apoptosis.	*Galderisi et al 1999 (871)*
Semaphorins are mediators of apoptosis in neuronal cells.	*Shirvan et al 1999 (872)*
Imprinting of 1p, *MYCN* and other loci in NB.	*Caron and Hogarty 2000 (665)*
Ferentinid-induced apoptosis in NB	*Lovat et al 2000, 2002 (873, 874)*
	Osone et al 2004 (875)
Nuclear exclusion of p53 is mediated by MDM2 and attenuates etoposide-induced apoptosis.	*Rodriguez-Lopez et al 2000 (876)*
IGF-1 receptor activation and Bcl-2 overexpression prevent early apoptosis in NB.	*Van Golen et al 2000 (877)*
NB vs. glioblastoma response to apoptosis.	*Bursztajn et al 2001 (633)*
Iron chelation and induction of apoptosis and inhibition of *MYCN* expression in NB.	*Fan et al 2001 (878)*
The role of Bcl-2 and BAX in apoptosis has been reviewed.	*Harris and Johnson 2001 (879)*
Lysosomal proteases as potential targets for induction of apoptosis in NB.	*Castino et al 2002 (880)*
Apoptotic regulators CASP9 and DFFA (located at 1p36.2) undergo changes in NB.	*Abel et al 2003, 2004 (881, 882)*
Partial proteosome inhibition triggers apoptosis in NB.	*Nahreini et al 2003 (883)*
Coexpression of insulin receptor-related receptor (IRRR) and ILGF-1 receptor are associated with enhanced apoptosis and differentiation in NB.	*Weber et al 2003 (884)*
SCA2 gene, encoding ataxin-2, sensitizes NB cells to apoptosis.	*Wiedemeyer et al 2003 (885)*
Guanine nucleotide deletion triggers apoptosis in NB cell lines.	*Messina et al 2004 (886)*
Caspase-8 and sensitization to drug-induced apoptosis in NB.	*Mühlethaler-Mottet et al 2004 (444)*
Mitochondrial pro- and antiapoptotic mediators in unfavorable NB.	*Abel et al 2005 (887)*
Induction of p19^{INK4} by ultraviolet light confers resistance to apoptosis in NB.	*Ceruti et al 2005 (888)*
Livin, an inhibitor of apoptosis, is a protein expressed in NB and NB cell lines and whose increased expression is an indicator of poor outcome in NB.	*D.K. Kim et al 2005 (889)*
Antisense Bcl-2 transfection up-regulates antiapoptotic and antioxidant thioredoxin in NB.	*Li et al 2005 (890)*
Activation of the phosphatidylinositol (PI3K)-Akt pathway plays a pivotal role in cellular proliferation and differentiation. The multiple components of this pathway are dysregulated in a wide spectrum of tumors, including NB. Akt is in effect an anti apoptotic protein and can serve as a target for therapy inhibiting its effects on apoptosis.	*Liu et al 2005 (891)*
p73 antisense-induced apoptosis.	*Simões-Wüst et al 2005 (892)*
Apigenin induction of apoptosis: caspase-dependent and p53- mediated.	*Torkin et al 2005 (893)*
BBC3, a member of the Bcl-2 family, mediates fenretinide- induced apoptosis in NB.	*Wei et al 2005 (894)*
Flavonoids can activate calpain and caspase cascades for mediation of apoptosis in SH-SY5Y NB cell line. Plant-derived flavonoids should be explored as possible therapy for NB.	*Das et al 2006 (895)*
LGI1 gene triggers apoptosis in NB.	*Gabellini et al 2006 (896)*
Caspase-dependent apoptosis demonstrated in unfavorable NB.	*Koizumi et al 2006 (423)*
Spontaneous regression likely reflects the activation of an apoptotic/differentiation process with p73 or survivin as factors. Caspase-8 plays a key role in apoptosis and its inactivation by methylation inhibits its transcription in some *MYCN*-amplified NB; other processes may inactivate caspase-8. Demethylation of the cellular genome in NB may up-regulate caspase-8.	*Li and Brattain 2006 (897)*
	Casciano et al 2004 (449, 450)
FLIP(L) expression and apoptosis.	*Flahaut et al 2006 (898)*
Resveratrol inhibits NB growth and mediates apoptosis by affecting the mitochondria.	*van Ginkel et al 2007 (898A)*

primary tumors (391–393). However, there did not seem to be a convincing correlation between prognosis and *BCL2* expression (45).

Several studies have shown that Bcl-2 is expressed in primary NB tumors (386, 387, 392, 393, 395, 395A, 396, 397, 397A) and cell lines (388, 390, 394). Bcl-2 is immuno-histochemically detected in 40–100% of primary NB before treatment (388, 390, 391, 394, 395, 397) and expressed in areas where TUNEL-positive cells are not distributed; double staining for Bcl-2 and TUNEL indicates that Bcl-2 is expressed in the cytoplasm of tumors cells that are TUNEL-negative (386, 387, 392, 395A, 398). This inverse relationship between Bcl-2 expression and TUNEL positivity suggests that Bcl-2 may function to regulate the process of apoptosis in NB cells. Although Bcl-2 is widely expressed in primary tumor specimens, there are conflicting reports regarding the relationship between Bcl-2 expression and other prognostic factors for NB. In cultured NB cells, Bcl-2 is primarily expressed in lines of chromaffin lineage, whereas Bcl-x$_L$ is expressed in both chromaffin and nonchromaffin lineage lines. *Bcl-x* is a gene that seems to function as a bcl-2 independent regulator of apoptosis and encodes two distinct species of mRNA, Bcl-x/*Bcl-x* and Bcl-xg/*Bcl-xg*. Cell lines that express high levels of one protein generally express low levels of the other (390,399). Immunohistochemical staining showed *BCL2* to be present in islets of treated NB (400). Response to therapy in NB may be related to *BCL2* and *BAX* expression; high expression of the former is associated with a poor response and vice versa for *BAX* (401).

Studies on the SK-N-SH NB cell line showed in inverse relation between *BCL2* (p26) level of expression and differentiation (389). Resistance to therapy in NB occurs via an increase in *BCL2* expression and alterations in apoptosis (402).

Drug-induced apoptosis in NB is mediated via CD95 (APO-1/Fas)-L-receptor system (403), independent of p53. Apoptosis also may be induced via caspases, probably by interference with mitochondrial function and independent of the CD95 pathway (404). Although there are differences in their mechanisms of action, caspase-3 and caspase-8 induce apoptosis in NB cells by perturbing mitochondrial function (403). A direct effect on mitochondria by drug-inducing apoptosis has been postulated (404).

Spontaneous Regression of NB

NB has the highest rate of spontaneous cures and regressions recorded in human cancer (45) and have been documented for many years (405). These cases have included extensive metastases (406–415). This regression usually consists of disappearance of the disease, although maturation to ganglioneuromas may occasionally take place; the cases involved consisted of stage 2 or 4S disease (416). Stage IVS NB rarely differentiate into benign glioneuromas, but they do regress completely in the majority of cases (410). Various hypotheses, especially those related to apoptosis, have been advanced as responsible for spontaneous regression of NB (396, 417–419); none of them has been rigorously proven or generally accepted.

NB that regress spontaneously do not contain del(1p36), *MYCN* amplification, or triploid content of DNA (414) or gain of 17q21-q25 (418). In contrast, caspases are induced in regressing NB cells, and members of the neurotrophin signaling pathways, especially the TrkA receptor, may play an important role in regression (418,420).

Maturation or regression of NB was found to be associated with diploidy in the neuronal and schwann cells and absence of 1p- (421). Abnormal genomic findings in neuroblastic and ganglion cells in NB suggest that the schwann cells are a reactive population of normal cells that invade the NB (421).

Studies have proposed (419) that spontaneous NB regression occurs via a caspase-independent, nonapoptotic mechanism (422). H-ras expression was found to be twice as frequent in tumors detected by mass screening (60.9%) than in high-stage tumors (29.2%) (419). Degenerating cells were fragmented without any evidence of apoptosis, but with features of autophagy. The authors (419) concluded that *Ras*-mediated nonapoptotic degeneration was responsible for spontaneous regression. In a subsequent study, neither of the aforementioned mechanisms was found to be involved in spontaneous regression. Instead, other mechanisms, perhaps related to tumor maturation, seemed more likely to be responsible for the regression phenomenon (423), supported by the correlation of the incidence of caspase-dependent apoptosis with indicators of poor prognosis in the NB examined (Shimada's unfavorable histology, *MYCN* amplification, and high mitotic index), but not with factors related to tumor regression, such as clinical stage and mass screening. Spontaneous regression of IV-S NB involves hyperexpression of N-*Ras* and initiation factor 4E (424).

Epigenetic Changes in NB

Epigenetic abnormalities, especially those resulting from DNA methylation alterations, are intimately involved in the development of various tumors. Aberrant methylation of promoter CpG islands causes inactivation of TSG. Thus, methylation of promoter CpG islands of the TSG *RASSF1A* and *BLU* in NB was shown to be associated with a poor prognosis, possibly due to the silencing of important genes as an underlying mechanism (425). Similar findings had been obtained by others (426A,426). Methylation of *RASSF1A* in NB had been described previosly (427,428). In Table 7.11 are shown some changes in NB related to methylation.

Caspases

Caspases (a family of cysteine acid proteases) are proteolytic enzymes responsible for the execution of the apoptotic signal, particularly caspase-3 (429). Studies have shown that an increased expression of interleukin 1β-converting enzyme (caspase-1) and other caspases in NB are associated with a favorable clinical outcome and histologic features (430–432). These findings are in keeping with the notion that NB prone to undergo apoptosis are more likely to spontaneously regress, respond favorably to cytotoxic agents (45), or both This suggests a key role for caspase-dependent apoptosis in the regression process.

Caspase-3, caspase-8, and caspase-10 deregulation or deficiency in NB leads to resistance to apoptosis (433, 434). Caspase-8 is deleted or silenced (by methylation) in NB with *MYCN* amplification (435). Loss of caspase-8 expression in NB correlates with resistance to tumor necrosis factor related

TABLE 7.11. Methylation in NB.

	References
Epigenetic abnormalities, especially aberrant methylation of promoter CpG islands, may cause inactivation of TSG in NB, either by hypomethylation or hypermethylation.	*Thoraval et al 1996* (899)
Gene silencing in NB.	*von Noesel et al 2003* (426A) *van Noesel and Versteeg 2004* (900) *Yang et al 2004* (901) *Nahreini et al 2004* (902)
Silencing of important genes through methylation in NB was shown to be associated with a poor prognosis.	*Abe et al 2005* (425)
Methylation status of some genes (RASSF1A and DCR2) is a more reliable predictor of NB progression than *MYCN* amplification. Furthermore, NB with single-copy *MYCN* showed a tendency to progression when associated with methylation of the above-mentioned genes.	*Banelli et al 2005* (903A, 903)
Methylation silenced nuclear receptor 112 gene in NB.	*Misawa et al 2005* (904)

apoptosis (436) and metastatic spread (437). Doxorubicin-induced cell death in NB is caspase-independent (438). Apoptosis in NB is not Fas-dependent, but it is activated by caspase-3, which is activated by Fas (439). Caspase-9 and apoptotic protease activating factor 1 (Apaf-1) are expressed in NB cell lines with LOH of 1p and *MYCN* amplification (440). Caspase-8 acts in the early stages of the apoptotic process and its silencing may have a role in tumor progression, including NB (435), by the induction of tumor cell resistance to apoptosis triggered by cytotoxic and other agents (441, 442). Induction of caspase-8 expression by gene transfer in NB cells restored sensitivity to death-receptor ligands and to their combinations with cytotoxic drugs (443, 444).

Interferon (IFN)-γ sensitizes cells to apoptosis including the induction of caspase-8 expression (445–450). The role of IFN-γ and other factors in conjunction with caspase-8 expression in the apoptotic process of NB cells has been described previously (442).

Cells of favorable NB usually express high levels of the high-affinity nerve growth factor (NGF) receptor gene *TrkA*, and such cells differentiate in vitro in response to NGF (420, 431). In NB, high levels of caspase-1 and caspase-3 are significantly correlated with a high level of *TrkA* expression, single copy of *MYCN*, younger age of patients, lower stages of disease, and improved prognosis (431, 451–453). Studies on the expression and cellular localization of these proteases indicate that they are more highly expressed in the nuclei of favorable NB than unfavorable tumors (43, 451–453). In addition, it is suggested that these enzymes may be translocated from the cytoplasm into the nuclei and play an important role in the process of apoptosis in favorable NB (430, 431, 452, 453).

Loss of caspase-8 protein expression was found in 7% of NB and was not restricted to advanced stages of the disease (455). No correlation was found between the loss of caspase-8 expression and *MYCN* amplification or with even-free or overall survival. The methylation status of caspase-8 has been linked to *MYCN* amplification in some studies (435, 436, 443, 448), but not in others (427, 433, 456).

Trk Receptors

Studies have shown a favorable clinical outcome of NB associated with high levels of TrkA (*NTRK*) in these tumors (279, 431, 452). Parallel with these studies have been in vitro demonstrations of the antineoplastic effects of TrkA (465A, 464–466).

An outline of the neurotrophin-signaling pathway includes NGF, brain-derived neurotrophic factor (BDNF) and Trk family of kinases and their receptors (TRKA, TRKB, and TRKC). The neurotrophin pathway is critical to the development of the sympathetic nervous system and hence plays and important role in the biology and clinical behavior of NB (466A).

The Trk family is a group of high-affinity receptor proteins that are critical in the transfer of extracellular signals from nerve growth and differentiation factors, or neurotrophins, into the intracellular milieu of neural cells (467–470,470A). Three relevant neurotrophin receptors, or Trk proteins, have been identified: TrkA, TrkB, and TrkC. TrkA and TrkC are differentiation-inducing proteins whose expression is inversely related to NB stage and *MYCN* amplification (471). TrkB, in contrast, is overexpressed in advanced tumors (45).

High levels of TrkA mRNA were present in tumors from patients with favorable-stage disease (Figure 7.10), whereas low-to-undetectable levels were observed in *MYCN*-amplified tumors. In contrast, coexpression of full-length TrkB (*NTRK2*) (there is also a truncated isoform lacking the tyrosine kinase) and its ligand BDNF was highly associated with *MYCN* amplification resistance to chemotherapy (472) and advanced-stage NB and may represent an autocrine survival pathway (468, 473). At variance with the aforementioned work, another study showed that an alternative-spliced transcript of TrkA is preferentially expressed in high-risk tumors, is resistant to the differentiation influence of its ligand NGF, and likely plays an oncogenic role in human NB (474). Thus, Trk overexpression may be a therapeutic target that can exploit the relatively limited normal tissue expression and theoretically provide a wide therapeutic index.

Differential expression of the neurotrophin receptors is strongly correlated with the biologic and clinical features of

NB. Some primary NB differentiate in vitro in the presence of NGF but die in its absence. Transfection of *TRKA* into a non-TrkA–expressing NB cell line restores its ability to differentiate in response to NGF(464, 465). *TRKA* expression is inversely related to disease-stage and *MYCN* amplification status (452). Thus, high *TRKA* expression is a marker of "favorable" NB, and it is correlated with an increased probability of good outcome (431, 457–459, 462, 463). *TRKB* is either expressed in low amounts, or as its truncated isoform, in biologically favorable tumors. *TRKC* is expressed in favorable NB, essentially all of which also express *TRKA* (Figure 7.11) (460, 461, 463).

An evolving model of neurotrophin ligand/receptor interactions in NB indicates that low-stage tumors, particularly those in infants, usually express high levels of TrkA (475). If NGF is present, TrkA-expressing cells will terminally differentiate. In contrast, if NGF is limiting, TrkA-expressing cells will enter a programmed cell death pathway (453). *MYCN*-amplified higher-stage tumors usually express TrkB and rely on autocrine production of BDNF to maintain a growth-promoting signal, even in the absence of exogenous neurotrophins. There is an inverse relation of *TrkA* expression and *MYCN* amplification in NB (431, 452).

Neurotrophins have been shown to have diverse effects on NB cell lines, depending on whether these lines expressed p75NTR or TrkA, and on the cell type (476). The coexpression of both low- and high-affinity NGF receptors is not only more efficient in the restoration of NGF-induced differentia-tion in NB cell lines but also leads to activation of the proapop-totic activity of low-affinity receptor (p75NTR) and causes the NB cells to become NGF-dependent and irreversibly differen-tiated (477).

A NB cell line in which the cells either expressed TrkA (NTRK1) or TrkB (NTRK2) was shown to be associated with up-regulation of proapoptotic genes and angiogenesis inhibitors in the former and with up-regulation of genes involved in invasion or therapy resistance in the latter (478).

Neurotrophin effects are induced by binding to two different classes of cell surface receptors, i.e., p75NTR (low affinity) and Trk receptors (high affinity). The latter belong to the large family of receptor tyrosine kinases, and p75NTR is structurally a member of the Fas/TNF-R family. The Trk receptors are mediators of neurotrophin action. The signaling pathways involved are complex; see reviews of this area for details (211)A.

The clinical and therapeutic aspects of the role of Trk recep-tors in NB have received much attention (470)A, particularly their prognostic effects, although they are not independent predictors of patient outcome (481). In one study (481), levels of TrkA expression did not add significant information to prognostic grouping, as defined by a combination of clinical stage, histopathology and *MYCN* status. There was a biolog-ically relevant correlation between molecular properties, i.e., TrkA expression and *MYCN* amplification (Figure 7.12), and histopathologic features of NB (481). Nevertheless, TrkA remains one of the important potential targets in NB research, if only as a guide related to developing of appropriate therapy and management of NB.

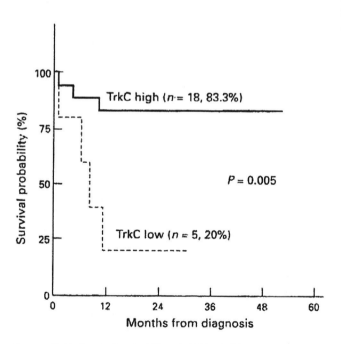

FIGURE 7.11. Survival probability of children with NB with interme-diate or high TrkC mRNA expression (*solid line*) versus those with low TrkC expression (*broken line*). High TrkC expression is asso-ciated with a significantly higher survival rate than cases with low expression. Similar findings have been associated with TrkA expres-sion (see text) (from ref. 460, with permission).

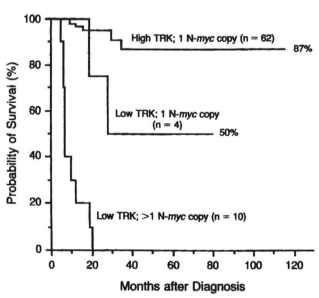

FIGURE 7.12. Cumulative survival of NB patients according to combined expression of Trk mRNA and *MYCN* amplification. Survival is high when Trk expression is high and no amplification of *MYCN* is present. Low survival is associated with low Trk expression and *MYCN* amplification (from ref. 431, with permission).

TrkB and BDNF can contribute to the chemoresistance of NB of advanced stages via the phosphatidylinosital 3′-kinase pathway (PI3K/c-Akt) (482). The complexity of the latter pathway and its role in cellular survival have been reviewed (482A). Table 7.12 presents cogent areas of information related to NB and the role of various neurotrophins in its genesis and biology.

Conditions Related to or Confused with NB

Olfactory Neuroblastoma (ONB)

ONB (esthesioneuroblastoma, olfactory neuroepithelioma, and olfactory neurocytoma) is a rare but distinctive tumor of the sinonasal area thought to originate from the neuroendocrine cells of the olfactory epithelium based on morphologic, immunophenotypic, and ultrastructural features (483–488). A neural derivation is indicated ultrastructurally by the presence of dense core neurosecretory granules and neuritic processes within tumor cells and immunohistochemically by expression of neurofilament proteins, neuron-specific enolase, and S-100 protein. Several origins of ONB have been suggested, including the sympathetic fibers of the anterior nasal cavity, the neuroectodermal cells of the olfactory placode, the sphenopalatine ganglion, and the organ of Jacobson (484,489). The proposed relationship to NB implied in the name is based on morphologic similarities to primitive forms of NB and a possible origin from sympathetic fibers of the nervus terminalis located in the anterior nasal cavity. Unlike NB, however, ONB rarely occurs in early childhood, tends to behave less aggressively than NB, and is rarely associated with catecholamine secretion (490, 491). Moreover, the location of ONB within the nasal vault favors an origin from neuroectodermally derived cells of the olfactory placode. These considerations have led to the alternative hypothesis that ONB is related to the pPNET family of primitive sarcomas (492). Survival and prognosis in ONB have been discussed previously (493).

An early report (494) of a t(11;22)(q24;q12) in one ONB and reports of a few other such tumors with this cytogenetic change (490, 492) led to the speculation about the relationship of these tumors to ES and pPNET, even though some cases studied cytogenetically, by FISH or by molecular techniques, were shown not to have evidence of the translocation (495–497). One ONB was shown to have +8 as the only change (496). Two established cell lines from an ONB showed rearrangements of EWS; both lines expressed EWS-FLI1 genomic fusion transcripts with in-frame junctions between exon 7 of EWS and exon 6 of FLI1 as described for ES and pPNET (492). Similar gene fusions were seen by these authors (492) in four of six primary ONB. None of the cases expressed tyrosine hydroxylases, indicating that ONB is not an NB but a member of the pPNET group of tumors.

In contrast, in one study (498) no evidence for the EWS-FLI1 fusion product (or MIC2 expression) in 11 ONB examined by RT-PCR was found. These findings were corroborated by another study (499), which examined 16 cases of ONB with MIC2 and found only one that was positive. When 13 of these 15 ONB were evaluated for the EWS-FLI1 fusion with FISH, distinct fusion signals were seen in only one tumor, indicating that ES and pPNET tumors of the sinonasal cavity occur rarely in that area and that the preponderance of these tumors is related to NB (500). A study of a single ONB revealed a complex hyperdiploid karyotype with none of the aberrations being similar to those previously described (501). The cytogenetic changes in one ONB were complex with a hyperdiploid karyotype and quite dissimilar from the changes in adenocarcinomas of the nasal area (260).

Examination for TrkA and p75 meurotrophin receptor (p75^{NTR}) expression in 10 ONB revealed both to be expressed but at different levels (502). The demonstration of such expression may assist in differentiating ONB from other tumors.

CGH studies of ONB have shown variable results. The first such study of a tumor from a 46-year-old woman showed many changes, including gain of whole chromosomes 4, 8, 11, and 14 and partial gains of 1q and 17q (100). Loss of entire chromosomes 16, 18, 19, and X also was observed and partial losses of 5q and 17p.

In another CGH study of three ONB, gain of the entire chromosome 19 and partial gains of 8q, 15q, and 22q were found (503). Loss of 8q also was observed. CGH studies of 12 ONB revealed deletions or gains in most of the chromosomes (504). Loss of 1p21-p31 was associated with a poor prognosis.

In a more extensive CGH study of 12 primary ONB and 10 metastases, loss of 3p and gain of 17q were found in all cases (505, 506). Other frequent changes (>80% of cases) were losses of 1p, 3p/q, 9p, and 10p/q and gains of 17p13, 20p, and 22q. Multiple other less frequent changes also were encountered in metastases.

A pigmented ONB showed dual differentiation in nude mice, with the melanotic part having a different chromosome pattern than the ONB (507). The latter was shown to have a modal chromosome number of 48, with many numerical and structural changes (507). The results indicate that ONB is a genetically distinct entity different from other conditions, e.g., loss of 1p was less common and that of 3p more common in ONB than in NB in general.

An ONB (grade III–IV) with intracranial and extracranial involvement was studied with several methodologies (e.g., cytogenetics, M-FISH, locus-specific FISH, and high-density SNP arrays) (508). A large number of structural chromosome changes were found, with del(2)(q37), del(6q), del(21)(q22), and del(22q) being the most common. The authors stress the usefulness of applying complementary methods for comprehensive cytogenetic analyses of ONB (508).

It would seem, then, that although a rare case of ONB may belong to the PNET group of tumors containing the

TABLE 7.12. Some aspects of the receptors for neurotrophins in NB.

	References
TrkA High-affinity receptor for NGF. Developing neurons differentiate in response to this ligand or enter an apoptotic pathway if NGF is withdrawn.	*Barbacid 1995 (466A)*
TrkB Binds with high-affinity both BDNF (and neurotrophin-4).	
TrkC High-affinity receptor for neurotrophin-3.	
A subset of NB with response to NGF may exist.	*Borello et al 1993 (905)*
γ-Interferon increases levels of Trk expression in NB cell line, while decreasing the *MYCN* levels.	*Shikata et al 1994 (906)*
NGF induces apoptosis in NB cell line SK-N-KC expressing p75.	*Kuner and Hertel 1998 (907)*
Neurotrophins and glial family ligands in CNS development and NB tumorigenesis.	*McConville and Forsyth 2003 (908)*
TrkA and TrkB have opposite effects in NB. The former is activated by its ligand NGF and the latter by BDNF. TrkA is a strong (positive) prognostic factor in NB without *MYCN* amplification, whereas NB with TrkB expression are usually associated with *MYCN* amplification and a poor prognosis. Aggressive NB cells have high BDNF and/or TrkB and escape control by NGF.	*Eggert et al 2000 (908–913)*
Some primary NB differentiate in vitro in the presence of NGF, but die in its absence. Transfection of TrkA into NB cell line lacking TrkA restores its ability to differentiate in response to NGF.	*Lavenius et al 1995 (464)*
The expression of TrkA, a high-affinity receptor for NGF, and $p75^{NTR}$, a low-affinity receptor for NGF, are important indicators of prognosis. A favorable prognosis is associated with high levels of TrkA expression and a poor prognosis with low levels of TrkA.	*Lucarelli et al 1997 (465)* *Nakagawara et al 1993 (431)*
TrkA expression is inversely related to disease stage and *MYCN* amplification status in NB. High TrkA expression is a favorable index in NB.	*Nakagawara and Brodeur 1997 (451)* *Nakagawara 1998, 2001, 2004 (420, 914, 915)* *Nakagawara et al 1993 (431)* *Kogner et al 1993 (457)* *Suzuki et al 1993 (458)* *Tanaka et al 1995 (459)* *Combaret et al 1997 (462)* *Svenson et al 1997 (463)*
High levels of TrkA expression (p140) and $p75^{NGFR}$ neuro- trophic receptors determined by immunohistochemistry are associated with favorable outcome in NB.	*Dominici et al 1997 (916)*
TrkA is frequently expressed in NB cell lines.	*Horii et al 1998 (465A)*
Up-regulation of TrkA in NB leads to neuronal differentiation by NGF.	*Shikata et al 2000 (917)*
Somewhat at variance with the information on TrkA in NB presented above are the findings in a study which showed that an alternative-spliced transcript of TrkA is preferentially expressed in high-risk NB; this transcript is resistant to the differentiation influence of its ligand NGF and likely plays an oncogenic role in NB.	*Tacconelli et al 2004 (474)*
Based on the findings of the above mentioned study, the authors challenged the concept of an exclusively tumor-suppressing role for TrkA in NB, by showing a novel hypoxia-regulated mechanism for oncogenic TrkA activation in NB cells.	*Tacconelli et al 2004 (474)*
TrkA induces apoptosis in NB through a p53-dependent mechanism.	*Lavoie et al 2005 (918)*
High levels of expression of EPHB6, EFNB2 and EFNB3 are associated with low-stage NB and high TrkA expression and prognostic significance of these parameters. NB cells with high TrkA and p75 have a favorable outlook. Depletion of NGF is also a factor.	*Tang et al 2000 (919,920)*
NGF is involved in differentiation of NB.	*Jensen et al 1992 (921)*
The BDNF/TrkB pathway may be important for growth and differentiation of NB with *MYCN* amplification.	*Nakagawara et al 1994 (473)*
TrkB (with tyrosine activity) preferentially expressed in advanced stage NB with *MYCN* amplification. These tumors usually express BDNF, thus establishing an autocrine pathway promoting cell growth. TRKB is expressed in low concentration or in its truncated (lacking tyrosine kinase) isoform in favorable NB.	*Matsumoto et al 1995 (468)*
PCR methodology for TrkB expression has been described.	*Combaret et al 1997 (674)* *Eggert et al 2000 (922)*
Autocrine and/or paracrine mechanisms involving BDNF may stimulate signal transduction via TrkB receptors in NB with unfavorable outcome, resulting in aberrant survival of tumor cells.	*Aoyama et al 2001 (923)*
TrkB cooperates with c-Met to promote NB invasiveness.	*Hecht et al 2005 (924)*
Akt serves as a mediator of TrkB effects in NB; possible role in therapy.	*Li et al 2005 (925)*
Activation of TrkB induces VEGF expression via hypoxia- inducible factor 1-α in NB.	*Nakamura et al 2006 (926)*
TrkC is expressed in favorable NB in which TrkA is also expressed.	*Ryden et al 1996 (460)* *Yamashiro et al 1996 (461)* *Svensson et al 1997 (463)*
Although NB are of neural origin, the tumors develop only in differentiated sympathetic ganglia or in the adrenal medulla. Genes important in the normal development of sympathetic neurons include *MYCN* (encoding a basic helix-loop-helix transcription factor), MASH1 and PHOX2A and PHOX2B. GDNF and its receptors (Ret, 6FRα1-3) and ligands (neurturin, catenin), pleotrophin (PTN), medkine (MK), all affect the development and physiology of sympathetic neurons, but their role in NB has not been established. The presence or absence of NGF signals strongly affect and regulate survival or death or normal sympathetic neurons.	
Three neuronal miRNAs were shown to control NB cell proliferation by repressing the truncated isoform of the neurotrophin receptor tropomyosin-related kinase C. These miRNAs were found to be downregulated in NB tumors.	*Laneve et al 2007 (463)A*

characteristic t(11;22), the bulk of ONB belong to the NB group of tumors, possibly showing features that differentiate ONB from the other NB tumors.

Glioneuroblastoma

GNB is generally regarded as an intermediate form in the maturation or regression process of NB, and it is associated with a better prognosis and survival rate than NB (509). Indicative of the better prognosis was the absence of *MYCN* amplification in the 16 GNB examined (509) and that the various prognostic indices in GNB and NB indicated a more favorable disease in the former than the latter. GNB is a relatively benign form of NB, and it consists of a mixture of fibrils, mature and maturing ganglion cells, and undifferentiated neuroblasts.

Systems for the classification and grading of GNB have been published previously (29, 31, 510, 511).

A primary site for GNB is often the thoracic region and less frequently the adrenal medulla (512, 513). In a report on thoracic NB (514), the better survival of these cases was emphasized, possibly related to scarcity of *MYCN* amplification in these tumors. Because the histology of these tumors was not mentioned, one wonders whether they may have been GNB.

The histologic heterogeneity of GNB is evidenced in the molecular heterogeneity demonstrated in these tumors (Figure 7.13) (515). A composite GNB was examined for *MYCN* copy status and DNA content at initial resection and 2 years later after progressive disease. In the latter sample, the more differentiated gangliomatous portion of the tumor was analyzed separately from the neuroblastic foci and shown to differ significantly (515).

Some reports have indicated that *MYCN* amplification was lower in GNB than in NB (516–518). Some of the variability may partly be due to institutional differences in the diagnostic criteria for GNB and its subtypes (31). In one report (287), *MYCN* amplification was found in two of 11 nodular type of GNB with undifferentiated cells, whereas no *MYCN* amplification was found in well-differentiated or intermixed GNB.

Negligible cytogenetic, FISH, and LOH information is available on GNB. It is possible that some GNB have been included in studies of NB without the authors mentioning their inclusion. Furthermore, in contrast to NB for which many established cell lines are available, no known cell line of GNB could be found in the literature.

Only a few CGH studies have been performed on GNB, the results being somewhat at variance, e.g., in one study (519) loss of 1p was present, whereas in another study (520) no such loss was observed; the same applies to gains of 1p, 2p, 4p15.1, 5p15.1-p15.3, and chromosome 7. When results were similar, the percentages showed wide variations. *MYCN* amplification and expression in localized GNB (stroma-rich NB) were associated with a good prognosis (521) (Figure 7.14).

A constitutional t(1;13)(q22;q12) in a patient with GNB has been described previously (522); the cells of this patient served as a source of establishing the breakpoint on chromosome 13 (523) and ultimately for the description of a gene, *WAVE3*, an actin-polymerization gene, which seems to be truncated and inactivated in GNB (524). Whether this gene,

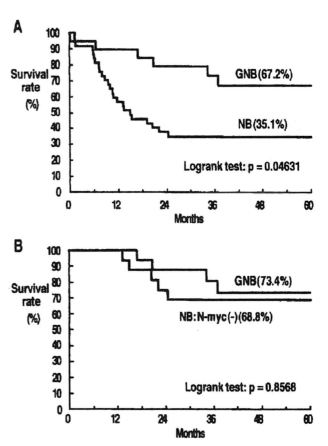

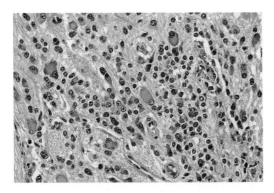

FIGURE 7.13. Histologic picture of a GNB composed of a mixture of neuroblasts and ganglion cells surrounded by a fibrillary stroma (from ref. 8, with permission).

FIGURE 7.14. Survival of patients with GNB versus that of NB cases. **(A)** General survival that was significantly different between the two groups. **(B)** Survival of GNB and NB cases without *MYCN* amplification (from ref. 509, with permission).

or the translocation described above, play a major role in the development of GNB is an open question, because 13q12 is not often affected by changes as determined cytogenetically or by CGH either in NB or GNB (519,520). A study compared NB with GNB and showed 1p36 deletions in the former but *not* in the latter (359). It also was found that the neuroblastic cells of these tumors differed from the schwann cells and that schwann cells have a less aggressive phenotype and play a role in tumor maturation (359).

Ganglioneuroma

Ganglioneuromas often arise in the posterior mediastinum rather than the abdomen (525). They may represent evidence of spontaneous maturation of previous NB (Figure 7.15), a process occasionally noted in NB patients. Alternatively, they may occur as metastatic, matured lesions in previously treated children. Maturation of NB to ganglioneuroma both in an extra-adrenal primary tumor and in local lymph node metastases has been reported previously (526).

Allelic status of 1p and 11p15 was evaluated in NB and ganglioneuromas (527) and loss of 1p found in one of the two ganglioneuromas examined and in three of 25 NB. Changes of 11p were observed only in NB. Cases of NB with loss of 1p do not always follow an aggressive clinical course and may differentiate into ganglioneuromas. The ultrastructural features of such maturation have been described previously (528).

The term "composite pheochromocytoma" is used for rare examples of pheochromocytoma that contain a significant histopathological component resembling ganglioneuroma, ganglioneuroblastoma, or NB (529). Patients with multiple endocrine neoplasia 2A with composite pheochromocytomas have been reported, one with ganglioneuroma (530) and another with ganglioneuroblastoma (531). Bilateral adrenal composite pheochromocytomas with ganglioneuroma in a 61-year-old woman with NF1 has been described previously (532).

A 58-year-old male with NF1, who died of necrotizing enteritis, was discovered to have a tumor of the adrenal gland

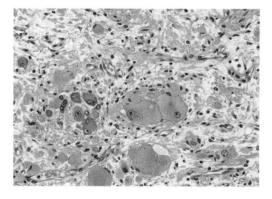

FIGURE 7.15. Histologic picture of a ganglioneuroma showing clusters of variably sized ganglion cells (from ref. 8, with permission).

consisting of pheochromocytoma and ganglioneuroma (533). The association of pheochromocytoma with ganglioneuroblastoma has been described in a 14-year-old girl with NF1 (529) and in a non-NF1 case (534). A pheochromocytoma with ganglioglioma was shown to produce catecholamines and various neuropeptides (535).

Table 7.13 show facets of GNB and ganglioneuroma not discussed in the text but cogent to these tumors.

The following sections present short descriptions of rare neural tumors that, on occasion, are to be differentiated from NB, ganglioneuroblastoma, and ganglioneuroma. These tumors include ganglioglioma, central neurocytoma, and cerebellar liponeurocytoma. Although cytogenetic and molecular genetic information on these tumors is sparse, considering their rarity, the findings are sufficiently informative to serve the purposes of diagnosis.

Ganglioglioma

Gangliogliomas are well-differentiated, slowly growing neuroepithelial tumors composed of neoplastic, mature ganglion (neural) cells, either alone (gangliocytoma), or in combination with neoplastic glial cells (ganglioglioma) (536).

Gangliogliomas show an additional neoplastic glial component, usually astrocytes, surrounded by a reticulin network. In contrast to normal neurons, the dysplastic neuronal cells of gangliogliomas show prominent neurosecretory activity (537). Gangliogliomas may occur throughout the CNS, including the cerebrum, brain stem, cerebellum, spinal cord, optic nerves, pituitary gland, and pineal gland. The majority are supratentorial and involve the temporal lobe (536–538). About three fourths of patients with gangliogliomas have seizures. Some of the chromosome changes in ganglioglioma are shown in Table 7.14.

A complex abnormal karyotype with three sublines showing inversions, translocations, and gains or losses of long or short chromosome arms and involving chromosomes 6, 7, 11, 13, 15, 16, 18, 19, and Y was observed in a rare case of ganglioglioma with malignant transformation (539). In addition, a ganglioglioma with an abnormal karyotype containing translocations and deletions of chromosomes (in order of frequency) 1, 5, 21, 12, 8, 13, 10, and X has been reported previously (540). Deletions on 17p were observed in an anaplastic ganglioglioma with diffuse leptomeningeal spread (541). Although the karyotypes described in gangliogliomas indicate a neoplastic origin (542), the cytogenetic data are too sparse to reach a conclusion regarding any recurrent changes.

A study based on CGH and FISH of five gangliogliomas revealed loss of material in the distal 9p and gain of chromosomes as the most common changes (543). An immunohistochemical study showed essentially absent expression of EGFR (located at 7p11~p13) in the five gangliogliomas examined (544).

TABLE 7.13. GNB and ganglioneuroma.

	References
Ganglioneuroma and NB in a child with NF1.	*Knudson and Amromin 1966* (576)
Clinical aspects of maturing NB and of GNB.	*McLaughlin and Urich 1977* (927)
Abnormal DNA content in NB and ganglioneuroma.	*Taylor et al 1988* (928)
Five ganglioneuromas showed *MYCN* expression but none of *MYC*.	*Slavc et al 1990* (310)
An ACTH-secreting GNB associated with opsomyoclonic encephalopathy and Cushing syndrome has been described.	*Clerico et al 1993* (929)
Composite GNB showed DNA content and *MYCN* copy number different in the neuroblastic vs. the more differentiated parts of the tumors.	*Schmidt et al 1993* (515)
No p53 abnormalities found in 14 GNB and 1 ganglioneuroma.	*Moll et al 1995* (930)
Cytopathology of NB, GNB and ganglioneuroma.	*Mondal 1995* (931)
Ganglioneuroma, GNB and NB in cases of opsoclonus- myoclonus syndrome.	*Blatt et al 1997* (932)
	Gambini et al 2003 (933)
A case of GNB and Wilms tumor has been reported in a 2-year- old child with NF1.	*Ito et al 1997* (934)
GNB may have areas with and without *MYCN* amplification as shown by FISH, findings of relevance to therapy.	*Lorenzana et al 1997* (935)
The p53 protein concentrations in adrenal PCC and ganglioneuromas were either normal or very low, in contrast to the high levels of aldosterone-producing adenomas.	*Adleff et al 1998* (1239)
A case of GNB and two of ganglioneuroma have been reported in NF1 patients.	*Geraci et al 1998* (936)
GNB and adrenal carcinoma in a child with Turner syndrome with germline mutation of p53 (Li-Fraumeni syndrome?).	*Pivnick et al 1998* (937)
Ganglioneuroma in a Turner syndrome case.	*Sasaki et al 2000* (938)
Melanin-concentrating hormone receptor mRNA present in GNB and NB.	*Takahashi et al 2001* (939)
Glioblastoma development in resected ganglioneuroma.	*N.R. Kim et al 2003* (940)
A developmentally regulated gene, TUBEDOWN, whose expression has been found to be related to tumor type, i.e., high in NB and low in GNB and ganglioneuroma. This may define tumor types more accurately than other methodologies.	*Martin et al 2007* (940A)

In a study of 6 gangliogliomas, microsatellite instability was not observed (542). Numerous polymorphisms and a novel mutation of the *TSC2* gene were found in gangliogliomas (545,546). These changes seemed to affect the glial component of the tumors. In another study, *TSC1* and *TSC2* genes were not inactivated (547).

The development of tumors (e.g., meningioma and astrocytoma) within gangliogliomas is an interesting phenomenon. These tumors have a dual component, well-differentiated neuronal and glial cells, often of the pilocytic type (536).

Although generally gangliogliomas have a low Ki-67 index (>1%), increases in the Ki-67 index and expression of p53 in the neoplastic cells may be related to tumor recurrence (537, 548, 549). No tumors had aberrant expression of p53 (548). Accumulation of p53 has been described in a ganglioglioma due to a mutation of *TP53* (550). The aberrant product was shown to be associated predominantly with the neuronal (neoplastic) cells.

RFLP study in one ganglioglioma with diffuse leptomeningeal involvement with probes for chromosomes 5, 6, 10, 12, 15, and 17 showed loss of an allele at 17p13.3 in the tumor tissue but not in the blood cells (541). Microsatellite instability was not found in six gangliogliomas examined (551).

An infantile ganglioglioma was characterized by telomere association leading to the formation of dicentric and other derivative chromosomes involving chromosomes 17, 19, 14, 11, 9, 5, and 22 in the order of frequency. About a year after resection of the original tumor, a recurrent tumor was found to have a normal 46,XY karyotype (552).

In a desmoplastic infantile ganglioglioma (553), nonclonal chromosome changes were found, but the basic karyotype was 46, XY. Interphase FISH was not informative. Molecular studies showed no changes of *TP53*, *EGFR*, or the *p14*, *p15*, and *p16* genes, although hypermethylation of the promoter

TABLE 7.14. Chromosome changes in ganglioglioma.

Karyotype	References
46,XY,del(6)(q25)	*Neumann et al 1993* (730)
47,XY,+r(1)	*Neumann et al 1993* (730)
49,XX,+5,+6,+7	*Neumann et al 1993* (730)
**46,X,inv(Y),t(6;13)(q14;q33),t(18;?)(q11.2;?)/46,X,inv(Y)	*Jay et al 1994* (539)
43,XX,der(1)t(1;5)(q21;q12),-5,der(8;13)(q10;q10),-9, i(10)(q10)	*Debiec-Rychter et al 1995* (941)
*49,XY,+4,+5,inv(7)(q22q34),+7	*Jay et al 1997* (942)

*Temporal lobe tumor which recurred with intracranial and spinal metastases.
**Anaplastic tumor with neural and astrocytic components.

siblings inherited the predisposition to NB through an abnormality in the paternal 1p36 region and the NB probably developed as a consequence of somatic loss of the maternal 1p36 allele.

There have been several reports of an association of NF1 and NB (576), but epidemiologic analyses have suggested that this is coincidental (577). Two different transcripts of the *NF1* gene in a NB tumor has been reported (578). Two studies have documented *NF1* gene mutations in some NB cell lines (213, 579), and a patient with NF1 and NB whose tumor had a homozygous deletion of the *NF1* gene was reported (580). However, familial NB is not linked to the *NF1* locus, even in a family in which the proband had both conditions (209). Thus, germline inactivation of *NF1* does not seem to predispose to NB, but somatically acquired inactivation may occur as a late event in tumor evolution in some cases. Patients with familial NB exhibited chromosomal fragile sites (especially at band 1p13.1), with unaffected members showing hypersensitivity to aphidocolin-induced fragile sites (581). These findings are of interest in view of the suggestion that interference RNA (RNAi) (582) is synthesized at chromosomal fragile sites.

Germline chromosome abnormalities have been found in a relatively small group of NB patients. Of interest are two cases of NB with germline interstitial deletions of 1p within band p36 (175, 192), a region that may contain a TSG important in the development of NB. Studies of a constitutional balanced translocation t(1;17)(p36;q12-q21) have failed to reveal the gene responsible for NB (174, 241, 250, 251, 583). All other constitutional rearrangements in NB lack a consistent pattern indicative that they may be coincidental rather than causative, although views to the contrary have been expressed (207). The latter authors (207) have opined that, based on the incidence of NB in trisomy 21 cases (Down syndrome) and in patients affected by X-chromosome anomalies, genes of importance in the biology of NB may reside on these chromosomes (207).

Table 7.16 shows combined data on the cases with constitutional disorders and NB.

Hereditary (Familial) NB

Hereditary NB is rare, and a family history of the disease is present in only 1–2% of NB cases (209, 281, 577, 584–586). Patients with familial NB are characterized by an earlier median age (9 months vs. 22 months) at diagnosis and a higher frequency of multifocal primary tumors (577,584,587). However, patients with hereditary NB show the same clinical heterogeneity seen in sporadic cases, including cases as disparate as those with spontaneous regression and others with relentless progression within individual families (577, 585, 587). Thus, the rarity of familial NB may be explained, at least in part, both by the incomplete penetrance and by the lethality of the phenotype.

Familial NB was described as a distinct clinical entity in 1945 (588), and a genetic hypothesis was provided in 1972 (584). The families described in one study (585) conformed to the predictions of the two-mutation model. The median

TABLE 7.16. Constitutional anomalies assoicated with NB.

	References
Partial trisomy 15	*Sanger et al 1984 (* 220A)
11q deletion and 12q duplication syndrome and NB	*Koiffmann et al 1995* (264)
The incidence of NB and ganglioneuroma is in higher Turner syndrome higher than in the population; 12 cases reported, including one with an i(X)	*Blatt et al 1997* (932)
	Satgé et al 2003 (207)
Developmental anomalies, including those of the brain, occur in children with NB	*Blatt and Hamilton 1998* (958)
Balanced translocations (10 cases) (2 of 1p; 3 of 11q)	*Kaneko and Cohn 2000* (959)
	Satgé et al 2003 (207)
Unbalanced anomalies (20 cases) (3 of 1p; 2 of 11q)	*Kaneko and Cohn 2000* (959)
	Satgé et al 2003 (207)
Turner syndrome (12 cases)	*Kaneko and Cohn 2000* (959)
	Satgé et al 2003 (207)
Although NB in XXX females or XXY males is rare, single cases of such have been reported	*Satgé et al 2001* (960)
	Honma et al 2006 (962)
	Gül et al 2003 (961)
Partial 2p trisomy (1 case)	*Dowa et al 2002* (963)
(5 cases)	*Satgé et al 2003* (207)
del(11q)	*Mosse et al 2003* (265)
Disseminated NB in NF1 patient with homozygous deletion of the NF1 gene.	*Origone et al 2003* (601)
Miscellaneous (2 cases) (+18,-22)	*Satgé et al 2003* (207)
Trisomy 13 (4 cases)	*Satgé et al 2003* (207)
Trisomy 21 (4 cases)	*Satgé et al 2003* (207)
NB has been described in patients with Fanconi anemia and Bloom syndrome: possibly related to fragile site at 1p31-1p32.	*Bakshi and Hamre 2004* (964)
17p11.2 deletion in patient with Smith-Magenis syndrome	*Hienonen et al 2005* (965)

age at diagnosis of 14 months (585) was lower than that reported in surveys of large populations in sporadic NB cases, where the median age has typically been reported to be 23 months (39, 587). In addition, one quarter of the patients with familial NB were documented to have bilateral adrenal NB or more than one primary tumor, whereas multifocality is only rarely observed in sporadic NB (39). The seven pedigrees studied (585) suggested incomplete disease penetrance with generation-skipping in at least one family. Therefore, the penetrance value of 63% calculated in a previous review of hereditary NB (584) is a reasonable estimate of the chance of manifesting the disease in those with a heritable predisposition.

Allelic loss at 16p12-p13 in sporadic NB was found in 13% of 470 tumors (589). This loss was associated with favorable NB, possibly involving at least two candidate loci within the region. In one study (590), inactivation of the *HNB1* located at 16p12-p13 was found and the authors suggested that this gene may contribute to the pathogenesis of familial and sporadic NB.

In a subsequent linkage analysis of seven families with two or more first-degree relatives affected by NB revealed a single interval at chromosome bands 16p12-p13 and was the only genomic region consistent with linkage (585). LOH of multiple 16p polymorphic loci was identified in five of 11 familial NB and in 68 of 336 nonfamilial NB (20.2%). Together, these data suggest that a hereditary NB predisposition gene (*HNB1*) is located at 16p12-p13 and that disruption of this gene may also contribute to the pathogenesis of some nonfamilial (sporadic) NB. The observation that two of seven available tumor specimens from affected and 16p-linked individuals in these families showed 16p12-p13 LOH and that in both cases the allele retained was from the parent predisposed to the disease, further supports the conclusion that NB predisposition occurs due to heritable mutation in a TSG located on the short arm of chromosome 16 (16p) (585).

Three additional families were available for analysis, but they were distinguished from the 16p12-p13-linked families in that affected individuals were of second-degree or greater relation (585). These three families consisted of two first-cousin pairs and one uncle-proband pair. In none of these families was NB predisposition linked to 16p12-p13. This suggests that hereditary NB is genetically heterogeneous, like other pediatric cancer predisposition syndromes, such as familial Wilms tumor, and Li-Fraumeni syndrome (591–594). Alternatively, it is possible that some or all of these NB pedigrees are actually examples of chance occurrence of a relatively rare disease within the same family.

LOH was shown at multiple 16p12-p13 polymorphic loci in 20% of a representative panel of sporadic NB (585). This suggests that somatic deletions in this region might be targeting the second allele of a gene harboring a hemizygous mutation, but proof of this awaits identification of *HNB1*.

The first observation of germline mutations transmitted along a pedigree in NB patients (595) was reported for the paired-like homeobox 2B (*PHOX2B*) gene, which is considered an essential regulator in the development of autonomic neural crest derivatives (596) and is related to the TrkA differentiation pathway in NB. Two different missense mutations of *PHOX2B* were found in a transversion in a family with NB and two ganglioneuromas, and a different transversion in a sporadic case of Hirschsprung disease (HSCR) associated with NB. Another family with recurrent NB did not show any anomaly of *PHOX2B* (595). Subsequently, a heterozygous single-base deletion was found in a complex pedigree with NB in which seven members of three generations had NB; two of these patients also had HSCR (597). The proband had NB, HSCR, and NF1.

PHOX2B germline mutations seem to be rare in familial NB and when such mutations are involved in the initiation of NB, they occur in patients with disorders of the autonomic nervous system, i.e., HSCR and congenital central hypoventilation syndrome (CCHS) (597).

The locus for *PHOX2B* at 4p12 is not known as a frequent site of allelic deletions in NB (45), and to date biallelic inactivation of *PHOX2B* has not been reported (597).

Verification of the presence of *PHOX2B* mutations in three families with recurrent NB and one family with GNB and isolated HSCR was attempted by a number of techniques (598). No mutations were detected in the three exons of *PHOX2B* by analyzing constitutional DNA of the 10 NB patients and the one affected with BNG and HSCR. No haplotype segregation was found at the genomic interval encompassing the gene or in its vicinity. Therefore, the findings (598) exclude *PHOX2B* as the NB susceptibility gene in the families analyzed.

The different results obtained (595, 598) confirm the remarkable genetic heterogeneity of NB and raise questions about the *PHOX2B* involvement in the genetic determination of the disease. Moreover, the two distinct missense mutations affecting the homeodomain in the mid-portion of the protein (595) are different from the polyalanine stretch expansions of the distal portion of the protein, observed in sporadic ganglioneuroma and NB associated with CCHS (599, 600), and from the stop codon mutation detected in a single patient affected with CCHS, HSCR, and NB (600). Thus, involvement of *PHOX2B* might rather be limited to a subset of patients sharing specific phenotypic features, e.g., multifocal tumors and syndromic NB (595, 599, 600).

This hypothesis is further supported by a mutation screening (597) that demonstrated a *PHOX2B* germline mutation in only one of three generations of a NB family, but did not show any evidence for mutation in the eight other pedigrees. The family carrying a *PHOX2B* mutation showed a complex pedigree that included seven patients: one with ganglioneuroma, four with multifocal NB, and two with NB in association with HSCR, one of whom had multifocal tumors and NF1 (585). As the proband of this family was also shown

TABLE 7.17. Familial NB.

	References
NB of adrenal medulla in siblings.	*Dodge and Benner 1945* (588)
A family with NB and GNB, no chromosomal changes were present in blood cells.	*Klein et al 1975* (966)
NB in father and son.	*Arenson et al 1976* (967)
Risk of familial NB is <6% in siblings or offspring of NB cases.	*Kushner et al 1986* (577)
Only 22 hereditary cases in 10 families (18 of NB and 4 of ganglioneuromas) found among nearly 3,000 patients with NB (<1% of all cases)	*Claviez et al 2004* (968)
Familial NB has a multigenic mode of inheritance with the existence of different NB loci genetically interacting to cause and/or modify the disease.	*Longo et al 2005* (969)

to have an inactivating mutation in *NF1*, a NB patient with a constitutional *NF1* mutation was also screened (601), as were his parents, without finding any *PHOX2B* mutations.

Six different frameshift *PHOX2B* mutations also have been identified in a series of 237 sporadic NB and 22 cell lines (2.3%). Two of these were de novo constitutional mutations detected in two NB patients, one of whom was diagnosed with CCHS and HSCR (602).

Combining the results obtained by *PHOX2B* mutation screenings (595, 597, 598, 602) with those achieved by linkage analysis and deletion mapping, it is unlikely that a single gene predisposes to familial NB. All these data further confirm the remarkable genetic and consequent phenotypic (histologic) heterogeneity of NB and reinforce the concept that an oligogenic mode of disease transmission, probably governed by a major susceptibility gene, is more likely to explain the genetic basis of familial NB (597, 603). *PHOX2B* is mutated in NB associated with high expression of delta-like-1 (Dlk1) (604).

Except for a possible gene on 16p, a number of other regions and genes as loci for putative TSG in familial NB have been examined; linkage analysis at candidate loci has excluded each of these regions, including 1p (209, 603, 605). LOH on 4p was shown in two tumors from a family with NB (132). Linkage for predisposition for NB located at 4p16 was found to be rather weak (132).

Table 7.17 presents information on familial NB.

Miscellaneous Aspects of NB

This section presents an encapsulization of some of the extensive, if not daunting, literature on the genetics and molecular biology and other facets of NB.

Angiogenesis

Angiogenesis, or the formation of new blood vessels from existing ones, is a fundamental requirement for tumor growth, invasion, and metastasis (606). Vascular endothelial growth factor (VEGF) together with its receptors has been shown to play a significant role in tumor-induced neovascularization. Disruption of the VEGF pathway has been shown in experimental models of tumorigenesis to inhibit angiogenesis, and,

consequently, tumor growth. This has been done in a variety of ways, including the administration of a soluble, truncated VEGF receptor that inhibits VEGF action by a "dominant-negative" mechanism (607).

Engineered expression of flk-1, a competitive inhibitor of VEGF, by tumor cells results in the production of an inhibitor of endothelial cell proliferation and migration that greatly restricts the growth of the tumor cells in vivo (607). Inhibition of growth of a xenograft of NB by an inhibitor of angiogenesis (TNP-470) has been reported (608, 609). Similar indications of possible effectiveness of antiangiogenetic agents, i.e., ABT-510 peptide derivative of natural angiogenic inhibitor thrombospondin-1 and valproic acid, against NB have been proposed (610). Gene therapy-mediated delivery of angiogenesis inhibitors may provide an alternative approach to treating refractory tumors such as advanced NB. Areas related to angiogenesis in NB and the role played by some factors in angiogenesis are shown in Table 7.18.

p16 Pathway

The p16-CDK/cyclin D-pRb pathway plays a critical role in cell cycle progression. Protein complexes of D-type cyclins and cyclin-dependent kinase (CDK) (CDK4 and CDK6) induce phosphorylation of pRb to promote the G1-S-phase transition. The phosphorylated pRb releases transcriptional factors such as E2F, which activate the expression of genes essential for S-phase entry. CDK inhibitor proteins, including p16, play critical roles in the G1-S cell cycle transition by inhibiting the cyclin D-CDK4/6-mediated pRb phosphorylation. Alterations of any component of the pathway, such as deletion/mutation of the *p16* gene, amplification/overexpression of CDK or D cyclins, and mutations to CDK that affect p16 binding, result in pRb phosphorylation and subsequent progression of G1 into S-phase. Similarly, alterations of pRb itself may also lead to the G1-S-phase transition. These alterations have been found frequently in various human tumors, including NB, suggesting that inactivation of the p16-CDK/cyclin D-pRb pathway may play an important role in their pathogenesis.

Deregulation of the p16-CDK/cyclin D-pRb pathway in NB has been reported (214, 332, 337, 611–617) in which almost all of the NB cell lines and primary tumors were shown to retain the wild-type *p16* gene. It was demonstrated (611) that seven

TABLE 7.18. Angiogenesis in NB.

	References‘
Angiogenesis inhibitor (TNP-470) reduces growth of NB in nude mice and in other systems.	Katzenstein et al 1999 (970) Kim et al 2002 (971)
Angiogenetic characteristics and facets of NB do not seen to influence prognostic aspects.	Cañete et al 2000 (972)
High expression of VEGF associated with advanced stage NB	Eggert et al 2000 (973)
Pigment epithelium-derived factor (PEDF), a potent inhibitor of angiogenesis and inducer of neural differentiation, is produced by ganglion cells and schwann cells, but not by primitive tumor cells. PEDF may serve as an antitumor agent by inhibiting angio- genesis, while promoting an increase in schwann cells and differentiated tumor cells that in turn produce PEDF.	Crawford et al 2001 (974)
Deletion of retSDR1, a regulator of vitamin A metabolism, in NB with MYCN amplification could compromise sensitivity of the cells to retinol, thereby contributing to NB development and progression.	Cerignoli et al 2002 (975)
VEGF and its receptor (FIK-1) have been demonstrated in NB; these not only affect angiogenesis but may act directly on the NB cells.	Fukuzawa et al 2002 (976)
	Langer et al 2002 (977)
Flavopiridol inhibits VEGF effects induced by hypoxia or picolinic acid in NB.	Rapella et al 2002 (978)
Arginine deiminase inhibits NB growth by its antiangiogenic effect.	Gong et al 2003 (979)
The effects of imatinib in NB are probably produced through suppression of PDGFR and KIT phosphorylation and inhibition of VEGF suppression.	Beppu et al 2004 (980)
VEGF levels in NB are regulated by autocrine and paracrine effects of growth factors (e.g., hypoxia-inducible factor-1α), a system that may be affected by some drugs.	Beppu et al 2005 (981)
Neuropilin-1 and VEGF are determinants of NB growth.	Marcus et al 2005 (982)
Poorly vascularized NB become immature and maintain an aggressive phenotype, possibly due to stabilization and activation of hypoxia-inducible factor-2α.	Nilsson et al 2005 (983)
BH3 is a potent activator of apoptosis in NB.	Goldsmith et al 2006 (984)
Glomeruloid microvascular proliferation (MVP) was significantly associated with schwannian stroma-poor histology and decreased survival in neuroblastoma.	Peddinti et al 2007 (1243)

of 19 NB cell lines displayed very high p16 expression at both the mRNA and protein levels. In addition, a preliminary study of six primary NB samples revealed elevated p16 expression in three samples, suggesting that the finding on cell lines may represent a general feature of NB in vivo (611). It also was demonstrated that in NB cell lines, alterations in components downstream of p16, which may negate the regulatory effect of p16, were limited to infrequent CDK4 gene amplifications and cyclin D2 expression. All 19 cell lines exhibited various degrees of phosphorylated pRb protein; p16 expression was independent of the pRb phosphorylation status (611). These results suggest that the elevated p16 protein may not be functioning properly in the regulation of the pathway and that p16 transcription is induced by a pRb-independent mechanism, in contrast to a pRb/p16 feedback regulatory loop that has been reported in other cancers (615).

Alterations of the downstream components of the p16 pathway in NB were infrequent. pRb was deregulated in the majority of samples investigated (20 of 33; 82%), 24 with hyperphosphorylated pRb, and three with no pRb protein (615). The phosphorylation status of pRb did not correlate with p16 protein expression, suggesting that the elevated p16 protein may not be functioning properly to regulate the pathway. Among patients of all stages of NB, p16 expression was significantly associated with a lower overall survival. There was no overexpression of MDM2, and loss

of $p14^{ARF}$ expression and p53 mutations were infrequent events. Together, these findings suggest that upregulated p16 expression may represent a unique feature of aggressive NB (615).

The $p16^{INK4A}$ locus has been found to also encode a second protein, $p14^{ARF}$, with a distinct reading frame. Studies have demonstrated that $p14^{ARF}$ regulates the cell cycle progression through an interaction with p53 and MDM2; it physically interacts with MDM2 and blocks both MCM2-induced p53 degradation and transcriptional silencing of p53. Similar to the p16-CDK/cyclin D-pRb pathway, the $p14^{ARF}$-MDM2-p53 pathway seems to also be inactivated in human cancers through alteration of various components, such as deletion and silencing of $p14^{ARF}$, amplification of MDM2, or p53 mutation. The deregulation of this novel cell cycle pathway in NB has not been explored yet.

NB Cell Lines

Neuroblastoma cell lines are transformed, neural crest-derived cells, capable of unlimited proliferation in vitro. These cell lines retain the ability to differentiate into neuronal cell types when exposed to various agents. Thus, the NB cell lines represent an excellent in vitro system for a wide range of studies, including the effectiveness and neurotoxicity of drugs as anticancer agents, including NB. The versatility of NB cell lines as an in vitro model in neurobiology and neuropathology has been appropriately emphasized (618–623).

The description of the genetic and molecular changes in NB cell lines is of importance in view of the widespread use of these cell lines as models for the study of the biology of NB (66). An application will be a comparison and interpretation of gene expression data with the available knowledge on the genomic abnormalities in these cell lines. FISH studies have revealed cryptic chromosome rearrangements, e.g., t(2;4), which was not ascertainable by CGH, and the latter confirmed the many genomic changes frequently found in NB (e.g., 1p-, 17q+, 11q-). Furthermore, established cell lines afford a readily available, diverse, biologically and genetically well-defined and manipulable source of NB cells for various studies (624–627). The latter have ranged from protein markers (628), gene expression analyses (629), response to NGF (630), cadherin expression (631), the relation of loss of 1p to cellular events (e.g., caspase-9 and APAF-1), genetics (632), apoptosis (633), effects of retinoic acid and vitamin A derivates (634), differentiation studies (635) and a wide range of studies on the *MYCN* gene, p53 effects and related molecular events (636–638).

The cytogenetic findings in such lines are shown in Table 7.18, and they date back more than three decades (639). In interpreting the findings shown in Table 7.18, it should be pointed out that somewhat different karyotypes may be obtained on the same cell line in different laboratories and in the same laboratory at different times of examination due to several factors, e.g., variability in culture conditions, basic instability or heterogeneity of the cell line, and the confusion regarding the provenance or identity of a particular cell line (640). A reproducible DNA fingerprinting or profiling method should be routinely adopted for cell lines to ensure proper authentication (640).

Early cytogenetic studies of established NB cell lines showed them to contain abnormal karyotypes, e.g., 47,XX, -22,+2mar in the SK-N-SH cell line and 46-47,XX,-2,-5, -16,+20,+2mar,+dmin in the SK-N-MC cell line (639), which were later rigorously defined with more advanced cytogenetic methodologies (641) (Table 7.18). In another study of three cell lines (CHP-100, CHP-126, and CHP-134), abnormal karyotypes (not based on banding) with marker chromosomes were present in each cell line (642). In a cell line (TR-14), an hsr was present in three copies without any normal chromosomes 1 being present (643). The modal chromosome number was 50, with four markers, one possibly being a t(1q;3q), and dmin. Another cell line was shown to have the karyotype, 49,XX,-12,der(5),+17,+2mar,+dmin (644).

Cell lines may show karyotypic instability and heterogeneity (645) probably related to in vitro effects and conditions not fully understood. An example is the SK-N-SH cell line and those (SH-SY5Y and SH-EP) derived from it. This karyotypic instability should be kept in mind when interpreting data based on established NB cell lines.

CGH studies of NB cell lines revealed findings often encountered in NB tumors, i.e., gain of 17q, *MYCN* amplification and loss of 1p (646). Cell lines resistant to several chemotherapeutic agents showed in addition gains at 13q14.1-q32 and 7q11.2-q31.3 and amplification of 7q21.1, consistent with *MDR1* amplification.

Early studies on the changes in the karyotypes of NB cell lines described abnormalities of the chromosome number and the presence of abnormal (marker) chromosomes, e.g., 7q+ (in SK-N-SH cell line), hsr in the short arm of chromosome 1 (IMR-32 cell line), two markers in cell line SK-N-MC, a marker with hsr in cell line LA-N-1 and an abnormal chromosome 1 in LA-N-2 (647).

In the SMS series, the seven lines examined had near-diploid chromosome numbers with many structural changes with del(1), del(11) and translocations involving chromosome 17 frequently present (648). All lines had dmin. No detailed karyotypes were presented. Two chromosomes 12-derived amplification units were found in the NFP cell line; the regions involved were 12q24 and 12q21 (649).

In a cell line derived from an advanced NB, the following karyotype was reasonably reflective of the findings described (no karyotypes shown in the publication): 49–54,XY,+1p,+2,+7,+8,+12,+20,+hsr(13). The latter was shown to contain *MYCN* (650).

Of the cell lines for which full karyotypes were available (Table 7.17), more than half (17 of 28) had either hsr (13 cell lines) or dmin (four cell lines) and *MYCN* amplification. One cell line (IGR-N-91) contained hsr in one cell population and dmin in another. Some of the cell lines without demonstrable hsr or dmin in the karyotypes were shown to have *MYCN* amplification.

In general, the chromosome changes in NB cell lines were more complicated and numerous than those in primary tumors (Tables 7.3 and 7.19). In a few cell lines, the karyotypes showed instability, although retaining some of the chromosome changes during the evolution of the sublines.

Retinoids and Neuroblastoma

Among retinoid agents, retinoic acid, particularly its isomers all-*trans*-RA (ATRA) and 9-*cis* (9CRA), have been found to have a significant therapeutic effect in acute promyelocytic leukemia, juvenile chronic myelocytic leukemia and to have antitumor effects against NB in vitro (651).

The role played by retinoids and their receptors, particularly their effects on cell growth by inducing different degrees either of apoptosis or differentiation in NB has been reviewed (653A, 651–653). The use of retinoic acid isomers in the treatment of NB continues to receive attention (654–656). Down-regulation of hASH1 via retinoic acid-induced differentiation of NB cell lines has been reported (657). Some information on retinoic acid in NB is supplied in Table 7.20.

p53

The protein products of the *TP53* gene, p53, has been described to be involved in a host of human tumors. Under normal conditions, p53, a phosphorprotein localized in the nucleus, plays a central role in the regulation of cell growth and tumor formation, because p53 operates as a cell cycle checkpoint and monitors DNA damage. Inactivation of p53 occurs in about 50% of all cancers, either by point mutation of one allele or by LOH of the reciprocal allele (658). The *TP53* gene is rarely mutated in NB (659–661). Mutation of p53 was found in a NB cell line established after the patient had received chemotherapy (662).

TABLE 7.19. Chromosome changes in neuroblastoma cell lines.

Cell line	Karyotype	References
MC-NB-1	44,XY,del(1)(p22:),-4,-7,+del(7)(q22:),-16,+t(7;16)(q22;q24),-17	*Casper et al 1983 (624)*
GI-LA-N	45,XY,-4,-10,del(1)(p32),t(1;11)(q43;q14),del(7)(q31),+mar	*Donti et al 1988 (232)*
GI-ME-N	90,XX,-X,-X,-1,-1,-2,-2,-4,-4,-10,-10,-11,-11,+13,-15,-19,-19,+der(1)t(1;?)(p36.2;?),dup(2)(p21p24)x2,dup(4)(q25q34)x2,del(6)(p21),del(6)(q13)x2,der(10)t(10;?)(p11;?)x2,der(11)t(11;17)(q13;q11)x2,+2mar	*Donti et al 1988 (232)*
N206	47~50,X,-1,+der(1)t(1;3)(q12;?),+t(1;7)(q?;p?),+add(2)(p14),+add(3)(p12),-4,add(5)(p13),-6,-7,add(7)(p15),-8,-9,add(10)(q26),add(11)(q23),+add(12)(q24),add(14)(q32),+add(15)(?),-17,-17,+5-6mar,hsr in 2 of the markers	*Versteeg et al 1990 (985)*
KP-N-NS	46,XX,-1,+der(1)t(1;?)(p36.1;?)	*Yoshihara et al 1993 (986)*
CLB-Ba	75,XXY,add(1)(p22)x2,del(2)(p16)x2,-7,+8,+add(11)(p14)x2,+17,+18,+22	*Combaret et al 1995 (753)*
CLB-Be	49~53,XY,del(1)(p2-p3?) or add(1)(p2?),hsr2(q11),hsr(6)(q12),+7,+9,+12, -14,add(15)(p13),+18,+3mar,inc	*Combaret et al 1995 (753)*
CLB-Br	87,XXX,-X,add(1)(p33),del(2)(p21),-3,-8,+9,-10,-11,-20	*Combaret et al 1995 (753)*
CLB-Ca	46,XY,der(1)t(1;8)(p21;p12),t(3;15)(p22;q21),t(4;5)(p15.1;p15.3),add(7)(p14),der(8)t(1;8)(p21p33;p12),t(12;22)(q14;q13),-14,+21,del(22)(q12),+r	*Combaret et al 1995 (753)*
CLB-Ga	45,X,-Y,der(1)t(1;17)(p31;q25),der(3)t(3;17)(p12;q22q24),add(4)(p15),t(14;17)(q23;q23),del(6q12),del(11)(q14),add(13)(q23),der(17)t(3;17)(p12;q22q24)	*Combaret et al 1995 (753)*
CLB-MA	46,XX,add(1)(p22)/46,idem,2-40dmin/46,idem,del(1)(q26)	*Combaret et al 1995 (753)*
CLB-Pe	56,XXX,del(6)(q21q26),-10,-10,-13,add(13)(q33),-19,-19,add(21)(p13),add(22)(q13),hsr(22)(q11),+2mar	*Combaret et al 1995 (753)*
NGP	47,XY,der(1)t(1;15)(p36.2;q24),hsr(4)(p16),der(6)t(6;17)(q?;q21),der(10)t(1;10)(q23;q26),der(11)t(2;11)(p14;q14),hsr(12)(q13),del(15)(q24),der(18)t(1;18)(p36.2),del(19)(q13),+19	*Muresu et al 1995 (987)* *Van Roy et al 2001 (66)*
MASS-NB-SCH1	42,XY,+1,1p-,+4,-10,-15,-16,-17,19q+,-22,-22	*Hiraiwa et al 1997 (988)*
KSC	45-47,XX,del(1)(p13),del(9)(p21),dmin	*Kao and Correa 1997 (989)*
MHH-NB-11	49-54,XY,+1p-,+1p-,+3,+7,+12,hsr(13q),+20	*Pietsch et al 1988 (650)*
SIMA	45,XY,t(1;17)(q35;q12),t(6;9)(p21;q33),?inv(12)(q?15-21q21-24),-17,der(22)t(17;22)(q22;p13); 22% of cells near-triploid	*Marini et al 1999 (990)*
IGR-N-331	46,XX,der(1)t(1;7)(p22;q11),der(5)t(5;17)(q35;q21),dmin	*Valent et al 1999 (784)*
IGR-N-337	52,XX,-1,+der(1)t(1;17)(p35;q21)x2,+2,+2,+4,+der(7)t(7;7)(p21;q21),-15,-16,+18,+20,+mar1,dmin/53,XX,idem,+mar1,+mar2,dmin	*Valent et al 1999 (784)*
IGR-N-91	43-44,XY,del(1)(q12),der(2)t(2;10)(p14;q11),der(4)t(1;4)(q12;p15),add(6)(q22),der(9)t(9;17)(q34;q21),-10,der(16)t(16;17)(q24;q22),-17,+der(?)t(?;2;10),+mar1,+mar2,hsr(14)(q32)/43-44,idem,+1/42-44,XY,-del(1)(q12),der(2)t(2;10;17)(p14;q11;q22),der(4)t(1;4)(q12;p15),del(9)(p13),-10,-13,-17,der(21)t(13;21)(q13;p13),+der(?)t(?;2;10),+mar1,dmin	*Valent et al 1999, 2001 (784,991)*
GICIN-1	53,XY,dup(1)(q21q32),t(2;4)(p23;q32),+der(2)t(2;4)(p23;q32)del(2)(q13),+der(6)t(6;7)(q15;p12),+der(10)t(3;10)(q13;q22),der(11)t(11;17)(p15;q12)t(11;17)(q22;q12),+12,+der(15)t(3;15)(?;p13),+20,+der(22)t(1;22)(q12;p13)	*Panarello et al 2000 (785)*
BCH-N-DR	44,X,-Y,der(1)t(1;11)(p32;q13)t(11;17)(q13;q21),inv(2)(p11q21),del(3)(p21p25),der(4)t(2;4)(p13;p16),dic(6;22)(q11;p11),der(11)t(11;17)(q13;q21),der(18)t(3;18)(p21;q21),der(20)t(2;20)(p?23;q13),der(21)t(17;21)(p13;p11),der(22)t(13;22)(q14;p11)	*McConville et al 2001 (786)*
BCH-N-JW	46,XY,i(1)(q10),t(2;17)(q33)q24~q25),der(3)t(3;17)(p21;q21),t(4;5)(q21;p15),+der(4)t(4;5),der(4)t(4;7)(p14;p13),der(11)t(11;17)(q13;q21)t(2;17)(q33;q24~q25),add(14)(q24),-17,add(21)(q22)	*McConville et al 2001 (786)*
G1-ME-N	78~82,XX,-X,-X,der(1)t(1;17)(p36;q11.2)x2,dup(2)(p21p24)x2,- 4,dup(4)(q25q34),der(5)t(4;5)(?;p14)x2,der(6)t(6;19)(p11;?)x2,der(6)t(6;10)(q12;p13)x2,-7,-9,-9,-10,der(10)t(6;10)(?;p13),der(11)t(11;17)(q13;q11.2)x2,-13,-15,-17,-17,der(17)t(17;22)(p12;q11),-19,-21,-22	*Van Roy et al 2001 (66)*
NBL-S	49,XY,der(1)t(1;1)(q32;q12),+2,t(2;4)(p24.3;q34.3),del(6)(p23),-11,+15,+19, +r(11)	*Van Roy et al 2001 (66)*

TABLE 7.19. Continued

Cell line	Karyotype	References
NLF	78~82,XXY,-Y,der(1)t(1;17)(p36;q12)x2,del(3)(p13),-4,der(4)t(4;13)(q34;q31), der(4)t dup(4)(q22q28),der(6)t(6;17)(q21;q21)x2,der(8)t(2;8) (p22;q23),hsr(8)(q12),hsr(8)(q?),der(8)t(8;19)(?;?),der(9)t(7;9)(p13;p21), -11,-13,del(13)(q31),+del(13)(q31),-14,der(16)(16pter→16q23::hsr::8?), -17,-18,-19,-20,-21,-22	*Van Roy et al 2001* (66)
NMB	81,XX,-X,-X,der(1)t(1;17)(p36;q12),der(2)t(2;10)(q13;q22),der(3)t(3;11)(?;q14), -4,-5,der(5)(2pter→2p12::5p14→5q32::2p12→2pter),+der(7)t(7;20)(q32;?), -9,-10,-11,der(12)t(1;12)(?;q?),hsr(13)(p11.2),+hsr(13)(p11.2),-13,-14, der(14)t(14;16)(q22;p12),-18,+del(20)(p11.2),-21,-22	*Van Roy et al 2001* (66)
SJNB-6	47,XY,der(1)t(1;2)(p31;p12),del(2)(p12),del(3)(p14),hsr(3)(p25),+der(11) t(11;17)(?;q12),der(16)hsr(16)(?),del(17)(p?12),der(19)t(17;19)(q21;q13.1), der(22)t(1;22)(p?;p?10)	*Van Roy et al 2001* (66)
SJNB-8	44,X,-X,der(1)t(1;17)(p33;q12)t(1;6)(q43;?),der(2)t(2;6)(p22;?),+der(2)t(2;6), der(5)t(5;15)(q34;q25),hsr(6)(p22),der(6)(8pter→8p12::6p12→6q16::15q24→ 15qter),der(8)t(1;8)(p33;q21),der(9)t hsr(9)(p24),19(9qter→9p24::hsr::19?), der(14)t(9;14)(?;q32),-18,-19,der(22)(?;p15;p?34)	*Van Roy et al 2001* (66)
SJNB-12	83,XXXX,-1,der(1)t(1;17)(p34;q11~12),+der(1)t(1;17),-2,der(2)t(2;14)(q36;q21), der(2)t(2;3)(p10;p12)x2,+del(2)(q24),-8,der(8)t(8;16)(q22;?),der(9)t(9;16) (p13?),-10,-11,-13,-14,-14,der(15;22)(q10;q10)x2,der(15)t(2;15)(q?;q10), der(16)t(16;19)(p?;?)x2,der(17)t(1;17)(q12;q24),-17,r(20),+dmin	*Van Roy et al 2001* (66)
SK-N-BE	66,XX,-1,der(1)t(1;2)(p22;p15)x2,-2,-2,add(2)(p24)x2,der(3)t(3;14)(p12;q12), der(3)t(3;17)(p21;q12),hsr(4)(q24),hsr(6)(p12),+7,-10,-11,+12,-13,der(13) t(13;15)(q10;q10),-14,-15,del(16)(?),der(17)t(11;17)(q12;q?12),der(17) (4?::8?::17q24→17pter),der(19)t(16;19)(q12;q13.3)x2	*Van Roy et al 2001* (66)
SK-N-FI	48,XY,del(1)(q42),del(6)(q13q16),+12,+13,der(14)t(7;14)(q32;q24),t(5;15) (q26;q11.2),t(17;19)(q11~12;q13)	*Van Roy et al 2001* (66)
STA-NB-12	47,XY,der(1)t(1;2)(p34;p15),del(6)(q23),der(10)t(7;10)(q22;q24),+12	*Van Roy et al 2001* (66)
IMR-32	48,XY,+1,der(1)t(1;17)(p32;q12)hsr(17)(q21)x2,+6,der(16)t(15;16)(q21;q23)	
SK-N-AS	42-45,XX,der(3)t(3;17)(p13;q21),der(4)t(4;8)(q32;q24.2),inv(5)(p15.3q11.2), der(6)t(4;6)(q25;q21),del(8)(q24.2),der(9)t(8;9)(q23;p21),der(11)t(11;17) (q13;q11),dup(13)(q12q23)ins(13;8)(q32;q22q24),der(14)inv(14)(q13q32), der(16)t(12;16)(q24;q24),del(22)(q13)	*Kim et al 2001* (783) *Spengler et al 2002* (641)
SK-N-MC	46,XX,t(2;3)(q23;q27),del(4)(p12),t(5;10)(q11;q21),+6,del(8)(p12),del(9)(q34), -10,der(11)add(11)(p15)add(11)(q23),-15,der(17)t(16;17)(p10;p10), der(21)t(21;22)(p11;q13.1),del(22)(q12),+mar	*Spengler et al 2002* (641)
SK-N-SH	47,XX,+7,der(9)t(7;9)(q22;q34),der(22)t(17;22)(q21;q13)	*Kim et al 2001* (783) *Spengler et al 2002* (641)
SK-N-SH	47,XX,dup(1)(q12q25),+7,t(7;8)(q34;q24.2),der(9)t(2;9)(p15;q34),der(22) t(17;22)(q21.3;q13)/47,XX,dup(1)(q12q25),+7,t(7;8)(q34;q24.2),der(9) t(9;18)(q34;?),del(12)(q?),der(18)t(18;12;2)(q23;q13q21;p15),der(22)t(17;22) (q21.3;q13)/47,XX,dup(1)(q12q25),+7,t(7;8)(q34;q24.2),der(9)t(9;18)(q34;?), del(12)(q?),der(18)t(18;12;11;2)(q23;q13q21;q14q23p15),der(22)t(17;22) (q21.3;q13)/47,XX,+der(7)t(1;7)(q12;q3?),t(7;8)(q34;q24.2),der(9)t(2;9) (p15;q34),der(22)t(17;22)(q21.3;q13)	*Cohen et al 2003* (992)
SH-SY5Y	47,XX,der(1)dup(1)(q25q11)dup(1)(q44q25),der(9)t(7;9)(q22;q34),der(22) t(17;22)(q21;q13)	*Kim et al 2001* (783) *Spengler et al 2002* (641)
SH-SY5Y	47,XX,dup(1)(q12q25),+7,t(7;8)(q34;q24.2),der(9)t(2;9)(p15;q34),der(22) t(17;22)(q21.3;q13)	*Cohen et al 2003* (992)
*SH-EP	47,X,-X,+i?(1)(q10),+7,t(7;8)(q34;q24.2),der(19)t(10;19)(?;?),der(22)t(17;22) (q21.3;q13)/49,XX,+i?(1)(q10),+del(5)(q1?),der(5)t(3;5)(?;p13),+7,t(7;8) (q34;q24.2),der(22)t(17;22)(q21.3;q13)	*Cohen et al 2003* (992)
*SH-EP	49,XX,+i(1)(q10),+del(7)(p11.2),t(7;8)(q34;q24.2),+der(8)t(7;8)(q34;q24.2), t(12;14)(p13;q12),der(22)t(17;22)(q21;q13)	*Cohen et al 2003* (992)
SMS-KCNR	46,XY,del(1)(p34),der(6)t(2;6)(p16;q16),der(20)t(17;20)(q21;q13)	*Spengler et al 2002* (641)
NB-1643	46~48,XY,der(1)t(1;17)(p?43;?q22),+der(1)t(1;17)(p?34;q?22),+7	*Yoon et al 2006* (781)
NB-1691	41~52,XX,der(3)hsr(3)(p26)hsr(3)(q29),hsr(3)(q29),+der(3)add(3)(q29), del(4)(q24),+7,der(14)add(14)(p11),+17,+17	*Yoon et al 2006* (781)

The SH-SY5Y and SH-EP cell lines were derived from the SK-N-SH cell line. However, the exact phenotypic (cellular) provenance of the two former cell lines has not been settled (993, 994). The two karyotypes of SH-EP shown in Table 7.18 were determined in two different laboratories (993). Cell line SK-N-MC may have originated from a Ewing type of tumor (640).

TABLE 7.20. Retinoic acid in NB.

	References
Retinoic acid and cell death in NB.	*Piacentini et al 1992* (995)
High levels of retinoic acid receptor-β reflect a good prognosis in in NB; work through effects on p21 expression and consequent G0/G1 cell cycle arrest.	*B. Cheung et al 1998* (996)
HMGI(Y) (at 6p21) and HMGI-C (at 12q13-q15) genes, expressed in NB and NB cell lines, may participate in the differentiation and growth-related progression of NB, possibly affecting response to retinoic acid.	*Giannini et al 1999* (997)
Retinoids in NB.	*Raschellà et al 2000* (651)
	Vertuani et al 2003 (653)
Retinoic acid induces apoptosis in NB cell lines and is associated with nuclear accumulation of p53.	*Takada et al 2001* (998)
retSDR1, a short-chain retinal reductase, is retinoic acid-inducible and frequently deleted in NB cell lines.	*Cerignoli et al 2002* (975)
Overexpression of RAR-γ increases apoptosis in NB cell lines in response to retinoic acid but not to fenretinide. Retinoid therapy in pediatric tumors, including decreased *MYCN* expression, precedes retinoic acid-induced NB differentiation.	*Goranov et al 2006* (999)
	Reynolds and Lemons 2001 (1000)
	Thiele et al 1985 (1001)
Activation of PI3/Akt signaling pathway by retinoic acid is required for neural differentiation of the SH-SY5Y NB cell line.	*López-Carballo et al 2002* (1002)
Fenretinide (a synthetic retinoid) as therapy in NB.	*Garaventa et al 2003* (656)
Synthetic retinoid (Ro 13-6307) inhibits NB growth and induces differentiation.	*Ponthan et al 2003* (1003)
Retinoic acid up-regulates Wnt-5a gene expression in NB.	*Blanc et al 2005* (1004)

A candidate TSG located at 1p36 with sequence and structural homology to the *TP53* gene is *TP73* (663). *TP73* may share functional similarities with *TP53*. In nine NB cell lines studied, no mutations of *TP73* were found (663). Several studies failed to demonstrate a key role for *TP73* in the tumorigenesis of NB (194, 212, 664–666). The changes and other information on p53 in NB are shown in Table 7.21.

H-*ras* and NB

Although the *ras* family of oncogenes plays a major role in the genesis of several tumors, their role in NB tumorigenesis has not been established (667). H-*ras* through its protein product p21, a member of the G protein family, located on the inner surface of the cell membrane, relays signals from transmembrane receptors to intracellular effector proteins (658). Also, p21 is highly expressed in neuronal cells and participates in the NGF signal transduction pathway. The role of H-*ras* in NB development remains unknown. Nevertheless, it has been promulgated that expression of H-*ras* (p21) in NB is associated with a favorable prognosis and can serve as an independent prognostic factor (658, 668) and as a more informative marker than *MYCN* amplification.

CD44 and NB

CD44 represents a heterogeneous group of cell surface and secreted glycoproteins. CD44 expression is widely distributed in many tissues and cells and has been related to a variety of physiological processes, including cell–cell and cell–extracellular matrix interactions (669). The CD44 hyaluronan receptor is differentially expressed (more highly) in low-grade versus high-grade NB (670, 671). Further information on the role of *MYCN*, the enzymatic systems involved in the regulation of CD44 glycosylation, and the consequences of nonfunctional CD44 receptor expression in NB growth and tumorigenesis should facilitate the understanding of the invasive mechanisms involved in NB progression and dissemination (669).

CD44 is repressed in NB through hypermethylation; reactivation of CD44 could be a target of NB therapy (672). Lack of CD44 expression is a prognostic marker in NB, associated with aggressive disease (462, 670, 673, 674).

CD44 is one of the most powerful factors for predicting clinical outcome in NB at the time of initial staging (674). Thus, CD44 expression is strongly correlated with favorable tumor stages and histology, younger patient age and normal *MYCN* copy numbers (675). The authors strongly recommend the determination of CD44 expression as an additional biological markers in the initial staging of NB (676, 677). Hypermethylation of CD44 may regulate its expression in NB (678).

Mass Screening of NB

Mass screening to detect aggressive NB has produced paradoxical results, in that it has revealed an increased incidence of infantile NB, whereas NB with unfavorable biological markers were rarely found in the screening-positive tumors (679, 680). It is possible that the mass screening systems tend to identify additional NB cases that would regress spontaneously or mature (679). The lack of *MYCN* amplification and the presence of its expression in NB detected by screening suggest that the role of *MYCN* differs in NB identified by screening vs. *MYCN*-amplified tumors (679). The general experience with mass screening has been discussed in previous publications (681–688).

TABLE 7.22. Continued

	References
mice. The authors suggested that ARID3B may play a key role in the malignant evolution of NB and may serve not only as a marker of malignancy but also as a potential target for cancer therapy of stage IV NB, for which there is currently no effective treatment available.	
Hypoglycemia-induced cell death in a NB cell line mediated through the induction of the transcription factor [EBP homology protein (CHOP)] and hypoxia counteracted this cell death via repression of CHOP and induction of VEGF.	*Kogel et al 2006* (1039)
Doppel-induced apoptosis, which may be counteracted by cellular prion protein, has been studied in NB cells.	*Qin et al 2006* (1040)
The successful use of a small-molecule MDM2 antagonist to disrupt the p53-MDM2 pathway holds promise as therapy for NB without imposing a genotoxic effect.	*Van Maerken et al 2006* (1041)
A highly significant correlation was found between a high mitotic karyotypic index and *MYCN* amplification in 43 NB. Neither 1p36 deletion nor p36 imbalance significantly correlated with the mitotic karyotypic index. Deletion of 1p36 was found to be a reliable parameter in determining unfavorable prognosis in NB.	*Altungöz et al 2007* (1042)
A novel attenuated oncolytic poliovirus has been suggested as a candidate for effective oncolytic treatment of NB (or other cancers) even in the presence of induced antipolio immunity, based on studies in novel poliovirus-susceptible mice	*Toyoda et al 2007* (1042)A
A possible role of ploidy heterogeneity in localized NB in relation to prognosis has been evaluated.	*Mora et al 2007* (1042B)
Synthetic triterpenoids were found to be inhibitors of growth and induce apoptosis in NB (and glioblastoma) cells through inhibition of the Akt, nuclear factor-kB, and Notch1 pathways.	*Gao et al 2007* (1244)
Pathology and biology guidelines for NB resection.	*Ambros et al 2001* (1244A)
A study of isolated neuroblasts revealed a profile compatible with NB originating from these cells, as well as revealing a number of molecular and genetic pathways of importance in NB biology.	*De Preter et al 2006* (1244B)
Clonal ploidy heterogeneity based on DNA content has been found in locoregional NB and related to prognostic facets.	*Mora et al 2007* (1244C)
Profiles of genomic DNA content of NB predict clinical phenotype and regional gene expression.	*Mosse et al 2007* (1244D)
Methylguanine-DNA methyltransferase is widely expressed in NB and may serve as a cogent target in the treatment of NB.	*Wagner et al 2007* (1244E)

TABLE 7.23. Molecular and genetic changes in NB.

	References
Differentiation of NB cells in vitro and suppression of tumorigenicity are controlled by many genes.	*Bader et al 1991* (176)
P-glycoprotein and multidrug resistance in NB.	*Chan et al 1991* (1044)
	Favrot et al 1991 (1045)
ILGF-II mediated proliferation in NB occurs with TrkA inactivation.	*El-Badry et al 1991* (1046)
	C.J. Kim et al 1999 (1047)
	Russo et al 2005 (1048)
NM23 overexpression, DNA amplification and mutation in aggressive NB.	*Leone et al 1993* (1049)
Pituitary adenylate cyclase activating polypeptide (PACAP)-like immunoreactivity in NB and GNB may play a role in these tumors.	*Takahashi et al 1993* (1050)
DNA ploidy and H-Ras (p21) in NB.	*Kusafuka et al 1995* (1051)
Coamplification of *DEAD* box gene and *MYCN* in NB; no *MYCN* amplification, no *DDXT* amplification.	*Manohar et al 1995* (1052)
No significant changes affect p16, p18 and p27 in NB.	*Kawamata et al 1996* (612)
Survey of allelotype in NB.	*Perri et al 1996* (1053)
Mutations of p16 do not play a part in NB.	*Castresana et al 1997* (613)
MEN-2B/RET enhances metastatic spread and activates Jun kinase in NB; *RET* proto-oncogene in NB.	*Marshall et al 1997* (1054)
	M. Takahashi et al 1991 (1055)
GDNF regulates growth and differentiation and apoptosis in NB.	*Hishiki et al 1998* (1056)
Serum levels of NSE parallel NB stage.	*Massaron et al 1998* (1057)
Metastatic sites of grades 4 and 4S NB: age, tumor biology, and survival.	*DuBois et al 1999* (1058)
Achaete-scute homologue-1 (HASH-1) is down-regulated in differentiating NB.	*Soderholm et al 1999* (1059)
	Cooper et al 2001 (1032)
Ganglioside expression and stratification in NB; may be used as a marker for NB cells in marrow.	*Hoon et al 2001* (1060)
	Hettmer et al 2003 (1061)
Gem, a small GTP-binding protein within the Ras family, promotes differentiation in NB.	*Leone et al 2001* (1062)
PTEN/MMAC1 alterations detected in NB.	*Moritake et al 2001* (1214)
Possible role for melanin-concentrating hormone receptor in GNB and NB.	*K. Takahashi et al 2001* (939)
SWI1 expression in NB.	*K. Takahashi et al 2001* (939)
Deletion of *CDKN2A* (p16 and p14) but not of 1p36 in NB.	*Thompson et al 2001* (135)
Novel products of the *HUD, HUC, NNP-1*, and α-internexin found in NB.	*Behrends et al 2002* (1191)
Galanin and its receptor in NB.	*Perel et al 2002* (1063)
ALX is hypermethylated in NB.	*Wimmer et al 2002* (1064)
ILGF binding protein 5 (IGFBP-5), a modulator of IGF-2, is a determinant of proliferation and metastases in NB.	*Cesi et al 2004* (1065)
SDHD not involved in NB.	*De Preter et al 2004* (1066)
Cyclooxygenase-2 (COX-2) is expressed in NB and may be a target for nonsteroidal agents as additional therapy.	*Johnsen et al 2004* (1067)
Tubulin tyrosine kinase in NB.	*Kato et al 2004* (1068)

TABLE 7.23. Continued

	References
KIT expressed in NB with a favorable prognosis and may be an area for therapeutic considerations.	*Krams et al 2004* (1069)
	Uccini et al 2005 (1070)
EWS/FLI1 may change NB to PNET/ES.	*Rorie et al 2004* (1071)
Authors describe genes (*EPHB6, EFNB2, EFNB3, NTRK1* and *CD44*) whose high level of expression predicts favorable NB outcome and forced expression of these genes inhibits growth of unfavorable NB. This area may have an impact on the development of therapeutic strategies for unfavorable NB.	*Tang et al 2000* (919,920)
CDK9 is expressed in more differentiated NB and correlates with tumor grade.	*De Falco et al 2005* (1072)
Effects of histone deacetylase inhibitor (BL1521) on gene expression in NB.	*de Ruijter et al 2005* (1073)
Growth of NB is mediated by EGFR, which can be inhibited by therapy.	*Ho et al 2005* (1074)
Delta-like gene is up-regulated in glioneuromas but not in NB or GNB.	*Hsiao et al 2005* (1075)
Calreticulin is a prognostic factor in NB, its expression correlates with differentiation.	*Hsu et al 2005* (1076)
Glucose-regulated protein 78 (GRP78), an endoplasmic reticulum protein, is essential for differentiation of NB and its presence correlates with a good prognosis.	*Hsu et al 2005* (1077)
Glutathione S-transferase polymorphisms not related to NB susceptibility.	*Lanciotti et al 2005* (1078)
NM23-H1 protein in the serum and clinical significance.	*Okabe-Kado 2005* (1079)
MDR-associated protein 1 gene in NB biology and outcome.	*Pajic et al 2005* (1080)
Protocol for obtaining NB specimens for various analyses.	*Qualman et al 2005* (1081)
The *PTEN* gene in NB.	*Qiao et al 2005* (1082)
Ras/MAPK signaling in NB.	*Stockhausen et al 2005* (1083)
Copy number defects of *MYCN, CDC42* and *NM23* and differentiation pathway in NB.	*Valentijn et al 2005* (1084)
Anti-proliferative effects of an agonist for PPARγ (peroxisone proliferators-activated receptor γ) in NB cell lines.	*Cellai et al 2006* (1085)
Polysialic acid is expressed in NB and is synthesized by sialyltransferase (STX), whose mRNA levels have potential utility as a molecular marker for metastases.	*Cheung et al 2006* (1086)
FCGR2A polymorphism and clinical outcome in NB.	*Cheung et al 2006* (1087)
Flavonoids can activate calpain and caspase cascades for mediation of apoptosis.	*Das et al 2006* (895)
Protein kinase B modulates the sensitivity of NB ILGF inhibition.	*Guerreiro et al 2006* (1088)
MRP1 (multidrug transporter gene-1) expression is a poor sign in NB.	*Haber et al 2006* (1089)
High expression of N-acetylglucosaminyltransferase V is a favorable sign in NB.	*Inamori et al 2006* (1090)
BMCC1 (has homology to *BCH*) expression is associated with favorable outcome in NB.	*Machida et al 2006* (1091)
In high-grade NB, caspase-8 levels correlate with those of STAT-1 (signal transducer and activator of transcription-1), but not in lower grades.	*Muscat et al 2006* (1092)
P2X$_7$ receptor, substance P-dependent mechanism and NB growth.	*Raffaghello et al 2006* (1093)
A study of genomic events in a NB cell line led the authors to propose a sequence of translocation-excision-deletion-amplification of oncogenes (involving *MYC* and *ATBF1*), rather than the breakage-fusion-bridge model previously favored.	*Van Roy et al 2006* (1094)
Integrative genomics points to multiple gene alterations in NB.	*Wang et al 2006* (1095)
The synergy between hypoxia and etoposide causing purging of the NB stem cell compartment in vitro may have applicability to spontaneous regression of NB.	*Marzi et al 2007* (1095A)
A protective effect of DMSG against H_2O_2-induced cell death in a NB cell line (SY5Y) via inhibition of both mRNA and protein expression levels of *BAX* and CASP3 has been described.	*Park et al 2007* (1241)
A CGH study of NB without *MYCN* amplification, although confirming previous findings, attempted to define those cases likely to relapse.	*Schleiermacher et al 2007* (1245)
LOH of 1p is an indicator of a poor prognosis in NB.	*Schleiermacher et al 1996* (1246)
Neuroblastic brain tumors containing abundant neuropil and true rosettes may constitute an entity with the PNET.	*Eberhart et al 2000* (1247)
High expression of the gene Skp2 characterizes high-risk NB independent of the *MYCN* status.	*Westermann et al 2007* (1249)
Recombinant human erythropoietin (Epo) and its receptor (EpoR) were expressed, but did not modify tumor cell proliferation in NB. High expression of EpoR was associated with an increased survival. The authors suggested that Epo may be used safely in children with NB.	*Sartelet et al 2007* (1250)
A study utilizing high-resolution oligonucleotide array CGH showed losses of whole chromosomes 3, 4, 10 and 16 and gains of 6, 7, 8, 13, 17, 18, and 20 in localized NB. In contrast, disseminated tumors showed gain of 17q and 3p and 11q loss among a number of structural chromosomal changes.	*Scaruffi et al 2007* (1251)
The existence and spatial distribution of neurotrophin receptors in NB lend supportive evidence that neurotrophic influences may be involved in tumor persistence or regression.	*Hoehner et al 1995* (1252)
T-Cadherin is a negative growth regulator of epidermal growth factor in NB cells.	*Jakeuchi et al*
Anaplastic lymphoma kinase (*ALK*) gene activation was responsible for SHH-C (a member of docking proteins) hyperphosphorylation in NB cell lines.	*Miyake et al 2002* (1252B)
The insulin-like growth factor (IGF) axis is frequently activated in NB and its cell lines. By using miRNA and siRNA to silence endogenous expression of IGF Binding Protein-5 in NB to cause mitochondrial apoptosis, which was characterized by release of cytochrome C into the cytoplasm and activation of caspase-9, Erk-1 and Erk-2 inhibition and up-regulation of pro-apoptotic proteins Bim and Bax.	*Tanno et al 2006* (1252C)
The complexity of changes in NB is reflected in a study utilizing microarrays and FISH in which more than 1000 genes were identified, whose expression was consistently altered due to genomic alterations (reflected in loss of 1p, 3p, 4p, 10q and 11q and gain of 2p, 17q and the *MYCN* amplicon). Of these genes, 84 correlated with survival, with 17q gain and 4p loss being significant. These genes are involved in RNA and DNA metabolism and in apoptosis.	*Łastkowska et al 2007* (1252D)

TABLE 7.24. Genes and related factors in NB.

	References
DNA loop attachment and replication sites.	*Smith et al 1984* (1096)
Neuron-specific enolase in the diagnosis of NB.	*Tsokos et al 1984* (1097)
A NB xenograft with loss of 1p and the presence of either dmin or hsr has been used for testing the effects of various agents.	*Makino et al 1986, 1993* (1098, 1099)
Procoagulant activity in NB cell lines and their growth.	*Esumi et al 1989* (1100)
MYCN/max system in NB.	*Wenzel et al 1991* (1101)
	Wenzel and Schwab 1995 (1102)
Achaete-scute homolog 1 in olfactory and autonomic neuron development.	*Guillemot et al 1993* (1103)
NM23-H1 mutations in NB.	*C.L. Chang et al 1994* (1198)
Stem cell factor and c-*kit* in NB.	*P.S. Cohen et al 1994* (1104)
ErbB2 expression in NB.	*Goji et al 1995* (1105)
Pleiotrophin and midkine in NB.	*Nakagawara et al 1995* (1106)
RET not mutated in sporadic and hereditary NB.	*Hofstra et al 1996* (1107)
Calbindin proteins in NB.	*Ishiguro et al 1996* (1108)
Expression of MDR gene and outcome in NB.	*Norris et al 1996* (1109)
NM23H1: poor prognostic factor independent of LOH of 1p in NB.	*Takeda et al 1996* (1110)
Effects of *RET* (MEN-2B gene) in NB.	*Marshall et al 1997* (1054)
Cerebellin and its mRNA in GNB and NB.	*Satoh et al 1997* (1111)
Mononucleotide repeat instability is infrequent in NB.	*Hogarty et al 1998* (1112)
When taken with other factors affecting the biology of NB, determination of bcl-2 protein levels can provide prognostic information.	*Mejia et al 1998* (1242)
Kinases and their effects on NB cell line.	*Encinas et al 1999* (1113)
EPHB6, EFNB2, and *EFNB3* expression with low tumor stage and high TrkA in NB.	*Tang et al 1999,2000* (919,920,1114)
Antigen expression and complement susceptibility in differentiated NB cells.	*S. Chen et al 2000* (1115)
Gene expression of neuronal nitric oxide synthase and adrenomedullin in NB.	*Dotsch et al 2000* (1116)
Integrins in NB.	*Erdreich-Epstein et al 2000* (1117)
Genomic imprinting of ILGF-2 in NB.	*Hattori et al 2000* (1118)
Gene encoding DNA fragmentation factor 40 (caspase-activated nuclease) gene in a TSG in NB.	*Judson et al 2000* (1119)
Cross-resistance of topoisomerase I and II inhibitors in NB cell lines.	*Keshelava et al 2000* (1120)
ALK gene in NB.	*Lamant et al 2000* (1121)
h*CDC10* and favorable prognosis in NB.	*Nagata et al 2000* (1122)
Drosophila Delta gene in NB.	*van Limpt et al 2000* (1123)
Composite pheochromocytoma with NB.	*Candanedo-Gonzalez et al 2001* (1124)
Inhibition of SK-N-MC cell line growth by y nucleotide phosphodiesterase inhibitors.	*Giorgi et al 2001* (1125)
Factors in NB differentiation.	*Han et al 2001* (1126)
EPB4.1 gene at 1p36.	*Huang et al 2001* (1127)
CD2 synthase: molecular marker for NB.	*Lo Piccolo et al 2001* (1128)
SMARCF1, truncated SWI1, in NB.	*Takeuchi et al 2001* (1129)
M-FISH in NB reveals targets for study.	*Van Roy et al 2001* (66)
HUD, HUC, NNP1, and α-internexin genes identified by autologous antibody screening of a pediatric NB library.	*Behrends et al 2002* (1191)
Ras and Seppuku in NB.	*Reynolds 2002* (422)
Survivin: Fas ratio is predictive of recurrent NB.	*Sandler et al 2002* (1130)
RASSF1A as TSG in NB.	*Agathanggelou et al 2003* (1131)
RT-PCR of tyrosine hydroxylase gene expression for MRD in NB.	*Lambooy et al 2003* (1132)
In vitro and *in vivo* effects of liposomal fenretinide on NB.	*Raffaghello et al 2003* (1133)
Dopamine decarboxylase expression in marrow and blood as NB marker.	*Bozzi et al 2004* (1134)
No alterations of topoisomerase in NB cells resistant to irinotecan.	*Calvet et al 2004* (1135)
ID2, CDKN1B and *CDKN2A* in NB.	*Gebauer et al 2004* (1136)
HGF/c-Met in experimental human NB.	*Hecht et al 2004* (1137)
PRAME, a tumor-associated antigen, associated with high-stage NB and poor prognosis.	*Oberthuer et al 2004* (1138)
EWS/FLI-1 effect on p53 in NB.	*Rorie and Weissman 2004* (1139)
Nestin in NB.	*Thomas et al 2004* (1140)
Galectins in NB growth and physiology.	*Andre et al 2005* (1141)
LM03 and HEN2 interaction: act as an oncogene in NB?	*Aoyama et al 2005* (1142)
DLK1-GTL2 alteration epigenetically.	*Astuti et al 2005* (1143)
Sperm protein 17 expression defines subsets of NB.	*Bumm et al 2005* (1144)
P244 receptor participates in the concomitant to differentiation and cell death in SH-SY5Y NB cells.	*Cavaliere et al 2005* (1197)
HDL blocks differentiation of NB by inhibiting EGFR.	*Evangelopoulos et al 2005* (1145)
cAMP enhances siRNA-mediated gene silencing in NB.	*Hanson et al 2005* (1146)
Gangliosides in NB cell lines.	*Hettmer et al 2005* (1147)
NB vasculature and matrix metalloproteinase-9.	*Jodele et al 2005* (1148)

TABLE 7.24. Genes and related factors in NB.

	References
No *KIT* mutations similar to GIST in NB.	*Korja et al 2005* (1149)
Neurokinin receptors and NB proliferation.	*Mukerji et al 2005* (1150)
MRP4/ABCC: marker of poor prognosis in NB.	*Norris et al 2005* (1017)
Doublecortin gene, a marker for MRD in NB.	*Oltra et al 2005* (1151)
Anaplastic lymphoma kinase in NB.	*Osajima-Hakomori et al 2005* (1152)
Cox-2 inhibitor effects on NB cell growth.	*Parashar et al 2005* (1152A)
Shc family expression in poor outcome NB.	*Terui et al 2005* (1153)
NPFF2 receptor in NB cell line	*Anko and Panula 2006* (1154)
Mutations of UBE4A and UBE4B ubiquitination factors in NB.	*Carén et al 2006* (1155)
WSB1 increased copy number correlates with increased survival in NB.	*Chen et al 2006* (1199)
PDMP, hyperdiploidy and sensitivity of NB cells to paclitaxel.	*Dijkhuis et al 2006* (1156)
Lactoferricin B is cytotoxic to NB.	*Eliassen et al 2006* (1157)
BMI1, target gene of E2F-1, strongly expressed in NB.	*Nowak et al 2006* (1158)
Risk estimation of localized NB based on loss of 1p and loss of 11q.	*Simon et al 2006* (1159)
The results obtained with gene-expression studies may be useful in the risk estimation of treatment in NB.	*Oberthuer et al 2006* (1159A)
A subset of NB in which enhanced expression of *DDX1* and low expression of *NAG*, consequent to *DDX1* amplification without that of *NAG*, may contribute to therapy response. *DDX1* (DEAD box polypeptide 1) and NAG (neuroblastoma amplified gene) are part of the *MYCN* amplicon.	*Kaneko et al 2007* (1159B)
Based on high-resolution arrayCGH results, classification of NB according to their genomic manifestations has been proposed.	*Michels et al 2007* (1159C)
1α-hydroxyvitamin D$_2$ inhibits growth of human NB cells in grafted mice.	*van Ginkel et al 2007* (1159D)
Overexpression of 2p was found in almost all NB cell lines with *MYCN* amplification. Using comparative expressed sequence hybridization, it was shown that expression peaks were present at 2pter, 2p24 (*MYCN* locus), 2p23.3-p23.2, and/or 2p23.1. Cell lines with enhanced peak at 2 p23.2-p23.3 had also amplified *ALK* gene; the latter is associated with a poor prognosis. The data indicate that besides *MYCN* amplification, other genes proximal and distal to it are highly expressed in NB.	*Stock et al 2008* (1159E)

TABLE 7.25. Some of the methodologies used for the determination of changes in NB.

	References
Methods for determining *MYCN* expression and amplification in NB have been described over the years and have included Southern blotting, chromogenic in situ hybridization, RT-PCR, dot blotting, and various FISH procedures. The latter approach seems to be favored by workers in the field. It should be stressed that these methodologies are also generally applicable to fixed tissues.	*Crabbe et al 1992* (1161A) *Hiyama et al 1999* (150A) *Barroca et al 2001* (1160) *Bhargava et al 2005* (1161) *de Cremoux et al 1997* (1162) *Layfield et al 2005* (1163) *Mathew et al 2001* (1164) *Melegh et al 2003* (1165) *Miyajima et al 1996* (1166) *Raggi et al 1999* (1167) *Sartelet et al 2002* (1168) *Squire et al 1996* (1169) *Thorner et al 2006* (1170)
SKY and M-FISH studies may reveal novel regions as genetic targets worthy of investigation in NB.	*Van Roy et al 2001* (1171)
Detection of LOH in NB can be accomplished with single- strand conformation polymorphism techniques.	*White et al 1993* (1172)
Detection of *MYCN* amplficiation with competitive PCR quantitation.	*Oude Luttikhuis et al 2000* (1173)
Methodologies for quantitation fo marrow disease in NB by PCR, as well as of GD2 synthase mRNA have been described.	*Cheung et al 2001,2002,2003* (1174–1176)
Laser-capture microdissection applied to the genetic profiling of NB has been reported.	*De Preter et al 2003* (1177) *Mehes et al 2001,2003* (1178, 1179)
The establishment of DNA copy number and levels in NB can be accomplished by several techniques.	*Mosse et al 2005* (1180, 1181) *Michels et al 2006* (1182)
The application of PCR for the detection of disseminated NB has been described.	*Swerts et al 2006* (1183)
Gene expression studies may serve as predictors of outcome in NB.	*Wei et al 2004* (381A)

TABLE 7.25. Continued

	References
Automatic detection anda genetic profiling of disseminated NB cells.	*Méhes et al 2001* (1178)
Quantitation of *MYCN*, DDX1 and NAG genes has been determined by PCR.	*De Preter et al 2002* (1184)
RT-PCR assay for prediction of the uptake of labeled iodobenzyl- guanidine by NB.	*Carlin et al 2003* (1185)
Laser-capture microdissection in the genetic analysis of NB cells and their precursors.	*De Preter et al 2003* (1177)
Measurement of circulating NB cells by quantitative RT-PCR and correlation with findings in marrow and other disease markers.	*Cheung et al 2004* (1186)
The detection of minimal residual disease in NB has been evaluated.	*Beiske et al 2005* (1187)
Multiplex real-time PCR appears to be a promising approach for reducing the complexity of information obtained from whole-arm genome array studies. Such an approach could be a clinically useful test to predict NB outcome.	*Schramm et al 2007* (1187A)

reflect the complexity of these facets associated with the process of tumorigenesis. No specific gene has been found to be consistently associated with NB. Even in conditions, e.g., CML, in which a specific translocation involving known genes is associated with the genesis of the disease, molecularly the pathways leading to the development and progression of CML, are complex. Hence, it is not surprising that NB, and many other tumors, are associated with a host of changes as seen in Tables 7.22 and 7.23.

Table 7.25 are presented some comments regarding the methodologies used to determine changes in NB and a listing of some of these techniques.

Closing Comments

Almost all of the genes, their products, and related substances mentioned in this chapter are involved in signaling and related pathways of great complexity, making the deciphering of the key and ensuing events responsible for the development and progression of NB a challenge yet to be met. In this NB is not unique, because the aforementioned statements apply to almost all cancers and leukemias.

References

1. Brodeur, G.M., and Ambros, P.F. (2000) Genetic and biological markers of prognosis in neuroblastoma. In *Neuroblastoma* (Brodeur, G.M., Sawada, T., Tsuchida, Y., and Voûte, P.A., eds.), Elsevier Science B.V., Amsterdam, The Nethrlands, pp. 355–369.
1A. Brodeur, G.M., Sawada, T., Tsuchida, Y., and Voûte, P.A., editors. (2000) *Neuroblastoma*. Elsevier Science B.V., Amsterdam, The Netherlands.
2. Brodeur, G.M. (1998) Clinical and biological aspects of neuroblastoma. In *The Genetic Basis of Human Cancer* (Vogelstein, B., and Kinzler, K.W., eds.), McGraw-Hill, New York, pp. 691–711.
3. Brodeur, G.M. (2003) Neuroblastoma: biological insights into a clinical enigma. *Nat. Rev.* **3**, 203–216.
4. Schor, N.F. (1999) Neuroblastoma as a neurobiological disease. *J. Neurooncol.* **41**, 159–166.
5. Alvarado, C.S., London, W.B., Look, A.T., Brodeur, G.M., Altmiller, D.H., Thorner, P.S., Joshi, V.V., Rowe, S.T., Nash, M.B., Smith, E.I., Castelberry,R.P., and Cohn, S.L. (2000) Natural history and biology of stage A neuroblastoma: A Pediatric Oncology Group Study. *J. Pediatr. Hematol. Oncol.* **22**, 197–205.
6. Schwab, M., Shimada, H., Joshi, V., and Brodeur, G.M. (2000) Neuroblastic tumours of adrenal gland and sympathetic nervous system. In *World Health Organization Classification of Tumours. Pathology and Genetics of Tumours of the Nervous System* (Kleihues, P., and Cavenee, W.K., eds.), IARC Press, Lyon, France, pp. 153–161.
7. Schwab, M., Westermann, F., Hero, B., and Berthold, F. (2003) Neuroblastoma: biology and molecular and chromosomal pathology. *Lancet Oncol.* **4**, 472–480.
8. Weiss, S.W., and Goldblum, J.R. (2001) *Enzinger and Weiss's Soft Tissue Tumors*, 4th ed. Mosby, St. Louis, MO.
9. Westermann, F., and Schwab, M. (2002) Genetic parameters of neuroblastomas. *Cancer Lett.* **184**, 127–147.
10. Miettinen, M. (2003) Neuroblastoma. Miettinen, M. (2003) Nerve sheath tumors. In *Diagnostic Soft Tissue Pathology*, Churchill Livingstone, Philadelphia, PA, pp.440–460.
11. Weinstein, J.L., Katzenstein, H.M., and Cohn, S.L. (2003) Advances in the diagnosis and treatment of neuroblastoma. *Oncologist* **8**, 278–292.
12. Kushner, B.H., and Cheung, N.-K.V. (1996) Allelic loss of chromosome 1p in neuroblastoma. *N. Engl. J. Med.* **334**, 1608.
13. Ambros, I.M., Amann, G., and Ambros, P.F. (2002) Correspondence re: J. Mora *et al.*, Neuroblastic and schwannian stromal cells of neuroblastoma are derived from a tumoral progenitor cell. *Cancer Res.* **61**, 6892–6898, 2001. *Cancer Res.* **62**, 2986–2989.
14. Brodeur, G.M. (1990) Neuroblastoma: clinical significance of genetic abnormalities. *Cancer Surv.* **9**, 673–688.
15. Shimada, H., Bonadio, J., and Sano, H. (2005) Neublatoma and related tumors. *Pathol. Case Rev.* **10**, 252–256.
16. Olshan, A.F., and Bunin, G.R. (2000) Epidemiology of neuroblastoma. In *Neuroblastoma* (Brodeur, G.M., Sawada, T., Tsuchida, Y., and Voûte, P.A., eds.), Elsevier Science B.V., Amsterdam, The Netherlands, pp. 33–39.

17. Gurney, J.G., Davis, S., Severson, R.K., Fang, J.V., Ross, J.A., and Robinson, L.L. (1996) Trends in cancer incidence among chidren in the U.S. *Cancer* **78,** 532–541.

18. Cotterill, S.J., Pearson, A.D., Pritchard, J., Foot, A.B., Roald, B., Kohler, J.A., and Imeson, J. (2000) Clinical prognostic factors in 1277 patients with neuroblastoma: results of The European Neuroblastoma Study Group 'Survey' 1982–1992. *Eur. J. Cancer* **36,** 901–908.

19. London, W.B., Boni, L., Simon, T., Berthold, F., Twist, C., Schmidt, M.L., Castleberry, R.P., Matthay, K.K., Cohn, S.L., and De Bernardi, B. (2005) The role of age in neuroblastoma risk stratification: the German, Italian, and Children's Oncology Group perspectives. *Cancer Lett.* **228,** 257–266.

20. London, W.B., Castleberry, R.P., Matthay, K.K., Look, A.T., Seeger, R.C., Shimada, H., Thorner, P., Brodeur, G., Maris, J.M., Reynolds, C.P., and Cohn, S.L. (2005) Evidence for an age cutoff greater than 365 days for neuroblastoma risk group stratification in the Children's Oncology Group. *J. Clin. Oncol.* **2 3,** 6459–6465.

21. McArdle, L., McDermott, M., Purcell, R., Grehan, D., O'Meara, A.O., Breatnach, F., Catchpoole, D., Culhane, A.C., Jeffery, I., Gallagher, W.M., and Stallings, R.L. (2004) Oligonucleotide microarray analysis of gene expression in neuroblastoma displaying loss of chromosome 11q. *Carcinogenesis* **25,** 1599–1609.

22. Brodeur, G.M., and Nakagawara, A. (1992) Molecular basis of clinical heterogeneity in neuroblastoma. *Am. J. Pediatr. Hematol. Oncol.* **14,** 111–116.

23. Brodeur, G.M. (1995) Molecular basis for heterogeneity in human neuroblastomas. *Eur. J. Cancer* **31A,** 505–510.

24. Brodeur, G.M. (2002) Significance of intratumoral genetic heterogeneity in neuroblastomas. *Med. Pediatr. Oncol.* **38,** 112–113.

25. Evans, A.E. (2000) Neuroblastoma: a historial perspective 1864–1998. In *Neuroblastoma* (Brodeur, G.M., Sawada, T., Tsuchida, Y., and Voûte, P.A., eds.), Elsevier Science B.V., Amsterdam, The Netherlands, pp. 1–7.

26. Brodeur, G.M., Pritchard, J., Berthold, F., Carlsen, N.L.T., Castel, V., Castleberry, R.P., De Bernardi, B., Evans, A.E., Favrot, M., Hedborg, F., Kaneko, M., Kemshead, J., Lampert, F., Lee, R.E.J., Look, A.T., Pearson, A.D.J., Philip, T., Roald, B., Sawada, T., Seeger, R.C., Tsuchida, Y., and Voute, P.A. (1993) Revisions of the international criteria for neuroblastoma diagnosis, staging, and response to treatment. *J. Clin. Oncol.* **11,** 1466–1477.

27. Shimada, H., Ambros, I.M., Dehner, L.P., Hata, J., Joshi, V.V., and Roald, B. (1999) Terminology and morphologic criteria of neuroblastic tumors. Recommendations by the International Neuroblastoma Pathology Committee. *Cancer* **86,** 349–363.

28. Shimada, H., Ambros, I.M., Dehner, L.P., Hata, J., Joshi, V.V., Roald, B., Stram, D.O., Gerbing, R.B., Lukens, J.N., Matthay, K.K., and Castleberry, R.P. (1999) The International Neuroblastoma Pathology Classification (the Shimada System). *Cancer* **86,** 364–372.

29. Joshi, V.V., Cantor, A.B., Altshuler, G., Larkin, E.W., Neill, J.S.A, Shuster, J.J., Holbrook, C.T., Hayes, F.A., and Castleberry, R.P. (1992) Recommendations for modification of terminology of neuroblastic tumors and prognostic significance of Shimada Classification. A clinicopathologic study

of 213 cases from the Pediatric Oncology Group. *Cancer* **69,** 2183–2196.

30. Joshi, V.V., Cantor, A.B., Altshuler, G., Larkin, E.W., Neill, J.S.A, Shuster, J.J., Holbrook, C.T., Hayes, F.A., Nitschke, R., Duncan, M.H., Shochat, S.J., Talbert, J., Smith, E.I., and Castleberry, R.P. (1992) Age-linked prognostic categorization based on a new histologic grading system of neuroblastomas. A clinicopathologic study of 211 cases from the Pediatric Oncology Group. *Cancer* **69,** 2197–2211.

31. Joshi, V.V., Rao, P.V., Cantor, A.B., Altshuler, G., Shuster, J.J., and Castleberry, R.P. (1996) Modified histologic grading of neuroblastomas by replacement of mitotic rate with mitosis karyorrhexis index. A clinicopathologic study of 223 cases from the Pediatric Oncology Group. *Cancer* **77,** 1582–1588.

32. Joshi, V.V. (2000) Peripheral neuroblastic tumors: pathologic classification based on recommendations of International Neuroblastoma Pathology Committee (modification of Shimada classification). *Pediatr. Dev. Pathol.* **3,** 184–199.

32A. Peuchmaur, M., d'Amore, E.S.G., Joshi, V.V., Hata, J., Roald, B., Dehner, L.P., Gerbing, R.B., Stram, D.O., Lukens, J.N., Matthay, K.K., and Shimada, H. (2003) Revision of the International Neuroblastoma Pathology Classification. Confirmation of favorable and unfavorable prognostic subsets in ganglioneuroblastoma, nodular. *Cancer* **98,** 2274–2281.

33. Brodeur, G.M., Seeger, R.C., Barrett, A., Berthold, F., Castleberry, R.P., D'Angio, G., De Bernardi, B., Evans, A.E., Favrot, M., Freeman, A.I., Haase, G., Hartmann, O., Hayes, F.A., Helson, L., Kemshead, J., Lampert, F., Ninane, J., Ohkawa, H., Philip, T., Pinkerton, C.R., Pritchard, J., Sawada, T., Siegel, S., Smith, E.I., Tsuchida, Y., and Voute, P.A. (1988) International criteria for diagnosis, staging, and response to treatment in patients with neuroblastoma. *J. Clin. Oncol.* **6,** 1874–1881.

34. Bowman, L.C., Castleberry, R.P., Cantor, A., Joshi, V., Cohn, S.L., Smith, E.I., Yu, A., Brodeur, G.M., Hayes, F.A., and Look, A.T. (1997) Genetic staging of unresectable or metastatic neuroblastoma in infants: a pediatric oncology group study. *J. Natl. Cancer Inst.* **89,** 373–380.

35. Matthay, K.K., Villablanca, J.G., Seeger, R.C., Stram, D.O., Harris, R.E., Ramsay, N.K., Swift, P., Shimada, H., Black, C.T., Brodeur, G.M., Gerbing, R.B., and Reynolds, C.P. for the Children's Cancer Group. (1999) Treatment of high-risk neuroblastoma with intensive chemotherapy, radiotherapy, autologous bone marrow transplantation, and 13-*CIS*-retinoic acid. *N. Engl. J. Med.* **341,** 1165–1173.

36. Perez, C.A., Matthay, K.K., Atkinson, J.B., Seeger, R.C., Shimada, H., Haase, G.M., Stram, D.O., Gerbing, R.B., and Lukens, J.N. (2000) Biologic variables in the outcome of stages I and II neuroblastoma treated with surgery as primary therapy: a Children's Cancer Group Study. *J. Clin. Oncol.* **18,** 18–26.

37. Simon, T., Spitz, R., Faldum, A., Hero, B., and Berthold, F. (2004) New definition of low-risk neuroblastoma using stage, age, and 1p and *MYCN* status. *J. Pediatr. Hematol. Oncol.* **26,** 791–796.

38. Pritchard, J., Cotterill, S.J., Germond, S.M., Imeson, J., de Kraker, J., and Jones, D.R. (2005) High dose melphalan in the treatment of advanced neuroblastoma: results of a

randomized trial (ENSG-1) by the European Neuroblastoma Study Group. *Pediatr. Blood Cancer* **44**, 348–357.

39. Brodeur, G.M., and Maris, J.M. (2002) Neuroblastoma. In *Principles and Practice of Pediatric Oncology*, 4th ed. (Pizzo, P.A., and Poplack, D.G., eds.), Lippincott Williams & Wilkins, Philadelphia, PA, pp. 865–937.

40. O'Hanlon, L.H. (2006) Genetic deletions help determine neuroblastoma prognosis. *Lancet Oncol.* **7**, 18.

41. Sandberg, A.A., and Bridge, J.A. (1994) *The Cytogenetics of Bone and Soft Tissue Tumors*. R.G. Landes Company, Austin/Georgetown.

42. Sandberg, A.A., and Bridge, J.A. (2000) Updates on cytogenetics and molecular genetics of bone and soft tissue tumors: Ewing sarcoma and peripheral primitive neuroectodermal tumors. *Cancer Genet. Cytogenet.* **123**, 1–26.

43. Oppenheimer, O., Alaminos, M., and Gerald, W.L. (2003) Genomic medicine and neuroblastoma. *Expert Rev. Mol. Diagn.* **3**, 39–54.

44. Maris, J.M. (2005) The biologic basis for neuroblastoma heterogeneity and risk stratification. *Curr. Opin. Pediatr.* **17**, 7–13.

44A. Evans, A.E., D'Angio, G.J., Propert, K., Anderson, J., and Hann, H.-W.L. (1987) Prognostic factors in neuroblastoma. *Cancer* **59**, 1853–1859.

45. Maris, J.M., and Matthay, K.K. (1999) Molecular biology of neuroblastoma. *J. Clin. Oncol.* **17**, 2264–2279.

46. Lastowska, M., Cotterill, S., Pearson, A.D.J., Roberts, P., McGuckin, A., Lewis, I., and Bown, N. On behalf of the U.K. Children's Cancer Study Group and the U.K. Cancer Cytogenetics Group. (1997) Gain of chromosome arm 17q predicts unfavourable outcome in neuroblastoma patients. *Eur. J. Cancer* **33**, 1627–1633.

47. Abel, F., Ejeskär, K., Kogner, P., and Martinsson, T. (1999) Gain of chromosome arm 17q is associated with unfavourable prognosis in neuroblastomas, but does not involve mutations in the somatostatin receptor 2 (SSTR2) gene at 17q24. *Br. J. Cancer* **81**, 1402–1409.

48. Bown, N., Cotterill, S., Lastowska, M., O'Neill, S., Pearson, A.D.J., Plantaz, D., Meddeb, M., Danglot, G., Brinkschmidt, C., Christiansen, H., Laureys, G., and Speleman, F. (1999) Gain of chromosome arm 17q and adverse outcome in patients with neuroblastoma. *N. Engl. J. Med.* **340**, 1954–1961.

49. Bown, N., Lastowska, M., Cotterill, S., O'Neill, S., Ellershaw, C., Roberts, P., Lewis, I., Pearson, A.D., U.K. Cancer Cytogenetics Group and the United Kingdom Children's Cancer Study Group. (2001) 17q gain in neuroblastoma predicts adverse clinical outcome. U.K. Cancer Cytogenetics Group and the U.K. Children's Cancer Study Group. *Med. Pediatr. Oncol.* **36**, 14–19.

50. Attiyeh, E.F., London, W.B., Mossé, Y.P., Wang, Q., Winter, C., Khazi, D., McGrady, P.W., Seeger, R.C., Look, A.T., Shimada, H., Brodeur, G.M., Cohn, S.L., Matthay, K.K., Maris, J.M., for the Children's Oncology Group. (2005) Chromosome 1p an 11q deletions and outcome in neuroblastoma. *N. Engl. J. Med.* **353**, 2243–2253.

51. Look, A.T., Hayes, F.A., Nitschke, R., McWilliams, N.B., and Green, A.A. (1984) Cellular DNA content as predictor of response to chemotherapy in infants with unresectable neuroblastoma. *N. Engl. J. Med.* **311**, 231–235.

52. Look, A.T., Hayes, F.A., Shuster, J.J., Douglass, E.C., Castleberry, R.P., Bowman, L.C., Smith, E.I., and Brodeur, G.M. (1991) Clinical relevance of tumor cell ploidy and N-*myc* gene amplification in childhood neuroblastoma: a Pediatric Oncology Group Study. *J. Clin. Oncol.* **9**, 581–591.

53. Hayashi, Y., Inaba, T., Hanada, R., and Yamamoto, K. (1988) Chromosome findings and prognosis in 15 patients with neuroblastoma found by VMA mass screening. *J. Pediatr.* **112**, 567–571.

54. Kaneko, Y., Kanda, N., Maseki, N., Sakurai, M., Tsuchida, Y., Takeda, T., Okabe, I., and Sakurai, M. (1987) Different karyotypic patterns in early and advance stage neuroblastomas. *Cancer Res.* **47**, 311–318 .

55. Kaneko, Y., Maseki, N., Sakurai, M., Okabe, I., Takeda, T., Sakurai, M., and Kanda, N. (1988) Chromosomes and screening for neuroblastoma. *Lancet* **1**, 174–175.

56. Ikeda, H., and Cohn, S.L. (2000) Ploidy and cytogenetics of neuroblastoma. In *Neuroblastoma* (Brodeur, G.M., Sawada, T., Tsuchida, Y., and Voûte, P.A., eds.), Elsevier Science B.V., Amsterdam, The Netherlands, pp. 41–56.

57. George, R.E., London, W.B., Cohn, S.L., Maris, J.M., Kretschmar, C., Diller, L., Brodeur, G.M., Castleberry, R.P., and Look, A.T. (2005) Hyperdiploidy plus nonamplified *MYCN* confers a favorable prognosis in children 12 to 18 months old without disseminated neuroblastoma: a Pediatric Oncology Group Study. *J. Clin. Oncol.* **23**, 6466–6473.

58. Hayashi, Y., Hanada, R., Yamamoto, K., and Bessho, F. (1987) Chromosome findings and prognosis in neuroblastomas. *Cancer Genet. Cytogenet.* **29**, 175–177.

59. Schwab, M., Praml, C., and Amler, L.C. (1996) Genomic instability in 1p and human malignancies. *Genes Chromosomes Cancer* **16**, 211–229.

60. Christiansen, H., and Lampert, F. (1988) Tumour karyotype discriminates between good and bad prognostic outcome in neuroblastoma. *Br. J. Cancer* **57**, 121–126.

61. Burchill, S.A., Wheeldon, J., Cullinane, C., and Lewis, I.J. (1997) EWS-FLI1 fusion transcripts identified in patients with typical neuroblastoma. *Eur. J. Cancer* **33**, 239–243.

62. Janoueix-Lerosey, I., Hupe, P., Maciorowski, Z., La Rosa, P., Schleiermacher, G., Pierron, G., Liva, S., Barillot, E., and Delattre, O. (2005) Preferential occurrence of chromosome breakpoints within early replicating regions in neuroblastoma. *Cell Cycle* **4**, 1842–1846.

63. Gehring, M., Berthold, F., Schwab, M., and Amler, L. (1994) Genetic stability of microsatellites in human neuroblastomas. *Int. J. Oncol.* **4**, 1043–1045.

64. Trakhtenbrot, L., Cohen, N., Rosner, E., Gipsh, N., Brok-Simoni, F., Mandel, M., Amariglio, N., and Rechavi, G. (1999) Coexistence of several unbalanced translocations in a case of neuroblastoma: the contribution of multicolor spectral karyotyping. *Cancer Genet. Cytogenet.* **112**, 119–123.

65. Glaser-Gabay, L., Jeison, M., Stark, B., Finkelstein, S., Katzin, N., Stein, J., Zaizov, R., Yaniv, I., and Bar-Am, I. (2001) Hidden translocations revealed by spectral karyotyping (SKY) in neuroblastoma tumors. *Cancer Genet. Cytogenet.* **128**, 68.

66. Van Roy, N., Van Limbergen, H., Vandesompele, J., Van Gele, M., Poppe, B., Salwen, H., Laureys, G., Manoel, N., De Paepe, A., and Speleman, F. (2001) Combined M-FISH and CGH analysis allows comprehensive description

of genetic alterations in neuroblastoma cell lines. *Genes Chromosomes Cancer* **32**, 126–135.

67. Stark, B., Jeison, M., Bar-Am, I., Glaser-Gabay, L., Mardoukh, J., Luria, D., Feinmesser, M., Goshen, Y., Stein, J., Abramov, A., Zaizov, R., and Yaniv, I. (2002) Distinct cytogenetic pathways of advanced-stage neuroblastoma tumors, detected by spectral karyotyping. *Genes Chromosomes Cancer* **34**, 313–324.

68. Stark, B., Jeison, M., Glaser-Gabay, L., Bar-Am, I., Mardoukh, J., Ash, S., Atias, D., Stein, J., Zaizov, R., and Yaniv, I. (2003) der(11)t(11;17): a distinct cytogenetic pathway of advanced stage neuroblastoma (NBL) detected by spectral karyotyping (SKY). *Cancer Lett.* **197**, 75–79.

69. Vernole, P., Concato, C., Pianca, C., Nicoletti, B., and Melino, G. (1988) Association of cytogenetic abnormalities in a neuroblastoma and fragile sites expression. *Br. J. Cancer* **58**, 287–291.

70. Cox, D., Yuncken, C., and Spriggs, A.I. (1965) Minute chromatin bodies in malignant tumours of childhood. *Lancet* ii, 55–58.

71. Levan, A., Manolov, G., and Clifford, P. (1968) The chromosomes of a human neuroblastoma: a new case with accessory minute chromosomes. *J. Natl. Cancer Inst.* **41**, 1377–1387.

72. Sandberg, A.A., Sakurai, M., and Holdsworth, R.N. (1972) Chromosomes and causation of human cancer and leukemia. VIII. DMS chromosomes in a neuroblastoma. *Cancer* **29**, 1671–1679.

73. Biedler, J.L., and Spengler, B.A. (1976) A novel chromosome abnormality in human neuroblastoma and antifolate-resistant Chinese hamster cell lines in culture. *J. Natl. Cancer Inst.* **57**, 683–695.

74. Biedler, J.L., and Spengler, B.A. (1976) Metaphase chromosome anomaly: association with drug resistance and cell-specific products. *Science* **191**, 185–187.

75. Schwab, M., Alitalo, K., Klempnauer, K.-H., Varmus, H.E., Bishop, J.M., Gilbert, F., Brodeur, G., Goldstein, M., and Trent, J. (1983) Amplified DNA with limited homology to *myc* cellular oncogene is shared by human neuroblastoma cell lines and a neuroblastoma tumour. *Nature* **305**, 245–248.

76. Schwab, M., Varmus, H.E., Bishop, J.M., Grzeschik, K.H., Naylor, S.L., Sakaguchi, A.Y., Brodeur, G., and Trent, J. (1984) Chromosome localization in normal human cells and neuroblastomas of a gene related to c-myc. *Nature* **308**, 288–291.

77. Amler, L.C., and Schwab, M. (1989) Amplified N-*myc* in human neuroblastoma cells is often arranged as clustered tandem repeats of differently recombined DNA. *Mol. Cell. Biol.* **9**, 4903–4913.

78. Amler, L.C., Shibasaki, Y., Savelyeva, L., and Schwab, M. (1992) Amplification of the N-myc gene in human neuroblastomas: tandemly repeated amplicons within homogeneously staining regions on different chromosomes with the retention of single copy gene at the resident site. *Mutat. Res.* **276**, 291–297.

79. Corvi, R., Savelyeva, L., and Schwab, M. (1997) Patterns of oncogene activation in human neuroblastoma cells. *J. Neurooncol.* **31**, 25–31.

80. Savelyeva, L., and Schwab, M. (2001) Amplification of oncogenes revisited: from expression profiling to clinical application. *Cancer Lett.* **167**, 115–123.

81. Moreau, L.A., McGrady, P., London, W.B., Shimada, H., Cohn, S.L., Maris, J.M., Diller, L., Look, A.T., and George, R.E. (2006) Does *MYCN* amplification manifested as homogeneously staining regions at diagnosis predict a worse outcome in children with neuroblastoma? A Chidlren's Oncology Group Study. *Clin. Cancer Res.* **12**, 5693–5697.

82. Ambros, I.M., Rumpler, S., Luegmayr, A., Hattinger, C.M., Strehl, S., Kovar, H., Gadner, H., and Ambros, P.F. (1997) Neuroblastoma cells can actively eliminate supernumerary MYCN gene copies by micronucleus formation—sign of tumour cell revertance? *Eur. J. Cancer* **33**, 2043–2049.

83. Valent, A., Bénard, J., Clausse, B., Barrois, M., Valteau-Couanet, D., Terrier-Lacombe, M.-J., Spengler, B., and Bernheim, A. (2001) *In vivo* elimination of acentric double minutes containing amplified *MYCN* from neuroblastoma tumor cells through the formation of micronuclei. *Am. J. Pathol.* **158**, 1579–1584.

84. Yoshimoto, M., Caminada de Toledo, S.R., Monteiro Caran, E.M., de Seixas, M.T., de Martino Lee, M.L., de Campos Vieira Abib, S., Rossi Vianna, S.M., Schettini S.T., and Duffles Andrade, J.A. (1999) MYCN gene amplification. Identificaiton of cell populations containing double minutes and homogeneously staining regions in neuroblastoma tumors. *Am. J. Pathol.* **155**, 1439–1443.

85. Brodeur, G.M., Green, A.A., Hayes, F.A., Williams, K.J., Williams, D., and Tsiatis, A.A. (1981) Cytogenetic features of human neuroblastomas and cell lines. *Cancer Res.* **41**, 4678–4686.

86. Solovei, I., Kienle, D., Little, G., Eils, R., Savelyeva, L., Schwab, M., Jäger, W., Cremer, C., and Cremer, T. (2000) Topology of double minutes (dmins) and homogeneously staining regions (HSRs) in nuclei of human neuroblastoma cell lines. *Genes Chromosomes Cancer* **29**, 297–308.

87. Manuelidis, L. (1985) Individual interphase chromosome domains revealed by in situ hybridization. *Hum. Genet.* **71**, 288–293.

88. Cremer, T., Kurz, A., Zirbel, R., Dietzel, S., Rinke, B., Schröck, E., Speicher, M.R., Mathieu, U., Jauch, A., Emmerich, P., Scherthan, H., Ried, T., Cremer, C., and Lichter P. (1993) Role of chromosome territories in the functional compartmentalization of the cell nucleus. *Cold Spring Harb. Symp. Quant. Biol.* **58**, 777–792.

89. Lichter, P., Cremer, T., Borden, J., Manuelidis, L., and Ward, D.C. (1988) Delineation of individual human chromosomes in metaphase and interphase cells by in situ suppression hybridization using recombinant DNA libraries. *Hum. Genet.* **80**, 224–234.

90. Pinkel, D., Landegent, J., Collins, C., Fuscoe, J., Segraves, R., Lucas, J., and Gray, J. (1988) Fluorescence in situ hybridization with human chromosome-specific libraries: detection of trisomy 21 and translocations of chromosome 4. *Proc. Natl. Acad. Sci. USA* **85**, 9138–9142.

91. Spector, D.L. (1996) Nuclear organization and gene expression. *Exp. Cell Res.* **229**, 189–197.

92. Lamond, A.I., and Earnshaw, W.C. (1998) Structure and function in the nucleus. *Science* **280**, 547–535.

93. Misteli, T., Caceres, J.F., and Spector, D.L. (1997) The dynamics of a pre-mRNA splicing factor in living cells. *Nature* **387**, 523–527.

94. Misteli, T., and Spector, D.L. (1998) The cellular organization of gene expression. *Curr. Opin. Cell. Biol.* **10**, 323–331.

95. Altura, R.A., Maris, J.M., Li, H., Boyett, J.M., Brodeur, G.M., and Look, A.T. (1997) Novel regions of chromosomal loss in familial neuroblastoma by comparative genomic hybridization. *Genes Chromosomes Cancer* **19**, 176–184.

96. Brinkschmidt, C., Christiansen, H., Terpe, H.J., Simon, R., Boecker, W., Lampert, F., and Stoerkel, S. (1997) Comparative genomic hybridization (CGH) analysis of neuroblastomas—an important methodololgical approach in paediatric tumour pathology. *J. Pathol.* **181**, 394–400.

97. Brinkschmidt, C., Poremba, C., Christiansen, H., Simon, R., Schäfer, K.L., Terpe, H.J., Lampert, F., Boecker, W., Dockhorn-Dworniczak, B. (1998) Comparative genomic hybridization and telomerase activity analysis identify two biologically different groups of 4s neuroblastoma. *Br. J. Cancer* **77**, 2223–2229.

98. Lastowska, M., Nacheva, E., McGuckin, A., Curtis, A., Grace, C., Pearson, A., and Bown, N. on behalf of the United Kingdom Children's Cancer Study Group. (1997) Comparative genomic hybridization study of primary neuroblastoma tumors. *Genes Chromosomes Cancer* **18**, 162–169.

99. Plantaz, D., Mohapatra, G., Matthay, K.K., Pellarin, M., Seeger, R.C., and Feuerstein, B.G. (1997) Gain of chromosome 17 is the most frequent abnormality detected in neuroblastoma by comparative genomic hybridization. *Am. J. Pathol.* **150**, 81–89.

100. Szymas, J., Wolf, G., Kowalczyk, D., Nowak, S., and Petersen, I. (1997) Olfactory neuroblastoma: detection of genomic imbalances by comparative genomic hybridization. *Acta Neurochir. (Wien)* **139**, 839–844.

101. Van Gele, M., Van Roy, N., Jauch, A., Laureys, G., Benoit, Y., Schelfhout, V., De Potter, C.R., Brock, P., Uyttebroeck, A., Sciot, R., Schuuring, E., Versteeg, R., and Speleman, F. (1997) Sensitive and reliable detection of genomic imbalances in human neuroblastomas using comparative genomic hybridisation analysis. *Eur. J. Cancer* **33**, 1979–1982.

102. Breen, C.J., O'Meara, A., McDermott, M., Mullarkey, M., and Stallings, R.L. (2000) Coordinate deletion of chromosome 3p and 11q in neuroblastoma detected by comparative genomic hybridization. *Cancer Genet. Cytogenet.* **120**, 44–49.

103. Cunsolo, C.L., Bicocchi, M.P., Petti, A.R., and Tonini, G.P. (2000) Numerical and structural aberrations in advanced neuroblastoma tumours by CGH analysis: survival correlates with chromosome 17 status. *Br. J. Cancer* **83**, 1295–1300.

104. Vandesompele, J., Van Roy, N., Van Gele, M., Laureys, G., Ambros, P., Heimann, P., Devalck, C., Schuuring, E., Brock, P., Otten, J., Gyselinck, J., De Paepe, A., and Speleman, F. (1998) Genetic heterogeneity of neuroblastoma studied by comparative genomic hybridization. *Genes Chromosomes Cancer* **23**, 141–152.

105. Hirai, M., Yoshida, S., Kashiwagi, H., Kawamura, T., Ishikawa, T., Kaneko, M., Ohkawa, H., Nakagawara, A., Miwa, M., and Uchida, K. (1999) 1q23 gain is associated with progressive neuroblastoma resistant to aggressive treatment. *Genes Chromosomes Cancer* **25**, 261–269.

106. Vettenranta, K., Aalto, Y., Wikström, S., Knuutila, S., Saarinen-Pihkala, U. (2001) Comparative genomic hybridization reveals changes in DNA-copy number in poor-risk neuroblastoma. *Cancer Genet. Cytogenet.* **125**, 125–130.

107. Plantaz, D., Vandesompele, J., Van Roy, N., Lastowska, M., Bown, N., Combaret, V., Favrot, M.C., Delattre, O., Michon, J., Bénard, J., Hartmann, O., Nicholson, J.C., Ross, F.M., Brinkschmidt, C., Laureys, G., Caron, H., Matthay, K.K., Feuerstein, B.G., and Speleman. F. (2001) Comparative genomic hybridization (CGH) analysis of stage 4 neuroblastoma reveals high frequency of 11q deletion in tumors lacking *MYCN* amplification. *Int. J. Cancer* **91**, 680–686.

108. Iehara, T., Hamazaki, M., and Sawada, T. (2002) Cytogenetic analysis of infantile neuroblastomas by comparative genomic hybridization. *Cancer Lett.* **178**, 83–89.

109. Schleiermacher, G., Janoueix-Lerosey, I., Combaret, V., Derré, J., Couturier, J., Aurias, A., and Delattre, O. (2003) Combined 24-color karyotyping and comparative genomic hybridization analysis indicates predominant rearrangements of early replicating chromosome regions in neuroblastoma. *Cancer Genet. Cytogenet.* **141**, 32–42.

110. Brodeur, G.M., and Hogarty, M.H. (1998) Oncogene amplification in human cancers. In: *The Genetic Basis of Human Cancer* (Vogelstein, B., and Kinzler K., eds.), McGraw Hill, New York, pp. 161–172.

111. Brinkschmidt, C., Christiansen, H., Terpe, H.J., Simon, R., Lampert, F., Boecker, W., and Dockhorn-Dworniczak, B. (2001) Distal chromosome 17 gains in neuroblastoma detected by comparative genomic hybridization (CGH) are associated with a poor clinical outcome. *Med. Pediatr. Oncol.* **36**, 11–13.

112. Morrison, C. (2006) Fluorescent in situ hybridization and array comparative genomic hybridization. Complementary techniques for genomic evaluation. *Arch. Pathol. Lab. Med.* **130**, 967–974.

113. Pollack, J.R., Perou, C.M., Alizadeh, A.A., Eisen, M.B., Pergamenschikov, A., Williams, C.F., Jeffrey, S.S., Botstein, D., and Brown, P.O. (1999) Genome-wide analysis of DNA copy-number changes using cDNA microarrays. *Nat. Genet.* **23**, 41–46.

114. Daigo. Y., Chin., S.-F., Gorringe, K.L., Bobrow, L.G., Ponder, B.A.J., Pharoah, P.D.P., and Caldas, C. (2001) Degenerate oligonucleotide primed-polymerase chain reaction-based array comparative genomic hybridization for extensive amplicon profiling of breast cancers. A new approach for the molecular analysis of paraffin-embedded cancaer tissue. *Am. J. Pathol.* **158**, 1623–1631.

115. Kashiwagi, H., Kanai, M., Yoshida, S., Miwa, M., and Uchida, K. (2000) Analysis of DNA copy number alterations in neuroblastomas by array-based CGH. *Cancer Genet. Cytogenet.* **123**, 81.

116. Yamanaka, Y., Hamazaki, Y., Sato, Y., Ito, K., Watanabe, K., Heike, T., Nakahata, T., and Nakamura, Y. (2002) Maturational sequence of neuroblastoma revealed by molecular analysis on cDNA microarrays. *Int. J. Oncol.* **21**, 803–807.

117. Hackett, C.S., Hodgson, J.G., Law, M.E., Fridlyand, J., Osoegawa, K., de Jong, P.J., Nowak, N.J., Pinkel, D., Albertson, D.G., Jain, A., Jenkins, R., Gray, J.W., and Weiss, W.A. (2003) Genome-wide array CGH analysis of murine neuroblastoma reveals distinct genomic aberrations which parallel those in human tumors. *Cancer Res.* **63**, 5266–5273.

118. Beheshti, B., Braude, I., Marrano, P., Thorner, P., Zielenska, M., and Squire, J.A. (2003) Chromosomal localization of DNA amplifications in neuroblastoma tumors using cDNA microarray comparative genomic hybridization. *Neopolasia* **5,** 53–62.

119. Fix, A., Peter, M., Pierron, G., Aurias, A., Delattre, O., and Janoueix-Lerosey, I. (2004) High-resolution mapping of amplicons of the short arm of chromosome 1 in two neuroblastoma tumors by microarray-based comparative genomic hybridization. *Genes Chromosomes Cancer* **40,** 266–270.

120. Chen, Q.-R., Bilke, S., Wei, J.S., Whiteford, C.C., Cenacchi, N., Krasnoselsky, A.L., Greer, B.T., Son, C.-G., Westermann, F., Berthold,, F., Schwab, M., Catchpoole, D., and Khan, J. (2004) cDNA array-CGH profiling identifies genomic alterations specific to stage and *MYCN*-amplification in neuroblastoma. *BMC Genomics* **5,** 70–82.

121. Chen, Q.R., Bilke, S., and Khan, J. (2005) High-resolution cDNA microarray-based comparative genomic hybridization analysis in neuroblastoma. *Cancer Lett.* **228,** 71–81.

122. Spitz, R., Oberthuer, A., Zapatka, M., Brors, B., Hero, B., Ernestus, K., Oestreich, J., Fischer, M., Simon, T., and Berthold, F. (2006) Oligonucleotide array-based genomic hybridization (aCGH) of 90 neuroblastomas reveals aberration patterns closely associated with relapse pattern and outcome. *Genes Chromosomes Cancer* **45,** 1130–1142.

123. Selzer, R.R., Richmond, T.A., Pofahl, N.J., Green, R.D, Eis, P.S., Nair, P., Brothman, A.R., and Stallings, R.L. (2005) Analysis of chromosome breakpoints in neuroblastoma at sub-kilobase resolution using fine-tiling oligonucleotide array CGH. *Genes Chromosomes Cancer* **44,** 305–319.

124. Lázcoz, P., Muñoz, J., Nistal, M., Pestaña, Á., Encío, I.J., and Castresana, J.S. (2007) Loss of heterozygosity and microsatellite instability on chromosome arm 10q in neuroblastoma. *Cancer Genet. Cytogenet.* **174,** 1–8.

125. Schwab, M. (2002) Amplified MYCN in human neuroblastoma: paradigm for the translation of molecular genetics to clinical oncology. *Am. N.Y. Acad. Sci.* **963,** 67–73.

126. Scott, D., Elsden, J., Pearson, A., and Lunec, J. (2003) Genes co-amplified with *MYCN* in neuroblastoma: silent passengers or co-determinants of phenotype? *Cancer Lett.* **197,** 81–86.

127. Scott, D.K., Board, J.R., Lu, X., Pearson, A.D., Kenyon, R.M., and Lunec, J. (2003) The neuroblastoma amplified gene, *NAG*: genomic structure and charaterisation of the 7.3 kb transcript predominantly expressed in neuroblastoma. *Gene* **307,** 1–11.

128. Saito-Ohara, F., Imoto, I., Inoue, J., Hosoi, H., Nakagawara, A., Sugimoto, T., and Inazawa, J. (2003) PPM1D is a potential target for 17q gain in neuroblastoma. *Cancer Res.* **63,** 1876–1883.

129. Guo, C., White, P.S., Weiss, M.J., Hogarty, M.D., Thompson, P.M., Stram, D.O., Gerbing, R., Matthay, K.K., Seeger, R.C., Brodeur, G.M., and Maris, J.M. (1999) Allelic deletion at 11q23 is common in MYCN single copy neuroblastomas. *Oncogene* **18,** 4948–4957.

130. Bauer, A., Savelyeva, L., Claas, A., Praml, C., Berthold, F., and Schwab, M. (2001) Smallest region of overlapping deletion in 1p36 in human neuroblastoma: a 1 Mbp cosmid and PAC contig. *Genes Chromosomes Cancer* **31,** 228–239.

131. Mora, J., Cheung, N.-K.V., Kushner, B.H., LaQuaglia, M.P., Kramer, K., Fazzari, M., Heller, G., Chen, L., and Gerald, W.L. (2000) Clinical categories of neuroblastoma are associated with different patterns of loss of heterozygosity on chromosome arm 1p. *J. Mol. Diagn.* **2,** 37–46.

132. Perri, P., Longo, L., Cusano, R., McConville, C.M., Rees, S.A., Devoto, M., Conte, M., Battista Ferrara, G., Seri, M., Romeo, G., and Tonini, G.P. (2002) Weak linkage at 4p16 to predisposition for human neuroblastoma. *Oncogene* **21,** 8356–8360.

133. Benedict, W.F., Murphree, A.L., Banerjee, A., Spina, C.A., Sparkes, M.C., and Sparkes, R.S. (1983) Patient with 13 chromosome deletion: evidence that the *retinoblastoma* gene is a recessive cancer gene. *Science* **219,** 973–975.

134. Kamb, A., Gruis, N.A., Weaver-Feldhaus, J., Liu, Q., Harshman, K., Tavitgian, S.V., Stockert, E., Day, R.S., III, Johnson, B.E., and Skolnick, M.H. (1994) A cell cycle regulatory potentially involved in genesis of many tumor types. *Science* **264,** 436–440.

135. Thompson, P.M., Maris, J.M., Hogarty, M.D., Seeger, R.C., Reynolds, C.P., Brodeur, G.M., and White P.S. (2001) Homozygous deletion of *CDKN2A* ($p16^{INK4a}/p14^{ARF}$) but not within 1p36 or at other tumor suppressor loci in neuroblastoma. *Cancer Res.* **61,** 679–686.

136. Spitz, R., Hero, B., Ernestus, K., and Berthold, F. (2003) Deletions in chromosome arms 3p and 11q are new prognostic markers in localized and 4s neuroblastoma. *Clin. Cancer Res.* **9,** 52–58.

137. Caron, H., van Sluis, P., Buschman, R., Pereira, do Tanque, R., Maes, P., Beks, L., de Kraker, J., Voûte, P., Vergnaud, G., Westerveld, A., Slater, R., and Versteeg. R. (1996) Allelic loss of the short arm of chromosome 4 in neuroblastoma suggests a novel tumour suppressor gene locus. *Hum. Genet.* **97,** 834–837.

138. Caron, H., van Sluis, P., de Kraker, J., Bökkerink, J., Egeler, M., Laureys, G., Slater, R., Westerveld, A., Voûte, P.A., and Versteeg. R. (1996) Allelic loss of chromosome 1p as a predictor of unfavourable outcome ini patients with neuroblastoma. *N. Engl. J. Med.* **334,** 225–230.

139. Mora, J., Cheung, N.-K.V., Chen, L., Qin, J., and Gerald, W. (2001) Loss of heterozygosity at 19q13.3 is associated with locally aggressive neuroblastoma. *Clin.Cancer Res.* **7,** 1358–1361.

140. Mora, J., Alaminos, M., de Torres, C., Illei, P., Qin, J., Cheung, N.-K., and Gerald, W.L. (2004) Comprehensive analysis of the 9p21 region in neuroblastoma suggests a role for genes mapping to 9p21–23 in the biology of favourable stage 4 tumours. *Br. J. Cancer* **91,** 1112–1118.

141. Thompson, P.M., Seifried, B.A., Kyemba, S.K., Jensen, S.J., Guo, C., Maris, J.M., Brodeur, G.M., Stram, D.O., Seeger, R.C., Gerbing, R., Matthay, K.K., Matise, T.C., and White, P.S. (2001) Loss of heterozygosity for chromosome 14q in neuroblastoma. *Med. Pediatr. Oncol.* **36,** 28–31.

142. Isobe, K., Yashiro, T., Omura, S., Kaneko, M., Kaneko, S., Kamma, H., Tatsuno, I., Takekoshi, K., Kawakami, Y., and Nakai, T. (2004) Expression of the human telomerase reverse transciptase in pheochromocytoma and neuroblastoma tissues. *Endocr. J.* **51,** 47–52.

143. Zhang, A., Zheng, C., Lindvall, C., Hou, M., Ekedahl, J., Lewensohn, R., Yan, Z., Yang, X., Henriksson, M.,

Blennow, E., Nordenskjöld, M., Zetterberg, A., Björkholm, M., Gruber, A., and Xu, D. (2000) Frequent amplification of the *Telomerase Reverse Transcriptase* gene in human tumors. *Cancer Res.* **60**, 6230–6235.

144. Ducrest, A.-L., Szutorisz, H., Lingner, J., and Nabholz, M. (2002) Regulation of the human telomerase reverse transcriptase gene. *Oncogene* **21**, 541–552.

145. Hiyama, E., Hiyama, K., Yokoyama, T., Matsuura, Y., Piatyszek, M.A., and Shay, J.W. (1995) Correlating telomerase activity levels with human neuroblastoma outcomes. *Nat. Med.* **1**, 249–255.

146. Hiyama, E., Hiyama, K., Nishiyama, M., Reynolds, C.P., Shay, J.W., and Yokoyama, T. (2003) Differential gene expression profiles between neuroblastomas with high telomerase activity and low telomerase activity. *J. Pediatr. Surg.* **38**, 1730–1734.

147. Krams, M., Claviez, A., Heidorn, K., Krupp, G., Parwaresch, R., Harms, D., and Rudolph, P. (2001) Regulation of telomerase activity by alternate splicing of human telomerase reverse transciptase mRNA in a subset of neuroblastomas. *Am. J. Pathol.* **159**, 1925–1932.

148. Krams, M., Hero, B., Berthold, F., Parwaresch, R., Harms, D., and Rudolph, P. (2003) Full-length telomerase reverse transcriptase messenger RNA is an independent prognostic factor in neuroblastoma. *Am. J. Pathol.* **162**, 1019–1026.

149. Streutker, C.J., Thorner, P., Fabricius, N., Weitzman, S., and Zielenska, M. (2001) Telomerase activity as a prognostic factor in neuroblastomas. *Pediatr. Dev. Pathol.* **4**, 62–67.

150. Hiyama, E., and Hiyama, K. (2002) Clinical utility of telomerase in cancer. *Oncogene* **21**, 643–649.

150A. Hiyama, E., Hiyama, K., Yokoyama, T., Fukuba, I., Yamaoka, H., Shay, J.W., and Matsuura, Y. (1999) Rapid detection of *MYCN* gene amplification and telomerase expression in neuroblastoma. *Clin. Cancer Res.* **5**, 601–609.

151. Kim, N.W., Piatyszek, M.A., Prowse, K.R., Harley, C.B., West, M.D., Ho, P.L.C., Coviello, G.M., Wright, W.E., Weinrich, S.L., and Shay, J.W. (1994) Specific association of human telomerase activity with immortal cells and cancer. *Science* **266**, 2011–2015.

152. Brodeur, G.M., Azar, C., Brother, M., Hiemstra, J., Kaufman, B., Marshall, H., Moley, J., Nakagawara, A., Saylors, R., Scavarda, N., Schneider, S., Wasson, J., White, P., Seeger, R., Look, T., and Castleberry R. (1992) Effect of genetic factors on prognosis and treatment. *Cancer* **70**, 1685–1694.

153. Reynolds, C.P., Zuo, J.J., Kim, N.W., Wang, H., Lukens, J.N., Matthay, K.K., and Seeger, R.C. (1997) Telomerase expression in primary neuroblastomas. *Eur. J. Cancer* **33**, 1929–1931.

154. Maitra, A., Yashima, K., Rathi, A., Timmons, C.F., Barton Rogers, B., Shay, J.W., and Gazdar A.F. (1999) The RNA component of telomerase as a marker of biologic potential and clinical outcome in childhood neuroblastic tumors. *Cancer* **85**, 741–749.

155. Hiyama, E., and Reynolds, C.P. (2000) Telomerase as a biological and prognostic marker in neuroblastoma. In *Neuroblastoma* (Brodeur, G.M., Sawada, T., Tsuchida, Y., and Voûte, P.A., eds.), Elsevier Science B.V., Amsterdam, The Netherlands, pp. 159–174.

156. Nozaki, C., Horibe, K., Iwata, H., Ishiguro, Y., Hamaguchi, M., and Takahashi, M. (2000) Prognostic impact of telomerase activity in patients with neuroblastoma. *Int. J. Oncol.* **17**, 341–345.

157. Poremba, C., Scheel, C., Hero, B., Christiansen, H., Schaefer, K.-L., Nakayama, J., Berthold, F., Juergens, H., Boecker, W., and Dockhorn-Dworniczak, B. (2000) Telomerase activity and telomerase subunits gene expression patterns in neuroblastoma: a molecular and immunohistochemical study establishing prognostic tools for fresh-frozen and paraffin-embedded tissues. *J. Clin. Oncol.* **18**, 2582–2592.

158. Poremba, C., Hero, B., Heine, B., Scheel, C., Schaefer, K.L., Christiansen, H., Berthold, F., Kneif, S., Stein, H., Juergens, H., Boecker, W., and Dockhorn-Dworniczak, B. (2000) Telomeras is a strong indicator for assessing the proneness to progression in neuroblastomas. *Med. Pediatr. Oncol.* **35**, 651–655.

159. Choi, L.M., Kim, N.W., Zuo, J.J., Gerbing, R., Stram, D., Lukens, J.N., Matthay, K.K., Seeger, R.C., and Reynolds, C.P. (2000) Telomerase activity by TRAP assay and telomerase RNA (hTR) expression are predictive of outcome in neuroblastoma. *Med. Pediatr. Oncol.* **35**, 647–650.

160. Sanborn, C.K., O'Connor, A., Sawin, R.S., Moore, K., Dehart, M.J., and Azarow, K.S. (2000) Comparison of telomerase levels before and after differentiation of two cell lines of human in neuroblastoma. *J. Surg. Res.* **93**, 206–210.

161. Bisoffi, M., Heaphy, C.M., and Griffith, J.K. (2006) Telomeres: prognostic markers for solid tumors. *Int. J. Cancer* **119**, 2255–2260.

162. Brodeur, G.M., Sekhon, G.S., and Goldstein, M.N. (1975) Specific chromosomal aberration in human neuroblastoma. *Am. J. Hum. Genet.* **28**, 20A.

163. Brodeur, G.M., Sekhon, G.S., and Goldstein, M.N. (1977) Chromosomal aberrations in human neuroblastomas. *Cancer* **40**, 2256–2263.

164. Gilbert, F., Balaban, G., Moorhead, P., Bianchi, D., and Schlesinger, H. (1982) Abnormalities of chromosome 1p in human neuroblastoma tumors and cell lines. *Cancer Genet. Cytogenet.* **7**, 33–42.

165. Kushner, B.H., and Cheung, N.-K.V. (2006) Neuroblastoma—from genetic profiles to clinical challenge. *N. Engl. J. Med.* **353**, 2215–2217.

166. Kushner, B.H., LaQuaglia, M.P., Cheung, N.K., Kramer, K., Hamelin, A.C., Gerald, W.L., and Ladanyi, M. (1999) Clinically critical impact of molecular genetic studies in pediatric solid tumors. *Med. Pediatr. Oncol.* **33**, 530–535.

167. Blatt, J. (2001) Tumor suppressor genes on the short arm of chromosome 1 in neuroblastoma. *Pediatr. Hematol. Oncol.* **18**, 3–5.

168. Fong, C., Dracopoli, N.C., White, P.S., Merrill, P.T., Griffith, R.C., Housman, D.E., and Brodeur, G.M. (1989) Loss of heterozygosity for the short arm of chromosome 1 in human neuroblastomas: correlation with N-*myc* amplification. *Proc. Natl. Acad. Sci. USA* **86**, 3753–3757.

169. Weith, A., Martinsson, T., Cziepluch, C., Bruderlein, S., Amler, L.C., Berthold, F., and Schwab, M. (1989) Neuroblastoma consensus deletion maps to 1p36.1–2. *Genes Chromosomes Cancer* **1**, 159–166.

170. White, P.S., Maris, J.M., Beltinger, C., Sulman, E., Marshall, H.N., Fujimori, M., Kaufman, B.A., Biegel, J.A., Allen, C., Hilliard, C., Valentine, M.B., Look, A.T., Enomoto, II., Sakiyama, S., and Brodeur, G.M. (1995) A region of consistent deletion in neuroblastoma maps within human chromosome 1p36.2–36.3. *Proc. Natl. Acad. Sci USA* **92**, 5520–5524.

171. Godfried, M.B., Veenstra, M., Valent, A., van Sluis, P., Voute, P.A., Versteeg, R., and Caron, H.N. (2002) Lack of interstitial chromosome 1p deletions in clinically-detected neuroblastoma. *Eur. J. Cancer* **38**, 1513–1519.

172. Hogarty, M.D., Liu, X., Guo, C., Thompson, P.M., Weiss, M.J., White, P.S., Sulman, E.P., Brodeur, G.M., and Maris, J.M. (2000) Identification of a 1-megabase consensus region of deletion at 1p36.3 in primary neuroblastomas. *Med. Pediatr. Oncol.* **35**, 512–515.

173. Ejeskär, K., Sjöberg, R.-M., Abel, F., Kogner, P., Ambros, P.F., and Martinsson, T. (2001) Fine mapping of a tumour suppressor candidate gene region in 1p36.2–3, commonly deleted in neuroblastomas and germ cell tumours. *Med. Pediatr. Oncol.* **36**, 61–66.

174. Laureys, G., Speleman, F., Opdenakker, G., Benoit, Y., and Leroy, J. (1990) Constitutional translocation t(1;17)(p36;1;q12–21) in a patient with neuroblastoma. *Genes Chromosomes Cancer* **2**, 252–254.

175. Biegel, J.A., White, P.S., Marshall, H.N., Fujimori, M., Zackai, E.H., Scher, C.D., Brodeur, G.M., and Emanuel, B.S. (1993) Constitutional 1p36 deletion in a child with neuroblastoma. *Am. J. Hum. Genet.* **52**, 176–182.

176. Bader, S.A., Fasching, C., Brodeur, G.M., and Stanbridge, E.J. (1991) Dissociation of suppression of tumorigenicity and differentiation *in vitro* effected by transfer of single human chromosomes into human neuroblastoma cells. *Cell Growth Differ.* **2**, 245–255.

177. Caron, H., van Sluis, P., van Hoeve, M., de Kraker, J., Bras, J., Slater, R., Mannens, M., Voute, P.A., Westerveld, A., and Versteeg, R. (1993) Allelic loss of chromosome 1p36 in neuroblastoma is of preferential maternal origin and correlates with N-myc amplification. *Nat. Genet.* **4**, 187–190.

178. Caron, H., Spieker, N., Godfried, M., Veenstra, M., van Sluis, P., de Kraker, J., Voûte, P., and Versteeg, R. (2001) Chromosome bands 1p35–36 contain two distinct neuroblastoma tumor suppressor loci, one of which is imprinted. *Genes Chromosomes Cancer* **30**, 168–174.

179. Hogarty, M.D., Winter, C.L., Liu, X., Guo, C., White, P.S., Look, A.T., Brodeur, G.M., and Maris, J.M. (2002) No evidence for the presence of an imprinted neuroblastoma suppressor gene within chromosome sub-band 1p36.3. *Cancer Res.* **62**, 6481–6484.

180. Gehring, M., Berthold, F., Edler, L., Schwab, M., and Amler, L.C (1995) The 1p deletion is not a reliable marker for the prognosis of patients with neuroblastoma. *Cancer Res.* **55**, 5366–5369.

181. Maris, J.M., White, P.S., Beltinger, C.P., Sulman, E.P., Castelberry, R.P., Shuster, J.J., Look, A.T., and Brodeur, G.M. (1995) Significance of chromosome 1p loss of heterozygosity in neuroblastoma. *Cancer Res.* **55**, 4664–4669.

182. Schleiermacher, G., Peter, M., Michon, J., Hugot, J.-P., Vielh, P., Zucker, J.-M., Magdelénat, H., Thomas, G., and Delattre, O. (1994) Two distinct deleted regions on the short arm of chromosome 1 in neuroblastoma. *Genes Chromosomes Cancer* **10**, 275–281.

183. Rubie, H., Delattre, O., Hartmann, O., Combaret, V., Michon, J., Bénard, J., Peyroulet, M.C., Plantaz, D., Coze, C., Chastagner, P., Baranzelli, M.C., Frappaz, D., Lemerle, J., and Sommelet, D. (1997) Loss of chromosome 1p may have a prognostic value in localized neuroblastoma: results of the French NBL 90 Study. Neuroblastoma Study Group of the Societe Française d'Oncologie Pediatrique (SFOP). *Eur. J. Cancer* **33**, 1917–1922.

184. Maris, J.M., Guo, C., Blake, D., White, P.S., Hogarty, M.D., Thompson, P.M., Rajalingam, V., Gerbing, R., Stram, D.O., Matthay, K.K., Seeger, R.C., and Brodeur, G.M. (2001) Comprehensive analysis of chromosome 1p deletions in neuroblastoma. *Med. Pediatr. Oncol.* **36**, 32–36.

185. Burgues, O., Navarro, S., Noguera, R., Pellin, A., Ruiz, A., Castel, V., and Llombart-Bosch, A. (2006) Prognostic value of the International Neuroblastoma Pathology Classification in Neuroblastoma (Schwannian stroma-poor) and comparison with other prognostic factors: a study of 182 cases from the Spanish Neuroblastoma Registry. *Virchows Arch.* **449**, 410–420.

186. Spitz, R., Hero, B., Westermann, F., Ernestus, K., Schwab, M., and Berthold, F. (2002) Fluorescence in situ hybridization analyses of chromosome band 1p36 in neuroblastoma detect two classes of alterations. *Genes Chromosomes Cancer* **34**, 299–305.

187. Maris, J.M., Hii, G., Gelfand, C.A., Varde, S., White, P.S., Rappaport, E., Surrey, S., and Fortina, P. (2005) Region-specific detection of neuroblastoma loss of heterozygosity at multiple loci simultaneously using a SNP-based tag-array platform. *Genome Res.* **15**, 1168–1176.

188. Maris, J.M., Weiss, M.J., Guo, C., Gerbing, R.B., Stram, D.O., White, P.S., Hogarty, M.D., Sulman, E.P., Thompson, P.M., Lukens, J.N., Matthay, K.K., Seeger, R.C., and Brodeur, G.M. (2000) Loss of heterozygosity at 1p36 independently predicts for disease progression but not decreased overall survival probability in neuroblastoma patients: a Children's Cancer Group Study. *J. Clin. Oncol.* **18**, 1888–1899.

189. Caron, H., Peter, M., van Sluis, P., Speleman, F., de Kraker, J., Laureys, G., Michon, J., Brugières, L., Voûte, P.A., Westerveld, A., Slater, R., Delattre, O., and Versteeg, R. (1995) Evidence for two tumour suppressor loci on chromosomal bands 1p35–36 involved in neuroblastoma: one probably imprinted, another associated with N-myc amplification. *Hum. Mol. Genet.* **4**, 535–539.

190. Martinsson, T., Sjöberg, R.-M., Hedborg, F., and Kogner, P. (1995) Deletion of chromosome 1p loci and microsatellite instability in neuroblastomas analyzed with short-tandem repeat polymorphisms. *Cancer Res.* **55**, 5681–5686

191. Martinsson, T., Sjöberg, R.M., Hallstensson, K., Nordling, M., Hedborg, F., and Kogner, P. (1997) Delimitation of a critical tumour suppressor region at distal 1p in neuroblastoma tumours. *Eur. J. Cancer* **33**, 1997–2001.

192. White, P.S., Maris, J.M., Sulman, E.P., Jensen, S.J., Kyemba, S.M., Beltinger, C.P., Allen, C., Kramer, D.L., Biegel, J.A., and Brodeur, G.M. (1997) Molecular analysis of the region

of distal 1p commonly deleted in neuroblastoma. *Eur J. Cancer* **33,** 1957–1961.

193. White, P.S., Thompson, P.M., Gotoh, T., Okawa, E.R., Igarashi, J., Kok, M., Winter, C., Gregory, S.G., Hogarty, M.D., Maris, J.M., and Brodeur, G.M. (2005) Definition and characterization of a region of 1p36.3 consistently deleted in neuroblastoma. *Oncogene* **24,** 2684–2694.

194. Ichimiya, S., Nimura, Y., Kageyama, H., Takada, N., Sunahara, M., Shishikura, T., Nakamura, Y., Sakiyama, S., Seki, N., Ohira, M., Kaneko, Y., McKeon, F., Caput, D., and Nakagawara, A. (1999) *p73* at chromosome 1p36.3 is lost in advanced stage neuroblastoma but its mutation is infrequent. *Oncogene* **18,** 1061–1066.

195. Spieker, N., Beitsma, M., van Sluis, P., Roobeek, I., den Dunnen, J.T., Speleman, F., Caron, H., and Versteeg, R. (2000) An integrated 5-Mb physical, genetic, and radiation hybrid map of a 1p36.1 region implicated in neuroblastoma pathogenesis. *Genes Chromosomes Cancer* **27,** 143–152.

196. Spieker, N., Beitsma, M., Van Sluis, P., Chan, A., Caron, H., and Versteeg, R. (2001) Three chromosomal rearrangements in neuroblastoma cluster within a 300-kb region on 1p36.1. *Genes Chromosomes Cancer* **31,** 172–181.

197. Krona, C., Ejeskär, K., Abel, F., Kogner, P., Bjelke, J., Björk, E., Sjöberg, R.-M., and Martinsson, T. (2003) Screening for gene mutations in a 500 kb neuroblastoma tumor suppressor candidate region in chromosome 1p; mutation and stage-specific expression in *UBE4B/UFD2*. *Oncogene* **22,** 2343–2351.

198. Amler, L.C., Corvi, R., Praml, C., Savelyeva, L., Le Paslier, D., and Schwab, M. (1995) A reciprocal translocation (1;15)(p36.2;q24) in a neuroblastoma cell line is accompanied by DNA duplication and may signal the site of a putative tumor suppressor-gene. *Oncogene* **10,** 1095–1011.

199. Cheng, N.C., Van Roy, N., Chan, A., Beitsma, M., Westerveld, A., Speleman, F., and Versteeg, R. (1995) Deletion mapping in neuroblastoma cell lines suggests two distinct tumor suppressor genes in the 1p35–36 region, only one of which is associated with N-myc amplification. *Oncogene* **10,** 291–297.

200. Takeda, O., Homma, C., Maseki, N., Sakurai, M., Kanda, N., Schwab, M., Nakamura, Y., and Kaneko, Y. (1994) There may be two tumor suppressor genes on chromosome arm 1p closely associated with biologically distinct subtypes of neuroblastoma. *Genes Chromosomes Cancer* **10,** 30–39.

201. Mora, J., Cheung, N.-K.V., Chen, L., Qin, J., and Gerald, W. (2001) Survival analysis in clinical, pathologic, and genetic features in neuroblastoma presenting as locoregional disease. *Cancer* **91,** 435–442.

202. Casciano, I., Marchi, J.V.M., Muresu, R., Volpi, E.V., Rozzo, C., Opdenakker, G., and Romani, M. (1996) Molecular and genetic studies on the region of translocation and duplication in the neuroblastoma cell line NGP at the 1p36.13 p36.32 chromosomal site. *Oncogene* **12,** 2101–2108.

203. Savelyeva, L., Corvi, R., and Schwab, M. (1994) Translocation involving 1p and 17q is a recurrent genetic alteration of human neuroblastoma cells. *Am.J. Genet.* **55,** 334–340.

204. Hunt, J.D., and Tereba, A. (1990) Molecular evaluation of abnormalities of the short arm of chromosome 1 in a neuroblastoma. *Genes Chromosomes Cancer* **2,** 137–146.

205. van der Drift, P., Chan, A., Laureys, G., van Roy, N., Sickmann, G., den Dunnen, J., Westerveld, A., Speleman, F., and Versteeg, R. (1995) Balanced translocation in a neuroblastoma patient disrupts a cluster of small nuclear RNA U1 and tRNA genes in chromosomal band 1p36. *Genes Chromosomes Cancer* **14,** 35–42.

206. Ohira, M., Kageyama, H., Mihara, M., Furuta, S., Machida, T., Shishikura, T., Takayasu, H., Islam, A., Nakamura, Y., Takahashi, M., Tomioka, N., Sakiyama, S., Kaneko, Y., Toyoda, A., Hattori, M., Sakaki, Y., Ohki, M., Horii, A., Soeda, E., Inazawa, J., Seki, N., Kuma, H., Nozawa, I., and Nakagawara, A. (2000) Identification and characterization of a 500-kb homozygously deleted region at 1p36.2-p36.3 in a neuroblastoma cell line. *Oncogene* **19,** 4302–4307.

207. Satgé, D., Moore, S.W., Stiller, C.A., Niggli, F.K., Pritchard-Jones, K., Bown, N., Bénard, J., and Plantaz, D. (2003) Abnormal constitutional karyotypes in patients with neuroblastoma: a report of four new cases and review of 47 others in the literature. *Cancer Genet. Cytogenet.* **147,** 89–98.

208. Barker, P.E., Savelyeva, L., and Schwab, M. (1993) Translocation junctions cluster at the distal short arm of chromosome 1 (1p36.1–2) in human neuroblastoma cells. *Oncogene* **8,** 3353–3358.

209. Maris, J.M., Kyemba, S.M., Rebbeck, T.R., White, P.S., Sulman, E.P., Jensen, S.J., Allen, C., Biegel, J.A., Yanofsky, R.A., Feldman, G.L., and Brodeur, G.M. (1996) Familial predisposition to neuroblastoma does not map to chromosome band 1p36. *Cancer Res.* **56,** 3421–3425.

210. Janoueix-Lerosey, I., Novikov, E., Monteiro, M., Gruel, N., Schleiermacher, G., Loriod, B., Nguyen, C., and Delattre, O. (2004) Gene expression profiling of 1p35–36 in neuroblastoma. *Oncogene* **23,** 5912–5922.

211. White, P.S., and Versteeg, R. (2000) Allelic loss and neuroblastoma suppressor genes. In *Neuroblastoma* (Brodeur, G.M., Sawada, T., Tsuchida, Y., and Voûte, P.A., eds.), Elsevier Science B.V., Amsterdam, The Netherlands, pp. 57–74.

212. White, P.S., Thompson, P.M., Seifried, B.A., Sulman, E.P., Jensen, S.J., Guo, C., Maris, J.M., Hogarty, M.D., Allen, C., Biegel, J.A., Matise, T.C., Gregory, S.G., Reynolds, C.P., and Brodeur, G.M. (2001) Detailed molecular analysis of 1p36 in neuroblastoma. *Med. Pediatr. Oncol.* **36,** 37–41.

213. The, I., Murthy, A.E., Hannigan, G.E., Jacoby, L.B., Menon, A.G., Gusella, J.F., and Bernards, A. (1993) Neurofibromatosis type 1 gene mutations in neuroblastoma. *Nat. Genet.* **3,** 62–66.

214. Diccianni, M.B., Chau, L.S., Batova, A., Vu, T.Q., and Yu, A.L. (1996) The *p16* and *p18* tumor suppressor genes in neuroblastoma: implications for drug resitance. *Cancer* **104,** 183–192.

215. Kaelin, W.G., Jr. (1999) The emerging *p53* gene family. *J. Natl. Cancer Inst.* **91,** 594–598.

216. Sharpless, N.E., and DePinho, R.A. (1999) The *INK4A/ARF* locus and its two gene product. *Curr. Opin. Genet. Dev.* **9,** 22–30.

217. Brodeur, G.M., Seeger, R.C., Schwab, M., Varmus, H.E., and Bishop, J.M. (1984) Amplifiation of N-myc in untrated human neuroblastomas correlates with advanced disease stage. *Science* **224,** 1121–1124.

218. Biedler, J., Ross, R.A., Shanske, S., and Spengler, B.A. (1980) Human neuroblastoma cytogenetics: search for significance of homogeneously staining regions and double minute chromosomes. In *Advances in Neuroblastoma Research* (Evans, A.E., ed.), Raven Press, New York, pp. 81–96.

219. Gilbert, F., Feder, M., Balaban, G., Brangman, D., Lurie, D.K., Podolsky, R., Rinaldt, V., Vinikoor, N., and Weisband, J. (1984) Human neuroblastomas and abnormalities of chromosomes 1 and 17. *Cancer Res.* **44,** 5444–5449.

220. Van Roy, N., Laureys, G., Cheng, N.C., Willem, P., Opdenakker, G., Versteeg, R., and Speleman, F. (1994) 1;17 translocations and other chromosome 17 rearrangements in human primary neuroblastoma tumors and cell lines. *Genes Chromosomes Cancer* **10,** 103–114.

220A. Sanger, W.G., Howe, J., Fordyce, R., and Purtilo, D.T. (1984) Inherited partial trisomy #15 complicated by neuroblastoma. *Cancer Genet. Cytogenet.* **11,** 153–159.

221. Van Roy, N., van der Drift, P., Van Gele, M., Chan, A., Laureys, G., Opdenakker, G., Maertens, L., Vandesompele, J., Versteeg, R., and Speleman, F. (1997) Chromosome 17 involvement in neuroblastoma: an overview. *Cancer Genet. Cytogenet.* **98,** 168.

222. Van Roy, N., Jauch, A., Van Gele, M., Laureys, G., Versteeg, R., De Paepe, A., Cremer, T., and Speleman, F. (1997) Comparative genomic hybridization analysis of human neuroblastomas: detection of distal 1p deletions and further molecular genetic characterization of neuroblastoma cell lines. *Cancer Genet. Cytogenet.* **135,** 135–142.

222A. Franke, F., Forster, W., Rudolph, B., and Lampert, F. (1985) Metastatic neuroblastoma in an infant: translocation (1;11), deletion (2) and double minute chromosomes. *Eur. J. Pediatr.* 143, 305–308.

222B. Franke, F., Rudolph, B., Christiansen, H., Harbott, J., and Lampert, F. (1986) Tumour karyotype may be important in the prognosis of human neuroblastoma. *J. Cancer Res. Clin. Oncol.* **111,** 26–272.

223. Brown, N., Mazzocco, K., Lastowska, M., Thompson, K., and Roberts, P. (2001) The cytogenetics of 17q in neuroblastoma: gain, multiple gain, involvement in LOH and interactions with MYCN. *Cancer Genet. Cytogenet.* **128,** 68.

223A. Hafez, M., El-Meadday, M., Sheir, M., Al-Tonbarry, Y., Nada, N., and El-Desoky, I. (1985) Chromosomal analysis of neuroblastomas. *Br. J. Cancer* **51,** 237–243.

224. Meddeb, M., Danglot, G., Chudoba, I., Vénuat, A.-M., Bénard, J., Avet-Loiseau, H., Vasseur, B., Le Paslier, D., Terrier-Lacombe, M.-J., Hartmann, O., and Bernheim, A. (1996) Additional copies of a 25 Mb chromosomal region originating from 17q23.1–17qter are present in 90% of high-grade neuroblastomas. *Genes Chromosomes Cancer* **17,** 156–165.

224A. Kaneko, Y., Tsuchida, Y., Maseki, N., Takasaki, N., Sakurai, M., and Saito, S. (1985) Chromosome findings in human neuroblastomas xenografted in nude mice. *Jpn. J. Cancer. Res.* **76,** 359–364.

225. Caron, H., van Sluis, P., van Roy, N., de Kraker, J., Speleman, F., Voûte, P.A., Westerveld, A., Slater, R., and Versteeg, R. (1994) Recurrent 1;17 translocations in human neuroblastoma reveal nonhomologous mitotic recombina-

tion during the S/G2 phase as a novel mechanism for loss of heterozygosity. *Am. J. Hum. Genet.* **55,** 341–347.

226. Lawce, H., Huang, M., and Magenis, R.E. (2000) Translocation of chromosomes 11 and 17 in neuroblastoma in G-banded and FISH studies from seven pediatric patients; does this rearrangement affect outcome? *Cancer Genet. Cytogenet.* **123,** 83.

227. Spitz, R., Hero, B., Ernestus, K., and Berthold, F. (2003) Gain of distal chromosome arm 17q is not associated with poor prognosis in neuroblastoma. *Clin. Cancer Res.* **9,** 4835–4840.

228. Lastowska, M., Roberts, P., Pearson, A.D.J., Lewis, I., Wolstenholme, J., and Bown, N. (1997) Promiscuous translocations of chromosome arm 17q in human neuroblastomas. *Genes Chromosomes Cancer* **19,** 143–149.

229. Lastowska, M., Van Roy, N., Bown, N., Speleman, F., Lunec, J., Strachan, T., Pearson, A.D.J., and Jackson, M.S. (1998) Molecular cytogenetic delineation of 17q translocation breakpoints in neuroblastoma cell lines. *Genes Chromosomes Cancer* **23,** 116–122.

230. Speleman, F., and Bown, N. (2000) 17q gain in neuroblastoma. In *Neuroblastoma* (Brodeur, G.M., Sawada, T., Tsuchida, Y., and Voûte, P.A., eds.), Elsevier Science B.V., Amsterdam, The Netherlands, pp. 113–124.

231. Bown, N. (2001) Neuroblastoma tumour genetics: clinical and biological aspects. *J. Clin. Pathol.* **54,** 897–910.

232. Donti, E., Longo, L., Tonini, G.P., Verdona, G., Melodia, A., Lanino, E., and Cornaglia-Ferraris, P. (1988) Cytogenetic and molecular study of two human neuroblastoma cell lines. *Cancer Genet. Cytogenet.* **30,** 225–231.

233. McRobert, T.L., Rudduck, C., Kees, U.R., and Garson, O.M. (1992) Detection of MYCN amplification in three neuroblastoma cell lines by non-radioactive chromosomal in situ hybridization. *Cancer Genet. Cytogenet.* **59,** 128–134.

234. Avet-Loiseau, H., Venuat, A.-M., Bénard, J., Leibovitch, M.-P., Hartmann, O., and Bernheim, A. (1995) Morphologic and molecular cytogenetics in neuroblastoma. *Cancer* **75,** 1694–1699.

235. Stallings, R.L., Carty, P., McArdle, L., Mullarkey, M., McDermott, M., Breatnach, F., and O'Meara, A. (2004) Molecular cytogenetic analysis of recurrent unbalanced t(11;17) in neuroblastoma. *Cancer Genet. Cytogenet.* **154,** 44–51.

236. Schleiermacher, G., Raynal, V., Janoueix-Lerosey, I., Combaret, V., Aurias, A., and Delattre, O. (2004) Variety and complexity of chromosome 17 translocations in neuroblastoma. *Genes Chromosomes Cancer* **39,** 143–150.

237. Islam, A., Kageyama, H., Hashizume, K., Kaneko, Y., and Nakagawara, A. (2000) Role of survivin, whose gene is mapped to 17q25, in human neuroblastoma and identification of a novel dominant-negative isoform, survivin-beta/2B. *Med. Pediatr. Oncol.* **35,** 550–553.

238. Islam, A., Kageyama, H., Takada, N., Kawamoto, T., Takayasu, H., Isogai, E., Ohira, M., Hashizume, K., Kobayashi, H., Kaneko, Y., and Nakagawara, A. (2000) High expression of *Survivin*, mapped to 17q25, is significantly associated with poor prognostic factors and promotes cell survival in human neuroblastoma. *Oncogene* **19,** 617–623.

239. Schleiermacher, G., Bourdeaut, F., Combaret, V., Picrron, G., Raynal, V., Aurias, A., Ribeiro, A., Janoueix-Lerosey, I.,

and Delattre, O. (2005) Stepwise occurrence of a complex unbalanced translocation in neuroblastoma leading to insertion of a telomere sequence and late chromosome 17q gain. *Oncogene* **24**, 3377–3384.

240. Van Roy, N., Cheng, N.C., Laureys, G., Opdenakker, G., Versteeg, R., and Speleman, F. (1995) Molecular cytogenetic analysis of 1;17 translocations in neuroblastoma. *Eur. J. Cancer* **31A**, 530–535.

241. Van Roy, N., Vandesompele, J., Berx, G., Staes, K., Van Gele, M., De Smet, E., De Paepe, A., Laureys, G., van der Drift, P., Versteeg, R., Van Roy, F., and Speleman, F. (2002) Localization of the 17q breakpoint of a constitutional 1;17 translocation in patient with neuroblastoma within a 25-kb segment located between the *ACCN1* and *TLK2* genes and near the distal breakpoints of two microdeletions in neurofibromatosis type 1 patients. *Genes Chromosomes Cancer* **35**, 113–120.

242. Aguzzi, A., Ellmeier, W., and Weith, A. (1992) Dominant and recessive molecular changes in neuroblastoma. *Brain Pathol.* **2**, 195–208.

243. Trakhtenbrot, L., Cohen, N., Betts, D.R., Niggli, F.K., Amariglio, N., Brok-Simoni, F., Rechavi, G., and Meitar, D. (2002) Interphase fluorescence in situ hybridization detection of chromosome 17 and 17q region gains in neuroblastoma: are they secondary events?. *Cancer Genet. Cytogenet.* **137**, 95–101.

244. Rudduck, C., Lukeis, R.E., McRobert, T.L., Chow, C.W., and Garson, O.M. (1992) Chromosomal localization of amplified N-*myc* in neuroblastoma cell using a biotinylated probe. *Cancer Genet. Cytogenet.* **58**, 55–59.

245. Gotoh, T., Sugiharta, H., Matsumara, T., Katsura, K., Takamatsu, T., and Sawada, T. (1998) Human neuroblastoma demonstrating clonal evolution in vivo. *Genes Chromosomes Cancer* **19**, 143–149.

246. Brodeur, G.M., and Fong, C.T. (1989) Molecular biology and genetics of human neuroblastoma. *Cancer Genet. Cytogenet.* **41**, 153–174.

247. Mitelman, F., Johansson, B., and Mertens, F. (1994) Catalog of chromosome aberrations in cancer, 5th ed. Wiley-Liss, New York.

248. Van Roy, N., Laureys, G., Van Gele, M., Opdenakker, G., Miura, R., van der Drift, P., Chan, A., Versteeg, R., and Speleman, F. (1997) Analysis of 1;17 breakpoints in neuroblastoma: implications for mapping neuroblastoma genes. *Eur. J. Cancer* **33**, 1974–1978.

249. Vandesompele, J., Speleman, F., Van Roy, N., Laureys, G., Brinkschmidt, C., Christiansen, H., Lampert, F., Lastowska, M., Bown, N., Pearson, A., Nicholson, J.C., Ross, F., Combaret, V., Delattre, O., Feuerstein, B.G., and Plantaz. D. (2001) Multicentre analysis of patterns of DNA gains and losses in 204 neuroblastoma tumors: how many genetic subgroups are there? *Med. Pediatr. Oncol.* **36**, 5–10.

250. Laureys, G., Versteeg, R., Speleman, F., van der Drift, P., Francke, U., Opdenakker, G., and Van Roy, N. (1995) Characterization of the chromosome breakpoints in a patient with a constitutional translocation t(1;17)(p36.31-p36.13;q11.2-q12) and neuroblastoma. *Eur. J. Cancer* **31A** 523–526.

251. Laureys, G., Speleman, F., Versteeg, R., van der Drift, P., Chan, A., Leroy, J., Francke, U., Opdenakker, G., and Van Roy, N.. (1995) Constitutional translocation t(1;17)(p36.31-p36.13;q11.2-q12.1) in a neuroblastoma patient. Establishment of somatic cell hybrids and identification of PND/A12M2 on chromosome 1 and NF1/SCYA7 on chromosome 17 as breakpoint flanking single copy markers. *Oncogene* **10**, 1087–1093.

252. van der Drift, P., Chan, A., Laureys, G., van Roy, N., Westerveld, A., Speleman, F., and Versteeg, R. (1996) Analysis of a constitutional balanced translocation t(1;17)(p36;q11.2–12.1) in a neuroblastoma patient. *Cancer Genet. Cytogenet.* **91**, 171.

253. Janoueix-Lerosey, I., Penther, D., Thioux, M., de Crémoux, P., Derré, J., Ambros, P., Vielh, P., Bénard, J., Aurias, A., Delattre, O. (2000) Molecular analysis of chromosome arm 17q gain in neuroblastoma. *Genes Chromosomes Cancer* **28**, 276–284.

254. Lastowska, M., Cullinane, C., Variend, S., Cotterill, S., Bown, N., O'Neill, S., Mazzocco, K., Roberts, P., Nicholson, J., Ellershaw, C., Pearson, A., and Jackson, M. (2001) Comprehensive genetic and histopathological analysis reveals three types of neuroblastoma tumours. *Cancer Genet. Cytogenet.* **128**, 68.

255. Lastowska, M., Cullinane, C., Variend, S., Cotterill, S., Bown, N., O'Neill, S., Mazzocco, K., Roberts, P., Nicholson, J., Ellershaw, C., Pearson, A.D.J., and Jackson, M.S. for the United Kingdom Children Cancer Study Group and the United Kingdom Cancer Cytogenetics Group. (2001) Comprehensive genetic and histopathologic study reveals three types of neuroblastoma tumors. *J. Clin. Oncol.* **19**, 3080–3090.

256. Lastowska, M., Van Roy, N., Bown, N., Speleman, F., Roberts, P., Lunec, J., Strachan, T., Pearson, A.D., and Jackson, M.S. (2001) Molecular cytogenetic definition of 17q translocation breakpoints in neuroblastoma. *Med. Pediatr. Oncol.* **36**, 20–23.

257. Lastowska, M., Cotterill, S., Bown, N., Cullinane, C., Variend, S., Lunec, J., Strachan, T., Pearson, A.D.J., and Jackson, M.S. (2002) Breakpoint position on 17q identifies the most aggressive neuroblastoma tumors. *Genes Chromosomes Cancer* **34**, 428–436.

258. Lastowska, M., Chung, Y.J., Cheng Ching, N., Haber, M., Norris, M.D., Kees, U.R., Pearson, A.D., and Jackson, M.S. (2004) Regions syntenic to human 17q are gained in mouse and rat neuroblastoma. *Genes Chromosomes Cancer* **40**, 158–163.

259. Bown, N.P., Reid, M.M., Pearson, A.D.J., Davison, E.V., Malcolm, A., and Craft, A.W. (1987) Cytogenetic investigations in the assessment of response to treatment in neuroblastoma. *J. Clin. Pathol.* **40**, 1334–1336.

259A. Caron, H. (1995) Allelic loss of chromosome 1 and additional chromosome 17 material are both unfavourable prognostic markers in neuroblastoma. *Med. Pediatr. Oncol.* **24**, 215–221.

260. Mertens, F., Johansson, B., Höglund, M., and Mitelman, F. (1997) Chromosomal imbalance maps to malignant solid tumors: a cytogenetic survey of 3185 neoplasms. *Cancer Res.* **57**, 2765–2780.

261. Van Roy, N., Laureys, G., Plantaz, D., Feuerstein, B.G., Brinkschmidt, C., Christiansen, H., Lastowska, M., Bown, N., Nicholson, J.C., Ross, F., and Speleman, F. (1999) Multicenter CGH analysis of 184 neuroblastomas reveals charac-

teristic gains and losses in localised tumors and high incidence of 11q loss in stage 4 tumors without *MYCN* amplification and/or 1p deletion. *Cancer Genet. Cytogenet.* **112,** 70.

262. Srivatsan, E.S., Ying, K.L., and Seeger, R.C. (1993) Deletion of chromosome 11 and of 14q sequences in neuroblastoma. *Genes Chromosomes Cancer* **7,** 32–37.

263. Takita, J., Hayashi, Y., Kohno, T., Shiseki, M., Yamaguchi, N., Hanada, R., Yamamoto, K., and Yokota, J. (1995) Allelotype of neuroblastoma. *Oncogene* **11,** 1829–1834.

264. Koiffmann, C.P., Gonzalez, C.H., Vianna-Morgante, A.M., Kim, C.A., Odone-Filho, V., and Wajntal, A. (1995) Neuroblastoma in a boy with MCA-MR syndrome, deletion 11q, and duplication 12q. *Am. J. Med. Genet.* **58,** 46–49.

265. Mosse, Y., Greshock, J., King, A., Khazi, D., Weber, B.L., and Maris, J.M. (2003) Identification and high-resolution mapping of a constitutional 11q deletion in an infant with multifocal neuroblastoma. *Lancet Oncol.* **4,** 769–771.

265A. Bown, N.P., Pearson, A.D., and Reid, M.M. (1993) High incidence of constitutional balanced translocations in neuroblastoma. *Cancer Genet. Cytogenet.* **69,** 166–167.

266. Hecht, F., Hecht, B.K., Northrup, J.C., Trachtenberg, N., Wood, T.S., and Cohen, J.D. (1982) Genetics of familial neuroblastoma: long range studies. *Cancer Genet. Cytogenet.* **7,** 227–230.

267. Luttikhuis, M.E.M., Powell, J.E., Rees, S.A., Genus, T., Chughtai, S., Ramani, P., Mann, J.R., McConville, C.M. (2001) Neuroblastomas with chromosome 11q loss and single copy MYCN comprise a biologically distinct group of tumours with adverse prognosis. *Br. J. Cancer* **85,** 531–537.

268. De Preter, K., Vandesompele J., Menten, B., Carr, P., Fiegler, H., Edsjö, A., Carter, N.P., Yigit, N., Waelput, W., Van Roy, N., Bader, S., Påhlman, S., and Speleman, F. (2005) Positional and functional mapping of a neuroblastoma differentiation gene on chromosome 11. *BMC Cancer* **6,** 97–206.

269. Wada, M., Seeger, R.C., Mizoguchi, H., and Koeffler, H.P. (1995) Maintenance of normal imprinting of *H19* and *IGF2* genes in neuroblastoma. *Cancer Res.* **55,** 3386–3388.

270. Luttikhuis, M.E., Powell, J.E., Rees, S.A., Genus, T., Chughtai, S., Ramani, P., Mann, J.R., and McConville, C.M. (2001) Neuroblastomas with chromosome 11q loss and single copy MYCN comprise a biologically distinct group of tumours with adverse prognosis. *Br. J. Cancer* **85,** 531–537.

271. Spitz, R., Hero, B., Ernestus, K., and Berthold, F. (2003) FISH analyses for alterations in chromosomes 1, 2, 3, and 11 define high-risk groups in neuroblastoma. *Med. Pediatr. Oncol.* **41,** 30–35.

272. Maris, J.M., Guo, C., White, P.S., Hogarty, M.D., Thompson, P.M., Stram, D.O., Gerbing, R., Matthay, K.K., Seeger, R.C., and Brodeur, G.M. (2001) Allelic deletion at chromosome bands 11q14–23 is common in neuroblastoma. *Med. Pediatr. Oncol.* **36,** 24–27.

273. Guo, C., White, P.S., Hogarty, M.D., Brodeur, G.M., Gerbing, R., Stram, D.O., and Maris, J.M. (2000) Deletion of 11q23 is a frequent event in the evolution of *MYCN* single-copy high-risk neuroblastomas. *Med. Pediatr. Oncol.* **35,** 544–546.

274. Stallings, R.L., Howard, J., Dunlop, A., Mullarkey, M., McDermott, M., Breatnach, F., and O'Meara, A. (2003) Are gains of chromosomal regions 7q and 11p important abnormalities in neuroblastoma? *Cancer Genet. Cytogenet.* **140,** 133–137.

274A. Flahaut, M., Mühlethaler-Mottet, A., Martinet, D., Fattet, S., Bourloud, K.B., Auderset, K., Meier, R., Besuchet Schmutz, N., Delattre, O., Joseph, J.-M., and Gross, N. (2006) Molecular cytogenetic characterization of doxorubicin-resistant neuroblastoma cell lines: evidence that acquired multidrug resistance results from a unique large amplification of the 7q21 region. *Genes Chromosomes Cancer* **45,** 495–508.

275. Katoh, M., and Katoh, M. (2003) Identification and characterization of *FLJ10737* and *CAMTA1* genes on the commonly deleted region of neuroblastoma at human chromosome 1p36.31-p36.23. *Int. J. Oncol.* **23,** 1219–1224.

276. Kohl, N.E., Kanda, N., Schreck, R.R., Bruns, G., Latt, S.A., Gilbert, F., and Alt, F.W. (1983) Transposition and amplification of oncogene-related sequences in human neuroblastomas. *Cell* **35,** 359–367.

277. Seeger, R.C., Wada, R., Brodeur, G.M., Moss, T.J., Bjork, R.L., Sousa, L., and Slamon, D.J. (1988) Expression of N-myc by neuroblastomas with one of multiple copies of the oncogene. *Prog. Clin. Biol. Res.* **271,** 41–49.

278. Brodeur, G.M., Hayes, F.A., Green, A.A., Casper, J.T., Wasson, J., Wallach, S., and Seeger, R.C. (1987) Consistent N-myc copy number in simultaneous of consecutive neuroblastoma samples from sixty individual patients. *Cancer Res.* **47,** 4248–4253.

279. Brodeur, G.M., Maris, J.M., Yamashiro, D.J., Hogarty, M.D., and White, P.S. (1997) Biology and genetics of human neuroblastomas. *J. Pediatr. Hematol./Oncol.* **19,** 93–101.

280. Seeger, R.C., Brodeur, G.M., Sather, H., Dalton, A., Siegel, S.E., Wong, K.Y., and Hammond, D. (1985) Association of multiple copies of the N-*myc* oncogene with rapid progression of neuroblastomas. *N. Engl. J. Med.* **313,** 1111–1116.

281. Berthold, F., Sahin, K., Hero, B., Christiansen, H., Gehring, M., Harms, D., Horz, S., Lampert, F., Schwab, M., and Terpe, J. (1997) The current contribution of molecular factors to risk estimation in neuroblastoma patients. *Eur. J. Cancer* **33,** 2092–2097.

282. Mora, J., Gerald, W.L., Qin, J., and Cheung, N.-K.V. (2001) Molecular genetics of neuroblastoma and the implications for clinical management: a review of the MSKCC experience. *Oncologist* **6,** 263–268.

283. Schmidt, M.L., Lukens, J.N., Seeger, R.C., Brodeur, G.M., Shimada, H., Gerbing, R.B., Stram, D.O., Perez, C., Haase, G.M., and Matthay, K.K. (2000) Biologic factors determine prognosis in infants with stage IV neuroblastoma: a prospective Children's Cancer Group Study. *J. Clin. Oncol.* **18,** 1260–1268.

284. Schmidt, M.L., Lal, A., and Dittmann, D. (2000) Differential expression in an antisense MYCN neuroblastoma model. *Med. Pediatr. Oncol.* **35,** 669–672.

285. Schmidt, M.L., Lal, A., Seeger, R.C., Maris, J.M., Shimada, H., O'Leary, M., Gerbing, R.B., and Matthay, K.K. (2005) Favorable prognosis for patients 12 to 18 months of age with stage 4 nonamplified *MYCN* neuroblastoma: a Children's Cancer Group Study. *J. Clin. Oncol.* **23,** 6474–6480.

286. Goto, S., Umehara, S., Gerbing, R.B., Stram, D.O., Brodeur, G.M. Seeger, R.C., Lukens, J.N., Matthay, K.K., and Shimada, H. (2001) Histopathology (International Neuroblastoma Pathology Classification) and MYCN status in

patients with peripheral neuroblastic tumors. A report from the Children's Cancer Group. *Cancer* **92**, 2699–2708.

287. Shimada, H., Stram, D.O., Chatten, J., Joshi, V.V., Hachitanda, Y., Brodeur, G.M., Lukens, J.N., Matthay, K.K., and Seeger, R.C. (1995) Identificaiton of subsets of neuroblastomas by combined histopathologic and N-myc analysis. *J. Natl. Cancer Inst.* **87**, 1470–1476.

288. Shimada, H., Umehara, S., Monobe, Y., Hachitanda, Y., Nakagawa, A., Goto, S., Gerbing, R.B., Stram, D.O., Lukens, J.N., and Matthay, K.K. (2001) Internatiuonal Neuroblastoma Pathology Classification for prognostic evaluation of patients with peripheral neuroblastic tumors. A report from the Children's Cancer Group. *Cancer* **92**, 451–561.

289. Schwab, M., Ellison, J., Busch, M., Rosenau, W., Varmus, H.E., and Bishop, J.M. (1984) Enhanced expression of the human gene N-*myc* consequent to amplification of DNA may contribute to malignant progression of neuroblastoma. *Proc. Natl. Acad. Sci. USA* **81**, 4940–4944.

290. Alaminos, M., Gerald, W.L., and Cheung, N.K. (2005) Prognostic value of *MYCN* and *ID2* overexpression in neuroblastoma. *Pediatr. Blood Cancer* **45**, 909–915.

291. Cohn, S.L., Look, A.T., Joshi, V.V., Holbrook, T., Salwen, H., Chagnovich, D., Chesler, L., Rowe, S.T., Valentine, M.B., Komuro, H., Castleberry, R.P., Bowman, L.C., Rao, P.V., Seeger, R.C., and Brodeur, G.M. (1995) Lack of correlation of N-*myc* gene amplification with prognosis in localized neuroblastoma: a Pediatric Oncology Group study. *Cancer Res.* **55**, 721–726.

292. Cohn, S.L., London, W.B., Huang, D., Katzenstein, H.M., Salwen, H.R., Reinhart, T., Madafiglio, J., Marshall, G.M., Norris, M.D., and Haber, M. (2000) *MYCN* expression is not prognostic of adverse outcome in advanced-stage neuroblastoma with nonamplified *MYCN*. *J. Clin. Oncol.* **18**, 3604–3613.

293. Tang, X.X., Zhao, H., Kung, B., Kim, D.Y., Hicks, S.L., Cohn, S.L, Cheung, N.-K., Seeger, R.C., Evans, A.E., and Ikegaki, N. (2006) The *MYCN* enigma: significance of *MYCN* expression in neuroblastoma. *Clin. Cancer Res.* **66**, 2826–2833.

294. Matthay, K.K. (2000) *MYCN* expression in neuroblastoma: a mixed message? *J. Clin. Oncol.* **18**, 3591–3594.

295. Hiyama, E., Hiyama, K., Yokoyama, T., and Ishii, T. (1991) Immunohistochemical analysis of N-myc protein expression in neuroblastoma: correlation with prognosis of patients. *J. Pediatr. Surg.* **26**, 838–843.

296. Nakada, K., Fujioka, T., Kitagawa, H., Takakuwa, T., and Yamate, N. (1993) Expressions of N-myc and ras oncogene products in neuroblastoma and their correlations with prognosis. *Jpn. J. Clin. Oncol.* **23**, 149–155.

297. Chan, H.S.L, Gallie, B.L., DeBoer, G., Haddad, G., Ikegaki, N., Dimitroulakos, J., Yeger, H., and Ling, V. (1997) MYCN protein expression as a predictor of neuroblastoma prognosis. *Clin. Cancer Res.* **3**, 1699–1706.

298. Bordow, S.B., Norris, M.D., Haber, P.S., Marshall, G.M., and Haber, M. (1998) Prognostic significance of MYCN oncogene expression in childhood neuroblastoma. *J. Clin. Oncol.* **16**, 3286–3294.

299. Hogarty, M.D. (2003) The requirement for evasion of programmed cell death in neuroblastomas with *MYCN* amplification. *Cancer Lett.* **197**, 173–179.

300. Weiss, W.A., Aldape, K., Mohapatra, G., Feuerstein, B.G., and Bishop, J.M. (1997) Targeted expression of *MYCN* causes neuroblastoma in transgenic mice. *EMBO J.* **16**, 2985–2995.

301. Shapiro, D.N., Valentine, M.B., Rowe, S.T., Sinclair, A.E., Sublett, J.E., Roberts, W.M., and Look, A.T. (1993) Detection of N-*myc* gene amplification by fluorescence *in situ* hybridization. Diagnostic utility for neuroblastoma. *Am. J. Pathol.* **142**, 1339–1346.

302. Tsuchida, Y., Hemmi, H., Inoue, A., Obana, K., Yang, H.W., Hayashi, Y., Kanda, N., and Shimatake, H. (1996) Genetic clinical markers of human neuroblastoma with special reference to N-myc oncogene: amplified or not amplified?—an overview. *Tumour Biol.* **17**, 65–74.

303. Boon, K., Caron, H.N., van Asperen, R., Valentijn, L., Hermus, M.-C., van Sluis, P., Roobeek, I., Weis, I., Voûte, P.A., Schwab, M., and Versteeg, R. (2001) N-*myc* enhances the expression of a large set of genes functioning in ribosome biogenesis and protein synthesis. *EMBO J.* **20**, 1383–1393.

304. Alaminos, M., Mora, J., Cheung, N.-K.V., Smith, A., Qin, J., Chen, L., and Gerald, W.L. (2003) Genome-wide analysis of gene expression associated with *MYCN* in human neuroblastoma. *Cancer Res.* **63**, 4538–4546.

305. Spieker, N., Van Sluis, P., Beitsma, M., Boon, K. van Schaik, B.D.C., van Kampen, A.H.C., Caron, H., and Versteeg, R. (2001) The MEIS1 oncogene is highly expressed in neuroblastoma and amplified in cell line IMR32. *Genomics* **71**, 214–221.

306. Lasorella, A., Boldrini, R., Dominici, C., Donfrancesco, A., Yokota, Y., Inserra, A., and Iavarone, A., (2002) Id2 is critical for cellular proliferation and is the oncogenic effector of N-Myc in human neuroblastoma. *Cancer Res.* **62**, 301–306.

307. Vandesompele, J., Edsjö, A., De Preter, K., Axelson, H., Speleman,F., and Påhlman, S. (2003) *ID2* expression in neuroblastoma does not correlate to *MYCN* levels and lacks prognostic value. *Oncogene* **22**, 456–460.

308. Wang, Q., Hii, G., Shusterman, S., Mosse, Y., Winter, C.L., Guo, C., Zhao, H., Rappaport, E., Hogarty, M.D., and Maris, J.M. (2003) ID2 expression is not associated with *MYCN* amplification or expression in human neuroblastomas. *Cancer Res.* **63**, 1631–1635.

309. Cohn, S.L., and Tweddle, D.A. (2004) MYCN amplification remains prognostically strong 20 years after its "clinical debut". *Eur. J. Cancer* **40**, 2639–2642.

310. Slavc, I., Ellenbogen, R., Jung, W.H., Vawter, G.F., Kretschmar, C., Grier, H., and Korf, B.R. (1990) Myc gene amplification and expression in primary human neuroblastoma. *Cancer Res.* **50**, 1459–1463.

311. Spitz, R., Hero, B., Skowron, M., Ernestus, K., and Berthold, F. (2004) MYCN-status in neuroblastoma: characteristics of tumours showing amplification, gain, and non-amplification. *Eur. J. Cancer* **40**, 2753–2759.

312. Amler, L.C., Schürmann, J., and Schwab, M. (1996) The *DDX1* gene maps within 400 kbp 5' to *MYCN* and is frequently coamplified in human neuroblastoma. *Genes Chromosomes Cancer* **15**, 134–137.

313. George, R.E., Kenyon, R.M., McGuckin, A.G., Malcolm, A.J., Pearson, A.D.J., and Lunec, J. on behalf of the United Kingdom Children's Cancer Study Group. (1996) Investigation of co-amplification of the candidate genes ornithine

decarboxylase, ribonucleotide reductase, syndecan-1 and a DEAD box gene, *DDX1*, with N-*myc* in neuroblastoma. *Oncogene* **12**, 1583–1587.

314. Wimmer, K., Zhu, X.X., Lamb, B.J., Kuick, R., Ambros, P.F., Kovar, H., Thoraval, D., Motyka, S., Alberts, J.R., and Hanash, S.M. (1999) Co-amplification of a novel gene, NAG, with the N-*myc* gene in neuroblastoma. *Oncogene* **18**, 233–238.

315. Shiloh, Y., Shipley, J., Brodeur, G.M., Bruns, G., Korf, B., Donlon, T., Schreck, R.R., Seeger, R., Sakai, K., and Latt, S.A. (1985) Differential amplification, assembly, and relocation of multiple DNA sequences in human neuroblastomas and neuroblastoma cell lines. *Proc. Natl. Acad. Sci. USA* **82**, 3761–3765.

316. Stallings, R.L., Carty, P., McArdle, L., Mullarkey, M., McDermott, M., O'Meara, A. , Ryan, E., Catchpoole, D., and Breatnach, F. (2004) Evolution of unbalanced gain of distal chromosome 2p in neuroblastoma. *Cytogenet. Genome Res.* **106**, 49–54.

317. Willatt, L.R., Pearson, J., and Green, A.J. (2001) Partial trisomy of 2p and neuroblastoma. *Am. J. Med. Genet.* **102**, 304–305.

318. Oberthuer, A., Skowron, M., Spitz, R., Kahlert, Y., Westermann, F., Mehler, K., Berthold, F., and Fischer, M. (2005) Characterization of a complex genomic alteration on chromosome 2p that leads to four alternatively spliced fusion transcripts in the neuroblastoma cell lines IMR-5, IMR-5/75 and IMR-32. *Gene* **363**, 41–50.

319. Jones, T.A., Flomen, R.H., Senger, G., Nizetic, D., and Sheer, D. (2000) The homeobox gene MEIS1 is amplified in IMR-32 and highly expressed in other neuroblastoma cell lines. *Eur. J. Cancer* **36**, 2368–2374.

320. Jones, T., Edwards, C., Flomen, R., Nizetic, D., and Sheer, D. (2001) Amplification and/or high expression of the homeobox gene meis 1 in neuroblastoma. *Cancer Genet. Cytogenet.* **128**, 91.

321. Geerts, D., Schilderink, N., Jorritsma, G., and Versteeg, R. (2003) The role of the MEIS homeobox genes in neuroblastoma. *Cancer Lett.* **197**, 87–92.

322. Geerts, D., Revet, I., Jorritsma, G., Schilderink, N., and Versteeg, R. (2005) MEIS homeobox genes in neuroblastoma. *Cancer Lett.* **228**, 43–50.

323. Takita, J., Yang, H.W., Chen, Y.Y., Hanada, R., Yamamoto, K., Teitz, T., Kidd, V., and Hayashi, Y. (2001) Allelic imbalance on chromosome 21 and alterations of the *caspase 8* gene in neuroblastoma. *Oncogene* **20**, 4424–4432.

324. Hallstensson, K., Thulin, S., Aburatani, H., Hippo, Y., and Martinsson, T. (1997) Representational difference analysis and loss of heterozygosity studies detect 3p deletions in neuroblastoma. *Eur. J. Cancer* **33**, 1966–1970.

325. Ejeskär, K., Aburatani, H., Abrahamsson, J., Kogner, P., and Martinsson, T. (1998) Loss of heterozygosity of 3p markers in neuroblastoma tumours implicate a tumour-suppressor locus distal to the FHIT gene. *Br. J. Cancer* **77**, 1787–1791.

326. Agathanggelou, A., Dallol, A., Zochbauer-Muller, S., Morrissey, C., Honorio, S., Hesson, L., Martinsson, T., Fong, K.M., Kuo, M.J., Yuen, P.W., Maher, E.R., Minna, J.D., and Latif, F. (2003) Epigenetic inactivation of the candidate 3p21.3 tumor suppressor gene in BLU in human cancers. *Oncogene* **22**, 1580–1588.

327. Nair, P.N., McArdle, L., Cornell, J., Cohn, S.L., and Stallings, R.L. (2007) High-resolution analysis of 3p deletion in neuroblastoma and differential methylation of the *SEMA3B* tumor suppressor gene. *Cancer Genet. Cytogenetics* **174**, 100–110.

328. Hoebeeck, J., Vandesompele, J., Nilsson, H., De Preter, K., Van Roy, N., De Smet, E., Yigit, N., De Paepe, A., Laureys, G., Påhlman, S., and Speleman, F. (2006) The von Hippel-Lindau tumor suppressor gene expression level has prognostic value in neuroblastoma. *Int. J. Cancer* **119**, 624–629.

329. Hoebeeck, J., Michels, E., Menten, B., Van Roy, N., Eggert, A., Schramm, A., De Preter, K., Yigit, N., De Smet, E., De Paepe, A., Laureys, G., Vandesompele, J., and Speleman, F. (2007) High resolution tiling-path BAC array deletion mapping suggests commonly involved 3p21-p22 tumor suppressor genes in neuroblastoma and more frequent tumors. *Int. J. Cancer* **120**, 533–538.

330. Meltzer, S.J., O'Doherty, S.P., Frantz, C.N., Smolinski, K., Yin, J., Cantor, A.B., Liu, J., Valentine, M., Brodeur, G.M., and Berg, P.E. (1996) Allelic imbalance on chromosome 5q predicts long-term survival in neuroblastomas. *Br. J. Cancer* **74**, 1855–1861.

331. Takita, J., Hayashi, Y., and Yokota, J. (1997) Loss of heterozygosity in neuroblastomas—an overview. *Eur. J. Cancer* **33**, 1971–1973.

332. Iolascon, A., Giordani, L., Moretti, A., Tonini, G.P., Lo Cunsolo, C., Mastropietro, S., Borriello, A., and Ragione, F.D. (1998) Structural and functional analysis of cyclin-dependent kinase inhibitor genes (*CDKN2A, CDKN2B,* and *CDKN2C*) in neuroblastoma. *Pediatr. Res.* **43**, 139–144.

333. Beltinger, C.P., White, P.S., Sulman, E.P., Maris, J.M., and Brodeur, G.M. (1995) No CDKN2 mutations in neuroblastoma. *Cancer Res.* **55**, 2053–2055.

334. Giordani, L., Iolascon, A., Servedio, V., Mazzocco, K., Longo, L., and Tonini, G.P. (2002) Two regions of deletion in 9p22~p24 in neuroblastoma are frequently observed in favorable tumors. *Cancer Genet. Cytogenet.* **135**, 42–47.

335. Marshall, B., Isidro, G., Martins, A.G., and Boavida, M.G. (1997) Loss of heterozygosity at chromosome 9p21 in primary neuroblastomas: evidence for two deleted reigons. *Cancer Genet. Cytogenet.* **96**, 134–139.

336. Takita, J., Hayashi, Y., Nakajima, T., Adachi, J., Tanaka, T., Yamaguchi, N., Ogawa, Y., Hanada, R., Yamamoto, K., and Yokota, J. (1998) The *p16* (*CDKN2A*) gene is involved in the growth of neuroblastoma cells and its expression is associated with prognosis of neuroblastoma patients. *Oncogene* **17**, 3137–3143.

337. Takita, J., Hayashi, Y., Kohno, T., Yamaguchi, N., Hanada, R., Yamamoto, K., and Yokota, J. (1997) Deletion map of chromosome 9 and *p16* (*CDKN2A*) gene alterations in neuroblastoma. *Cancer Res.* **57**, 907–912.

338. Muñoz, J., and Castresana, J.S. (2006) Silencing of DMBT1 in neuroblastoma cell lines is not due to methylation of CCWGG motifs on its promoter. *Neoplasma* **53**, 15–18.

339. Muñoz, J., Lázcoz, P., del Mar Inda, M., Nistal, M., Pestaña, A., Encío, I.J., and Castresana, J.S. (2004) Homozygous deletion and expression of PTEN and DMBT1 in human primary neuroblastoma and cell lines. *Int. J. Cancer* **109**, 673–679.

340. Su, W.T., Alaminos, M., Mora, J., Cheung, N.-K., La Quaglia, M.P., and Gerald, W.L. (2004) Positional gene

expression analysis identifies 12q overexpression and amplification in a subset of neuroblastomas. *Cancer Genet. Cytogenet.* **154**, 131–137.

341. Petković, I., Nakic, M., and Cepulic, M. (1989) Direct cytogenetic analysis of primary neuroblastoma. *Cancer Genet. Cytogenet.* **42**, 147–152.

342. Petković, I., and Cepulic, M. (1991) Cytogenetic analysis of primary neuroblastoma with del(1), del(14), hsr, and dmin chromosomes. *Cancer Genet. Cytogenet.* **55**, 231–234.

343. Petković, I., Nakic, M., Cepulic, M., and Konja, J. (1993) Chromosomal analysis of two neuroblastomas. *Cancer Genet. Cytogenet.* **65**, 167–169.

344. Theobald, M., Christiansen, H., Schmidt, A., Melekian, B., Wolkewitz, N., Christiansen, N.M., Brinkschmidt, C., Berthold, F., and Lampert, F. (1999) Sublocalization of putative tumor suppressor gene loci on chromosome arm 14q in neuroblastoma. *Genes Chromosomes Cancer* **26**, 40–46.

345. Hoshi, M., Otagiri, N., Shiwaku, H.O., Asakawa, S., Shimizu, N., Kaneko, Y., Ohi, R., Hayashi, Y., and Horii, A. (2000) Detailed deletion mapping of chromosome band 14q32 in human neuroblastoma defines a 1.1-Mb region of common allelic loss. *Br. J. Cancer* **82**, 1801–1807.

346. Hoshi, M., Shiwaku, H.O., Hayashi, Y., Kaneko, Y., and Horii, A. (2000) Deletion mapping of 14q32 in human neuroblastoma defines an 1,100-kb region of common allelic loss. *Med. Pediatri. Oncol.* **35**, 522–525.

347. Suzuki, T., Yokota, J., Mugishima, H., Okabe, K., Ookuni, M., Sugimura, T., and Terada, M. (1989) Frequent loss of heterozygosity on chromosome 14q in neuroblastoma. *Cancer Res.* **49**, 1095–1098.

348. Fong, C., White, P.S., Peterson, K., Sapienza, C., Cavenee, W.K., Kern, S.E., Vogelstein, B., Cantor, A.B., Look, A.T., and Brodeur, G.M. (1992) Loss of heterozygosity for chromosomes 1 or 14 defines subsets of advanced neuroblastomas. *Cancer Res.* **52**, 1780–1785.

349. Srivatsan, E.S., Murali, V., and Seeger, R.C. (1991) Loss of heterozygosity for alleles on chromosomes 11q and 14q in neuroblastoma. *Prog. Clin. Biol. Res.* **366**, 91–98.

350. Takayama, H., Suzuki, T., Mugishima, H., Fujisawa, T., Ookuni, M., Schwab, M., Gehring, M., Nakamura, Y., Sugimura, T., Terada, M., and Yokota. J. (1992) Deletion mapping of chromosomes 14q and 1p in human neuroblastoma. *Oncogene* **7**, 1185–1189.

351. Takita, J., Hayashi, Y., Takei, K., Yamaguchi, N., Hanada, R., Yamamoto, K., and Yokota, J. (2000) Allelic imbalance on chromosome 18 in neuroblastoma. *Eur. J. Cancer* **36**, 508–513.

352. Reale, M.A., Reyes-Mugica, M., Pierceall, W.E., Rubinstein, M.C., Hedrick, L., Cohn, S.L., Nakagawara, A., Brodeur, G.M., and Fearon, E.R. (1996) Loss of *DCC* expression in neuroblastoma is associated with disease dissemination. *Clin. Cancer Res.* **2**, 1097–1102.

353. Kong, X.-T., Choi, S.H., Inoue, A., Xu, F., Chen, T., Takita, J., Yokota, J., Bessho, F., Yanagisawa, M., Hanada, R., Yamamoto, K., and Hayashi, Y. (1997) Expression and mutationl analysis of the *DCC, DPC4,* and *MADR2JV18–1* genes in neuroblastoma. *Cancer Res.* **57**, 3772–3778.

354. laminos, M., Dávalos, V., Ropero, S., Setién, F., Paz, M.F., Herranz, M., Fraga, M.F., Mora, J., Cheung, N.-K.V., Gerald, W.L., and Esteller, M. (2005) *EMP3*, a myelin-related gene located in the critical 19q13.3 region, is epigenetically silenced and exhibits features of a candidate tumor suppressor in glioma and neuroblastoma. *CancerRes.* **65**, 2565–2571.

355. Alaminos, M., Dávalos, V., Cheung, N.-K.V., Gerald, W.L., and Esteller, M. (2004) Clustering of gene hypermethylation associated with clinical risk groups in neuroblastoma. *J. Natl. Cancer Inst.* **96**, 1208–1219.

356. Stallings, R.L., Nair, P., Maris, J.M., Catchpoole, D., McDermott, M., O'Meara, A., and Breatnach, F. (2006) High-resolution analysis of chromosomal breakpoints and genomic instability identifies *PTPRD* as a candidate tumor suppressor gene in neuroblastoma. *Cancer Res.* **66**, 3673–3680.

356A. Alaminos, M., Dávalos, V., Ropero, S., Setién, F., Paz, M.F., Herranz, M., Fraga, M.F., Mora, J., Cheung, N.-K., V., Gerald, W.L, and Esteller, M. (2005) *EMP3*, a myelin-related gene located in the critical 19q13.3 region, is epigenetically silenced and exhibits features of a candidate tumor suppressor in glioma and neuroblastoma. *Cancer Res.* **65**, 2565–2571.

357. Ohira, M., Shishikura, T., Kawamoto, T., Inuzuka, H., Morohashi, A., Takayasu, H., Kageyama, H., Takada, N., Takahashi, M., Sakiyama, S., Suzuki, Y., Sugano, S., Kuma, H., Nozawa, I., and Nakagawara, A. (2000) Hunting the subset-specific genes of neuroblastoma: expression profiling and differential screening of the full-length-enriched oligo-capping cDNA libraries. *Med. Pediatr. Oncol.* **35**, 547–549.

358. Fischer, M., Oberthuer, A., Brors, B., Kahlert, Y., Skowron, M., Voth, H., Warnat, P., Ernestus, K., Hero, B., and Berthold, F. (2006) Differential expression of neuronal genes defines subtypes of disseminated neuroblastoma with favorable and unfavorable outcome. *Clin. Cancer Res.* **12**, 5118–5128.

359. Coco, S., Defferrari, R., Scaruffi, P., Cavazzana, A., Di Cristofano, C., Longo, L., Mazzocco, K., Perri, P., Gambini, C., Moretti, S., Bonassi, S., and Tonini, G.P. (2005) Genome analysis and gene expression profiling of neuroblastoma and ganglioneuroblastomas reveal differences between neuro- blastic and Schwannian stromal cells. *J. Pathol.* **207**, 346–357.

360. Gilbert, J., Haber, M., Bordow, S.B., Marshall, G.M., and Norris, M.D. (1999) Use of tumor-specific gene expression for the differential diagnosis of neuroblastoma from other pediatric small round-cell malignancies. *Am. J. Pathol.* **155**, 17–21.

361. Speleman, F., Van Roy, N., Berx, G., Strumane, K., De Paepe, A., Van Roy, F., and Vandesompele, J. (2001) Gene expression profiling in high-stage neuroblastomas. *Cancer Genet. Cytogenet.* **128**, 68.

362. Truckenmiller, M.E., Vawter, M.P., Cheadle, C., Coggiano, M., Donovan, D.M., Freed, W.J., and Becker, K.G. (2001) Gene expression profile in early stage of retinoic acid-induced differentiation of human SH-SY5Y neuroblastoma cells. *Restor. Neurol. Neurosci.* **18**, 67–80.

363. Sardi, I., Tintori, V., Marchi, C., Veltroni, M., Lippi, A., Tucci, F., Tamburini, A., Bernini, G., and Faulkner, L. (2002) Molecular profiling of high-risk in neuroblastoma by cDNA array. *Int. J. Mol. Med.* **9**, 541–545.

364. Ohira, M., Morohashi, A., Inuzuka, H., Shishikura, T., Kawamoto, T., Kageyama, H., Nakamura, Y., Isogai, E., Takayasu, H., Sakiyama, S., Suzuki, Y., Sugano, S., Goto, T., Sato, S., and Nakagawara, A. (2003) Expression profiling and characterization of 4200 genes cloned from primary neuroblastoma: identification of 305 genes differentially expressed between favorable and unfavorable subsets. *Oncogene* **22**, 5525–5536.

365. Ohira, M., Oba, S., Nakamura, Y., Isogai, E., Kaneko, S., Nakagawa, A., Hirata, T., Kubo, H., Goto, T., Yamada, S., Yoshida, Y., Fuchioka, M., Ishii, S., and Nakagawara, A. (2005) Expression profiling using tumor-specific cDNA microarray predicts the prognosis of intermediate risk neuroblastomas. *Cancer Cell* **7**, 337–350.

366. Ohira, M., Oba, S., Nakamura, Y., Hirata, T., Ishii, S., and Nakagawara, A. (2005) A review of DNA microarray analysis of human neuroblastomas. *Cancer Lett.* **228**, 5–11.

367. Takita, J., Ishii, M., Tsutsumi, S., Tanaka, Y., Kato, K., Toyoda, Y., Hanada, R., Yamamoto, K., Hayashi, Y., and Aburatani, H. (2004) Gene expression profiling and identification of novel prognostic marker genes in neuroblastoma. *Genes Chromosomes Cancer* **40**, 120–132.

368. Fan, J., Chen, Y., Chan, H.M., Tam, P.K.H., and Ren, Y. (2005) Removing intensity effects and identifying significant genes for Affymetrix arrays in macrophage migration inhibitory factor-suppressed neuroblastoma cells. *Proc. Natl. Acad. Sci. USA* **102**, 17751–17756.

369. Fischer, M., Skowron, M., and Berthold, F. (2005) Reliable transcript quantification by real-time reverse transcriptase-polymerase chain reaction in primary neuroblastoma using normalization to averaged expression levels of the control genes *HPRT1* and *SDHA*. *J. Mol. Diagn.* **7**, 89–96.

370. Krause, A., Combaret, V., Iacono, I., Lacroix, B., Compagnon, C., Bergeron, C., Valsesia-Wittmann, S., Leissner, P., Mougin, B., and Puisieux, A. (2005) Genome-wide analysis of gene expression in neuroblastomas detected by mass screening. *Cancer Lett.* **225**, 111–120.

371. Scaruffi, P., Valent, A., Schramm, A., Astrahantseff, K., Eggert, A., and Tonini, G.P. (2005) Application of microarray-based technology to neuroblastoma. *Cancer Lett.* **228**, 13–20.

372. Schramm, A., Schulte, J.H., Klein-Hitpass, L., Havers, W., Sieverts, H., Berwanger, B., Christiansen, H., Warnat, P., Brors, B., Eils, J., Eils, R., and Eggert, A. (2005) Prediction of clinical outcome and biological characterization of neuroblastoma by expression profiling. *Oncogene* **24**, 7902–7912.

373. Simon, R. (2006) Development and evaluation of therapeutically relevant predictive classifiers using gene expression profiling. *J. Natl. Cancer Inst.* **98**, 1169–1171.

374. Asgharzadeh, S., Pique-Regi, R., Sposto, R., Wang, H., Yang, Y., Shimada, H., Matthay, K., Buckley, J., Ortega, A., and Seeger, R.C. (2006) Prognostic significance of gene expression profiles of metastatic neuroblastomas lacking MYCN gene amplficiation. *J. Natl. Cancer Inst.* **98**, 1193–1203.

375. Khan, J., Wei, J.S., Ringnér, M., Saal, L.H., Ladanyi, M., Westermann, F., Berthold, F., Schwab, M., Antonescu, C.R., Peterson, C., and Meltzer, P.S. (2003) Classification and diagnostic prediction of cancers using gene expression profiling and artificial neural networks. *Nat. Med.* **7**, 673–679.

376. Mora, J., Cheung, N.-K.V., Oplanich, S., Chen, L., and Gerald, W.L. (2002) Novel regions of allelic imbalance identified by genome-wide analysis of neuroblastoma. *Cancer Res.* **62**, 1761–1767.

377. Mora, J., Gerald, W.L., and Cheung, N.K. (2003) Evolving significance of prognostic markers associated with new treatment strategies in neuroblastoma. *Cancer Lett.* **197**, 119–124.

378. Hiyama, E., Hiyama, K., Yamaoka, H., Sueda, T., Reynolds, C.P., and Yokoyama, T. (2004) Expression profiling of unfavorable and unfavorable neuroblastomas. *Pediatr. Surg. Int.* **20**, 33–38.

379. Fransson, S., Martinsson, T., and Ejeskär, K. (2007) Neuroblastoma tumors with favorable and unfavorable outcomes: significant differences in mRNA expression of genes mapped at 1p36.2. *Genes Chromosomes Cancer* **46**, 45–52.

380. Krasnoselsky, A.L., Whiteford, C.C., Wei, J.S., Bilke, S., Westermann, F., Chen, Q.-R., and Khan, J. (2005) Altered expression of cell cycle genes distinguishes aggressive neuroblastoma. *Oncogene* **24**, 1533–1541.

381. Vasudevan, S.A., Nuchtern, J.G., and Shohet, J.M. (2005) Gene profiling of high risk neuroblastoma. *World J. Surg.* **29**, 317–324.

381A. Wei, J.S., Greer, B.T., Westermann, F., Steinberg, S.M., Son, C.-G., Chen, Q.-R., Whiteford, C.C., Bilke, S. Krasnoselsky, A.L., Cenacchi, N., Catchpoole, D., Berthold, F., Schwab, M., and Khan, J. (2004) Prediction of clinical outcome using gene expression profiling and artificial neural networks for patients with neuroblastoma. *Cancer Res.* **64**, 6883–6891.

382. Carr, J., Bown, N.P., Case, M.C., Hall, A.G., Lunec, J., and Tweddle, D.A. (2007) High-resolution analysis of allelic inbalance in neuroblastoma cell lines by single nucleotide polymorphism arrays. *Cancer Genet. Cytogenet.* **172**, 127–138.

383. Berwanger, B., Hartmann, O., Bergmann, E., Bernard, S., Nielsen, D., Krause, M., Kartal, A., Flynn, D., Wiedemeyer, R., Schwab, M., Schäfer, H., Christiansen, H., and Eilers, M. (2002) Loss of a *FYN*-regulated differentiation and growth arrest pathway in advanced stage neuroblastoma. *Cancer Cell* **2**, 377–386.

384. Kerr, J.F.R., Wyllie, A.H., and Currie, A.R. (1972) Apoptosis: a basic biologial phenomenon with wide-ranging implications in tissue kinetics. *Br. J. Cancer* **26**, 239–257.

385. Ikeda, H., and Castle, V.P. (2000) Apoptosis in neuroblastoma and pathways involving bcl-2 and caspases. In *Neuroblastoma* (Brodeur, G.M., Sawada, T., Tsuchida, Y., and Voûte, P.A., eds.), Elsevier Science B.V., Amsterdam, The Netherlands, pp. 197–205.

386. Ikeda, H., Hirato, J., Akami, M., Suzuki, N., Takahashi, A., Kuroiwa, M., and Matsuyama, S. (1996) Massive apoptosis detected by in situ DNA nick end labeling in neuroblastoma. *Am. J. Surg. Pathol.* **20**, 649–655.

387. Hoehner, J.C., Gestblom, C., Olsen, L., and Påhlman, S. (1997) Spatial association of apoptosis-related gene expression and cellular death in clinical neuroblastoma. *Br. J. Cancer* **75**, 1185–1194.

388. Reed, J.C., Meister, L., Tanaka, S., Cuddy, M., Yum, S., Geyer, C., and Pleasure, D. (1991) Differential expres-

sion of *bcl2* protooncogene in neuroblastoma and other human tumor cell lines of neural origin. *Cancer Res.* **541**, 6529–6538.

389. Hanada, M., Krajewski, S., Tanaka, S., Cazals-Hatem, D., Spengler, B.A., Ross, R.A., Biedler, J.L., and Reed, J.C. (1993) Regulation of Bcl-2 oncoprotein levels with differentiation of human neuroblastoma cells. *Cancer Res.* **53**, 4978–5086.

390. Dole, M.G., Jasty, R., Cooper, M.J., Thompson, C.B., Nuñez, G., and Castle, V.P. (1995) Bcl-x$_L$ is expressed in neuro- blastoma cells and modules chemotherapy-induced apoptosis. *Cancer Res.* **55**, 2576–2582.

391. Castle, V.P., Heidelberger, K.P., Bromberg, J., Ou, X., Dole, M., and Nuñez, G. (1993) Expression of the apoptosis-suppressing protein *bcl-2*, in neuroblastoma is associated with unfavorable histology and N-*myc* amplification. *Am. J. Pathol.* **143**, 1543–1550.

392. Hoehner, J.C., Hedborg, F., Jernberg Wiklund, H., Olsen, L., and Påhlman, S. (1995) Cellular death in neuroblastoma: *in situ* correlation of apoptosis and bcl-2 expression. *Int. J. Cancer* **62**, 19–24.

393. Romani, M., Scaruffi, P., Casciano, I., Mazzocco, K., Lo Cunsolo, C., Cavazzana, A., Gambini, C., Boni, L., De Bernardi, B., and Tonni, G.P. (1999) Stage-independent expression and genetic analysis of *tp73* in neuroblastoma. *Int. J. Cancer* **84**, 365–369.

394. Dole, M., Nuñez, G., Merchant, A.K., Maybaum, J., Rode, C.K., Bloch, C.A., and Castle, V.P. (1994) Bcl-2 inhibits chemotherapy-induced apoptosis in neuroblastoma. *Cancer Res.* **54**, 3253–3259.

395. Ikegaki, N., Katsumata, M., Tsujimoto, Y., Nakagawara, A., and Brodeur, G.M. (1995) Relationship between bcl-2 and myc gene expression in human neuroblastoma. *Cancer Lett.* **91**, 161–168.

395A. Koizumi, H., Wakisaka, M., Nakada, K., Takakuwa, T., Fujioka, T., Yamate, N., and Uchikoshi, T. (1995) Demonstration of apoptosis in neuroblastoma and its relationship to tumour regression. *Virchows Arch.* **427**, 167–173.

396. Oue, T., Fukuzawa, M., Kusafuka, T., Kohmoto, Y., Imura, K., Nagahara, S., and Okada, A. (1996) In situ detection of DNA fragmentation and expression of *bcl*-2 in human neuro- blastoma: relation to apoptosis and spontaneous regression. *J. Pediatr. Surg.* **31**, 251–257.

397. Ramani, P., and Lu, Q.L. (1994) Expression of bcl-2 gene product in neuroblastoma. *J. Pathol.* **172**, 273–278.

397A. Castle, V.P., Heidelberger, K.P., Bromberg, J., Ou, X., Dole, M., Nuñez, G. (1993) Expression of the apoptosis-suppressing protein bcl-2, in neuroblastoma is associated with unfavorable histology and N-myc amplification. *Am.J. Pathol.* **143**, 1543–1550.

398. Ikeda, H., Hirato, J., Akami, M., Matsuyama, S., Suzuki, N., Takahashi, A., and Kuroiwa, M. (1995) Bcl-2 oncoprotien expression and apoptosis in neuroblastoma. *J. Pediatr. Surg.* **30**, 805–808.

399. Dole, M.G., Clarke, M.F., Holman, P., Benedict, M., Lu, J., Jasty, R., Eipers, P., Thompson, C.B., Rode, C., Bloch, C., Nuñez, G., and Castle, V.P. (1996) Bcl-x$_S$ enhances adenoviral vector-induced apoptosis in neuroblastoma cells. *Cancer Res.* **56**, 5734–5740.

400. Krajewski, S., Chatten, J., Hanada, M., and Reed, J.C. (1995) Immunohistochemical analysis of the Bcl-2 oncoprotein in human neuroblastomas. Comparisons with tumor cell differentiation and N-Myc protein. *Lab. Inves.* **72**, 42–54.

401. Gallo, G., Giarnieri, E., Bosco, S., Cappelli, C., Alderisio, M., Giovagnoli, M.R., Giordano, A., and Vecchione, A. (2003) Aberrant bcl-2 and bax protein expression related to chemotherapy response in neuroblastoma. *Anticancer Res.* **23**, 777–784.

402. Lasorella, A., Iavarone, A., and Israel, M.A. (1995) Differentiation of neuroblastoma enhances bcl-2 expression and induces alterations of apoptosis and drug resistance. *Cancer Res.* **55**, 4711–4716.

403. Fulda, S., Sieverts, H., Friesen, C., Herr, I., and Debatin, K.-M., (1997) The CD95 (APO-1/Fas) system mediates drug-induced apoptosis in neuroblastoma cells. *Cancer Res.* **57**, 3823–3829.

404. Fulda, S., Friesen, C., Los, M., Scaffidi, C., Mier, W., Benedict, M., Nuñez, G., Krammer, P.H., Peter, M.E., and Debatin, K.-M., (1997) Betulinic acid triggers CD95 (APO-1/Fas)- and p53-independent apoptosis via activation of caspases in neuroectodermal tumors. *Cancer Res.* **57**, 4956–4964.

405. Rutiloni, C., and Ammaniti, M. (1967) Considerazioni sul decorso e prognosi del neuroblastoma neonatale. *Min. Ped.* **19**, 1773–1778.

406. Palásthy, V.G. (1973) Die spontane regression des neuroblastoms. Ein überblick über die biolgischen und immunologischen ursahen der heilung. In *Kinderärztliche Praxis*, Leipzig Georg Thieme Verlag, pp. 23–30.

407. Takeda, T., and Sotooka, T. (1977) Spontaneous regression of neuroblastoma. *Hokkaido Igaku Zasshi.* **52**, 12–15.

408. Rangecroft, L., Lauder, I., and Wagget, J. (1978) Spontaneous maturation of stage IV-S neuroblastoma. *Arch. Dis. Child.* **53**, 815–817.

409. Eklof, O., Sandstedt, B., Thonell, S., and Ahström, L. (1983) Spontaneous regression of stage IV neuroblastoma. *Acta Paediatr. Scand.* **72**, 473–476.

410. Haas, D., Ablin, A.R., Miller, C., Zoger, S., and Matthay, K.K. (1988) Complete pathologic maturation and regression of Stage IVS neuroblastoma withtout treatment. *Cancer* **62**, 818–825.

411. Carlsen, N.L. (1990) How frequent is spontaneous remission of neuroblastomas? Implications for screening. *Br. J. Cancer* **61**, 441–446.

412. Kushner, B.H., Cheung, N.-K.V., LaQuaglia, M.P., Ambros, P.F., Ambros, I.M., Bonilla, M.A., Gerald, W.L., Ladanyi, M., Gilbert, F., Rosenfield, N.S., and Yeh, S.D.J. (1996) Survival from locally invasive or widespread neuroblastoma without cytotoxic therapy. *J. Clin. Oncol.* **14**, 373–381.

413. Yamamoto, K., Hanada, R., Kikuchi, A., Ichikawa, M., Aihara, T., Oguma, E., Moritani, T., Shimanuki, Y., Tanimura, M., and Hayashi, Y. (1998) Spontaneous regression of localized neuroblastoma detected by mass screening. *J. Clin. Oncol.* **16**, 1265–1269.

414. Ambros, P.F., and Brodeur, G.M. (2000) Concept of tumorigenesis and regression. In *Neuroblastoma* (Brodeur. G.M., Sawada, T., Tsuchida, Y., and Voûte, P.A., eds.), Elsevier Science B.V., pp. 21–32.

415. Nishio, N., Mimaya, J., Horikoshi, Y., Okada, N., Nara, T., Takashima, Y., Urushihara, N., Hasegawa, S., Aoki, K., and Hamasaki, M. (2006) Spontaneous regression of metastases including meningeal metastasis after gross resection of primary tumor in an infant with stage 4 neuroblastoma. *J. Pediatr. Hematol. Oncol.* **28**, 537–539.

416. Evans, A.E., Gerson, J., and Schnaufer, L. (1976) Spontaneous regression of neuroblastoma. *Natl. Cancer Inst. Monogr.* **44**, 49–54.

417. Pritchard, J., and Hickman, J.A. (1994) Why does stage 4s neuroblastoma regress spontaneously? *Lancet* **344**, 869–870.

418. Nakagawara, A. (1998) Molecular basis of spontaneous regression of neuroblastoma: role of neurotrophic signals and genetic abnormalities. *Hum.Cell* **11**, 115–124.

419. Kitanaka, C., Kato, K., Ijiri, R., Sakurada, K., Tomiyama, A., Noguchi, K., Nagashima, Y., Nakagawara, A., Momoi, T., Toyoda, Y., Kigasawa, H., Nishi, T., Shirouzu, M., Yokoyama, S., Tanaka, Y., and Kuchino, Y. (2002) Increased Ras expression and caspase-independent neuroblastoma cell death: possible mechanism of spontaneous neuroblastoma regression. *J. Natl. Cancer Inst.* **94**, 358–368.

420. Nakagawara, A. (1998) The NGF story and neuroblastoma. *Med. Pediatr. Oncol.* **31**, 113–115.

421. Ambros, I.M., Zellner, A., Roald, B., Amann, G., Ladenstein, R., Printz, D., Gadner, H., and Ambros, P.F. (1996) Role of ploidy, chromosome 1p, and Schwann cells in the maturation of neuroblastoma. *N. Engl. J. Med.* **334**, 1505–1511.

422. Reynolds, C.P. (2002) Ras and Seppuku in neuroblastoma. *J. Natl. Cancer Inst.* **94**, 319–321.

423. Koizumi, H., Hamano, S., Doi, M., Tatsunami, S., Nakada, K., Shinagawa, T., and Tadokoro, M. (2006) Increased occurrence of caspase-dependent apoptosis in unfavorable neuroblastomas. *Am. J. Surg. Pathol.* **30**, 249–257.

424. Fowler, C.L., Zimmer, C.C., and Zimmer, S.G. (2000) Spontaneous progression of stage IV-S human neuroblastoma cell line involves the increased expression of the protooncogenes ras and eukaryotic initiation factor 4E. *Pediatr. Pathol. Mol. Med.* **19**, 433–447.

425. Abe, M., Ohira, M., Kaneda, A., Yagi, Y., Yamamoto, S., Kitano, Y., Takato, T., Nakagawara, A., and Ushijima, T. (2005) CpG island methylator phenotype is a strong determinant of poor prognosis in neuroblastomas. *Cancer Res.* **65**, 828–834.

426. Yang, Q., Zage, P., Kagan, D., Tian, Y., Seshadri, R., Salwen, H.R., Liu, S., Chlenski, A., and Cohn, S.L. (2004) Association of epigenetic inactivation of *RASSF1A* with poor outcome in human neuroblastoma. *Clin. Cancer Res.* **10**, 8493–8500.

426A. van Noesel, M.M., van Bezouw, S., Voûte, P.A., Herman, J.G., Pieters, R., and Versteeg, R. (2003) Clustering of hypermethylated genes in neuroblastoma. *Genes Chromosomes Cancer* **38**, 226–233.

427. Harada, K., Toyooka, S., Maitra, A., Maruyama, R., Toyooka, K.O., Timmons, C.F., Tomlinson, G.E., Mastrangelo, D., Hay, R.J., Minna, J.D., and Gazdar, A.F. (2002) Aberrant promoter methylation and silencing of the *RASSF1A* gene in pediatric tumors and cell lines. *Oncogene* **21**, 4345–4349.

428. Astuti, D., Agathanggelou, A., Honorio, S., Dallol, A., Martinsson, T., Kogner, P., Cummins, C., Neumann, H.P.H., Voutilainen, R., Dahia, P., Eng, C., Maher, E.R., and Latif, F. (2001) *RASSF1A* promoter region CpG island hypermethylation in phaeochromocytomas and neuroblastoms tumours. *Oncogene* **20**, 7573–7577.

429. Abraham, M.C., and Shaham, S. (2004) Death with caspases, caspases without death. *Trends Cell Biol.* **14**, 184–193.

430. Ikeda, H., Nakamura, Y., Hiwasa, T., Sakiyama, S., Kuida, K., Su, M.S., and Nakagawara, A. (1997) Interleukin-1β converting enzyme (ICE) is preferentially expressed in neuroblastomas with favourable prognosis. *Eur. J. Cancer* **33**, 2081–2083.

431. Nakagawara, A., Arima-Nakagawara, M., Scavarda, N.J., Azar, C.G., Cantor, A.B., and Brodeur, G.M. (1993) Association between high levels of expression of the Trk gene and favorable outcome in human neuroblastoma. *N. Engl. J. Med.* **328**, 847–854.

432. Posmantur, R., McGinnis, K., Nadimpalli, R., Gilbertsen, R.B., and Wang, K.K. (1997) Characterization of CPP32-like protease activity following apoptotic challenge in SH-SY5Y neuroblastoma cells. *J. Neurochem.* **68**, 2328–2337.

433. Harada, K., Toyooka, S., Shivapurkar, N., Maitra, A., Reddy, J.L., Matta, H., Miyajima, K., Timmons, C.F., Tomlinson, G.E., Mastrangelo, D., Hay, R.J., Chaudhary, P.M., and Gazdar, A.F. (2002) Deregulation of *Caspase 8* and *10* expression in pediatric tumors and cell lines. *Cancer Res.* **62**, 5897–5901.

434. Iolascon, A., Borriello, A., Giordani, L., Cucciolla, V., Moretti, A., Monno, F., Criniti, V., Marzullo, A., Criscuolo, M., and Della Ragione, F. (2003) Caspase 3 and 8 deficiency in human neuroblastoma. *Cancer Genet. Cytogenet.* **146**, 41–47.

435. Teitz, T., Wei, T., Valentine, M.B., Vanin, E.F., Grenet, J., Valentine, V.A., Behm, F.G., Look, A.T., Lahti, J.M., and Kidd, V.J. (2000) Caspase 8 is deleted or silenced preferentially in childhood neuroblastomas with amplification of *MYCN. Nat. Med.* **6**, 529–535.

436. Hopkins-Donaldson, S., Bodmer, J.-L., Bourloud, K.B., Brognara, C.B., Tschopp, J., and Gross, N. (2000) Loss of caspase-8 expression in highly malignant human neuroblastoma cells correlates with resistance to tumor necrosis factor-related apoptosis-inducing ligand-induced apoptosis. *Cancer Res.* **60**, 4315–4319.

437. Stupack, D.G., Teitz, T., Potter, M.D., Mikolon, D., Houghton, P.J., Kidd, V.J., Lahti, J.M., Cheresh, D.A. (2006) Potentiation of neuroblastoma metastasis by loss of caspase-8. *Nature* **439**, 95–99.

438. Hopkins-Donaldson, S., Yan, P., Bourloud, K.B., Mühlethaler, A., Bodmer, J.-L., and Gross, N. (2002) Doxorubicin-induced death in neuroblastoma does not involve death receptors in S-type cells and is caspase-independent in N-type cells. *Oncogene* **21**, 6132–6137.

439. Koizumi, H., Ohkawa, I., Tsukahara, T., Momoi, T., Nakada, K., and Uchikoshi, T. (1999) Apoptosis in favourable neuroblastomas is not dependent on Fas (CD95/APO-1) expression but on activated caspase 3 (CPP32). *J. Pathol.* **189**, 410–415.

440. Teitz, T., Wei, T., Liu, D., Valentine, V., Valentine, M., Grenet, J., Lahti, J.M., and Kidd, V.J. (2002) Caspase-9 and Apaf-1 are expressed and functionally active in human neuroblastoma tumor cell lines with 1p36 LOH and amplified *MYCN*. *Oncogene* **21**, 1848–1858.

441. Eggert, A., Grotzer, M.A., Zuzak, T.J., Wierrodt, B.R., Ho, R., Ikegaki, N., and Brodeur, G.M. (2001) Resistance to tumor necrosis factor-related apoptosis-inducing ligand (TRAIL)-induced apoptosis in neuroblastoma cells correlates with a loss of caspase-8 expression. *Cancer Res.* **61**, 1314–1319.

442. De Ambrosis, A., Casciano, I., Croce, M., Pagnan, G., Radic, L., Banelli, B., Di Vinci, A., Allemanni, G., Tonini, G.P., Ponzoni, M., Romani, M., and Ferrini, S. (2006) An interferon-sensitive response element is involved in constitutive caspase-8 gene expression in neuroblastoma cells. *Int. J. Cancer* **120**, 39–47.

443. Fulda, S., Kufer, M.U., Meyer, E., van Valen, F., Dockhorn-Dworniczak, B., and Debatin, K.M. (2001) Sensitization for death receptor- or drug-induced apoptosis by re-expression of caspase-8 through demethylation or gene transfer. *Oncogene* **20**, 5865–5877.

444. Mühlethaler-Mottet, A., Bourloud, K.B., Auderset, K., Joseph, J.-M., and Gross, N. (2004) Drug-mediated sensitization to TRAIL-induced apoptosis in caspase-8-complemented neuroblastoma cells proceeds via activation of intrinsic and extrinsic pathways and caspase-dependent cleavage of XIAP, Bcl-x$_L$ and RIP. *Oncogene* **23**, 5415–5425.

445. Annicchiarico-Petruzzelli, M., Bernassola F., Lovat, P.E., Redfern, C.P., Pearson, A.D., and Melino, G. (2001) Apoptosis in neuroblastomas induced by interferon-γ involves the CD95/CD95L pathway. *Med. Pediatr. Oncol.* **36**, 115–117.

446. Banelli, B., Casciano, I., Croce, M., Di Vinci, A., Gelvi, I., Pagnan, G., Brignole, C., Allemanni, G., Ferrini, S., Ponzoni, M., and Romani, M. (2002) Expression and methylation of CASP8 in neuroblastoma: identification of a promoter region. *Nat. Med.* **8**, 1333–1335.

447. Fulda, S., and Debatin, K.M. (2002) IFNγ sensitizes for apoptosis by upregulating caspase-8 expression through the Stat1 pathway. *Oncogene* **21**, 2295–2308.

448. Yang, X., Merchant, M.S., Romero, M.E., Tsokos, M., Wexler, L.H., Kontny, U., Mackall, C.L., and Thiele, C.J. (2003) Induction of caspase 8 by interferon-γ renders some neuroblastoma (NB) cells sensitive to tumor necrosis factor-related apoptosis-inducing ligand (TRAIL) but reveals that a lack of membrane TR1/TR2 also contributes to TRAIL resistance in NB. *Cancer Res.* **63**, 1122–1129.

449. Casciano, I., Banelli, B., Croce, M., De Ambrosis, A., di Vinci, A., Gelvi, I., Pagnan, G., Brignole, C., Allemanni, G., Ferrini, S., Ponzoni, M., and Romani, M. (2004) Caspase-8 gene expression in neuroblastoma. *Am. N Y Acad. Sci.* **1028**, 157–167.

450. Casciano, I., De Ambrosis, A., Croce, M., Pagnan, G., Di Vinci, A., Allemanni, G., Banelli, B., Ponzoni, M., Romani, M., and Ferrini, S. (2004) Expression of the caspase-8 gene in neuroblastoma cells is regulated through an essential interferon-sensitive response element (ISRE). *Cell Death Differ.* **11**, 131–134.

451. Nakagawara, A., and Brodeur, G.M. (1997) Role of neurotrophins and their receptors in human neuroblastomas: a primary culture study. *Eur. J. Cancer* **33**, 2050–2053.

452. Nakagawara, A., Arima, M., Azar, C.G., Scavarda, N.J., and Brodeur, G.M. (1992) Inverse relationship between *trk* expression and N-*myc* amplification in humam neuroblastomas. *Cancer Res.* **52**, 1364–1368.

453. Nakagawara, A., Nakamura, Y., Ikeda, H., Hiwasa, T., Kuida, K., Su, M.S.-S., Zhao, H, Cnaan, A., and Sakiyama, S. (1997) High levels of expression and nuclear localization of interleukin-1 β converting enzyme (ICE) and CPP32 in favorable human neuroblastomas. *Cancer Res.* **57**, 4578–4584.

454. Mao, P.-L., Jiang, Y., Wee, B.Y., and Porter, A.G. (1998) Activation of caspase-1 in the nucleus requires nuclear translocation of pro-caspase-1 mediated by its prodomain. *J. Biol. Chem.* **273**, 23621–23624.

455. Fulda, S., Poremba, C., Berwanger, B., Häcker, S., Eilers, M., Christiansen, H., Hero, B., and Debatin, K.-M. (2006) Loss of Caspase-8 expression does not correlate with *MYCN* amplification, aggressive disease, or prognosis in neuroblastoma. *Cancer Res.* **66**, 10016–10023.

456. Nakagawara, A., and Ohira, M. (2004) Comprehensive genomics linking between neural development and cancer: neuroblastoma as a model. *Cancer Lett.* **204**, 213–224.

457. Kogner, P., Barbany, G., Dominici, C., Castello, M.A., RaschellàG., and Persson, H. (1993) Coexpression of messenger RNA for *TRK* protooncogene and low affinity nerve growth factor receptor in neuroblastoma with favorable prognosis. *Cancer Res.* **53**, 2044–2050.

458. Suzuki, T., Bogenmann, E., Shimada, H., Stram, D., Seeger, R.C. (1993) Lack of high-affinity nerve growth factor receptors in aggressive neuroblastomas. *J. Natl. Cancer Inst.* **85**, 377–384.

459. Tanaka, T., Hiyama, E., Sugimoto, T., Sawada, T., Tanabe, M., and Ida, N. (1995) trk A gene expression in neuroblastoma. The clinical significance of an immunohistochemical study. *Cancer* **76**, 1086–1095.

460. Ryden, M., Sehgal, R., Dominici, C., Schilling, F.H., Ibañez, C.F., and Kogner, P. (1996) Expression of mRNA for the neurotrophin receptor trkC in neuroblastomas with favourable tumour stage and good prognosis. *Br. J. Cancer* **74**, 773–779.

461. Yamashiro, D.J., Nakagawara, A., Ikegaki, N., Liu, X., and Brodeur, G.M. (1996) Expression of TrkC in favorable human neuroblastomas. *Oncogene* **12**, 37–41.

462. Combaret, V., Gross, N., Lasset, C., Balmas, K., Bouvier, R., Frappaz, D., Beretta-Brognara, C., Philip, T., Favrot, M.C., and Coll, J.L. (1997) Clinical relevance of TRKA expression on neuroblastoma: comparison with N-MYC amplification and CD44 expression. *Br. J. Cancer* **75**, 1151–1155.

463. Svensson, T., Ryden, M., Schilling, F.H., Dominici, C., Sehgal, R., Ibanñez, C.F., and Kogner, P. (1997) Coexpression of mRNA for the full-length neurotrophin receptors? trk-C and trk-A in favourable neuroblastoma. *Eur. J. Cancer* **33**, 2058–2063.

463A. Laneve, P., Di Marcotullio, L., Gioia, U., Fiori, M.E., Ferretti, E., Gulino, A., Bozzoni, I., and Caffarelli, E. (2007)

The interplay between microRNAs and the neurotrophin receptor tropomyosin-related kinase C controls proliferation of human neuroblastoma cells. *Proc Natl Acad Sci USA* **104**, 7957–7962.

464. Lavenius, E., Gestblom, C., Johansson, I., Nanberg, E., and Påhlman, S. (1995) Transfection of TRK-A into human neuroblastoma cells restores their ability to differentiate in response to nerve growth factor. *Cell Growth Differ.* **6**, 727–736.

465. Lucarelli, E., Kaplan, D., and Thiele, C.J. (1997) Activation of trk-A but not trk-B signal transduction pathway inhibits growth of neuroblastoma cells. *Eur. J. Cancer* **33**, 2068–2070.

465A. Horii, Y., Kuroda, H., and Sugimoto, T. (1998) Frequent detection of TrkA expression in human neuroblastoma cell lines. *Acta Paediatr. Jpn.* **40**, 644–646.

466. Edsjö, A., Hallberg, B., Fagerström, S., Larsson, C., Axelson, H., and Påhlman, S. (2001) Differences in early and late responses between neurotrophin-stimulated *trkA*- and *trkC*-transfected SH-SY5Y neuroblastoma cells. *Cell Growth Differ.* **12**, 39–50.

466A. Barbacid, M. (1995) Neurotrophic factors and their receptors. *Curr. Opin. Cell. Biol.* **7**, 148–155.

467. Kaplan, D.R., Hempstead, B.L., Martin-Zanca, D., Chao, M.V., and Parada, L.F. (1991) The trk proto-oncogene product: a signal transducing receptor for nerve growth factor. *Science* **252**, 554–558.

468. Matsumoto, K., Wada, R.K., Yamashiro, J.M., Kaplan, D.R., and Thiele, C.J. (1995) Expression of brain-derived neurotrophic factor and p145TrkB affects survival, differentiation, and invasiveness of human neuroblastoma cells. *Cancer Res.* **55**, 1798–1806.

469. Poluha, W., Poluha, D.K., and Ross, A.H. (1995) TrkA neurogenic receptor regulates differentiation of neuroblastoma cells. *Oncogene* **10**, 185–189.

470. Sugimoto, T., Kuroda, H., Horii, Y., Moritake, H., Tanaka, T., and Hattori, S. (2001) Signal transduction pathways through TRK-A and TRK-B receptors iin human neuroblastoma cells. *Jpn J. Cancer Res.* **92**, 152–160.

470B. Eggert, A., Ikegaki, N., Liu, X., Chou, T.T., and Brodeur, G.M. (2001) TrkA signal transduction pathways in neuroblastoma. *Med. Pediatr. Oncol.* **36**, 108–119.

471. Tanaka, T., Sugimoto, T., and Sawada, T. (1998) Prognostic discrimination among neuroblastomas according to Ha-*ras/trk A* gene expression. A comparison of the profiles of neuroblastomas detected clinically and those detected through mass screening. *Cancer* **83**, 1626–1633.

472. Ho, R., Eggert, A., Hishiki, T., Minturn, J.E., Ikegaki, N., Foster, P., Camoratto, A.M., Evans, A.E., and Brodeur, G.M. (2002) Resistance to chemotherapy mediated by TrkB in neuroblastomas. *Cancer Res.* **62**, 6462–6466.

473. Nakagawara, A., Azar, C.G., Scavarda, N.J., and Brodeur, G.M. (1994) Expression and function of *TRK-B* and *BDNF* in human neuroblastomas. *Mol. Cell. Biol.* **14**, 759–767.

474. Tacconelli, A., Farina, A.R., Cappabianca, L., DeSantis, G., Tessitore, A., Vetuschi, A., Sferra, R., Rucci, N., Argenti, B., Screpanti, I., Gulino, A., and Mackay, A.R. (2004) TrkA alternative splicing: a regulated tumor-promoting switch in human neuroblastoma. *Cancer Cell* **6**, 347–360.

475. Brodeur, G.M., Nakagawara, A., Yamashiro, D.J., Ikegaki, N., Liu, X.G., Azar, C.G., Lee, C.P., and Evanas, A.E. (1997) Expression of TrkA, TrkB and TrkC in human neuroblastoma. *J. Neurooncol.* **31**, 49–55.

476. Evangelopoulos, M.E., Weis, J., and Krüttgen, A. (2004) Neurotrophin effects on neuroblastoma cells: correlation with Trk and p75NTR expression and influence of Trk receptor bodies. *J. Neurooncol.* **66**, 101–110.

477. Chen, J., and Zhe, X. (2003) Cotransfection of TrkA and p75(NTR) in neuroblastoma cell line (IMR-32) promotes differentiation and apoptosis of tumor cells. *Chin. Med. J. (Engl.)* **116**, 906–912.

478. Schulte, J.H., Schramm, A., Klein-Hitpass, L., Klenk, M., Wessels, H., Hauffa, B.P., Eils, J., Eils, R., Brodeur, G.M., Schweigerer, L., Havers, W., and Eggert, A. (2005) Microarray analysis reveals differential gene expression patterns and regulation of single target genes contributing to the opposing phenotype of TrkA- and TrkB-expressing neuroblastomas. *Oncogene* **24**, 165–177.

479. Patapoutian, A., and Reichardt, L.F. (2001) Trk receptors: mediators of neurotrophin action. *Curr. Opin. Neurobiol.* **11**, 272–280.

480. Hempstead, B.L., Martin-Zanca, D., Kaplan, D.R., Parada, L.F., and Chao, M.V. (1991) High-affinity NGF binding requires coexpression of the trk proto-oncogene and the low-affinity NGF receptor. *Nature (Lond.)* **350**, 678–683.

481. Shimada, H., Nakagawa, A., Peters, J., Wang, H., Wakamatsu, P.K., Lukens, J.N., Matthay, K.K., Siegel, S.E., and Seeger, R.C. (2004) *TrkA* expression in peripheral neuroblastic tumors. Prognostic significance and biological relevance. *Cancer* **101**, 1873–1881.

482. Jaboin, J., Kim, C.J., Kaplan, D.R., and Thiele, C.J. (2002) Brain-derived neurotrophic factor activation of TrkB protects neuroblastoma cells form chemotherapy-induced apoptosis via phosphatidylinositol 3'-kinase pathway. *Cancer Res.* **62**, 6756–6763.

482A. Datta, S.R., Brunet, A., and Greenberg, M.E. (1999) Cellular survival: a play in three Akts. *Genes Dev.* **13**, 2905–2927.

483. Hirose, T., Scheithauer, B.W., Lopes, M.B., Gerber, H.A., Altermatt, H..J., Harner, S.G., and VandenBerg, S.R. (1995) Olfactory neuroblastoma: an immunohistochemical, ultrastructural, and flow cytometric study. *Cancer* **76**, 4–19.

484. Finkelstein, S.D., Hirose, T., and VandenBerg, S.R. (2000) Olfactory neuroblastoma. In *World Health Organization Classi- fication of Tumours. Pathology and Genetics of Tumours of the Nervous System* (Kleihues, P., and Cavenee, W.K., eds.), IARC Press, Lyon, France, pp. 150–152.

485. Resto, V.A., Eisele, D.W., Forastiere, A., Zahurak, M., Lee, D.J., and Westra, W.H. (2000) Esthesioneuroblastoma: the Johns Hopkins experience. *Head Neck* **22**, 550–558.

486. Dulguerov, P., Allal, A.S., and Calcaterra, T.C. (2001) Esthesioneuroblastoma: a meta-analysis and review. *Lancet Oncol.* **2**, 683–690.

487. Oskouian, R.J., Jane, J.A., Dumont, A.S., Sheehan, J.M., Laurent, J.J., and Levine, P.A. (2002) Esthesioneuroblastoma: clinical presentation, radiological, and pathological features, treatment, review of the literature, and the University of Virginia experience. *Neurosurg. Focus* **12**, 1–9.

488. Mahooti, S., and Wakely, P.E., Jr. (2006) Cytopathologic features of olfactory neuroblastoma. *Cancer (Cancer Cytopathol.)* **108**, 86–92.

489. Enzinger, F.M., and Weiss, S.W. (1983) *Soft Tissue Tumors.* CV Mosby Co., St. Louis, MO, pp. 657–675.

490. Cavazzana, A.O., Navarro, S., Noguera, R., Reynolds, P.C., and Triche, T.J. (1988) Olfactory neuroblastoma is not a neuroblastoma but is related to primitive neuroectodermal tumor (PNET). *Prog. Clin. Biol. Res.* **271**, 463–473.

491. Broich, G., Pagliari, A., and Ottaviani, F. (1997) Esthesioneuroblastoma: a general review of the cases published since the discovery of the tumour in 1924. *Anticancer Res.* **17**, 2683–2706.

492. Sorensen, P.H.B., Wu, J.K., Berean, K.W., Lim, J.F., Donn, W., Frierson, H.F., Reynolds, C.P., López-Terrada, D., and Triche, T.J. (1996) Olfactory neuroblastoma is a peripheral primitive neuroectodermal tumor related to Ewing sarcoma. *Proc. Natl. Acad. Sci. USA* **93**, 1038–1043.

493. Hwang, S.-K., Paek, S.H., Kim, D.G., Jeon, Y.K., Chi, J.G., and Jung, H.W. (2002) Olfactory neuroblastomas: survival rate and prognostic factor. *J. Neurooncol.* **59**, 217–226.

494. Whang-Peng, J., Freter, C.E., Knutsen, T., Nanfro, J.J., and Gazdar, A. (1987) Translocation t(11;22) in esthesioneuroblastoma. *Cancer Genet. Cytogenet.* **29**, 155–157.

495. Castaneda, V.L, Cheah, M.S., Saldivar, V.A., Richmond, C.M., and Parmley, R.T. (1991) Cytogenetic and molecular evaluation of clinically aggressive esthesioneuroblastoma. *Am. J. Pediatr. Hematol. Oncol.* **13**, 62–70.

496. VanDevanter, D.R., George, D., McNutt, M.A., Vogel, A., Luthardt, F. (1991) Trisomy 8 in primary esthesioneuroblastoma. *Cancer Genet. Cytogenet.* **57**, 133–136.

497. Mezzelani, A., Tornielli, S., Minoletti, F., Pierotti, M.A., Sozzi, G., and Pilotti, S. (1999) Esthesioneuroblastoma is not a member of the primitive peripheral neuroectodermal tumour-Ewing's group. *Br. J. Cancer* **81**, 586–591.

498. Argani, P., Perez-Ordoñez, B., Xiao, H., Caruana, S.M., Huvos, A.G., and Ladanyi, M. (1998) Olfactory neuroblastoma is not related to the Ewing family of tumors: absence of EWS/FLI1 gene fusion and MIC2 expression. *Am. J. Surg. Pathol.* **22**, 391–398.

499. Kumar, S., Perlman, E., Pack, S., Davis, M., Zhang, H., Meltzer, P., and Tsokos, M. (1999) Absence of *EWS/FLI1* fusion in olfactory neuroblastomas indicates these tumors do not belong to the Ewing's sarcoma family. *Hum. Pathol.* **30**, 1356–1360.

500. Nelson, R.S., Perlman, E.J., and Askin, F.B. (1995) Is esthesioneuroblastoma a peripheral neuroectodermal tumor? *Hum. Pathol.* **26**, 639–641.

501. Jin, Y., Mertens, F., Arheden, K., Mandahl, N., Wennerberg, J., Dictor, M., Heim, S., and Mitelman, F. (1995) Karyotypic features of malignant tumors of the nasal cavity and paranasal sinuses. *Int. J. Cancer* **60**, 637–641.

502. Zhao, S.-P., and Zhou, X.-F. (2002) Co-expression of trkA and p75 neurotrophine receptor in extracranial olfactory neuroblastoma cells. *Neuropathol. Appl. Neurobiol.* **28**, 301–307.

503. Riazimand, S.H., Brieger, J., Jacob, R., Welkoborsky, H.-J., and Mann, W.J. (2002) Analysis of cytogenetic aberrations in esthesioneuroblastomas by comparative genomic hybridization. *Cancer Genet. Cytogenet.* **136**, 53–57.

504. Petersen, I., Arps, H., Draf, W., and Bockmühl, U. (2005) Cytogenetic aberrations of esthesioneuroblastoma studied by comparative genomic hybridization. *Clin. J. Oncol.* **27**, 16–21.

505. Bockmühl, U., You, X., Pacyna-Gengelbach, M., Arps, H., Draf, W., and Petersen, I. (2004) CGH pattern of esthesioneuroblastoma and their metastases. *Brain Pathol.* **14**, 158–163.

506. You, X.J., Petersen, I., Arps, H., Draf, W., and Bockmuhl,U. (2005) Cytogenetic aberrations of esthesioneuroblastoma studied by comparative genomic hybridization. *Zhonghua Zhong Liu Za Zhi* **27**, 16–21.

507. Llombart-Bosch, A., Carda, C., Peydro-Olaya, A., Noguera, R., Boix, J., Pellín, A. (1989) Pigmented esthesioneuroblastoma showinig dual differentiation following transplantation in nude mice. An immunohistochemical, electron microscopical, and cytogenetic analysis. *Virchows Arch. A Pathol. Anat. Histopathol.* **414**, 199–208.

508. Holland, H., Koschny, R., Krupp, W., Meixensberger, J., Bauer, M., Kirsten, H., and Ahnert, P. (2007) Comprehensive cytogenetic characterization of an esthesioneuroblastoma. *Cancer Genet. Cytogenet.* **173**, 89–96.

509. Kubota, M., Suita, S., Tajiri, T., Shono, K., and Fujii, Y. (2000) Analysis of the prognostic factors relating to better clinical outcome in ganglioneuroblastoma. *J. Pediatr. Surg.* **35**, 92–95.

510. Bove, K.E., and McAdams, A.J. (1981) Composite ganglioneuroblastoma: an assessment of the significance of histological maturation in neuroblastoma diagnosed beyond infancy. *Arch. Pathol. Lab. Med.* **105**, 325–330.

511. Shimada, H., Chatten, J., Newton, W.A., Jr., Sachs, N., Hamoudi, A.B., Chiba, T., Marsden, H.B., and Misugi, K. (1984) Histopathologic prognostic factors in neuroblastic tumors: definition of subtypes of ganglioneuroblastoma and an age-linked classification of neuroblastomas. *J. Natl. Cancer Inst.* **73**, 405–416.

512. Adam, A., and Hochholzer, L. (1981) Ganglioneuroblastoma of the posterior mediastinum: a clinicopathologic review of 80 cases. *Cancer* **47**, 373–381.

513. Suita, S., Zaizen, Y., Kaneko, M., Uchino, J., Takeda, T., Iwafuchi, M., Utsumi, J., Takahashi, H., Yokoyama, J., Nishihira, H., Okada, A., Kawa, K., Nagahara, N., Yano, H., and Tsuchida, Y. (1994) What is the benefit of aggressive chemotherapy for advanced neuroblastoma with N-myc amplification? A report from the Japanese Study Group for the Treatment of Advanced Neuroblastoma. *J. Pediatr. Surg.* **29**, 746–750.

514. Morris, J.A., Schochat, S.J., Smith, E.I., Look, A.T., Brodeur, G.M., Cantor, A.B., and Castleberry, R.P. (1995) Biological variables in thoracic neuroblastoma: a Pediatric Oncology Group study. *J. Pediatr. Surg.* **30**, 296–303.

515. Schmidt, M.L., Salwen, H.R., Chagnovich, D., Bauer, K.D., Crawford, S.E., and Cohn, S.L. (1993) Evidence for molecular heterogeneity in human ganglioneuroblastoma. *Pediatr. Pathol.* **13**, 787–796.

516. Hedborg, F., Lindgren, P.G., Johansson, I., Kogner, P., Samuelsson, B.O., Bekassy, A.N., Olson, L., Kreuger, A., and Påhlman, S. (1992) N-myc gene amplification in neuroblastoma: a clinical approach using ultrasound guided

cutting needle biopsies collected at diagnosis. *Med. Pediatr. Oncol.* **20,** 292–300.

517. Barontini, M., Gutierrez, M.I., Levin, G., Mur, N., and Diez, B. (1993) N–myc oncogene and urinary catecholamines in children with neuroblastoma. *Med. Pediatr. Oncol.* **21,** 499–504.

518. Tsuda, T., Obara, M., Hirano, H., Gotoh, S., Kubomura, S., Higashi, K., Kuroiwa, A., Nakagawara, A., Nagahara, N., and Shimizu, K. (1987) Analysis of N-myc amplification in relation to disease stage and histologic types in human neuroblastomas. *Cancer* **60,** 820–826.

519. Keser, I., Burckhardt, E., Ortaç, R., Tunali, N., Çaglar, M., and Alkan, M. (1998) Detection of gains and losses of DNA sequences in five ganglioneuroblastomas by comparative genomic hybridization using do uble step DOP-PCR. *Eur. J. Hum. Genet.* **6,** 99.

520. Toraman, A.D., Keser, I., Lüleci, G., Tunali, N., and Gelen, T. (2002) Comparative genomic hybridization in ganglioneuroblastomas. *Cancer Genet. Cytogenet.* **132,** 36–40.

521. Fabbretti, G., Valenti, C., Loda, M., Brisigotti, M., Cozzutto, C., Tonini, G.P., and Callea, F. (1993) N-*myc* gene amplification/expression in localized stroma-rich neuroblastoma (ganglioneuroblastoma). *Hum. Pathol.* **24,** 294–297.

522. Michalski, A.J., Cotter, F.E., and Cowell, J.K. (1992) Isolation of chromosome-specific DNA sequences from an Alu polymerase chain reaction library to define the breakpoint in a p atient with a constitutional translocation t(1;13)(q22;q12) in ganglioneuroblastoma. *Oncogene* **7,** 1595–1602.

523. Michalski, A.J., and Cowell, J.K. (1994) Constructing a physical map around a constitutional t(1;13)(q22;q12) breakpoint in a patient with a ganglioneuroblastoma. *Prog. Clin. Biol. Res.* **385,** 79–85.

524. Sossey-Alaoui, K., Su, G., Malaj, E., Roe,B., and Cowell, J.K. (2002) *WAVE2*, an actin-polymerization gene, is truncated and inactivated as a result of a constitutional t(1;13)(q21;q12) chromosome translocation in a patient with ganglioneuroblastoma. *Oncogene* **21,** 5967–5974.

525. Stout, A.P. (1947) Ganglioneuroma of the sympathetic nervous system. *Surg. Gynecol. Obstet.* **84,** 101–110.

526. MacMillan, R.W., Blanc, W.B., and Santulli, T.V. (1976) Maturation of neuroblastoma to ganglioneuroma in lymph nodes. *J. Pediat. Surg.* **11,** 461–462.

527. Kurahashi, H., Oue, T., Akagi, K., Fukuzawa, M., Okada, A., Tawa, A., Okada, S., and Nishisho, I. (1995) Allelic status on 1p and 11p15 in neuroblastoma and benign ganglioneuroma. *Int. J. Oncol.* **6,** 669–674.

528. Hicks, M.J., and Mackay, B. (1995) Comparison of ultrastructural features among neuroblastic tumors: maturation from neuroblastoma to ganglioneuroma. *Ultratruct. Pathol.* **19,** 311–322.

529. Nakagawara, A., Ikeda, K., Tsuneyoshi, M., Daimaru, Y., and Enjoji, M. (1985) Malignant pheochromocytoma with ganglioneuroblastoma elements in a patient with von Recklinghausen's disease. *Cancer* **55,** 2794–2798.

530. Brady, S., Lechan, R.M., Schwaitzberg, S.D., Dayal, Y., Ziar, J., and Tischler, A.S. (1997) Composite pheochromocytoma/ganglioneuroma of the adrenal gland associated with multiple endocrine neoplasia 2A: case report with immunohistochemical analysis. *Am. J. Surg. Pathol.* **21,** 102–108.

531. Matias-Guiu, X., and Garrastazu, M.T. (1998) Composite phaeochromo- cytoma-ganglioneuroblastoma in a patient with multiple endocrine neoplasia type IIA. *Histopathology* **32,** 281–282.

532. Chetty, R., and Duhig, J.D. (1993) Bilateral pheochromocytoma-ganglioneuroma of the adrenal in type 1 neurofibromatosis. *Am. J. Surg. Pathol.* **17,** 837–841.

533. Choutet, P., Benatre, A., Cosnay, P., Huten, N., Alison, D., Ginies, G., Perrotin, D., and Lamisse, F. (1981) Tumor of the adrenal gland associating pheochromocytoma and ganglioneuroma. *Sem.Hop.* **57,** 590–592.

534. Donadio, F., La Ganga, V., Vajo, M., Campanella, G., and Coverlizza, S. (1982) Ganglioneuroblastoma with areas of pheochromocytoma of the adrenal gland. Case report. *Minerva Chir.* **37,** 551–558.

535. Salmi, J., Pelto-Huikko, M., Auvinen, O., Karvonen, A.L., Saaristo, J., Paronen, I., Poyhonen, L., and Seppanen, S. (1988) Adrenal pheochromocytoma-ganglioneuroma producing catecholamines and various neuropeptides. *Acta Med. Scand.* **224,** 403–408.

536. Nelson, J.S., Bruner, J.M., Wiestler, O.D., and VandenBerg, S.R. (2000) Ganglioglioma and gangliocytoma. In *World Health Organization Classification of Tumours. Pathology and Genetics of Tumours of the Nervous System* (Kleihues, P., and Cavenee, W.K., eds.), IARC Press, Lyon, France, pp. 96–98.

537. Hirose, T., Scheithauer, B.W., Lopes, M.B.S., Gerber, H.A., Altermatt, H..J., and VandenBerg, S.R. (1997) Ganglioglioma. An ultrastructural and immunohistochemical study. *Cancer* **79,** 989–1003.

538. Jallo, G.I., Freed, D., and Epstein, F.J. (2004) Spinal cord gangliogliomas: a review of 56 patients. *J. Neurooncol.* **68,** 71–77.

539. Jay, V., Squire, J., Becker, L.E., and Humphreys, R. (1994) Malignant transformation in a ganglioglioma with anaplastic neuronal and astrocytic components. Report of a case with flow cytometric and cytogenetic analysis. *Cancer* **73,** 2862–2868.

540. Kordek, R., Klimek, A., Karpinska, A., Alwasiak, J., Diebiec-Rychter, M., and Liberski, P. (1996) The immunohistochemistry and ultrastructure of ganglioglioma with chromosomal alterations: a case report. *Pol. J. Pathol.* **47,** 37–39.

541. Wacker, M.R., Cogen, P.H., Etzell, J.E., Daneshvar, L., Davis, R.L., and Prados, M.D. (1992) Diffuse leptomeningeal involvement by a ganglioglioma in a child. Case report. *J. Neurosurg.* **77,** 302–306.

542. Zhu, J.J, Leon, S.P., Folkerth, R.D., Guo, S.-Z., Wu J.K., and Black, P.M. (1997) Evidence for clonal origin of neoplastic neuronal and glial cells in gangliomas. *Am. J. Pathol.* **151,** 565–571.

543. Yin, X.L., Hui, A.B., Pang, J.C.., Poon, W.S., and Ng, H.K. (2002) Genome-wide survey for chromosomal imbalances in ganglioglioma using comparative genomic hybridization. *Cancer Genet. Cytogenet.* **134,** 71–76.

544. Yin, X.L., and Ng, H.K. (2005) Molecular genetic studies on ganglioglioma. *Zhonghua Bing Li Xue Za Zhi.* **34,** 147–149.

545. Platten, M., Meyer-Puttlitz, B., Blumcke, I., Waha, A., Wolf, H.K., Nothen, M.M., Louis, D.N., Sampson, J.R., and von Deimling, A. (1997) A novel splice site associated polymorphism in the tuberous sclerosis 2 (TSC2) gene may predispose to the development of sporadic gangliogliomas. *J. Neuropathol. Exp. Neurol.* **56,** 806–810.

546. Becker, A.J., Löbach, M., Klein, H., Normann, S., Nöthen, M.M., von Deimling, A., Mizuguchi, M., Elger, C.E., Schramm, J., Wiestler, O.D., and Blümcke, I. (2001) Mutational analysis of *TSC1* and *TSC2* in gangliogliomas. *Neuropathol. Appl. Neurobiol.* **27,** 105–114.

547. Parry, L., Maynard, J.H., Patel, A., Hodges, A.K., von Deimling, A., Sampson, J.R., and Cheadle, J.P. (2000) Molecular analysis of the *TSC1* and *TSC2* tumour suppressor genes in sporadic glial and glioneuronal tumours. *Hum. Genet.* **107,** 350–356.

548. Wolf, H.K., Müller, M.B., Spänle, M., Zentner, J., Schramm, J., and Wiestler, O.D. (1994) Ganglioglioma: a detailed histopathological and immunohistochemical analysis of 61 cases. *Acta Neuropathol.* **88,** 166–173.

549. Prayson, R.A., Khajavi, K., and Comair, Y.G. (1995) Cortical architectural abnormalities and MIB1 immnoreactivity in gangliogliomas: a study of 60 patients with intracranial tumors. *J. Neuropathol. Exp. Neurol.* **54,** 513–520.

550. Fukushima, T., Katayama, Y., Watanabe, T., Yoshino, A., Komine, C., and Yokoyama, T. (2005) Aberrant TP53 protein accumulation in the neuronal component of ganglioglioma. *J. Neurooncol.* **72,** 103–106.

551. Zhu, J., Guo, S.-Z., Beggs, A.H., Maruyama, T., Santarius, T., Dashner, K., Olsen, N., Wu J.K., and Black, P. (1996) Microsatellite instability analysis of primary human brain tumors. *Oncogene* **12,** 1417–1423.

552. Park, J.P., Dossu, J.R., and Rhodes, C.H. (1996) Telomere associations in desmoplastic infantile ganglioglioma. *Cancer Genet. Cytogenet.* **92,** 4–7.

553. Cerdá-Nicolás, M., López-Ginés, C., Gil-Benso, R., Donat, J., Fernandez-Delgado, R., Pellin, A., Lopez-Guerrero, J.A., Roldan, P., and Barbera, J. (2006) Desmoplastic infantile ganglioglioma. Morphological, immunohistochemical and genetic features. *Histopathology* **48,** 605–626.

553A. Prayson, R.A. (1999) Bcl-2 and Bcl-X expression in gangliogliomas. *Hum. Pathol.* **30,** 701–705.

554. Keyvani, K., Rickert C.H., von Wild, K., and Paulus, W. (2001) Rosetted glioneuronal tumor: a case with proliferating neuronal nodules. *Acta. Neuropathol.* **101,** 525–528.

554A. Rickert, C.H., Jasper, M., Sepehrnia, A., and Jeibmann, A. (2006) Rosetted glioneuronal tumour of the spine: clinical, histological and cytogenetic data. *Acta Neuropathol.* **112,** 231–233.

555. von Deimling, A., Kleihues, P., Saremaslani, P., Yasargil, M.G., Spoerri, O., Sudhof, T.C., and Wiestler, O.D. (1991) Histogenesis and differentiation potential of central neurocytomas. *Lab Invest.* **64,** 585–591.

555A. Figarella-Branger, D., Söylemezoglu, F., Kleihues, P., and Hassoun, J. (2000) Central neurocytoma. In *World Health Organization Classification of Tumours. Pathology and Genetics of Tumours of the Nervous System* (Kleihues, P., and Cavenee, W.K., eds.), IARC Press, Lyon, France, pp. 107–109.

555B. Schmidt, M.H., Gottfried, O.N., von Koch, C.S., Chang, S.M., and McDermott, M.W. (2004) Central neurocytoma: a review. *J. Neurooncol.* **66,** 377–384.

555C. Rades D., and Schild, S.E. (2006) Treatment recommendations for the various subgroups of neurocytomas. *J. Neurooncol.* **77,** 305–309.

556. Cerdá-Nicolás, M., López-Ginés, C., Peydro-Olaya, A., and Llombart-Bosch, A. (1993) Central neurocytoma: a cytogenetic case study. *Cancer Genet. Cytogenet.* **65,** 173–174.

557. Taruscio, D., Danesi, R., Montaldi, A., Cerasoli, S., Cenacchi, G., and Giangaspero, F. (1997) Nonrandom gain of chromosome 7 in central neurocytomas: a chromosomal analysis and fluorescence in situ hybridization study. *Virchows Arch.* **430,** 47–51.

558. Yin, X.-L., Pang, J.C.-S., Hui, A.B.-Y., and Ng, H.-K. (2000) Detection of chromosomal imbalances in central neurocytomas by using comparative genomic hybridization. *J. Neurosurg.* **93,** 77–81.

558A. Korshunov, A., Sycheva, R., and Golanov, A. (2007) Recurrent cytogenetic aberrations in central neurocytomas and their biological relevance. *Acta Neuropathol.* **113,** 303–312.

558B. Knoepfler, P.S., Cheng, P.F., and Eisenman, R.N. (2002) N-*myc* is essential during neurogenesis for the rapid expansion of progenitor cell populations and the inhibition of neuronal differentiation. *Genes Dev.***16,** 2699–2712.

558C. Lachyankar, M.B., Sultana, N., Schonhoff, C.M., Mitra, P., Poluha, W., Lambert, S., Quesenberry, P.J., Litofsky, N.S., Recht, L.D., Nabi, R., Miller, S.J., Ohta, S., Neel, B.G., and Ross, A.H. (2000) The role for nuclear PTEN in neuronal differentiation. *J. Neurosci.* **20,** 1404–1413.

559. Englund, C., Alvord, E.C., Jr., Folkerth, R.D., Silbergeld, D., Born, D.E., and Hevner, R.F. (2005) NeuN expression correlates with reduced mitotic index of neoplastic cells in central neurocytomas. *Neuropathol. Appl. Neurobiol.* **31,** 429–438.

560. Rades, D., Schild, S.E., and Fehlauer, F. (2004) Prognostic value of the MIB-1 labeling index for central neurocytomas. *Neurology* **62,** 987–989.

561. Christov, C., Adle-Biassette, H., and Le Guerinel, C. (1999) Recurrent central neurocytoma with marked increase in MIB-1 labelling index. *Br. J. Neurosurg.* **13,** 496–499.

562. Kanamori, M., Kumabe, T., Shimizu, H., and Yoshimoto, T. (2002) ^{201}Tl-SPECT, ^{1}H-MRS, and MIB-1 labeling index of central neurocytomas: three case reports. *Acta Neurochir. (Wien)* **144,** 157–163.

563. Kim, D.G., Kim, J.S., Chi, J.G., Park, S.H., Jung, H.-W., Choi, K.S., and Han, D.H. (1996) Central neurocytoma: proliferative potential and biological behavior. *J. Neurosurg.* **84,** 742–747.

564. Fujimaki, T., Matsuno, A., Sasaki, T., Toyoda, T., Matsuura, R., Ogai, M., Kitanaka, C., Asai, A., Matsutani, M., and Kirino, T. (1997) Proliferative activity of central neurocytoma: measurement of tumor volume doubling time, MIB-1 staining index and bromodeoxyuridine labeling index. *J. Neurooncol.* **32,** 103–109.

565. Sharma, M.C., Rathore, A., Karak, A.K., and Sarkar, C. (1998) A study of proliferative markers in central neurocytoma. *Pathology* **30,** 355–359.

566. Uro-Coste, E., Bousquet, P., Arrue, P., and Delisle, M.B. (1999) Central neurocytoma. Immunohistochemical study:

MIB1, p53 and bcl-2. Report of 5 cases. *Arch. Anat. Cytol. Pathol.* **47**, 13–18.

567. Ogawa, Y., Sugawara, T., Seki, H., and Sakuma, T. (2006) Central neurocytomas with MIB-1 labeling index over 10% showing rapid tumor growth and dissemination. *J. Neurooncol.* **79**, 211–216.

568. Tong, C.Y., Ng, H.K., Pang, J.C., Hu, J., Hui, A.B., and Poon, W.S. (2000) Central neurocytomas are genetically distinct from oligodendrogliomas and neuroblastomas. *Histopathology* **37**, 160–165.

569. Ohgaki, H., Eibl, R.H., Schwab, M., Reichel, M.B., Mariani, L., Gehring, M., Petersen, I., Holl, T., Wiestler, O.D., and Kleihues, P. (1993) Mutations of the p53 tumor suppressor gene in neoplasms of the human nervous system. *Mol. Carcinog.* **8**, 74–80.

570. Jay, V., Edwards, V., Hoving, E., Rutka, J., Becker, L., Zielenska, M., and Teshima, I. (1999) Central neurocytoma: morphological, flow cytometric, polymerase chain reaction, fluorescence in situ hybridization, and karyotypic analyses. *J. Neurosurg.* **90**, 348–354.

571. Mena, H., Morrison, A.L., Jones, R.V., and Gyure, K.A. (2001) Central neurocytomas express photoreceptor differentiation. *Cancer* **91**, 136–143.

572. Nozaki, M., Tada, M., Matsumoto, R., Sawamura, Y., Abe, H., and Iggo, R.D. (1998) Rare occurrence of inactivating p53 gene mutations in primary non-astrocytic tumors of the central nervous system: reappraisal by yeast functional assay. *Acta Neuropathol.* **95**, 291–296.

573. Horstmann, S., Perry, A., Reifenberger, G., Giangaspero, F., Huang, H., Hara, A., Masuoka, J., Rainov, N.G., Bergmann, M., Heppner, F.L., Brandner, S., Chimelli, L., Montagna, N., Jackson, T., Davis, D.G., Markesbery, W.R., Ellison, D.W., Weller, R.O., Taddei, G.L., Conti, R., Del Bigio, M.R., González-Cámpora, R., Radhakrishnan, V.V., Söylemezoglu, F., Uro-Coste, E., Qian, J., Kleihues, P., and Ohgaki, H. (2004) Genetic and expression profiles of cerebellar liponeurocytomas. *Brain Pathol.* **14**, 281–289.

574. Kleihues, P., Chimelli, L., and Giangaspero, F. (2000) Cerebellar liponeurocytoma. In *World Health Organization Classification of Tumours. Pathology and Genetics of Tumours of the Nervous System* (Kleihues, P., and Cavenee, W.K., eds.), IARC Press, Lyon, France, pp. 110–111.

575. Lo Cunsolo, C., Iolascon, A., Cavazzana, A., Cusano, R., Strigini, P., Mazzocco, K., Giordani, L., Massimo, L., De Bernardi, B., Conte, M., and Tonini, G.P. (1999) Neuroblastoma in two siblings supports the role of 1p36 deletion in tumor development. *Cancer Genet. Cytogenet.* **109**, 126–130.

576. Knudson, A.G., Jr., and Amromin, G.D. (1966) Neuroblastoma and ganglioneuroma in a child with multiple neurofibromatosis: implications for the mutational origin of neuroblastoma. *Cancer* **19**, 1032–1037.

577. Kushner, B.H., Gilbert, F., and Helson, L. (1986) Familial neuroblastoma. Case reports, literature review, and etiologic considerations. *Cancer* **57**, 1887–1893.

578. Girgert, R., and Schweizer, P. (1997) Regulation of expression of two different transcripts of the NF-1 gene in neuroblastoma. *J. Neurooncol.* **31**, 91–97.

579. Johnson, M.R., Look, A.T., DeClue, J.E., Valentine, M.B., and Lowy, D.R. (1993) Inactivation of the *NF1* gene in human melanoma and neuroblastoma cell lines without impaired regulation of GTP-Ras. *Proc. Natl. Acad. Sci. USA* **90**, 5539–5543.

580. Martinsson, T., Sjöberg, R.-M., Hedborg, F., and Kogner, P. (1997) Homozygous deletion of the neurofibromatosis-1 gene in the tumor of a patient with neuroblastoma. *Cancer Genet. Cytogenet.* **95**, 183–189.

581. Ankathil, R., Kusumakumary, P., Priyakumary, T., and Krishnan Nair, M. (2002) Expression of folate sensitive and apidicolin induced chromosomal fragile sites in familial neuroblastoma. *J. Exp. Clin. Cancer Res.* **21**, 383–388.

582. Calin, G.A., Dumitru, C.D., Shimizu, M., Bichi, R., Zupo, S., Noch, E., Aldler, H., Rattan, S., Keating, M., Rai, K., Rassenti, L., Kipps, T., Negrini, M., Bullrich, F., and Croce, C.M. (2002) Frequent deletions and down-regulation of micro-RNA genes miR15 and MiR16 at 13q14 in chronic lymphocytic leukemia. *Proc. Natl. Acad. Sci. USA* **99**, 15524–15529.

583. Laureys, G.G. (1995) Identification of the breakpoint-flanking markers on chromosomes 1 and 17 of a constitutional translocation t(1;17)(p36;q12–21) in a patient with neuroblastoma. *Verh. K. Acad. Geneeskd. Belg.* **57**, 389–422.

584. Knudson, A.G., Jr., and Strong, L.C. (1972) Mutation and cancer: neuroblastoma and pheochromocytoma. *Am. J. Hum. Genet.* **24**, 514–532.

585. Maris, J.M., Weiss, M.J., Mosse, Y., Hii, G., Guo, C., White, P.S., Hogarty, M.D., Mirensky, T., Brodeur, G.M., Rebbeck, T.R., Urbanek, M., and Shusterman, S. (2002) Evidence for a hereditary neuroblastoma predisposition locus at chromosome 16p12–13. *Cancer Res.* **62**, 6651–6658.

586. Chompret, A., de Vathaire, F., Brugieres, L., et al. (1998) Excess of cancers in relatives of patients with neuroblastoma. *Med. Pediatr. Oncol.* **31**, 211.

587. Maris, J.M., and Tonini, G.P. (2000) Genetics of familial neuroblastoma. In *Neuroblastoma* (Brodeur. G.M., Sawada, T., Tsuchida, Y., and Voûte, P.A., eds.), Elsevier Science B.V., Amsterdam, The Netherlands, pp. 125–135.

588. Dodge, H.J., and Benner, M.C. (1945) Neruoblastoma of the adrenal medulla in siblings. *Rocky Mt. Med.* **42**, 35–38.

589. Furuta, S.,. Ohira, M., Machida, T., Hamano, S., and Nakagawara, A. (2000) Analysis of loss of heterozygosity at 16p12-p13 (familial neuroblastoma locus) in 470 neuroblastomas including both sporadic and mass screening tumors. *Med. Pediatr. Oncol.* **35**, 531–533.

590. Weiss, M.J., Guo, C., Shusterman, S., Hii, G., Mirensky, T.L., White, P.S., Hogarty, M.D., Rebbeck, T.R., Teare, D., Urbanek, M., Brodeur, G.M., and Maris, J.M. (2000) Localization of a hereditary neuroblastoma predisposition gene to 16p12-p13. *Med. Pediatr. Oncol.* **35**, 526–530.

591. Malkin, D., Li, F.P., Strong, L.C., Fraumeni, J.F., Jr., Nelson, C.E., Kim, D.H., Kassel, J., Gryka, M.A., Bischoff, F.Z., Tainsky, M.A., and Friend, S.H. (1990) Germ line p53 mutations in a familial syndrome of breast cancer, sarcomas, and other neoplasms. *Science (Wash. DC)* **250**, 1233–1238.

592. Bell, D.W., Varley, J.M., Szydlo, T.E., Kang, D.H., Wahrer, D.C., Shannon, K.E., Lubratovich, M., Verselis, S.J., Isselbacher, K.J., Fraumeni, J.F., Birch, J.M., Li, F.P., Garber, J.E., and Haber, D.A. (1999) Heterozygous germ line hCHK2 mutations in Li-Fraumeni syndrome. *Science (Wash. DC)* **286**, 2528–2531.

593. Huff, V., Amos, C.I., Douglass, E.C., Fisher, R., Geiser, C.F., Krill, C.E., Li, F.P., Strong, L.C., and McDonald, J.M. (1997) Evidence for genetic heterogeneity in familial Wilms' tumor. *Cancer Res.* **57,** 1859–1862.

594. McDonald, J.M., Douglass, E.C., Fisher, R., Geiser, C.F., Krill, C.E., Strong, L.C., Virshup, D., and Huff, V. (1998) Linkage of familial Wilms' tumor predisposition to chromosome 19 and a two-loucs model for the etiology of familial tumors. *Cancer Res.* **58,** 1387–1390.

595. Trochet, D., Bourdeaut, F., Janoueix-Lerosey, I., Deville, A., de Pontual, L., Schleiermacher, G., Coze, C., Philip, N., Frébourg, T., Munnich, A., Lyonnet, S., Delattre, O., and Amiel, J. (2004) Germline mutations of the paired-like homeobox 2B (*PHOX2B*) gene in neuroblastoma. *Am. J. Hum. Genet.* **74,** 761–764.

596. Pattyn, A., Morin, X., Cremer, H., Goridis, C., and Brunet, J.F. (1999) The homeobox gene Phox2b is essential for the development of autonomic neural crest derivates. *Nature* **399,** 366–370.

597. Mosse, Y.P., Laudenslager, M., Khazi, D., Carlisle, A.J., Winter C.L., Rappaport, E., and Maris, J.M. (2004) Germline *PHOX2B* mutation in hereditary neuroblastoma. *Am. J. Hum. Genet.* **75,** 727–730.

598. Perri, P., Bachetti, T., Longo, L., Matera, I., Seri, M., Tonini, G.P., and Ceccherini, I. (2005) *PHOX2B* mutations and genetic predisposition to neuroblastoma. *Oncogene* **24,** 3050–3053.

599. Amiel, J., Laudier, B., Attie-Bitach, T., Trang, H., de Pontual, L., Gener, B., Trochet, D., Etchevers, H., Ray, P., Simonneau, M., Vekemans, M., Munnich, A., Gaultier, C., and Lyonnet, S. (2003) Polyalanine expansion and frameshift mutations of the paired-like box gene *PHOX2B* in congenital central hypoventilation syndrome. *Nat. Genet.* **33,** 459–461.

600. Weese-Mayer, D.E., Berry-Kravis, E.M., Zhou, L., Maher, B.S., Silvestri, J.M., Curran, M.E., and Marazita, M.L. (2003) Idiopathic congenital central hypoventilation syndrome: analysis of genes pertinent to early autonomic nervous system embryologic development and identification in PHOX2b. *Am. J. Med. Genet.* **A 123,** 267–278.

601. Origone, P., Defferrari, R., Mazzocco, K., Lo Cunsolo, C., De Bernardi, B., and Tonini, G.P. (2003) Homozygous inactivation of the *NF1* gene in a patient with familial NF1 and disseminated neuroblastoma. *Am. J. Med. Genet.* **118A,** 309–313.

602. van Limpt, V., Schramm, A., van Lakeman, A., Sluis, P., Chan, A., van Noesel, M., Baas, F., Caron, H., Eggert, A., and Versteeg, R. (2004) The Phox2B homeobox gene is mutated in sporadic neuroblastomas. *Oncogene* **23,** 9280–9288.

603. Tonini, G.P., Longo, L., Coco, S., and Perri, P. (2003) Familial neuroblastoma: a complex heritable disease. *Cancer Lett.* **197,** 41–45.

604. van Limpt, V., Chan, A., Schramm, A., Eggert, A., and Versteeg, R. (2005) Phox2B mutations and the Delta-Notch pathway in neuroblastoma. *Cancer Lett.* **228,** 59–63.

605. Tonini, G.P., Lo Cunsolo, C., Cusano, R., Iolascon, A., Dagnino, M., Conte, M., Milanaccio, C., De Bernardi, B., Mazzocco, K., and Scaruffi, P. (1997) Loss of heterozy-gosity for chromosome 1p in familial neuroblastoma. *Eur. J. Cancer* **33,** 1953–1956.

606. Folkman, J. (1992) The role of angiogenesis in tumor growth. *Semin. Cancer Biol.* **3,** 65–71.

607. Davidoff, A.M., Leary, M.A., Ng, C.Y.C., and Vanin, E.F. (2001) Gene therapy-mediated expression by tumor cells of the angiogenesis inhibitor flk-1 results in inhibition of neuroblastoma growth in vivo. *J. Pediatr. Surg.* **36,** 30–36.

608. Shusterman, S., Grupp, S.A., and Maris, J.M. (2000) Inhibition of tumor growth in a human neuroblastoma xenograft model with TNP-470. *Med. Pediat. Oncol.* **35,** 673–676.

609. Shusterman, S., Grupp, S.A., Barr, R., Carpentieri, D., Zhao, H., and Maris, J.M. (2001) The angiogenesis inhibitor TNP-470 effectively inhibits human neuroblastoma xenografts growth, especially in the setting of subclinical disease. *Clin. Cancer Res.* **7,** 977–984.

610. Yang, Q., Tian, Y., Liu, S., Zeine, R., Chlenski, A., Salwen, H.R., Henkin, J., and Cohn, S.L. (2007) Thrombospondin-1 peptide ABT-510 combined with valproic acid is an effective antiangiogenesis strategy in neuroblastoma. *Cancer Res.* **67,** 1716–1724.

611. Diccianni, M.B., Omura-Minamisawa, M., Batova, A., Le T., Bridgeman, L., and Yu, A.L. (1999) Frequent deregulation of p16 and the p16/G1 cell cycle-regulatory pathway in neuroblastoma. *Int. J. Cancer* **80,** 145–154.

612. Kawamata, N., Seriu, T., Koeffler, H.P., and Bartram, C.R. (1996) Molecular analysis of the cyclin-dependent kinase inhibitor family: p16(CDKN2/MTS1/INK4A), p18(INK4C) and p27(Kip1) genes in neuroblastomas. *Cancer* **77,** 570–575.

613. Castresana, J.S., Gomez, L., Garcia-Miguel, P., Queizan, A., and Pestaña, A. (1997) Mutational analysis of the p16 gene in human neuroblastomas. *Mol. Carcinog.* **18,** 129–133.

614. Easton, J., Wei, T., Lahti, J.M., and Kidd, V.J. (1998) Disruption of the cyclin D/cyclin-dependent kinase/INK4/retinoblastoma protein regulatory pathway in human neuroblastoma. *Cancer Res.* **58,** 2624–2632.

615. Omura-Minamisawa, M., Diccianni, M.B., Chang, R.C., Batova, A., Bridgeman, L.J., Schiff, J., Cohn, S.L., London, W.B., and Yu, A.L. (2001) p16/p14^{ARF} cell cycle regulatory pathways in primary neuroblastoma: p16 expresion is associated with advanced stage disease. *Clin. Cancer Res.* **7,** 3481–3490.

616. Molenaar, J.J., van Sluis, P., Boon, K., Versteeg, R., and Carson, H.N. (2003) Rearrangements and increased expression of cyclin D1 (*CCND1*) in neuroblastoma. *Genes Chromosomes Cancer* **36,** 242–249.

617. Seo, M., Kim, Y., Lee, Y.-I., Kim, S.-Y., Ahn, Y.-M., Kang, U.G., Roh, M.-S., Kim, Y.-S., and Juhnn, Y.-S. (2006) Membrane depolarization stimulates the proiliferation of SH-SY5Y human neuroblastoma cells by increasing retinoblastoma protein (RB) phosphorylation through the activation of cyclin-dependent kinase 2 (Cdk2). *Neurosci. Lett.* **404,** 87–92.

618. Nojima, T., Abe, S., Furuta, Y., Nagashima, K., Alam, A.F., Takada, N., Sasaki, F., and Hata, Y. (1991) Morphological and cytogenetic characterization and N-myc oncogene analysis of a newly established neuroblastoma cell line. *Acta. Pathol. Jpn.* **41,** 507–515.

619. Lambert, D.G. (1995) Characterisation of a new human neuroblastoma cell line, NAL-GT. *Neurosci. Lett.* **183,** 4–7.

620. Fowler, C.L., Zimmer, C.C., Ruktanouchai, D., and Zimmer, S.G. (1998) Two new human neuroblastoma cell lines exhibiting tumorigenesis and metastasis in the nude mouse model. *Anticancer Res.* **18,** 1393–1398.

621. Yasuno, T., Matsumura, T., Shikata, T., Inazawa, J., Sakabe, T., Tsuchida, S., Takahata, T., Miyairi, S., Naganuma, A., and Sawada, T. (1999) Establishment and characterization of a cisplatin-resistant human neuroblastoma cell line. *Anticancer Res.* **19,** 4049–4057.

622. Shastry, P., Basu, A., and Rajadhyaksha, M.S. (2001) Neuroblastoma cell lines—a versatile in vitro model in neurobiology. *Int. J. Neurosci.* **108,** 109–126.

623. Walton, J.D., Kattan, D.R., Thomas, S.K., Spengler, B.A., Guo, H.-F., Biedler, J.L., Cheung, N.-K.V., and Ross, R.A. (2004) Characteristics of stem cells from human neuroblastoma cell lines and in tumors. *Neoplasia* **6,** 838–845.

624. Casper, J.T., Trent, J.M., Harb, J., Piaskowski, V., Helmsworth, M., Finlan, J., and Von Hoff, D.D. (1983) Distinguishing characteristics of a new neuroblastoma cell line. *Cancer Genet. Cytogenet.* **10,** 177–186.

625. Carachi, R., Raza, T., Robertson, D., Wheldon, T.W., Wilson, L., Livingstone, A., van Heyningen, V., Spowart, G., Middleton, P., Gosden, J.R., Kemshead, J.T., and Clayton, J.P. (1987) Biologial properties of a tumour cell line (NB1-G) derived from human neuroblastoma. *Br. J. Cancer* **55,** 407–411.

626. Bettan-Renaud, L., Bayle, C., Teyssier, J.R., and Bénard, J. (1989) Stability of phenotypic and genotypic traits during the establishment of a human neuroblastoma cell line, IGR-N-835. *Int. J. Cancer* **44,** 460–466.

627. Lutz, W., Stohr, M., Schurmann, J., Wenzel, A., Lohr, A., and Schwab, M. (1996) Conditional expression of N-myc in human neuroblastoma cells increases expression of alpha-prothymosin and ornithine decarboxylase and accelerates progression into S-phase early after mitogenic stimulation of quiescent cells. *Oncogene* **13,** 803–812.

628. Escobar, M.A., Hoelz, D.J., Sandoval, J.A., Hickey, R.J., Grosfeld, J.L., and Malkas, L.H. (2005) Profiling of nuclear extract proteins from human neuroblastoma cell lines: the search for fingerprints. *J. Pediatr. Surg.* **40,** 349–358.

629. Fischer, M., and Berthold, F. (2003) Characterization of the gene expression profile of neuroblastoma cell line IMR-5 using serial analysis of gene expression. *Cancer Lett.* **190,** 79–87.

630. Oe, T., Sasayama, T., Nagashima, T., Muramoto, M., Yamazaki, T., Morikawa, N., Okitsu, O., Nishimura, S., Aoki, T., Katayama, Y., and Kita, Y. (2005) Differences in gene expression profile among SH-SY5Y neuroblastoma subclones with different neurite outgrowth responses to nerve growth factor. *J. Neurochem.* **94,** 1264–1276.

631. Shimono, R., Matsubara, S., Takamatsu, H., Fukushige, T., and Ozawa, M. (2000) The expression of cadherins in human neuroblastoma cell lines and clinical tumors. *Anticancer Res.* **20,** 917–923.

632. Tumilowicz, J.J., Nichols, W.W., Cholon, J.J., and Greene, A.E. (1970) Definition of a continuous human cell line derived from neuroblastoma. *Cancer Res.* **30,** 2110–2118.

633. Bursztajn, S., Feng, J.-J., Nanda, A., and Berman, S.A. (2001) Differential responses to human neuroblastoma and glioblastoma to apoptosis. *Mol. Brain Res.* **91,** 57–72.

634. Stio, M., Celli, A., and Treves, C. (2001) Synergistic anti-proliferative effects of vitamin derivatives and 9-cis retinoic acid in SH-SY5Y human neuroblastoma cells. *J. Steroid Biochem. Mol. Biol.* **77,** 213–222.

635. Messina, E., Gazzaniga, P., Micheli, V., Barile, L., Lupi, F., Aglianò, A.M., and Giacomello, A. (2004) Low levels of mycophenolic acid induce differentiation of human neuroblastoma cell lines. *Int. J. Cancer* **112,** 352–354.

636. Tweddle, D.A., Malcolm, A.J., Cole, M., Pearson, A.D.J., and Lunec, J. (2001) p53 cellular localization and function in neuroblastoma. Evidence for defective G_1 arrest despite *WAF1* induction in *MYCN*-amplified cells. *Am. J. Pathol.* **158,** 2067–2077.

637. Nakamura, Y., Ozaki, T., Koseki, H., Nakagawara, A., and Sakiyama, S. (2006) Accumulation of $p27^{KIP1}$ is associated with BMP2-induced growth arrest and neuronal differentiation of human neuroblastoma-derived cell lines. *Biochem. Biophys. Res. Commun.* **307,** 206–213.

638. Chesler, L., Schlieve, C., Goldenberg, D.D., Kenney, A., Kim, G., McMillan, A., Matthay, K.K., Rowitch, D., and Weiss, W.A. (2006) Inhibition of phosphatidylinositol 3-kinase destabilizes Mycn protein and blocks malignant progression in neuroblastoma. *Cancer Res.* **66,** 8139–8146.

639. Biedler, J.L., Helson, L., and Spengler, B.A. (1973) Morphology and growth, tumorigenicity, and cytogenetics of human neuroblastoma cells in continuous culture. *Cancer Res.* **33,** 2643–2652.

640. Vandesompele, J., and Speleman, F. (2002) A brief commentary on "chromosomal aberrations in neuroblastoma cell lines identified by cross species color banding and chromosome painting". *Cancer Genet. Cytogenet.* **135,** 196.

641. Spengler, B.A., Biedler, J.L., and Ross, R.A. (2002) A corrected karyotype for the SH-SY5Y human neuroblastoma cell line. *Cancer Genet. Cytogenet.* **138,** 177–178.

642. Schlesinger, H.R., Gerson, J.M., Moorhead, P.S., Maguire, H., and Hummeler, K. (1976) Establishment and characterization of human neuroblastoma cell lines. *Cancer Res.* **36,** 3094–3100.

643. Cowell, J.K., and Rupniak, H.T. (1983) Chromosome analysis of human neuroblastoma cell line TR14 showing double minutes and an aberration involving chromosome 1. *Cancer Genet. Cytogenet.* **9,** 273–280.

644. Rudolph, G., Schilbach-Stuckle, K., Handgretinger, R., Kaiser, P., and Hameister, H. (1991) Cytogenetic and molecular characterization of a newly established neuroblastoma cell line in LS. *Hum. Genet.* **86,** 562–566.

645. Inoue, A., Yokomori, K., Tanabe, H., Mizusawa, H., Sofuni, T., Hayashi, Y., Tsuchida, Y., and Shimatake, H. (1997) Extensive genetic heterogeneity in the neuroblastoma cell line NB(TU)1. *Int. J. Cancer* **72,** 1070–1077.

646. Bedrnicek, J., Vicha, A., Jarosova, M., Holzerova, M., Cinatl, J., Jr., Michaelis, M., Cinatl, J., Eckschlager, T. (2005) Characterization of drug-resistant neuroblastoma cell lines by comparative genomic hybridization. *Neoplasma* **52,** 415–419.

647. Seeger, R.C., Rayner, S.A., Banerjee, A., Chung, H., Laug, W.E., Neustein, H.B., and Benedict, W.F. (1977) Morphology, growth, chromosomal pattern, and fibrinolytic activity of two new human neuroblastoma cell lines. *Cancer Res.* **37,** 1364–1371.

648. Reynolds, C.P., Biedler, J.L., Spengler, B.A., Reynolds, D.A., Ross, R.A., Frenkel, E.P., and Smith, R.G. (1986) Characterization of human neuroblastoma cell lines established before and after therapy. *J. Natl. Cancer Inst.* **76,** 375–387.

649. Van Roy, N., Forus, A., Myklebost, O., Cheng, N.C., Versteeg, R., and Speleman, F. (1995) Identification of two distinct chromosome 12-derived amplification units in neuroblastoma cell line NGP. *Cancer Genet. Cytogenet.* **82,** 151–154.

650. Pietsch, T., Göttert, E., Meese, E., Blin, N., Feickert, H.-J., Riehm, H., and Kovacs, G. (1988) Characaterization of a continuous cell line (MHH-NB-11) derived from advanced neuroblastoma. *Anticancer Res.* **8,** 1329–1334.

651. Raschellà, G., Melino, G., and Thiele, C.J. (2000) Retinoids and neuroblastoma: differentiation or apoptosis? In *Neuroblastoma* (Brodeur, G.M., Sawada, T., Tsuchida, Y., and Voûte, P.A., eds.), Elsevier Science B.V., pp. 183–195.

652. Melino, G., Thiele, C.J., Knight, R.A., and Piacentini, M. (1997) Retinoids and the control of growth/death decisions in human neuroblastoma cell lines. *J. Neurooncol.* **31,** 65–83.

653. Vertuani, S., De Geer, A., Levitsky, V., Kogner, P., Kiessling, R., and Levitskaya, J. (2003) Retinoids act as multistep modulators of the major histocompatibility class I presentation pathway and sensitize neuroblastomas to cytotoxic lymphocytes. *Cancer Res.* **63,** 8006–8013.

653A. Dufner-Beattie, J., Lemons, R.S., and Thorburn, A. (2001) Retinoic acid-induced expression of autotoxin in N-myc-amplified neuroblastoma cells. *Mol. Carcinog.* **30,** 181–189.

654. Reynolds, C.P., and Lie, S.O. (2000) Retinoid therapy of neuroblastoma. In *Neuroblastoma* (Brodeur, G.M., Sawada, T., Tsuchida, Y., and Voûte, P.A., eds.), Elsevier Science B.V., Amsterdam, The Nethrlands, pp. 519–539.

655. Garaventa, A., Gambini, C., Villavecchia, G., Di Cataldo, A., Bertolazzi, L., Pizzitola, M.R., De Bernardi, B., and Haupt, R. (2003) Second malignancies in children with neuroblastoma after combined treatment with ^{131}I-metaiodobenzylguanidine. *Cancer* **97,** 1332–1338.

656. Garaventa, A., Luksch, R., Lo Piccolo, M.S., Cavadini, E., Montaldo, P.G., Pizzitola, M.R., Boni, L., Ponzoni, M., Decensi, A., De Bernardi, Fossati Bellani, F., and Formelli, F. (2003) Phase I trial and pharmacokinetics of fenretinide in children with neuroblastoma. *Clin. Cancer Res.* **9,** 2032–2039.

657. Ichimiya, S., Nimura, Y., Seki, N., Ozaki, T., Nagase, T., and Nakagawara, A. (2001) Downregulation of hASH1 is associated with the retinoic acid-induced differentiation of human neuroblastoma cell lines. *Med. Pediatr. Oncol.* **36,** 132–134.

658. Tanaka, T., and Bénard, J. (2000) Expression of Ha-*ras,* *TP53,* and *TP73* in neuroblastoma. In *Neuroblastoma* (Brodeur, G.M., Sawada, T., Tsuchida, Y., and Voûte, P.A., eds.), Elsevier Science B.V., Amsterdam, The Netherlands, pp. 175–182.

659. Imamura, J., Bartram, C.R., Berthold, F., Hasrms, D., Nakamura, H., and Koeffler, H.P. (1993) Mutation of the *p53* gene in neuroblastoma and its relationship with the N-*myc* amplification. *Cancer Res.* **53,** 4053–4058.

660. Vogan, K., Bernstein, M., Leclerc, J.-M., Brisson, L., Brossard, J., Brodeur, G.M., Pelletier, J., and Gros, P. (1993) Absence of *p53* gene mutations in primary neuroblastomas. *Cancer Res.* **53,** 5269–5273.

661. Hosoi, G., Hara, J., Okamura, T., Osugi, Y., Ishihara, S., Fukuzawa, M., Okada, A., Okada, S., and Tawa, A. (1994) Low frequency of the p53 gene mutations in neuroblastoma. *Cancer* **73,** 3087–3093.

662. Tweddle, D.A., Malcolm, A.J., Bown, N., Pearson, A.D.J., and Lunec, J. (2001) Evidence for the development of *p53* mutations after cytotoxic therapy in a neuroblastoma cell line. *Cancer Res.* **61,** 8–13.

663. Kaghad, M., Bonnet, H., Yang, A., Creancier, L., Biscan, J.-C., Valent, A., Minty, A., Chalon, P., Lelias, J.-M., Dumont, X., Ferrara, P., McKeon, F., and Caput, D. (1997) Monoallelically expressed gene related to p53 at 1p36, a region frequently deleted in neuroblastoma and other human cancers. *Cell* **90,** 809–819.

664. Kovalev, S., Marchenko, N., Swendeman, S., LaQuaglia, M., and Moll, U.M. (1998) Expression level, allelic origin and mutation analysis of the *TP53* gene in neuroblastoma tumors and cell lines. *Cell Growth Differ.* **9,** 897–903.

665. Caron, H.N., and Hogarty, M.D. (2000) Imprinting of 1p, *MYCN* and other loci. In *Neuroblastoma* (Brodeur, G.M., Sawada, T., Tsuchida, Y., and Voûte, P.A., eds.), Elsevier Science B.V., Amsterdam, The Netherlands, pp. 101–112.

666. Norris, M.D., Gilbert, J., Smith, S.A., Marshall, G.M., Salwen, H., Cohn, S.L., and Haber, M. (2001) Expression of the putative tumour suppressor gene, *p73,* in neuroblastoma and other childhood tumours. *Med. Pediatr. Oncol.* **36,** 48–51.

667. Moley, J.F., Brother, M.B., Wells, S.A., Spengler, B.A., Biedler, J.L., and Brodeur, G.M. (1991) Low frequency of *ras* gene mutations in neuroblastomas, pheochromocytomas, and medullary thyroid cancers. *Cancer Res.* **51,** 1596–1599.

668. Tanaka, T., Slamon, D.J., Shimada, H., Shimoda, H., Fujisawa, T., Ida, N., and Seeger, R.C. (1991) A significant association of Ha-*ras* p21 in neuroblastoma cells with patient prognosis. A retrospective study of 103 cases. *Cancer* **68,** 1296–1302.

669. Favrot, M.C., and Gross, N. (2000) Adehsion molecules pexression in neuroblastoma. In *Neuroblastoma* (Brodeur, G.M., Sawada, T., Tsuchida, Y., and Voûte, P.A., eds.), Elsevier Science B.V., Amsterdam, The Netherlands, pp. 217–228.

670. Favrot, M.C., Combaret, V., Lasset, C., and Centre Léon Bérard (1993) CD44—A new prognostic marker for neuroblastoma. *N. Engl. J. Med.* **329,** 1965.

671. Tang, X.X., Robinson, M.E., Riceberg, J.S., Kim, D.Y., Kung, B., Titus, T.B., Hayashi, S., Flake, A.W., Carpentieri, D., and Ikegaki, N. (2004) Favorable neuroblastoma genes and molecular therapeutics of neuroblastoma. *Clin. Cancer Res.* **10,** 5837–5844.

672. Shtivelman, E., and Bishop, J.M. (1991) Expression of *CD44* is repressed in neuroblastoma cells. *Mol. Cell. Biol.* **11,** 5446–5453.

673. Combaret, V., Coll, J.L., and Favrot, M.C. (1994–1995) Expression of integrin and CD44 adhesion molecules on neuroblastoma: the relation to tumor aggressiveness and embryonic neural-crest differentiatin. *Invasion Metastasis* **14**, 156–163.

674. Combaret, V., Gross, N., Lasset, C., Frappaz, D., Beretta-Brognara, C., Philip, T., Beck, D., and Favrot, M.C. (1997) Clinical relevance of CD44 cell surface expression and MYCN gene amplification in neuroblastoma. *Eur. J. Cancer* **33**, 2101–2105.

675. Combaret, V., Gross, N., Lasset, C., Frappaz, D., Peruisseau G., Philip, T., Beck, D., and Favrot, M.C. (1996) Clinical relevance of CD44 cell surface expression and N-myc gene amplification in a multicentric analysis of 121 pediatric neuroblastomas. *J. Clin. Oncol.* **14**, 25–34.

676. Combaret, V., Lasset, C., Bouvier, R., Frappaz, D., Thiesse, P., Rebillard, A.C., Philip, T., and Favrot, M.C. (1995) CD44 expression: a new prognostic factor in neuroblastoma.. *Bull. Cancer* **82**, 131–136.

677. Combaret, V., Lasset, C., Frappaz, D., Bouvier, R., Thiesse, P., Rebillard, A.C., Philip, T., and Favrot, M.C. (1995) Evaluation of CD44 prognostic value in neuroblastoma: comparison with the other prognostic factors. *Eur. J. Cancer* **31A**, 545–549.

678. Yan, P., Mühlethaler, A., Balmas Bourloud, K., Nenadov Beck, M., and Gross, N. (2003) Hypermethylation-mediated regulation of *CD44* gene expression in human neuroblastoma. *Genes Chromosomes Cancer* **36**, 129–138.

679. Matsunaga, T., Shirasawa, H., Hishiki, T., Yoshida, H., Kouchi, K., Ohtsuka, Y., Kawamura, K., Etoh, T., and Ohnuma, N. (2000) Enhanced expression of N-*myc* messenger RNA in neuroblastomas found by mass screening. *Clin. Cancer Res.* **6**, 3199–3204.

680. Woods, W.G., Gao, R.-N., Shuster, J.J., Robison, L.L., Bernstein, M., Weitzman, S., Bunin, G., Levy, I., Brossard, J., Dougherty, G., Tuchman, M., and Lemieux, B. (2002) Screening of infants and mortality due to neuroblastoma. *N. Engl. J. Med.* **346**, 1041–1046.

681. Sawada, T., Nakata, T., Takasugi, N., Maeda, K., Hanawa, Y., Shimizu, K., Hirayama, M., Takeda, T., Mori, T., Koide, R., Tsunoda, A., Nagahara, N., and Yamamoto, K. (1984) Mass screening for neuroblastoma in infants in Japan. Interim Report of a Mass Screening Study Group. *Lancet* **2**, 271–273.

682. Schilling, F.H., Erttmann, R., Ambros, P.F., Strehl, S., Christiansen, H., Kovar, H., Kabisch, H., and Treuner, J. (1994) Screening for neuroblastoma. *Nature* **344**, 1157–1158.

683. Schilling, F.H., Spix, C., Berthold, F., Erttmann, R., Fehse, N., Hero, B., Klein, G., Sander, J., Schwarz, K., Treuner, J., Zorn, U., and Michaelis, J. (2002) Neuroblastoma screening at one year of age. *N. Engl. J. Med.* **346**, 1047–1053.

684. Ater, J.L., Gardner, K.L., Foxhall, L.E., Therrell, B.L., and Bleyer, W.A. (1998) Neuroblastoma screening in the United States. Results of the Texas Outreach Program for Neuroblastoma Screening. *Cancer* **82**, 1593–1602.

685. Matsumura, T., Gotoh, T., Sawada, T., and Sugimoto, T. (1998) A neuroblastoma cell line derived from a case detected through a mass screening system in Japan: a case report including the biologic and phenotypic characteristics of the cell line. *Cancer* **82**, 1416–1417.

686. Tajiri, T., Suita, S., Sera, Y., Takamatsu, H., Mizote, H., Nagasaki, A., Kurosaki, N., Handa, N., Hara, T., Okamura, J., Miyazaki, S., Sugimoto,T., Kawakami, K., Eguchi, H., Tsuneyoshi, M., and the Committee for Pediatric Solid Malignant Tumors in the Kyushu Area (2001) Clinical and biologic characteristics for recurring neuroblastoma at mass screening cases in Japan. *Cancer* **92**, 349–353.

687. Kerbel, R., Urban, C.E., Ambros, I.M., Dornbusch, H.J., Schwinger, W., Lackner, H., Ladenstein, R., Strenger, V., Gadner, H., and Ambros, P.F. (2003) Neuroblastoma mass screening in late infancy: insights into the biology of neuroblastic tumors. *J. Clin. Oncol.* **21**, 4228–4234.

688. Que, T., Inoue, M., Yoneda, A., Kubota, A., Okuyama, H., Kawahara, H., Nishikawa, M., Nakayama, M., and Kawa, K. (2005) Profile of neuroblastoma detected by mass screening, resected after observation without treatment: results of the Wait and See pilot study. *J. Pediatr. Surg.* **40**, 359–363.

689. Calin, G.A., Sevignani, C., Dumitru, C.D., Hyslop, T., Noch, E., Yendamuri, S., Shimizu, M., Rattan, S., Bullrich, F., Negrini, M., and Croce, C.M. (2004) Human MiRNA genes are frequently located at fragile sites and genomic regions involved in cancers. *Proc. Natl. Acad. Sci. USA* **101**, 2999–3004.

690. Calin, G.A., Liu, C.G., Sevignani, C., Ferracin, M., Felli, N., Dumitru, C.D., Shimizu, ., Cimmino, A., Zupo, S., Dono, M., Dell'Aquila, M.L., Aldler, H., Rassenti, L., Kipps, T.J., Bullrich, F., Negrini, M., and Croce, C.M. (2004) MicroRNA profiling reveals distinct signatures in B cell chronic lymphocytic leukemia. *Proc. Natl. Acad. Sci. USA* **101**, 11755–11760.

691. Michael, M.Z., O'Connor, S.M., van Holst Pellekaan, N.G., Young, G.P., and James, R.J. (2003) Reduced accumulation of specific microRNAs in colorectal neoplasia. *Mol. Cancer Res.* **1**, 882–891.

692. Eis, P.S., Tam, W., Sun, L., Chadburn, A., Li, Z., Gomez, M.F., Lund, E., and Dahlberg, J.E. (2005) Accumulation of miR-155 and BIC RNA in human B cell lymphomas. *Proc. Natl. Acad. Sci. USA* **102**, 3627–3632.

693. Lu, J., Getz, G., Miska, E.A., Alvarez-Saavedra, E., Lamb, J., Peck, D., Sweet-Cordero, A., Ebert, B.L., Mak, R.H., Ferrando, A.A., Downing, J.R., Jacks, T., Horvitz, H.R., and Golub, T.R. (2005) MicroRNA expression profiles classify human cancers. *Nature* **435**, 834–838.

694. Tanno, B., Cesi, V., Vitali, R., Sesti, F., Giuffrida, M.L., Mancini, C., Calabretta, B., and Raschellà, G. (2005) Silencing of endogenous IGFBP-5 by micro RNA interference affects proliferation, apoptosis and differentiation of neuroblastoma cells. *Cell Death Differ.* **12**, 213–223.

695. Tanno, B., Vitali, R., De Arcangelis, D., Mancini, C., Eleuteri, P., Dominici, C., and Raschellà, G. (2006) Bim-dependent apoptosis follows IGFBP-5 down-regulation in neuroblastoma cells. *Biochem. Biophys. Res. Commun.* **351**, 547–552.

696. Kang, J.H., Rychahou, P.G., Ishola, T.A., Qiao, J., Evers, B.M., and Chung, D.H. (2006) *MYCN* silencing induces differentiation and apoptosis in human neuroblastoma cells. *Biochem. Biophys. Res. Commun.* **351**, 192–197.

697. Chen, Y., and Stallings, R.L. (2007) Differential patterns of microRNA expression in neuroblastoma are correlated with

prognosis, differentiation and apoptosis. *Cancer Res.***67,** 976–983.

698. Mora, J., Cheung, N.-K.V., Juan, G., Illei, P., Cheung, I., Akram, M., Chi, S., Ladanyi, M., Cordon-Cardo, C., and Gerald, W.L. (2001) Neuroblastic and schwannian stromal cells of neuroblastoma are derived from a tumoral progenitor cell. *Cancer Res.* **61,** 6892–6898.

699. Ross, R.A., and Spengler, B.A. (2004) The conundrum posed by cellular heterogeneity in analysis of human neuroblastoma. *J. Natl. Cancer Inst.* **96,** 1192–1193.

700. Tonini, G.P. (1993) Neuroblastoma: a multiple biological disease. *Eur. J. Cancer* **29A,** 802–804.

701. Tonini, G.P., and Romani, M. (2003) Genetic and epigenetic alterations in neuroblastoma. *Cancer Lett.* **197,** 69–73.

702. Combaret, V., Delattre, O., Bénard, J., and Favrot, M.C. (1998) Biological markers for the prognosis of neuroblastoma: proposal of a method of analysis. *Bull. Cancer* **85,** 262–266.

703. Riley, R.D., Heney, D., Jones, D.R., Sutton, A.J., Lambert, P.C., Abrams, K.R., Young, B., Wailoo, A.J., and Burchill, S.A. (2004) A systematic review of molecular and biological tumor markers in in neuroblastoma. *Clin. Cancer Res.* **10,** 4–12.

704. Rudie Hovland, A., Nahreini, P., Andreatta, C.P., Edwards-Praasad, J., and Prasad, K.N. (2001) Identifying genes involved in regulating differentiation of in neuroblastoma cells. *J. Neurosci. Res.* **64,** 302–310.

705. Tomioka, N., Kobayashi, H., Kageyama, H., Ohira, M., Nakamura, Y., Sasaki, F., Todo, S., Nakagawara, A., and Kaneko, Y. (2003) Chromosomes that show partial loss or gain in near-diploid tumors coincide with chromosomes that show whole loss or gain in near-triploid tumors: evidence suggesting the involvement of the same genes in the tumorigenesis of high- and low-risk neuroblastomas. *Genes Chromosomes Cancer* **36,** 139–150.

706. Melegh, Z., Csernák, E., Tóth, E., Veleczki, Z., Magyarosy, E., Nagy, K., and Szentirmay, Z. (2005) DNA content heterogeneity in neuroblastoma analyzed by means of image cytometry and its potential significance. *Virchows Arch.* **446,** 517–524.

707. Vandesompele, J., Baudis, M., De Preter, K., Van Roy, N., Ambros, P., Bown, P., Brinkschmidt, C., Christiansen, H., Combaret, V., Lastowska, M., Nicholson, J., O'Meara, A., Plantaz, D., Stallings, R., Brichard, B., Van den Broecke, C., De Bie, S., De Paepe, A., Laureys, G., and Speleman, F. (2005) Unequivocal delineation of clinicogenetic subgroups and development of a new model for improved outcome prediction in neuroblastoma. *J. Clin. Oncol.* **23,** 2280–2299.

708. Feder, M.K., and Gilbert, F. (1983) Clonal evolution in a human neuroblastoma. *J. Natl. Cancer Inst.* **70,** 1051–1056.

709. Shen, L., Wu, Z., Kuroda, T., Honna, T., Tanaka, Y., and Miyauchi, J. (2005) Clonality analysis of childhood neuroblastoma by polymerase chain reafction for the human androgen receptor gene. *Int. J. Oncol.* **26,** 1329–1335.

710. Schwab, M. (1993) Amplification of N-myc as a prognostic marker for patients with neuroblastoma. *Semin. Cancer Biol.* **4,** 13–18.

711. Schwab, M. (1999) Human neuroblastoma: from basic science to clinical debut of cellular oncogenes. *Naturwissenschaften* **86,** 71–78.

712. Schwab, M. (1999) Oncogene amplification in solid tumors. *Semin. Cancer Biol.* **9,** 319–325.

713. Schwab, M., and Evans, A. (2003) Neuroblastoma – developmental and molecular biology meet therapy design. *Cancer Lett.* **197,** 1.

714. Sawada, T., and Matsumura, T. (2000) DNA ploidy and prognosis of neuroblastoma. *Pediatr. Hematol. Oncol.* **17,** 1–3.

715. Tajiri, T., Shono, K., Tanaka, S., and Suita, S. (2002) Evaluation of genetic heterogeneity in neuroblastoma. *Surgery* **131,** S283–S287.

716. Goldsby, R.E., and Matthay, K.K. (2004) Neuroblastoma: evolving therapies for a disease with many faces. *Paediatr. Drugs* **6,** 107–122,

717. Bilke, S., Chen, Q.-R., Westerman, F., Schwab, M., Catchpoole, D., and Khan, J. (2005) Inferring a tumor progression model for neuroblastoma from genomic data. *J. Clin. Oncol.* **23,** 7322–7331.

718. Douglass, E.C., Poplack, D.G., and Whang-Peng. J. (1980) Involvement of chromosome no. 22 in neuroblastoma. *Cancer Genet. Cytogenet.* **2,** 287–291.

719. Haag, M.M., Soukup, S.W., and Neeley, J.E. (1981) Chromosome analysis of a human human neuroblastoma. *Cancer Res.* **41,** 2995–2999.

720. O'Malley, D.P., Cousineau, A.J., Kulkarni, R., and Higgins, J.V. (1986) Double minute chromosomes. A bone marrow indicator of neuroblastoma metastasis and relapse: two case reports. *Cancer* **57,** 2158–2161.

721. Christiansen, H., and Lampert, F. (1989) Tumor cytogenetics and prognosis in neuroblastoma. *Monatsschr. Kinderheilkd.* **137,** 666–671.

722. Pearson, A.D.J., Reid, M.M., Davison, E.V., Bown, N., Malcolm, A.J., and Craft, A.W. (1988) Cytogenetic investigations of solid tumours of children. *Arch. Dis. Child.* **63,** 1012–1015.

723. Vernole, P., and Melino, G. (1988) Genetic and cytogenetic characteristics of human neuroblastoma. *Clin. Ter.* **124,** 279–285.

724. Hayashi, Y., Kanda, N., Inaba, T., Hanada, R., Nagahara, N., Muchi, H., and Yamamoto, K. (1989) Cytogenetic findings and prognosis in neuroblastoma with emphasis on marker chromosome 1. *Cancer* **63,** 126–132.

725. Skinnider, L., Wang, S.Y., Wang, H.C., and Krespin, H.I. (1990) A translocation (1;10)(p32;q24) in a neuroblastoma cell line derived from marrow metastases. *Cancer Genet. Cytogenet.* **45,** 131–135.

726. Carlsen, N.L., Mortensen, B.T., Andersson, P.K., Larsen, J.K., Nielsen, O.H., and Tommerup, N. (1991) Chromosome abnormalities, cellular DNA content, oncogene amplification and growth pattern in agar culture of human neuroblastomas. *Anticancer Res.* **11,** 353–358.

727. Fletcher, J.A., Kozakewich, H.P., Hoffer, F.A., Lage, J.M., Weidner, N., Tepper, R., Pinkus, G.S., Morton, C.C., and Corson, J.M. (1991) Diagnostic relevance of clonal cytogenetic aberrations in malignant soft-tissue tumors. *N. Engl. J. Med.* **324,** 436–443.

728. Dressler, L.G., Duncan, M.H., Varsa, E.E., and McConnell, T.S. (1993) DNA content measurement can be obtained using archival material for DNA flow cytometry. A compar-

ison with cytogenetic analysis in 56 pediatric solid tumors. *Cancer* **72**, 2033–2041.

729. Sainati, L., Stella, M., Montaldi, A., Bolcato, S., Guercini, N., Bonan, F., Ninfo, V., and Zanesco, L. (1992) Value of cytogenetic in the differential diagnosis of the small round cell tumors of childhood. *Med. Pediatr. Oncol.* **20**, 130–135.

730. Neumann, E., Kalousek, D.K., Norman, M.G., Steinbok, P., Cochrane, D.D., and Goddard, K. (1993) Cytogenetic analysis of 109 pediatric central nervous system tumors. *Cancer Genet. Cytogenet.* **71**, 40–49.

731. Bown, N.P., Reid, M.M., Malcolm, A.J., Davison, E.V., Craft, A.W., and Pearson, A.D. (1994) Cytogenetic abnormalities of small round cell tumours. *Med. Pediatr. Oncol.* **23**, 124–129.

732. Chen, Z., Morgan, R., Stone, J.F., and Sandberg, A.A. (1994) Appreciation of the significance of cytogenetic and FISH analysis of bone marrow in clinical oncology. *Cancer Genet. Cytogenet.* **78**, 10–14.

733. Fujii, Y., Hongo, T., and Hayashi, Y. (1994) Chromosome analysis of brain tumors in childhood. *Genes Chromosomes Cancer* **11**, 205–215.

734. Lo, R., Perlman, E., Hawkins, A.L., Hayashi, R., Wechsler, D.S., Look, A.T., and Griffin, C.A. (1994) Cytogenetic abnormalities in two cases of neuroblastoma. *Cancer Genet. Cytogenet.* **74**, 30–34.

735. Hattinger, C.M., Rumpler, S., Ambros, I.M., Strehl, S., Lion, T., Zoubek, A., Gadner, H., and Ambros, P.F. (1996) Demonstration of the translocation der(16)t(1;16)(q12;q11.2) in interphase nuclei of Ewing tumors. *Genes Chromosomes Cancer* **17**, 141–150.

736. Mitchell, E.L., McNally, K., and Kelsey, A. (1996) Involvement of chromosomes 1 and 17 in a case of neuroblastoma. *J. Pediatr. Hematol.* **13**, 457–461.

737. Sawyer, J.R., Roloson, G.J., Bell, J.M., Thomas, J.R., Teo, C., and Chadduck, W.M. (1996) Telomeric associations in the progression of chromosome aberrations in pediatric solid tumors. *Cancer Genet. Cytogenet.* **90**, 1–13.

738. Cowan, J.M., Dayal, Y., Schwaitzberg, S., and Tischler, A.S. (1997) Cytogenetic and immunohistochemical analysis of an adult anaplastic neuroblastoma. *Am. J. Surg. Pathol.* **21**, 957–963.

739. Kaneko, Y., and Knudson, A.G. (2000) Mechanism and relevance of ploidy in neuroblastoma. *Genes Chromosomes Cancer* **29**, 89–95.

740. Rechavi, G., and Meitar, D. (2001) Detection of unidentified chromosome abnormalities in human neuroblastoma by spectral karyotyping (SKY). *Genes Chromosomes Cancer* **31**, 201–208.

741. Udayakumar, A.M., Sundareshan, T.S., Goud, T.M., Biswas, S., Devi, M.G., Kumari, B.S., and Appaji, J. (2001) Cytogenetics of neuroblastoma: a study using fine needle aspiration cultures. *Ann. Genet.* **44**, 161–166.

742. Leon, M.E., Hou, J.,S., Galindo, L.M., and Garcia, F.U. (2004) Fine-needle aspiration of adult small-round-cell tumors studied with flow cytometry. *Diagn. Cytopathol.* **31**, 147–154.

743. Betts, D.R., Cohen, N., Leibundgut, K.E., Kuhne, T., Caflisch, U., Greiner, J., Trakhtenbrot, L., and Niggli, F.K. (2005) Characterization of karyotypic events and evolution in neuroblastoma. *Pediatr. Blood Cancer* **44**, 147–157.

744. Shapira, S.K., McCaskill, C., Northrup, H., Spikes, A.S., Elder, F.F.B., Sutton, V.R., Korenberg, J.R., Greenberg, F., and Shaffer, L.G. (1997) Chromosome 1p36 deletions: the clinical phenotype and molecular characterization of a common newly delineated syndrome. *Am. J. Hum. Genet.* **61**, 642–650.

745. Slavotinek, A., Shaffer, L.G., and Shapira, S.K. (1999) Monosomy 1p36. *J. Med. Genet.* **36**, 657–663.

746. Avigad, S., Tamir, Y., Yaniv, I., Goshen, J., Kaplinsky, C., Cohen, I.J., and Zaizov, R. (1994) Genetic alterations involving chromosome 1p in children with neuroblastoma. *Isr. J. Med. Sci.* **30**, 639–641.

747. Enomoto, H., Ozaki, T., Takahashi, E., Nomura, N., Tabata, S., Takahashi, H., Ohnuma, N., Tanabe, M., Iwai, J., Yoshida, H., Matsunaga, T., and Sakiyama, S. (1994) Identification of human *DAN* gene, mapping to the putative neuroblastoma tumor suppressor locus. *Oncogene* **9**, 2785–2791.

748. Lahti, J.M., Valentine, M., Xiang, J., Jones, B., Amann, J., Grenet, J., Richmond, G., Look, A.T., and Kidd, V.J. (1994) Alterations in the PITSLRE protein kinase gene complex on chromosome 1p36 in childhood neuroblastoma. *Nat. Genet.* **7**, 370–375.

749. Amler, L.C., Bauer, A., Corvi, R., Dihlmann, S., Praml, C., Cavenee, W.K., Schwab, M., and Hampton, G.M. (2000) Identification and characterization of novel genes located at the t(1;15)(p36.2;q24) translocation breakpoint in the neuroblastoma cell line NGP. *Genomics* **64**, 195–202.

750. Cheng, N.C., Van Roy, N., Chan, A., Beitsma, M., Westerveld, A., Speleman, F., and Versteeg, R. (1995) Deletion mapping in neuroblastoma cell lines suggests two distinct tumor suppressor genes in the 1p35–36 region, only one of which is associated with N-myc amplification. *Oncogene* **10**, 291–297.

751. Scaruffi, P., Parodi, S., Mazzocco, K., Defferrari, R., Fontana, V., Bonassi, S., and Tonini, G.P. (2004) Detection of *MYCN* amplification and chromosome 1p36 loss in neuroblastoma by cDNA microarray comparative genomic hybridization. *Mol. Diagn.* **8**, 93–100.

752. Strehl, S., and Ambros, P.F. (1993) Fluorescence in situ hybridization combined with immunohistochemistry for highly sensitive detection of chromosome 1 aberrations in neuroblastoma. *Cytogenet. Cell Genet.* **63**, 24–28.

753. Combaret, V., Turc-Carel, C., Thiesse, P., Rebillard, A.-C., Frappaz, D., Haus, O., Philip, T., and Favrot, M.C. (1995) Sensitive detection of numerical and structural aberrations of chromosome 1 in neuroblastoma by interphase fluorescence *in situ* hybridization. Comparison with restriction fragment length polymorphism and conventional cytogenetic analyses. *Int. J. Cancer* **61**, 185–191.

754. Peter, M., Michon, J., Vielh, P., Neuenschwander, S., Nakamura, Y., Sonsino, E., Zucker, J.M., Vergnaud, G., Thomas, G., and Delattre, O. (1992) PCR assay for chromosome 1p deletion in small neuroblastoma samples. *Int. J. Cancer* **52**, 544–548.

755. Anderson, J., Gibson, S., Williamson, D., Rampling, D., Austin, C., Shipley, J., Sebire, N., and Brock, P. (2006) Rapid and accurate determination of *MYCN* copy number and 1p deletion in neuroblastoma by quantitative PCR. *Pediatr. Blood Cancer* **46**, 820–824.

756. Leong, P.K., Thnorner, P., Yeger, H., Ng, K., Zhang, Z., and Squire, J. (1993) Detection of MYCN gene amplification and deletions of chromosome 1p in neuroblastoma by in situ hybridization using routine histologic sections. *Lab. Invest.* **69,** 43–50.

757. Mead, R.S., and Cowell, J.K. (1995) Molecular characterization of a (1;10)(p22;q21) constitutional translocation from a patient with neuroblastoma. *Cancer Genet. Cytogenet.* **81,** 151–157.

758. Roberts, T., Chernova, O., and Cowell, J.K. (1998) Molecular characterization of the 1p22 breakpoint region spanning the constitutional translocation breakpoint in a neuroblastoma patient with a t(1;10)(p22;q21). *Cancer Genet. Cytogenet.* **100,** 10–20.

759. Versteeg, R., Caron, H., Cheng, N.C., van der Drift, P., Slater, R., Westerveld, A., Voûte, P.A., Delattre, O., Laureys, G., Van Roy, N., and Speleman, F. (1995) 1p36: every subband a suppressor? *Eur. J. Cancer* **31A,** 538–541.

760. Grenet, J., Valentine, V., Kitson, J., Li, H., Farrow, S.N., and Kidd, V.J. (1998) Duplication of the *DR3* gene on human chromosome 1p36 and its deletion in human neuroblastoma. *Genomics* **49,** 385–393.

761. Kaneko, Y., Kobayashi, H., Maseki, N., Nakagawara, A., and Sakurai, M. (1999) Disomy 1 with terminal 1p deletion is frequent in mass-screening-negative/late-presenting neuroblastomas in young children, but not in mass-screening-positive neuroblastomas in infants. *Int. J. Cancer* **80,** 54–59.

762. Jogi, A., Abel, F., Sjöberg, R.M., Toftgard, R., Zaphiropoulos, P.G., Påhlman, S., Martinsson, T., and Axelson, H. (2000) Patched 2, located in 1p32–34, is not mutated, in high stage neuroblastoma tumors. *Int. J. Oncol.* **16,** 943–949.

763. Mora, J., Gerald, W.L., Qin, J., and Cheung, N.-K.V. (2000) Evolving significance of prognostic markers associated with treatment improvement in patients with stage 4 neuroblastoma. *Cancer* **94,** 2756–2765.

764. Nakagawara, A., Ohira, M., Kageyama, H., Mihara, M., Furuta, S., Machida, T., Takayasu, H., Islam, A., Nakamura, Y., Takahashi, M., Shishikura, T., Kaneko, Y., Toyoda, A., Hattori, M., Sakaki, Y., Ohki, M., Horii, A., Soeda, E., Inazawa, J., Sedki, N., Kuma, H., Nozawa, I., and Sakiyama, S. (2000) Identification of the homozygously deleted region at chromosome 1p36.2 in human neuroblastoma. *Med. Pediatr. Oncol.* **35,** 516–521.

765. Ambros, P.F., Ambros, I.M., Kerbl, R., Luegmayr, A., Rumpler, S., Ladenstein, R., Amann, G., Kovar, H., Horcher, E., De Bernardi, B., Michon, J., and Gadner, H. (2001) Intratumoural heterogeneity of 1p deletions and MYCN amplification in neuroblastomas. *Med. Pediatr. Oncol.* **36,** 1–4.

766. 766. Chen, Y.Z., Soeda, E., Yang, H.W., Takita, J., Chai, L., Horii, A., Inazawa, J., Ohki, M., and Hayashi, Y. (2001) Homozygous deletion in a neuroblastoma cell line defined by a high-density STS map spanning human chromosome band 1p36. *Genes Chromosomes Cancer* **31,** 326–332.

767. De Toledo, M., Coulon, V., Schmidt, S., Fort, P., and Blangy, A. (2001) The gene for a new brain specific RhoA exchange factor maps to the highly unstable chromosomal region 1p36.2–1p36.3. *Oncogene* **20,** 7307–7317.

768. Hiyama, E., Hiyama, K., Ohtsu, K., Yamaoka, H., Fukuba, I., Matsuura, Y., and Yokoyama, T. (2001) Biological characteristics of neuroblastoma with partial deletion in the short arm of chromosome 1. *Med. Pediatr. Oncol.* **36,** 67–74.

769. Hogarty, M.D., Maris, J.M., White, P.S., Guo, C., and Brodeur, G.M. (2001) Analysis of genomic imprinting at 1p35–36 in neuroblastoma. *Med. Pediatri. Oncol.* **36,** 52–55.

770. Yang, H.W., Chen, Y.Z., Takita, J., Soeda, E., Piao, H.Y., and Hayashi, Y. (2001) Genomic structure and mutational analysis of the human *KIF1B* gene which is homozygously deleted in neuroblastoma at chromosome 1p36.2. *Oncogene* **20,** 5075–5083.

771. Chen, Y.Y., Takita, J., Chen, Y.Z., Yang, H.W., Hanada, R., Yamamoto, K., and Hayashi, Y. (2003) Genomic structure and mutational analysis of the human *KIF1Bα* gene located at 1p36.2 in neuroblastoma. *Int. J. Oncol.* **23,** 737–744.

772. Gonzalez-Gomez, P., Bello, M.J., Lomas, J., Arjona, D., Alonso, M.E., Amiñoso, C., Lopez-Marin, I., Anselmo, N.P., Sarasa, J.L., Gutierrez, M., Casartelli, C., and Rey, J.A. (2003) Aberrant methylation of multiple genes in neuroblastic tumours: relationship with *MYCN* amplification and allelic status at 1p. *Eur. J.Cancer.* **39,** 1478–1485.

773. Krona, C., Ejeskär, K., Caren, H., Abel, F., Sjöberg, R.M., and Martinsson, T. (2004) A novel 1p36.2 located gene, *APITD1*, with tumour-suppressive properties and a putative binding domain, shows low expression in neuroblastoma tumours. *Br. J. Cancer* **91,** 1119–1130.

774. Mathysen, D., Van Roy, N., Van Hul, W., Laureys, G., Ambros, P., Speleman, F., and Wuyts, W. (2004) Molecular analysis of the putative tumour-suppressor gene *EXTL1* in neuroblastoma patients and cell lines. *Eur. J. Cancer* **40,** 1255–1261.

775. Carén, H., Ejeskär, K., Fransson, S., Hesson, L., Latif, F., Sjöberg, R.-M., Krona, C., and Martinsson, T. (2005) A cluster of genes located in 1p36 are down-regulated in neuroblastomas with poor prognosis, but not due to CpG island methylation. *Mol. Cancer* **4,** 10–18.

776. Ejeskär, K., Krona, C., Carén, H., Zaibak, F., Li, L., Martinsson, T., and Iaonnou, P.A. (2005) Introduction of *in vitro* transcribed *ENO1* mRNA into neuroblastoma cells induces cell death. *BMC Cancer* **5,** 161–174.

777. Henrich, K.-O., Fischer, M., Mertens, D., Benner, A., Wiedemeyer, R., Brors, B., Oberthuer, A., Berthold, F., Wei, J.S., Khan, J., Schwab, M., and Westermann, F. (2006) Reduced expression of *CAMTA1* correlates with adverse outcome in neuroblastoma patients. *Clin. Cancer Res.* **12,** 131–138.

777A. Wang, Q., Diskin, S., Rappaport, E., Attiyeh, E., Mosse, Y., Shue, D., Seiser, E., Jagannathan, J., Shusterman, S., Bansal, M., Khazi, D., Winter, C., Okawa, E., Grant, G., Cnaan, A., Zhao, H., Cheung, N.-K., Gerald, W., London, W., Matthay, K.K., Brodeur, G.M., and Maris, J.M. (2006) Integrative genomics identifies distinct molecular classes of neuroblastoma and shows that multiple genes are targeted by regional alterations in DNA copy number. *Cancer Res.* **66,** 6050–6062.

777B. Fransson, S., Martinsson, T., and Ejeskär, K. (2007) Neuroblastoma tumors with favorable and unfavorable outcomes: Significant differences in mRNA expression of genes mapped to 1 p36.2 *Genes Chromosomes Cancer* **46,** 45–52.

777C. Welch, C., Chen, Y., and Stallings, R.L. (2007) MicroRNA-34a functions as a potential tumor suppressor by inducing apoptosis in neuroblastoma cells. *Oncogene* **26**, 5017–5022.

778. O'Neill, S., Ekstrom, L., Lastowska, M., Roberts, P., Brodeur, G.M., Kees, U.R., Schwab, M., and Bown, N. (2001) *MYCN* amplification and 17q in 4302–4307: evidence for structural association. *Genes Chromosomes Cancer* **30**, 87–90.

779. Godfried, M.B., Veenstra, M., van Sluis, P., Boon, K., van Asperen, R., Hermus, M.C., van Schaik, B.D.C., Voute, T.P.A., Schwab, M., Versteeg, R., and Caron, H.N. (2002) The N-*myc* and c-*myc* downstream pathways include the chromosome 17q genes nm23-H1 and nm23-H2. *Oncogene* **21**, 2097–2101.

780. Morowitz, M., Shusterman, S., Mosse, Y., Hii, G., Winter, C.L., Khazi, D., Wang, Q., King, R., and Maris, J.M. (2003) Detection of single-copy chromosome 17q gain in human neuroblastomas using real-time quantitative polymerase chain reaction. *Mod. Pathol.* **16**, 1248–1256.

781. Yoon, K.J., Danks, M.K., Ragsdale, S.T., Valentine, M.B., and Valentine, V.A. (2006) Translocations of 17q21~qter in neuroblastoma cell lines infrequently include the topoisomerase IIα gene. *Cancer Genet. Cytogenet.* **167**, 92–94.

781A. Vandesompele, J., Michels, E., De Preter, K., Menten, B., Schramm, A,, Eggert, A., Ambros, P.F., Combaret, V., Francotte, N., Antonacci, F., De Paepe, A., Laureys, G., Speleman, F., and Van Roy, N. (2008) Identification of 2 putative critical segments of 17q gain in neuroblastoma through integrative genomics. *Int. J. Cancer* **122**, 1177–1182.

782. Raetz, E.A., Kim, M.K.H., Moos, P., Carlson, M., Bruggers, C., Hooper, D.K., Foot, L., Liu, T., Seeger, R., and Carroll, W.L. (2003) Identification of genes that area regulated transcriptionally by Myc in childhood tumors. *Cancer* **98**, 841–853.

783. Kim, G.-J., Park, S.-Y., Kim, H., Chun, Y.-H., and Park, S.-H. (2001) Chromosomal aberrations in neuroblastoma cell lines identified by cross species color banding and chromosome painting. *Cancer Genet. Cytogenet.* **129**, 10–16.

784. Valent, A., Bénard, J., Vénuat, A.-M., Da Silva, J., Duverger, A., Duarte, N., Hartmann, O., Spengler, B.A., and Bernheim, A. (1999) Phenotypic and genotypic diversity of human neuroblastoma studied in three IGR cell line models derived from bone marrow metastases. *Cancer Genet. Cytogenet.* **112**, 124–129.

785. Panarello, C., Morerio, C., Russo, I., Pasquali, F., Rapella, A., Corrias, M.V., Morando, A., and Rosanda, C. (2000) Full cytogenetic characterization of a new neuroblastoma cell line with a complex 17q translocation. *Cancer Genet. Cytogenet.* **116**, 124–132.

786. McConville, C.M., Dyer, S., Rees, S.A., Oude Luttikhuis, M.E.M., McMullan, D.J., Vickers, S.J., Ramani, P., Redfern, D., and Morland, B.J. (2001) Molecular cytogenetic characterization of two non-MYCN amplified neuroblastoma cell lines with complex t(11;17). *Cancer Genet. Cytogenet.* **130**, 133–140.

787. Ross, R.A., Spengler, B.A., and Biedler, J.L. (1983) Coordinate morphological and biochemical interconversion of human neuroblastoma cells. *J. Natl Cancer Inst.* **71**, 741–747.

788. Spitz, R., Hero, B., Simon, T., and Berthold, F. (2006) Loss in chromosome 11q identifies tumors with increased risk for metastatic relapses in localized and 4S neuroblastoma. *Clin. Cancer Res.* **12**, 3368–3373.

789. Boensch, M., Oberthuer, A., Fischer, M., Skowron, M., Oestreich, J., Berthold, F., and Spitz, R. (2005) Quantitative real-time PCR for quick simultaneous determination of therapy-stratifying markers MYCN amplification, deletion 1p and 11q. *Diagn. Mol. Pathol.* **14**, 177–182.

789A. George, R.E., Attiyeh, E.F., Li, S., Moreau, L.A., Neuberg, D., Li, C., Fox, E.A., Meyerson, M., Diller, L., Fortina, P., Look, A.T., and Maris, J.M. (2007) Genome-wide analysis of neuroblastomas using high-density single nucleotide polymorphism arrays. *PLoS ONE* **2**, e255.

790. Bernards, R., Dessain, S.K., and Weinberg, R.A. (1986) N-*myc* amplification causes down-regulation of MHC class 1 antigen expression in neuroblastoma. *Cell* **47**, 667–674.

791. Slamon, D.J., Boone, T.C., Seeger, R.C., Keith, D.E., Chazin, V., Lee, H.C., and Souza, L.M. (1986) Identification and characterization of the protein encoded by the human N-myc oncogene. *Science* **232**, 768–772.

792. Cohen, P.S., Seeger, R.C., Triche, T.J., and Israel, M.A. (1988) Detection of N-myc gene expression in neuroblastoma tumors by in situ hybridization. *Am. J. Pathol.* **131**, 391–397.

793. Misra, D.N., Dickman, P.S., and Yunis, E.J. (1995) Fluorescence in situ hybridization (FISH) detection of MYCN oncogene amplification in neuroblastoma using paraffin-embedded tissues. *Diagn. Mol. Pathol.* **4**, 128–135.

794. Hachitanda, Y., Saito, M., Mori, T., and Hamazaki, M. (1997) Application of fluorescaence in situ hybridization to detect N-myc (MYCN) gene amplification on paraffin-embedded tissue sections of neuroblastomas. *Med. Pediatr. Oncol.* **29**, 135–138.

795. Noguchi, M., Hirohashi, S., Tsuda, H., Nakajima, T., Hara, F., and Shimosato, Y. (1988) Detection of amplified N-myc gene in neuroblastoma by in situ hybridization using the single-step silver enhancement method. *Mod. Pathol.* **1**, 428–432.

796. Nisen, P.D., Waber, P.G., Rich, M.A., Pierce, S., Garvin, J.R., Jr., Gilbert, F., and Lanzkowsky, P. (1988) N-myc oncogene RNA expression in neuroblastoma. *J. Natl. Cancer Inst.* **80**, 1633–1637.

797. Tsuda, H., Shimosato, Y., Upton, M.P., Yokota, J., Terada, M., Ohira, M., Sugimura, T., and Hirohashi, S. (1988) Retrospective study on amplification of N-myc and c-myc genes in pediatric solid tumors and its association with prognosis and tumor differentiation. *Lab. Invest.* **59**, 321–327.

798. Bourhis, J., Bénard, J., Hartmann, O., Boccon-Gibod, L., Lemerle, J., and Riou, G. (1989) Correlation of MDR1 gene expression with chemotherapy in neuroblastoma. *J. Natl. Cancer Inst.* **81**, 1401–1405.

799. Breit, S., and Schwab, M. (1989) Suppression of MYC by high expression of NMYC in human neuroblastoma cells. *Neurosci. Res.* **24**, 21–28.

800. Cohn, S.L., Salwen, H., Quasney, M.W., Ikegaki, N., Cowan, J.M., Herst, C.V., Kennett, R.H., Rosen, S.T., DiGiuseppe, J.A., and Brodeur, G.M. (1990) Prolonged N-myc protein half-life in a neuroblastoma cell line lacking N-myc amplification. *Oncogene* **5**, 1821–1827.

shRNA.ribozyme, and shRNA.antisense in neuroblastoma cells. *Cell Mol. Neurobiol.* **24,** 781–792.

903. Banelli, B., Di Vinci, A., Gelvi, I., Casciano, I., Allemanni, G., Bonassi, S., and Romani, M. (2005) DNA methylation in neuroblastic tumors. *Cancer Lett.* **228,** 37–41.

903A. Banelli, B., Gelvi, I., Di Vinci, A., Scaruffi, P., Casciano, I., Allemanni, G., Bonassi, S., Tonini, G.P., and Romani, M. (2005) Distinct CpG methylation profiles characterize different clinical groups of neuroblastic tumors. *Oncogene* **24,** 5619–5628.

904. Misawa, A., Inoue, J., Sugino, Y., Hosoi, H., Sugimoto, T., Hosoda, F., Ohki, M., Imoto, I., and Inazawa, J. (2005) Methylation-associated silencing of the *nuclear receptor 112* gene in advanced-type neuroblastomas, identified by bacterial artificial chromosome array-based methylated CpG island amplification. *Cancer Res.* **65,** 10233–10242.

905. Borrello, M.G., Bongarzone, I., Pierotti, M.A., Luksch, R., Gasparini, M., Collini, P., Pilotti, S., Rizzetti, M.G., Mondellini, P., De Bernardi, B., Di Martino, D., Garaventa, A., Brisigotti, M., and Tonini, G.P. (1993) *trk* and *ret* proto-oncogene expression in human neuroblastoma specimens: high frequency of *trk* expression in non-advanced stages. *Int. J. Cancer* **54,** 540–545.

906. Shikata, A., Sugimoto, T., Hosoi, H., Sotozono, Y., Shikata, T., Sawada, T., and Parada, L.F. (1994) Increased expression of trk proto-oncogene by gamma-interferon in human neuroblastoma cell lines. *Jpn. J. Cancer Res.* **85,** 122–126.

907. Kuner, P., and Hertel, C. (1998) NGF induces apoptosis in a human neuroblastoma cell line expressing the neurotrophin receptor p75NTR. *J. Neurosci. Res.* **54,** 465–474.

908. McConville, C.M., and Forsyth, J. (2003) Neuroblastoma – a developmental perspective. *Cancer Lett.* **197,** 3–9.

909. Eggert, A., Ikegaki, N., Liu, X., and Brodeur, G.M. (2000) Prognostic and biological role of neurotrophin-receptor TrkA and TrkB in neuroblastoma. *Klin. Pediatr.* **212,** 200–205.

910. Eggert, A., Grotzer, M.A., Ikegaki, N., Liu, X., Evans, A.E., and Brodeur, G.M. (2000) Expression of neurotrophin receptor TrkA inhibits angiogenesis in neuroblastoma. *Med. Pediatr. Oncol.* **35,** 569–572.

911. Eggert, A., Sieverts, H., Ikegaki, N., and Brodeur, G.M. (2000) p75 mediated apoptosis in neuroblastoma cells is inhibited by expression of TrkA. *Med. Pediatr. Oncol.* **35,** 573–576.

912. Eggert, A., Ho, R. Ikegaki, N., Liu, X., and Brodeur, G.M. (2000) Different effects of TrkA expression in neuroblastoma cell lines with or without *MYCN* amplification. *Med. Pediatr. Oncol.* **35,** 623–627.

913. Eggert, A., Ikegaki, N., Liu, X., Chou, T.T., Lee, V.M., Trojanowski, J.Q., and Brodeur, G.M. (2000) Molecular dissection of TrkA signal transduction pathways mediating differentiation in human neuroblastoma cells. *Oncogene* **19,** 2043–2051.

914. Nakagawara, A. (2001) Trk receptor tyrosine kinases: a bridge between cancer and neural development. *Cancer Lett.* **169,** 107–114.

915. Nakagawara, A. (2004) Neural crest development and neuroblastoma: the genetic and biological link. *Prog. Brain Res.* **146,** 233–242.

916. Domenici, C., Nicotra, M.R., Alemà, S., Bosman, C., Castello, M.A., Donfrancesco, A., Gallo, P., McDowell, H., and Natali, P.G. (1997) Immunohistochemical detection of p140^{trkA} and p75^{LNGFR} neurotrophin receptors in neuroblastoma. *J. Neurooncol.* **31,** 57–64.

917. Shikata, A., Shikata, T., Sotozono, Y., Hosoi, H., Matsumura, T., Sugimoto, T., and Sawada, T. (2000) Neuronal differentiation in human neuroblastoma cells by nerve growth factor following TrkA up-regulation by interferon-gamma. *Med. Pediatr. Oncol.* **34,** 394–401.

918. Lavoie, J.-F., LeSauteur, L., Kohn, J., Wong, J., Furtoss, O., Thiele, C.J., Miller, F.D., and Kaplan, D.R. (2005) TrkA induces apoptosis of neuroblastoma cells and does so via a p53-dependent mechanism. *J. Biol. Chem.* **280,** 29199–29207.

919. Tang, X.X., Zhao, H., Robinson, M.E., Cnaan, A., London, W., Cohn, S.L., Cheung, N.-K., Brodeur, G.M., Evans, A.E., and Ikegaki, N. (2000) Prognostic significance of *EPHB6*, *EFNB2*, and *EFNB3* expressions in neuroblastoma. *Med. Pediatr. Oncol.* **35,** 656–658.

920. Tang, X.X., Zhao, H., Robinson, M.E., Cohn, B., Cnaan, A., London, W., Cohn, S.L., Cheung, N.-K.V., Brodeur, G.M., Evans, A.E., and Ikegaki, N. (2000) Implications of *EPHB6*, *EFNB2*, and *EFNB3* expressions in human neuroblastoma. *Proc. Natl. Acad. Sci. USA* **97,** 10936–10941.

921. Jensen, L.M., Zhang, Y., and Shooter, E.M. (1992) Steady-state polypeptide modulations associated with nerve growth factor (NGF)-induced terminal differentiation and NGF deprivation-induced apoptosis in human neuroblastoma cells. *J. Biol. Chem.* **267,** 19325–19333.

922. Eggert, A., Brodeur, G.M., and Ikegaki, N. (2000) Relative quantitative RT-PCR protocol for *TrkB* expression in neuroblastoma using *GAPD* as an internal control. *BioTechniques* **28,** 681–691.

923. Aoyama, M., Asai, K., Shishikura, T., Kawamoto, T., Miyachi, T., Yokoi, T., Togari, H., Wada, Y., Kato, T., and Nakagawara, A. (2001) Human neuroblastomas with unfavorable biologies express high levels of brain-derived neurotrophic factor mRNA and a variety of its variants. *Cancer Lett.* **164,** 51–60.

924. Hecht, M., Schulte, J.H., Eggert, A., Wilting, J., and Schweigerer, L. (2005) The neurotrophin TrkB cooperates with c-Met in enhancing neuroblastoma invasiveness. *Carcinogenesis* **26,** 2105–2115.

925. Li, Z., Jaboin, J., Dennis, P.A., and Thiele, C.J. (2005) Genetic and pharmacologic identification of Akt as a mediator of brain-derived neurotrophic factor/TrkB rescue of neuroblastoma cells from c hemotherapy-induced cell death. *Cancer Res.* **65,** 2070–2075.

926. Nakamura, K., Martin, K.C., Jackson, J.K., Beppu, K., Woo, C.-W., and Thiele, C.J. (2006) Brain-derived neurotrophic factor activation of TrkB induces vascular endothelial growth factor expression via hypoxia-inducible factor-1α in neuroblastoma cells. *Cancer Res.* **66,** 4249–4255.

927. McLaughlin, J.E., and Urich, H. (1977) Maturing neuroblastoma and ganglioneuroblastoma: a study of four cases with long survival. *J. Pathol.* **121,** 19–26.

928. Taylor, S.R., Blatt, J., Costantino, J.P., Roederer, M., and Murphy, R.F. (1988) Flow cytometric DNA anal-

ysis of neuroblastoma and ganglioneuromas. A 10-year retrospective study. *Cancer* **62**, 749–754.

929. Clerico, A., Tenore, A., Bartolozzi, S., Remotti, D., Ruco, L., Dominici, C., Properzi, E., and Castello, M.A. (1993) Adrenocorticotropic hormone-secreting ganglioneuroblastoma associated with opsomyoclonic encephalopathy: a case report with immunohistochemical study. *Med. Pediatr. Oncol.* **21**, 690–694.

930. Moll, U.M., LaQuaglia, M., Bénard, J., and Riou, G. (1995) Wild-type p53 protein undergoes cytoplasmic sequestration in undifferentiated neuroblastomas but not in differentiated tumors. *Proc. Natl. Acad. Sci. USA* **92**, 4407–4411.

931. Mondal, A. (1995) Cytopathology of neuroblastoma, ganglioneuroblastoma and ganglioneuroma. *J. Indian Med. Assoc.* **93**, 340–343.

932. Blatt, J., Olshan, A.F., Lee, P.A., and Ross, J.L. (1997) Neuroblastoma and related tumors in Turner's syndrome. *J. Pediatr.* **133**, 666–670.

933. Gambini, C., Conte, M., Bernini, G., Angelini, P., Pession, A., Paolucci, P., Donfrancesco, A., Veneselli, E., Mazzocco, K., Tonini, G.P., Raffaghello, L., Dominici, C., Morando, A., Negri, F., Favre, A., de Bernardi, B., and Pistoia, V. (2003) Neuroblastic tumors associated with opsoclonus-myoclonus syndrome: histological, immunohistochemical and molecular features of 15 Italian cases. *Virchows Arch.* **442**, 555–562.

934. Ito, F., Watanabe, Y., and Ito, T. (1997) Synchronous occurrence of Wilms tumor and ganglioneuroblastoma in a child with neurofibromatosis. *Eur. J. Pediatr. Surg.* **7**, 308–310.

935. Lorenzana, A.N., Zielenska, M., Thorner, P., Gerrie, B., Weitzman, S., and Squire, J. (1997) Heterogeneity of MYCN amplification in a child with stroma-rich neuroblastoma (ganglioneuroblastoma). *Pediatr. Pathol. Lab. Med.* **17**, 875–883.

936. Geraci, A.P., de Csepel, J., Shlasko, E., and Wallace, S.A. (1998) Ganglioeuroblastoma and ganglioneuroma in association with neurofibromatosis type I: report of three cases. *J. Child Neurol.* **13**, 356–358.

937. Pivnick, E.K., Furman, W.L., Velagaleti, G.V.N., Jenkins, J.J., Chase, N.A., and Ribeiro, R.C. (1998) Simultaneous adrenocortical carcinoma and ganglioneuroblastoma in a child with Turner syndrome and germline p53 mutation. *J. Med. Genet.* **35**, 328–332.

938. Sasaki, Y., Nakayama, H., and Ikeda, M. (2000) Turner syndrome and ganglioneuroma. *J. Pediatr. Hematol. Oncol.* **22**, 89–90.

939. Takahashi, K., Totsune, K., Murakami, O., Sone, M., Satoh, F., Kitamuro, T., Noshiro, T., Hayashi, Y., Sasano, H., and Shibahara, S. (2001) Expression of melanin-concentrating hormone receptor messenger ribonucleic acid in tumor tissues of pheochromocytoma, ganglioneuroblastoma, and neuroblastoma. *J. Clin. Endocrinol. Metab.* **86**, 369–374.

940. Kim, N.R., Wang, K.C., Bang, J.S., Choe, G., Park, Y., Kim, S.K., Cho, B.K., and Chi, J.G. (2003) Glioblastomatous transformation of ganglioglioma: a case report with reference to molecular genetic and flow cytometric analysis. *Pathol. Int.* **53**, 874–882.

941A. Martin, D.T., Gendron, R.L., Jarzembowski, J.A., Perry, A., Collins, M.H., Pushpanathan, C., Miskiewicz, E., Castle, V.P., and Paradis, H. (2007) Tubedown expression correlates with the differentiation status and aggressiveness of neuroblastic tumors. *Clin. Cancer Res.* **13**, 1480–1487.

941. Debiec-Rychter, M., Liberski, P.P., Alwasiak, J., and Klimek, A. (1995) Chromosomal alterations in a case of ganglioglioma. *Cancer Genet. Cytogenet.* **85**, 155–156.

942. Jay, V., Squire, J., Blaser, S., Hoffman, H.J., and Hwang, P. (1997) Intracranial and spinal metastases from a ganglioglioma with unusual cytogenetic abnormalities in a patient with complex partial seizures. *Childs Nerv. Syst.* **13**, 550–555.

943. Kalyan-Raman, U.P., and Olivero, W.C. (1987) Ganglioglioma: a correlative clinicopathological and radiological study of ten surgically treated cases with follow-up. *Neurosurgery* **20**, 428–433.

944. Lang, F.F., Epstein, F.J., Ransohoff, J., Allen, J.C., Wisoff, J., Abbott, R., and Miller, D.C. (1993) Central nervous system gangliogliomas. Part 2: clinical outcome. *J. Neurosurg.* **79**, 867–873.

945. Miller, D.C., Lang, F.L., and Epstein, F.J. (1993) Central nervous system gangliogliomas. Part 1: pathology. *J. Neurosrug.* **79**, 859–866.

946. Miller, D.C., Kricheff, I.I., Patel, U., Pinto, R.S., Epstein, F.J., and Rorke, L.B. (1999) Synaptophysin staining for ganglioglioma. *Am. J. Neuroradiol.* **20**, 527–529.

947. Nishio, S., Morioka, T., Hamada, Y., Fukui, M., and Nakagawara, A. (1998) Immunohistochemical expression of tyrosine kinase (Trk) receptor proteins in mature neuronal cell tumors of the central nervous system. *Clin. Neuropathol.* **17**, 123–130.

948. Quinn, B. (1998) Synaptophysin staining in normal brain. Importance for diagnosis of ganglioglioma. *Am. J. Surg. Pathol.* **22**, 550–556.

949. Quinn, B. (1999) Synaptophysin staining for ganglioglioma. *Am. J. Neuroradiol.* **20**, 526–527.

950. Rumana, C.S., Valadka, A.B., and Contant, C.F. (1999) Prognostic factors in supratentorial ganglioglioma. *Acta Neurochir. (Wien)* **141**, 63–69.

951. Sawin, P.D., Theodore, N., and Rekate, H.L. (1999) Spinal cord ganglioglioma in a child with neurofibromatosis type 2. Case report and literature review. *J. Neurosurg.* **90**, 231–233.

952. Wierzba-Bobrowicz, T., Schmidt-Sidor, B., Gwiazda, E., and Bertrand, E. (1999) The significance of immunocytochemical markers, synaptophysin and neurofilaments in diagnosis of ganglioglioma. *Folia Neuropathol.* **37**, 156–161.

953. Bannykh, S.I., Perry, A., Powell, H.C., Hill, A., and Hansen, L.A. (2002) Malignant rhabdoid meningioma arising in the setting of preexisting ganglioglioma: a diagnosis supported by fluorescence in situ hybridization. *J. Neurosurg.* **97**, 1450–1455.

954. Becker, A.J., Klein, H., Baden, T., Aigner, L., Normann, S., Elger, C.E., Schramm, J., Wiestler, O.D., and Blümcke, I. (2002) Mutational and expression analysis of the reelin pathway components CDK5 and doublecortin in gangliogliomas. *Acta Neuropathol.* **104**, 403–408.

955. Aronica, E., Gorter, J.A., van Vliet, E.A., Spliet, W.G.M., van Veelen, C.W.M., van Rijen, P.C., Leenstra, S., Ramkema, M.D., Scheffer, G.L., Scheper, R.J., Sisodiya, S.M., and Troost, D. (2003) Overexpresion of the human

major vault protein in gangliogliomas. *Epilepsia* **44**, 1166–1175.

956. Knapp, J., Olson, L., Tye, S., Bethard, J.R., Welsh, C.A., Rumbolt, Z., Takacs, I., and Maria, B.L. (2005) Case of desmoplastic infantile ganglioglioma screeting ceruloplasmin. *J. Child Neurol.* **20**, 920–924.

957. Resta, N., Lauriola, L., Puca, A., Susca, F.C., Albanese, A., Sabatino, G., Di Giacomo, M.C., Gessi, M., and Guanti, G. (2006) Ganglioglioma arising in a Peutz-Jeghers patient: a case report with molecular implications. *Acta Neuropathol.* **112**, 106–111.

958. Blatt, J., and Hamilton R.L. (1998) Neurodevelopmental anomalies in children with neuroblastoma. *Cancer* **82**, 1603–1608.

959. Kaneko, Y., and Cohn, S.L. (2000) Ploidy and cytogenetics of neuroblastoma. In *Neuroblastoma* (Brodeur, G.M., Sawada, T., Tsuchida, Y., and Voûte, P.A., eds.), Elsevier Science B.V., Amsterdam, The Netherlands, pp. 41–56.

960. Satgé, D., Sasco, A.J., Plantaz, D., Bénard, J., and Vekemans, M.J. (2001) Abnormal number of X chromosomes and neuroblastic tumors. *J. Pediatr. Hematol. Oncol* .**23**, 331–332.

961. Gül, D., Akin, R., Kismet, E., and Köseoglu, V. (2003) Neuroblastoma in a patient with 47,XXX karyotype. *Cancer Genet. Cytogenet.* **146**, 84–85.

962. Honma, T., Kanegane, H., Shima, T., Otsubo, K., Nomura, K., and Miyawaki, T. (2006) Neuroblastoma in an XYY male. *Cancer Genet. Cytogenet.* **168**, 84–84.

963. Dowa, Y., Yamamoto, T., Abe, Y., Kobayashi, M., Hoshino, R., Tanaka, K., Aida, N., Take, H., Kato, K., Tanaka, Y., Ariyama, J., Harada, N., Matsumoto, N., and Kurosawa, K. (2006) Congenital neuroblastoma in a patient with parital trisomy of 2p. *J. Pediatr. Hematol. Oncol.* **28**, 379–382.

964. Bakhshi, S., Hamre, M.H. (2004) Neuroblastoma in chromosomal breakage syndromes. *Cancer Genet. Cytogenet.* **148**, 85.

965. Hienonen, T., Sammalkorpi, H., Isohanni, P., Versteeg, R., Karikoski, R., and Aaltonen, L.A. (2005) A 17p11.2 germline deletion in a patient with Smith-Magenis syndrome and neuroblastoma. *J. Med. Genet.* **42**, e31–e36.

966. Klein, H., Plochl, E., and Lampert, F. (1975) Familial neuroblastoma: cytogenetic investigation of the peripheral blood. *Humangenetik.* **28**, 217–220.

967. Arenson, E.B., Jr., Hutter, J.J., Jr., Restuccia, R.D., and Holton, C.P. (1976) Neuroblastoma in father and son. *JAMA* **235**, 727–729.

968. Claviez, A., Lakomek, M., Ritter, J., Suttorp, M., Kremens, B., Dickerhoff, R., Harms, D., Berthold, F., and Hero, B. (2004) Low occurrence of familial neuroblastomas and ganglioneuromas in five consectuvie GPOH neuroblastoma treatment studies. *Eur. J. Cancer* **40**, 2760–2765.

969. Longo, L., Tonini, G.P., Ceccherini, I., and Perri, P. (2005) Oligogenic inheritance in neuroblastoma. *Cancer Lett.* **228**, 65–69.

970. Katzenstein, H.M., Rademaker, A.W., Senger, C., Salwen, H.R., Nguyen, N.N., Thorner, P.S., Litsas, L., and Cohn, S.L. (1999) Effectiveness of the angiogenesis inhibitor TNP-470 in reducing the growth of human neuroblastoma in nude mice inversely correlates with tumor burden. *Clin. Cancer Res.* **5**, 4273–4278.

971. Kim, E.S., Soffer, S.Z., Huang, J., McCrudden, K.W., Yokoi, A., Manley, C.A., Middlesworth, W., Kandel, J.J., and Yamashiro, D.J. (2002) Distinct response of experimental neuroblastoma to combination antiangiogenic strategies. *J. Pediatr. Surg.* **37**, 518–522.

972. Cañete, A., Navarro, S., Bermúdez, J., Pellín, A., Castel, V., and Llombart-Bosch, A. (2000) Angiogenesis in neuroblastoma: relationship to survival and other prognostic factors in a cohort of neuroblastoma patients. *J. Clin. Oncol.* **18**, 27–34.

973. Eggert, A., Ikegaki, N., Kwiatkowski, J., Zhao, H., Brodeur, G.M., and Himelstein, B.P. (2000) High-level expression of angiogenic factors is associated with advanced tumor stage in human neuroblastomas. *Clin. Cancer Res.* **6**, 1900–1908.

974. Crawford, S.E., Stellmach, V., Ranalli, M., Huang, X., Huang, L., Volpert, O., De Vries, G.H., Abramson, L.P., and Bouch, N. (2001) Pigment epithelium-derived factor (PEDF) in neuroblastoma: a multifunctional mediator of Schwann cell antitumor activity. *J. Cell Sci.* **114**, 4421–4428.

975. Cerignoli, F., Guo, X., Cardinali, B., Rinaldi, C., Casaletto, J., Frati, L., Screpanti, I., Gudas, L.J., Gulino, A., Thiele, C.J., and Giannini, G. (2002) retSDR1, a short-chain retinol dehydrogenase/reductase, is retinoic acid-inducible and frequently deleted in human neuroblastoma cell lines. *Cancer Res.* **62**, 1196–1204.

976. Fukuzawa, M., Sugiura, H., Koshinaga, T., Ikeda, T., Hagiwara, N., and Sawada, T. (2002) Expression of vascular endothelial growth factor and its receptor Flk-1 in human neuroblastoma using in situ hybridization. *J. Pediatr. Surg.* **37**, 1747–1750.

977. Langer, I., Vertongen, P., Perret, J., Fontaine, J., Atassi, G., and Robberecht, P. (2002) Expression of vascular endothelial growth factor (VEGF) and VEGF receptors in human neuroblastomas. *Med. Pediatr. Oncol.* **34**, 386–393.

978. Rapella, A., Negrioli, A., Melillo, G., Pastorino, S., Varesio, L., and Bosco, M.C. (2002) Flavopiridol inhibits vascular endothelial growth factor production induced by hypoxia or picolinic acid in human neuroblastoma. *Int. J. Cancer* **99**, 658–664.

979. Gong, H., Pöttgen, C., Stüben, G., Havers, W., Stuschke, M., and Schweigerer, L. (2003) Arginine deiminase and other antiangiogenic agents inhibit unfavorable neuroblastoma growth: potentiation by irradiation. *Int. J. Cancer* **106**, 723–728.

980. Beppu, K., Jaboine, J., Merchant, M.S., Mackall, C.L., and Thiele, C.J. (2004) Effect of imatinib mesylate on neuroblastoma tumorigenesis and vascular endothelial growth factor expression. *J. Natl. Cancer Inst.* **96**, 46–55.

981. Beppu, K., Nakamura, K., Linehan, W.M., Rapisarda, A., and Thiele, C.J. (2005) Topotecan blocks hypoxia-inducible factor-1α and vascular endothelial growth factor expression induced by insulin-like growth factor-1 in neuroblastoma cells. *Cancer Res.* **65**, 4775–4781.

982. Marcus, K., Johnson, M., Adam, R.M., O'Reilly, M.S., Donovan, M., Atala, A., Freeman, M.R., and Soker, S. (2005) Tumor cell-associated neuropilin-1 and vascular endothelial growth factor expression as determinants of tumor growth in neuroblastoma. *Neuropathology* **25**, 178–187.

983. Nilsson, H., Jogi, A., Beckman, S., Harris, A.L., Poelliger, L., and Pahlman, S. (2005) HIF-2alpha expression in human fetal paraganglia and neuroblastoma: relation to sympathetic differentiation, glucose deficiency, and hypoxia. *Exp. Cell Res.* **303**, 447–456.

984. Goldsmith, K.C., Liu, X., Dam, V., Morgan, B.T., Shabbout, M., Cnaan, A., Letai, A., Korsmeyer, S.J., and Hogarty, M.D. (2006) BH3 peptidomimetics potently activate apoptosis and demonstrate single agent efficacy in neuroblastoma. *Oncogene* **25**, 4525–4533.

985. Versteeg, R., van der Minne, C., Plomp, A., Sijts, A., van Leeuwen, A.A.D., and Schrier, P. (1990) N-*myc* expression switched off and class I human leukocyte antigen expression switched on after somatic cell fusion of neuroblastoma cells. *Mol. Cell Biol.* **10**, 5416–5423.

986. Yoshihara, T., Ikushima, S., Hibi, S., Misawa, S., and Imashuku, S. (1993) Establishment and characterization of a human neuroblastoma cell line derived from a brain metastatic lesion. *Hum. Cell* **6**, 210–217.

987. Muresu, R., Casciano, I., Volpi, E.V., Siniscalco, M., and Romani, M. (1995) Cytogenetic and molecular studies on the neuroblastoma cell line NGP: identification of a reciprocal t(1;15) involving the "consensus region" 1p36.1. *Genes Chromosomes Cancer* **13**, 66–71.

988. Hiraiwa, H., Hamazaki, M., Takata, A., Kikuchi, H., and Hata, J. (1997) A neuroblastoma caell line derived from a case detected through a mass screening system in Japan. A case report including the biologic and phenotypic characteristics of the cell line. *Cancer* **79**, 2036–2044.

989. Kao, Y.S., and Correa, H. (1997) Characterization of a new human neuroblastoma cell line. *Cancer Genet. Cytogenet.* **98**, 171.

990. Marini, P., MacLeod, R.A.F., Treuner, C., Bruchelt, G., Böhm, W., Wolburg, H., Schweizer, P., and Girgert, R. (1999) SiMa, a new neuroblastoma cell line combining poor prognostic cytogenetic markers with high adrenergic differentiation. *Cancer Genet. Cytogenet.* **112**, 161–164.

991. Valent, A., Vénuat, A.-M., Danglot, G., Da Silva, J., Duarte, N., Bernheim, A., and Bénard, J. (2001) Stromal cells and human malignant neuroblasts derived from bone marrow metastasis may share common karyotypic abnormalities: the case of IGR-N-91 cell line. *Med. Pediatr. Oncol.* **36**, 100–103.

992. Cohen, N., Betts, D.R., Rechavi, G., Amariglio, N., and Trakhtenbrot, L. (2003) Clonal expansion and not cell interconversion is the basis for the neuroblast and nonneuronal types of the SK-N-SH neuroblastoma cell line. *Cancer Genet. Cytogenet.* **143**, 80–84.

993. Cohen, N., Betts, D.R., Rechavi, G., Amariglio, N., and Trakhtenbrot, L. (2004) Authors' response to Goranov, B.B., and Redfern, C.P.F. (2004) Cytogenetic differences and neuroblastoma cell interconversion. *Cancer Genet. Cytogenet.* **149**, 175.

994. Goranov, B.B., and Redfern, C.P.F. (2004) Cytogenetic differences and neuroblastoma cell interconversion. *Cancer Genet. Cytogenet.* **149**, 173–174.

995. Piacentini, M., Annicchiarico-Petruzzelli, M., Oliverio, S., Piredda, L., Biedler, J.L., and Melino, G. (1992) Phenotype-specific "tissue" transglutaminase regulation in human neuroblastoma cells by response to retinoic acid: correlation with cell death by apoptosis. *Int. J. Caner* **52**, 271–278.

996. Cheung, B., Hocker, J.E., Smith, S.A., Norris, M.D., Haber, M., and Marshall, G.M. (1998) Favorable prognostic significance of high-level retinoic acid receptor β expression in neuroblastoma mediated by effects on cell cycle regulation. *Oncogene* **17**, 751–759.

997. Giannini, G., Di Marcotullio, L., Ristori, E., Zani, M., Crescenzi, M., Scarpa, S., Piaggio, G., Vacca, A., Peverali, F.A., Diana, F., Screpanti, I., Frati, L., and Gulino, A. (1999) *HMGI(Y)* and *HMGI-C* genes are expressed in neuroblastoma cell lines and tumors and affect retinoic acid responsiveness. *Cancer Res.* **59**, 2484–2492.

998. Takada, N., Isogai, E., Kawamoto, T., Nakanishi, H., Todo, S., and Nakagawara, A. (2001) Retinoic acid-induced apoptosis of the CHP134 neuroblastoma cell line is associated with nuclear accumulation of p53 and is rescued by the GDNF/Ret signal. *Med. Peidatr. Oncol.* **36**, 122–126.

999. Goranov, B.B., Campbell Hewson, Q.D., Pearson, A.D., and Redfern, C.P. (2006) Overexpression of RARgamma increases death of SH-SY5Y neuroblastoma cells in response to retinoic acid but not fenretinide. *Cell Death Differ.* **13**, 676–679.

1000. Reynolds, C.P., and Lemons, R.S. (2001) Retinoid therapy of childhood cancer. *Hematol. Oncol. Clin. North Am.* **15**, 867–910.

1001. Thiele, C.J., Reynolds, C.P., and Israel, M.A. (1985) Decreased expression of N-myc precedes retinoic acid-induced morphological differentiation of human neuroblastoma. *Nature* **313**, 404–406.

1002. López-Carballo, G., Moreno, L., Masiá, S., Pérez, P., and Barettino, D. (2002) Activation of the phosphatidylinositol 3-kinase/Akt signaling pathway by retinoic acid is required for neural differentiation of SH-SY5Y human neuroblastoma cells. *J. Biol. Chem.* **277**, 25297–25304.

1003. Ponthan, F., Johnsen, J.I., Klevenvall, L., Castro, J., and Kogner, P. (2003) The synthetic retinoid Ro 13–6307 induces neuroblastoma differentiation *in vitro* and inhibits neuroblastoma tumour growth *in vivo*. *Int. J. Cancer* **104**, 418–424.

1004. Blanc, E., Le Roux, G., Bénard, J., and Raguénez, G. (2005) Low expression of *Wnt-5a* gene is associated with high-risk neuroblastomas. *Oncogene* **24**, 1277–1283.

1005. Bill, A.H., Jr. (1968) The regression of neuroblastoma. *J. Pediatr. Surg.* **3**, 103–106.

1006. Komuro, H., Hayashi, Y., Kawamura, M., Hayashi, K., Kaneko, Y., Kamoshita, S., Hanada, R., Yamamoto, K., Hongo, T., Yamada, M., and Tsuchida, Y. (1993) Mutations of the *p53* gene are involved in Ewing's sarcomas but not in neuroblastomas. *Cancer Res.* **53**, 5284–5288.

1007. Adesina, A.M., Nalbantoglu, J., and Cavenee, W.K. (1994) *p53* gene mutations and *mdm2* gene amplification are uncommon in medulloblastoma. *Cancer Res.* **54**, 5649–5651.

1008. Goldman, S.C., Chen, C.-Y., Lansing, T.J., Gilmer, T.M., and Kastan, M.B. (1996) The p53 signal transduction pathway is intact in human neuroblastoma despite cytoplasmic localization. *Am. J. Pathol.* **148**, 1381–1385.

1009. Ostermeyer, A.G., Runko, E., Winkfield, B., Ahn, B., and Moll, U.M. (1996) Cytoplasmically sequestered wild-type p53 protein in neuroblastoma is relocated to the nucleus by a C-terminal peptide. *Proc. Natl. Acad. Sci. USA* **93,** 15190–15194.

1010. Isaacs, J.S., Hardman, R., Carman, T.A., Barrett, J.C., and Weissman, B.E. (1998) Differential subcellular p53 localization and function in N- and S-type neuroblastoma cell lines. *Cell Growth Differ.* **9,** 545–555.

1011. Manhani, R., Cristofani, L.M., Odone Filho, V., and Bendit, I. (1997) Concomitant p53 mutation and MYCN amplification in neuroblastoma. *Med. Pediatr. Oncol.* **29,** 206–207.

1012. Ejeskär, K., Sjöberg, R.M., Kogner, P., and Martinsson, T. (1999) Variable expression and absence of mutations in p73 in primary neuroblastoma tumors argues against a role in neuroblastoma development. *Int. J. Mol. Med.* **3,** 585–589.

1013. Yang, H.W., Piao, H.Y., Chen, Y.Z., Takita, J., Kobayashi, M., Taniwaki, M., Hashizume, K., Hanada, R., Yamamoto, K., Taki, T., Bessho, F., Yanagisawa, M., and Hayaashi, Y. (2000) The p53 gene is less involved in the development but involved in the progression of neuroblastoma. *Int. J. Mol. Med.* **5,** 379–384.

1014. Barrios, M., Eychenne, M.K., Terrier-Lacombe, M.J., Duarte, N., Dubourg, C., Douc-Rasy, S., Chompret, A., Khagad, M., Hartmann, O., Caput, D., and Benard, J. (2001) Genomic and allelic expression status of the p75 gene in human neuroblastoma. *Med. Pediatr. Oncol.* **36,** 45–47.

1014A. Casciano, I., Mazzocco, K., Boni, L., Pagnan, G., Banelli, B., Allemanni, G., Ponzoni, M., Tonini, G.P., and Romani, M. (2002) Expression of DeltaNp73 is a molecular marker for adverse outcome in neuroblastoma patients. *Cell Death Differ.* **9,** 246–251.

1015. Ichimiya, S., Nimura, Y., Kageyama, H., Takada, N., Sunahara, M., Shishikura, T., Nakamura, Y., Sakiyama, S., Seki, N., Ohira, M., Kaneko, Y., McKeon, F., Caput, D., and Nakagawara, A. (2001) Genetic analysis of *p73* localized at chromosome 1p36.3 is primary neuroblastomas. *Med. Pediatr. Oncol.* **36,** 42–44.

1015A. Casciano, I., Banelli, B., Croce, M., Allemanni, G., Ferrini, S., Tonini, G.P., Ponzoni, M., and Romani, M. (2002) Role of methylation in the control of DeltaNp73 expression in neuroblastoma. *Cell Death Differ.* **9,** 343–345.

1016. Matos, P., Isidro, G., Vieira, E., Lacerda, A.F., Martins, A.G., and Boavida, M.G. (2001) p73 expression in neuroblastoma: a role in the biology of advanced tumors? *Pediatr. Hematol. Oncol.* **18,** 37–46.

1016A. Banelli, B., Casciano, I., and Romani, M. (2000) Methylation-independent silencing of the *p73* gene in neuroblastoma. *Oncogene* **19,** 4553–4556.

1017. Norris, M.D., Smith, J., Tanabe, K., Tobin, P., Flemming, C., Scheffer, G.L., Wielinga, P., Cohn, S.L., London, W.B., Marshall, G.M., Allen, J.D., and Haber, M. (2005) Expression of multidrug transporter *MRP4/ABCC4* is a marker of poor prognosis in neuroblastoma and confers resistance to irinotecan *in vitro. Mol. Cancer Ther.* **4,** 547–553.

1017A. Wolff, A., Technau, A., Ihling, C., Technau-Ihling, K., Erber, R., Bosch, F.X., and Brandner, G. (2001) Evidence that wild-type p53 in neuroblastoma cells is in a conformation refractory to integration into the transcriptional complex. *Oncogene* **20,** 1307–1317.

1018. Takada, N., Ozaki, T., Ichimiya, S., Todo, S., and Nakagawara, A. (1999) Identification of a trtansactivation

activity in the COOH-terminal region of p73 which is impaired in the naturally occurring mutants found in human neuroblastomas. *Cancer Res.* **59,** 2810–2814.

1018A. Kopitz, J., André, S., von Reitzenstein, C., Versluis, K., Kaltner, H., Pieters, R.J., Wasano, K., Kuwabara, I., Liu, F.-T., Cantz, M., Heck, A.J.R., and Gabius, H.-J. (2003) Homodimeric galectin-7 (p53-induced gene 1) is a negative growth regulator for human neuroblastoma cells. *Oncogene* **22,** 6277–6288.

1019. Goldschneider, D., Million, K., Meiller, A., Haddada, H., Puisieux, A., Bénard, J., May, E., and Douc-Rasy, S. (2005) The neurogene $BTG2^{TIS21/PC3}$ is transactivated by ΔNp73α via p53 specifically in neuroblastoma cells. *J. Cell Sci.* **118,** 1245–1253.

1019A. Xue, C., Haber, M., Flemming, C., Marshall, G.M., Lock, R.B., MacKenzie, K.L., Gurova, K.V., Nonis, M.D., and Gudkov, A.V. (2007) p53 determines multidrug sensitivity of childhood neuroblastoma. *Cancer Res.* **67,** 10351–10360.

1020. Schuster, V., and Kreth, H.W. (1987) Molecular biology aspects of neuroblastoma. *Monatsschr. Kinderheilkd.* **135,** 672–676.

1021. Ambros, P.F., Ambros, I.M., Strehl, S., Bauer, S., Luegmayr, A., Kovar, H., Ladenstein, R., Fink, F.-M., Horcher, E., Printz, G., Mutz, I., Schilling, F., Urban, C., and Gadner, H. (1995) Regression and progression in neuroblastomas. Does genetics predict tumour behaviour? *Eur. J. Cancer* **31A,** 510–515.

1022. Castelberry, R.P., Pritchard, J., Ambros, P., Berthold, F., Brodeur, G.M., Castel, V., Cohn, S.L., De Bernardi, B., Dicks-Mireaux, C., Frappaz, D., Haase, G.M., Haber, M., Jones, D.R., Joshi, V.V., Kaneko, M., Kemshead, J.T., Kogner, P., Lee, R.E.J., Matthay, K.K., Michon, J.M., Monclair, R., Roald, B.R., Seeger, R.C., Shaw, P.J., Shimada, H., and Shuster, J.J. (1997) The International Neuroblastoma Risk Groups (INRG): a preliminary report. *Eur. J. Cancer* **33,** 2113–2116.

1023. Rudolph, P., Lappe, T., Hero, B., Berthold, F., Parwaresch, R., Harms, D., and Schmidt, D. (1997) Prognostic significance of the proliferative activity in neuroblastoma. *Am. J. Pathol.* **150,** 133–145.

1024. Castelberry, R.P. (1999) Predicting outcome in neuroblastoma. *N. Engl. J. Med.* **340,** 1992–1993.

1025. Katzenstein, H.M., Bowman, L.C., Brodeur, G.M., Thorner, P.S., Joshi, V.V., Smith, E.I., Look, A.T., Rowe, S.T., Nash, M.B., Holbrook, T., Alvarado, C., Rao, P.V., Castleberry, R.P., and Cohn, S.L. (1998) Prognostic significance of age, *MYCN* oncogene amplification, tumor cell ploidy, and histology in 110 infants with stage D(S) neuroblastoma— The Pediatric Oncology Group Experience—A Pediatric Oncology Group Study. *J. Clin. Oncol.* **16,** 2007–2017.

1026. Kaneko, M., Nishihira, H., Mugishima, H., Ohnuma, N., Nakada, K., Kawa, K., Fukuzawa, M., Suita, S., Sera, Y., and Tsuchida, Y. (1998) Stratification of treatment of stage 4 neuroblastoma patients based on N-myc amplification status. Study Group of Japan for Treatment of Advanced neuroblastoma, Tokyo, Japan. *Med. Pediatr. Oncol.* **31,** 1–7.

1027. Ladenstein, R., Philip, T., Lasset, C., Hartmann, O., Garaventa, A., Pinkerton, R., Michon, J., Pritchard, J., Klingebiel, T., Kremens, B., Pearson, A., Coze, C., Paolucci, P., Frappaz, D., Gadner, H., and Chauvin, F. (1998) Multivariate analysis of risk factors in stage 4 neuroblastoma patients over the age of one year treated with megatherapy and stem-cell transplantation: a report from the European Bone Marrow Transplantation Solid Tumor Registry. *J. Clin. Oncol.* **16,** 953–965.

1028. Laprie, A., Michon, J., Hartmann, O., Munzer, C., Leclair, M.-D., Coze, C., Valteau-Couanet, D., Plantaz, D., Carrie, C., Habrand, J.-L., Bergeron, C., Chastagner, P., Défachelles, A.-S., Delattre, O., Combaret, V., Bénard, J., Pérel, Y., Gandemer, V., Rubie, H., for the Neuroblastoma Study Group of the French Society of Pediatric Oncology. (2004) High-dose chemotherapy followed by locoregional irradiation improves the outcome of patients with International Neuroblastoma Staging System stage II and III neuroblastoma with *MYCN* amplification. *Cancer* **101,** 1081–1089.

1029. Powell, J.E., Esteve, J., Mann, J.R., Parker, L., Frappaz, D., Michaelis, J., Kerbl, R., Mutz, I.D., and Stiller, C.A. (1998) Neuroblastoma in Europe: differences in the pattern of disease in the UK. SENSE. Study group for the Evaluation of Neuroblastoma Screening in Europe. *Lancet* **352,** 682–687.

1030. De Bernardi, B. (2005) Why does neuroblastoma attract us so much? *Cancer Lett.* **228,** 3–4.

1031. Niizuma, H., Nakamura, Y., Ozaki, T., Nakanishi, H., Ohira, M., Isogai, E., Kageyama, H., Imaizumi, M., and Nakagawara, A. (2006) Bcl-2 is a key regulator for the retinoic acid-induced apoptotic cell death in neuroblastoma. *Oncogene* **25,** 5046–5055.

1032. Cooper, R., Khakoo, Y., Matthay, K.K., Lukens, J.N., Seeger, R.C., Stram, D.O., Gerbing, R.B., Nakagawa, A., and Shimada, H. (2001) Opsoclonus-myoclonus-ataxia syndrome in neuroblastoma: histopathologic features—a report from the Children's Cancer Group. *Med. Pediatr. Oncol.* **36,** 623–629.

1033. Chang, B.H., Koch, T., Hopkins, K., and Malempati, S. (2006) Neuroblastoma found in a 4-year-old after rituximab therapy for opsoclonus-myoclonus. *Pediatr. Neurol.* **35,** 213–215.

1034. Ladenstein, R., Ambros, I.M., Potschger, U., Amann, G., Urban, C., Fink, F.M., Schmitt, K., Ones, R., Slociak, M., Schilling, F., Ritter, J., Berthold, F., Gadner, H., and Ambros, P.F. (2001) Prognostic significance of DNA di-tetraploidy in neuroblastoma. *Med. Pediatr. Oncol.* **36,** 83–92.

1035. Bourdeaut, F., Trochet, D., Janoueix-Lerosey, I., Ribeiro, A., Deville, A., Coz, C., Michiels, J.F., Lyonnet, S., Amiel, J., and Delattre, O. (2005) Germline mutations of the paired-like homeobox 2B (PHOX2B) gene in neuroblastoma. *Cancer Lett.* **228,** 51–58.

1036. Carr, J., Bell, E., Pearson, A.D.J., Kees, U.R., Beris, H., Lunec, J., and Tweddle, D.A. (2006) Increased frequency of aberrations in the p53/MDM2/p14^{ARF} pathway in neuroblastoma cell lines established at relapse. *Cancer Res.* **66,** 2138–2145.

1037. Hussein, D., Estlin, E.J., Dive, C., and Makin, G.W. (2006) Chronic hypoxia promotes hypoxia-inducible factor-1alpha-dependent resistance to etoposide and vincristine in neuroblastoma cells. *Mol. Cancer Ther.* **5,** 2241–2250.

1038. Kobayashi, K., Era, T., Takebe, A., Jakt, L.M., and Nishikawa, S.-I. (2006) ARID3B induces malignant transformation of mouse embryonic fibroblasts and is strongly associated with malignant neuroblastoma. *Cancer Res.* **66,** 8331–8336.

1039. Kogel, D., Svensson, B., Copanaki, E., Anguissola, S., Bonner, C., Thurow, N., Gudorf, D., Hetschko, H., Muller, T., Peters, M , Konig, H.G., Prehn, J.H. (2006) Induction of transcription factor CEBP homology protein mediates hypoglycaemia-induced necrotic cell death in human neuroblastoma cells. *J. Neurochem.* **99,** 952–964.

1040. Qin, K., Zhao, L., Tang, Y., Bhatta, S., Simard, J.M., and Zhao, R.Y. (2006) Doppel-induced apoptosis and counteraction by cellular prion protein in neuroblastoma and astrocytes. *Neuroscience* **141,** 1375–1388.

1041. Van Maerken, T., Speleman, F., Vermeulen, J., Lambertz, I., De Clercq, S., De Smet, E., Yigit, N., Coppens, V., Philippé, J., De Paepe, A., Marine, J.-C., and Vandesompele, J. (2006) Small-molecule MDM2 antagonists as a new therapy concept for neuroblastoma. *Cancer Res.* **66,** 9646–9655.

1042. Altungoz, O., Aygun, N., Tumer, S., Ozer, E., Olgun, N., and Sakizli, M. (2007) Correlation of modified Shimada classification with *MYCN* and 1p36 status detected by fluorescence in situ hybridization in neuroblastoma. *Cancer Genet. Cytogenet.* **172,** 113–119.

1042A. Toyoda, H., Yin, J., Mueller, S., Wimmer, E., and Cello, J. (2007) Oncolytic treatment and cure of neuroblastoma by a novel attenuated poliovirus in a novel poliovirus-susceptible animal mode. *Cancer Res.* **67,** 2857–2864

1042B. Mora, J., Lavarino, C., Alaminos, M., Cheung, N.-K.V., Ríos, J., de Torres, C., Illei, P., Juan, G., and Gerald, W.L. (2007) Comprehensive analysis of tumoral DNA content reveals clonal ploidy heterogeneity as a marker with prognostic significance in locoregional neuroblastoma. *Genes Chromosomes Cancer* **46,** 385–396.

1043. Bader, P., Schilling, F., Schlaud, M., Girgert, R., Handgretinger, R., Klingebiel, T., Treuner, J., Liu, C., Niethammer, D., and Beck, J.F. (1999) Expression analysis of multidrug resistance associated genes in neuroblastomas. *Oncol. Rep.* **6,** 1143–1146.

1044. Chan, H.S.L., Haddad, G., Thorner, P.S., DeBoer, G., Lin, Y.P., Ondrusek, N., Yeger, H., and Ling, V. (1991) P-glycoprotein expression as a predictor of the outcome of therapy for neuroblastoma. *N. Engl. J. Med.* **325,** 1608–1614.

1045. Favrot, M., Combaret, V., Goillot, E., Wagner, J.P., Bouffet, E., Mazingue, F., Thyss, A., Bordigoni, P., Delsol, G., Bailly, C., Fontaniere, B., and Philip, T. (1991) Expression of P-glycoprotein restricted to normal cells in neuroblastoma biopsies. *Br. J. Cancer* **64,** 233–238.

1046. El-Badry, O.M., Helman, L.J., Chatten, J., Steinberg, S.M., Evans, A.E., and Israel, M.A. (1991) Insulin-like growth factor II-mediated proliferation of human neuroblastoma. *J. Clin. Invest.* **87,** 648–657.

1047. Kim, C.J., Matsuo, T., Lee, K.-H., and Thiele, C.J. (1999) Up-regulation of insulin-like growth factor-II expression is a feature of TrkA but not TrkB activation in SH-SY5Y neuroblastoma cells. *Am. J. Pathol.* **155,** 1661–1670.

1048. Russo, V.C., Schutt, B.S., Andaloro, E., Ymer, S.I., Hoeflich, A., Ranke, M.B., Bach, L.A., and Werther, G.A. (2005) Insulin-like growth factor binding protein-2 binding to extracellular matrix plays a critical role in neuroblastoma cell proliferation, migration, and invasion. *Endocrinology* **146,** 4445–4455.

1049. Leone, A., Seeger, R.C., Hong, C.M., Hu, Y.Y., Arboleda, M.J., Brodeur, G.M., Stram, D., Slamon, D.J., and Steeg, P.S. (1993) Evidence for *nm23* RNA overexpression, DNA amplification and mutation in aggressive childhood neuroblastomas. *Oncogene* **8,** 855–865.

1050. Takahashi, K., Totsune, K., Murakami, O., Sone, M., Itoi, K., Hayashi, Y., Ohi, R., and Mouri, T. (1993) Pituitary adenylate cyclase activating polypeptide (PACAP)-like immunoreactivity in ganglioneuroblastoma and neuroblastoma. *Regul. Pept.* **49,** 19–24.

1051. Kusafuka, T., Nagahara, N., Oue, T., Imura, K., Nakamura, T., Kobayashi, Y., Komoto, Y., Fukuzawa, M., Okada, A., and Nakayama, M. (1995) Unfavorable DNA ploidy and Ha-*ras* p21 findings in neuroblastomas detected through mass screening. *Cancer* **76,** 695–699.

1052. Manohar, C.F., Salwen, H.R., Brodeur, G.M., and Cohn, S.L. (1995) Co-amplification and concomitant high levels of expression of a DEAD box gene with MYCN in human neuroblastoma. *Genes Chromosomes Cancer* **14,** 196–203.

1053. Perri, P., Pession, A., Mazzocco, K., Strigini, P., Iolascon, A., Basso, G., and Tonini, G.P. (1996) Peculiar allelotype associated with susceptibility to neuroblastoma. *Genes Chromosomes Cancer* **17,** 60–63.

1054. Marshall, G.M., Peaston, A.E., Hocker, J.E., Smith, S.A., Hansford, L .M., Tobias, V., Norris, M.D., Haber, M., Smith, D.P., Lorenzo, M.J., Ponder, B.A.J., and Hancock, J.F. (1997) Expression of multiple endocrine neoplasia 2B RET in neuroblastoma cells alters cell adhesion *in vitro*, enhances metastatic behavior *in vivo*, and activates Jun kinase. *Cancer Res.* **57,** 5399–5405.

1055. Takahashi, M., Buma, Y., and Taniguchi, M. (1991) Identification of the *ret* proto-oncogene products in neuroblastoma and leukemia cells. *Oncogene* **6,** 297–301.

1056. Hishiki, T., Nimura, Y., Isogai, E., Kondo, K., Ichimiya, S., Nakamura, Y., Ozaki, T., Sakiyama, S., Hirose, M., Seki, N., Takahashi, H., Ohnuma, N., Tanabe, M., and Nakagawara, A., (1998) Glial cell line-derived neurotrophic factor/neurturin-induced differentiation and its enhancement by retinoic acid in primary human neuroblastomas expressing c-*Ret*, *GFRα-1*, and *GFRα-2*. *Cancer Res.* **58,** 2158–2165.

1057. Massaron, S., Seregni, E., Luksch, R., Casanova, M., Botti, C., Ferrari, L., Martinetti, A., Molteni, S.N., Bellani, F.F., and Bombardieri, E. (1998) Neuron-specific enolase evaluation in patients with neuroblastoma. *Tumour Biol.* **19,** 261–268.

1058. DuBois, S.G., Kalika, Y., Lukens, J.N., Brodeur, G.M., Seeger, R.C., Atkinson, J.B., Haase, G.M., Black, C.T., Perez, C., Shimada, H., Gerbing, R., Stram, D.O., and Matthay, K.K. (1999) Metastatic sites in stage IV and IVS neuroblastoma correlate with age, tumor biology, and survival.. *Pediatr. Hematol. Oncol.* **21,** 181–189.

1059. Soderholm, H., Ortoft, E., Johnansson, I., Ljungberg, J., Larsson, C., Axelson, H., and Påhlman, S. (1999) Human achaete-scute homologue 1 (HASH-1) is downregulated in differentiating neuroblastoma cells. *Biochem. Biophys. Res. Commun.* **256,** 557–563.

1060. Hoon, D.S.B., Kuo, C.T., Wen, S., Wang, H., Metelitsa, L., Reynolds, C.P., and Seeger, R.C. (2001) Ganglioside GM2/GD2 synthetase mRNA is a marker for detection of infrequent neuroblastoma cells in bone marrow. *Am. J. Pathol.* **159,** 493–500.

1061. Hettmer, S., Malott, C., Woods, W., Ladisch, S., and Kaucic, K. (2003) Biological stratification of human neuroblastoma by complex "B" pathway ganglioside expression. *Cancer Res.* **63,** 7270–7276.

1062. Leone, A., Mitsiades, N., Ward, Y., Spinelli, B., Poulaki, V., Tsokos, M., and Kelly, K. (2001) The Gem GTP-binding protein promotes morphological differentiation in neuroblastoma. *Oncogene* **20,** 3217–3225.

1063. Perel, Y., Amrein, L., Dobremez, E., Rivel, J., Daniel, J.Y., and Landry, M. (2002) Galanin and galanin receptor expression in neuroblastic tumours: correlation with their differentiation status.*Br. J. Cancer* **86,** 117–122.

1064. Wimmer, K., Zhu, X.X., Rouillard, J.M., Ambros, P.F., Lamb, B.J., Kuick, R., Eckart, M., Weinhäusl, A., Fonatsch, C., and Hanash, S.M. (2002) Combined restriction landmark genomic scanning and virtual genome scans identify a novel human homeobox gene, *ALX3*, that is hypermethylated in neuroblastoma. *Genes Chromosomes Cancer* **33,** 285–294.

1065. Cesi, V., Vitali, R., Tanno, B., Giuffrida, M.L., Sesti, F., Mancini, C., Raschellà, G. (2004) Insulin-like growth factor binding protein 5: contribution to growth and differentiation of neuroblastoma cells. *Am. N Y Acad. Sci.* **1028,** 59–68.

1066. De Preter, K., Vandesompele J., Hoebeeck, J., Vandenbroecke, C., Smet, J., Nuyts, A., Laureys, G., Combaret, V., Van Roy, N., Roels, F., Van Coster, R., Praet, M., De Paepe, A., and Speleman, F. (2004) No evidence for involvement of *SDHD* in neuroblastoma. *BMC Cancer* **4,** 55–69.

1067. Johnsen, J.I., Lindskog, M., Ponthan, F., Pettersen, I., Elfman, L., Orrego, A., Sveinbjörnsson, B., and Kogner, P. (2004) Cyclooxygenase-2 is expressed in neuroblastoma, and non- steroidal anti-inflammatory drugs induce apoptosis and inhibit tumor growth *in vivo*. *Cancer Res.* **64,** 7210–7215.

1068. Kato, C., Miyazaki, K., Nakagawa, A., Ohira, M., Nakamura, Y., Ozaki, T., Imai, T., and Nakagawara, A. (2004) Low expression of human tubulin tyrosine ligase and suppressed tubulin tyrosination/detyrosination cycle are associated with impaired neuronal differentiation in neuroblastomas with poor prognosis. *Int. J. Cancer* **112,** 365–375.

1069. Krams, M., Parwaresch, R., Sipos, B., Heidorn, K., Harms, D., and Rudolph, P. (2004) Expression of the c-kit receptor characterizes a subset of neuroblastomas with favorable prognosis. *Oncogene* **23,** 588–595.

1070. Uccini, S., Mannarino, O., McDowell, H.P., Pauser, U., Vitali, R., Natali, P.G., Altavista, P., Andreano, T., Coco, S., Boldrini, R., Bosco, S., Clerico, A., Cozzi, D., Donfrancesco, A., Inserra, A., Kokai, G., Losty, P.D., Nicotra, M.R., Raschellà, G., Tonini, G.P., and Dominici, C. (2005) Clinical and molecular evidence for c-kit receptor as a therapeutic target in neuroblastic tumors. *Clin. Cancer Res.* **11,** 380–389.

1071. Rorie, C.J., Thomas, V.D., Chen, P., Pierce, H.H., O'Bryan, J.P., and Weissman, B.E. (2004) The *Ews/Fli-1* fusion gene switches the differentiation program of neuroblastomas to Ewing sarcoma/peripheral primitive neuroectodermal tumors. *Cancer Res.* **64,** 1266–1277.

1072. De Falco, G., Bellan, C., D'Amuri, A., Angeloni, G., Leucci, E., Giordano, A., and Leoncini, L. (2005) Cdk9 regulates neural differentiation and its expression correlates with the differentiation grade of neuroblastoma and PNET tumors. *Cancer Biol. Ther.* **4,** 277–281.

1073. de Ruijter, A.J., Meinsma, R.J., Bosma, P., Kemp, S., Caron, H.N., and van Kuilenburg, A.B. (2005) Gene expression profiling in response to the histone deacetylase inhibitor BL1521 in neuroblastoma. *Exp. Cell Res.* **309,** 451–467.

1074. Ho, R., Minturn, J.E., Hishiki, T., Zhao, H., Wang, Q., Cnaan, A., Maris, J., Evans, A.E., and Brodeur, G.M. (2005) Proliferation of human neuroblastomas mediated by the epidermal growth factor receptor. *Cancer Res.* **65,** 9868–9875.

1075. Hsiao, C.-C., Huang, C.-C., Sheen, J.-M., Tai, M.-H., Chen, C.-M., Huang, L.L.H., and Chuang, J.-H. (2005) Differentiation expression of delta-like gene and protein in neuroblastoma, ganglioneuroblastoma and ganglioneuroma. *Mod. Pathol.* **18,** 656–662.

1076. Hsu, W.M., Hsieh, F.J., Jeng, Y.M., Kuo, M.L., Chen, C.N., Lai, D.M., Hsieh, L.J., Wang, B.T., Tsao, P.N., Lee, H., Lin, M.T., Lai, H.S., and Chen, W.J. (2005) Calreticulin expression in neuroblastoma—a novel independent prognostic factor. *Ann. Oncol.* **16,** 314–321.

1077. Hsu, W.-M., Hsieh, F.J., Jeng, Y.-M., Kuo, M.-L., Tsao, P.-N., Lee, H., Lin, M.-T., Lai, H.-S., Chen, C.-N., Lai, D.-M., and Chen, W.-J. (2005) GRP78 expression correlates with histologic differentiation and favorable prognosis in neuroblastoma tumors. *Int. J. Cancer* **113,** 920–927.

1078. Lanciotti, M., Coco, S., Di Michele, P., Haupt, R., Boni, L., Pigullo, S., Dufour, C., Garaventa, A., and Tonini, G.P. (2005) Glutathione S-transferase polymorphisms and susceptibility to neuroblastoma. *Pharmacogenet. Genomics* **15,** 423–426.

1079. Okabe-Kado, J., Kasukabe, T., Honma, Y., Hanada, R., Nakagawara, A., and Kaneko, Y. (2005) Clinical significance of serum NM23-H1 protein in neuroblastoma. *Cancer Sci.* **96,** 653–660.

1080. Pajic, M., Norris, M.D., Cohn, S.L., and Haber, M. (2005) The role of the multidrug resistance-associated protein 1 gene in neuroblastoma biology and clinical outcome. *Cancer Lett.* **228,** 241–246.

1081. Qualman, S.J., Bowen, J., Fitzgibbons, P.L., Cohn, S.L., and Shimada, H. for the Members of the Cancer Committee, College of American Pathologists. (2005) Protocol for the examination of specimens from patients with neuroblastoma and related neuroblastic tumors. *Arch. Pathol. Lab. Med.* **129,** 874–883.

1082. Qiao, J., Kang, J., Cree, J., Evers, B.M., and Chung, D.H. (2005) Gastrin-releasing peptide-induced down-regulation of tumor suppressor protein PTEN (*P*hosphatase and *T*ensin homolog deleted on chromosome t*EN*) in neuroblastomas. *Ann. Surg.* **241,** 684–692.

1083. Stockhausen, M.T., Sjölund, J., and Axelson, H. (2005) Regulation of the Notch target gene Hes-1 by TGFalpha induced Ras/MAPK signaling in human neuroblastoma cells. *Exp. Cell Res.* **310,** 218–228.

1084. Valentijn, L.J., Koppen, A., van Asperen, R., Root, H.A., Haneveld, F., and Versteeg, R. (2005) Inhibition of a new differentiation pathway in neuroblastoma by copy number defects of *N-myc*, *Cdc42*, and *nm23* genes. Cancer Res. **65,** 3136–3145.

1085. Cellai, I., Benvenuti, S., Luciani, P., Galli, A., Ceni, E., Simi, L., Baglioni, S., Muratori, M., Ottanelli, B., Serio, M., Thiele, C.J., and Peri, A. (2006) Antineoplastic effects of rosiglitazone and PPARγ transactivation in neuroblastoma cells. *Br. J. Cancer* **95,** 879–888.

1086. Cheung, I.Y., Vickers, A., and Cheung, N.-K.V. (2006) Sialyltransferase STX (ST8SiaII): a novel molecular marker of metastatic neuroblastoma. *Int. J. Cancer* **119,** 152–156.

1087. Cheung, N.-K.V. Sowers, R., Vickers, A.J., Cheung, I.Y., Kushner, B.H., and Gorlick, R. (2006) *FCGR2A* polymorphism is correlated with clinical outcome after immunotherapy of neuroblastoma with anti-GD2 antibody and granulocyte macrophage colony-stimulating factor. *J. Clin. Oncol.* **24,** 2885–2890.

1088. Guerreiro, A.S., Boller, D., Shalaby, T., Grotzer, M.A., and Arcaro, A. (2006) Protein kinase B modulates the sensitivity of human neuroblastoma cells to insulin-like growth factor receptor inhibition. *Int. J. Cancer* **119,** 2527–2538.

1089. Haber, M., Smith, J., Bordow, S.B., Flemming, C., Cohn, S.L., London, W.B., Marshall, G.M., and Norris, M.D. (2006) Association of high-level *MRP1* expression with poor clinical outcome in a large prospective study of primary neuroblastoma. *J. Clin. Oncol.* **24,** 1546–1553.

1090. Inamori, K., Gu, J., Ohira, M., Kawasaki, A., Nakamura, Y., Nakagawa, T., Kondo, A., Miyoshi, E., Nakagawara, A., and Taniguchi, N. (2006) High expression of N-acetylglucosaminyltransferase V in favorable neuroblastomas: involvement of its effect on apoptosis. *FEBS Lett.* **580,** 627–632.

1091. Machida, T., Fujita, T., Ooo, M.L., Ohira, M., Isogai, E., Mihara, M., Hirato, J., Tomotsune, D., Hirata, T., Fujimori, M., Adachi, W., and Nakagawara, A. (2006) Increased expression of proapoptotic *BMCC1*, a novel gene with the *BNIP2* and *Cdc42GAP* homology (BCH) domain, is associated with favorable prognosis in human neuroblastomas. *Oncogene* **25,** 1931–1942.

1092. Muscat, A., Hawkins, C., and Ashley, D.M. (2006) Caspase-8 levels correlate with the expression of signal transducer and activator of transcription 1 in high-grade but not lower grade neuroblastoma. *Cancer* **107,** 824–831.

1093. Raffaghello, L., Chiozzi, P., Falzoni, S., Di Virgilio, F., and Pistoia, V. (2006) The P2X$_7$ receptor sustains the growth of human neuroblastoma cells through a substance P-dependent mechanism. *Cancer Res.* **66,** 907–914.

1094. Van Roy, N., Vandesompele, Menten, B., Nilsson, H., De Smet, E., Rocchi, M., De Paepe, A., Påhlman, S., and Speleman, F. (2006) Translocation-excision-deletion-amplification mechanism leading to nonsyntenic coamplification of *MYC* and *ATBF1*. *Genes Chromosomes Cancer* **45,** 107–117.

1095. Wang, Q., Diskin, S., Rappaport, E., Attiyeh, E., Mosse, Y., Shue, D., Seiser, E., Jagannathan, J., Shusterman, S., Bansal, M., Khazi, D., Winter, c., Okawa, E., Grant, G., Cnaan, A., Zhao, H., Cheung, N.-K., Gerald, W., London, W., Matthay, K.K., Brodeur, G.M., and Maris, J.M. (2006)

Integrative genomics identifies distinct molecular classes of neuroblastoma and shows that multiple genes are targeted by regional alterations in DNA copy number. *Cancer Res.* **66,** 6050–6062.

1095A. Marzi, I., D'Amico, M., Biagiotti, T., Giunti, S., Carbone, M.V., Fredducci, D., Wanke, E., and Olivotto, M. (2007) Purging of the neuroblastoma stem cell compartment and tumor regression on exposure to hypoxia or cytotoxic treatment. *Cancer Res.* **67,** 2402–2407.

1096. Smith, H.C., Puvion, E., Buchholz, L.A., and Berezney, R. (1984) Spatial distribution of DNA loop attachment and replicational sites in the nuclear matrix. *J. Cell Biol.* **99,** 1794–1802.

1097. Tsokos, M., Linnoila, R.I., Chandra, R.S., and Triche, T.J. (1984) Neuron-specific enolase in the diagnosis of neuroblastoma and other small, round-cell tumors in children. *Hum. Pathol.***15,** 575–584.

1098. Makino, S., Tsuchida, Y., Nakajo, T., Kanda, N., and Kaneko, Y. (1986) Effects of cyclophosphamide, cisplatinum, nitrosourea (ACNU), melphalan, and dacarbazine on a cytogenetically highly malignant neuroblastoma xenograft. *Med. Pediatr. Oncol.* **14,** 36–40.

1099. Makino, S., Kashii, A., Kanazawa, K., and Tsuchida, Y. (1993) Effects of newly introduced chemotherapeutic agents on a cytogenetically highly malignant neuroblastoma, xenotransplanted in nude mice. *J. Pediatr. Surg.* **28,** 612–616.

1100. Esumi, N., Imashuku, S., Tsunamoto, K., Todo, S., Misawa, S., Goto, T., and Fujisawa, Y. (1989) Procoagulant activity of human neuroblastoma cell lines, in relation to cell growth, differentiation and cytogenetic abnormalities. *Jpn. J. Cancer Res.* **80,** 438–443.

1101. Wenzel, A., Cziepluch, C., Hamann, U., Schürmann, J., and Schwab, M. (1991) The N-Myc oncoprotein is associated *in vivo* with the phosphoprotein Max(p20/22) in human neuroblastoma cells. *EMBO J.* **10,** 3703–3712.

1102. Wenzel, A., and Schwab, M. (1995) The mycN/max protein complex in neuroblastoma. Short review. *Eur. J. Cancer* **31A,** 516–519.

1103. Guillemot, F., Lo, L.-C., Johnson, J.E., Auerbach, A., Anderson, D.J., and Joyner, A.L. (1993) Mammalian *achaete-scute* homolog 1 is required for the early development of olfactory and autonomic neurons. *Cell* **75,** 463–476.

1104. Cohen, P.S., Chan, J.P., Lipkunskaya, M., Biedler, J.L., Seeger, R.C., and The Children's Cancer Group (1994) Expression of stem cell factor and c-*kit* in human neuroblastoma. *Blood* **84,** 3465–3472.

1105. Goji, J., Nakamura, H., Ito, H., Mabuchi, O., Hashimoto, K., and Sano, K. (1995) Expression of c-ErbB2 in human neuroblastoma tissues, adrenal medulla adjacent to tumor, and developing mouse neural crest cells. *Am. J. Pathol.* **146,** 660–672.

1106. Nakagawara, A., Milbrandt, J., Muramatsu, T., Deuel, T.F., Zhao, H., Cnaan, A., and Brodeur, G.M. (1995) Differential expression of pleiotrophin and midkine in advanced neuroblastomas. *Cancer Res.* **55,** 1792–1797.

1107. Hofstra, R.M.W., Cheng, N.C., Hansen, C., Stulp, R.P., Stelwagen, T., Clausen, N., Tommerup, N., Caron, H., Westerveld, A., Versteeg, R., and Buys, C.H.C.M. (1996) No mutations found by *RET* mutation scanning in sporadic and hereditary neuroblastoma. *Hum. Genet.* **97,** 362–364.

1108. Ishiguro, Y., Kato, K., Akatsuka, H., Iwata, H., Ito, F., Watanabe, Y., and Nagaya, M. (1996) Comparison of calbindin D-28k and S-100 protein B in neuroblastoma as determined by enzyme immunoassay. *Jpn. J. Cancer Res.* **87,** 62–67.

1109. Norris, M.D., Bordow, S.B., Marshall, G.M., Haber, P.S., Cohn, S.L., and Haber, M. (1996) Expression of the gene for multidrug-resistance-associated protein and outcome in patients with neuroblastoma. *N. Engl. J. Med.* **334,** 231–238.

1110. Takeda, O., Handa, M., Uehara, T., Maseki, N., Sakashita, A., Sakurai, M., Kanda, N., Arai, Y., and Kaneko, Y. (1996) An increased *NM23H1* copy number may be a poor prognostic factor independent of LOH on 1p in neuroblastomas. *Br. J. Cancer* **74,** 1620–1626.

1111. Satoh, F., Takahashi, K., Murakami, O., Totsune, K., Ohneda, M., Mizuno, Y., Sone, M., Miura, Y., Takase, S., Hayashi, Y., Sasano, H., and Mouri, T. (1997) Cerebellin and cerebelin mRNA in the human brain, adrenal glands and the tumour tissues of adrenal tumour, ganglioneuroblastoma and neuroblastoma. *J. Endocrinol.* **154,** 27–34.

1112. Hogarty, M.D., White, P.S., Sulman, E.P., and Brodeur, G.M. (1998) Mononucleotide repeat instability is infrequent in neuroblastoma. *Cancer Genet. Cytogenet.* **106,** 140–143.

1113. Encinas, M., Iglesias, M., Llecha, N., and Comella, J.X. (1999) Extracellular-regulated kinases and phosphatidylinositol 3-kinase are involved in brain-derived neurotrophic factor-mediated survival and neuritogenesis of the neuroblastoma cell line SH-SY5Y. *J. Neurochem.* **73,** 1409–1421.

1114. Tang, X.X., Evans, A.E., Zhao, H., Cnaan, A., London, W., Cohn, S.L., Brodeur, G.M., and Ikegaki, N. (1999) High-level expression of *EPHB6*, *EFNB2*, and *EFNB3* is associated with low tumor stage and high *TrkA* expression in human neuroblastomas. *Clin. Cancer Res.* **5,** 1491–1496.

1115. Chen, S., Caragine, T., Cheung, N.-K.V., and Tomlinson, S. (2000) Surface antigen expression and complement susceptibility of differentiated neuroblastoma clones. *Am. J. Pathol.* **156,** 1085–1091.

1116. Dotsch, J., Harmjanz, A., Christiansen, H., Hanze, J., Lampert, F., and Rasher, W. (2000) Gene expression of neuronal nitric oxide synthase and adrenomedullin in human neuroblastoma using real-time PCR. *Int. J. Cancer* **88,** 172–175.

1117. Erdreich-Epstein, A., Shimada, H., Groshen, S., Liu, M., Metelitsa, L.S., Kim, K.S., Stins, M.F., Seeger, R.C., and Durden, D.L. (2000) Integrins $\alpha_v\beta_3$ and $\alpha_v\beta_5$ are expressed by endothelium of high-risk neuroblastoma and their inhibition is associated with increased endogenous ceramide. *Cancer Res.* **60,** 712–721.

1118. Hattori, H., Matsuzaki, A., Suminoe, A., Ihara, K., Eguchi, M., Tajiri, T., Suita, S., Ishii, E., and Hara, T. (2000) Genomic imprinting of insulin-like growth factor-2 in infant leukemia and childhood neuroblastoma. *Cancer* **88,** 2372–2377.

1119. Judson, H., von Roy, N., Strain, L., Vandesompele, J., Van Gele, M., Speleman, F., and Bonthron, D.T. (2000) Structure and mutation analysis of the gene encoding DNA fragmentation factor 40 (caspase-activated nuclease), a candidate neuroblastoma tumour suppressor gene. *Hum. Genet.* **106,** 406–413.

1120. Keshelava, N., Groshen, S., and Reynolds, C.P. (2000) Cross-resistance of topoisomerase I and II inhibitors in human neuroblastoma cell lines. *Cancer Chemother. Pharmacol.* **45**, 1–8.

1121. Lamant, L., Pulford, K., Bischof, D., Morris, S.W., Mason, D.Y., Delsol, G., and Mariamé, B. (2000) Expression of the ALK tyrosine kinase gene in neuroblastoma. *Am. J. Pathol.* **156**, 1711–1721.

1122. Nagata, T., Takahashi, Y., Asai, S., Ishii, Y., Mugishima, H., Suzuki, T., Chin, M., Harada, K., Koshinaga, S., and Ishikawa, K. (2000) The high level of h*CDC10* gene expression in neuroblastoma may be associated with favorable characteristics of the tumor. *J. Surg. Res.* **92**, 267–275.

1123. van Limpt, V., Chan, A., Carón, H., Van Sluis, P., Boon, K., Hermus, M.-C., and Versteeg, R. (2000) SAGE analysis of neuroblastoma reveals a high expression of the human homologue of the *Drosophila Delta* gene. *Med. Pediatr. Oncol.* **35**, 554–558.

1124. Candanedo-Gonzalez, F.A., Alvarado-Cabrero, I., Gamboa-Dominguez, A., Cerbulo-Vazquez, A., Lopez-Romero, R., Bornstein-Quevedo, L.,, and Salcedo-Vargas, M. (2001) Sporadic type comoposite pheochromocytoma with neuroblastoma: clinicomorphologic, DNA content and reet gene analysis. *Endocr. Pathol.* **12**, 343–350.

1125. Giorgi, M., Leonetti, C., Citro, G., and Augusti-Tocco, G. (2001) *In vitro* and *in vivo* inhibition of SK-N-MC neuroblastoma growth using cyclic nucleotide phosphodiesterase inhibitors. *J. Neurooncol.* **51**, 25–31.

1126. Han, S., Wada, R.K., and Sidell, N. (2001) Differentiation of human neuroblastoma by phenylacetate is mediated by peroxisome proliferator-activated receptor γ^1. *Cancer Res.* **61**, 3998–4002.

1127. Huang, S., Lichtenauer, U.D., Pack, S., Wang, C., Kim, A.C., Lutchman, M., Koch, C.A., Torres-Cruz, J., Huang, S.C. Benz, E.J., Jr., Christiansen, H., Dockhorn-Dworniczak, B., Poremba, C., Vortmeyer, A.O., Chisti, A.H., and Zhuang, Z. (2001) Reassignment of the EPB4.1 gene to 1p36 and assessment of its involvement in neuroblastomas. *Eur. J. Clin. Invest.* **31**, 907–914.

1128. Lo Piccolo, M.S., Cheung, N.K.V., and Cheung, I.Y. (2001) CD2 synthase. A new molecular marker for detecting neuroblastoma. *Cancer* **92**, 924–931.

1129. Takeuchi, T., Nicole, S., Misaki, A., Furihata, M., Iwata, J., Sonobe, H., and Ohtsuki, Y. (2001) Expression of SMARCF1, a truncated form of SWI1, in neuroblastoma. *Am. J. Pathol.* **158**, 663–672.

1130. Sandler, A., Scott, D., Azuhata, T., Takamizawa, S., and O'Dorisio, S. (2002) The survivin:Fas ratio is predictive of recurrent disease in neuroblastoma . *J. Pediatr. Surg.* **37**, 507–511.

1131. Agathanggelou, A., Bièche, I., Ahmed-Choudhury, J., Nicke, B., Dammann, R., Baksh, S., Gao, B., Minna, J.D., Downward, J., Maher, E.R., and Latif, F. (2003) Identification of novel gene expression targets for the Ras association domain family 1 (*RASSF1A*) tumor suppressor gene in non-small cell lung cancer and neuroblastoma. *Cancer Res.* **63**, 5344–5351.

1132. Lambooy, L.H.J., Gidding, C.E.M., van den Heuvel, L.P., Hulsbergen-van de Kaa, C.A., Ligtenberg, M., Bökkerink, J.P.M., and De Abreu, R.A. (2003) Real-time analysis of *tyrosine hydroxylase* gene expression: a sensitive semiquantitative marker for minimal residual disease detection of neuroblastoma. *Clin. Cancer Res.* **9**, 812–819.

1133. Raffaghello, L., Pagnan, G., Pastorino, F., Cosimo, E., Brignole, C., Marimpietri, D., Montaldo, P.G., Gambini, C., Allen, T.M., Bogenmann, E., and Ponzoni, M. (2003) *In vitro* and *in vivo* antitumor activity of liposomal fenretinide targeted to human neuroblastoma. *Int. J. Cancer* **104**, 559–567.

1134. Bozzi, F., Luksch, R., Collini, P., Gambirasio, F., Barzanò, E., Polastri, D., Podda, M., Brando, B., and Fossati-Bellani, F. (2004) Molecular detection of dopamine decarboxylase expression by means of reverse transriptase and polymerase chain reaction in bone marrow and peripheral blood. Utility as a tumor marker for neuroblastoma. *Diagn. Mol. Pathol.* **13**, 135–143.

1135. Calvet, L., Santos, A., Valent, A., Terrier-Lacombe, M.J., Opolon, P., Merlin, J.L., Aubert, G., Morizet, J., Schellens, J.H., Benard, J., and Vassal, G. (2004) No topoisomerase I alteration in a neuroblastoma model with in vivo acquired resistance to irinotecan. *Br. J. Cancer* **91**, 1205–1212.

1136. Gebauer, S., Yu, A.L., Omura-Minamisawa, M., Batova, A., and Diccianni, M.B. (2004) Expression profiles and clinical relationships of *ID2*, *CDKN1B*, and *CDKN2A* in primary neuroblastoma. *Genes Chromosomes Cancer* **41**, 297–308.

1137. Hecht, M., Papoutsi, M., Tran, H.D., Wilting, J., and Schweigerer, L. (2004) Hepatocyte growth factor/c-Met signaling promotes the progression of experimental human neuroblastomas. *Cancer Res.* **64**, 6109–6118.

1138. Oberthuer, A., Hero, B., Spitz, R., Berthold, F., and Fischer, M. (2004) The tumor-associated antigen *PRAME* is universally expressed in high-stage neuroblastoma and associated with poor outcome. *Clin. Cancer Res.* **10**, 4307–4313.

1139. Rorie, C.J., and Weissman, B.E., (2004) The Ews/Fli-1 fusion gene changes the status of p53 in neuroblastoma tumor cell lines. *Cancer Res.* **64**, 7288–7295.

1140. Thomas, S.K., Messam, C.A., Spengler, B.A., Biedler, J.L., and Ross, R.A. (2004) Nestin is a potential mediator of malignancy in human neuroblastoma cells. *J. Biol. Chem.* **279**, 27994–27999.

1141. Andre, S., Kaltner, H., Lensch, M., Russwurm, R., Siebert, H.-C., Fallsehr, C., Tajkhorshid, E., Heck, A.J.R., von Knebel Doeberitz, M., Gabius, H.-J., and Kopitz, J. (2005) Determination of structural and functional overlap/divergence of five proto-type galectins by analysis of the growth-regulatory interaction with ganglioside GM_1 *in silico* and *in vitro* on human neuroblastomas cells. *Int. J. Cancer* **114**, 46–57.

1142. Aoyama, M., Ozaki, T., Inuzuka, H., Tomotsune, D., Hirato, J., Okamoto, Y., Tokita, H., Ohira, M., and Nakagawara, A. (2005) LM03 interacts with neuronal transcription factor, HEN2, and acts as an oncogene in neuroblastoma. *Cancer Res.* **65**, 4587–4597.

1143. Astuti, D., Latif, F., Wagner, K., Gentle, D., Cooper, W.N., Catchpoole, D., Grundy, R., Ferguson-Smith, A.C., and Maher, E.R. (2005) Epigenetic alteration at the *DLK1-GTL2* imprinted domain in human neoplasia: analysis of neuroblastoma, phaeochromocytoma and Wilms' tumour. *Br. J. Cancer* **92**, 1574–1580.

1144. Bumm, K., Grizzi, F., Franceschini, B., Koch, M., Iro, H., Wurm, J., Ceva-Grimaldi, G., Dimimler, A., Cobos, E., Dioguardi, N., Sinha, U.K., Kast, W.M., and Chiriva-Internati, M. (2005) Sperm protein 17 expression defines 2 subsets of primary neuroblastoma. *Hum. Pathol.* **36**, 1289–1293.

1145. Evangelopoulos, M.E., Weis, J., and Krüttgen, A. (2005) Signalling pathways leading to neuroblastoma differentiation after serum withdrawal: HDL blocks neuroblastoma differentiation by inhibition of EGFR. *Oncogene* **24**, 3309–3318.

1146. Hanson, A.J., Nahreini, P., Andreatta, C., Yan, X.-D., and Prasad, K.N. (2005) Role of the adenosine 3',5'-cyclic monophosphate (cAMP) in enhancing the efficacy of siRNA-mediated gene silencing in neuroblastoma cells. *Oncogene* **24**, 4149–4154.

1147. Hettmer, S., Ladisch, S., and Kaucic, K. (2005) Low complex ganglioside expression characterizes human neuroblastoma cell lines. *Cancer Lett.* **225**, 141–149.

1148. Jodele, S., Chantrain, C.F., Blavier, L ., Lutzko, C., Crooks, G.M., Shimada, H., Coussens, L.M., and DeClerck, Y.A. (2005) The contribution of bone marrow-derived cells to the tumor vasculature in neuroblastoma is matrix metalloproteinase-9 dependent. *Cancer Res.* **65**, 3200–3208.

1149. Korja, M., Finne, J., Salmi, T.T., Haapasalo H., Tanner, M., and Isola, J. (2005) No GIST-type c-kit gain of function mutations in neuroblastic tumours. *J. Clin. Pathol.* **58**, 762–765.

1150. Mukerji, I., Ramkissoon, S.H., Reddy, K.K.R., and Rameshwar, P. (2005) Autocrine proliferation of neuroblastoma cells is partly mediated through neurokinin receptors: relevance to bone marrow metastases. *J. Neurooncol.* **71**, 91–98.

1151. Oltra, S., Martinez, F., Orellana, C., Grau, E., Fernandez, J.M., Cañete, A., and Castel, V. (2005) The doublecortin gene, a new molecular marker to detect minimal residual disease in neuroblastoma. *Diagn. Mol. Pathol.* **14**, 53–57.

1152. Osajima-Hakomori, Y., Miyake, I., Ohira, M., Nakagawara, A., Nakagawa, A., and Sakai, R. (2005) Biological role of anaplastic lymphoma kinase in neuroblastoma. *Am. J. Pathol.* **167**, 213–222.

1152A. Parashar, B., Shankar, S.L., O'Guin, K., Butler, J., Vikram, B., and Shafit-Zagardo, B. (2005) Inhibition of human neuroblastoma cell growth by CAY10404, a highly selective Cox-2 inhibitor. *J. Neurooncol.* **71**, 141–148.

1153. Terui, E., Matsunaga, T., Yoshida, H., Kouchi, K., Kuroda, H., Hishiki, T., Saito, T., Yamada, S., Shirasawa, H., and Ohnuma, N. (2005) Shc family expression in neuroblastoma: high expression of *ShcC* is associated with a poor prognosis in advanced neuroblastoma. *Clin. Cancer Res.* **11**, 3280–3287.

1154. Anko, M.L., and Panula, P. (2006) Regulation of endogenous human NPFF2 receptor by neuropeptide FF in SK-N-MC neuroblastoma cell line. *J. Neurochem.* **96**, 573–584.

1155. Carén, H., Holmstrand, A., Sjöberg, R.M., and Martinsson, T. (2006) The two human homologues of yeast UFD2 ubiquitination factor, UBE4A and UBE4B, are located in common neuroblastoma deletion regions and are subject to mutations in tumours. *Eur. J. Cancer* **42**, 381–387.

1156. Dijkhuis, A.J., Klappe, K., Jacobs, S., Kroesen, B.J., Kamps, W., Sietsma, H., and Kok, J.W. (2006) PDMP sensitizes neuroblastoma to paclitaxel by inducing aberrant cell cycle progression leading to hyperdiploidy. *Mol. Cancer Ther.* **5**, 593–601.

1157. Eliassen, L.T., Berge, G., Leknessund, A., Wikman, M., Lindin, I., Lokke, C., Ponthan, F., Johnsen, J.I., Sveinbjornsson, B., Kogner, P., Flaegstad, T., and Rekdal, Ø. (2006) The antimicrobial peptide, Lactoferricin B, is cytotoxic to neuroblastoma cells *in vitro* and inhibits xenograft growth *in vivo*. *Int. J. Cancer* **119**, 493–500.

1158. Nowak, K., Kerl, K., Fehr, D., Kramps, C., Gessner, C., Killmer, K., Samans, B., Berwanger, B., Christiansen, H., and Lutz, W. (2006) *BMI1* is a target gene of E2F-1 and is strongly expressed in primary neuroblastomas. *Nucleic Acids Res.* **34**, 1745–1754.

1159. Simon, T., Spitz, R., Hero, B., Berthold, F., and Faldum, A. (2006) Risk estimation in localized unresectable single copy MYCN neuroblastoma by the status of chromosomes 1p and 11q. *Cancer Lett.* **237**, 215–222.

1159A. Oberthuer, A., Berthold, F., Warnat, P., Hero, B., Kahlert, Y., Spitz, R., Enlestus, K., König, R., Haas, S., Eils, R., Schwab, M., Brors, B., Westermann, F. (2006) Custoinized oligonucleotide microarray gene expression-based classification of neuroblastoma patients outperforms current clinical risk stratification. *J. Clin. Oncol.* **24**, 5070–5078.

1159B. Kaneko, S., Ohira, M., Nakamura, Y., Isogai, E., Nakagawara, A., and Kaneko, M. (2007) Relationship of *DDX1* and *NAG* gene amplification/overexpression to the prognosis of patients with *MYCN*-amplified neuroblastoma. *J. Cancer Res. Clin. Oncol.* **133**, 185–192.

1159C. Michels, E., Vandesompele, J., De Preter, K., Hoebeeck, J., Veermeulen, J., Schramm, A., Molenaar, J.J., Menten, B., Marques, B., Stallings, R.L., Combaret, V., Devalck, C., De Paepe, A., Versteeg, R., Eggert, A., Laureys, G., Van Roy, N., and Speleman, F. (2007) ArrayCGH-based classification of neuroblastoma into genomic subgroups. *Genes Chromosomes Cancer* **46**, 1098–1108.

1159D. van Ginkel, P.R., Yang, W., Marcet, M.M., Chow, C.C., Kulkarni, A.D., Darjatmoko, S., Lindstrom, M.J., Lokken, J., Bhattacharya, S., and Albert, D.M. (2007) 1α-hydroxyvitamin D$_2$ inhibits growth of human neuroblastoma. *J. Neurooncol.* **85**, 255–262.

1159E. Stock, C., Bozsaky, E., Watzinger, F., Poetschger, U., Orel, L., Lion, T., Kowalska, A., and Ambros, P.F. (2008) Genes proximal and distal to *MYCN* are highly expressed in human neuroblastoma as visualized bu comparative expressed sequence hybridization. *Am. J. Pathol.* **172**, 203–214.

1160. Barroca, H., Carvalho, J.L., Gil da Costa, M.J., Cirnes, L., Seruca, R., and Schmitt, F.C. (2001) Detection of *N-myc* amplification in neuroblastomas using Southern blotting on find needle aspirates. *Acta Cytol.* **45**, 169–172.

1161. Bhargava, R., Oppenheimer, O., Gerald, W., Jhanwar, S.C., and Chen, B. (2005) Identification of MYCN gene amplification in neuroblastoma using chromogenic in situ hybridization (CISH). An alternative and practical method. *Diagn. Mol. Pathol.* **14**, 72–76.

1161A. Crabbe, D.C.G., Peters, J., and Seeger, R.C. (1992) Rapid detection of MYCN gene amplification in neuroblastomas using the polymerase chain reaction. *Diagn. Mol. Pathol.* **1,** 229–234.

1162. de Cremoux, P., Thioux, M., Peter, M., Vielh, P., Michon, J., Delattre, O., and Magdelenat, H. (1997) Polymerase chain reaction compared with dot blotting for the determination of N-myc gene amplification in neuroblastoma. *Int. J. Cancer* **72,** 518–521.

1163. Layfield, L.J., Willmore-Payne, C., Shimada, H., and Holden, J.A. (2005) Assessment of NMYC amplification: a comparison of FISH, quantitative PCR monoplexing and traditional blotting methods used with formalin-fixed, paraffin-embedded neuroblastomas. *Anal. Quant. Cytol. Histol.* **27,** 5–14.

1164. Mathew, P., Valentine, M.B., Bowman, L.C., Rowe, S.T., Nash, M.B., Valentine, V.A., Cohn, S.L., Castleberry, R.P., Brodeur, G.M., and Look, A.T. (2001) Detection of *MYCN* gene amplification in neuroblastoma by fluorescence *in situ* hybridization: a Pediatric Oncology Group Study. *Neoplasia* **3,** 105–109.

1165. Melegh, Z., Bálint, I., Tóth, E., Csernák, E., Szentirmay, Z., and Nagy, K. (2003) Detection of N-*myc* gene amplification in neuroblastoma by comparative, in situ, and real-time polymerase chain reaction. *Pediatr. Pathol. Mol. Med.* **22,** 213–222.

1166. Miyajima, Y., Horibe, K., Fukuda, M., Matsumoto, K., Numata, S., Mori, H., and Kato, K. (1996) Sequential detection of tumor cells in the peripheral blood and bone marrow of patients with stage IV neuroblastoma by the revaerse transcription-polymerase chain reaction for tyrosine hydroxylases mRNA. *Cancer* **77,** 1214–1219.

1167. Raggi, C.C., Bagnoni, M.L., Tonini, G.P., Maggi, M., Vona, G., Pinzani, P., Mazzocco, K., De Bernardi, B., Pazzagli, M., and Orlando, C. (1999) Real-time quantitative PCR for the measurement of *MYCN* amplification in human neuroblastoma with the TaqMan detection system. *Clin. Chem.* **45,** 1918–1924.

1168. Sartelet, H., Grossi, L., Pasquier, D., Combaret, V., Bouvier, R., Ranchere, D., Plantaz, D., Munzer, M., Philip, T., Birmebaut, P., Zahm, J.M., Bergeron, C., Gaillard, D., and Pasquier, B. (2002) Detection of N-myc amplification by FISH in immature areas of fixed neuroblastomas: more efficient than Southern blot/PCR. *J. Pathol.* **198,** 83–91.

1169. Squire, J.A., Thorner, P., Marrano, P., Parkinson, D., Ng, Y.K., Gerrie, B., Chilton-Macneill, S., and Zielenska, M. (1996) Identification of MYCN copy number heterogeneity by direct FISH analysis of neuroblastoma preparations. *Mol. Diagn.* **1,** 281–289.

1170. Thorner, P.S., Ho, M., Chilton-MacNeill, S., and Zielenska, M. (2006) Use of chromogenic in situ hybridization to identify *MYCN* gene copy number in neuroblastoma using routine tissue sections. *Am. J. Surg. Pathol.* **30,** 635–642.

1171. Van Roy, N., Van Limbergen, H., Vandesompele, J., Van Gele, M., Poppe, B., Laureys, G., De Paepe, A., and Speleman, F. (2001) M-FISH analysis reveals novel chromosomal regions as targets for genetic investigation in neuroblastoma. *Cancer Genet. Cytogenet.* **128,** 69.

1172. White, P.S., Kaufman, B.A., Marshall, H.N., and Brodeur, G.M. (1993) Use of the single-strand conformation polymorphism technique to detect loss of heterozy-gosity in neuroblastoma. *Genes Chromosomes Cancer* **7,** 102–108.

1173. Oude Luttikhuis, M.E., Iyer, V.K., Dyer, S., Ramani, P., and McConville, C.M. (2000) Detection of MYCN amplification in neuroblastoma using competitive PCR quantitation. *Lab. Invest.* **80,** 271–273.

1174. Cheung, I.Y., and Cheung, N.-K.V. (2001) Quantitation of marrow disease in neuroblastoma by real-time reverse transcription-PCR. *Clin. Cancer Res.* **7,** 1698–1705.

1175. Cheung, I.Y., Lo Piccolo, M.S., Collins, N., Kushner, B.H., and Cheung, N.-K.V. (2002) Quantitation of GD2 synthase mRNA by real-time reverse transcription-polymerase chain reaction. *Cancer* **94,** 3042–3048.

1176. Cheung, I.Y., Lo Piccolo, M.S., Kushner, B.H., Kramer, K., and Cheung, N.-K.V. (2003) Quantitation of GD2 synthase mRNA by real-time reverse transcriptase polymerase chain reaction: clinical utility in evaluating adjuvant therapy in neuroblastoma. *J. Clin. Oncol.* **21,** 1087–1093.

1177. De Preter, K., Vandesompele J., Heimann, P., Kockx, M.M., Van Gele, M., Hoebeeck, J., De Smet, E., Demarche, M., Laureys, G., Van Roy, N., De Paepe, A., and Speleman, F. (2003) Application of laser capture microdissection in genetic analysis of neuroblastoma and neuroblastoma precursor cells. *Cancer Lett.* **197,** 53–61.

1178. Méhes, G., Luegmayr, A., Hattinger, C.M., Lorch, T., Ambros, I.M., Gadner, H., and Ambros, P.F. (2001) Automatic detection and genetic profiling of disseminated neuroblastoma cells. *Med. Pediatr. Oncol.* **36,** 205–209.

1179. Méhes, G., Luegmayr, A., Kornmüller, R., Ambros, I.M., Ladenstein, R., Gadner, H., and Ambros, P.F. (2003) Detection of disseminated tumor cells in neuroblastoma. 3 log improvement in sensitivity by automatic immunofluorescence plus FISH (AIPF) analysis compared with classical bone marrow cytology. *Am. J. Pathol.* **163,** 393–399.

1180. Mosse, Y.P., Greshock, J., Weber, B.L., and Maris, J.M. (2005) Measurement and relevance of neuroblastoma DNA copy number changes in the post-genome era. *Cancer Lett.* **228,** 83–90.

1181. Mosse, Y.P., Greshock, J., Margolin, A., Naylor, T., Cole, K., Khazi, D., Hii, G., Winter, C., Shahzad, S., Usman Asziz, M., Biegel, J.A., Weber, B.L., and Maris, J.M. (2005) High-resolution detection and mapping of genomic DNA alterations in neuroblastoma. *Genes Chromosomes Cancer* **43,** 390–403.

1182. Michels, E., Vandesompele, J., Hoebeeck, J., Menten, B., De Preter, K., Laureys, G., Van Roy, N., and Speleman, F. (2006) Genome wide measurement of DNA copy number changes in neuroblastoma: dissecting amplicons and mapping losses, gains and breakpoints. *Cytogenet. Genome Res.* **115,** 273–282.

1183. Swerts, K., De Moerloose, B., Dhooge, C., Vandesompele, J., Hoyoux, C., Beiske, K., Benoit, Y., Laureys, G., and Philippe, J. (2006) Potential application of ELAVL4 real-time quantitative reverse transcription-PCR for detection of disseminated neuroblastoma cells. *Clin. Chem.* **52,** 438–445.

1184. De Preter, K., Speleman, F., Combaret, V., Lunec, J., Laureys, G., Eussen, B.H.J., Francotte, N., Board, J., Pearson, A.D.J., De Paepe, A., Van Roy, N., and Vandesompele J. (2002) Quantification of *MYCN, DDX1,* and

NAG gene copy number in neuroblastoma using a real-time quantitative PCR assay. *Mod. Pathol.* **15,** 159–166.

1185. Carlin, S., Mairs, R.J., McCluskey, A.G.., Tweddle, D.A., Sprigg, A., Estlin, C., Board, J., George, R.E., Ellershaw, C., Pearson, A.D.J., Lunec, J., Montaldo, P.G., Ponzoni, M., van Eck-Smit, B.L., Hoefnagel, C.A., van den Brug, M.D., Tytgat, G.A.M., and Caron, H.N. (2003) Development of real-time polymerase chain reaction assay for prediction of the uptake of meta-[131 I]iodobenzylguanidine by neuroblastoma tumors.*Clin. Cancer Res.* **9,** 3338–3344.

1186. Cheung, I.Y., Sahota, A., and Cheung, N.-K.V. (2004) Measuring circulating neuroblastomas cells by quantitative reverse transcriptase-polymerase chain reaction analysis. Correlation with its paired bone marrow and standard disease markers. *Cancer* **101,** 2303–2308.

1187. Beiske, K., Ambros, P.F., Burchill, S.A., Cheung, I.Y., and Swerts, K. (2005) Detecting minimal residual disease in neuroblastoma patients–the present state of the art. *Cancer Lett.* **228,** 229–240.

1187A. Schramm, A., Vandesompele, J., Schulte, J.H., Dreesmann, S., Kaderali, L., Brors, B., Eils, R., Speleman, F., and Eggert, A. (2007) Translating expression profiling into a clinically feasible test to predict neuroblastoma outcome. *Clin. Cancer Res.* **13,** 1459–1465.

1188. Amitay, R., Nass, D., Meitar, D., Goldberg, I., Davidson, B., Trakhtenbrot, L., Brok-Simoni, F., Ben-Ze'ev, A., Rechavi, G., and Kaufmann, Y. (2001) Reduced expression of plakoglobin correlates with adverse outcome in patients with neuroblastoma. *Am. J. Pathol.* **159,** 43–49.

1189. Anglesio, M.S., Lestou, V.S., Lim, J.E., Peters, J., and Sorensen, P.H.B. (2001) Identification of a novel gene, *N-COG,* co-amplified with *MYC* in aggressive childhood neuroblastoma. *Cancer Genet. Cytogenet.* **128,** 60.

1190. Bannasch, D., Weis, I., and Schwab, M. (1999) Nmi protein interacts with regions that differ between MycN and Myc and is localized in the cytoplasm of neuroblastoma cells in contrast to nuclear MycN. *Oncogene* **18,** 6810–6817.

1191. Behrends, U., Jandl, T., Golbeck, A., Lechner, B., Müller-Weihrich, S., Schmid, I., Till, H., Berthold, F., Voltz, R., and Mautner, J.M. (2002) Novel products of the *HUD, HUC, NNP-1* and β-internexin genes identified by autologous antibody screening of a pediatric neuroblastoma library. *Int. J. Cancer* **100,** 669–677.

1192. Bergmann, E., Wanzel, M., Weber, A., Shin, I., Christiansen, H., and Eilers, M. (2001) Expression of p27[Kip1] is prognostic and independent of *MYCN* amplification in human neuroblastoma. *Int. J. Cancer (Pred. Oncol.)* **95,** 176–183.

1193. Blanc, E., Goldschneider, D., Ferrandis, E., Barrois, M., Le Roux, G., Leonce, S., Douc-Rasy, S., Bénard J., and Raguénez, G. (2003) MYCN enhances P-gp/*MDR1* gene expression in the human metastatic neuroblastoma IGR-N-91 model. *Am. J. Pathol.* **163,** 321–331.

1194. Bordow, S.B., Haber, M., Madafiglio, J., Cheung, B., Marshall, G.M., and Norris, M.D. (1994) Expression of the multidrug resistance-associated protein (*MRP*) gene correlates with amplification and overexpression of the N-*myc* oncogene in childhood neuroblastoma. *Cancer Res.* **54,** 5036–5040.

1195. Breit, S., Rössler, J., Fotsis, T., and Schweigerer, L. (2000) N-*myc* down-regulates activin A. *Biochem. Biophys. Res. Commun.* **274,** 405–409.

1196. Breit, S., Ashman, K., Wilting, J., Rössler, J., Hatzi, E., Fotsis, T., and Schweigerer, L. (2000) The N-*myc* oncogene in human neuroblastoma cells: down-regulation of an angiogenesis inhibitor identified as activin A[1]. *Cancer Res.* **60,** 4596–4601.

1197. Cavaliere, F., Nestola, V., Amadio, S., D'Ambrosi, N., Angelini, D.F., Sancesario, G., Bernardi, G., and Volonte, C. (2005) The metabotropic P2Y4 receptor participates in the commitment to differentiation and cell death of human neuroblastoma SH-SY5Y cells. *Neurobiol. Dis.* **18,** 100–109.

1198. Chang, C.L., Zhu, X.-x., Thoraval, D.H., Ungar, D., Rawwas, J., Hora, N., Strahler, J.R., Hanash, S.M., and Radany, E. (1994) *nm23-H1* mutation in neuroblastoma. *Nature* **370,** 335–336.

1199. Chen, Q.-R., Bilke, S., Wei, J.S., Greer, B.T., Steinberg, S.M., Westermann, F., Schwab, M., and Khan, J. (2006) Increased *WSB1* copy number correlates with its over-expression which associates with increased survival in neuroblastoma. *Genes Chromosomes Cancer* **45,** 856–862.

1200. Corvi, R., Savelyeva, L., Breit, S., Wenzel, A., Handgretinger, R., Barak, J., Oren, M., Amler, L., and Schwab, M. (1995) Non-syntenic amplification of *MDM2* and *MYCN* in human neuroblastoma. *Oncogene* **10,** 1081–1086.

1201. De Preter, K., Speleman, F., Combaret, V., Lunec, J., Board, J., Pearson, A., De Paepe, A., Van Roy, N., Laureys, G., and Vandesompele J. (2005) No evidence for correlation of *DDX1* gene amplification with improved survival probability in patients with *MYCN*-amplified neuroblastomas. *J. Clin. Oncol.* **23,** 3167–3168.

1202. Edsjö, A., Lavenius, E., Nilsson, H., Hoehner, J.C., Simonsson, P., Culp, L.A., Martinsson, T., Larsson, C., and Påhlman, S. (2003) Expression of *trkB* in human neuroblastoma in relation to *MYCN* expression and retinoic acid treatment. *Lab. Invest.* **83,** 813–823.

1203. Ejeskär, K. Abel, F.,, Sjöberg, R., Backstrom, J., Kogner, P., and Martinsson, T. (2000) Fine mapping of the human preprocortistatin gene (CORT) to neuroblastoma consensus deletion region 1p36.3→p36.2, but absence of mutations in primary tumors. *Cytogenet. Cell Genet.* **89,** 62–66.

1204. Gilbert, J., Norris, M.D., Haber, M., Kavallaris, M., Marshall, G.M., and Stewart, B.W. (1993) Determination of N-myc gene amplification in neuroblastoma by differential polymerase chain reaction. *Mol. Cell. Probes* **7,** 227–234.

1205. Hailat, N., Keim, D.R., Melhem, R.F., Zhu, X.-x., Eckerskorn, C., Brodeur, G.M., Reynolds, C.P., Seeger, R.C., Lottspeich, F., Strahler, J.R., and Hanash, S.M. (1991) High levels of p19/nm23 protein in neuroblastoma are associated with advanced stage disease and with N-*myc* gene amplification. *J. Clin. Invest.* **88,** 341–345.

1206. Hatzi, E., Murphy, C., Zoephel, A., Rasmussen, H., Morbidelli, L., Ahorn, H., Kunisada, K., Tontsch, U., Klenk, M., Yamauchi-Takihara, K., Ziche, M., Rofstad, E.K., Schweigerer, L., and Fotsis, T. (2002) N-*myc* oncogene overexpression down-regulates IL-6; evidence that IL-6 inhibits angiogenesis and suppresses neuroblastoma tumor growth. *Oncogene* **21,** 3552–3561.

1207. Hogarty, M.D., Liu, X., Thompson, P.M., White, P.S., Sulman, E.P., Maris, J.M., and Brodeur, G.M. (2000) BIN1 inhibits colon formation and induces apoptosis in neuroblastoma cell lines with *MYCN* amplification. *Med. Pediatri. Oncol.* **35**, 559–562.

1208. Keshelava, N., Zuo, J.J., Chen, P., Waidyaratne, S.N., Luna, M.C., Gomer, C.J., Triche, T.J., and Reynolds, C.P. (2001) Loss of p53 function confers high-level multidrug resistance in neuroblastoma cell lines. *Cancer Res.* **61**, 6185–6193.

1209. Kong, X.-T., Valentine, V.A., Rowe, S.T., Valentine, M.B., Ragsdale, S.T., Jones, B.G., Wilkinson, D.A., Brodeur, G.M., Cohn, S.L., and Look, A.T. (1999) Lack of homozygously inactivated *p73* in single-copy *MYCN* primary neuroblastomas and neuroblastoma cell lines. *Neoplasia* **1**, 80–89.

1210. Krams, M., Hero, B., Berthold, F., Parwaresch, R., Harms, D., and Rudolph, P. (2002) Proliferation marker Ki-S5 discriminates between favorable and adverse prognosis in advanced stages of neuroblastoma with and without *MYCN* amplification. *Cancer* **94**, 854–861.

1211. Lazarova, D.L., Spengler, B.A., Biedler, J.L., and Ross, R.A. (1999) HuD, a neuronal-specific RNA-binding protein, is a putative regulator of N-*myc* pre-mRNA processing/stability in malignant human neuroblasts. *Oncogene* **18**, 2703–2710.

1212. Manohar, C.F., Bray, J.A., Salwen, H.R., Madafiglio, J., Cheng, A., Flemming, C., Marshall, G.M., Norris, M.D., Haber, M., and Cohn, S.L. (2004) MYCN-mediated regulation of the *MRP1* promoter in human neuroblastoma. *Oncogene* **23**, 753–762.

1213. Misawa, A., Hosoi, H., Arimoto, A., Shikata, T., Akioka, S., Matsumura, T., Houghton, P.J., and Sawada, T. (2000) N-Myc induction stimulated by insulin-like growth factor I through mitogen-activated protein kinase signaling pathway in human neuroblastoma cells. *Cancer Res.* **60**, 64–69.

1214. Moritake, H., Horii, Y., Kuroda, H., and Sugimoto, T. (2001) Analysis of *PTEN/MMAC1* alteration in neuroblastoma. *Cancer Genet. Cytogenet.* **125**, 151–155.

1215. Noguchi, T., Akiyama, K., Yokoyama, M., Kanda, N., Matsunaga, T., and Nishi, Y. (1996) Amplification of a DEAD box gene (*DDX1*) with the *MYCN* gene in neuroblastoma as a result of cosegregation of sequences flanking the *MYCN* locus. *Genes Chromosomes Cancer* **15**, 129–133.

1216. Noujaim, D., van Golen, C.M., van Golen, K.L., Grauman, A., and Feldman, E.L. (2002) N-Myc and Bcl-2 coexpression induces MMP-2 secretion and activation in human neuroblastoma cells. *Oncogene* **21**, 4549–4557.

1217. Reiter, J.L., and Brodeur, G.M. (1996) High-resolution mapping of a 130-kb core region of the MYCN amplicon in neuroblastomas. Genomics **32**, 97–103.

1218. Reiter, J.L., and Brodeur, G.M. (1998) *MYCN* is the only highly expressed gene from the core amplified domain in human neuroblastomas. *Genes Chromosomes Cancer* **23**, 134–140.

1219. Rozzo, C., Chiesa, V., and Ponzoni, M. (1997) Integrin upregulation as marker of neuroblastoma cell differentiation: correlation with neurite extension. *Cell Death Differ.* **4**, 713–724.

1220. Schwab, M. (1990) Amplification of the *MYCN* oncogene and deletion of putative tumour suppressor gene in human neuroblastomas. *Braini Pathol.* **1**, 41–46.

1221. Shohet, J.M., Hicks, M.J., Plon, S.E., Burlingame, S.M., Stuart, S., Chen, S.-Y., Brenner, M.K., and Nuchtern, J.G. (2002) Minichromosome maintenance protein *MCM7* is a direct target of the MYCN transcription factor in neuroblastoma. *Cancer Res.* **62**, 1123–1128.

1222. Squire, J.A., Thorner, P.S., Weitzman, S., Maggi, J.D., Dirks, P., Doyle, J., Hale, M., and Godbout, R. (1995) Co-amplification of *MYCN* and a DEAD box gene (*DDX1*) in primary neuroblastoma. *Oncogene* **10**, 1417–1422.

1223. Tajiri, T., Shono, K., Fujii, Y., Noguchi, S., Kinoshita, Y., Tsuneyoshi, M., and Suita, S. (1999) Highly sensitive analysis for N-*myc* amplification in neuroblastoma based on fluorescence in situ hybridization. *J. Pediatr. Surg.* **34**, 1615–1619.

1224. Tajiri, T., Liu, X., Thompson, P.M., Tanaka, S., Suita, S., Zhao, H., Maris, J.M., Prendergast, G.C., and Hogarty, M.D. (2003) Expression of a *MYCN*-interacting isoform of the tumor suppressor *BIN1* is reduced in neuroblastomas with unfavorable biological features. *Clin. Cancer Res.* **9**, 3345–3355.

1225. Tsai, H.-Y., Hsi, B.-L., Hung, I.-J., Yang, C.-P., Lin, J.-N., Chen, J.-C., Tsai, S.-F., and Huang, S.-F. (2004) Correlation of MYCN amplification with MCM7 protein expression in neuroblastomas: a chromogenic in situ hybridization study in paraffin sections. *Hum. Pathol.* **35**, 1397–1403.

1226. Valsesia-Wittmann, S., Magdeleine, M., Dupasquier, S., Garin, E., Jallas, A.-C., Combaret, V., Krause, A., Leissner, P., and Puisieux, A. (2004) Oncogenic cooperation between *H-Twist* and *N-Myc* overrides failsafe programs in cancer cells. *Cancer Cell* **6**, 625–630.

1227. Vitali, R., Cesi, V., Nicotra, M.R., McDowell, H.P., Donfrancesco, A., Mannarino, O., Natali, P.G., Raschellà, G., and Dominici, C. (2003) c-KIT is preferentially expressed in *MYCN*-amplified neuroblastoma and its effect on cell proliferation is inhibited *in vitro* by STI-571. *Int. J. Cancer* **106**, 147–152.

1228. Wallick, C.J., Gamper, I., Thorne, M., Feith, D.J., Takasaki, K.Y., Wilson, S.M., Seki, J.A., Pegg, A.E., Byus, C.V., and Bachmann, A.S. (2005) Key role for p27^{Kip1}, retinoblastoma protein Rb, and MYCN in polyamine inhibitor-induced G_1 cell cycle arrest in *MYCN*-amplified human neuroblastoma cells. *Oncogene* **24**, 5606–5618.

1229. Weber, A., Imisch, P., Bergmann, E., and Christiansen. H. (2004) Coamplification of *DDX1* with an improved survival probability in children with *MYCN*-amplified human neuroblastoma. *J. Clin. Oncol.* **22**, 2681–2690.

1230. Wimmer, K., Zhu, X.X., Lamb, B.J., Kuick, R., Ambros, P., Kovar, H., Thoraval, D., Elkahloun, A., Meltzer, P., and Hanash, S.M. (2001) Two-dimensional DNA electrophoresis identifies novel CpG islands frequently coamplified with MYCN in neuroblastoma. *Med. Pediatr. Oncol.* **36**, 75–79.

1231. Yaari, S., Jacob-Hirsch, J., Amariglio, N., Haklai, R., Rechavi, G., and Kloog, Y. (2005) Disruption of cooperation between Ras and MycN in human neuroblastoma cells promotes growth arrest. *Clin. Cancer Res.* **11**, 4321–4330.

1232. Schmidt, M.H., Gottfried, O.N., von Koch, C.S., Chang, S.M., and McDermott, M.W. (2004) Central neurocytoma: a review. *J. Neurooncol.* **66**, 377–384.

1233. Taruscio, D., Danesi, R., Montaldi, A., Cerasoli, S., Cenacchi, G., and Giangaspero, F. (1997) Nonrandom gain of chromosome 7 in central neurocytoma: a chromosomal analysis and fluorescence in situ hybridization study. *Virchows Arch.* **430,** 47–51.

1234. Fujisawa, H., Marukawa, K., Hasegawa, M., Tohma, Y., Hayashi, Y., Uchiyama, N., Tachibana, O., and Yamashita, J. (2002) Genetic differences between neurocytoma and dysembryoplastic neuroepithelial tumor and oligodendroglial tumors. *J. Neurosurg.* **97,** 1350–1355.

1235. Buxton, N., Ho, K.J., Macarthur, D., Vloeberghs, M., Punt, J., and Robertson, I. (2001) Neuroendoscopic third ventriculosotomy for hydrocephalus in adults: report of a single unit's experience with 63 cases. *Surg., Neurol.* **55,** 74–78.

1236. Nesbit, C.E., Tersak, J.M., and Prochownik, E.V. (1999) *MYC* oncogenes and human neoplastic disease. *Oncogene* **18,** 3004–3016.

1237. Knoepfler, P.S., Cheng, P.F., and Eisenman, R.N. (2002) N-*myc* is essential during neurogenesis for the rapid expansion of progenitor cell poulations and the inhibition of neuronal differentiation. *Genes Dev.* **16,** 2699–2712.

1238. Goodman, L.A., Liu, B.C., Thiele, C.J., Schmidt, M.L., Cohn, S.L., Yamashiro, J.M., Pai, D.S., Ikegaki, N., and Wada, R.K. (1997) Modulation of N-myc expression alters the invasiveness of neuroblastoma. *Clin. Exp. Metastasis* **15,** 130–139.

1239. Adleff, V., Rácz, K., Tóth, M., Varga, I., Bezzegh, A., and Gláz, E. (1998) p53 protein and its messenger ribonucleic acid in human adrenal tumors. *J. Endocrinol. Invest.* **21,** 753–777.

1240. Kros, J.M., Delwel, E.J., de Jong, T.H.R., Tanghe, H.L.J., van Run, P.R.W.A., Vissers, K., and Alers, J.C. (2002) Desmoplastic infantile astrocytoma and ganglioglioma: a search for genomic characteristics. *Acta Neuropathol.* **104,** 144–148.

1241. Park, H.-J., Kim, Y., Kim, H., Ha, E., Park, H., Yoon, S., Kim, M.-J., Choi, S., Ryu, Y., Park, H.-K., and Hong, M. (2007) Neuroprotective effect by Dammishimgyu-herbal acupuncture against H_2O_2-induced apoptosis in human neuroblastoma, SH-SY5Y cells. *Neurol. Res.* **29,** 593–597.

1242. Mejia, M.C., Navarro, S., Pellin, A., Castel, V., and Llombart-Bosch, A. (1998) Study of bcl-2 protein expression and the apoptosis phenomenon in neuroblastoma. *Anticancer Res.* **18,** 801–806.

1243. Peddinti, R., Zeine, R., Luca, D., Seshadri, R., Chlenski, A., Cole, K., Pawel, B., Salwen, H.R., Maris, J.M., and Cohn, S.L. (2007) Prominent microvascular proliferation in clinically aggressive neuroblastoma. *Clin. Cancer Res.* **13,** 3499–3506.

1244. Gao, X., Deeb, D., Jiang, H., Liu, Y., Dulchavsky, S.A., and Gautam, S.C. (2007) Synthetic triterpenoids inhibit growth and induce apoptosis in human glioblastoma and neuroblastoma cells through inhibition of prosurvival Akt, NF-κB and Notch1 signaling. *J. Neurooncol.* **84,** 147–157.

1244A. Ambros, P.F., Ambros, I.M.; SIOP Europe Neuroblastoma Pathology, Biology, and Bone Marrow Group (2001) Pathology and biology guidelines for resectable and unresectable neuroblastic tumors and bone marrow examination guidelines. *Med. Pediatr. Oncol.* **37,** 489–504.

1244B. De Preter, K., Vandesompele, J., Heimann, P., Yigit, N., Beckman, S., Schramm, A., Eggert, A., Stallings, R.L., Benoit, Y., Renard, M., De Paepe, A., Laureys, G., Påhlman,

S., and Speleman. F. (2006) Human fetal neuroblast and neuroblastoma transcriptome analysis confirms neuroblast origin and highlights neuroblastoma candidate genes. Genome Biol. **7,** R84.

1244C. Mora, J., Lavarino, C., Alaminos, M., Cheung, N.-K.V., Rios, J., de Torres, C., Illei, P., Juan, G., and Gerald, W.L. (2007) Comprehensive analysis of tumoral DNA content reveals clonal ploidy heterogeneity as a marker with prognostic significance in locaregional neuroblastoma. *Genes Chromosomes Cancer* **46,** 385–396.

1244D. Mosse, Y.P., Diskin, S.J., Wasserman, N., Rinaldi, K., Attiyeh, E.F., Cole, K., Jagannathan, J., Bhambhani, K., Winter, C., and Maris, J.M. (2007) Neuroblastomas have distinct gneomic DNA profiles that predict clinical phenotype and regional gene expression. *Genes Chromosomes Cancer* **46,** 936–949.

1244E. Wagner, L.M., McLendon, R.E., Yoon, K.J., Weiss, B.D., Billups, C.A. and Danks, M.K. (2007) Targeting methylguaning-DNA methyltransferase in the treatment of neuroblastoma. *Clin Cancer Res.* **13,** 5418–5425.

1245. Schleiermacher, G., Michon, J., Huon, I., Dubois d'Enghien, C., Klijanienko, J., Brisse, H., Ribeiro, A., Mosseri, V., Rubie, H., Munzer, C., Thomas, C., Valteau-Couanet, D., Auvrignon, A., Plantaz, D., Delattre, O., and Couturier, J. on behalf of the Société Française des Cancers de l'Enfant (SFCE). (2007) Chromosomal CGH identifies patients with a higher risk of relapse in neuroblastoma without *MYCN* amplification. *Br. J. Cancer* **97,** 238–246.

1246. Schleiermacher, G., Delattre, O., Peter, M., Mosseri, V., Delonlay, P., Vielh, P., Thomas, G., Zucker, J.M., Magdelénat, M., and Michon, J. (1996) Clinical relevance of loss of heterozygosity of the short arm of chromosome 1 in neuro- blastoma: a single-institution study. *Int. J. Cancer* **69,** 73–78.

1247. Eberhart, C.G., Brat, D.J., Cohen, K.J., and Burger, P.C. (2000) Pediatric neuroblastic brain tumors containing abundant neuropil and true rosettes. *Pediatr. Dev. Pathol.* **3,** 346–352.

1248. Stallings, R.L., Yoon, K., Kwek, S., and Ko, D. (2007) The origin of chromosome imbalances in neuroblastoma. *Cancer Genet. Cytogenet.* **176,** 28–34.

1249. Westermann, F., Henrich, K.-O., Wei, J.S., Lutz, W., Fischer, M., König, R., Wiedemeyer, R., Ehemann, V., Brors, B., Ernestus, K., Leuschner, I., Benner, A., Khan, J., and Schwab, M. (2007) High *Skp2* expression characterizes high-risk neuroblastomas independent of *MYCN* status. *Clin. Cancer Res.* **13,** 4696–4703.

1250. Sartelet, H., Fabre, M., Castaing, M., Bosq, J., Racu, I., Lagonotte, E., Scott, V., Lecluse, Y., Barette, S., Michiels, S., and Vassal, G. (2007) Expression of erythropoietin and its receptor in neuroblastomas. *Cancer* **110,** 1096–1106.

1251. Scaruffi, P., Coco, S., Cifuentes, F., Albino, D., Nair, M., Defferrari, R., Mazzocco, K., and Tonini, G.P. (2007) Identification and characterization of DNA imbalances in neuro- blastoma by high-resolution oligonucleotide array comparative genomic hybridization. *Cancer Genet. Cytogenet.* **177,** 20–29.

1252. Hoehner, J.C., Olsen, L., Sandstedt, B., Kaplan, D.R., and Påhlman, S. (1995) Association of neurotrophin receptor expression and differentiation in human neuroblastoma. *Am. J. Pathol.* **147,** 102–113.

1252A. Takeuchi, T., Misaki, A., Liang, S.-B., Tachibana, A., Hayashi, N., Sonobe, H., and Ohtsuki, Y. (2000) Expression of T-cadherin (CDH13, H-cadherin) in human brain and its characteristics as a negative grouth regularor of epidermal growth factor in neuroblastoma cells. *J. Neurochem.* **74,** 1489–1497.

1252B. Miyake, I., Hakomori, Y., Shinohara, A., Gamou, T., Saito, M., Iwamatsu, A., and Sakai, R. (2002) Activation of anaplastic lymphoma kinase is responsible for hyperphosphorylation of ShcC in neuroblastoma cell lines. *Oncogene* **21,** 5823–5334.

1252C. Tanno, B., Vitali, R., De Arcangelis, D., Mancini, C., Eleuteri, P., Dominici, C., and Raschellà, G. (2006) Bim-dependent apoptosis follows IGFBP-5 downregulation in neuroblastoma cells. *Biochem. Biophys. Res. Commun.* **351,** 547–552.

1252D. Łastowska, M., Viprey, V., Santibanez-Koref, M., Wappler, I., Peters, H., Cullinane, C., Roberts, P., Hall, A.G., Tweddle, D.A., Pearson, A.D.J., Lewis, I., Burchill, S.A., and Jackson, M.S. (2007) Identification of candidate genes involved in neuroblastoma progression by combining genomic and expression microarrays with survival data. *Oncogene* **26,** 7432–7444.

8

Medulloblastoma, Primitive Neuroectodermal Tumors, and Pineal Tumors

Summary   Chapter 8 addresses three different tumors: medulloblastoma, primitive neuroectodermal tumors, and pineal tumors. The genetic and molecular studies in medulloblastoma are legion; yet, the basic changes responsible for the development of these tumors have remained evasive. Nevertheless, cytogenetically about one third of the medulloblastomas contain loss of 1p or i(17q), and less common anomalies of chromosomes 1 and 7, loss of chromosome 7, and loss of chromosome 22 (-22) and other changes. Of interest is the presence of double minutes in some medulloblastomas (approx 7%). The aforementioned changes have been supplemented with those obtained by spectral karyotyping, fluorescence in situ hybridization, comparative genomic hybridization (CGH), and loss of heterozygosity (LOH) results. Possible candidate genes playing a role in medulloblastoma development may reside at 10q or 8p. The changes of *MYCN* (amplification) in approximately 10% of medulloblastomas are of interest. Other genes investigated in these tumors include *TP53* and *ARF*, *IGF*, epidermal growth factor receptor (*EGFR*)-*ErbB*, *PDFGFα*, *RAS/MAPK*, and *PAX* and *ENI*. The sonic hedgehog (SHH) and Wingless signaling pathways may play a role in medulloblastoma biology, and also neurotrophins and other factors (related to apoptosis), such as somatostatin receptors, Notch signaling, epigenetic events, the *HIC* gene, *caspase8* gene, *RASSF1A* gene, and telomerase activity. The hereditary and genetic susceptibilities to medulloblastoma development are discussed. Chapter 8 also addresses primitive neuroectodermal tumors (PNET) cytogenetically; the changes in PNET show a variable pattern, as do the changes obtained with CGH and LOH. Molecular studies of PNET have included p53, Notch signaling, and the SHH pathway. Approaches similar to those used for the study of PNET have been applied to pineal tumors with the results being variable and not specific for any of the tumors.

Keywords   medulloblastoma · PNET · pineal tumors · genetics · molecular biology.

Introduction

Chapter 8 has been divided into three main sections: medulloblastoma (posterior fossa or infratentorial tumors), primitive neuroectodermal tumors (PNET) (anterior fossa or supratentorial tumors or cerebral neuoblastomas) and pineal tumors (primarily pineoblastoma). Although there is biologic and histologic overlap among these tumors (1, 2), we have followed "a rule of thumb" in which we treat all infratentorial (posterior fossa) tumors, essentially those of the cerebellum, as medulloblastomas, and those of supratentorial (anterior fossa) location as PNET, including cerebral neuroblastomas. Although problems arose in occasional cases with this approach, in the preponderant number of tumors the correlations seemed to be valid and informative. Generally, the genetic changes (particularly cytogenetic) in these tumors complemented their anatomic location, histology and

behavior, lending the approach not only practicality but also a correlative meaning.

Medulloblastoma

Clinical Aspects and Pathology of Medulloblastoma

Medulloblastoma is a malignant, poorly differentiated, highly aggressive, invasive embryonal tumor of the cerebellum primarily affecting children, showing predominantly neuronal differentiation, and an inherent tendency to disseminate via the cerebral spinal fluid (CSF) pathways (3–5,5A). Adult cases, although rare, have been reported, and their manifestations summarized (5B). Information regarding the incidence, epidemiology, and histoclinical aspects of medulloblastoma is given in Table 8.1.

From: A. A. Sandberg and J. F. Stone: *The Genetics and Molecular Biology of Neural Tumors*
© 2008 Humana Press, Totowa, NJ

TABLE 8.1. Some features of medulloblastoma incidence, histology, and clinical aspects.

	References
Medulloblastoma is the most common CNS tumor in children.	*Kaplan and Meier 1958* (625)
Medulloblastomas account for 20–25% of all childhood brain tumors, and only for 1% in adults.	*Fuller 1996* (626)
Incidence of medulloblastoma is 1–2 cases per million and 1:150,000 children with a peak of patient age of 5–7 years.	*Lantos et al 1996* (627)
	Packer et al 1999 (628)
Approximately 65% of patients with medulloblastoma are male. Over 70% of medulloblastomas occur in patients younger than 16 years.	*Arseni and Ciurea 1981* (628A)
	Roberts et al 1991 (629)
In adults, 80% of medulloblastomas arise in patients 21–40-years-old; these tumors are rare beyond the 5th decade of life.	*Hubbard et al 1989* (629A)
	Giordana et al 1999 (630)
In general, more than 80% of medulloblastomas are infratentorial (posterior fossa) and arise primarily in the cerebellum. PNET are located supratentorially (anterior fossa). Spinal medulloblastomas are very rare.	*Dirks et al 1996* (547)
	Girschick et al 2001 (536)
Progress and future challenges in the therapy of medulloblastoma have been addressed. At present, 5-year survival rates of patients with medulloblastoma range from 40–80% and 10-year rates 40–50%.	*Packer 1999* (697)
Medulloblastomas are believed to originate from cerebellar granule cell precursors that transform and fail to undergo differentiation, based on expression of the neurotrophin receptor p75 NTR. However, alternative origins may exist, e.g., subependymal cells of the medullary velum region and possibly the internal granule cell layer.	*Bühren et al 2000* (385)
	Provias and Becker 1996 (230)
The bulk of childhood medulloblastomas are in the vermis (arising from the ventricular matrix) and project into the fourth ventricle. Involvement of cerebellar hemisphere increases with age of the patient.	*Giangaspero et al 2000* (4)
In adults, medulloblastomas occur more often in the cerebellum than the vermis and are usually desmoplastic.	*Greenberg et al 2001* (630A)
Medulloblastomas are the most frequent malignant brain tumors. They are composed of primitive neural cells with some medulloblastomas showing signs of differentiation along neuronal, glial, myogenic and rarely melanocytic or retinal lineages.	*Rubinstein 1985* (631)
	Giangaspero et al 2000 (4)
	Kleihues et al 2002 (6)
Medulloblastomas express neuronal intermediate filaments, synaptic vesicle proteins, growth factor receptors and adhesion molecules, suggesting that these tumors in general arise from neuroblasts that escape terminal differentiation.	*Trojanowski et al 1992* (632)
Medulloblastomas in adults (often desmoplastic) often do not require the aggressive therapy needed in pediatric tumors (often classic). The differences in adult vs. pediatric medulloblastoma are also reflected in the clinicopathologic aspects, proliferation index (MIB-1) and apoptotic index.	*Sarkar et al 2002* (414)
Clinical, histologic, diagnostic markers and prognosis of medulloblastoma.	*Gajjar et al 2004* (742)
	Gilbertson 2002 (743)
	McLendon et al 1999 (744)
Stratification of medulloblastomas on the basis of histopathological grading and the problem of establishing strict criteria for large cell/anaplastic medulloblastoma have been discussed.	*Min et al 2006* (739)
Low capillary permeability and blood flow were demonstrated in medulloblastoma, findings which could explain the limited response rates of these tumors to chemotherapy.	*Warnke et al 2006* (740)
Genome-wide expression profiles of medulloblastomas can partition large tumor cohorts into subgroups that are enriched for specific genetic alterations, an approach which may ultimately assist in the selection of patients for clinical trials of molecularly-targeted therapies.	*Thompson et al 2006* (777)
Understanding the various biological and molecular aspects of medulloblastoma, advances in the clinical management of these cases may be made.	*Frühwald and Plass 2002* (780)
HLA class I-related APM component defects have been described in pediatric medulloblastomas. These findings may be useful in the development of T-cell immunotherapy of medulloblastoma using autologous tumor-specific CTL.	*Raffaghello et al 2007* (778)
Histologic grading of adult medulloblastomas, based on the presence of anaplasia and large-cell morphology has been proposed.	*Rodriguez et al 2007* (779).

Medulloblastoma are not simple tumors histologically. The World Health Organization (WHO) classification (6) lists four variants in addition to the classic type (i.e., desmoplastic, large-cell, medullomyoblastic, and melanotic). Other types have been described and reviewed previously (1). The WHO classification of brain tumors (6, 7) has been generally accepted, with some sobering but mostly approving commentaries regarding the WHO classification (8–10).

Chapter 8 provides the genetic information for those medulloblastoma types for which such information is available. Not infrequently, there is no clear-cut dividing line existing in the histologic features of medulloblastomas, particularly between classic and large-cell types, and this may account for some of the variability of results for these tumors presented in the literature.

The salient histologic features of the various types of medulloblastoma applicable to the interpretation of the genetic and molecular data presented here summarized in Figures 8.1 and 8.2. Concise descriptions of the histologic aspects of medulloblastoma are presented in Table 8.2.

Classic medulloblastomas

Classic medulloblastomas account for about 80% of all medulloblastomas (1), and they grow as sheets of cells with a high nuclear:cytoplasmic ratio and a capacity to invade adjacent brain and leptomeninges. Classic medulloblastomas are composed of densely packed cells with round-to-oval or carrot-shaped highly hyperchromatic nuclei surrounded by scanty cytoplasm. Neuroblastic rosettes are a typical but not a constant feature (observed in <40% of tumors) (4). Round

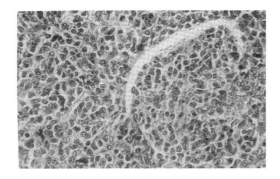

FIGURE 8.1. Histologic appearance of a medulloblastoma in an adult showing sheets of undifferentiated cells without glial or neuronal differentiation (637).

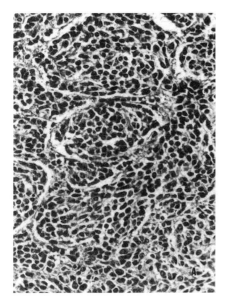

FIGURE 8.2. Histologic presentation of a desmoplastic medulloblastoma with a lobulated pattern and high cellularity.

cells with chromatin of varying density are frequently intermingled, and occasionally they form the main population of classic medulloblastomas. Occasional ganglion cells may be seen. Mitoses are infrequent and seen in about 25% of the cases (11,12). Apoptosis is frequent, whereas areas of necrosis are less common.

Large-Cell Medulloblastoma

Large-cell medulloblastoma, a highly malignant tumor, has emerged as a variant with significant clinical and molecular associations, such as a poor prognosis and early dissemination through the CSF pathways (13). Large-cell medulloblastomas represent approximately 4% of all medulloblastomas (13, 14). The cells of these tumors have large, round and/or pleomorphic nuclei with prominent nucleoli and a more abundant cytoplasm than that found in classic medulloblastoma. Large areas of necrosis, high mitotic activity, and high apoptotic rate are common findings in large-cell medulloblastomas. This variant resembles atypical teratoid/rhabdoid tumors (AT/RT) of the cerebellar region, but it differs on the basis of immunophenotype and cytogenetic findings (14), e.g., deletion of 22q or monosomy of chromosome 22 often seen in AT/RT. Most large-cell (and some classic) medulloblastomas contain regions where the cells show an anaplastic immunophenotype, increased nuclear pleomorphism and a distinctive molding of cells in one another; these tumors have been labeled as anaplastic. Large-cell/anaplastic medulloblastoma has been a label applied to these tumors (15). Although reported cases of large-cell/anaplastic medulloblastoma are few, the age of manifestation ranged from 13 months to 4-years, and all the patients were male. These neoplasms are usually located in the cerebellar vermis, and they are characterized by highly aggressive behavior. *MYC* amplification is seen more often in large-cell medulloblastoma than other types (14), as are double minutes and an isochromosome 17q (16, 17).

Classic, anaplastic, and large-cell medulloblastomas are not distinguished by their immunophenotype; reactivities signifying variable degrees of differentiation along neuroepithelial lines can be evident (or not) in each variant.

TABLE 8.2. Histologic highlights of medulloblastomas.

Type	References
Classic medulloblastomas: cells contain small round nuclei which are generally arranged in sheets.	*Frank et al 2004* (207)
Desmoplastic medulloblastomas: contain nodules of tumor cells that frequently demonstrate neurocytic differentiation and are surrounded by collagen-rich tissue.	*Frank et al 2004* (207)
Desmoplastic medulloblastomas have astroglial differentiation as shown by reactivity for glial fibrillary acidic protein (GFAP), in comparison with the classic type.	*Herpers and Budka 1985* (633)
Large-cell/anaplastic medulloblastomas: contain large cells with pleomorphic nuclei, prominent nucleoli and abundant cytoplasm Regions of anaplasia also characterize large-cell/anaplastic medulloblastomas, a tumor type associated with a poor prognosis.	*Eberhart et al 2002* (5) *McManamy et al 2003* (28)
Medulloblastomas with extensive nodularity: These tumors may show considerable neuronal differentiation and are characterized by intranodular nuclear uniformity. These tumors have a favorable outcome.	*Pearl and Takei 1981* (22) *Burger and Scheithauer 1994* (12) *Giangaspero et al 2000* (4)

Desmoplastic Medulloblastomas

Desmoplastic medulloblastomas (constitute 15% of all medulloblastomas) are prevalent in adult patients (18) and show distinctive architectural features, combining reticulin-free nodules and reticulin-rich internodular regions (Figure 8.2). These nodular, reticulin-free zones ("pale islands") are surrounded by densely packed, highly proliferative cells that produce a dense intercellular reticulin fiber network (12, 16, 19). The nodules have reduced cellularity, a rarified fibrillar matrix and marked nuclear uniformity. Medulloblastomas showing only an increased amount of collagenous and reticulin fibers without the nodular pattern are not classified as a desmoplastic variant. The mitotic count and growth fraction, as assessed by the Ki-67 immunolabeling index, are higher in internodular regions; yet, apoptotic bodies tend to be more frequent in the nodules (20, 21). This biphasic phenotype is reflected in immunohistochemical patterns of protein expression. Signs of neuronal differentiation, such as synaptophysin reactivity, tend to be greater in the nodular than internodular regions.

Medulloblastomas with Extensive Nodularity (MBEN)

MBEN show advanced neuronal differentiation, intranodular nuclear uniformity, and cell streaming in a fine fibrillary background. Large nodules, some looking ovoid or reniform, dominate the histopathology of MBEN. These tumors had been labeled in the past as cerebellar neuroblastomas (12, 22, 23). MBEN represents the most differentiated form of the nodular/desmoplastic medulloblastomas. Cells in the nodules show neurocytic differentiation, adopt a laminar pattern, and they are set against a neuropil-like background. A neuronal immunophenotype is characteristic. Mature ganglion cells may be observed. These neoplasms occur predominantly in children less than 3 years of age, and they seem to be associated with a favorable prognosis. Neoplasms of this type occasionally undergo maturation to more differentiated ganglion cell tumors, i.e., gangliomas (24, 25). It has been suggested that they constitute a pathogenetically separate group (26).

Despite the current emphasis on molecular classification of tumors, including medulloblastoma, the histopathological classification of medulloblastomas continues to evolve (1, 27). The poor prognosis associated with the anaplastic variant, has now been demonstrated in two large cohorts of children treated in clinical trials (5, 14, 28). Large-cell/anaplastic medulloblastomas seem to have an aggressive course, as reflected in the increased cellular proliferation (28), and, hence, an unfavorable prognosis compared with the other types of medulloblastomas (1, 5, 14).

Immunohistochemistry of Medulloblastoma

The common immunohistochemical (IMHC) features of medulloblastoma are shown in Table 8.3. The application of appropriate immunohistochemical markers can be quite helpful diagnostically in these tumors (4, 29, 30). In contrast, the application of immunohistochemical data to the prognosis of medulloblastoma has been criticized (31) on the basis that the data of various immunohistochemical assays in medulloblastoma are heterogeneous and that universally applicable markers for verification of medulloblastomas are not yet a reality. Furthermore, these authors (31) do not consider any immunohistochemical variable to be a weighty prognostic indicator, which, in addition to clinical factors, is useful for the further definition of medulloblastoma risk groups and for choice of optimal medulloblastoma treatment. Perhaps *MYC* and tyrosine kinase (Trk)C mRNA levels could be more useful for clinical purposes, but to introduce these biomolecular markers in clinical protocols, its distinct prognostic significance needs to be proved by prospective studies (31). Reviews on the clinical, histologic, and epidemiologic aspects of medulloblastoma besides those already referred to may be found in a number of publications (32–37).

Cytogenetic Findings in Medulloblastoma

In the following sections, results obtained with individual techniques, e.g., cytogenetics, comparative genomic hybridization (CGH), fluorescence in situ hybridization (FISH), and loss of heterozygosity (LOH), are presented; however, a more informative and panoramic view of the genetic changes in medulloblastoma should be based on

TABLE 8.3. IMHC features of medulloblastomas.

Positivity for synaptophysin is prominent in tumor nodules and in the centers of neuroblastic rosettes.
Nestin expression is not unique for medulloblastoma.
Vimentin is strongly reactive in classic medulloblastomas.
GFAP-positive cells are present in classic and desmoplastic medulloblastomas; such cells are not present in large-cell medulloblastoma.
Photoreceptor proteins (rhodopsin and retinal S-antigen) may be focally expressed in medulloblastoma.
Neural cell adhesion molecules (N-CAM) expression may be seen in medulloblastoma, but is not specific, for such expression may be seen in other tumor types.
High-affinity NGFR (nerve growth factor receptor), also known as Trk (A–C), is comomonly expressed in medulloblastoma.
Neuron-specific enolase and neurofilament protein are markers of differentiation in medulloblastoma (741)
Apoptosis, based on histologic criteria of involved nuclei, is present in classic and desmoplastic medulloblastoma, though the distribution of apoptotic cells is uneven and tends to occur in areas with a high mitotic index and the presence Ki-67–positive cells.

Based on Giangaspero et al. (4), where further details and references dealing with the IMHC features may be found.

a composite of the findings obtained with all techniques. Although each of the methodologies used for studying the genetic aspects of medulloblastoma tends to cover some of the same ground, each affords nevertheless different or additional views and details sufficient to warrant separate sections within this chapter.

A detailed chromosome analysis of medulloblastomas was first reported in 1970 (38) in which four tumors were examined, including one with a metastasis. For two of these tumors, prebanding karyotypes were presented. The modal chromosome numbers for the four tumors were 45, 60, 46, and 83, respectively. Three of the tumors had cell populations in the triploid-tetraploid range, and two tumors had distinct marker chromosomes.

Subsequent studies have demonstrated that the most common aberration detected by chromosomal banding was loss of 17p, which has been detected in 30–50% of the cases (39–45). This was substantiated in a large interphase FISH study in which deletions of chromosomal arm 17p were seen in 44% and an i(17q) in 30% of the cases (46). In another study based on CGH, gain of only 17q or only loss of 17p in separate groups of patients was observed (47), suggesting that these two abnormalities may involve different tumorigenic pathways. In a small number of medulloblastomas, partial or complete loss of 17p may occur through several mechanisms, e.g., variable interstitial deletions (48), monosomy 17 (−17), or an unbalanced translocation (49).

Abnormal karyotypes of all published cases of medulloblastoma, PNET, and pineal tumors are shown in Table 8.4. Non-clonal changes are not included in Table 8.4.

The published cytogenetic changes in >160 medulloblastomas shown in Table 8.4 by and large reveal that most of these tumors have many chromosomal alterations, and some were associated with very complicated karyotypes. The chromosome number in medulloblastomas ranged from hypodiploidy (<46 chromosomes) to >150 chromosomes (Figures 8.3–8.6). Approximately one third of the tumors had a pseudodiploid (46 chromosomes but with an abnormal karyotype), and the remaining two thirds were equally distributed in the hypodiploid (<46 chromosomes), hyperdiploid (from 47 to >60 chromosomes) and triploid-tetraploid (69–92 or more chromosomes) ranges.

In keeping with the heterogeneity of the histological picture in medulloblastoma, it is possible that cytogenetically defined groups represent subtypes of medulloblastoma at the genetic and consequently at the histologic and clinical level, and although some studies point in that direction, this hypothesis remains to be more rigorously determined in appropriate studies.

The most common specific abnormality in medulloblastomas, which is present in approximately one third to one half of reported cases, is an isochromosome 17q [i(17q)] (39, 40), with the breakpoint in the proximal portion of the p-arm, so that the resultant structure is dicentric. In Table 8.4, containing the karyotypes of published medulloblastomas,

one third had an i(17q). In about 5% of these tumors, an i(17q) was the sole cytogenetic change (Figure 8.3). This is a unique phenomenon, because an i(17q) generally occurs as a secondary change in leukemias and solid tumors (50, 51). This observation has been confirmed by restriction fragment length polymorphism (RFLP) and microsatellite analyses.

In a review of isochromosomes (51) in human neoplasia, 6% of "malignant neurogenic neoplasms" were found to have an i(17q), most of these occurring in patients <15 years old (20 of 24) and having medulloblastomas or PNET. In five of 24 cases, an i(17q) was the only anomaly, as reported previously (41, 52). Although other karyotypic changes were present, both structural and numerical, rearrangements of chromosome 17 were found in 25% of medulloblastomas, with most of these tumors showing the net result of loss of 17p rather than gain of 17q (49). The authors of a review on i(17q) (51) mention that only two of 37 tumors had been described to have TP53 rearrangements (49, 53, 54).

Changes of chromosome 1, often translocations and deletions, were present in 20% of medulloblastomas (39, 40, 44), and next to chromosome 17, was the most commonly involved chromosome in medulloblastoma. However, the types of abnormalities were variable, including unbalanced translocations, deletions, and duplications (39, 40). Although there is no consistency with regard to specific breakpoints or type of rearrangements of chromosome 1, these alterations often result in an extra copy of all or part of 1q through an unbalanced translocation, gain, of a deleted chromosome (55).

Trisomy 7 (+7) was present in 15% of medulloblastomas, and except for one case in which it was the only change, +7 was part of complex karyotypes and often accompanied the i(17q) (40, 41, 47, 55, 56). In addition to +7, four tumors had loss of chromosome 7 (−7) and nine displayed structural anomalies of this chromosome. The potential oncogenic contribution of +7 remains unclear inasmuch as trisomy 7 also has been detected in normal tissue adjacent to neoplasms lacking this change (57).

Examination of the changes of chromosome 10 in medulloblastoma revealed more than 10–12% of these tumors to have a deletion or loss of this chromosome (−10). Only one tumor had +10 in addition to other abnormalities in the karyotype. Structural changes (translocations or inversions) of chromosome 10 were seen in a small number of cases. No medulloblastoma with an anomaly of chromosome 10 as the sole karyotypic change has been reported to date.

Nearly 10% of the medulloblastomas had monosomy 22 (−22) as the sole karyotypic change (Table 8.4), a finding that may indicate that a subgroup exists within medulloblastomas whose genetic background is different from those with i(17q). In some studies as much as 25% of medulloblastomas showed −22 as the sole cytogenetic abnormality (55, 56, 58). In addition to those cases with −22 as the sole anomaly, >10% of the medulloblastomas had −22 or loss of 22q as part of more complicated karyotypes (59, 60), again pointing to a possible

TABLE 8.4. Chromosome changes in medulloblastoma, PNET, and pineoblastoma.

Karyotype	References
Medulloblastoma	
47,XY,8q+,+11,17p+,20q+	*Friedman et al 1985 (269)*
47,X,-Y,+2mar/52,X,-Y,inc	*Latimer et al 1987 (634)**
47,XY,+11,i(17q),der(8)t(8;?)(q13;?),der(20)t(5;20)(q11;q13)	*Bigner et al 1988 (39)*
46,XX,der(20)t(?13;20)(?q13;q13)	*Bigner et al 1988 (39)*
49,XY,+del(1)(p13),+6,+8,i(17)(q10),+18,-22,dmin	*Bigner et al 1988 (39)*
87-XXYY,del(1)(p22),+del(1)(q24),-2,-8,-8,-9,del(11)(q22),-13,-14,i(17)(q10)x2	*Bigner et al 1988 (39)*
49,XY,+,6+8,-22,+del(1)(p13),+i(17q),+dmin	*Friedman et al 1988 (113)*
45,XY,-6/45,idem,del(3)(q?)	*Griffin et al 1988 (40)*
46,XY,add(1)(q44)	*Griffin et al 1988 (40)*
46,XX,?del(6)(q?)	*Griffin et al 1988 (40)*
47,XY,+mar	*Griffin et al 1988 (40)*
78-80,XXY,+Y,add(1)(p3?), der(4)t(1;4)(q21;q35), add(5)(p13), inv(7)(p13q11), del(9)(p13p21), del(11)(p13),i(17)(q10), inc	*Griffin et al 1988 (40)*
46,XY,dmin	*Pearson et al 1988 (635)**
45,XY,der(5)t(5;11)(q35;q11),add(8)(p23),-11,i(17)(q10)	*Biegel et al 1989 (41)*
45,XX,-6	*Biegel et al 1989 (41)*
45,X,-X,add(5)(q35),del(11)(q14),add(22)(q13)	*Biegel et al 1989 (41)*
45,X,-X,i(17)(q10)	*Biegel et al 1989 (41)*
45,X,-Y,i(17)(q10)	*Biegel et al 1989 (41)*
46,XY,del(1)(p21), +del(1)(q31), der(3)add(3)(p?)add(3)(q?),add(4)(q?), add(4)(q35), add(5)(q35),-6,-6,-12,-12, add(16)(q12-q21), add(17)(p13),+18,-19, del(22)(q11),+4mar	*Biegel et al 1989 (41)*
46,XY,del(9)(q22),dmin	*Biegel et al 1989 (41)*
46,XY,del(16)(q11)	*Biegel et al 1989 (41)*
46,XY,der(10)t(4;10)(q23-q25;q23-24),dmin	*Biegel et al 1989 (41)*
46,X,-X,+i(17)(q10)	*Biegel et al 1989 (41)*
46,XY,t(16;20)(q13-q22;q13), i(17)(q10)/46, idem, del(2)(q2?)	*Biegel et al 1989 (41)*
53,XY,+4,+4,+5,+7,+7,+12,+13,i(17)(q10),dmin	*Biegel et al 1989 (41)*
69,XXX,add(11)(q23),add(16)(q11-q12),inc	*Biegel et al 1989 (41)*
83-89,XXYY,-3,-4,+7,+der(7)t(7;11)(p11;q11),-8,-10,-11,-11,-13,del(16)(q11),i(17)(q10)x2, -20,+mar,dmin	*Biegel et al 1989 (41)*
84,XX,-X,-X,-3,-8,-10,-13,-15,i(17)(q10),-20,add(21)(q22)	*Biegel et al 1989 (41)*
89,XXX,-X,+del(1)(p13p32-p35)x2,-4,-4,-5,+7,+8,+8,add(11)(p15),-14,-14,-16,-17,- 17,-18,-18,add(19)(q13)x2,+3mar	*Biegel et al 1989 (41)*
91-101,XXYY,+13,+17	*Biegel et al 1989 (41)*
49,XX,+6,+7,-8,+der(11)t(8;11)(q11;p11),+i(17)(q10)/50,idem,+18	*Callen et al 1989 (636)*
61-66,XY,-X,-3,+add(5)(q35),+del(6)(q21q25),+der(6)(6;?8)(q13;q13),-7,-8,-8,+9,+der(11) t(8;11)(q11;p11),-13,-14,+15,+add(16)(p?),+i(17)(q10),+18,+19,-21,+mar	*Callen et al 1989 (636)*
46,XY,t(1;15)(p36;q11), der(4)/46,XY, t(2;12)(q21;q23), t(3;9)(q27;q22), del(4)(p14),t(7;19) (p11;p11),t(13;15)(q32;q22)	*Chadduck et al 1991 (550)*
46,XY,t(3;12)(p21;q13), t(3;17)(p13;p13), t(5;8;10)(q34;q24;q24)/46,XY,t(1;1)(q32;p15), del(4)(p?), add(9)(q?), der(11)	*Chadduck et al 1991 (550)*
46,XY, del(9)(q12), del(11)(q22)	*Lòpez-Ginés et al 1991 (637)*
48,XY,+8,+8,der(14)t(1;14)(q11;p11),i(17)(q10)/51, idem,+5,+6,+20/51, idem,+5,+6,+20, dmin	*Sawyer et al 1991 (109)*
46,XX,der(6)t(1;6)(q21;q13)	*Stratton et al 1991 (638)*
46,XX,del(7)(q32)	*Karnes et al 1992 (639)*
46,XX,del(7)(q33)	*Karnes et al 1992 (639)*
46,XY,-11,+mar	*Karnes et al 1992 (639)*
48,X,-Y, der(1)t(1;11)(q31;q21), dup(2)(q11q31), del(6)(q12),-8,-9,der(11)t(1;11)(q32;q12), add(12)(q21),+14, der(17)i(17)(q10)t(11;17)(q13;q22),+21,+3mar/90,XX,-Y,-Y,-6,+11,del(11)(q11)x2, +der(11)t(11;12)(p11;p13),+12,+t(16;17)(q?;q?)x2,i(17)(q10)x2,-18,-18, -19,-19	*Karnes et al 1992 (639)*
92,XXYY,+4,+5,+7,+7,-8,-8,-10,+11,-14,-14,-16,i(17)(q10),+18,-19,+21	*Karnes et al 1992 (639)*
45,XX,-22	*Vagner-Capodano et al 1992 (52)*
45,XY,-22 (3 cases)	*Vagner-Capodano et al 1992 (52)*
45,XY,-17	*Vagner-Capodano et al 1992 (52)*
46,XY,add(10)(q?),der(11),-17,+mar	*Vagner-Capodano et al 1992 (52)*
46,XY,i(17)(q10)	*Vagner-Capodano et al 1992 (52)*
46,XX,t(1;8)(q25;q22),add(3)(q29),del(9)(p12),dup(12)(q12q23),-17,+22	*Vagner-Capodano et al 1992 (52)*
46,XY,t(1;14)(p22;q31)	*Vagner-Capodano et al 1992 (52)*
46,XX,inv(2)(p13q37),del(13)(q14)	*Dressler et al 1993 (68)*
45,XX,t(9;19)(q22;q13),-22	*Neumann et al 1993 (43)*
45,XY,-6/45,idem,?del(17)(p?)	*Neumann et al 1993 (43)*
46,Y,del(X)(p?)	*Neumann et al 1993 (43)*
46,XY,t(12;13)(p13;p11)/46,XY,del(1)(q11)	*Neumann et al 1993 (43)*

Table 8.4. Continued

Karyotype	References
46,XY,-11,+mar	*Neumann et al 1993 (43)*
50,Y,der(X)t(X;1)(p22;q21),+5,+6,+8,+8,i(17)(q10)	*Neumann et al 1993 (43)*
49,XX,+5,+6,+8, add(11)(q?)x2,i(17)(q10), add(22)(p?)/50,XX,+5,+6,+8, add(11),+13,i(17)(q10), add(22)	*Stuart et al 1993 (640)*
45,XY,-22	*Fort et al 1994 (172)*
44-46,XY,+16,-18	*Fujii et al 1994 (44)*
44-46,XX,+20	*Fujii et al 1994 (44)*
45,X,-X,i(17)(q10)/45,idem,+2,-8/89-94,XX,-X,-X,i(17)(q10)x2	*Fujii et al 1994 (44)*
51-54,XY,+4,+5,+10,+11,+14,+15,i(17)(q10),+18,3-5dmin/102-108,XXYY,i(17)(q10)x2,5-10dmin	*Fujii et al 1994 (44)*
54-60,XX,+1,+2,+4,+5,+10,+11,+12,+14,+15,+17,+18,+21,+22	*Fujii et al 1994 (44)*
62,XY,-X,-3,-5,-6,-8,-9,-10/115,XXYY,inc	*Fujii et al 1994 (44)*
78-91,XXY,-Y,-1,-6,-8,-10,-14,-15,-16,i(17)(q10)x2	*Fujii et al 1994 (44)*
90-93,XXY,-Y,-3,-10,-15,+18,+21	*Fujii et al 1994 (44)*
148-153,X?,inc	*Fujii et al 1994 (44)*
68-78,XXY,+Y,+Y,-2,-3,-4,+add(5)(q11),-8,-9,-10,-11,add(12)(p13),t(?15;16)(q13;p13), +add(16)(q13),+17,add(17)(p11)x2,add(18)(q12),+20,-22,+5-15mar,dmin	*Tomlinson et al 1994 (108)*
43-45,XX,-22	*Vagner-Capodano et al 1994 (93)*
44-50,XY,+21	*Vagner-Capodano et al 1994 (93)*
45,XX,add(9)(p24),-10	*Vagner-Capodano et al 1994 (93)*
45,XX,-22 (2 cases)	*Vagner-Capodano et al 1994 (93)*
45,X,-X,i(17)(q10)	*Vagner-Capodano et al 1994 (93)*
46,XY,add(10)(q26),-17,+mar	*Vagner-Capodano et al 1994 (93)*
46,XX,i(17)(q10)	*Vagner-Capodano et al 1994 (93)*
46,XY,-22	*Vagner-Capodano et al 1994 (93)*
48-54,XX,+1,+11	*Vagner-Capodano et al 1994 (93)*
49,XX,+7,+11,+12	*Vagner-Capodano et al 1994 (93)*
49-52,XY,+1,i(17)(q10)	*Vagner-Capodano et al 1994 (93)*
36-91,XX, del(1)(p36), add(2)(q33),t(3;6)(p21;q12),add(7)(q22), add(8)(q24),der(12;16)(q10;p10)	*Agmanolis and Malone 1995 (554)*
40-88,X,-X,-X,-X,t(2;15)(q37;q15),add(5)(p15),t(5;6)(q13;q21),add(11)(p15)x2,add(14)(p11), add(21)(p11),+3-6mar	*Agmanolis and Malone 1995 (554)*
80-91,XX,-X,-X,-3,+7,-8,-13,-13,i(17)(q10) x2,+18	*Agmanolis and Malone 1995 (554)*
45,XY,-8,i(17)(q10)	*Biegel et al 1995 (46)*
46,XY,add(11)(p11),i(17)(q10)/47,idem,+7	*Biegel et al 1995 (46)*
46,XY,del(3)(q?),del(6)(q?)	*Biegel et al 1995 (46)*
46,XX,del(5)(q3?),add(6)(p22),del(6)(q?)	*Biegel et al 1995 (46)*
46,XY,i(17)(q10),dmin	*Biegel et al 1995 (46)*
46,XY,t(1;6)(p21;q11-q13)	*Biegel et al 1995 (46)*
47,XX,+mar	*Biegel et al 1995 (46)*
45-46,XX,?del(17)(p?),inc	
47,XY,+19	*Biegel et al 1995 (46)*
70-73,XX,-X,+3,-4,+6,+7,+7,+7,+9,-14,+19,-20,-22	*Biegel et al 1995 (46)*
74-90,XXYY,-3,+7,+7,-8,-11,-14,-14,-15,-16,i(17)(q10)x2,+2mar	*Biegel et al 1995 (46)*
82-92,XXY?Y,-2,+7,-8,-10,-11,-13,i(17)(q10)x2,inc	*Biegel et al 1995 (46)*
90,XXX,-X,i(17)(q10)x2,-21	*Biegel et al 1995 (46)*
93-96,XXYY,-3,-3,+7,-8,-8,i(11)(q10)x2,+13,-17,i(17)(q10),inc	*Biegel et al 1995 (46)*
45,XY,add(1)(p32),del(6)(q21),t(7;13)(q11;q34),add(9)(q?),del(11)(q23),-16,der(20)t(1;20) (p13;q13)	*Sainati et al 1996 (641)*
45,X,-Y,i(17)(q10)	*Sainati et al 1996 (641)*
46,XY,add(8)(q?),i(17)(q10)/46,idem,dmin	*Sainati et al 1996 (641)*
46,XY,add(1)(p?)/46,XY,der(3)t(1;3)(q23;p25)	*Sainati et al 1996 (641)*
46,XX,t(17;17)(p?;p?)	*Sainati et al 1996 (641)*
48,XY,+add(1)(p13),add(4)(q?),+8,+10,dup(11)(q13q23),+12,-16,-20	*Sainati et al 1996 (641)*
84-88,XXYY,i(17)(q10)x2,inc	*Sainati et al 1996 (641)*
44,X,-X,i(17)(q10), dmin/90, idemx2/50-95,idemx2,+5,+6,+7/106, idemx2,+X,+X,+5,+5,+6,+6,+7,+7,+7,+8,+9,+10,+12,+13,+19,+20	*Bhattacharjee et al 1997 (56)*
45,XY,-8,del(11)(q21q23),add(18)(p11)/47,idem,+7,+8,add(20)(q13)	*Bhattacharjee et al 1997 (56)*
45,XY,-8,i(17)(q10)/67-93,idem,x2,-3,+7,+7,-11	*Bhattacharjee et al 1997 (56)*
45,XX,-22 (2 cases)	*Bhattacharjee et al 1997 (56)*
45,XX,-22/44,idem,-10	*Bhattacharjee et al 1997 (56)*
45-52,XY,add(3)(q13),+5,+6,+7, add(14)(q32),i(17)(q10),+18,+19, +mar	*Bhattacharjee et al 1997 (56)*
46,XY,add(8)(p11)	*Bhattacharjee et al 1997 (56)*
46,XY,inv(3)(p21q27),del(10)(q22q26)	*Bhattacharjee et al 1997 (56)*
46,XY,+7,i(17)(q10),+mar	*Bhattacharjee et al 1997 (56)*
46,XX,-22,+mar/46,idem,i(7)(q10)	*Bhattacharjee et al 1997 (56)*
46-92,XXYY, del(9)(p22), add(13)(q?34)x2, add(16)(q?13)x2,i(17)(q10)x2	*Bhattacharjee et al 1997 (56)*

TABLE 8.4. Continued

Karyotype	References
57-129,XXYY,i(17)(q10),+mar,inc	*Bhattacharjee et al 1997* (56)
60,XX,-?X,-2,-3,+5,-8,-11,-12,-15,-20,-21,-22/114-116,idemx2	*Bhattacharjee et al 1997* (56)
63-98,XY,-X,i(1)(q10),inc	*Bhattacharjee et al 1997* (56)
66,XY,-X,+Y,-2,-3,+5,-6,+9,-10,+11,+12,-16,-18,+20,-21,-22	*Bhattacharjee et al 1997* (56)
67,XXY,t(1;3)(p13;p13),-2,+3,+5,+7,-8,-10,-11,-13,der(16)t(1;16)(q21;q24),+17,+18,-21,-22	*Bhattacharjee et al 1997* (56)
67-84,XXY,+1,add(2)(q21),+3,+4,+6,+7,+8,+9,+11,+12-13,-14,-15,-16,-17,add(19)(q13),+21,+7-13 mar	*Bhattacharjee et al 1997* (56)
85-96,XXYY,+7,+8,i(17)(q10),+18	*Bhattacharjee et al 1997* (56)
46,XY,i(17)(q10)	*Biegel and Wentz 1997* (100)
51,XX,+6,+8,+8,+13,i(17)(q10),+19	*Biegel and Wentz 1997* (100)
44,X,-Y,+del(1)(p13),-16,i(17)(q10),-21,dmin/45-46,idem,+6,+8,+18,+mar	*Bigner et al 1997* (55)
45,XY,+del(1)(p22p34),del(3)(q12),-22,-22/46,XY,+der(1)t(1;19)(q11;q13),+3,del(3)x2,+9,-18,-19,-22,-22,+mar	*Bigner et al 1997* (55)
45,XY,-8,i(17)(q10)	*Bigner et al 1997* (55)
45,XX,-22 (2 cases)	*Bigner et al 1997* (55)
46,XX,der(13)t(1;13)(p11;q34),i(17)(q10),dmin/50,idem,+3,+6,+18,+21	*Bigner et al 1997* (55)
46,XY,t(1;4)(q31;q35)	*Bigner et al 1997* (55)
47,XY,+7	*Bigner et al 1997* (55)
53,XX,der(1)t(1;1)(p36;q21),+6,+7,+17,+18,+18,+19,+21	*Bigner et al 1997* (55)
56,XX,+der(1)t(1;1)(p36;q25), add(4)(q35),+6,+7,+8,+13, add(16)(q24),+18,+20,+20, +2mar	*Bigner et al 1997* (55)
79,X,-X,-Y,+1,+1,+2,+2,+3,-6,+7,+7,-10,+11,+12,-13,-13,-15,add(15)(p11)x2,-16,-17,-18,+20,+7-8mar,dmin,inc	*Bigner et al 1997* (55)
87,XXYY,add(1)(p36)x2,i(17)(q10)x2,dmin,inc	*Bigner et al 1997* (55)
94-96,XX,-X,-X,del(3)(p13p25),i(17)(q10)x2,inc	*Bigner et al 1997* (55)
91,XXXX,del(9)(q13),-12,add(12)(q24),-14,der(15)t(3;15)(q11;p11),-17,i(17)(q10),+2mar	*Lpez-Gins et al 1997* (642)
46,XX,add(16)(q24),der(19)t(1;19)(q23;q13)/65-68,XX,-X,+der(19)t(1;19)x2	*Vagner-Capodano et al 1997* (643)
39-49,X,-X(or -Y),add(6)(p22 or p23),add(8)(q23 or q24),add(16)(q10),i(17)(q10) add(20)(q13)	*Bruggers et al 1998* (185)
45,XY,del(10)(q22), der(13)t(1;13)(q11;p11), der(16)t(7;16)(q21;q24), idic(17)(p11),-21	*Scheurlen et al 1999* (643A)
43-46,X,der(Y)t(Y;15)(q?;q?),-5,+8,+11,-12,-15,-21	Bayani et al 2000 (85)
45,XX,der(1)t(1;7)(p36;?),der(15)t(15;21),-21,-22,+der(?)t(?;1;7)/42,XX,+der(4),-6,	*Bayani et al 2000* (85)
-11,der(12)t(12;19)(p?;q?),-15,-19,-20/46,XX,der(8)t(8;21),der(15)t(15;21),der(20)t(13;20),dmin	
45,X,der(3)(p?),der(7)(q?),-10,t(17;18)(p11;q11),-22,+mar,60dmin,inc	*Bayani et al 2000* (85)
45-46,XY,der(1),-22,?der(22)	*Bayani et al 2000* (85)
46,XY,add(14)(q24)/47,idem,+7	*Bayani et al 2000* (85)
46,XY,inv(7)(p11q34),inv(10)(q12q22), add(18)(p11),-20, +mar/46,XY,del(8)(p21), +mar/46,XY, del(5)(p13),	*Bayani et al 2000* (85)
t(11;13)(p13;q14),t(18;22)(q23;q11)/46,XY,t(1;3)(q32;q27),inv(10)(p12q25),add(16)(p?)/46,XY,del(3)(q?)x2,	
t(6;14)(q27;q11),-18,+2mar	
46,XY,-3,+der(5)t(5;7), der(10)t(10;14)(q2?3;q3?1), del(11)(p?)/41,XY, der(3)t(3;7)(q11;q11),-8, der(10)t(10;14),	*Bayani et al 2000* (85)
-11,-17,-21,-22/45,XY,der(3)t(3;7),+i(7)(q10),del(11),-22,inc/46,XY,-3,+der(5)x2,-10,del(11)(p?)	
46,XX,-6	*Bayani et al 2000* (85)
46,XX,-10,+r	*Bayani et al 2000* (85)
47,XY,add(5)(q?),?del(6)(q?),-9,-10,+13,+14,?i(17)(q10), ?add(22)(q?), +mar	*Bayani et al 2000* (85)
92,X?,100-200dmin,inc	*Bayani et al 2000* (85)
95,XXYY,+6,+8,-13,-14,-15, del(17)(p11),+20,+21,+2mar	*Bayani et al 2000* (85)
99-105,XXX,-X,+1,-3,-4,+7,-8,-10,-11,+12,-13,+14,+15,-16,+17,+18,+18,+19,+20,-21,-22	*Bayani et al 2000* (85)
45,XY,-7,add(8)(q24.1), add(16)(q13), i(17q)(q10), der(20)t(1;20)(q12;q13.3)	*Gilhuis et al 2000* (117)
46,XX,del(10)(q22q24)	*Gilhuis et al 2000* (117)
80-90,XXXX,-1,-9,-12,-14,-16,del(17)(p11),-18,+mar	*Gilhuis et al 2000* (117)
46,XY,add(8)(p?),-10,add(11)(p13),-15,del(15)(q12),?del(17)(p?),+2mar/92,idemx2	*Roberts et al 2001* (593)
46,XY,?add(17)(q?)	*Roberts et al 2001* (593)
46,XY,der(2)add(2)(p24)add(2)(q33),del(15)(q1?q21)	*Roberts et al 2001* (593)
46-47,XX,+2,+mar/89-92,XX,-X,-X,+add(1)(q?),+2,+add(2)(p?),-6,-6,-7,-7,+8,-10,-14,-14,-16,-17,-17,	*Roberts et al 2001* (593)
-17,-20,+4mar	
80-86,XYY,add(X)(p22),i(3)(q10)x2,der(7)t(7;11)(q22;q13),i(9)(p10),-11,add(11)(q23),-14,-14,-16,-17,20,-22	*Roberts et al 2001* (593)
93,XX,-Y,Y,+X,-2,-3,-5,-8,-8,+9,+9,+9,-10,-10,-11,-11,-13,-13,-	*Roberts et al 2001* (593)
15,+17,+add(17)(p11)x2,i(17)(q10)x2,+18,+18,+18,+18,+20,+20,+20,-21,+22,+2mar	
46,XY,ins(1;10)(q31;q23q26)	*DeChiara et al 2002* (644)
47,XY,+4,i(17)(q10)/92-93, idemx2/88-92, idemx2,der(8)t(8;17)(p21;q2?3)x2/48, idem,+7, der(8)t(8;17)/48,	*Cohen et al 2004* (551)
idem,+7, der(8)t(8;17)der(19)t(12;19)(p12;q13)/47,idem,dup(7)(q?),der(19)t(12;19)/92-93,	
idemx2,dup(7)(q?)x2,der(19)t(12;19)x2/49,idem,der(8)t(8;17),der(19)t(12;19),+r	
86-90,XX,-X,-X,der(11)t(11;14)(p11;q11)x2,-12,der(13)t(X;13)(?;p11)x2,-14,-14,der(16) t(X;16)(p11;q22)x2,	*Cohen et al 2004* (551)
add(19)(q13)x2, der(22)t(2;22)(?;q13)x2,+r/88, idem,10-200dmin/88, idem,t(10;16)	

TABLE 8.4. Continued

Karyotype	References
45-46,XY7add(1)(p31.l),t(2;8)(p21;q24.1),add(3)(q25),t(9;15)(q22;q13),add(12)(p11.2), +1 -2mar,inc **(lingual metastasis)**	*Donner et al 2005 (746)*
40-46,XY,add(1)(p36.l),add(2)(q33),t(2;8)(p21;q24.1),-3,t(3;5)(q21;p15.1), add(4)(p14),-5,de1(5) (q 15q33),de1(6)(p2l.l),add(8)(q13),-9,add(9)(p22),t(9;15)(q22;q13), add(10)(q22),add(11)(p13),-12,add(12) (p11.2),de1(12)(q15),-14,add(l7)(q25),add(18)(q21), add(20)(q13. 1),-22,+1-5mar.inc **(vertebral metastasis)**	
36-47,X,-X,+der(1)t(1;14)(p22;q11),del(2)(q23),ins(3;?)(p21;?),-4,del(5)(q22),del(6)(p21), der(7)t(7;15)(q22;q15), add(8)(p23), add(11)(q23), add(12)(q24),-13,-14, der(16)t(8;16)(p11;p13),-17, der(20)t(2;20)(q21;q13), add(22)(q13),+r,inc	*Glanz et al 2007 (807)*
45,XX,-22 (2 cases)	*Glanz et al 2007 (807)*
46,XY,der(1;8)(q10;q10),i(7)(q10),add(17)(q21),der(18)t(1;18)(p13;q21)	*Glanz et al 2007 (807)*
88,XXYY,+1,+1,-3,-3,-11,-11,-17,-17	*Glanz et al2007 (807)*
PNET	
46-47,XY,+3,del(3)(q13)x2,+del(5)(p11)x2,-10,del(11)(p13),?i(13)(q10),-18	*Bigner et al 1988 (39)*
85-87,XXYY,del(1)(p21)x2, del(1)(q11)x2, add(2)(q37), del(3)(p11), ?i(3)(p10), i(17)(q10)x2, dmin,inc	*Bigner et al 1988 (39)*
45,XY,-22	*Chadduck et al 1991 (550)*
43-44,XY,+2,-6,der(10)t(10;11)(q26;q21),-12,-13,+mar/84-90,XXYY,der(10)t(10;11)(q26;q21)der(10) t(10;11)(q26;q21),del(11)(q21),del(11)(q21),+2mar	*Fujii et al 1994 (44)*
46,XY,+i(1)(q10),-9,t(9;11)(q34;q13)	*Bhattacharjee et al 1997 (556)*
46,XY,t(6;19)(q21;q13),del(10)(q22)/45,idem,t(11;13)(q15;q11),-13	*Bigner et al 1997 (55)*
49,XX,add(3)(q23 or q24),+5,+8, dup(11)(q12q22.3 or q13q23), del(16)(q22q24), add(19)(p13), +21	*Burnett et al 1997 (99)*
90,XX,add(X)(p22)x2,dic(1;9)(q42;p21)x2,dic(4;9)(q3?5;p2?2),-6,add(6)(p24),-9,add(11)(p15)x2,- 15x2,add(16)(q2?2),+mar1,+mar2	*Burnett et al 1997 (96)*
46,X,?der(X),?der(10)(p?), ?der(14)(q?), add(19)(q?), ?add(22)(q?), del(22)(q?)	*Bayani et al 2000 (85)*
46,XY,t(6;13)(q25;q14)	*Bayani et al 2000 (85)*
55-75,XX,-X,del(1)(p22),i(4)(p10),-5,+6,+add(7)(q36),add(9)(p21),-11,-13,-17,add(18)(q23),-19,- 19,+13mar,dmin	*Bayani et al 2000 (85)*
70-103,X?,+2r,dmin,inc	*Bayani et al 2000 (85)*
46,XX,del(2)(p22p23),del(5)(q33q35)/46,idem,del(17)(q21q21)	*Roberts et al 2001 (593)*
56-59,**Xc**,+X,+1,+add(1)(p?), +add(1)(q?),+del(2)(p24),+7, +add(8)(q?), der(9;15)(q10;q10)x2,+11,+13,+18,+20,+20	*Roberts et al 2001 (593)*
69-75,XX,-X,add(1)(q42)x2,-4,-4,add(4)(q3?),-10,-11,del(11)(q2?),-13,-16,-18,+7-13mar,dmin	*Roberts et al 2001 (593)*
52,XX,+1,+1,add(3)(q25),+7,+7,add(11)(q25)x2,+21,+21	*Uematsu et al 2002 (645)*
49,XX,+9,der(11)del(11)(q?)t(1;11)(p?;q?),+13,+der(18)t(11;18)(q?;q?)	*Batanian et al 2003 (552)*
39-42,X,-X,-1,ins(1)(q21q42q12), t(1;8)(q1?;q2?), der(2)t(2;21)(q37;q10), der(3)t(3;13)(q27;?), der(4)t(4;21)(q1?;q10),der(5)t(5;21)(q?;?),der(6)t(2;6)(?;q26),der(7)t(7;13;4)(q22;?;q2?), der(11)t(11;14)(p13;q11),-13,-14,-21,-21/39-42,idem,der(2)t(2;15)(p11;q?),-6/39-42,X,-X,del(1)(q12),t(1;8), der(1)t(1;14)(q12;?),der(2)t(2;16),der(3)t(3;13)(q27;?),der(4)t(4;21)(q1?;q10),der(5)t(5;21)(q?;?),der(6)t(2;6)(?;q26), der(7)t(7;13;4),der(9)t(2;9)(?;q34),der(11)t(11;14),-13,-14,-14,-16,-19,-21,-21 39-42,idem,der(2)t(2;21)(q37;q10),-der(2)t(2;16),-11	*Cohen et al 2004 (551)*
Pineoblastoma	
42-46,XX,add(1)(p36), del(1)(p13p21), inc	*Griffin et al 1988 (40)*
45-47,XY,add(1)(q44),Dq+,inc	*Grififn et al 1988 (40)*
46,XY,del(11)(q13.1q13.5)	*Sreekantaiah et al 1989 (646)*
45,XY,-22	*Fort et al 1994 (172)*
Near-diploid,i(17q)	*Kees et al 1994 (594)**
72-86,add(Xp)x2,-3,-9,-10,-12,-13,add(13q),-17,i(17q),dmin	*Kees et al 1994 (594)**
46,XX,+14,-22/46,idem,-20,+mar	*Bigner et al 1997 (55)*
46,XY,der(10)t(10;17)(q21;q22-q23),der(16)t(1;16)(q12;q11.2)	*Kees et al 1998 (595)*
45,XX,-22	*Roberts et al 2001 (593)*
63-82,XXX,+X,+2,+3,+4,+5,+7,+9,+11,+12,-14,+15,+16,+17,+18,+20,+21,+22	*Roberts et al 2001 (593)*
*42,XX,+1, dup(1)(q11q25)x2,-8, der(11)t(11;17)(p11.2;q11), der(13)t(13;?;17)(p11;?;q11),-16,-17,-18,-19,-20,-21, +2mar/42, idem, add(2)(p21)	*Brown et al 2001 (744)*
*54~56,XX,t(1;16)(p13;q13),+add(1)(p21)x2,-15,+19,+19,del(19)(q13),+21,+21,+2-4mar	*Brown et al 2001 (744)*
*80-83<4n>,XXXX,-1,-2,-3,-4,-6,-9,add(12)(p13),-13,-16,-20,-21,+mar	*Brown et al 2006 (744)*
*93-99,XXXX,+9,-13,+14,i(17)(q10),+19,+19,+20,+0-5mar/93 99,idem,der(1)t(1;11)(p36;q14),-6,+12	*Brown et al 2006 (744)*

*See text.

gain of chromosome 17 (47). The loss of chromosome 11 was found to be more frequent in patients who had localized disease (47).

As with other malignant tumors, examination of the cytogenetic changes in medulloblastomas reveals varying numbers of normal karyotypes in some tumors, particularly after short-term culture, and the meaning of the normal karyotypes has been a subject of controversy, i.e., whether they belong to the tumor or are of normal cell origin with growth advantages over the tumor cells in short-term cultures.

An example is the study (69) of nine medulloblastomas with normal karyotypes after culture of tumor samples for 4–9 days. Centromeric FISH probes for chromosomes 3 and 17 were applied to touch preparations of frozen tumor tissues and revealed seven of nine of the medulloblastomas to contain significant numbers of cells with three to four signals (indicative of triploidy or tetraploidy), although in four of the tumors the majority of the cells had 2 normal signals (diploidy) for chromosome 3 or 17 (two tumors showed two signals for both chromosomes). DNA studies on paraffin-embedded tumor showed six of the tumors to contain 3-4n values. The results are somewhat baffling, particularly in the tumor with a 2n DNA content and yet containing 70% of cells with three signals for chromosome 3 and 18% for chromosome 17, and tumors with preponderantly disomic signals for chromosomes 3 and 17 and with 3-4n DNA content. It would seem that in the nine tumors with normal karyotypes, the normal diploid cells had a growth advantage over the polyploid cells, if in fact the latter were tumors cells.

Cytogenetic Aspects of Chromosome 17 in Medulloblastoma

Much attention has been given to changes of chromosome 17 in medulloblastomas, due to its frequent involvement in rearrangements in these tumors, i.e., as an iso(17q) (Figures 8.3– 8.5) and variable loss of the short arm of chromosome 17 (17p), and the strong possibility that this arm harbors a gene (other than *TP53)* causally related to medulloblastoma (41, 70–72, 72A).

The key role of i(17q) in the pathogenesis of medulloblastoma has already been mentioned, particularly because cases with this karyotypic change as the sole anomaly have been reported (Table 8.4). Deletion of 17p may occur independently of the i(17q) formation and vice versa. Other structural changes of chromosome 17, including nonreciprocal translocations have been reported (55, 73, 74). LOH and CGH studies have shown variable loss of 17p in 30–50% of medulloblastomas (47, 59, 60, 75–87). Although loss of 17p usually involves substantial parts of the chromosome arm, small distal deletions have been described previously (49, 76). Mutations of *TP53* (located at 17p13.1) occur in <10% of medulloblastomas (75, 88–91), and cogent genes on 17p related to medulloblastoma causation are yet to be identified. The effects of an isochromosome (17q) in medulloblastoma are mediated by gene-dosage effects of genes on 17p or 17q rather than by the disruption or deregulation of a "breakpoint" gene (91A).

Most medulloblastomas with 17p loss have breakpoints at 17p11.2 (78). This chromosomal region is deleted or duplicated in the Smith–Magenis syndrome (92), and instability in this region is a plausible explanation for the tendency for 17p loss to occur through isochromosome formation. Even though the incidence of medulloblastoma is not increased in this syndrome, the underlying mechanism that produces chromosomal disruption and recombination within this region of chromosome 17 might be relevant to medulloblastoma genesis (78). Isochromosome 17q has been demonstrated in interphase of medulloblastoma nuclei using FISH (46, 93, 94). Although i(17q) is the most common mechanism for 17p loss in medulloblastomas, in some cases, partial or complete loss of 17p occurs through interstitial deletions, unbalanced translocations, or monosomy 17 (49).

Because loss of the short arm of chromosome 17 is a most common cytogenetic and molecular change in medulloblastoma, much effort has been expended in studies attempting to delineate the area on 17p in which is located a gene possibly responsible for the development of medulloblastoma. The failure to find *TP53* (located at 17p13.1) mutations or involvement in a significant percentage of medulloblastomas (95–98), has led to the suggestion that a gene distal to *TP53* is the likely candidate.

Loss at 17p13.3 is not parentally related (99), nor is the origin of the i(17q) (100). Attempts to identify a possible gene on 17q that may play a role in medulloblastoma have so far not been successful (101). Deletions of 17p lead to loss of *REN* effects, particularly as a SHH antagonist, thus removing a checkpoint of SHH-dependent events during cerebellar development and tumorigenesis (101A).

LOH studies of 17p13.3-p13.2, although revealing loss in 50% of medulloblastomas, excluded the involvement of the *TP53* and *ABR* genes (102). Loss of 17p in medulloblastoma has been reported to be associated with an unfavorable outcome (Figure 8.7) (90, 103, 104); yet the presence of an i(17q) seems not to influence the outcome (105). Examination of Table 8.4 reveals that loss of a sex chromosome is not infrequent in medulloblastoma, i.e., -Y in 5% of the males and loss of an X in >10% of the tumors, mostly of female origin. These findings are reflected by the not infrequent loss of the X-chromosome revealed by CGH and the rather infrequent loss of the Y-chromosome.

dmin in Medulloblastoma

Table 8.4 reveals that approximately 7% of medulloblastomas have dmin as part of their karyotypes (Figure 8.5). These dmin are usually associated with gene amplification (e.g., *MYC* or *MYCN*), they and are present in a majority of medulloblastoma cell lines and xenografts (39), where they may

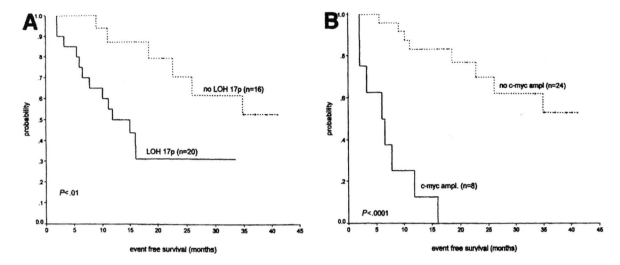

FIGURE 8.7. Survival of patients with medulloblastoma is related to LOH of 17p and *MYC* amplification. Survival was shortest for the patients with amplification of *MYC* in their tumors. LOH of 17p carried with it a poor survival outcome (from ref. (79) with permission).

assume the morphologic form of homogeneously staining regions (hsr).

Amplification of *MYC* in medulloblastoma (90, 106, 107) and less often of *MYCN* (108), as mentioned, seem to be related to the presence of dmin (85, 109). An association of *MYC* amplification with the presence of dmin and an unfavorable clinical course in patients with medulloblastoma has been reported previously (78, 110).

A study (111) investigated *MYC* amplification in 77 cases of paraffin-embedded medulloblastomas and found four positive tumors (5.2%). This is in keeping with the general experience regarding the incidence of dmin in medulloblastoma, which has ranged from 5 to 10%.

In a study (106) of three medulloblastoma cell lines and four xenografts, four of seven were shown to contain dmin, and *MYC* amplification. No amplification was present in the specimens without dmin, although in one tumor *MYC* was rearranged. A medulloblastoma cell line with a complex karyotype contained dmin (112). The presence of dmin is almost always associated with gene amplification, but the latter may occur in the absence of dmin.

Dmin are more commonly seen in medulloblastoma cell lines than in primary tumors (113–115).

It is of interest that the incidence of dmin in medulloblastoma (approx 7%) observed cytogenetically is similar to the percentage of tumors with *MYC* amplification.

SKY and FISH Studies in Medulloblastoma

FISH analysis, for which a large number of specific probes are available, can identify chromosomal anomalies not only in metaphases, but most importantly in interphase cells. Thus, dividing cells obtained on fresh tissues and cells are not always necessary for FISH analysis, which can be performed on archival and fixed specimens. For example, i(17q) was

identified in interphase nuclei of medulloblastomas by FISH (46,93). In fact, the dicentric nature of the i(17q) in these specimens was thus established (46).

Recent advances in molecular cytogenetic procedures have allowed the recognition of the provenance of abnormal chromosomes (usually labeled as "markers") and for detection of subtle chromosomal rearrangements that usually are not recognized by conventional G-banding analysis. SKY and multiplex-FISH (M-FISH) and related techniques are multicolored FISH procedures that allow for the simultaneous identification of all chromosomes, both normal and abnormal. SKY has become a powerful tool in detecting complex chromosomal rearrangements, including the chromosomal origins of marker chromosomes, and in elucidating structural rearrangements that may provide more insight into the nature of the disease. Together with other methods, such as G-banding, CGH, FISH, and molecular assays, the amount of information that can be generated from one specimen has greatly increased. These procedures are complementary and when used in combination they can identify abnormalities not detectable by conventional cytogenetic analysis (85).

One study (85) analyzed 24 pediatric cases in a retrospective manner, including 19 medulloblastomas and five PNET, by using conventional G-banding, CGH, SKY, FISH, and semiquantitative polymerase chain reaction (PCR), either alone or in combination. The goal of the study (85) was to screen a series of tumors to ascertain the chromosomal changes that occur both numerically and structurally, and to determine whether there are consistent rearrangements or translocations specific to medulloblastomas and PNET.

The combined results, although in line with those reported previously, e.g., gain of 17q and chromosome 7 in 60% of the tumors, revealed a higher frequency of rearrangements than that observed by any single technique (85). For example, a normal CGH pattern in one case was shown to be abnormal by cytogenetics and loss or gain or chromosomal material was more evident in some tumors upon CGH analysis than by cytogenetics. Yet, loss of 17p, the commonest alteration in medulloblastoma, was observed in only one of 10 tumors by CGH, with no indications that the other methodologies pointed to such loss.

CGH Studies in Medulloblastoma

A succinct and yet comprehensive review of CGH findings in central nervous system (CNS) tumors, including medulloblastoma, PNET, and pineoblastoma has been published (36). Studies using quantitative comparative genomic hybridization (CGH) (116) have demonstrated a greater degree of genomic imbalance in medulloblastoma than previously recognized, confirming loss of 17p with gain of 17q as the most frequent findings (47, 59, 60, 117). Table 8.5 and Figure 8.8 show the common changes (e.g., gains, losses, and amplifications) of chromosomal material as determined by CGH in medulloblastoma. In general, little variation was present among the more frequent findings in the various CGH studies, although some authors pointed to special features or unique observations related to these tumors, some of which are mentioned in the following paragraphs.

Not all chromosomal gains and losses in medulloblastoma determined by CGH are shown in Table 8.5, primarily because they were encountered in a small number of these tumors. Yet, some of these infrequent changes may be related to small subsets of medulloblastoma associated with if not caused by genes, e.g., the *PTCH* gene and loss of 9q and the *DCC* gene and loss of 18q and the minimally deleted region of 13q14-q22 and genes *RB1*, *DBM1*, and *BRCA2*. Some detailed features of CGH findings in medulloblastoma are listed in Table 8.6.

TABLE 8.5. Common CGH findings in medulloblastoma.

Losses: 8p, 10q, 11q, 16q, 17p
Gains: 1q, 2p, chromosome 7, 8q, 12q, 17q
Amplifications at: 17q, 7q, 8q24, 2p21, 5p15.3, 11q22.3 (in order of their frequency)
Minimal deletion regions: 1q31-q32, 10q25, 11p13-p15.1, 16q24.1-q24.3

The data represent our consensus of the most common CGH changes in the 150 medulloblastomasthe available in the literature [*Brown et al 2000* (14), *Reardon et al 1997, 2000* (17, 59), *Avet-Loiseau et al 1999* (47), *Russo et al 1999* (82), *Rienstein et al 2000* (83), *Aldosari et al 2000* (84), *Bayani et al 2000* (85), *Nishizaki et al 1999* (86), *Langdon et al 2006* (110), *Gilhuis et al 2000* (117), *Rickert and Paulus 2004* (118), *Eberhart et al 2002* (120), *Michiels et al 2002* (121), *Rossi et al 2006* (122),*Tong et al 2004* (126),*Jay et al 1999* (180), *Inda et al 2004* (562), *Frhwald et al 2000* (563), and *Hui et al 2005* (647).

The introduction of array-CGH methodologies with higher sensitivity (at 1 Mbp resolution) *Rossi et al 2006* (122), *Hui et al 2005* (647), *Lo et al 2007* (648)], than that of previous methods (10 Mbp) will undoubtedly lead to modifications and expansion of the data shown in the Table.

Although one CGH study (86) suggested an association between the number of chromosomal changes in medulloblastoma and an unfavorable prognosis on the basis of only six cases, larger studies (59, 80, 81, 117) failed to identify specific genetic abnormalities or patterns associated with prognosis. In one of these studies (80), an association of loss of chromosome 22 (–22) and a poor prognosis in classic medulloblastomas possibly existed. The classic tumors with –22 tended to occur in young patients with a median age of 8 months (80). Loss of 22 (–22) was found in 12% of medulloblastomas in another study (82). These findings point to a possible relationship between these medulloblastomas and AT/RT in which –22 is common (118). Although an association between LOH on 17p and an adverse prognosis has been reported previously (119), these findings could not be supported by CGH findings (59, 80, 86).

Loss of 17p by CGH is more frequent in classic than desmoplastic medulloblastomas (80), and it has been reported to occur almost exclusively in anaplastic tumors (120) in which there is a strong association between *MYC* and *MYCN* amplification and loss of 17p. Gain of 17q did not influence survival (80).

Gain of 12q was encountered in 16–20% of medulloblastomas based on CGH (59, 80, 120), possibly involving the *EGFR* pathway (80). Gain of the distal chromosome arm 2p (117) is of interest in medulloblastomas because it contains the *MYCN* gene, amplification of which has been detected in about 7% of medulloblastomas and it may be associated with a poor prognosis (121).

A cell line studied by CGH showed amplification on the distal end of chromosome arm 8q, which corresponds to the 20-fold amplification of *MYC* established previously by Southern hybridization (117). Thus, the gains of 2p and 8q in medulloblastomas, as detected by CGH, probably represent amplifications of *MYCN* and *MYC*, respectively.

Amplification of *MYC* (located on 8q24) and *MYCN* (located on 2p24) in some medulloblastomas based on CGH studies (121) are associated with a poor prognosis (60, 85, 117, 120). An aggressive medulloblastoma with both *MYC* and *MYCN* amplification has been reported (107). In a CGH study of 27 medulloblastomas (81), eight of 27 were found to have an i(17q), an additional nine tumors had either loss of 17p or gain of 17q and 10 tumors were apparently balanced. Concomitant results with FISH were concordant with those of CGH in 21/27 medulloblastomas.

Loss of 10q (10q22-10qter, possibly 10q25.1) determined by CGH (122) was present in 18–42% of medulloblastomas (relevant tumor suppressor genes located on that chromosomal band are *DMBT* and *PTEN*) (47, 59, 83, 117).

Medulloblastomas with +7 or gain of 7q also had –8 (82). About 80% of medulloblastomas with loss of 8p also had gain of 17q. The latter did not influence survival (80).

Oncogene amplifications as determined by CGH are relatively rare in medulloblastoma (59, 90, 114, 123), when they occur they involve *MYC* (at 8q24), *MYCN* (at 2p24.1) and *EGFR* (at 7p12) (85, 107, 108, 114, 124, 125).

Medulloblastoma   357

FIGURE 8.8. CGH results of chromosomal gains (*bars to the right*) and losses (*bars to the left*) in medulloblastoma (*upper part*) and in PNET (*lower part*). Loss of 17p and gain of 17q were the most common changes in medulloblastoma. In this series of cases, loss of the X chromosomes was frequent. An extra chromosome 7 was seen in both types of tumors. Note absence of chromosome 17 involvement in PNET. The thick bars, denoting amplifications, indicate such changes in 1p, 3p, 4p, and 6p in medulloblastoma and none in PNET (from ref. (82) with permission).

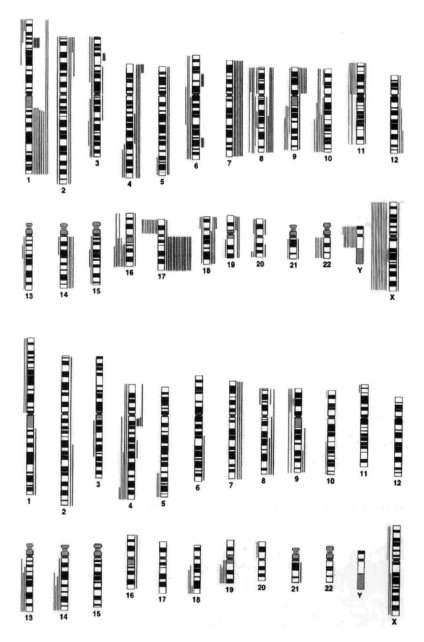

Amplification based on CGH studies of oncogenes included *PGY1* (located at 7q21.1), *MDM2* (at 12q14.1-q15), and *ERBB2* (at 17q21.2) (126) in nearly 40% of the medulloblastomas studied. The results were confirmed with FISH studies.

Digital karyotyping revealed amplification at 14q22.3 in 2/5 medulloblastomas, which included the *OTX2* gene, a homeobox gene of importance in brain embryogenesis (127).

The genesis and nature of the i(17q) seen in medulloblastoma, besides its apparent dicentric morphology, may be more complicated in some cases. Thus, in a study of a medulloblastoma cell line (84), an apparent i(17q) was, in fact, an unbalanced translocation between two copies of chromosome 17, with breakpoints at p12 and q11.1, an explanation that reconciled the cytogenetic and LOH findings (84). This observa-

tion suggests that the chromosomal breakpoints may play a more important role than the imbalance of genetic material on the short or long arm of chromosome 17 in medulloblastoma (122). Besides the results on chromosome 17, CGH (122) revealed losses of 8p, 10q, 16q, and 20p and gains of 2p, 4p, and chromosomes 7 and 19, findings in keeping with those shown in Table 8.5. Fine-resolution mapping provided by arrayCGH regarding small deletions included 1q23.3-q24.2, 2q13.12, 6q25-qter, 8p23.1, 10q25.1, and 12q13.12-q13.2. Amplifications were rare, the most common involving *MYC* in 16% of the medulloblastomas (122).

Based on arrayCGH, the most frequent aberration was the coincident loss of part of 17p and gain of 17q through formation of an isochromosome 17q. The breakpoint on 17p in this study (122) occurred consistently within a small interval,

TABLE 8.6. Highlights of CGH studies in medulloblastoma.

	References
Desmoplastic and large-cell medulloblastomas exhibited higher karyotypic heterogeneity, amplification, and aneusomy than the classic type, as shown by CGH and SKY.	*Bayani et al 2000* (85)
CGH revealed amplicons at 2p24-p25, 2q12-q22 and 17p11, and losses at 11q and chromosome 18, and low level gains at 3q, 11p, 13q, and 14q in a large-cell medulloblastoma. *MYCN* amplification was present, but not of p53.	*Reardon et al 2000* (17)
In a CGH study of eight high-grade intraventricular medulloblastomas, loss of 4q was found in six of the tumors and of 6q in four; losses of 10q and 16q were seen in three and two of the tumors, respectively.	*Shlomit et al 2000* (797)
Desmoplastic medulloblastomas had gains of 4q and 13q, changes not found in other types of medulloblastomas. Desmoplastic medulloblastomas were also characterized by loss of 9q in 25% of cases, whereas gain of 17q was seen in only 10% of these tumors. Desmoplastic tumors had fewer DNA copy number changes than the other subtypes of medulloblastoma.	*Eberhart et al 2002* (120) *Rickert and Paulus 2004* (118)
Large-cell/anaplastic medulloblastomas had more alterations (6.8 per tumor, especially gains) than other types of medulloblastoma (3.3 per tumor). A high percentage (50%) of these tumors showed either *MYC* or *MYCN* amplification in conjunction with loss of 17p.	*Eberhart et al 2002* (120) *Lamont et al 2004* (15)
In classic medulloblastomas gain of 17q was the single most common alteration in 45–61% of the tumors. Deletions of 17p was the single most common loss in 30–50% of medulloblastomas, often resulting from the formation of an i(17q).	*Rickert and Paulus 2004* (118) *Nicholson et al 2000* (81)
The combination of gain of 17q, as a result of i(17q) formation, and loss of 17p is the most common imbalance in large-cell/ anaplastic medulloblastomas, occurring in 33 and 62% of cases, respectively.	*Rickert and Paulus 2004* (118)
Trisomy 7 (+7), often seen in tumors with an i(17q), was the second most common abnormality among classic and desmoplastic medulloblastomas.	*Rickert and Paulus 2004* (118)
Utilizing array-CGH homozygous deletion was demonstrated in a medulloblastoma cell line and single copy loss in 30.3% of medulloblastoma tumors.	*Hui et al 2005* (647)
Metaphase and array CGH studies of medulloblastoma in South American patients yielded results somewhat at variance from those of other geographical areas. The association of *MYCN* and *TERT* amplification with a poor prognosis in medulloblastoma was stressed.	*Yoshimoto et al 2006* (743A)
Utilizing array-CGH, losses of 1p12-p22.1 and gains of 19p were shown to be more frequent in PNET than in medulloblastoma. By means of FISH, deletions of 9p21 were common in PNET, some of them being homozygous. All 9p21 deletions were associated with loss of CDKN2A protein expression, a rare finding in medulloblastoma. Gains of 19p were more common in PNET than in medulloblastoma, whereas gains of 17q showed an opposite picture. The authors suggested that genetic differences exist between PNET and medulloblastomas, and it will be an important line of future research to identify the cells of origin for both tumor entities and molecular pathways in their development. This may also provide novel options in the treatment strategy of these pediatric neoplasms.	*Pfister et al 2007* (808)

with high-resolution analysis, suggesting that the breakpoint occurred 5′ of the *EPN2* gene (122). Whether the genetic event that gives rise to this loss of 17p material specifically affects this gene, rather than exposing recessive mutation in as-yet-undetermined genes on 17p, is not known (122). The *EPN2* gene (128) is expressed in the cerebellum and interacts with proteins in the endocytic machinery related to protein transport within the cell.

The reasons for some of the discordant results of CGH versus cytogenetics are likely due to weaknesses of both procedures. Cytogenetic analysis only evaluates a limited number of clones, i.e., those in division within each tumor. Dividing cells (metaphases) generally constitute a minor population of tumor cells, compared with the dominance of interphase cells, which primarily contribute to CGH results. Hence, routine cytogenetic examination may not provide a complete overview of all genetic aberrations within a tumor. Culturing of solid tumors is not always successful, and the cultures may be overgrown by normal cells. CGH only detects copy number changes involved. Balanced translocations and inversions cannot be detected by CGH, because equal amounts of tumor and control DNA are used in CGH experiments.

Although a substantial amount of genetic information has been accumulated using CGH, as mentioned, a more complete picture of medulloblastoma can be obtained by correlating

data obtained by a combination of CGH, FISH, and LOH (110). For example, >40% of medulloblastomas were shown to have genetic losses by LOH without evidence of DNA copy number alterations detected by CGH and FISH (110). This indicates that genetic alterations in medulloblastoma arise through a complexity of established mechanisms.

LOH Studies in Medulloblastoma

The methodologies used to ascertain LOH in medulloblastoma have included RFLP, single-strand conformation polymorphism (SSCP), and microsatellite analysis combined with reverse transcription-polymerase chain reaction (RT-PCR).

These methodologies are more sensitive (84) than CGH, cytogenetics, or FISH in establishing certain genetic changes in tumors, although for optimal results a combination of these methodologies should be used (110).

LOH studies in medulloblastoma (49, 70, 84) dealt with the common loss of 17p in up to 50% of these tumors, particularly the region 17p12-p13.1 (49, 70, 75, 78, 79, 110, 129, 130), probably containing a gene of significance in medulloblastoma development (70, 78, 79, 90, 96, 103, 110, 119, 129, 131–134). Addressed was the possible involvement of *TP53*; however, these studies showed that the LOH area was more distal to and distinct from *TP53*, although occasional mutations

(5–10% of medulloblastomas) of this gene were encountered (49,70,88,135,136), suggesting that the *TP53* gene alterations may contribute to the pathogenesis of only a small subset of tumors and that a second tumor suppressor gene may be located on the short arm of chromosome 17. In support of the latter hypothesis, deletion mapping has indicated the presence of another locus distal to *TP53* (90, 129). Using pulsed-field gel electrophoresis, an active *BCR*-related (*ABR*) gene locus in this region that is deleted in approximately 90% of informative medulloblastomas has been reported previously (129). A deletion mapping study localized a common chromosomal disruption within a more centromeric region, 17p11.2, in roughly 35% of cases (78).

Loss of 17p has been reported to be indicative of a poor prognosis (70, 90), although others disagree (103, 104, 132). However, it has been emphasized that patients with LOH of 17p usually had metastases at the time of diagnosis (79).

Critical loss of 17p may occur through a number of mechanisms: often as a result of i(17q) formation, terminal deletions, breakage of the i(17q), unbalanced translocations, or homologous recombinations (84).

Next in LOH frequency in medulloblastoma was involvement of 10q (minimal region 10q21.1-q26.3), observed in approximately 20–25% of the tumors. Loss at 10q25 was thought to be the most frequent (137). LOH of 16q22 (minimal region 16q24.1-q24.3) involved about 20% of medulloblastomas (137). Less frequent (10–15%) LOH has been shown for both arms of chromosome 11 (11q24-qter), although conflicting reports have been published (76). For example, in one study of 19 medulloblastomas, no LOH of 11p was found (76), whereas others have shown LOH changes in both 11p and 11q in a substantial percentage of these tumors (13–30%) (79,138–141). The LOH on 11p has been localized to 11p15.5-pter (139) or 11p13-p15.1 (137). The same state-

ment applies to LOH of 22q (131, 141). Loss of 9q22 was primarily seen in the nodular/desmoplastic variant of medulloblastoma, whereas loss of 10q24 was seen in all histologic types (15).

LOH at 17p13, a frequent finding in medulloblastoma, was explored as to the possible genes involved (142). The gene *KCTD11* (*REN*) was found to be expressed at a lower level in medulloblastoma than in adult cerebellum and higher than in neural stem cells. Another gene mapping to 7p13, *HIC1*, was not involved in medulloblastoma pathogenesis.

Less commonly LOH affected chromosomal sites at 9q, 8p, 2p, and chromosome 1. Involvement of 9q could be implicated in changes of the *PTCH* gene (located at 9q22.3-q31) and that of chromosome 1 in the *PTCH2* gene (located at 1p32.1-p32.3), genes that have been shown to have mutations in sporadic medulloblastoma (143–146, 146A).

LOH in chromosomes and chromosome regions in medulloblastoma is shown in Table 8.7, including the frequently affected chromosomes and chromosome arms and more defined chromosome regions affected by LOH.

Discrepancies in LOH findings in medulloblastoma may be explained (141) by the small number of tumors examined in some studies (123, 138–140) or by the small number of markers used (76, 79, 123, 138). LOH studies in medulloblastoma have been often restricted to specific regions, e.g., 9q, 11p, 17p (76), chromosome 11 and 22q (141) and 10q, and chromosomes 11 and 16 (137).

Molecular Genetics of Medulloblastoma

Candidate Genes on Chromosome Arm 10q

Losses of genetic material on 10q have been observed in approximately 20–40% of medulloblastomas (59, 79, 134,140). The frequency of these alterations varied

TABLE 8.7. Chromosomes and chromosomal regions (exclusive of 17p) affected by LOH in medulloblastoma.

	References
Chromosome arms most frequently affected by LOH: 10q, 11p, 16q, 6q, 9p	
Chromosome arms less frequently affected by LOH: 1q, 11q, 22q (1p, 5q, 8p, 16p)	*James et al 1990* (138)
	Thomas and Raffel 1991 (131)
	Batra et al 1995 (90)
	Blaeker et al 1996 (140)
	Albrecht et al 1994 (76)
	Kraus et al 1996 (134)
	Slavc et al 1997 (649)
	Scheurlen et al 1998 (79)
	Lescop et al 1999 (141)
	Yin et al 2001, 2002 (137, 148)
	Zakrzewska et al 2004 (559)
Chromosome regions affected by LOH: 1q31, 9q22, 10q21-q26.3, 11p13-p15.1, 11q24-qter, 16q22, 16q24.1-q24.3	*Blaeker et al 1996* (140)
	Scheurlen et al 1998 (79)
	Yin et al 2001, 2002 (137, 148)
	Langdon et al 2006 (110)
Homozygous deletion by LOH: 8p22-p23.1	*Yin et al 2001, 2002* (137, 148)

Not shown in this table are LOH data on chromosome 17, which affects approximately 50% of medulloblastomas and is the most common genetic change in these tumors (see text).
LOH of 9q were rare and not seen at 11p in one study (76).

between different studies, and the target genes were not always identified with certainty. Two candidate genes located on 10q have been studied in some detail. The tumor suppressor gene *PTEN* located on 10q23 and encoding a dual-specific phosphatase was found to be frequently affected in glioblastomas, but only in a single medulloblastoma (59). Using representational difference analysis, a homozygous deletion at 10q25.3-q26.1 was identified in a medulloblastoma cell line and a novel gene, *DMBT1*, spanning this deletion was cloned (147). *DMBT1* shows homology to the scavenger receptor cysteine-rich (SRCR) superfamily. Intragenic homozygous deletions have been detected in two of 20 medulloblastomas (147). Therefore, *DMBT1* may represent a putative tumor suppressor gene implicated in the pathogenesis of a small subset of medulloblastoma. Refined mapping of the region 10q23.3-10q25.3 in medulloblastomas and cell lines was undertaken (147A). Allelic loss on 10q was found in five of 32 of the former and in two of eight of the latter. Changes of three TSG were evaluated (*MXI1*, *SUFU*, and *BTRC*); except for one *MXI1* mutation, no other changes were found.

Allelic Losses at 8p in Medulloblastoma

A high-resolution genome-wide allelotype analysis was performed on 12 medulloblastomas (148), with an aim to characterize putative tumor suppressor loci involved in this malignant neoplasm. With this methodology, nonrandom allelic losses somewhat at variance with those obtained with the usual LOH techniques (Table 8.7), were identified on 8p (66.7%), 16q (58.3%), and 17p (58.3%), strongly indicating that alterations of these chromosome arms play critical roles in medulloblastoma tumorigenesis. Further detailed mapping showed 8p22-p23.1 to be a region of homozygous deletion.

The aforementioned findings, especially loss of 8p (148), were compared with those of previous investigations; genetic data from six genome-wide studies covering a total of 107 medulloblastomas were reviewed (47, 59, 80, 117, 140, 149) and the summarized results revealed that deletions of 17p (39%), 10q (35%), 8p (32%), 11p (28%), 11q (28%), and 16q (28%) were the most frequent genetic abnormalities.

The most striking finding in the study on allelic losses in medulloblastoma (148), although based on only 12 tumors, was the high incidence (67%) of allelic deletion on 8p. This result is in agreement with that of previous low-resolution allelotyping studies, but it was about twofold higher than those determined by CGH studies (47, 59, 80, 117, 140). The use of high-density markers in the microsatellite analysis (148) probably explains the increased sensitivity of detecting allelic loss. Frequent deletion of 8p has also been detected in a variety of tumors, such as prostate, breast, lung, colorectal, bladder, head and neck, liver, ovarian, and gastric cancers (148).

Deletion mapping has identified several subchromosomal regions on 8p that are important for cancer development. These regions are localized to 8p12, 8p21, 8p22, and 8p23, although the region encompassing an interval of 1.8 cM of

homozygous deletion at 8p22-p23.1 seems to be particularly affected in medulloblastoma (148). However, the exact gene(s), located in this region and responsible for medulloblastoma pathogenesis has remained elusive (148, 149A).

PinX1, a gene located at 8p23.1, a region often deleted in medulloblastoma through its protein product, is a potent inhibitor of telomerase activity. Studies in medulloblastomas and PNET revealed no suppression of *PinX1* function, although telomerase activation was present in these tumors (150).

In a study reflecting the general experience, involving a large number (approx 150) of medulloblastomas (15), the frequency of molecular cytogenetic abnormalities in pathologic variants of these tumors revealed the following: except for nodular desmoplastic, i(17q) and loss at 17p13.3 involved all other tumor types, loss at 9q22 was seen primarily in nodular desmoplastic and large-cell/anaplastic medulloblastomas and loss of 10q24 affected all types.

Hereditary Syndromes and Medulloblastoma

Highlights of the three most common familial syndromes with a high (but variable) incidence of medulloblastoma are outlined in Table 8.8. These syndromes are associated with mutations of the genes shown in Table 8.8.

There is an association between medulloblastoma and Gorlin syndrome, or nevoid basal cell nevus syndrome. This is an autosomal dominant syndrome associated with jaw cysts, skeletal abnormalities, and multiple basal cell carcinomas. Population-based studies, however, have suggested that the association is not as strong as was originally thought, with medulloblastomas being seen in perhaps 2% of all patients with Gorlin syndrome and in 5% of those under age 3 (151, 152).

A strong association between medulloblastoma and certain forms of Turcot syndrome may exist. Turcot syndrome may be genetically diverse, with germline mutations in either the adenomatous polyposis coli (*APC*) gene on 5q or in DNA mismatch repair genes referred to as hMLH-1 and hPM52. In the *APC* form of Turcot syndrome (Table 8.8), seven medulloblastomas were seen in 10 patients in one study, suggesting that the *APC* gene may be an important oncogene in sporadic medulloblastomas; however, initial mutational analyses have been negative (152).

Germline mutations in the *TP53* gene have been shown to be present in at least 50% of families with Li-Fraumeni syndrome (153–156). However, not all patients who carry a *TP53* germline mutation fulfill the criteria for Li-Fraumeni syndrome, which are development of a sarcoma before the age of 45, at least one first–degree relative with any tumor before the age of 45, and a second (or first)-degree relative with cancer before age 45 or a sarcoma at any age (156, 157). In a review of 91 families with germline mutations in the *TP53* gene, only 45 (49%) met these criteria (158). Thus, the definition of Li-Fraumeni syndrome includes only a subset of

TABLE 8.8. Autosomal dominant familial syndromes with high incidence of medulloblastoma.

	References
Nevoid basal cell carcinoma syndrome (NBCCS or Gorlin syndrome) Approximately 3–5% of patients develop medulloblastomas, usually desmoplastic. Responsible gene is *PTCH* located at 9q22.	*Gorlin 1987* (650) *Lacombe et al 1990* (651) *Evans et al 1993* (151) *Hahn et al 1996* (276) *Johnson et al 1996* (277) *Ragge et al* (277A)
Turcot syndrome (Type 2) Patients with Turcot syndrome (type 2) develop medulloblastoma in the setting of familial adenomatous polyps of the colon (20% of which become malignant). Responsible gene is *APC* located at 5q21.	*Hamilton et al 1995* (368) *Hahn et al 1996* (276) *Paraf et al 1997* (652)
Li-Fraumeni syndrome Nearly 15% of the tumors in families with the Li-Fraumeni syndrome are brain tumors (astrocytomas, ependymomas, choroid plexus tumors, schwannomas, meningiomas), with medulloblastomas constituting about 10% of these tumors. The dominant tumors in Li-Fraumeni families are sarcomas and carcinomas of the breast, lung, and gastrointestinal tract. Responsible gene is *TP53* located at 17p13.	*Li et al 1988* (157) *Malkin et al 1990* (154)

families with *TP53* germline mutations. In contrast, not all Li-Fraumeni families have been found to carry *TP53* mutations (153, 158). It is unclear whether these families have genomic alterations outside the commonly screened *TP53* coding regions, e.g., in the promoter region or in intronic gene sequence, or carry germline mutations in other, not yet identified, genes. In addition, it remains to be clarified whether the definition of Li-Fraumeni syndrome should be modified in order to allow inclusion of those families with *TP53* germline mutation that for genetic or epigenetic reasons happen to present with a different spectrum of malignancies. Thus, although the incidence of medulloblastoma is higher in patients with the Li-Fraumeni syndrome than in the general population, the number of cases described so far has not been very high.

About 12% of the tumors that occur in patients with *TP53* germline mutation are located in the central nervous system, and more than two thirds of these are astrocytic gliomas (158). Only five histologically proven medulloblastomas have been reported in patients carrying a *TP53* germline mutation (158).

Rubinstein–Taybi syndrome, a complex malformation syndrome is caused by germline mutations in the gene for the transcriptional coactivator CGP (159, 160) and may be associated with medulloblastoma (161).

Only a small number of reports of familial medulloblastomas have been published. These included medulloblastomas in monozygotic twins (162–164) and in dizygotic twins and siblings (165–167, 167A). In one report on desmoplastic medulloblastomas occurring simultaneously in monozygotic twins, the tumors had been investigated by molecular-genetic techniques (163). The authors (163) found no evidence of tumor-associated LOH on 17p and 9q, suggesting that alterations of the *TP53* gene on 17p13.1 and the *PTCH* gene on 9q22 were not involved in the pathogenesis of these tumors.

Genetic Susceptibility to Medulloblastoma

Table 8.9 shows unique cases of medulloblastoma demonstrating the complexity, including genetic, that may be associated with these tumors. Associations of medulloblastoma with other brain tumors (168) and extraneural malignancies, including Wilms tumor (169, 170) and renal malignant rhabdoid tumor (171, 172), also have been observed. Occasionally, medulloblastomas develop within the setting of complex malformations, e.g., intestinal malrotation, omphalocele, and

TABLE 8.9. Unique cases of medulloblastoma.

	References
A 5-year-old girl with NF-1 developed a Wilms tumor, T-acute lymphoblastic leukemia, medulloblastoma, and then acute myelocytic leukemia in the sequence shown.	*Perilango et al 1993* (653)
Molecular characterization of critical region and breakpoint cluster on 17p for the Smith-Magenis syndrome, findings possibly applicable to medulloblastoma.	*Wilgenbus et al 1997* (654)
Identical twin infants with medulloblastoma.	*Chidambaram et al 1998* (737)
Histological evidence that a medulloblastoma in an adult woman originated in the fetal external granular layer of the cerebellum.	*Miyata et al 1998* (655)
Two patients with Gorlin syndrome developed BCC at the site of radiation for medulloblastoma; another patient developed a glioblastoma and still another a meningioma after radiation of medulloblastomas.	*Stavrou et al 2001* (656)
An 11-year-old girl with Gorlin syndrome had a cerebellar medulloblastoma resected, which recurred with CSF seeding and followed by a "spontaneous" cure.	*Su et al 2003* (275)

bladder entrophy (173), and Coffin–Siris syndrome (mental retardation, deficiency of postnatal growth, joint laxity, and brachydactyly of the fifth digit with absence of the nail bed) (174). Some studies suggest that relatives of patients with medulloblastomas have an increased risk of developing other childhood malignancies, particularly leukemia and lymphoma (175, 176).

MYC *and* MYCN *Genes in Medulloblastoma*

As with other genes described in this chapter, so it is with *MYC* and *MYCN*, i.e., much has been written about them and much is known about them, yet much is unknown about them, particularly as it applies to medulloblastoma.

Anomalies of *MYC* and *MYCN* genes in medulloblastoma are shown in Table 8.10.

TABLE 8.10. Some features of *MYC* and *MYCN* changes in medulloblastoma.

	References
Two medulloblastomas with *MYCN* amplification exhibited neuronal differentiation.	*Rouah et al 1989* (124)
Infrequent *MYC* and *MYCN* amplification in medulloblastomas; more common in cell lines of these tumors.	*Bigner et al 1990* (106)
	Fuller and Bigner 1992 (125)
	Batra et al 1995 (90)
Expression of *MYC* is elevated in a much higher proportion of medulloblastomas than is *MYCN* and is not necessarily related to *MYC* amplification; however, hyperexpression of MYC may indicate a poor prognosis.	*MacGregor and Ziff 1990* (179)
	Brown et al 2000 (14)
	Herms et al 2000 (115)
	Grotzer et al 2001 (188)
	Eberhart et al 2004 (189)
Amplification and overexpression of *MYC* have an unfavorable impact on medulloblastoma.	*Badiali et al 1991* (107)
	Liu et al 2001 (657)
	Aldosari et al 2002 (182)
	Ellison 2002 (1)
The high frequency of *MYC* expression (>50%) in medulloblastomas cannot be explained by gene amplification alone (4% incidence in medulloblastoma) and therefore other factors are responsible for this overexpression.	*Badiali et al 1991* (107)
	Herms et al 2000 (115)
	Eberhart et al 2004 (189)
	Misaki et al 2005 (658)
Amplifications in 5–15% of *MYCN* and *MYC*, respectively, occur in medulloblastomas, usually associated with anaplastic/large-cell phenotype and aggressive behavior.	*Valery et al 1991* (659)
	Jay et al 1995, 1999 (180, 660)
	Scheurlen et al 1998 (79)
	Brown et al 2000 (14)
	Gilbertson et al 2001 (661)
	Taipale and Beachy 2001 (336)
	Eberhart et al 2002 (120)
	Ellison 2002 (1)
	Ellison et al 2003, 2005 (324, 662)
	Frank et al 2004 (207)
	Lamont et al 2004 (15)
MYC may be activated in the nucleus by β-catenin in some medulloblastomas. However, β-catenin binding sites in the *MYC* promoter were not required for elevated *MYC* expression In two medulloblastoma cell lines.	*He et al 1998* (354)
	Siu et al 2003 (663)
Since *MYC* is located downstream of the *Wnt* signal pathway, activating mutations of upstream components including *APC*, *Axin* and *β-catenin* genes have been described, but only in approximately 5% of medulloblastomas for each of these genes. Thus, these findings cannot account for *MYC* overexpression.	*He et al 1998* (354)
	Siu et al 2003 (663)
	Zurawel et al 1998 (346)
	Eberhart et al 2000 (371)
	Huang et al 2000 (361)
	Koch et al 2001 (365)
	Yokota et al 2002 (370)
	Baeza et al 2003 (366)
	Misaki et al 2005 (658)
Amplification of *MYC* and *MYCN* are more likely to be associated with large-cell/anaplastic than other types of medulloblastomas; these tumors tend to be aggressive.	*Brown et al 2000* (14)
	Reardon et al 2000 (59)
	Leonard et al 2001 (181)
	Eberhart et al 2002 (120)
Not all medulloblastomas with an increased *MYC* copy number have elevated expression of this gene.	*Grotzer et al 2001* (188)
MYCN amplification, seen often in nodular and anaplastic medulloblastomas, is associated with a trend to a poor outcome; reports in the past held that it is indicative of a poor prognosis.	*Eberhart et al 2004* (198)
	Tomlinson et al 1994 (108)
	Eberthart et al 2001 (20)
	Reardon et al 2000 (59)
MYCN can substitute for *ILGF* signaling in a murine model of SHH-induced medulloblastoma.	*Brown et al 2006* (738)
MYC overexpression causes anaplasia in medulloblastoma.	*Stearns et al 2006* (185A)

MYC, the protein encoded by the *MYC* oncogene, is a transcription factor. Normally, MYC is induced after mitogenic stimulation, it is required for cell cycle entry, and it promotes cell proliferation, providing that growth factors are available. However, MYC may also induce apoptosis when insulin-like growth factor (IGF) and platelet-derived growth factor (PDGF), which normally suppress MYC-induced apoptosis, are not available. It has been proposed (177, 178) that when MYC is active, there is simultaneous induction of cell proliferation and apoptosis. This apparently paradoxical coupling may be regarded as a safeguard against neoplasia. Successful cell proliferation requires a dual signal, one of which overrides the apoptotic pathway and another, which simultaneously promotes growth. IGF, PDGF, and *BCL2* all inhibit MYC-induced apoptosis independent of cell proliferation. The precise mechanisms of MYC-induced apoptosis are not clear, and p53 may not be necessarily involved.

In interpreting results on amplification, it should be kept in mind that the range in *MYC* and *MYCN* amplification in medulloblastoma may be affected by, among other factors, the methodology utilized, i.e., FISH, Southern blotting, or CGH. For example, studies of *MYC* and *MYCN* amplification in medulloblastoma based on FISH showed variable results both in terms of the pattern of FISH signals in individual nuclei and the proportion of nuclei demonstrating amplification (15). Indicative of a wide range of amplification of *MYC* and *MYCN* is the failure to find any amplification of *MYCN* in 27 medulloblastomas (107), a study in which only one tumor was found to have *MYC* amplification associated with the presence of dmin and an aggressive clinical course.

MYC and *MYCN* amplifications have been reported in approx 10 and 5%, respectively, of medulloblastomas (13, 15, 90, 107, 108, 114, 115, 123, 125, 179–182). As mentioned, *MYCN* gene amplification is usually found in the form of dmin by cytogenetics (17, 84, 107, 108, 114) and may be associated with a poor response to therapy (Figure 8.7) (78, 79, 99). All patients with evidence of *MYC* amplification in their tumors died within months of diagnosis, compared with only 50% of patients who died within the same period with nonamplified tumors (182). These data support the observation that *MYC* amplification may serve as a marker of poor prognosis in medulloblastoma. The prognostic significance of *MYCN* amplification is uncertain, although it has been reported in three aggressive tumors (14, 17, 108).

CGH studies showed high copy number for regions on 8q24 and 2p24, suggesting amplification of the *MYC* and *MYCN* genes in 10–22% of cases, respectively (60, 117). In contrast, a report using the same method, none of these regions was amplified (47, 79). Two of 20 tumors showing *MYCN* amplification also exhibited neuronal differentiation (124).

MYC amplification is relatively high in the large-cell variant of medulloblastoma (14, 15) and established cell lines of such tumors (113, 114, 183), although the association is not absolute (17, 180, 184). *MYC* amplification is seldom seen in desmoplastic medulloblastoma (15, 23). The association between

MYC amplification and large-cell/anaplastic phenotype was seen when the extent of amplification was high, whereas when it was low, the medulloblastomas generally had a classic phenotype (15).

In contrast to amplification, *MYC* expression is common in medulloblastomas with elevated levels of the protein being present in the absence of *MYC* amplification (185). *MYC* overexpression appears to play an important role in the development of anaplasia in medulloblastoma (185A) and may be a useful marker of aggressive disease. Increased expression of *MYCN* may be an indicator for poor prognosis in medulloblastoma and PNET (186). Low levels of *MYC* protein expression in medulloblastoma were associated with a relatively long disease-free interval (187, 188). Overexpression of *MYC* has been reported in >40–90% of medulloblastomas (111, 185), such overexpression being associated with an unfavorable outcome (Figure 8.9) (115, 188).

The role of *MYCN* in medulloblastoma pathogenesis is not well established. *MYCN* expression seems to be regulated by Hedgehog signaling (see Section 2.15) in embryogenesis and medulloblastomas, providing a link between development and neoplasia. However, several unanswered questions remain, including whether *MYCN* expression is linked to clinical

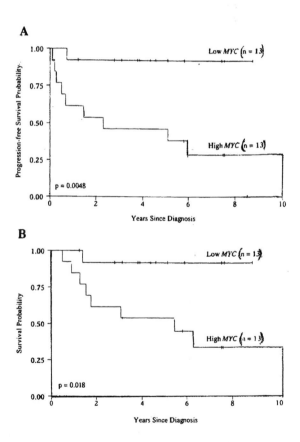

FIGURE 8.9. Progression-free survival and survival probabilities in patients with medulloblastoma as related to *MYC* expression. High expression carried with it a poor prognosis (from reference 402 with permission).

outcome and to the anaplastic medulloblastoma phenotype. To address these issues, *MYCN* and *MYC* expressions were analyzed in a large group of medulloblastomas using in situ hybridization, correlating mRNA levels with tumor subtype and clinical outcome (189). The study (189) examined several mechanisms, such as gene amplification, Wnt signaling and *MXI1* (a regulator of *MYC*) mutation, that might account for increased *MYC* activity in medulloblastomas.

SHH signaling positively regulates *MYCN* expression in cerebellar granule cells, providing a mechanism for the increased *MYCN* expression in all six nodular medulloblastomas examined (189A). *MYCN* expression may not have the prognostically unfavorable effect as it does in non-nodular tumors with elevated SHH activity. In the study referred to above (189), all six patients with nodular tumors were long-term survivors despite expression of *MYCN*. However, analysis of non-nodular medulloblastoma alone did not reveal a significant association between *MYCN* expression and survival.

The data (189) support the value of *MYC* mRNA level as a prognostic marker in medulloblastoma patients and indicate that increased *MYC* expression, in addition to gene amplification, is associated with the large-cell/anaplastic subtype. *MYCN*'s potential as a prognostic marker is less clear, perhaps due to its expression in more-differentiated tumors as a consequence of Hedgehog signaling. Regulation of *MYC* transcription in the medulloblastomas examined was apparently not affected by Wnt signaling or changes in the *MXI1* gene.

A screening study of 20 medulloblastomas by using a panel of known oncogenes, including *ERBB1*, *GLI*, *NEU*, *RAS*, *SIS*, *FOS*, and *SRC*, showed amplification in only a single tumor; the amplified gene was *ERBB1* (114). Immunohistochemical expression of the *ERBB2* product has been detected in approximately 84% of tumors studied. In a separate study the combined incidence for *MYC*, *MYCN*, and *ERBB1* amplification in medulloblastomas was reported as approximately 10% (125). Amplification of *EGFR* and *MDM2* genes is very rare, and it does not seem to represent a major mechanism of oncogene activation in medulloblastomas (90, 125).

TP53/p53 and ARF Pathways in Medulloblastoma

Among its many diverse functions (190A), the p53 protein acts as a transcription factor that upregulates or represses the activity of a range of genes, including several that modulate apoptosis. The p53 protein normally prevents DNA replication in cells that have sustained DNA damage by reversibly maintaining the cell in G1 for "DNA repairs." If the DNA damage is irreparable, the cell undergoes apoptosis. In cells lacking functional p53, there is reduced susceptibility to apoptosis induced by radiation and cytotoxic drugs. Cells with inactivated p53 might therefore survive abnormally and allow further genetic damage to accumulate. Missense mutations in *TP53* are very commonly encountered in many cancer types, and p53-deficient tumors show reduced apoptosis, thereby facilitating tumor growth. However, this is not supported by experimental data or by observations on spontaneous human tumors. One p53 function is to inhibit malignant transformation resulting from DNA damage or ectopic oncogenic expression, by activation of genes involved in cell cycle arrest and apoptosis (190).

Using array-CGH genetic instability was shown in medulloblastomas of mice resulting from inactivation of p53 together with deregulation of proliferation induced by Rb loss leading to neoplastic transformation with acquisition of additional genetic changes mainly affecting *MYCN* and *PTCH2* genes. The p53 loss influences molecular mechanisms that cannot be mimicked by the loss of p19^{ARF}, p21, or ATM (190B). Some features of p53 changes in medulloblastoma are presented in Table 8.11.

Due to its role in controlling cell growth, p53 function is tightly regulated by the oncogene *MDM2* (12q14), a ubiquitin ligase that facilitates degradation of p53 by proteasomes (Figure 8.10). Loss of p15^{ARF}, a critical upstream regulator of p53 (190C), also can represent a common mechanism of functional p53 loss in tumors. The different expression of the isoforms of p53 in human tumors may explain the difficulties in correlating p53 status to the biological properties of the tumors and drug sensitivity (190).

TABLE 8.11. Some findings related to p53 changes in medulloblastoma.

	References
High expression of p53 may be seen in medulloblastoma cell lines, compared with the low levels in primary tumors.	*Loda et al 1992* (664)
Low levels of expression of p53 are common in medulloblastomas, whereas overexpression is much less common (apprx 10%) and carries with it an unfavorable prognosis. Expression of p53 has been shown in 18% of all medulloblastomas, in 45% of anaplastic tumors, in 67% of AT/RT and in 88% of PNET.	*Jaros et al 1993* (572) *Mirabell et al 1999* (254) *Jaros et al 1993* (572) *Gudkov and Kamarova 2003* (206) *Ray et al 2004* (665)
The absence of germline and somatic aberrations of p53 in medulloblastoma has been stressed.	*Portwine et al 2001* (666)
Intense nuclear staining for p53 may be a most important prognostic indicator of non-metastatic medulloblastoma. No correlation was obtained for disease-free interval and immuno-histochemical staining for KI-67 index, glial fibrillary acidic protein (GFAP) or synaptophysin.	*Woodburn et al 2001* (203)
Elevated p53 levels were found in an anaplastic medulloblastoma, but not in the classic or nodular types. JCV was not detected in any of these tumors.	*Eberhart et al 2005* (524)

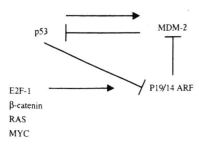

FIGURE 8.10. Schematic presentation of a partial loop of the p53/ARF pathway (from ref. (191), with permission). The activation and enhancement of the effects of p53 by a number of oncogenes (*E2F1*, β-*catenin*, *RAS*, *MYC*) are mediated in part by a positive feedback loop that results in the transcription of *p19* or *p14*ARF (196A,523A,195–197,523), which in turn inhibits MDM2 ubiquitin ligase that is responsible for inactivating p53 (523B), resulting in an increased level of p53 in the cell.

In addition to the complexity in interpreting findings for p53 introduced by the p53 positive and negative feedback loops (191), the description of a family of p53 proteins further complicates this situation (192). The protein product of the murine double minute-2 gene (*MDM2*) has a controlling effect on p53. For example, the proapoptotic activity of p53 can be nullified by the amplification or overexpression of the MDM2 protein by forming complexes with the p53 protein. Amplification of *MDM2* has not been found in pediatric (89, 90) and adult medulloblastomas (193). From a clinical viewpoint, this relationship is important, because the response of medulloblastoma to radiation depends on an active apoptotic system in which p53 is an essential component. Loss of p53 function can be the consequence of *TP53* mutations or the amplification and/or overexpression of the *MDM2* gene.

In a study of 51 medulloblastomas the immunohistochemical expression of MDM2 protein and p53 were mutually exclusive, however, both being associated with a short survival. This association may be due to a disruption of the p53 apoptotic pathway by two distinctive but still unclarified events resulting in the accumulation of wildtype p53 and overexpression of the MDM2 protein, thus reducing response to radiation (193).

In a study of 56 pediatric medulloblastomas, LOH of 17p was present in eight of 18 (29%) of the tumors examined (90), and were associated with a poor prognosis. Mutations of *TP53* were found in only two of 46 tumors and *MYC* amplification in three of 43. None of the medulloblastomas showed amplification of the *MYCN*, *EGFR*, or *MDM2* genes (90). Similar findings had been obtained in an earlier study (89). In a subsequent study (193), an analysis of 51 adult medulloblastomas revealed no amplification of *MDM2*, although 10 showed expression of MDM2, only one had a *TP53* mutation and overexpression of p53 were shown in 16. Only two tumors showed both MDM2 and p53 expression. The authors concluded that the overexpression of MDM2 protein and the accumulation of wild-type p53 are unrelated in medulloblas-

toma, but both may result in a reduced apoptotic response after radiotherapy and contribute to a shortened survival.

The TP53-ARF tumor suppressor pathway is disrupted in the majority of human cancers (194). This pathway includes ARF, one of two cell cycle inhibitors encoded by the *INK/ARF* locus (Figure 8.11). Activated oncogenes, including *MYC*, induce *ARF* expression via E2F-1 and thus regulate p53-dependent apoptosis (195–197). In turn, *ARF* stabilizes *TP53* by sequestering the *TP53*-negative regulator *MDM2*, thereby diverting cells to either cell cycle arrest or apoptosis (198). Disruption of this tumor suppressor pathway in cancer can occur directly by inactivating *TP53* mutations, or though mutually exclusive *p14*ARF inactivation (by hypermethylation or deletion) or by amplification of *MDM2* (194). Alterations in *TP53* have been implicated in the formation of at least a subset of medulloblastomas. For example, loss of p53 accelerated medulloblastoma development in genetic mouse models (199–202), and humans carrying germline mutations of *TP53* (Li-Fraumeni syndrome) are predisposed to

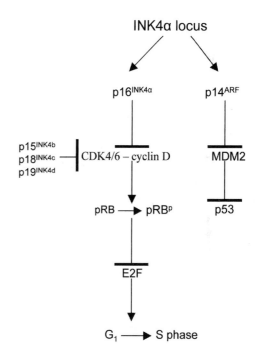

FIGURE 8.11. Simplified schematic presentation of the regulation of the pRB and p53 based on Li et al *(2005)* (360). INK4α/ARF tumor suppressor pathway that may be deregulated in medulloblastoma and PNET. The INK4α locus encodes p16INK4a and ARF through the use of alternate reading frames. The p16INK4a and related INK4 proteins (INK4b–d) inhibit CDK4/6 activity with consequent hypophosphorylation of pRB. Hypophosphorylated pRB acts with E2F proteins to repress transcription of genes necessary for the G1-S phase transition. The p14ARF regulates p53 activity by inhibiting MDM2-mediated degradation of p53, allowing p53 to induce cell cycle arrest. Thus, this pathway plays a key role in controlling cell growth by integrating multiple mitogenic and antimitogenic stimuli, which may serve as targets for therapy (from ref. (523C) with permission).

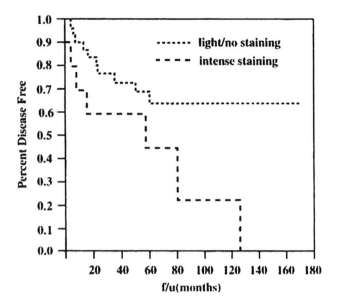

FIGURE 8.12. Disease-free survival of patients with medulloblastoma according to the intensity of staining for p53 in the tumors. Intense staining, reflective of overexpression of p53, carries with it a poor outcome (from ref. (203) with permission).

developing medulloblastoma. *TP53* mutations also have been identified in 10% of sporadic human medulloblastomas (1), and high-level expression of p53 in medulloblastoma is associated with a poor clinical outcome (Figure 8.12) (203, 204).

The identification of TP53-ARF pathway alterations may have important implications for the treatment of medulloblastoma. This pathway plays a critical role in modulating tumor radio- and chemosensitivity (205, 206); thus, the finding that this pathway is disrupted in large-cell medulloblastoma (207) may explain, at least in part, why this tumor subtype is highly resistant to treatment (1, 5, 207). A study of the *TP53-ARF* tumor suppressor pathway (207) showed three of 29 (10%) of large-cell medulloblastomas to have mutated *TP53*, but a further 10% of these tumors had mutually exclusive alterations in the *INK4A/ARF* pathway indicating that the *TP53/ARF* pathway is likely to be disrupted in about 20% of large-cell medulloblastomas. Five of these tumors were large-cell/anaplastic medulloblastomas or tumors with significant anaplasia. The authors (207) suggested that the alterations within the *TP53-ARF* pathway may contribute to the development of aggressive forms of medulloblastoma.

Gene Expression by Microarrays in Medulloblastoma

The technology of DNA microarrays and similar methods have been used for gene expression profiling and prediction of clinical outcome in medulloblastoma (208, 209).

Gene expression studies in medulloblastoma of mice and humman identified a group of genes that may be central to medulloblastoma tumorigenesis, including high expression of *MYC* and *ERBB2* (210).

A gene expression profiling study showed a large number of genes to be either up-regulated or downregulated in medulloblastoma (211). Of interest was the up-regulation of matrix metalloproteinases 11 and 12. Down-regulated genes included *TrkC*. Matrix metalloproteinases were expressed in varying extents in four medulloblastomas tested (including a desmoplastic type) (212). Metalloproteinases were expressed in different patterns in gliomas and ependymomas versus medulloblastomas (212A). The aforementioned studies represent the search for biological markers helpful to investigate the embryological origin of medulloblastoma and for markers relevant to the prognosis and therapy of these tumors (213, 214).

A microarray study showed cohorts of genes that exhibited differences from control cerebellum specimens (215). The most accurate outcome in medulloblastoma is best predicted through microarray gene expression profiling, an opinion voiced in a review of risk stratification based on new directions in this area (217A). Genome-wide expression profiles are capable of partitioning tumor groups into subgroups associated with specific genetic alterations; the latter may ultimately be of high utility in the selection of appropriate therapy (217B).

Other Genes and Factors in Medulloblastoma

Survivin, encoded by the *BIRC5* gene, located at 17q25, a chromosomal region frequently amplified in medulloblastoma (55), is a member of the family of inhibitors of apoptosis and plays an important role in the control of cell division (216, 217). Its expression is upregulated in most cancers, including those of the nervous system (218). Generally, survivin is undetectable in most adult tissues, including the cerebellum, but *BIRC5* is strongly expressed in embryonic and fetal organs and in virtually every tumor studied (218). *BIRC5* was found to be overexpressed in all 23 medulloblastomas examined, both pediatric and adult (219), but not in adult cerebellum or brain. Similar to pigment epithelium-derived factor (PEDF), survivin has been advocated as a therapeutic target in medulloblastoma (219, 220).

The *MATH1* gene encodes a basic helix-loop-helix transcription factor specifically expressed in the granule cell precursors and essential for genesis of cerebellar granule neurons (221). Medulloblastomas expressing *MATH1* originate from the external granule layer and have high expression of Trk-C and PEDF (located at 17p13.1) and occur in the cerebellar hemispheres, whereas medulloblastomas that do not express *MATH1* originate in the ventricular matrix and are located in the vermis (219). Furthermore, *MATH1* controls cerebellar granule cell differentiation by regulating multiple components of the Notch signaling pathway (222). Overexpression of *MATH1* may disrupt this differentiation (223).

PEDF is a potent and broadly acting neurotrophic factor that protects neurons in the CNS against insults and damage (224). Thus, cerebellar cells in culture use PEDF as a survival factor (225, 226). PEDF is also a strong antiangiogenic factor and activates a number of pathways which could be appropriate therapeutic targets (224).

The *MXI1* gene, whose protein product is a member of the MAD family of proteins, capable of repressing the effects of *MYC* (227), was studied for mutations in 22 medulloblastomas (189). None was observed. In another study the *MNT* gene, located at 17p13.3, was studied in medulloblastomas of which 49% of informative cases had allelic loss of that chromosomal region (228). The results of the study indicated the presence and functional integrity of *MNT* and that *MNT* is not likely to be a gene involved in the pathogenesis of medulloblastoma. *MNT* is thought to be a TSG encoding a *MAX*-interacting nuclear protein with transcriptional-repressor activity (228). *MAX*, in contrast, heterodimerizes with *MYC* to regulate gene expression (227).

Prognostic Factors in Medulloblastoma

The prognostic aspects of medulloblastoma have been addressed from clinical, histologic, cytogenetic and molecular viewpoints, and although consensus may exist about some of them, the sizable literature in this area is witness to the uncertainty of the prognostic value of some of these aspects. Findings in medulloblastoma that generally have a favorable prognostic impact are shown in Table 8.12.

Unfavorable prognostic factors in medulloblastoma are shown in Table 8.13 essentially in their effective order, although unanimity does not exist regarding the prognostic impact of each of the factors. The interpretation of the prognostic elements involved in medulloblastoma is still complicated by the difficulty in the critical interpretation of the clinical, therapeutic and histologic factors involved (1, 15).

Prognosis in medulloblastoma has been linked to histological variants, with the desmoplastic type tending to have a more favorable outcome than classic medulloblastomas (229, 230). However, other studies (80, 117) found no difference in survival between patients with desmoplastic and nondesmoplastic tumors. The need for more practical changes in medulloblastomas applicable to the clinical situation and then to the prognosis of this tumor has been voiced (230A). For further discussions and evaluations of prognostic aspects of medulloblastoma, the following publications should be consulted (32, 231–247).

Mitotic Indices in Medulloblastoma

Data on dividing cells within tumors, so-called proliferation or mitotic indices, have been established for almost all tumors, medulloblastoma included. Ki-67 is a nuclear antigen expressed in all phases of the cell cycle, excluding the G0 phase. Initially selected as a measure of dividing cells was the reaction of an antibody with Ki-67, which had the major disadvantage that its immunohistochemical application required fresh or frozen tissue; hence, retrospective studies on paraffin-embedded tissues were not possible (248). Subsequently, other Ki-67 analog antibodies raised to bacterially expressed Ki-67 peptide sequences were introduced. These reacted well in formalin-fixed tissues. The most widely used of these antibodies is MIB-1 (248). Results obtained with MIB-1 immunostaining have generally shown that high scores were related to tumor grade and prognosis, including findings in medulloblastoma. Even with a caveat (249), the Ki-67 mitotic index remains a valuable prognostic marker in tumors.

Another index of mitotic activity is that demonstrated by the proliferating cell nuclear antigen (PCNA), which is an auxiliary protein of DNA polymerase-δ necessary for DNA replication; its expression is highest in the G1/S cell cycle phase. Results obtained with anti Ki-67 antibodies are less

TABLE 8.12. Favorable prognostic factors in medulloblastoma.

	References
Sex-female	*Roberts et al 1991* (629)
	Weil et al 1998 (667)
Gross total resection	*Zeltzer et al 1999* (668)
No metastases present	*Zeltzer et al 1999* (668)
Desmoplastic histology	*Chatty et al 1971* (669)
Increased apoptotic index (A1)	*Haslam et al 1998* (401)
Hyperdiploidy (DNA levels)	*Gajjar et al 1993* (265)
High *TRKC* expression	*Segal et al 1994* (670)
	Kim et al 1999 (671)
	Grotzer et al 2000 (184, 263)
	Pomeroy et al 2002 (208)
	Ray et al 2004 (665)
β-catenin expression	*Ellison et al 2005* (665)
	Misaki et al 2005 (658)
Expression of genes characteristic of cerebellar differentiation β-*NAP*, *NSCL1*, sodium channels/and genes encoding extracellular matrix proteins (PLDD lysyl hydroxylases, collagen type V and I elastin)*NSCL1*, sodium channels/and genes encoding extracellular matrix proteins (PLDD lysyl hydroxylases, collagen type V and I elastin)	*MacDonald et al 2001* (129)
	Pomeroy et al 2002 (208)
	Fernandez-Teijeiro et al 2004 (672)

Table based on Fisher et al. 2004 (214).

TABLE 8.13. Unfavorable prognostic factors in medulloblastoma.

	References
Young age of patients	*Gurney et al 1999* (673)
	McNeil et al 2002 (674)
Subtotal resection	*Zeltzer et al 1999* (668)
Metastases present	*Zeltzer et al 1999* (668)
	Ray et al 2004 (665)
Large-cell/anaplastic histology of medulloblastoma	*Giangaspero et al 1992* (13)
	Eberhart et al 2002 (120)
	McManamy et al 2003 (28)
Elevated Ki-67/MIB-1 proliferative index	*Grotzer et al 2001* (251)
Aneuploidy (DNA level)	*Zerbini et al 1993* (266)
Increased *ERBB2* expression	*Gilbertson et al 1998, 2001* (426,661)
	Gajjar et al 2004 (742A)
	Ray et al 2004 (665)
Increased expression or amplification of *MYC*	*Scheurlen et al 1998* (79)
	Brown et al 2000 (14)
	Herms et al 2000 (115)
	Grotzer et al 2001 (188)
	Aldosari et al 2002 (182)
Upregulation of PDGFR	*Scheurlen et al 1998* (79)
	MacDonald et al 2001 (129)
Overexpression of calbindin-D$_{28K}$	*Pelc et al 2002* (675)
Genes related to cell proliferation and metabolism (*MYBL2*), enolase 1, *LDH, HMG1(Y)*, cytochrome *c* oxidase, and multidrug resistance (sorcin)	*Pomeroy et al 2002* (208)
	Fernandez-Teijeiro et al 2004 (672)
High MIB index and anaplasia	*Miralbell et al 1999* (254)
	Eberhart et al 2000 (371)
	Ellison 2002 (1)
	Ray et al 2004 (665)
Active apoptosis	*Korshunov et al 2002* (31)
Value of apoptotic rate as prognostic indicator has been questioned.	*Pizem et al 2005* (676)
17p loss in medulloblastoma*	*Cogen et al 1992* (70)
	Batra et al 1995 (90)
	Scheurlen et al 1998 (79)
	Gilbertson et al 2001 (661)
	Lamont et al 2004 (15)
Isolated 17p LOH in medulloblastoma	*Bigner et al 1988* (39)
	Biegel et al 1997 (104)
	Gilbertson et al 2001 (661)
Amplifications of *EGFR* (*HER2* and *HER4*), *MYCN, ERBB2* and *ERBB4* genes	*Gilbertson et al 1997, 2001* (425,661)
	Herms et al 2000 (115)
Note that an increased apoptotic index (AI) has been listed as a favorable prognostic factor. Two studies indicate either an unfavorable or lack of prognostic value to the AI. No prognostic value was found for the KI-67 index or for the reaction to GFAP or synaptophysin.	*Grotzer et al 2001* (402)
	Ray et al 2004 (665)
	Miralbell et al 1999 (254)
	Woodburn et al 2001 (203)
Loss of caspase-8 protein expression	*Pingoud-Meier et al 2003* (483)
Iso(17q) in poor-risk childhood medulloblastoma	*Pan et al 2005* (677)

This table is based partially on an editorial on biologic risks in medulloblastoma (214). The factors are listed in the order of their significance.
*Failure to find an association between 17p loss and prognosis has been described (103,104).

variable than those obtained with PCNA; thus, Ki-67 seems to be a more specific marker of proliferation. A high Ki-67 index (>20%) has been considered to be an unfavorable sign in medulloblastoma (and in PNET(250–252)); other found no such association (253,254). The high Ki-67 indices found in the above studies are in keeping with those of previous reports (253, 255–258). Medulloblastoma is one of the most rapidly proliferating primary intracranial tumors and, hence, MIB-1 (Ki-67) indices are high (≥20%) (256,258).

Molecular Causation and Progression in Medulloblastoma

Evidence of molecular progression in medulloblastomas and other CNS embryonal tumors is reflected in and associated with sequential genetic alterations of the type seen in most other tumors (27). The strongest support comes from cases in which tumoral material was available for molecular analysis on more than one occasion. For example, CGH analysis of material from both a primary nodular medulloblastoma and an anaplastic recurrence removed 3 years later (27) showed that the genetic alterations in the recurrent tumor involved the same four chromosomes as in the primary tumor and six additional chromosomes as well.

A study (259) compared gene expression profiles of 23 metastatic and nonmetastatic medulloblastomas using oligonucleotide arrays and identified increased signaling through the MAP kinase pathway via up-regulation of PDGFα in tumors that spread from the cerebellum, suggesting this pathway also may be involved in malignant progression and metastasis.

Although many details of molecular and histological progression in medulloblastoma are still unclear, a rough framework can be suggested (Figure 8.13). The increased incidence of medulloblastomas in patients inheriting mutations that lead to activation of Wnt and SHH signaling pathways indicates that these pathways act early in medulloblastoma formation, with the latter pathway especially important in the pathogenesis of nodular lesions (143, 208).

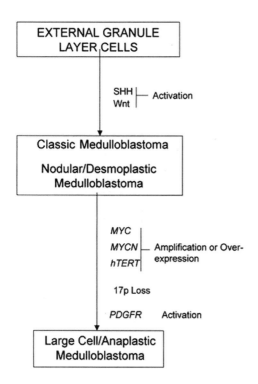

FIGURE 8.13. Schematic presentation of molecular progression of medulloblastoma (from ref. (27), with permission). As suggested by these authors, the increased incidence of medulloblastoma in patients inheriting mutations that lead to activation of the Wnt and SHH signaling indicates that these pathways act early in some medulloblastoma formation, with the latter pathway especially important in the pathogenesis of nodular lesions (77, 208), and as supported by animal studies (260, 261). Undoubtedly other genetic events involving pathways other than Wnt an SHH are involved in early medulloblastoma causation. Subsequent events, including amplifications, overexpressions or mutations of genes may be responsible for progression of the medulloblastomas.

Animal models also support this concept (260, 261). Additional genetic hits, including 'initiation' events involving pathways other than hedgehog or Wnt, almost certainly occur as well. All medulloblastoma subtypes can then progress histologically and molecularly via mutation or amplification of oncogenes (MYC, MYCN, hTERT), upregulation of various signaling pathways (PDGF) or loss of as yet uncharacterized tumor suppressor loci especially on i(17p) (27).

Future studies will map the genetic and histological course of medulloblastoma progression more fully. Primary and recurrent lesions should be compared both histologically and molecularly to determine what molecular abnormalities and gene expression changes that are associated with recurrence and progression. Similar analyses of microdissected anaplastic and nonanaplastic regions should yield additional information. A better understanding of how molecular and microscopic events interrelate will hopefully facilitate greater precision and efficacy in the treatment of patients with medulloblastoma.

Molecular and Cytogenetic Correlations with Medulloblastoma Phenotype

The molecular alteration most strongly associated with the large-cell/anaplastic medulloblastoma subtype is MYC oncogene amplification found in an early study, which was prompted by the presence of numerous dmin in a large-cell medulloblastoma (13). MYC is well known to promote cellular proliferation, increased cell size and apoptosis, all features of large-cell/anaplastic medulloblastomas. MYCN or MYC amplification has been detected in the majority of large-cell/anaplastic medulloblastomas examined (14, 27, 85, 120, 180, 181, 262). Increased MYC mRNA levels predict an unfavorable clinical outcome (115, 263).

Several groups have detected increased aneuploidy in large-cell/anaplastic medulloblastomas, suggesting that additional genetic changes associated with anaplasia are yet to be discovered (85, 120). On average, there are more than twice as many chromosomal copy number alterations in tumors with anaplasia than in their non-anaplastic counterparts (27). Examination of additional large-cell anaplastic medulloblastomas may ultimately reveal specific chromosomal loci commonly altered in these tumors. Expression profiling, which has already been used to discriminate between classic and nodular lesions (208), should prove useful in identifying genes contributing to the large-cell/anaplastic medulloblastoma phenotype.

In a study of 87 medulloblastomas using FISH to detect chromosome 17 abnormalities and losses of 9q22 and 10q24 and amplification of MYC and MYCN (15), the frequency of abnormalities was essentially in keeping with results previously published (16, 79, 90, 111, 120, 264). In toto, these findings suggest an association between pathological variants and genetic abnormalities, e.g., an association between diploid status and desmoplastic phenotype, a high incidence of i(17q) and loss of 17p in large-cell tumors and the lack of correlation between ploidy status and clinical outcome (15). However, conflicting data in the literature exist on the latter issue (264–266).

Established Medulloblastoma Cell Lines

Established medulloblastoma cell lines whose studies have yielded information of direct interest to this chapter date back several decades (267). In general, the findings in cell lines essentially limn those in primary tumors, though some differences exist, e.g., complex karyotypes observed in some of the cell lines (112), although often containing changes common in primary tumors, e.g., 10q-, i(17q), –22 (72,111,113), a higher incidence of dmin (182,268) and hence *MYC* amplification in cell lines (106).

The use of medulloblastoma cell lines and their xenografts (269–274) for the study of various genetic, metabolic, and therapeutic parameters has been advocated.

Increased *MYC* expression was shown to exist in medulloblastoma cell lines (275) without the presence of amplification. The authors suggested that a number of mechanisms exist for increased *MYC* expression.

The two pathways to be discussed next, i.e., SHH and Wnt pathways, although they may account for the development of only a small percentage of medulloblastomas, particularly of the sporadic type, the understanding of these pathways and the information they have yielded on medulloblastoma tumorigenesis will certainly help to decipher the processes involved in the development of these tumors.

SHH Signaling Pathway in Medulloblastoma

To simplify annotations referring to genes and other factors in human tumors, we have used the following annotations for the human homologs of those of the mouse (276,277), i.e., SHH to refer to the sonic hedgehog, PTCH to signify patched, and SMOH to refer to smoothened. Table 8.14 shows some of the changes affecting members of the SHH signaling pathway in medulloblastoma.

Interest in some of the genes involved in the SHH signaling pathway and their possible role in medulloblastoma development stems from a number of observations. The most cogent is related to mutations of the *PTCH* gene responsible for the nevoid basal cell carcinoma (Gorlin) syndrome (NBCCS) (Table 8.8), which is associated with a high incidence of medulloblastoma (277), especially of the desmoplastic type (26). The gene responsible for NBCCS acts as a tumor suppressor gene (TSG) in medulloblastoma (278).

Another observation is the spontaneous development of medulloblastomas in mice with mutations of the *PTCH* gene or other anomalies of the SHH pathway (279–284).

The key role played by the SHH signaling pathway in cerebellar development and especially as a potent mitogen for the granular cell precursor(s) is another source of interest (285).

Studies of medulloblastoma in mice defective for *PTCH* (260, 261, 286–290) have thrown much light on events in the human, though also revealing important differences (290A). Puzzling has been the observation that only approximately15% of these mice develop medulloblastoma in the first year of life, akin to the incidence in the human (290B,C). Hence, other factors must be implicated as causative agents for medulloblastoma tumorigenesis, e.g., IGF has been reported as a required factor in medulloblastoma formation in mice (284A, 289). Incidentally, DNA microarray studies have shown increased expression of IGF in desmoplastic medulloblastoma (208).

SHH, produced by the Purkinje neurons of the cerebellum, regulates the proliferation of cerebellar granule cell precursors (CGCP) (285, 291, 291A, 292) The SHH antagonist, *REN*^KCTD11^, regulates proliferation and apoptosis in developing granule cell precursors (292A). Mutational inactivation of *PTCH* may result in inappropriate or sustained proliferation of these progenitor cells in the external granular layer of the cerebellum (291, 292), suggesting that developmental control genes play an important role in the pathogenesis of medulloblastoma (285).

Activity of SHH proteins is required for expansion of specific neuronal precursor populations. SHH signaling upgrades expression of *MYCN* and cultured CGCP and thus may play a role in medulloblastoma tumorigenesis (293).

The SHH signaling pathway is a very complicated one with many constituents and functions, (Figure 8.14) the latter required for normal regulation of cell proliferation and differentiation during a diverse array of patterning events, including neural tube differentiation in vertebrates (294, 295). Aberrant activity of the SHH pathway is associated with the initiation and growth of a variety of human malignant tumors, including medulloblastoma (296). The already existing complexity of the SHH pathway is further compounded by the significant gaps remaining in our understanding of SHH signal transduction (297). In this chapter, only those aspects of the SHH pathway shown to be involved in medulloblastoma will be presented.

SHH-PTCH-SMOH signaling has a novel signal transduction mechanism (298, 299). In the absence of the SHH signal, the transmembrane protein PTCH represses the constitutive signaling pathway of a second membrane protein, SMOH, by virtue of its ability to form PTCH-SMOH complexes. The interaction of SHH with PTCH or mutations or the *PTCH* gene result in failure of PTCH to inhibit SMOH and leads to constitutive activity of the SHH pathway and to transcription of SHH target genes and to cancer (e.g., basal cell carcinoma), including medulloblastoma (144, 277, 285, 300–302). The secreted protein SMOH becomes activated, leading to post-transcriptional modification and nuclear translocation of the *GLI* family of transcriptional factors. Once in the nucleus,

TABLE 8.14. Some changes in the SHH pathway in medulloblastoma.

	References
Inactivating mutations of *PTCH* identified in approximately 8–12% of sporadic medulloblastomas, although the percentages have been much lower or much higher in some of the studies.	*Schofield et al 1995 (26)* *Pietsch et al 1997 (143)* *Raffel et al 1997 (144)* *Vorechovsky et al 1997 (317)* *Wolter et al 1997 (145)* *Zurawel et al 2000 (290B)*
One study suggested that such mutations occur preferentially in desmoplastic medulloblastomas.	*Pietsch et al 1997 (143)*
Mutations of *PTCH* (located at 9q22) and loss of this locus primarily occur in desmoplastic medulloblastoma, but may also occur in other variants of medulloblastoma.	*Schofield et al 1995 (26)* *Pietsch et al 1997 (143)* *Lamont et al 2004 (15)*
A point mutation of *SHH* in exon 2 was found in 1/14 medulloblastoma. However, this mutation was not found in a larger series of cases.	*Oro et al 1997 (320)* *Wicking et al 1998 (321)* *Zurawel et al 2000 (290B)*
Activation of the SHH pathway occurs primarily in nodular sub-types of medulloblastoma.	*Pietsch et al 1997 (143)* *Pomeroy et al 2002 (208)*
SSCP analysis showed *PTCH* mutations in 10–15% of medulloblastomas.	*Pietsch et al 1997 (143)* *Raffel et al 1997 (144)* *Berman et al 2002 (318)* *Di Marcotullio et al 2004 (305)* *Oliver et al 2005 (290A)*
LOH of *PTCH* in 5/24 medulloblastomas; in three of these, mutations of remaining allele were identified and located on exon 17.	*Raffel et al 1997 (144)*
In two separate studies, LOH of 9q22-q23, locus for *PTCH*, or mutations of the gene were found in 6/50 medulloblastomas, in 3/10 PNET and in 2/9 BCC.	*Vorechovsky et al 1997 (317)* *Wolter et al 1997 (145)*
Mutations of *PTCH* found in 2/15 classical and in 2/17 desmoplastic medulloblastomas and in 1/5 PNET.	*Wolter et al 1997 (145)*
PTCH mutations (nonsense) found in 2/14 sporadic medulloblastomas.	*Xie et al 1997 (315A)*
Missense mutations of *SMOH* found in two medulloblastomas. One of these mutations was of the activating variety, as described in basal cell carcinoma (BCC).	*Reifenberger et al 1998 (304)* *Lam et al 1999 (298)* *Xie et al 1998 (315)*
No *SHH* mutations found in 55 medulloblastoma samples from 52 patients; nine were desmoplastic.	*Wicking et al 1998 (321)*
SSCP and heteroduplex analyses showed inactivation of one allele of *PTCH2* (located at 1p32-p34) in 1/92 samples of medulloblastomas.	*Smyth et al 1999 (146)*
LOH of *PTCH* found in 2/2 medulloblastomas from patients with NBCCS and in only 1/33 sporadic tumors.	*Vortmeyer et al 1999 (316)*
A method for detecting *PATCHED* mutations in medulloblastoma has been described, utilizing direct sequencing. The results did not differ from those obtained by other methods, e.g., SSCP, i.e., about 2% of the tumors contained mutations of *PATCHED*.	*Dong et al 2000 (781)*
The down-regulation of the SHH pathway in cultured medulloblastoma cells and reversal of the effects of mutations in this pathway as they relate to possible therapeutic approaches have been described.	*Taipale et al 2000 (678)* *Sasai et al 2006 (679)*
The *SHH* pathway in development and disease.	*Villavicencio et al 2000 (740)*
Whereas, *PTCH* is believed to have tumor suppressor activity, both ligand-encoding genes *SHH* and *SMOH* are candidates for putative oncogenes.	*Zurawel et al 2000 (290B)*
PTCH involved in the transcriptional regulation of the *WNT* and *TGFβ* gene families.	*Zurawel et al 2000 (290B)*
PTCH mutations occurred in 3/37 medulloblastomas; *SHH* and *SMOH* were not mutated.	*Zurawel et al 2000 (290B)*
The *SUFU* gene (located at 10q24.3) may be affected by loss of the locus in all medulloblastoma variants. Mutations of *SUFU* occur in low frequency in medulloblastoma.	*Rubin and Rowitch 2002 (680)* *Taylor et al 2002 (308)*
A caveat regarding the use of cultured medulloblastoma cell lines for testing inhibitors of the SHH pathway was issued. The authors indicate that the propagation of tumor cells in culture inhibits SMOH activity, raising concern about the use of such cells to test the efficacy of SHH pathway inhibitors as anticancer therapies.	*Gumireddy et al 2003 (681)* *Hallahan et al 2003 (682)* *Sasai et al 2006 (679)*
Reviews of some of the molecular pathways, e.g., hedgehog/PTCH, which may serve as a target for appropriate therapy in medulloblastoma have appeared.	*Pomeroy and Sturla 2003 (782)* *Gilbertson 2004 (783)* *Newton 2004 (784)*
Low-level SHH signaling in medulloblastoma is associated with the selective maintenance of high cytoplasmic domain expression of the CYT-1 isoform of ErB-4, a phenomenon that exerts a tumor-promoting effect.	*Ferretti et al 2006 (741)*
From results in mice and literature data, it was proposed that SHH signaling might synergize simultaneously with Notch and Wnt signaling in medulloblastoma development by controlling Notch and Wnt pathway ligand and/or target gene expression.	*Dakubo et al 2006 (742)*

GLI proteins bind to DNA and increase the transcription of specific target genes (303).

Of the known components of the SHH signaling pathway, the only gene mutated with notable frequency in sporadic medulloblastoma is *PTCH* (located at 9q22.2) (290B). These mutations affect medulloblastomas, more often, but not exclusively, of the desmoplastic type. Mutations of the *SHH*, *SMOH* and *SUFU* genes appear to be rare events in medulloblastoma

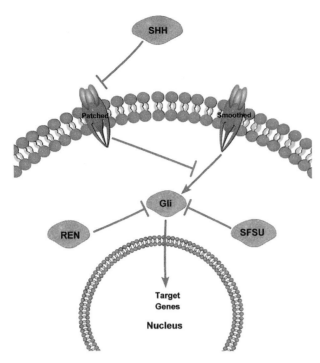

FIGURE 8.14. SHH signaling pathway. PTCH, a membrane protein, is a receptor for SHH ligand, a secreted signaling protein. SMOH is another membrane protein inhibited by Patched, in the absence of an SHH signal. When SHH signaling inhibits PTCH, SMOH becomes active and transduces signals that convert Gli proteins into transcriptional activators of target genes. Inactivating mutations of PTCH or activating mutations of SMOH may be involved in medulloblastoma tumorigenesis. The gist of the pathway involves *PTCH* which encodes a transmembrane glycoprotein that functions as a negative regulator of the SHH signaling pathway. Binding of the SHH ligand to the PTCH receptor liberates SMOH activity, which leads to release and nuclear translocation of activator and repressor forms of the *GLI* family of zinc protein transcription factors (from refs. (329) and (330), with permission). Inactivation of *PTCH* via loss-of-function mutation leads to constitutive pathway activation due to unrestrained SMOH activity (310,328,331,332,335), leading to overexpression of the oncogene *GLI1* and some members of the *WNT* and *TGFβ* gene families (284A).

(290B, C, 298,304). For example, a point mutation of the *SHH* gene was seen in one of 14 tumors, of *SMOH* in 1/21, and of *SHH* in none of >50 samples.

Information regarding the SHH pathway was enlarged by studies of the *REN* gene (305). The *REN* gene, located at 17p13.2, was given attention since deletions at that chromosomal region occur in nearly 40% of sporadic medulloblastomas and LOH in up to 50% of these tumors (1, 102) and might involve the *REN* gene. The *REN* gene is involved in neural cell proliferation and differentiation in the mouse (306), in which it consistently induces growth arrest and differentiation of neural cells.

REN displays allelic deletion as well as significantly reduced expression in medulloblastoma (142, 305). Furthermore, *REN* inhibits medulloblastoma cell proliferation and

colony formation in vitro and suppresses xenograft tumor growth in vivo. *REN* seems to inhibit medulloblastoma growth by negatively regulating the SHH by a number of mechanisms: it antagonizes the *GLI*-mediated transactivation of *SHH* target genes, by affecting *GLI1* nuclear transfer, and by its growth inhibitory activity being impaired by *GLI1* inactivation. Therefore, *REN* seems to be a suppressor of SHH signaling and its inactivation might lead to a deregulation of the tumor-promoting SHH pathway in medulloblastoma.

Another gene involved in the SHH pathway is *SUFU* located at 10q24.3, a band often affected by deletions in medulloblastoma. SUFU (suppressor of fused) affects the functions of the SHH pathway, including functional repression of *GLI1* (307). A subset of children with medulloblastoma was found to carry germline and somatic mutations of *SUFU*, accompanied by LOH of the wildtype allele (308). Mutations of *SUFU* in four of 46 (9%) medulloblastomas have been reported (308). This is comparable to the frequency seen for *PTCH* (9%) and *CTNNβ1* (encoding β-catenin) of 5% (310). The truncated proteins of *SUFU* are unable to export the GLI transcription factor from the nucleus to the cytoplasm, thus activating the SHH signal (308). The authors concluded that *SUFU* functions as a TSG in a subset of desmoplastic medulloblastomas. Interestingly, when patients with NBCCS but without *PTCH* mutations were examined for possible *SUFU* mutations, none was found (308).

As already mentioned above, mutations of *PTCH, SMOH,* and *SUFU* genes lead to abnormal SHH pathway activation in a proportion of syndrome-associated and sporadic medulloblastomas (309). Mutations of *PTCH* (located at 9q22.2) and related genes (e.g., *SMOH*) are the cause of NBCCS or Gorlin syndrome in >50% of these cases (276,277,310–313); such mutations also are seen in one of three of sporadic basal cell carcinomas (BCC) (145,298,304,308,314–316) and in 10–15% of medulloblastomas, more often of the desmoplastic type than others (26, 76, 143–145, 290A, 290A, 304, 305, 308, 315A, 317, 318). Thus, it seems that inactivation of *PTCH* function is involved in at least a subset of sporadic medulloblastomas. This is in keeping with observed losses on 9q in 10–18% of medulloblastomas (26, 76, 319), including the locus for *PTCH* (276).

Rare mutations of the *SHH* gene in medulloblastoma have been described (320), although more extensive studies have failed to find such mutations (290B, C, 321, including desmoplastic medulloblastomas.

Proliferation of cerebellar granule neuron precursors (CGNP) is largely driven by basic helix-loop-helix transcription factors in the HES family and the helix-loop-helix zip transcription factor, *MYCN* (290A, 293, 322). Notch signaling induces HES family members and SHH signaling induces *GLI1* and *MYCN*. Both pathways regulate *Cyclin D1*, a possible mediator of CGNP proliferation (288). *MYCN* is a direct target of the SHH pathway functioning as a regulator of cell cycle progression in cerebellar granule neuron precursor(s) (290A, 293).

Analysis of neurodevelopmental pathways in medulloblastoma provides the foundation for new therapies that may specifically block mitogenic signal transduction (285, 292, 323). The combined mutations causing increased SHH pathway activity have been identified in up to 20% of medulloblastoma (144, 298, 308, 317, 324, 325). These include inactivating mutations of *PTCH1* or *SUFU*, genes that encode negative regulators of SHH signaling or activating mutations of *SMOH*, a gene that encodes the SHH-mediated SMOH. Agents have been shown to block SHH signaling in medulloblastoma cell cultures, thus causing extensive cell death and could serve as a basis for therapy (318, 323).

SHH and Notch pathways were activated in 16/24 medulloblastomas, including all desmoplastic tumors. Of nondesmoplastic medulloblastomas, 50% (nine of 18) had elevated *GLI1*, indicating that SHH pathway activation is much more common than previously known. *GLI2* was elevated in *all* desmoplastic medulloblastomas and in 39% of others. *MYCN*, a direct target of SHH in CGNP, was elevated in 88% (21 of 24) medulloblastomas. *PTCH1* expression was present in 33% (2/6) of desmoplastic and 11% (2/18) of nondesmoplastic tumors. *PTCH2* paralleled that of *PTCH1* (288).

The *BTRC* gene (ubiquitin ligase involved in SHH and Wnt signaling), located at 10q24.3, distal to the *PTEN* gene, located at 10q23.31, showed no mutations in 21 and 36 medulloblastomas, respectively (308).

Allelic deletion of *PTCH* is infrequent in sporadic classic medulloblastoma and a major role for *PTCH* in sporadic medulloblastoma appears unlikely (316). In the two medulloblastomas (both desmoplastic) of patients with NBCCS, LOH of the *PTCH* locus was present. In sporadic BCC, *PTCH* acts as a TSG (311) and is found to be mutated in more than 30% of BCC (314). *SMOH*, when mutated, can function as an oncogene in BCC (315), and it has been found to be mutated in 20% of BCC (298).

An additional putative tumor suppressor gene that is a homolog of the *PTCH* gene has been isolated, the *PTCH2* gene (146), which maps to 1p32.1-p32.3. Using SSCP and heteroduplex analysis, inactivation of one allele of this gene was demonstrated in one of 92 medulloblastoma tissue samples and cell lines studied. It remains a possibility that mutation of *PTCH2* may be involved in the development of some medulloblastomas (146). *Patched-1*, a TSG, containing exon 12b may be expressed in some medulloblastoma tissues and cell lines, possibly playing a role in the pathogenesis of this tumor (146A). Appropriate information may be found in the literature dealing with the broad facets of the SHH pathway in embryogenesis and cancer (301, 303, 326–339, 339A).

Wnt Signaling Pathway in Medulloblastoma

The cellular functions of the Wnt pathway have been comprehensively reviewed (339B, 340–342), including its possible role in medulloblastoma tumorigenesis (343).

Wnt proteins, named after the *Drosophila* protein "wingless" and the mouse protein "Int-1", represent a large family of secreted cysteine-rich glycosylated proteins (344). Members of the Wnt family of secreted signaling proteins induce subsets of target genes at defined distances from their source by activating conserved transduction cascades (341). Aberrant activation of these pathways can lead to various types of cancers, possibly including medulloblastoma (336, 342A).

Table 8.15 shows features of β-catenin changes and effects in medulloblastoma.

The Wnt pathway controls transcription of target genes via its central component β-catenin, which serves as an interaction hub for a complex array of membrane-associated, cytosolic and nuclear proteins (Figure 8.15) (343A). At the cell membrane, β-catenin associates with E-cadherin in adherens junctions to provide a cell adhesion, signal transduction, cell proliferation and transcriptional control functions (344A, 345–347). In the absence of Wnt signaling, the remaining cytoplasmic β-catenin protein is constitutively marked for proteasomal degradation by glycogen synthase kinase 3β (GSK3β)-mediated phosphorylation (344A, 340, 348–351). GSK3β is part of a degradation complex comprising Axin and APC. APC and Axin proteins form a complex with β-catenin in the cytoplasm and act together with the GSK3 kinase to promote β-catenin phosphorylation and degradation (352, 353). Upon binding of an extracellular Wnt ligand to members of Frizzled receptors, the degradation complex is disrupted (345). As a result, the levels of free cytoplasmic β-catenin rise and β-catenin enters the nucleus where it interacts with Lef/Tcf DNA binding proteins to initiate transcription of Wnt-responsive genes, e.g., *MYC*, *Cyclin D1* and others (354, 355, 355A). Like the SHH pathway, the Wnt pathway is also very complicated with gaps remaining at various levels, from the structure of the core complex formed by β-catenin to the mechanisms controlling transcription of Wnt targets (341). The available information on the Wnt pathway in medulloblastoma must be viewed with the above background in perspective.

Activation of the Wnt/Wg signaling pathway occurs in up to 25% of medulloblastomas and is associated with a favorable outcome. Medulloblastomas showing evidence of Wnt/Wg activation (*CTNNB1* mutation, β-catenin nuclear stabilization, or both) were exclusively associated with a distinct genomic signature involving loss of an entire copy of chromosome 6 with few other changes (341A). In contrast, Wnt/Wg-negative medulloblastomas clustered into an unrelated subgroup characterized by multiple genomic defects common in these tumors (loss of 17p, chromosomes 8, 10, and 16; gain of chromosome 7 and 17q). Activation of the Wnt/Wg pathway was independent of chromosome 17 aberrations. Wnt/Wg active

TABLE 8.15. Some features of β-catenin in medulloblastoma.

	References
β-catenin interacts in the nucleus with T-cell factor 4 (Tcf-4), which leads to the overexpression of *MYC*, which, may be related to tumor growth.	*Korinek et al 1997 (682A)*
	Morin et al 1997 (649)
β-Catenin mutations not present in medulloblastoma with *PTCH* mutations.	*Raffel et al 1997 (144)*
	Zurawel et al 1998 (346)
β-catenin expression and mutation in sporadic medulloblastomas. Mutations present in approximately 5% of tumors.	*Zurawel et al 1998 (346)*
	Eberhart et al 2000 (371)
	Huang et al 2000 (361)
	Koch et al 1996 (130)
	Yokota et al 2002 (370)
β-Catenin expression binding to SOX proteins modulates Wnt signaling.	*Zorn et al 1999 (683)*
β-catenin upregulates *Cyclin D1* rather than *MYC*, whereas *γ-catenin* activates *MYC*.	*Barker and Clevers 2000 (684)*
	Kolligs et al 2000 (685)
β-Catenin was found in the nuclei of 9/50 (18%) of sporadic medulloblastomas by immunohistochemistry and in a tumor from a patienat with Turcot syndrome. The nuclear localization of *β-catenin* expression points to activation of the Wnt pathway.	*Eberhart et al 2000 (371)*
β-Catenin activity can be repressed by being dislocated from the nucleus through binding to the SUFU protein (Meng et al 2001).	*Meng et al 2001 (686)*
β-Catenin and *γ-catenin* function in the Wnt signal pathway in the nucleus through binding to the SUFU protein in a similar manner.	*Misaki et al 2005 (658)*
β-Catenin was expressed in nearly 80% of medulloblastomas (19/24), *γ-catenin* in 37% (9/24) and *Cyclin D1* in 25% (6/24). No amplification of these genes was found.	*Misaki et al 2005 (658)*
γ-Catenin expression and accumulation has been evaluated in medulloblastomas and found to be a favorable prognostic index. The reverse was true of *Cyclin D1*.	*Misaki et al 2005 (658)*

tumors could not be distinguished on the basis of clinical or pathological features.

APC is a crucial part of this degradation complex which plays a role in controlling the subcellular localization, accumulation and activity of β-catenin (356–358). As already mentioned, the latter functions as a downstream transcriptional activator of the Wnt signaling pathway and also as a submembrane component of a cell-cell adhesion system (357, 359, 360).

Of relevance to medulloblastoma is that β-catenin may assume oncogenic potential through several mechanisms including mutations of *APC* which ablate its ability to degrade β-catenin, through mutations of β-catenin which prevent its down-regulation by wild-type *APC*, or through mutations of the *AXIN* gene resulting in the activation of Wnt receptors (347, 361, 262).

The higher incidence of medulloblastoma in patients with Turcot syndrome (363), who have germline mutations of the *APC* gene (located at 5q21) as an underlying cause of their disease (364) (Table 8.8) has received much attention as to the possible role this gene and the Wnt pathway may play in medulloblastoma development. In a series of 173 sporadic medulloblastomas, 3–5% had Gorlin syndrome (151).

Malfunctions of the Wnt signaling pathway constitute a hallmark of many aggressive cancers (340, 365). Disruption of the Wnt pathway through either *β-catenin, APC* or *AXIN* mutations or GSK3β dysfunction is involved in the development of a small but significant fraction of medulloblastomas (10–15%) (316, 362, 366).

Involvement of the Wnt pathway in the evolution of sporadic medulloblastomas was first indicated by the presence of β-catenin mutations in three of 67 (4.5%) of these tumors

(346). This raised the question whether sporadic medulloblastomas also contain *APC* mutations.

In the absence of Turcot syndrome, somatic mutations of the *APC* gene were shown in early studies to be very rare in sporadic medulloblastomas (316, 367–369). In one study (367), for example, no mutations of *APC* were found in 41 sporadic medulloblastoma and another group (369) observed no LOH at the gene locus on 5q in 27 sporadic medulloblastomas. No LOH of the *APC* locus was demonstrated in 35 medulloblastomas (including two from NBCCS cases), although amplification of *APC* was present in 32/35 of these tumors, a finding to which the authors attached no significance (316).

However, in a study using SSCP and direct DNA sequencing, three miscoding *APC* point mutations were found in two of 46 (4.3%) medulloblastoma (361). In addition, miscoding β-catenin mutations (in codons 33 and 37) were detected in four additional tumors (8.7%). *APC* and β-catenin mutations were mutually exclusive and occurred in a total of six of 46 cases, suggesting involvement of the Wnt pathway in approximately 13% of sporadic medulloblastomas (361).

A study of the *APC* gene in three subjects with Turcot syndrome (a rare and probably autosomal recessive disorder) with colonic tumors and two with medulloblastoma, showed germline mutations in the three patients and a somatic mutation of *APC* in 3 colonic adenomas of one patient (367). Examination of the medulloblastomas from the two cases of Turcot syndrome, and 91 archival medulloblastomas, failed to reveal any mutations of *APC*.

The study referred to in the preceding paragraph seems to be the only one in which the mutated germline genes in one of the syndromes described in Table 8.8 have been examined in

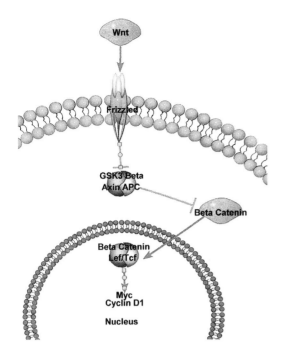

FIGURE 8.15. Schematic presentation of the Wnt signaling pathway. Secreted proteins of the Wnt family act by binding to transmembrane receptors called Frizzled and by regulating the levels of intracellular β-catenin. In the absence of Wnt signal, β-catenin is phosphorylated and degraded by a complex of proteins that includes GSK3β, the APC tumor suppressor protein and the scaffolding protein Axin. Binding of Wnt to Frizzled inactivates GSK3β, reducing the phosphorylation and degradation of β-catenin. This allows β-catenin to translocate to the nucleus where it binds to transcription factors of the Lef/Tcf family, converting them from repressors to activators. The β-catenin-Lef/Tcf complex induces expression of target genes such as *MYC* and *Cyclin D1*, which promote cell cycle progression. Mutations inactivating APC or Axin render β-catenin resistant to degradation are associated with a variety of tumor types, including medulloblastoma.

the tumors of affected individuals. The lack of other informative studies on such tumors in all three syndromes leaves a gap in our understanding of the pathogenesis of medulloblastoma in these conditions.

One study (346) examined the status of the *β-catenin* and *GSK3β* genes in 67 sporadic medulloblastomas. Mutations of the *β-catenin* gene were observed in a small proportion (three of 67) of the tumors, suggesting that this gene may be involved in the development of a limited subset of medulloblastomas. Although LOH of the *GSK3β* locus was detected in some tumors (seven of 32), none had an alteration in the remaining *GSK3β* allele, indicating that this gene is not a significant TSG for medulloblastoma.

In five of 23 sporadic medulloblastomas accumulation of β-catenin was present by immunohistochemical staining; two cases were positive for Wnt-1 and another showed either mutation of *AXIN1* or *CTNNB1*. The *AXIN1* mutation involved exon 3, site of GSK3β binding and that of *CTNNB1* (β-catenin gene) at exon 3, corresponding to its phosphoryla-

tion site. These events could upregulate the Wnt signaling and accumulation of β-catenin, followed by cell proliferation and medulloblastoma development (370). Although the frequency of β-catenin mutations is low in medulloblastoma, pathways leading to its activation may be relevant to medulloblastoma biology.

AXIN1 gene, a component of the Wnt pathway, was screened in 39 medulloblastomas, but only two tumors had missense mutations (366). Large deletions of *AXIN1* in 5/12 medulloblastomas may not have significance, because they were seen in similar frequency in normal brain tissue and could also be due to RT-PCR errors (366). Alterations of *AXIN1* (Axis inhibition protein-1) were seen in 12% of 86 medulloblastomas and 11 cell lines and manifested often as large deletions (362), but they have been pointed to as possible methodological artifacts of RT-PCR (366).

About 5–10% of sporadic medulloblastomas have inactivating and usually exclusive point mutations of the *APC* and *β-catenin* (*CTNNB1*) genes (343, 346, 361, 371). Codons 33 and 37 of the *β-catenin* gene constitute mutational hotspots in medulloblastoma.

Table 8.16 shows some salient findings in the three pathways (SHH, Wnt, and p53) in medulloblastoma.

Neurotrophins in Medulloblastoma

Cerebellar granule cell neuron growth and differentiation have been shown to occur in coordination with the expression of the neurotrophin family of growth factors (372). The members of this family include nerve growth factor (NGF), brain-derived neurotrophic factor (BDNF), neurotrophin (NT)-3, NT-4/5, and NT-6 (373–380). Medulloblastomas express one or more of the neurotrophins, and their receptors (381).

Two classes of transmembrane glycoproteins serve as receptors for the neurotrophins. The first class, represented by the p75 low-affinity nerve growth factor receptor, has no tyrosine kinase activity. It binds NGF, BDNF, and NT-3. NGF, when bound to the p75 receptor, may promote the survival of normal and neoplastic neurons by blocking apoptosis (378,382), although this is an area that is not totally clear (383). The activity of p75 is critical for cerebellar development under pathologic conditions where neurotrophin levels are reduced (384); p75 is present in cerebellar granule cells and in Purkinje cells (384, 385).

The second class of neurotrophin receptors is represented by members of the Trk family of tyrosine kinases, homologous with a previously described oncogene, tropomyoses receptor kinase, or *Trk* (375, 378). These glycoproteins function as selective high-affinity neurotrophin receptors whose tyrosine kinase is activated by neurotrophin binding. Activated kinases, in turn, autophosphorylate tyrosine residues in the cytoplasmic domain of the receptors. The first member of this family, TrkA, preferentially binds NGF. Another member, TrkB, specifically binds BDNF and NT-4/5, whereas TrkC binds only NT-3. The biological effects of NT4/5 and NT6 are

TABLE 8.16. Some of findings in the SHH, Wnt, and p53 pathways in medulloblastoma.

Constitutive activation of the SHH signaling pathway occurs in approximately 25% of cases overall in medulloblastoma, by mutually exclusive mutations of the *PTCH1*, *SUFU*, and *SMO* genes (462).

Activation of the Wnt signaling pathway occurs in 10–30% of medulloblastomas by mutations of *β-catenin* (*CTNNB1*), *APC*, or *AXIN1* genes (462).

Defects in the *TP53* tumor suppressor pathway in approximately 20% of medulloblastomas are due to *TP53* mutations and *p14*^{ARF} deletions (462).

Critical evaluation of the functional and clinical relationship of the genetic defects just mentioned substantiate their roles in medulloblastoma tumorigenesis and suggest that their differential involvement may indicate divergent mechanisms of pathogenesis and clinical behavior in subsets of medulloblastoma (462).

Despite the progress reflected by the aforementioned findings in medulloblastoma, a genetic basis for the majority of medulloblastomas remains to be discovered (462).

A review of the role of the Wnt signaling pathway in cancer has been publsihed (802).

A study demonstrated that wild-type *AXIN2* antagonized Wnt signaling in medulloblastoma cells (785), whereas a truncated *AXIN2* increased T-cell factor (TCF) activity. The findings indicated that mutations of *AXIN2* can lead to an oncogenic activation of the Wnt pathway in medulloblastomas. The authors *(785)*, stated that the presence of *β-catenin*, *APC*, *AXIN1*, and *AXIN2* mutations in medulloblastomas described in other studies suggests that disruption of the Wnt signaling pathway by different mechanisms is involved in the molecular pathogenesis of these tumors and the findings *(785)* underline those of others (662,777) that activating mutations within the Wnt pathway define a subset of medulloblastomas.

Activation of AXIN2 expression by β-catenin-T cell factor has been described (786).

not as well delineated at this time, but NT4/5 binds to both TrkA and TrkB (386,387). Both BDNF and NT-3 are found in the immature cerebellum. Early cells in the external granule cell layer of the cerebellum respond to BDNF, whereas more mature granule cells respond to NT-3 but not BDNF.

The neurotrophin receptor TrkC, whose expression in medulloblastoma is generally considered to be a favorable indicator, is given some attention in Table 8.17, in which its mechanisms of action are listed. During development, TrkC is expressed in the most mature granule cells of the cerebellum (381, 388), cells from which arise at least some of the medulloblastomas. Thus, the favorable outcome in cases expressing TrkC may be related to the above observation (Figure 8.16) (389). Some of the characteristics of the receptor TrkC are presented in Table 8.17. Some characteristics of TrkC in medulloblastoma are shown in Figure 8.16A.

Two studies on neurotrophins and their expression in medulloblastomas (and a few PNET), performed by essentially the same group of investigators (390, 391), found that expression for TrkA was seen in 25–27% of the tumors, for TrkB in 40–62%, and for TrkC in 48–85%. The presence of NGF, BDNF, NT-3, and NT-4/5 was observed in 15–40% of medulloblastomas. These studies imply that signal transduction pathways mediated by neurotrophins, their receptors, or both influence the induction or progression of medulloblastomas.

Neurotrophins act through their cognate receptors to promote the differentiation, survival, or both of neuronal

progenitor cells, immature neurons, and other cells (392). The best-studied representative of neurotrophins to date has been NGF. Studies have indicated that a NGF/TrkA signal transduction pathway could activate apoptosis in medulloblastoma (392).

In a medulloblastoma cell line, NGFR expression could be inhibited by NGF (393). In a study based on the expression of p75, the authors indicated that progenitor cells of the external granule cell layer of the cerebellum served as precursors of medulloblastoma, particularly of the desmoplastic type (385).

The expression of neurotrophins and neurotrophin receptors by PNET (medulloblastoma?) cell lines was variable. The presence of activated Trk receptors may be required for rapid growth (394).

NGF, which has the potential to induce cellular differentiation, was found to be present in four of 10 medulloblastomas (272A) and confined to 5–8% of the cells, an observation previously made by others (395); attempts to increase NGF in an in vitro system were essentially negative. Thus, it seems unlikely that NGF will be a useful therapeutic agent for medulloblastoma (272A).

An early study (396) showed >50% of medulloblastomas to be positive for NGF receptors, particularly those tumors with glial features. Although the NGF receptor-positive tumors may be responsive to the effects of NGF, other reports (397, 397A) indicate that the mere expression of NGF receptors does not guarantee that NGF will induce differentiation of such tumors.

TABLE 8.17. Some characteristics of the receptor TrkC in medulloblastoma.

	References
Expressed in most mature granule cells of cerebellum and may underlie possible reason for favorable outcome in medulloblastoma expressing TrkC.	*Segal et al 1992* (389)
	Pomeroy et al 1997(208)
Inhibits the growth of medulloblastoma xenografts in mice.	*Kim et al 1999* (671)
TrkC expression highly correlated with apoptosis in medulloblastoma.	*Kim et al 1999* (671)
TrkC-mediated NT3 signaling promotes apoptosis by activating multiple parallel signaling pathways and by inducing early expression of *JUN* and *FOS* genes. In toto, the biological actions of TrkC activation affect medulloblastoma outcome by inhibiting tumor growth through the promotion of apoptosis.	*Kim et al 1999* (671)

FIGURE 8.16. Cumulative survival of patients with medulloblastoma according to expression levels of *MYC* and *TrkC*. Survival was the best in the group with high levels of *TrkC* expression and with low *MYC* expression, whereas patients with low levels of *TrkC* expression, regardless of those of *MYC* had the worst survival (from ref. (188), with permission).

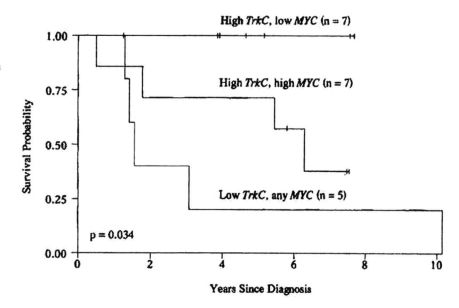

The expression of neurotrophins and their high-affinity Trk receptors in medulloblastoma is variable, in part due to inconsistencies in laboratory techniques and growth conditions (394).

A review of the neurotrophin signal transduction and applicability to the development of agents against medulloblastoma has been published (398).

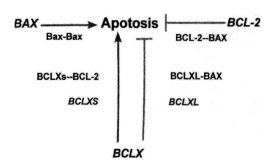

Apoptosis in Medulloblastoma

The term *apoptosis* is derived from the Greek and means literally "to drop off," as in leaves falling from a tree (190A). As used in cell biology, apoptosis refers to genetically mediated individual cell death affected by a host of genes and factors, e.g., *BCL2*, p53, *MYC*, *BAX*, *FOS*, *RB1*, and caspases, to name a few (Table 8.18 and Figure 8.17). These genes and their products constantly interact, thereby potentiating or repressing one another's apoptosis-regulating effects (399). The literature on apoptosis is vast and hence here we shall confine ourselves to its role in medulloblastoma.

TABLE 8.18. Selected oncogenes and TSG initiators or stimulators of apoptosis.

Stimulators of Apoptosis	Inhibitors of Apoptosis
MYC	BCL2
p53	BCLXL
BAX	MCL1
BAK	ABL
BAD	RB1
BCLXS	
FOS	
JUN	

Based on Staunton and Gaffney (190A), where appropriate references may be found.

FIGURE 8.17. Schematic presentation of the parts played by the *BAX* and *BCL* genes and their protein products in the control of apoptosis. Increased expression of *BAX* promotes apoptosis and increased expression of *BCL2* suppresses apoptosis. *BCL2* expression has been shown to confer resistance to cell death and thus promote tumor formation by prolonging cell survival inappropriately (736B). Alternative splicing of *BCLX* gene produces two transcripts with opposing effects, i.e., *BCLXS* promotes apoptosis, whereas *BCLXL* suppresses it. The proteins encoded by these genes form dimers with competing effects on apoptosis. Thus, the BAX-BAX homodimer promotes apoptosis, whereas the BCL2-BAX and BCLXL-BAX heterodimers suppress apoptosis. The BCLXS-BCL2 dimer may promote apoptosis by blocking the formation of the antiapoptotic BCL2-BAX heterodimer. The relative expression of the genes shown appears to determine whether a cell will undergo apoptotic death. Figure based on the composite information in Boise et al. 1993 (737B), Oltavi et al. 1993 (409), Sedlak et al. 1995 (738B), Kroemer 1997 (739A), Levine 1997 (740B), Bruggers et al. 1999 (410), and Cory et al. 2003 (736B).

Apoptosis (programmed cell death) is a crucial regulator of normal development and tissue homoeostasis, and the abrogation of apoptotic pathways is a frequent event in tumor development. Apoptosis requires energy in the form of ATP, indicating that programmed cell death, as opposed to necrosis, is an energy-dependent active physiological and pathophysiological phenomenon (400).

Apoptosis in medulloblastoma can be identified by light microscopy, electron microscopy, agarose gel electrophoresis, flow cytometry, in situ labeling, and time-lapse video microscopy (190A). Each of these methods assesses a somewhat different aspect of apoptosis, hence the use of more than one technique yields more cogent and meaningful information.

The number of cells involved by the apoptotic process, i.e., the so-called apoptotic index (AI) can be established by a number of methodologies, some of which are far from optimal.

The methodological inconsistencies in determining the AI (apoptotic index) in medulloblastomas have led to contrasting results (401, 402) and, hence, the interpretation of findings based on AI must take these inconsistencies into consideration.

The use of techniques other than terminal deoxynucleotidyl transferase dUTP nick-end labeling (TUNEL) to evaluate apoptosis has been advocated as more reliable in answering important questions about tumor biology and treatment (402, 403).

Classic and desmoplastic medulloblastomas have a $1.3 \pm 0.5\%$ apoptotic nuclei (258). The distribution of apoptotic nuclei in medulloblastoma was uneven, i.e., they were generally more frequent in areas of the tumor with a high mitotic index and higher numbers of Ki-67-positive cells (404). No definable differences were observed in the number of apoptotic cells among types of medulloblastoma; also, the AI varied from tumor to tumor and in different areas of the same tumor (21), nor does it correlate with prognosis (402), although it is higher in recurrent tumors. The AI was higher in medulloblastomas of male than female patients (31) and had a negative relation to patient age (21).

An AI of $>1.5\%$ was associated with an unfavorable outcome (31, 405). This contrasted with the limited prognostic value of *MYC* and *TrkC* expression. Inactivation of *p53* causes decreased levels of radiation-induced apoptosis in medulloblastoma (406). A positive correlation between apoptosis and expression of cell-cycle inhibitor p27^{KIP-1} in medulloblastoma has been reported (407, 408).

NGF acting through TrkA may induce apoptosis in medulloblastoma (392), although the process and mechanism of this effect are rather complex.

Undoubtedly, apoptosis plays an important role in medulloblastoma tumorigenesis; however, the information on the genes and pathways involved (except for bcl-2 and p53) is relatively scanty.

Deregulated cell proliferation, together with apoptosis, operates to support neoplastic progression (178, 190A). Thus, there is little doubt that apoptosis is an integral part of the tumorigenic process in medulloblastoma, but what has remained controversial is the meaning of the quantitative and qualitative aspects of apoptosis biologically and prognostically in these cases (21). The primary role that apoptosis plays in neural development, oncogenesis, and therapeutic response suggests that aberrantly expressed apoptotic genes could be molecular determinants of tumorigenesis in the brain.

The *BCL2* gene product is a potent inhibitor of apoptosis, thus prolonging cell survival. *BCL2* is considered a downstream regulator of apoptosis because it inhibits apoptosis induced by a variety of stimuli, including gamma radiation, hyperthermia, chemotherapeutic agents, and steroids. Thus, *BCL2* expression is one of several defined mechanisms of chemoresistance and radioresistance in medulloblastoma.

BCL2 and its homologs interact closely in the regulation of apoptosis. The ratio of *BCL2/BAX* and that of their protein products is a critical determinant for the induction of apoptosis; when *BAX* is present in excess, apoptosis is stimulated, and when *BCL2* is present in excess, apoptosis is inhibited (409). The complexity of the apoptotic process is reflected in the *BAX* and *BCL* and their related genes effects and interactions, where the formation of dimers of the protein products of the genes either increase or decrease apoptosis (410).

In three medulloblastoma cell lines, expression of *BCL2* and *BCLXS* was not present, whereas *BCLXL* and *BAX* were expressed (410). In a PNET cell line, *BCL2*, *BCLXL*, and *BAX* were expressed, but not *BCLXS*. The findings in these cell lines and those of other brain tumors, and in tumor specimens, led the authors to conclude that coexpression of proapoptotic and antiapoptotic genes in these tumors supports the hypothesis that the relative expression of competing genes determines the apoptotic threshold (410).

Positivity for *BCL2* expression in medulloblastoma has varied among studies, e.g., in one study, *BCL2* was not expressed in any medulloblastoma cell line or tumor specimen (411, 412). However, 25–30% of medulloblastomas showed positivity in other studies (407, 413). In another study (414), only 7.8% (7/90) of medulloblastomas were positive for *BCL2*, three were classical and four desmoplastic.

One group (415) reported increased levels of BCL2 protein expression in a significant proportion of medulloblastomas, possibly associated with an unfavorable outcome. Others (254) did not find such a correlation (416, 417). It has been suggested (254) that medulloblastomas being undifferentiated tumors, it is not surprising that *BCL2* expression is frequently absent in these tumors.

In a study based on RT-PCR analysis for expression of *BCL2*, *BCLXL*, *BCLXS*, and *BAX*, genes affecting apoptosis, three medulloblastomas and one PNET cell line were examined (410). All cell lines *BCLXL* and *BAX* and only the PNET cell line also expressed *BCL2*. The coexpression of proapoptotic *(BCLXL)* and antiapoptotic *(BAX)* genes in the same tumors is to be noted. Medulloblastomas expressing *BCL2* have been found to show features of neuronal differentiation (407).

Growth Factors and Their Receptors in Medulloblastoma

In an article published a little over a decade ago (418), the author referred to the fact that growth factors and their receptors had been extensively studied in glial tumors, but not in medulloblastoma, and the few reports that were published (418A,419,420) did not shed much light in this area. However, since 1996 several studies on these growth factors in medulloblastoma have benn published and are summarized in the following paragraphs.

IGF in Medulloblastoma

Insulin like growth factor (IGF) may play a role in medulloblastoma development (289). For example, DNA microarray studies have shown increased expression of IGF2, mainly in desmoplastic medulloblastomas (208). Evidence for the role of IGF2 is reflected in murine models (284A). The effects of IGF2 are mediated by the IGF1 receptor (IGF1R), a receptor tyrosine kinase. Immunohistochemistry studies (421) have shown expression of insulin receptor substrate 1 (IRS-1) and phosphorylated (activated) IGF1R in medulloblastomas. The authors of one study (289) indicated that the combined activation of the SHH and IGF signaling pathways is an important mechanism in medulloblastomas pathogenesis. IGF1R was shown to be present in a single medulloblastoma by quantitative autoradiography (418A).

In a study using an established medulloblastoma cell line (422), stimulation of IGF1 led to phosphorylation of its receptor (IGF1R) and resulted in cell proliferation. Antibodies to IGF1R blocked the effects of IGF1, as did an inhibitor of the mitogen-activated protein kinase (MAPK) pathway (422). The results point to an autocrine IGF1/IGF1R loop and to IGF1 promoting proliferation via the MAPK pathway.

IGF1R is a multifunctional tyrosine kinase receptor that regulates cellular growth by a number of different ways, i.e., by being mitogenic, by protecting normal and tumor cells from apoptosis, by controlling neuronal differentiation and by maintenance of the transformed phenotype (423).

Positive staining for IGF-1R in 6 medulloblastoma biopsies and overexpression in 6 medulloblastoma cell lines led the authors (423) to suggest that IGF1R may contribute to the growth of medulloblastoma.

Similar to the synergism of *MYC* with SHH in inducing medulloblastoma formation from nestin-expressing neural progenitors in mice (424), IGF likewise cooperated with SHH in tumor induction (289). The IGF1R has been proposed as a potential therapeutic target in medulloblastoma (424A).

EGFR-ErbB *Oncogenes in Medulloblastoma*

This group of genes belongs to the RTK-I (receptor tyrosine kinase) signaling family which includes ErbB1-4 and is associated with a complex network of ligands of the neuregulin (NRG) family (425–427).

The malignant phenotype of medulloblastoma (including PNET) is maintained by the expression of the *EGFR* family of genes (428). Very high expression of *erbB4* was found in 73% of the 48 tumors examined. Transcript levels of *erbB2* and *erbB3* exceeding 10-fold that of the cerebellum were found in 35% of the cases. Only two tumors had *erbB1* expression at these levels. The results suggest that various forms of stimulation through these receptors may contribute to the malignant phenotype of medulloblastoma (428).

In isolated immunohistochemical studies of medulloblastoma, EGFR has been detected in a small proportion of the tumors, but survival analyses have not shown any statistical significance (425). In a study *erbB2* expression was found in the majority of the 55 medulloblastomas examined and was associated with an unfavorable outcome (429). In a subsequent study (429A), expression of erbB2 was much less common, particularly in tumors of children <3 years old. Possible reasons for this discrepancy were discussed (105). The results of the studies of these authors (105) showed that overexpression of erbB2 was neither evident nor correlative with the presence of i(17q). Coexpression of *erbB2* (HER2) and *erbB4* (HER4) demonstrated prognostic significance (425), in contrast to expression of individual receptors or NRG expression.

PDGFRα *and* RAS/MAPK *Oncogenes in Medulloblastoma*

In a study of expression profiles of 23 medulloblastomas (259), PDGFRα and members of the downstream RAS/MAPK signal transduction pathway were up-regulated in tumors with metastases, as evidenced by overexpression of PDGFRα in another set of tumors examined immunohistochemically. In vitro studies (259) showed that PDGFRα can enhance medulloblastoma aggressiveness and increase downstream phosphorylation of MAPK components; the latter could be blocked by antibodies to PDGFRα. The PDGFRα-RAS/MAPK pathway may be associated with medulloblastoma development in certain cases. No evidence of oncogenic mutations of *KRAS, NRAS, HRAS, BRAF,* or *PDGFRB* were found (259A).

In a previous study (420) on expression of PDGF in 5 medulloblastomas, none of the tumors had the ligand and/or receptor for either the PDGFRα or PDGFRβ autocrine loop (420).

Basic fibroblast growth factor expression was not found in an immunohistochemical study (419). The vascular endothelial growth factor (VEGF), which is of demonstrated importance in glial tumorigenesis, has not been studied sufficiently in medulloblastoma. An investigation of the angiogenic profile of medulloblastoma showed that VEGF may not be the only determining factor in the vascularization of these tumors (430), neither was PEDF one of the factors tested.

Other Genes and Factors in Medulloblastoma

OTX *Genes and Expression*

The transcription factors OXT1 and OTX2 are essential for normal development. It has been found that the *OTX2* gene is amplified in medulloblastoma (127, 431, 432) and may reflect a major role for OTX2 in medulloblastoma pathogenesis, though it is unknown how OTX2 relates to the clinical and histopathologic features of medulloblastoma. Expression of *OTX1* and *OTX2* has been used as a means of defining layers and regions in the developing cerebral cortex and cerebellum (433).

Almost all medulloblastomas 114 of 152 (75%) and cell lines expressed either OTX1 or OTX2 or both; the *OTX2* gene was highly amplified in a medulloblastoma cell line (432). OTX2 was expressed in 29/34 classic medulloblastomas and OTX1 in nine of 11 of the nodular/desmoplastic type. The more frequent *OTX2* expression in tumors (127) could, therefore, result from analysis of a different histologic subset. No expression of either OTX1 or OTX2 was detected in postnatal normal cerebellum. IMHC analysis showed OTX2 expression in 83 of 107 (75%) classic medulloblastomas (432). These tumors were preferentially localized in the vermis, whereas the negative cases were frequently localized in the cerebellar hemispheres. Amplified *OTX2* was found in eight of 42 medulloblastoma tumors and in two of 11 medulloblastoma cell lines (127). No *OTX2* amplifications were seen in 44 analyzed tumors, and only in three of 13 medulloblastoma cell lines (431). Taken in toto, the findings strongly suggest that amplification of *OTX2* is not a frequent or consistent event in medulloblastomas but rather is found in medulloblastoma cell lines (431, 432).

Serial analysis of gene expression (SAGE) in a medulloblastoma of an 11-year-old girl and fetal cerebellum (433A) revealed high expression of *OXT2* in the tumor, and in the cerebellar germinal layers, supporting the hypothesis that medulloblastomas arise from the germinal layer of the cerebellum. Analysis by suppression subtraction hybridization of a medulloblastoma cell line library revealed upregulation of the homeobox *OTX2* (434).

Interference with *OTX2* expression inhibited medulloblastoma growth in vitro; ATRA repressed *OTX2* expression and induced apoptosis only in cell lines that expressed *OTX2* (431). The authors stated that *OTX2* seems essential for the pathogenesis of anaplastic medulloblastoma and that tumors expressing *OXT2* may be candidates for ATRA therapy.

PAX *and* EN *Genes in Medulloblastoma*

Members of the paired box-containing *(PAX)* gene family code for transcription factors expressed in specific regions of the brain, e.g., *PAX6* is expressed in the granule cells of the developing cerebellum and is implicated in the control of cell fate specification, proliferation, and/or migration of neuroectodermal precursor cells during development (435).

In addition, *PAX* genes have shown the potential of oncogenes, e.g., the involvement of *PAX3* and *PAX7* genes in specific translocations in rhabdomyosarcoma (436); hence, their possible relationship to medulloblastoma (435). The homeobox-containing genes *EN1* and *EN2* are transcribed in the embryonic brain region giving rise to the cerebellum (437); hence, their possible role in medulloblastoma development.

Consistent expression of *PAX5*, *PAX6*, *EN1*, and *EN2* ranging from 60 to 80% in medulloblastomas has been demonstrated (435). *EN1*, *EN2*, and *PAX6* are expressed in normal cerebellum, adding evidence that medulloblastomas originate from the external granular layer of the developing cerebellum (435). *PAX5* was not expressed in neonatal cerebellum and its deregulated expression in medulloblastoma indicates its possible role in the development of this tumor and correlates positively with cell proliferation and inversely with neuronal differentiation in desmoplastic medulloblastomas. *PAX2* and *PAX3*, which are expressed in cell types other than the granular in the cerebellum, are rarely transcribed in medulloblastoma.

In another study (438), the results suggested a regulatory loop between *PAX6* and *EN* in medulloblastoma. The *PAX3* and p53 genes were examined by several molecular techniques in 58 medulloblastomas and 50 paraffin-embedded medulloblastoma tissues, respectively (439). Four mutations of p53 were encountered, whereas only one polymorphism was identified in the *PAX3* gene.

Somatostatin Receptors in Medulloblastoma

Somatostatin (SST) is a neuropeptide found in different central and peripheral nervous system structures and in tissues involved in endocrine, digestive and immune functions (440). SST has a number of active forms which play a role in central nervous system functions, although of interest to this chapter is their participation as growth regulators of the developing cerebellum, and their antiproliferative activity (440, 441), which involves a number of membrane receptors for SST (440–442), including some in the cerebellum (443, 444).

The receptors for SST are expressed in a high proportion of medulloblastomas (445–447). Some types of SST, e.g., SSTR2, are more noticeably expressed than others (446, 447); this opens some interesting prospects for the diagnosis and therapy of medulloblastoma. Clinically, somatostatin receptor scintigraphy has been used in the diagnosis of medulloblastoma and in the follow-up of pediatric cases (445).

Notch Signaling Pathway in Medulloblastoma

Notch signaling is involved in the differentiation of cerebellar granule neuron precursors and other aspects of cerebellar development (322, 448); hence, it bears cogently on medulloblastoma tumorigenesis. The role of *notch* signaling in cancer development has been documented, although its specific role in medulloblastoma pathogenesis remains unknown (449, 450).

In embryonal brain tumors, Notch1 and Notch2 have opposing effects on growth that parallel their functions in normal development (448). These strikingly antagonistic changes are highly unusual, and further characterization of the immediate effectors and downstream targets of Notch1 and Notch2 in these tumors should provide insight into the growth control functions of the Notch pathway. Using RNA interference and inhibition of γ-secretase, an ongoing requirement for *Notch2* and its transactivation of gene targets, e.g., *Hes1* (451) in medulloblastoma growth was documented (448), suggesting that modulation of Notch signaling may be a therapeutically useful approach (452).

The demonstration that malignant cells with activated *notch1* signaling become chemoresistant through inhibition of the p53 pathway (453) may have therapeutic and clinical implications in medulloblastoma.

The best-characterized Notch effector, Hes1, negatively regulates basic helix-loop-helix neurogenic transcription activators such as the human Achaete Scute 1 gene (*hASH1*) (360). Expression of the Notch pathway target gene, *Hes1*, in medulloblastoma was associated with an unfavorable prognosis (448). Notch1 was more frequently overexpressed in medulloblastoma than was Notch2; Hes1 and Hes2 were shown to be Notch effectors in medulloblastoma. Several ligands of the Notch receptor have been characterized (454).

Much has to be learned about how the Notch pathway may contribute to the development of cancer, including medulloblastoma. Inappropriate activation or inactivation of the Notch pathway will derail cells from their normal course of differentiation (454), suggesting that the Notch pathway might offer a novel route for therapeutic intervention.

Epigenetic Events in Medulloblastoma

Epigenetics is the study of stable alterations in gene expression patterns that are not attributable to mutations of or base changes in primary DNA sequences. One focus of epigenetics is DNA methylation. Cytosines located 5′ to guanines, called CpG dinucleotides, are found clustered (CpG islands) in the regulatory regions of approximately 50% of all human genes. In normal cells, the cytosines in the CpG dinucleotides are generally unmethylated or methylated at low frequency.

Aberrant methylation of normally unmethylated CpG islands has been documented as a relatively frequent event in immortalized and transformed cells and is associated with transcriptional inactivation of defined TSG in human cancers.

Genes methylated in medulloblastoma are shown in (Table 8.19). The genes have been arranged in three categories: consistently methylated genes, less frequently methylated genes and possibly methylated genes. The cellular effects of these genes are also shown. Genes not methylated in medulloblastoma are shown in Table 8.20.

TABLE 8.19. Genes affected by methylation in medulloblastoma.

Consistently affected in >60% of medulloblastomas	
RASSF1A	(microtubule stabilization; regulation of mitosis)
CASP8	(regulation of apoptosis)
HIC1	(regulation of transcription)
*p16*INK4A	(cell cycle regulator)
Consistent evidence of hypermethylation	
MGMT	(DNA repair)
CDH1	(regulation of cell adhesion)
TIMP3	(regulation of extracellular matrix)
GSTP1	
DAPK	(regulation of extracellular matrix)
Possibly methylated	
*p14*ARF	(cell cycle regulator)
TP53	(cell cycle regulator)
TP73	(cell cycle regulator)
EDNRB	
MCJ	(response to chemotherapy)
RB1	(cell cycle regulator)
THBS1	

Frühwald et al 2001 (87), Harada et al 2002 (466, 467), Lusher et al 2002 (468) Rood et al 2002, 2004 (469, 470), Zuzak et al 2002 (471), Gonzalez-Gomez et al 2003 (472, 473, 687), Horiguchi et al 2003 (474), Rathi et al 2003 (475), Waha et al 2003 (476), Ebinger et al 2004 (477), Chang et al 2005 (688), Inda et al 2006 (689), Lindsey et al 2004, 2006 (478, 690).

The information in the table findings in medulloblastoma tumors and tumor cell lines. Because the latter show a higher frequency of methylation of CpG islands, the information should be viewed with that in mind.

Based on Lindsey et al. 2005 (462).

TABLE 8.20. Nonmethylated genes in medulloblastoma.

	Reference
APC	*Harada et al 2002 (467)*
CDHI	*Harada et al 2002 (467)*
DLC1	*Pang et al 2005 (479)*
FHIT	*Chang et al 2005 (688)*
*p15*INK4B	*Frühwald et al 2001 (465), Lindsey et al 2004 (478)*
RARβ	*Harada et al 2002 (467)*
RIZ1	*Lindsey et al 2004 (478)*
sFRP1	*Chang et al 2005 (688)*
TSLC1	*Lindsey et al 2004 (478)*
VHL	*Frühwald et al 2001 (87, 465)*

A large body of evidence now implicates a major role for epigenetic mechanisms in cancer development (455–460), particularly hypermethylation which affects specific genes and leads to their epigenetic inactivation and thus to play more specific functional roles in cancer pathogenesis (459). The latter mechanism represents the most widely investigated epigenetic events in cancer, and typically occur somatically at CpG islands, resulting in an altered chromatin structure and transcriptional silencing.

Aberrant DNA hypermethylation events affecting a large number of TSG have been described for most tumor types, including medulloblastoma, and characterize genes involved across the spectrum of tumor development processes (455,

459,461,462). Epigenetic expression (through methylation) of *RASSF1A* but not *CASP8* occurs in PNET and AT/RT (462A).

Somatic methylation of CpG islands in cancer is accompanied by associated histone modifications, transcriptional silencing, growth advantage, and subsequent clonal selection (463,464).

The epigenetic events affecting TSG by promoter hypermethylation have become the focus of intense investigations in multiple tumor types, including medulloblastoma (462). These investigations have led to the identification of common gene-specific events that can occur either independently of or in conjunction with other genetic defects to disrupt genes involved in tumor pathogenesis. This area has been compressively reviewed, as it applies to medulloblastoma (462).

The list of genes examined for evidence of CpG island methylation in medulloblastoma contains >24 genes, with more genes expected to be added to the list as studies are expanded in this area. Tables 8.19 and 8.20 contain information regarding some of these methylated genes (87, 462, 465–479). It has been suggested that the variation in the frequencies of methylation between studies may be related to small cohorts analyzed in some studies and to different methods used in others (462).

In the following paragraphs, we present some information regarding the most consistently affected genes by hypermethylation in the bulk of medulloblastomas examined.

HIC-1 *Gene (Located at 17p13.3)*

The chromosomal location of this tumor suppressor gene is frequently affected by abnormalities as demonstrated by cytogenetic studies (Table 8.4) (324). *HIC-1* methylation is observed in a significant number of medulloblastomas (469, 475, 476). Mutations of *HIC-1* in medulloblastoma are very rare. Findings to-date indicate that inactivation of *HIC-1* arises either through biallelic epigenetic mechanisms or epigenetic mechanisms together with chromosomal deletions (480). The protein product of *HIC-1* functions as a transcriptional repressor. Accumulated genetic, epigenetic and functional evidence implicates *HIC-1* as an epigenetically inactivated TSG and as a strong candidate for a TSG in medulloblastoma (462). Expression of *HIC-1* is activated by p53 (480A, 481); however, much remains to be clarified on *HIC-1* as a TSG.

CASP8 *Gene (Located at 2q33-q34)*

Cysteine-aspartyl-protease-8 (caspase-8) encodes an initiator caspase involved in death receptor-mediated apoptosis and other facets of apoptosis (462, 481A, 481B). A significant proportion of medulloblastomas show evidence of complete methylation of *CASP8* (478), although deletion of 2q33-q34 is rare in these tumors, compatible with an epigenetic mechanism for inactivation of this gene. *CASP8* is a candidate for a TSG in medulloblastoma and inactivation of *CASP8* by CpG

island methylation may play a role in abrogation of apoptotic pathways involved in medulloblastoma tumorigenesis. Methylation of *CASP8* was present in 81% of medulloblastomas (15 of 18) and highly correlated with methylation of the TSG *RUSSF1A* (located at 3p21.3). *CASP10* (466), a related but functionally independent gene, also was methylated.

Studies have shown activated caspase-3, caspase-7 and poly(ADP-ribose) polymerase 1 (PARP1) cleavage in medulloblastoma (482). Cleaved caspase-3 and PARP expression parallel apoptosis. Some cleavage of caspase-7 was also found.

Ras-Associated Domain Family 1 Isoform A (RASSF1A) *Gene (Located at 3p21.3)*

Inactivation of this gene occurs by biallelic hypermethylation with no involvement of genetic mutation or deletion in medulloblastoma (468). That the expression of *CASP8* may have clinical implications is shown in Figure 8.18. Essentially, weak expression of caspase-8 (? due to hypermethylated gene?) was associated with an unfavorable outcome compared with cases with a moderate or strong expression (483). *RASSF1A* is a TSG (462, 484–486). Among its functions are microtubule stabilization, effects on apoptosis and cell cycle arrest mediated through both the G1/S-phase and G2/M-phase mitotic checkpoints (462). Methylation of *RASSF1A* was found in two of three of medulloblastomas (467). Aberrant methylation of normally unmethylated CpG islands located in the promoter regions is associated with transcriptional inactivation of defined TSG in human tumors, thereby contributing to the pathogenesis and progression of malignant states. Cell lines are more likely to show methylation than primary tumors.

Telomerase in Medulloblastoma

Telomerase is an enzyme synthesizing the repetitive nucleotide sequence TTAGGG in chromosome telomeres. Human telomerase consists of an RNA subunit (hTR), which is widely expressed, and a protein subunit, telomerase reverse transcriptase (hTERT), whose expression is more tightly regulated. The *hTERT* gene is located at 5p15.33; its expression is repressed in normal human somatic cells, but it may be reactivated in most tumors (487). In many neoplasms, increased telomerase activity is associated with a poor clinical outcome (487). It has been reported (488) that telomerase activity is controlled through *MYC*-dependent inhibition and alternative splicing of *hTERT*.

High-level gains of chromosomal material of the 5p15 region have been detected in some medulloblastomas (489–491), suggesting that the *hTERT* gene could be amplified in these tumors (59, 60, 121), although data on telomerase levels in medulloblastoma are sparse.

A study (491) showed that the *hTERT* oncogene is amplified in a significant proportion of medulloblastomas and other

FIGURE 8.18. Progression-free survival and survival probabilities in patients with medulloblastoma based on the expression of caspase-8 in their tumors. Moderate to strong caspase 8 expression was associated with a much better outcome than that of patients whose tumors showed a weak caspase-8 expression (from ref. (483), with permission).

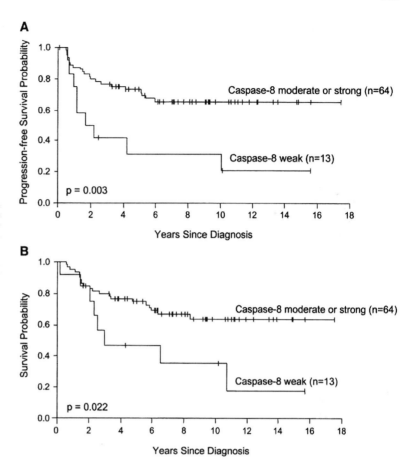

CNS embryonal neoplasms. This gene amplification correlates with increased expression of hTERT mRNA. Measurement of the latter in medulloblastoma showed that 38 of 50 (76%) had up-regulated expression. The finding indicated a possible role of telomerase in the pathogenesis of medulloblastoma and the presence of alternative lengthening of telomeres in subsets of medulloblastoma and PNET, which could be addressed therapeutically (492).

A study indicated that hTERT may enhance tumorigenesis and regulate the expression of genes that control cell growth, activities which cannot be explained by telomere stabilization per se (493). The correlation of hTERT expression with aggressive tumor behavior suggests that hTERT levels could be a useful molecular prognostic marker (491). The role of telomerase and telomere anomalies in tumorigenesis have been extensively reviewed (487, 494–498).

Basic Helix-Loop-Helix Genes in Medulloblastoma

Escape of cells from terminal differentiation is apparently related to the NeuroD family of basic helix-loop-helix (bHLH) proteins, NeuroD and other *achaete scute* transcription factors regulate activation of genes that commit progenitors to a neural fate and induce terminal neuronal differentiation (499). *NeuroD* and *achaete scute* genes are expressed in overlapping but distinct populations of neuroblasts during neuronal development. *NeuroD1*, *NeuroD2*, and *NeuroD3* (*neurogenin*) are expressed throughout cerebellar development and in the adult cerebellum, primarily in the granule cell layer (500–502).

NeuroD family members may be expressed in medulloblastomas as markers of a neurogenic lineage. Because distinct populations of neurons express different neurogenic bHLH transcription factors during development, it was further hypothesized that bHLH patterns would vary between tumors (medulloblastoma, PNET) arising from different neuronal lineages (499).

NeuroD1 (located at 2q32) was expressed in all of the 12 medulloblastomas examined, and in two cell lines (499), whereas *NeuroD2* (located at 17q12) and *NeuroD3* (*neurogenin*) (located at 5q23-q31) were expressed in partly overlappig subsets of medulloblastoma and in all tumors with distant metastases. The human *achaete scute* homolog, *HASH1*, was not expressed in medulloblastoma, but it was expressed in 3/5 PNET (499). These findings may contribute to or reflect biological differences that arise during malignant transformation in medulloblastoma and PNET (499). However, the study of mammalian neurogenic bHLH transcription factors is complex and their prognostic implications are yet to be established.

Human Polyomavirus (HPV) and Medulloblastoma

HPV, and especially the JC virus (JCV), have been implicated in the pathogenesis of medulloblastoma. JCV, a neurotrophic HPV, is the cause of progressive multifocal leucoencephalopathy (503) and is ubiquitously present in the world population, infecting >70% of humans (503). The virus has been implicated in brain tumor development, including medulloblastomas, based on experimental animal studies (504–513) and the presence of virus DNA sequences in tumors (508, 514). However, studies in which viral sequences could not be found in medulloblastomas and other brain tumors also have been reported (515–520). The possible role of products of JCV in medulloblastoma has received support from studies on agnoprotein, a product of JCV (513, 521). The finding of agnoprotein expression in the absence of T-antigen expression, suggests a potential role for agnoprotein in pathways involved in the development of JCV-associated medulloblastomas (513, 521). This area continues to be controversial and its solution is not likely to occur soon (520, 522).

Adenovirus *E1A* oncogene activates p53 through a signaling pathway involving the Rb protein and the tumor suppressor gene p19^{ARF} (523). Increased expression of p53, as shown by immunoreactivity, in anaplastic medulloblastomas and PNET is not caused by JC virus (524).

Of clinical (therapeutic) interest is an in vitro study (525) on JCV T-antigen–positive mouse medulloblastoma cell line in which IRS-1 when translocated to the nucleus where it coo-localized with Rad51, one of homologous recombination-directed DNA repair proteins, sensitized the medulloblastoma cells to the genotoxic effects of cisplatin and γ-radiation.

Comments on the Genetic Changes in Medulloblastoma

Table 8.21 presents genes that may or may not be involved in the biology of medulloblastoma or are rarely involved in these tumors. The genes listed complement those already discussed above.

TABLE 8.21. Changes affecting some genes and factors in medulloblastoma.

	References
p16^{INK4A} and *p15^{ARF}* (located at 9p21), are adjacent genes whose proteins are suppressors of certain CDKs. They also regulate pRB ad p53. In one study, deletions of these genes were found in glioblastomas, but not in medulloblastomas. In a subsequent study, 1/20 medulloblastomas was associated with p16 inactivation. Rare homozygous deletions of p16 have been reported, but no point mutations.	*Jen et al 1994* (691) *Petronio et al 1996* (692) *Sato et al 1996* (693) *Barker et al 1997* (694) *Roussel 1999* (695)
Up to 25% of large-cell and aggressive medulloblastomas show genetic or epigenetic lesions in the *INK4/ARF/p53* pathway.	*Frank et al 2004* (207)
ZIC (located at 3q24), is expressed in the nuclei of the cerebellar granule cells; its protein was present in 26/29 medulloblastomas, but not in PNET or other tumors, a finding supporting the origin of medulloblastoma from the germinal layer of the cerebellum. *ZIC* could serve as a potential biomarker for medulloblastoma (similar to *OXT2*).	*Yokota et al 1996* (434A) *Michiels et al 1999* (433A)
DMBT1 (located on 10q), is a possible TSG with homology to the scavenger receptor cysteine-rich (SRCR) superfamily. Homozygous deletions were present in 2/20 medulloblastomas in one study and in 7/31 medulloblastomas and 1/5 PNET in another study. *DMBT1* may play a role in the pathogenesis of these tumors.	*Mollenhauer et al 1997* (147) *Inda et al 2004* (562)
INI1/hSNF5 (located at 22q11.2), is a gene whose involvement is frequent in atypical teratoid/rhabdoid tumors (AT/RT) and whose deletions are characteristic of these tumors and thus can be used to separate these tumors from confusing medulloblastomas, PNET and pineoblastomas.	*Rorke et al 1997* (532) *Versteege et al 1998* (696) *Taylor et al 2000* (572A) *Kraus et al 2002* (63) *Judkins et al 2004* (621) *Imbalzano and Jones 2005* (698)
PEDF (pigment epithelium-derived factor), is located at 17p13; LOH of *PEDF* was found in 10/20 medulloblastomas, but with no evidence that it is a TSG.	*Slavc et al 1997* (649)
Waf1/p21 is a cyclin-dependent kinase (CDK) inhibitor gene. A silent mutation of this gene was described in 1/6 medulloblastomas.	*Tsumanuma et al 1997* (620)
The Fas ligand/Fas system appears to participate in ATRA-associated chemosensitivity in medulloblastoma. ATRA inhibited medulloblastoma cell proliferation and induced apoptosis in part by activating caspase-3/poly(ADP-ribose) polymerase 1 pathway.	*Liu et al 2000* (699A, 699) *Gumireddy et al 2003* (681)
Insulin-like growth factor II and its receptor are not expressed in medulloblastoma (and central neuroblastoma) and this can be used to differentiate it from other lesions and tumors, e.g., AT/RT, choroid plexus papilloma and anaplastic ependymoma.	*Ogino et al 2001* (738A)
SOX proteins are transcription factors involved in neural development with a possible role in medulloblastoma. *SOX4* and *SOX11* expression is strong in classic medulloblastomas.	*Lee et al 2002* (700)
BTRC (located at 10q24.3) (β-transducin repeat-containing protein), is a negative regulator of both Wnt and SHH signaling pathways. *BTRC* encodes a ubiquitin ligase affecting these pathways. No mutations of *BTRC* were found in 62 medulloblastomas and four PNET, though it was expressed in all the tumors. *BTRC* probably does not play a role in the pathogenesis of medulloblastoma.	*Taylor et al 2002* (308) *Wolter et al 2003* (561)

TABLE 8.21. Continued

	References
DLC1 gene (deleted in liver cancer) (located at 8p22-p23.1), has been analyzed in medulloblastomas without showing homozygous deletions or somatic mutations. Loss of expression of *DLC1* was found in 1/9 medulloblastomas and in 1/3 PNET. Except for the one PNET, no hypermethylation of *DLC1* was found in a large series of medulloblastomas, PNET and their cell lines.	*Yin et al 2002* (148) *Pang et al 2005* (479)
ERBB2 overexpression in medulloblastoma is associated with a poor prognosis and metastases. In a study it was shown that *ERBB2* up-regulates the expression of a number of pro-metastatic genes, including *S100A4*, via a pathway involving phosphatidylinositol 3-kinase, AKT1, and extracellular kinase 1/2. Some activity can be blocked by the ERBB tyrosine kinase inhibitor OSI-774.	*Hernan et al 2003* (787)
ATM gene (located at 11q), a possible TSG, was not involved in 19 pediatric medulloblastomas.	*Liberzon et al 2003* (701)
A large-cell medulloblastoma with an i(17q) cytogenetic change produced arrestin-like protein not related to visual arrestin.	*Vincent et al 2003* (560A)
KIT is a receptor tyrosine kinase oncogene involved in a number of signal transduction pathways. All 10 medulloblastomas examined showed *KIT* expression, but no mutations.	*Chilton-Macneill et al 2004* (702)
PTEN gene (located at 10q23), is a possible TSG. Homozygous losses were found in 7/23 medulloblastomas and none in 9 PNET and 7 PNET cell lines. Rare mutations of *PTEN* have been described in medulloblastoma.	*Inda et al 2004* (562) *Rasheed et al 1997* (703) *Kraus et al 2002* (560)
Blockade of apoptosis by Bcl-2 cooperates with SHH-stimulated proliferation to transform cerebellar progenitor cells and induce medulloblastoma formation in mice. Clinically, a correlation between Bcl-2 expression and prognosis could not be established suggesting to the authors that Bcl-2 might not be required for the maintenance of established tumors.	*McCall et al 2007* (788) *Schüller et al 2004* (789) *McCall et al 2007* (788)
The cytogenetic and other differences among the medulloblastoma cell lines may explain their different responses to differentiation aspects.	*Li et al 2004* (809)
ERBB2 (*HER2/neu*) may be involved in a prometastatic pathway in medulloblastoma and possibly provide a target for therapeutic intervention in these tumors.	*Hernan et al 2003* (810)
Utilizing medulloblastoma cell lines, it was demonstrated that ATRA (all-*trans*-retinoic acid) inhibited cell proliferation and induced apoptotic cell death in part by activating the casepase-3/poly(ADP-ribose) polymerase 1 effector pathway. The latter leads to cleavage of PARP.	*Gumireddy et al 2003* (811)
Retinoic acids and novel retinoids may be potential agents in the therapy of medulloblastoma. These agents may modulate fas expression and enhance chemosensitivity of medulloblastoma.	*Liu et al 2000* (812)
Syngeneic fibroblasts transfected with IL-2 and IL-12 mediate therapeutic effects against established tumors, possibly related to generating immunological memory. Such an approach may provide an immunotherapeutic strategy for the treatment of medulloblastoma.	*Barker et al 2007* (813)
Utilizing an established medulloblastoma cell line (DAOY), a study showed that cell cycle arrest, rather than apoptosis, is the main mechanism of retinoic acid (RA) inhibition of cell proliferation in DAOY cells. Inhibition of Cyclin D1 or C-*myc* which regulates the G1 S-phase transition and are the targets of b-catenin-LEF/TCF pathway may be involved in the mechanism of retinoid retardation of cell cycle progression. RARb is likely to play a role in RA and b-catenin-LEF/TCF pathways. Therefore, ATRA may have potential clinical value in the management of medulloblastoma.	*Chang et al 2007* (814)
Members of the *S100* gene family of calcium-binding proteins show divergent patterns of cell and tissue-specific expression, and the expression of specific family members is disrupted in a range of disease including cancer. Overexpression of several of the S100 proteins is thought to be involved in tumor progression, such as S100A4 and S100A, which is upregulated in many cancers including medulloblastoma and promotes angiogenesis and metastasis. Other S100 proteins are downregulated in tumors and have putative tumor suppressor roles, including S100A2 and S100A6.	*Lindsey et al 2007* (815) *Hernan et al 2003* (810) *Lindsey et al 2007* (815)
A study showed complex patterns of somatic methylation affecting *S100* genes in the normal cerebellum and changes in these genes in medulloblastoma vis-à-vis those of the cerebellum, i.e., S100A6 and S100A10 were hypermethylated and associated with aggressive large-cell/anaplastic medulloblastoma, whereas S100A4, a pro metastatic factor, was hypomethylated.	*Lindsey et al 2007* (815)
Overexpression of DNA topoisomerase II a and b mRNA was found in medulloblastoma with the levels showing a correlation with etoposide sensitivity	*Uesaka et al 2007* (816)
INI1 protein was found in 152/158 medulloblastomas (classic, nodular/desmoplastic, large cell types); the 6 negative tumors were of the classic type.	*Haberler et al 2006* (793)

Changes affecting various genes in medulloblastoma are shown in Table 8.22, in categories of their mechanisms of action in the cell, i.e., cell cycle genes, signaling pathways, and transcription factors.

Table 8.23 Some highlights of findings of a number of aspects of medulloblastoma (some of which have been discussed in the text of the chapter). The information presented in Tables 8.21–8.24 is in some respects a recapit-

ulation of the content presented above for medulloblastoma. What appears clear from Tables 8.21–8.23 and the findings presented in this chapter is that medulloblastoma is generally an embryonic tumor whose genesis is due to developmentally uncontrolled events affecting a number of different and diverse genes and their products and leading to abnormal growth and accounting for the clinical, histologic, molecular and biologic heterogeneity of medulloblastoma (199, 526,

TABLE 8.22. Genetic molecular anomalies in medulloblastoma.

	References
Cell cycle genes	
TP53 mutations (<10%)*	*Cogen et al 1990 (75), Badiali et al 1993 (705A)*
	Ohgaki et al al 1993 (88), Raffel et al 1993 (569),
	Batra et al 1995 (90), Felix et al 1995 (704),
	Nozaki et al 1998 (571), Orellana et al 1998 (91),
	Adesina et al 2000 (705)
TP53 increased expression (<40%)*	*Jaros et al 1993 (572), Adesina et al 1994 (89),*
	Giordana et al 1998 (94), Woodburn et al 2001 (203),
	Burns et al 2002 (204)
MDM2 expression (<20%)	*Batra et al 1995 (90), Giordana et al 2002 (193),*
	Adesina et al 2000 (705)
p15/p16 mutation or deletion (rare)	*Jen et al 1994 (691), Petronio et al 1996 (692),*
	Sato et al 1996 (693), Barker et al 1997 (694),
	Ebinger et al 2004 (477)
CDK4/cyclin D amplification (rare)	*Petronio et al 1996 (692), Sato et al 1996 (693)*
Signaling pathways	
PTCH mutations (<20%)**	*Pietsch et al 1997 (143), Raffel et al 1997 (144),*
	Vořechovskŷ et al 1997 (317), Wolter et al 1997 (145),
	Vortmeyer et al 1999 (316),
	Zurawel et al 2000 (290B, C)
SMOH mutations (rare)	*Reifenberger et al 1998 (304)*
APC mutations (<5%)	*Mori et al 1994 (367), Huang et al 2000 (361),*
	Baeza et al 2003 (366)
β-catenin mutations (<15%)	*Huang et al 2000 (361), Zurawel et al 2000 (290C),*
	Baeza et al 2003 (366)
Axin1 mutations (rare)	*Dahmen et al 2001 (361), Baeza et al 2003 (366)*
ErbB2/HER2 expression (<39%)*	*Gilbertson et al 1995, 1998, 2001 (426,429,661)*
TrkC overexpression (<75%)§	*Goumnerova 1996 (418)*
	Grotzer et al 2000, 2001 (263,418)
Transcription factors	
PAX5 abnormal expression (?70%)§	*Kozmik et al 1995 (435)*
NeuroD3 abnormal expression (?60%)*	*Rostomily et al 1997 (499)*
Zic overexpression (?90%)	*Yokota et al 1996 (434A)*
MYC amplification (<10%)†	*Badiali et al 1991 (107), Fuller and Bigner 1992 (125),*
	Batra et al 1995 (90), Scheurlen et al 1997 (78),
	Giangaspero et al 1999 (23), Aldosari et al 2000 (84),
	Eberhart et al 2002 (120)
MYC overexpression (<50%)*	*Debiec-Rychter et al 1994 (564), Bruggers et al 1998 (185),*
	Grotzer et al 2001 (188), Herms et al 2000 (115)
MYCN amplification (<10%)	*Badiali et al 1991 (107), Fuller and Bigner 1992 (125),*
	Batra et al 1995 (90), Aldosari et al 2000 (84),
	Lamont et al 2004 (15)
FOXG1 deregulation is frequently found in medulloblastoma based on arrayCGH studies.	*Adesina et al 2007 (185C)*
No methylation of *PTCH1* was found in medulloblastoma.	*Pritchard and Olson 2008 (186A)*

*Associated with aggressive disease.
**Desmoplastic medulloblastoma.
§Favorable outcome.
†Anaplastic medulloblastoma.
Table based on Ellison 2002 *(1)* with some modifications.

527). This heterogeneity is reflected by the large number of genes involved in the pathogenesis of medulloblastoma.

Developments in the genetic aspects of medulloblastoma have been related to two areas which have shed much light on this field. One of these areas is the identification of genes responsible for hereditary conditions associated with a much higher incidence of medulloblastoma than in the general population (Table 8.8). The results of investigations of these genes in sporadic medulloblastoma are presented in this chapter. Another area with an extensive literature based on experimental results is that of medulloblastoma development in genetically affected or modified animals, particularly the

TABLE 8.23. Some molecular biologic features of medulloblastoma.

	References
Desmoplastic medulloblastomas are associated with inactivating mutations of the *PTCH* gene, activating mutations of its interaction partners, and an activation of the SHH-Patched signaling pathway.	*Pietsch et al 1997* (143) *Raffel et al 1997* (144) *Wolter et al 1997* (145) *Zurawel et al 1998* (290B) *Reifenberger et al 1998* (304)
It has also been shown that PI3K/AKT signaling enhances the effects of SHH in these tumors and their cerebellar precursors and may, in fact, represent an indispensable mediator for the realization of the SHH signal.	*Hartmann et al 2005* (750) *Rubin et al 2003* (765) *Klein et al 2001* (766)
In contrast, the classic subtype of medulloblastoma has been less characterized molecularly, though there is evidence for PI3K/AKT signaling playing a part in these tumors, i.e., several growth factor receptors involving PI3K/AKT signal transduction are expressed in classic medulloblastoma, e.g., ILGF1-R,TrkB, PDGFR, CXCR4, ErbB2, and c-Kit.	*Patti et al 2000* (422) *Wang et al 2001* (423) *Washiyama et al 1996* (390) *MacDonald et al 2001* (259) *Gilbertson and Clifford 2003* (767) *Barbero et al 2002* (768) *Schüller et al 2005* (754) *Ray et al 2004* (665) *Chilto-Macneill* (702)
Also, classic medulloblastoma often show allelic losses at10q23 where the *PTEN* gene is located; this gene is a regulator of the PI3K/AKT signaling pathway.	*Li et al 1997* (769)
Thus, a recent study confirmed much of the above and provided evidence that PI3K/AKT signaling may play an important role in human medulloblastoma, in particular in context with further signaling pathways involved in its pathogenesis, e.g., Hedgehog and Wnt signaling. The data also indicate that PTEN may be involved in the activation of the PI3K/AKT pathway in medulloblastoma, possibly representing an alternative or additional molecular step to abundant membrane receptor activation or mutations in PI3KCA. Additional studies will be required to further characterize the role of the PI3K/AKT/PTEN system in medulloblastoma.	*Hartmann et al 2006* (770) *Broderick et al 2004* (771)
A study of 69 medulloblastomas frozen samples for the status of a number of TSG revealed no significant association between amplification of *MYC* and clinical outcome and a high level of methylation of caspase-8. No evidence was found for *GAS7* as a TSG in medulloblastoma.	*Ebinger et al 2006* (745)
A case of classic medulloblastoma differentiating into a metastatic large cell medulloblastoma with focal rhabdoid differentiation led the author to conclude that some AT/RT of the cerebellum and medulloblastoma may be histogenetically related.	*Donner 2005* (746)
In keeping with the findings previously obtained in a number of studies, it was shown in a mouse model that genetic alterations in medulloblastoma genesis from primary cerebellar granule neuron precursors involves the functional interplay between a network of specific genes that recurrently contribute to this genesis.	*Zindy et al 2007* (747)
The presence of OVCA2, a putative serine-hydrolase, was shown to be present in high abundance in medulloblastoma and its cell lines.	*Azizi et al 2006* (748)
The minichromosome maintenance protein 4, required for initiation and elongation of chromosomal DNA, ensuring that DNA replication takes place only once, was found to be present in a medulloblastoma cell line (DAOY) but not in 11 other cell lines.	*Gruber-Olipitz et al 2006* (749)
Medulloblastomas were shown to depend on IGF-II) for control of cellular proliferation in both cerebellar precursors and medulloblastoma cells.	*Hartmann et al 2005* (750)
Ku80, a factor in the maintenance of the genome by repairing DNA double strand breaks, in conjunction with p53-mediated neuronal apoptosis, suppresses neoplasia caused by non-specific DNA damage in a mouse model.	*Holcomb et al 2006* (751)
JPO2, a novel binding protein to *MYC* NH$_2$-terminal domain, is specifically expressed in human and murine medulloblastoma and may play a role in the neoplastic process.	*Huang et al 2005* (752)
Elevated expression of survivin mRNA, a member of the inhibitor of apoptosis protein family, was common in medulloblastoma but not in normal brain. Overexpression of survivin-deltaEx3 was predictive of a poor clinical outcome, and may serve as a therapeutic target in medulloblastoma. Similar observations have been made in other types of brain tumors.	*Li et al 2007* (753) *Pizem et al 2005* (753A) *Body et al 2004* (753B) *Fangusaro et al 2005* (753C) *Yamada et al 2003* (753D)
The chemokine receptor gene *CXCR4* is essential for normal differentiation of the cerebellar cortex. Very low levels of *CXCR4* were found in classic medulloblastomas; high amounts were expressed in desmoplastic and nodular medulloblastomas. A significant correlation was found between the high *CXCR4* expression and the presence of neurotrophin receptor p75^{NTR} or expression of *ATOH1* and *GLI1*. The data suggest that *CXCR4* may be involved in the pathogenesis of medulloblastoma.	*Schüller et al 2005* (754)
Expression of the neurotrophin-3 receptor tyrosine kinase C (TrkC) is associated with a favorable prognosis in medulloblastoma, probably due to increased tumor apoptosis induced by TrkC. The latter action may be inhibited by the MAPK. A study of these parameters demonstrated an apoptosis pathway independent of TrkC and neurotrophin-3 and involving extracellular signaling regulated kinase-5 (MAP/ERK5) and myosis enhancer factor-2 (MEF-2).	*Sturla et al 2005* (755)
Histone deacetylase inhibitors (HDI) are a promising class of anti-tumor agents due to their ability to induce apoptosis and growth arrest of cancer cells. Studies have suggested that HDI may be useful as a monotherapy in medulloblastoma and particularly in combination with ionizing radiation, appropriate cytostatics and TRAIL.	*Sonnemann et al 2006* (756)

TABLE 8.24. Continued

	References
O^6-methylguanine-DNA-methyltransferase (MGMT) is a DNA repair protein which confers resistance to some chemotherapeutic agents. The demonstration of high levels of MGMT in medulloblastoma may explain such resistance in these tumors. Similarly, apurinic/apyrimidinic endonuclease activity in medulloblastoma and PNET promotes resistance to radiation and chemotherapy. These observations provide a potential target for inhibition of the parameters mentioned.	*Bobola et al 2001 (723)* *Bobola et al 2005 (724)*
Insulin-like growth factor-I (IGF-1) and its signal transduction substrates (IRS-1 and PI3-K) were detected in the majority of medulloblastomas (10/17).	*Del Valle et al 2002 (421)*
Gene expression profiles of medulloblastomas though revealing a restricted set of activated genes, yet none of them is thought to be involved in early causation of these tumors; the authors suggest these genes as potential targets for immunotherapy.	*Boon et al 2003 (752)*
The *HIC1* (hypermethylated in cancer-1) gene (at 17p13.3) was found to be hypermethylated at the 5' UTR and/or at the second exon of the gene in 33/39 (85%) of medulloblastomas and in 7/8 (88%) of medulloblastoma cell lines. The authors suggest that epigenetic silencing of the *HIC1* gene may contribute to the pathogenesis of most medulloblastomas.	*Waha et al 2003 (476)*
A synthetic small molecule (CUR61414) capable of inhibiting the SHH signaling pathway may be a valid therapeutic agent for treating basal cell carcinoma and by extension some of the medulloblastomas.	*Williams et al 2003 (726)*
Telomerase activation is involved in medulloblastoma, but *PINX1*, a gene whose product is a potent inhibitor of telomerase, does not play a role in the oncogenesis of medulloblastoma.	*Chang et al 2004 (150)*
Amplification and overexpression of *OTX2* gene in anaplastic medulloblastoma, which may be targeted by therapy, such as ATRA.	*Di et al 2005 (431)*
Amplified regions at 17q23.2 and 7q21.2 contain the *PPM1D* and *CDK6* genes, respectively. Strong expression of the former in 88% of medulloblastomas and of the latter in 30% of the tumors has been described. Overexpression of *CDK6* is associated with a poor prognosis. The increased protein levels of the above genes may have an effect on the *TP53* and *RB1* pathways in medulloblastoma.	*Mendrzyk et al 2005 (727)*
Depletion of mutant p53 in cancer cell lines reduces their aggressiveness.	*Bossi et al 2006 (727A)*
Human neural stem cells have been shown to be an effective delivery system to target and disseminate therapeutic agents in medulloblastoma.	*Kim et al 2006 (737A)*
In 1/8 medulloblastomas a mutation of *PTPN11* gene was seen. The authors, however, stated that this gene played a minor role in solid tumors.	*Martinelli et al 2006 (728)*
Two-hit model for progression of medulloblastoma in mice was demonstrated.	*Pazzaglia et al 2006 (736A)*
Cytogenetic findings in two pineocytomas have been reported. Both tumors had complex karyotypic changes.	*Aparecida Rainho et al 1992 (798)* *Bello et al 1993 (799)*
Methylation status of $p16^{INK4a}$, $p14^{ARF}$, *TIMP3* and *CDH1* in medulloblastoma and PNETs is at bst of low level.	*Mühlisch et al 2007 (801)*
A proteome-wide analysis of medulloblastomas and PNET showed overexpression of stathmin, a protein belonging to a family of phosphoproteins involved in microtubule dynamics, whereas annexin A1 and calcyphosine were significantly overexpressed in ependymomas.	*de Bont et al 2007 (803)*
Reviews on the molecular cytogeneics of medulloblastoma, PNET, neuroblastomas and other brain tumors have appeared.	*Bigner and Schröck 1997 (804)*

Figures 8.19 and 8.20 show schematic suggestions for the genetic pathways that may be involved in the formation of medulloblastomas of varying histology. Thus, as is true of most malignant tumors in the human, medulloblastoma is characterized by a stepwise and orchestrated process of genetic and molecular changes and by the strong likelihood that medulloblastoma consists of a number of distinct entities, each defined by its unique genetic and molecular background.

Table 8.23 are presented some recent observations related to the molecular biology of medulloblastoma, demonstrating again the wide range of changes present in these tumors.

Primitive Neuroectodermal Tumors

PNET are small round blue cell tumors that are relatively common in children and are mainly intracranial in location. Primary spinal PNET is a rare condition in adults and even more so in children (530B). PNET can be either central (cPNET), i.e., located in the CNS, or peripheral (pPNET), i.e., located in sites other than the CNS, in origin. cPNET are morphologically indistinguishable neoplasms from medulloblastoma, but they are located elsewhere in the CNS outside the posterior fossa.

Although medulloblastoma and PNET are histologically similar (531–533), they are markedly different clinically and biologically (47, 82, 534) with PNET representing a much more aggressive tumor group than the medulloblastomas (535, 536). This difference is reflected in the cytogenetic and molecular aspects of these two groups of tumors, to be presented below, and raises the problem of limitations to the histologic aspects in the differentiation of medulloblastoma from PNET or to past proposals for the common histogenesis of these tumors (531, 532, 537). The enigma of what is or is not a PNET even at present, is reflected in a conversational vignette between a neuro-oncologist and a pathologist (538), which leaves the reader in an undecided state as well.

FIGURE 8.19. Gene expression profiles during cerebellar development and medulloblastoma nodule formation. The cerebellar external granule layer (EGL) contains mitotically active neuroblasts histologically similar to the pleomorphic internodular medulloblastoma cells, whereas the differentiated neurons of the internal granule cell layer (IGL) are similar to the tumor cells in the nodules of medulloblastoma. Thus, BCL2, p75^{NTR}, and p53 are expressed in both the cerebellar EGL and medulloblastoma nodules, whereas synaptophysin (Syn), TrkC and p27^{KIP1} are expressed in both the cerebellar IGL and medulloblastoma nodules. These facets of the developing cerebellum and nodular and internodular regions of medulloblastoma reflect their similarity (from ref. (20), with permission).

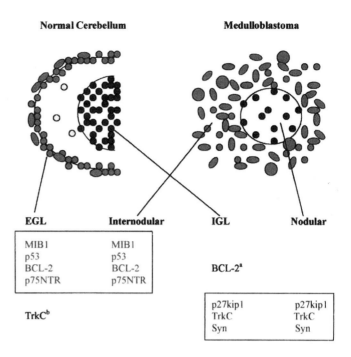

In Table 8.25 are listed some clinical, epidemiologic and genetic aspects of PNET.

The primitive neuroectodermal tumor (PNET) concept (531, 532, 539) has been controversial for some years, largely due to the uncertainties regarding the histogenesis of medulloblastoma and its relationship to histologically similar

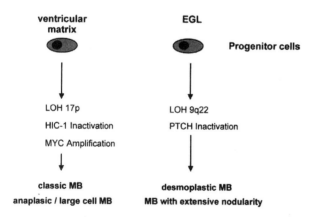

FIGURE 8.20. Hypothetical molecular pathways leading to the development of histologically different types of medulloblastoma. Classic medulloblastoma, the most common variant, is characterized by frequent alterations of chromosome 17, whereas the desmoplastic subtype is associated with frequent activation of the SHH pathway through *PTCH* inactivation (see text). Desmoplastic medulloblastoma is thought to arise from EGL precursors. The cellular origin of the classic subtype, which is usually located in the midline, is uncertain but may be the ventricular matrix. The anaplastic variant, presenting a continuum of the classic type, has *MYC* amplification. Medulloblastoma with extensive nodularity, represents a variant of the desmoplastic type, usually occurs in young children and is associated with good outcome (from ref. (522), with permission).

neoplasms at other sites. However, over the years a number of aspects characterizing these tumors has emerged, including the unique origin of medulloblastoma from the external granular cell layer of the cerebellum, supported by the occasional presence of *PTCH* mutations in medulloblastoma (144), *PTCH* being the part of the SHH pathway which controls cell proliferation of this layer during embryonal development (291). The presence of mutations of *β-catenin* and *APC* genes (361), and the pattern of DNA methylation (87, 465) in some medulloblastomas and their absence in PNET and pineoblastoma differentiates these tumors. Differences in the clinical behavior of medulloblastoma versus PNET may be a reflection of the above differences (533, 540, 541). The challenging diagnostic aspects and the histologic variability of PNET have been discussed in a number of publications (532, 542–545).

In evaluating the various findings in PNET, the problem one faces consists of several elements. Some authors continue to use the term PNET to include medulloblastoma, although in most publications the nature of the tumors can be deduced from the clinical, histologic, anatomic and source descriptions; however, in some publications this cannot be done, though by their much higher prevalence it can be assumed that the tumors described consist primarily of medulloblastomas. Yet, this remains a serious problem in evaluating the literature on PNET. Even when authors at one point separate medulloblastomas from PNET, at another point the results often do not clearly specify as to which tumor type these results apply to (546).

Primitive neuroectodermal tumors (PNET) are embryonal tumors of the cerebrum and/or supratentorial region composed of undifferentiated or poorly-differentiated neuroepithelial cells which have the capacity for differentiation along neuronal, astrocytic, ependymal, muscular or melanocytic

TABLE 8.25. Miscellaneous genetic and clinical features of PNET.

	References
The age range for cerebral PNET is 4 weeks to 10 years with a mean of 5.5 years.	*Rorke et al 2000* (535)
65% of PNET occur in patients less than 5 years of age. There is no sex predominance, although a 2:1 male:female ratio has been shown for cerebral PNET. PNET are very rare in adults in whom the clinical course may be more favorable.	*Rorke et al 2000* (535) *Kim et al 2002* (729) *Kouyialis et al 2005* (730)
PNET is 1/10th as common as medulloblastoma.	*Kim et al 2002* (729)
PNET account for <2% of pediatric intracranial tumors	*Rickert and Paulus 2001* (36)
PNET may be secondary to therapy for other neoplasias (leukemia, lymphoma) and other intracranial tumors.	*Leung et al 2001* (730A)
The clinical and treatment aspects of PNET in adults (most of them in the 3rd decade) do not differ significantly from those in children.	*Kim et al 2002* (729) *Kouyialis et al 2005* (730)
Prognosis of PNET is related to age of patient, location of tumor and tumor stage.	*Krischer et al 1991* (730B) *Chan et al 2000* (731) *Packer et al 2003* (732)
Animal models for PNET have been described.	*Fung and Trojanowski 1995* (379)
PNET (and medulloblastoma) is differentiated from AT/RT by changes in the *INI1* gene and del(22q) seen in the latter but not in the former tumors.	*Biegel et al 2001* (62) *Bruch et al 2001* (733)
No aberrant CpG methylation at 17p11.2 was present in PNET in contrast to medulloblastoma.	*Frühwald et al 2001* (87)
Promoter hypermethylation of the *RASSF1A* gene (located at 3p21.3) was present in 6/9 PNET and in all 25 medulloblastomas. *FHIT* (fragile histidine triad) and *sFRP1* (secreted frizzled-related protein 1) genes were affected in a fraction of PNET (11% and 22%, respectively); medulloblastoma was not affected.	*Chang et al 2005* (688)
bHLH (basic helix-loop-helix) transcription factor *NEUROD1* was expressed in 1/4 of the PNET analyzed, in contrast to all 12 medulloblastomas expressing this gene; 3/5 PNET expressed *HASH1* (human *achaete scute1*), a gene not found to be expressed in the medulloblastomas.	*Rostomily et al 1997* (499)
Loss of 17p was reported in a PNET of the right temporal lobe of a 33-year-old female patient.	*James et al 1990* (138)
Abnormalities of hTERT (human telomerase reverse transcriptase) and ALT (alternative lengthening of telomeres) may play a role in the pathogenesis of PNET.	*Didiano et al 2004* (492)
One PNET (the only one examined) was shown to be *BCL2*, BCLX$_L$, BCLX$_S$ and *BAX* positive, genes involved in apoptosis (pro and anti) and whose relative expression determines the apoptotic threshold.	*Bruggers et al 1999* (410)
An analysis of LOH in one PNET of a 1-year-old girl showed loss at 22q11.2.	*Blaeker et al 1996* (140)
The *hSNF5/INI1* product is a component of the SW1/SNF chromatin remodeling complex and although its alterations are more common in other tumor types (e.g., AT/RT), a few PNET have shown changes of this complex. Thus, 2/17 PNET showed nonsense mutations resulting in a truncated *INI1* product and a constitutional *INI1* mutation was found in a PNET in a patient who also had a renal rhabdoid tumor.	*Imbalzano and Jones 2005* (698) *Sévenet et al 1999* (734,735)
An analysis of TP53 mutations in PNET revealed them to be more common in tumors with neuronal differentiation than in non-neuronal tumors and more common in adult cases than in pediatric ones. The *TP53* status did not affect survival, nor did the neuronal differentiation.	*Ho et al 1996* (570)
Mutually exclusive mutations of p53, methylation of p14ARF or deletion of *INK4/ARF* have been demonstrated in up to 25% of medulloblastomas.	*Frank et al 2004* (207)
Examination of formalin-fixed, paraffin-embedded PNET samples for quantitative mRNA expression analysis of TrkC and MYC significantly correlated with the findings obtained from matching fresh-frozen tissues.	*Kunz et al 2006* (794) *Rutkowski et al 2007* (795)
Utilizing a number of molecular techniques, it was shown that PNET cell lines characterized by deletions of 10q, 16q and 17p and to a lesser extent of Xq, 1p, 7p and 13q. The prevalence of hemizygous loss in these tumors suggests that haploinsufficiency affecting multiple TSG may play a fundamental role in their pathogenesis.	*Dallas et al 2005* (796)
PNET harbors deletions of the *CDKN2A* locus and other aberrations distinct from medulloblastomas.	*Pfister et al 2007* (797A)

lines. Tumors with a distinct neuronal differentiation are termed cerebral neuroblastomas (Figure 8.21) or, if ganglion cells are also present, ganglioneuroblastoma (532, 535, 545). Light microscopic features of PNET are basically very similar to those of medulloblastoma, thus presenting a problem in the histologic diagnosis of at least some of these tumors.

A microarray-based study in which expression profiles on PNET were compared with those of normal cerebellum and medulloblastomas showed the PNET profile to be distinctly different from those of the normal cerebellum and medulloblastomas (208). This study also supported the granule cell origin of medulloblastoma but not of PNET. The latter may arise from cells of periventricular germinal matrix or subependymal cells of the diencephalon.

A PNET most commonly arises in the cerebrum, and is seen less frequently in deep paraventricular or midline locations such as the diencephalon and basal ganglia (547); rarely, PNET can present in the leptomeninges without evidence of a primary tumor in the supra- or infratentorial compartments (548). These are proposed to arise from heterotopic glial nests within the subarachnoid space (549); however, their true histogenesis and their relationship to either medulloblastoma or PNET is unknown.

Cytogenetic Changes in PNET

Abnormal karyotypes in only 18 PNET have been published (Table 8.4), reflecting their relative uncommon incidence.

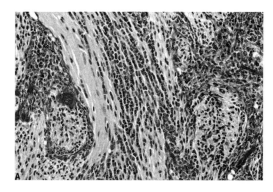

FIGURE 8.21. Histologic appearance of a PNET with advanced neuronal differentiation (cerebral neuroblastoma) exhibiting a nodular architecture with typical streaming of tumor cells (from ref. (535), with permission).

The chromosome number ranged from 39 to 90; two were hypodiploid, five were pseudodiploid (46 chromosomes but with abnormalities), five were hyperdiploid, and five triploid-tetraploid.

Examination of the 18 abnormal karyotypes observed in PNET (Table 8.4) does not reveal two similar karyotypes. Of interest is one PNET with two copies of i(17q) (39), one with −22 as the sole anomaly (550) and another with an extremely complex karyotype (551). These findings should be viewed against the high incidence of i(17q) in medulloblastoma and −22 in RT/AT tumors, pineoblastomas and meningiomas. The most frequently affected chromosome was 11, with changes varying from translocations, losses or gains. In one publication (552), a case of PNET with an unusual structural aberration involving one copy of chromosome 11 was studied by FISH, the genesis of the abnormal chromosome was clarified (Table 8.4), and its significance was discussed.

PNET generally show complex karyotypes with many structural alterations. The latter include deletions, additions and translocations. However, none of these changes is sufficiently recurrent to be characteristic of PNET. For example, 10q22-q26 and 6q21-q25 have been described to be involved

in translocations, but only in two to four cases with variable bands affected. In one case (85) a t(6;13)(q25;q24) as the sole karyotypic abnormality, confirmed by SKY, in the PNET of a 2-year-old child without a history of retinoblastoma may imply involvement of the *RB1* locus at 13q14. However, no other PNET cases with changes at 13q14 have been reported.

Four tumors had dmin, a finding usually associated with gene amplification. No hsr have been described in PNET (41, 553).

Normal karyotypes in PNET have been reported (44, 55, 554), but these probably represent normal cells that were in division at the expense of the abnormal tumor cells, as is evident from CGH studies.

As mentioned, the t(11;22)(q24;q12) characterizes pPNET, and it is associated with the genesis of a chimeric gene *EWS/FLI1* (436). These tumors are also characterized by the expression of the MIC2 glycoprotein. A study (555) of PNET and medulloblastoma showed negativity for MIC2. The demonstration of the aforementioned translocation differentiated a metastatic peripheral PNET to the spinal epidural space from a primary CNS PNET (556).

CGH Studies in PNET

Features of CGH studies in PNET are shown in Table 8.26. CGH data on 17 cases have been presented by 4 groups (47, 80, 82, 85) and have revealed almost all of them to have abnormal profiles in contrast to the cytogenetic studies in which a significant percentage (35–40%) has been found to be diploid (normal). This discrepancy is probably due to the overgrowth of the tumor cells by normal cells present in the tumor in cytogenetic preparations. Compared with the relatively large number of medulloblastomas that has been examined by CGH, that of PNET is small and, hence, generalizations may be spurious. Nevertheless, some of the findings are worthy of emphasis. No losses of 17p or the presence of i(17q) were seen in any PNET, findings commonly present in most medulloblastomas. The observations of one study (82), which

TABLE 8.26. CGH Studies in PNET.

	References
PNET in 2 siblings (brother and sister) with germline *TP53* mutations were studied. One tumor had 17 changes by CGH and the other had 14 changes. Five alterations were common to both tumors: gains of chromosomes 1 and 7 and 21q, and loss of chromosome 3 and 13q.	*Reifenberger et al 1998* (567)
A PNET showed gains of 1q, chromosomes 3 and 17, but no losses. Another PNET showed loss of 9pterq22 and 11q23-qter.	*Avet-Loiseau et al 1999* (47)
No abnormalities of chromosome 17 were found in 4 PNET; 3 cases showed loss of 3p12.3-p14	*Nicholson et al 1999* (80)
CGH of 10 PNET (including 3 pineal) showed loss of 4q, 9p, 13q, 14q, 19q, and gain of chromosome 7 and 8q. One tumor had amplification at 4q12-q13.	*Russo et al 1999* (82)
Results on 5 PNET were combined with those of 19 medulloblastomas, thus comprehensive results on individual PNET were not available except for the following: one PNET had normal results with CGH, though cytogenetically and by SKY it had a balanced t(6;13), one tumor had loss of 17p but no reciprocal gain of 17q and the other two PNET had normal results. Amplification of 2p24 was found.	*Bayani et al 2000* (85)
Gain of 1q was a frequent change by CGH in 6 PNET studied. Losses affected 16p and 19p. An i(17q) and loss of chromosome 10, changes common in medulloblastoma, were not seen in 6 PNET.	*Inda et al 2005* (706)

revealed losses of 4q, 13q, 14q, and 19q and gains of chromosome 7 and 8q in 21–32% of the tumors (based on 10 tumors, 3 pineal) and loss of 3p12.3-p14 in 3 PNET in another study (80) await confirmation on more cases, so that their meaning and significance can be evaluated with some certainty.

Moreover, a comparative study of 43 medulloblastomas and 10 PNET found statistically significant differences in copy number aberrations for chromosomes 14 (40% of PNET but 0% of medulloblastomas), 17 (0% of PNET but 37% of medulloblastomas), and 19 (40% of PNET but only 2% of medulloblastomas) (82). Consequently, i(17q), which is frequently involved in medulloblastomas, does not seem to play a role among PNET.

In a series of 17 pediatric PNET, in which high-resolution CGH was used (557), genomic imbalance was reported in 70% of the tumors, with a median of four changes per tumor. Most changes (63%) involved whole chromosomes, with partial chromosome imbalance seen in 37% of the tumors. Using high-resolution array-based CGH, the breakpoints associated with i(17q) in medulloblastoma were more clearly delineated (558), and differences were shown that could be applied to the diagnosis of supratentorial PNET versus medulloblastoma.

LOH Studies in PNET

LOH studies (131) of 6q, 16q, 17p, and chromosome 22, although revealing anomalies in medulloblastomas, were negative in the four PNET examined.

In an analysis of LOH of regions thought to harbor genes possibly involved in medulloblastoma and PNET tumorigenesis, five pediatric PNET were examined (559). One PNET showed LOH of 9q and 17p and also contained an i(17q) by FISH, findings more compatible with medulloblastoma than PNET. Of three PNET examined for LOH of 17p, chromosomes 6, 9, and 16 (78), one displayed LOH at 17p11.2 and another of chromosome 9. LOH on 22q was observed in 2/5 PNET (559), with the region of deletion covering the *hSNF5/INI1* gene locus. Otherwise, allelotyping studies in >40 PNET have not disclosed loss of 17p (85, 96, 123, 139, 140, 560).

Other regions of loss observed at significant frequency include chromosome 9 (centering around distal 9p), and 13q and 14q. Less frequently, losses involving all or part of chromosomes 5, 6, 10q, 12, 18q, and 19q have been reported previously (360). An isolated 3p21 interstitial deletion has been reported in one tumor (80); it remains to be determined whether the *hMLH1* DNA repair gene mapping to this site was involved. Losses of 10q observed in approximately 25% of PNET may rarely involve tumor suppressors *BTRC* (561), *PTEN*, or *DMBT1*. Only one mutation of *PTEN* in one of 21 PNET (560, 562) and one of nine of *DMBT1* (562) have been reported.

Similar to observations in medulloblastomas (47, 104), gains at chromosome 7 (six of 16), followed by 1q, occur most frequently in PNET. Gains of 9qter and chromosomes 13 and 17 have been reported in more than one case. Consistent with detection of dmin in karyotypic studies, CGH analyses of PNET have identified high-level DNA amplification at 2p24, 4q12-q13, 7q21.3, and 7p11.2. Additional amplicons at 1q;6p, 12q21, 6p, and 20q also have been reported (557). With the exception of the 2p24 amplicon, which has been shown to be related to *MYCN* gene amplification (85), the majority of putative oncogenic loci indicated by cytogenetic findings in PNET remain unknown. Studies with higher-resolution single gene mapping tools (such as the recently developed array-based CGH technology) (563) and with larger tumor numbers will be very valuable for defining the true spectrum and frequency of genomic imbalances, and to locate oncogenes and TSG that underlie PNET development.

Using several techniques (e.g., representational difference analysis, microsatellite mapping, and quantitative PCR) previously described deletions of 10q, 16q, and 17p were confirmed in PNET, but it also revealed deletions of regions not commonly described, i.e., Xq, 1p, 7p, and 13q (563A). The authors suggest that hemizygous loss in PNET affecting multiple TSG may play a role in their pathogenesis.

Molecular Studies in PNET

Because of the rarity of PNET, studies of only small numbers of PNET have been reported to date. Clearly, comprehensive knowledge of the molecular genetic features of PNET will be essential to advance our knowledge regarding the diagnosis, therapy, and prognosis of this rare but challenging tumor. Existing molecular studies of PNET have been reviewed (360), with a view to improving our current understanding of putative signaling defects and molecular abnormalities in these tumors.

Using a method of detecting low levels of amplification, one group (563B) found *MYCN* amplification in a PNET occurring 2 years after chemotherapy and radiation for a primary medulloblastoma in a 7-year-old girl. Overexpression of *MYC* was present in 10% of cerebral PNET (564), but it is not associated with amplification of this gene. Neither was *MYCN* or *erbB1* amplified. In a study aimed at determining the status of the *PTCH* gene, somatic mutations were found in 4/33 medulloblastomas, but in none of the 4 PNET examined, even though two of them had LOH of 9q (317), where the *PTCH* gene is located. Mutation of *PTCH* was seen in one of five PNET, although all these tumors expressed *PTCH* (145).

α-Synuclein, a presynaptic nerve terminal protein, was found to be highly expressed in the one PNET in which it was examined (565), indicative of neuronal differentiation. The status of neurotrophin receptor proteins (TrkA, TrkB, and TrkC) in PNET is uncertain, because only a few tumors have been examined (390).

Prognostic aspects and clinical outcome of patients with PNET, though small in number, have been reported (263, 502, 566). Reviews of various facets of PNET have been published (462, 522).

Scattered reports of other genetic abnormalities in these tumors include identification of *MYCN* expression (186), germline mutation of *TP53* (567), expression of the Neuro D family of bHLH transcription factors, and *achaete scute*, another neurogenic transcription factor with homology to *Neuro D* genes (499). This latter gene was actually expressed in three of five PNET but not in medulloblastoma. Thus, although the number of studies of PNET is small, it seems that genetic events associated with development of PNET are essentially different from those in medulloblastoma.

p53 in PNET

Although involvement of the locus on 17p for the *TP53* gene is very uncommon in PNET, the product of this gene, p53, has been investigated in this tumor on the basis of its frequent alterations in a host of other types of cancers. Molecular-genetic investigation of two children with PNET revealed that the mother, who in the past had an operation to remove an ovarian carcinoma, had transmitted a *TP53* germline mutation at codon 213 to both children (567). LOH analysis at microsatellite markers located proximal and distal to the *TP53* locus at 17p13.1 showed loss of the paternal alleles in both PNET. Analysis of the PNET by CGH revealed multiple chromosomal imbalances, indicating substantial genomic instability in both tumors. Five changes were common to the tumors, i.e., gains of chromosomes 1 and 7 and 21q and losses of chromosome 3 and 13q (567). Mutation of *TP53* and lack of wildtype p53 in one of these PNET led to an accumulation of centrosomes which the authors (568) propose as resulting in uneven chromosome segregation in the tumor.

PNET may be present in the Li-Fraumeni syndrome (568).

Mutations of p53 not found in high incidence (two of 28) in PNET of children (96, 560, 569). In 14 adult and pediatric PNET, six of 11 of the adult tumors had p53 mutations, primarily in neuronal tumors; no mutations were present in the pediatric cases (570). This led to the suggestion (360) that pediatric and adult PNET may differ genetically.

Although p53 mutations are rare in PNET (571), overexpression of p53 is not uncommon (due to p53 dysfunction) (203, 204, 524, 572). JCV was not detected in any of the tumors; hence, it is not the cause of the elevated p53 (524). Amplification of MDM2 is not a common mechanism for p53 loss-of-function in PNET.

Notch Signaling in PNET

Notch signaling studies have demonstrated activation of the pathway in PNET and differential expression of Notch pathway components in medulloblastoma and PNET, perhaps reflecting the distinct cells of origin for these tumors. Thus, in a study of 13 medulloblastomas and 5 PNET, hASH1 expression was seen in three of five PNET but in none of the medulloblastomas (499). Expression of *NeuroD* was more prevalent in the medulloblastomas.

High Notch2 expression was present in PNET when compared with medulloblastoma (448), i.e., 50% of PNET (including a cell line) expressed extremely high levels of Notch2, whereas 2 medulloblastomas and 2 medulloblastoma cell lines expressed Notch2 and Notch1 at levels comparable to those of control cerebellar cells. Three of the PNET with high Notch2 expression had genomic amplification or copy number gains of the Notch2 locus at 1p11-p13 (448). Signaling pathways, essential for normal brain development may be dysregulated by mutations in some medulloblastoma (161, 572A), i.e., SHH-Gli, Notch and less so Wnt pathways, but they also may be involved in PNET development (360). A PNET in a newborn male was shown by FISH not to have *TP53* involved (536) or amplified, an example that inactivation of *TP53* probably does not play a role in the pathogenesis of PNET.

An example is one study (96) in which 8 PNET and 35 medulloblastomas were examined for 17p loss by RFLP. None of the PNET had such loss, whereas 13 of 35 medulloblastomas showed this loss. In two of the latter tumors mutations of *TP53* were present, but none in PNET (96).

A mutation of *TP53* in the PNET with a favorable course in a 9-year-old girl has been reported previously (573).

A mutation of *TP53* was found in one of 12 PNET without the tumor showing 17p allelic loss (560). No changes (deletions or amplifications) were found in the *CDKN2A*, *EGFR*, cyclin-dependent kinase (*CDK*)4, or *MDM2* genes. A highly malignant PNET was shown to harbor two distinct copies of *MYC* without any hint of amplification (79).

Secondary PNET have been reported in the context of cancer predisposition syndromes in two patients. One child presented with a PNET 5 years after treatment for unilateral retinoblastoma (574); in another child with neurofibromatosis type 1, PNET developed after radiation treatment for a brainstem astrocytoma (575). PNET also can occur as secondary malignancies in both children and adults without genetic background. Tumors occurring up to 18 years after radiation treatment for low-grade primary intracranial neoplasms, such as cerebellar pilocytic astrocytoma and low-grade astrocytoma, have been described (575, 576).

The current literature contains reports suggesting that metastatic disease, extent of surgery, and age at diagnosis may be of prognostic relevance in PNET (540, 577). Based on more aggressive clinical features, PNET have traditionally received therapies designed for metastatic, high-risk medulloblastomas, which involves more intensive chemotherapy and higher doses of radiation to the head and spine. Several retrospective (540,578) and prospective studies (577,579,580) have consistently reported significantly inferior outcomes for even localized PNET treated with high-risk medulloblastoma therapy. To date, the overall survival rate for PNET is substantially lower than that for medulloblastomas, with an expected 3-year progression-free survival of approximately 50% for localized PNET (540, 577). These observations suggest intrinsic biological differences between PNET and medulloblastoma.

From the studies presented in this section on PNET (cytogenetic, CGH, LOH, and molecular), it is clear that PNET and medulloblastoma are separate and distinct tumors and that we should alleviate the confusion regarding these neoplastic entities. In essence, it is the features that characterize medulloblastoma, i.e., presence of an i(17q), loss of 17p, anomalies of the SHH and Wnt pathways, and the pattern of DNA methylation, features that are absent in PNET (and pineoblastoma) that differentiate these two groups of tumors.

Established PNET Cell Lines

Only a few established cell lines of PNET with information appropriate for this chapter have been reported in the literature. One was of cerebral origin (581) and another of spinal origin (268). Sublines of one of these lines (581) showed complex karyotypes with some variability among the sublines. Thus, one subline showed a –22, not an uncommon finding in PNET, whereas other sublines did not. In contrast, a translocation (17;22)(q21;q13.1) was present in two of the sublines. Two other sublines had +8 as the only change. The other PNET cell line had the karyotype: 50,del(3)(q13.2-q21),+8,t(11;22)(q24;q12),+12,+14,+21 (268).

SHH Pathway in PNET

Increased expression of PTCH, SMOH, and Gli mRNA has been reported in PNET (three of five) (304). Based on immunohistochemistry (186), three PNET were reported to express *MYCN*, a downstream target of SHH signaling. Missense mutation and LOH involving the *PTCH* locus at 9q22.3 has been reported in three of eight pediatric PNET (145, 317).

These limited studies indicate that SHH pathway aberrations are likely to have a pathogenetic role in at least some PNET, the scope of such involvement and whether it correlates with specific histopathological features await more extensive studies (360). Due to the rarity of PNET, the status of the SHH signaling pathway (and other pathways) remains incompletely investigated, and findings generally have been reported collectively with those on medulloblastoma (365)., resulting in β-catenin nuclear accumulation and inappropriate Wnt activation.

hASH1 expression was found only in PNET and *not* in medulloblastoma (499). Expression of *NeuroD* genes was more prevalent in medulloblastoma. Examination of 12 PNET revealed significantly higher levels of Notch1 and Hes1 mRNA than in medulloblastoma (448). Notch2 levels tended to be higher in PNET than in medulloblastoma. More than 50% of PNET (including a cell line) expressed very high levels of Notch2 mRNA. Medulloblastomas (including cell lines) expressed Notch1 and Notch2 at levels comparable with control cerebellar cells. Three of six PNET with high levels

of Notch2 mRNA had genomic amplification or copy number gains of the *Notch2* locus at 1p11-p13 (448). Knockdown of Notch 2 in a PNET cell line (581) resulted in inhibition of cell growth, supporting the role of Notch in PNET development (360).

An early arising, highly malignant supratentorial PNET (79) harbored two distinct copies of the *MYC* oncogene without any hint for amplification. A previously reported molecular analysis of medulloblastomas occurring simultaneously in monozygotic twins also lacked aberrations of 17p and amplifications of *MYC* (163).

The gene *DLC1* (deleted in liver cancer) is located at 8p22, a region deleted in medulloblastoma and PNET. Of 21 medulloblastomas, nine PNET and six PNET cell lines, only one of nine medulloblastoma and one of three PNET showed loss of *DLC1* by gene expression (479). In the PNET, the gene was silenced by promoter hypermethylation. Based on this finding, an additional 20 medulloblastomas, eight PNET, four medulloblastoma cell lines, and two PNET cell lines were examined for hypermethylation of the *DLC1* locus, with negative results (479).

A family in which three children were affected by brain tumors, two with PNET has been described (479A). The family was heavily consanguineous with cosegregation of learning difficulties, café-au-lait spots and early onset of brain tumors inherited in a recessive manner. One of the PNET from a 14-year-old girl had a near-tetraploid karyotype, loss of one chromosome 13, and an unbalanced rearrangement resulting in 10q loss. Missense mutation of exon 14 of the *PMS2* gene was found in the affected cases. The authors suggested that *PMS2* mutations are more likely to result in a recessively inherited childhood cancer syndrome than in the autosomally dominant Turcot syndrome.

Pineoblastoma

Pineal parenchymal tumors (PPT) are rare neoplasms derived from pineocytes or their embryonal precursors and span a wide spectrum of differentiation that is paralleled by differences in biological behavior, clinical outcome, and molecular genetics (533, 582–584A). They account for <2% of pediatric brain tumors.

Pineoblastoma is the most common of these tumors, with pineocytoma and pineal parenchymal tumor of intermediate differentiation (PPTID) being less common (585–587). Heritable cases of pineoblastoma are shown in Table 8.27.

Pineoblastomas make up 45–50% of pineal parenchymal tumors, with variable reactivity to neuronal and glial markers (586, 588). Like PNET, pineoblastoma is mainly a childhood disease, with cases only occasionally described in adults (518, 585). Pineoblastomas have been collectively analyzed with PNET in most therapeutic and molecular studies, comprising 20–50% of reported PNET in some series (540, 547, 578, 580, 589, 590). Nevertheless, there is some suggestion that in very

TABLE 8.27. Cases with heritable pineoblastoma.

	References
Discordant tumors in identical twins: one with a pineoblastoma and the other with a medulloblastoma.	*Waldbaur al 1976* (707)
Mother and daughter with pineoblastoma.	*Lesnick et al 1985* (708)
Pineoblastoma and astrocytoma in the same patient.	*Brockmeyer et al 1997* (709)
Pineoblastoma with familial adenomatous polyposis.	*Ikeda et al 1998* (710)
Congenital pineoblastoma and parameningeal rhabdomyosarcoma in an infant.	*Çorapcíoğlu et al 2006* (711)

(a)

(b)

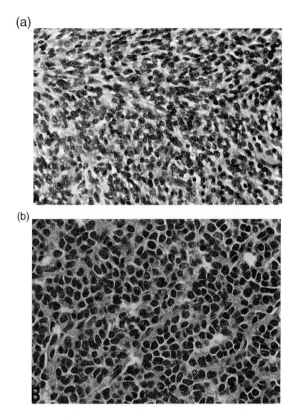

FIGURE 8.22. Histologic view (*top*) of a pineoblastoma showing high cellularity with numerous mitotic figures. Histologic picture (*bottom*) of another pineoblastoma with highly cellular features showing undifferentiated small cell histology (from ref. (4), with permission).

young children (<18 months), pineal tumors may occur at a higher frequency than nonpineal lesions (25).

Pineoblastomas are highly cellular tumors. Small, undifferentiated cells proliferate diffusely without lobular architecture, mimicking medulloblastoma and PNET arising in other sites (Figure 8.22). Tumor cells have round to irregular, hyperchromatic nuclei and scant cytoplasm (591, 592). The fibrillary matrix is poor, and there are occasionally neuroblastic rosettes (Homer Wright rosettes) and Flexner–Wintersteiner rosettes. Fleurettes, a bundle of cytoplasmic processes showing bulbous expansion at their distal portion, are interpreted as evidence of photoreceptor differentiation. Necrosis is common, and mitotic activity varies from case to case but is commonly brisk. Silver carbonate impregnation discloses a few short, thin processes.

Cytogenetic Changes in Pineoblastoma

Only 10 cases (including three cell lines) of pineoblastoma with abnormal karyotypes have been described (Table 8.4). Two of these were associated with trilateral retinoblastoma (TRB). In eight of these tumors, the chromosome number was near 46. Loss of chromosome 22 was the only anomaly in two cases (172, 593) and associated with other changes in another pineoblastoma (55). An i(17q), a common karyotypic change in medulloblastoma, was found in two cell lines derived from the pineoblastoma of a treated patient (594). Two unbalanced translocations, der(10)t(10;17) and der(16)t(1;16), were present in the same tumor (595). This tumor had increased *MYC* and *MYCN* expression in the absence of amplification. Increased expression of *MYC* and *MYCN*, downstream β-catenin targets, has been reported in pineoblastoma cell lines (594).

Although pineoblastomas have been linked to hereditary retinoblastoma as part of the TRB complex, none of the sporadic cases of pineoblastoma exhibited a visible change in the region of the *RB1* locus at 13q14. Karyotypic changes in three pineocytomas have been published (596–598). Two of these tumors had complex changes (596, 597), whereas the third had only numerical alterations (598).

CGH Findings in Pineal Tumors

A CGH study (118, 599, 600) of three specimens for each tumor type revealed no changes in pineocytomas and more than five changes for pineoblastoma and PPTID. The most frequent DNA copy number changes among PPTID and pineoblastoma were gains of 12q (50%), 4q, 5p, and 5q (33% each) and losses of 22 (67%), 9q, and 16q (33% each). Among PPTID, the most common imbalances were gains of 4q and 12q, and −22 (67% each), the latter of which was also prominent in pineoblastomas (67%). In accordance with previous cytogenetic results, −22 seems to play a role in PPT, whereas other reported changes, such as gain of 17q or loss of chromosome 11 and 12q , were not found in any of the tumors. In contrast, gains of 12q were found in three of the six higher grade PPT. Moreover, regarding alterations in the *MYC* and *TP53* gene regions, only one gain of 8q and no loss of 17p were encountered; however, in view of the role of the *RB1* gene mapping to 13q14 in the familial TRB, it is intriguing to speculate about its involvement in one of the three pineoblastomas showing a loss for 13q. Thus, PPTID are cytogenetically more similar to pineoblastomas than pineocytomas.

Furthermore, imbalances in higher-grade PPT consisted of gain of 12q and -22 (600).

Pineoblastomas may exhibit histopathological features indicative of photosensory differentiation, ranging from focal expression of retinal S-antigen and proteins to characteristic Flexner-Wintersteiner rosettes and fleurettes reminiscent of retinoblastic differentiation (586, 601, 602). Embryologically, the pineal gland develops from primitive neuroepithelial cells, which line the diencephalic roof (603) and have features of photoreceptor organs, with photosensory and neuroendocrine differentiation. These cells are thought to be ontogenetically linked to the photoreceptor cells of the retina. The histological features of pineal tumors and PNET arising in the context of trilateral retinoblastoma (TRB) (360) (see below), can bear exceptionally close resemblance to retinoblastoma, and this tumor type has been labeled by some as "primary ectopic intracranial retinoblastoma" (601, 604). Thus, at least a proportion of pineoblastomas may share ontogenic aspects with precursor cells targeted for transformation during retinoblastoma development.

Pineoblastoma can occur, as just mentioned, in the setting of heritable and sporadic retinoblastoma, so-called TRB, a condition first described in 1971 (605). TRB is characterized by intracranial neuroblastic tumors that typically occur in the setting of germline *RB1* gene mutations and heritable disease that manifests as bi- or multifocal disease. Less frequently, TRB can manifest with unilateral ocular retinoblastoma (606). Rare germline mutations of *RB* manifest as "forme fruste" retinoblastoma, with only intracranial pineoblastoma or PNET, and no ocular tumors (607). TRB with intracranial disease that precede the development of retinoblastoma are exceptional (601, 608). Up to 6% of patients with bilateral retinoblastoma and 10% of patients with family history of these lesions have been reported to harbor an intracranial pineoblastoma or PNET (609). TRB develops within a median duration of 21–23 months from the retinoblastoma diagnosis, has a poor prognosis, and a 6-month median survival duration following diagnosis of the intracranial tumor (604, 607).

More than 80% of tumors associated with retinoblastomas are pineal in location, whereas supra- or parasellar tumors comprise <20% of intracranial tumors with TRB. Rare cases of this disease with cerebellar/fourth ventricular PNET or medulloblastoma have been reported (608, 610). Only one case of cerebral PNET in association with retinoblastoma has been reported (574), which occurred in a child in whom a temporal-parietal supratentorial PNET developed 5 years after radiation therapy for a unilateral retinoblastoma. The atypical location and exceptionally long lag time to development of the intracranial lesions suggests a therapy-related secondary tumor rather than a manifestation of TRB.

The observed differences in histological features and location of TRB-associated PNET suggest that pineal/suprasellar PNET and nonpineal cerebral supratentorial PNET may arise from distinct precursor cell populations. In this respect, it may

be relevant that pineal PNET presenting in very young children are reported to have worse outcomes that cerebral supratentorial PNET (25, 590). Several therapeutic studies have reported opposite trends in older children, with higher survival rates and lower incidences of metastatic disease in pineal compared with nonpineal supratentorial PNET (547,579,580). Whether these observed differences reflect true age-related differences in tumor biology or result from earlier clinical manifestations and hence detection of pineal region tumors is not known.

Mutations of the *RB1* gene (located at 13q14) must play a role in pineoblastoma development in TRB, considering the genetic background of this syndrome (611,612).

It is not known whether direct mutations of *RB1* or alternate epigenetic mechanisms of *RB1* gene inactivation or functional pRB loss occur in sporadic pineoblastoma. Observations show significantly worse outcomes in patients with pineoblastoma associated with constitutional RB1 defects compared with sporadic pineoblastoma (613) and suggest possible biological differences between these diseases. The more aggressive disease course in TRB-associated pineoblastoma may reflect the greater genomic instability reported to occur with constitutional pRB loss (614). It is not known whether changes in 1q31 and 6p22, which have been reported in 50% of human retinoblastomas (615,616), occur in pineoblastomas or PNET associated or not associated with retinoblastoma.

In a microarray study, an attempt was made to identify genes that are differentially expressed between pineoblastoma and fetal pineal gland; however, the small number of tumors analyzed (two) significantly limits the interpretation of this study (617). Based on a study with oligonucleotide arrays of pineal tumors (618), differences in the expression of a number of genes in PPT, i.e., pineocytoma and pineoblastoma versus PTPR were demonstrated and may be of diagnostic value.

Single-gene investigations of pineoblastoma have included five pineal tumors with negative results for p53 (619) and Waf1/p21 (620) gene aberrations. A missense *INI1* mutation in a 7-month-old infant with a histologically confirmed pineoblastoma without rhabdoid features and monosomy 22 as the only abnormality has been reported (62). It is not known whether *INI1* mutations are more likely to occur in pineoblastomas presenting in very young children with particularly aggressive disease. The development of *INI1*-specific antibodies (621) should facilitate a broader scope of investigations into the genetic and functional role of the *INI1* locus in the development of pineoblastoma.

Studies on somatostatin in normal pineal gland and in pineal tumors have been reported (622), and reports on the molecular genetics in pineal tumors have been published (611, 623). A tumor of the pineal gland diagnosed as a PNET but also consistent with the diagnosis of pineoblastoma in a 6-month-old infant with a constitutional t(16;22)(p13.3;q22.2~12) has been described (624). Cytogenetic and SKY studies of the tumor revealed only the presence of the translocation.

References

1. Ellison, D. (2002) Classifying the medulloblastoma: insights from morphology and molecular genetics. *Neuropathol. Appl. Neurobiol.* **28**, 257–282.
2. Collins, V.P. (2004) Brain tumours: Classification and genes. *J. Neurol. Neurosurg. Psychiatry* **75(Suppl. II),** ii2–ii11.
3. Reddy, A.T., and Packer, R.J. (1999) Medulloblastoma. *Neurology* **12,** 681–685.
4. Giangaspero, F., Bigner, S.H., Kleihues, P., Pietsch, T., and Trojanowski, J.Q. (2000) Medulloblastoma. In *World Health Organization Classification of Tumours* (Kleihues, P., and Cavenee, W.K., eds.), IARC Press, Lyon, France, pp.129–137.
5. Eberhart, C.G., Kepner, J.L., Goldthwaite, P.T., Kun, L.E., Duffner, P.K. Friedman, H.S., Strother, D.R., and Burger, P.C. (2002) Histopathologic grading of medulloblastomas. A Pediatric Oncology Group Study. *Cancer* **94**, 552–560.
5A. Sarkar, C., Deb, P., and Sharma, M.C. (2005) Recent advances in embryonal tumours of the central nervous system. *Childs Nerv. Syst.* **21**, 272–293.
5B. Sheikh, B.Y., and Kanaan, I.N. (1994) Medulloblastoma in adults. *J. Neurosrug. Sci.* **38**, 229–234.
6. Kleihues, P., Louis, D.N., Scheithauer, B.W., Rorke, L.B., Reifenberger, G., Burger, P.C., and Cavenee, W.K. (2002) The WHO Classification of tumors of the nervous system. *J. Neuropathol. Exp. Neurol.* **61**, 215–225.
7. Kleihues, P., and Cavenee, W.K., eds. (2000) World Health Organization Classification of Tumours: *Pathology and Genetics of Tumours of the Nervous System.* IARC Press, Lyon, France.
8. Daumas-Duport, C. (2002) Commentary on the WHO classification of tumors of the nervous system. *J. Neuropathol. Exp. Neurol.* **61**, 226–227.
9. Kepes, J.J. (2002) Comments on the new WHO classification of tumors of the nervous system. *J. Neuropathol. Exp. Neurol.* **61**, 227–228.
10. Rorke, L.B. (2002) Commentary on the WHO classification of tumors of the nervous system. *J. Neuropathol. Exp. Neurol.* **61**, 228–229.
11. Burger, P.C., Grahmann, F.C., Bliestle, A., and Kleihues, P. (1987) Differentiation in the medulloblastoma. A histological and immunohistochemical study. *Acta Neuropathol. (Berl.)* **73**, 115–123.
12. Burger, P.C., and Scheithauer, B.W. (1994) *Tumors of the Central Nervous System.* Armed Forces Institute of Pathology: Washington.
13. Giangaspero, F., Rigobello, L., Badiali, M., Loda, M., Andreini, L., Basso, G., Zorzi, F., and Montaldi, A. (1992) Large-cell medulloblastomas. A distinct variant with highly aggressive behavior. *Am. J. Surg. Pathol.* **16**, 687–693.
14. Brown, H.G., Kepner, J.L., Perlman, E.J., Friedman, H.S., Strother, D.R., Duffner, P.K., Kun, L.E., Goldthwaite, P.T., and Burger, P.C. (2000) "Large cell/anaplastic" medulloblastomas: A Pediatric Oncology Group Study. *J. Neuropathol. Exp. Neurol.* **59**, 857–865.
15. Lamont, J.M., McManamy, C.S., Pearson, A.D., Clifford, S.C., and Ellison, D.W. (2004) Combined histopathological and molecular cytogenetic stratification of medulloblastoma patients. *Clin. Cancer Res.* **10**, 5482–5493.
16. Giangaspero, F., Chieco, P., Ceccarelli, C., Lisignoli, G., Pozzuoli, R., Gambacorta, M., Rossi, G., and Burger, P.C. (1991) "Desmoplastic" versus "classic" medulloblastoma: comparison of DNA content, histopathology and differentiation. *Virchows Arch A Pathol. Anat. Histopathol.* **418**, 207–214.
17. Reardon, D.A., Jenkins, J.J., Sublett, J.E., Burger, P.C., and Kun, L.E. (2000) Multiple genomic alterations including N-*myc* amplification in a primary large cell medulloblastoma. *Pediatr. Neurosurg.* **32**, 187–191.
18. Giordana, M.T., Cavalla, P., Dutto, A., Borsotti, L., Chiò, A., and Schiffer, D. (1997) Is medulloblastoma the same tumor in children and adults? *J Neurooncol.* **35**, 169–176.
19. Katsetos, C.D., Herman, M.M., Frankfurter, A., Gass, P., Collins, V.P., Walker, C.C., Rosemberg, S., Barnard, R.O., and Rubinstein, L.J. (1989) Cerebellar desmopolastic medulloblastomas. A further immunohistochemical characaterization of the reticulin-free pale islands. *Arch. Pathol. Lab. Med.* **113**, 1019–1029.
20. Eberhart, C.G., Kaufman, W.E., Tihan, T., and Burger, P.C. (2001) Apoptosis, neuronal maturation, and neurotrophin expression within medulloblastoma nodules. *J. Neuropathol. Exp. Neurol.* **60**, 462–469.
21. Schubert, T.E.O., and Cervos-Navarro, J. (1998) The histopathological and clinical relevance of apoptotic cell death in medulloblastoma. *J. Neuropathol. Exp. Neurol.* **57**, 10–15.
22. Pearl, G.S., and Takei, Y. (1981) Cerebellar 'neuroblastoma': nosology as it relates to medulloblastoma. *Cancer* **47**, 772–779.
23. Giangaspero, F., Perilongo, G., Fondelli, M.P., Brisigotti, M., Carollo, C., Burnelli, R., Burger, P.C., and Garre, M.L. (1999) Medulloblastoma with extensive nodularity: a variant with favorable prognosis. *J. Neurosurg.* **91**, 971–977.
24. de Chadarevian, J.P., Montes, J.L., O'Gorman, A.M., and Freeman, C.R. (1987) Maturation of cerebellar neuroblastoma into ganglioneuroma with melanosis. A histologic, immunocyto-chemical, and ultrastructural study. *Cancer* **59**, 69–76.
25. Geyer, J.R., Zeltzer, P.M., Boyett, J.M., Rorke, L.B., Stanley, P., Albright, A.L., Wisoff, J.H., Milstein, J.M., Allen, J.C., Finlay, J.L., Ayers, G.D., Shurin, S.B., Stevens, K.R., and Bleyer, W.A. (1994) Survival of infants with primitive neuroectodermal tumors or malignant epemdymomas of the CNS treated with eight drugs in 1 day: A report from the Childrens Cancer Group. *J. Clin. Oncol.* **12**, 1607–1615.
26. Schofield, D., West, D.C., Anthony, D.C., Marshal, R., and Sklar, J. (1995) Correlation of loss of heterozygosity at chromosome 9q with histological subtype in medulloblastomas. *Am. J. Pathol.* **146**, 472–480.
27. Eberhart, C.G., and Burger, P.C. (2003) Anaplasia and grading in medulloblastomas. *Brain Pathol.* **13**, 376–385.
28. McManamy, C.S., Lamont, J.M., Taylor, R.E., Cole, M., Pearson, A.D.J., Clifford, S.C., and Ellison, D.W. (2003) Morphophenotypic variation predicts clinical behavior in childhood non-desmoplastic medulloblastomas. *J. Neuropathol. Exp. Neurol.* **62**, 627–632.
29. Molenaar, W.M., Jansson, D.S., Gould, V.E., Rorke, L.B., Franke, W.W., Lee, V.M., Packer, R.J., and Trojanowski, J.Q.

(1989) Molecular markers of primitive neuroectodermal tumors and other pediatric central nervous system tumors. Monoclonal antibodies to neuronal and glial antigens distinguish subsets of primitive neuroectodermal tumors. *Lab. Invest.* **61,** 635–643.

30. Jennings, M.T., Jennings, D.L., Ebrahim, S.A., Johnson, M.D., Turc-Carel, C., Philip, T., Philip, I., Lapras, C., and Shapiro, J.R. (1992) In vitro karyotypic and immunophenotypic characterization of primitive neuroectodermal tumours: similarities to malignant gliomas. *Eur. J. Cancer* **28A,** 762–766.

31. Korshunov, A., Savostikova, M., and Ozerov, S. (2002) Immunohistochemical markers for prognosis of average-risk pediatric medulloblastomas. The effect of apoptotic index, TrkC, and c-myc expression. *J. Neurooncol.* **58,** 271–279.

32. Packer, R.J., Sutton, L.N., Rorke, L.B., Littman, P.A., Sposto, R., Rosenstock, J.G., Bruce, D.A., and Schut, L. (1984) Prognostic importance of cellular differentiation in medulloblastoma of childhood. *J. Neurosurg.* **61,** 296–301.

33. Schofield, D. (1995) Diagnostic histopathology, cytogenetics, and molecular markers of pediatric brain tumors. *Neurosurg. Clin. N. Am.* **3,** 723–738.

34. Packer, R.J. (1999) Childhood medulloblastoma: progress and future challenges. *Brain Dev.* **21,** 75–81.

35. Packer, R.J. (1999) Primary central nervous system tumors in children. *Curr. Treat. Options Neurol.* **1,** 395–408.

36. Rickert, C.H., and Paulus, W. (2001) Epidemiology of central nervous system tumors in childhood and adolescence based on the new WHO classification. *Childs Nerv. Syst.* **17,** 503–511.

37. MacDonald, T.J., Rood, B.R., Santi, M.R., Vezina, G., Bingaman, K., Cogen, P.H., and Packer, R.J. (2003) Advances in the diagnosis, molecular genetics, and treatment of pediatric embryonal CNS tumors. *Oncologist* **8,** 174–186.

38. Mark, J. (1970) Chromosomal characteristics of neurogenic tumors in children. *Acta Cytologica* **14,** 510–518.

39. Bigner, S.H., Mark, J., Friedman, H.S., Biegel, J.A., and Bigner, D.D. (1988) Structural chromosomal abnormalities in human medulloblastoma. *Cancer Genet. Cytogenet.* **30,** 91–101.

40. Griffin, C.A., Hawkins, A.L, Packer, R.J., Rorke, L.B., and Emanuel, B.S. (1988) Chromosome abnormalities in pediatric brain tumors. *Cancer Res.* **48,** 175–180.

41. Biegel, J.A., Rorke, L.B., Packer, R.J., Sutton, L.N., Schut, L., Bonner, K., and Emanuel, B.S. (1989) Isochromosome 17q in primitive neuroectodermal tumors of the central nervous system. *Genes Chromosomes Cancer* **1,** 139–147.

42. Rasheed, B.K., and Bigner, S.H. (1991) Genetic alterations in glioma and medulloblastoma. *Cancer Metastasis Rev.* **10,** 289–299.

43. Neumann, E., Kalousek, D.K., Norman, M.G., Steinbok, P., Cochrane, D.D., and Goddard, K. (1993) Cytogenetic analysis of 109 pediatric central nervous system tumors. *Cancer Genet. Cytogenet.* **71,** 40–49.

44. Fujii, Y., Hongo, T., and Hayashi, Y., (1994) Chromosome analysis of brain tumors in childhood. *Genes Chromosomes Cancer* **11,** 205–215.

45. Vagner-Capodano, A.M., Zattara-Cannoni, H., Quilichini, B., Giocanti, G., and Groupe Français de Cytogénétique

Oncologique (2003) From cytogenetics to cytogenomics of brain tumors: 1. Medulloblastoma. *Bull. Cancer* **90,** 315–318.

46. Biegel, J.A., Rorke, L.B., Janss, A.J., Sutton, L.N., and Parmiter, A.H. (1995) Isochromosome 17q demonstrated by interphase fluorescence in situ hybridization in primitive neuroectodermal tumors of the central nervous system. *Genes Chromosomes Cancer* **14,** 85–96.

47. Avet-Loiseau, H. Vénuat, A.-M., Terrier-Lacombe, M.-J., Lellouch-Tubiana, A., Zerah, M., and Vassal, G. (1999) Comparative genomic hybridization detects many recurrent imbalances in central nervous system primitive neuroectodermal tumours in children. *Br. J. Cancer* **79,** 1843–1847.

48. Margaret, E.B., Eileen, C.W., Sharon, S., Markus, S., and Philip, C. (1997) Chromosome arm 17p deletion analysis reveals molecular genetic heterogeneity in supratentorial and infratentorial primitive neuroectodermal tumors of the central nevous system. *Cancer Genet. Cytogenet.* **97,** 25–31.

49. Biegel, J.A., Burk, C.D., Barr, F.G., and Emanuel, B.S. (1992) Evidence for a 17p tumor related locus distinct from p53 in pediatric primitive neuroectodermal tumors. *Cancer Res.* **52,** 3391–3395.

50. Sandberg, A.A. (1990) *The Chromosomes in Human Cancer and Leukemia*, 2nd ed., Elsevier, New York.

51. Mertens, F., Johansson, B., and Mitelman, F. (1994) Isochromosomes in neoplasia. *Genes Chromosomes Cancer* **10,** 221–230.

52. Vagner-Capodano, A.M., Gentet, J.C., Gambarelli, D., Pellissier, J.F., Gouzien, M., Lena, G., Genitori, L., Choux, M., and Raybaud, C. (1992) Cytogenetic studies in 45 pediatric brain tumors. *Pediatr. Hematol./Oncol.* **9,** 223–235.

53. Ohgaki, H., Eibl, R.H., Wiestler, O.D., Gazi Yasargil, M., Newcomb, E.W., and Kleihues, P. (1991) *p53* mutations in nonastrocytic human brain tumors. *Cancer Res.* **51,** 6202–6205.

54. Saylors, R.L., Sidransky, D., Friedman, H.S., Bigner, S.H., Bigner, D.D., Vogelstein, B., and Brodeur, G.M. (1991) Infrequent *p53* gene mutations in a medulloblastoma. *Cancer Res.* **51,** 4721–4723.

55. Bigner, S.H., McLendon, R.E., Fuchs, H., McKeever, P.E., and Friedman, H.S. (1997) Chromosomal characteristics of childhood brain tumors. *Cancer Genet. Cytogenet.* **97,** 125–134.

56. Bhattacharjee, M.B., Armstrong, D.D., Vogel, H., and Cooley, L.D. (1997) Cytogenetic analysis of 120 primary pediatric brain tumors and literature review. *Cancer Genet. Cytogenet.* **97,** 39–53.

57. Johansson, B., Heim, S., Mandahl, N., Mertens, F., and Mitelman, F. (1993) Trisomy 7 in nonneoplastic cells. *Genes Chromosomes Cancer* **6,** 199–205.

58. Vagner-Capodano, A.M., Gentet, J.C., Choux, M., Lena, G., Gambarelli, D., Bernard, J.L., and Raybaud, C. (1988) Chromosome abnormalities in sixteen pediatric brains tumors. *Pediatr Neurosci.* **14,** 159–160.

59. Reardon, D.A., Michalkiewicz, E., Boyett, J.M., Sublett, J.E., Entrekin, R.E., Ragsdale, S.T., Valentine, M.B., Behm, F.G., Li, H., Heideman, R.L., Kun, L.E., Shapiro, D.N., and Look, A.T. (1997) Extensive genomic abnormalities in childhood medulloblastoma by comparative genomic hybridization. *Cancer Res.* **57,** 4042–4047.

60. Schütz, B.R., Scheurlen, W., Krauss, J., du Manoir, S., Joos, S., Bentz, M., and Lichter, P. (1996) Mapping of chromosomal gains and losses in primitive neuroectodermal tumors by comparative genomic hybridization. *Genes Chromosomes Cancer* **16**, 196–203.

61. Biegel, J.A., Rorke, L.B., Packer, R.J., and Emanuel, B.S. (1990) Monosomy 22 in rhabdoid or atypical tumors of the brain. *J. Neurosurg.* **73**, 710–714.

62. Biegel, J.A., Fogelgren, B., Zhou, J.-Y., James, C.D., Janss, A.J., Allen, J.C., Zagzag, D., Raffel, C., and Rorke, L.B. (2000) Mutations of the *INI1* rhabdoid tumor suppressor gene in medulloblastomas and primitive neuroectodermal tumors of the central nervous system. *Clin. Cancer Res.* **6**, 2759–2763.

63. Kraus, J.A., Oster, C., Sörensen, N., Berthold, F., Schlegel, U., Tonn, J.C., Wiestler, O.D., and Pietsch. T. (2002) Human medulloblastomas lack point mutations and homozygous deletions of the *hSNF5/INI1* tumour suppressor gene. *Neuropathol. Appl.. Neurobiol.* **28**, 136–141.

64. Burger, P.C., Scheithauer, B.W., and Vogel, F.S. (2002) Surgical Pathology of the Central Nervous System and Its Coverings, 4th ed. Churchill Livingstone, Philadelphia, PA.

65. Burger, P.C., Yu, I.T., Tihan, T., Friedman, H.S., Strother, D.R., Kepner, J.L., Duffner, P.K., Kun, L.E., and Perlman, E.J. (1998) Atypical teratoid/rhabdoid tumor of the central nervous system: a highly malignant tumor of infancy and childhood frequently mistaken for medulloblastoma: a Pediatric Oncology Group study. *Am. Surg. Pathol.* **22**, 1082–1092.

66. Jay, V., Pienkowska, M., Becker, L., and Zielenska, M. (1995) Primitive neuroectodermal tumors of the cerebrum and cerebellum: absence of t(11;22) translocation by RT-PCR analysis. *Mod. Pathol.* **8**, 488–491.

67. Goussia, A.C., Bruner, J.M., Kyritsis A.P., Agnantis, N.J, and Fuller, G.N. (2000) Cytogenetic and molecular genetic abnormalities in primitive neuroectodermal tumors of the central nervous system. *Anticancer Res.* **20**, 65–74.

68. Dressler, L.G., Duncan, M.H., Varsa, E.E., and McConnell, T.S. (1993) DNA content measurement can be obtained using archival material for DNA flow cytometry. *Cancer* **72**, 2033–2041.

69. Rajcan-Separovic, E., Hendson, G., Tang, S., Seto, E., Thomson, T., Phillips, D., and Kalousek, D. (2002) Interphase fluorescence in situ hybridization and DNA flow cytometry analysis of medulloblastomas with a normal karyotype. *Cancer Genet. Cytogenet.* **133**, 94–97.

70. Cogen, P.H., Daneshvar, L., Metzger, A.K., Duyk, G., Edwards, M.S.B., and Sheffield, V.C. (1992) Involvement of multiple chromosome 17p loci in medulloblastoma tumorigenesis. *Am. J. Hum. Genet.* **50**, 584–589.

71. Cogen, P.H., and McDonald, J.D. (1996) Tumor suppressor genes and medulloblastoma. *J. Neurooncol.* **29**, 103–112.

72. Bigner, S.H., and Schröck, E. (1997) Molecular cytogenetics of brain tumors. *J. Neuropathol. Exp. Neurol.* **56**, 1173–1181.

72A. Raffel, C. (1999) Molecular biology of the primitive neuroectodermal tumor: a review. *Neurosurg. Focus* **7**, e2.

73. Biegel, J.A. (1997) Genetics of pediatric central nervous system tumors. *J. Pediatr. Hematol./Oncol.* **19**, 492–501.

74. Biegel, J.A. (1999) Cytogenetic and molecular genetics of childhood brain tumors. *J. Neurooncol.* **1**, 139–151.

75. Cogen, P.H., Daneshvar, L., Metzger, A.K., and Edwards, M.S.B. (1990) Deletion mapping of the medulloblastoma locus on chromosome 17p. *Genomics* **8**, 279–285.

76. Albrecht, S., von Deimling, A., Pietsch, T., Giangaspero, F., Brandner, S., Kleihuest, P., and Wiestler, O.D. (1994) Microsatellite analysis of loss of heterozygosity on chromosomes 9q, 11p and 17p in medulloblastomas. *Neuropathol. Appl. Neurobiol.* **20**, 74–81.

77. Pietsch, T., Koch, A., and Wiestler, O.D. (1997) Molecular genetic studies in medulloblastomas: evidence for tumor suppressor genes at the chromosomal regions 1q31–32 and 17p13. *Klin. Padiatr.* **209**, 150–155.

78. Scheurlen, W.G., Seranski, P., Mincheva, A., Kühl, J., Sörensen, N., Krauss, J., Lichter, P., Poustka, A., and Wilgenbus, K.K. (1997) High-resolution deletion mapping of chromosome arm 17p in childhood primitive neuroectodermal tumors reveals a common chromosomal disruption within the Smith-Magenis region, an unstable region in chromosome band 17p11.2. *Genes Chromosomes Cancer* **18**, 50–58.

79. Scheurlen, W.G., Schwabe, G.C., Joos, S., Mollenhauer, J., Sörensen, N., and Kühl, J. (1998) Molecular analysis of childhood primitive neuroectodermal tumors defines markers associated with poor outcome. *J. Clin. Oncol.* **16**, 2478–2485.

80. Nicholson, J.C., Ross, F.M., Kohler, J.A., and Ellison, D.W. (1999) Comparative genomic hybridization and histological variation in primitive neuroectodermal tumours. *Br. J. Cancer* **80**, 1322–1331.

81. Nicholson, J., Wickramasinghe, C., Ross, F., Crolla, J., and Ellison, D. (2000) Imbalances of chromosome 17 in medulloblastomas determined by comparative genomic hybridisation and fluorescence in situ hybridisation. *J. Clin. Pathol:Mol. Pathol.* **53**, 313–319.

82. Russo, C., Pellarin, M., Tingby, O., Bollen, A.W., Lamborn, K.R., Mohapatra, G., Collins, V.P., and Feuerstein, B.G. (1999) Comparative genomic hybridization in patients with supratentorial and infratentorial primitive neuroectodermal tumors. *Cancer* **86**, 331–339.

83. Rienstein, S., Aviram-Goldring, A., Daniely, M., Amerigleo, N., Simoni, F., Goldman, B., Barkai, G., Rechavi, G., and Constantini, S. (2000) Gains and losses of DNA sequences in childhood brain tumors analyzed by comparative genomic hybridization. *Cancer Genet. Cytogenet.* **121**, 67–72.

84. Aldosari, N., Ahmed Rasheed, B.K., McLendon, R.E., Friedman, H.S., Bigner, D.D., and Bigner, S.H. (2000) Characterization of chromosome 17 abnormalities in medulloblastomas. *Acta Neuropathol.* **99**, 345–351.

85. Bayani, J., Zielenska, M., Marrano, P., Ng Y.K., Taylor, M.D., Jay, V., Rutka, J.T., and Squire, J.A. (2000) Molecular cytogenetic analysis of medulloblastomas and supratentorial primitive neuroectodermal tumors by using conventional banding, comparative genomic hybridization, and spectral karyotyping. *J. Neurosurg.* **93**, 437–448.

86. Nishizaki, T., Harada, K., Kubota, H., Harada, K., Ozaki, S., Ito, H., and Sasaki, K. (1999) Genetic alterations in pediatric medulloblastomas detected by comparative genomic hybridization. *Pediatr. Neurosurg.* **31**, 27–32.

87. Frühwald, M.C., O'Dorisio, M.S., Dai, Z., Rush, L.J., Krahe, R., Smiraglia, D.J., Pietsch, T., Elsea, S.H., and Plass, C. (2001) Aberrant hypermethylation of the major breakpoint

cluster region in 17p11.2 in medulloblastomas but not supratentorial PNETs. *Genes Chromosomes Cancer* **30,** 38–47.

88. Ohgaki, H., Eibl, R.H., Schwab, M., Reichel, M.B., Mariani, L., Gehring, M., Petersen, I., Holl, T., Wiestler, O.D., and Kleihues, P. (1993) Mutations of the *p53* tumor suppressor gene in neoplasms of the human nervous system. *Mol. Carcinog.* **8,** 74–80.

89. Adesina, A.M., Nalbantoglu, J., and Cavenee, W.K. (1994) *p53* gene mutation and *mdm2* gene amplification are uncommon in medulloblastoma. *Cancer Res.* **54,** 5649–5651.

90. Batra, S.K., McLendon, R.E., Koo, J.S., Castelino-Prabhu, S., Fuchs, H.E., Krischer, J.P. Friedman, H.S., Bigner, D.D., and Bigner, S.H. (1995) Prognostic implications of chromosome 17p deletions in human medulloblastomas. *J. Neurooncol.* **24,** 39–45.

91. Orellana, C., Hernandez-Martí, M., Martínez, F., Castel, V., Millán, J.M., Alvarez-Garijo, J.A., Prieto, F., and Badía, L. (1998) Pediatric brain tumors: Loss of heterozygosity at 17p and TP53 gene mutations. *Cancer Genet. Cytogenet.* **102,** 93–99.

91A. Mendrzyk, F., Korshunov, A., Toedt, G., Schwarz, F., Korn, B., Joos, S., Hochhaus, A., Schoch, C., Lichter, P., and Radlwimmer, B. (2006) Isochromosome breakpoints on 17p in medulloblastoma are flanked by different classes of DNA sequence repeats. *Genes Chromosomes Cancer* **45,** 401–410.

92. Greenberg, F., Guzzetta, V., Montes de Oca-Luna, R., Magenis, R.E., Smith, A.C., Richter, S.F., Kondo, I., Dobyns, W.B., Patel, P.I., and Lupski, J.R. (1991) Molecular analysis of the Smith-Magenis syndrome: a possible contiguous-gene syndrome associated with del(17)(p11.2). *Am. J. Hum. Genet.* **49,** 1207–1218.

93. Vagner-Capodano, A.M., Zattara-Cannoni, H., Gambarelli, D., Gentet, J.C., Genitori, L., Lena, G., Graziani, N., Raybaud, C., Choux, M., and Grisoli, F. (1994) Detection of i(17q) chromosome by fluorescent in situ hybridization (FISH) with interphase nuclei in medulloblastoma. *Cancer Genet. Cytogenet.* **78,** 1–6.

94. Giordana, M.T., Migheli, A., and Pavanelli, E. (1998) Isochromosome 17q is a constant finding in medulloblastoma. An interphase cytogenetic study on tissue sections. *Neuropathol. Appl. Neurobiol.* **24,** 223–238.

95. Phelan, C.M., Liu, L., Ruttledge, M.H., Muntzning, K., Ridderheim, P.A., and Collins, V.P. (1995) Chromosome 17 abnormalities and lack of TP53 mutations in paediatric central nervous system tumors. *Hum. Genet.* **96,** 684–690.

96. Burnett, M.E., White, E.C., Sih, S., von Haken, M.S., and Cogen, P.H. (1997) Chromosome arm 17p deletion analysis reveals molecular genetic heterogeneity in supratentorial and infratentorial primitive neuroectodermal tumors of the central nervous system. *Cancer Genet. Cytogenet.* **97,** 25–31.

97. Haataja, L., Raffel, C., Ledbetter, D.H., Tanigami, A., Petersen, D., Heisterkamp, N., and Groffen, J. (1997) Deletion with the *D17S34* locus in a primitive neuroectodermal tumor. *Cancer Res.* **57,** 32–34.

98. Smith, J.S., Tachibana, I., Allen, C., Chiappa, S.A., Lee, H.K., McIver, B., Jenkins, R.B., and Raffel, C. (1999) Cloning of a human ortholog (RPH3AL) of (RNO)Rph3al from a candidate 17p13.3 medulloblastoma tumor suppressor locus. *Genomics* **59,** 97–101.

99. Scheurlen, W.G., Krauss, J., and Kühl, J. (1995) No preferential loss of one parental allele of chromosome 17p13.3 in childhood medulloblastoma. *Int. J. Cancer* **63,** 372–374.

100. Biegel, J.A., and Wentz, E. (1997) No preferential parent of origin for the isochromosome 17q in childhood primitive neuroectodermal tumor (medulloblastoma). *Genes Chromosomes Cancer* **18,** 143–146.

101. Lastowska, M., Cotterill, S., Bown, N., Cullinane, C., Variend, S., Lunec, J., Strachan, T., Pearson, A.D.J., and Jackson, M.S. (2002) Breakpoint position on 17q identifies the most aggressive neuroblastoma tumors. *Genes Chromosomes Cancer* **34,** 428–436.

101A. Ferretti, E., De Smaele, E., Di Marcotullio, L., Srepanti, I., and Gulino, A. (2005) Hedgehog checkpoints in medulloblastoma: the chromosome 17p deletion paradigm. *Trends Mol. Med.* **11,** 537–545.

102. Steichen-Gersdorf, E., Baumgartner, M., Kreczy, A., Maier, H., and Fink, F.-M. (1997) Deletion mapping on chromosome 17p in a medulloblastoma. *Br. J. Cancer* **76,** 1284–1287.

103. Emadian, S.M., McDonald, J.D., Gerken, S.C., and Fults, D. (1996) Correlation of chromosome 17p loss with clinical outcome in medulloblastoma. *Clin. Cancer Res.* **2,** 1559–1564.

104. Biegel, J.A., Janss, A.J., Raffel, C. Sutton, L., Rorke, L.B., Harper, J.M., Phillips, P.C. (1997) Prognostic significance of chromosome 17p deletions in childhood primitive neuroectodermal tumors (medulloblastomas) of the central nervous system. *Clin. Cancer Res.* **3,** 473–478.

105. Nam, D.-H., Wang, K.-C., Kim, Y.-M., Chi, J.G., Kim, S.-K., and Cho, B.-K. (2000) The effect of isochromosome 17q presence, proliferative and apoptotic indices, expression of c-erbB-2, bcl-2 and p53 proteins on the prognosis of medulloblastoma. *J. Korean Med. Sci.* **15,** 452–456.

106. Bigner, S.H., Friedman, H.S., Vogelstein, B., Oakes, W.J., and Bigner, D.D. (1990) Amplification of the c-*myc* gene in human medulloblastoma cell lines and xenografts. *Cancer Res.* **50,** 2347–2350.

107. Badiali, M., Pession, A., Basso, G., Andreini, L., Rigobello, L., Galassi, E., and Giangaspero, F. (1991) N-myc and c-myc oncogenes amplification in medulloblastomas. Evidence of particularly aggressive behavior of a tumor with c-myc amplification. *Tumori* **77,** 118–121.

108. Tomlinson, F.H., Jenkins, R.B., Scheithauer, B.W., Keelan, P.A., Ritland, S., Parisi, J.E., Cunningham, J., and Olsen, K.D. (1994) Aggressive medulloblastoma with high-level N-myc amplification. *Mayo Clin. Proc.* **69,** 359–365.

109. Sawyer, J.R., Swanson, C.M., Roloson, G.J., Longee, D.C., Boop, F.A., and Chadduck, W.M. (1991) Molecular cytogenetic analysis of a medulloblastoma with isochromosome 17 and double minutes. *Cancer Genet. Cytogenet.* **57,** 181–186.

110. Langdon, J.A., Lamont, J.M., Scott, D.K., Dyer, S., Prebble, E., Bown, N., Grundy, R.G., Ellison, D.W., and Clifford, S.C. (2006) Combined genome-wide allelotyping and copy number analysis identify frequent genetic losses without copy number reduction in medulloblastoma. *Genes Chromosomes Cancer* **45,** 47–60.

111. Aldosari, N., Wiltshire, R.N., Dutra, A., Schröck, E., McLendon, R.E., Friedman, H.S., Bigner, D.D., and Bigner, S.H. (2002) Comprehensive molecular cytogenetic

investigation of chromosomal abnormalities in human medulloblastoma cell lines and xenografts. *J. Neurooncol.* **4,** 75–85.

112. Jacobsen, P.F., Jenkyn, D.J., and Papadimitriou, J.M. (1985) Establishment of a human medulloblastoma cell line and its heterotransplantation into nude mide. *J. Neuropathol. Exp. Neurol.* **44,** 472–485.

113. Friedman, H.S., Burger, P.C., Bigner, S.H., Trojanowski, J.Q., Brodeur, G.M., He, X., Wikstrand, C.J., Kurtzberg, J., Berens, M.E., Halperin, E.C., and Bigner, D.D. (1988) Phenotypic and genotypic analysis of a human medulloblastoma cell line and transplantable xenografts (D341 Med) demonstrating amplification of c-*myc. Am. J. Pathol.* **130,** 472–484.

114. Wasson, J.C., Saylors, R.L., III, Zeltzer, P., Friedman, H.S., Bigner, S.H., Burger, P.C., Bigner, D.D., Look, A.T., Douglass, E.C., and Brodeur, G.M. (1990) Oncogene amplification in pediatric brain tumors. *Cancer Res.* **50,** 2987–2990.

115. Herms, J., Neidt, I., Luscher, B., Sommer, A., Schurmann, P., Schröder, T., Bergmann, M., Wilken, B., Probst-Cousin, S., Hernaiz-Driever, P., Behnke, J., Hanefeld, F., Pietsch, T., and Kretzschmar, H.A. (2000) C-MYC expression in medulloblastoma and its prognostic value. *Int. J. Cancer* **89,** 395–402.

116. du Manoir, S., Schröck, E., Bentz, M., Speicher, M.R., Joos, S., Ried, T., Lichter, P., and Cremer, T. (1995) Quantitative analysis of comparative genomic hybridization. *Cytometry* **19,** 27–41.

117. Gilhuis, H.J., Anderl, K.L., Boerman, R.H., Jeuken, J.M., James, C.D., Raffel, C., Scheithauer, B.W., and Jenkins, R.B. (2000) Comparative genomic hybridization of medulloblastomas and clinical relevance: eleven new cases and a review of the literature.*Clin. Neurol. Neurosurg.* **102,** 203–209.

118. Rickert, C.H., and Paulus, W. (2004) Comparative genomic hybridization in central and peripheral nervous system tumors of childhood and adolescence. *J. Neuropathol. Exp. Neurol.* **63,** 399–417.

119. Cogen, P.H. (1991–92) Prognostic significance of molecular genetic markers in childhood brain tumors. *Pediatr. Neurosurg.* **17,** 245–250.

120. Eberhart, C.G., Kratz, J.E., Schuster, A., Goldthwaite, P., Cohen, K.J., Perlman, E.J., and Burger, P.C. (2002) Comparative genomic hybridization detects an increased number of chromosomal alterations in large cell/anaplastic medulloblastomas. *Brain Pathol.* **12,** 36–44

121. Michiels, E.M.C., Weiss, M.M. Hoovers, J.M.N., Baak, J.P.A., Voûte, P.A., Baas, F., and Hermsen, M.A.J.A. (2002) Genetic alterations in childhood medulloblastoma analyzed by comparative genomic hybridization. *J. Pediatr. Hematol./Oncol.* **24,** 205–210.

122. Rossi, M.R., Conroy, J., McQuaid, D., Nowak, N.J., Rutka, J.T., and Cowell, J.K. (2006) Array CGH analysis of pediatric medulloblastomas. *Genes Chromosomes Cancer* **45,** 290–303.

123. Raffel, C., Gilles, F.E., and Weinberg, K.I. (1990) Reduction to homozygosity and gene amplification in central nervous system primitive neuroectodermal tumors of childhood. *Cancer Res.* **50,** 587–591.

124. Rouah, E., Wilson, D.R., Armstrong, D.L., and Darlington, G.J. (1989) N-*myc* amplification and neuronal differentiation in human primitive neuroectodermal tumors of the central nervous system. *Cancer Res.* **49,** 1797–1801.

125. Fuller, G.N., and Bigner, S.H. (1992) Amplified cellular oncogenes in neoplasms of the human central nervous system. *Mutat. Res.* **276,** 299–306.

126. Tong, C.Y.K., Hui, A.B.Y., Yin, X.-L., Pang, J.C.S., Zhu, X.-L., Poon, W.-S., and Ng, H.-K. (2004) Detection of oncogene amplifications in medulloblastomas by comparative genomic hybridization and array-based comparative genomic hybridization. *J. Neurosurg.* **100,** 187–193.

127. Boon, K., Eberhart, C.G., and Riggins, G.J. (2005) Genomic amplification of *orthodenticle homologue 2* in medulloblastomas. *Cancer Res.* **65,** 703–707.

128. Rosenthal, J.A., Chen., H., Slepnev, V.I., Pellegrini, L., Salcini, A.E., Di Fiore, P.P., and DeCamilli, P. (1999) The epsins define a family of proteins that interact with components of the clathrin coat and contain a new protein module. *J. Biol. Chem.* **274,** 33959–33965.

129. McDonald, J.D., Daneshvar, L., Willert, J.R., Matsumura, K., Waldman, F., and Cogen, P.H. (1994) Physical mapping of chromosome 17p13.3 in the region of a putative tumor suppressor gene important in medulloblastoma. Genomics **23,** 229–232.

130. Koch, A., Tonn, J., Kraus, J.A., Sörensen, N., Albrecht, S., Wiestler, O.D., and Pietsch, T. (1996) Molecular analysis of the lissencephaly gene 1 (*LIS-1*) in medulloblastomas. *Neuropathol. Appl. Neurobiol.* **22,** 233–242.

131. Thomas, G.A., and Raffel, C. (1991) Loss of heterozygosity on 6q, 16q, and 17p in human central nervous system primitive neuroectodermal tumors. *Cancer Res.* **51,** 639–643.

132. Jung, H.L., Wang, K.-C., Kim, S.-K., Sung, K.W., Koo, H.H., Shin, H.Y., Ahn, H.S., Shin, H.J., and Cho, B.-K. (2004) Loss of heterozygosity analysis of chromosome 17p13.1–13.3 and its correlation with clinical outcome in medulloblastoma. *J. Neurooncol.* **67,** 41–46.

133. Jung, H.L., and Lau, C.C. (1998) Loss of heterozygosity studies on chromosomes 9, 22 and 17p in 12 various pediatric brain tumors. *Korean J Pediatr. Hematol. Oncol.* **5,** 293–303.

134. Kraus, J.A., Koch, A., Albrecht, S., von Deimling, A., Wiestler, O.D., and Pietsch, T. (1996) Loss of heterozygosity at locus F13B on chromosome 1q in human medulloblastoma. *Int. J. Cancer* **67,** 11–15.

135. Barnes, L., and Kapadia, S.B. (1994) The biology and pathology of selected skull base tumors. *J. Neurooncol.* **20,** 213–240.

136. Orellana, C., Martínez, F., Hernandez-Martí, M., Castel, V., Cañete, A., Prieto, F., and Badía, L. (1998) A novel TP53 germ-line mutation identified in a girl with a primitive neuroectodermal tumor and her father. *Cancer Genet. Cytogenet.* **105,** 103–108.

137. Yin, X.-L., Pang, J.C., Liu, Y.-H., Chong, E.Y., Cheng, Y., Poon, W.-S., and Ng, H.-K. (2001) Analysis of loss of heterozygosity on chromosomes 10q, 11, and 16 in medulloblastoma. *J. Neurosurg.* **94,** 799–805.

138. James, C.D., He, J., Carlbom, E., Mikkelsen, T., Ridderheim, P-A., Cavenee, W.K., and Collins, V.P. (1990) Loss of genetic information in central nervous system tumors common to children and young adults. *Genes Chromosomes Cancer* **2,** 94–102.

139. Fults, D., Petronio, J., Noblett, B.D., and Pedone, C.A. (1992) Chromosome 11p15 deletions in human malignant astrocy-

tomas and primitive neuroectodermal tumors. *Genomics* **14,** 799–801.

140. Blaeker, H., Ahmed Rasheed, B.K. McLendon, R.E., Friedman, H.S., Batra, S.K., Fuchs, H.E., and Bigner, S.H. (1996) Microsatellite analysis of childhood brain tumors. *Genes Chromosomes Cancer* **15,** 54–63.

141. Lescop, S., Lellouch-Tubiana, A., Vassal, G., and Besnard-Guerin, C. (1999) Molecular genetic studies of chromosome 11 and chromosome 22q DNA sequences in pediatric medulloblastomas. *J. Neurooncol.* **44,** 119–127.

142. Zawlik, I., Zakrzewska, M., Witusik, M., Golanska, E., Kulczycka-Wojdala, D., Szybka, M., Piaskowski, S., Wozniak, K., Zakrzewski, K., Papierz, W., Liberski, P.P., and Rieske, P. (2006) *KCDT11* expression in medulloblastoma is lower than in adult cerebellum and higher than in neural stem cells. *Cancer Genet. Cytogenet.* **170,** 24–28.

143. Pietsch, T., Waha, A., Koch, A., Kraus, J., Albrecht, S., Tonn, J., Sörensen, N., Berthold, F., Henk, B., Schmandt, N., Wolf, H.K., von Deimling, A., Wainwright, B., Chenevix-Trench, G., Wiestler, O.D., and Wicking, C. (1997) Medulloblastomas of the desmoplastic variant carry mutations of the human homologue of *Drosophila patched. Cancer Res.* **57,** 2085–2088.

144. Raffel, C., Jenkins, R.B., Frederick, L., Hebrink, D., Alderete, B., Fults, D.W., and James, C.D. (1997) Sporadic medulloblastomas contain *PTCH* mutations. *Cancer Res.* **57,** 842–845.

145. Wolter, M., Reifenberger, J., Sommer, C., Ruzicka, T., and Reifenberger, G. (1997) Mutations in the human homologue of the *Drosophila* segment polarity gene *patched (PTCH)* in sporadic basal cell carcinomas of the skin and primitive neuroectodermal tumors of the central nervous system. *Cancer Res.* **57,** 2581–2585.

146. Smyth, I., Narang, M.A., Evans, T., Heimann, C., Nakamura, Y., Chenevix-Trench, G., Pietsch, T., Wicking, C., and Wainwright, B.J. (1999) Isolation and characterization of human *Patched 2 (PTCH2)*, a putative tumour suppressor gene in basal cell carcinoma and medulloblastoma on chromosome 1p32. *Hum. Mol. Genet.* **8,** 291–297.

146A. Uchikawa, H., Toyoda, M., Nagao, K., Miyauchi, H., Nishikawa, R., Fujii, K., Kohno, Y., Yamada, M., and Miyashita, T. (2006) Brain- and heart-specific Patched-1 containing exon 12b is a dominant negative isoform and is expressed in medulloblastoma. *Biochem. Biophys. Res. Commun.* **349,** 277–283.

147. Mollenhauer, J., Wiemann, S., Scheurlen, W., Korn, B., Hayashi, Y., Wilgenbus, K.K., von Deimling, A., and Poustka, A. (1997) DMBT1, a new member of the SRCR superfamily, on chromosome 10q25.3–26.1 is deleted in malignant brain tumours. *Nat. Genet.* **17,** 32–39.

147A. Scott, D.K., Straughton, D., Cole, M., Bailey, S., Ellison, D.W., and Clifford, S.C. (2006) Identification and analysis of tumor suppressor loci at chromosome 10q23.3–10q25.3 in medulloblastoma. *Cell Cycle* **5,** 2381–2389.

148. Yin, X.-L., Pang, J.C., and Ng, H.-K. (2002) Identification of a region of homozygous deletion on 8p22–23.1 in medulloblastoma. *Oncogene* **21,** 1461–1468.

149. von Deimling, A., Fimmers, R., Schmidt, M.C., Bender, B., Fassbender, F., Nagel, J., Jahnke, R., Kaskel, P., Duerr, E.-M., Koopmann, J., Maintz, D., Steinbeck, S., Wick, W.,

Platten, M., Müller, D.J., Przkora, R., Waha, A., Blümcke, B., Wellenreuther, R., Meyer-Puttlitz, B., Schmidt, O., Mollenhauer, J., Poustka, A., Stangl, A.P., Lenartz, D., von Ammon, K., Henson, J.W., Schramm, J., Louis, D.N., and Wiestler, O.D. (2000) Comprehensive allelotype and genetic analysis of 466 human nervous system tumors. *J. Neuropathol. Exp. Neurol.* **59,** 544–558.

149A. Wang, J.C., Radford, D.M., Holt, M.S., Helms, C., Goate, A., Brandt, W., Parik, M., Phillips, N.J., DeSchryver, K., Schuh, M.E., Fair, K.L., Ritter, J.H., Marshall, P., and Donis-Keller, H. (1999) Sequence-ready contig for the 1.4-carcinoma in situ loss of heterozygosity region on chromosome 8p22-p23. *Genomics* **60,** 1–11.

150. Chang, Q., Pang, J.C., Li, J., Hu, L., Kong, X., and Ng, H.K. (2004) Molecular analysis of PinX1 in medulloblastoma. *Int. J. Cancer* **109,** 309–314.

151. Evans, D.G.R., Farndon, P.A., Burnell, L.D., Gattamaneni, R., and Birch, J.M. (1991) The incidence of Gorlin syndrome in 173 consecutive cases of medulloblastoma. *Br. J. Cancer* **64,** 959–961.

152. Louis, D.N., and von Deimling, A. (1995) Hereditary tumor syndromes of the nervous system: Overview and rare syndromes. *Brain Pathol.* **5,** 145–151.

153. Malkin, D. (1994) p53 and the Li-Fraumeni syndrome. *Biochim. Biophys. Acta* **1198,** 197–213.

154. Malkin, D., Li, F.P., Strong, L.C., Fraumeni, J.G., Nelson, C., Kim, D.H., Kassel, J., Gryka, M.A., Bishoff, F.Z., Taisky, M.A., and Friend, S.H. (1990) Germline p53 mutations in a familial syndrome of breast cancer, sarcomas, and other neoplasms. *Science* **250,** 1222–1228.

155. Srivastava, S., Zou, Z.Q., Pirollo, K., Blattner, W., and Chang, E.H. (1990) Germ-line transmission of a mutated *p53* gene in a cancer-pronte family with Li-Fraumeni syndrome. *Nature* **348,** 747–749.

156. Garber, J.E., Goldstein, A.M., Kantor, A.F., Dreyfus, M.G., Fraumeni, J.F., Jr., and Li, F.P. (1991) Follow-up study of twenty-four families with Li-Fraumeni syndrome. *Cancer Res.* **51,** 6094–6097.

157. Li, F.P., Fraumeni, J.F., Jr., Mulvihill, J.J., Blattner, W.A., Dreyfus, M.G., Tucker, M.A., and Miller, R.W. (1988) A cancer family syndrome in twenty-four kindreds. *Cancer Res.* **48,** 5358–5362.

158. Kleihues, P., Schauble, B., zur Hausen, A., Esteve, J., and Ohgaki, H. (1997) Tumors associated with *p53* germline mutations: A synopsis of 91 families. *Am. J. Pathol.* **150,** 1–13.

159. Miller, R.W., and Rubinstein, J.H. (1995) Tumors in Rubinstein-Taybi syndrome. *Am. J. Med. Genet.* **56,** 112–115.

160. Petrij, F., Giles, R.H., Dauwerse, H.G., Saris, J.J., Hennekam, R.C.M., Masuno, M., Tommerup, N., van Ommen, G.-J.B., Goodman, R.H., Peters, D.J.M., and Breuning, M.H. (1995) Rubinstein-Taybi syndrome caused by mutations in the transcriptional co-activator CBP. *Nature* **376,** 348–351.

161. Taylor, M.D., Mainprize, T.G., and Rutka, J.T. (2000) Molecular insight into medulloblastoma and central nervous system primitive neuroectodermal tumor biology from hereditary syndromes. A review. *Neurosurgery* **47,** 888–901.

162. Griepentrog, F., and Pauly, H. (1957) Intra- und extracranielle, frühmanifeste Medulloblastome bei erbgleichen Zwillingen. *Zentralbl. Neurochir.* **17,** 129–140.

163. Scheurlen, W.G., Sorensen, N., Roggendorf, W., and Kühl, J. (1996) Molecular analysis of medulloblastoma occurring simultaneously in monozygotic twins. *Eur. J. Pediatr.* **155**, 880–884.

164. Chidambaram, B., Santhosh, V., and Shankar, S.K. (1998) Identical twins with medulloblastoma occurring in infancy. *Childs Nerv. Syst.* **14**, 421–425.

165. Bickerstaff, E.R., Connolly, R.C., and Woolf, A.L. (1967) Cerebellar medulloblastoma occurring in brothers. *Acta Neuropathol. (Berl.)* **8**, 104–107.

166. Yamashita, Y., Handa, H., and Toyama, M. (1975) Medulloblastoma in two brothers. *Surg. Neurol.* **4**, 225–227.

167. Hung, K.L., Wu, C.M., Huang, J.S., and How, S.W. (1990) Familial medulloblastoma in siblings Report in one family and review of the literature. *Surg. Neurol.* **33**, 341–346.

167A. Tijssen, C.C., Halprin, M.R., and Endtz, L.J. (1982) Familial brain tumors. A commented register. Matinus Nijhoff Publishers, The Hague, The Nethrlands, pp. 10–73.

168. Farwell, J.R., Dohrmann, G.J., and Flannery, J.T. (1984) Medulloblastoma in childhood: an epidemiological study. *J. Neurosurg.* **61**, 657–664.

169. Olson, J.M., Breslow, N.E., and Barce, J. (1993) Cancer in twins of Wilms tumor patients. *Am. J. Med. Genet.* **47**, 91–94.

170. Rainov, N.G., Lübbe, J., Renshaw, J., Pritchard-Jones, K., Lüthy, A.R., and Aguzzi, A. (1995) Association of Wilms' tumor with primary brain tumor in siblings. *J. Neuropathol. Exp. Neurol.* **54**, 214–223.

171. Bonnin, J.M., Rubinstein, L.J., Palmer, N.F., and Beckwith, J.B. (1984) The association of embryonal tumors originating in the kidney and in the brain. *Cancer* **54**, 2137–2146.

172. Fort, D.W., Tonk, V.S., Tomlinson, G.E., Timmons, C.F., and Schneider, N.R. (1994) Rhabdoid tumor of the kidney with primitive neuroectodermal tumor of the central nervous system: associated tumors with different histologic, cytogenetic, and molecular findings. *Genes Chromosomes Cancer* **11**, 146–152.

173. Ogasawara, H., Inagawa, T., Yamamoto, M., Kamiya, K., Yano, T., and Utsunomiya, H. (1988) Medulloblastoma in infancy associated with omphalocele, malrotation of the intestine, and extrophy of the bladder. *Childs Nerv. Syst.* **4**, 108–111.

174. Rogers, L., Pattisapu, J., Smith, R.R., and Parker, P. (1988) Medulloblastoma in association with the Coffin-Siris syndrome. *Child. Nerv. Syst.* **4**, 41–44.

175. Farwell, J., and Flannery, J.T. (1984) Cancer in relatives of children with central-nervous-system neoplasms. *N. Engl. J. Med.* **311**, 749–753.

176. Kuijten, R.R., Strom, S.S., Rorke, L.B., Boesel, C.P., Buckley, J.D., Meadows, A.T., and Bunin, G.R. (1993) Family history of cancer and seizures in young children with brain tumors: a report from the Childrens Cancer Group (United States and Canada). *Cancer Causes Control* **4**, 455–464.

177. Evan, G.I., Wyllie, A.H., Gilbert, C.S., Littlewood, T.D., Land, H., Brooks, M., Waters, C.M., Penn, L.Z., and Hancock, D.C. (1992) Induction of apoptosis in fibroblasts by c-myc protein. *Cell* **69**, 119–128.

178. Evan, G.I., and Vousden, K.H. (2001) Proliferation, cell cycle and apoptosis in cancer. *Nature* **411**, 342–348.

179. MacGregor, D.N., and Ziff, E.B. (1990) Elevated c-*myc* expression in childhood medulloblastoma. *Pediatr. Res.* **28**, 63–68.

180. Jay, V., Squire, J., Bayani, J., Alkhani, A.M., Rutka, J.T., and Zielenska, M. (1999) Oncogene amplification in medulloblastoma: analysis of a case by comparative genomic hybridization and fluorescence in situ hybridization. *Pathology* **31**, 337–344.

181. Leonard, J.R., Cai, D.X., Rivet, D.J., Kaufman, B.A., Park, T.S., Levy, B.K., and Perry, A. (2001) Large cell/anaplastic medulloblastomas and medullomyoblastoma: clinicopathological and genetic features. *J. Neurosurg.* **95**, 82–88.

182. Aldosari, N., Bigner, S.H., Burger, P.C., Becker, L., Kepner, J.L., Friedman, H.S., and McLendon, R.E. (2002) *MYCC* and *MYCN* oncogene amplification in medulloblastoma. *Arch. Pathol. Lab. Med.* **126**, 540–544.

183. Bigner, S.H., and Vogelstein, B. (1990) Cytogenetics and molecular genetics of malignant gliomas and medulloblastoma. *Brain Pathol.* **1**, 12–18.

184. Grotzer, M.A., Janss, A.J., Fung, K.-M., Biegel, J.A., Sutton, L.N., Rorke, L.B., Zhao, H., Cnaan, A., Phillips, P.C., Lee, V.M.-Y., and Trojanowski, J.Q. (2000A) TrkC expression predicts good clinical outcome in primitive neuroectodermal brain tumors. *J. Clin. Oncol.* **18**, 1027–1035.

185. Bruggers, C.S., Tai, K.-F., Murdock, T., Sivak, L., Le, K., Perkins, S.L., Coffin, C.M., and Carroll, W.L. (1998) Expression of the c-Myc protein in childhood medulloblastoma. *J. Pediatr. Hematol./Oncol.* **20**, 18–25.

185A. Stearns, D., Chaudhry, A., Abel, T.W., Burger, P.C., Dang, C.V., and Eberhart, C.G. (2006) c-Myc overexpression causes anaplasia in a medulloblastoma. *Cancer Res.* **66**, 673–681.

185B. Pession, A., and Tonelli, R. (2005) The *MYCN* oncogene as a specific and selective drug target for peripheral and central nervous system tumors. *Curr. Cancer Drug Targets* **5**, 273–283.

185C. Adesina, A.M., Nguyen, Y., Mehta, V., Takei, H., Strangeby, P., Crabtree, S., Chintagumpala, M., and Gumerlock, M.K. (2007) *FOXG1* dysregulation is a frequent event in medulloblastoma. *J. Neurooncol.* **85**, 111–112.

186. Moriuchi, S., Shimizu, K., Miyao, Y., and Hayakawa, T. (1996) An immunohistochemical analysis of medulloblastoma and PNET with emphasis on N-myc protein expression. *Anticancer Res.* **16**, 2687–2692.

186A. Pritchard, J.I. and Olson, J.M. (2008) Methylation of *PTCH1*, the Patched-1 gene, in a panel of primary medulloblastomas. *Cancer Genet. Cytogenet.* **180**, 47–50.

187. Garson, J.A., Pemberton, L.F., Sheppard, P.W., Varndell, I.M., Coakham, H.B., and Kemshead, J.T. (1989) N-myc gene expression and oncoprotein characterization in medulloblastoma. *Br. J. Cancer* **59**, 889–894.

188. Grotzer, M.A., Hogarty, M.D., Janss, A.J., Liu, X., Zhao, H., Eggert, A., Sutton, L.N., Rorke, L.B., Brodeur, G.M., and Phillips, P.C. (2001) *MYC* messenger RNA expression predicts survival outcome in childhood primitive neuroectodermal tumor/medulloblastoma. *Clin. Cancer Res.* **7**, 2425–2433.

189. Eberhart, C.G., Kratz, J., Wang, Y., Summers, K., Stearns, D., Cohen, K., Dang, C.V., and Burger, P.C. (2004) Histopathological and molecular prognostic markers in medulloblastoma: c-my, N-myc, TrkC, and anaplasia. *J. Neuropathol. Exp. Neurol.* **63**, 441–449.

189A. Olsen, C.L., Hsu, P.-P., Glienke, J., Rubanyi, G.M., and Brooks, A.R. (2004) Hedgehog-interacting protein is highly expressed in endothelial cells but down-regulated during angiogenesis and in several human tumors. *BMC Cancer* **4**, 4–14.

190. Bourdon, J.-C., Fernandes, K., Murray-Zmijewski, F., Liu, G., Diot, A., Xirodimas, D.P., Saville, M.K., and Lane, D.P. (2005) p53 isoforms can regulate p53 transcriptional activity. *Genes Dev.* **19**, 2122–2137.

190A. Staunton, M.J., and Gaffney, E.F. (1998) Apoptosis. Basic concepts and potential significance in human cancer. *Arch. Pathol. Lab. Med.* **122**, 310–319.

190B. Shakhova, O., Leung, C., van Montfort, E., Berns, A., and Marino, S. (2006) Lack of Rb and p53 delays cerebellar development and predisposes to large cell anaplastic medulloblastoma through amplification of *N-Myc* and *Ptch2*. *Cancer Res.* **66**, 5190–5200.

190C. Sherr, C.J. (2001) The INK4a/ARF network in tumour suppression. *Nat. Rev. Mol. Cell Biol.* **2**, 731–737.

191. Harris, S.L., and Levine, A.J. (2005) The p53 pathway: positive and negative feedback loops. *Oncogene* **24**, 2899–2908.

192. Irwin, M.S., and Kaelin, W.G., Jr. (2001) Role of the newer p53 family proteins in malignancy. *Apoptosis* **6**, 17–29.

193. Giordana, M.T., Duó, D., Gasverde, S., Trevisan, E., Boghi, A., Morra, I., Pradotto, L., Mauro, A., and Chiò, A. (2002) MDM2 overexpression is associated with short survival in adults with medulloblastoma. *J. Neurooncol.* **4**, 115–122.

194. Sherr, C.J., and McCormick, F. (2002) The RB and p53 pathways in cancer. *Cancer Cell* **2**, 103–112.

195. Bates, S., Phillips, A.C., Clark, P.A., Stott, F., Peters, G., Ludwig, R.L., and Vousden, K.H. (1998) p14ARF links to tumour suppresors RB and p53. *Nature* **395**, 124–125.

196. Palmero, I., Pantoja, C., and Serrano, M. (1998) p19ARF links the tumour suppressor p53 to Ras. *Nature* **395**, 125–126.

196A. Stott, F.J., Bates, S., James, M.C., McConnell, B.B., Starborg, M., Brookes, S., Palmero, I, Ryan, K., Hara, E., Vousden, K.H., and Peters, G. (1998) The alternative product from the human *CDKN2A* locus, p14^{ARF}, participates in a regulatory feedback loop with p53 and MDM2. *EMBO J.* **17**, 5001–5014.

197. Zindy, F., Eischen, C.M., Randle, D.H., Kamijo, T., Cleveland, J.L., Sherr, C.J., and Roussel, M.F. (1998) Myc signaling via the ARF tumor suppressor regulates p53-dependent apoptosis and immortalization. *Genes Dev.* **12**, 2424–2433.

198. Pomerantz, J., Schreiber-Agus, N., Liegeois, N.J., Silverman, A., Alland, L., Chin, L., Potes, J., Chen, K., Orlow, I., Lee, H.W., Cordon-Cardo, C., and DePinho, R.A. (1998) The Ink4a tumor suppressor gene product, p19Arf, interacts with MDM2 and neutralizes MDM2's inhibition of p53. *Cell* **92**, 713–723.

199. Marino, S. (2005) Medulloblastoma: developmental mechanisms out of control. *Trends Mol. Med.* **11**, 17–22.

200. Wetmore, C., Eberhart, D.E., and Curran, T. (2001) Loss of *p53* but not *ARF* accelerates medulloblastoma in mice heterozygous for *patched*. *Cancer Res.* **61**, 513–516.

201. Lee, Y., and McKinnon, P.J. (2002) DNA ligase IV suppresses medulloblastoma formation. *Cancer Res.* **62**, 6395–6399.

202. Tong, W.-M., Ohgaki, H., Huang, H., Granier, C., Kleihues, P., and Wang, Z.-Q. (2003) Null mutation of DNA strand break-binding molecule poly(ADP-ribose) polymerase causes medulloblastomas in p53$^{-/-}$ mice. *Am. J. Pathol.* **162**, 343–352.

203. Woodburn, R.T., Azzarelli, B., Montebello, J.F., and Goss, I.E. (2001) Intense p53 staining is a valuable prognostic indicator for poor prognosis in medulloblastoma/central nervous system primitive neuroectodermal tumors. *J. Neurooncol.* **52**, 57–62.

204. Burns, A.S., Jaros, E., Cole, M., Perry, R., Pearson, A.J., and Lunec, J. (2002) The molecular pathology of p53 in primitive neuroectodermal tumours of the central nervous system. *Br. J. Cancer* **86**, 1117–1123.

205. Schmidt, C.A., Fridman, J.S., Yang, M., Lee, S., Baranov, E., Hoffman, R.M., and Lowe, S.W. (2002) A senescence program controlled by p53 and p16INK4a contributes to the outcome of cancer therapy. *Cell* **109**, 335–346.

206. Gudkov, A.V., and Kamarova, E.A. (2003) The role of p53 in determining sensitivity to radiotherapy. *Nat. Rev. Cancer* **3**, 117–129.

207. Frank, A.J., Hernan, R., Hollander, A., Lindsey, J.C., Lusher, M.E., Fuller, C.E., Clifford, S.C., and Gilbertson, R.J. (2004) The TP53-ARF tumor suppressor pathway is frequently disrupted in large/cell anaplastic medulloblastoma. *Mol. Brain Res.* **121**, 137–140.

208. Pomeroy, S.L., Tamayo, P., Gaasenbeek, M., Sturla, L.M., Angelo, M., McLaughlin, M.E., Kim, J.Y.H., Goumnerova, L.C., Black, P.M., Lau, C., Allen, J.C., Zagzag, D., Olson, J.M., Curran, T., Wetmore, C., Biegel, J.A., Poggio, T., Mokherjee, S., Rifkin, R., Califano, A., Stolovitzky, G., Louis, D.N., Mesirov, J.P., Lander, E.S., and Golub, T.R. (2002) Prediction of central nervous system embryonal tumour outcome based on gene expression. *Nature* **415**, 436–442.

209. Chopra, A., Brown, K.M., Rood, B.R., Packer, R.J., and MacDonald, T.J. (2003) The use of gene expression analysis to gain insights into signaling mechanisms of metastatic medulloblastoma. *Pediatr. Neurosurg.* **39**, 68–74.

210. Lee, Y., Miller, H.L., Jensen, P., Hernan, R., Connelly, M., Wetmore, C., Zindy F., Roussel, M.F., Curran, T., Gilbertson, R.J., and McKinnon, P.J. (2003) A molecular fingerprint for medulloblastoma. *Cancer Res.* **63**, 5428–5437.

211. Ramachandran, C., Khatib, Z., Escalon, E., Rodriguez, S., Raveendran Nair, P.K., Jhabvala, P., and Melnick, S.J. (2004) Gene expression profiling of pediatric medulloblastomas by microarray hybridization. *J. Neurooncol.* **6**, 413.

212. Lampert, K., Machein, U., Machein, M.R., Conca, W., Peter, H.H., and Volk, B. (1998) Expression of matrix metalloproteinases and their tissue inhibitors in human brain tumors. *Am. J. Pathol.* **153**, 429–437.

212A. Vince, G.H., Herbold, C., Klein, R., Kühl, J., Pietsch, T., Franz, S., Roosen, K., Tonn, J.-C. (2001) Medulloblastoma displays distinct regional matrix metalloprotease expression. *J. Neurooncol.* **53**, 99–106.

213. Brandes, A.A., Paris, M.K., and Basso, U. (2003) Medulloblastomas: do molecular and biologic markers indicate different prognoses and treatments? *Expert Rev. Anticancer Ther.* **3**, 615–630.

214. Fisher, P.G., Burger, P.C., and Eberhart, C.G. (2004) Biologic risk stratification of medulloblastoma: the real time is now. *J. Clin. Oncol.* **22**, 971–974.

215. Park, P.C., Taylor, M.C., Mainprize, T.G., Becker, L.E., Ho, M., Dura, W.T., Squire, J., and Rutka, J.T. (2003) Transcriptional profiling of medulloblastoma in children. *J. Neurosurg.* **99,** 534–541.

216. Altieri, D.C. (2003) Validating survivin as a cancer therapeutic target. *Nat. Rev. Cancer* **3,** 46–54.

217. Altieri, D.C. (2003) Survivin, versatile modulation of cell division and apoptosis in cancer. *Oncogene* **22,** 8581–8589.

217A. Sarkar, C., Deb, P., and Sharma, M.C., (2006) Medulloblastomas: new directions in risk stratification. *Neurol. India* **54,** 16–23.

217B. Thompson, M.C., Fuller, C., Hogg, T.L., Dalton, J., Finkelstein, D., Lau, C.C., Chintagumpala, M., Adesina, A., Ashley, D.M., Kellie, S.J., Taylor, M.D., Curran, T., Gajjar, A., and Gilbertson, R.J. (2006) Genomics identifies medulloblastoma subgroups that are enriched for specific genetic alterations. *J. Clin. Oncol.* **24,** 1924–1931.

218. Sasaki, T., Beatriz, M., Lopes, S., Hankins, G.R., and Helm, G.A. (2002) Expression of survivin, an inhibitor of apoptosis protein, in tumors of the nervous system. *Acta Neuropathol.* **104,** 105–109.

219. Salsano, E., Pollo, B., Eoli, M., Giordana, M.T., and Finocchiaro, G. (2004) Expression of *MATH1*, a marker of cerebellar granule cell progenitors, identifies different medulloblastoma sub-types. *Neurosci. Lett.* **370,** 180–185.

220. Araki, T., Taniwaki, T., Becerra, S.P., Chader, G.J., and Schwartz, J.P. (1998) Pigment epithelium-derived factor (PEDF) differentially protects immature but not mature cerebellar granule cells against apoptotic cell death. *J. Neurosci. Res.* **53,** 7–15.

221. Ben-Arie, N., Bellen, H.J., Armstrong, D.L., McCall, A.E., Gordadze, P.R.., Guo, Q., Matzuk, M.M., and Zoghbi, H.Y. (1997) *Math1* is essential for genesis of cerebellar granule neurons. *Nature* **390,** 169–172.

222. Gazit, R., Krizhanovsky, V., and Ben-Arie, N. (2003) *Math1* controls cerebellar granule cell differentiation by regulating multiple components of the Notch signaling pathway. *Development* **131,** 903–913.

223. Helms, A.W., Gowan, K., Abney, A., Savage, T., and Johnson, J.E. (2001) Overexpression of MATH1 disrupts the coordination of neural differentiation in cerebellum development. *Mol. Cell Neurosci.* **17,** 671–682.

224. Tombran-Tink, J., and Barnstable, C.J. (2003) PEDF: a multifaceted neurotrophic factor. *Nat. Rev. Neurosci.* **4,** 628–636.

225. Taniwaki, T., Becerra, S.P., Chader, G.J., and Schwartz, J.P. (1995) Pigment epithelium-derived factor is a survival factor for cerebellar granule cells in culture. *J. Neurochem.* **64,** 2509–2517.

226. Taniwaki, T., Hirashima, N., Becerra, S.P., Chader, G.J., Etcheberrigaray, R., and Schwartz, J.P. (1997) Pigment epithelium-derived factor protects cultured cerebellar granule cells against glutamate-induced neurotoxicity. *J. Neurochem.* **68,** 26–32.

227. Dang, C.V., Resar, L.M., Emison, E., Kim, S., Li, Q., Prescott, J.E., Wonsey, D., and Zeller, K. (1999) Function of the c-Myc oncogenic transcription factor. *Exp. Cells Res.* **253,** 63–77.

228. Sommer, A., Koch, A., Tonn, J., Sorensen, N., Hurlin, P., Luscher, B., and Pietsch, T. (1999) Analysis of the Max binding protein MNT in human medulloblastoma. *Int. J. Cancer* **82,** 810–816.

229. Peterson, K., and Walker, R.W. (1995) Medulloblatoma/primitive neuroectodermal tumor in 45 adults. *Neurology* **45,** 440–442.

230. Provias, J.P., and Becker, L.E. (1996) Cellular and molecular pathology of medulloblastoma. *J. Neuro-Oncol.* **29,** 35–43.

230A. Gilbertson, R.J., and Gajjar, A. (2005) Molecular biology of medulloblastoma: will it ever make a difference to clinical management? *J. Neurooncol.* **75,** 273–278.

231. Caputy, A.J., McCullough, D.C., Manz, H.J., Patterson, K., and Hammock, M.K. (1987) A review of the factors influencing the prognosis of medulloblastoma. *J. Neurosurg.* **66,** 80–87.

232. Goldberg-Stern, H., Gadoth, N., Stern, S., Cohen, I.J., Zaizov, R., and Sandbank, U. (1991) The prognostic significance of glial fibrillary acidic protein staining in medulloblastoma. *Cancer* **68,** 568–573.

233. Schofield, D.E., Yunis, E.J., Geyer, J.R., Albright, A.L., Berger, M.S., and Taylor, S.R. (1992) DNA content and other prognostic features in childhood medulloblastoma: proposal of a scoring system. *Cancer* **69,** 1307–1314.

234. Tait, D.M., Eeles, R.A., Carter, R., Ashley, S., and Ormerod, M.G. (1993) Ploidy and proliferative index in medulloblastoma: Useful prognostic factors? *Eur. J. Cancer* **29A,** 1383–1387.

235. Carrie, C., Lasset, C., Alapetite, C., Haie-Meder, C., Hoffstetter, S., Demaille, M.-C., Kerr, C., Wagner, J.-P., Lagrange, J.-L., Maire, J.-P., Seng, S.-H., Chong Tao Kong Man, Y.O., Murraciole, X., and Pinto, N. (1994) Multivariate analysis of prognostic factors in adult patients with medulloblastoma. *Cancer* **74,** 2352–2360.

236. Albright, A.L., Wisoff, J.H., Zeltzer, P., Boyett, J., Rorke, L.B., Stanley, P., Geyer, J.R., and Milstein, J.M. (1995) Prognostic factors in children with supratentorial (nonpineal) primitive neuroectodermal tumors. *Pediatr. Neurosurg.* **22,** 1–7.

237. Albright, A.L., Wisoff, J.H., Zeltzer, P.M., Boyett, J.M., Rorke, L.B., and Stanley, P. (1996) Effects of medulloblastoma resections on outcome in children: a report from the Children's Cancer Group. *Neurosurgery* **38,** 265–271.

238. Giordana, M.T., Cavalla, P., Chiò, A., Marino, S., Soffietti, R., Vigliani, M.C., and Schiffer, D. (1995) Prognostic factors in adult medulloblastoma. A clinico-pathologic study. *Tumori* **81,** 338–346.

239. Giordana, M.T., Schiffer, P., and Schiffer, D. (1998) Prognostic factors in medulloblastoma. *Child's Nerv. Syst.* **14,** 256–262.

240. Giordana, M.T., D'Agostino, C., Pollo, B., Silvani, A., Ferracini, R., Paiolo, A., Ghiglione, P., and Chiò, A. (2005) Anaplasia is rare and does not influence prognosis in adult medulloblastoma. *J. Neuropathol. Exp. Neurol.* **64,** 869–874.

241. Sure, U., Berghorn, W.J., Bertalanffy, H., Wakabayashi, T., Yoshida, J., Sugita, K., and Seeger, W. (1995) Staging, scoring and grading of medulloblastomas. A postoperative prognosis predicting system based on the cases of a single institute. *Acta Neurochir (Wien)* **132,** 59–65.

242. Janss, A.J., Yachnis, A.T., Silber, J.H., Trojanowski, J.Q., Lee, V.M., Sutton, L.N., Perilongo, G., Rorke, L.B., and Phillips, P.C. (1996) Glial differentiation predicts poor

clinical outcome in primitive neuroectodermal brain tumors. *Ann. Neurol.* **39,** 481–489.

243. Gilbertson, R.J., Jaros, E., Perry, R.H., Kelly, P.J., Lunee, J., and Pearson, A.D. (1997) Mitotic percentage index: a new prognostic factor for childhood medulloblastoma. *Eur. J. Cancer* **33,** 609–615.

244. Jenkin, D., Shabanah, M.A., Shail, E.A., Gray, A., Hassounah, M., Khafaga, Y., Kofide, A., Mustafa, M., and Schultz, H. (2000) Prognostic factors for medulloblastoma. *Int. J. Radiat. Oncol. Biol. Phys.* **47,** 573–584.

245. Neben, K., Korshunov, A., Benner, A., Wrobel, G., Hahn, M., Kokocinski, F., Golanov, A., Joos, S., and Lichter, P. (2004) Microarray-based screening for molecular markers in medulloblastoma revealed *STK15* as independent predictor for survival. *Cancer Res.* **64,** 3103–3111.

246. Ozer, E., Sarialioglu, F., Cetingoz, R., Yuceer, N., Cakmakci, H., Ozkal, S., Olgun, N., Uysal, K., Corapcioglu, F., and Canda, S. (2004) Prognostic significance of anaplasia and angiogenesis in childhood medulloblastoma: A Pediatric Oncology Group Study. *Pathol. Res. Pract.* **200,** 501–509.

247. Rickert, C.H. (2004) Prognosis-related molecular markers in pediatric central nervous system tumors. *J. Neuropathol. Exp. Neurol.* **63,** 1211–1224.

248. Ross, W., and Hall, P.A. (1995) KI67 – From antibody to molecule to understanding. *J. Clin. Pathol-Clin. Mol. Pathol.* **48,** M113–M117.

249. Jalava, P., Kuopio, T., Juntti-Patinen, L., Kotkansalo, T., Kronqvist, P., and Collan, Y. (2006) Ki67 immunohistochemistry: a valuable marker in prognostication but with a risk of misclassification: proliferation subgroups formed based on Ki67 immunoreactivity and standardized mitotic index. *Histopathology* **48,** 674–682.

250. Ito, S., Hoshino, T., Prados, M.D., and Edwards, M.S.B. (1992) Cell kinetics of medulloblastomas. *Cancer* **70,** 671–678.

251. Grotzer, M.A., Geoerger, B., Janss, A.J., Zhao, H., Rorke, L.B., and Phillips, P.C. (2001) Prognostic significance of Ki-67 (MIB-1) proliferation index in childhood primitive neuroectodermal tumors or the central nervous system. *Med. Pediatr. Oncol.* **36,** 268–273.

252. Kayaselçuk, F., Zorludemir, S., Gümürdülü, D., Zeren, H., and Erman, T. (2002) PCNA and Ki-67 in central nervous system tumors: correlation with the histological type and grade. *J. Neurooncol.* **57,** 115–121.

253. Jay, V., Parkinson, D., Becker, L., and Chan, F.W. (1994) Cell kinetic analysis in pediatric brain and spinal tumors: a study of 117 cases with Ki-67 quantitation and flow cytometry. *Pediatr. Pathol.* **14,** 253–276.

254. Miralbell, R., Tolnay, M., Bieri, S., Probst, A., Sappino, A.-P., Berchtold, W., Pepper, M.S., and Pizzolato, G. (1999) Pediatric medulloblastoma: prognostic value of p53, bcl-2, Mib-1, and microvessel density. *J. Neuro-Oncol.* **45,** 103–110.

255. Karamitopoulou, E., Perentes, E., Melachrinou, M., and Maraziotis, T. (1993) Proliferating cell nuclear antigen immunoreactivity in human central nervous system neoplasm. *Acta Neuropathol.* **85,** 316–322.

256. Karamitopoulou, E., Perentes, E., Diamantis, I., and Maraziotis, T. (1994) Ki-67 immunoreactivity in human central nervous system tumors: a study with MIB 1 monoclonal antibody on archival material. *Acta Neuropathol.* **87,** 47–54.

257. Shibuya, M., Satoyuki, I., Miwa, T., Davis, R.L., Wilson, C.B., and Hoshino, T. (1993) Proliferation potential of brain tumors, analyses with Ki-67 and anti-DNA polymerase alpha monoclonal antibodies, bromodeoxyuridine labeling, and nucleolar organizer region counts. *Cancer* **71,** 199–205.

258. Schiffer, D., Cavalla, P., Chiò, A., Giordana, M.T., Marino, S., Mauro, A., and Migheli, A. (1994) Tumor cell proliferation and apoptosis in medulloblastoma. *Acta Neuropathol.* **87,** 362–370.

258A. Onda, K., Davis, R.L., and Edwards, M.S.B. (1996) Comparison of bromodeoxyuridine uptake and MIB 1 immunoreactivity in medulloblastomas determined with single and double immunohistochemical staining methods. *J. Neuro-Oncol.* **29,** 129–136.

259. MacDonald, T.J., Brown, K.M., LaFleur, B., Peterson, K., Lawlor, C., Chen, Y., Packer, R.J., Cogen, P., and Stephan, D.A. (2001) Expression profiling of medulloblastoma: PDGFRA and the RAS/MAPK pathway as therapeutic targets for metastatic disease. *Nat. Genet.* **29,** 143–152.

259A. Gilbertson, R.J., Langdon, J.A., Hollander, A., Hernan, R., Hogg, T.L., Gajjar, A., Fuller, C., and Clifford, S.C. (2006) Mutation analysis of *PDGFR-RAS/MAPK* pathway activation in childhood medulloblastoma. *Eur. J. Cancer* **42,** 646–649.

260. Goodrich, L.V., Milenkovic, L., Higgins, K.M., and Scott, M.P. (1997) Altered neural cell fates and medulloblastoma in mouse *patched* mutants. *Science* **277,** 1109–1113.

261. Hahn, H., Wojnowski, L., Zimmer, A.M., Hall, J., Miller, G., Zimmer, A. (1998) Rhabdomyosarcomas and radiation hypersensitivity in a mouse model of Gorlin syndrome. *Nat. Med.* **4,** 619–622.

262. Freeman, C.R., Taylor, R.E., Kortmann, R.D., and Carrie, C. (2002) Radiotherapy for medulloblastoma in children: a perspective on current international clinical research efforts. *Med. Pediatr. Oncol.* **39,** 99–108.

263. Grotzer, M.A., Janss, A.J., Phillips, P.C., and Trojanowski, J.Q. (2000) Neurotrophin receptor TrkC predicts good clinical outcome in medulloblastoma and other primitive neuroectodermal brain tumors. *Klin. Padiatr.* **212,** 196–199.

264. Yasue, M., Tomita, T., Engelhard, H., Gonzalez-Crussi, F., McLone, D.G., and Bauer, K.D. (1989) Prognostic importance of DNA ploidy in medulloblastoma of childhood. *J. Neurosurg.* **70,** 385–391.

265. Gajjar, A.J., Heideman, R.L., Douglass, E.C., Kun, L.E., Kovnar, E.H., Sanford, R.A., Fairclough, D.L., Ayers, D., and Look, A.T. (1993) Relation of tumor-cell ploidy to survival in children with medulloblastoma. *J. Clin. Oncol.* **11,** 2211–2217.

266. Zerbini, C., Gelber, R.D., Weinberg, D., Sallan, S.E., Barnes, P., Kupsky, W., Scott, R.M., and Tarbell, N.J. (1993) Prognostic factors in medulloblastoma, including DNA ploidy. *J. Clin. Oncol.* **11,** 616–622.

266A. Tomita, T., Yasue, M., Engelhard, H.H., McLone, D.G., Gonzalez-Crussi, F., and Bauer, K.D. (1988) Flow cytometric DNA analysis of medulloblastoma. Prognostic implication of aneuploidy. *Cancer* **61,** 744–749.

267. McAllister, R.M., Isaacs, H., Rongey, R., Peer, M., Au, W., Soukup, S.W., and Gardner, M.B. (1977) Establishment of a human medulloblastoma cell line. *Int. J. Cancer* **20,** 206–212.

268. Pietsch, T., Scharmann, T., Fonatsch, C., Schmidt, D., Öckler, R., Freihoff, D., Albrecht, S., Wiestler, O.D., Zeltzer, P., and Riehm, H. (1994) Characterization of five new cell lines derived from human primitive neuroectodermal tumors of the central nervous system. *Cancer Res.* **54**, 3278–3287.

269. Friedman, H.S., Burger, P.C., Bigner, S.H., Trojanowski, J.Q., Wikstrand, C.J., Halperin, E.C., and Bigner, D.D. (1985) Establishment and characterization of the human medulloblastoma cell line and transplantable xenograft D283 Med. *J. Neuropathol. Exp. Neurol.* **6**, 592–605.

270. He, X.M., Wikstrand, C.J., Friedman, H.S., Bigner, S.H., Pleasure, S., Trojanowski, J.Q., and Bigner D.D. (1991) Differentiation characteristics of newly established medulloblastoma cell lines (D384 Med, D425 Med, and D458 Med) and their transplantable xenografts. *Lab. Invest.* **64**, 833–843.

271. He, X., Ostrowski, L.E., von Wronski, M.A., Friedman, H.S., Wikstrand, C.J., Bigner, S.H., Rasheed, A., Batra, S.K., Mitra, S., Brent, T.P., and Bigner, D.D. (1992) Expression of O^6-methylguanine-DNA methyltransferase in six human medulloblastoma cell lines. *Cancer Res.* **52**, 1144–1148.

272. Keles, G.E., Berger, M.S., Schofield, D., and Bothwell, M. (1993) Nerve growth factor receptor expression in medulloblastomas and the potential role of nerve growth factor as a differentiating agent in medulloblastoma cell lines. *Neurosurgery* **32**, 274–280.

272A. Keles, G.E., Berger, M.S., Srinivasan, J., Kolstoe, D.D., Bobola, M.S., and Silber, J.R. (1995) Establishment and characterization of four human medulloblastoma-derived cell lines. *Oncol. Res.* **7**, 493–503.

273. Srinivasan, J., Berger, M.S., and Silber, J.R. (1996) *p53* expression in four human medulloblastoma-derived cell lines. *Child's Nerv. Syst.* **12**, 76–80.

274. Macaulay, R.J.B., Wang, W., Dimitroulakos, J., Becker, L.E., and Yeger, H. (1999) Lovastatin-induced apoptosis of human medulloblastoma cell lines *in vitro*. *J. Neurooncol.* **42**, 1–11.

275. Su, C.-W., Lin, K.-L., Hou, J.-W., Jung, S.-M., and Zen, E.-C. (2003) Spontaneous recovery from a medulloblastoma by a female with Gorlin-Goltz syndrome. *Pediatr. Neurol.* **28**, 231–234.

276. Hahn, H., Wicking, C., Zaphiropoulos, P.G., Gailani, M.R., Shanley, S., Chidambaram, A., Vorechovsky, I., Holmberg, E., Unden, A.B., Gillies, S., Negus, K., Smyth, I., Pressman, C., Leffell, D.J., Gerrard, B., Goldstein, A.M., Dean, M., Toftgard, R., Chenevix-Trench, G., Wainwright, B., and Bale, A.E. (1996) Mutations of the human homolog of drosophila *patched* in the nevoid basal cell carcinoma syndrome. *Cell* **85**, 841–851.

277. Johnson, R.L., Rothman, A.L., Xie, J., Goodrich, L.V., Bare, J.W., Bonifas, J.M., Quinn, A.G., Myers, R.M., Cox, D.R., Epstein, E.H. Jr., and Scott, M.P. (1996) Human homolog of *patched*, a candidate gene for the basal cell nevus syndrome. *Science* **272**, 1668–1671.

277A. Ragge, N.K., Sait, A., Collin, J.R., Michalski, A., and Farndon, P.A. (2005) Gorlin syndrome: the *PTCH* gene links ocular developmental defects and tumour formation. *Br. J. Ophthalmol.* **89**, 988–991.

278. Cowan, R., Hoban, P., Kelsey, A., Birch, J.M., Gattamaneni, R., and Evans, D.G. (1997) The gene for the naevoid basal cell carcinoma syndrome acts as tumour-suppressor gene in medulloblastoma. *Br. J. Cancer* **76**, 141–145.

279. Holland, E.C. (2001) Brain tumor animal models: importance and progress. *Curr. Opin. Oncol.* **13**, 143–147.

280. Weiner, H.L., Bakst, R., Hurlbert, M.S., Ruggiero, J., Ahn, E., Lee, W.S., Stephen, D., Zagzag, D., Joyner, A.L., and Turnbull, D.H. (2002) Induction of medulloblastomas in mice by Sonic hedgehog, independent of *Gli1*. *Cancer Res.* **62**, 6385–6389.

281. Zindy, F., Nilsson, L.M., Nguyen, L., Meunier, C., Smeyne, R.J., Rehg, J.E., Eberhart, C., Sherr, C.J., and Roussel, M.F. (2003) Hemangiosarcomas, medulloblastomas, and other tumors in *Ink4c/p53*-null mice. *Cancer Res.* **63**, 5420–5427.

282. Dyer, M.A. (2004) Mouse models of childhood cancer of the nervous system. *J. Clin. Pathol.* **57**, 561–576.

283. Fults, D.W. (2005) Modeling medulloblastoma with genetically engineered mice. *Neurosurg. Focus* **19**(5)E7, 1–8.

284. Kimura, H., Stephen, D., Joyner, A., anda Curran, T. (2005) Gli1 is important for medulloblastoma formation in $Ptc1^{+/--}$ mice. *Oncogene* **24**, 4026–4036.

284A. Hahn, H., Wojnowski, L., Specht, K., Kappler, R., Calzada-Wack, J., Potter, D., Zimmer, A., Müller, U., Samson, E., Quintanilla-Martinez, L., and Zimmer, A. (2000) *Patched* target *Igf2* is indispensable for the formation of medulloblastoma and rhabdomyosarcoma. *J. Biol. Chem.* **275**, 28341–28344.

285. Dahmane, N., Sánchez, P., Gitton, Y., Palma, V., Sun, T., Beyna, M., Weiner, H., Ruiz i Altaba, A. (2001) The Sonic Hedgehog-Gli pathway regulates dorsal brain growth and tumorigenesis. *Development* **128**, 5201–5212.

286. Wetmore, C., Eberhart, D.E., and Curran, T. (2000) The normal *patched* allele is expressed in medulloblastomas from mice with heterozygous germ-line mutation of *patched*. *Cancer Res.* **60**, 2239–2246.

287. Kim, J.Y., Nelson, A.L., Algon, S.A., Graves, O., Sturla, L.M., Goumnerova, L.C., Rowitch, D.H., Segal, R.A., and Pomeroy, S.L. (2003) Medulloblastoma tumorigenesis diverges from cerebellar granular cell differentiation in patched heterozygous mice. *Dev. Biol.* **263**, 50–66.

288. Hallahan, A.R., Pritchard, J.I., Hansen, S., Benson, M., Stoeck, J., Hatton, B.A., Russell, T.L., Ellenbogen, R.G., Bernstein, I.D., Beachy, P.A., and Olson, J.M. (2004) The SmoA1 mouse model reveals that Notch signaling is critical for the growth and survival of Sonic Hedgehog-induced medulloblastomas. *Cancer Res.* **64**, 7794–7800.

289. Rao, G., Pedone, C.A., Del Valle, L., Reiss, K., Holland, E.C., and Fults, D.W. (2004) Sonic hedgehog and insulin-like growth factor signaling synergize to induce medulloblastoma formation from nestin-expressing neural progenitors in mice. *Oncogene* **23**, 6156–6162.

290. Romer, J.T., Kimura, H., Magdaleno, S., Sasai, K., Fuller, C., Baines, H., Connelly, M., Stewart, C.F., Gould, S., Rubin, L.L., and Curran, T. (2004) Suppression of the Shh pathway using a small molecule inhibitor eliminates medulloblastoma in $Ptc1^{+/-}p53^{-/-}$ mice. *Cancer Cell* **6**, 229–240.

290A. Oliver, T.G., Read, T.A., Kessler, J.D., Mehmeti, A., Wells, J.F., Huynh, T.T.T., Lin, S.M., and Wechsler-Reya, R.J. (2005) Loss of *patched* and disruption of granule cell development in a pre-neoplastic stage of medulloblastoma. *Development* **132**, 2425–2439.

290B. Zurawel, R.H., Allen, C., Chiappa, S.A., Cato, W., Biegel, J., Cogen, P., de Sauvage, F., and Raffel, C. (2000) Analysis of

PTCH/SMO/SHH pathway genes in medulloblastoma. *Genes Chromosomes Cancer* **27,** 44–51.

290C. Zurawel, R.H., Allen, C., Wechsler-Reya, R., Scott, M.P., and Raffel, C. (2000) Evidence that haploinsufficiency of *PTCH* leads to medulloblastoma in mice. *Genes Chromosomes Cancer* **28,** 77–81.

291. Wechsler-Reya, R.J., and Scott, M.P. (1999) Control of neuronal precursor proliferation in the cerebellum by Sonic hedgehog. *Neuron* **22,** 103–114.

291A. Wallace, V.A. (1999) Purkinje-cell-derived Sonic hedgehog regulates granule neuron precursor cell proliferation in the developing mouse cerebellum. *Curr. Biol.* **9,** 445–448.

292. Wechsler-Reya, R., and Scott, M.P. (2001) The developmental biology of brain tumors. *Annu. Rev. Neurosci.* **24,** 385–428.

292A. Argenti, B., Gallo, R., Di Marcotullio, L., Ferretti, E., Napolitano, M., Canterini, S., De Smaele, E., Greco, A., Fiorenza, M.T., Maroder, M., Screpanti, I., Alesse, E., andGulino, A. (2005) Hedgehog antagonist REN^{KCTD11} regulates proliferation and apoptosis of developing granule cell prognosis. *J. Neurosci.* **25,** 8338–8346.

293. Kenney, A.M., Cole, M.D., and Rowitch, D.H. (2003) *Nmyc* upregulation by sonic hedgehog signaling promotes proliferation in developing cerebellar granule neuron precursors. *Development* **130,** 15–28.

294. Jessell, T.M. (2000) Neuronal specification in the spinal cord: inductive signals and transcriptional codes. *Nat. Rev. Genet.* **1,** 20–29.

295. McMahon, A.P., Ingham, P.W., and Tabin, C.J. (2003) Developmental roles and clinical significance of hedgehog signaling. *Curr. Top. Dev. Biol.* **53,** 1–114.

296. Beachy, P.A., Karhadkar, S.S., and Berman, D.M. (2004) Tissue repair and stem cell renewal in carcinogenesis. *Nature* **432,** 324–331.

297. Yao, S., Lum, L., and Beachy, P. (2006) The ihog cell surface proteins bind Hedgehog and mediate pathway activation. *Cell* **125,** 343–357.

298. Lam, C.W., Xie, J., To, K.-F., Ng, H.-K., Lee, K.-C., Yuen, N.W.-F., Lim, P.-L., Chan, L.Y.-S., Tong, S.-F., and McCormick, F. (1999) A frequent activated smoothened mutation in sporadic basal cell carcinomas. *Oncogene* **18,** 833–836.

299. Walterhouse, D.O., Yoon, J.W., and Iannoccone, P.M. (1999) Developmental pathways: Sonic hedgehog-Patched-GLI. *Environ. Heatlh Perspect.* **107,** 167–171.

300. Alcedo, J., and Noll, M. (1997) Hedgehog and its patched-smoothened receptor complex: a novel signaling mechanism at the cell surface. *Biol. Chem.* **378,** 583–590.

301. Hahn, H., Wojnowski, L., Miller, G., and Zimmer, A. (1999) The patched signaling pathway in tumorigenesis and development: lessons from animal models. *J. Mol. Med.* **77,** 459–468.

302. Ingham, P.W., and McMahon, A.P. (2001) Hedgehog signaling in animal development: paradigms and principles. *Genes Dev.* **15,** 3059–3087.

303. Dean, M. (1996) Polarity, proliferation and the hedgehog pathway. *Nat. Genet.* **14,** 245–247.

304. Reifenberger, J., Wolter, M., Weber, R.G., Megahed, M., Ruzicka, T., Lichter, P., and Reifenberger, G. (1998) Missense mutations in *SMOH* in sporadic basal cell carcinomas of the skin and primitive neuroectodermal tumors of the central nervous system. *Cancer Res.* **58,** 1798–1803.

305. Di Marcotullio, L., Ferretti, E., De Smaele, E., Argenti, B., Minicione, C., Zazzeroni, F., Gallo, R., Masuelli, L., Napolitano, M., Maroder, M., Modesti, A., Giangaspero, F., Screpanti, I., Alesse, E., and Gulino, A. (2004) REN^{KCTD11} is a suppressor of Hedgehog signaling and is deleted in human medulloblastoma. *PNAS* **101,** 10833–10838.

306. Gallo, R., Zazzeroni, F., Alesse, E., Mincione, C., Borello, U., Buanne, P., D'Eugenio, R., Mackay, A.R., Argenti, B., Gradini, R., Russo, M.A., Maroder, M., Cossu, G., Frati, L., Screpanti, I., and Gulino, A. (2002) *REN*: a novel, developmentally regulated gene that promotes neural cell differentiation. *J. Cell Biol.* **158,** 731–740.

307. Dunaeva, M., Michelson, P., Kogerman, P., and Toftgard, R. (2003) Characterization of the physical interaction of Gli proteins with SUFU proteins. *J. Biol. Chem.* **278,** 5116–5122.

308. Taylor, M.D., Liu, L., Raffel, C., Huiu, C.-c., Mainprize, T.G., Zhang, X., Agatep, R., Chiappa, S., Gao, L., Lowrance, A., Hao, A., Goldstein, A.M., Stavrou, T., Scherer, S.W., Dura, W.T., Wainwright, B., Squire, J.A., Rutka, J.T., and Hogg, D. (2002) Mutations of *SUFU* predispose to medulloblastoma. *Nat. Genet.* **31,** 306–310.

309. Taylor, M.D., Mainprize, T.G., Rutka, J.T., Becker, L., Bayani, J., and Drake, J.M. (2001) Medulloblastoma a child with Rubenstein-Taybi syndrome: Case report and review of the literature. *Pediatr. Neurosurg.* **35,** 235–238.

310. Unden, A.B., Holmbert, E., Lundh-Rozell, B., Stähle-Bäckdahl, M., Zaphiropoulos, P.G., Toftgärd, R., and Vorechovsky, I. (1996) Mutations in the human homologue of *Drosophila patched* (*PTCH*) in basal cell carcinomas and the Gorlin syndrome: Different *in vivo* mechaniasms of *PTCH* inactivation. *Cancer Res.* **56,** 4562–4565.

311. Stone, D.M., Hynes, M., Armanini, M., Swanson, T.A., Gu, Q., Johnson, R.L., Scott, M.P., Pennica, D., Goddard, A., Phillips, H., Noll, M., Hooper, J.E., de Sauvage, F., and Rosenthal, A. (1996) The tumour-suppressor gene *patched* encodes a candidate receptor for Sonic hedgehog. *Nature* **384,** 129–134.

312. Wicking, C., and Bale, A.E. (1997) Molecular basis of the nevois basal cell carcinoma syndrome. *Curr. Opin. Pediatr.* **9,** 630–635.

313. Wicking, C., Gilles, S., Smyth, I., Shanley, S., Fowles, L., Ratcliffe, J., Wainwright, B., and Chenevix-Trench, G. (1997) De novo mutations of the Patched gene in nevoid basal cell carcinoma syndrome help to define the clinical phenotype. *Am. J. Med. Genet.* **73,** 304–307.

314. Gailani, M.R., Stahle-Backdahl, M., Leffell, D.J., Glynn, M., Zaphiropoulos, P.E., Pressman, C., Unden, A.B., Dean, M., Brash, D.E., Bale, A.E., and Toftgard, R. (1996) The role of the human homologue of *Drosophila patched* in sporadic basal cell carcinomas. *Nat. Genet.* **14,** 7–8.

315. Xie, J., Murone, M., Luoh, S.-M., Ryan, A., Gu, Q., Zhang, C., Bonifas, J.M., Lam, C.-W., Hynes, M., Goddard, A., Rosenthal, A., Epstein, E.H., Jr., and de Sauvage, F.J. (1998) Activating *Smoothened* mutations in sporadic basal-cell carcinoma. *Nature* **391,** 90–92.

315A. Xie, J., Johnson, R.L., Zhang, X., Bare, J.W., Waldman, F.M., Cogen, P.H., Menon, A.G., Wararen, R.S., Chen, L.-C., Scott, M.P., and Epstein, E.H., Jr. (1997) Mutations of the *PATHCED* gene in several types of sporadic extracutaneous tumors. *Cancer Res.* **57,** 2369–2372.

316. Vortmeyer, A.O., Stavrou, T., Selby, D., Li, G., Weil, R.J., Park, W.-S., Moon, Y.-W., Chandra, R., Goldstein, A.M., and Zhuang, Z. (1999) Deletion analysis of the adenomatous polyposis coli and *PTCH* gene loci in patients with sporadic and nevoid basal cell carcinoma syndrome-associated medulloblastoma. *Cancer* **85**, 2662–2667.

317. Vorechovsky, I., Tingby, O., Hartman, M., Strömberg, B., Nister, M., Collins, V.P., and Toftgård, R. (1997) Somatic mutations in the human homologue of *Drosophila patched* in primitive neuroectodermal tumours. *Oncogene* **15**, 361–366.

318. Berman, D.M., Karhadkar, S.S., Hallahan, A.R., Pritchard, J.I., Eberhart, C.G., Watkins, D.N., Chen, J.K., Cooper, M.K., Taipale, J., Olson, J.M., and Beachy, P.A. (2002) Medulloblastoma growth inhibition by Hedgehog pathway blockade. *Science* **297**, 1559–1561.

319. Scheurlen, W., and Kühl, J. (1998) Current diagnostic and therapeutic management of CNS metastasis in childhood primitive neuroectodermal tumors and ependymomas. *J. Neurooncol.* **38**, 181–185.

320. Oro, A.E., Higgins, K.M., Hu, Z., Bonifas, J.M., Epstein, E.H., Jr., and Scott, M.P. (1997) Basal cell carcinomas in mice overexpressing Sonic hedgehog. *Science* **276**, 817–821.

321. Wicking, C., Evans, T., Henk, B., Hayward, N., Simms, L.A., Chenevix-Trench, G., Pietsch, T., and Wainwright, B. (1998) No evidence for the H133Y mutation in *SONIC HEDGEHOG* in a collection of common tumour types. *Oncogene* **16**, 1091–1093.

322. Solecki, D.J., Liu, XL, Tomoda, T., Fang, Y., and Hatten, M.E. (2001) Activated *Notch2* signaling inhibits differentiation of cerebellar granule neuron precursors by maintaining proliferation. *Neuron* **31**, 557–568.

323. Romer, J., and Curran, T. (2005) Targeting medulloblastoma: Small-molecule inhibitors of the Sonic Hedgehog pathway as potential cancer therapeutics. *Cancer Res.* **65**, 4975–4978.

324. Ellison, D.W., Clifford, S.C., Gajjar, A., and Gilbertson, R.J. (2003) What's new in neuro-oncology? Recent advances in medulloblastoma. *Eur. J. Paediatr. Neurol.* **7**, 53–66.

325. Corcoran, R.B., and Scott, M.P. (2001) A mouse model for medulloblastoma and basal cell nevus syndrome. *J. Neurooncol.* **53**, 307–318.

326. Piedimonte, L.R., Wailes, I.K., and Weiner, H.L. (2005) Medulloblastoma: mouse models and novel targeted therapies based on the Sonic hedgehog pathway. *Neurosurg. Focus* **19**, E8.

327. Pazzaglia, S., Tanori, M., Mancuso, M., Rebessi, S., Leonardi, S., Di Majo, V., Covelli, V., Atkinson, M.J., Hahn, H., and Saran, A. (2006) Linking DNA damage to medulloblastoma tumorigenesis in patched heterozygous knockout mice. *Oncogene* **25**, 1165–1173.

328. Aza-Blanc, P., Ramírez-Weber, F.-A., Laget, M.-P., Schwarz, C., and Kornberg, T.B. (1997) Proteolysis that is inhibited by Hedgehog targets cubitus interruptus protein to the nucleus and converts it to a repressor. *Cell* **89**, 1043–1053.

329. Goodrich, L.V., and Scott, M.P. (1998) Hedgehog and Patched in neural development and disease. *Neuron* **21**, 1243–1257.

330. Ingham, P.W. (1998) Transducing Hedgehog: the story so far. *EMBO J.* **17**, 3505–3511.

331. Chen, C.-H., von Kessler, D.P., Park, W., Wang, B., Ma, Y., and Beachy, P.A. (1999) Nuclear trafficking of cubitus interruptus in the transcriptional regulation of Hedgehog target gene expression. *Cell* **98**, 305–316.

332. Méthot, N., and Basler, K. (1999) Hedgehog controls limb development by regulating the activities of distinct transcriptional activator and repressor forms of cubitus interruptus. *Cell* **96**, 819–831.

333. Murone, M., Rosenthal, A., and de Sauvage, F.J. (1999) Sonic hedgehog signaling by the patched-smoothened receptor complex. *Curr. Biol.* **9**, 76–84.

334. Toftgard, R. (2000) Hedgehog signaling in cancer. *Cell Mol. Life Sci.* **57**, 1720–1731.

335. Wang, B., Fallon, J.F., and Beachy, P.A. (2000) Hedgehog-regulated processing of Gli3 produces an anterior/posterior repressor gradient in the developing vertebrate limb. *Cell* **100**, 423–434.

336. Taipale, J., and Beachy, P.A. (2001) The Hedgehog and Wnt signaling pathways in cancer. *Nature* **411**, 349–354.

337. Nybakken, K., and Perrimon, N. (2002) Hedgehog signal transduction: recent findings. *Curr. Opin. Genet. Dev.* **12**, 503–511.

338. Ruiz i Altaba, A., Sanchez, P., and Dahmane, N. (2002) Gli and hedgehog in cancer: tumours, embryos and stem cells. *Nat. Rev. Cancer* **2**, 361–372.

339. Wetmore, C. (2003) Sonic hedgehog in normal and neoplastic proliferation: insight gained from human tumors and animal models. *Curr. Opin. Genet. Dev.* **13**, 34–42.

339A. Adolphe, C., Hetherington, R., Ellis, T., and Wainwright, B. (2006) Patched1 functions as a gatekeeper by promoting cell cycle progression. *Cancer Res.* **66**, 2081–2088.

339B. Moon, R.T., Kohn, A.D., De Ferrari, G.V., and Kaykas, A. (2004) WNT and beta-catenin signaling: diseases and therapies. *Nat. Rev. Genet.* **5**, 691–701.

340. Reya, T., and Clevers, H. (2005) Wnt signaling in stem cells and cancer. *Nature* **434**, 843–850.

341. Mosimann, C., Hausmann, G., and Basler, K. (2006) Parafibromin/Hyrax activates Wnt/Wg target gne transcription by direct association with β-catenin/Armadillo. *Cell* **125**, 327–341.

341A. Clifford, S.C., Lusher, M.E., Lindsey, J.C., Langdon, J.A., Gilbertson, R.J., Straughton, D., and Ellison, D.W. (2006) Wnt/Wingless pathway activation and chromosome 6 loss characterize a distinct molecular sub-group of medulloblastomas associated with a favorable prognosis. *Cell* Cycle **5**, 2666–2670.

342. Sierra, J., Yoshida, T., Joazeiro, C.A., and Jones, K.A. (2006) The APC tumor suppressor counteracts β-catenin activation and H3K4 methylation at Wnt target genes. *Genes Dev.* **20**, 586–600.

342A. Gregorieff, A., and Clevers, H. (2005) Wnt signaling in the intestinal epithelium: from endoderm to cancer. *Genes Dev.* **19**, 877–890.

343. Ohgaki, H., Baeza, N., Masuoka, J., and Kleihues, P. (2003) Involvement of the Wnt pathway in the development of medulloblastomas. *J. Neuropathol. Exp. Neurol.* **62**, 581.

343A. Daniels, D.L., Eklof Spink, K., and Weis, W.I. (2001) Beta-catenin: molecular plasticity and drug design. *Trends Biochem. Sci.* **26**, 672–678.

344. Chong, Z.Z., Li, F., and Maiese, K. (2005) Oxidative stress in the brain: novel cellular targets that govern survival during neurodegenerative disease. *Prog. Neurobiol.* **75**, 207–246.

344A. Rubinfeld, B., Albert, I., Porfiri, E., Fiol, C., Munemitsu, S., and Polakis, P. (1996) Binding of GSK3β to the APC-β-catenin complex and regulation of complex assembly. *Science* **272,** 1023–1026.

345. Bullions, L.C., and Levine, A.J. (1998) The role of beta-catenin in cell adhesion, signal transduction, and cancer. *Curr. Opin. Oncol.* **10,** 81–87.

346. Zurawel, R.H., Chiappa, S.A., Allen, C., and Raffel, C. (1998) Sporadic medulloblastomas contain oncogenic β-catenin mutations. *Cancer Res.* **58,** 896–899.

347. Polakis, P. (1999) The oncogenic activation of beta-catenin. *Curr. Opin. Genet. Dev.* **9,** 15–21.

348. Yost, C., Torres, M., Miller, J.R., Huang, E., Kimelman, D., and Moon, R.T. (1996) The axis-inducing activity, stability, and subcellular distribution of β-catenin is regulated in *Xenopus* embryos by glycogen synthase kinase 3. *Genes Dev.* **10,** 1443–1454.

349. Morin, P.J., Sparks, A.B., Korinek, V., Barker, N., Clevers, H., Vogelstein, B., and Kinzler, K.W. (1997) Activation of β-catenin-Tcf signaling in colon cancer by mutations in β-catenin or APC. *Science* **275,** 1787–1790.

350. Hart, M.J., de los Santos, R., Albert, I.N., Rubinfeld, B., and Polakis, P. (1998) Downregulation of beta-catenin by hum;an Axin and its association with the APC tumor suppressor, beta-catening and GSK3 beta. *Curr. Biol.* **8,** 573–581.

351. Peifer, M., and Polakis, P. (2000) Wnt signaling in oncogenesis and embryogenesis—a look outside the nucleus. *Science* **287,** 1606–1609.

352. Hamada, F., Tomoyasu, Y., Takatsu, Y., Nakamura, M., Nagai, S., Suzuki, A., Fujita, F., Shibuya, H., Toyoshima, K., Ueno, N., and Akiyama, T. (1999) Negative regulation of Wingless signaling by D-axin, a *Drosophila* homolog of axin. *Science* **283,** 1739–1742.

353. Willert, K., Shibamoto, S., and Nusse, R. (1999) Wnt-induced dephosphorylations of Axin releases β-catenin from the Axin complex. *Genes Dev.* **13,** 1768–1773.

354. He, T.-C., Sparks, A.B., Rago, C., Hermeking, H., Zawel, L., da Costa, L.T., Morin, P.J., Vogelstein, B., and Kinzler, K.W. (1998) Identification of c-*MYC* as a target of the APC pathway. *Science* **281,** 1509–1512.

355. Tetsu, O., and McCormick, F. (1999) β-catenin regulates expression of cyclin D1 in colon carcinoma cells. *Nature* **398,** 422–426.

355A. Nelson, W.J., and Nusse, R. (2004) Convergence of Wnt, β-catenin, and cadherin pathways. *Science* **303,** 1483–1487.

356. Behrens, J., von Kries, J.P., Kuhl, M., Bruhn, L., Wedlich, D., Grosschedl, R., and Birchmeier, W. (1996) Functional interaction of beta-catenin and the transcription factor LEF-1. *Nature* **382,** 638–642.

357. Barth, A.I., Nathke, I.S., and Nelson, W.J. (1997) Cadherins, catenins and APC protein: interplay between cytoskeletal complexes and signaling pathways. *Curr. Opin. Cell. Biol.* **9,** 683–690.

358. Xiong, Y., and Kotake, Y. (2006) No exit strategy? No problem: APC inhibits β-catenin inside the nucleus. *Genes Dev.* **20,** 637–642.

359. Hirohashi, S. (1998) Inactivation of the E-cadherin-mediated cell adhesion system in human cancers. *Am. J. Pathol.* **153,** 333–339.

360. Li, M.H., Bouffet, E., Hawkins, C.E., Squire, J.A., and Huang, A. (2005) Molecular genetics of supratentorial primitive neuroectodermal tumors and pineoblastoma. *Neurosurg. Focus* **19**.

361. Huang, H., Mahler-Araujo, B.M., Sankila, A., Chimelli, L., Yonekawa, Y., Kleihues, P., and Ohgaki, H. (2000) *APC* mutations in sporadic medulloblastomas. *Am. J. Pathol.* **156,** 433–437.

362. Dahmen, R.P., Koch, A., Denkhaus, D., Tonn, J.C., Sörensen, N., Berthold, F., Behrens, J., Birchmeier, W., Wiestler, O.D., and Pietsch, T. (2001) Deletions of *AXIN1*, a component of the *WNT/wingless* pathway in sporadic medulloblastomas. *Cancer Res.* **61,** 7039–7043.

363. Itoh, H., Ohsato, K., Yao, T., Iida, M., and Watanabe, H. (1979) Turcot's syndrome and its mode of inheritance. *Gut.* **20,** 414–419.

364. Van Meir, E.G. (1998) "Turcot's syndrome": phenotype of brain tumors, survival and mode of inheritance. *Int. J. Cancer* **75,** 162–164.

365. Koch, A., Waha, A., Tonn, J.C., Sorensen, N., Berthold, F., Wolter, M., Reifenberger, J., Hartmann, W., Friedl, W., Reifenberger, G., Wiestler, O.D., and Pietsch, T. (2001) Somatic mutations of WNT/wingless signaling pathway components in primitive neuroectodermal tumors. *Int. J. Cancer* **93,** 445–449.

366. Baeza, N., Masuoka, J., Kleihues, P., and Ohgaki, H. (2003) *AXIN1* mutations but not deletions in cerebellar medulloblastomas. *Oncogene* **22,** 632–636.

367. Mori, T., Nagase, H., Horii, A., Miyoshi Y., Shimano, T., Nakatsuru, S., Aoki, T., Arakawa, H., Yanagisawa, A., Ushio, Y., Takano, S., Ogawa, M., Nakamura, M., Shibuya, M., Nishikawa, R., Matsutani, M., Hayashi, Y., Takahashi, H., Ikuta, F., Nishihara, T., Mori, S., and Nakamura, Y. (1994) Germ-line and somatic mutations of the *APC* gene in patients with Turcot syndrome and analysis of *APC* mutations in brain tumors. *Genes Chromosomes Cancer* **9,** 168–172.

368. Hamilton, S.R., Liu, B., Parsons, R.E., Papadopoulos, N., Jen, J., Powell, S.M., Krush, A.J., Berk, T., Cohen, Z., Tetu, B., Burger, P.C., Wood, P.A., Taqi, F., Booker, S.V., Petersen, G.M., Offerhaus, G.J.A., Tersmette, A.C., Giardiello, F.M., Vogelstein, B., and Kinzler, K.W. (1995) The molecular basis of Turcot's syndrome. *N. Engl. J. Med.* **332,** 839–847.

369. Yong, W.H., Raffel, C., von Deimling, A., and Louis, D.N. (1995) The *APC* gene in Turcot's syndrome. *N. Engl. J. Med.* **332,** 839–847.

370. Yokota, N., Nishizawa, S., Ohta, S., Date, H., Sugimura, H., Namba, H., and Maekawa, M. (2002) Role of Wnt pathway in medulloblastoma oncogenesis. *Int. J. Cancer* **101,** 198–201.

371. Eberhart, C.G., Tihan, T., and Burger, P.C. (2000) Nuclear localization and mutation of β-catenin in medulloblastomas. *J. Neuropathol. Exp. Neurol.* **59,** 333–337.

372. Bates, B., Rios, M., Trumpp, A., Chen, C., Fan, G., Bishop, J.M., and Jaenisch, R. (1999) Neurotrophin-3 is required for proper cerebellar development. *Nat. Neurosci.* **2,** 115–117.

373. Korsching, S. (1993) The neurotrophic factor concept: a reexamination. *J. Neurobiol.* **13,** 2739–2748.

374. Miranda, R.C., Sohrabji, F., and Toran-Allerand, C.D. (1993) Neuronal colocalization of mRNAs for neurotrophins and their receptors in the developing central nervous system suggests a potential for autocrine interactions. *Proc. Natl. Acad. Sci. USA* **90,** 6439–6443.

375. Barbacid, M. (1994) The Trk family of neurotrophin receptors. *J. Neurobiol.* **25**, 1386–1403.

376. Davies, A.M. (1994) The role of neurotrophins in the developing nervous system. *J. Neurobiol.* **25**, 1334–1348.

377. Kaplan, D.R., and Stephens, R.M. (1994) Neurotrophin signal transduction by the Trk receptor. *J. Neurobiol.* **25**, 1404–1417.

378. Chao, M.V., and Hempstead, B.L. (1995) p75 and Trk: a two receptor system. *Trends Neurosci.* **7**, 321–326.

379. Fung, K.-M., and Trojanowski, J.Q. (1995) Animal models of medulloblastomas and related primitive neuroectodermal tumors. A review. *J. Neuropathol. Exp. Neurol.* **54**, 285–296.

380. Pello, J.M., Guate, J.L., Naves, F.J., Escaf, S., and Vega, J.A. (1999) Neurotrophins and neurotrophin receptors in some neural crest-derived tumours (ganglioneuroma, phaeochromocytoma and paraganglioma). *Histopathology* **34**, 216–225.

381. Pomeroy, S.L., Sutton, M.E., Goumnerova, L.C., and Segal, R.A. (1997) Neurotrophins in cerebellar granule cell development and medulloblastoma. *J. Neuro-Oncol.* **35**, 347–352.

382. Rabizadeh, S., Oh, J., Zhong, L.-T., Yang, J., Bitler, C.M., Butcher, L.L., and Bredesen, D.E. (1993) Induction of apoptosis by the lower affinity NGF receptor. *Science* **261**, 345–348.

383. Barrett, G.L. (2000) The p75 neurotrophin receptor and neuronal apoptosis. *Prog. Neurobiol.* **61**, 205–229.

384. Carter, A.R., Berry, E.M., and Segal, R.A. (2003) Regional expression of p75NTR contributes to neurotrophin regulation of cerebellar patterning. *Mol. Cell Neurosci.* **22**, 1–13.

385. Bühren, J., Christoph, A.H.A., Buslei, R., Albrecht, S., Wiestler, O.D., and Pietsch, T. (2000) Expression of the neurotrophin receptor p75^{NTR} in medulloblastomas is correlated with distinct histopathological and clinical features: evidence for a medulloblastoma subtype derived from the external granule cell layer. *J. Neuropathol. Exp. Neurol.* **59**, 229–240.

386. Berkemeier, L.R., Winslow, J.W., Kaplan, D.R., Nikolics, K., Goeddel, D.V., and Rosenthal, A. (1991) Neurotrophin-5: a novel neurotrophic factor that actaivates trk and trk B. *Neuron* **7**, 857–866.

387. Götz, R., Köster, R., Winkler, C., Raulf, F., Lottspeich, F., Schartl M., and Thoenen, H. (1994) Neurotrophin-6 is a new member of the nerve growth factor family. *Nature* **372**, 266–269.

388. Minichiello, L., and Klein, R. (1996) TrkB and TrkC neurotrophin receptors cooperate in promoting survival of hippocampal and cerebellar granule neurons. *Genees & Dev.* **10**, 2849–2858.

389. Segal, R.A., Takahashi, H., and McKay, R.D.G. (1992) Changes in neurotrophin responsiveness during the development of cerebellar granule neurons. *Neuron* **9**, 1041–1052.

390. Washiyama, K., Muragaki, Y., Rorke, L.B., Lee, V.M.-Y., Feinstein, S.C., Radeke, M.J., Blumberg, D., Kaplan, D.R., and Trojanowski, J.Q. (1996) Neurotrophin and neurotrophin receptor proteins in medulloblastomas and other primitive neuroectodermal tumors of the pediatric central nervous system. *Am. J. Pathol.* **148**, 929–940.

391. Tajima, Y., Molina, R.P., Rorke, L.B., Kaplan, D.R., Radeke, M., Feinstein, S.C., Lee, V.M.-Y., and Trojanowski, J.Q. (1998) Neurotrophins and neuronal versus glial differentiation in medulloblastomas and other pediatric brain tumors. *Acta. Neuropathol.* **95**, 325–332.

392. Muragaki, Y., Chou, T.T., Kaplan, D.R., Trojanowski, J.Q., and Lee, V.M.-Y. (1997) Nerve growth factor induces apoptosis in human medulloblastoma cell lines that express TrkA receptors. *J. Neurosci.* **17**, 530–542.

393. Kokunai, T., Sawa, H., Tatsumi, S., and Tamaki, N. (1994) Expression of nerve growth factor receptor by human primitive neuroectodermal tumors. *Neurol. Med. Chir. (Tokyo)* **34**, 523–529.

394. Chiappa, S.A., Chin, L.S., Zurawel, R.H., and Raffel, C. (1999) Neurotrophins and Trk receptors in primitive neuroectodermal tumor cell lines. *Neurosurgery* **45**, 1148–1155.

395. Prior, R., Reifenberger, G., and Wechsler, W. (1989) Nerve growth factor receptor in tumours of the human nervous system. Immunohistochemical analysis of receptor expression and tumour growth function. *Pathol. Res. Pract.* **185**, 332–338.

396. Baker, D.L., Reddy, U.R., Pleasure, S., Hardy, M., Williams, M., Tartaglione, M., Biegel, J.A., Emanuel, B.S., Lo Presti, P., Kreider, B., Trojanowski, J.Q., Evans, A., Roy, A., Venkatakrishnan, G., Chen, J., Ross, A.H., and Pleasure, D. (1990) Human central nervous system primitive neuroectodermal tumor expressing nerve growth factor receptors: CHP707m. *Am. Neurol.* **28**, 136–145.

397. Azar, C.G., Scavarda, N.J., Reynolds, C.P., and Brodeur, G.M. (1990) Multiple defects of the nerve growth factor receptor in human neuroblastoma. *Cell Growth Diff.* **1**, 421–428.

397A. Pleasure, S.J., Reddy, U.R., Venkatakrishnan, G., Roy, A.K., Chen, J., Ross, A.H., Trojanowski, J.Q., Pleasure, D.E., and Lee, V.M.-Y. (1990) Introduction of nerve growth factor (NGF) receptors into a medulloblastoma cell line results in expression of high- and low-affinity NGF receptors but not NGF-mediated differentiation. *Proc. Natl. Acad. Sci. USA* **87**, 8496–8500.

398. Chou, T.T., Trojanowski, J.Q., and Lee, V.M.-Y. (1997) Neurotrophin signal transduction in medulloblastoma. *J. Neurosci. Res.* **49**, 522–527.

399. Hengartner, M.O. (2000) The biochemistry of apoptosis. *Nature* **407**, 770–776.

400. Kaiser, H.E., and Bodey, B. (2000) The role of apoptosis in normal ontogenesis and solid human neoplasms. *In Vivo* **14**, 789–803.

401. Haslam, R.H.A, Lamborn, K.R., Becker, L.E., and Israel, M.A. (1998) Tumor cell apoptosis present at diagnosis may predict treatment outcome for patients with medulloblastoma. *J. Pediatr. Hematol./Oncol.* **20**, 520–527.

402. Grotzer, M.A., Janss, A.J., Fung, K.-M., Sutton, L.N., Zhao, H., Trojanowski, J.Q., Rorke, L.B., and Phillips, P.C. (2001) Abundance of apoptotic neoplastic cells in diagnostic biopsy samples is not a prognostic factor in childhood primitive neuroectodermal tumors or the central nervous system. *J. Pediatr. Hematol./Oncol.* **23**, 25–29.

403. Walker, J.A., and Quirke, P. (2001) Viewing apoptosis through a 'TUNEL'. *J. Pathol.* **195**, 275–276.

404. Schiffer, D., Cavalla, P., Migheli, A., Chiò, A., Giordana, M.T., Marino, S., and Attanasio, A. (1995) Apoptosis and cell proliferation in human neuroepithelial tumors. *Neurosci. Lett.* **195**, 81–84.

405. Korshunov, A., Golanov, A., Ozerov, S., and Sycheva, R. (1999) Prognostic value of tumor-associated antigens immunoreactivity and apoptosis in medulloblastomas. An analysis of 73 cases. *Brain Tumor Pathol.* **16**, 37–44.

406. Dee, S., Haas-Kogan, D.A., and Israel, M.A. (1995) Inactivation of p53 is associated with decreased levels of radiation-induced apoptosis in medulloblastoma cell lines. *Cell Death Differ.* **2**, 267–275.

407. Schiffer, D., Cavalla, P., Migheli, A., Giordana, M.T., and Chiado-Piat, L. (1996) bcl-2 distribution in neuroepithelial tumors: an immunohistochemical study. *J. Neuro-Oncol.* **27**, 101–109.

408. Schiffer, D., Bortolotto, S., Bosone, I., Cancelli, I., Cavalla, P., Schiffer, P., and Piva, R. (1999) Cell-cycle inhibitor p27/Kip-1 expression in non-astrocytic and non-oligodendrocytic human nevous system tumors. *Neurosci. Lett.* **264**, 29–32.

409. Oltvai, Z.N., Milliman, C.L., and Korsmeyer, S.J. (1993) Bcl-2 heterodimerizes in vivo with a conserved homolog, Bax, that accelerates programmed cell death. *Cell* **74**, 609–619.

410. Bruggers, C.S., Fults, D., Perkins, S.L., Coffin, C.M., and Carroll, W.L. (1999) Coexpression of genes involved in apoptosis in central nervous system neoplasms. *J. Pediatr.Hematol./Oncol.* **21**, 19–25.

411. Reed, J.C., Meister, L., Tanaka, S., Cuddy, M., Yum, S., Geyer, C., and Pleasure, D. (1991) Differential expression of *bcl2* protooncogene in neuroblastoma and other human tumor cell lines of neural origin. *Cancer Res.* **51**, 6529–6538.

412. Hara, A., Hirose, Y., Yoshimi, N., Tanaka, T., and Mori, H. (1997) Expression of Bax and bcl-2 proteins, regulators of programmed cell death, in human brain tumors. *Neurol. Res.* **19**, 623–628.

413. Krajewski, S., Krajewska, M., Ehrmann, J., Sikorska, M., Lach, B., Chatten, J., and Reed, J.C.I. (1997) Immunohisto-chemical analysis of bcl-2, bcl-x, mcl-1, and bax in tumors of central and peripheral nervous system origin. *Am. J. Pathol.* **150**, 805–814.

414. Sarkar, C., Pramanik, P., Karak, A.K., Mukhopadhyay, P., Sharma, M.C., Singh, V.P., and Mehta, V.S. (2002) Are childhood and adult medulloblastomas different? A comparative study of clinicopathological features, proliferation index and apoptotic index. *J. Neurooncol.* **59**, 49–61.

415. Heck, K., Pyle, P., and Bruner, J.H. (1994) bcl-2 expression in cerebellar medulloblastomas. *Brain Pathol.***4**, 438.

416. Nakasu, S., Nakasu, Y., Nioka, H., Nakajima, M., and Handa, H. (1994) bcl-2 protein expression in tumors of the central nervous system. *Acta Neuropathol.* **88**, 520–526.

417. Ballantyne, E.S., McDowell, H., May, P.L., and Gosden, C. (1996) Childhood medulloblastoma: bcl-2 expression and survival. *J. Neurooncol.* **30**, 99.

418. Goumnerova, L.C. (1996) Growth factor receptors and medulloblastoma. *J. Neuro-Oncol.* **29**, 85–89.

418A. Kurihara, M., Ochi, A., Tokunaga, Y., Kawaguchi, T., Niwa, M., and Mori, K. (1989) Expression of insulin-like growth factor I (IGF-I) and epidermal growth factor (EGF) receptors in primary non-glial human brain tumors. *No To Shinkei.* **41**, 1127–1133.

419. Brem, H., and Klagsbrun, M. (1992) The role of fibroblast growth factors and related oncogenes in tumor growth. *Cancer Treat. Res.* **63**, 211–231.

420. Black, P., Carroll, R., and Glowacka, D. (1996) Expression of platelet-derived growth factor transcripts in medulloblastomas and ependymomas. *Pediatr. Neurosurg.* **24**, 74–78.

421. Del Valle, L., Enam, S., Lassak, A., Wang, J.Y., Croul, S., Khalili, K., and Reiss, K. (2002) Insulin-like growth factor I receptor activity in human medulloblastomas. *Clin. Cancer Res.* **8**, 1822–1830.

422. Patti, R., Reddy, C.D., Geoerger, B., Grotzer, M.A., Raghunath, M., Sutton, L.N., and Phillips, P.C. (2000) Autocrine secreted insulin-like growth factor-I stimulates MAP kinase-dependent mitogenic effects in human primitive neuroectodermal tumor/medulloblastoma. *Int. J. Oncol.* **16**, 577–584.

423. Wang, J.Y., Del Valle, L., Gordon, J., Rubini, M., Romano, G., Croul, S., Peruzzi, F., Khalili, K., and Reiss, K. (2001) Activation of the IGF-IR system contributes to malignant growth of human and mouse medulloblastomas. *Oncogene* **20**, 3857–3868

424. Rao, G., Pedone, C.A., Coffin, C.M., Holland, E.C., and Fults, D.W. (2003) c-Myc enhances sonic Hedgehog-induced medulloblastoma formation from nestin-expressing neural progenitors in mice. *Neoplasia* **5**, 198–204.

424A. Reiss, K. (2002) Insulin-like growth factor-I receptor – a potential therapeutic target in medulloblastomas. *Expert Opin. Ther. Targets* **6**, 539–544.

425. Gilbertson, R.J., Perry, R.H., Kelly, P.J., Pearson, A.D.J., and Lunec, J. (1997) Prognostic significance of HER2 and HER4 coexpression in childhood medulloblastoma. *Cancer Res.* **57**, 3272–3280.

426. Gilbertson, R.J., Clifford, S.C., MacMeekin, W., Wright, C., Perry, R.H., Kelly, P., Pearson, A.D.J., and Lunec, J. (1998) Expression of the ErbB-neuregulin signaling network during human cerebellar development: Implications for the biology of medulloblastoma. *Cancer Res.* **58**, 3932–3941.

427. Pinkas-Kramarski, R., Eilam, R., Alroy, I., Levkowitz, G., Lonai, P., and Yarden, Y. (1997) Differential expression of NDF/neuregulin receptors ErbB-3 and ErbB-4 and involvement in inhibition of neuronal differentiation. *Oncogene* **15**, 2803–2815.

428. Collins, V.P., Ichimura, K., and Liu, L. (2002) The expression of the EGF receptor family and their ligands in human medulloblastomas. *Neuropathol. Appl. Neurobiol.* **28**, 156.

429. Gilbertson, R.J., Pearson, A.D., Perry, R.H., Jaros, E., and Kelly, P.J (1995) Prognostic significance of the c-erbB-2 oncogene product in childhood medulloblastoma. *Br. J. Cancer* **71**, 473–477.

429A. Herms, J.W., Behnke, J., Bergmann, M., Christen, H.-J., Kolb, R., Wilkening, M., Markakis, E., Hanefeld, F., and Kretzschmar, H.A. (1997) Potential prognostic value of C-erbB-2 expression in medulloblastomas in very young children. *J. Pediatr. Hematol./Oncol.* **19**, 510–515.

430. Huber, H., Eggert, A., Janss, A.J., Wierwrodt, R., Zhao, H., Sutton, L.N., Rorke, L.B., Phillips, P.C., and Grotzer, M.A. (2001) Angiogenic profile of childhood primitive neuroectodermal brain tumours/medulloblastoma. *Eur. J. Cancer* **37**, 2064–2072.

431. Di, C., Liao, S., Adamson, D.C., Parrett, T.J., Broderick, D.K., Shi, Q., Lengauer, C., Cummins, J.M., Velculescu, V.E., Fults, D.W., McLendon, R.E., Bigner, D.D., and Yan, H. (2005) Identification of *OXT2* as a medulloblastoma

oncogene whose product can be targeted by all-*trans* retinoic acid. *Cancer Res.* **65,** 919–924.

432. de Haas T., Oussoren, E., Grajkowska, W., Perek-Polnik, M., Popovic, M., Zadravec-Zaletel, L., Perera, M., Corte, G., Wirths, O., van Sluis, P., Pietsch, T., Troost, D., Baas, F., Versteeg, R., and Kool, M. (2006) OTX1 and OTX3 expression correlates with the clinicopathologic classification of medulloblastomas. *J. Neuropathol. Exp. Neurol.* **65,** 176–186.

433. Frantz, G.D., Weimann, J.M., Levin, M.E., and McConnell, S.K. (1994) Otx1 and Otx2 define layers and regions in developing cerebral cortex and cerebellum. *J. Neurosci.* **14,** 5725–5740.

433A. Michiels, E.M.C., Oussoren, E., van Groenigen, M., Pauws, E., Bossuyt, P.M.M., Voûte, P.A., and Baas, F. (1999) Genes differentially expressed in medulloblastoma and fetal brain. *Physiol. Genomics* **1,** 83–91.

434. Yokota, N., Mainprize, T.G., Taylor, M.D., Kohata, T., Loreto, M., Ueda, S., Dura, W., Grajkowska, W., Kuo, J.S., and Rutka, J.T. (2004) Identification of differentially expressed and developmentally regulated genes in medulloblastoma using suppression subtraction hybridization. *Oncogene* **23,** 3444–3453.

434A. Yokota, N., Aruga, J., Takai, S., Yamada, K., Hamazaki, M., Iwase, T., Sugimura, H., and Mikoshiba, K. (1996) Predominant expression of human Zic in cerebellar granule cell lineage and medulloblastoma. *Cancer Res.* **56,** 377–383.

435. Kozmik, Z., Sure, U., Rüedi, D., Busslinger, M., and Aguzzi, A. (1995) Deregulated expression of *PAX5* in medulloblastoma. *Proc. Natl. Acad. Sci. USA* **92,** 5709–5713.

436. Sandberg, A.A., and Bridge, J.A. (1996) The Cytogenetics of Bone and Soft Tissue Tumors. R.G. Landes Co., Austin, 1994.

437. Davis, C.A., and Joyner, A.L. (1988) Expression patterns of the homeo box-containing genes En-1 and En-2 and the proto-oncogene int-1 diverge during mouse development. *Genes Dev.* **2,** 1736–1744.

438. Vincent, S., Turque, N., Plaza, S., Dhellemmes, P., Hladky, J.P., Assaker, R., Ruchoux, M. M., and Saule, S. (1996) Differential expression between PAX-6 and EN proteins in medulloblastoma. *Int. J. Oncol.* **8,** 901–910.

439. Wang, W., Kumar, P., Wang, W., Whalley, J., Schwarz, M., Malone, G., Haworth, A., and Kumar, S. (1998) The mutation status of PAX3 and p53 genes in medulloblastoma. *Anticancer Res.* **18,** 849–853.

440. Patel, Y.C. (1999) Somatostatin and its receptor family. *Frontiers Neuroendo.* **20,** 157–198.

441. Pollak, M.N., and Schally, A.V. (1998) Mechanisms of antineoplastic action of somatostatin analogs. *Proc. Soc. Exp. Biol. Med.* **217,** 143–152.

442. Reisine, T., and Bell, G.I. (1995) Molecular biology of somatostatin receptors. *Endocr. Rev.* **16,** 427–442.

443. Thoss, V.S., Perez, J., Probat, A., and Hoyer, D. (1996) Expression of five somatostatin receptor mRNAs in the human brain and pituitary. *Arch. Pharmacol.* **354,** 411–419.

444. Schindler, M., Holloway, S., Humphrey, P.P.A., Waldvogel, H., Faull, R.L.M., Berger, W., and Emson, P.C. (1998) Localization of the somatostatin SST2A receptor in human cerebral cortex, hippocampus and cerebellum. *Neuro. Rep.* **9,** 521–525.

445. Müller, H.L., Frühwald, M.C. Scheubeck, M., Rendl, J., Warmuth-Metz, M., Sörensen, N., Kühl, J., and Reubi, J.C. (1998) A possible role for somatostatin receptor scintigraphy in the diagnosis and follow-up of children with medulloblastoma. *J. Neurooncol.* **38,** 27–40.

446. Frühwald, M.C., O'Dorisio, M.S., Pietsch, T., and Reubi, J.C. (1999) High expression of somatostatin receptor subtype 2 (sst) in medulloblastoma: Implications for diagnosis and therapy. *Pediatr. Res.* **45,** 697–708.

447. Guyotat, J., Champier, J., Saint Pierre, G., Jouvet, A., Bret, P., Brisson, C., Belin, M.F., Signorelli, F., and Montange, M.F. (2001) Differential expression of somatostatin receptors in medulloblastoma. *J. Neurooncol.* **51,** 93–103.

448. Fan, X., Mikolaenko, I., Elhassan, I., Ni, XZ., Wang, Y., Ball, D., Brat, D.J., Perry, A., and Eberhart, C.G. (2004) *Notch1* and *Notch2* have opposite effects on embryonal brain tumor growth. *Cancer Res.* **64,** 7787–7793.

449. Allenspach, E.J., Maillard, I., Aster, J.C., and Pear, W.S. (2002) Notch signaling in cancer. *Cancer Biol. Ther.* **1,** 466–476.

450. Radtke, F., and Raj, K. (2003) The role of Notch in tumorigenesis: oncogene or tumour suppressor? *Nat. Rev. Cancer* **3,** 756–767.

451. Iso, T., Kedes, L., and Hamamori, Y. (2003) HES and HERP families: multiple effectors of the Notch signaling pathway. *J. Cell Physiol.* **194,** 237–255.

452. Nickoloff, B.J., Osborne, B.A., and Miele, L. (2003) Notch signaling as a therapeutic target in cancer: a new approach to the development of cell fate modifying agents. *Oncogene* **22,** 6598–6608.

453. Mungamuri, S.K., Yang, X., Thor, A.D., and Somasundaram, K. (2006) Survival signaling by Notch1: Mammalian target of rapamycin (mTOR)-dependent inhibition of p53. *Cancer Res.* **66,** 4715–4724.

454. Gray, G.E., Mann, R.S., Mitsiadis, E., Henrique, D., Carcangiu, M.-L., Banks, A., Leiman, J., Ward, D., Ish-Horowitz, D., and Artavanis-Tsakonas, S. (1999) Human ligands of the Notch receptor. *Am. J. Pathol.* **154,** 785–794.

455. Jones, P.A., and Laird, P.W. (1999) Cancer epigenetics comes of age. *Nat. Genet.* **21,** 163–167.

456. Baylin, S.B., Esteller, M., Rountree, M.R., Bachman, K.E., Scheubel, K., and Herman, J.G. (2001) Aberrant patterns of DNA methylation, chromatin formation and gene expression in cancer. *Hum. Mol. Genet.* **10,** 687–692.

457. Esteller, M., Corn, P.G., Baylin, S.B., and Herman, J.G. (2001) A gene hypermethylation profile of human cancer. *Cancer Res.* **61,** 3225–3229.

458. Ehrlich, M. (2002) DNA methylation in cancer: too much, but also too little. *Oncogene* **21,** 5400–5413.

459. Jones, P.A., and Baylin, S.B. (2002) The fundamental role of epigenetic events in cancer. *Nat. Rev.* **3,** 415–428.

460. Egger, G., Liang, G., Aparicio, A., and Jones, P.A. (2004) Epigenetics in human disease and prospects for epigenetic therapy. *Nature* **429,** 457–463.

461. Jones, P.A. (1999) The DNA methylation paradox. *Trends Genet.* **15,** 34–37.

462. Lindsey, J.C., Anderton, J.A., Lusher, M.E., and Clifford, S.C. (2005) Epigenetic events in medulloblastoma development. *Neurosurg. Focus* **19,** E10.

462A. Mühlisch, J., Schwering, A., Grotzer, M., Vince, G.H., Roggendorf, W., Hagemann, C., Sörensen, N., Rickert, C.H., Osada, N., Jürgens, H., and Frühwald, M.C. (2006) Epigenetic repression of *RASSF1A* but not *CASP8* in supratentorial PNET (sPNET) and atypical teratoid/rhaboid tumors (AT/RT) of childhood. *Oncogene* **25**, 1111–1117.

463. Stirzaker, C., Song, J.Z., Davidson, B., and Clark, S.J. (2004) Transcriptional gene silencing promotes DNA hypermethylation through a sequential change in chromatin modifications in cancer cells. *Cancer Res.* **64**, 3871–3877.

464. Ushijima, T., and Okochi-Takada, E. (2005) Aberrant methylations in cancer cells: where do they come from? *Cancer Sci.* **96**, 206–211.

465. Frühwald, M.C., O'Dorisio, M.S., Dai, Z., Tanner, S.M., Balster, D.A., Gao, X., Wright, F.A., and Plass, C. (2001) Aberrant promoter rmethylation of previously unidentified target genes is a common abnormality in medulloblastomas— implications for tumor biology and potential clinical utility. *Oncogene* **20**, 5033–5042.

466. Harada, K., Toyooka, S., Shivapurkar, N., Maitra, A., Reddy, J.L., Matta, H., Miyajima, K., Timmons, C.F., Tomlinson, G.E., Mastrangelo, D., Hay, R.J., Chaudhary, P.M., and Gazdar, A.F. (2002) Deregulation of *Caspase 8* and *10* expression in pediatric tumors and cell lines. *Cancer Res.* **62**, 5897–5901.

467. Harada, K., Toyooka, S., Maitra, A., Maruyama, R., Toyooka, K.O., Timmons, C.F., Tomlinson, G.E., Mastrangelo, D., Hay, R.J., Minna, J.D., and Gazdar, A.F. (2002) Aberrant promoter methylation and silencing of the *RASSF1A* gene in pediatric tumors and cell lines. *Oncogene* **21**, 4345–4349.

468. Lusher, M.E., Lindsey, J.C., Latif, F., Pearson, A.D.J., Ellison, D.W., and Clifford, S.C. (2002) Biallelic epigenetic inactivation of the *RASSF1A* tumor suppressor gene in medulloblastoma development. *Cancer Res.* **62**, 5906–5911.

469. Rood, B.R., Zhang, H., Weitman, D.M., and Cogen, P.H. (2002) Hypermethylation of *HIC-1* and 17p allelic loss in medulloblastoma. *Cancer Res.* **62**, 3794–3797.

470. Rood, B.R., Zhang, H., and Cogen, P.H. (2004) Intercellular heterogeneity of expression of the *MGMT* DNA repair gene in pediatric medulloblastoma. *J. Neurooncol.* **6**, 200–207.

471. Zuzak, T.J., Steinhoff, D.F., Sutton, L.N., Phillips, P.C., Eggert, A., and Grotzer, M.A. (2002) Loss of caspase-8 mRNA expression is common in childhood primitive neuroectodermal brain tumour/ medulloblastoma. *Eur. J. Cancer* **38**, 83–91.

472. Gonzalez-Gomez, P., Bello, M.J., Alonso, M.E., Arjona, D., Lomas, J., de Campos, J.M., Isla, A., and Rey, J.A. (2003) CpG island methylation status and mutation analysis of the *RB1* gene essential promoter region and protein-binding pocket domain in nervous system tumors. *Br. J. Cancer* **88**, 109–114.

473. Gonzalez-Gomez, P., Bello, M.J., Inda, M.M., Alonso, M.E., Arjona, D., Aminoso, C., Lopez-Marin, I., de Campos, J.M., Sarasa, J.L., Castresana, J.S., and Rey, J.A. (2003) Deletion and aberrant CpG island methylation of caspase 8 gene in medulloblastoma. *Oncol. Rep.* **12**, 663–666.

474. Horiguchi, K., Tomizawa, Y., Tosaka, M., Ishiuchi, S., Kurihara, H., Mori, M., and Saito, N. (2003) Epigenetic inactivation of *RASSF1A* candidate tumor suppressor gene at 3p21.3 in brain tumors. *Oncogene* **22**, 7862–7865.

475. Rathi, A., Virmani, A.K., Harada, K., Timmons, C.F., Miyajima, K., Hay, R.J., Mastrangelo, D., Maitra, A., Tomlinson,

G.E., and Gazdar, A.F. (2003) Aberrant methylation of the *HIC1* promoter is a frequent event in specific pediatric neoplasms. *Clin. Cancer Res.* **9**, 3674–3678.

476. Waha, A., Waha, A., Koch, A., Meyer-Puttlitz, B., Weggen, S., Sörensen, N., Tonn, J.C., Albrecht, S., Goodyer, C.G., Berthold, F., Wiestler, O.D., and Pietsch, T. (2003) Epigenetic silencing of the *HIC-1* gene in human medulloblastomas. *J. Neuropathol. Exp. Neurol.* **62**, 1192–1201.

477. Ebinger, M., Senf, L., Wachowski, O., and Scheurlen, W. (2004) Promoter methylation pattern of caspase-8, p16^{INK4A}, MGMT, TIMP-3, and E-cadherin in medulloblastoma. *Pathol. Oncol. Res.* **10**, 17–21.

478. Lindsey, J.C., Lusher, M.E., Anderton, J.A., Bailey, S., Gilbertson, R.J., Pearson, A.D.J., Ellison, D.W., and Clifford, S.C. (2004) Identification of tumour-specific epigenetic events in medulloblastoma development by hypermethylation profiling. *Carcinogenesis* **25**, 661–668.

479. Pang, J.C.-S., Chang, Q., Chung, Y.F., Teo, J.G.C., Poon, W.S., Zhou, L.-F., Kong, X., and Ng, H.-K. (2005) Epigenetic inactivation of DLC-1 in supratentorial primitive neuroectodermal tumor. *Hum. Pathol.* **36**, 36–43.

479A. De Vos, M., Hayward, B.E., Picton, S., Sheridan, E., and Bonthron, D.T. (2004) Novel *PMS2* pseudogenes can conceal recessive mutations causing a distinctive childhood cancer syndrome. *Am. J. Hum. Genet.* **74**, 954–964.

480. Chen, D., Livne-bar, I., Vanderluit, J.L., Slack, R.S., Agochiya, M., and Bremner, R. (2004) Cell-specific effects of RB or RB/p107 loss on retinal development implicate an intrinsically death-resistant cell-of-origin in retinoblastoma. *Cancer Cell* **5**, 539–551.

480A. Wales, M.M., Biel, M.A., el Deiry, W., Nelkin, B.D., Issa, J.P., Cavenee, W.K., Kuerbitz, S.J., and Baylin, S.B. (1995) p53 activates expression of HIC-1, a new candidate tumour suppressor gene on 17p13.3. *Nat. Med.* **1**, 570–577.

481. Guerardel, C., Deltour, S., Pinte, S., Monte, D., Begue, A., Godwin, A.K., and Leprince, D. (2001) Identification in the human candidate tumor suppressor gene *HIC-1* of a new major alterantive TATA-less promoter positively regulated by p53. *J. Biol. Chem.* **276**, 3078–3089.

481A. Chen, M., and Wang, J. (2002) Initiation capsases in apoptosis signaling pathways. *Apoptosis* **7**, 313–319.

481B. Salvesen, G.S. (2002) Caspases and apoptosis. *Essays Biochem.* **38**, 9–19.

482. Puig, B., Tortosa, A., and Ferrer, I. (2001) Cleaved caspase-3, caspase-7 and poly (ADP-ribose) polymerase are complementarily but differentially expressed in human medulloblastomas. *Neurosci. Lett.* **306**, 85–88.

483. Pingoud-Meier, C., Lang, D., Janss, A.J., Rorke, L.B., Phillips, P.C., Shalaby, T., and Grotzer, M.A. (2003) Loss of caspase-8 protein expression correlates with unfavorable survival outcome in childhood medulloblastoma. *Clin. Cancer Res.* **9**, 6401–6409.

484. Dammann, R., Li, C., Yoon, J.-H., Chin, P.L., Bates, S., and Pfeifer, G.P. (2000) Epigenetic inactivation of a RAS association domain family protein from the lung tumour suppressor locus 3p21.3. *Nat. Genet.* **25**, 315–319.

485. Dammann, R., Schagdarsurengin, U., Seidel, C., Strunnikova, M., Rastetter, M., Baier, K., and Pfeifer, G.P. (2005) The tumor suppressor *RASSF1A* in human carcinogenesis: an update. *Histol. Histopathol.* **20**, 645–663.

486. Tommasi, S., Dammann, R., Zhang, Z., Wang, Y., Liu, L., Tsark, W.M., Wilczynski, S.P., Li, J., You, M., and Pfeifer, G.P. (2005) Tumor susceptibility of *Rassf1a* knockout mice. *Cancer Res.* **65,** 92–98.

487. Hiyama, E., and Hiyama, K. (2002) Clinical utility of telomerase in cancer. *Oncogene* **21,** 643–649.

488. Cerezo, A., Kalthoff, H., Schuermann, M., Schäfer, B., and Boukamp, P. (2002) Dual regulation of telomerase activity through c-Myc-dependent inhibition and alternative splicing of hTERT. *J. Cell Sci.* **115,** 1305–1312.

489. Zhang, A., Zheng, C., Lindvall, C., Hou, M., Ekedahl, J., Lewensohn, R., Yan, Z., Yang, X., Henriksson, M., Blennow, E., Nordenskjöld, M., Zetterberg, A., Björkholm, M., Gruber, A., and Xu, D. (2000) Frequent amplification of thet *Telomerase Reverse Transcriptase* gene in human tumors. *Cancer Res.* **60,** 6230–6235.

490. Desmaze, C., Soria, J.C., Freulet-Marriere, M.A., Mathieu, N., and Sabatier, L. (2003) Telomere-driven genomic instability in cancer cells. *Cancer Lett.* **194,** 173–182.

491. Fan, X., Wang, Y., Kratz, J., Brat, D.J., Robitaille, Y., Moghrabi, A., Perlman, E.J., Dang, C.V., Burger, P.C., and Eberhart, C.G. (2003) *hTERT* gene amplification and increased mRNA expression in central nervous system embryonal tumors. *Am. J. Pathol.* **162,** 1763–1769.

492. Didiano, D., Shalaby, T., Lang, D., and Grotzer, M.A. (2004) Telomere maintenance in childhood primitive neuroectodermal brain tumors. *J. Neurooncol.* **6,** 1–8.

493. Chung, H.K., Cheong, C., Song, J., and Lee, H.W. (2005) Extratelomeric functions of telomerase. *Curr. Mol. Med.* **5,** 233–241.

494. Hackett, J.A., and Greider, C.W. (2002) Balancing instability: dual roles for telomerase and telomere dysfunction in tumorigenesis. *Oncogene* **21,** 619–626.

495. Hardy, C.B. (2002) Telomerase is not an oncogene. *Oncogene* **21,** 494–502.

496. Mu, J., and Wei, L.X. (2002) Telomere and telomerase in oncology. *Cell Res.* **12,** 1–7.

497. Neumann, A.A., and Reddel, R.R. (2002) Telomere maintenance and cancer – Look, no telomerase. *Nat. Rev. Cancer* **2,** 879–884.

498. Shay, J.W., and Wright, W.E. (2002) Telomerase: a target for cancer therapeutics. *Cancer Cell.* **2,** 257–265.

499. Rostomily, R.C., Bermingham-McDonogh, O., Berger, M.S., Tapscott, S.J., Reh, T.A., and Olson, J.M. (1997) Expression of neurogenic basic helix-loop-helix genes in primitive neuroectodermal tumors. *Cancer Res.* **57,** 3526–3531.

500. Lee, J.E., Hollenberg, S.M., Snider, L., Turner, D.L., Lipnick, N., and Weintraub, H. (1995) NeuroD, a new basic helix-loop-helix protein can convert *Xenopus* ectoderm into neurons. *Science* **268,** 836–844.

501. Ma, Q., Kimmer, C., and Anderson, D.J. (1996) Identification of *neurogenin*, a verbebrate neuronal determination gene. *Cell* **87,** 43–52.

502. McCormick, M.B., Tamini, R.M., Snider, L., Asakura, A., Bergstrom, D., and Tapscott, S.J. (1996) *NeuroD2* and *neuroD3*: distinct expression patterns and transcriptional activation potentials within the *neuroD* gene family. *Mol. Cell Biol.l* **16,** 5792–5800.

503. Lammie, G.A., Beckett, A., Courtney, R., and Scaravilli, F. (1994) An immunohistochemical study of p53 and prolifer-

ating cell nuclear antigen expression in progressive multifocal leucoencephalopathy. *Acta Neuropathol.* **88,** 465–471.

504. Zu Rhein, G.M., and Varakis, J.N. (1979) Perinatal induction of medulloblastomas in Syrian golden hamsters by a human polyoma virus (JC). *Natl. Cancer Inst. Monogr.* **51,** 205–208.

505. Matsuda, M., Yasui, K., Nagashima, K., and Mori, W. (1987) Origin of the medulloblastoma experimentally induced by human polyomavirus JC. *J. Natl. Cancer Inst.* **79,** 585–591.

506. Nagashima, K., Yasui, K., Kimura, J., Washizu, M., Yamaguchi, K., and Mori, W. (1984) Induction of brain tumors by a newly isolated JC virus (Tokyo-1 strain). *Am. J. Pathol.* **116,** 455–463.

507. Khalili, K., Krynska, B., Del Valle, L., Katsetos, C.D., and Croul, S. (1999) Medulloblastomas and the human neurotrophic polyomavirus JC virus. *Lancet* **353,** 1152–1153.

508. Khalili, K., Del Valle, L., Otte, J., Weaver, M., and Gordon, J. (2003) Human neurotrophic polyomavirus, JCV, and its role in carcinogenesis. *Oncogene* **22,** 5181–5191.

509. Krynska, B., Otte, J., Franks, R., Khalili, K., and Croul, S. (1999) Human ubiquitous JCV(CY) T-antigen gene induces brain tumors in experimental animals. *Oncogene* **18,** 39–46.

510. Krynska, B., Del Valle, L., Croul, S., Gordon, J., Katsetos, C.D., Carbone, M., Giordano, A., and Khalili, K. (1999) Detection of human neurotrophic JC virus DNA sequence and expression of the viral oncogenic protein in pediatric medulloblastomas. *Proc. Natl. Acad. Sci. USA* **96,** 11519–11524.

511. Caldarelli-Stefano, R., Boldorini, R., Monga, G., Meraviglia, E., Zorini, E.O., and Ferrante, P. (2000) JC virus in human glial-derived tumors. *Hum. Pathol.* **31,** 394–395.

512. Del Valle, L., Gordon, J., Assimako+poulou, M., Enam, S., Geddes, J.F., Varakis, J.N., Katsetos, C.D., Croul, S., and Khalili, K. (2001) Detection of JC virus DNA sequences and expression of the viral regulatory protein T-antigen in tumors of the central nervous system. *Cancer Res.* **61,** 4287–4293.

513. Del Valle, L., Enam, S., Lara, C., Ortiz-Hidalgo, C., Katsetos, C.D., and Khalili, K. (2002) Detection of *JC polyomavirus* DNA sequences and cellular localization of T-antigen and agnoprotein in oligodendrogliomas. *Clin. Cancer Res.* **8,** 3332–3340.

514. Huang, H., Reis, R., Yonekawa, Y., Lopes, J.M., Kleihues, P., and Ohgaki, H. (1999) Identification in human brain tumors of DNA sequences specific for SV40 large T antigen. *Brain Pathol.* **9,** 33–44.

515. Arthur, R.R., Grossman, S.A., Ronnett, B.M., Bigner, S.H., Vogelstein, B., and Shah, K.V. (1994) Lack of association of human polyomaviruses with human brain tumors. *J. Neurooncol.* **20,** 55–58.

516. Weggen, S., Bayer, T.A., von Deimling, A., Reifenberger, G., von Schweinitz, D., Wiestler, O.D., and Pietsch, T. (2000) Low frequency of SV40, JC and BK polyomavirus sequences in human medulloblastomas, meningiomas and ependymomas. *Brain Pathol.* **10,** 85–92.

517. Hayashi, H., Endo, S., Suzuki, S., Tanaka, S., Sawa, H., Ozaki, Y., Sawamura, Y., and Nagashima, K. (2001) JC virus large T protein transforms rodent cells but is not involved in human medulloblastoma. *Neuropathology* **21,** 129–137.

518. Kim, J.Y.H., Koralnik, I.J., LeFave, M., Segal, R.A., Pfister, L.-A., and Pomeroy, S.L. (2002) Medulloblastomas and primitive neuroectodermal tumors rarely contain polyomavirus DNA sequences. *J. Neurooncol.* **4,** 165–170.

519. Rollison, D.E., Utaipat, U., Ryschkewitsch, C., Hou, J., Goldthwaite, P., Daniel, R., Helzlsouer, K.J., Burger, P.C., Shah, K.V., and Major, E.O. (2005) Investigation of human brain tumors for the presence of polyomavirus genome sequences by two independent laboratories. *Int. J. Cancer* **113**, 769–774.

520. Muñoz-Mármol, A.M., Mola, G., Ruiz-Larroya, T., Fernández-Vasalo, A., Vela, E., Mate, J.L., and Ariza, A. (2006) Rarity of JC virus DNA sequences and early proteins in human gliomas and medulloblastomas: the controversial role of JC virus in human neurooncogenesis. *Neuropathol. Appl. Neurobiol.* **32**, 131–140.

521. Del Valle, L., Gordon, J., Enam, S., Delbue, S., Croul, S., Abraham, S., Radhakrishnan, S., Assimakopoulou, M., Katsetos, C.D., and Khalili, K. (2002) Expression of human neurotrophic polyomavirus JCV late gene product agnoprotein in human medulloblastoma. *J. Natl. Cancer Inst.* **94**, 267–273.

522. Pietsch, T., Taylor, M.D., and Rutka, J.T. (2004) Molecular pathogenesis of childhood brain tumors. *J. Neurooncol.* **70**, 203–215.

523. de Stanchina, E., McCurrach, M.E., Zindy, F., Shieh, S.-Y., Ferbeyre, G., Samuelson, A.V., Prives, C., Roussel, M.F., Sherr, C.J., and Lowe, S.W. (1998) E1A signaling to p53 involve sthe p19^{ARF} tumor suppressor. *Genes Dev.* **12**, 2434–2442.

523A. Deshpande, A., Sicinski, P., and Hinds, P.W. (2005) Cyclins and cdks in development and cancer: a perspective. *Oncogene* **24**, 2909–2915.

523B. Honda, R., and Yasuda, H. (1999) Association of p19^{ARF} with Mdm2 inhibits ubiquitin ligase activity of Mdm2 for tumor suppressor p53. *EMBO J.* **18**, 22–27.

523C. Ortega, S., Malumbres, M., and Barbacid, M. (2002) Cyclin D-dependent kinases, INK4 inhibitors and cancer. *Biochim. Biophys. Acta* **1602**, 73–87.

524. Eberhart, C.G., Chaudhry, A., Daniel, R.W., Khaki, L., Shah, K.V., and Gravitt, P.E. (2005) Increased p53 immunoreactivity in anaplastic medulloblastoma and supratentorial PNET is not caused by JC virus. *BMC Cancer* **5**, 1–8.

525. Trojanek, J., Ho, T., Croul, S., Wang, J.Y., Chintapalli, J., Koptyra, M., Giordano, A., Khalili, K., and Reiss, K. (2006) IRS-1-Rad51 nuclear interaction sensitizes JCV T-antigen positive medulloblastoma cells to genotoxic treatment. *Int. J. Cancer* **119**, 539–548.

526. Adesina, A.M. (1999) Molecular heterogeneity in medulloblastoma with implications for differing tumor biology. *J. Child. Neurol.* **14**, 411–417.

527. Raffel, C. (2004) Medulloblastoma: molecular genetics and animal models. *Neoplasia* **6**, 310–322.

528. Stewart, C.L., Soria, A.M., and Hamel, P.A. (2001) Integration of the pRB and p53 cell cycle control pathways. *J. Neurooncol.* **51**, 183–204.

529. Newton, H.B. (2001) Review of the molecular genetics and chemotherapeutic treatment of adult and paediatric medulloblastomas. *Expert Opin. Investig. Drugs* **10**, 2089–2104.

530. Schneider, N.R. (1993) Cytogenetic evaluation of childhood neoplasms. *Arch. Pathol. Lab. Med.* **117**, 1220–1224.

530A. Falls, D.L. (2003) Neuregulins: functions, forms, and signaling strategies. *Exp. Cell Res.* **284**, 14–30.

530B. Perry, R., Gonzales, I., Finlay, J., and Zacharoulis, S. (2007) Primary peripheral primitive neuroectodermal tumors of the spinal cord: report of two cases and review of the literature. *J. Neurooncol.* **81**, 259–264.

531. Rorke, L.B. (1983) The cerebellar medulloblastoma and its relationship to primitive neuroectodermal tumors. *J. Neuropathol. Exp. Neurol.* **42**, 1–15.

532. Rorke, L.B., Trojanowski, J.Q., Lee, V.M.-Y., Zimmerman, R.A., Sutton, L.N., Biegel, J.A., Goldwein, J.W., and Packer, R.J. (1997) Primitive neuroectodermal tumors of the central nervous system. *Brain Pathol.* **7**, 765–784.

533. Jakacki, R.I. (1999) Pineal and nonpineal supratentorial primitive neuroectodermal tumors. *Childs Nerv. Syst.* **15**, 586–591.

534. Dai, A.I., Backstrom, J.W., Burger, P.C., and Duffner, P.K. (2003) Supratentorial primitive neuroectodermal tumors of infancy: Clinical and radiologic findings. *Pediatr. Neurol.* **29**, 430–434.

535. Rorke, L.B., Hart, M.N., and McLendon, R.E. (2000) Supratentorial primitive neuroectodermal tumour (PNET). In *World Health Organization Classification of Tumours. Pathology and Genetics of Tumours of the Nervous* System (Kleihues, P., and Cavenee, W.K., eds.), IARC Press, Lyon, France, pp. 141–144.

536. Girschick, H.J., Klein, R., Scheurlen, W.G., and Kühl, J. (2001) Cytogenetic and histopathologic studies of congenital supratentorial primitive neuroectodermal tumors: a case report. *Pathol. Oncol. Res.* **7**, 67–71.

537. McLendon, R.E., and Burger, P.C. (1987) The primitive neuroectodermal tumor: A cautionary view. *J. Pediatric Neurosci.* **3**, 1–8.

538. Burger, P.C. (2006) Supratentorial primitive neuroectodermal tumor (sPNET). *Brain Pathol.* **16**, 86.

539. Moss, S.D., Haines, S.J., Leonard, A.S., and Dehner, L.P. (1986) Congenital supratentorial and infratentorial peripheral neurogenic tumor: A clinical, ultrastructural, and immunohistochemical study. *Neurosurgery* **19**, 426–433.

540. Reddy, A.T., Janss, A.J., Phillips, P.C., Weiss, H.L., and Packer, R.J. (2000) Outcome for children with supratentorial primitive neuroectodermal tumors treated with surgery, radiation and chemotherapy. *Cancer* **88**, 2189–2193.

541. Jakacki, R.I. (2005) Treatment strategies for high-risk medulloblastoma and supratentorial primitive neuroectodermal tumors. *J. Neurosurg. (Pediatrics 1)* **102**, 44–52.

542. Becker, L.E., and Hinton, D. (1983) Primitive neuroectodermal tumors of the central nervous system. *Hum. Pathol.* **14**, 538–550.

543. Berger, M.S., Edwards, M.S.B., Wara, W.M., Levin, V.A., and Wilson, C.B. (1983) Primary cerebral neuroblastoma. Long-term follow-up review and therapeutic guidelines. *J. Neurosurg.* **59**, 418–423.

544. Kleihues, P., and Zuulch, K.J. (1986) *Brain Tumours: Their Biology and Pathology*, ed. 3. Springer-Verlag, Berlin, Germany.

545. Rorke, B. (1997) Atypical teratoid/rhabdoid tumours: In *Pathology and Genetics – Tumours of the Nervous System* (Kleihues, P., and Cavenee, W.K., eds.), IARC Press, Lyon, France, pp. 110–111.

546. Hongeng, S., Brent, T.P., Sanford, R.A., Li, H., Kun, L.E., and Heideman, R.L. (1997) O^6-methylguanine-DNA methyl-

transferase protein levels in pediatric brain tumors. *Clin. Cancer Res.* **3**, 2459–2463.

547. Dirks, P.B., Harris, L., Hoffman, H.J., Humphreys, R.P., Drake, J.M., and Rutka, J.T. (1996) supratentorial primitive neuroectodermal tumors in children. *J. Neurooncol.* **29**, 75–84.

548. Begemann, M., Lyden, D., Rosenblum, M.K., Lis, E., Wolden, S., Antunes, N.L., and Dunkel, I.J. (2003) Primary leptomeningeal primitive neuroectodermal tumor. *J. Neurooncol.* **63**, 299–03.

549. Cooper, I.S., and Kernohan, J.W. (1951) Heterotopic glial nests in the subarachnoid space: histopathologic characteristics, mode of origin and relation to meningeal gliomas. *J. Neuropathol. Exp. Neurol.* **10**, 16–29.

550. Chadduck, W.M., Boop, F.A., and Sawyer, J.R. (1991) Cytogenetic studies of pediatric brain and spinal cord tumors. *Pediatr. Neurosurg.* **92**, 57–65.

551. Cohen, N., Betts, D.R., Tavori, U., Toren, A., Tam, T., Constantini, S., Grotzer, M.A., Amariglio, N., Rechavi, G., and Trakhtenbrot, L. (2004) Karyotypic evolution pathways in medulloblastoma/primitive neuroectodermal tumor determined with a combination of spectral karyotyping, G-banding, and fluorescence in situ hybridization. *Cancer Genet. Cytogenet.* **149**, 44–52.

552. Batanian, J.R., Havlioglu, N., Huang, Y., and Gadre, B. (2003) Unusual aberration involving the short arm of chromosome 11 in an 8-month-old patient with a supratentorial primitive neuroectodermal tumor. *Cancer Genet. Cytogenet.* **141**, 143–147.

553. Benner, S.E., Wahl, G.M., and Von Hoff, D.D. (1991) Double minute chromosomes and homogeneously staining regions in tumors taken directly from patients versus in human tumor cell lines. *Anti-Cancer Drugs* **2**, 11–25.

554. Agamanolis, D.P., and Malone, J.M. (1995) Chromosomal abnormalities in 47 pediatric brain tumors. *Cancer Genet. Cytogenet.* **81**, 125–134.

555. Ishii, N., Hiraga, H., Sawamura, Y., Shinohe, Y., and Nagashima, K. (2001) Alternative EWS-FLI1 fusion gene and MIC2 expression in peripheral and central primitive neuroectodermal tumors. *Neuropathology* **21**, 40–44.

556. Izycka-Swieszewska, E., Stefanowicz, J., Debiec-Rychter, M., Rzepko, R., and Borowska-Lehman, J. (2001) Peripheral primitive neuroectodermal tumor within the spinal epidural space. *Neuropathology* **21**, 218–221.

557. Prebble, E., Dyer, S., Brundler, M.-A., Ellison, D., Davison, V., and Grundy, R. (2004) Genomic imbalances in supratentorial primitive neuroectodermal tumours (STPNETs). *J. Neurooncol.* **6**, 413.

558. McCabe, M.G., Ichimura, K., Liu, L., Plant, K., Bäcklund, L.M., Pearson, D.M., and Collins, V.P. (2006) High-resolution array-basead comparative genomic hybridization of medulloblastomas and supratentorial primitive neuroectodermal tumors. *J. Neuropathol. Exp. Neurol.* **65**, 549–561.

559. Zakrzewska, M., Rieske, P., Debiec-Rychter, M., Zakrzewski, K., Polis, L., Fiks, T., and Liberski, P.P. (2004) Molecular abnormalities in pediatric embryonal brain tumors – analysis of loss of heterozygosity on chromosome, 1, 5, 9, 10, 11, 16, 17 and 22. *Clin. Neuropathol.* **23**, 209–217.

560. Kraus, J.A., Felsberg, J., Tonn, J.C., Reifenberger, G., and Pietsch, T. (2002) Molecular genetic analysis of the *TP53, PTEN, CDKN2A, EGFR, CDK4* and *MDM2* tumour-associated genes in supratentorial primitive neuroectodermal tumours and glioblastomas of childhood. *Neuropathol. Appl. Neurobiol.* **28**, 325–333.

560A. Vincent, S., Mirshari, M., Nicolas, C., Adenis, C., Dhellemmes, P., Soto, Ares, G., Maurage, C.A., Baranzelli, M.C., Giangaspero, F., and Ruchoux, M.M. (2003) Large-cell medulloblastoma with arrestin-like protein expression. *Clin. Neuropathol.* **22**, 1–9.

561. Wolter, M., Scharwächter, C., Reifenberger, J., Koch, A., Pietsch, T., and Reifenberger, G. (2003) Absence of detectable alterations in the putative tumor suppressor gene *BTRC* in cerebellar medulloblastomas and cutaneous basal cell carcinomas. *Acta. Neuropathol.* **106**, 287–290.

562. Inda, M.M., Mercapide, J., Muñoz, J., Coullin, P., Danglot, G., Tunon, T., Martinez-Penuela, J.M., Rivera, J.M., Burgos, J.J., Bernheim, A., and Castresana, J.S. (2004) PTEN and DMBT1 homozygous deletion and expression in medulloblastomas and supratentorial primitive neuroectodermal tumors. *Oncol. Rep.* **12**, 1341–1347.

563. Davies, J.J., Wilson, I.M., and Lam, W.L. (2005) Array CGH technologies and their applications to cancer genomes. *Chromosome Res.* **13**, 237–248.

563A. Dallas, P.B., Terry, P.A., and Kees, U.R. (2005) Genomic deletions in cell lines derived form primitive neuroectodermal tumors of the central nervous system. *Cancer Genet. Cytogenet.* **159**, 105–113.

563B. Frühwald, M.C., O'Dorisio, M.S., Rush, L.J., Reiter, J.L., Smiraglia, D.J., Wenger, G., Costello, J.F., White, P.S., Krahe, R., Brodeur, G.M., and Plass, C. (2000) Gene amplification in PNETs/medulloblastomas: mapping of a novel amplified gene within the *MYCN* amplicon. *J. Med. Genet.* **37**, 501–509.

564. Debiec-Rychter, M., Alwasiak, J., Papierz, W., and Liberski, P.P. (1994) Expression of N-myc, c-myc and c-erbB-1 proto-oncogenes in cerebral primitive neuroectodermal tumors (PNET). *Folia Neuropathol.* **32**, 229–231.

565. Kawashima, M., Suzuki, S.O., Doh-ura, K., and Iwaki, T. (2000) α-Synuclein is expressed in a variety of brain tumors showing neuronal differentiation. *Acta.Neuropathol.* **99**, 154–160.

566. Ironside, J.W., Moss, T.H., Louis, D.N., Lowe, J.S., and Weller, R.D., eds. (2002) Embryonal tumours. In *Diagnostic Pathology of Nervous System Tumours*, Churchill Livingstone, London, UK, pp. 185–215.

567. Reifenberger, J., Janssen, G., Weber, R.G., Boström, J., Engelbrecht, V., Lichter, P., Borchard, F., Göbel, U., Lenard, H.-G., and Reifenberger, G. (1998) Primitive neuroectodermal tumors of the cerebral hemispheres in two siblings with *TP53* germline mutation. *J. Neuropathol. Exp. Neurol.* **57**, 179–187.

568. Weber, R.G., Bridger, J.M., Benner, A., Weisenberger, D., Ehemann, V., Reifenberger, G., and Lichter, P. (1998) Centrosome amplification as a possible mechanism for numerical chromosome aberrations in cerebral primitive neuroectodermal tumors with TP53 mutations. *Cytogenet. Cell Genet.* **83**, 266–269.

569. Raffel, C., Thomas, G.A., Tishler, D.M., Lassoff, S., and Allen, J.C. (1993) Absence of p53 mutations in childhood central nervous system primitive neuroectodermal tumors. *Neurosurgery* **33**, 301–306.

570. Ho, Y.S., Hsieh, L.L., Chen, J.S., Chang, C.N., Lee, S.T., Chiu, L.L., Chin, T.Y., and Cheng, S.C. (1996) p53 gene mutation in cerebral primitive neuroectodermal tumors in Taiwan. *Cancer Lett.* **104**, 103–113.

571. Nozaki, M., Tada, M., Matsumoto, R., Sawamura, Y., Abe, H., and Iggo, R.D. (1998) Rare occurrence of inactivating p53 gene mutations in primary non-astrocytic tumors of the central nervous system: reappraisal by yeast functional assay. *Acta Neuropathol.* **95**, 291–296.

572. Jaros, E., Lunec, J., Perry, R.H., Kelly, P.J., and Pearson, A.D. (1993) p53 protein overexpression identifies a group of central primitive neuroectodermal tumours with poor prognosis. *Br. J. Cancer* **68**, 801–807.

572A. Taylor, M.D., Gokgoz, N., Andrulis, I.L., Mainprize, T.G., Drake, J.M., and Rutka, J.T. (2000) Familial posterior fossa brain tumors of infancy secondary to germline mutation of the *hSNF5* gene. *Am. J. Hum. Genet.* **66**, 1403–1406.

573. Postovsky, S., Ben Arush, M.W., Elhasid, R., Davidson, S., Leshanski, L., Vlodavsky, E., Guilburd, J.N., and Amikam, D. (2003) A novel case of a CAT to AAT transversion in codon 179 of the p53 gene in a supratentorial primitive neuroectodermal tumor harbored by a young girl. Case report and review of the literature. *Oncology* **65**, 46–51.

574. Dorfmuller, G., Wurtz, F.G., Kleinert, R., and Lanner, G. (1997) Cerebral primitive neuro-ectodermal tumour following treatment of a unilateral retinoblastoma. *Acta Neurochir. (Wien)* **139**, 749–755.

575. Chen, A.Y., Lee, H., Hartman, J., Greco, C., Ryu, J.K., O'Donnell, R., and Boggan, J. (2005) Secondary supratentorial primitive neuroectodermal tumor following irradiation in a patient with low-grade astrocytoma. *AJNR Am. J. Neuroradiol.* **26**, 160–162.

576. Hader, W.J., Drovini-Zis, K., and Maguire, J.A. (2003) Primitive neuroectodermal tumors in the central nervous system following cranial irradiation. *Cancer* **97**, 1072–1076.

577. Hong, T.S., Mehta, M.P., Boyett, J.M., Donahue, B., Rorke, L.B., Yao, M.S., and Zeltzer, P.M. (2004) Patterns of failure in supratentorial primitive neuroectodermal tumors treated in Children's Cancer Group Study 921, a phase III combined modality study. *Int. J. Radiat. Oncol. Biol. Phys.* **60**, 204–213.

578. Paulino, A.C., and Melian, E. (1999) Medulloblastoma and supratentorial primitive neuroectodermal tumors. An institutional experience. *Cancer* **86**, 142–148.

579. Jakacki, R.I., Zeltzer, P.M., Boyett, J.M., Albright, A.L., Allen, J.C., Geyer, J.R., Rorke, L.B., Stanley, P., Stevens, K.R., Wisoff, J.H., McGuire-Cullen, P.L., Milstein, J.M., Packer, R.J., and Finlay, J.L. (1995) Survival and prognostic factors following radiation and/or chemotherapy for primitive neuroectodermal tumors of the pineal region in infants and children: a report from the Childrens Cancer Group. *J. Clin. Oncol.* **13**, 1377–1383.

580. Cohen, B.H., Zeltzer, P.M., Boyett, J.M., Geyer, J.R., Allen, J.C., Finlay, J.L., McGuire-Cullen, P., Milstein, J.M., Rorke, L.B., Stanley, P., Stehbens, J.A., Shurin, S.B., Wisoff, J., Stevens, K.R., and Albright, A.L. (1995) Prog-nostic factors and treatment results for supratentorial primitive neuroectodermal tumors in children using radiation and chemotherapy: A Children's Cancer Group Randomized Trial. *J. Clin. Oncol.* **13**, 1687–1696.

581. Fults, D., Pedone, C.A., Morse, H.G., Rose, J.W., and McKay, R.D.G. (1992) Establishment and characterization of a human primitive neuroectodermal tumor cell line from the cerebral hemisphere. *J. Neuropathol. Exp. Neurol.* **51**, 272–280.

582. Tsumanuma, I., Tanaka, R., and Washiyama, K. (1999) Clinicopathologic study of pineal parenchymal tumors: correlation between histopathological features, proliferative potential, and prognosis. *Brain Tumor Pathol.* **16**, 61–68.

583. Hinkes, B.G., von Hoff, K., Deinlein, F., Warmuth-Metz, M., Soerensen, N., Timmermann, B., Mittler, U., Urban, C., Bode, U., Pietsch, T., Schlegel, P.G., Kortmann, R.D., Kuehl, J., and Rutkowski, S. (2007) Childhood pineoblastoma: experience from the prospective multicenter trials HIT-SKK87, HIT-SKK92 and HIT91. *J. Neurooncol.* **81**, 217–223.

584. Hirato, J., and Nakazato, Y. (2001) Pathology of pineal region tumors. *J. Neurooncol.* **54**, 239–249.

584A. Mena, H., Rushing, E.J., Ribas, J.L., Delahunt, B., and McCarthy, W.F. (1995) Tumors of pineal parenchymal cells: A correlation of histological features, including nucleolar organizer regions, with survival in 35 cases. *Hum. Pathol.* **26**, 20–30.

585. Mena, H., Rushing, E.J., Ribas, J.L., Delahunt, B., and McCarthy, W.F. (1995) Tumors of the pineal parenchymal cells: A correlation of histological features, including nucleolar organizer regions, with survival in 35 cases. *Hum. Pathol.* **26**, 20–30.

586. Mena, H., Nakazato, Y., Jouvet, A., and Scheithauer, B.W. (2000) Pineoblastoma. In *World Health Organization Classification of Tumours. Pathology and Genetics of Tumours of the Nervous* System (Kleihues, P., and Cavenee, W.K., eds.), IARC Press, Lyon, France, pp. 116–120.

587. Muller, M., Hubbard, S.L., Fukuyama, K., Dirks, P., Matsuzawa, K., and Rutka, J.T. (1995) Characterization of a pineal region malignant rhabdoid tumor. *Pediatr. Neurosurg.* **22**, 204–209.

588. Yamane, Y., Mena, H., and Nakazato, Y. (2002) Immunohistochemical characterization of pineal parenchynmal tumors using novel monoclonal antibodies to the pineal body. *Neuropathology* **22**, 66–76.

589. Marec-Berard, P., Jouvet, A., Thiesse, P., Kalifa, C., Doz, F., and Frappaz, D. (2002) Supratentorial embryonal tumors in children under 5 years of age: an SFOP study of treatment with postoperative chemotherapy alone. *Med. Pediatr. Oncol.* **38**, 83–90.

590. Timmermann, B., Kortmann, R.-D., Kühl, J., Meisner, C., Dieckmann, K., Pietsch, T., and Bamberg, M. (2002) Role of radiotherapy in the treatment of supratentorial primitive neuroectodermal tumors in childhood: Results of the Prospective German Brain Tumor Trials HIT 88/89 and 91. *J. Clin. Oncol.* **20**, 842–849.

591. Jouvet, A., Fèvre-Montange, M., Besançon, R., Derrington, E., Saint-Pierre, G., Belin, M.F., Pialat, J., and Lapras, C. (1994) Structural and ultrastructural characteristics of human pineal gland, and pineal parenchymal tumors. *Acta Neuropathol.* **88**, 334–348.

592. Jouvet, A., Saint-Pierre, G., Fauchon, F., Privat, K., Bouffet, E., Ruchoux, M.-M., Chauveinc, L., and Fèvre-Montange, M. (2000) Pineal parenchymal tumors: A correlation of histological features with prognosis in 66 cases. *Brain Pathol.* **10**, 49–60.

593. Roberts, P., Chumas, P.D., Picton, S., Bridges, L., Livingstone, J.H., and Sheridan, E. (2001) A review of cytogenetics of 58 pediatric brain tumors. *Cancer Genet. Cytogenet.* **131**, 1–12.

594. Kees, U.R., Biegel, J.A., Ford, J., Ranford, P.R., Peroni, S.E., Hallam, L.A., Parmiter, A.H., Willoughby, M.L.N., and Spagnolo, D. (1994) Enhanced *MYCN* expression and isochromosome 17q in pineoblastoma cell lines. *Genes Chromosomes Cancer* **9**, 129–135.

595. Kees, U.R., Spagnolo, D., Hallam, L.A., Ford, J., Ranford, P.R., Baker, D.L., Callen, D.F., and Biegel, J.A. (1998) A new pineoblastoma cell line, PER-480, with der(10t(10;17),der(16)t(1;16), and enhanced *MYC* expression in the absence of gene amplification. *Cancer Genet. Cytogenet.* **100**, 159–164.

596. Rainho, C.P., Rogatto, S.R., Correa de Moraes, L., and Barbieri-Neto, J. (1992) Cytogenetic study of a pineocytoma. *Cancer Genet. Cytogenet.* **64**, 127–132.

597. Bello, M.J., Rey, J.A., de Campos, J.M., and Kusak, M.E. (1993) Chromosomal abnormalities in a pineocytoma. *Cancer Genet. Cytogenet.* **71**, 185–186.

598. Dario, A., Cerati, M., Taborelli, M., Finzi, G., Pozzi, M., and Dorizzi, A. (2000) Cytogenetic and ultrastructural study of a pineocytoma case report. *J. Neurooncol.* **48**, 131–134.

599. Rickert, C.H., Simon, R., Bergmann, M., Dockhorn-Dworniczak, B., and Paulus, W. (2000) Comparative genomic hybridization in pineal germ cell tumors. *J. Neuropathol. Exp. Neurol.* **59**, 815–821.

600. Rickert, C.H., Simon, R., Bergmann, M., Dockhorn-Dworniczak, B., and Paulus, W. (2001) Comparative genomic hybridization in pineal parenchymal tumors. *Genes Chromosomes Cancer* **30**, 99–104

601. Lopes, M.B.S., Gonzalez-Fernandez, F., Scheithauer, B.W., and VandenBerg, S.R. (1993) Differential expression of retinal proteins in a pineal parenchymal tumor. *J. Neuropathol. Exp. Neurol.* **52**, 516–524.

602. Bader, J.L., Meadows, A.T., Zimmerman, L.E., Rorke, L.B., Voute, P.A., Champion, L.A.A., and Miller, R.W. (1982) Bilateral retinoblastoma with ectopic intracranial retinoblastoma: Trilateral retinoblastoma *Cancer Genet. Cytogenet.* **5**, 203–213.

603. Russell, D.S., and Rubenstein, L.J., editors. (1989) *Pathology of Tumors of the Nervous System*, ed. 5. Williams and Wilkins, Baltimore, MD, pp. 380–394.

604. Kivelä, T. (1999) Trilateral retinoblastoma: A meta-analysis of hereditary retinoblastoma associated with primary ectopic intracranial retinoblastoma. *J. Clin. Oncol.* **17**, 1829–1837.

605. Jensen, R.D., and Miller, R.W. (1971) Retinoblastoma: Epidemiologic characteristics. *N. Engl. J. Med.* **285**, 307–316.

606. Whittle, I.R., McClellan, K., Martin, F.J., and Johnston, I.H. (1985) Concurrent pineoblastoma and unilateral retinoblastoma forme fruste of trilateral retinoblastoma? *Neurosurgery* **17**, 500–505.

607. Paulino, A.C. (1999) Trilateral retinoblastoma. Is the location of the intracranial tumor important? *Cancer* **86**, 135–141.

608. Finelli, D.A., Shurin, S.B., and Bardenstein, D.S. (1995) Trilteral retinoblastoma: Two variations. *Am. J. Neuroradiol.* **16**, 166–170.

609. Blach, L.E., McCormick, B., Abramson, D.H., and Ellsworth, R. (1994) Trilateral retinoblastoma—incidence and outcome: A decade of experience. *Int. J. Radiat. Oncol. Biol. Phys.* **29**, 729–733.

610. Elias, W.J., Lopes, B.S., Golden, W.L., Jane, J.A., and Gonzalez-Fernandez, F. (2001) Trilateral retinoblastoma variant indicative of the relevance of the retinoblastoma tumor-suppressor pathway to medulloblastomas in humans. *J. Neurosurg.* **95**, 871–878.

611. Taylor, M.D., Mainprize, T.G., Squire, J.A., and Rutka, J.T. (2001) Molecular genetics of pineal region neoplasms. *J. Neurooncol.* **54**, 219–238.

612. Skrypnyk, C., and Bartsch, O. (2004) Retinoblastoma, pinealoma, and mild overgrowth in a boy with a deletion of *RB1* and neighbor genes on chromosome 13q14. *Am. J. Med. Genet.* **124**, 397–401.

613. Plowman, P.N., Pizer, B., and Kingston, J.E. (2004) Pineal parenchymal tumours: II. On the aggressive behaviour of pineoblastoma in patients with an inherited mutation of the *RB1* gene. *Clin. Oncol. (R. Coll. Radiol.)* **16**, 244–247.

614. Amare Kadam, P.S., Ghule, P., Jose, J., Bamne, M., Kurkure, P., Banavali, S., Sarin, R., and Advani, S. (2004) Constitutional genomic instability, chromosome aberrations in tumor cells and retinoblastoma. *Cancer Genet. Cytogenet.* **150**, 33–43.

615. Chen, D., Gallie, B.L., and Squire, J.A. (2001) Minimal regions of chromosomal imbalance in retinoblastoma detected by comparative genomic hybridization. *Cancer Genet. Cytogenet.* **129**, 57–63.

616. Chen, WY., Cooper, T.K., Zahnow, C.A., Overholtzer, M., Zhao, Z., Ladanyi, M., Karp, J.E., Gokgoz, N., Wunder, J.S., Andrulis, I.L., Levine, A.J., Mankowski, J.L., and Baylin, S.B. (2004) Epigenetic and genetic loss of *Hic1* function accentuates the role of *p53* in tumorigenesis. *Cancer Cell* **6**, 387–398.

617. Champier, J., Jouvet, A., Rey, C., Brun, V., Bernard, A., and Fèvre-Montange, M. (2005) Identification of differentially expressed genes in human pineal parenchymal tumors by microarray analysis. *Acta Neuropathol.* **109**, 306–313.

618. Fèvre-Montange, M., Champier, J., Szathmari, A., Wierinckx, A., Mottolese, C., Guyotat, J., Fiagarella-Branger, D., Jouvet, A., and Lachuer, J. (2006) Microarray analysis reveals differential gene expression patterns in tumors of the pineal region. *J. Neuropathol. Exp. Neurol.* **65**, 675–684.

619. Tsumanuma, I., Sato, M., Okazaki, H., Tanaka, R., Washiyama, K., Kawasaki, T., Kumanishi, T. (1995) The analysis of p53 tumor suppressor gene in pineal parenchymal tumors. *Noshuyo Byori* **12**, 39–43.

620. Tsumanuma, I., Tanaka, R., Abe, S., Kawasaki, T., Washiyama, K., and Kumanishi, T. (1997) Infrequent mutation of Waf1/p21 gene, a CDK inhibitor gene, in brain tumors. *Neurol. Med. Chir. (Tokyo)* **37**, 150–156.

621. Judkins, A.R., Mauger, J., Rorke, L.B., and Biegel, J.A. (2004) Immunohistochemical analysis of hSNF5/INI1 in pediatric CNS neoplasms. *Am. J. Surg. Pathol.* **28**, 644–650.

622. Champier, J., Jouvet, A., Rey, C., Guyotat, J., and Fevre-Montange, M. (2003) Differential somatostatin receptor

subtype expression in human normal pineal gland and pineal parenchymal tumors. *Cell. Mol. Neurobiol.* **23,** 875.

623. Tsumanuma, I., Tanaka, R., Ichikawa, T., Washiyama, K., and Kumanishi, T. (2000) Demonstration of hydroxyindole-O-methyltransferase (HIOMT) mRNA expression in pineal parenchymal tumors: histochemical in situ hybridization. *J. Pineal Res.* **28,** 203–209.

624. Sawyer, J.R., Sammartino, G., Husain, M., and Linskey, M.E. (2003) Constitutional t(16;22)(p13.3;q11.2ˉ12) in a primitive neuroectodermal tumor of the pineal region. *Cancer Genet. Cytogenet.* **142,** 73–76.

625. Kaplan, E.L., and Meier, P. (1958) Nonparametric estimation from incomplete observations. *J. Am. Stat. Assoc.* **53,** 457–481.

626. Fuller, G.N. (1996) Central nervous system tumors. In: *Pediatric Neoplasia: Morphology and Biology* (Parham, D.M., ed.), Lippincott-Raven Publishers, Philadelphia, PA, pp. 153–202.

627. Lantos, P.L., Vandenberg, S.R., and Kleihues, P. (1996) Tumours of the nervous system. In *Greenfield's Neuropathology,* 6th ed., vol. 2 (Graham, D.I., and Lantos, P.L., eds.), Edward Arnold, London, UK, pp. 583–879.

628. Packer, R.J. Cogen, P., Vezina, G., and Rorke, L.B. (1999) Medulloblatoma: clinical and biologic aspects. *Neuro-Oncol.* **1,** 232–250.

628A. Arseni, C., and Ciurea, A.V. (1981) Statistical survey of 276 cases of medulloblastoma (1935–1978). *Acta Neurochir. (Wien)* **57,** 159–162.

629. Roberts, R.O., Lynch, C.F., Jones, M.P., and Hart, M.N. (1991) Medulloblastoma: a population-based study of 532 cases. *J. Neuropathol.I Exp. Neurol.* **50,** 134–144.

629A. Hubbard, J.L., Scheithauer, B.W., Kispert, D.B., Carpenter, S.M., Wick, M.R., and Laws, E.R., Jr. (1989) Adult cerebellar medulloblastomas: the pathological, radiographic, and clinical disease spectrum. *J. Neurosurg.* **70,** 534–544.

630. Giordana, M.T., Schiffer, P., Lanotte, M., Girardi, P., and Chiò, A. (1999) Epidemiology of adult medulloblastoma. *Int. J. Cancer* **80,** 689–692.

630A. Greenberg, F., Guzzetta, V., Montes de Oca-Luna, R., Magenis, R.E., Smith, A.C., Richter, S.F., Kondo, I., Dobyns, W.B., Patel, P.I., and Lupski, J.R. (1991) Molecular analysis of the Smith-Magenis syndrome: a possible contiguous-gene syndrome associated with del(17)(p11.2). *Am. J. Hum. Genet.* **49,** 1207–1218.

631. Rubinstein, L.J. (1985) Embryonal central neuroepithelial tumors and their differentiating potential. A cytogenetic view of a complex neuro-oncological problem. *J. Neurosurg.* **62,** 795–805.

632. Trojanowski, J.Q., Tohyama, T., and Lee, V.M. (1992) Medulloblastomas and related primitive neuroectodermal brain tumors of childhood recapitulate molecular milestones in the maturation of neuroblasts. *Mol. Chem. Neuropathol.* **17,** 121–135.

633. Herpers, M.J., and Budka, H. (1985) Primitive neuroectodermal tumors including the medulloblastoma: glial differentiation signaled by immunoreactivity for FAP is restricted to the pure desmoplastic medulloblastoma ("arachnoidal sarcoma of the verebellum"). *Clin. Neuropathol.* **4,** 12–18.

634. Latimer, F.R., Al Saadi, A.A., and Robbins, T.O. (1987) Cytogenetic studies of human brain tumors and their clinical significance. *J. Neurooncol.* **4,** 287–291.

635. Pearson, A.D., Reid, M.M., Davison, E.V., Bown, N., Malcolm, A.J., and Craft, A.W. (1988) Cytogenetic investigations of solid tumours of children. *Arch Dis. Child.* **63,** 1012–1015.

636. Callen, D.F., Cirocco, L., and Moore, L. (1989) A der(11)t(8;11) in two medulloblastomas. A possible nonrandom cytogenetic abnormality. *Cancer Genet. Cytogenet.* **38,** 255–260.

637. López-Ginés, C., Cerdá-Nicolás, M., and Llombart-Bosch, A. (1991) A new case of medulloblastoma in an adult. *Cancer Genet. Cytogenet.* **57,** 235–237.

638. Stratton, M.R., Darling, J., Cooper, C.S., and Reeves, B.R. (1991) A case of cerebellar medulloblastoma with a single chromosome abnormality. *Cancer Genet. Cytogenet.* **53,** 101–103.

639. Karnes, P.S., Tran, T.N., Cui, M.Y., Raffel, C., Gilles, F.H., Barranger, J.A., and Ying, K.L. (1992) Cytogenetic analysis of 39 pediatric central nervous system tumors. *Cancer Genet. Cytogenet.* **59,** 12–19.

640. Stuart, A.G., Pearson, A.D., Emslie, J., Lennard, A., Davison, E.V., Perry, R.H., and Crawford, P.J. (1993) Cytogenetic abnormalities in a disseminated medulloblastoma. *Med. Pediatr. Oncol.* **21,** 295–298.

641. Sainati, L., Bolcato, S., Montaldi, A., Celli, P., Stella, M., Leszl, A., Silvestro, L., Perilongo, G., Cordero di Montezemolo, L., and Basso, G. (1996) Cytogenetics of pediatric central nervous system tumors. *Cancer Genet. Cytogenet.* **91,** 13–27.

642. López-Ginés, C., Cerdá-Nicolás, M., Gil-Benso, R., Barcia-Salorio, J.L., and Llombart-Bosch, A. (1997) Involvement of the long arm of chromosome 9 in medulloblastoma in an adult. *Cancer Genet. Cytogenet.* **81,** 81–84.

643. Vagner-Capodano, A.M., Mugneret, F., Zattara-Cannoni, H., Gabert, J., Favre, B., Figarella-Branger, D., Gentet, J.C., Lena, G., and Bernard, J.L. (1997) Translocation 1;19 in two brain tumors. *Cancer Genet. Cytogenet.* **97,** 1–4.

643A. Scheurlen, W.G., Schwabe, G.C., Seranski, P., Joos, S., Harbott, J., Metzke, S., Dohner, H., Poustka, A., Wilgenbus, K., and Haas, O.A. (1999) Mapping of the breakpoints on the short arm of chromosome 17 in neoplasms with an i(17q). *Genes Chromosomes Cancer* **25,** 230–240.

644. DeChiara, C., Borghese, A., Fiorillo, A., Genesio, R., Conti, A., D'Amore, R., Pettinato, G., Varone, A., and Maggi, G. (2002) Cytogenetic evaluation of isochromosome 17q in posterior fossa tumors of children and correlation with clinical outcome in medulloblastoma. Detection of a novel chromosomal abnormality. *Child's Nerv. Syst.* **18,** 380–384.

645. Uematsu, Y., Takehara, R., Shimizu, M., Tanaka, Y., Itakura, T., and Komai, N. (2002) Pleomorphic primitive neuroectodermal tumor with glial and neuronal differentiation: clinical, pathological, cultural, and chromosomal analysis of a case. *J. Neuro-Oncol.* **59,** 71–79.

646. Sreekantaiah, C., Jockin, H., Brecher, M.L., and Sandberg, A.A. (1989) Interstitial deletion of chromosome 11q in a pineoblastoma. *Cancer Genet. Cytogenet.* **39,** 125–131.

647. Hui, A.B.Y., Takano, H., Lo, K.-W., Kuo, W.-L., Lam, C.N.Y., Tong, C.Y.K., Chang, Q., Gray, J.W., and Ng, H.-K. (2005)

Identification of a novel homozygous deletion region at 6q23.1 in medulloblastomas using high-resolution array comparative genomic hybridization analysis. *Clin. Cancer Res.* **11,** 4707–4716.

647A. Ehrbrecht, A., Muller, U., Wolter, M., Hoischen, A., Koch, A., Radlwimmer, B., Actor, B., Mincheva, A., Pietsch, T., Lichter, P., Reifenberger, G., and Weber, R.G. (2006) Comprehensive genomic analysis of desmoplastic medulloblastomas: identification of novel amplified genes and separate evaluation of the different histologic components. *J. Pathol.* **208,** 554–563.

648. Lo, K.C., Rossi, M.R., Burkhardt, T., Pomeroy, S.L., and Cowell, J.K. (2007) Overlay analysis of the oligonucleotide array gene expression profiles and copy number abnormalities as determined by array comparative genomic hybridization in medulloblastomas. *Genes Chromosomes Cancer* **46,** 53–66.

649. Slavc, I., Rodriguez, I.R., Mazuruk, K., Chader, G.J., and Biegel, J.A. (1997) Mutation analysis and loss of heterozygosity of PEDF in central nervous system primitive neuroectodermal tumors. *Int. J. Cancer* **72,** 277–282.

650. Gorlin, R.J. (1987) Nevoid basal-cell carcinoma syndrome. *Medicine* **66,** 96–113.

651. Lacombe, D., Chateil, J.F., Fontan, D., and Battin, J. (1990) Medulloblastoma in the nevoid basal-cell carcinoma syndrome: case reports and review of the literature. *Genet. Couns.* **1,** 273–277.

652. Paraf, F., Jothy, S., and Van Meir, E.G. (1997) Brain tumor-polyposis syndrome: two genetic diseases? *J. Clin. Oncol.* **15,** 2744–2758.

653. Perilongo, G., Felix, C.A., Meadows, A.T., Nowell, P., Biegel, J., and Lange, B.J. (1993) Sequential development of Wilms tumor, T-cell acute lymphoblastic leukemia, medulloblastoma and myeloid leukemia in a child with type 1 neurofibromatosis: a clinical and cytogenetic case report. *Leukemia* **7,** 912–915.

654. Wilgenbus, K.K., Seranski, P., Brown, A., Leuchs, B., Mincheva, A., Lichter, P., and Poustka, A. (1997) Molecular characterization of a genetically unstable region containing the SMS critical area and a breakpoint cluster for human PNETs. *Genomics* **42,** 1–10.

655. Miyata, H., Ikawa, E., and Ohama, E. (1998) Medulloblastoma in an adult suggestive of external granule cells as its origin: a histological and immunohistochemical study. *Brain Tumor Pathol.* **15,** 31–35.

656. Stavrou, T., Bromley, C.M., Nicholson, H.S., Byrne, J., Packer, R.J., Goldstein, A.M., and Reaman, G.H. (2001) Prognostic factors and secondary malignancies in childhood medulloblastoma. *J. Pediatr. Hematol./Oncol.* **23,** 431–436.

656A. Campbell, R.M., Mader, R.D., and Dufresne, R.G., Jr. (2005) Meningiomas after medulloblastoma irradiation treatment in a patient with basal cell nevus syndrome. *J. Am. Acad. Dermatol.* **53,** S256–259.

657. Liu, Y., Mao, B., Hokeung, N., Chen, Y., and Gao, L. (2001) Expression of C-myc and N-myc protein in adulthood and childhood medulloblastomas and prognostic analysis. *Hua Xi Yi Ke Da Xue Xue Bao* **32,** 120–122.

658. Misaki, K., Marukawa, K., Hayashi, Y., Fukusato, T., Minamoto, T., Hasegawa, M., Yamashita, J., and Fujisawa, H.

(2005) Correlation of γ-catenin expression with good prognosis in medulloblastomas. *J. Neurosrug. (Pediatrics 2)* **102,** 197–206.

659. Valery, C., Vassal, G., Lelouche-Tubiana, A., Hirsh, J.F., Kalifa, C., Riou, G., and Lemerle, J. (1991) Expression of c-myc gene in medulloblastoma. *Pediatr. Oncol.* **19,** 344.

660. Jay, V., Squire, J., Zielenska, M., Gerrie, B., and Humphreys, R. (1995) Molecular and cytogenetic analysis of a cerebellar primitive neuroectodermal tumor with prominent neuronal differentiation: detection of MYCN amplification by differential polymerase chain reaction and Southern blot analysis. *Pediatr. Pathol. Lab. Med.* **15,** 733–744.

661. Gilbertson, R., Wickramasinghe, C., Hernan, R., Balaji, V., Hunt, D., Jones-Wallace, D., Crolla, J., Perry, R., Lunec, J., Pearson, A., and Ellison, D. (2001) Clinical and molecular stratification of disease risk in medulloblastoma. *Br. J. Cancer* **85,** 705–712.

662. Ellison, D.W., Onilude, O.E., Lindsey, J.C., Lusher, M.E., Weston, C.L., Taylor, R.E., Pearson, A.D., and Clifford, S.C. (2005) β-catenin status predicts a favorable outcome in childhood medulloblastoma: The United Kingdom Children's Cancer Study Group Brain Tumour Committee. *J. Clin. Oncol.* **23,** 7951–7957.

663. Siu, I.-M., Lal, A., Blankenship, J.R., Aldosari, N., and Riggins, G.J. (2003) c-Myc promoter activation in medulloblastoma. *Cancer Res.* **63,** 4773–4776

664. Loda, M., Giangaspero, F., Baldini, M., Capodieci, P., and Pession. A. (1992) p53 gene expression in medulloblastoma by quantitative polymerase chain reaction. *Diagn. Mol. Pathol.* **1,** 36–44.

665. Ray, A., Ho, M., Ma, J., Parkes, R.K., Mainprize, T.G., Ueda, S., McLaughlin, J., Bouffet, E., Rutka, J.T., and Hawkins, C.E. (2004) A clinicobiological model predicting survival in medulloblastoma. *Clin. Cancer Res.* **10,** 7613–7620.

666. Portwine, C., Chilton-MacNeill, S., Brown, C., Sexsmith, E., McLaughlin, J., and Malkin, D. (2001) Absence of germline and somatic *p53* alterations in children with sporadic brain tumors. *J. Neurooncol.* **52,** 227–235.

667. Weil, M.D., Lamborn, K., Edwards, M.S.B., and Wara, W.M. (1998) Influence of a child's sex on medulloblastoma outcome. *J.A.M.A.* **279,** 1474–1476.

668. Zeltzer, P.M., Boyett, J.M., Finlay, J.L., Albright, A.L., Rorke, L.B., Milstein, J.M., Allen, J.C., Stevens, K.R., Stanley, P., Li, H., Wisoff, J.H., Geyer, J.R., McGuire-Cullen, P., Stehbens, J.A., Shurin, S.B., and Packer, R.J. (1999) Metastasis stage, adjuvant treatment, and residual tumor are prognostic factors for medulloblastoma in children: Conclusions from the Children's Cancer Group 921 randomized phase III study. *J. Clin. Oncol.* **17,** 832–845.

669. Chatty, E.M., and Earle, K.M. (1971) Medulloblastoma: A report of 201 cases with emphasis on the relationship of histologic variants to survival. *Cancer* **28,** 977–983.

670. Segal, R.A., Goumnerova, L.C., Kwon, Y.K., Stiles, C.D., and Pomeroy, S.L. (1994) Expression of the neurotrophin receptor TrkC is linked to a favorable outcome in medulloblastoma. *Proc. Natl. Acad. Sci. USA* **91,** 12867–12871.

671. Kim, J.Y.H., Sutton, M.E., Lu, D.J., Cho, T.A., Goumnerova, L.C., Goritchenko, L., Kaufman, J.R., Lam, K.K., Billet, A.L., Tarbell, N.J., Wu, J., Allen, J.C., Stiles, C.D.,

Segal, R.A., and Pomeroy, S.L. (1999) Activation of neurotrophin-3 receptor TrkC induces apoptosis in medulloblastomas. *Cancer Res.* **59**, 711–719.

671A. Sinnappah-Kang, N.D., Mrak, R.E., Paulsen, D.B., and Marchetit, D. (2006) Heparanase expression and TrkC/p75NTR ratios in human medulloblastoma. *Clin. Exp. Metastasis* **23**, 55–63.

671B. Sinnappah-Kang, N.D., Kaiser, A.J., Blust, B.E., Mrak, R.E., and Marchetit, D. (2005) Heparanase, TrkC, and p75NTR: their functional involvement in human medulloblastoma cell invasion. *Int. J. Oncol.* **27**, 617–626.

672. Fernandez-Teijeiro, A., Betensky, R.A., Sturla, L.M., Kim, J.Y.H., Tamayo, P., and Pomeroy, S.L. (2004) Combining gene expression profiles and clinical parameters for risk stratification in medulloblastomas. *J. Clin. Oncol.* **22**, 994–998.

673. Gurney, J.G., Smith, M.A., and Bunin, G.R. (1999) CNS and miscellaneous intracranial and intraspinal neoplasms. In *Cancer Incidence and Survival among Children and Adolescents. United States SEER Program 1975–1995*, (Ries, L.A.G., Smith, M.A., Gurney, J.G., Linet, M., Tamra, T., Young, J.L., and Bunin, G.R, eds.) National Cancer Institute, SEER Program. NIH Pub. No. 99–4649. National Cancer Institute, Bethesda, MD, pp. 51–63.

674. McNeil, D.E., Cote, T.R., Clegg, L., and Rorke, L.B. (2002) Incidence and trends in pediatric malignancies medulloblastoma/primitive neuroectodermal tumor: a SEER update. Surveillance Epidemiology and End Results. *Med. Pediatr. Oncol.* **39**, 190–194.

675. Pelc, K., Vincent, S., Ruchoux, M.-M., Kiss, R., Pochet, R., Sariban, E., Decaestecker, C. Heizmann, C.W., (2002) Calbindin-D . A marker of recurrence for medulloblastomas. *Cancer* **95**, 410–419.

676. Pizem, J., Cor, A., Zadravec Zalatel, L., and Popovic, M. (2005) Prognostic significance of apoptosis in medulloblastoma. *Neurosci. Lett.* **381**, 69–73.

677. Pan, E., Pellarin, M., Holmes, E., Smirnov, I., Misra, A., Eberhart, C.G., Burger, P.C., Biegel, J.A., and Feuerstein, B.G. (2005) Isochromosome 17q is a negative prognostic factor in poor-risk childhood medulloblastomas patients. *Clin. Cancer Res.* **11**, 4733–4840.

678. Taipale, J., Chen, J.K., Cooper, M.K., Wang, B., Mann, R.K., Milenkovic, L., Scott, M.P., and Beachy, P.A. (2000) Effects of oncogenic mutations in *Smoothened* and *Patched* can be reversed by cyclopamine. *Nature* **406**, 1005–1009.

679. Sasai, K., Romer, J.T., Lee, Y., Finkelstein, D., Fuller, C., McKinnon, P.J., and Curran, T. (2006) Shh pathway activity is down-regulated in cultured medulloblastoma cells: Implications for preclinical studies. *Cancer Res.* **66**, 4215–4222.

680. Rubin, J.B., and Rowitch, D.H. (2002) Medulloblastoma: a problem of developmental biology. *Cancer Cell.* **2**, 7–8.

681. Gumireddy, K., Sutton, L.N., Phillips, P.C., and Reddy, C.D. (2003) All-*trans*-retinoic acid-induced apoptosis in human medulloblastoma: Activation of caspase-3/poly(ADP-ribose) polymerase 1 pathway. *Clin. Cancer Res.* **9**, 4052–4059.

682. Hallahan, A.R., Pritchard, J.I., Chandraratna, R.A.S., Ellenbogen, R.G., Geyer, J.R., Overland, R.P., Strand, A.D., Tapscott, S.J., and Olson, J.M. (2003) BMP-2 mediates retinoid-induced apoptosis in medulloblastoma cells through a paracrine effect. *Nat. Med.* **9**, 1033–1038.

682A. Korinek, V., Barker, N., Morin, P.J., van Wichen, D., de Weger, R., Kinzler, K.W., Vogelstein, B., and Clevers, H. (1997) Constitutive transcriptional actiation by a β-catenin-Tcf complex in APC$^{--/--}$ colon carcinoma. *Science* **275**, 1784–1787.

683. Zorn, A.M., Barish, G.D., Williams, B.O., Lavender, P., Klymkowsky, M.W., and Varmus, H.E. (1999) Regulation of Wnt signaling by Sox proteins: XSox17 alpha/beta and XSox3 physically interact with beta-catenin. *Mol. Cell.* **4**, 487–498.

684. Barker, N., and Clevers, H. (2000) Catenins, Wnt signaling and cancer. *Bioessasy* **22**, 961–965.

685. Kollligs, F.T., Kolligs, B., Hajra, K.M., Hu, G., Tani, M., Cho, K.R., and Fearon, E.R. (2000) γ-Catenin is regulated by the APC tumor suppressor and its oncogenic activity is distinct from that of β-catenin. *Genes Dev.* **14**, 1319–1331.

686. Meng., X., Poon, R., Zhang, X., Cheah, A., Ding, Q., Hui, C., and Alman B. (2001) Suppressor of fused negatively regultes β-catenin signaling. *J. Biol. Chem.* **276**, 40113–40119.

687. Gonzalez-Gomez, P., Bello, M.J., Lomas, J., Arjona, D., Alonso, M.E., Aminoso, C., de Campos, J.M., Vaquero, J., Sarasa, J.L., Casartelli, C., and Rey, J.A. (2003) Epigenetic changes in pilocytic astrocytomas and medulloblastomas. *Int. J. Mol. Med.* **11**, 655–660.

688. Chang, Q., Pang, J.C.-S., Li, K.K.W., Poon, W.S., Zhou, L., and Ng, H.-K. (2005) Promoter hypermethylation profile of RASSF1A, FHIT, and sFRP1 in intracranial primitive neuroectodermal tumors. *Hum. Pathol.* **36**, 1265–1272.

689. Inda, M.M., Muñoz, J., Coullin, P., Fauvet, D., Danglot, G., Tuñón, T., Bernheim, A., and Castresana, J.S. (2006) High promoter hypermethylation frequency of *p14/ARF* in supratentorial PNET but not in medulloblastoma. *Histopathology* **48**, 579–587.

690. Lindsey, J.C., Lusher, M.E., Strathdee, G., Brown, R., Gilbertson, R.J., Bailey, S., Ellison, D.W., and Clifford, S.C. (2006) Epigenetic inactivation of MCJ (DNAJD1) in malignant paediatric brain tumours. *Int. J. Cancer* **118**, 346–352

691. Jen, J., Harper, J.W., Bigner, S.H., Bigner, D.D., Papadopoulos, N., Markowitz, S., Willson, J.K.V., Kinzler, K.W., and Vogelstein, B. (1994) Deletion of *p16* and *p15* genes in brain tumors. *Cancer Res.* **57**, 6353–6358.

692. Petronio, J., He, J., Fults, D., Pedone, C., James, C.D., and Allen, J.R. (1996) Common alternative gene alterations in adult malignant astrocytomas, but not in childhood primitive neuroectodermal tumors: *P16*^{ink4} homozygous deletions and *CDK4* gene amplifications. *J. Neurosurg.* **84**, 1020–1023.

693. Sato, K., Schauble, B., Kleihues, P., and Ohgaki, H. (1996) Infrequent alterations of the p15, p16, CDK4 and cyclin D1 genes in non-astrocytic human brain tumors. *Int. J. Cancer* **66**, 305–308.

694. Barker, F.G., Chen, P., Furman, F., Aldape, K.D., Edwards, M.S., and Israel, M.A. (1997) P16 deletion and mutation analysis in human brain tumors. *J. Neurooncol.* **31**, 17–23.

695. Roussel, M.F. (1999) The INK4 family of cell cycle inhibitors in cancer. *Oncogene* **18**, 5311–5317.

696. Versteege, I., Sévenet, N., Lange, J., Rousseau-Merck, M.-F., Ambros, P., Handgretinger, R., Aurias, A., and Delattre, O. (1998) Truncating mutations of hSNF5/INI1 in aggressive paediatric cancer. *Nature* **394**, 203–206.

697. Packer, R.J. (1999) Childhood medulloblastoma: progress and future challenges. *Brain Dev.* **21,** 75–81.

698. Imbalzano, A.N., and Jones, S.N. (2005) Snf5 tumor suppressor couples chromatin remodeling, checkpoint control, and chromosomal stability. *Cancer Cell* **7,** 294–295.

699. Liu, J., Guo, L., Luo, Y., Li, J.W., and Li, H. (2000) All trans-retinoic acid suppresses in vitro growth and down-regulates LIF gene expression as well as telomerase activity of human medulloblastoma cells. *Anticancer Res.* **20,** 2659–2664.

699A. Liu, J., Guo, L.,., Jun-Wei, L., Liu, N., and Li, H. (2000) All trans-retinoic acid modulates fas expression and enhances chemosensitivity of human medulloblastoma cells. *Int. J. Mol. Med.* **5,** 145–149.

700. Lee, C.-J., Appleby, V.J., Orme, A.T., Chan, W.-I., and Scotting, P.J. (2002) Differential expression of *SOX4* and *SOX11* in medulloblastoma. *J. Neurooncol.* **57,** 201–214.

701. Liberzon, E., Avigad, S., Cohen, I.J., Yaniv, I., Michovitz, S., and Zaizov, R. (2003) *ATM* gene mutations are not involved in medulloblastoma in children. *Cancer Genet. Cytogenet.* **146,** 167–169.

702. Chilton-Macneill, S., Ho, M., Hawkins, C., Gassas, A., Zielenska, M., and Baruchel, S. (2004) C-kit expression and mutational analysis in medulloblastoma. *Pediatr. Dev. Pathol.* **7,** 493–498.

703. Rasheed, B.K.A., Stenzel, T.T., McLendon, R.E., Parsons, R., Friedman, A.H., Friedman, H.S., Bigner, D.D., and Bigner, S.H. (1997) *PTEN* gene mutations are seen in high-grade but not in low-grade gliomas. *Cancer Res.* **57,** 4187–4190.

704. Felix, C.A., Slavc, I., Dunn, M., Strauss, E.A., Phillips, P.C., Rorke, L.B., Sutton, L., Bunin, G.R., and Biegel, J.A. (1995) p53 gene mutations in pediatric brain tumors. *Med. Pediatr. Oncol.* **25,** 431–436.

705. Adesina, A.M., Dunn, S.T., Moore, W.E., and Nalbantoglu, J. (2000) Expression of p27^{kip1} and p53 in medulloblastoma: relationship with cell proliferation and survival. *Pathol. Res. Pract.* **196,** 243–250.

705A. Badiali, M., Iolascon, A., Loda, M., Scheithauer, B.W., Basso, G., Trentini, G.P., and Giangaspero, F. (1993) p53 gene mutations in medulloblastoma. Immunohistochemistry, gel shift analysis, and sequencing. *Diagn. Mol. Pathol.* **2,** 23–28.

706. Inda, M.M., Perot, C., Guillaud-Bataille, M., Danglot, G., Rey, J.A., Bello, M.J., Fan, X., Eberhart, C., Zazpe, I., Portillo, E., Tuñón, T., Martínez-Peñuela, J.M., Bernheim, A., and Castresana, J.S. (2005) Genetic heterogeneity in supratentorial and infratentorial primitive neuroectodermal tumours of the central nervous system. *Histopathology* **47,** 631–637.

707. Waldbaur, H., Gottschaldt, M., Schmidt, H., and Neuhauser, G. (1976) Medulloblastoma and pineoblastoma in monozygous twins. *Klin. Paeiatr.* **188,** 366–371.

708. Lesnick, J.E., Chayt, K.J., Bruce, D.A., Rorke, L.B., Trojanowski, J., Savino, P.J., and Schatz, N.J. (1985) Familial pineoblastoma. Report of two cases. *J. Neurosurg.* **62,** 930–932.

709. Brockmeyer, D.L., Walker, M.L., Thompson, G., and Fults, D.W. (1997) Astrocytoma and pineoblastoma arising sequentially in the fourth ventricle of the same patient. *Pediatr. Neurosurg.* **26,** 36–40.

710. Ikeda, J., Sawamura, Y., and Van Meir, E.G. (1998) Pineoblastoma presenting in familial adenomatous polyposis (FAP): random association, FAP variant or Turcot syndrome? *Br. J. Neurosurg.* **12,** 576–578.

711. Çorapcíoglu, F., Özek, M.M., Sav, A., and Üren, D. (2006) Congenital pineoblastoma and parameningeal rhabdomyosarcoma: concurrent two embryonal tumors in a young infant. *Childs Nev. Syst.* **22,** 533–538.

712. Hoshino, T. (1993) Proliferativfe potential of brain tumors, analyses with Ki-67 and anti-DNA polymerase alpha monoclonal antibodies, bromodeoxyuridine labeling, and nucleolar organizer region counts. *Cancer* **71,** 199–205.

713. Hatakeyama, M., and Weinberg, R.A. (1995) The role of RB in cell cycle control. *Prog. Cell Cycle Res.* **1,** 9–19.

714. Korf, H.-W., Götz, W., Herken, R., Theuring, F., Gruss, P., and Schachenmayr, W. (1990) S-antigen and rod-opsin immunoreactions in midline brain neoplasms of transgenic mice: similarities to pineal cell tumors and certain medulloblastomas in man. *J. Neuropathol. Exp. Neurol.* **49,** 424–437.

715. Maraziotis, T., Perentes, E., Karamitopoulou, E., Nakagawa, Y., Gessaga, E.C., Probst, A., and Frankfurter, A. (1992) Neuron-associated class III β-tubulin isotype, retinal S-antigen, synaptophysin, and glial fibrillary acidic protein in human medulloblastomas: a clinicopathological analysis of 36 cases. *Acta Neuropathol.* I **84,** 355–363.

716. Jennings, M.T., Kaariainen, I.T., Gold, L., Maciunas, R.J., and Commers, P.A. (1994) TGFβ1 and TGFβ2 are potential growth regulators for medulloblastomas, primitive neuroectodermal tumors, and ependymomas: Evidence in support of an autocrine hypothesis. *Hum. Pathol.* **25,** 464–475.

717. Katsetos, C.D., Herman, M.M., Krishna, L., Vender, J.R., Vinores, S.A., Agamanolis, D.P., Schiffer, D., Burger, P.C., and Urich, H. (1995) Calbindin-D in subsets of medulloblastoma and in the human medulloblastoma cell line D283 OR D28K? Med. *Arch. Pathol.I Lab. Med.* **119,** 734–743.

718. Rosenfeld, M.R., Meneses, P., Dalmau, M., Cordon-Cardo, C., and Kaplitt, M.G. (1995) Gene transfer of wild-type *p53* results in restoration of tumor suppressor function in a medulloblastoma cell line. *Neurology* **45,** 1533–1539.

719. Vassal, G., Terrier-Lacombe, M.J., Lellouch-Tubiana, A., Valery, C.A., Sainte-Rose, C., Morizet, J., Ardouin, P., Riou, G., Kalifa, C., and Gouyette, A. (1996) Tumorigenicity of cerebellar primitive neuro-ectodermal tumors in athymic mice correlates with poor prognosis in children. *Int. J. Cancer* **69,** 146–151.

720. Lee, S.E., Johnson, S.P., Hale, L.P., Li, J., Bullock, N., Fuchs, H., Friedman, A., McLendon, R., Bigner, D.D., Modrich, P., and Friedman, H.S. (1998) Analysis of DNA mismatch repair proteins in human medulloblastoma. *Clin. Cancer Res.* **4,** 1415–1419.

721. Iantosca, M.R., McPherson, C.E., Ho, S.Y., and Maxwell, G.D. (1999) Bone morphogenetic proteins-2 and -4 attenuate apoptosis in a cerebellar primitive neuroectodermal brain tumor cell line. *J. Neurosci. Res.* **56,** 248–258.

722. Bodey, B., Bodey, B., Jr., Siegel, S.E., and Kaiser, H.E. (2000) Immunocytochemical detection of the homeobox B3, B4, and C6 gene products in childhood medulloblastoma/primitive neuroectodermal tumors. *Anticancer Res.* **20,** 1769–1780.

723. Bobola, M.S., Berger, M.S., Ellenbogen, R.G., Roberts, T.S., Geyer, J.R., and Silber, J.R. (2001) O^6-methylguanine-DNA methyltransferase in pediatric primary brain tumors: Relation

757. Su, X., Gopalakrishnan, V., Stearns, D., Aldape, K., Lang, F.F., Fuller, G., Snyder, E., Eberhart, C.G., and Majumder, S. (2006) Abnormal expression of REST/NRSF and Myc in neural stem/progenitor cells causes cerebellar tumors by blocking neuronal differentiation. *Mol. Cell Biol.* **26**, 1666–1678.

758. Uziel, T., Zindy, F., Xie, S., Lee, Y., Forget, A., Magdaleno, S., Rehg, J.E., Calabrese, C., Solecki, D., Eberhart, C.G., Sherr, S.E., Plimmer, S., Clifford, S.C., Hatten, M.E., McKinnon, P.J., Gilbertson, R.J., Curran, T., Sherr, C.J., and Roussel, M.F. (2005) The tumor suppressors *Ink4c* and *p53* collaborate independently with *Patched* to suppress medulloblastoma formation. *Genes Dev.* **19**, 2656–2667.

759. Werbowetski-Ogilvie, T.E., Sayed Sadr, M., Jabado, N., Angers-Loustau, A., Agar, N.Y.R., Wu, J., Bjerkvig, R., Antel, J.P., Faury, D., Rao, Y., and Del Maestro, R.F. (2006) Inhibition of medulloblastoma cell invasion by Slit. *Oncogene* **25**, 5103–5112.

760. Hernan, R., Fasheh, R., Calabrese, C., Frank, A.J., Maclean, K.H., Allard, D., Barraclough, R., and Gilbertson, R.J. (2003) *ERBB2* up-regulated *S100A4* and several other prometastatic genes in medulloblastoma. *Cancer Res.* **63**, 140–148.

761. Nakamura, M., Rieger, J., Weller, M., Kim, J., Kleihues, P., and Ohgaki, H. (2000) APO2L/TRAIL expression in human brain tumors. *Acta Neuropathol.* **99**, 1–6.

762. Fulda, S., and Debatin, K.-M. (2006) 5-Aza-2'-deoxycytidine and IFN-γ cooperate to sensitize for *TRAIL*-induced apoptosis by upregulating caspase-8. *Oncogene* **25**, 5125–5133.

763. Gilbertson, R.J. (2004) Medulloblastoma: signaling a change in treatment. *Lancet Oncol.* **5**, 209–218.

764. Leung, C., Lingbeek, M., Shakhova, O., Liu, J., Tanger, E., Saremaslani, P., van Lohulzen, M., and Marino, S. (2004) Bmi1 is essential for cerebellar development and is overexpressed in human medulloblastomas. *Nature* **428**, 337–341.

765. Rubin, J.B., Kung, A.L., Klein, R.S., Chan, J.A., Sun, Y., Schmidt, K., Kieran, M.W., Luster, A.D., and Segal, R.A. (2003) A small-molecule antagonist of CXCR4 inhibits intracranial growth in primary brain tumors. *Proc. Natl. Acad. Sci. USA* **100**, 13513–13518.

766. Klein, R.S., Rubin, J.B., Gibson, H.D., DeHaan, E.N., Alvarez-Hernandez, X., Segal, R.A., and Luster, A.D. (2001) SDF-1 alpha induces chemotaxis and enhanaes Sonic hedgehog-induced proliferation of cerebellar granule cells. *Development* **128**, 1971–1981.

767. Gilbertson, R.J., and Clifford, S.C. (2003) *PDGFRB* is overexpressed in metastatic medulloblastoma. *Nat. Genet.* **35**, 197–198.

768. Barbero, S., Bajetto, A., Bonavia, R., Porcile, C., Piccioli, P., Pirani, P., Ravetti, J.L., Zona, G., Spaziante, R., Florio, T., Schettini, G. (2002) Expression of the chemokine receptor CXCR4 and its ligand stromal cell-derived factor 1 in human brain tumors and their involvement in glial proliferation in vitro. *Ann. N Y Acad. Sci.* **973**, 60–69.

769. Li, J., Yen, C., Liaw, D., Podsypanina, K., Bose, S., Wang, S.I., Puc, J., Miliaresis, C., Rodgers, L., McCombie, R., Bigner, S.H., Giovanella, B.C., Ittmann, M., Tycko, B., Hibshoosh, H., Wigler, M.H., and Parsons, R. (1997) *PTEN*, a putative protein tyrosine phosphatase gene mutated in human brain, breast, and prostate cancer. *Science* **275**, 1943–1947.

770. Hartmann, W., Digon-Söntgerath, B., Koch, A., Waha, A., Endl, E., Dani, I., Denkhaus, D., Goodyer, C.G., Sörensen, N., Wiestler, O.D., and Pietsch, T. (2006) Phosphatidylinositol 3—kinase/AKT signaling is activated in medulloblastoma cell proliferation and is associated with reduced expression of *PTEN*. *Clin. Cancer Res.* **12**, 3019–3027.

771. Broderick, D.K., Di, C., Parrett, T.J., Samuels, Y.R., Cummins, J.M., McLendon, R.E., Fults, D.W., Velculescu, V.E., Bigner, D.D., and Yan, H. (2004) Mutations of the *PIK3CA* in anaplastic oligodendrogliomas, high-grade astrocytomas, and medulloblastomas. *Cancer Res.* **64**, 5048–5050.

772. Damodar Reddy, C., Marwaha, S., Patti, R., Raghunath, M., Duhaime, A.C., Sutton, L., and Phillips, P.C. (2001) Role of MAP kinase pathways in primitive neuroectodermal tumors. *Anticancer Res.* **21**, 2733–2738.

773. Cheng, Y.C., Lee, C.J., Badge, R.M., Orme, A.T., and Scotting, P.J. (2001) Sox8 gene expression identifies immature glial cells in developing cerebellum and cerebellar tumours. *Brain Res. Mol. Brain Res.* **92**, 193–200.

774. Scotting, P.J., Thompson, S.L., Punt, J.A.G., and Walker, D.A. (2000) Paediatric brain tumours: an embryological perspective. *Childs Nerv. Syst.* **16**, 261–268.

775. Lee, C.J., Chan, W.I., Cheung, M., Cheng, Y.C., Appleby, V.J., Orme, A.T., and Scotting, P.J. (2002) *CIC*, a member of a novel subfamily of the HMG-box superfamily, is transiently expressed in developing granule neurons. *Brain Res. Mol. Brain Res.* **106**, 151–156.

776. Lee, C.J., Chan, W.I., and Scotting, P.J. (2005) *CIC*, a gene involved in cerebellar development and ErbB signaling, is significantly expressed in medulloblastomas. *J. Neurooncol.* **73**, 101–108.

777. Thompson, M.C., Fuller, C., Hogg, T.L., Dalton, J., Finkelstein, D., Lau, C.C., Chintagumpala, M., Adesina, A., Ashley, D.M., Kellie, S.J., Taylor, M.D., Curran, T., Gajjar, A., and Gilbertson, R.J. (2006) Genomics identifies medulloblastoma subgroups that ar enriched for specific genetic alterations. *J. Clin. Oncol.* 24, 1924–1931.

778. Raffaghello, L., Nozza, P., Morandi, F., Camoriano, M., Wang, X., Garrè, M.L., Cama, A., Basso, G., Ferrone, S., Gambini, C., and Pistoia, V. (2007) Expression and functional analysis of human leukocyte antigen class I antigen-processing machinery in medulloblastoma. *Cancer Res.* **67**, 5471–5478.

779. Rodriguez, F.J., Eberhart, C., O'Neill, B.P., Slezak, J., Burger, P.C., Goldthwaite, P., Wu, W., and Giannini, C. (2007) Histopathologic grading of adult medulloblastomas. *Cancer* **109**, 2557–2565.

780. Frühwald, M.C., and Plass, C. (2002) Metastatic medulloblastoma—therapeutic success through molecular target identification? *Pharmacogenomics J.* **2**, 7–10.

781. Dong, J., Gailani, M.R., Pomeroy, S.L., Reardon, D., and Bale, A.E. (2000) Identification of PATCHED mutations in medulloblastomas by direct sequencing. *Hum. Mutat.* **16**, 89–90.

782. Pomeroy, S.L., and Sturla, L.-M. (2003) Molecular biology of medulloblastoma therapy. *Pediatr. Neurosurg.* **39**, 299–304.

783. Gilbertson, R.J. (2004) Medulloblastoma: signaling a change in treatment. *Lancet Oncol.* **5**, 209–218.

784. Newton, H.B. (2004) Moleuclar neuro-oncology and development of targeted therapeutic strategies for brain tumors. Part 2: PI3K/Akt/PTEN, mTOR, SHH/PTCH and angiogenesis. *Expert Rev. Anticancer Ther.* **4,** 105–128.

785. Koch, A., Hrychyk, A., Hartmann, W., Waha, A., Mikeska, T., Waha, A., Schüller, U., Sörensen, N., Berthold, F., Goodyer, C.G., Wiestler, O.D., Birchmeier, W., Behrens, J., and Pietsch, T. (2007) Mutations of the Wnt antagonist *AXIN2* (*Conductin*) result in TCF-dependent transcription in medulloblastomas. *Int. J. Cancer* **121,** 284–291.

786. Leung, J.Y., Kolligs, F.T., Wu, R., Zhai, Y., Kuick, R., Hanash, S., Cho, K.R., and Fearon, E.R. (2002) Activation of AXIN2 expression by β-catenin-T cell factor. A feedback repressor pathway regulating Wnt signaling. *J. Biol. Chem.* **277,** 21657–21665.

787. Hrnan, R., Fasheh, R., Calabrese, C., Frank, A.J., Maclean, K.H., Allard, D., Barraclough, R., and Gilbertson, R.J. (2003) ERBB2 up-regulates *S100A4* and several other prometastatic genes in medulloblastoma. *Cancer Res.* **63,** 140–148.

788. McCall, T.D., Pedone, C.A., and Fults, D.W. (2007) Apoptosis suppression by somatic cell transfer of Bcl-2 promotes Sonic hedgehog-dependent medulloblastoma formation in mice. *Cancer Res.* **67,** 5179–5185.

789. Schüller, U., Schober, F., Kretzschmar, H.A., and Herms, J. (2004) Bcl-2 expression inversely correlates with tumour cell differentiation in medulloblastoma. *Neuropathol. Appl. Neurobiol.* **10,** 513–521.

790. Jaffey, P.B., To, G.T., Xu, H.-J., Hu, S.-X., Benedict, W.F., Donoso, L.A., and Campbell, G.A. (1995) Retinoblastoma-like phenotype expressed in medulloblastoma. *J. Neuropathol. Exp. Neurol.* **54,** 664–672.

791. Lee, W.-H., Bookstein, R., Hong, F., Young, L.-J., Shew, J.-Y., Lee, E.Y.-H.P. (1987) Human retinoblastoma susceptibility gene: cloning, identification, and sequence. *Science* **235,** 1394–1399.

792. Lawinger, P., Venugopal, R., Guo, Z.S., Immaneni, A., Sengupta, D., Lu, W., Rastelli, L., Marin Dias Carneiro, A., Levin, V., Fuller, G.N., Echelard, Y., and Majumder, S. (2000) The neuronal repressor REST/NRSF is an essential regulator in medulloblastoma cells. *Nat. Med.* **6,** 826–831.

793. Haberler, C., Laggner, U., Slavc, I., Czech, T., Ambros, I.M., Ambros, P.F., Budka, H., and Hainfellner, J.A. (2006) Immunohistochemical analysis of INI1 protein in malignant pediatric CNS tumors: Lack of INI1 in atypical teratoid/rhabdoid tumors and in a fraction of primitive neuroectodermal tumors without rhabdoid phenotype. *Am. J. Surg. Pathol.* **30,** 1462–1468.

794. Kunz, F., Shalaby, T., Lang, D., von Bueren, A., Hainfellner, J.A., Slavc, I., Tabatabai, G., and Grotzer, M.A. (2006) Quantitative mRNA expression analysis of neurotrophin-receptor TrkC and oncogene c-MYC from formalin-fixed, paraffin-embedded primitive neuroectodermal tumor samples. *Neuropathology* **26,** 393–399.

795. Rutkowski, S., von Bueren, A., von Hoff, K., Hartmann, W., Shalaby, T., Deinlein, F., Warmuth-Metz, M., Soerensen, N., Emser, A., Bode, U., Mittler, U., Urban, C., Benesch, M., Kortmann, R.D., Schlegel, P.G., Kuehl, J., Pietsch, T., and Grotzer, M. (2007) Prognostic relevance of clinical and biological risk factors in childhood medulloblastoma: Results of patients treated in the prospective multicenter trial HIT'91. *Clin. Cancer Res.* **13,** 2651–2657.

796. Dallas, P.B., Terry, P.A., and Kees, U.R. (2005) Genomic deletions in cell lines derived from primitive neuroectodermal tumors of the central nerovus system. *Cancer Genet. Cytogenet.* **159,** 105–113.

797. Shlomit, R., Ayala, A.-G., Michal, D., Ninett, A., Frida, S., Boleslaw, G., Gad, B., Gideon, R., and Shlomi, C. (2000) Gains and losses of DNA sequences in childhood brain tumors analyzed by comparative genomic hybridization. *Cancer Genet. Cytogenet.* **121,** 67–72.

797A. Pfister, S., Remke, M., Toedt, G., Werft, W., Benner, A., Mendrzyk, F., Wittmann, A., Devens, F., von Hoff, K., Rutkowski, S., Kulozik, A., Radlwimmer, B., Scheurlen, W., Lichter, P., Korshunov, A. (2007) Supratentorial primitive neuroectodermal tumors of the central nervous system frequently harbor deletions of the *CDKN2A* locus and other genomic aberrations distinct from medulloblastomas. *Genes Chromosomes Cancer* **46,** 839–851.

798. Aparecida Rainho, C., Rogatto, S.R., Correa de Moraes, L., and Barbieri-Neto, J. (1992) Cytogenetic study of a pineocytoma. *Cancer Genet. Cytogenet.* **64,** 127–132.

799. Bello, M.J., Rey, J.A., de Campos, J.M., and Kusak, M.E. (1993) Chromosomal abnormalities in a pineocytomas. *Cancer Genet. Cytogenet.* **71,** 185–186.

800. Holland, H., Koschny, R., Krupp, W., Meixensberger, J., Bauer, M., Schober, R., Kirsten, H., Walczak, H., and Ahnert, P. (2007) Cytogenetic and molecular biological characterization of an adult medulloblastoma. *Cancer Genet. Cytogenet.* **in press.**

801. Mühlisch, J., Bajanowski, T., Rickert, C.H., Roggendorf, W., Würthwein, G., Jürgens, H., and Frühwald, M.C. (2007) Frequent but borderline methylation of $p16^{INK4a}$ and *TIMP3* in medulloblastoma and sPNET revealed by quantitative analysis. *J. Neurooncol.* **83,** 17–29.

802. Giles, R.H., van Es, J.H., and Clevers, H. (2003) Caught up in a Wnt storm: Wnt signaling in cancer. *Biochim. Biophys. Acta.* **1653,** 1–24.

803. de Bont, J.M., den Boer, M.L., Kros, J.M., Passier, M.M.C.J., Reddingius, R.E., Sillevis Smitt, P.A.E., Luider, T.M., and Pieters, R. (2007) Identification of novel biomarkers in pediatric primitive neuroectodermal tumors and ependymomas by proteome-wide analysis. *J. Neuropathol. Exp. Neurol.* **66,** 505–516.

804. Bigner, S.H., and Schröck, E. (1997) Molecular cytogenetics of brain tumors. *J. Neuropathol. Exp. Neurol.* **56,** 1173–1181.

804A. Waha, A., Koch, A., Hartmann, W., Milde, U., Felsberg, J., Hübner, A., Mikeska, T., Goodyer, C.G., Sörensen, N., Lindberg, I., Wiestler, O.D., Pietsch, T., and Waha, A. (2007) *SGNE1/7B2* is epigenetically altered and transcriptionally downregulated in human medulloblastomas. *Oncogene* **26,** 5662–5668.

805. Bayani, J., Pandita, A., and Squire, J.A. (2005) Moleuclar cytogenetic analysis in the study of brain tumors: findings and applications. *Neurosurg. Focus* **19,** E1.

806. Ahmed, N., Ratnayake, M., Savoldo, B., Perlaky, L., Dotti, G., Wels, W.S., Bhattacharjee, M.B., Gilbertson, R.J., Shine, H.D., Weiss, H.L., Rooney, C.M., Heslop, H.E., and Gottschalk, S. (2007) Regression of experimental

(a)   **T-loop**

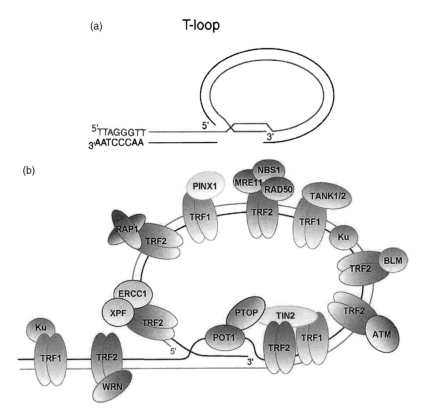

(b)

FIGURE 9.1. (a) The primary and secondary structure of the human telomere showing the G strand overhang forming the G-loop. (b) Simplified diagram of the protein–DNA complexes at the telomere (from refs. (5) and (7), with permission).

also would exist in humans (e.g., somatic strand-specific imprinting and segregation) (14).

The DNA of higher eukaryotes is methylated almost exclusively at cytosine residues whish are located 5′ to a guanosine residue in a CpG dinucleotide. CpG occurs less frequently in the mammalian genome than predicted by chance, and these dinucleotides are clustered in the promoter regions of many genes known as CpG islands. Methylation of the cytosines in CpG islands is associated with the transcriptional silencing of the adjacent gene. Methylation of the CpG islands may show a distinct tissue-specific pattern, is associated with imprinting of genes, and is involved with the differential transcriptional silencing of genes on the X chromosome in its inactivation in mammalian females.

DNA methylation is effected by a family of DNA methyltransferases (DNAMTs). DNAMT1 is the primary enzyme involved with maintaining methylation, and it has high affinity for hemimethylated, newly synthesized DNA (15). DNAMT3a and DNAMT3b are involved with de novo DNA methylation, and they participate in embryogenesis (16).

A technique called methylations specific-polymerase chain reaction (MSP) has been developed to examine methylation of specific CpG islands (17). This technique is based on the chemical conversion of cytosine to uracil by exposure to the bisulfite ion. Cytosine, which is methylated, is protected from the action of bisulfite. Thus, when treated with bisulfite, the base sequence of the DNA is converted from cytosine to uracil at every unmethylated site. PCR primers can then be

designed to recognize either the methylated (unconverted) or the unmethylated (converted) sequences. Primers are designed such that amplification products from the methylation-specific primers are of a different size from those specific for the unmethylated sequence. The fragment sizes are then resolved by gel electrophoresis. MSP is currently widely used for the analysis of CpG island methylation in both research and clinical settings.

Histone modifications occur primarily at the amino terminus and involve acetylation, methylation, and ADP ribosylation. The most studied histone modifications are of acetylation and methylation of the amino tails of histones H3 and H4 (18). In general, histone acetylation is associated with transcriptionally active chromatin, whereas methylation is associated with transcriptionally silent chromatin. An exception is the methylation of lysine 4 (K4) of histone H3, which is found in active chromatin and actually promotes acetylation of other lysine residues in H3 (19).

Histone acetyltransferases and histone deacetylases (HDACs) mediate histone modification. These enzymes affect the binding of transcription factors and are, in turn, recruited to regions of chromatin by transcription factors. In addition, the N-terminal domain of DNAMT1 binds to HDACs and inhibits transcription by increasing HDAC activity (20). DNAMT3a colocalizes with proteins associated with methylated histones (21). Therefore, epigenetic regulation of genes expression involves a complex interaction between histones,

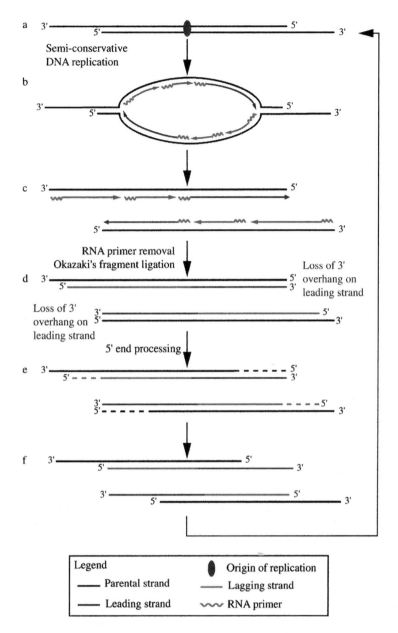

a 3' ——————————————●——————————— 5'
          5'   3'

Semi-conservative
DNA replication

b
3' ——————   —————— 5'
          5'   3'

c 3' ——————————————————————— 5'
       wwww————→ wwww————→ wwww————→

                            ←———— wwww ←———— wwww ←————
          5'   3'

RNA primer removal
Okazaki's fragment ligation   Loss of 3'
                                                overhang on
d 3' ——————————————————— 5'   leading strand
          5'   3'

Loss of 3'   3' ——————————————— 5'
overhang on   5'   3'
leading strand

5' end processing

e 3' ——————————————————— - - - 5'
          5' - - - - ————————————— 3'

          3' —————————————— - - - - 5'
          5' - - - - ——————————————— 3'

f 3' ————————————————— 5'
          5'   3'

          3' ————————————————— 5'
          5'   3'

Legend   ● Origin of replication
—— Parental strand   —— Lagging strand
—— Leading strand   www RNA primer

FIGURE 9.2. Diagrammatic representation of the shortening of telomere length with each cycle of semiconservative replication (from ref. (9), with permission).

DNA, the enzymes that modify chromatin and specific chromatin binding factors such as transcription factors.

An aberrant pattern of methylation is often observed in CpG islands of specific genes in human cancers. The pattern is usually that of a gene whose CpG island is unmethylated (transcriptionally active) in normal tissue and methylated (transcriptionally inactive) in the tumor. The class of genes showing this pattern is one in which transcriptional inactivation also can occur by deletion of the gene sequence. Because the loss of function of these genes is associated with the disruption of the normal cell cycle in cancer, as a class they are referred to as tumor suppressor genes. This phenomenon has been described for many different types of human tumors, including carcinomas and adenocarcinomas,

bone and soft tissue tumors, benign tumors, and both malignant and premalignant hematological disorders (reviewed in (22–24)). At least in some instances, changes in methylation are thought to be one of the initiating steps in tumorigenesis (25–27). Methylation status of the CpG islands of particular genes also has been shown to have prognostic value in the clinical evaluation of cancer patients (28). Studies of cultured cancer cells (28–34) and of treated patients (35) have shown potential for demethylating agents as therapeutic candidates. Recently, the appearance of methylated CpG islands of tumor suppressor genes in DNA from serum, bronchial lavage and urine from cancer patients has been demonstrated and may portend recurrence, and, it is hoped, possibly serve as a screen for the early detection of cancer (36–40).

TP53: The Cell's Sentry

The *TP53* gene, localized to 17p13.1, encodes a protein of 393 amino acids whose primary function seems to be surveillance for cellular damage events and promoting cell cycle arrest, apoptosis and senescence; recent studies have shown p53 to also be involved in regulating glycolysis, autophagy, repair of genotoxic damage, regulation of oxidative stress, invasion and motility, angiogenesis, differentiation, and bone remodeling (41). Disruption of p53 by mutation or deletion is associated with tumor progression, as would be expected if its cell cycle arresting and apoptotic induction functions are compromised. In addition, other gene products can diminish the function of p53. Amplification of *MDM2* promotes degradation of p53, and the HPV E6 protein specifically inhibits p53 (42).

TP53 transcription is induced by cellular stresses such as DNA damage, oncogene activation, hypoxia, nutrient deprivation, telomere erosion, and ribosomal stress. The p53 protein interacts with the Wilms' tumor protein, WT1, and, depending on the degree of stress, either induces inhibitors of cyclin-dependent kinases (CDKs) such as p21, or proapoptotic genes such as PUMA or NOXA. Inhibition of CDK2 and CDK4 arrests the cell cycle in late G1. In the apoptotic pathway, the p53–WT1 complex activates proapoptotic genes whose protein products inhibit BCL2, which inhibit the apoptosis-promoting BAX protein (both of which are described in more detail below). This pathway is shown schematically in Figure 9.3.

pRB Pathway

The other major pathway regulating the cell cycle is that mediated by the retinoblastoma protein (RB1). The protein is a 928 amino acid phosphoprotein coded for by a gene localized to 13q14.1-13q14.2. In its unphosporylated form, it complexes with and inactivates E2F transcription factors. Phosphorylated pRB is unable to bind E2A. Free E2F binds to DNA and promotes transcription of genes necessary for the G1⇒S transition. Therefore, binding of E2A by pRB represents a major point at which control of the cell cycle is exerted. pRB is itself regulated by CDKs that promote phosphorylation and inactivation of pRB. The interface with CDKs connects the p53 and pRB pathways. Although p53-mediated cell cycle control is primarily induced by cellular stress, the pRB control is constitutive. The pRB pathway is shown schematically in Figure 9.4. Most solid tumors have now been shown to have disruptions of either the p53 or the pRB pathway, and it is possible that such disruption is required for tumor progression.

Cell Cycle

The cell cycle is critical in carcinogenesis, because its dysregulation results in the uncontrolled cell division associated with tumor proliferation. As already mentioned, pRB and p53 are the primary controllers of whether the cell cycle proceeds. They mediate the effect of stage-specific cycle

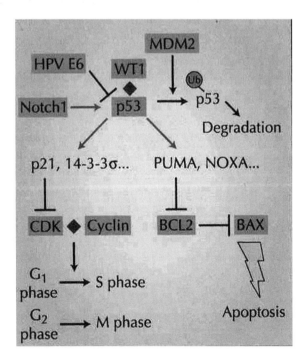

FIGURE 9.3. Simplified diagram of the p53 pathway. See text for details (from ref. (42), with permission).

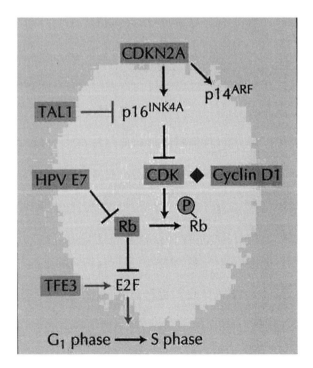

FIGURE 9.4. Simplified diagram of the pRB pathway. See text for details (from ref. (42), with permission).

progression factors, promoting entry into and continuation of the cell cycle. The various phases of the cell cycle are entered into by promotion by the complex of cyclins and CDKs. As shown in Figure 9.5, CyclinD1/CDK4/6 promotes entry into and progression through G1. CyclinE/CDK2 is associated with the transition from G1 to S, whereas cyclinA/CDK2 is expressed through S and early G2. The progression through G2 into M is mediated by cyclinA/CDK1 and cyclinB/CDK1 (55,56).

TP53-mediated checkpoints exist in G1 and G2, during which the integrity of the genome before replication (G1 checkpoint) and after replication, but before chromosome segregation and cell division (G2 checkpoint). It is at these points that, in response to genome damage, p53 can induce DNA repair mechanisms or an apoptotic response (vide infra) depending on the nature of the damage. p53 inhibits the cell cycle via cyclin-dependent kinase inhibitors (e.g., Ink4a–d, Cip1, and Kip1/2), arresting the cell cycle at either checkpoint (57).

Overexpression of cyclins might be expected to promote carcinogenesis by stimulating the cell cycle. Cyclin D1 (also called PRAD1) is amplified in several solid tumors (47–49). Cyclin D1 can also be activated by an inversion and translocation (49, 50). In addition to stimulating the cell cycle directly, overexpression of cyclin D1 also up-regulates fibroblast growth factor receptor and promotes hyperphosphorylation of pRB, thus upregulating E2F-1 (51). A truncated form of cyclin E is overexpressed in breast cancer, and it seems to induce the G1⇒S transition more rapidly than the full-length form (52). CDK4 is amplified in gliomas (53).

Apoptosis

Programmed cell death (PCD) is an important process in embryological development and normal tissue turnover. The 10 known pathways leading to PCD can be classified as apoptotic and nonapoptotic. Nonapoptotic death includes

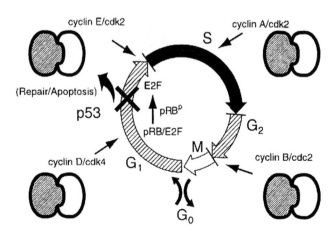

FIGURE 9.5. Simplified representation of the cell cycle and the genes affecting it. See text for details (from ref. (43), with permission).

autophagic cell death, necrosis, mitotic cell death, and senescence. Apoptosis includes caspase-dependent processes of "classic" or type I apoptosis, anoikis (death due to detachment from substrate or other cells) and the caspase-independent cell death pathway (CICD) (54).

Type I apoptosis is characterized by membrane blebbing, cytoplasmic shrinkage, and reduced cell volume (pyknosis), chromatin condensation, and nuclear fragmentation (karyorrhexis), which all lead to the formation of apoptotic bodies, the characteristic cytological feature of apoptosis. Several genes govern the apoptotic process, some of which are mutated or otherwise dysregulated in tumors. The actual destruction of cellular and biochemical targets is mediated by caspases, a family of zymogenic proteases. The range of targets and effects of caspase cleavage is shown in Figure 9.6.

Type I apoptosis can be induced by either intrinsic or extrinsic factors. The extrinsic pathway is triggered by the binding of an extracellular ligand (e.g., FasL and tumor necrosis factor [TNF]-α) to its cell surface receptor (e.g., Fas/CD95 and TNF-R1), which recruits cytosolic adaptor proteins (e.g., Fas-associated death domain) and subsequent activation of caspases-8 and -10. These caspases, in turn, activate the effector caspases-3, -6, and -7, which initiate the cellular degradation processes associated with apoptosis. In addition, caspases-8 and -10 induce the mitochondria to release cytochrome c through permeabilization of the mitochondrial outer membrane. Cytochrome c then complexes with caspase-9 and apoptotic protease activating factor-1 (Apaf-1), in a structure called the apoptosome to activate the effector caspases.

The intrinsic pathway is triggered by intracellular stresses such as DNA damage cytoskeletal damage, endoplasmic reticulum dysfunction, loss of cell adhesion, and growth factor starvation. Under such conditions, the mitochondria release cytochrome c by the outer membrane permeabilization (MOMP) initiating the apoptosome-mediated process described above. The mitochondria also release several other proapoptotic molecules, such as Smac/Diablo (second mitochondrial activator of caspases/direct inhibitor of apoptosis (IAP) binding protein with low pH) and Omi/HtrA2 (high temperature requirement factor), which inactivate various IAPs. IAPs bind to activated caspase-3, -7, and 9, inhibiting their activities. Thus, Smac/Diablo and Omi/HtrA2 potentiate apoptosis by blocking IAP caspase inhibition.

The p53 tumor suppressor protein positively regulates both the extrinsic and intrinsic pathways. p53 expression transactivates genes for the proapoptotic members of the Bcl-2 family (e.g., Bax, Bak, Bid, Puma, and Noxa), the disruptor of mitochondrial function (p53AIP), Apaf-1, effector caspases, and p53-induced protein with death domain among others. p53 represses genes for antiapoptotic proteins such as survivin. p53 also translocates to the mitochondria in response to apoptotic stimuli, where it neutralizes antiapoptotic proteins (e.g., Bcl-xL) and activates Bax and Bak, resulting in the inhibition of the antiapoptotic activity of Bcl-2.

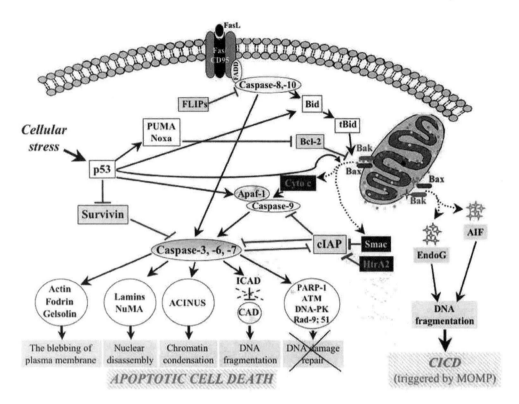

FIGURE 9.6. Diagram of the CCID pathways of apoptosis. See text for details (from ref. (47), with permission).

The CICD pathway is induced when MOMP has been triggered, but caspase activation is inhibited. Mitochondrial function is lost, and the proteins of the intermembrane space are released into the cytoplasm. Two of these proteins, apoptosis-inducing factor and endonuclease G, directly effect DNA fragmentation without caspase activation.

It can be appreciated from this complexity that there are many targets for disruption of the apoptotic process in tumorigenesis. Loss of proper apoptotic function can allow damaged (mutated) cells to survive, continuing to acquire the genetic changes associated with tumor progression. Loss or diminution of the apoptotic response has been proposed as a mechanism of resistance to chemotherapy (55), but this hypothesis has recently been questioned (56).

Signaling Pathways

The signaling pathways mediated by the Notch, sonic hedgehog (SHH), Wingless (Wnt), and the receptor tyrosine kinase (RTK) cell surface receptors play important roles in neurogenesis and in neural tumors.

Notch

The Notch pathway affects the development of neural potential and regulates whether these cells differentiate into neural or glial cells (57). Notch and its ligands. Delta and Jagged, are transmembrane proteins with large extracellular domains of epidermal growth factor (EGF)-like repeats. After ligand activation and binding to Notch, the Notch intracellular domain (Ncid) is cleaved by ADAM-family metalloproteinases and by γ-secretase, and then it is translocated to the nucleus where it effects its signaling function (the intricacies of ligand and receptor activation are reviewed in (58–62)).

Once inside the nucleus, Ncid forms a trimeric complex with the DNA-binding protein CSL [CBF1, Su(H), LAG-1] and the coactivator Mastermind. This complex recruits histone acetylases and chromatin remodeling complexes, promoting transcription of the target genes. The bound Ncid is modified by kinases such as CDK8 and SEL 10, making it a substrate for proteosomal degradation. In the absence of Ncid, corepressors associate with CSL and recruit histone deactylases, resulting in transcriptional repression of the target genes. The cycle of Notch target gene activation and repression is shown in Figure 9.7.

Sonic Hedgehog

The sonic hedgehog family is a member of a family of proteins involved in embryonic patterning. Desert hedgehog (Dhh) is involved with male germ cell development. Indian hedgehog (Ihh) functions in the development of cartilage (63). The SHH signaling pathway has a key role in the differentiation of ventral cell in the central nervous system, and in axis patterning in the developing limb, lung morphogenesis, angiogenesis, and bone and cartilage formation (64, 65). Loss of

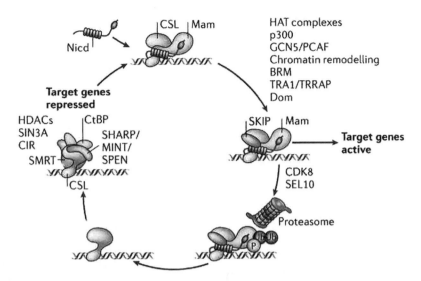

FIGURE 9.7. Notch signaling pathway cycle showing activation of target genes by the Nicd and repression of target genes upon notch degradation. See text for details (adapted from ref. (51), with permission).

function in the various components of the SHH pathway is associated with developmental anomalies (66). Unrepressed SHH signaling is associated with several cancers, including those discussed in previous chapters (67–69).

Hedgehog proteins are extracellular signaling molecules. SHH binds to its membrane receptor, Patched1 (Ptch1), which, in turn, allows a second membrane protein, Smoothed (Smo), to act as a G-protein coupled receptor. G_1 proteins and arrestin are recruited to Smo. G proteins are second messengers that in the SHH pathway activate phosphoinositide 3-kinase (PI3K), which, in turn phosphorylates the Akt protein. Activated Akt inhibits glycogen synthase kinase-3β (GSK-3β). GSK-3β phosphorylates the mycn transcription factor leading to its degradation. Therefore, inactivation of GSK-3β leads to unrepressed mycn-driven transcription. GSK-3β also phosphorylates the Gli transcription factors, causing them to remain cytoplasmic. The suppressor of fused (SuFu) protein in concert with the Smo-G_1-arrestin complex facilitates the GSK-3β inactivation of Gli. The Smo-G_1-arrestin complex also inhibits adenyl cyclase (AC). AC activates cAMP-dependent protein kinase, which also phosphorylates and inactivates GSK-3β. This complex, interwoven set of pathways is shown in Figure 9.8.

Wingless

Wnts are a family of secreted lipid-modified glycoproteins (70), which, under normal conditions, are involved in maintaining stem cell-like fates in various renewing tissues, such as skin, intestinal epithelium, and hematopoietic cells (71–73). With its cellular self-renewal function, it is not surprising that activation of the Wnt signaling pathway occurs in numerous human cancers (74–76).

Wnt signaling occurs in two generalized ways: one that cause stabilization and nuclear translocation of β-catenin

(the canonical pathway) and another that does not affect β-catenin. In the canonical pathway (reviewed in (77, 78)), Wnt complexes with the cell surface receptor Frizzled (Fz) and low-density lipoprotein receptor-relate protein (LRP). The Wnt-LRP-Fz complex binds the Disheveled protein (Dsh) and the Axin-Adenomatous polyposis coli (APC)-GSK-3 complex, phosphorylating Axin and Dsh. Unbound Axin-APC-GSK-3 binds to β-catenin, marking it for ubiquitylation and destruction by the proteosome. With the Axin-APC-GSK-3 complex inactivated by the Wnt-LRP-Fz complex, β-catenin is free to enter the nucleus, where it displaces transcriptional repressors, such as Groucho, allowing transcription factors, such as T-cell specific transcription factor/lymphoid enhancer protein factor 1 to effect

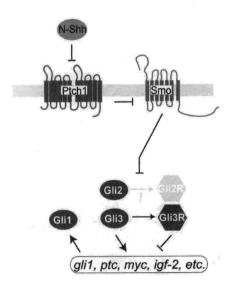

FIGURE 9.8. The Shh signaling pathway. See text for details (from ref. (69), with permission).

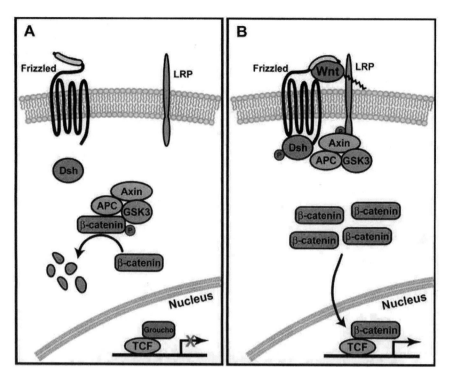

FIGURE 9.9. The canonical Wnt pathway. See text for details (from ref. (74), with permission).

transcription of target genes. This pathway is shown diagrammatically in Figure 9.9.

Noncanonical Wnt signaling acts independently of β-catenin, initiating a variety of cellular responses (reviewed in (79–81)) In the best understood branch of the noncanonical pathway, Wnt binds to Fz in the presence of RTK, specifically the Ryk family, instead of LRP. This initiates calcium influx, activating calcium-sensitive kinase II and protein kinase C (PKC). It is also thought that another, noncanonical pathway involves the activation of Rho GTPases. In this scheme, the Formin homology protein Daam1 binds to both Dsh and Rho, activating it as a GTPase, in turn activating Rho kinase, one of

its effectors. Activation of cdc42 by the noncanonical pathway also has been suggested. The noncanonical pathway is shown diagrammatically in Figure 9.10.

Receptor Tyrosine Kinases

RTKs are cell surface molecules that, on binding of ligand, activate its intracellular tyrosine kinase domain, resulting in autophosphorylation and recruitment of downstream effector molecules (reviewed in (82)). RTKs are a superfamily of receptors that include the epidermal growth factor receptor

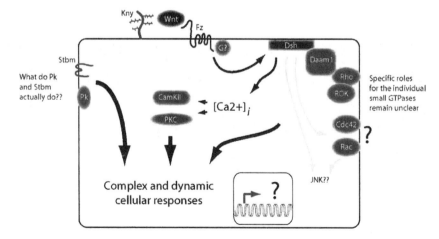

FIGURE 9.10. The noncanonical Wnt pathway showing calcium-mediated kinase activation and Rho kinase activation. See text for details (from ref. (72), with permission).

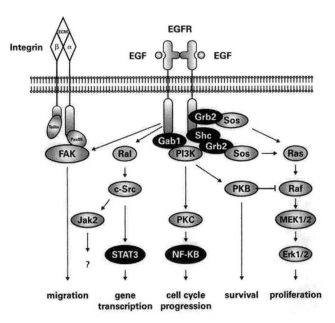

FIGURE 9.11. The canonical EGFR signal transduction pathway. See text for details (from ref. (76), with permission).

(EGFR), fibroblast growth factor, and platelet-derived growth factor families, and various other cytokine receptors.

The best understood and most studied of the RTK families is EGFR. Most cancers have aberrant EGFR-mediated signalling (83). This group consists of EGFR (also known as HER1/ErbB-1), HER2/neu (ErbB-2), HER3 (ErbB3) and HER4 (ErbB4). Ligands for these receptors include EGF, transforming growth factor-α, and amphiregulin, which are specific to EGFR, β-cellulin, epiregulin, and heparin-binding-EGF, which bind to both EGFR and HER4, and neuregulin, which binds to both HER3 and HER4 and their heterodimer.

Ligand binding induces dimerization of EGFR and subsequent activation of its intracellular tyrosine kinase domain. This results in phosphorylation of critical tyrosine residues in the EGFR cytoplasmic domain that act as docking sites for specific effector and adaptor proteins, which are then activated by the kinase activity. Downstream effects then can follow several pathways. The best-understood effect is through the mitogen-activate protein kinase (MAPK) known as mitogen-activated protein kinase kinase (MEK). Phosphorylated EGFR recruits adaptor proteins Shc and Grb2, which through a third factor, Sos, activate the Ras protein. Ras activates the Raf protein, which then activates MEK. MEK then activates a

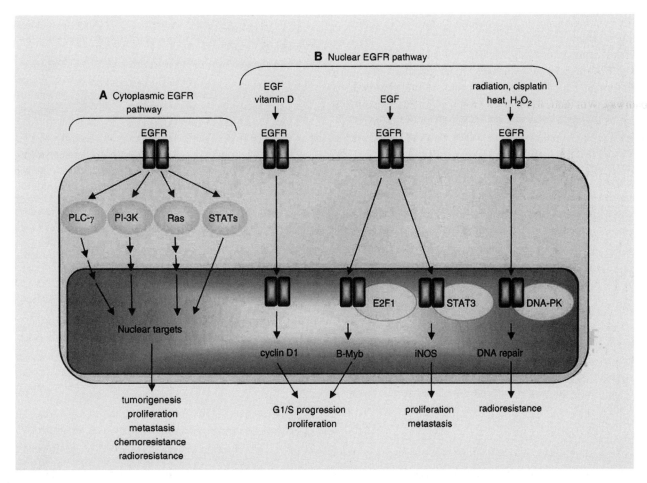

FIGURE 9.12. The noncanonical, nuclear EGFR pathway. See text for details (from ref. (77), with permission).

second MAPK, extracellular signal-regulated kinase, which in turn, induces expression of transcription factors. The various downstream pathways are shown diagrammatically in Figure 9.11.

EGFR also stimulates cell division and survival by way of activation of PI3K, mediated by the adaptor protein Gab1. PI3K then activates the PKC-NF-κB pathway as already discussed. EGFR also can act through c-Src activation, which, in turn, activates the signal transducer and activator of transcription (STAT) family of transcription factors. Finally, the EGFR tyrosine kinase inactivates focal adhesion kinase, a key component of integrin mediated cell–cell and cell–matrix interactions.

EGFR has recently been shown to have noncanonical pathways in which, upon ligand binding, the dimerized EGFR is translocated directly into the nucleus, where it up-regulates cyclin D1, transcription factors E2F1 and STAT3, and activates DNA-mediated protein kinase, which, in turn, activates various transcription factors (84). The noncanonical pathway for EGFR signaling is shown in Figure 9.12.

Complex Systems Modeling

Signal transduction involves very complex interactions between components of the various pathways. In addition, many proteins are involved in several different types of pathway. This complexity implies that an understanding of the effects of one pathway on cellular metabolism cannot be fully understood without taking into account these effects on other pathways. The field of complex systems modeling has emerged from such an appreciation for this complexity (85, 86). By creating algorithms that account for all of the possible branches in all of the pathways, and using the power of advanced computing technology, this field hopes to be able to predict the specific effect of mutational changes in pathway components and the effect of specific therapeutic compounds on the entirety of cellular metabolism, thereby creating better diagnosis, prognosis, and therapeutic regimens.

Chromosome Abnormalities

For descriptions of chromosome nomenclature, details of banding techniques and band identification, fluorescence in situ hybridization, spectral karyotyping, and related techniques, comparative genomic hybridization methodology and the genesis, recognition, and meaning of cytogenetic changes in tumors (e.g., translocations, inversions, insertions, terminal and interstitial deletions, and additions), see (87–93). The central dogma of the translation scheme of DNA producing messenger RNA (mRNA) which is then responsible for protein production has been "revolutionized" by the discovery of microRNAs (miRNAs) and other small noncoding interference RNAs (ncRNAs) (94–96).

The miRNAs of which hundreds or possibly > 1000 are present in human cells, are singlestranded and *ca* 22 nucleotides in size and make up a small percentage of cellular RNA. The main function of miRNA is to guide the regulation of mRNA translation by interacting with the 3'-untranslated regions of target mRNA. A single miRNA may recognize a large number of different mRNAs (possibly > 1000). The miRNAs, although noncoding in nature, play a role in cell proliferation, apoptosis and differentiation. Thus, they act in the nature of tumor preventors and not surprisingly miRNAs are very often up- or downregulated in a variety of tumors examined to-date.

A vast and burgeoning literature has been and is being developed on miRNA changes in cancers and leukemias (97–99). However, miRNA data on the tumors presented in this book (except for a few studies in NB; see Chapter 7) are still to be realized, but will undoubtedly shed light on hitherto undeciphered genetic and molecular facets of the tumors presented. The role of miRNA in cancer biology, the significance of their cellular effects, the consequences of changes in miRNA and their applicability to therapeutic approaches remain essentially to be determined (100–105).

References

1. Mefford, H. C., and Trask, B.J. (2002) The complex structure and dynamic evolution of human subtelomeres. *Nat. Rev. Genet.* **3,** 91–102.
2. Reithman, H., Ambrosini, A., and Paul, S. (2005) Human subtelomere structure and variation. *Chromosome Res.* **13,** 505–515.
3. Wright, W.E., Tesmer, V.M., Huffman, K.E., Levene, S.D., and Shay, J.W. (2006) Normal human chromosomes have long G-rich telomeric overhangs at one end. *Genes Dev.* **11,** 2801–2809.
4. Griffeth, J.D., Corneau, L., Rosenfeld, S., Stansel, R.M., Bianchi, A., Moss, H, and de Lange, T. (1999) Mammalian telomere end in a large duplex loop. *Cell* **97,** 503–514.
5. Olaussen, K.A., Dubrana, K., Domont, J., Spano, J-P., Sabatier, L., and Soria, J.-C. (2006) Telomeres and telomerase as targets for anticancer drug development. *Crit. Rev. Oncol/hematol.* **57,** 191–214.
6. Ferreira, M.G., Miller, K.M., and Cooper, J. P. (2004) Indecent exposure: when telomeres become uncapped. *Mol. Cell* **13,** 7–18.
7. Stewart, S.A., and Weinberg, R.A. (2006) Telomeres: cancer to human aging. *Annu. Rev. Cell Dev. Biol.* **22,** 531–557.
8. Harley, C.B. (1997) Human ageing and telomeres. *Ciba Found. Symp.* **211,** 129–139.
9. Hug, N., and Lingner, J. (2006) Telomere length homeostasis. *Chromosoma* **115,** 413–425.
10. Collins, K. (2006) the biogenesis and regulation of telomerase haloenzymes. *Nat. Rev. Mol. Cell. Biol.* **7,** 484–494.
11. Stewart, S.A., and Weinberg, R.A. (2006) Telomeres: cancer to human aging. *Annu. Rev. Cell. Dev. Biol.* **22,** 531–557.
12. Marciniak, R.A., Cavazos, D., Montellano, R. Chen, Q., Guarente, L., and Johnson, F.B. (2005) A novel telomere structure in a human alternative lengthening of telomeres cell line. *Cancer Res.* **65,** 2730–2737.

13. Muntoni, A., and Reddel, R.R. (2005) The first molecular details of ALT in human tumor cells. *Hum. Mol. Genet.* **14,** R191–R196.

14. Klar, A.J. (1999) Genetic models for handedness, brain lateralization, schizophrenia, and manic-depression. *Schizophr. Res.* **39,** 207–218.

15. Robertson, K.D. (2001) DNA methylation, methyltransferases and cancer. *Oncogene* **20,** 3139–3155.

16. Okano, M., Bell, D.W., Haber, D.A., and Li, E. (1999) Dnmt3a and Dnmt3b are essential for *de novo* methylation and mammalian development. *Cell* **99,** 247–257.

17. Herman, J.G., Graff, J.R., Myöhänen, S, Nelkin, B., and Baylin, S.B. (1996) Methylation-specific PCR: a novel PCR assay for methylation status of CpG islands. *Proc. Natl. Acad. Sci. USA* **93,** 9821–9826.

18. Fischle, W, Wang, Y., and Allis, C.D. (2003) Histone and chromatin cross-talk. *Curr. Opin. Cell Biol.* **15,** 172–183.

19. Wang, H., Cao, R., Xia, L., Erdjument-Bromage, H., Borchers, C., Tempst, P., and Zhang, Y. (2001) Purification and functional characterization of a histone H3 lysine-4-specific methyltransferase. *Mol. Cell* **8,** 1207–1217.

20. Roundtree, M.R., Bachman, K.E., and Baylin, S.B. (2000) DNAMT1 binds HDAC2 and a new co-repressor, DMAP1, to form a complex at replication foci. *Nat. Genet.* **25,** 269–277.

21. Bachman, K.E., Rountree, M.R., and Baylin, S.B. (2001) Dnamt3a and Dnamt3b are transcriptional repressors that exhibit unique localization properties to heterochromatin. *J. Biol. Chem.* **276,** 32282–32287.

22. Melki, J.R., and Clark, S.J. (2002) DNA methylation changes in leukemia. *Cancer Biol.* **12,** 347–57.

23. Baylin, S.B, Esteller, M., Rountree, M.R., Bachman, K.E., Scheubel, K., and Merman, J.G. (2003) Aberrant patterns of DNA methylation, chromatin formation and gene expression in cancer. *Hum. Mol. Genet.* **10,** 687–692.

24. Herman, J.G., and Baylin, S.B. (2003) Mechanisms of disease: gene silencing in cancer in association with promoter hypermethylation. *N. Engl. J. Med.* **349,** 2042–2054.

25. Lotem, J., and Sachs, L. (2002) Epigenetics wins over genetics: induction of differentiation in tumor cells. *Semin. Cancer Biol.* **12,** 339–346.

26. Jaffe, L.F. (2003) Epigenetic theories of cancer initiation. *Adv. Cancer Res.* **90,** 209–230.

27. Wajed, S., Laird, P.W., and DeMeester, T.R. (2001) DNA methylation: an alternative pathway to cancer. *Ann. Surg.* **234,** 10–20.

28. Santini, V., Kanterajian, H.M., and Issa J.P. (2001) Changes in DNA methylation in neoplasia: pathophysiology and therapeutic implications. *Ann. Intern. Med.* **134,** 573–586.

29. Izbicka, E., MacDonald, J.R., Davidson, K., Lawrence R.A., Gomez, L., and VonHoff, D.D. (1999) 5,6-Dihydro-5′-azacytidine (DHAC) restores androgen responsiveness in androgen-insensitive prostate cancer cells. *Anticancer Res.* **19,** 1285–1291.

30. Fulda, S., Küker, M.U., Meyer, E., van Valen, F., Dockham-Dworniczak, and Debatin, K.-M. (2001) Sensitization for death receptor- or drug-induced apoptosis by re-expression of caspase-8 through demethylation or gene transfer. *Oncogene* **20,** 5865–5877.

31. Worm, J., Kirkin, A.F., Dzhandzhugazyan, K.N., and Gulberg, P. (2001) Methylation-dependent silencing of the reduced folate carrier gene in inherently methotrexate-resistant human breast cancer cells. *J. Biol. Chem.* **276,** 39990–40000.

32. Szyf, M., Pakneshan, P., and Rabbinic, S.A. (2004) DNA methylation and breast cancer. *Biochem. Pharmacol.* **68,** 1187–1197.

33. Murakami, J., Asumi, J-i., Maki, Y., Tsujigiwa, H., Kuroda, M., Nagai, N., Yanagi, Y., Inoue, I., Kumasaki, S, Tanaka, N., Matsubara, N., and Kishi, K. (2004) Effects of demethylating agent 5-aza-2('-)-deoxycytidine and histone deactylase inhibitor FR901228 on maspin gene expression in oral cancer cell lines. *Oral Oncol.* **40,** 597–603.

34. Pulukuri, S.M., and Rao, J.S. (2005) Activation of p53/p21 Waf1/Cip1 pathway by 5-aza-2'-deoxycytidine inhibits cell proliferation, induces pro-apoptotic genes and mitogen-activated protein kinases in human prostate cancer cells. *Int. J. Oncol.* **26,** 863–871.

35. Lubbert, M. (2000) DNA methylation inhibitors in the treatment of leukemias, myelodysplasic syndromes and hemoglobinopathies: clinical results and possible mechanisms of action. *Curr. Top. Microbiol. Immunol.* **249,** 135–164.

36. Jerónimo, H., Usadel, H., Henrique, R., Silva, C., Oliviera, J., Lopez. C., and Sidransky, D. (2002) Quantitative GSTP1 hypermethylation in bodily fluids of patients with prostate cancer. *Urology* **60,** 1131–2002.

37. Nayayama, Hibi, K., Taguchi, M., Takuse, T., Yamazaki, T., Kasai, U., Ito, K., Akiyama, S., and Nakao, A. (2002) Molecular detection of *p16* promoter methylation in the serum of colorectal cancer patients. *Cancer Lett.* **188,** 115–119.

38. Yamaguchi, S., Asao, T., Nakamura, J-I., Ide, M., and Kuwano, H. (2003) High frequency of *DAP-kinase* gene promoter methylation in colorectal cancer specimens and its identification in serum. *Cancer Lett.* **194,** 99–105.

39. Balaña, C., Ramirez, J.L., Taron, M., Roussos, Y., Ariza, A. Ballester, R., Sarries, C., Mendez, P., Sanchez, J.J., and Rosell, R. (2003) O^6-methyl-guanine-DNA methyltransferase methylation in serum and tumor DNA predicts response to 1,3-bis (2-chloroethyl)-1-nitrosurea but not temozolamide plus cisplatin in glioblastoma multiforme. *Clin. Cancer Res.* **9,** 1461–1468.

40. Wong, I.H.N., Zhang, J., Lai, P.B.S., Lau, W.L., and Lo, Y.M.D. (2003) Quantitative analysis of tumor-derived methylated *p16INK44a* sequences in plasma, serum and blood cells of hepatocellular carcinoma patients. *Clin. Cancer Res.* **9,** 1047–1052.

41. Vouseden, K.H., and Lane, D.P. (2007) p53 in health and disease. *Nat. Rev. Mol. Cell Biol.* **8,** 275–283.

42. Vogelstein, B., and Kinzler, K.W. (2004) Cancer genes and the pathways they control. *Nat. Med.* **10,** 789–799.

43. Cordon-Cardo, C. (1995) Mutation of cell cycle regulators: biological and clinical implications for human neoplasia. *Am. J. Pathol.* **147,** 545–560.

44. Mainprize, T.G., Taylor, M.D., Rutka, J.T., and Dirks, P.B. (2001) Cip/Kip cell-cycle inhibitors: a neuro-oncological perspective. *J. Neorooncol.* **51,** 205–218.

45. Tashiro, E., Tsuchiya, A., and Imoto, M. (2007) Functions of cyclin D1 as an oncogene and regulation of cyclin D1 expression. *Cancer Sci.* **98,** 629–635.

46. Keyomarsi, K., and Pardee, A.B. (1993) Redundant cyclin over expression and gene amplification in breast cancer cells. *Proc. Natl. Acad. Sci. USA* **90,** 1112–1116.

47. Schauer, I.E., Siriwardana, S. Langan, T.A., and Scalfani, R.A. (1994) Cyclin D1 overexpression vs. retinoblastoma inacti-

vation: implications for growth control evasion in non-small cell and small cell lung cancer. *Proc. Natl. Acad. Sci. USA* **91,** 7827–7831.

48. Gong, J., Ardelt, B., Traganos, F., and Darzynkiewicz, Z. (1994) Unscheduled expression of cyclin B1 and cyclin E in several leukemic and solid tumor cell lines. *Cancer Res.* **54,** 4285–4288.

49. Arnold, A. (1995) The cyclin D1/PRAD oncogene in neoplasia. *J. Investig. Med.* **43,** 543–549.

50. Reifenberger, G, Reifenberger, J, Ichimura, K., Meltzer, P.S., and Collins, V.P. (1994) Amplification of multiple genes from chromosomal region 12q13-q14 in human malignant gliomas: preliminary mapping of amplicons shows preferential involvement of CDK4, SAS, and MDM2. *Cancer Res.* **54,** 4299–4303.

51. Tsujimoto, Y., Yunis, J., Onorato-Showe, L. Erickson, J. Nowell, P.C., and Croce, C.M. (1984) Molecular cloning of the chromosome breakpoint of B-cell lymphomas and leukemias with the t(11;14) chromosome translocation. *Science* **224,** 1403–1406.

52. Porter, D.C., Zhang, N, Danes, C., McGahren, M.J., Harwell, R.M., Faruki, S., and Keyomarsi, K. (2001) Tissue-specific proteolytic processing of cyclin E generates hyperactive lower-molecular-weight forms. *Mol. Cell Biol.* **21,** 6254–6259.

53. Blank, M., and Shiloh, Y. (2007) Programs for cell death: apoptosis is only one way to go. *Cell Cycle* **6,** 686–695.

54. Lowe, S.W., Ruley, H.E., Jacks, T., and Housman, D.E. (1993) p-53 dependent apoptosis modulates the cytotoxicity of anticancer agents. *Cell* **74,** 957–967.

55. Brown, J.M., and Attardi, L.D. (2005) The role of apoptosis in cancer development and treatment response. *Nat. Rev. Cancer* **5,** 231–237.

56. Louvi, A., and Artavanis-Tsakonas, S. (2006) Notch signaling in vertebrate development. *Nature Rev. Neurosci.* **7,** 91–102.

57. Bray, S.J. (2006) Notch signalling: a simple pathway becomes complex. *Nat. Rev. Mol. Cell Biol.* **7,** 678–689.

58. Radtke, F., and Raj, K. (2003) The role of Notch in tumorigenesis: oncogene or tumour suppressor? *Nature Rev. Cancer* **3,** 756–767.

59. Fortini, M.E. (2002) γ-Secretase-mediated proteolysis in cell-surface receptor signalling. *Nat. Rev. Mol. Cell Biol.* **3,** 673–684.

60. Selkoe, D., and Kopan, R. (2003) Notch and presenilin: regulated intramembrane proteolysis links development and degeneration. *Annu. Rev. Neurosci.* **26,** 565–597.

61. Mumm, S.J., and Kopan, R. (2000) Notch signaling: from the inside out. *Dev. Biol.* **228,** 151–165.

62. Hammerschmidt, M., Brook, A., and McMahon, A.P. (1997) The world according to hedgehog. *Trends Genet.* **13,** 14–21.

63. Ingraham, P.W., and McMahon, A.P. (2001) Hedgehog signaling in animal development: paradigms and principles. *Genes Dev.* **15,** 3059–3087.

64. Ingraham, P.W., and Placzek, M. (2006) Orchestrating ontogenesis: variations on a theme by sonic hedgehog. *Nat. Rev. Genet.* **7,** 841–850.

65. Ming, J.E., Roessler, E., and Meunke, M. (1998) Human developmental disorders and the sonic hedgehog pathway. *Mol. Med. Today* **4,** 343–349.

66. Taipale, J., and Beachy, P. (2001) The hedgehog and Wnt signaling pathways in cancer. *Nature* **411,** 349–354.

67. Grimmer, M.R., and Weiss, W.A. (2006) Childhood tumors of the nervous system as disorders of normal development. *Curr. Opin. Pediat.* **18,** 634–638.

68. Riobo, N.A., and Manning D.R. (2007) Pathways of signal transduction employed by vertebrate Hedgehogs. *Biochem. J.* **403,** 369–379.

69. Willert, K., Brown, J.D., Danenberg, E., Duncan, A.W., Weissman, I.L., Reya, T., Yates, J.R., 3rd, and Nusse, R. (2003) Wnt proteins are lipid-modified and can act as stem cell growth factors. *Nature* **423,** 448–452.

70. Lowry, W.E., Blanpain, C., Nowak, J.A., Gausch, G., Lewis, L., and Fuchs, E. (2005) Defining the impact of beta-catenin/Tcf transactivation on epithelial stem cells. *Genes Dev.* **19,** 1596–1611.

71. Pinto, D., and Clevers, H. (2005) Wnt, stem cells and cancer in the intestine. *Biol. Cell* **97,** 185–196.

72. Reya, T., Duncan, A.W., Ailles, L., Domen, J., Scherer, D.C., Willert, K., Hintz, L. Nusse, R., and Weissman, I.L. (2003) A role for Wnt signaling in self-renewal of haematopoeitic stem cells. *Nature* **423,** 409–414.

73. Polakis, P. (2000) Wnt signaling and cancer. *Genes Dev.* **14,** 1837–1851.

74. van Es, J.H., Barker, N., and Clevers, H. (2003) You Wnt some, you lose some: oncogenes in the Wnt signaling pathway. *Curr. Opin. Genet. Dev.* **13,** 28–33.

75. Gregorieff, A., and Clevers, H. (2005) Wnt signaling in the intestinal epithelium: from endoderm to cancer. *Genes Dev.* **19,** 877–890.

76. Logan, C.Y., and Nusse, R. (2004) The Wnt signaling pathway in development and disease. *Annu. Rev. Cell Dev. Biol.* **20,** 781–810.

77. Cadigan, K.M., and Liu, Y.I. (2006) Wnt signallng: complexity at the core. *J. Cell Sci.* **119,** 395–402.

78. Veeman, M.T., Axelrod, J.D., and Moon, R. T. (2003) A second canon: functions and mechanisms of β-catenin-independent Wnt signaling. *Dev. Cell* **5,** 367–377.

79. Kohn, A.D., and Moon, R.T. (2005) Wnt and calcium signaling: β-catenin-independent pathways. *Cell Calcium* **38,** 439–446.

80. Gordon, M.D., and Nusse, R. (2006) Wnt signaling: pathways, multiple receptors, and multiple transcription factors. *J. Biol. Chem.* **281,** 22429–22433.

81. Vivekanand, P., and Rebay, I. (2006) Intersection of signal transduction pathways and development. *Annu. Rev. Genet.* **40,** 139–157.

82. Pal, S.K., and Pegram, M. (2005) Epidermal growth factor receptor and signal transduction: potential targets for anticancer therapy. *Anticancer Drugs* **16,** 483–494.

83. Lo, H.-W., and Hung, M.-C. (2006) Nuclear EGFR signalling network in cancers: linking EGFR pathway to cell cycle progression, nitric oxide pathway and patient survival. *Br. J. Cancer* **94,** 184–188.

84. Khalil, I.G., and Hill, C. (2005) Systems biology for cancer. *Curr. Opin. Oncol.* **17,** 444–48.

85. Hlavacek, W.S., Faeder, J.R., Blinov, M.L., Posner, R.G, Hucka, M., and Fontana, W. (2006) Rules for modeling signal-transduction systems. *Sci. STKE* **344,** re6.

86. Sandberg, A.A. (1990) *The Chromosomes in Human Cancer and Leukemia*, 2nd ed. Elsevier Science Publishing Co., Inc., New York.

87. Sandberg, A.A., and Chen, Z. (2001) Cytogenetic analysis. In *Hematologic Malignancies. Methods and Techniques* (Faguet, G.B., ed.), Humana Press, Totowa, NJ, pp. 3–18.

88. Sandberg, A.A., and Chen, Z. (2001) FISH Analysis. In *Hematologic Malignancies. Methods and Techniques* (Faguet, G.B., ed.), Humana Press, Totowa, NJ, pp. 19–42.

89. Baudis, M., and Bentz, M. (2001) Comparative Genomic Hybridization for the Analysis of Leukemias and Lymphomas. In *Hematologic Malignancies. Methods and Techniques* (Faguet, G.B., ed.), Humana Press, Totowa, NJ, pp. 43–64.

90. Hilgenfeld, E., Padilla-Nash, H., Haas, O.A., Serve, H., Schröck, E., and Ried, T. Sandberg, A.A., and Chen, Z. (2001) Spectral Karyotyping (SKY) of Hematologic Malignancies. In *Hematologic Malignancies. Methods and Techniques* (Faguet, G.B., ed.), Humana Press, Totowa, NJ, pp. 65–79.

91. Sandberg, A.A., Stone, J.F., and Chen, Z. (2005) Karyotyping. In *Medical Biomethods Handbook* (Walker, J.M., and Rapley, R., eds.), Humana Press, Totowa, NJ, pp. 447–463.

92. ISCN 2005. An International System for Human Cytogenetic Nomenclature (Shaffer, L.G., and Tommerup, N., eds.), S. Karger, Basel, Switzerland.

93. ISCN 1995. An International System for Human Cytogenetic Nomenclature. (Mitelman, F., ed.), S. Karger, Basel, Switzerland.

94. Nelson, P.T. and Keller, J.N. (2007) RNA in brain disease: No longer just "the messenger in the middle". *J. Neuropathol. Exp. Neurol.* **66**, 461–468.

95. Meltzer, P.S. (2005) Small RNAs with big impacts. *Nature* **435**, 745–746.

96. Rychahou, P.G., Jackson, L.N., Farrow, B.J., and Evers, B.M. (2006) RNA interference: Mechanisms of action and therapeutic consideration. *Surgery* **140**, 719–725.

97. Calin, G.A. and Croce, C.M. (2007) Chromosomal rearrangements and microRNAs: a new cancer link with clinical implications. *J. Clin. Invest.* **117**, 2059–2066.

98. Tricoli, J.V. and Jacobson, J.W. (2007) MicroRNA: Potential for cancer detection, diagnosis, and prognosis. *Cancer Res.* **67**, 4553–4555.

99. Esquela-Kerscher, A. and Slack, F.J. (2006) Oncomirs—microRNAs with a role in cancer. *Nat. Rev. Cancer* **6**, 259–269.

100. Yang, N., Coukos, G., Zhang, L. (2008) MicroRNA epigenetic alterations in human cancer: One step forward in diagnosis and treatment. *Int. J. Cancer* **122**, 963–968.

101. Barbarotto, E., Schmittgen, T.D., and Calin, G.A. (2008) MicroRNAs and cancer: Profile, profile, profile. *Int. J. Cancer* **122**, 969–977.

102. Zhang, W., Dahlberg, J.E., and Tam, W. (2007) MicroRNAs in tumorigenesis. A primer. *Am. J. Pathol.* **171**, 728–738.

103. Wu, W., Sun, M., Zou, G.-M., and Chen, J. (2006) MicroRNA and cancer. Current status and prospective. *Int. J. Cancer* **120**, 953–960.

104. Chuang, J.C. and Jones, P.A. (2007) Epigenetics and microRNAs. *Pediatr. Res.* **61**, 24R–29R.

105. Wijnhoven, B.P.L., Michael, M.Z., and Watson, D.I. (2007) MicroRNAs and cancer. *Br. J. Surg.* **94**, 23–30

Index